About the authors

Kym Anderson is George Gollin Professor Emeritus of Economics at the University of Adelaide and founding Executive Director of its Wine Economics Research Centre. He is also an Honorary Professor at the Australian National University's Crawford School of Public Policy, and Vice-President of the American Association of Wine Ecoomists and Co-Editor of its *Journal of Wine Economics*. He has worked also at the GATT (now WTO) Secretariat in Geneva (1990-92) and at the World Bank as Lead Economist (Trade Policy) during 2004-07. He served during 2000-05 as a non-executive Director on the Board of Australia's Grape and Wine Research and Development Corporation. Since graduating from the University of Chicago and Stanford University he has published more than 400 articles and 40 books. His latest wine books are *The International Economics of Wine* (World Scientific, 2020), *Wine Globalization: A New Comparative History* (Cambridge University Press, 2018 in English and 2020 in Chinese) and *Global Wine Markets, 1860 to 2016: A Statistical Compendium* (University of Adelaide Press, 2017). The latter two were co-winners of the 2018 prize for the best wine economics books awarded by the Paris-based Organisation Internationale de la Vigne et du Vin (OIV). The first edition of the present book won the 2014 OIV Prize in the best viticulture books category.

Signe Nelgen is a researcher at Geisenheim University's Institute for Wine and Beverage Business Research in Germany. Following doctoral studies and a post-doctoral fellowship at the University of Adelaide, she worked as an agricultural economist at the International Livestock Research Institute in Nairobi and in Washington DC (affiliated with IFPRI) and then as a policy analyst at the UN Food and Agriculture Organization in Rome. Her books include *Global Wine Markets, 1860 to 2016: A Statistical Compendium* (with K. Anderson and V. Pinilla, University of Adelaide Press, 2017), *Distortions to Agricultural Markets: Trends and Fluctuations, 1955-2010* (SVH-Verlag, 2012) and *Global Wine Markets, 1961 to 2009: A Statistical Compendium* (with K. Anderson, University of Adelaide Press, 2011).

Which Winegrape Varieties are Grown Where?

A Global Empirical Picture

Revised edition

Kym Anderson and Signe Nelgen

This volume, which is freely available as an ebook at **www.adelaide.edu.au/press/titles/winegrapes**, provides winegrape area data for the major winegrape regions of the world. It complements a volume on global wine markets which was published in 2017 by University of Adelaide Press as:

K. Anderson, S. Nelgen and V. Pinilla, ***Global Wine Markets, 1860 to 2016: A Statistical Compendium***, Adelaide: University of Adelaide Press, 2017 (also freely available as an ebook, at **www.adelaide.edu.au/press/titles/ global-wine-markets**).

The data underlying both books are also freely available in Excel spreadsheets at https://economics.adelaide.edu.au/wine-economics/databases

The authors welcome comments on how to improve the quality and coverage of data and the way they have been summarized. Please send feedback to:

Professor Kym Anderson, Executive Director
Wine Economics Research Centre
School of Economics
University of Adelaide
Adelaide SA 5005 Australia
Phone (+61 8) 8313 4712
kym.anderson@adelaide.edu.au
https://economics.adelaide.edu.au/wine-economics/

Which Winegrape Varieties are Grown Where?

A Global Empirical Picture

Revised edition

Kym Anderson and Signe Nelgen

Wine Economics Research Centre
School of Economics
University of Adelaide
Adelaide, South Australia

THE UNIVERSITY *of* ADELAIDE

UNIVERSITY OF
ADELAIDE PRESS

Published in Adelaide by

University of Adelaide Press
University of Adelaide
North Terrace
Adelaide SA 5005 Australia
press@adelaide.edu.au
www.adelaide.edu.au/press

The University of Adelaide Press publishes externally refereed scholarly books by staff of the University of Adelaide. It aims to maximise access to the University's best research by publishing works through the internet as free downloads and for sale as high quality printed volumes.

For the full Cataloguing-in-Publication data please contact the National Library of Australia: cip@nla.gov.au

ISBN (paperback) 978-1-925261-86-8
ISBN (ebook) 978-1-925261-87-5

Cover image: iStockphoto Cover design: Emma Spoehr
Production: John Emerson Publishing Solutions

Paperback printed by Ingram Lightning Source worldwide locations

Table of contents

List of charts

B. Top 25 wine-producing countries: 20 leading varieties

C. Top 25 varieties: 20 leading countries

List of tables

III. Winegrape areas for world's top varieties, by country

VII. Location and climate indicators of the world's wine regions

VIII. Regional Varietal Intensity Indexes for world's top varieties

IX. Varietal National and Regional Similarity Indexes (VSI)

X. Climatic National Similarity Index and a premium climate indicator

XI. National winegrape area, world's top 300 varieties, 2016

Authors' preface and acknowledgements

Over the past two decades the University of Adelaide has provided numerous editions of a global statistical compendium of annual time series data and various key indicators of national markets for grape wines. The ninth version was published by the University of Adelaide Press in 2017 as a book and e-book (www.adelaide.edu.au/press/titles/global-wine-markets) and the data are freely available at the website of the University's Wine Economics Research Centre (https://economics.adelaide.edu.au/wine-economics/databases). However, very little of the data in that compendium relate to the grapes that are the key ingredient in winemaking. Nor are data included by wine region within each of the countries covered. One reason is space: that compendium is already almost 600 pages long, so subdividing each country's area and production data into regions would have turned the volume into a brick. Also, the readily available annual data for grapes do not distinguish winegrapes from grapes for fresh consumption or drying, and most countries release detailed winegrape area data by region and variety (cultivar) irregularly and rarely annually.[1]

Another reason for our global wine markets compendium including little information on winegrapes is that the relatively scant data on bearing area (and the even scanter data on winegrape production, yield and price) refer to varieties that have different names in different countries – and sometimes in different regions within countries – even though they may have the same DNA. This challenge has been reduced greatly, however, thanks to new DNA research. In particular, the well-known Robinson/Harding/Vouillamoz 2012 book called *Wine Grapes* (hereafter also referred to as RHV) provides a detailed guide to 1368 commercially grown 'prime' varieties, and it also identifies their various synonyms. The 'prime' name is chosen by those authors according to the name used in what they consider its country or region of origin. In addition, the Julius Kühn-Institut for Grapevine Breeding at the Federal Research Centre for Cultivated Plants in Geilweilerhof, Germany maintains a *Vitis International Variety Catalogue* (www.vivc.de, hereafter referred to as VIVC). That very comprehensive resource provides additional DNA-based varietal information. As in RHV, the present volume defines prime names of varieties using Cyrillic letters if that is what is used by the country of origin. VIVC, by contrast, uses transliterated versions of those names, based on rules adopted by the journal *Chemical Abstracts*, thereby omitting any language-specific symbol (accent, cedilla, tilde, dieresis) in a prime name. Such transliterated spellings appear among the listed synonyms to our primes, in Table 15. Since there is still uncertainty about the country of origin of some varieties, it is unsurprising that VIVC considers a subset of our synonyms to be primes and nominates a different country of origin

[1] For several decades Australia was among the exceptions. Its regional and varietal time series data are compiled in Anderson, K. (with the assistance of N.R. Aryal), *Growth and Cycles in Australia's Wine Industry: A Statistical Compendium, 1843 to 2013*, University of Adelaide Press, 2015. That volume is freely available as an e-book at www.adelaide.edu.au/press/titles/austwine, and in Excel format at https://economics.adelaide.edu.au/wine-economics/databases), and further details that include winegrape crush tonnages and prices are available in Excel at the bottom of that webpage ("Section V: Regional varietal area, production and price data, 1999 to 2013").

for some varieties. These are shown in Tables C to E of the Sources of Data part of these front pages.

With the far greater capacity now available to avoid spuriously indicating diversity of winegrape varieties across regions and countries, and with the European Union publishing decadal census data on bearing area by variety and region circa 2000 and 2010 for most of its winegrape-producing member countries, the first edition of our global compendium of data on bearing area by variety and region (and hence also by country) was produced in 2013. It is a supplement to *Wine Grapes* and *The World Atlas of Wine* (or similar atlases) for readers seeking an idea of the relative importance of the world's wine regions and varieties, at least as reflected in winegrape bearing area data since the turn of the century (with more-limited data on national and global varietal totals for 1990 and on a few nations' data back to 1960).

The volume's popularity surprised us. Not counting sales of hard copies, it took less than five years following its publication in 2013 for 100,000 copies of the 700-page ebook version to be downloaded.

Unexpectedly, the European Union recently published data on bearing area by variety and region for the 2015 vintage for the EU's winegrape-producing member countries. We therefore decided to update our global winegrape database and ebook to circa 2016 (that is, using late 2015 data from the northern hemisphere and early 2016 data from the southern hemisphere, or as close to those vintages as possible). That also gave us the opportunity to revise the database for earlier years in numerous minor ways, such as using the exact spelling for each variety as used in its nominated country of origin, ensuring the same names are used across the years for each region, and concording regional and super-regional names where aggregations within countries varied across time.

Assembling and 'cleaning' those new data has been once again an extremely time-consuming task, but it would have taken much longer (and in some cases been impossible) without the generous assistance of a large number of people in numerous countries. First and foremost, grateful thanks go to Jancis Robinson MW, Julia Harding MW and José Vouillamoz for promptly responding to emailed questions and for sharing their vast knowledge by reacting to drafts of numerous tables, as well as for providing 2010 data for such countries as China, Japan, Russia and Ukraine. Also extremely helpful for the latest revision and update were comments on varietal names/spellings and countries of origin by Domen Presern of the University of Oxford (and President of its Blind Tasting Society). As well, Erika Maul provided detailed information on VIVC prime names and synonyms. Peter Dry of the Australian Wine Research Institute and Gregory Jones of Linfield University in Oregon cast their eyes over our listings and provided many additional helpful comments.

Professor Gregory Jones also provided invaluable guidance on the nature and limitations of regional location and climate data. That he has helped us compile and report new data in Section VII of this volume. Our thanks too to Germán Puga, an Argentinean PhD student at the University of Adelaide, who provided excellent research assistance in concording regional names across years with those used in the 8th edition of *The World Atlas of Wine,* in assembling those data for Section VI, in finding the latitude and longitude of a representative town or city for each region, in aggregating these climate data in various ways as reported in Section VII, and in generating the Climatic National Similarity Index numbers in Section X.

We remain grateful also to the late Patrick Fegan of the Chicago Wine School, whose 2003 book *The Vineyard Handbook: Appellations, Maps and Statistics* helped with its 1990 varietal data and also circa 2000 data for several small wine-producing countries, for which we otherwise would have had only 2010 and 2016 information. The data for Italy for 1990 (and 1970) were greatly expanded to include many minor varieties, thanks to Table 3 of Ian D'Agata's 2014 book, *Native Wine Grapes of Italy* (University of California Press).

At the risk of accidentally omitting some names (for which we humbly apologize), our sincere thanks for providing or leading us to the following national data up to 2010 go to, in author alphabetical order, Julian Alston, Kate Fuller and Sandro Steinbach (California and Washington States, USA), Georgi Apkhazava (Georgia), Peter Bailey, Sheralee Davies, Alan Nankivell and Mark Rowley (Australia), Stefan Bojnec (Croatia, Serbia and Slovenia), Bruce Bordelon (Indianan, USA), Jasna Čačić (Croatia), Mark Chien (Pennsylvania, USA), Donald Cyr (Canada), Dominique Desbois (for carefully assembling French survey data for 2009, pending the publication of the official census data which have yet to be released), Christy Eckstein (Ohio, USA), Denis Gastin (Thailand), Anatassios Haniotis and Kargarita Koumanioti (for advance access to the 2009 Greek census data), Giulia Meloni (EUROSTAT data plus Brazil, Italy, Peru), Jimena Estrella, Javier Merino and Germán Puga (Argentina), Taner Öğütoğlu (Turkey), Sergey Oleichenko, Dauren Oshakbaev and Alfinura Sharafeyeva (Kazakhstan), Bruce Reisch (New York State, USA), Jorge Tenotio (Mexico), Gabriel Tinguely (Switzerland), Áron Török (for advance access to the 2010 Hungarian census data), Angeliki Tsiolo of the OIV (for contacts in various countries), and last but definitely not least, the trio of Annalisa Zezza, Roberta Sardone and Eugenio Pomarici (for advance access to and heroic efforts to polish the 2010 Italian census data).

Thanks also to many of those same people who helped us secure circa 2016 data, along with (alphabetically) Irina Bîstriţchi (Moldova), Bruce Bordelon (Indiana), Doug Caspey (Colorado), Umay Çeviker (Turkey), Elodie Comby and Océane Gex (Switzerland), Giorgio Delgrosso and Barbara Iasiello (OIV, Paris), Yulin Fang, Demei Li and Huiqin Ma (China), Bruno Gabersek (Slovenia), Carolyn Gilby (Bulgaria, Moldova, Romania and Serbia), Antonio Graca (Portugal), Sandy Hathaway (Australia), Simone Heidinger and Florian Schütky (Austria), Nelli Hovhannisyan and Aramayis Mkrtchyan (Armenia), Tomislav Ivanovic and Igor Lukovic (Serbia), Ronald Jackson, L. Kittmer and Bruce Wright (Canada), Erik Lindås (Norway), Dan McLaughlin (North Carolina), Diana Mereanu (Romania), Elena Miloshevska (North Macedonia), Vicente Pinilla (Spain), Germán Puga (Argentina), Venelin Roychev (Bulgaria), Steffen Schindler and Eberhard Abele (Germany), Áron Török and Gabriella Szmilkó (Hungary), Michael White (Iowa) and Doniella Winchell and Maria Smith (Ohio).

We acknowledge and thank Australia's Grape and Wine Research and Development Corporation (GWRDC) for assisting with funding the research project that produced the initial data collection. We are grateful also to Lachlan Deer and Claire Hollweg for earlier research assistance with circa 2000 winegrape varietal data compilation for a dozen key countries that provided a prototype for the present much more comprehensive study (see Anderson, K., "Varietal Intensities and Similarities of the World's Wine Regions", *Journal of Wine Economics* 5(2): 270-309, Winter 2010). GWRDC has since been absorbed into Wine Australia which, together with the University of Adelaide's Faculty of Business, Economics and Law and its School of Agriculture, Food and Wine, has helped finance the current revision and update of this database and book. We also wish to acknowledge the very helpful advice of our research project's Industry Reference Group, comprising Peter Hayes (former

President of OIV), Brett McKinnon (Global Operations Director of Pernod Ricard Winemakers) and Marc Soccio (formerly of the beverage section of Rabobank's Food and Agribusiness Research division).

The previous revision of the database, following the publication of the 2013 book, was July 2014, in which two dozen very minor varietal names were corrected, the datasets for four countries (France, Italy, Germany and Romania) were expanded, and Ethiopia was added. For France, more-detailed regional and varietal data were provided for 2010 (its original 45 regions were further divided into 72 regions and there were 272 prime varieties, up from 96 previously). For Italy, more-detailed regional data were provided for 2010 (its original 20 regions were subdivided into 110 sub-regions, similar to the 103 regions for 2000). For Germany, the 'Other varieties' category was shrunk from around one-third to just 1% of the total bearing area, and there were 91 instead of 48 varieties shown in 2010 and 68 instead of 57 in 2000. For Romania, the 'Other varieties' category for 2010 was shrunk only a little but the bearing areas of more minor varieties were included, raising the total number of its varieties from 25 to 99. We are grateful to Norbert Tischelmayer of Austria for pointing to errors in some minor varietal names or colours, to Patrick Aigrain of France, Eugenio Pomarici of Italy, Steffen Schindler of Germany and Diana Mereanu of Romania for providing the more-detailed data for their countries, to Etsegenet Kiflu of the DebreZeit Agricultural Research Center for providing 2011/12 data for Ethiopia's emerging wine industry, to Patrick Fegan's 2003 *Vintage Handbook* for data for Israel and Taiwan for 2000 (assumed to be the same in 2010), and to www.vivaioenotria.com for 2011 data for Korea and for the newly added data for pre-2000 years for several countries. Together these revisions expanded the number of regions in the 2010 global dataset by 23% and the number of prime varieties by 14%.

For the 2016 collection, several EU countries provided less-detailed data than in their full census years of 2010 and 2000, while some other countries provided more-detailed data. That makes it difficult in some cases to compare across time the varietal mix of regions, or the importance of minor varieties (which, in less-detailed collections, get absorbed in the residual 'Other varieties' categories). We have also revised the spelling of numerous more-obscure varieties and added a little to the earlier and scanter national data for 1960, 1970, 1980 and 1990. Together these revisions have expanded the global dataset to more than 1700 prime varieties. In addition, there are more than 1350 synonyms for those prime varieties in the complete dataset. As for the number regions, there are 629 in 2016, a little below the 706 in 2010 but above the 489 in 2000. The extent of regional disaggregation varies over the past two decades though, such that there more than 800 regions and sub-regions listed in those three years for the countries with such disaggregated data, as reported in Tables 67 to 69. For the convenience of the reader, regions within each state are identified for Australia, Italy and the United States in Table 70, some concordances of regions in a few other countries over time are provided in Table 71, and in Tables 72 to 74 concordances are provided between the regions in this database and the regions identified in the 8th edition (2019) of the Hugh Johnson/Jancis Robinson *World Atlas of Wine.*

The making of wine is intimately related to the climate in which its grapes are grown. The climate of each winegrowing region is a critical determinant of the suitability of the region to particular winegrape varieties, their potential wine style, and the region's overall productivity and profitability. Thus as climates change and planting areas expand in the various wine regions, so will the quality, productivity and profitability of the wines produced there. Hence the effort made to include in this new edition at least some key climate variables for each of our identified wine regions. The geographic location of each region and its climate variables are shown in Section VII. By classifying each region as either cool, temperate, warm or hot as suggested by Gregory Jones, according to its growing season average temperature, we have been able to estimate shares of the national and global winegrape area of each variety growing in cool, temperate, warm and hot climates.

While we have made every effort to ensure the accuracy and currency of information within this compendium, we cannot accept responsibility for information that may later prove to be misrepresented or inaccurate, or for any reliance placed on the information by readers. We warmly welcome comments on the raw data and the indicators derived from them, and we would gratefully receive any new databases for omitted countries or updated, expanded or revised databases for those countries already included.

Kym Anderson and Signe Nelgen
Adelaide, South Australia
August 2020

Sources of data

The most important source of winegrape bearing area data for this compendium is EUROSTAT, because it provides data by region for the European Union's member countries for the two most-recent decadal censuses, which were circa 2000 and 2010, and the update for 2015. They can be found at http://epp.eurostat.ec.europa.eu/portal/page/portal/statistics/search_database [In the Data Navigation Tree, click on "Agriculture, forestry and fisheries" then "Agriculture" then "Structure of orchards and vineyards" then "Vineyard " and then "Basic vineyard survey".] Since that source provides data for a large share of the world's winegrape production, those years are the ones targeted for all other countries. For the majority of the EU countries the census dates were a year earlier, so 1999 and 2009 were the vintages targeted for other Northern Hemisphere countries while 2000 and 2010 (plus 2016) were targeted for Southern Hemisphere countries – bearing in mind that harvesting is late in the calendar year in the north and early in the calendar year in the south.

Not all EU-28 countries had their latest census data uploaded on that EUROSTAT website initially, so we approached government officials in the missing member countries (France, Greece, Hungary, Italy) to secure advance copies of the circa 2010 data that were yet to be uploaded in those countries, as we did also for Hungary and Romania for more-detailed 2015 data.

The national and regional data sources, and the exact years to which they relate, are listed in Table A of this section.

The choice of countries to include was determined primarily by national shares of global wine production, in addition to availability of data. The 53 countries for which data are available for circa 2016 account for 99% of global grapewine output. The only other country producing more than 0.1% of the world's wine in 2010 is North Macedonia (0.3%), for which we were unable to locate varietal area data until 2016. Of the 44 countries reported for 2010, we were unable to secure reliable data for 2000 for 9 of them (China, Japan, Kazakhstan, Mexico, Myanmar, Peru, Thailand, Turkey, and Ukraine). The combined share of global wine production of those 9 countries in 2000 was only 1.6% (compared with 5.1% in 2010), but to retain their unusual varietal contributions we included them as a group (called "Missing 9 in 2000") by assuming each of them had (i) the same varietal mix then as in 2010 and (ii) a national acreage in 2000 that was the same fraction of its 2010 acreage as was its national wine production volume. Of the extra countries added for 2016, four were included in our July 2014 update, and the other five represent only 0.2% of global wine production.

In many of the country tables we also include a 'World' total. In section I's Tables 2 and 3 that includes an estimate for 'Rest of the world'. Since that residual comprises no more than 1% of global wine production, we assume that share is also its share of the global winegrape area. Being so small, we ignore that residual in all other tables, where 'World' refers to the sum of the shown countries. This total is sub-divided into 'Old World' and 'New World', as summarized in column B of Table 1. 'Old World' refers to traditional winegrape-growing countries of Europe, the former Soviet Union, the Levant, and France's former North African colonies. All others ('New World' countries) grow winegrapes in and for newer markets and include, unusually, Norway and the United Kingdom (although their winegrape areas are so tiny that it makes little difference which sub-section includes them) plus Asian countries other than those of Central Asia that had been part of the Soviet Union. Prior to World War II, the 'Old World' accounted for all but 7% of global wine production, and as recently as the 1990s its share still exceeded 80%.

In addition to aggregate national data, bearing area data by variety are available for regions within 29 of our 53 countries in 2016, for 29 of 48 countries in 2010, and for 12 of 38 countries in 2000. In aggregate there are 629 unique within-country regions represented in 2016, 706 in 2010 and 489 in 2000. The degree of regional disaggregation of any one country varies through time though, as is evident in Tables 67-69. For the convenience of the reader, Table 71 reveals how some of them aggregate to super-regions, providing a total of 813 regions in the entire dataset. A concordance between these and the regions identified in the 8th edition of the Hugh Johnson/Jancis Robinson *World Atlas of Wine* is provided in Tables 72-74. For three countries, Table 70 shows in which State/Province each region is located.

As acknowledged in the Preface, our key source for identifying DNA-identical varieties and their synonyms is the Robinson/Harding/Vouillamoz book called *Wine Grapes: A Complete Guide to 1,368 Vine Varieties, Including their Origins and Flavours* (London: Allen Lane, 2012). Supplementing that source is the *Vitis International Variety Catalogue* (www.vivc.de), which also provides DNA-based varietal information. The RHV book's prime varieties account for most of the global winegrape area. VIVC accounts for much of the rest, and for a larger share of the number of minor varieties' prime names, as summarized in Tables B to E at the end of this section, with thanks to Erika Maul of VIVC for assisting with prime identifications. We also adopt RHV and VIVC berry colours, although we simplify their five categories to just three: the darkest two we call red, the lightest two we call white, and the middle colour we call 'grey' (which accounts for just 2.4% of the global area in 2016, 2.1% in 2010, and 1.4% in 2000, almost half of which is Pinot Gris/Grigio.[2]

There are two exceptions to our use of RHV prime names. One concerns Pinot, which is thought to have existed for two millennia and which therefore has many clones. Until recently the most popular clones – which include all three of our colour categories – were thought to be distinct varieties, and have been promoted separately to different niches in the market. For that reason we retain separately the following five, each of which has several synonyms identified by RHV: Pinot Blanc, Pinot Gris, Pinot Meunier, Pinot Noir, and Pinot Noir Précoce. The other exception is Garnacha, which also has both red and white mutations. In that case we retain separately the following four, each of which has several synonyms identified by RHV: Garnacha Blanca, Garnacha Peluda, Garnacha Roja, and Garnacha Tinta.

There are no official data on China's winegrape area by variety and region, so reliance has been on estimates by well-informed individuals. For the latest numbers we drew on national estimates by Professor Yulin Fang, Dean of the College of Enology, Northwest Agricultural and Forestry University, who in turn drew on data for 2017 from the China Grape & Wine Industry Network (http://www.chngw.net/Default.aspx). The total area is consistent with the volume of wine produced from domestic grapes in China.[3] Senior staff at Pernod Ricard Asia in China agreed these were the best estimates available, but they felt the

[2] Numerous countries have an 'other varieties' category for each region, only some of which sub-divide that category according to berry colour. When no sub-division is provided, we assume the proportions of 'other varieties' that are red, white and grey are the same as the proportions for the named varieties for that region.

[3] As estimated by Anderson, K and K. Harada (2018), 'How Much Wine is *Really* Produced and Consumed in China, Hong Kong and Japan?', *Journal of Wine Economics* 13(2): 199-220.

share of Cabernet Sauvignon in that total was probably twice the actual share (as producers of various other red grape varieties try to suggest their vines are this noble variety). We therefore halved the area listed for that variety and included the other half in 'Other red varieties'.

Australia's official area data had been compiled by the Australian Bureau of Statistics (ABS) until 2015, but ceased thereafter. Hence we had to assume the varietal mix in 2016 was the same as in 2015, and lowered each varietal area by 1.4% so they summed to the 2016 total bearing area. Those ABS data include only the largest 40 or so varieties. However, annual estimates of the winegrape crush by every variety and source region have been assembled by Wine Australia. We calculated for 2015-17 (to help even out seasonal variations), for each of the smaller red and white varieties not separately showing in the 2015 ABS data, their share of the total tonnes of missing red and white winegrape varieties. We then assumed the average winegrape yield per hectare was the same for each variety within the missing red and white winegrape variety categories in each region. That allowed us to use those shares to divide the ABS's aggregate areas of 'Other red' and 'Other white' for each region into estimates of bearing area for minor varieties, thereby expanding the number of varieties from 40 to 146. While those added varieties account for only 1.25% of the total bearing area for Australia in 2016, they provide an insight into the relative importance of what Darby Higgs calls the next 100 'Rare Ozzies'.[4]

We were unable to get 2015/16 data for several countries that were included in the 2010 dataset: Algeria, Ethiopia, Georgia, Kazakhstan, Korea, Mexico, Peru, Russia and Taiwan. Hence we assumed their varietal mix and total area were the same as in 2010. These countries accounted for just 2% of the global bearing area in 2010. In the case of Russia, though, we have added the region of Crimea (part of southeast Ukraine prior to its annexation by Russia in 2014) by assuming that region was half of Ukraine's winegrapes area[5] and, less plausibly but in the absence of further evidence, that the varietal mix in Crimea was the same as in the rest of Ukraine in 2016.

For six important EU wine-producing countries – France, Greece, Italy, Portugal and Romania – the number of varieties reported in Eurostat was far smaller in 2016 than in 2010, as only major varieties were shown separately. So too was the number of regions, but a concordance between the more-detailed and less-detailed sets was easy to compile (Tables 71-74). So for each of these five countries we have added the minor varieties that were separated out in 2010 but not in 2016 by assuming that for each of those red, white and grey varieties not separately showing in the 2016 data, their 2016 share of total area for each region was the same as in 2010. We then subtracted the sum of those newly added areas of red or white or grey varieties from the 2016 'Other red' or 'Other white' or 'Other grey' area for that region. Similar to Australia, those added varieties account for only a small fraction of the total bearing area for each country in 2016, but they help to indicate the relative importance of those minor varieties in the global bearing area.

---

[4] Higgs, D. (2019), *Rare Ozzies: A Hundred Rare Australian Grape Varieties*, self-publised in Willimastown, Victoria (see www.vinodiversity.com/rareozzies.html). See also Halliday, J. (2018), *Varietal Wines: A Guide to 140 Varieties Grown in Australia and their Place in the International Wine Landscape,* London: Hardie Grant Books.

[5] Tsymbliuk, K. and Y. Larina (2017), 'The Current State of the Vitiviniculture Sector in Ukraine', *Baltic Journal of Economic* Studies 3(5): 431-36.

Apart from the bearing area data, various other variables are included in some of the tables and charts. Their sources are as follows:

Tables 2 and 3: FAOSTAT data for total grapevine area, total grape production, grape yield per hectare, and agricultural land (arable land and land used for permanent crops) (http://faostat.fao.org).

Table 4: In our 2013 book, varietal bearing area data for 1990 are from Fegan, P.W, *The Vineyard Handbook: Appellations, Maps and Statistics,* revised edition (Springfield IL: Phillips Brothers, 2003). We estimated the global winegrape bearing area in 1990 based on total and winegrape production data and taking into account that the global winegrape yield per hectare averages 15% less than the total grape yield, based on data in Anderson, K. and D. Norman, *Global Wine Production, Consumption and Trade, 1961 to 2001: A Statistical Compendium* (Adelaide: Centre for International Economic Studies, 2003).

Table 29: The list of fungal-resistant hybrid varieties (a collective whose German name is Pilzwiderstandsfähige Sorten, popularly known by the acronym PIWI), includes all those varieties listed at http://www.zukunft-weinbau.de/forschung/piwi-liste/ that are reported in our global database. There may be others in production that are hidden in the 'Other varieties' residual catch-all for each country, but in almost all cases their bearing area is very minor as of 2016.

Section VII provides climate indicators for individual locations within or adjacent to the world's wine regions. Specifically, Table 75 provides the latitude and longitude of the region's most representative town or city and its elevation above sea level (from a Google Earth Pro app downloadable at https://www.google.com/earth/versions/#earth-pro) plus the growing season average temperature and precipitation. For convenience of readers, the page number of each region in the 8th edition of the Hugh Johnson/Jancis Robinson *World Atlas of Wine* also is provided in that table. Additional climate variables are provided in Table 76, taken from TerraClimate (see http://www.climatologylab.org/terraclimate.html) and updated to 2019 by Gregory Jones.[6] TerraClimate generate high-resolution (1/24°, 4-km) monthly data averaged over the period 1958-2019. This dataset includes the mean value for each longitude/latitude location for the following climate variables:

AnnP - annual precipitation (mm)

GSP - growing season precipitation (mm); growing season: April-October in the Northern Hemisphere, October-April in the Southern Hemisphere

RipeP - ripening month precipitation (mm); September in the Northern Hemisphere, March in the Southern Hemisphere

AnnT - annual average temperature (°C)

GST - growing season average temperature (°C); growing season: April-October in the Northern Hemisphere, October-April in the Southern Hemisphere

RPT - ripening period average temperature (°C); August and September in the Northern Hemisphere, February and March in the Southern Hemisphere

GDD - growing degree days (C° units); growing season: April-October in the Northern Hemisphere, October-April in the Southern Hemisphere[7]

---

[6] Abatzoglou, J., S. Dobrowski, S. Parks and K.C. Hegewisch (2018), 'TerraClimate, a High-Resolution Global Dataset of Monthly Climate and Climatic Water Balance from 1958–2015', *Scientific Data* 5: 170-191, 2018 (www.nature.com). https://doi.org/10.1038/sdata.2017.191

[7] GDD is defined for temperatures above a base temperature such as 10°C and with no upper cut off. A varient is BEDD (biologically effective degree days), which is defined as for GDD but with a maximum of 9°C above

GSDTR - growing season diurnal temperature range (°C); growing season April-October in the Northern Hemisphere, October-April in the Southern Hemisphere
RPDTR - ripening period diurnal temperature range (°C); August-September in the Northern Hemisphere, February-March in the Southern Hemisphere

For the convenience of readers and to stimulate discussion, we have classified each region as either cool, temperate, warm or hot, based on their GST and as defined by Gregory Jones in Chart 29 (reproduced with his permission). Of course the elevation and climate of each vineyard vary within each region,[8] and their median values would not be exactly the same as for the town we have chosen to represent the region. Hence these data are necessarily just a proxy for the climate of the region's vineyards. More specifically, since those locations are all urban, and many are at a lower elevation than nearby vineyards, they are probably warmer than the median growing season average temperature in the wine region they are purporting to represent. However, the world has warmed since the start of the period covered by those data (1958), so the average GST in regions today may not be below that 1958-2019 average. When we class each location according to the following GST range, it divides our 813 regions and generates GST global averages as follows:[9]

Class name	**GST range (°C)**	**Global average[a] GST (°C)**	**Share of global area[a] (%)**	**Share of regions (%)**
Cool	<15	14.5	3	11
Temperate	15-17	16.3	18	22
Warm	17-19	18.0	35	24
Hot	>19	20.4	44	42
All regions		**18.6**	**100**	**100**

[a] Weighted average across regions with weights based on regional winegrape bearing areas in 2016.

Having classified each region as either cool, temperate, warm or hot, based on their GST, Table 98 in Section X uses that information to generate an index of national climate similarity between pairs of countries' winegrape regions and relative to the world's. And having determined an estimated GST for each region, Table 99 in Section X reports an indicator of the extent to which the bearing area of the 21 premium winegrape varieties in Chart 29 fall within the ideal GST range for premium winegrape production, by country and globally.

the base. See Hall, A. and G.V. Jones (2009), 'Effect of Potential Atmospheric Warming on Temperature-Based Indices Describing Australian Winegrape Growing Conditions', *Australian Journal of Grape and Wine Research* 15(2): 97-119, June.

[8] Jones, G.V., R. Reid, and A. Vilks (2012), 'Climate, Grapes, and Wine: Structure and Suitability in a Variable and Changing Climate', pp. 109-133 in *The Geography of Wine: Regions, Terroir, and Techniques*, edited by P. Dougherty, New York: Springer.

[9] For the sake of comparison, we also derived a four-way classification of the locations from an optimum solution in a cluster analysis using kmeans, and it happened to divide the dataset into almost the same four GST ranges as in the table above.

Table A: Sources of national winegrape bearing area data[a]

Country	*Actual years*	*Data sources* (latest, then earlier date)
Algeria	2001, 2009, 2015	Fegan (2003); OIV (2018) for 2015; assumed same for 2009
Argentina	2002, 2011, 2016	Argentinean Instituto Nacional de Vitivinicultural (INV): provides more-detailed data in 2002 and 2011 than in 2000 and 2010, from http://www.inv.gov.ar
Armenia	2001, 2015	Fegan (2003); assumed unchanged for 2010; Nelli Hovhannisyan and Aramayis Mkrtchyan, personal communications
Australia	2001, 2010, 2015	http://www.wineaustralia.com; more-detailed series begins in 2001; mix assumed same for 2016, and augmented with minor varieties as described in text above.
Austria	1999, 2009, 2015	EUROSTAT and https://www.oesterreichwein.at
Brazil	2000, 2010, 2016	Fegan (2003; http://vitibrasil.cnpuv.embrapa.br/index.php?opcao=opt_03
Bulgaria	2001, 2009, 2015	Fegan (2003);_EUROSTAT; Venelin Roychev, personal communication (based on Roycheva, A.V., *Efficiency and Competitiveness of Bulgarian Viticulture*, PhD Dissertation, Plovdiv, 2019 and Roychev, A.V., *Ampelography,* Agricultural University Press, Plovdiv, 2012.)
Cambodia	2015	Denis Gastin, personal communication
Canada	2001, 2009, 2015	Fegan (2003); www.grapegrowersofontario.com/sites/default/files/2012%20annual%20report.pdf, British Columbia data are 2011, not 2009, from www.winebc.com; British Colombia Grapegrowers' Association, Grapegrowers of Ontario, Grapegrowers' Association of Nova Scotia, Vignerons Indépendants du *Québec*
Chile	2000, 2009, 2016	www.odepa.gob.cl/odepaweb/servicios-informacion/publica/catastro-vides-2009.pdf; Ministerio de Agricultura, *Catastro Viticola Nacional 2016*
China	2009, 2015	*China Agricultural Yearbook 2010*, via Julia Harding, personal communication for 2009; see text above re. 2015
Croatia	2001, 2011, 2015	Fegan (2003); Croatian Ministry of Agriculture; EUROSTAT
Cyprus	2000, 2009, 2015	Fegan (2003);_EUROSTAT
Czechia	2001,2009, 2015	Fegan (2003);_EUROSTAT
Ethiopia	2009, 2015	Ministry of Agriculture for 2009; assumed unchanged for 2015
France	1999, 2010, 2015	EUROSTAT and personal communication with Dominique Desbois
Georgia	2004, 2009, 2015	Georgian Wine Association for 2004; assumed mix unchanged for 2000, small growth in area by 2009 and 2015[b]
Germany	1999, 2009, 2015	EUROSTAT and www.deutscheweine.de
Greece	1999, 2009, 2015	EUROSTAT and Hellenic Statistical Authority
Hungary	2000, 2010, 2015	Fegan (2003); Hungarian Ministry of Agriculture; Áron Török and Gabriella Szmilkó (www.hnt.hu), personal communications
India	2015	OIV (2018)
Israel	1999, 2009, 2015	Fegan (2003); OIV (2018) for 2015 and assumed same for 2009
Italy	2000, 2010, 2015	EUROSTAT; Italian Ministry of Agriculture
Japan	2009, 2015	Julia Harding, personal communication; Kimie Harada, person communication

Kazakhstan	2007, 2015	Dauren Oshakbaev, personal communication; assumed mix unchanged for 2010 and 2015
Korea, Rep.	1999, 2011, 2015	www.vivaioenotria.com for 2011; assumed unchanged for 1999 and 2015
Lebanon	2015	OIV (2018)
Luxembourg	1999, 2009, 2015	EUROSTAT for 1999; OIV (2018) for 2015 and assumed same for 2009
Mexico	2011, 2015	Mexican Ministry of Agriculture, www.siap.gob.mx; assumed unchanged for 2015
Moldova	2009, 2015	Moldovan Ministry of Agriculture; Carolyn Gilby and Irina Bîstriţchi, personal communications
Morocco	1999, 2009, 2015	http://agriculture.ovh.org; OIV (2018) for 2015; assumed same for 2009
Myanmar	2012, 2015	http://redmountain-estate.com/varieties.html; assumed unchanged for 2015
New Zealand	2000, 2009, 2016	http://wineinf.nzwine.com/statistics_outputs.asp?id=89&cid=6&type=n; survey not conducted in 2010
North Macedonia	2015	Wines of Macedonia https://winesofmacedonia.mk; Elena Miloshevska, personal communication
Norway	2019	Erik Lindås, personal communication
Peru	2008, 2015	www.minag.gob.pe/portal/download/pdf/herramientas/boletines/DocumentoFinalVid.pdf; assumed unchanged for 2015
Portugal	1999, 2009, 2015	EUROSTAT
Romania	2001, 2009, 2015	Fegan (2003); EUROSTAT; Diana Mereanu (Oficiul Naţional Al Viei Şi Produselor Vitivinicole), personal communication
Russia	2000, 2009, 2015	Fegan (2003); Julia Harding, personal communication; assumed mix unchanged for 2015 except Crimea transferred from Ukraine
Serbia	2001, 2009, 2015	Fegan (2003); assumed unchanged for 2009; National Office of Statistics, National Winery Register
Slovakia	2000, 2009, 2015	Fegan (2003); EUROSTAT
Slovenia	2000, 2009, 2015	Fegan (2003); EUROSTAT; Bruno Gabersek, person communication
SouthAfrica	2002, 2011, 2016	http://www.wosa.co.za/sa/stats_sawis_annual.php
Spain	1999, 2009, 2015	EUROSTAT
Switzerland	1999, 2009, 2015	http://www.blw.admin.ch/themen/00013/00084/00344/index.html?lang=de and www.swisswine.ch
Taiwan	1999, 2009, 2015	Denis Gastin, personal communication
Thailand	2010, 2015	Thailand Grape Vine Survey, from Denis Gastin, personal communication
Tunisia	2000, 2009, 2015	Fegan (2003); assumed 6.3 kl/ha as in Algeria for 2015 and assumed same for 2009
Turkey	2010, 2015	Taner Öğütoğlu of Wines of Turkey, personal communication; Umay Çeviker, personal communication
Ukraine	2009, 2015	Ukrainian Ministry of Agriculture via Julia Harding, personal communication; assumed unchanged for 2015 except Crimea transferred to Russia
United Kingdom	1999, 2009, 2015	Wines of Great Britain
United States[c]	1999, 2009, 2015	www.nass.usda.gov and see footnote c for individual states

Uruguay	2000, 2012, 2016	http://www.inavi.com.uy/categoria/38-estada-sticas-de-via-edos.html; Robinson, J. *The Oxford Companion to Wine,* 3rd edition, 2006, p. 723; Departamento Registro de Vinedos, *Estadisticas de Vinedos 2016*

[a] EUROSTAT data are available at http://epp.eurostat.ec.europa.eu/portal/page/portal/statistics/search_database. In addition, some data for 1990 are provided from Fegan, P.W. (2003), *The Vineyard Handbook: Appellations, Maps and Statistics,* revised edition, Springfield IL: Phillips Brothers, for the Chicago Wine School. Some data for 2015 are from OIV (2018), *Grapevine Varieties Area by Country,* Paris: Organisation Internationale de la Vigne et du Vin, March. www.oiv.org

[b] More-detailed Georgian data are also available for 1953, from page 18 of Ketskhoveli, N., M. Ramishvili and D. Tabidze (2012), *Georgian Ampelopgraphy,* Tbilisi: Exclusive Print, Ltd.

[c] Dates for the various US states vary according to availability:
Arizona-2008, 2015, Arkansas-2009, 2015, California-1991, 1999, 2009, 2015, Colorado-2009, 2015, Georgia-2009, 2015, llinois-2011, 2015, Indiana-2011, 2013, Iowa-2006, 2015, Kentucky-2010, 2015, Michigan-2002, 2011, 2018, Minnesota-2007, 2015, Missouri-2010, 2015, New York-1990, 2001, 2011, 2015, North-Carolina-2009, 2015, Ohio-2008, 2015, Oregon-1990, 2000, 201, 2015, Pennsylvania-2008, 2015, Texas-2010, 2017, Virginia-2008, 2015, Washington-1990, 1999, 2011, 2016.

Table B: Sources of prime varietal names and their synonyms

Source ↓	No. of primes	%	2016 global area ('000 ha)	%
RHV	1091	64	4087	91
VIVC	549	32	50	8
Other	65	4	346	1
TOTAL	1705	100	4483	100

Table C: RHV prime names for which VIVC has one of our synonyms as their prime

RHV Prime ↓	**RHV origin**	**RHV colour**	**VIVC Prime**	*Where different:* **VIVC origin**	**VIVC colour**
Cabral	Portugal	R	Malvasia Cabral		G
Damaschino	Italy	W	Planta Fina	Spain	
Enantio	Italy	R	Lambrusco a Foglia Frastagliata		

Godello	Spain	W	Gouveio	Portugal
Graševina	Croatia	W	Welschriesling	Italy
Grechetto di Orvieto	Italy	W	Grechetto Bianco	
Hondarribi Beltza	Spain	R	Ondarrabi Beltza	
Juan Garcia	Spain	R	Mouraton	
Királyleányka	Hungary	W	Fetească Neagră	Romania
Leányka	Hungary	W	Fetească Albă	Romania
Listain de Huelva	Spain	W	Manteudo	Portugal
Macabeo	Spain	W	Viura	
Malvasia di Lipari	Italy	W	Malvasia Dubrovacka	
Mammolo	Italy	R	Sciaccarello	
Negramoll	Spain	R	Mollar Cano	Portugal
Norton	USA	R	Cynthiana	
Preto Martinho	Portugal	R	Amostrinha	
Rouge de Fully	Switzerland	R	Durize	
Seyval Noir	Canada	R	Seyve Villard 5-247	France
Tribidrag	Croatia	R	Primitivo	
Zametovka	Slovenia	R	Kavcina Crna	

Table D: RHV prime names for which VIVC has an alternative country of origin

RHV Prime ↓	RHV origin	RHV colour	VIVC Prime	VIVC origin	VIVC colour
Albana	Italy	W	Albana	Spain	
Blauer Portugieser	Austria	R	Portugieser Blau	Slovenia	
Cabernet Blanc	Switzerland	W	Cabernet Blanc	Austria	
Chasselas (R)	Switzerland	W	Chasselas Rose	France	G
Fetească Albă	Moldova	W	Fetească Albă	Romania	
Fetească Neagră	Moldova	R	Fetească Neagră	Romania	

Frühroter Veltliner	Austria	R	Veltliner Frühroter	Italy	G
Gouveio Real	Portugal	W	Gouveio Real	France	G
Grüner Veltliner	Austria	W	Veltliner Gruen	Italy	
Kalina	Switzerland	R	Kalina	Serbia	W
Loureiro	Portugal	W	Loureiro Blanc	Spain	
Malvasia	Greece	W	Malvasia	Portugal	
Roter Veltliner	Austria	G	Veltliner Rot	Italy	
Trajadura	Portugal	W	Trajadura	Spain	
Voskeat	Armenia	W	Voskeat	Azerbaijan	

Table E: RHV prime names for which VIVC has an alternative prime not in our database

				Where different:	
RHV Prime ↓	**RHV origin**	**RHV colour**	**VIVC Prime**	**VIVC origin**	**VIVC colour**
Bariadorgia	Italy	W	Carcajolo Blanc	France	
Carica l'Asino	Italy	W	Barbassese Bianco		
Casetta	Italy	R	Lambrusco del Casetta		
Çavuş	Turkey	W	Chaouch Blanc		
Chelva	Spain	W	Montua		
Completer	Switzerland	W	Malanstraube		
Debit	Croatia	W	Ruzevina		
Dona Branca	Portugal	W	Branda		
Eyholzer Rote	Switzerland	R	Hibou Rouge	France	
Gibi	Spain	W	Heben		
GR 7	USA	R	Geneva Red		
Incrocio Bianco Fedit 51	Italy	W	Dorona Veneziana		
Karasakiz	Turkey	R	Sakiz Kara		

Kraljevina	Croatia	W	Koenigstraube Weiss	Austria	
Lairen	Spain	W	Santa Magdalena		
Madeleine × Angevine 7672	Germany	W	Madeleine Angevine	France	
Mandon	Spain	R	Garro		
Menoir	Hungary	R	Medoc Noir		
Menu Pineau	France	W	Arbois Blanc		W
Mézes Fehér	Hungary	W	Honigler		
Moscato Rosa del Trentino	Italy	R	Tamjanika Crna	Balkans	
Muscat Swenson	USA	W	Osceola Muscat		
Muskat Moravsky	Czech Rep.	W	Mopr		
Pampanuto	Italy	W	Lagorthi	Greece	
Paolina	Italy	W	Montagna	Germany	
Picardan	France	W	Araignan		
Piculit Neri	Italy	R	Picolit Nero		
Prie	Italy	W	Agostenga		
Prosecco	Italy	W	Glera		
Prosecco Lungo	Italy	W	Bela Dinka	Serbia	
Räuschling	Germany	W	Raeuschling Rot	Switzerland	R
Royal de Alloza	Spain	R	Derechero		
Sauvignonasse	France	W	Friulano	Italy	
Silvaner (R)	Austria	W	Silvaner Rot		R
Termarina Rossa	Italy	G	Termarina		R
Yamabudo	Japan	R	Vitis Coignetiae Pulliat		

Technical notes

This section provides definitions of the units used in, and the various indicators generated from, the raw data in this compendium.

Definitions of unit measures

Variable	*Unit (per year)*
Grape vine bearing area	'000ha
Volume of grape production	kt
Grape yield	t/ha
Volume of wine production	ML
Volume of wine consumption	ML
Volume of wine exports and imports	ML
Value of wine exports and imports	current $USm
Unit value of wine exports and imports	current $US/L

Explanations of unit measures

Abbreviation	*Definition*	*Conversion*
ha	hectare	10,000 square metres or 2.471 acres
t	tonne	1,000 kilograms or 2,205 pounds
kt	kilotonne	1,000 tonnes
L	litre	1,000 millilitres or 0.2642 US gallons
ML	megalitre	1 million litres

Definitions of two indexes of internationalization

Two indexes of internationalization are presented in Tables 20 and 21: one of the relative importance of each country as a supplier of prime winegrape varieties to the world, and the other of the extent to which each nation's winegrape area is made up of exotic prime varieties (i.e., ones whose origin is another country). Specifically:

- The Index of Internationalization of Prime Varieties is defined for each country as the share of prime varieties originating from that nation in the global area of winegrapes divided by the share of that country in the global area of all winegrapes; and
- The Index of Internationalization of National Varietal Choice is defined for each country as the share of that nation's winegrape area made up of the rest of the world's prime varieties (i.e., ones whose origin is another country) divided by the share of prime varieties originating from all other nations in the global area of winegrapes.

Definitions of two indexes of varietal intensity

A Varietal Intensity Index has been defined by Anderson (2010) as a variety's share of a region's winegrape area divided by that variety's share of the global winegrape bearing area. This Varietal Intensity Index is thus a complement to share information in that it indicates the importance of a variety in a region not relative to other varieties in that region but rather relative to that variety in the world.

Specifically, define f_{im} as the proportion of bearing area of grape variety m in the total winegrape bearing area in region or country i such that the proportions fall between zero and one and sum to one (i.e., there is a total of M different grape varieties across the world, and $0 \leq f_{im} \leq 1$ and $\Sigma_m f_{im} = 1$). For the world as a whole, f_m is the bearing area of grape variety m as a proportion of the total global winegrape area, and $0 \leq f_m \leq 1$ and $\Sigma_m f_m = 1$. Then the Varietal Intensity Index, VII^m_i for variety m in region i, is:

$$(1) \qquad VII^m_i = f_{im}/f_m$$

This VII is simple to calculate and explain, but it is not ideal for comparing across varieties within a country, across countries or across time. For example, the maximum value of VII^m_i is the inverse of f_i, the proportion of the total global winegrape area that is in country i. That means (a) it does not vary between that country's unique varieties, and (b) small wine-producing countries with unique varieties (i.e., not grown elsewhere) will have a higher common maximum VII than will larger wine-producing countries with unique varieties. As well, it has a value of one if that variety's importance is the same in country i as in the world as a whole, but its range is just from zero to one if variety m less important in country i than globally but it can range from one to the inverse of f_i if variety m is more important in country i than globally.

These weaknesses in the VII are similar to those for all such ratio-of-shares indexes. A solution has been suggested for the revealed comparative advantage index[10] that 'normalizes' for country size. It can be adapted for present purposes as follows to generate a normalized VII, call it NVII. Consider a neutral situation in which variety m is equally important in country i as it is globally ($f_{im} = f_m$). In that case VII is one, but NVII will be zero. If variety m is more (less) important in country i than globally, NVII will be positive (negative). More than that, the extent of its divergence from zero for other varieties will be equally large for less-popular varieties as for more-popular varieties in country i, the only difference being that the former is negative while the latter is positive. To see that, define V as the global winegrape bearing area, V_i as the area in country i, V^m as the global area of variety m, and V^m_i as the area of variety m in country i. Then a neutral situation for variety m in country i is where its area equals the global area for that variety times country i's share of the total global area, that is, $V^m_i = V^m(V_i/V)$. The difference between that neutral situation and the actual situation for any variety, $V^m_i - V^m(V_i/V)$, is a measure of the intensity of that variety for country i. Normalizing that difference by dividing by V provides the definition of NVII, as follows:

$$(2) \qquad NVII^m_i = V^m_i/V - V^m(V_i/V)/V$$

The NVII's have the properties that they are in the range -1 to +1 and they are additive across varieties. In particular, their sum over all countries for a particular variety m is zero (Σ_i

[10] Yu, R., J. Cai and P. Leung (2009), 'The Normalized Revealed Comparative Advantage Index', *Annals of Regional Science* 43(1): 267–282, March.

$NVII^m_i = 0$), and their sum over all varieties for a particular country i is zero ($\Sigma_m NVII^m_i = 0$). Furthermore,

$$(3) \qquad NVII^m_i = (VII^m_i - 1)(V^m/V)(V_i/V) \equiv (VII^m_i - 1)(f_m)(f_i)$$

That is, in addition to f_{im} and f_m used to generate VII, f_i also impacts on NVII. Hence rankings using NVII in principle will differ from those using VII because the former does not include f_i, just as the rankings using VII in principle will differ from those using shares because of the inclusion of f_m in VII.

For practical reporting purposes and without loss of generality, the $NVII^m_i$ generated in this report are multiplied by 1000 to avoid the need for lots of decimals.

Definition of the Varietal National and Regional Similarity Indexes

An Index of Varietal National or Regional Similarity has been defined by Anderson (2010) to measure the extent to which the varietal mix of one region or country matches that of another region or country or the world. It can also be used to compare the varietal mix of a region or country over time. In defining the index, Anderson (2010) borrows and adapts an approach introduced by Jaffe (1986) and Griliches (1979). That approach has been used subsequently by Jaffe (1989), and by others including Alston, Norton and Pardey (1998) and Alston et al. (2010, Ch. 4), to measure inter-firm or inter-industry or inter-regional technology spillover potential.[11]

The mix of grape varieties is a form of revealed preference or judgement by vignerons about what is best to grow in their region. That judgement is affected by not only terroir but also past and present economic considerations, including current expectations about future price and yield trends and fluctuations plus the sunk cost that would be involved in grafting new varieties onto existing rootstocks or grubbing out and replacing existing varieties.

The vector of grape varietal shares defined above, $f_{im} = (f_{i1}, \ldots, f_{iM})$, locates region i in M-dimensional space. Noting that proximity is defined by the direction in which the f-vectors are pointing, but not necessarily their length, Jaffe (1989) proposes a measure called the angular separation of the vectors, which is equal to the cosine of the angle between them. If there were just two varieties, m and n, and region i had, say, 80 percent of its total vine area planted to variety m whereas only 40 percent of region j was planted to variety m, then their index of regional similarity is the cosine of the arrowed angle between the two vectors. When there are M varieties, this measure is defined as:

---

[11] Alston, J.M., M.A. Andersen, J.S. James and P.G. Pardey (2010), *Persistence Pays: U.S. Agricultural Productivity Growth and the Benefits from Public R&D Spending,* New York: Springer.
Alston, J.M., G.W. Norton and P.G. Pardey (1998), *Science under Scarcity: Principles and Practice for Agricultural Research Evaluation and Priority Setting,* London: CAB International.
Anderson, K. (2010), 'Varietal Intensities and Similarities of the World's Wine Regions', *Journal of Wine Economics* 5(2): 270-309, Winter.
Griliches, Z. (1979), 'Issues in Assessing the Contribution of R&D to Productivity Growth', *Bell Journal of Economics* 10: 92-116.
Jaffe, A.B. (1986), 'Technological Opportunity and Spillovers of R&D: Evidence from Firms' Patents Profits and Market Value', *American Economic Review* 76(5): 984-1001,
Jaffe, A.B. (1989), 'Real Effects of Academic Research', *American Economic Review* 79(5): 957-70.

$$(4) \qquad \omega_{ij} = \frac{\sum_{m=1}^{M} f_{im} f_{jm}}{\left(\sum_{m=1}^{M} f_{im}^2\right)^{1/2} \left(\sum_{m=1}^{M} f_{jm}^2\right)^{1/2}},$$

where again f_{im} is the area of plantings of grape variety m as a proportion of the total grape plantings in region i such that these proportions fall between zero and one and sum to one (i.e., there is a total of M different grape varieties across the world, and $0 \leq f_{im} \leq 1$ and $\Sigma_m f_{im} = 1$). This makes it possible to indicate the degree of varietal mix "similarity" of any pair of regions. The index also can be generated for each region relative to the average of the world's N regions, call it ω. In short, ω_{ij} measures the degree of overlap of f_i and f_j. The numerator of equation (4) will be large when i's and j's varietal mixes are very similar. The denominator normalizes the measure to be unity when f_i and f_j are identical. Hence, ω_{ij} will be zero for pairs of regions with no overlap in their grape varietal mix, and one for pairs of regions with an identical varietal mix. For cases in between those two extremes, $0 < \omega_{ij} < 1$. It is conceptually similar to a correlation coefficient. Like a correlation coefficient, it is completely symmetric in that $\omega_{ij} = \omega_{ji}$ and $\omega_{ii} = 1$. Thus the results can be summarized in a symmetric matrix with values of 1 on the diagonal, plus a vector that reports the index for each region (or nation) relative to the global varietal mix.

Definition of the Climatic National Similarity Index

Equation (4) can also be used to define a Climatic National Similarity Index, where f_{im} is the area of plantings in climate zone m (m = Cool, Temperate, Warm or Hot) as a proportion of the total winegrape plantings in country i such that these proportions fall between zero and one and sum to one (i.e., there is a total of M (in this case 4) different climate zones in each country, and $0 \leq f_{im} \leq 1$ and $\Sigma_m f_{im} = 1$). The index also can be generated for each country relative to the world's N (in this case 53) countries. The numerator of equation (4) will be large when country i's and country j's mix of climates among their wine regions are very similar. Hence, ω_{ij} will be zero for pairs of countries with no overlap in their climate mix, and one for pairs of countries with an identical mix of wine region climates. For cases in between those two extremes, $0 < \omega_{ij} < 1$. Again it is conceptually similar to a correlation coefficient and, like a correlation coefficient, it is completely symmetric in that $\omega_{ij} = \omega_{ji}$ and $\omega_{ii} = 1$. Thus the results can be summarized in a symmetric matrix with values of 1 on the diagonal, plus a vector that reports the index for each country or set of countries relative to the global winegrape-growing climate mix.

A premium climate indicator

Chart 29 reproduces what Gregory Jones has found to be the ideal GST ranges for premium winegrape production for 21 varieties. Using that information, it is possible to generate the share of the bearing area of each of those varieties in each country that are grown within that ideal GST range. We report those shares by country in Table 99 and for the world as a whole in Chart 37.

Where in the world various winegrape varieties are grown

The dramatic globalization of the world's wine markets over the past three decades (Anderson 2004; Anderson and Pinilla 2018) has brought forth many new producers and opened up new wine regions, and at the same time generated countless new wine consumers. This has added to both opportunities and competitive challenges for producers seeking to differentiate their product, or join the bandwagon of successful marketing of others, so as to attract the attention of consumers. Consumers in turn are always looking for new types and styles of wines, and more so as wines within at least the lower-priced product ranges become more homogeneous with multinationalization of both wineries and wine retailers.

One strategy for producers to attract consumer attention has been to display names of (especially popular) grape varieties on wine bottle labels. Its success, particularly for lower-priced New World wines, has led to demands in the European Union for freeing up labelling laws so as to allow such labelling there also. Meanwhile, producers in the New World are increasingly realizing the marketing value of going beyond country of origin to regional labelling as another form of product differentiation – something that has long been practiced by Europe's traditional producers, and for good reasons as outlined recently by Patterson and Buechsenstein (2018) and in recent surveys of the value of geographical indications (e.g., Ay 2020; Haeck, Meloni and Swinnen 2019; Raimondi et al. 2019).

In addition to striving to differentiate their product, producers are well aware also of the impact climate changes (higher temperatures, more extreme weather events) are having on their winegrape quality and vineyard yields. Adaptation strategies include switching to warmer-climate or more-resilient grape varieties, and re-locating to a region at a higher latitude or elevation to retain the current mix of grape varieties. Especially in the New World, where regions are still trying to identify their varietal comparative advantages and where regulations do not restrict varietal choice, winegrowers are continually on the lookout for attractive alternative varieties that do well in climates similar to what they expect theirs to become in the decades ahead. Moreover, the biotechnology revolution is providing breeders with new opportunities, which is increasing the interest in exploring traits of little-known varieties.

These marketing and climate adaptation needs are generating a rapidly growing demand for information on which winegrape varieties are grown in the world's various wine regions. Since 1971 *The World Atlas of Wine* has provided a great deal of information about where winegrapes are grown (the 8th Edition is by Johnson and Robinson 2019). That has been complemented by the seminal book called *Wine Grapes*, by Robinson, Harding and Vouillamoz (2012), which provides a detailed guide to 1368 commercially grown 'prime' varieties – and to their various synonyms, based on the latest DNA research. (The authors chose each 'prime' name according to the name used in its country or region of origin.) Neither of those books provides comprehensive global data on the bearing areas of winegrapes by region and variety. That is why we decided to fill that lacuna with the first

(2013) edition of this book, following a preliminary and partial attempt three years earlier (Anderson 2010).

Another reason for compiling a comprehensive global matrix of winegrape bearing areas by variety and region is because concern has been expressed that the diversity of winegrapes is narrowing to a few 'international' varieties. Johnson and Robinson (2013, page 8) note that vignerons are at last beginning to respond by reverting to neglected local varieties in the Old World and by exploring alternatives to the main 'international' varieties in the New World.[1] But how severe is the current concentration compared with earlier times; and how different is the concentration in the Old World compared with the New World? Answering that question requires first re-naming synonyms with their prime, to avoid overstating the degree of diversification.[2] That task is now possible, thanks to the book by Robinson, Harding and Vouillamoz (2012),[3] which in turn has been made possible by the DNA profiling of recent years that has added hugely to traditional ampelography (identification and description based on physical characteristics of the vine's appearance) as part of ampelology (grapevine science, see D'Agata 2014, p. 15).[4]

What's been added in this revised edition?

When a collection of 2015 data on winegrape varieties by region within EU countries became available recently (Eurostat 2019), we were stimulated to update our global database by compiling data for other Northern Hemisphere countries as of late 2015 and Southern Hemisphere data as of early 2016.[5] The combined dataset is referred to here as 2016, consistent with our treatment of data for our 2000 and 2010 compilations.

This provided an opportunity to revise and expand our presentation of the global time series in several other ways as well. First, we have moved to the precise spelling of varieties according to their country of origin, and listed transliterated spellings among our synonyms.

[1] As defined in the front pages under Sources of Data, the 'Old World' refers here to traditional winegrape-growing countries of Europe, the former Soviet Union, the Levant, and France's former North African colonies. All others ('New World' countries) grow winegrapes in and for newer markets and include, unusually, Norway and the United Kingdom plus Asian countries other than those of Central Asia that had been part of the Soviet Union. Prior to the World War II, the 'Old World' accounted for all but 7% of global wine production, and as recently as the 1990s its share still exceeded 80% (Anderson, Nelgen and Pinilla 2017).

[2] Some varieties are not as rare as previously believed. For example, Zinfandel is genetically identical not only to Pimitivo (in Puglia) but also to Tribidrag (in Croatia). Also identical are the two 'varieties' in Italy's Liguria region, near Genoa, of Pigato and Vermentino – which are also genetically identical to Favorita (in Italy's Piedmont) and Rolle (in southern France). Their prime name, according to Robinson et al., is Vermentino.

[3] The Vitis listing compiled by JKI (2020) for European countries provides another list of primes and synonyms. There is also a list of varieties and their 2015 bearing areas maintained by the OIV (2017, 2018) but, because OIV is an inter-governmental organization, it uses only the names adopted by each member country. Nor does it have within-country regional data.

[4] Scientific publications from that vine profiling began in South Australia in 1993 and in California at UC Davis in 1997, and have surged ahead since then. When one parent is missing, it is still possible for DNA profiling to identify parent-offspring relationships. And even when both parents are unknown, a probabilistic approach can be used to detect siblings, grandparents or grandchildren. The latter has been done for Syrah, for example: its parents were discovered to be Mondeuse Blanche and Dureza, its great grandparent is very likely Pinot (according to Vouillamoz and Grando 2006), and it is either a grandchild or a half-sibling of both Mondeuse Noire and Viognier. Undoubtedly further DNA profiling will reduce this uncertainty and add to our stock of knowledge of these and the other 10,000 or so grape varieties currently available globally.

[5] The EU's next decennial census data will not be published before the latter half of 2022, see https://ec.europa.eu/eurostat/web/agriculture/census-2020

Second, we have homogenized the spelling across the years of names of within-country winegrape-growing regions and, where the degree of regional disaggregation varied through time, we now provide a table to show the regions within each super-region so that trends in the latter are more-easily discernible. We also provide in Section VI a concordance between the names of regions in France, Italy and Spain as listed in the original national data source and the (sometimes more common) names of regions adopted in Johnson and Robinson's 8th *World Atlas of Wine*.

Third, we have introduced two new indexes of internationalization of varieties. One shows the extent to which various countries' native winegrape varieties have been adopted abroad; the other shows the extent to which each nation's varietal choice is focused on exotic (non-native) varieties.

Fourth, five additional countries have been added: Cambodia, India, Lebanon, North Macedonia and Norway. While these are all small producers of wine, they range from very low to very high Northern latitudes and so add to the diversity of climates and thus varieties.

Most importantly, this edition of the book has a new set of tables (Section VII) focused on the climate of each of the world's 800+ winegrape-growing regions. Based on the location (latitude and longitude) of the region's main town, nine climate variables have been extracted by Gregory Jones from the 1958-2019 records of its nearest weather station. His research over the past quarter-century has found that growing season average temperature (GST) is the best single indicator of viticultural relevance, and his cluster analysis has determined that the world's winegrape regions can be usefully divided into four climate classifications: cool, temperate, warm and hot. The GST ranges over which Professor Jones finds key winegrape varieties thrive in premium wine regions are shown in Chart 29. After allocating each region to one of those four classifications, we have been able to determine the weighted average GST for each super-region (e.g. Bourgogne, made up of Cote-d'Or, Nievre, Saone-et-Loire and Yonne), by using regional winegrape bearing areas as weights. Those same weights have allowed us to estimate the national and global shares of each winegrape variety's bearing area in each of our four climate classifications. We report estimates for 2000 and 2016, to see to what extent if any vignerons have re-located their vines this century to cooler regions in response to global warming.

Drawing on our newly revised, updated and expanded global database (Anderson and Nelgen 2020), we have again estimated numerous indicators that capture changes over the first two decades of this century in the varietal mix and regional distribution of the world's vineyards. It builds directly on an earlier study that examined data for circa 2000 for just 12 countries (Anderson 2010) and the first edition of this book (Anderson 2013). Those earlier publications defined two helpful indicators: a varietal intensity index, which captures the degree of each region's specialization in certain varieties; and a varietal-based regional similarity index, which captures the degree of similarity of each region's varietal mix with that of any other region (or of the nation or world). Those and several other indicators are used in the present study too.

The years chosen correspond to the agricultural census periods of the European Union, which were 1999 or 2000 and 2009 or 2010, and an exceptional semi-decadal compilation of 2015. For the non-EU countries, data have been sought for the earlier year in the Northern Hemisphere and the latter year (and 2016) in the Southern Hemisphere, so they refer to vintages that were less than 6 months apart. Inevitably not all countries or regions had data for exactly those vintages, but those exceptions account for a very small fraction of the data (see Table A in the Sources of Data section at the front of this volume).

The database on which this volume draws thus involves the three years circa 2000, 2010 and 2016 (plus data for a subset of countries and an estimate of global data for the world's most important 50 or so varieties for 1990) and covers up to 813 regions (in 53 countries), and around 1705 prime varieties and 1350 synonyms. Such a large three-dimensional database of 4.2 million cells (many of which are zeros) is difficult to digest in the form of large spreadsheets, hence the present volume's attempt to summarize the data in numerous ways including though calculating various shares and indexes.

This Introduction provides a guide to the summary charts and tables, and is structured as follow. The next two sections describe the country and varietal coverage of the database in more detail. The following section reports on a key empirical indicator derived from the share data: an index of internationalization of each variety, to highlight its spread from its country of origin. Another indicator is then employed (the index of varietal intensity) to see how the degree of concentration on certain varieties varies across countries. The focus then turns to varieties, and then to regions within countries. The regional coverage of each country is first revealed, followed by location and climate indicators of those regions. The final sections report a series of regional climactic similarity indexes, used to distinguish between regions and countries according to their overall mix of varieties and climates.

Coverage of countries

Data on bearing area of winegrapes are available by variety and region for most key wine-producing countries. In the case of the European Union countries, plantings are available from one source (Eurostat 2019), while for other countries they are typically available online from a national wine industry body or the national statistical agency. The United States and Canada are key exceptions, where data are collected at the state/provincial level and only for those with significant wine production. The raw data have been compiled, and varietal synonyms have been changed to prime varietal names, by Anderson and Nelgen (2020).

Table 1 lists the 53 countries included in the dataset and shows the number of regions and prime varieties in each country in 2000, 2010 and 2016. The relative importance of those countries in global winegrape area and global wine production is reported in Table 2, which also shows another 20 countries reported to be producing wine. The latter set collectively account for just 0.8% of global wine output. Indeed, our sample covers 99% of the world' winegrapes in all three years. A more-extensive set of key indicators of grape production in those three years is provided in Table 3.

The **vast differences between countries in their winegrape bearing areas** are depicted in Chart 1. The biggest three, Spain, France and Italy, account for just over half of the world's winegrape vineyard area. The next biggest is the United States, but its share was just 5.3% in 2016, up from 3.6% in 2000. The United States accounts for 11% of global wine production, thanks to its relatively high yields. Likewise, while Australia was 10th in terms of its share of the world's vineyard area in 2016, it was 5th in terms of wine production. More generally, Table 2 reveals that the **differences between countries are greater in global wine production volume than they are in winegrape area.** Moreover, there are huge differences in the winegrape intensity of national cropland usage. Chart 2 shows that **the shares of national cropland under winegrapes are 6-12% in the seven countries where this indicator is highest (Chile, Georgia, Portugal, Italy, Slovenia, Moldova and New Zealand),** but that share is just 0.6% in China, 0.4% in Australia and just over 0.25% in the United States (Table 3).

China is the country whose winegrape vineyard area had grown fastest during 2000-16, in both absolute and percentage terms. The next biggest rise in national hectares is in the United States, followed by Chile and New Zealand; but the next biggest in percentage terms are for three cool-climate countries: New Zealand (bearing area almost quadrupled), the United Kingdom (more than doubled) and Canada (increased 62%). Despite those increases, **the global area of winegrapes has declined by 8% over this period.** The biggest falls were in Spain (13%), France, and several countries in southeastern Europe and North Africa (Chart 3 and Table 4). That overall decline continues an earlier trend: **the global winegrape area fell 8% in the final decade of the 20th century** (Anderson 2013, 2014).

Coverage of varieties

Drilling down from total winegrape area to the area under different varieties, Table 5 lists alphabetically all the prime varieties in the dataset in 2000, 2010 and 2016, while Table 6 ranks all but the smallest of them according to their 2016 shares of global area. The data for 1990 for the top 50 varieties are in Table 7.

The extent of varietal concentration in the world's vineyard has increased non-trivially between 2000 and 2016. Half the world's plantings were accounted for by 21 varieties in 2000 but, by 2010, that total had dropped to 15 varieties and it rose only by one, to 16, in 2016 (Chart 4). This **varietal concentration is more apparent in New World countries**, where in 2010 the top seven varieties accounted for over half of all plantings, whereas 16 varieties were needed in the Old World then to get to the half-way point (Anderson 2014, Figure 2).

Those changes in varietal concentration in the world's vineyard are reflected in the marked changes in the global rankings of varieties over the period since 1990. **Cabernet Sauvignon and Merlot have more than doubled their shares to take them from 8th and 7th to 1st and 2nd places, and Tempranillo and Chardonnay have more than trebled their shares to take 3rd and 5th places, while Syrah has jumped from 36th to 6th. Sauvignon Blanc and Pinot Noir are the other two to move into the top ten.** These have all been at the expense of Airén, which has fallen from 1st to 4th, Garnacha Tinta from 2nd to 7th, and Trebbiano Toscana from 5th to 9th. These changes ensure that the chart of the world's top 40 varieties as ranked in 1990 shows a quite different mix and rank ordering to the comparable chart for 2016 (Charts 5 and 6).

The decline in varietal diversity in the world's vineyard since 1990 was due in part to the large fall in the importance of the six most-common winegrape varieties in 1990 (especially low-quality Airén and Sultaniye) plus Monastrell and Boba. Airén was planted to huge areas of the low-rainfall La Mancha region because of its drought tolerance, and was destined mostly to brandy production. But with the decline in popularity of brandy relative to wine, plus the change in regulations in 1996 to allow irrigation to be used in Spain, this variety was removed or replaced by more-profitable varieties. Apart from tiny areas in North

Africa, Spain is the only country to grow this variety. Sultaniye is grown predominantly as a table grape and for drying. In the past it was used for making low-quality wine in such places as California (where it is known as Thompson Seedless) and Australia when the relative price of winegrapes was high enough, but its use in the wine industry has all but disappeared now thanks to the greater emphasis on quality upgrading and varietal labelling on bottles. Those varieties have been dramatically superseded by the rise in importance of several others. In order of increased hectares they are Tempranillo, Cabernet Sauvignon, Syrah, Sauvignon Blanc, Chardonnay, Merlot and Pinot Noir. These changes occurred as regions sought to improve both the quality of their winegrapes and the popularity of their wines. The world's most-expanded and most-contracted varietal plantings are depicted in Charts 7 and 8.

As part of that transformation in the varietal mix, **the share of red varieties in the global winegrape bearing area has risen considerably: from 46% to 56% between 1990 and 2016,** as white's share fell from 53% to 41% and grey's share trebled to 2.4% (Chart 9). The very considerable changes in the rankings of the top 30 red and top 30 white varieties are shown in Charts 10 and 11. Details for all three colours are in Tables 8 to 11.

The share of reds is higher in the New World than in the Old World, and has risen more in the New World: from 57% in 2000 to 65% in 2016, compared with a rise from 48% to 53% in the Old World. In 2016 that share was very high in China (86%) and very low in New Zealand (22%) but even lower in Kazakhstan, Georgia, and Luxembourg (Chart 12 and Table 9). Less than one-fifth of countries have raised the white varieties' share of their bearing area (Chart 13).

Within the red and white winegrape categories, the **varietal concentration is slightly greater for reds than whites, and it has increased for reds while changing little for white winegrapes over the 2000 to 2016 period**. For reds, 80% of the world's plantings were accounted for by 28 varieties in 2000 but, by 2016, that number had dropped to 24 varieties, whereas for whites it took the top 35 varieties to account for 80% of the world's plantings in both 2000 and 2016 (Chart 14).

Internationalization of varieties

Other ways to explore the varietal diversity issue involve examining how internationalized varieties have become. One way is to look at what share of the global area is devoted to varieties by their country of origin. In 2000, French and Spanish varieties dominated the global landscape, accounting for almost three-fifths of the world's winegrape vineyard area, with Italian varieties boosting that share to 70%. By 2016 that share had risen slightly to 72%, but France now dominates much more at the expense of Spain. Specifically, **between 2000 and 2016 the global winegrape share devoted to French varieties rose from 29% to 39%, while Spain's share fell from 29% to 21%** (Chart 15).

In terms of the number of prime varieties planted globally, Italy is the dominant country of origin followed by France, Portugal and Spain. Together those 'big 4' countries contributed 67% of all the prime varieties in production in 2000, and by 2016 their share was still 56% (Chart 15). The next-ranked countries of origin in terms of number of prime varieties are the United States and Hungary, bringing the total to 70% in 2016 (Table 12). Details of the varieties involved, including their synonyms, are shown in Tables 13 to 15.

How important are the varieties from the various countries of origin in each wine-producing country is revealed by shares in Tables 16 to 18 and by the normalized varietal intensity index in Tables 19 to 21.

Tables 22 to 24 reveal for each country of origin the distribution of the globally planted area of its varieties across the various wine-producing countries. It is not surprising that the varieties of the three biggest countries of origin dominate globally, because those countries are by far the world's three biggest producers of winegrapes (Table 3).[6] Yet France differs greatly from Spain and Italy in the spread of their native varieties to other countries. Chart 16 shows there are seven other countries where French varieties dominate more than they do in their native France, whereas Charts 17 and 18 show that the vast majority of the world's vine area devoted to Spanish and Italian varieties are in their country of origin. That is even more the case for varieties originating in Portugal, with no more than 9% of their global area being outside of Portugal in the past decade (Tables 23 and 24).

Particularly striking in Chart 16 is the **high and increasing dominance of winegrapes of French origin in the New World's vineyards: that share averaged 68% in 2016, up from 59% in 2000 as Spain's share fell from 5% to 3% and Italy's remained at just 2%. French varieties' dominant and rising shares in the New World compare with a similar increase for the Old World's vineyards but from a much lower base: from 21% to 29%** (Tables 16 and 18).

Yet the winegrape variety that has expanded its share of global plantings most since 2000 is Spain's Tempranillo (followed by France's Cabernet Sauvignon, Syrah, Sauvignon Blanc and Chardonnay, see Chart 7); and this is despite Spain being the country that has lost by far the most vineyard area this century (Chart 3). **The explanation for Tempranillo's expansion is not that it has become more international, but rather that it has largely replaced Airén as Spain's most planted grape variety**: between 2000 and 2016, Spain's area of Airén shrank 184,000 ha while its area of Tempranillo increased by 114,000 ha. Chart 19 makes clear that more than 80% of the world's plantings of this variety are still in Spain, and the only other significant area is next door in Portugal.

By contrast, **Syrah is more important in Australia and South Africa than in its native France, and the next three countries where it is a significant variety are also New World countries** (Chart 20). Its importance has grown substantially in at least 15 countries, so much so that even though its shares of Australian and French vineyards were higher in 2016 than in 2000, those increases were smaller than in the rest of the world and so the Syrah's Varietal Intensity Index (its national share divided by its global share of all vineyards) has fallen for both Australia and France while rising for Argentina, Chile, the United States, Spain and Italy (Chart 21). For Tempranillo, however, its Varietal Intensity Index has increased for Spain, from 3.5 to 4.5, because all but one-tenth of the increased planting of this variety since 2000 has been in Spain and most of the rest has been in Portugal. Note, though, from the lines in Chart 21 that the normalized varietal intensity index (NVII) has risen since 2000 for Australia, in contrast to the fall in its VII. This divergence is possible because of the rising importance of Syrah in the global bearing area.[7]

---

[6] In terms of number of varieties, Portugal appears to have a large global share but that is because it has introduced a particularly detailed reporting system that by 2010 captured many of its native varieties that are planted to a small fraction of 1% of its total plantings.

[7] See equation (3) in the Technical Notes at the front of this book.

The propensity for prime varieties to spread beyond their country of origin is captured for each country of origin in Chart 22 and for each country of planting in Chart 23. Not surprisingly, **those with the smallest share of exotic varieties in their national bearing area are all Old World countries**. By contrast, in most New World countries their local varieties account for less than 10% of their vineyard area.

To normalize for the size of each country of origin, Table 25 presents an Index of Internationalization of Prime Varieties that is defined for each country as the share of prime varieties originating from that nation in the global area of winegrapes divided by the share of that country in the global area of all winegrapes. By this measure, **the varieties of three smaller countries are more internationalized than those of France (Croatia, the UK and Greece), and three other countries' varieties are almost as internationalized as those of France, namely Georgia, Austria and Germany**. Only after those seven do Spain and Italy appear in that ranking. Most of the remaining varieties are from Southeastern Europe and the countries surrounding the Black and Caspian seas (Chart 24).

A complementary Index of Internationalization of National Varietal Choice is defined for each country as the share of that nation's winegrape area made up of the rest of the world's prime varieties (i.e., ones whose origin is another country) divided by the share of prime varieties originating from all other nations in the global area of winegrapes (Table 26). This Index, which is similar to the indicator in Chart 23 except for large providers of prime varieties, indicates that the vast majority of countries are heavily dependent on exotic winegrape varieties.

Yet another indicator of internationalization of varietal choices is the proportion of all the countries in our database that are growing particular varieties. Chart 25 depicts that for the world's top 30 varieties: in the most-recent decade, **Cabernet Sauvignon and Chardonnay were grown in at least three-quarters of our sample countries, and the next three (Merlot, Sauvignon Blanc and Pinot Noir) were grown in at least two-thirds of countries**. Table 27 extends that list to the top 100 varieties and shows the shares of our total number of intra-country regions that grow those varieties. In 2016, at least one-quarter of our 629 regions for that year grew those same five varieties plus Syrah. The next-most popular by this measure are Riesling, Cabernet Franc and Pinot Gris, which are grown in at least one-fifth of regions (Chart 26).

Finally in this section, Table 28 shows trends in the share of the national area taken by the top ten varieties. While the top variety's share on average for the world has not increased since 2000 (Chart 27), the shares of the top 3 and top 10 varieties nationally have increased slightly. This adds further weight to the argument that **varietal choices are becoming more concentrated even while they are becoming more internationalized** – notwithstanding efforts to breed new varieties to overcome vine pests and diseases (Table 29). The next section focuses on this from the viewpoint of the world's top 60 varieties, rather than countries.

National winegrape areas for world's top varieties

To see which countries have been most important in the growing of the top varieties, Tables 30 to 32 rank countries for the top 24 reds, top 24 whites and top 8 grey varieties. For Cabernet Sauvignon, for example, France, Chile, the United States and Australia have been the top four producers, although China may have nudged into 4th place by 2016 (Chart 65). For Merlot, it is France, Italy and the United States that hold the top three places, with China again nudging into 4th place by 2016. Spain and Chile have risen this past decade to be next in line (Chart 72).

Leaving aside Spain's dominant variety for producing brandy (Airén), Chardonnay is the world's top white variety. The ranking of the top five nations producing this variety was unchanged between 2000 and 2016: France, the United States, Australia, Italy and Chile, with South Africa and Spain close behind in recent years (Table 31 and Chart 67). As for grey varieties, Pinot Gris dominates. Again the ranking of the top five countries this past decade remains unchanged at Italy, the United States, Germany, Australia and France, but New Zealand is now close to joining that top group (Table 32 and Chart 76).

Tables 33-41 show the area of each of the top 30 red varieties, the national shares of the global area of each of those varieties, and their share of each nation's winegrape area, while Table 42 shows the change in their national areas since 2000. Tables 43 to 52 show similar data for the world's top 30 white varieties; and Tables 53-56 likewise for the few grey varieties.

Winegrape areas and Varietal Intensity Indexes for national top 15 varieties

The Varietal Intensity Index is defined by Anderson (2010) as a variety's share of a region's total winegrape area divided by that variety's share of the global winegrape area. This index is thus a complement to national share information in that it indicates the importance of a variety in a region not just relative to other varieties in that region but also taking account of that variety's importance in the world. It also complements information on a country's share of the global area for a variety: like that share, the VII can change for a region – even if its area remains unchanged – when that variety's area in the rest of the world changes.

That complementarity is exposed in Tables 57 and 58, which show the top 15 varieties for each country ranked in order of their national share in 2000 or 2016, and alongside that is shown the country's global share of that variety and its Varietal Intensity Index. For example, France's varietal mix altered relatively little over the period 2000 to 2016, yet its VIIs altered considerably. On the one hand, the **VIIs for France's four biggest varieties of French origin (Merlot, Syrah, Chardonnay and Cabernet Sauvignon) each fell by 10% or more, in each case because bearing areas of those varieties expanded considerably in the rest of the world**. On the other hand, France's VIIs for one of its three biggest varieties of non-French origin (Garnacha Tinta) rose by about 10%, in that case because its bearing area fell much more in the rest of the world than in France. Mazuelo was the big exception: its area in France fell by 67%, compared with a fall of 63% globally, and France's total winegrape area fell by 6% between 2000 and 2016, so its VII for that variety fell (from 4.2 to 3.7).

By contrast, **the global area of each of Spain's seven biggest varieties apart from Tempranillo contracted, and so even though the Spanish areas of each of those seven also contracted, the contractions were smaller in Spain than globally and Spain's total area shrunk by one-quarter, hence its VIIs rose for almost all of them** (the exception being Garnacha Tinta, whose VII for Spain fell slightly).

Another example of global interest relates to Argentina, where Côt (main synonym: Malbec) was the country's 3rd biggest variety in 2000 but its biggest in 2016 (19.6% of the national winegrape area), when it accounted for 77% of the world's Côt plantings. Since that variety represented only 1.17% of the global area of all varieties in that year, **Argentina's Varietal Intensity Index for Côt was 16.8 in 2016. But that was slightly smaller than its VII of 17.2 in 2000, because over that period the global area of Côt doubled.** Note also that for Argentina, Côt is not even ranked in the top 15 varieties in terms of VIIs in 2016 (Table 58), because there are numerous varieties that are unique to Argentina and that therefore have the even higher VII of 21.7 (see Chart 36: when a variety is grown only in one country, its VII is necessarily the inverse of the proportion of the global winegrape area accounted for by that country – and so is identical for each unique variety in that country and year).

To further illustrate the difference between the national share of a variety and its VII, consider again the national shares and the VII's for Syrah (main synonym: Shiraz). This is the most important variety in Australia, and its share of Australia's total winegrape area rose from 22% to 29% in the period 2000 to 2016. However, Charts 20 and 21 reveal that Syrah has become more important in numerous other countries as well since 2000. Thus its share of the global vineyard area rose, from 2.1% in 2000 to 4.0% in 2016. As a result, **Australia's share of Syrah's global area has fallen from 29% to 21% (Chart 20) and so Syrah's VII for Australia has fallen from 10.7 to 7.3 over that period** (Charts 22 and 37).

The fall in the VII for Australia is not unique to Syrah. Table 57 shows that of all top 14 varieties for which there were more than 1000 hectares in Australia in 2016, there are only two whose VII has risen since 2000 (Sauvignon Blanc and Colombard). Only a small fraction of that can be explained by Australia's share of the global area becoming larger, since its share has risen only marginally over that period (from 2.6% to 2.9%). The much more important reason for the VII falling for most of the key varieties in Australia is that the country's mix of varieties is becoming more similar to the global average.

Table 59 to 66 show by country the VIIs and NVIIs for each of 48 key varieties. And in sections B and C of the Charts at the front of this book, charts are provided for each of 25 key wine-producing countries to show the share and VII of the top 20 varieties in that country, and for each of the world's top 25 varieties to show the share and VII of the top 20 countries growing that variety.

Regional coverage of each country

Interest in winegrapes and wine tend to be even more focused on regions than on whole countries. For that reason we have assembled bearing area data by variety also at the regional level wherever possible. The total area of each region's winegrapes are reported in Tables 67 to 69.

Needless to say, the average area of vineyard regions within countries vary across countries, but it also varies a lot across time for some countries because of variations in the degree of regional aggregation in each year's published data. How those regions fit into states/provinces/departments are shown for Australia, Italy and the United States in Table 70, and Table 71 shows how they concord over time between smaller and residual regions of various countries. How they concord with the regions depicted in the latest (8th) edition of the

World Atlas of Wine (Johnson and Robinson 2019) are shown for France, Italy and Spain in Tables 72 and 73. Table 74 shows the precise regions for which varietal area data are available in each of 2000, 2010 and 2016 and how the more-disaggregated regions fit into less-disaggregated ones for all the countries for which this is an issue. It also provides the page numbers in the *World Atlas of Wine* for each region. The VIIs and NVIIs for the top regions for the world's top varieties are provided in Tables 80 to 87.

Location and climate indicators of the world's wine regions

The making of wine is intimately related to the climate in which its grapes are grown. The climate of each winegrowing region is a critical determinant of the suitability of the region to particular winegrape varieties. Hence vignerons' keen interest in climate change. As climates alter and planting areas expand or contract in the various wine regions, so will the quality, styles, productivity and profitability of the wines produced there. In Section VII some key climate variables are reported for each of the 813 regions in our database, after locating (in terms of latitude and longitude plus elevation above sea level) a representative town within or adjacent to each region. This is not perfect of course, given the variability of terroir in each region, but it provides a start to understanding how different are the climates of the world's various wine regions from each other and from national and global averages. It also allows one to see the range of climates over which the various winegrape varieties are grown in each country and globally. We summarize these patterns by classifying each region as either cool, temperate, warm or hot, as suggested by Gregory Jones (see Chart 29). That enables us to estimate shares of the national and global winegrape area of each variety that are growing in each of those climate zones.

Table 75 provides the latitude and longitude of each region's representative town and its elevation above sea level plus the average growing season temperature (GST) and precipitation (GSP). (For convenience of readers, the page number of each region in the 8th edition of the Hugh Johnson/Jancis Robinson *World Atlas of Wine* also is provided in that table.) These are from TerraClimate's high-resolution (1/24°, 4-km) monthly data, kindly averaged by Gregory Jones over the period 1958-2019. Seven additional climate variables from that same source are provided in Table 76. They are the mean value for each location of the following climate variables: annual average temperature, ripening period average temperature, growing season diurnal temperature range, ripening period diurnal temperature range, growing degree days, annual precipitation, and ripening month precipitation.

Jones and many others have found that GST is the most representative single indicator of climate insofar as it affects viticulture, so that is used here to summarize regional climates by classifying regions as either cool, temperate, warm or hot. True, towns are usually warmer than surrounding vineyards, whose elevation and climate can vary considerably within each region (Jones, Reid, and Vilks 2012). However, the world has warmed over the past six decades and so most regions' GST today would be higher than the representative location's 1958-2019 average.[8]

[8] For example, the 1958-2019 average GSTs for towns representing Australian regions were compared with those recently generated for a new Climate Atlas by Remenyi et al. (2019). The latter are for the regions themselves, from gridded data for the more-recent 1997-2017 period. Instead of it being cooler than the towns' GSTs, its unweighted average was 0.4^{0}C warmer across Australia's 64 GI regions.

The GST variable suggests that **44% of the world's vineyard area in 2016 was located in hot climates and only 3% in cool climates**. The hot share is down from 51% in 2000, indicating that vine-pulls this century have been concentrated in the hottest regions, particularly in the New World (Chart 30). Even so, **the share of New World vines in cool regions rose from 0.3% to 1.1% between 2000 and 2016, while in the Old World it has risen from 2.6% to 3.2%**. The cool regions are smaller in area than warmer regions, so their share of the number of regions in the world is somewhat higher, but still only 11%. These low shares are consistent with the fact that growing winegrapes in cool regions is more risky and perhaps less profitable today – although possibly not so in the future as those historically cool regions become warmer.

Table 77 ranks the world's top 120 varieties, in terms of bearing area, by their regional area-weighted average of the GSTs in which they were grown in 2016. Chart 31 depicts the climate zonal range for the world's top 40 varieties (again in terms of bearing area), revealing that **Riesling and Pinot Noir are in the coolest areas, Merlot and Chardonnay are on average in the temperate range, Côt and Syrah (Synonyms: Malbec and Shiraz) are on average in the warm range, and Tribidrag (Synonym: Zinfandel) and Colombard are closer to the hot end of the spectrum**. Similar data are shown in Table 77 also for the Old World and for the New World, while in Table 78 these same types of shares are shown for each of the top 25 wine-producing countries. That makes it possible to spot countries where the varieties shown in Chart 29 are being grown in what may be less-than-perfect climates for premium production. Australia, for example, grows cool-climate Pinot Gris and Gerwütztraminer in climates that have area-weighted average GSTs of 19.7°C and 20.6°C, respectively, when according to Chart 29 they do best where GSTs are in the 13-15°C range. Similarly, Merlot and Syrah do best in regions where GSTs are in the 16-19°C range but in Australia their area-weighted average GSTs are 19.3°C and 19.6°C, respectively.

Conversely, Table 79 shows the distribution of each country's plantings over those four climate zones. The top 40 of those countries, in terms of bearing area, are depicted in Chart 32, again ranked by average GST from the coolest (Switzerland, Germany, Canada) to the warmest (such as Greece, Cyprus, Turkey).

Varietal Similarity Indexes

While the Varietal Intensity Index is helpful in indicating the extent of specialization of a region or country in any particular variety vis-à-vis the rest of the world, a measure of how similar or different a region's overall mix of varieties is to that of other regions or the world also can be helpful. For that purpose an index of similarity of varietal mix between regions or countries or over time has been developed. As defined and explained in the Technical Notes at the front of this volume, this Varietal Similarity Index provides an indication of how closely the shares of different varieties in the winegrape area in one location match the shares in another location or in the world, or in that same location in another time period. The closer (further away) that match, the closer the index is to one (zero). That is, the index will be zero for pairs of regions with no overlap in their winegrape varietal mix, and one for pairs of regions with an identical varietal mix. The index is conceptually similar to a correlation coefficient in that it is completely symmetric so the results can be summarized in a symmetric matrix with values of 1 on the diagonal, plus a vector that reports the index for each region relative to the global varietal mix.

Various questions can be addressed with the help of this Varietal Similarity Index (VSI), given the heterogeneity across regions and even countries in their winegrape varietal mixes. The most obvious is: how similar is each country to the global average mix of varieties? The range of national-world VSI's is quite wide (Tables 88 to 90 and Chart 33), with a handful of countries above 0.55 and another handful below 0.15. Not surprisingly, the varietal mix in France is closest to the global mix, but there have been major changes since 2000: that for France and numerous other countries are now closer to the world average (i.e., VSI closer to 1), reflecting the fact that many other countries have adopted more French varieties over the past two decades. That global move toward French varieties has also contributed to the sharp rise in the VSI for the United States – and the small drop for Spain. Australia's VSI has risen in part because so many other countries have expanded their plantings of Australia's most-popular variety, namely Syrah.

The fact that the VSI with the world rose between 2000 and 2016 for New World countries except Canada and for two of the three biggest Old World countries (Chart 33) is a further reflection of the recent increase in varietal concentration in the world's vineyard over that decade. Meanwhile, the VSIs for many of the former communist countries of the Old World (but not Serbia, Bulgaria or Moldova) have fallen substantially since 2000 as those countries continue to restructure their vineyards and move toward more-profitable (including local) varieties.

The VSI is useful also for indicating, for any one region or country, how close its varietal mix in 2016 is to what it was in 2000 (Table 91). Chart 34 lays that out for each country for which there are comparable data for the two periods. While some countries have an across-time VSI close to one (Switzerland 0.98, Austria and Romania 0.97), others are below 0.5 which reflects considerable changes in their varietal mix of bearing areas over that decade.

The main use of the VSI is in examining the extent to which a region or country has a varietal mix similar to that of other regions or countries. **In both 2000 and 2016, the New World countries have varietal mixes closest to other New World countries, whereas the varietal mixes of Old World countries are closest to one of their neighbours** (Table 92 to 94). The latter is especially the case among the countries of Eastern Europe and the former Soviet Union. This shows up in Chart 34, which ranks countries according to their VSI with the country that has the closest varietal mix to theirs in 2016: half of the first 20 countries are former communist countries of the Old World, and their closest-matched country is also from that region – as are several of their other closest-matched countries shown in Table 94. So even though those countries tend to have varietal mixes very different from the world average (they are biased toward the right-hand side of Chart 33), those mixes are very similar to each other. By contrast, **several West European countries have no other country with a similar varietal mix, notably Greece, Portugal, Spain and Italy** (Chart 35). Such varietal distinctiveness may or may not be a good thing economically, depending on how unique their terroir is and how valued their local varieties are by consumers.

There are of course considerable differences in varietal mixes between regions within each country as well. That information may be helpful for producers thinking of altering their varietal mix or re-locating to a region with a higher latitude or elevation so as to maintain their firm's current varietal mix in the wake of global warming. Tables 95 to 97 show that for some countries such as Italy, the regions with the closest mix to theirs are neighbouring Italian regions, whereas the closest matches for many French regions are in other countries.

Climatic Similarity Index

The Climatic National Similarity Index draws on the shares of each country's winegrape area that are in Cool, Temperate, Warm and Hot climate zones. As explained in the Technical Notes at the front of this book, it is designed to see how similar the climate mix of one country's wine regions is to the climate mix of other countries or of the world. Table 98 reports those indexes as between each of the 53 countries in our sample, as well as with the Old World, the New World, and the world as a whole. Chart 36 ranks the top 35 wine-producing countries' climate mixes according to their proximity to the world's average for winegrape-growing regions. It suggests the United States, Australia and Portugal are the closest to that global average, and Canada, Germany and Switzerland are the countries most distant from that average.

A premium climate indicator

Chart 29 suggests the ideal GST ranges for premium winegrape production for 21 key varieties. Using that information, Table 99 provides the share of the bearing area of each of those varieties in each country that is grown within that ideal GST range. Chart 37 summarizes those data for the world. It indicates that two Italian varieties (Dolcetto and Nebbiolo) are ranked highest by this measure. That is perhaps because all but 6-7% of the area planted to those varieties globally are in Italy. By contrast, less than 40% of Chardonnay, Pinot Gris and Tribidrag globally is growing in ideal climates.

Between 2000 and 2016, as winegrapes globally have both internationalized and become more concentrated, 12 of the 21 varieties have seen a decline in the share of their area in ideal climates. Perhaps surprisingly, however, in the New World over that same period, only 4 of the 21 varieties have seen a decline in the share of their area in ideal climates (Table 99).

Summary data for each key variety and country

To assist readers wishing to focus on a particular variety or country, Chart sections B and C at the front of this book provide, in alphabetical order, charts for the top 25 wine-producing countries and for the world's top 25 varieties. In Table 100 a complete matrix of data is provided on how many hectares of each of the world's top 300 varieties are planted in each of our 53 various countries as of 2016 (and in the Old World, New World, and world as a whole).

Final word

While this volume provides a great deal of information about which winegrapes have been grown in various regions during the first two decades of the 21st century, it leaves open the question of *why* those varieties have been produced where they are. Is it driven mainly by what grows best in each location (the terroir explanation)? Gergaud and Ginsgurgh (2008) argue that even in Bordeaux that has not been the main explanation. Is the increasing concentration on major French varieties because non-French producers – particularly in newly expanding wine-producing countries – find it easier to market them because of France's strong reputation with those varieties? Might part of the explanation be that those varieties do well in a wide range of growing environments (including in the warmest regions of New World countries where popular commercial wines are produced by huge firms), or have been found to be desirable for blending with the traditional varieties of a region? Is the increasing concentration on red varieties simply because of China's rising demand for (overwhelmingly red) imported wines? These and other centripetal forces during the first decades of this century apparently have dominated the possible centrifugal forces mentioned at the start of this Introduction. It remains to be seen whether the latter will be strong enough to dominate the former over the next decade or so. Regardless, climate changes are going to encourage producers to consider a wider range of varieties in the future. Thankfully there are thousands of varieties from which to choose, even if one only focused on Italy (D'Agata 2014), Iberia, Eastern Europe (Gilby 2018) and the Caucasus (Ketskhoveli 2012, Maghradze et al. 2012, Yesayan et al. 2018, and Serdyuk 2020).

References (including those cited in earlier pages of this book)

Abatzoglou, J., S. Dobrowski, S. Parks and K.C. Hegewisch (2018), 'TerraClimate, a High-Resolution Global Dataset of Monthly Climate and Climatic Water Balance from 1958–2015', *Scientific Data* 5: 170-191.

Alston, J.M., M.A. Andersen, J.S. James and P.G. Pardey (2010), *Persistence Pays: U.S. Agricultural Productivity Growth and the Benefits from Public R&D Spending,* New York: Springer.

Alston, J., K. Anderson and O. Sambucci (2015), 'Drifting Towards Bordeaux? The Evolving Varietal Emphasis of U.S. Wine Regions', *Journal of Wine Economics* 10(3): 349-78.

Alston, J.M., G. W. Norton and P.G. Pardey (1998), *Science under Scarcity: Principles and Practice for Agricultural Research Evaluation and Priority Setting,* London: CAB International.

Anderson, K. (ed.) (2004), *The World's Wine Markets: Globalization at Work*, London: Edward Elgar.

Anderson, K. (2010), 'Varietal Intensities and Similarities of the World's Wine Regions', *Journal of Wine Economics* 5(2): 270-309, Winter.

Anderson, K. (with the assistance of N.R. Aryal) (2013), *Which Winegrape Varieties are Grown Where? A Global Empirical Picture*, Adelaide: University of Adelaide Press.

Anderson, K. (2014), 'Changing Varietal Distinctiveness of the World's Wine Regions: Evidence from a New Global Database', *Journal of Wine Economics* 9(3): 249-72.

Anderson, K. (with the assistance of N.R. Aryal) (2015), *Growth and Cycles in Australia's Wine Industry: A Statistical Compendium, 1843 to 2013*, Adelaide: University of Adelaide Press. Also freely available as an e-book at www.adelaide.edu.au/press/titles/austwine, and in Excel format at https://economics.adelaide.edu.au/wine-economics/databases

Anderson, K. (2017), 'How Might Climate Changes and Preference Changes Affect the Competitiveness of the World's Wine Regions?' *Wine Economics and Policy* 6(2): 23-27, June.

Anderson, K and K. Harada (2018), 'How Much Wine is *Really* Produced and Consumed in China, Hong Kong and Japan?', *Journal of Wine Economics* 13(2): 199-220.

Anderson, K. and S. Nelgen (2020), *Database of Regional, National and Global Winegrape Bearing Areas by Variety, 1960 to 2016,* Wine Economics Research Centre, University of Adelaide, June. Freely available at https://economics.adelaide.edu.au/wine-economics/databases.

Anderson, K., S. Nelgen and V. Pinilla (2017), *Global Wine Markets, 1860 to 2016: A Statistical Compendium,* Adelaide: University of Adelaide Press, freely available as an e-book at www.adelaide.edu.au/press/titles/global-wine-markets and in Excel files at https://economics.adelaide.edu.au/wine-economics/databases

Anderson, K. and V. Pinilla (eds.) (2017), *Wine Globalization: A New Comparative History*, Cambridge and New York: Cambridge University Press.

Ay, J.-S. (2020), 'The Informational Content of Geographical Indications', *American Journal of Agricultural Economics* (forthcoming).

D'Agata, I. (2014), *Native Wine Grapes of Italy,* Berkeley CA: University of California Press.

Eurostat (2019), *Basic Vineyard Survey*, accessible by navigating ["Agriculture, forestry and fisheries "□"Agriculture"□"Structure of Orchards and vineyards"□ Vineyard"□"Main vine varieties by age group"] at https://appsso.eurostat.ec.europa.eu/nui/show.do?dataset=vit_t4&lang=en

Fegan, P.W. (2003), *The Vineyard Handbook: Appellations, Maps and Statistics,* revised edition, Springfield IL: Phillips Brothers, for the Chicago Wine School.

Gergaud, O. and V. Ginsgurgh (2008), 'Endowments, Production Technologies and the Quality of Wines in Bordeaux: Does Terroir Matter?', *The Economic Journal* 118: F142-57. Reprinted in *Journal of Wine Economics* 5: 3-21, 2010.

Gilby, C. (2018), *The Wines of Bulgaria, Romania and Moldova,* Oxford: Infinite Ideas Ltd.

Griliches, Z. (1979), 'Issues in Assessing the Contribution of R&D to Productivity Growth', *Bell Journal of Economics* 10: 92-116.

Haeck, C., Meloni, G. and J. Swinnen (2019), 'The Value of Terroir. A Historical Analysis of the Bordeaux and Champagne Geographical Indications,'' *Applied Economic Perspectives and Policy* 41(4): 589–619.

Hall, A. and G.V. Jones (2009), 'Effect of Potential Atmospheric Warming on Temperature-Based Indices Describing Australian Winegrape Growing Conditions', *Australian Journal of Grape and Wine Research* 15(2): 97-119, June.

Halliday, J. (2018), *Varietal Wines: A Guide to 140 Varieties Grown in Australia and their Place in the International Wine Landscape,* London: Hardie Grant Books.

Higgs, D. (2019), *Rare Ozzies: A Hundred Rare Australian Grape Varieties,* self-publised in Willimastown, Victoria (see www.vinodiversity.com/rareozzies.html).

Jaffe, A.B. (1986), 'Technological Opportunity and Spillovers of R&D: Evidence from Firms' Patents Profits and Market Value', *American Economic Review* 76(5): 984-1001,

Jaffe, A.B. (1989), 'Real Effects of Academic Research', *American Economic Review* 79(5): 957-70.

JKI (Julius Kühn-Institut) (2020). *Vitis International Variety Catalogue.* Institute for Grapevine Breeding, Federal Research Centre for Cultivated Plants, Geilweilerhof. www.vivc.de

Johnson, H. and J. Robinson (2013), *World Atlas of Wine,* 7th edition, London: Mitchell Beasley.

Johnson, H. and J. Robinson (2019), *World Atlas of Wine,* 8th edition, London: Mitchell Beasley.

Jones, G.V. (2006), 'Climate and Terroir: Impacts of Climate Variability and Change on Wine', in *Fine Wine and Terroir: The Geoscience Perspective*, edited by R.W. Macqueen and L.D. Meinert, Geoscience Canada Reprint Series Number 9, Geological Association of Canada, St. John's, Newfoundland.

Jones, G.V. (2008), 'Climate Change and the Global Wine Industry', pp. 91-98 in the *Proceedings of the 13th Australian Wine Industry Technical Conference, Adelaide, 28 July-2 August 2007*, edited by R. Blair, P. Williams and S. Pretorius, Adelaide: Australian Wine Research Institute.

Jones, G.V., R. Reid, and A. Vilks (2012), 'Climate, Grapes, and Wine: Structure and Suitability in a Variable and Changing Climate', pp. 109-133 in *The Geography of Wine: Regions, Terroir, and Techniques*, edited by P. Dougherty, New York: Springer.

Ketskhoveli, N., M. Ramishvili and D. Tabidze (2012), *Georgian Ampelopgraphy,* Tbilisi: Exclusive Print, Ltd.

Maghradze, D., L. Rustioni, A. Scienza, J. Turok and O. Failla (eds.) (2012), *Caucasus and Northern Black Sea Region Ampelography,* Geilweilerhof: Julius Kühn-Institut.

OIV (2017), *Distribution of the World's Grapevine Varieties,* Paris: Organisation Internationale de la Vigne et du Vin (International Organisation of Vine and Wine). www.oiv.org

OIV (2018), *Grapevine Varieties Area by Country,* Paris: Organisation Internationale de la Vigne et du Vin (International Organisation of Vine and Wine), March. www.oiv.org

Patterson, T. and J. Buechsenstein (2018), *Wine and Place: A Terroir Reader,* Berkeley CA: University of California Press.

Peyrefitte, E. (2020), 'Small California Wine Producers in 2019: An Industry Note' *Wine Business Case Research Journal* 4(1): 1-17.

Raimondi, V., Falco, C., Curzi, D. and A. Olper (2019), 'Trade Effects of Geographical Indication Policy: The EU Case', *Journal of Agricultural Economics* 71(2): 330–356.

Remenyi, T.A., D.A. Rollins, P.T. Love, N.O. Earl, N.L. Bindoff and R.M.B. Harris (2019), *Australia's Wine Future: A Climate Atlas*, Hobart, University of Tasmania. Accessible at https://www.wineaustralia.com/growing-making/environment-and-climate/climate-atlas

Robinson, J., J. Harding and J. Vouillamoz (2012), *Wine Grapes: A Complete Guide to 1,368 Vine Varieties, Including their Origins and Flavours,* London: Allen Lane.

Serdyuk, I. (2020), 'Russia Explores its Grape Treasures', *Meninînger's Wine Business International* 15(3): 10-11, June-July.

Tsymbliuk, K. and Y. Larina (2017), 'The Current State of the Vitiviniculture Sector in Ukraine', *Baltic Journal of Economic* Studies 3(5): 431-36.

Vouillamoz, J.F. (2020), *Swiss Grapes: History and Origin,* self-published in Sion, Switzerland and available at https://www.amazon.com/Swiss-Grapes-History-Jose-Vouillamoz/dp/1729157440/ref=sr_1_1?dchild=1&qid=1591525400&refinements=p_27%3AJose+F+Vouillamoz&s=books&sr=1-1&text=Jose+F+Vouillamoz

Vouillamoz, J.F. and M.S. Grando (2006), 'Genealogy of Wine Grape Cultivars: Pinot is Related to Syrah', *Heredity* 97(2): 102-10.

Yesayan, A.A., A.A. Bobokhyan, M.V. Dallakyan, S.G. Hobosyan, B.Z. Gasparyan and N.A. Hovhannisyan (2018), *Armenian Vines and Wines,* Bonn: GIZ.

Yu, R., J. Cai and P. Leung (2009), "The Normalized Revealed Comparative Advantage Index", *Annals of Regional Science* 43(1): 267–282, March.

Charts

A. Overview

World's winegrape varieties at a glance

1. National shares of global winegrape area, 2000 and 2016 (%)

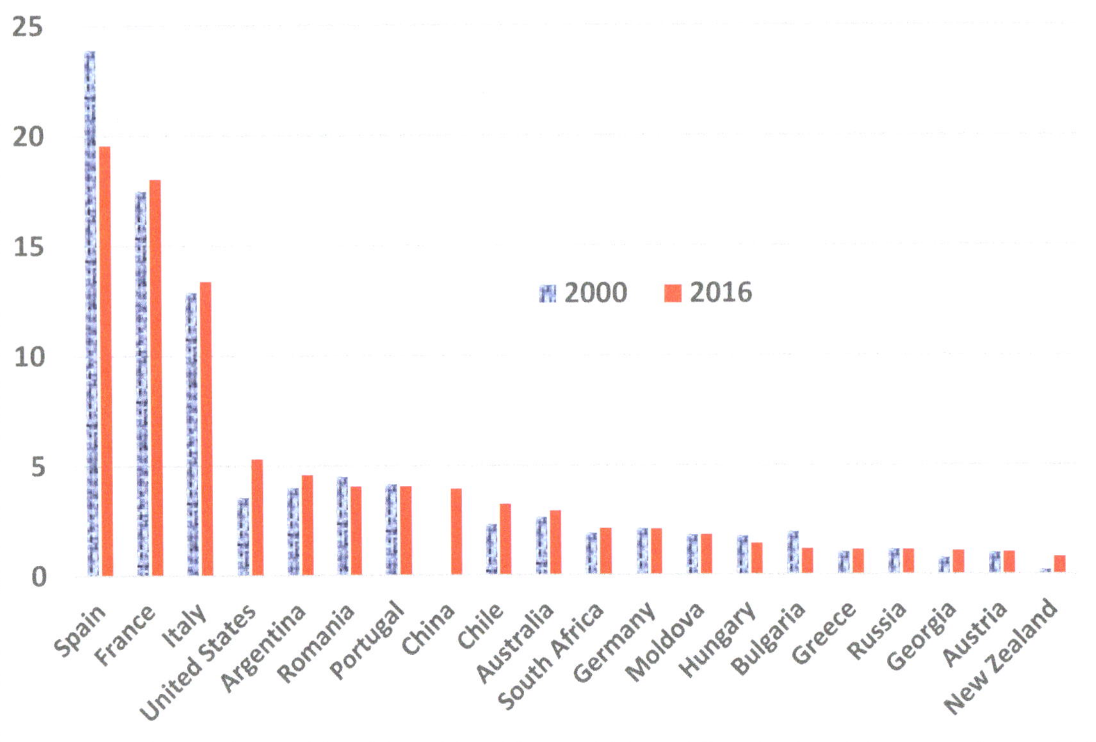

2. Share of national agricultural crop area under winegrapes, 2016 (%)

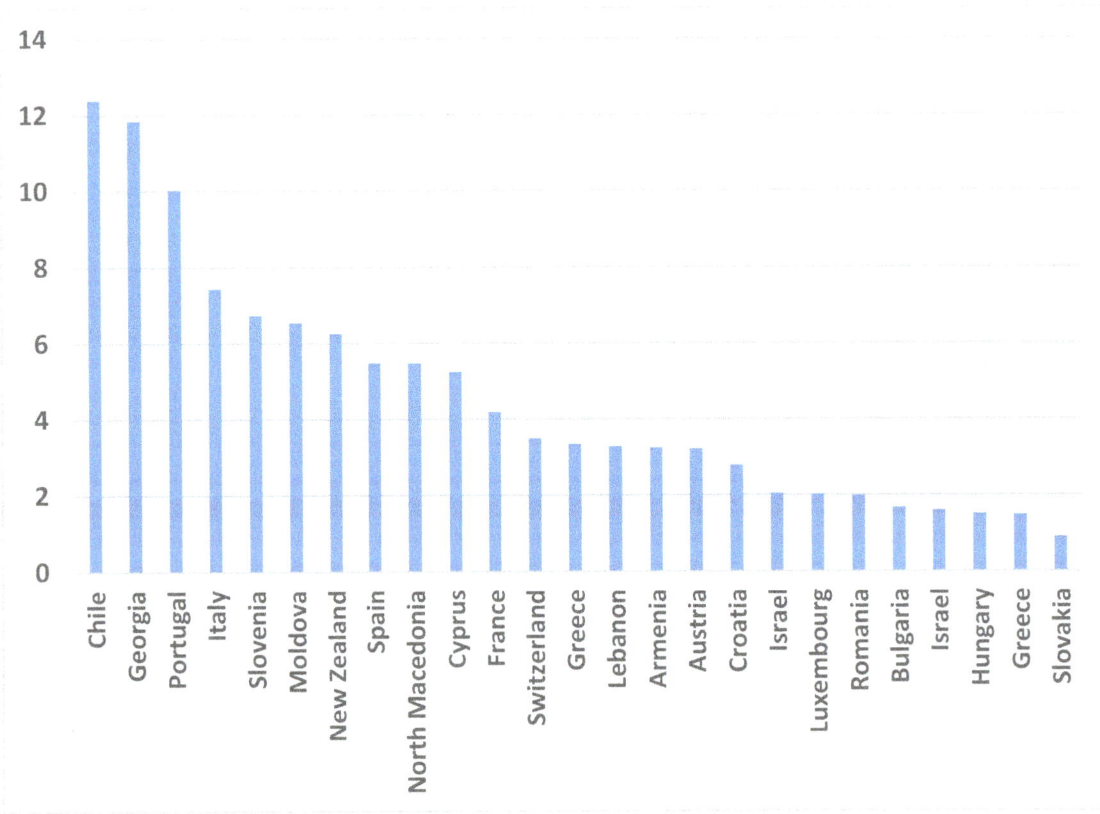

3. Largest increases and decreases in national winegrape bearing area, 2000* to 2016 ('000 ha)

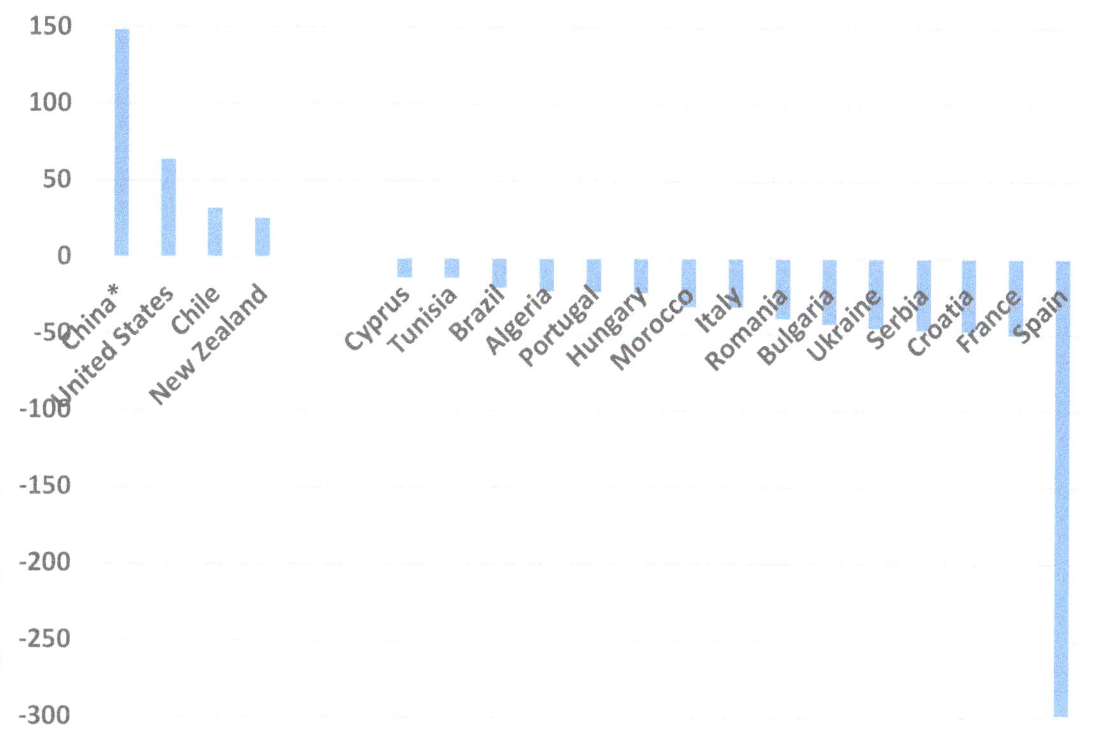

*For China, the increase shown is from 2010 to 2016, since the 2000 area is unknown

4. Cumulative varietal shares of global winegrape area, 2000 and 2016 (%)

5. World's top 40 varieties in 1990 compared with 2000 and 2016 ('000 ha)

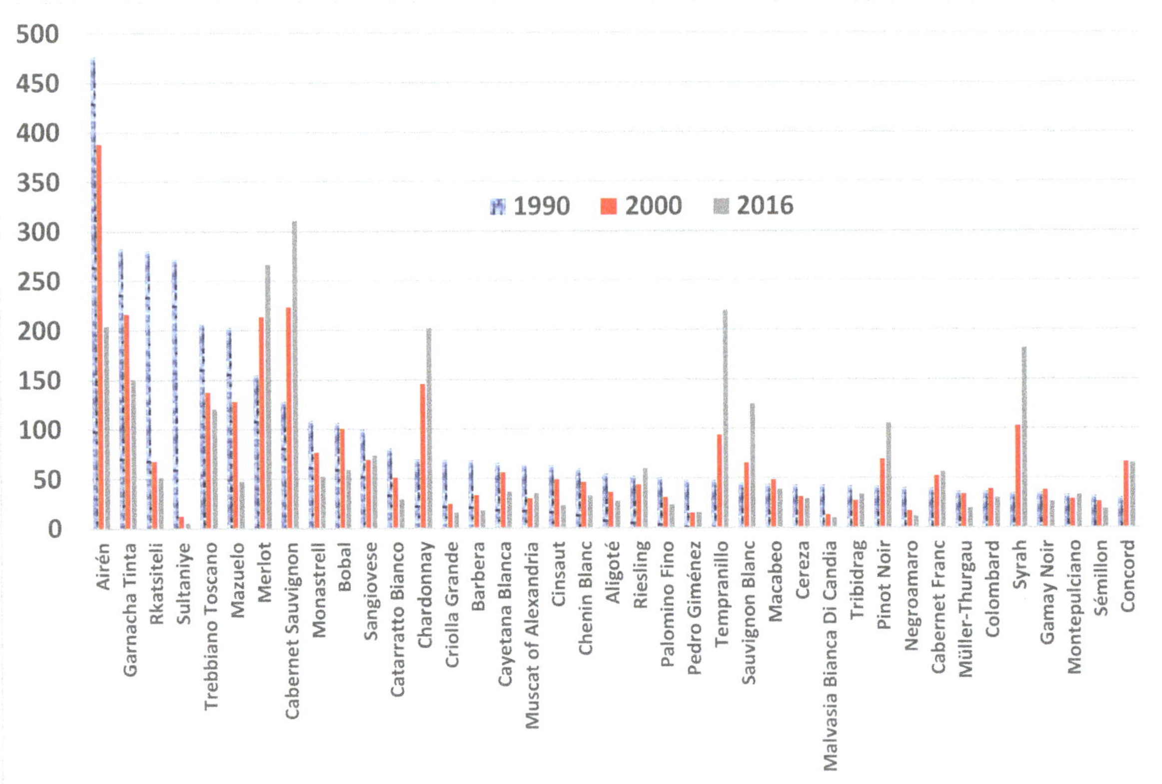

6. World's top 40 varieties in 2016 compared with 1990 and 2000 ('000 ha)

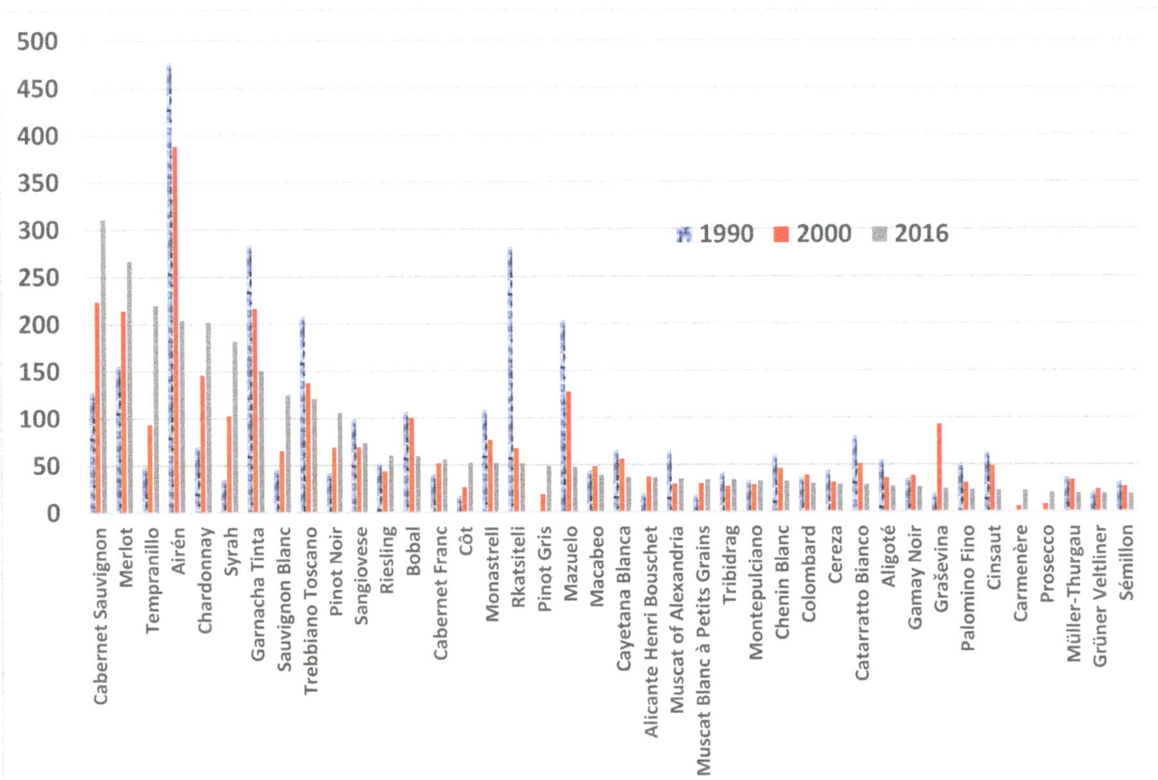

7. **World's 30 winegrape varieties with most-expanded areas, 2000 to 2016 ('000 ha)**

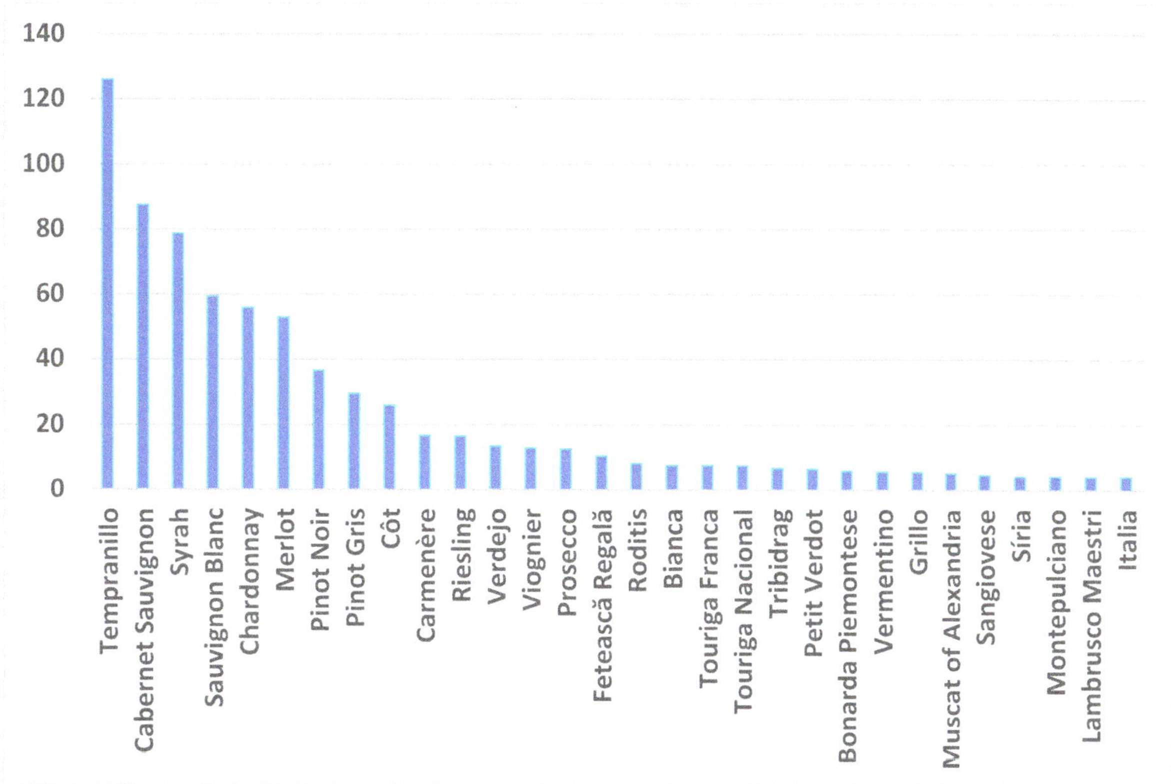

8. **World's 30 winegrape varieties with most-contracted areas, 2000 to 2016 ('000 ha)**

9. **Red, white and grey varietal shares of global winegrape area, 1990, 2000 and 2016 (%)**

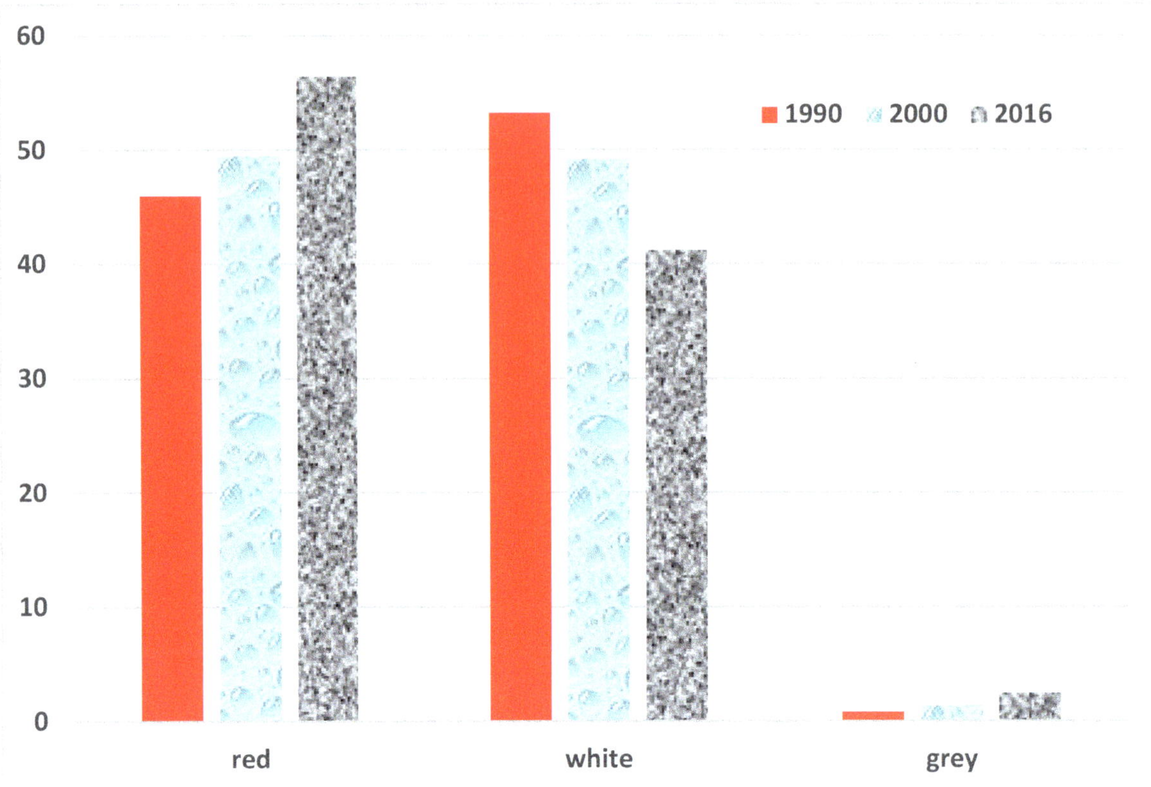

10. **Top 30 red varieties' shares of global wine area, 2000 and 2016 (%)**

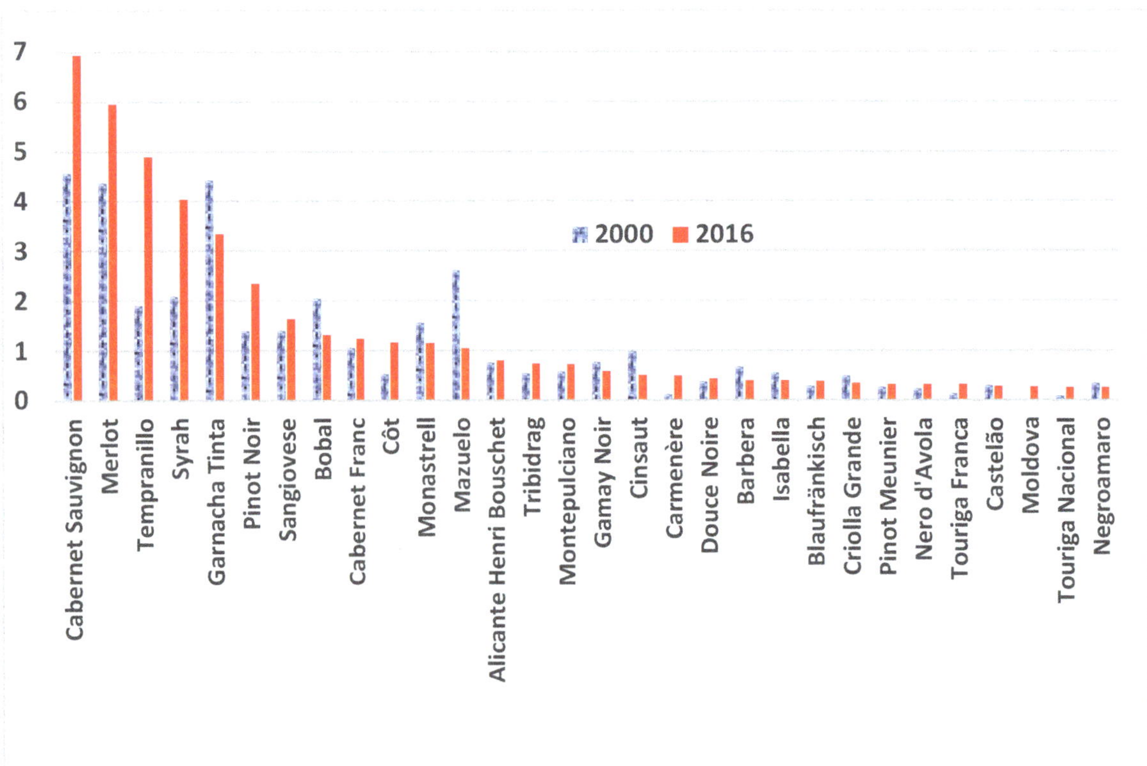

11. Top 30 white varieties' shares of global wine area, 2000 and 2016 (%)

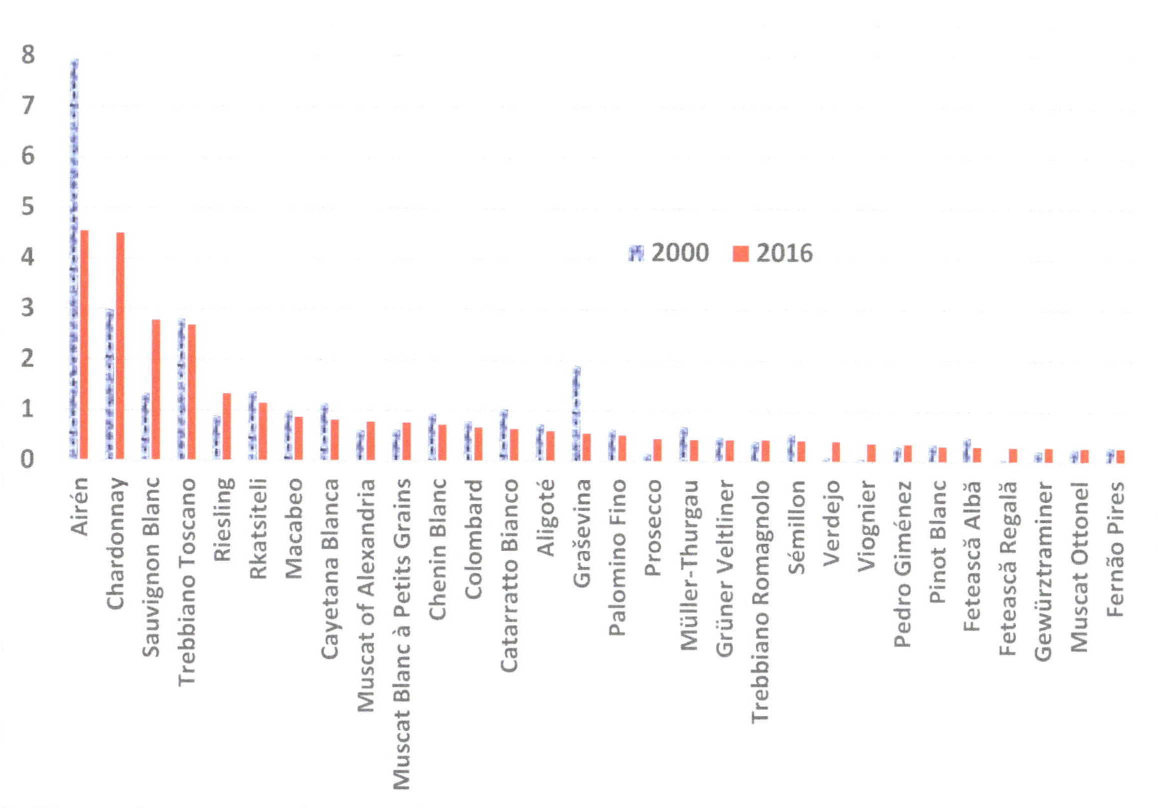

12. Share of red varieties in national winegrape area, 2000 and 2016

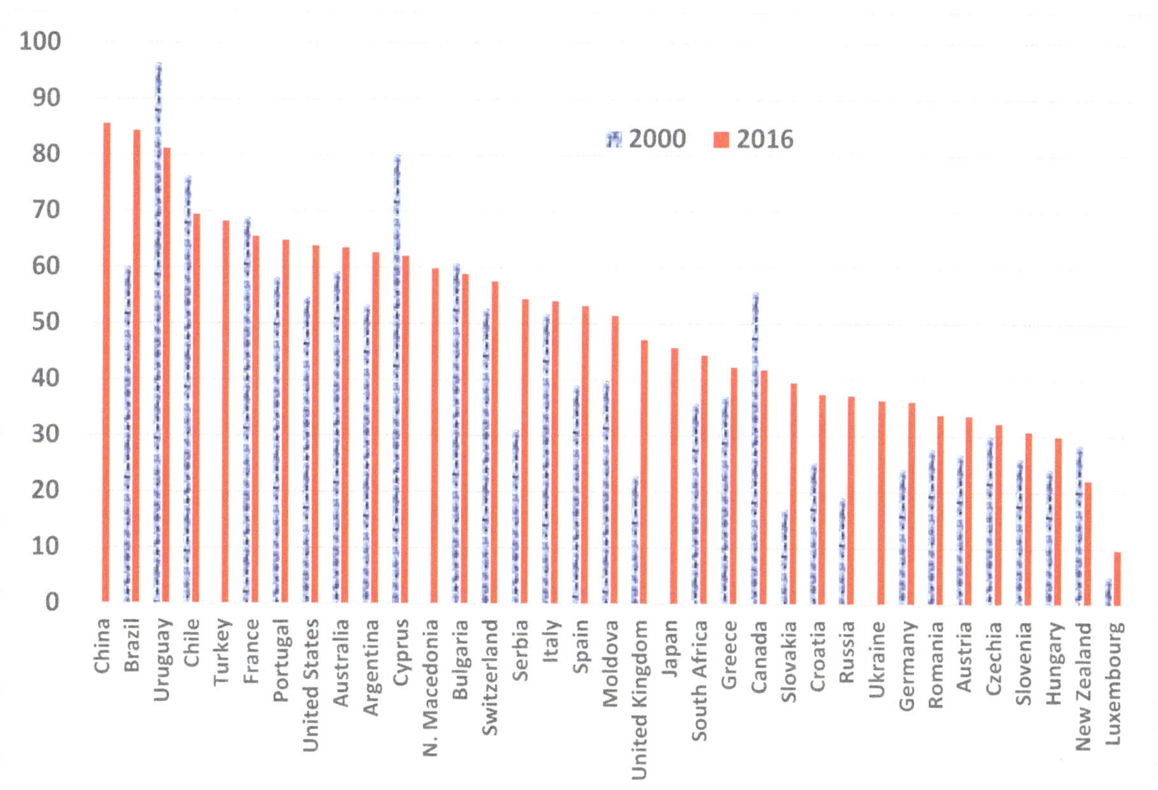

13. Percentage point changes in shares of red and white varieties in national winegrape area, 2000 to 2016

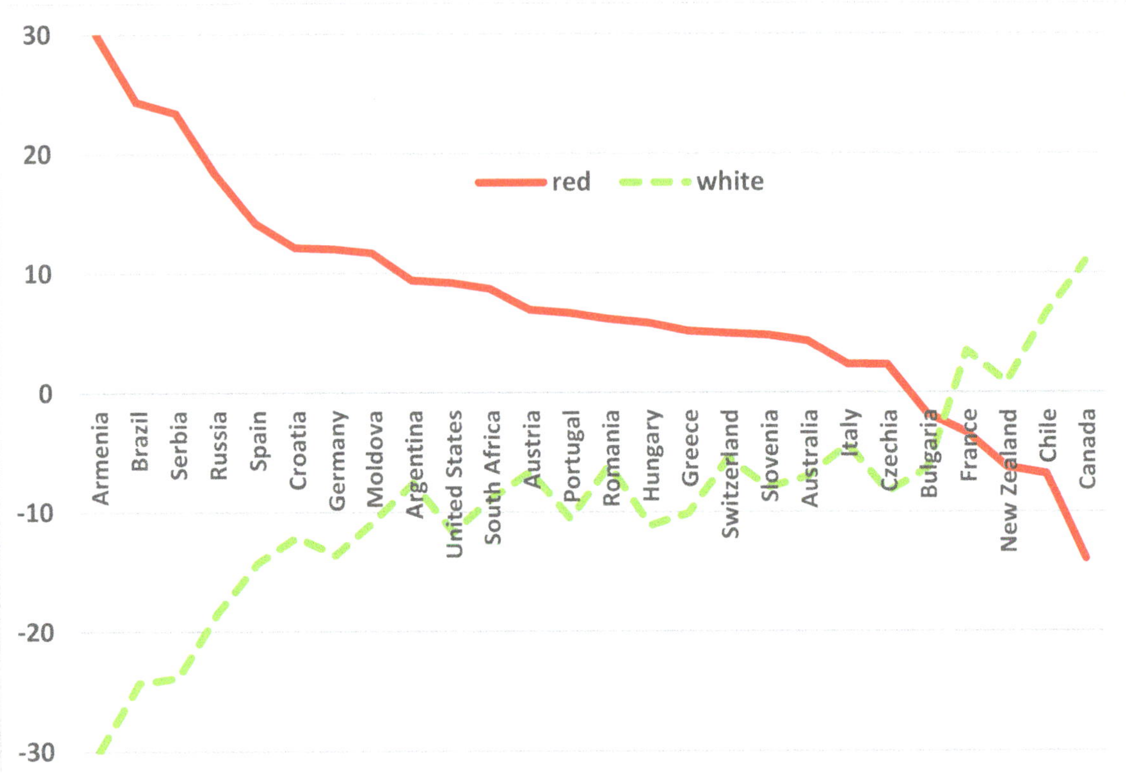

14. Cumulative varietal shares of world's red and white winegrape areas, 2000 and 2016 (%)

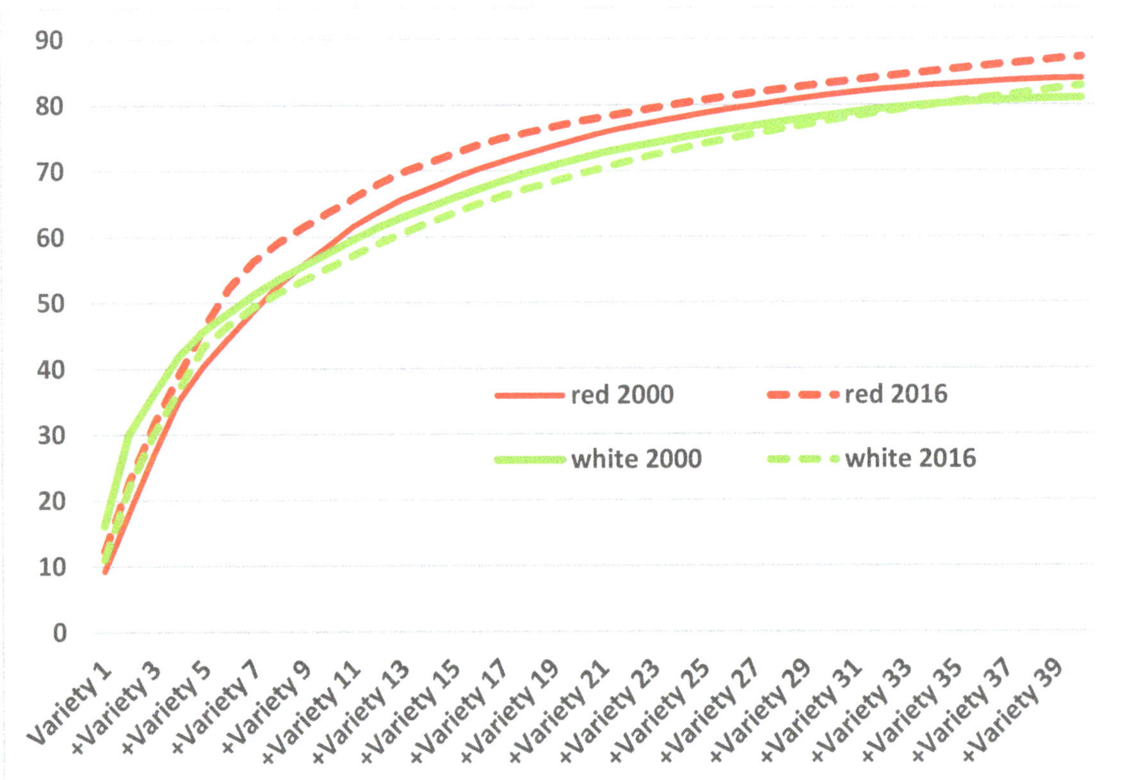

15. Number of prime varieties and their share of global bearing area, by country of origin of primes, 2000 and 2016 (# and %)

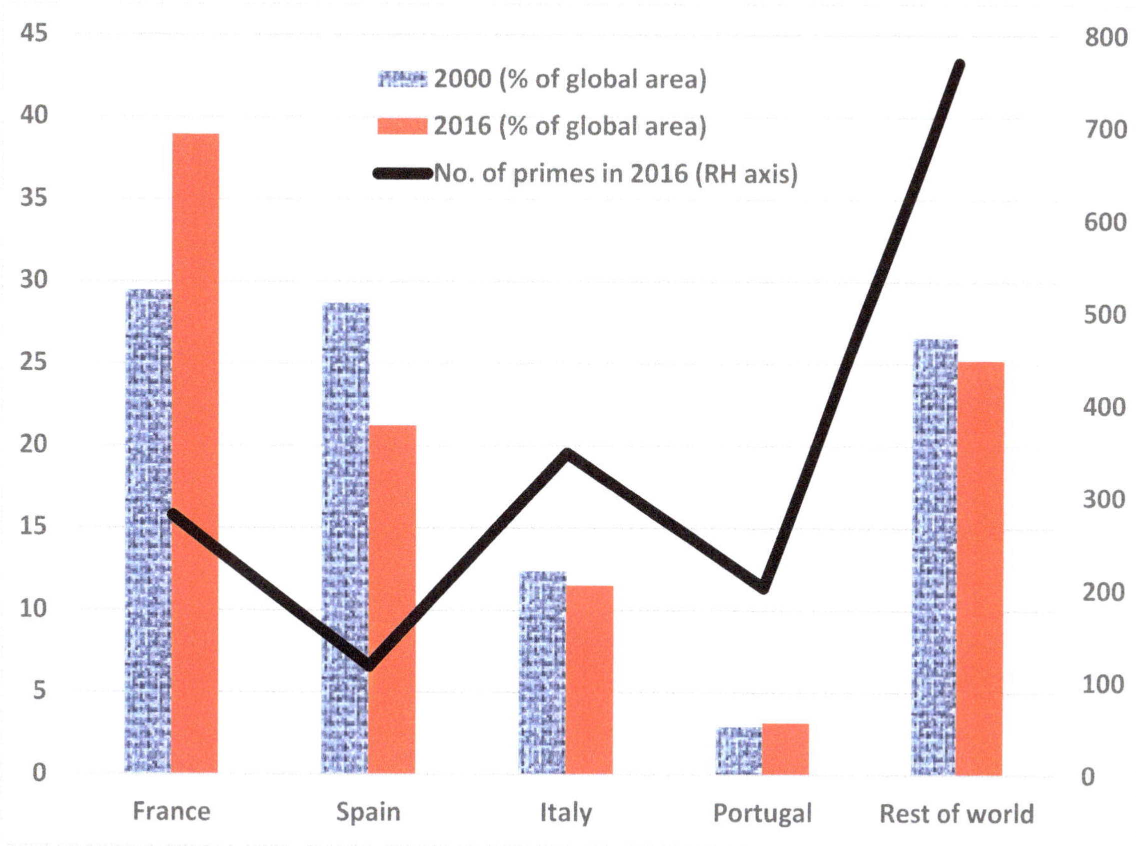

16. Shares of French varieties in national winegrape areas, 2000 and 2016 (%)

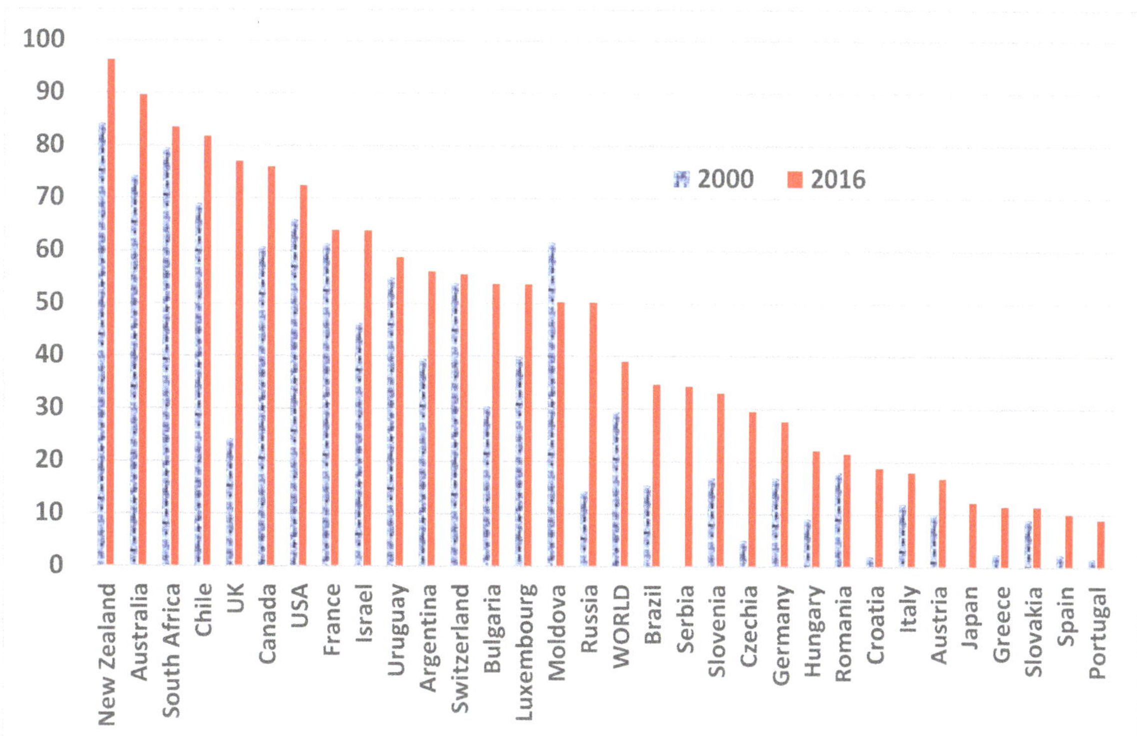

17. Shares of Spanish varieties in national winegrape areas, 2000 and 2016 (%)

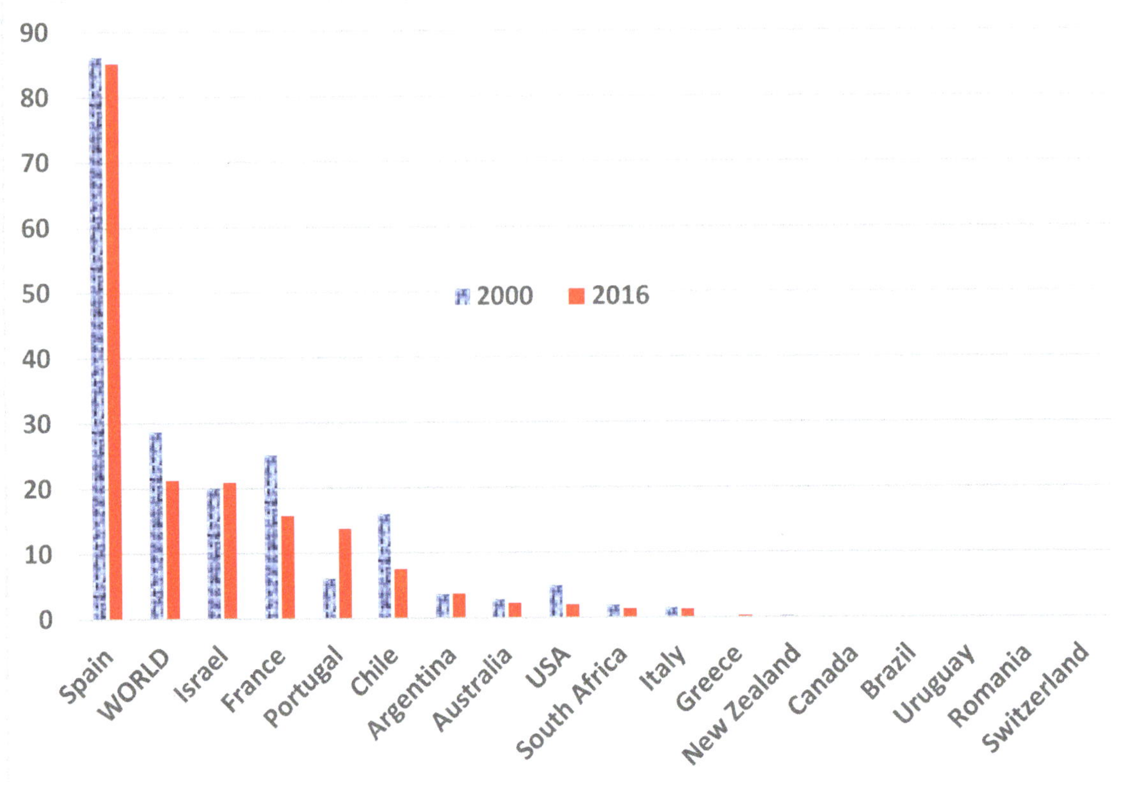

18. Shares of Italian varieties in national winegrape areas, 2000 and 2016 (%)

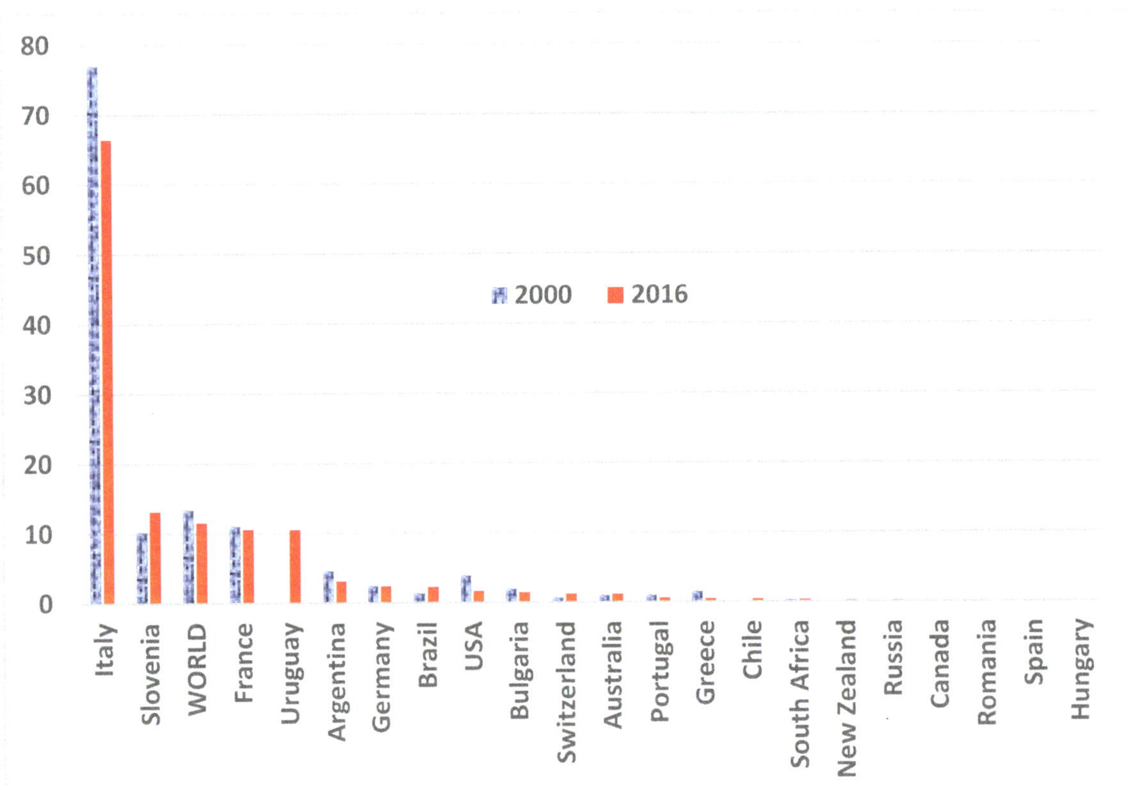

19. **Shares of Tempranillo in national winegrape area and national shares of global Tempranillo area, 2000 and 2016 (%)**

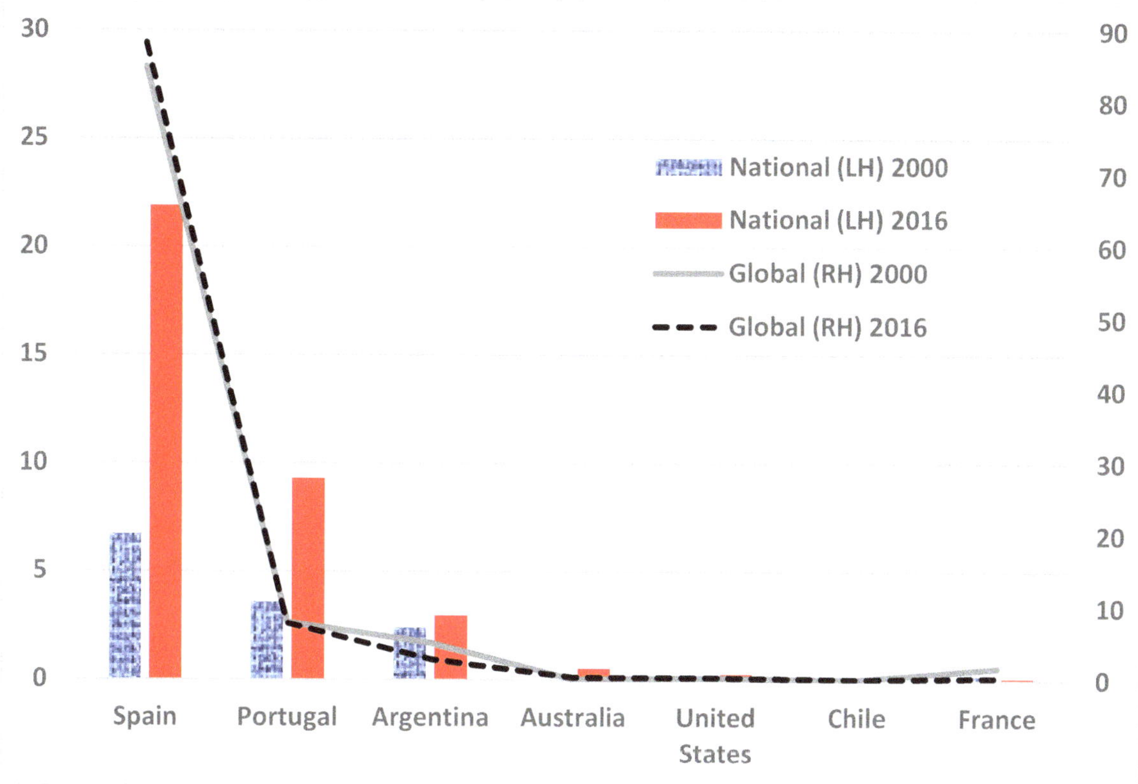

20. **Shares of Syrah in national winegrape area and national shares of global Syrah area, 2000 and 2016 (%)**

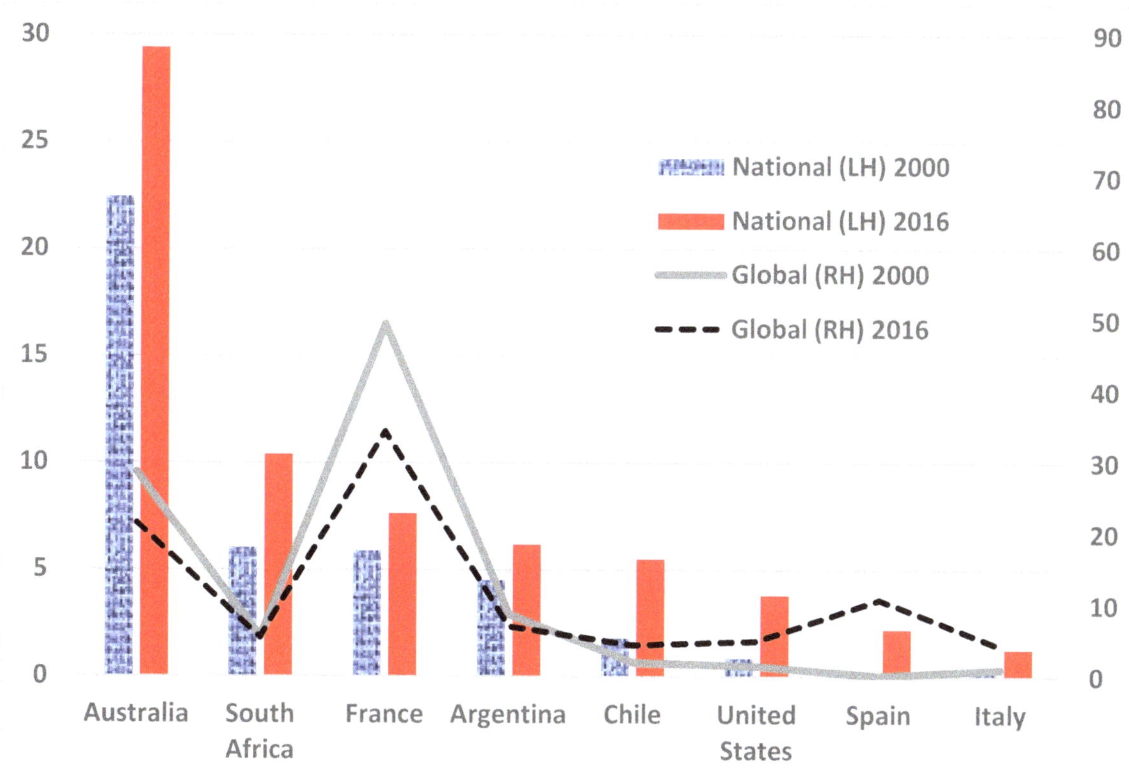

21. Varietal Intensity Indexes (VII and NVII), Syrah, key producing countries, 2000 and 2016

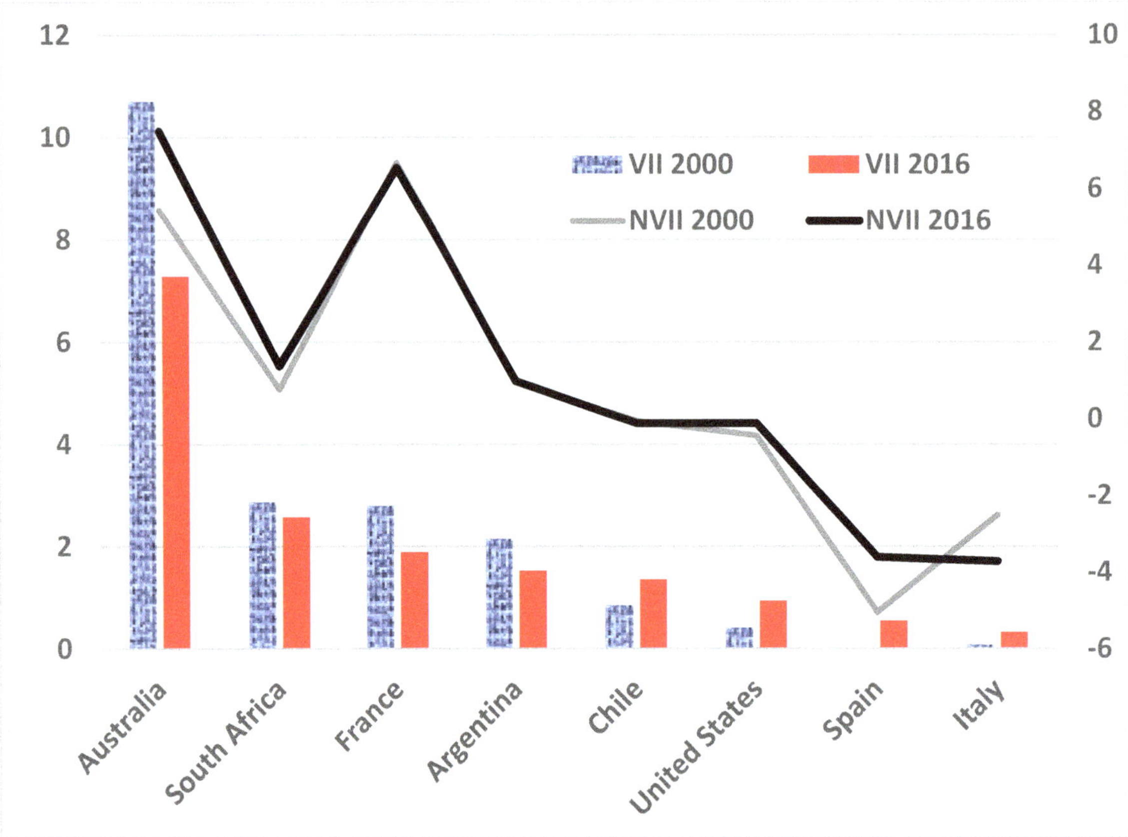

22. Share of global bearing area of prime varieties that is outside the country of origin, by country of origin, 2000 and 2016 (%)

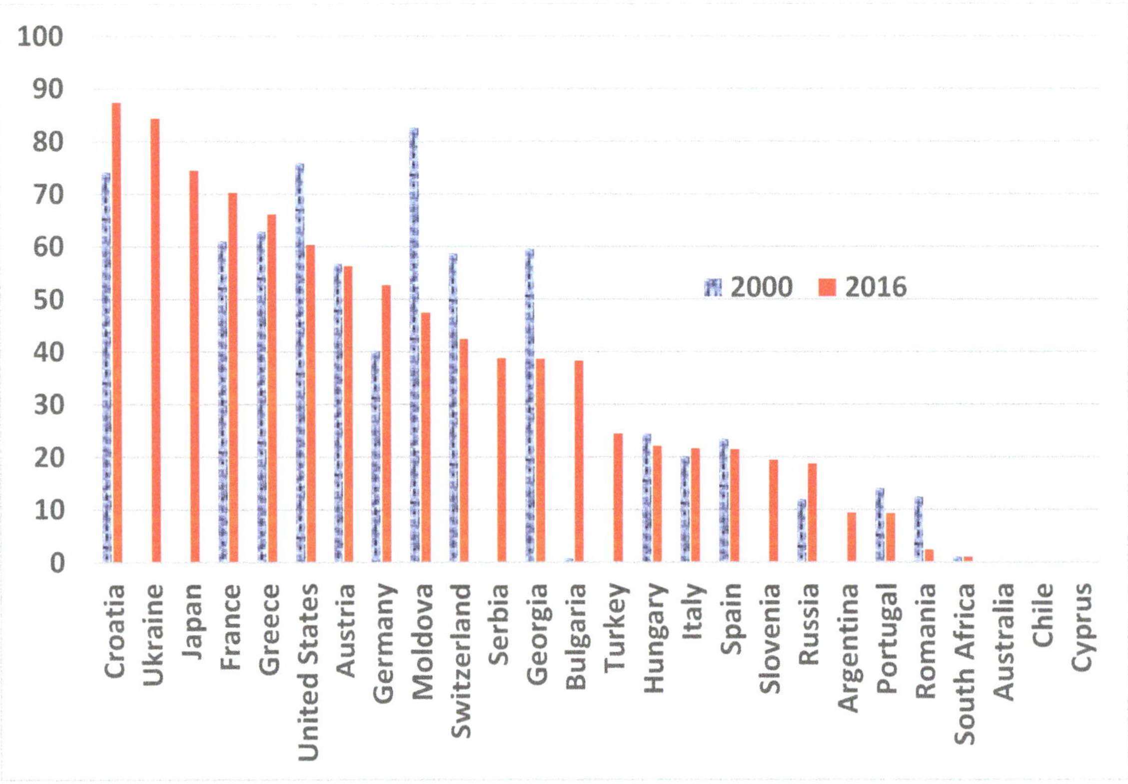

23. Share of national bearing area that is own countries' prime varieties, by country of planting,[a] 2000 and 2016 (%)

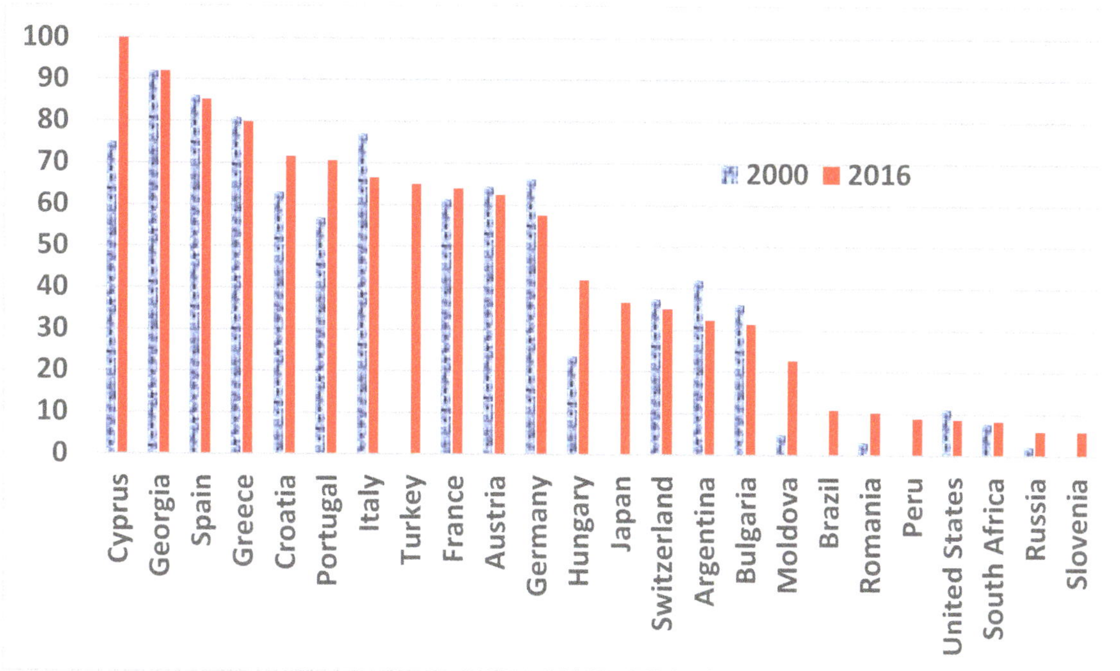

[a] All other countries are <5%.

24. Index of internationalization of prime varieties, by country of origin, 2000 and 2016

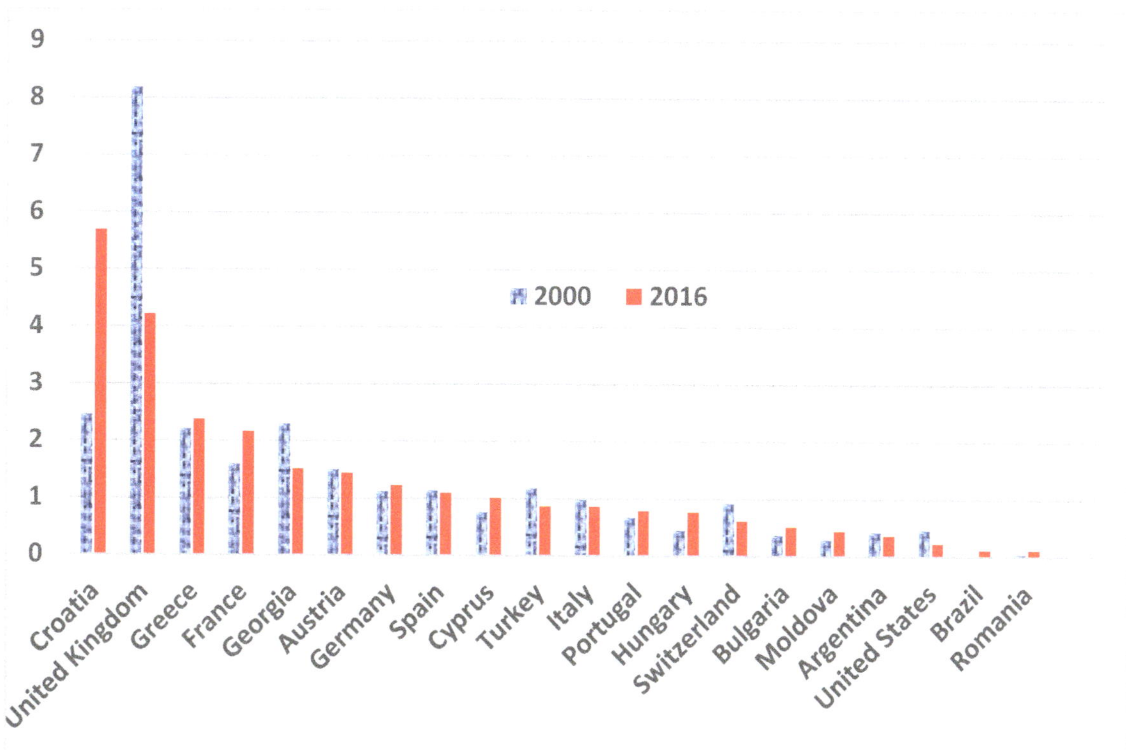

25. Share of database's countries growing top 30 varieties, 2000, 2010 and 2016 (%)

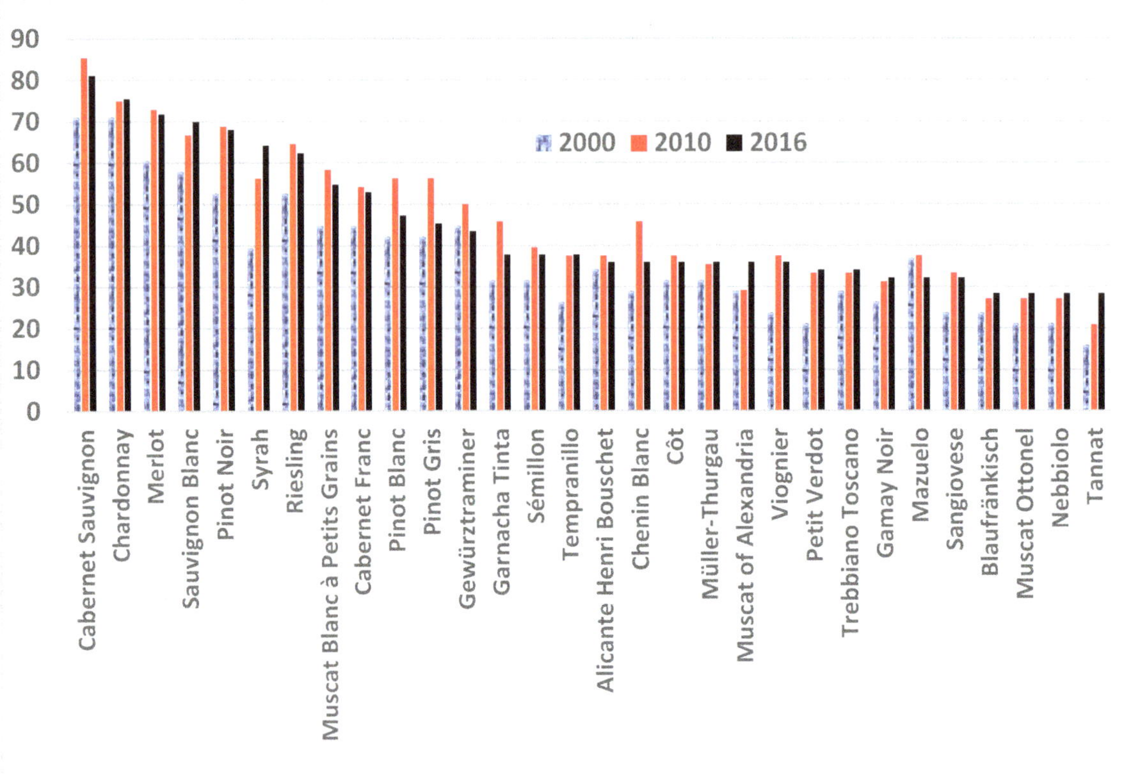

26. Share of database's regions growing top 30 varieties, 2000, 2010 and 2016 (%)

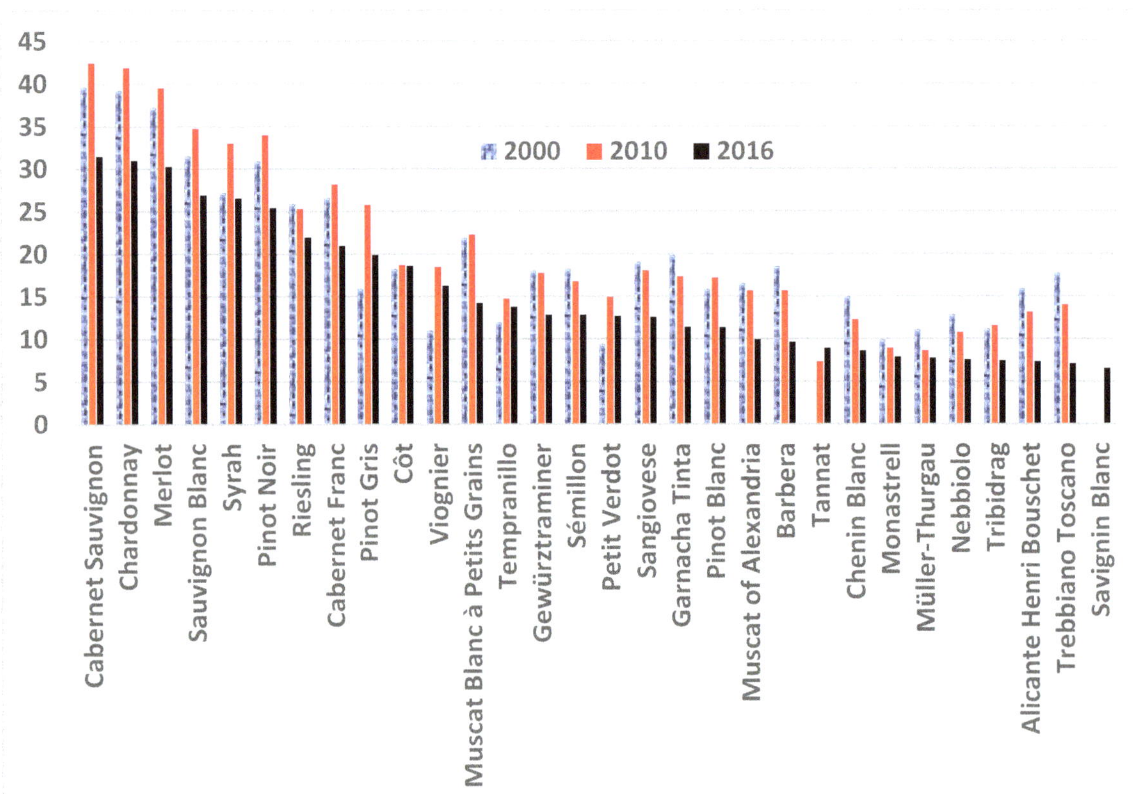

27. Share of top variety in national winegrape area, 2000 and 2016 (%)

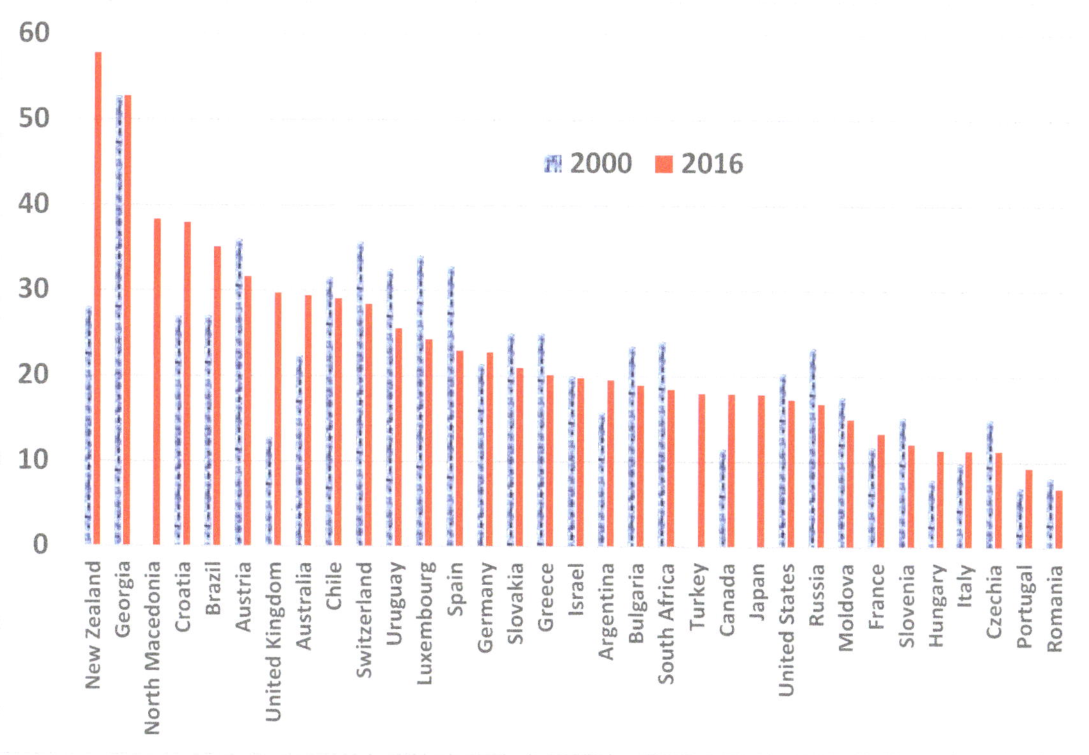

28. Highest national NVII[a] and corresponding VII for each of world's top 15 varieties, 2016

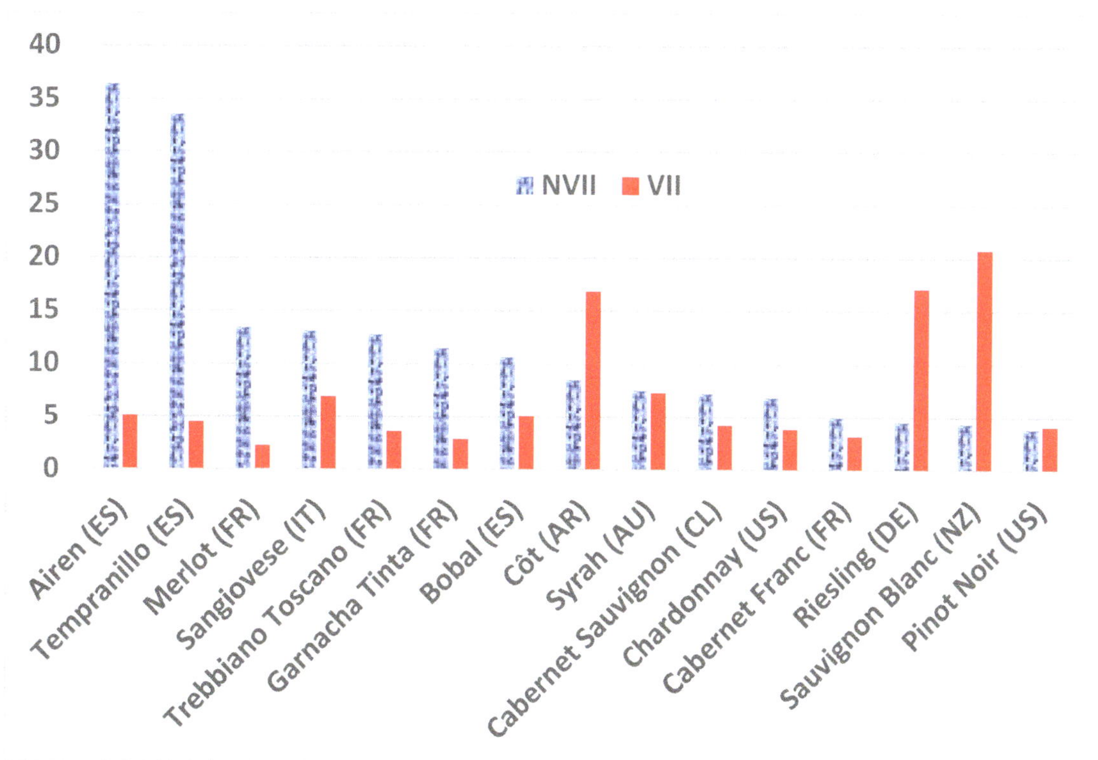

[a] NVII has been multiplied by 1000

29. Average growing season ripening temperature ranges for key winegrape varieties in premium regions (°C and °F)

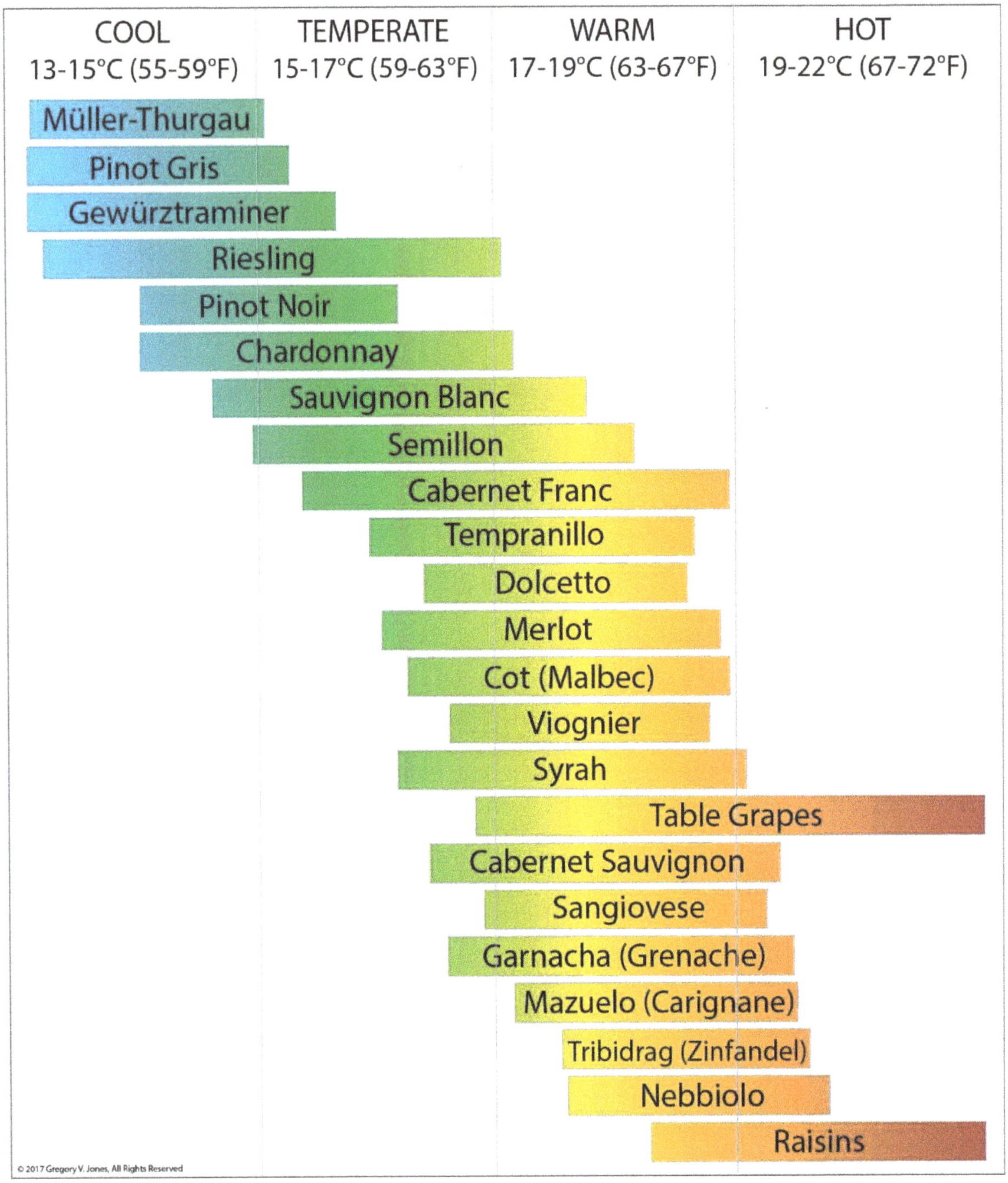

Note: Northern Hemisphere April-October, Southern Hemisphere October-April. Horizontal bars represent the range of temperatures that each variety is known to ripen and produce premium-quality wine in the world's benchmark regions. Adjustments may occur as more research and data become available, but changes of more than +/-0.5°C for any variety are highly unlikely. The figure is used with permission by its author, Professor Gregory V. Jones.

Sources: Jones, G.V. (2006), "Climate and Terroir: Impacts of Climate Variability and Change on Wine", in *Fine Wine and Terroir: The Geoscience Perspective*, edited by R.W. Macqueen and L.D. Meinert, Geoscience Canada Reprint Series Number 9, Geological Association of Canada, St. John's, Newfoundland; and Jones, G.V., R. Reid and A. Vilks (2012), "Climate, Grapes, and Wine: Structure and Suitability in a Variable and Changing Climate", pp. 109-133 in *The Geography of Wine: Regions, Terrior, and Techniques*, edited by P. Dougherty, New York: Springer.

30. Shares of winegrape area in cool, temperate, warm and hot climate regions, New World, Old World and World, 2000 and 2016 (%)

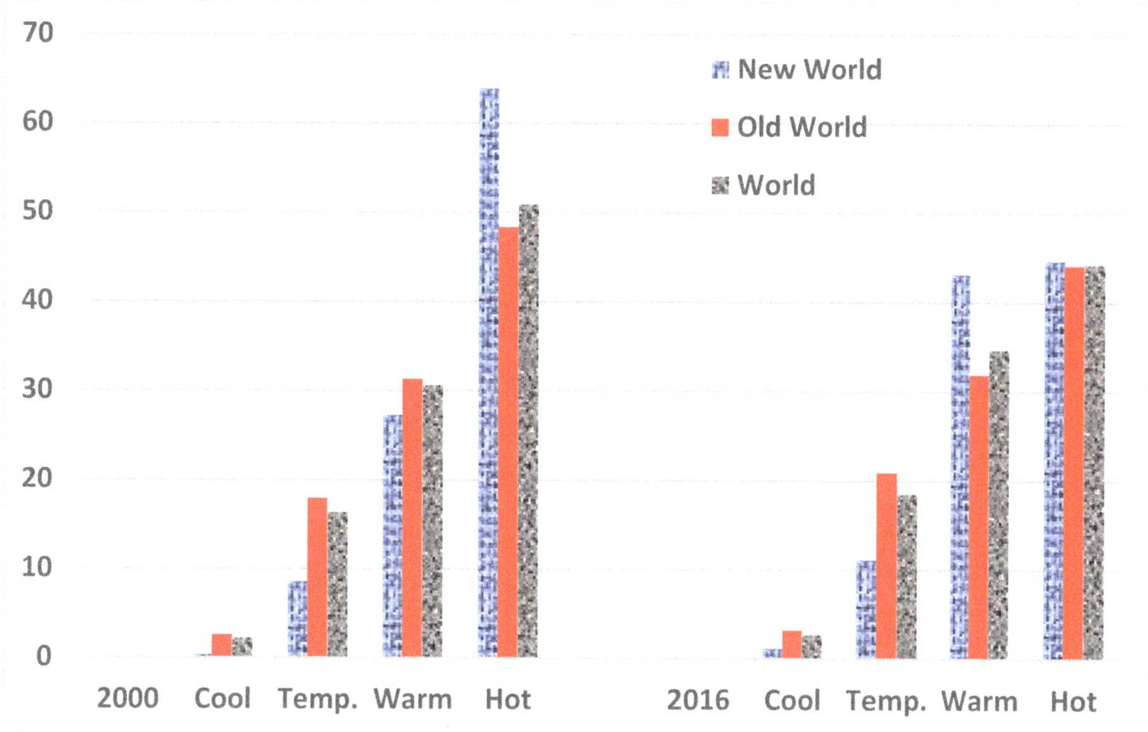

31. Shares of winegrape area in cool, temperate, warm and hot climate regions, top 40 varieties and all countries, 2016 (%)

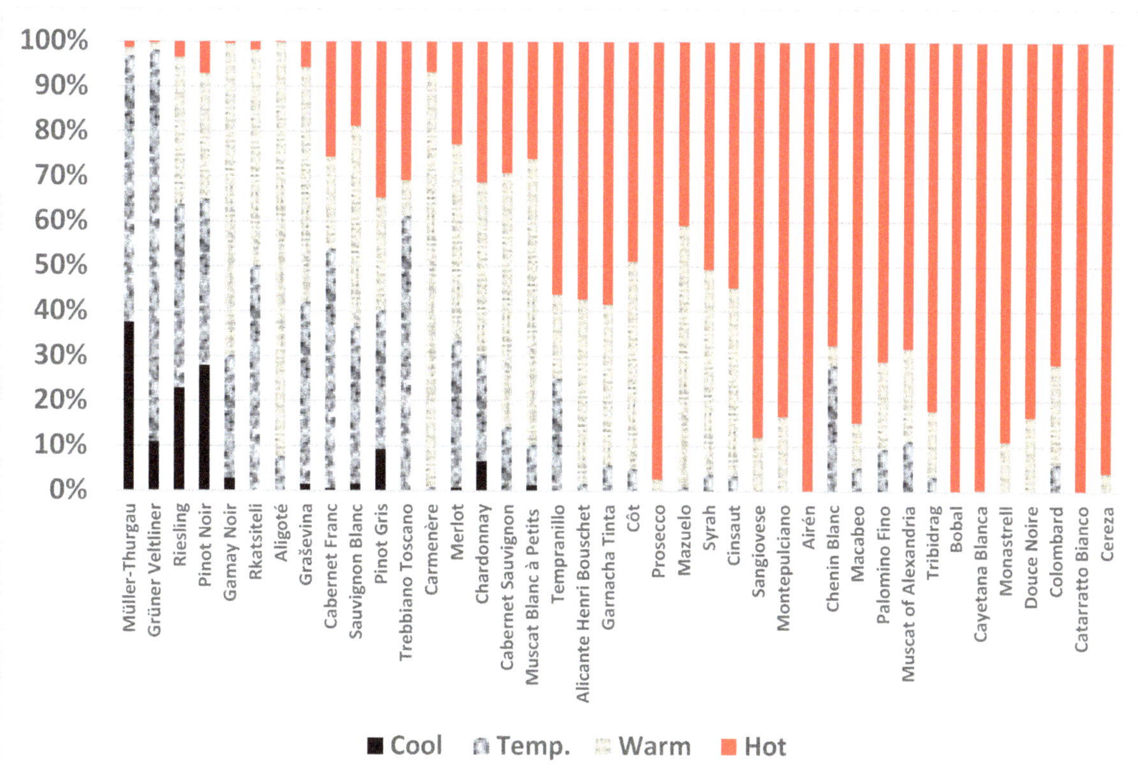

32. Shares of winegrape area in cool, temperate, warm and hot climate regions, top 40 countries and all varieties, 2016 (%)

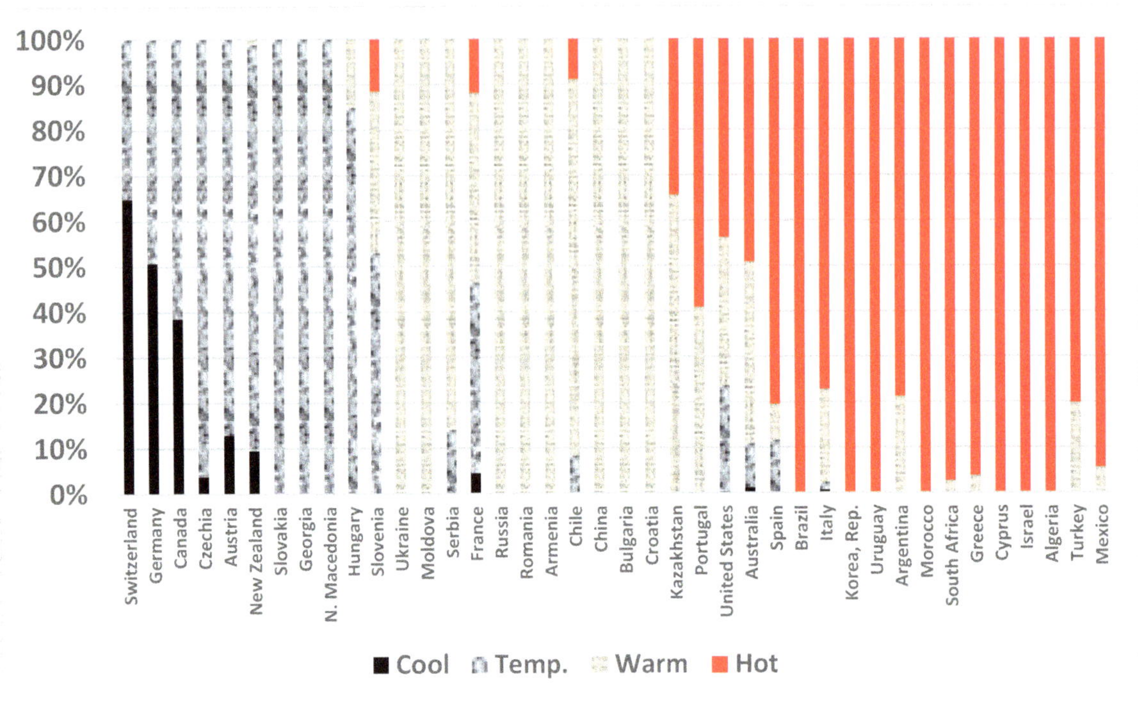

33. Index of Varietal Similarity of each country with the world, 2000 and 2016

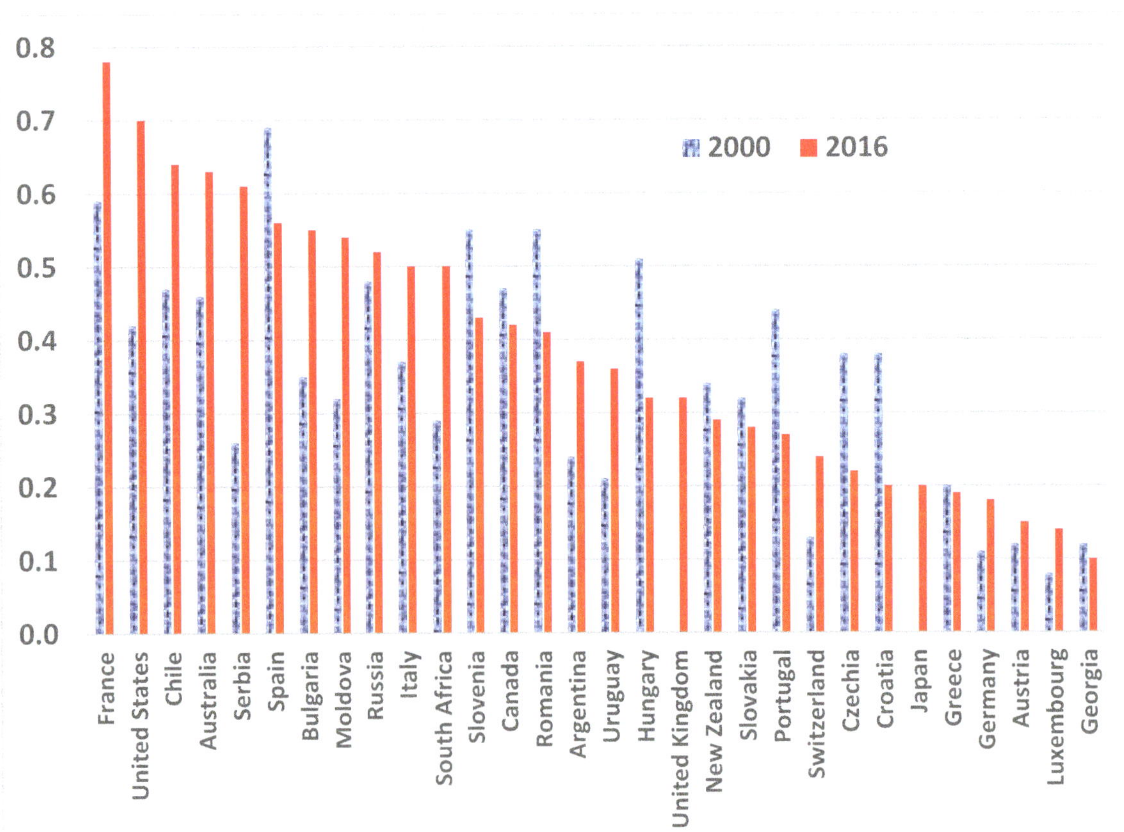

34. Index of Varietal Similarity between 2000 and 2016 for each country

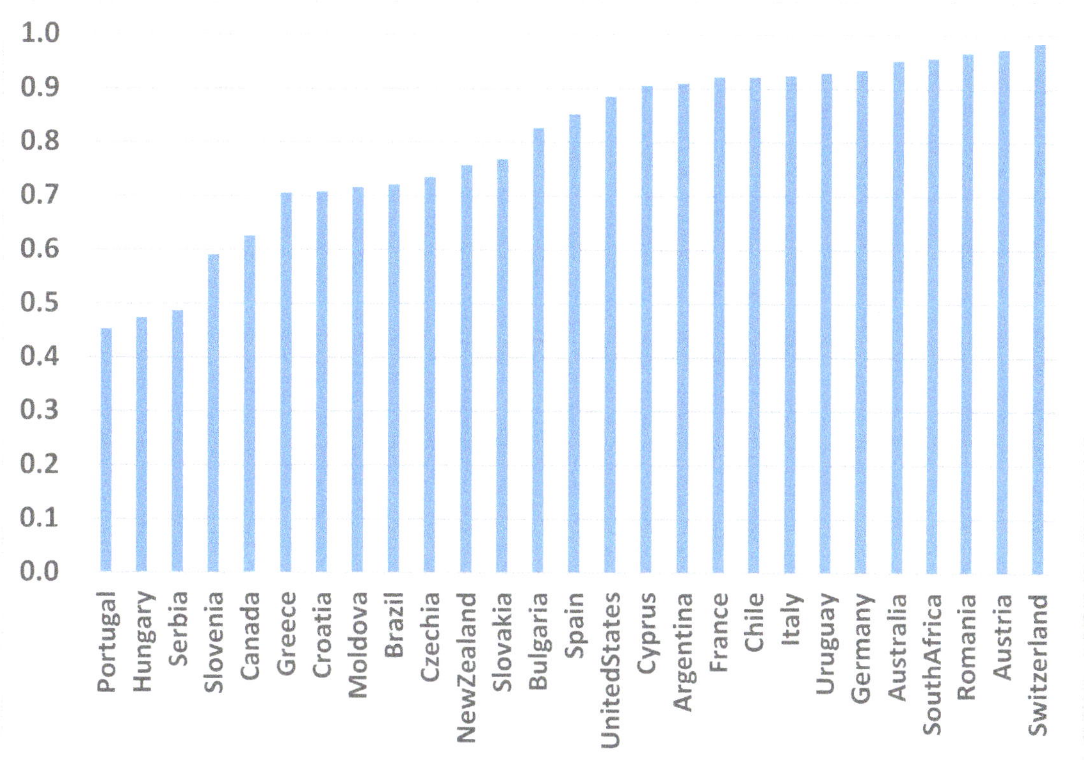

35. Index of Varietal Similarity of each country with the country with closest varietal mix, 2016

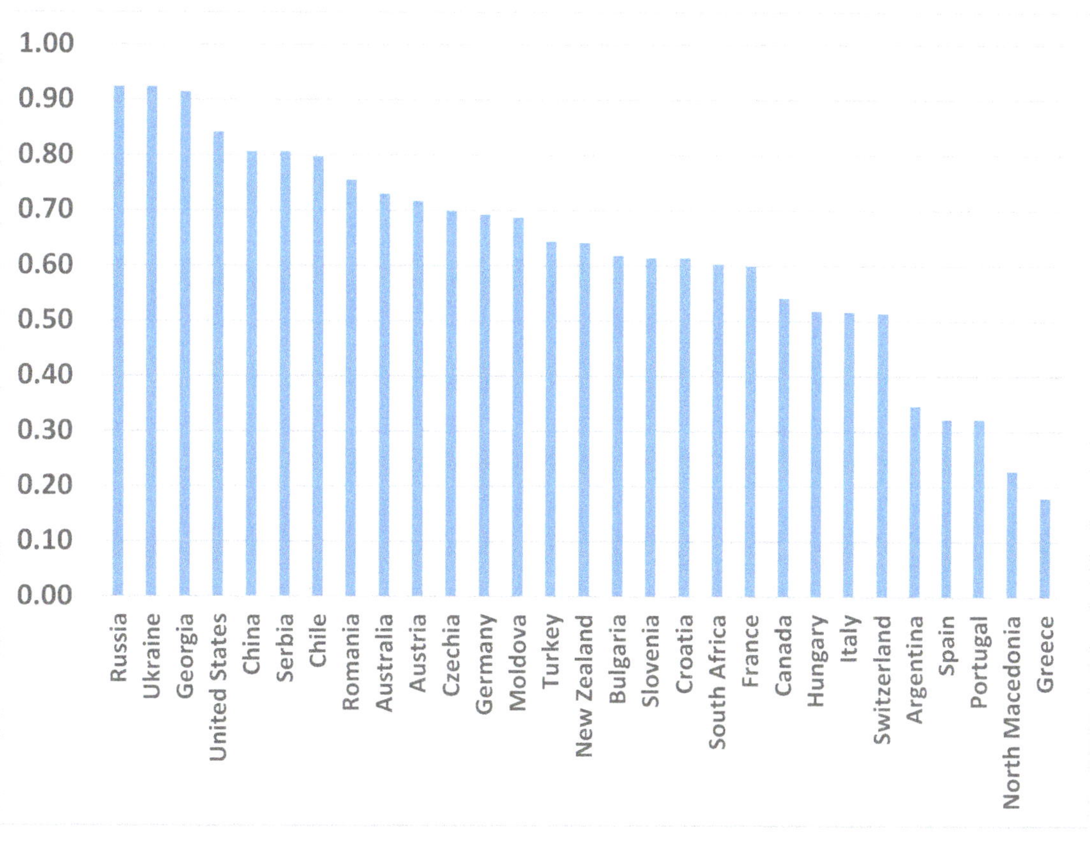

36. Index of Climatic Similarity of each country with the world, top 35 countries, 2016

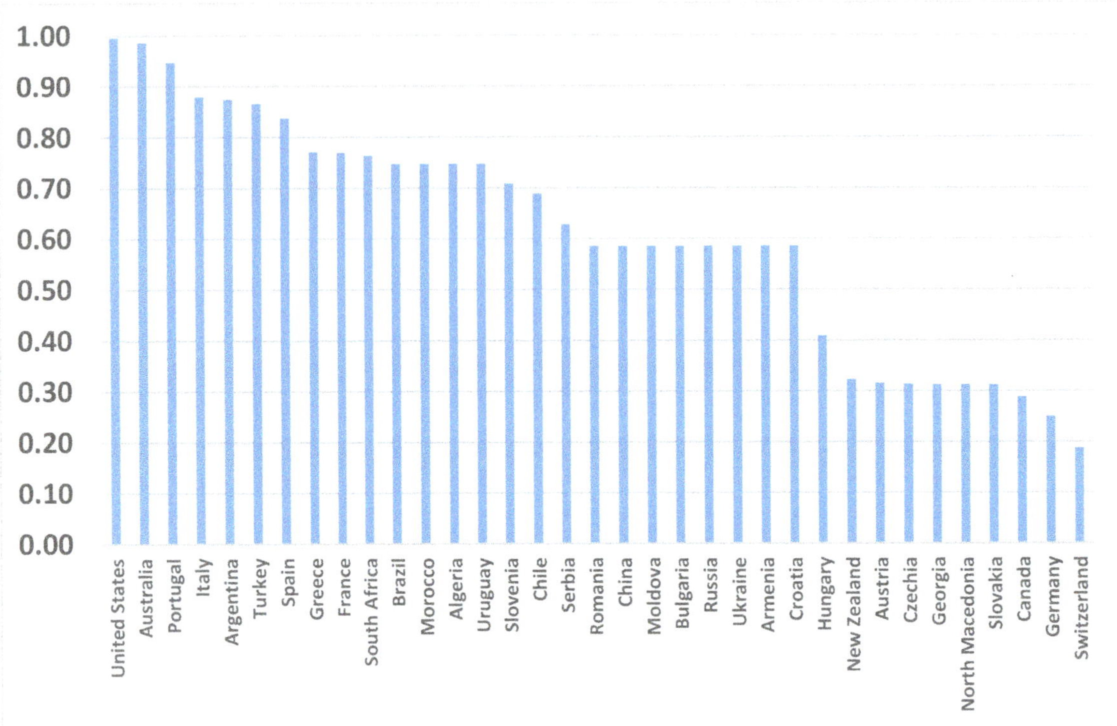

37. Global shares of selected varieties' bearing area within the ideal GST range for premium winegrapes, 2000 and 2016 (%)

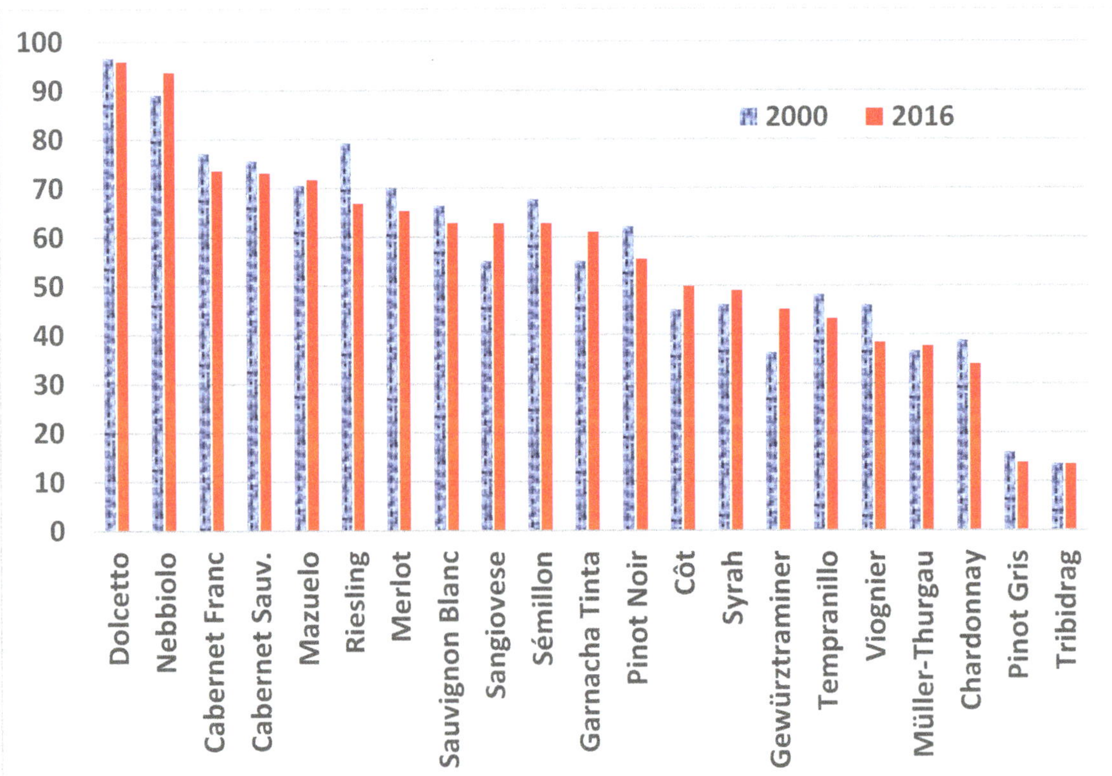

B. Top 25 wine-producing countries

20 leading varieties

38. Argentina

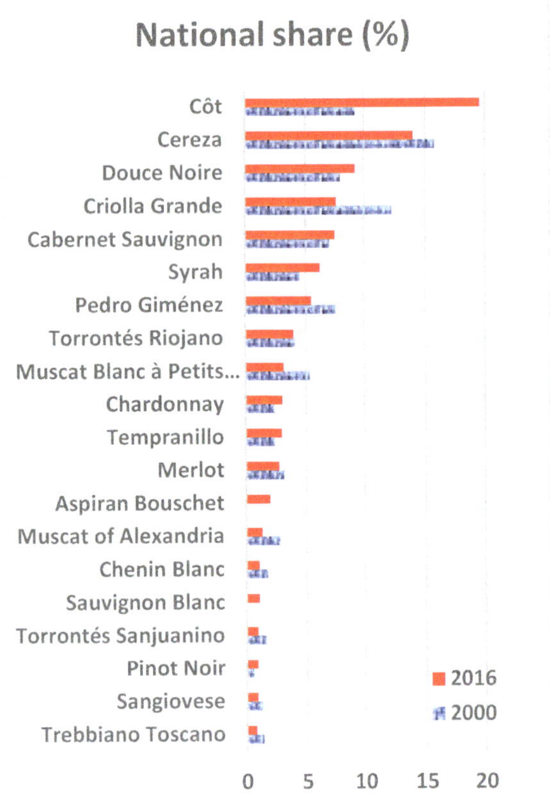

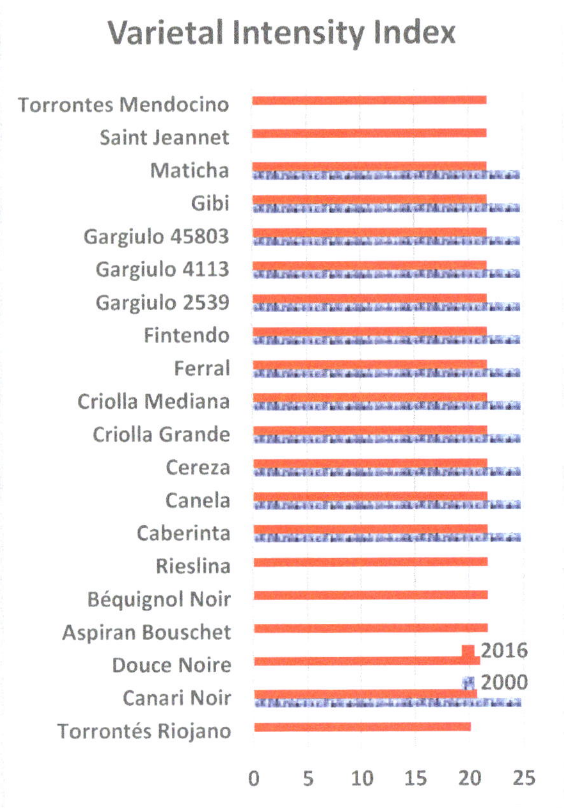

39. Australia

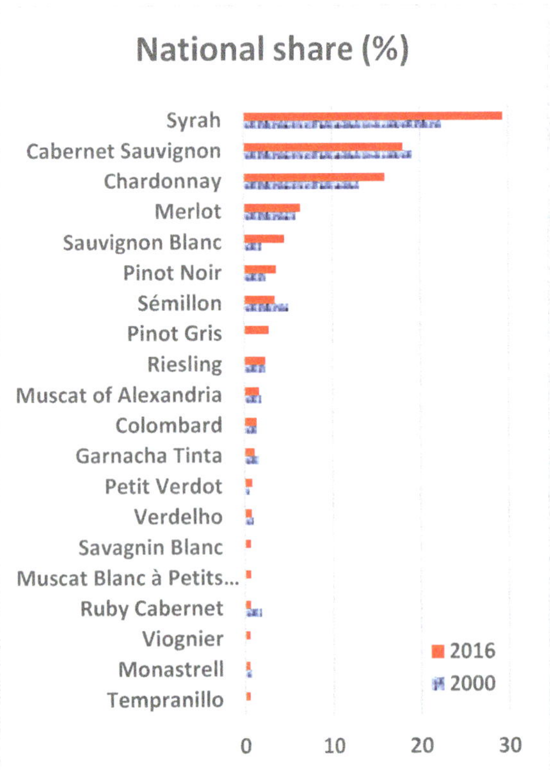

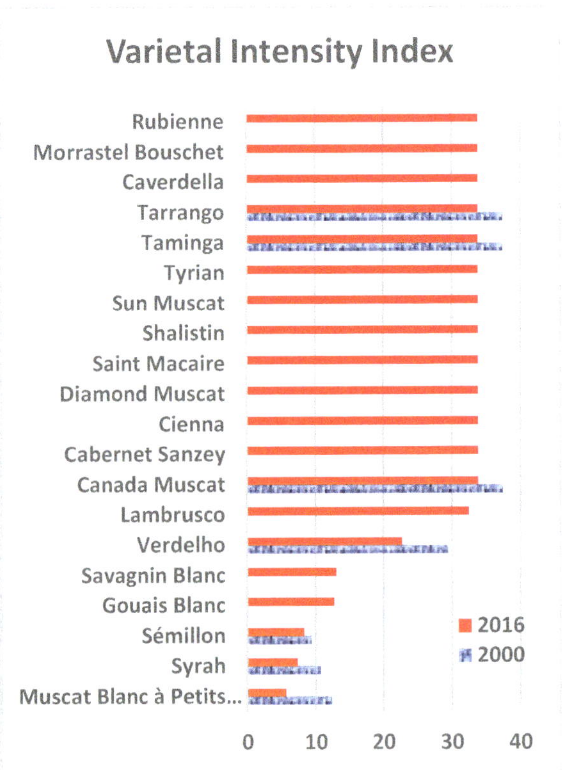

40. Austria

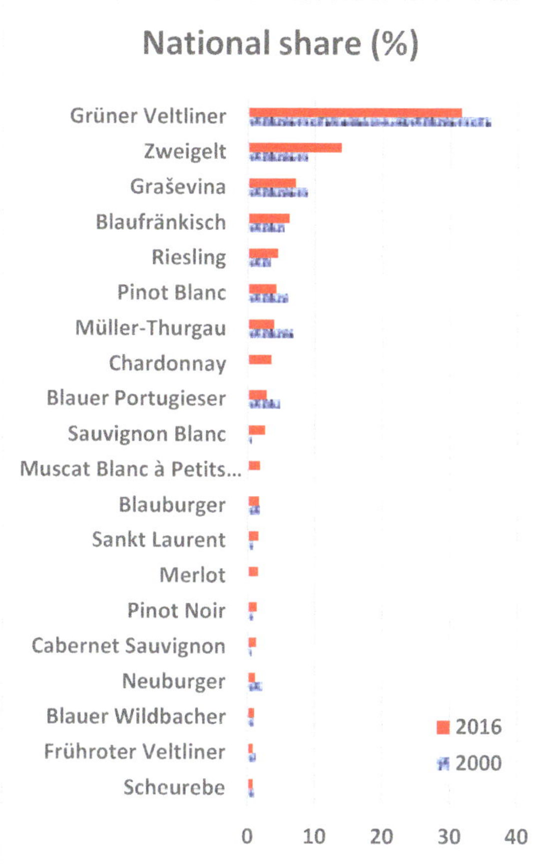

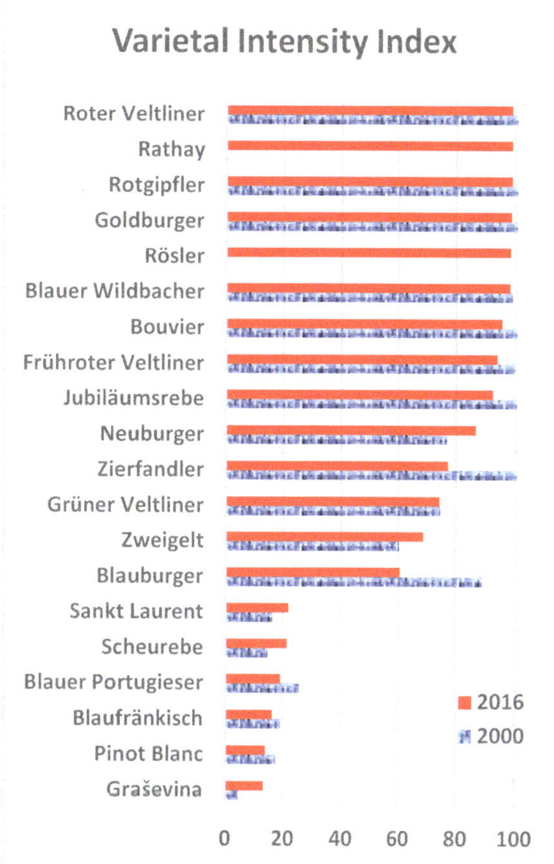

41. Brazil

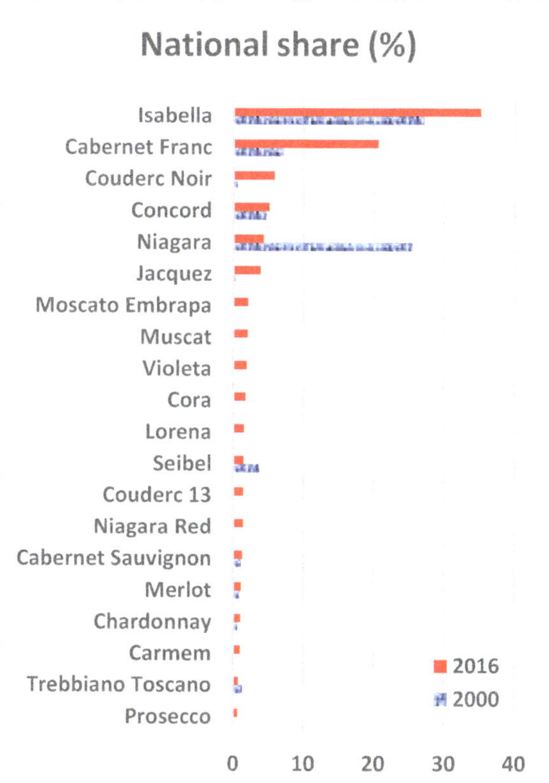

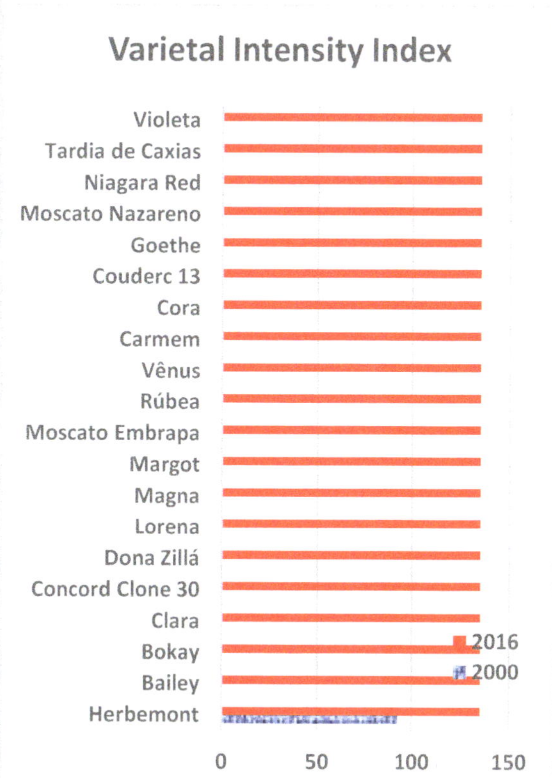

42. Bulgaria

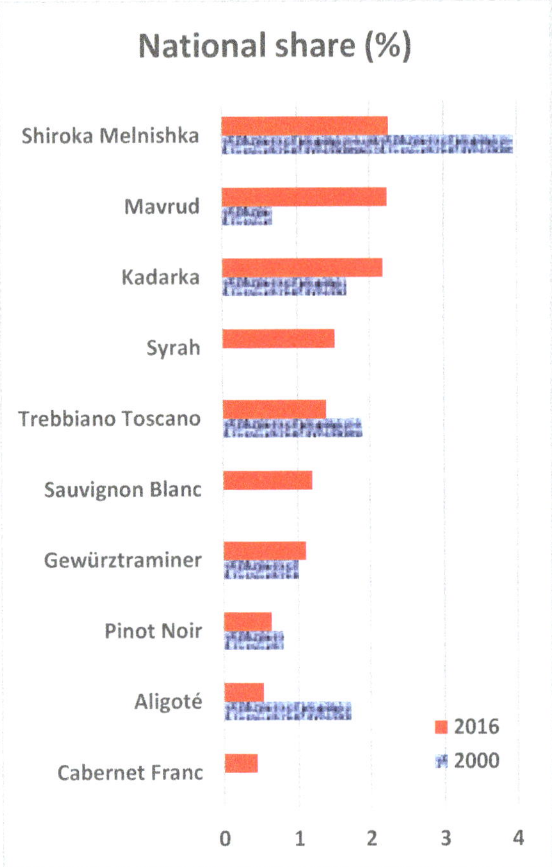

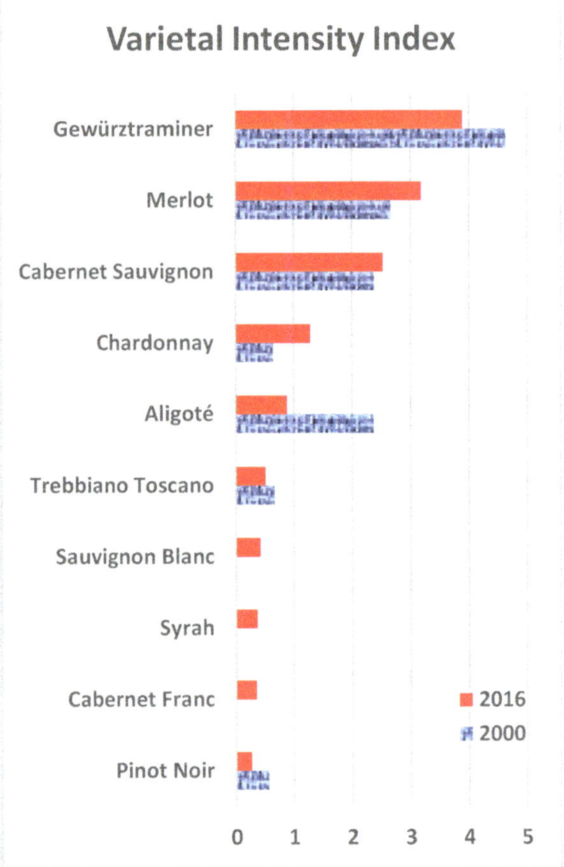

43. Canada

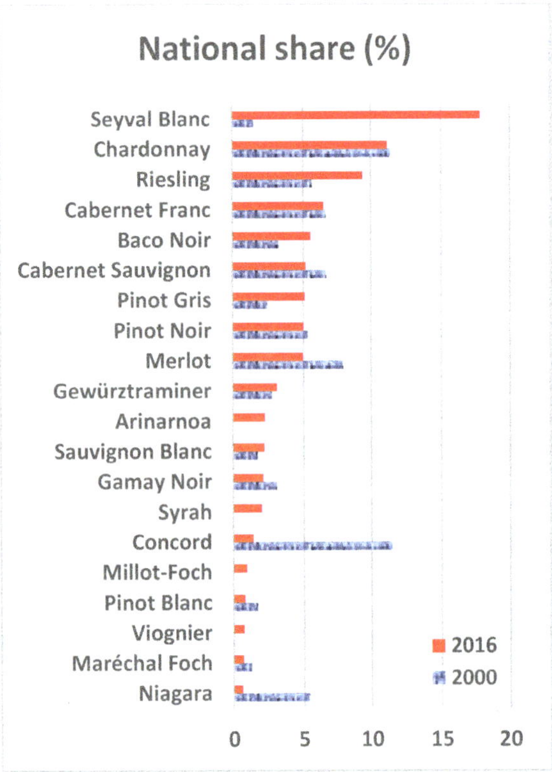

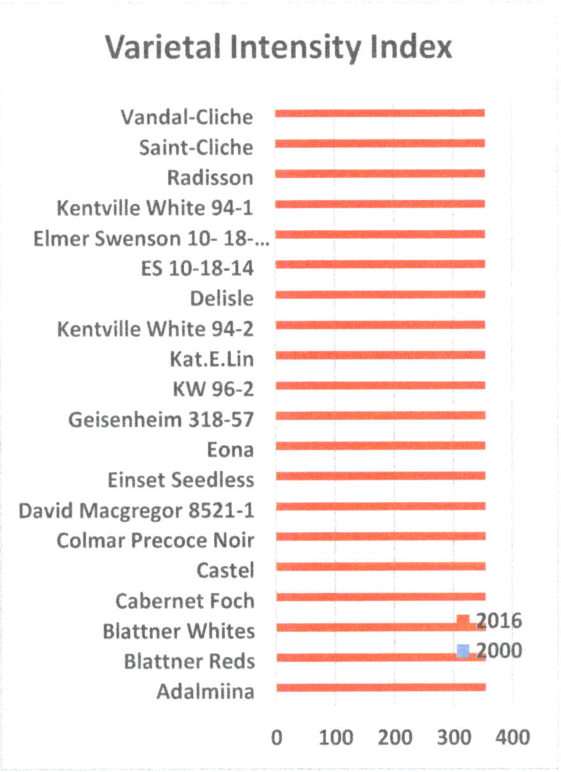

44. Chile

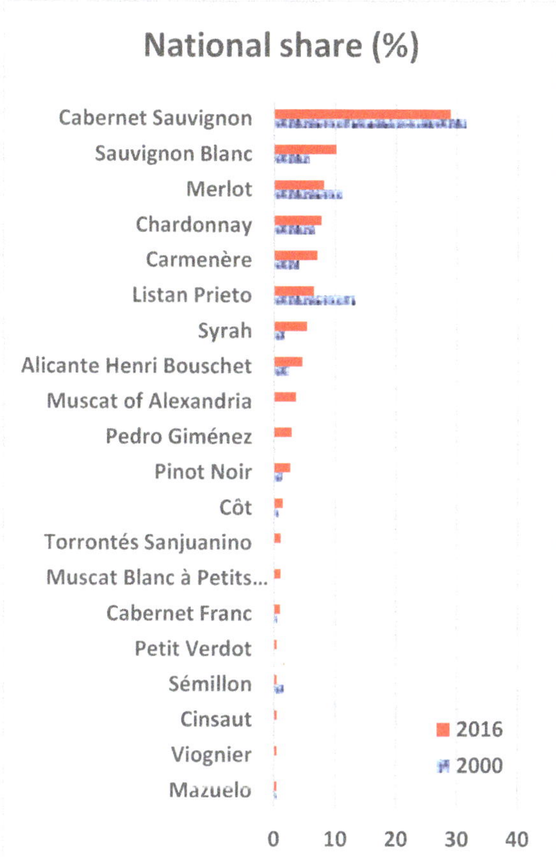

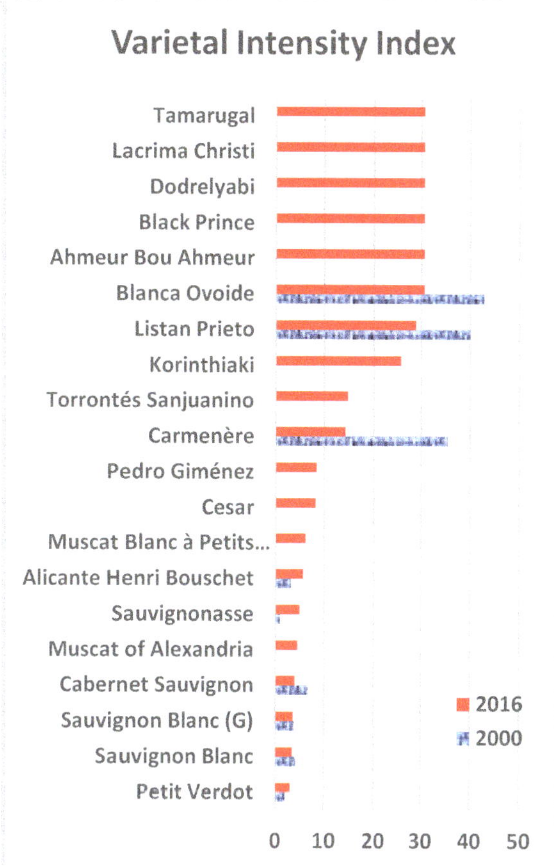

45. Croatia

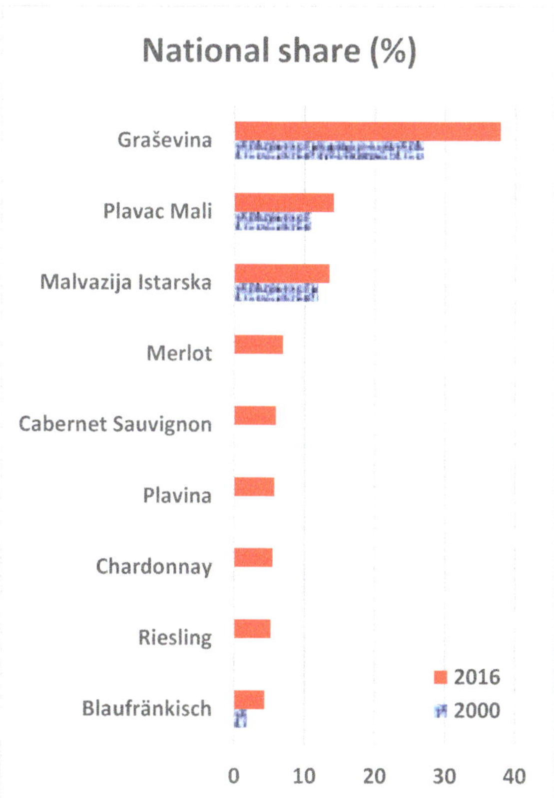

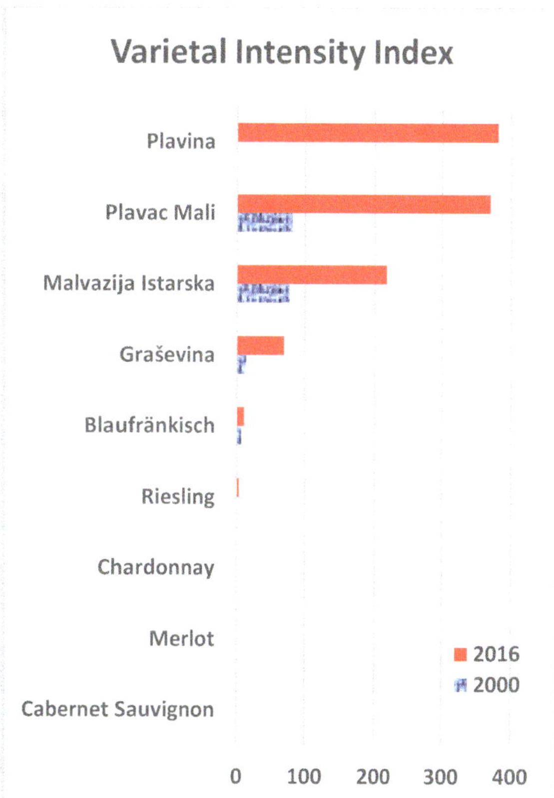

46. Czechia

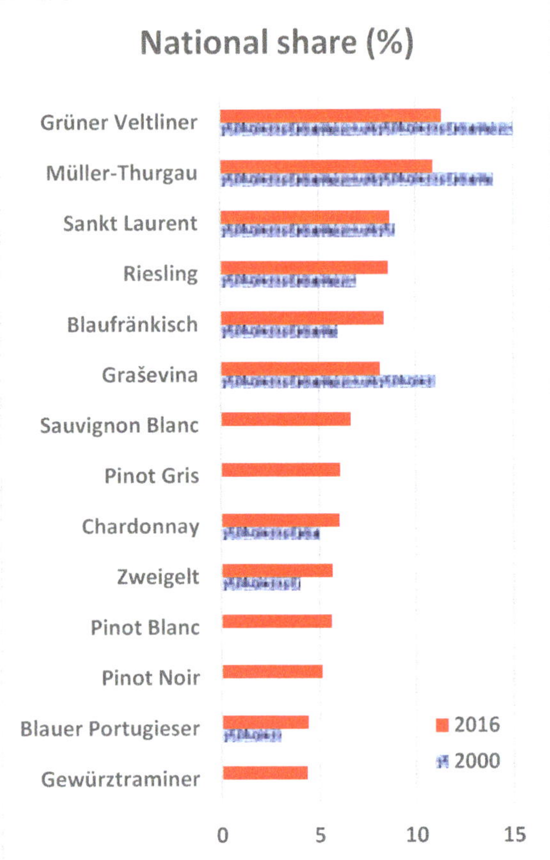

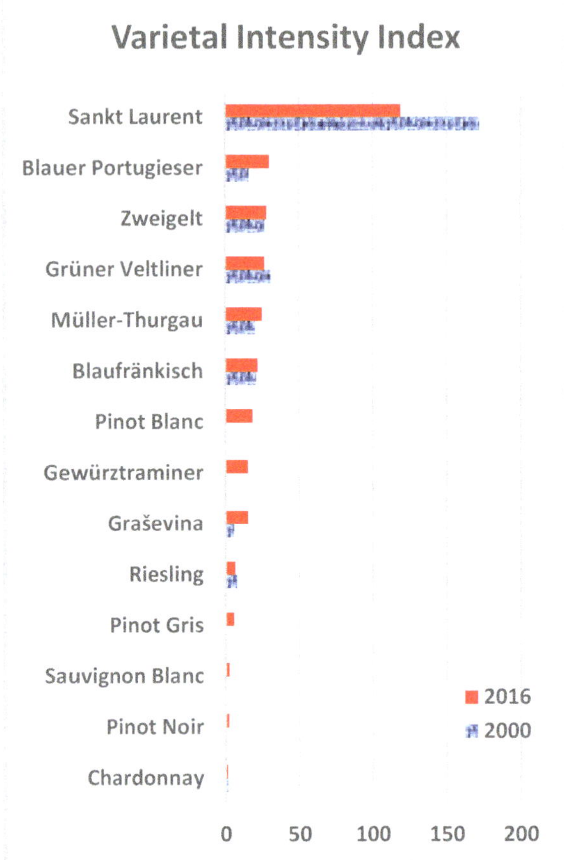

47. France

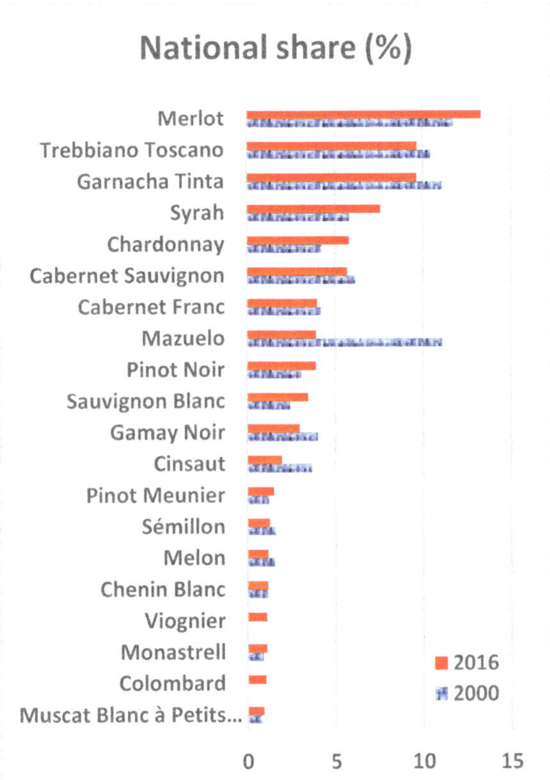

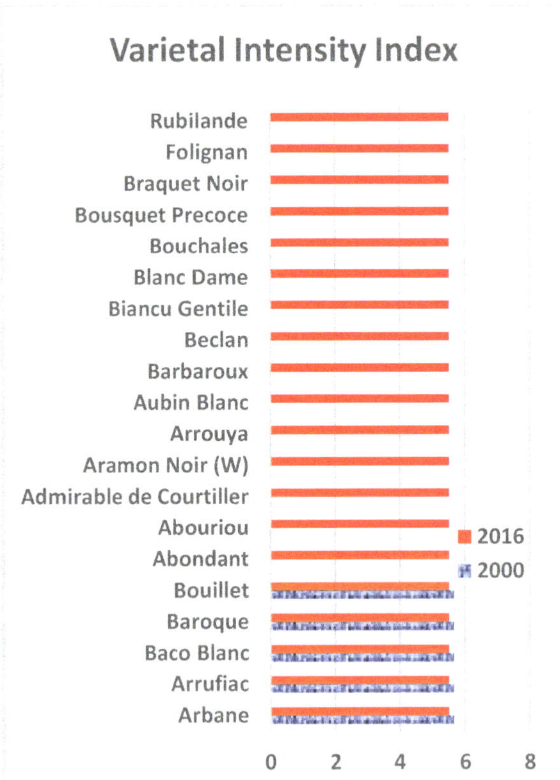

48. Germany

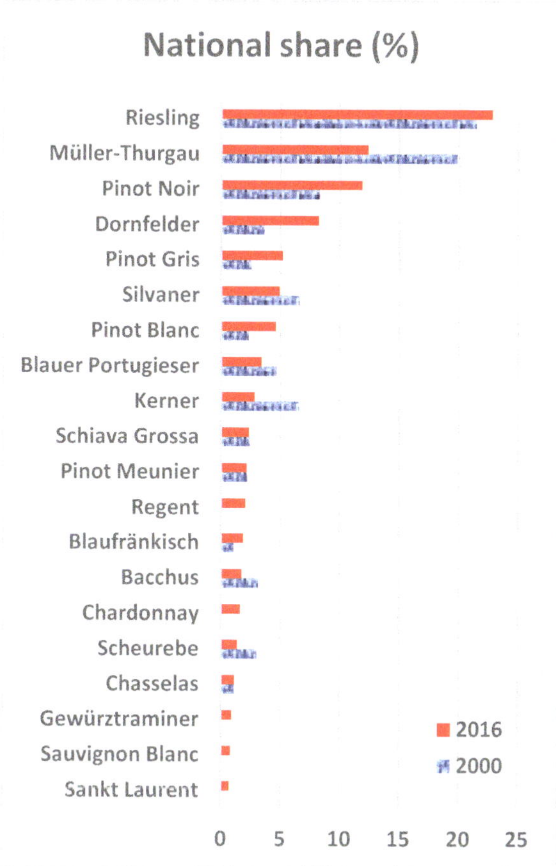

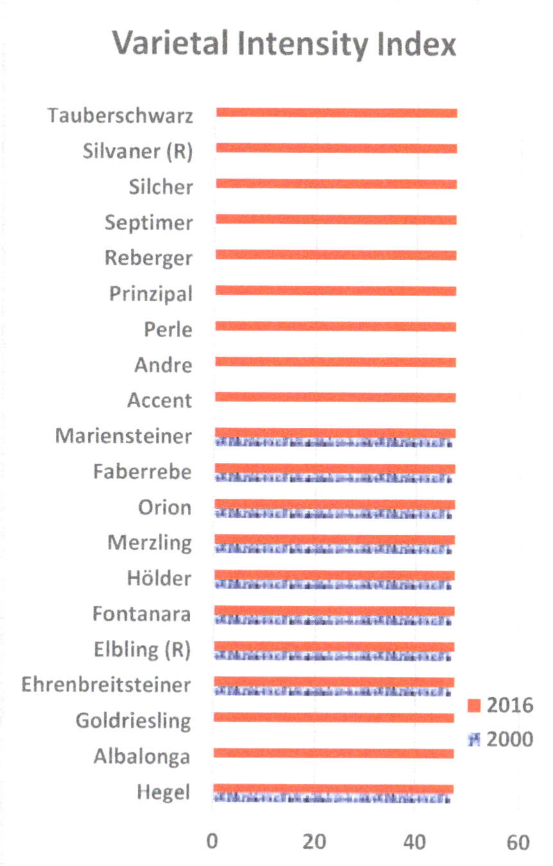

49. Greece

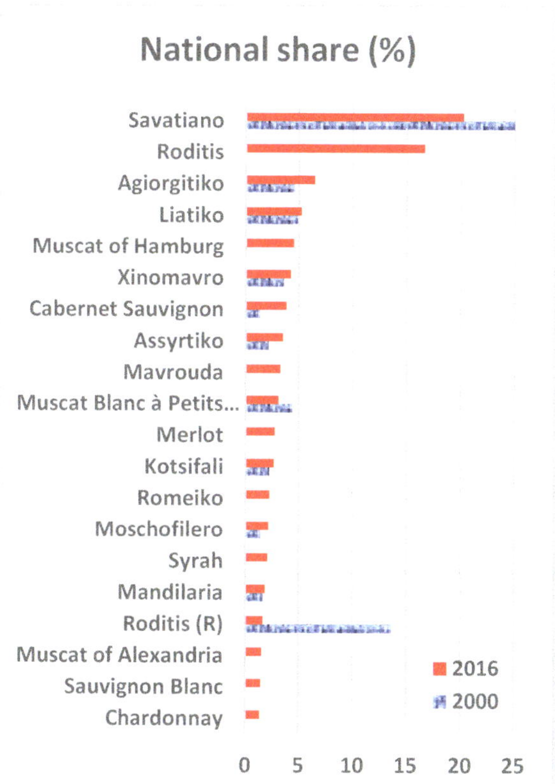

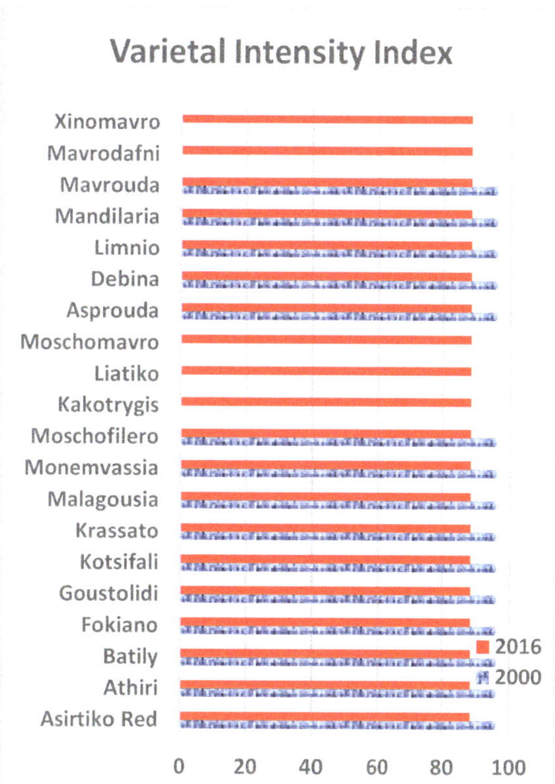

50. Hungary

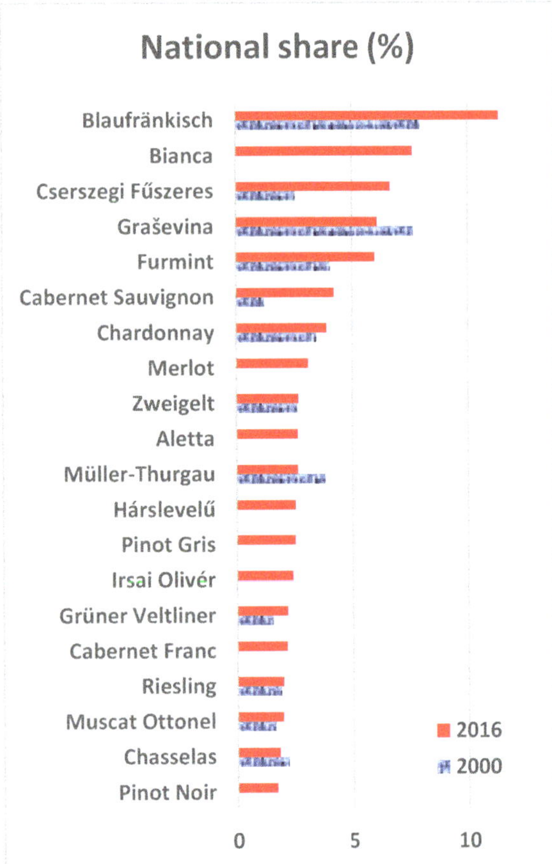

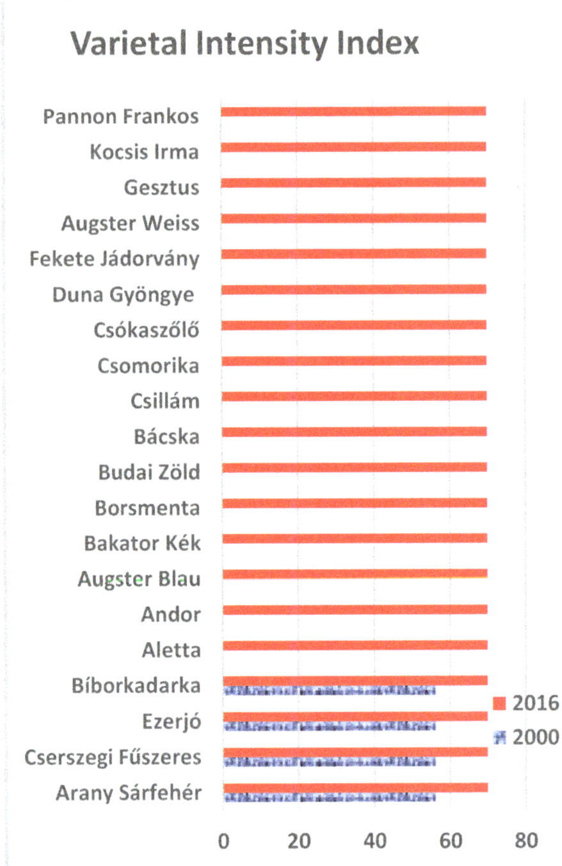

51. Italy

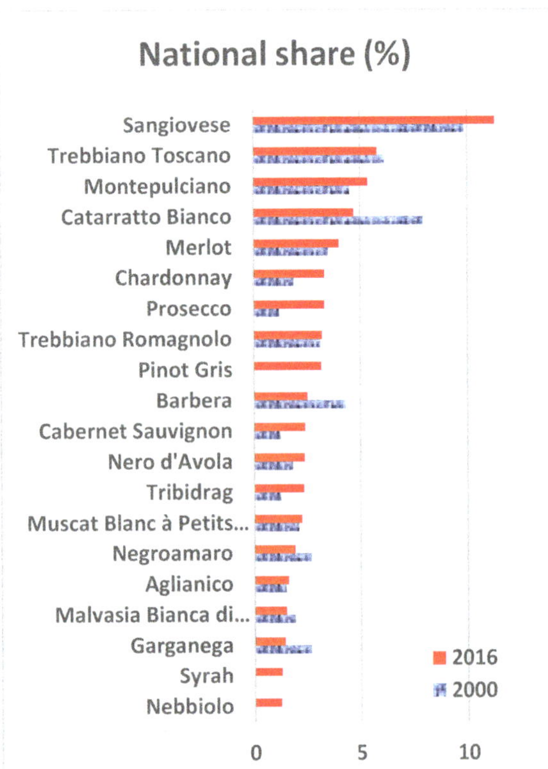

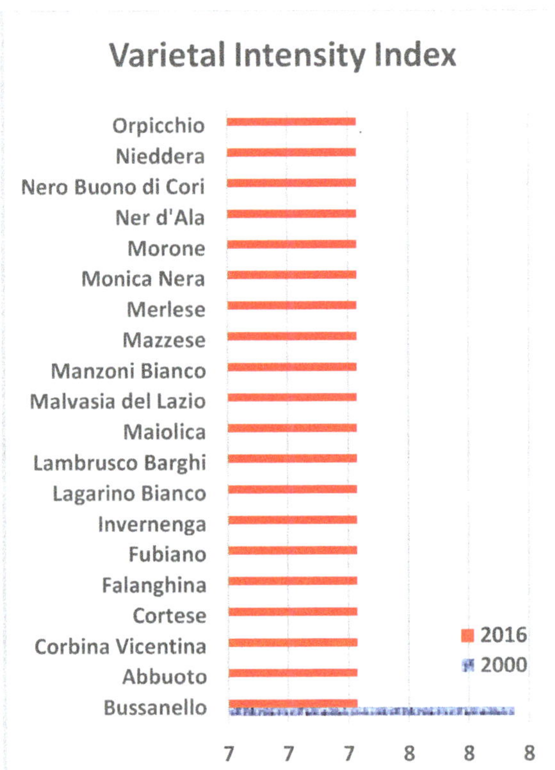

52. Moldova

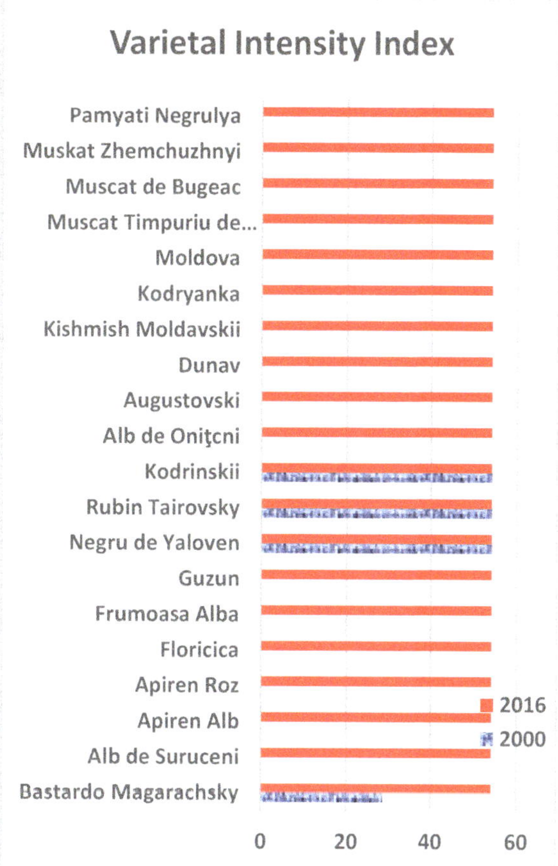

53. New Zealand

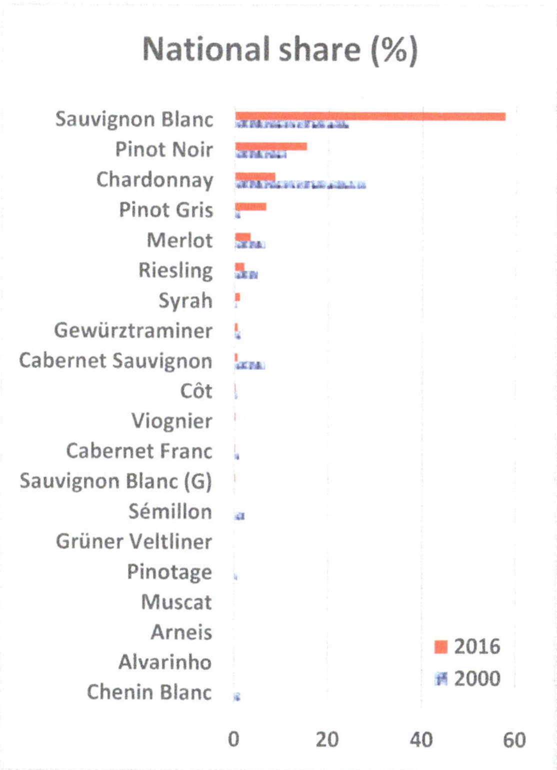

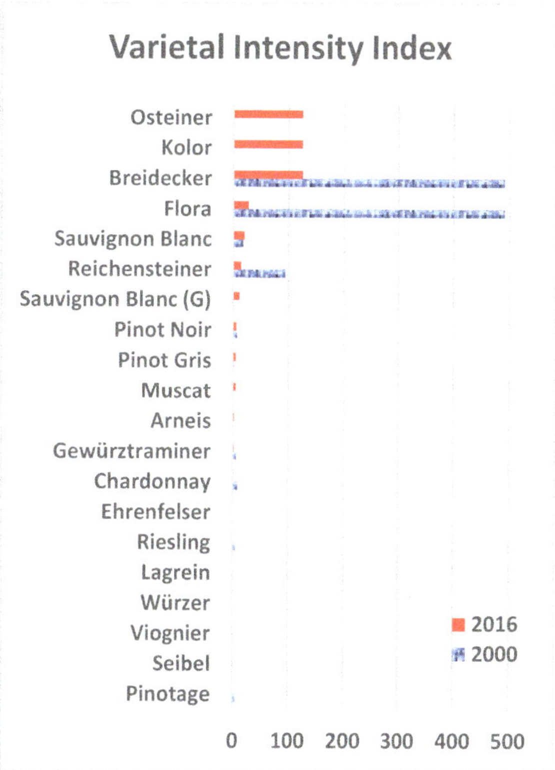

54. Portugal

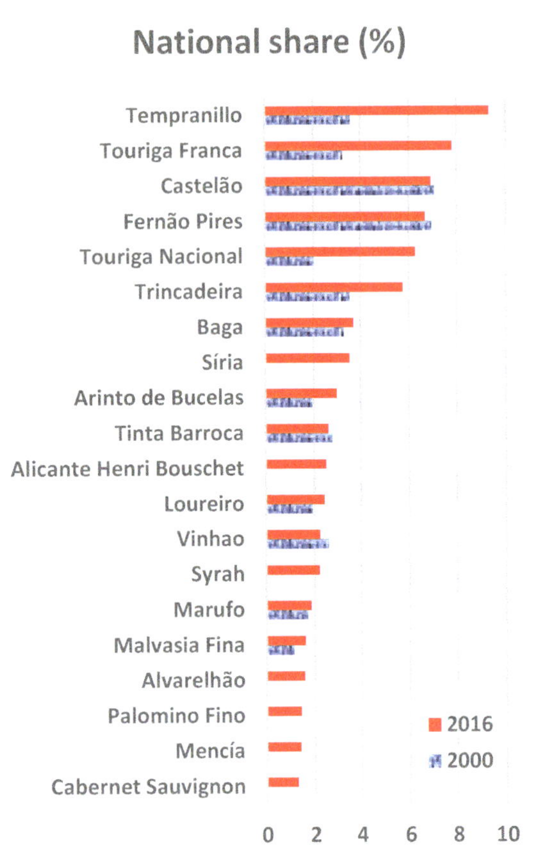

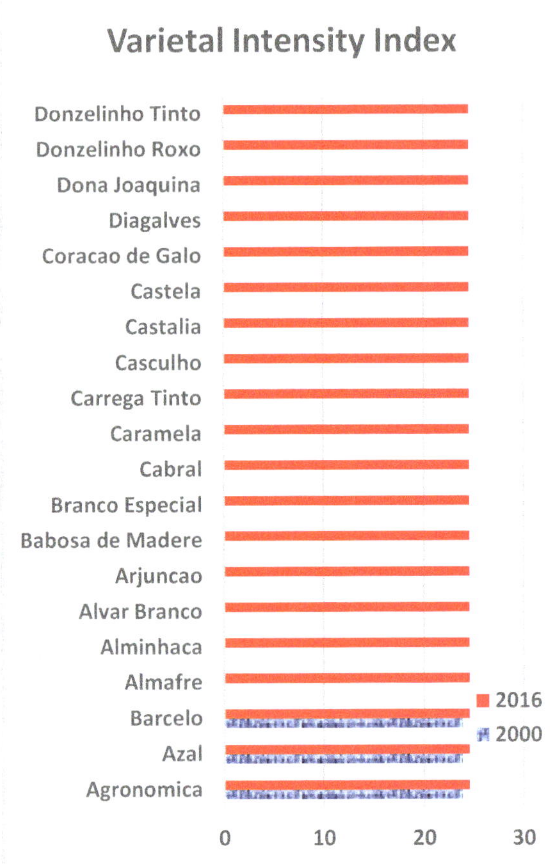

55. Romania

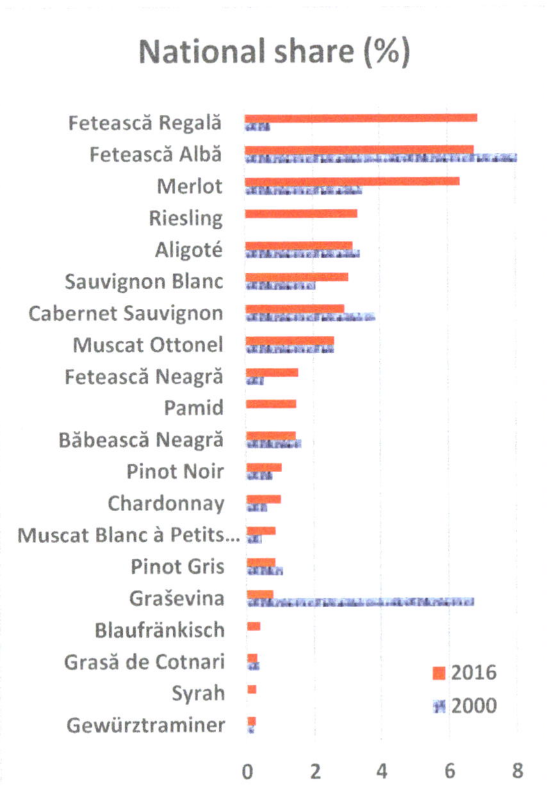

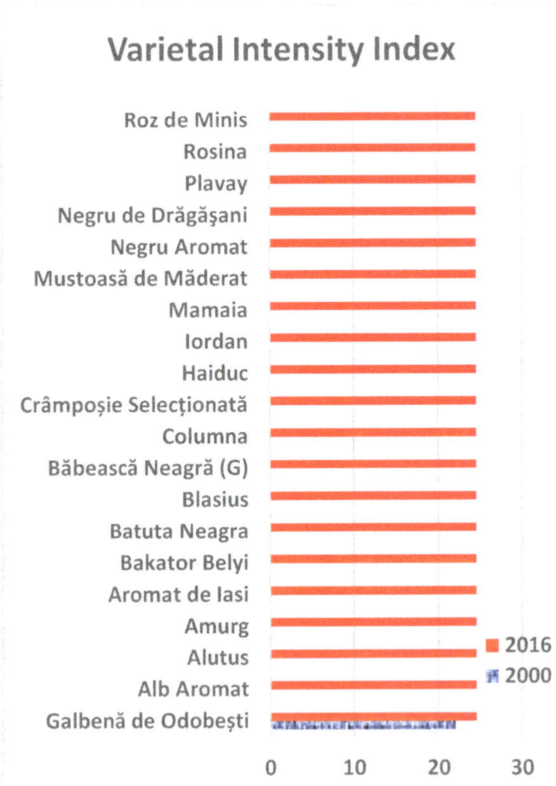

56. Russia

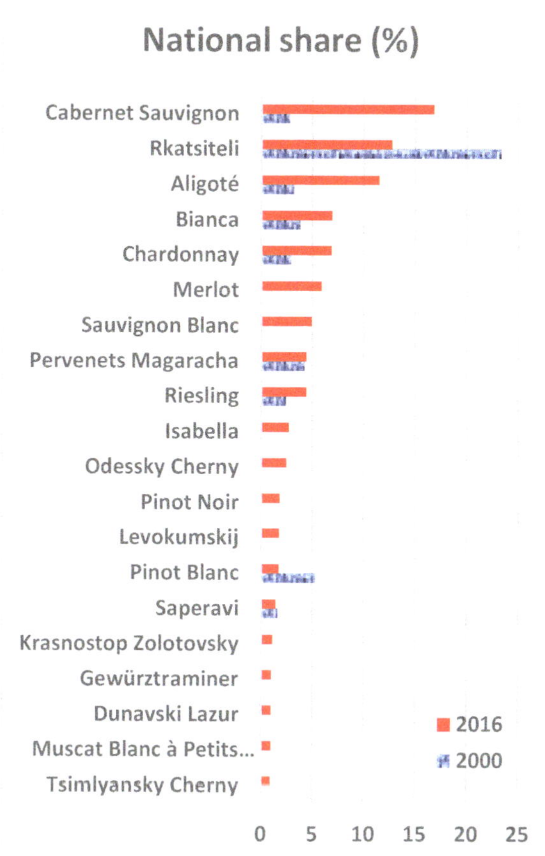

57. Slovenia

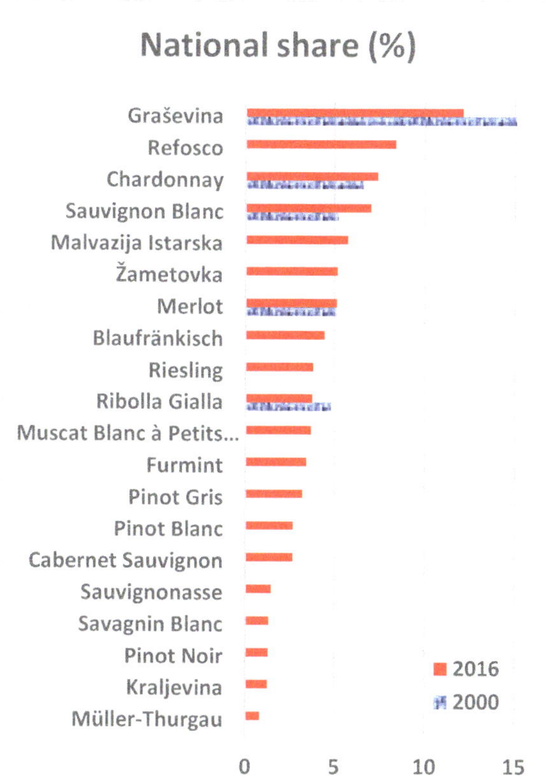

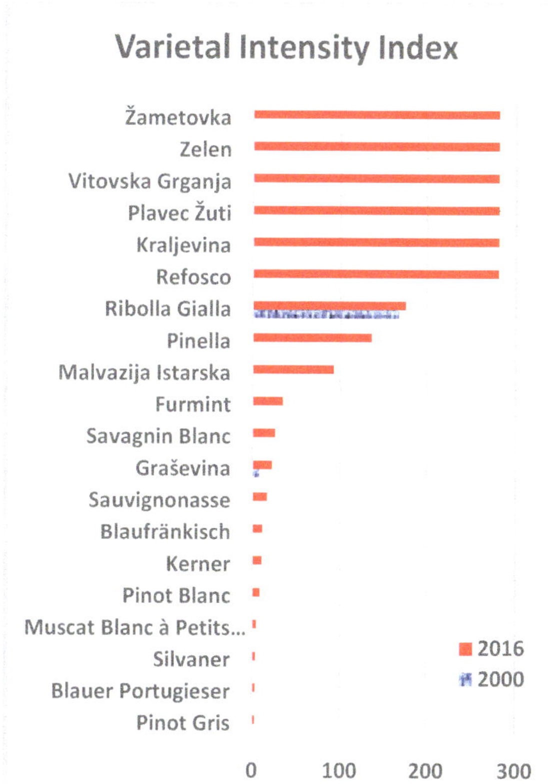

58. South Africa

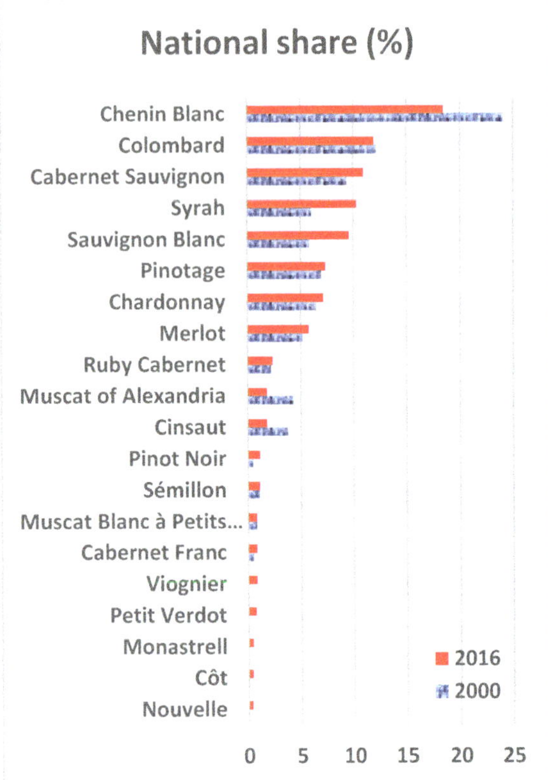

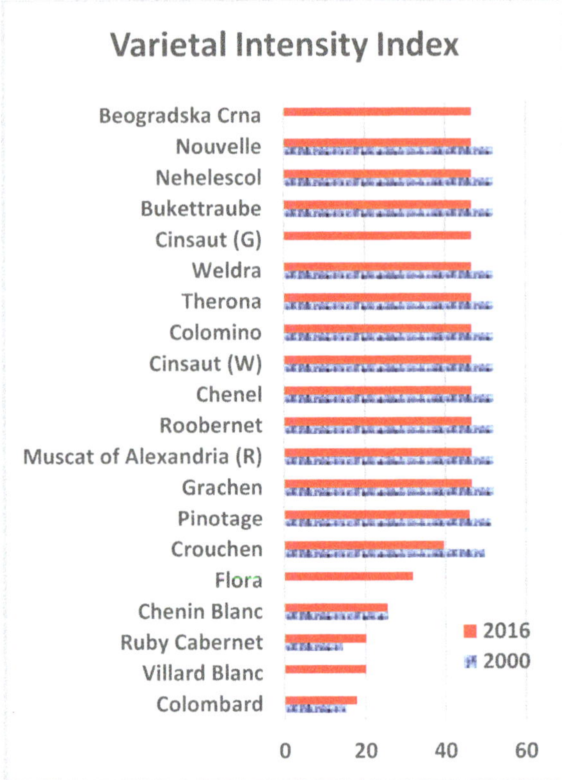

59. Spain

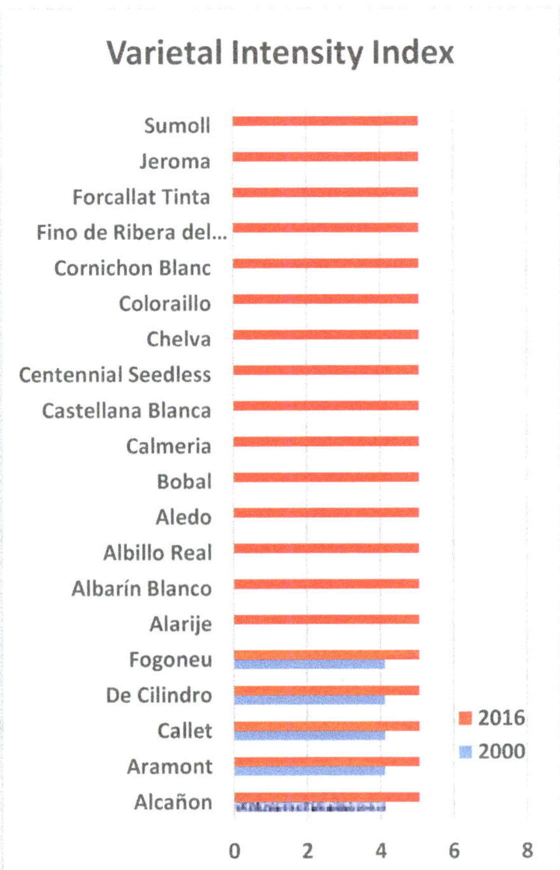

60. Switzerland

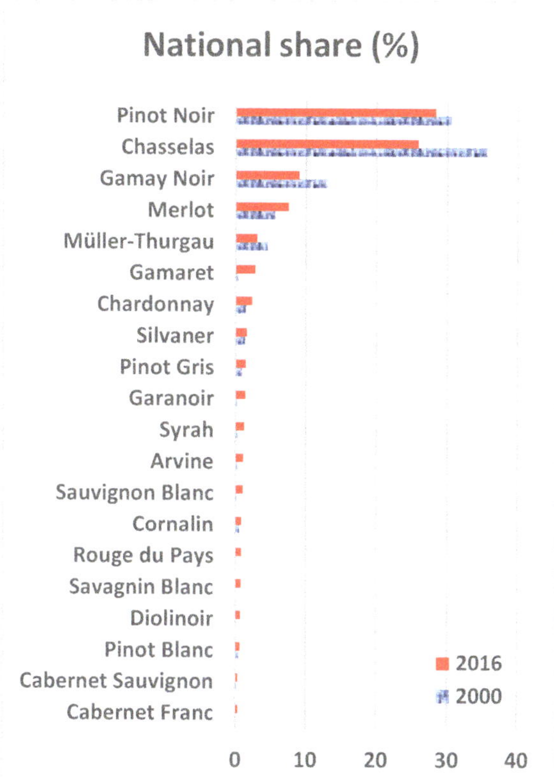

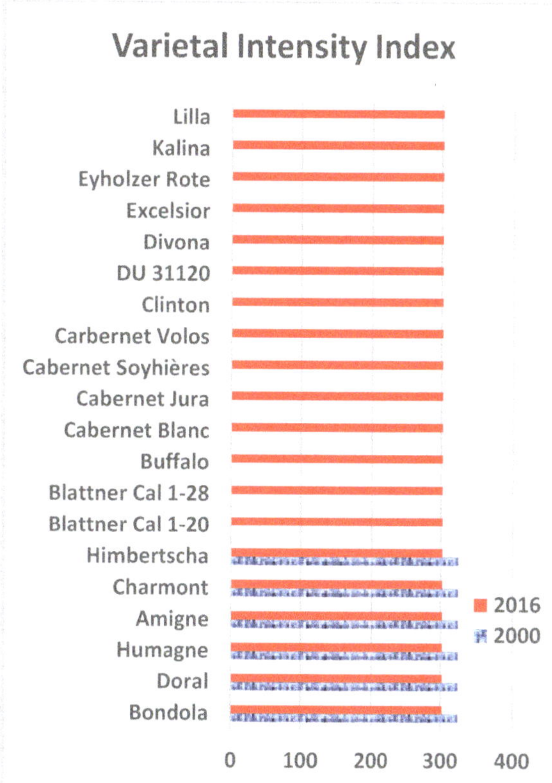

61. United States

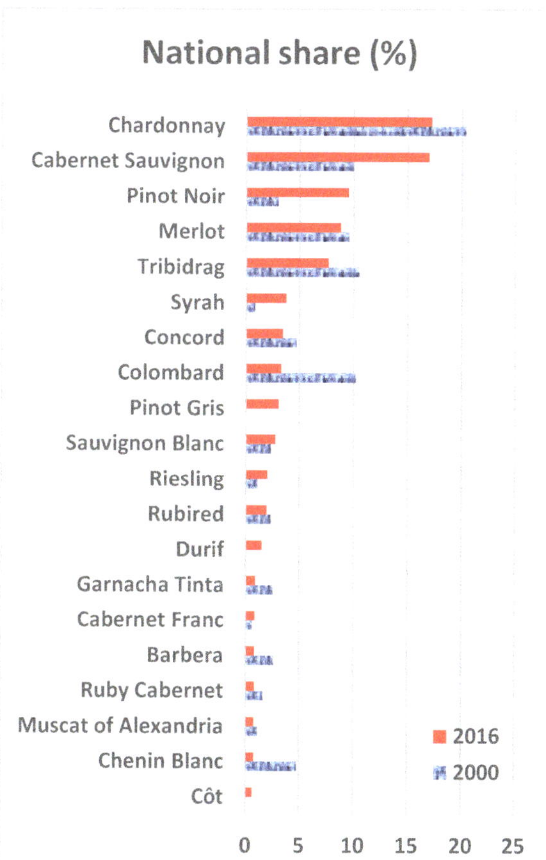

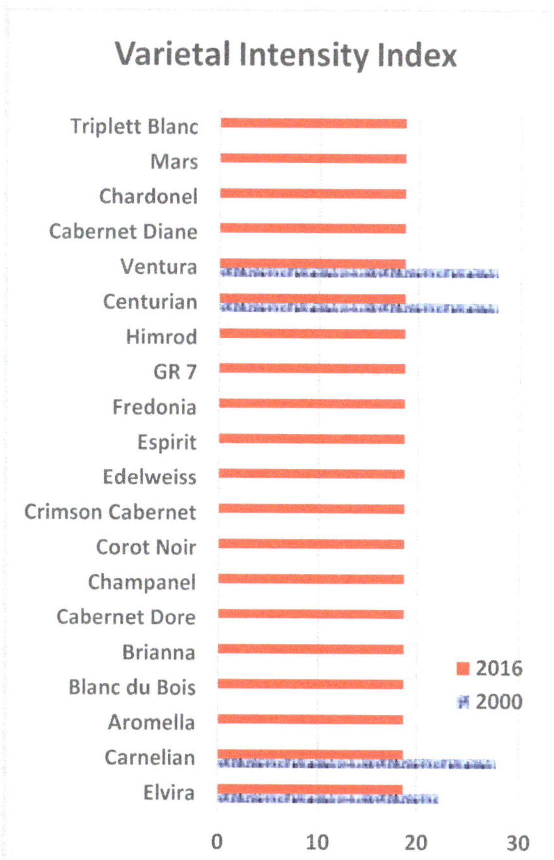

62. Uruguay

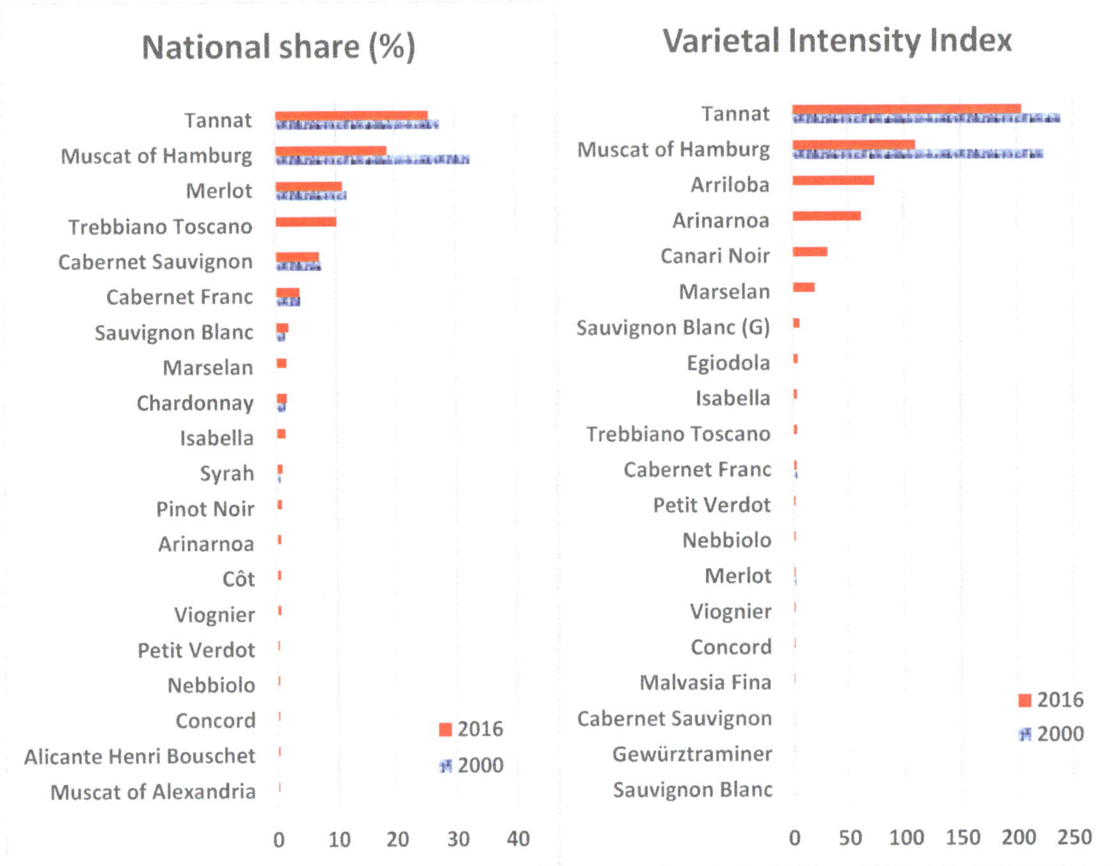

National share (%)
Tannat
Muscat of Hamburg
Merlot
Trebbiano Toscano
Cabernet Sauvignon
Cabernet Franc
Sauvignon Blanc
Marselan
Chardonnay
Isabella
Syrah
Pinot Noir
Arinarnoa
Côt
Viognier
Petit Verdot
Nebbiolo
Concord
Alicante Henri Bouschet
Muscat of Alexandria
0
10
20
30
40
2016
2000
Varietal Intensity Index
Tannat
Muscat of Hamburg
Arriloba
Arinarnoa
Canari Noir
Marselan
Sauvignon Blanc (G)
Egiodola
Isabella
Trebbiano Toscano
Cabernet Franc
Petit Verdot
Nebbiolo
Merlot
Viognier
Concord
Malvasia Fina
Cabernet Sauvignon
Gewürztraminer
Sauvignon Blanc
0
50
100
150
200
250
2016
2000

C. Top 25 varieties

20 leading countries

63. Airén

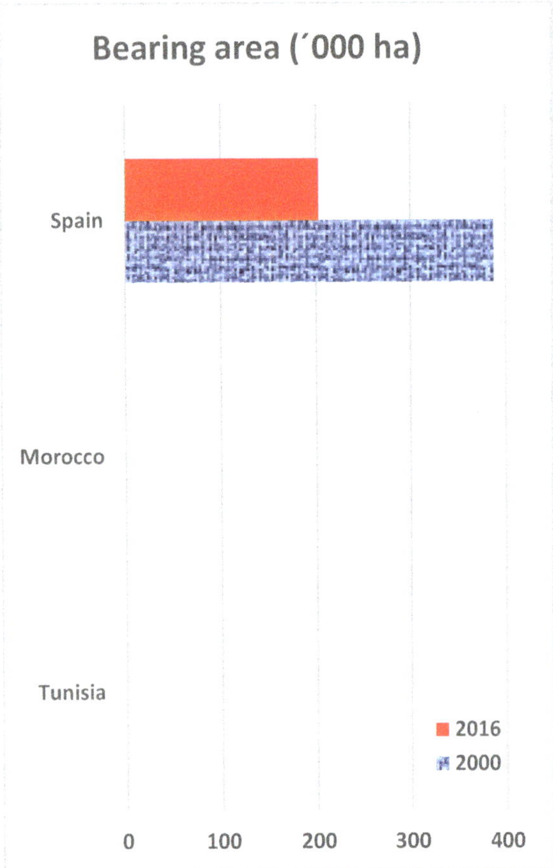

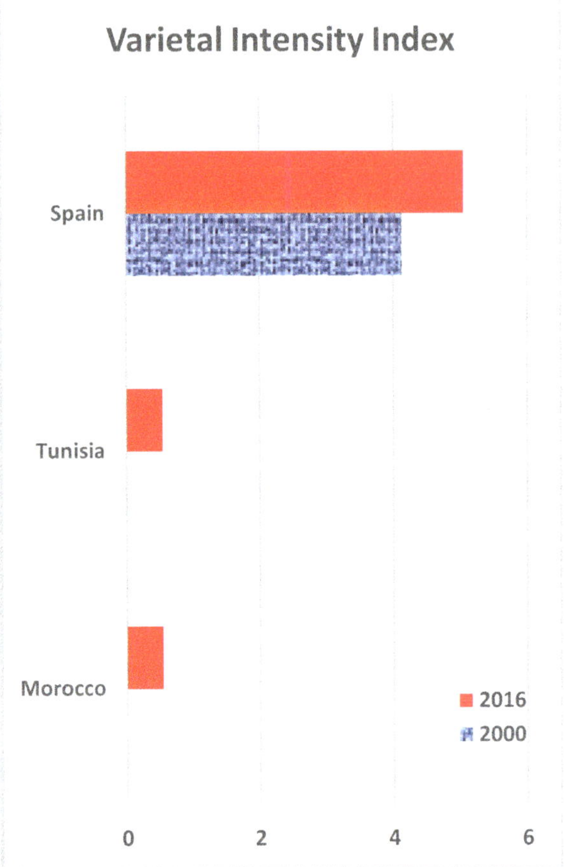

64. Alicante Henri Bouschet

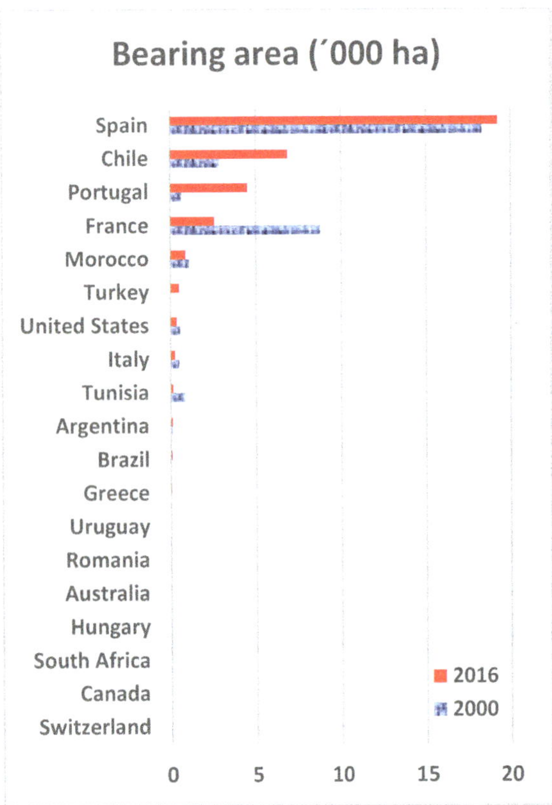

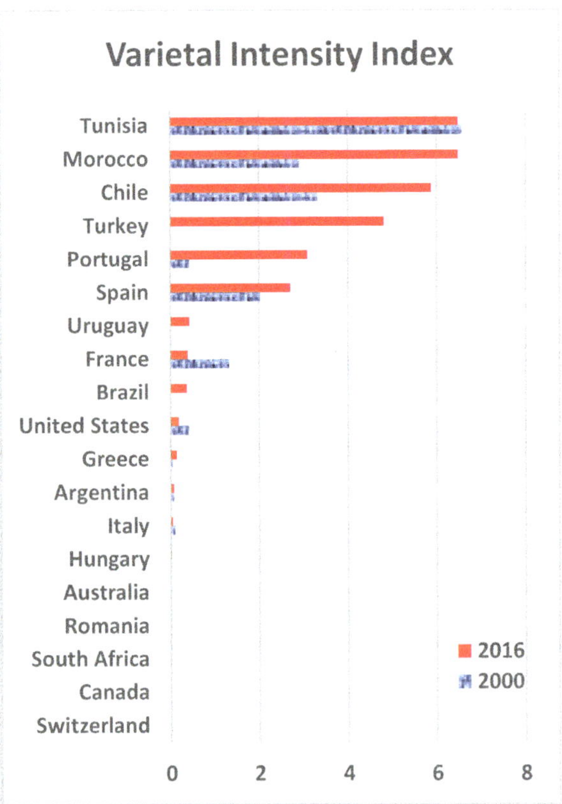

65. Bobal

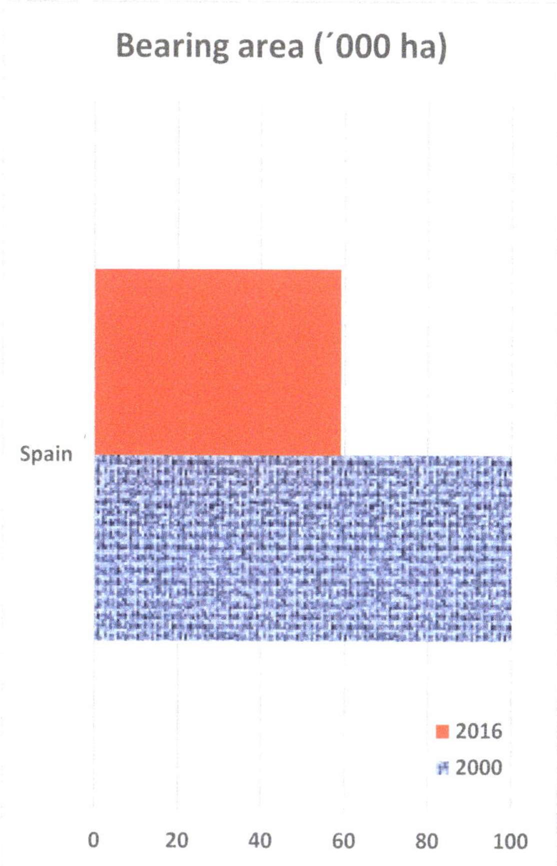

66. Cabernet Franc

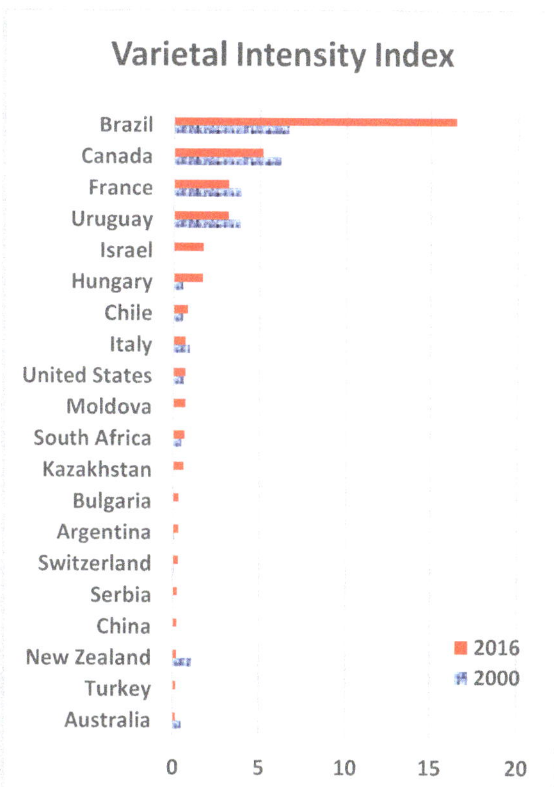

67. Cabernet Sauvignon

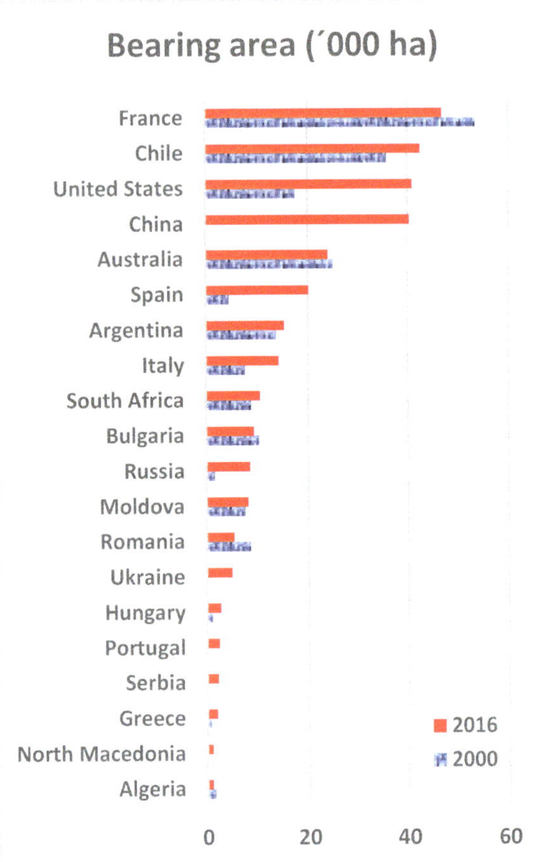

68. Cayetana Blanca

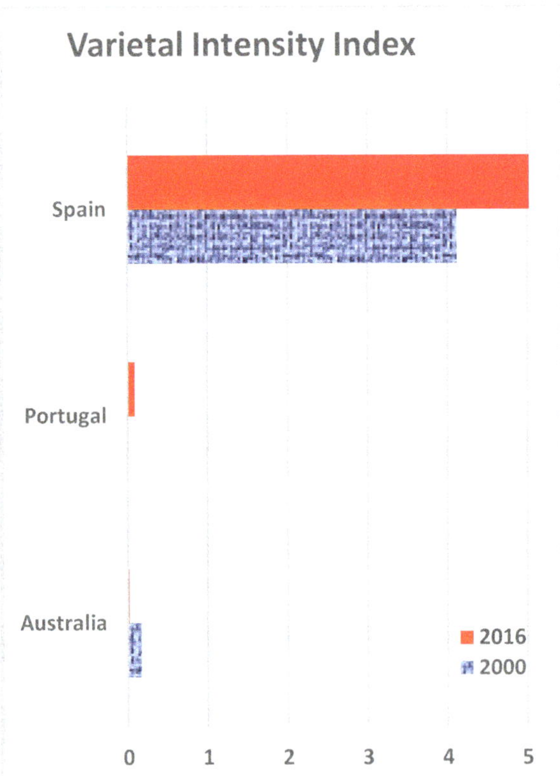

69. Chardonnay

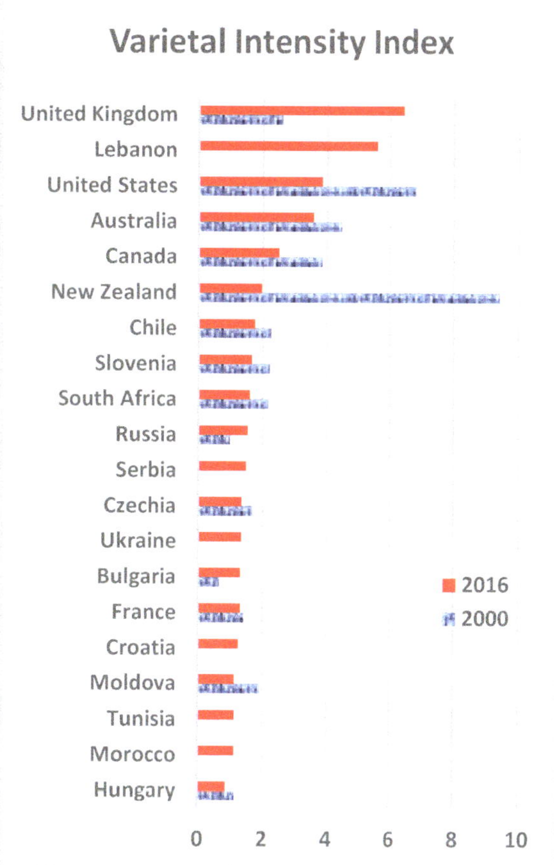

70. Côt

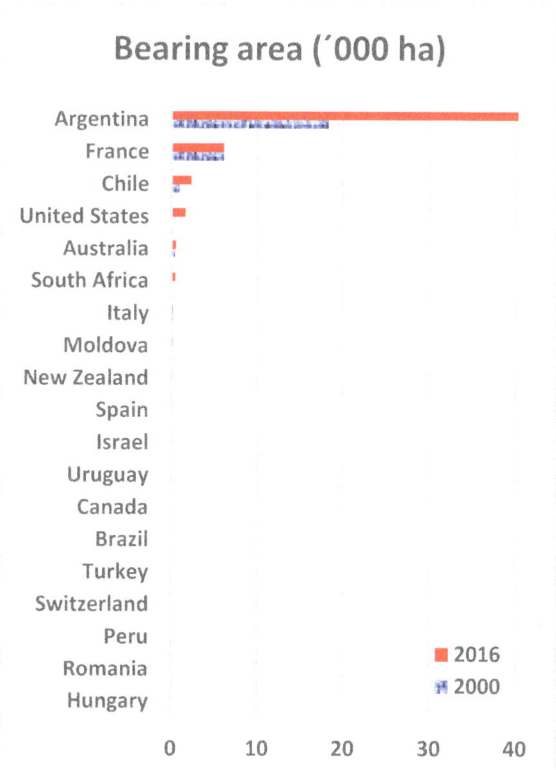

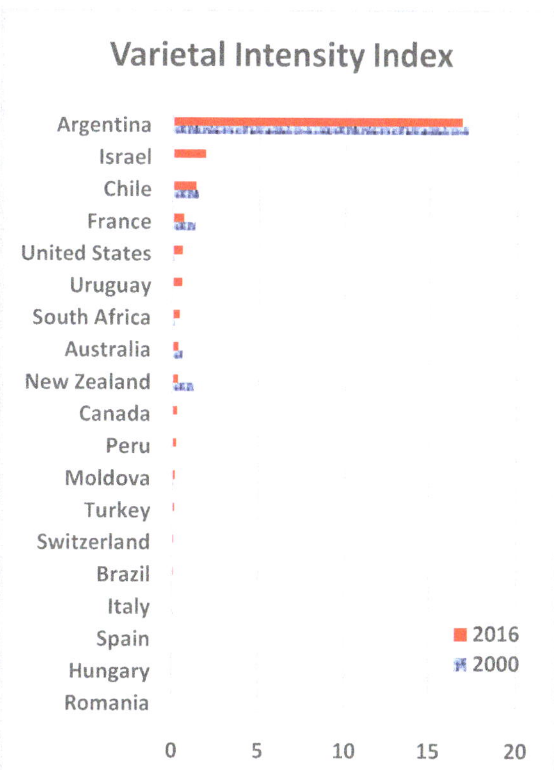

71. Garnacha Tinta

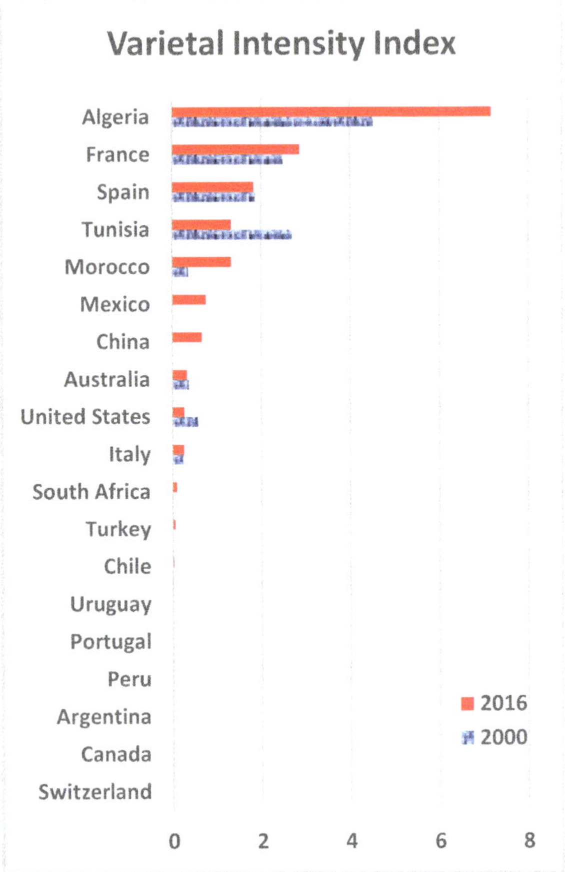

72. Macabeo

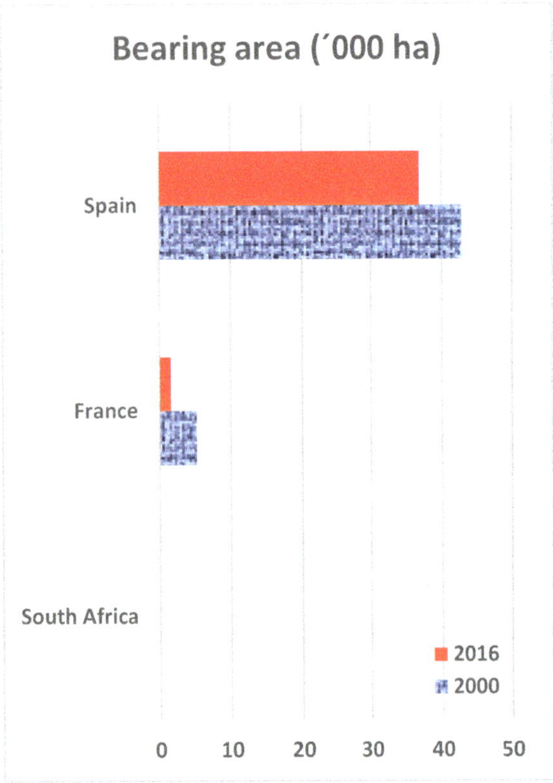

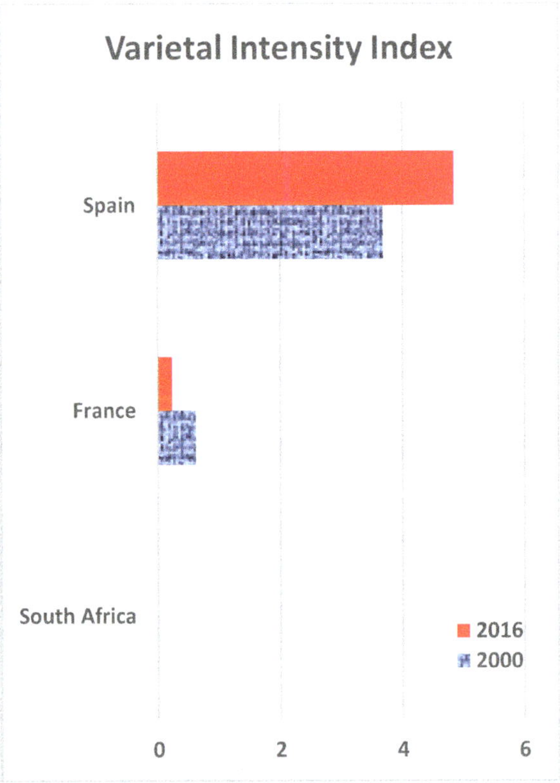

73. Mazuelo

74. Merlot

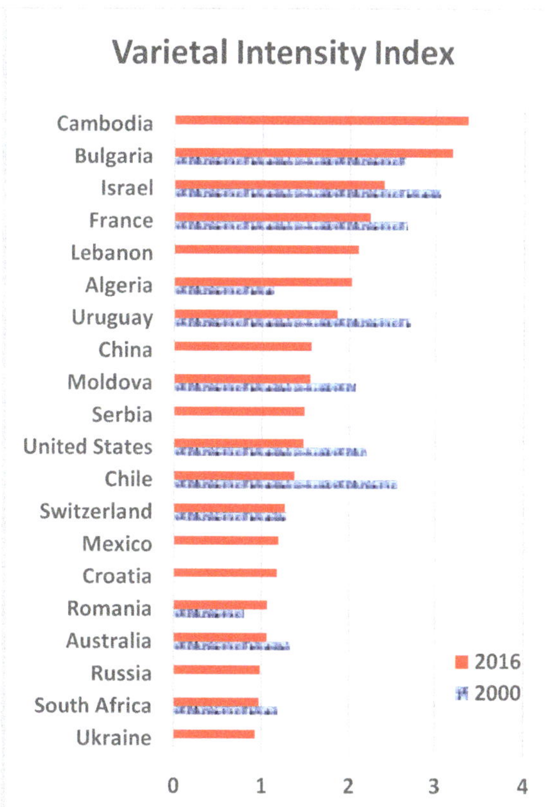

75. Monastrell

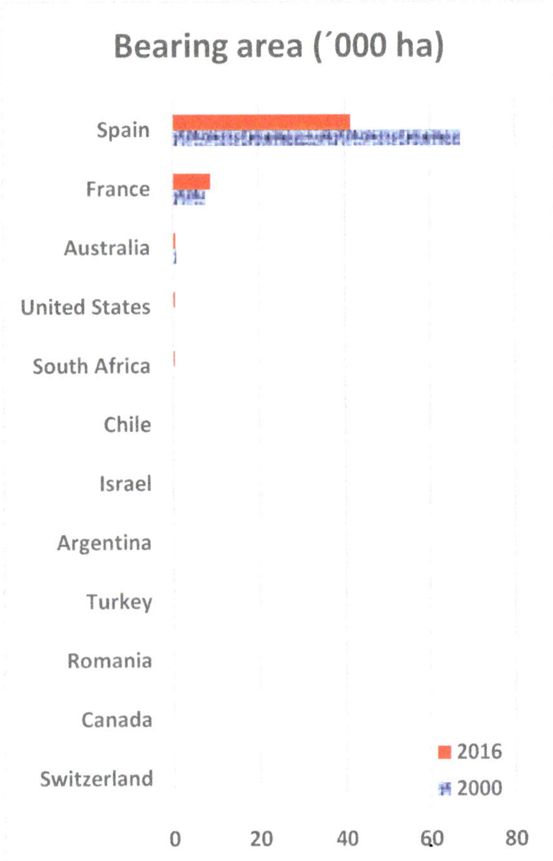

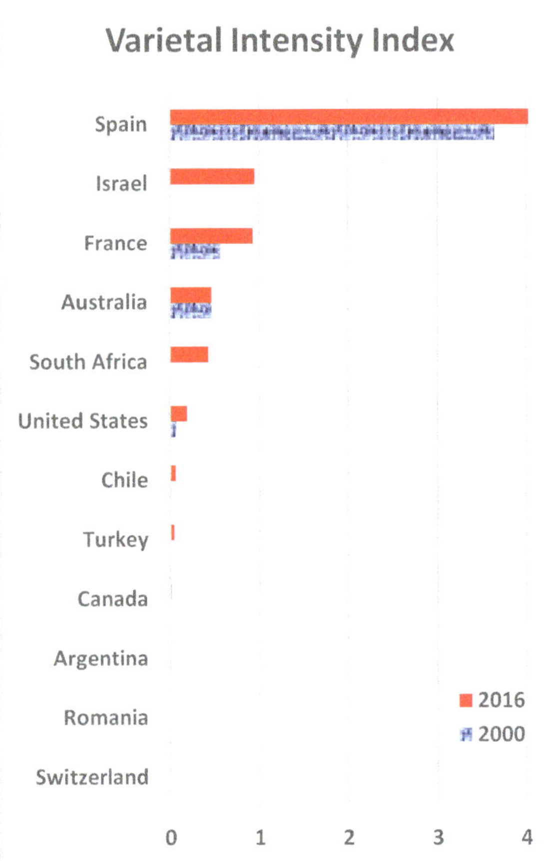

76. Muscat Blanc à Petits Grains

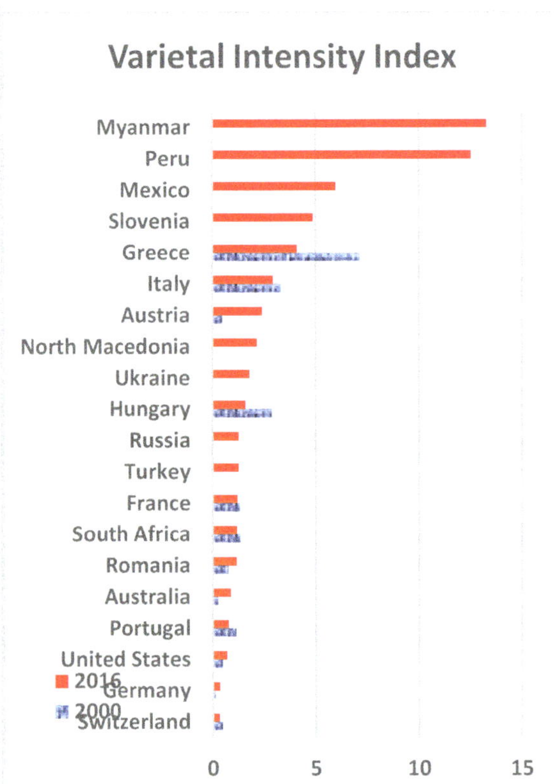

77. Muscat of Alexandria

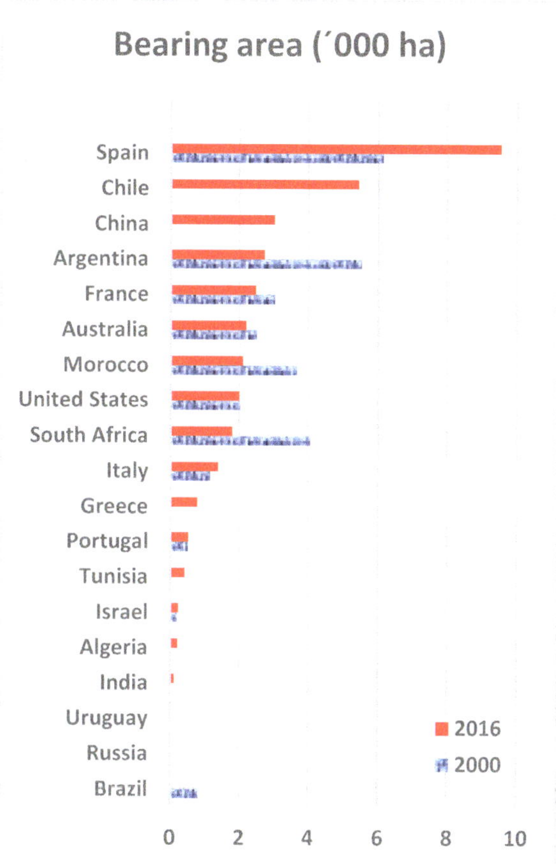

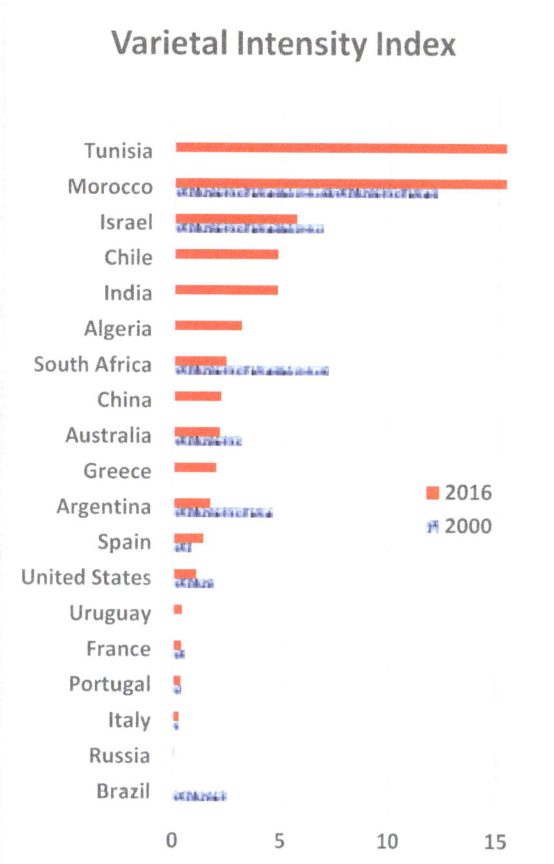

78. Pinot Gris

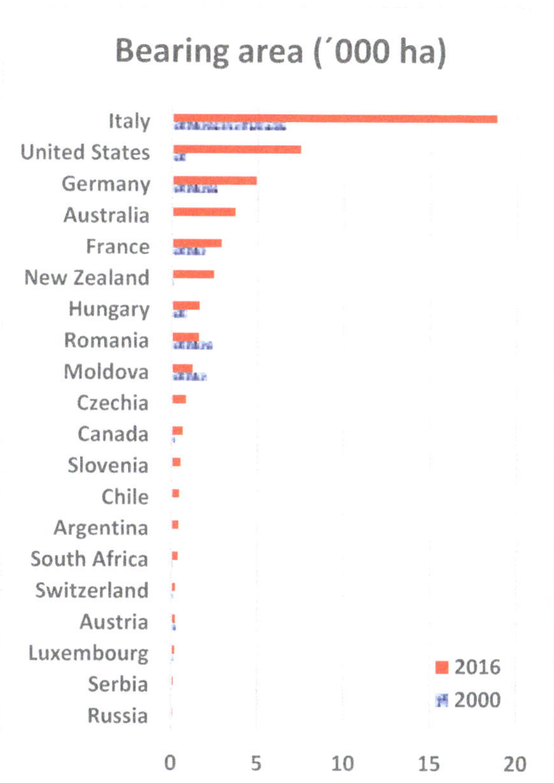

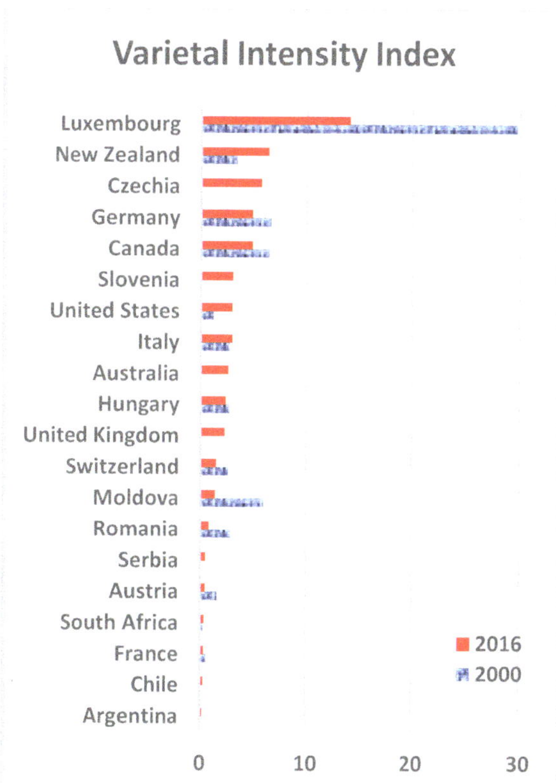

79. Pinot Noir

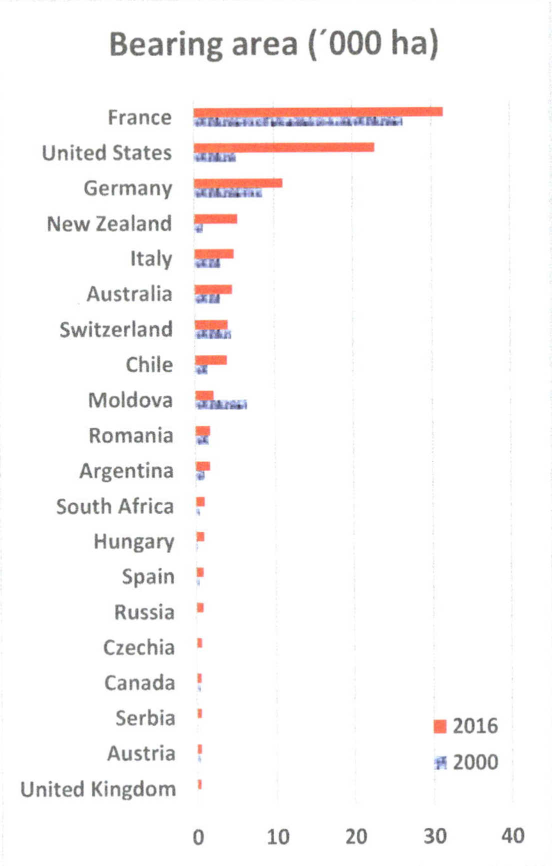

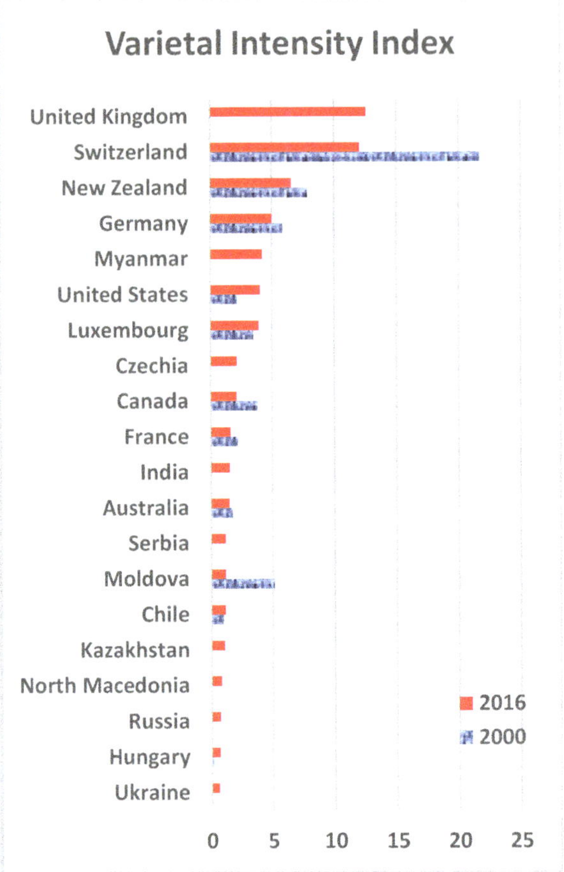

80. Riesling

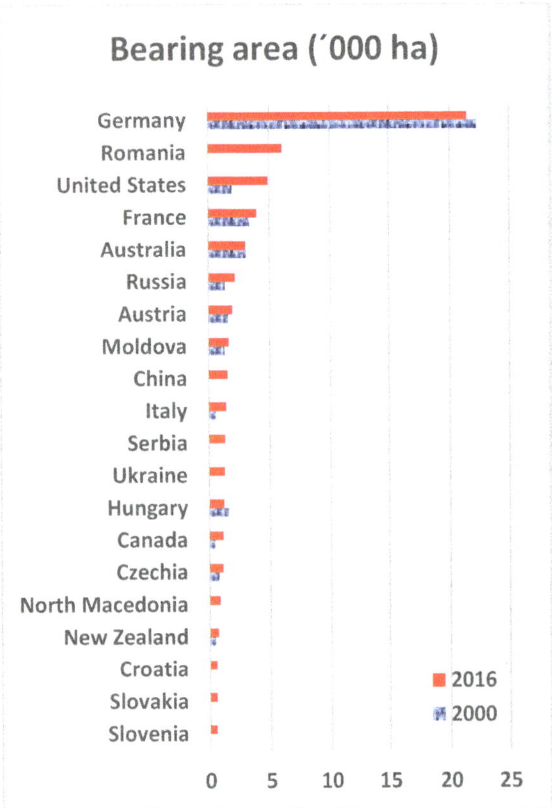

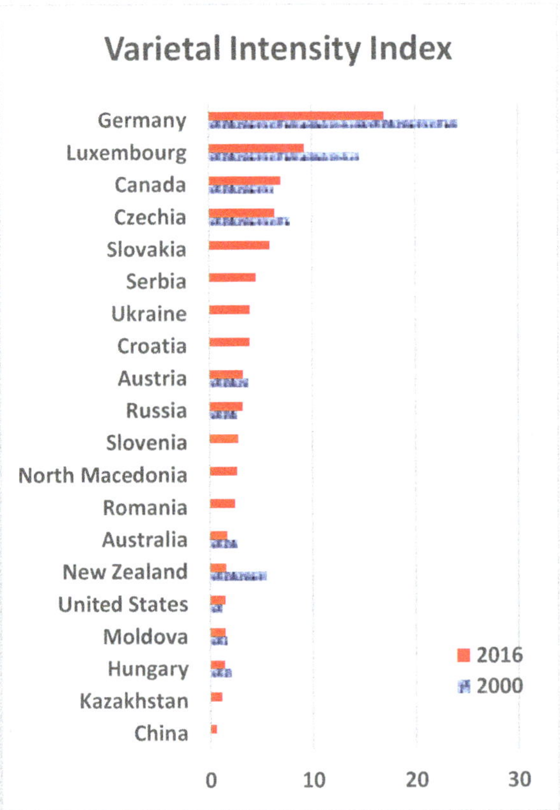

81. Rkatsiteli

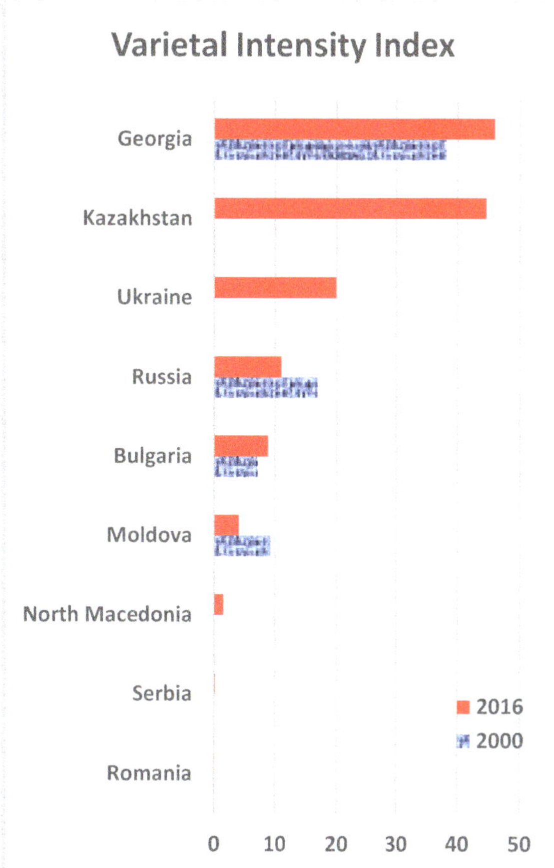

82. Sangiovese

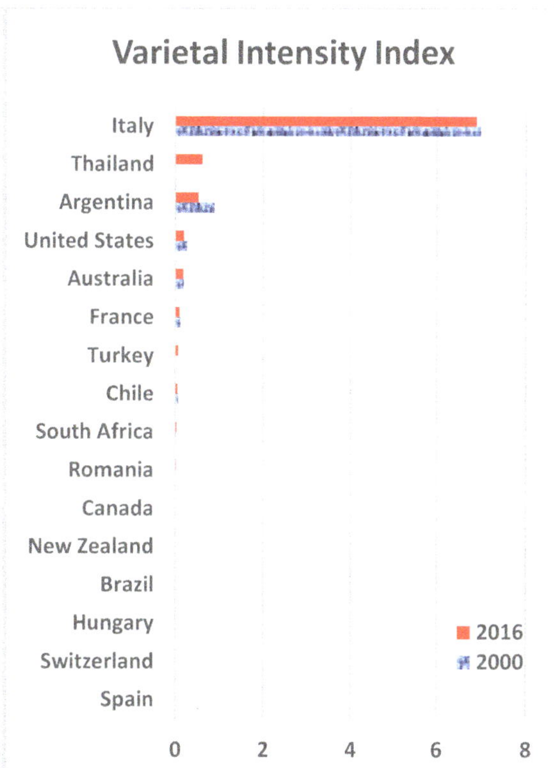

83. Sauvignon Blanc

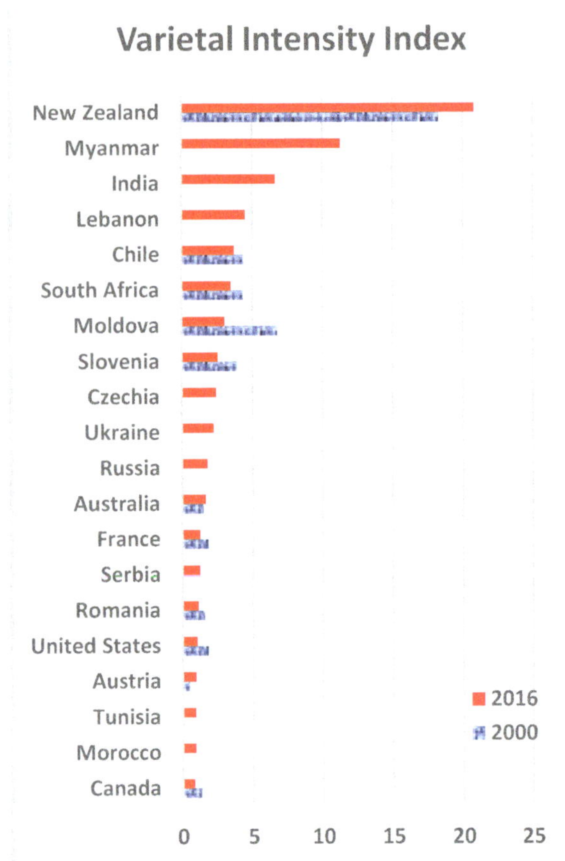

84. Syrah

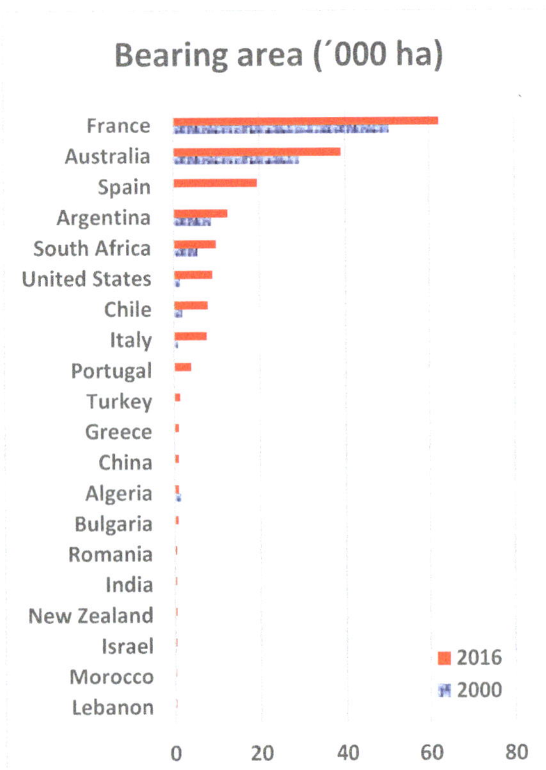

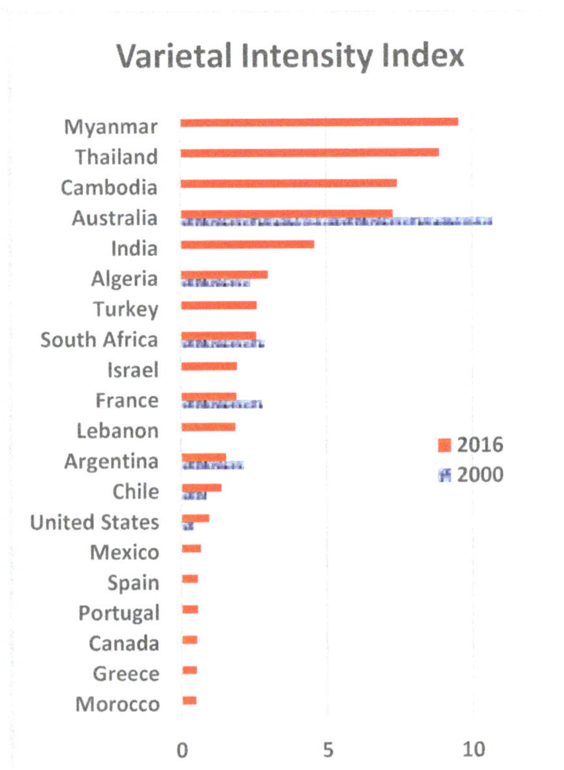

85. Tempranillo

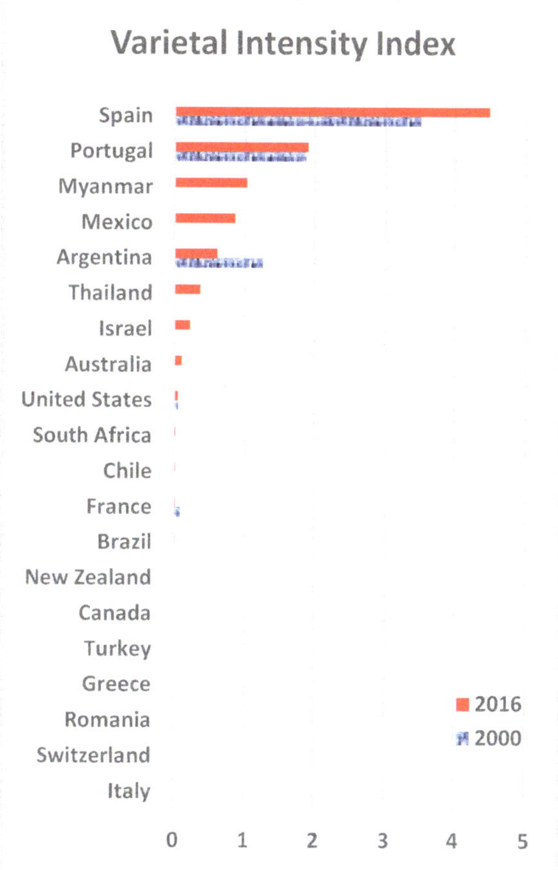

86. Trebbiano Toscano

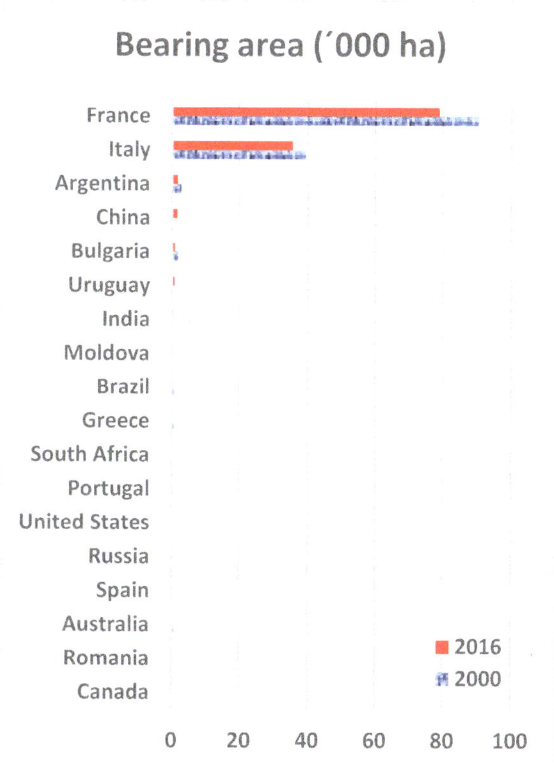

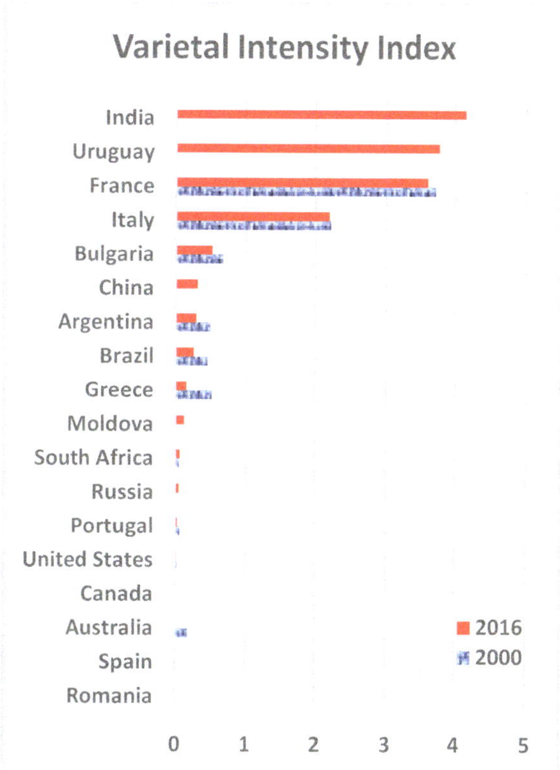

87. Tribidrag

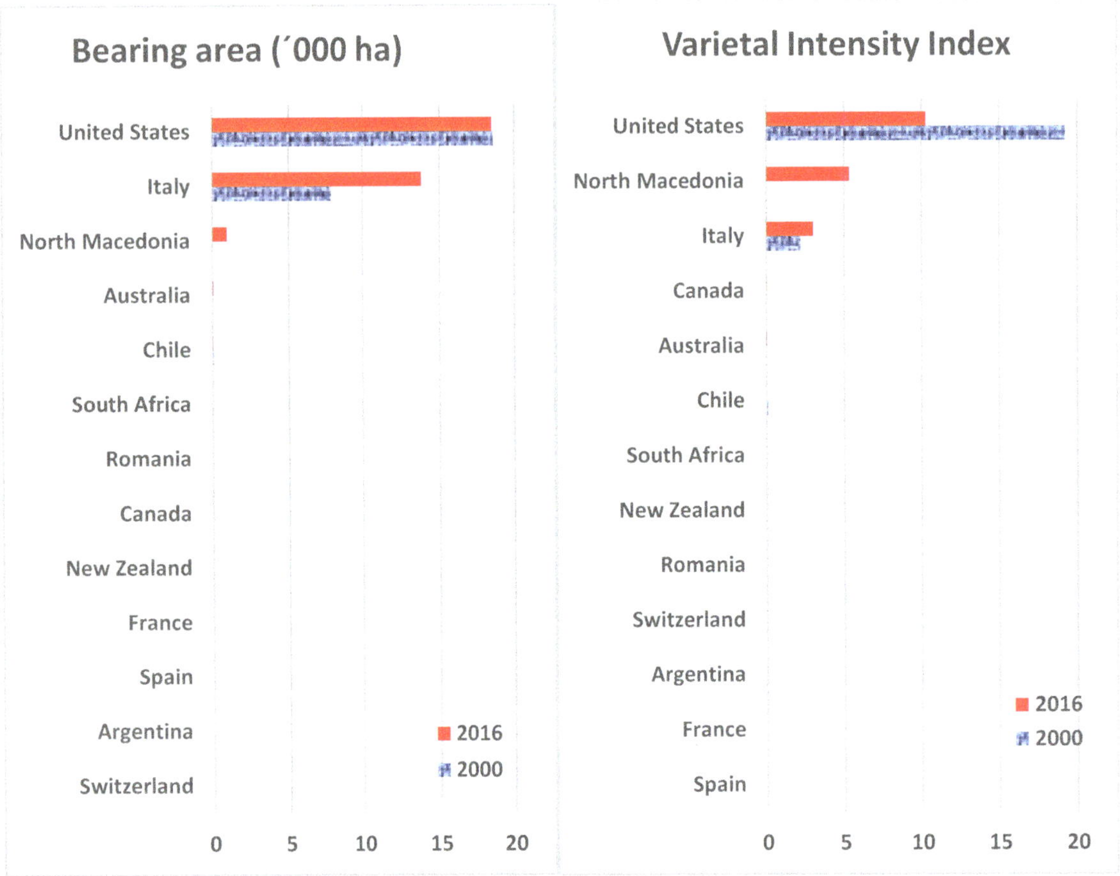

Tables

I. Country coverage

Table 1: Number of prime varieties and regions, by country, 2000, 2010 and 2016

Country		Code	2000 No. of prime varieties	2000 No. of regions	2010 No.of prime varieties	2010 No. of regions	2016 No.of prime varieties	2016 No. of regions
Algeria	OW	DZ	8	1	8	1	7	1
Argentina	NW	AR	111	75	117	92	107	123
Armenia	OW	AM	6	1	6	1	2	1
Australia	NW	AU	43	76	40	94	146	94
Austria	OW	AT	31	4	35	4	35	19
Brazil	NW	BR	18	1	98	1	86	1
Bulgaria	OW	BG	21	1	16	6	18	6
Cambodia	NW	KH					4	1
Canada	NW	CA	20	1	76	2	137	5
Chile	NW	CL	35	8	53	9	72	12
China	NW	CN			17	10	18	1
Croatia	OW	HR	7	1	71	13	9	2
Cyprus	OW	CY	2	1	15	1	2	1
Czechia	OW	CZ	10	1	32	2	14	5
Ethiopia	NW	ET			6	1	6	1
France	OW	FR	273	29	266	72	266	18
Georgia	OW	GE	21	1	21	1	21	1
Germany	OW	DE	67	13	74	13	103	13
Greece	OW	EL	53	13	52	13	54	13
Hungary	OW	HU	34	1	137	22	157	22
India	NW	IN					8	1
Israel	OW	IL	10	1	10	1	16	1
Italy	OW	IT	321	103	392	110	393	21
Japan	NW	JP			15	5	28	7
Kazakhstan	OW	KZ			14	6	14	6
Korea, Rep.	NW	KR	4	1	4	1	4	1
Lebanon	OW	LB					5	1
Luxembourg	OW	LU	11	1	10	1	11	1
Mexico	NW	MX			17	5	17	5
Moldova	OW	MD	39	1	39	1	87	1
Morocco	OW	MA	8	1	8	1	15	1
Myanmar	NW	MM			11	1	7	1
New Zealand	NW	NZ	22	10	45	11	51	11
North Macedonia	OW	MK					19	1
Norway	NW	NO					2	1
Peru	NW	PE			30	4	30	4
Portugal	OW	PT	78	9	252	9	253	7
Romania	OW	RO	18	1	99	8	102	1
Russia	OW	RU	11	1	53	2	55	3
Serbia	OW	RS	4	1	4	1	31	20
Slovakia	OW	SK	11	1	35	6	8	4
Slovenia	OW	SI	6	1	21	10	34	9
South Africa	NW	ZA	66	9	68	9	98	10
Spain	OW	ES	154	36	144	36	162	17
Switzerland	OW	CH	51	18	58	18	222	29
Taiwan	NW	TW	4	1	4	1	3	1
Thailand	NW	TH			13	1	15	1
Tunisia	OW	TN	9	1	9	1	15	1
Turkey	OW	TR			35	7	35	7
Ukraine	OW	UA			22	1	13	1
United Kingdom	NW	UK	9	1	43	1	16	1
United States	NW	US	83	61	129	89	142	98
Uruguay	NW	UY	8	1	40	1	38	15
"Missing 9 in 2000"		M9	100	1				
Total			**1001**	**489**	**1384**	**706**	**1558**	**629**

Table 2: National shares of global winegrape area and global wine production volume, 2000, 2010 and 2016 (%)

Included wine-producing countries	*Share (%) of global winegrape area*			*Share (%) of global wine production*		
	2000	2010	2016	2000	2010	2016
Algeria	0.61	0.65	0.18	0.16	0.18	0.19
Argentina	3.99	4.59	4.57	4.60	6.19	3.49
Armenia	0.23	0.24	0.33	0.02	0.02	0.02
Australia	2.64	3.26	2.93	3.15	4.39	4.86
Austria	0.98	0.98	1.01	0.86	0.66	0.72
Brazil	1.07	1.06	0.73	1.10	1.29	0.48
Bulgaria	1.94	1.21	1.17	0.67	0.57	0.45
Cambodia			0.00	0.00	0.00	0.00
Canada	0.17	0.22	0.28	0.16	0.21	0.20
Chile	2.31	2.40	3.23	2.09	3.54	3.72
China		0.64	3.94	1.09	3.62	3.92
Croatia	1.20	0.45	0.26	0.69	0.18	0.17
Cyprus	0.37	0.19	0.11	0.20	0.04	0.04
Czechia	0.23	0.35	0.30	0.19	0.17	0.17
Ethiopia		0.00	0.00	0.00	0.00	0.00
France	17.50	17.96	18.03	21.10	17.26	16.84
Georgia	0.76	1.03	1.06	0.26	0.39	0.33
Germany	2.11	2.20	2.09	3.61	2.63	3.34
Greece	1.03	1.17	1.13	1.35	1.28	0.93
Hungary	1.76	1.50	1.41	1.58	0.69	1.05
India			0.06	0.01	0.05	0.08
Israel	0.10	0.10	0.11	0.00	0.00	0.02
Italy	12.88	13.45	13.38	19.83	16.54	16.95
Japan		0.08	0.09	0.06	0.05	0.06
Kazakhstan		0.15	0.15	0.00	0.06	0.07
Korea, Rep.	0.11	0.12	0.12	0.10	0.10	0.12
Lebanon			0.09	0.00	0.05	0.05
Luxembourg	0.03	0.03	0.03	0.11	0.05	0.06
Mexico		0.12	0.12	0.38	0.13	0.14
Moldova	1.82	1.93	1.83	0.41	0.48	0.56
Morocco	1.00	1.05	0.39	0.11	0.13	0.13
Myanmar		0.00	0.00	0.00	0.00	0.00
New Zealand	0.20	0.69	0.78	0.22	0.72	1.16
North Macedonia			0.55	0.34	0.27	0.26
Norway			0.00	0.00	0.00	0.00
Peru		0.08	0.08	0.20	0.22	0.23
Portugal	4.15	3.51	4.04	2.34	2.63	2.23
Romania	4.50	3.66	4.04	2.00	1.25	1.22
Russia	1.14	0.55	1.12	1.13	2.90	2.45
Serbia	1.40	1.48	0.49	0.59	0.78	0.72
Slovakia	0.32	0.27	0.17	0.16	0.10	0.11
Slovenia	0.47	0.35	0.35	0.14	0.09	0.07
South Africa	1.89	2.17	2.12	3.07	3.75	3.91
Spain	23.91	22.10	19.55	15.10	13.46	14.83
Switzerland	0.30	0.32	0.33	0.47	0.39	0.41
Taiwan	0.06	0.06	0.00	0.00	0.00	0.00
Thailand		0.00	0.00	0.00	0.00	0.00
Tunisia	0.34	0.36	0.08	0.15	0.08	0.08
Turkey		0.28	0.30	0.09	0.11	0.16
Ukraine		1.12	0.56	0.47	1.14	0.45
United Kingdom	0.02	0.03	0.04	0.01	0.01	0.01
United States	3.55	4.90	5.30	8.55	10.09	11.68
Uruguay	0.18	0.16	0.15	0.33	0.25	0.28
"Missing 9 in 2000"	1.62					
Rest of world	*1.11*	*0.79*	*0.80*	*1.11*	*0.79*	*0.80*
Total	**100.00**	**100.00**	**100.00**	**100.00**	**100.00**	**100.00**

Non-sampled wine-producing countries	*Share (%) of global wine prodn, 2016*
Turkmenistan	0.13
Uzbekistan	0.12
Belarus	0.10
Albania	0.06
Montenegro	0.05
Colombia	0.04
Cuba	0.04
Estonia	0.04
Azerbaijan	0.03
Bolivia	0.03
Madagascar	0.03
Bosnia and Herzegc	0.03
Lithuania	0.02
Egypt	0.02
Belgium	0.01
Latvia	0.01
Malta	0.01
Zimbabwe	0.01
Kyrgyzstan	0.01
Paraguay	0.01
	0.80

Table 3: Key indicators of national grape area and production, 2000, 2010 and 2016

2000	*Total grapevine area harvested*[a]*('000ha)*	*Share of world grapevine area harvested (%)*	*Total wine-grape area ('000ha)*	*Share of world wine-grape area (%)*	*Share of wine-grapes in total grapevine area (%)*	*Share (%) of national agr. crop area under grapevine*	*Grape yield (tonnes per ha.)*	*Total grape prod'n (kt)*	*Share of world grape prod'n (%)*
Algeria	51	0.7	30.2	0.6	59	0.4	3.8	193	0.3
Argentina	201	2.7	197.4	4.0	98	0.7	12.2	2461	3.9
Armenia	15	0.2	11.2	0.2	76	2.3	7.9	116	0.2
Australia	137	1.9	130.6	2.6	95	0.3	10.0	1374	2.2
Austria	48	0.7	48.5	1.0	100	3.3	6.9	332	0.5
Brazil	61	0.8	52.8	1.1	87	0.1	16.5	1005	1.6
Bulgaria	117	1.6	96.0	1.9	82	2.5	3.6	419	0.7
Canada	8	0.1	8.5	0.2	100	0.0	8.1	69	0.1
Chile	160	2.2	114.0	2.3	71	5.4	11.0	1758	2.8
Croatia	59	0.8	59.4	1.2	100	5.9	6.2	369	0.6
Cyprus	18	0.2	18.3	0.4	100	13.0	5.6	103	0.2
Czechia	12	0.2	11.3	0.2	97	0.3	5.8	67	0.1
France	865	11.8	864.8	17.5	100	4.4	8.9	7676	12.3
Georgia	62	0.9	37.4	0.8	60	3.5	3.1	193	0.3
Germany	104	1.4	104.2	2.1	100	0.9	13.6	1415	2.3
Greece	126	1.7	50.9	1.0	40	1.3	9.8	1242	2.0
Hungary	90	1.2	86.9	1.8	97	1.8	7.7	688	1.1
Israel	7	0.1	4.9	0.1	69		14.6	102	0.2
Italy	870	11.9	636.7	12.9	73	5.6	10.4	9073	14.5
Korea, Rep.	29	0.4	5.4	0.1	18	1.5	5.6	163	0.3
Luxembourg	1	0.0	1.3	0.0	100	3.2	9.2	12	0.0
Moldova	146	2.0	89.8	1.8	61	4.1	3.8	558	0.9
Morocco	50	0.7	49.6	1.0	100	0.5	5.6	277	0.4
New Zealand	10	0.1	9.9	0.2	97	0.6	7.5	77	0.1
Portugal	235	3.2	205.0	4.1	87	8.6	4.3	1014	1.6
Romania	248	3.4	222.2	4.5	90	2.3	4.8	1178	1.9
Russia	61	0.8	56.3	1.1	93	0.0	4.2	254	0.4
Serbia	69	0.9	69.0	1.4	100	1.8	0.5	34	0.1
Slovakia	18	0.2	15.6	0.3	88	1.0	3.7	65	0.1
Slovenia	23	0.3	23.5	0.5	100	11.6	4.7	111	0.2
South Africa	108	1.5	93.7	1.9	87	0.7	13.4	1446	2.3
Spain	1182	16.1	1181.8	23.9	100	6.5	4.9	5806	9.3
Switzerland	15	0.2	15.0	0.3	100	3.4	10.8	163	0.3
Taiwan		0.0	2.8	0.1		0.0			0.0
Tunisia	28	0.4	16.8	0.3	60	0.3	4.9	139	0.2
United Kingdom	1	0.0	0.9	0.0	100	0.0	1.6	1	0.0
United States	376	5.1	175.7	3.6	47	0.1	16.5	6197	9.9
Uruguay	9	0.1	8.9	0.2	97	0.6	13.3	122	0.2
"Missing 9 in 2000"	995	13.6	80.2	1.6	8		7.9	7889	12.6
Rest of the world	744	10.2	54.9	1.1	7	0.0	11.4	8509	13.6
World	**7323**	**100.0**	**4942.5**	**100.0**	**67**	**0.3**	**8.5**	**62408**	**100.0**

Table 3 (cont.) Key indicators of national grape area and production, 2000, 2010 and 2016

2010	*Total grapevine area harvested[a] ('000ha)*	*Share of world grapevine area harvested (%)*	*Total wine-grape area ('000ha)*	*Share of world wine-grape area (%)*	*Share of wine-grapes in total grapevine area (%)*	*Share (%) of national agr. crop area under grapevine*	*Grape yield (tonnes per ha.)*	*Total grape prod'n (kt)*	*Share of world grape prod'n (%)*
Algeria	70	1.0	30.2	0.6	43	0.4	6.9	485	0.7
Argentina	223	3.1	213.4	4.6	96	0.5	11.3	2516	3.7
Armenia	14	0.2	11.2	0.2	77	2.3	15.2	220	0.3
Australia	158	2.2	151.8	3.3	96	0.3	11.0	1732	2.5
Austria	46	0.6	45.5	1.0	100	3.2	6.7	307	0.5
Brazil	82	1.2	49.4	1.1	60	0.1	17.2	1421	2.1
Bulgaria	88	1.2	56.1	1.2	64	1.7	2.9	252	0.4
Canada	11	0.2	10.1	0.2	94	0.0	6.7	72	0.1
Chile	200	2.8	111.5	2.4	56	6.3	14.4	2884	4.2
China	547	7.7	29.5	0.6	5	0.0	15.6	8519	12.5
Croatia	33	0.5	20.8	0.4	63	2.1	6.2	206	0.3
Cyprus	9	0.1	8.6	0.2	100	7.3	3.2	28	0.0
Czechia	16	0.2	16.2	0.3	100	0.5	4.2	69	0.1
Ethiopia	2	0.0	0.2	0.0	10		2.9	5	0.0
France	847	11.9	835.6	18.0	99	4.4	7.3	6162	9.1
Georgia	49	0.7	48.0	1.0	99	8.8	2.9	143	0.2
Germany	102	1.4	102.1	2.2	100	0.8	11.2	1146	1.7
Greece	100	1.4	54.4	1.2	54	1.5	9.4	938	1.4
Hungary	75	1.1	69.7	1.5	93	1.5	5.7	432	0.6
Israel	6	0.1	4.9	0.1	85		16.6	95	0.1
Italy	768	10.8	625.7	13.4	81	6.6	10.0	7715	11.4
Japan	18	0.3	3.7	0.1	21	0.1	10.3	187	0.3
Kazakhstan	10	0.1	6.9	0.1	72	0.0	5.8	56	0.1
Korea, Rep.	18	0.2	5.4	0.1	31	1.0	17.4	306	0.4
Luxembourg	1	0.0	1.3	0.0	100	2.1	12.9	17	0.0
Mexico	27	0.4	5.5	0.1	20	0.0	10.8	288	0.4
Moldova	132	1.9	89.8	1.9	68	4.3	4.4	587	0.9
Morocco	49	0.7	49.0	1.1	100	0.5	7.2	355	0.5
Myanmar		0.0	0.1	0.0		0.0			0.0
New Zealand	33	0.5	32.0	0.7	97	5.8	7.3	238	0.4
Peru	15	0.2	3.8	0.1	25	0.1	18.5	281	0.4
Portugal	180	2.5	163.5	3.5	91	8.8	4.6	826	1.2
Romania	179	2.5	170.3	3.7	95	1.8	4.9	870	1.3
Russia	43	0.6	25.6	0.6	59	0.0	8.0	345	0.5
Serbia	69	1.0	69.0	1.5	100	1.9	5.2	362	0.5
Slovakia	13	0.2	12.6	0.3	100	0.9	3.0	37	0.1
Slovenia	16	0.2	16.4	0.4	100	8.3	7.0	114	0.2
South Africa	113	1.6	101.0	2.2	89	0.8	11.5	1305	1.9
Spain	1028	14.4	1028.3	22.1	100	6.0	5.7	5817	8.6
Switzerland	15	0.2	14.8	0.3	100	3.5	9.0	134	0.2
Taiwan		0.0	2.8	0.1					0.0
Thailand	4	0.1	0.1	0.0	3	0.0	17.5	77	0.1
Tunisia	29	0.4	16.8	0.4	58	0.3	3.5	103	0.2
Turkey	476	6.7	12.9	0.3	3	0.1	9.0	4272	6.3
Ukraine	69	1.0	52.3	1.1	76	0.2	6.7	466	0.7
United Kingdom	1	0.0	1.2	0.0	100	0.0	0.8	1	0.0
United States	385	5.4	227.9	4.9	59	0.1	17.4	6721	9.9
Uruguay	8	0.1	7.7	0.2	96	0.4	13.5	108	0.2
Rest of the world	748	10.5	36.8	0.8	5	0.0	12.2	9142	13.5
World[a]	**7128**	**100.0**	**4652.6**	**100.0**	**65**	**0.3**	**9.5**	**67955**	**100.0**

[a]Non-sample countries, which have a high (mostly non-wine) grape area harvested in 2009-11, are (in '000ha): Iran-230, Uzbekistan-108, India-991, Egypt-64, Afghanistan-61, Syria-51, Tajikistan-35, Macedonia-20, Turkmenistan-18, Korea-18, Pakistan-15, Yemen-14, Azerbaijan-11, Saudi Arabia-11, Lebanon-11

Table 3 (cont.) Key indicators of national grape area and production, 2000, 2010 and 2016

2016	*Total grapevine area harvested ('000ha)*	*Share of world grapevine area harvested (%)*	*Total wine-grape area ('000ha)*	*Share of world wine-grape area (%)*	*Share of wine-grapes in total grapevine area (%)*	*Share (%) of national agr. crop area under grapevine*	*Grape yield (tonnes per ha.)**	*Total grape prod'n (kt)**	*Share of world grape prod'n (%)**
Algeria	66	0.8	8.3	0.2	13	0.8	8.2	571	0.7
Argentina	224	2.9	206.3	4.6	92	0.6	7.9	1758	2.0
Armenia	16	0.2	14.7	0.3	90	3.2	10.9	179	0.2
Australia	136	1.7	132.4	2.9	97	0.4	13.0	1773	2.0
Austria	45	0.6	45.4	1.0	100	3.2	5.7	260	0.3
Brazil	77	1.0	33.2	0.7	43	0.1	14.4	1113	1.3
Bulgaria	53	0.7	53.0	1.2	100	1.5	4.0	211	0.2
Cambodia	na	0.0	0.0	0.0	na	0.0			0.0
Canada	na	0.0	12.6	0.3	na	0.0	9.0	117	0.1
Chile	214	2.7	145.9	3.2	68	12.4	10.3	2200	2.5
China	799	10.2	178.0	3.9	22	0.6	16.8	12629	14.5
Croatia	26	0.3	11.7	0.3	46	2.8	5.4	124	0.1
Cyprus	7	0.1	5.1	0.1	78	5.2	2.9	18	0.0
Czechia	16	0.2	13.6	0.3	86	0.6	4.8	76	0.1
Ethiopia	2	0.0	0.2	0.0	7	0.0	2.2	5	0.0
France	815	10.4	814.9	18.0	100	4.2	7.7	6247	7.2
Georgia	57	0.7	48.0	1.1	85	11.8	2.8	159	0.2
Germany	100	1.3	94.5	2.1	95	0.8	12.3	1226	1.4
Greece	110	1.4	50.8	1.1	46	3.3	11.9	1083	1.2
Hungary	72	0.9	63.9	1.4	88	1.6	7.0	476	0.5
India	120	1.5	2.7	0.1	2	0.1	21.2	2590	3.0
Israel	8	0.1	5.0	0.1	62	2.0	8.2	66	0.1
Italy	673	8.6	604.6	13.4	90	7.4	12.3	8202	9.4
Japan	17	0.2	3.9	0.1	23	0.4	10.5	179	0.2
Kazakhstan	15	0.2	6.9	0.2	47	0.0	5.1	75	0.1
Korea, Rep.	15	0.2	5.4	0.1	35	0.9	14.9	229	0.3
Lebanon	8	0.1	4.0	0.1	47	3.3	8.3	70	0.1
Luxembourg	1	0.0	1.3	0.0	100	2.0	8.5	11	0.0
Mexico	28	0.4	5.5	0.1	19	0.1	13.0	337	0.4
Moldova	129	1.6	82.6	1.8	64	6.5	4.8	616	0.7
Morocco	45	0.6	17.6	0.4	39	0.5	7.9	365	0.4
Myanmar	na	0.0	0.1	0.0	na	0.0			0.0
New Zealand	40	0.5	35.5	0.8	88	6.3	10.6	436	0.5
North Macedonia	25	0.3	24.8	0.5	100	5.5	14.5	333	0.4
Norway	na	0.0	0.0	0.0	na	0.0			0.0
Peru	28	0.4	3.8	0.1	14	0.6	24.7	690	0.8
Portugal	183	2.3	182.6	4.0	100	10.0	4.2	774	0.9
Romania	183	2.3	182.8	4.0	100	2.0	4.0	737	0.8
Russia	63	0.8	50.8	1.1	81	0.1	8.7	601	0.7
Serbia	22	0.3	22.0	0.5	100	0.8	6.6	146	0.2
Slovakia	9	0.1	7.7	0.2	88	0.6	4.3	38	0.0
Slovenia	16	0.2	16.0	0.4	100	6.7	5.9	95	0.1
South Africa	120	1.5	95.8	2.1	79	1.0	16.3	1966	2.3
Spain	941	12.0	883.6	19.6	94	5.5	6.5	6103	7.0
Switzerland	15	0.2	14.8	0.3	100	3.5	9.2	136	0.2
Taiwan	3	0.0	0.1	0.0	5	0.4	27.1	80	0.1
Thailand	5	0.1	0.2	0.0	5	0.0	17.7	80	0.1
Tunisia	22	0.3	3.4	0.1	15	0.4	6.7	134	0.2
Turkey	462	5.9	13.7	0.3	3	1.9	9.2	4000	4.6
Ukraine	50	0.6	25.2	0.6	50	0.1	8.8	378	0.4
United Kingdom	2	0.0	1.8	0.0	100	0.0	1.1	1	0.0
United States	414	5.3	239.6	5.3	58	0.3	17.1	6983	8.0
Uruguay	7	0.1	6.7	0.1	99	0.3	15.4	105	0.1
Rest of the world	1320	16.9	36.1	0.8	3	0.2	6.2	20018	23.1
World	**7824**	**100.0**	**4519.2**	**100.0**	**58**	**0.5**	**11.3**	**86799**	**100**

* These final three columns refer to 2014

Table 4: National winegrape areas, 2000, 2010 and 2016, and changes from 2000 to 2016 and 2010 to 2016 (ha and %)

Country	*Area in 2000 (hectares)*	*Area in 2010 (hectares)*	*Area in 2016 (hectares)*	*Hectare change, 2000 to 2016*	*Hectare change, 2010 to 2016*	*% change, 2000-16*	*% change, 2010-16*
Algeria	30200	30200	8300	-21,900	-21,900	-73	-73
Argentina	197418	213372	206342	8,924	-7,030	5	-3
Armenia	11206	11206	14705	3,499	3,499	31	31
Australia	130602	151788	132435	1,833	-19,353	1	-13
Austria	48496	45533	45439	-3,057	-94	-6	0
Brazil	52840	49412	33205	-19,635	-16,207	-37	-33
Bulgaria	95997	56133	52974	-43,023	-3,159	-45	-6
Cambodia			10				
Canada	8498	10096	12600	4,102	2,504	48	25
Chile	113966	111525	145873	31,907	34,348	28	31
China		29545	178000				502
Croatia	59448	20754	11746	-47,702	-9,008	-80	-43
Cyprus	18282	8608	5133	-13,149	-3,475	-72	-40
Czechia	11331	16242	13600	2,269	-2,642	20	-16
Ethiopia		169	169				0
France	864846	835554	814882	-49,964	-20,672	-6	-2
Georgia	37419	48001	48000	10,581	-1	28	0
Germany	104233	102135	94501	-9,732	-7,634	-9	-7
Greece	50915	54390	50845	-70	-3,545	0	-7
Hungary	86886	69715	63881	-23,006	-5,835	-26	-8
India			2700				
Israel	4851	4851	5000	149	149	3	3
Italy	636662	625700	604551	32,111	21,149	-5	-3
Japan		3715	3869				4
Kazakhstan		6938	6938				0
Korea, Rep.	5400	5400	5400	0	0	0	0
Lebanon			4000				
Luxembourg	1348	1304	1300	-48	-4	-4	0
Mexico		5465	5465				0
Moldova	89844	89844	82600	-7,244	-7,244	-8	-8
Morocco	49600	49000	17590	-32,010	-31,410	-65	-64
Myanmar		75	70				-7
New Zealand	9942	31964	35463	25,521	3,500	257	11
North Macedonia			24777				
Norway			13				
Peru		3831	3831				0
Portugal	205003	163522	182649	-22,354	19,127	-11	12
Romania	222173	170292	182762	-39,411	12,470	-18	7
Russia	56332	25628	50794	-5,538	25,166	-10	98
Serbia	68999	68999	22014	-46,985	-46,985	-68	-68
Slovakia	15580	12637	7748	-7,832	-4,889	-50	-39
Slovenia	23472	16354	15989	-7,483	-365	-32	-2
South Africa	93656	101016	95775	2,120	-5,241	2	-5
Spain	1181806	1028258	883558	-298,248	-144,700	-25	-14
Switzerland	15042	14820	14793	-249	-27	-2	0
Taiwan	2833	2833	149	-2,684	-2,684	-95	-95
Thailand		149	208				39
Tunisia	16836	16836	3400	-13,436	-13,436	-80	-80
Turkey		12856	13704				7
Ukraine		52293	25166				-52
United Kingdom	873	1198	1839	966	641	111	54
United States	175693	227949	239632	63,938	11,683	36	5
Uruguay	8880	7657	6743	-2,137	-914	-24	-12
"Missing 9 in 2000"	80221						
Old World subtotal	**4006807**	**3658602**	**3373338**	**-570,272**	**-285,264**	-16	-8
New World subtotal	**800601**	**957160**	**1109790**	**245,992**	**152,631**	39	16
World total	**4887629**	**4615761**	**4483128**	**-404,501**	**-132,633**	-8	-3

II. Winegrape varietal coverage

Table 5: Global winegrape area, share of global area, and global ranks of each prime variety (alphabetic), 2000, 2010 and 2016 (ha and %)

		2000			2010			2016		
Prime vairiety	*Col*	*Global area (ha)*	*Global share %*	*Global rank*	*Global area (ha)*	*Global share %*	*Global rank*	*Global area (ha)*	*Global share %*	*Global rank*
Abbou	R	2375	0.05	187	2375	0.05	170			
Abbuoto	R	696	0.01	332	37	0.00	789	18	0.00	869
Abondant	W	0	0.00	974	0	0.00	1317	0	0.00	1401
Abouriou	R	419	0.01	401	329	0.01	437	310	0.01	417
Abrusco	R	399	0.01	406	423	0.01	395	215	0.00	473
Accent	R							1	0.00	1297
Acolon	R				490	0.01	379	477	0.01	362
Adakarası	R	48	0.00	648	69	0.00	691	89	0.00	602
Adalmiina	W							5	0.00	1076
Adirondac	R							24	0.00	830
Admirable de Courtiller	W				28	0.00	841	27	0.00	808
Afus Ali	W	1837	0.04	212	381	0.01	409	211	0.00	477
Agadai	W	1265	0.03	261						
Agasfark	R							0	0.00	1519
Agiorgitiko	R	2320	0.05	190	2905	0.06	153	3272	0.07	139
Aglianico	R	9346	0.19	78	9995	0.22	72	9734	0.22	73
Aglianicone	R	148	0.00	514	62	0.00	709	30	0.00	781
Agni	R				6	0.00	1061			
Agronomica	R	19	0.00	742	327	0.01	439	299	0.01	422
Agua Santa	R				78	0.00	671	76	0.00	633
Ahmeur Bou Ahmeur	R	3	0.00	867	1	0.00	1247	0	0.00	1407
Airén	W	387978	7.94	1	252364	5.47	3	203801	4.55	4
Aladasturi	R	46	0.00	650	59	0.00	718	59	0.00	665
Alarije	W	1686	0.03	220	1726	0.04	199	4407	0.10	126
Alb Aromat	W				24	0.00	863	24	0.00	832
Alb de Ialoveni	W	2	0.00	903	2	0.00	1175			
Alb de Suruceni	W							780	0.02	286
Albalonga	W	57	0.00	631	15	0.00	934	12	0.00	935
Albana	W	2487	0.05	183	1523	0.03	210	782	0.02	285
Albanello	W	117	0.00	541	18	0.00	911	2	0.00	1181
Albaranzeuli Bianco	W	72	0.00	601	7	0.00	1046	2	0.00	1164
Albaranzeuli Nero	R	40	0.00	665	49	0.00	754	28	0.00	800
Albarín Blanco	W				23	0.00	878	48	0.00	706
Albarola	W	4090	0.08	130	197	0.00	522	95	0.00	587
Albarossa	R	5	0.00	851	80	0.00	664	70	0.00	645
Albillo Mayor	W	5	0.00	848	1319	0.03	232	1152	0.03	241
Albillo Real	W	3368	0.07	148	861	0.02	296	601	0.01	329
Alcañon	W	54	0.00	637	60	0.00	716	27	0.00	804
Aleatico	R	458	0.01	391	346	0.01	429	165	0.00	512
Aledo	W				7	0.00	1043	2	0.00	1170
Aleksandrouli	R	219	0.00	476	281	0.01	463	281	0.01	436
Aletta	W				723	0.02	318	1676	0.04	196
Alfrocheiro	R	523	0.01	380	1188	0.03	249	1216	0.03	232
Alicante Henri Bouschet	R	37157	0.76	28	38462	0.83	24	36031	0.80	24
Aligoté	W	35668	0.73	29	36120	0.78	26	26929	0.60	33
Alionza	W	41	0.00	662	11	0.00	986	9	0.00	988
Almafra	W				0	0.00	1335	0	0.00	1495
Alphonse Lavallée	R	15	0.00	764	862	0.02	294	634	0.01	319
Altesse	W	294	0.01	444	359	0.01	419	227	0.01	464

Table 5 (cont.): Global winegrape area, share of global area, and global ranks of each prime variety (alphabetic), 2000, 2010 and 2016 (ha and %)

		2000			2010			2016		
Prime vairiety	*Col*	*Global area (ha)*	*Global share %*	*Global rank*	*Global area (ha)*	*Global share %*	*Global rank*	*Global area (ha)*	*Global share %*	*Global rank*
Alutus	R				2	0.00	1197	2	0.00	1217
Alvar Branco	W				0	0.00	1329	0	0.00	1457
Alvar Roxo	G				2	0.00	1192	2	0.00	1208
Alvarelhão	R	5272	0.11	117	5701	0.12	107	2910	0.06	147
Alvarelhao Ceitao	R				0	0.00	1381	0	0.00	1539
Alvarinho	W							5545	0.12	106
Amaral	R	582	0.01	362	92	0.00	634	93	0.00	589
Amigne	W	21	0.00	732	43	0.00	771	42	0.00	722
Amur	R				146	0.00	563	146	0.00	523
Amurg	R				3	0.00	1133	3	0.00	1119
Ancellotta	R	4405	0.09	122	4681	0.10	115	2739	0.06	154
Andor	W							1	0.00	1349
Andre	W				477	0.01	382	5	0.00	1073
Antao Vaz	W	376	0.01	412	1252	0.03	239	1768	0.04	190
Apiren Alb	W							1	0.00	1247
Apiren Roz	R							0	0.00	1376
Ar110	G	1	0.00	950	1	0.00	1244	1	0.00	1299
Ar99	G	3	0.00	864	5	0.00	1085	5	0.00	1071
Aramon Bouschet	R	10	0.00	802						
Aramon Noir	R	9157	0.19	80	2601	0.06	163	1181	0.03	238
Aramon Noir (W)	R	43	0.00	654	15	0.00	937	14	0.00	912
Aramont	R	1	0.00	941				0	0.00	1454
Aranel	W	22	0.00	729	5	0.00	1080	5	0.00	1060
Arany Sárfehér	W	2914	0.06	164	1133	0.02	255	586	0.01	331
Arbane	W	1	0.00	930	2	0.00	1190	1	0.00	1256
Arcas	R				1	0.00	1260	1	0.00	1306
Argaman	R	202	0.00	483	202	0.00	518	275	0.01	437
Ariana	R				3	0.00	1136			
Arinarnoa	R	150	0.00	513	189	0.00	528	486	0.01	358
Arinto de Bucelas	W	3966	0.08	134	4482	0.10	121	5409	0.12	107
Arkadia	W							303	0.01	420
Arneis	W	738	0.02	327	1122	0.02	256	1179	0.03	239
Arnsburger	W	3	0.00	871	30	0.00	824	29	0.00	786
Aromat de Iasi	W				62	0.00	710	66	0.00	652
Aromella	W							3	0.00	1145
Arriloba	W	59	0.00	627	55	0.00	729	54	0.00	677
Arrouya	R				0	0.00	1382	0	0.00	1542
Arrufiac	W	126	0.00	530	80	0.00	663	9	0.00	987
Arvesiniadu	W	147	0.00	516	30	0.00	829	13	0.00	927
Arvine	W	61	0.00	622	172	0.00	540	192	0.00	492
Asirtiko Red	R	22	0.00	730	5	0.00	1088	5	0.00	1067
Aspiran Bouschet	R	433	0.01	398	2245	0.05	176	4088	0.09	130
Asprouda	W	433	0.01	397	113	0.00	598	120	0.00	556
Assaraky	W				1	0.00	1235	0	0.00	1364
Assyrtiko	W	1106	0.02	279	902	0.02	286	1770	0.04	189
Astra	W				0	0.00	1332	0	0.00	1432
Athiri	W	1350	0.03	251	748	0.02	312	577	0.01	335
Aubin Blanc	W	2	0.00	913	1	0.00	1258	1	0.00	1335
Aubun	R	1411	0.03	243	553	0.01	358	537	0.01	344
Augster Blau	R							0	0.00	1487
Augster Weiss	W				1	0.00	1277	1	0.00	1336
Augustovski	W							0	0.00	1409
Aurelius	W				70	0.00	689			

Table 5 (cont.): Global winegrape area, share of global area, and global ranks of each prime variety (alphabetic), 2000, 2010 and 2016 (ha and %)

		2000			2010			2016		
Prime vairiety	*Col*	*Global area (ha)*	*Global share %*	*Global rank*	*Global area (ha)*	*Global share %*	*Global rank*	*Global area (ha)*	*Global share %*	*Global rank*
Aurore	W	299	0.01	439	268	0.01	473	255	0.01	446
Auxerrois	W	2302	0.05	191	2785	0.06	156	2853	0.06	149
Avana	R	53	0.00	641	28	0.00	842	18	0.00	870
Avarengo	R	1453	0.03	238	987	0.02	274	153	0.00	518
Avesso	W	636	0.01	344	685	0.01	328	699	0.02	301
Azal	W	3302	0.07	150	1072	0.02	267	1443	0.03	215
Băbească Neagră	R	3722	0.08	141	3122	0.07	146	2696	0.06	156
Băbească Neagră (G)	R				328	0.01	438	297	0.01	424
Babić	R	1189	0.02	271	359	0.01	420			
Babica	R				18	0.00	910			
Babosa de Madere	W				2	0.00	1170	2	0.00	1227
Bacchus	W	3374	0.07	147	2113	0.05	182	1759	0.04	191
Baco Blanc	W	2137	0.04	203	739	0.02	314	528	0.01	348
Baco Noir	R	397	0.01	407	475	0.01	383	735	0.02	293
Bácska	G							7	0.00	1011
Baga	R	6730	0.14	101	4108	0.09	125	6750	0.15	97
Bailey	R				34	0.00	803	49	0.00	702
Bakator Belyi	W				11	0.00	987	11	0.00	944
Bakator Kék	R				3	0.00	1154	2	0.00	1166
Bakator Roz	R				16	0.00	928	16	0.00	887
Balada	R				0	0.00	1297	0	0.00	1368
Baleille	W	1	0.00	923						
Baratuciat	W				2	0.00	1191	2	0.00	1209
Barbarossa	G	16	0.00	757						
Barbaroux	G	79	0.00	586	30	0.00	828	29	0.00	787
Barbera	R	33041	0.68	31	24366	0.53	38	17824	0.40	46
Barbera Bianca	W	251	0.01	464	181	0.00	534	114	0.00	563
Barbera Sarda	R	326	0.01	430	84	0.00	651	70	0.00	644
Barcelo	W	34	0.00	682	23	0.00	880	26	0.00	809
Bariadorgia	W	0	0.00	972						
Barkhatnyi	W				30	0.00	827	30	0.00	783
Baron	R							1	0.00	1274
Baroque	W	169	0.00	499	94	0.00	631	83	0.00	613
Barreto de Semente	R				3	0.00	1158	3	0.00	1161
Barsaglina	R	20	0.00	734	17	0.00	918	9	0.00	989
Bastardo Branco	W				15	0.00	942	14	0.00	914
Bastardo Magarachsky	R	1969	0.04	209	2370	0.05	171	180	0.00	500
Batili	W	27	0.00	703						
Batily	R	38	0.00	671	54	0.00	734	1	0.00	1277
Batoca	W	80	0.00	584	11	0.00	974	8	0.00	999
Batuta Neagra	R				3	0.00	1150	3	0.00	1143
Bayanshira	W	451	0.01	394	645	0.01	336	645	0.01	316
Beba	W	8250	0.17	84	6524	0.14	96	2556	0.06	161
Beclan	R	0	0.00	985	0	0.00	1324	0	0.00	1504
Beibinghong	R							1600	0.04	206
Bellandais	R	2	0.00	917						
Bellone	W	1315	0.03	256	511	0.01	370	184	0.00	497
Beogradska Crna	W							0	0.00	1419
Bequignol Gris	W				3	0.00	1149	2	0.00	1175
Béquignol Noir	R	1083	0.02	285	891	0.02	289	616	0.01	325
Bianca	W	2180	0.04	201	6462	0.14	97	9766	0.22	72
Biancame	W	1330	0.03	253	2599	0.06	164	1336	0.03	223
Bianchetta Trevigiana	W	53	0.00	639	13	0.00	952	12	0.00	940

Table 5 (cont.): Global winegrape area, share of global area, and global ranks of each prime variety (alphabetic), 2000, 2010 and 2016 (ha and %)

		2000			2010			2016		
Prime vairiety	*Col*	*Global area (ha)*	*Global share %*	*Global rank*	*Global area (ha)*	*Global share %*	*Global rank*	*Global area (ha)*	*Global share %*	*Global rank*
Bianco d'Alessano	W	941	0.02	300	419	0.01	398	39	0.00	728
Biancolella	W	385	0.01	409	164	0.00	547	23	0.00	843
Biancone di Portoferraio	W	67	0.00	616	78	0.00	670	34	0.00	761
Biancu Gentile	W	1	0.00	940	9	0.00	1018	5	0.00	1058
Bíborkadarka	R	202	0.00	482	136	0.00	575	109	0.00	573
Bical	W	912	0.02	305	924	0.02	281	1076	0.02	251
Birstaler Muskat	W							1	0.00	1288
Black Prince	R							1	0.00	1344
Black Queen	R	566	0.01	368	713	0.02	321	143	0.00	528
Blanc Dame	W	0	0.00	980	0	0.00	1338	0	0.00	1453
Blanc du Bois	W				28	0.00	838	81	0.00	620
Blanca Ovoide	W	107	0.00	551	40	0.00	778	44	0.00	718
Blanqueiro	W	1	0.00	958						
Blasius	W				14	0.00	946	15	0.00	899
Blattner Cal 1-15	R							0	0.00	1496
Blattner Cal 1-20	R							0	0.00	1369
Blattner Cal 1-22	R							0	0.00	1391
Blattner Cal 1-28	R							2	0.00	1216
Blattner Cal 1-31	R							0	0.00	1508
Blattner Cal 1-36	R							1	0.00	1252
Blattner Reds	R				39	0.00	781	8	0.00	995
Blattner Whites	W				25	0.00	855	7	0.00	1016
Blauburger	R	1002	0.02	292	1339	0.03	230	1223	0.03	231
Blauer Portugieser	R	9156	0.19	81	8027	0.17	87	6590	0.15	98
Blauer Wildbacher	R	472	0.01	388	368	0.01	415	437	0.01	377
Blaufränkisch	R	13997	0.29	59	17890	0.39	46	17180	0.38	48
Blush Seedless	R	10	0.00	799						
Boal Barreiro	W				1	0.00	1283	0	0.00	1418
Boal Vencedor	W				2	0.00	1161	2	0.00	1185
Bobal	R	100128	2.05	11	80120	1.74	13	59189	1.32	15
Boğazkere	R	773	0.02	321	1106	0.02	259	1436	0.03	218
Bogdanuša	W				48	0.00	759			
Boiziau	R	1	0.00	929						
Bokay	W							4	0.00	1094
Bombino Bianco	W	2903	0.06	165	1239	0.03	240	1147	0.03	242
Bombino Nero	R	1156	0.02	275	1201	0.03	247	865	0.02	275
Bonamico	R	336	0.01	427	233	0.01	497	149	0.00	521
Bonarda Grande	R	538	0.01	375						
Bonarda Piemontese	R	23	0.00	718	6	0.00	1064	5926	0.13	103
Bonda	R	3	0.00	873	7	0.00	1047	7	0.00	1008
Bondola	R	17	0.00	752	13	0.00	959	11	0.00	956
Borraçal	R	2654	0.05	177	683	0.01	329	512	0.01	350
Borsmenta	W							1	0.00	1292
Bosco	W	88	0.00	574	82	0.00	657	50	0.00	697
Bouchales	R	108	0.00	547	95	0.00	628	93	0.00	591
Bouillet	R	2	0.00	896	1	0.00	1265	1	0.00	1324
Bourboulenc	W	772	0.02	322	585	0.01	349	501	0.01	354
Bousquet Precoce	W	16	0.00	759	6	0.00	1071	0	0.00	1501
Bouvier	W	365	0.01	416	250	0.01	487	224	0.00	467
Bracciola Nera	R	89	0.00	572	26	0.00	850	4	0.00	1109
Brachetto del Piemonte	R	1534	0.03	232	1460	0.03	217	1694	0.04	194
Branco Escola	W				2	0.00	1189	2	0.00	1203
Branco Gouvaes	W				36	0.00	793	34	0.00	757

Table 5 (cont.): Global winegrape area, share of global area, and global ranks of each prime variety (alphabetic), 2000, 2010 and 2016 (ha and %)

		2000			2010			2016		
Prime vairiety	*Col*	*Global area (ha)*	*Global share %*	*Global rank*	*Global area (ha)*	*Global share %*	*Global rank*	*Global area (ha)*	*Global share %*	*Global rank*
Branco Sr. Joao	W				0	0.00	1299	0	0.00	1443
Branco Valente	W				0	0.00	1330	0	0.00	1471
Brandam	W				312	0.01	448	291	0.01	428
Braquet Noir	R	8	0.00	817	12	0.00	968	12	0.00	936
Breidecker	W	28	0.00	696	7	0.00	1042	0	0.00	1371
Brianna	W				12	0.00	964	21	0.00	853
Bric	R	21	0.00	733	2	0.00	1165	1	0.00	1242
Brocada	W	2	0.00	890						
Bronner	W				9	0.00	1017	6	0.00	1034
Brun Argente	R	14	0.00	772	11	0.00	978	11	0.00	954
Budai Zöld	W				6	0.00	1056	6	0.00	1045
Buffalo	R							0	0.00	1535
Bukettraube	W	280	0.01	447	71	0.00	685	54	0.00	676
Burdin	W	0	0.00	988	0	0.00	1365	0	0.00	1522
Bussanello	W	8	0.00	824	12	0.00	973	3	0.00	1132
Busuioacă de Bohotin	G				268	0.01	472	343	0.01	400
BX 81-83	R							1	0.00	1352
Cabaret Noir	R							3	0.00	1158
Caberinta	R	85	0.00	577	69	0.00	690	38	0.00	734
Cabernet Blanc	W							6	0.00	1033
Cabernet Cantor	R							1	0.00	1244
Cabernet Carbon	R							11	0.00	955
Cabernet Carol	R							6	0.00	1037
Cabernet Cortis	R							38	0.00	735
Cabernet Cubin	R				60	0.00	715	62	0.00	659
Cabernet Diane	R				0	0.00	1301	0	0.00	1374
Cabernet Dore	W				1	0.00	1219	1	0.00	1260
Cabernet Dorio	R				36	0.00	797	34	0.00	756
Cabernet Dorsa	R	43	0.00	655	252	0.01	486	272	0.01	438
Cabernet Early	R							0	0.00	1492
Cabernet Foch	R							2	0.00	1207
Cabernet Franc	R	51974	1.06	20	61295	1.33	17	56052	1.25	16
Cabernet Jura	R				19	0.00	899	27	0.00	802
Cabernet Malbec	R	34	0.00	681						
Cabernet Mitos	R				322	0.01	442	312	0.01	414
Cabernet Moravia	R				212	0.00	508			
Cabernet Sanzey	R							0	0.00	1475
Cabernet Sauvignon	R	223074	4.56	3	290083	6.28	1	310671	6.93	1
Cabernet Soyhières	R							1	0.00	1303
Cabernet x Maréchal Foch	R							0	0.00	1479
Cabertin	R							2	0.00	1223
Cabinda	R				362	0.01	418	355	0.01	395
Cabral	R				2	0.00	1186	2	0.00	1197
Caddiu	R	978	0.02	295	309	0.01	451	83	0.00	612
Caíño Blanco	W	69	0.00	609	128	0.00	586	77	0.00	630
Caladoc	R	1427	0.03	242	3675	0.08	132	5258	0.12	111
Calagrano	W	8229	0.17	85	4794	0.10	114			
Calitor Noir	R	85	0.00	579	26	0.00	849	26	0.00	817
Çalkarası	R	436	0.01	396	625	0.01	343	806	0.02	279
Callet	R	151	0.00	511	154	0.00	554	138	0.00	533
Calmeria	W							0	0.00	1404
Caloria	R	129	0.00	526	108	0.00	606	45	0.00	715
Calrao	R				1	0.00	1233	1	0.00	1305

Table 5 (cont.): Global winegrape area, share of global area, and global ranks of each prime variety (alphabetic), 2000, 2010 and 2016 (ha and %)

		2000			2010			2016		
Prime vairiety	*Col*	*Global area (ha)*	*Global share %*	*Global rank*	*Global area (ha)*	*Global share %*	*Global rank*	*Global area (ha)*	*Global share %*	*Global rank*
Camaralet de Lasseube	W	691	0.01	333	520	0.01	365	306	0.01	419
Campanario	R				2	0.00	1185	2	0.00	1200
Campbell Early	R	43	0.00	656	61	0.00	713	238	0.01	456
Canada Muscat	W	49	0.00	645				120	0.00	555
Canadice	R				0	0.00	1326	0	0.00	1373
Canaiolo Nero	R	2418	0.05	185	1068	0.02	268	1033	0.02	253
Canari Noir	R	218	0.00	479	163	0.00	549	84	0.00	611
Canela	G	0	0.00	984	1	0.00	1213	3	0.00	1146
Canelon	R	8	0.00	820	8	0.00	1024			
Canner Seedless	W				0	0.00	1368	0	0.00	1509
Canorroyo	W	157	0.00	509						
Capolongo	W				5	0.00	1079	0	0.00	1362
Caracol	W	14	0.00	773	33	0.00	814	33	0.00	766
Caramela	W				0	0.00	1292	0	0.00	1375
Carbernet Volos	R							0	0.00	1489
Cardinal	R	3870	0.08	136	536	0.01	361	1660	0.04	198
Carica l'Asino	W	299	0.01	438	17	0.00	913	5	0.00	1057
Carignan Bouschet	R	16	0.00	762	1	0.00	1203	1	0.00	1243
Carla	R				0	0.00	1358	0	0.00	1480
Carmem	R							328	0.01	405
Carmenère	R	5711	0.12	114	11366	0.25	65	22486	0.50	38
Carmine	R	10	0.00	805						
Carminoir	R				10	0.00	992	11	0.00	945
Carnelian	R	625	0.01	348	316	0.01	445	123	0.00	547
Carrega Branco	W				507	0.01	373	512	0.01	351
Carrega Tinto	R				17	0.00	917	17	0.00	880
Carricante	W	252	0.01	463	205	0.00	517	35	0.00	753
Cartouche	W	31	0.00	688						
Casavecchia	R				136	0.00	574	92	0.00	593
Cascade	R							22	0.00	848
Casculho	R				267	0.01	474	269	0.01	440
Casetta	R				12	0.00	962	14	0.00	922
Castalia	W				0	0.00	1333	0	0.00	1494
Castel	R	0	0.00	992	2	0.00	1169	3	0.00	1139
Castela	R				8	0.00	1026	7	0.00	1020
Castelão	R	14424	0.30	58	11088	0.24	66	12580	0.28	59
Castelão Branco	W				37	0.00	787	18	0.00	872
Castelino	R				147	0.00	560	144	0.00	525
Castellana Blanca	W							1	0.00	1272
Castelo Branco	W				5	0.00	1097	2	0.00	1176
Castets	R	0	0.00	998	0	0.00	1377	0	0.00	1545
Castiglione	R	83	0.00	581	18	0.00	906	4	0.00	1088
Castonotal	W	0	0.00	975						
Catalanesca	W				54	0.00	735	7	0.00	1019
Catanese Nero	R	76	0.00	592	15	0.00	941	7	0.00	1018
Catarratto Bianco	W	50711	1.04	21	34863	0.76	29	28613	0.64	32
Catawba	R	635	0.01	346	633	0.01	339	626	0.01	321
Caverdella	W							5	0.00	1066
Cavrara	R				23	0.00	877	1	0.00	1240
Çavuş	W							3	0.00	1151
Cayetana Blanca	W	55776	1.14	19	39781	0.86	23	36401	0.81	23
Cayuga White	W	108	0.00	549	212	0.00	509	217	0.00	470
Cellerina	R				2	0.00	1195	2	0.00	1222

Table 5 (cont.): Global winegrape area, share of global area, and global ranks of each prime variety (alphabetic), 2000, 2010 and 2016 (ha and %)

		2000			2010			2016		
Prime vairiety	*Col*	*Global area (ha)*	*Global share %*	*Global rank*	*Global area (ha)*	*Global share %*	*Global rank*	*Global area (ha)*	*Global share %*	*Global rank*
Centennial Seedless	W	2	0.00	888				1	0.00	1263
Centesimino	R				24	0.00	871	25	0.00	824
Centurian	R	134	0.00	520	34	0.00	805	33	0.00	764
Cep Rouge	R	0	0.00	996						
Cerceal Branco	W	597	0.01	358	379	0.01	410	261	0.01	443
Cereza	G	31113	0.64	32	29934	0.65	36	28887	0.64	31
Cesanese	R	1024	0.02	291	679	0.01	331	446	0.01	373
Cesar	R	8	0.00	815	10	0.00	989	15	0.00	909
Cetinka	W				35	0.00	802			
Chambourcin	R	257	0.01	460	1097	0.02	260	968	0.02	259
Champanel	R							1	0.00	1251
Chancellor	R	27	0.00	702	49	0.00	753	38	0.00	739
Chaouch Blanc	W	2	0.00	911	3	0.00	1156			
Chardonel	W				144	0.00	565	90	0.00	599
Chardonnay	W	145543	2.98	7	199743	4.33	5	201649	4.50	5
Chardoris	W							0	0.00	1448
Charmont	W	7	0.00	831	10	0.00	1003	10	0.00	968
Chasan	W	914	0.02	304	749	0.02	311	549	0.01	342
Chasselas	W	13318	0.27	60	13119	0.28	59	7377	0.16	93
Chasselas (R)	W	11	0.00	791	95	0.00	629	90	0.00	601
Chasselas Sabor	W				0	0.00	1370	0	0.00	1528
Chatus	R	15	0.00	765	79	0.00	667	71	0.00	640
Chelois	R				1	0.00	1202	2	0.00	1236
Chelva	W	10877	0.22	73	6168	0.13	102	5029	0.11	113
Chenanson	R	636	0.01	345	466	0.01	388	452	0.01	370
Chenel	W	339	0.01	425	79	0.00	666	33	0.00	765
Chenin Blanc	W	45761	0.94	24	35703	0.77	27	32221	0.72	29
Chenivesse	R	1	0.00	922						
Chinuri	W	955	0.02	298	1225	0.03	242	1225	0.03	230
Chkhaveri	G	20	0.00	736	26	0.00	852	26	0.00	815
Cianorie	R				2	0.00	1159	1	0.00	1255
Cidreiro	R				0	0.00	1344	0	0.00	1459
Cienna	R							70	0.00	642
Ciliegiolo	R	2527	0.05	182	1830	0.04	195	897	0.02	269
Cinsaut	R	48428	0.99	22	34751	0.75	30	22926	0.51	37
Cinsaut (G)	R							2	0.00	1237
Cinsaut (W)	R	41	0.00	660	7	0.00	1039	0	0.00	1442
Cinsaut Seedless	R	9	0.00	809	13	0.00	955			
Citronny Magarach	W				307	0.01	453	307	0.01	418
Cividin	W				4	0.00	1102	4	0.00	1091
Clairette	W	4359	0.09	124	3057	0.07	147	2420	0.05	165
Clara	W							0	0.00	1523
Clarin	W	11	0.00	792	6	0.00	1063	6	0.00	1046
Claverie	W	3	0.00	865	1	0.00	1264	1	0.00	1340
Clinton	R							0	0.00	1424
Coarnă Neagră	R							114	0.00	564
Cococciola	W	887	0.02	308	983	0.02	275	1671	0.04	197
Coda di Volpe Bianca	W	980	0.02	294	586	0.01	348	77	0.00	631
Codană	R				24	0.00	866	26	0.00	813
Codega de Larinho	W	4058	0.08	131	629	0.01	342	455	0.01	369
Codivarta	W	3	0.00	886	2	0.00	1167	1	0.00	1250
Colmar Precoce Noir	R							0	0.00	1426
Colobel	R	3	0.00	874	9	0.00	1007	8	0.00	1002

Table 5 (cont.): Global winegrape area, share of global area, and global ranks of each prime variety (alphabetic), 2000, 2010 and 2016 (ha and %)

Prime vairiety	Col	2000 Global area (ha)	2000 Global share %	2000 Global rank	2010 Global area (ha)	2010 Global share %	2010 Global rank	2016 Global area (ha)	2016 Global share %	2016 Global rank
Colombana Nera	R	126	0.00	528	38	0.00	784	16	0.00	886
Colombard	W	38632	0.79	26	32944	0.71	31	29996	0.67	30
Colomino	W	16	0.00	758	5	0.00	1096	2	0.00	1178
Coloraillo	R	614	0.01	350	374	0.01	412	109	0.00	574
Columna	W				24	0.00	870	26	0.00	818
Completer	W	2	0.00	899	3	0.00	1137	5	0.00	1078
Complexa	R	6	0.00	834	103	0.00	617	103	0.00	580
Concieira	R				52	0.00	742	53	0.00	685
Concord	R	11816	0.24	68	12238	0.27	61	10544	0.24	67
Concord Clone 30	R							196	0.00	489
Cora	R							570	0.01	336
Coracao de Galo	R				1	0.00	1257	1	0.00	1318
Corbina Vicentina	R				12	0.00	965	12	0.00	931
Cordenossa	R				5	0.00	1077	2	0.00	1211
Cornalin	R	93	0.00	568	256	0.01	481	147	0.00	522
Cornarea	R	22	0.00	731	13	0.00	953	8	0.00	997
Cornichon Blanc	W	4	0.00	859				1	0.00	1358
Cornifesto	R	259	0.01	458	499	0.01	377	509	0.01	352
Corot Noir	R				27	0.00	848	11	0.00	948
Cortese	W	3113	0.06	161	2953	0.06	152	2405	0.05	166
Corvina Veronese	R	4800	0.10	120	7496	0.16	91	6240	0.14	100
Corvinone	R	88	0.00	573	930	0.02	279	1140	0.03	244
Côt	R	26285	0.54	39	38158	0.83	25	52233	1.17	17
Couderc 13	W							474	0.01	363
Couderc Noir	R	614	0.01	351	3517	0.08	139	2136	0.05	174
Counoise	R	638	0.01	343	408	0.01	400	418	0.01	382
Courbu Blanc	W	47	0.00	649	43	0.00	770	32	0.00	769
Courbu Noir	R	2	0.00	915	1	0.00	1207	1	0.00	1254
Cove	W	56	0.00	633	6	0.00	1051	6	0.00	1039
Crâmpoşie Selecţionată	W				409	0.01	399	18	0.00	875
Crimposie	W				453	0.01	390	450	0.01	372
Crimson Cabernet	R				1	0.00	1218	1	0.00	1261
Crimson Seedless	R	1	0.00	927	8	0.00	1031	8	0.00	1006
Criolla Grande	R	24264	0.50	41	20745	0.45	42	15596	0.35	50
Criolla Mediana	R	1	0.00	920	3	0.00	1120	7	0.00	1017
Croatina	R	3116	0.06	160	5700	0.12	108	2695	0.06	157
Crouchen	W	2259	0.05	194	725	0.02	316	319	0.01	412
Crovassa	R	2	0.00	918	0	0.00	1325	0	0.00	1423
Cruciulita	W				0	0.00	1372	0	0.00	1510
Crystal	W	1	0.00	926	175	0.00	538	175	0.00	506
Csaba Gyöngye	W				89	0.00	638	175	0.00	505
Cserszegi Fűszeres	G	2185	0.04	200	3609	0.08	135	4299	0.10	128
Csillám	W				20	0.00	897	25	0.00	827
Csókaszőlő	R				2	0.00	1166	2	0.00	1232
Csomorika	W				0	0.00	1295	0	0.00	1367
Dakapo	R				51	0.00	747	68	0.00	649
Dalkauer	W	100	0.00	561						
Damaschino	W	3187	0.07	154	2171	0.05	178	1622	0.04	203
Danam	W				0	0.00	1322	0	0.00	1420
Danlas	W				255	0.01	482	203	0.00	482
Danuta	W				2	0.00	1171	2	0.00	1193
Dattier de St. Vallier	W				0	0.00	1298	0	0.00	1378
David Macgregor 8521-1	R							2	0.00	1214

Table 5 (cont.): Global winegrape area, share of global area, and global ranks of each prime variety (alphabetic), 2000, 2010 and 2016 (ha and %)

		2000			2010			2016		
Prime vairiety	*Col*	*Global area (ha)*	*Global share %*	*Global rank*	*Global area (ha)*	*Global share %*	*Global rank*	*Global area (ha)*	*Global share %*	*Global rank*
Dawn Seedless	W	19	0.00	740						
De Chaunac	R	186	0.00	487	91	0.00	636	102	0.00	581
De Cilindro	W	15	0.00	763				0	0.00	1540
Debina	W	455	0.01	393	239	0.01	495	14	0.00	917
Debit	W				403	0.01	404			
Deckrot	R	30	0.00	690	2	0.00	1172	12	0.00	934
Dedo de Dama	W				1	0.00	1259	1	0.00	1315
Dekabrskii	R				78	0.00	669	78	0.00	629
Delaware	G	234	0.00	470	227	0.00	503	421	0.01	380
Delhro	R				0	0.00	1334	0	0.00	1450
Deliciosa	R				0	0.00	1356	0	0.00	1477
Delisle	W							0	0.00	1484
Devin	W				133	0.00	582			
Diagalves	W	1088	0.02	283	1156	0.03	254	1090	0.02	248
Diamond Muscat	W							2	0.00	1173
Dimrit	R	602	0.01	354	863	0.02	293	704	0.02	299
Dimyat	W	7740	0.16	88	2401	0.05	169	9696	0.22	74
Dindarella	R	9	0.00	814	7	0.00	1045	5	0.00	1070
Diolinoir	R	31	0.00	687	114	0.00	596	122	0.00	551
Dišeća Ranina	W				2	0.00	1188			
Divico	R							10	0.00	967
Divona	W							0	0.00	1458
Docal	R				0	0.00	1341	0	0.00	1451
Dodrelyabi	R							0	0.00	1389
Doina	R				227	0.00	502	227	0.01	463
Dolcetto	R	7197	0.15	96	6333	0.14	99	4545	0.10	123
Dolciame	W	6	0.00	835	11	0.00	988	6	0.00	1043
Domina	R	187	0.00	486	407	0.01	401	375	0.01	390
Dominga	W	0	0.00	981	1	0.00	1242			
Dona Branca	W	296	0.01	441	276	0.01	466	204	0.00	481
Dona Joaquina	W				24	0.00	873	11	0.00	949
Dona Zillá	R							1	0.00	1308
Donaris	W				1	0.00	1254	1	0.00	1287
Donzelinho Branco	W	59	0.00	628	65	0.00	701	64	0.00	657
Donzelinho Roxo	R				0	0.00	1353	0	0.00	1468
Donzelinho Tinto	R				33	0.00	811	34	0.00	760
Doral	W	3	0.00	876	27	0.00	846	35	0.00	747
Dorinto	W				115	0.00	595	70	0.00	643
Dornfelder	R	3766	0.08	139	8182	0.18	84	7871	0.18	89
Dostoinyi	R				65	0.00	702	65	0.00	654
Douce Noire	R	18323	0.37	47	19630	0.43	43	19733	0.44	40
Doukkali	R	16557	0.34	52	16557	0.36	50			
Doux d'Henry	R	26	0.00	704	9	0.00	1005	6	0.00	1038
Droujba	W	3	0.00	869	3	0.00	1135			
Drupeggio	W	617	0.01	349	286	0.01	460	81	0.00	622
DU 31120	R							0	0.00	1491
Duna Gyöngye	R				63	0.00	707	45	0.00	714
Dunaj	R				46	0.00	764			
Dunav	R							11	0.00	946
Dunavski Lazur	W				483	0.01	380	483	0.01	359
Dunkelfelder	R	280	0.01	448	356	0.01	422	291	0.01	429
Dunze	R	1	0.00	954						
Duras	R	972	0.02	297	892	0.02	288	785	0.02	282

Table 5 (cont.): Global winegrape area, share of global area, and global ranks of each prime variety (alphabetic), 2000, 2010 and 2016 (ha and %)

		2000			2010			2016		
Prime vairiety	*Col*	*Global area (ha)*	*Global share %*	*Global rank*	*Global area (ha)*	*Global share %*	*Global rank*	*Global area (ha)*	*Global share %*	*Global rank*
Durella	W	599	0.01	357	470	0.01	385	480	0.01	361
Durif	R	1197	0.02	269	3557	0.08	137	4807	0.11	116
Edelweiss	W				32	0.00	820	16	0.00	891
Ederena	R	0	0.00	964	1	0.00	1261	1	0.00	1313
Egiodola	R	315	0.01	433	349	0.01	426	285	0.01	434
Ehrenbreitsteiner	W	13	0.00	782	10	0.00	994	8	0.00	1001
Ehrenfelser	W	289	0.01	446	113	0.00	599	82	0.00	614
Einset Seedless	G							0	0.00	1403
Ekigaina	R	5	0.00	841	4	0.00	1118	3	0.00	1114
Ekim Kara	R	19	0.00	746	27	0.00	847			
Elbling	W	1208	0.02	268	935	0.02	277	972	0.02	258
Elbling (R)	W	4	0.00	855	9	0.00	1014	10	0.00	966
Elmer Swenson 10- 18- 30	W							0	0.00	1482
Elvira	W	344	0.01	423	263	0.01	478	231	0.01	458
Emerald Riesling	W	937	0.02	301	508	0.01	371	177	0.00	503
Emerald seedless	W	16	0.00	761						
Emir	W	480	0.01	387	688	0.01	326	89	0.00	603
Enantio	R	1062	0.02	287	724	0.02	317	178	0.00	502
Encruzado	W	291	0.01	445	282	0.01	462	132	0.00	536
Enfarine Noir	R				0	0.00	1376	0	0.00	1534
Ensanyo Tintas	R				27	0.00	844			
Ensayo Blancas	W				1	0.00	1251			
Eona	W							2	0.00	1215
Erbaluce	W	329	0.01	429	319	0.01	444	316	0.01	413
Erbamat	W				24	0.00	865	2	0.00	1235
Ervi	R	5	0.00	840	4	0.00	1100	3	0.00	1112
ES 10-18-14	W							0	0.00	1485
Esganacao Preto	R				0	0.00	1380	0	0.00	1538
Esganinho	W				0	0.00	1351	0	0.00	1500
Espadeiro	R	1682	0.03	221	469	0.01	386	357	0.01	394
Espadeiro Mole	R				0	0.00	1308	0	0.00	1388
Espirit	W							2	0.00	1204
Esther	R							0	0.00	1558
Estreito Macio	W				3	0.00	1146	1	0.00	1253
Etraire de l'Adui	R	8	0.00	826	5	0.00	1089	5	0.00	1086
Exalta	W				4	0.00	1116	3	0.00	1126
Excelsior	W							0	0.00	1416
Eyholzer Rote	R							0	0.00	1435
Ezerfürtű	W	405	0.01	404	406	0.01	402	295	0.01	426
Ezerjó	W	3157	0.06	159	1074	0.02	265	636	0.01	318
Faberrebe	W	1586	0.03	230	554	0.01	357	331	0.01	402
Falanghina	W	1658	0.03	222	3037	0.07	148	323	0.01	410
Falanghina Flegrea	W							3634	0.08	136
Favorit	W							3	0.00	1142
Fenile	W				5	0.00	1073	1	0.00	1316
Fepiro	R				0	0.00	1383	0	0.00	1552
Fer	R	1626	0.03	228	1854	0.04	193	1686	0.04	195
Fernão Pires	W	14545	0.30	57	9609	0.21	75	12211	0.27	62
Ferral	R	48	0.00	646	30	0.00	823	31	0.00	776
Fertilia	R	13	0.00	777	3	0.00	1124	2	0.00	1183
Fetească Albă	W	23828	0.49	42	17469	0.38	47	13382	0.30	56
Fetească Neagră	R	1214	0.02	266	1719	0.04	200	3248	0.07	142

Table 5 (cont.): Global winegrape area, share of global area, and global ranks of each prime variety (alphabetic), 2000, 2010 and 2016 (ha and %)

		2000			2010			2016		
Prime vairiety	*Col*	*Global area (ha)*	*Global share %*	*Global rank*	*Global area (ha)*	*Global share %*	*Global rank*	*Global area (ha)*	*Global share %*	*Global rank*
Fetească Regală	W	2578	0.05	180	13136	0.28	58	12991	0.29	57
Feunate	R	0	0.00	1000						
Fiano	W	758	0.02	323	1377	0.03	226	2187	0.05	171
Fiesta	W	161	0.00	508	230	0.00	499	230	0.01	459
Fino de Ribera del Fresno	W	332	0.01	428	45	0.00	767	8	0.00	1005
Fintendo	R	144	0.00	518	118	0.00	591	185	0.00	496
Fioletovy Ranny	R				50	0.00	750	50	0.00	695
Flame Seedless	R	548	0.01	371	55	0.00	726	55	0.00	671
Flavis	W	12	0.00	785	3	0.00	1130	3	0.00	1156
Fleurtai	W							0	0.00	1498
Flora	G	6	0.00	838	8	0.00	1030	12	0.00	938
Florental	R	1	0.00	956	26	0.00	854	11	0.00	943
Floricica	W							14	0.00	921
Fogarina	R				5	0.00	1078	3	0.00	1138
Foglia Tonda	R	40	0.00	663	101	0.00	620	68	0.00	648
Fogoneu	R	36	0.00	678	35	0.00	799	15	0.00	910
Fokiano	R	162	0.00	507	262	0.01	479	212	0.00	475
Fokiano (W)	R	57	0.00	630						
Folgasao	W	409	0.01	403	182	0.00	533	162	0.00	517
Folgasao Roxo	R				18	0.00	904	18	0.00	876
Folha de Figueira	W				3	0.00	1121	2	0.00	1221
Folignan	W				51	0.00	746	51	0.00	690
Folle Blanche	W	2648	0.05	178	1803	0.04	196	1574	0.04	207
Fontan	R	0	0.00	983						
Fontanara	W	2	0.00	904				1	0.00	1296
Fonte Cal	W	355	0.01	418	111	0.00	602	52	0.00	688
Forastera	W	543	0.01	373	208	0.00	514	8	0.00	991
Forcallat Tinta	R	2690	0.06	176	1163	0.03	253	535	0.01	345
Forgiarin	R	2	0.00	912	4	0.00	1106	3	0.00	1147
Forsellina	R	9	0.00	807	7	0.00	1037	7	0.00	1027
Fortana	R	1252	0.03	262	642	0.01	338	469	0.01	365
Francavidda	W	86	0.00	576	13	0.00	960	2	0.00	1229
Frâncușă	W				621	0.01	344	365	0.01	393
Frappato	R	784	0.02	318	752	0.02	310	580	0.01	333
Fredonia	R	86	0.00	575	37	0.00	791	28	0.00	794
Freisa	R	1450	0.03	240	1054	0.02	270	519	0.01	349
Freisamer	W	17	0.00	754	8	0.00	1022	6	0.00	1042
Frontenac	R				135	0.00	576	212	0.00	474
Frontenac (G)	R				59	0.00	720	92	0.00	594
Frontenac (W)	R							26	0.00	811
Frühroter Veltliner	R	632	0.01	347	856	0.02	298	388	0.01	388
Frumoasa Alba	W							8	0.00	993
Fubiano	W	2	0.00	891	9	0.00	1009	2	0.00	1168
Fuella Nera	R	4	0.00	857	20	0.00	894	20	0.00	861
Fumin	R	73	0.00	598	31	0.00	821	25	0.00	823
Furmint	W	3481	0.07	145	5276	0.11	109	4435	0.10	125
Gaglioppo	R	3592	0.07	143	4214	0.09	123	4626	0.10	122
Gaillard	R	0	0.00	994	0	0.00	1362	0	0.00	1533
Galbenă de Odobești	W	546	0.01	372	385	0.01	407	417	0.01	383
Galego Dourado	W	51	0.00	643	16	0.00	927	7	0.00	1013
Galotta	R				13	0.00	954	35	0.00	755
Gamaret	R	71	0.00	604	405	0.01	403	441	0.01	374
Gamay Noir	R	37798	0.77	27	31927	0.69	34	26221	0.58	34

Table 5 (cont.): Global winegrape area, share of global area, and global ranks of each prime variety (alphabetic), 2000, 2010 and 2016 (ha and %)

		2000			2010			2016		
Prime vairiety	*Col*	*Global area (ha)*	*Global share %*	*Global rank*	*Global area (ha)*	*Global share %*	*Global rank*	*Global area (ha)*	*Global share %*	*Global rank*
Gamay Teinturier de Bouze	R	318	0.01	432	278	0.01	464	255	0.01	447
Gamay Teinturier de Chaudenay	R	267	0.01	453	157	0.00	553	142	0.00	529
Gamay Teinturier Freaux	R	132	0.00	521	104	0.00	614	79	0.00	627
Gamba Rossa	R				1	0.00	1279	0	0.00	1366
Gänsfüsser	R				0	0.00	1374	0	0.00	1548
Ganson	R	28	0.00	695	3	0.00	1127	3	0.00	1127
Garandmak	W	931	0.02	302	931	0.02	278			
Garanoir	R	76	0.00	594	216	0.00	506	229	0.01	460
Garganega	W	16553	0.34	53	15397	0.33	52	8554	0.19	82
Gargiulo 14260	R	1	0.00	939	1	0.00	1236			
Gargiulo 2539	R	69	0.00	610	49	0.00	757	29	0.00	788
Gargiulo 26189	R	6	0.00	839						
Gargiulo 26879	W	4	0.00	852	3	0.00	1157			
Gargiulo 4113	R	7	0.00	827	6	0.00	1058	3	0.00	1157
Gargiulo 45803	W	4	0.00	860	8	0.00	1023	10	0.00	960
Garnacha Blanca	W	10821	0.22	74	7398	0.16	92	7409	0.17	91
Garnacha Peluda	R	2024	0.04	207	1206	0.03	245	898	0.02	268
Garnacha Roja (Gris)	G	2761	0.06	173	2366	0.05	173	1462	0.03	213
Garnacha Tinta	R	216349	4.43	4	181553	3.93	7	150096	3.35	8
Garonnet	R	7	0.00	832	14	0.00	945	10	0.00	974
Garrido Fino	W	174	0.00	495	59	0.00	719	54	0.00	682
Gascon	R	0	0.00	1003	1	0.00	1252	1	0.00	1301
Gateta	R	7	0.00	830	2	0.00	1176			
Gegić	W				11	0.00	979			
Geilweilerhof Ga- 48- 12	W							1	0.00	1283
Geisenheim 318-57	W				106	0.00	611	14	0.00	920
Generosa	W	9	0.00	808	107	0.00	609	328	0.01	404
Gesztus	W				0	0.00	1320	0	0.00	1414
Gewürztraminer	W	10670	0.22	75	14355	0.31	55	12823	0.29	58
Gibi	W	1227	0.03	264	1074	0.02	266	785	0.02	283
Ginestra	W				4	0.00	1112	1	0.00	1359
Giro Nero	R	537	0.01	376	200	0.00	520	144	0.00	526
GM 322	R				17	0.00	919			
Göcseji Zamatos	W				55	0.00	725	50	0.00	698
Godello	W	1489	0.03	236	1332	0.03	231	1406	0.03	219
Goethe	R							20	0.00	859
Goldburger	W	309	0.01	434	140	0.00	570	98	0.00	586
Golden Muscat	W	1190	0.02	270	1191	0.03	248	50	0.00	694
Goldriesling	W	10	0.00	800	21	0.00	890	24	0.00	833
Goldtraminer	W				9	0.00	1012	5	0.00	1077
Golia	W				0	0.00	1361	0	0.00	1486
Golubok	R	50	0.00	644	87	0.00	641	37	0.00	743
Goncalo Pires	R				1	0.00	1270	1	0.00	1331
Gorgollasa	R							5	0.00	1064
Goruli Mtsvane	W	224	0.00	475	287	0.01	458	287	0.01	431
Gosen	R				1	0.00	1212	0	0.00	1493
Gouais Blanc	W	1	0.00	933				1	0.00	1270
Gouget Noir	R	1	0.00	948	10	0.00	998	3	0.00	1116
Goustolidi	W	112	0.00	544	68	0.00	693	19	0.00	866
Gouveio Preto	R				0	0.00	1315	0	0.00	1393
Gouveio Real	W				582	0.01	350	581	0.01	332

Table 5 (cont.): Global winegrape area, share of global area, and global ranks of each prime variety (alphabetic), 2000, 2010 and 2016 (ha and %)

		2000			2010			2016		
Prime vairiety	*Col*	*Global area (ha)*	*Global share %*	*Global rank*	*Global area (ha)*	*Global share %*	*Global rank*	*Global area (ha)*	*Global share %*	*Global rank*
GR 7	R				32	0.00	818	13	0.00	924
Grachen	W	3	0.00	875	2	0.00	1164	0	0.00	1525
Graciano	R	1960	0.04	210	3123	0.07	145	2910	0.06	146
Graisse	W	22	0.00	725	14	0.00	949	1	0.00	1268
Gramon	R	14	0.00	776	3	0.00	1144	3	0.00	1149
Grand Manchen	W				8	0.00	1025			
Grand Noir	R	949	0.02	299	955	0.02	276	707	0.02	297
Grangeal	R				1	0.00	1225	1	0.00	1267
Granho	W				0	0.00	1323	0	0.00	1422
Grapariol	W				2	0.00	1168	2	0.00	1169
Grasă de Cotnari	W	850	0.02	313	685	0.01	327	632	0.01	320
Graševina	W	92306	1.89	13	61200	1.33	18	24384	0.54	35
Grassen	R	1	0.00	942	0	0.00	1366	0	0.00	1502
Grechetto di Orvieto	W	1177	0.02	273	1501	0.03	212	1824	0.04	185
Grechetto Rosso	R	111	0.00	545	49	0.00	756	35	0.00	752
Greco	W	1325	0.03	254	158	0.00	552	21	0.00	851
Greco Bianco	W	660	0.01	336	1604	0.03	206	2050	0.05	176
Greco Nero	R	3229	0.07	153	1256	0.03	238	437	0.01	376
Grignolino	R	1353	0.03	250	915	0.02	283	911	0.02	267
Grillo	W	1803	0.04	214	6295	0.14	100	7383	0.16	92
Gringet	W	74	0.00	597	25	0.00	856	15	0.00	898
Grolleau Noir	R	3006	0.06	163	2759	0.06	157	1949	0.04	179
Groppello di Mocasina	R	120	0.00	536	81	0.00	661	24	0.00	828
Groppello di Revo	R				12	0.00	967	12	0.00	937
Groppello Gentile	R	219	0.00	478	326	0.01	440	78	0.00	628
Gros Manseng	W	2160	0.04	202	2960	0.06	151	3069	0.07	144
Grossa	R				73	0.00	677	54	0.00	679
Grüner Veltliner	W	23604	0.48	43	18834	0.41	44	19118	0.43	42
Gualarido	W							18	0.00	871
Guardavalle	W	168	0.00	501	33	0.00	809	16	0.00	892
Guillemot	W	0	0.00	995						
Gutenborner	W				2	0.00	1181			
Guzun	W							3	0.00	1153
Gyöngyrizling	W				23	0.00	875	16	0.00	897
Haiduc	R				3	0.00	1143	3	0.00	1130
Hajnalka	W				0	0.00	1373	0	0.00	1521
Hamvas	W							0	0.00	1430
Hárslevelű	W	1296	0.03	257	1856	0.04	192	1618	0.04	204
Hasansky Sladky	R							1	0.00	1258
Hegel	R	10	0.00	803	9	0.00	1015	7	0.00	1014
Helfensteiner	R	26	0.00	706	19	0.00	902	14	0.00	918
Helios	W							6	0.00	1032
Herbemont	R	1453	0.03	239	764	0.02	309	112	0.00	567
Heroldrebe	R	199	0.00	484	134	0.00	580	112	0.00	568
Heuréka	W							0	0.00	1449
Hibernal	W							20	0.00	854
Himbertscha	W	0	0.00	986				0	0.00	1429
Himrod	W				0	0.00	1293	1	0.00	1360
Hölder	W	13	0.00	781	5	0.00	1081	2	0.00	1186
Hondarribi Beltza	R	11	0.00	793	53	0.00	736	15	0.00	900
Hondarribi Zuri	W							624	0.01	322
Hrvatica	R	245	0.01	466	116	0.00	593	53	0.00	684
Humagne	W	9	0.00	811	30	0.00	825	29	0.00	790

Table 5 (cont.): Global winegrape area, share of global area, and global ranks of each prime variety (alphabetic), 2000, 2010 and 2016 (ha and %)

		2000			2010			2016		
Prime vairiety	*Col*	*Global area (ha)*	*Global share %*	*Global rank*	*Global area (ha)*	*Global share %*	*Global rank*	*Global area (ha)*	*Global share %*	*Global rank*
Huxelrebe	W	1289	0.03	259	630	0.01	340	466	0.01	368
Ignea	G				1	0.00	1267	1	0.00	1325
Ilichevskii Rannii	R	5	0.00	847	5	0.00	1082			
Imperial Napoleon	R				12	0.00	972	0	0.00	1526
Imperial Seedless	G	1	0.00	951	1	0.00	1243	1	0.00	1294
Impigno	W	60	0.00	625	7	0.00	1038	2	0.00	1213
Incrocio Bianco Fedit 51	W	11	0.00	790	5	0.00	1092	1	0.00	1314
Incrocio Bruni 54	W	12	0.00	784	12	0.00	966	3	0.00	1110
Incrocio Manzoni 2.15	R	166	0.00	506	86	0.00	648	72	0.00	639
Incrocio Terzi 1	R	65	0.00	618	44	0.00	769	12	0.00	933
Invernenga	W	32	0.00	686	7	0.00	1050	5	0.00	1075
Inzolia	W	9259	0.19	79	6133	0.13	103	4740	0.11	118
Iordan	W				315	0.01	446	311	0.01	415
IRAC 1933	R							0	0.00	1527
Irsai Olivér	W				1414	0.03	221	1790	0.04	188
Isa	W				9	0.00	1004	5	0.00	1084
Isabella	R	27450	0.56	37	32494	0.70	33	17813	0.40	47
Italia	W	1076	0.02	286	1463	0.03	216	5188	0.12	112
Italica	W	178	0.00	492	367	0.01	416	47	0.00	708
Ives	R	23	0.00	723	16	0.00	926			
Jacquere	W	1086	0.02	284	1014	0.02	272	621	0.01	323
Jacquez	R	226	0.00	474	2368	0.05	172	1443	0.03	216
Jádorvány	W							0	0.00	1402
Jampal	W	127	0.00	527	71	0.00	683	34	0.00	759
Jaoumet	W				0	0.00	1355	0	0.00	1472
Jardovany Fekete	R							0	0.00	1505
Jázmin	W							2	0.00	1198
Jeroma	R							0	0.00	1529
Joannes Seyve	R	3	0.00	887	1	0.00	1255	1	0.00	1309
Johanniter	W				86	0.00	646	111	0.00	569
Joubertin	R	2	0.00	916	1	0.00	1205			
Juan García	R	2077	0.04	205	1707	0.04	202	1545	0.03	209
Jubiläumsrebe	G	30	0.00	689	14	0.00	950	7	0.00	1026
Jubileum 75	R				194	0.00	525	91	0.00	595
Juhfark	W				186	0.00	530	195	0.00	490
Juliana	W				0	0.00	1327			
Jurançon Blanc	W	24	0.00	717	7	0.00	1040	2	0.00	1195
Jurançon Noir	R	1294	0.03	258	663	0.01	333	605	0.01	327
Jurie	W	0	0.00	1002						
Juwel	W	42	0.00	657	22	0.00	884	15	0.00	902
K.35	W				0	0.00	1337			
Kabar	W				18	0.00	909	30	0.00	782
Kadarka	R	2630	0.05	179	1181	0.03	251	1625	0.04	202
Kakotrygis	W				103	0.00	615	28	0.00	795
Kalecik Karası	R	601	0.01	355	861	0.02	297	704	0.02	298
Kalina	R							0	0.00	1382
Kangun	W	850	0.02	314	850	0.02	300			
Kanzler	W	53	0.00	640	33	0.00	813	28	0.00	796
Karalahna	R	3	0.00	883	4	0.00	1110	4	0.00	1097
Karasakiz	R	3	0.00	884	4	0.00	1109	4	0.00	1098
Karát	W				50	0.00	749	44	0.00	719
Kármin	R				36	0.00	794	24	0.00	837
Kat.E.Lin	R							0	0.00	1440

Table 5 (cont.): Global winegrape area, share of global area, and global ranks of each prime variety (alphabetic), 2000, 2010 and 2016 (ha and %)

		2000			2010			2016		
Prime vairiety	*Col*	*Global area (ha)*	*Global share %*	*Global rank*	*Global area (ha)*	*Global share %*	*Global rank*	*Global area (ha)*	*Global share %*	*Global rank*
Kay Gray	W				1	0.00	1217	2	0.00	1205
Kéknyelű	W				43	0.00	773	50	0.00	700
Kentville White 94-1	W							0	0.00	1415
Kentville White 94-2	W							1	0.00	1342
Kerner	W	7129	0.15	97	4093	0.09	127	2891	0.06	148
Khikhvi	W	5	0.00	844	6	0.00	1052	6	0.00	1028
Királyleányka	W				855	0.02	299	784	0.02	284
Kishmish Luchistyi	G							55	0.00	672
Kishmish Moldavskii	R							28	0.00	799
Kisi	W	20	0.00	737	26	0.00	853	26	0.00	816
Knipperlé	W	2	0.00	901	0	0.00	1352	0	0.00	1461
Kocsis Irma	W				11	0.00	984	2	0.00	1172
Kodrinskii	R	5	0.00	843	5	0.00	1084	229	0.01	462
Kodryanka	R							1143	0.03	243
Kokur Bely	W	641	0.01	342	918	0.02	282			
Kolor	R				2	0.00	1173	7	0.00	1025
Königin der Weingärten	W	750	0.02	325	61	0.00	712	70	0.00	641
Korinthiaki	R	834	0.02	316	54	0.00	732	106	0.00	576
Korona	W				1	0.00	1284	1	0.00	1302
Koshu	G	118	0.00	539	168	0.00	543	690	0.02	302
Kosmopolita	G							1	0.00	1334
Kotsifali	R	1148	0.02	277	2330	0.05	174	1338	0.03	222
Kövidinka	G	1214	0.02	265	1076	0.02	264	658	0.01	312
Kozma CS. 2	R				0	0.00	1364			
Kozmopoliten	W							0	0.00	1445
Krakhuna	W	36	0.00	676	46	0.00	762	46	0.00	710
Kraljevina	W				447	0.01	392	199	0.00	484
Krasnostop Zolotovsky	R				562	0.01	355	562	0.01	340
Krassato	R	38	0.00	670	52	0.00	744	5	0.00	1087
Kreaca	W				29	0.00	831	30	0.00	780
Kuban	R							32	0.00	772
Kujundžuša	W				206	0.00	515			
Kuldzhinskii	G	269	0.01	452	385	0.01	406	385	0.01	389
Kunbarát	W				9	0.00	1021	0	0.00	1379
Kunleány	W	1376	0.03	248	1211	0.03	244	974	0.02	257
Kurucvér	R				0	0.00	1343	0	0.00	1478
KW 96-2	W							0	0.00	1439
Kyoho (4N)	R	4003	0.08	132	4003	0.09	129	2762	0.06	152
L´Acadie Blanc	W							65	0.00	655
La Crescent	W				77	0.00	672	94	0.00	588
La Crosse	W				25	0.00	857	26	0.00	814
Labrusco	R				81	0.00	660	79	0.00	626
Lacoste	R	1	0.00	937						
Lacrima Christi	R				85	0.00	649	226	0.01	465
Lacrima di Morro d'Alba	R	652	0.01	339	421	0.01	397	252	0.01	448
Lado	W	1	0.00	953	1	0.00	1241	2	0.00	1201
Lafnetscha	W	1	0.00	928				2	0.00	1234
Lagarino Bianco	W				23	0.00	882	6	0.00	1053
Lagrein	R	471	0.01	389	718	0.02	319	251	0.01	450
Lairen	W	298	0.01	440	214	0.00	507	351	0.01	398
Lakhegyi Mézes	W	567	0.01	367	306	0.01	454	145	0.00	524
Lambrusca di Alessandria	R	888	0.02	307	137	0.00	573	58	0.00	667
Lambrusco	R	42	0.00	659	45	0.00	765	54	0.00	674

Table 5 (cont.): Global winegrape area, share of global area, and global ranks of each prime variety (alphabetic), 2000, 2010 and 2016 (ha and %)

Prime vairiety	Col	2000			2010			2016		
		Global area (ha)	*Global share %*	*Global rank*	*Global area (ha)*	*Global share %*	*Global rank*	*Global area (ha)*	*Global share %*	*Global rank*
Lambrusco Barghi	R				18	0.00	905	3	0.00	1154
Lambrusco di Sorbara	R	1409	0.03	244	1606	0.03	205	858	0.02	276
Lambrusco Grasparossa	R	1720	0.04	218	2734	0.06	159	954	0.02	261
Lambrusco Maestri	R	1513	0.03	233	2312	0.05	175	5657	0.13	104
Lambrusco Marani	R	2280	0.05	193	1394	0.03	224	1074	0.02	252
Lambrusco Montericco	R	262	0.01	455	70	0.00	686	25	0.00	821
Lambrusco Oliva	R				112	0.00	600	104	0.00	578
Lambrusco Salamino	R	4147	0.08	129	5003	0.11	111	6228	0.14	101
Lambrusco Viadanese	R	277	0.01	451	240	0.01	493	59	0.00	666
Lameiro	W				0	0.00	1305	0	0.00	1413
Landal	R	16	0.00	760	43	0.00	772	37	0.00	741
Landot Noir	R	0	0.00	976	3	0.00	1128	3	0.00	1128
Lariao	W				4	0.00	1114	5	0.00	1080
Lasina	R				14	0.00	944			
Laska	R	1	0.00	935				0	0.00	1513
Laurot	R				6	0.00	1062			
Lauzet	W	1	0.00	938	3	0.00	1152	3	0.00	1160
Leányka	W				838	0.02	303	719	0.02	295
Lecinaro	R				1	0.00	1274	0	0.00	1370
Leira	W				1	0.00	1281	0	0.00	1417
Len de l'El	W	734	0.02	328	629	0.01	341	603	0.01	328
Léon Millot	R	17	0.00	751	102	0.00	619	85	0.00	609
Leopoldo III	R	2	0.00	910						
Levokumskij	R				890	0.02	290	890	0.02	270
Liatiko	R	2476	0.05	184	1211	0.03	243	2633	0.06	159
Liatiko (W)	R	70	0.00	608						
Liliorila	W	3	0.00	885	4	0.00	1119	1	0.00	1257
Lilla	W							0	0.00	1559
Limnio	R	95	0.00	565	372	0.01	413	176	0.00	504
Limnio (W)	R	27	0.00	698						
Listain de Huelva	W	596	0.01	359	350	0.01	425	466	0.01	367
Listan Negro	R	3291	0.07	151	2666	0.06	162	2847	0.06	150
Listán Prieto	R	16589	0.34	51	4985	0.11	112	10267	0.23	70
Listrao Roxo	R	10	0.00	801						
Lival	R				101	0.00	621	99	0.00	585
Lomanto	R							3	0.00	1118
Longyan	R							1000	0.02	256
Lora	W							35	0.00	751
Lorena	W				519	0.01	366	500	0.01	356
Louise Swenson	W				3	0.00	1138	5	0.00	1063
Loureiro	W	4392	0.09	123	4054	0.09	128	4696	0.10	119
Lourela	R				0	0.00	1328	0	0.00	1431
Lucie Kuhlmann	R							21	0.00	850
Luisa Blanca	W							80	0.00	625
Lumassina	W	111	0.00	546	98	0.00	626	37	0.00	745
Lusitano	R				0	0.00	1354	0	0.00	1470
Luzidio	W				0	0.00	1321	0	0.00	1469
Lyana	W							41	0.00	723
Macabeo	W	48128	0.98	23	40864	1	22	38625	1	22
Maceratino	W	122	0.00	532	177	0.00	536	39	0.00	731
Madeleine × Angevine 7672	W				52	0.00	745	48	0.00	705
Madeleine Royale	W				3	0.00	1125	3	0.00	1137

Table 5 (cont.): Global winegrape area, share of global area, and global ranks of each prime variety (alphabetic), 2000, 2010 and 2016 (ha and %)

		2000			2010			2016		
Prime vairiety	*Col*	*Global area (ha)*	*Global share %*	*Global rank*	*Global area (ha)*	*Global share %*	*Global rank*	*Global area (ha)*	*Global share %*	*Global rank*
Madeleines	W				7	0.00	1044	7	0.00	1024
Madrasa	R	20	0.00	739	28	0.00	840	28	0.00	797
Magaracha Rannii	R							884	0.02	272
Magliasina	R							0	0.00	1532
Magliocco Canino	R	579	0.01	364	539	0.01	360	679	0.02	308
Magliocco Dolce	R	243	0.00	467	87	0.00	644	51	0.00	689
Magna	R							30	0.00	779
Magyarfrankos	R				0	0.00	1302	0	0.00	1412
Maiolica	R	70	0.00	605	26	0.00	851	13	0.00	926
Maiolina	R				1	0.00	1201	0	0.00	1384
Maiskii Chernyi	R	77	0.00	591	110	0.00	605	110	0.00	572
Malaga Blanc	W	11	0.00	789	16	0.00	922	54	0.00	678
Malagousia	W	23	0.00	721	182	0.00	532	126	0.00	542
Malbo Gentile	R	106	0.00	552	211	0.00	510	219	0.00	469
Malegue	W	1	0.00	936						
Malingre Precoce	W	3	0.00	880	0	0.00	1294	0	0.00	1372
Maliverne	W	0	0.00	969						
Malvarisco	R				3	0.00	1129	3	0.00	1122
Malvasia	W	61	0.00	624	45	0.00	766	2184	0.05	172
Malvasia Bianca di Basilicata	W	875	0.02	310	210	0.00	511	32	0.00	774
Malvasia Bianca di Candia	W	12889	0.26	64	9351	0.20	76	9685	0.22	75
Malvasia Bianca Lunga	W	3937	0.08	135	2544	0.06	166	1247	0.03	229
Malvasia Branca de Sao Jorge	W	55	0.00	634	110	0.00	603	110	0.00	571
Malvasia del Lazio	W	2366	0.05	188	590	0.01	347	680	0.02	307
Malvasia di Candia Aromatica	W	1754	0.04	217	927	0.02	280	1208	0.03	233
Malvasia di Casorzo	R	98	0.00	563	107	0.00	607	99	0.00	584
Malvasia di Lipari	W	516	0.01	381	310	0.01	450	113	0.00	566
Malvasia di Sardegna Rosada	G				3	0.00	1141	3	0.00	1140
Malvasia di Schierano	R	181	0.00	490	89	0.00	639	82	0.00	617
Malvasia Fina	W	7102	0.15	98	3501	0.08	140	3282	0.07	138
Malvasia Fina Roxa	G				25	0.00	858	24	0.00	831
Malvasia Moscata	W				554	0.01	356			
Malvasia Nera di Basilicata	R	754	0.02	324	114	0.00	597	39	0.00	727
Malvasia Nera di Brindisi	R	3174	0.06	156	1314	0.03	233	1264	0.03	226
Malvasia Nera Lunga	R				38	0.00	783	14	0.00	919
Malvasia Parda	W				4	0.00	1099	4	0.00	1095
Malvasia Preta	R	2210	0.05	197	1903	0.04	191	1933	0.04	182
Malvasia Romana	W				0	0.00	1340	0	0.00	1467
Malvazija Istarska	W	7559	0.15	89	2740	0.06	158	2788	0.06	151
Malvia	W				0	0.00	1331	0	0.00	1434
Mamaia	R				0	0.00	1313	0	0.00	1387
Mammolo	R	777	0.02	320	841	0.02	301	911	0.02	266
Mandilaria	R	845	0.02	315	885	0.02	291	932	0.02	263
Mandon	R	261	0.01	456				1	0.00	1346
Manseng Noir	R	10	0.00	797	32	0.00	819	29	0.00	791
Manteudo Preto	R				16	0.00	921	11	0.00	953
Manto Negro	R	470	0.01	390	273	0.01	467	311	0.01	416
Manzoni Bianco	W	8290	0.17	83	382	0.01	408	339	0.01	401

Table 5 (cont.): Global winegrape area, share of global area, and global ranks of each prime variety (alphabetic), 2000, 2010 and 2016 (ha and %)

Prime vairiety	*Col*	2000			2010			2016		
		Global area (ha)	*Global share %*	*Global rank*	*Global area (ha)*	*Global share %*	*Global rank*	*Global area (ha)*	*Global share %*	*Global rank*
Manzoni Moscato	R				20	0.00	896	19	0.00	863
Manzoni Rosa	G				29	0.00	832	23	0.00	845
Mara	R							10	0.00	969
Maratheftiko	R				152	0.00	557			
Maréchal Foch	R	173	0.00	496	356	0.01	421	229	0.01	461
Marfal	W	35	0.00	679	2	0.00	1177	1	0.00	1269
Margot	R							2	0.00	1220
Mariensteiner	W	9	0.00	810				2	0.00	1189
Marmajuelo	W	37	0.00	672	24	0.00	867	20	0.00	857
Marquette	R				88	0.00	640	166	0.00	510
Marquinhas	W				11	0.00	983	5	0.00	1069
Marquis	W							1	0.00	1264
Mars	R				2	0.00	1182	2	0.00	1218
Marsanne	W	1512	0.03	234	1763	0.04	198	1838	0.04	184
Marselan	R	176	0.00	494	2731	0.06	160	3941	0.09	132
Marufo	R	6339	0.13	107	6579	0.14	95	4683	0.10	120
Marzemina Bianca	W	78	0.00	589	54	0.00	733	55	0.00	673
Marzemino	R	994	0.02	293	1091	0.02	262	785	0.02	281
Maticha	W	354	0.01	419	311	0.01	449	257	0.01	444
Mátrai Muskotály	W				67	0.00	697	49	0.00	701
Maturana Blanca	W				18	0.00	912	13	0.00	925
Mauzac Blanc	W	3310	0.07	149	1933	0.04	189	1526	0.03	210
Mauzac Noir	R	10	0.00	804						
Mavro	R	10969	0.22	72	3575	0.08	136	3187	0.07	143
Mavro Messenikola	R	3	0.00	868						
Mavrodafni	R	537	0.01	377	345	0.01	432	324	0.01	409
Mavrouda	R	349	0.01	421	520	0.01	364	1658	0.04	199
Mavrud	R	647	0.01	341	1296	0.03	235	1193	0.03	237
Mayolet	R	4	0.00	858	7	0.00	1034	6	0.00	1047
Mazuelo	R	127692	2.61	9	75716	1.64	15	47312	1.06	21
Mazuelo (G)	R	25	0.00	713						
Mazuelo (W)	R	1035	0.02	289	3016	0.07	149	355	0.01	396
Mazzese	R	80	0.00	585	76	0.00	673	57	0.00	669
Mechta	R							19	0.00	865
Mecle de Bourgoin	R	2	0.00	897						
Medina	R				159	0.00	551	124	0.00	546
Melara	W	13	0.00	783	3	0.00	1147	1	0.00	1238
Melhorio	R				0	0.00	1367	0	0.00	1506
Melon	W	13253	0.27	61	12306	0.27	60	9551	0.21	76
Mencía	R	13138	0.27	62	10658	0.23	68	11052	0.25	66
Menoir	R				65	0.00	699	61	0.00	662
Menu Pineau	W	380	0.01	411	205	0.00	516	197	0.00	487
Mergeritar	W							0	0.00	1437
Merille	R	131	0.00	523	44	0.00	768	42	0.00	721
Merlese	R				14	0.00	943	8	0.00	1003
Merlot	R	213368	4.37	5	267888	5.80	2	266440	5.94	2
Merlot Blanc	W	176	0.00	493	46	0.00	761	44	0.00	717
Merlot Khorus	R							0	0.00	1446
Merseguera	W	7460	0.15	91	3946	0.09	130	2373	0.05	167
Merzling	W	5	0.00	842				1	0.00	1293
Meslier Saint-Francois	W	55	0.00	635	15	0.00	933	13	0.00	928
Meszi Kadarka	R							0	0.00	1365
Mézes Fehér	W				2	0.00	1163	1	0.00	1348

Table 5 (cont.): Global winegrape area, share of global area, and global ranks of each prime variety (alphabetic), 2000, 2010 and 2016 (ha and %)

		2000			2010			2016		
Prime vairiety	*Col*	*Global area (ha)*	*Global share %*	*Global rank*	*Global area (ha)*	*Global share %*	*Global rank*	*Global area (ha)*	*Global share %*	*Global rank*
Michele Palieri	R	24	0.00	714	1	0.00	1240	1	0.00	1332
Michurinets	R				0	0.00	1300	0	0.00	1377
Miguel del Arco	R	1267	0.03	260	468	0.01	387			
Milgranet	R	2	0.00	894	1	0.00	1208	1	0.00	1259
Milia	R				1	0.00	1210			
Millot-Foch	R							126	0.00	541
Mindelo	R				1	0.00	1287	1	0.00	1355
Minella Bianca	W	73	0.00	599	65	0.00	700	19	0.00	864
Minnesota Muscat	W							0	0.00	1398
Miorita	W				7	0.00	1041	8	0.00	1007
Mireille	W				0	0.00	1347	0	0.00	1462
Misket	W	3764	0.08	140						
Misket Cherven	G				4159	0.09	124	4349	0.10	127
Misket Varnenski	W				336	0.01	434			
Moldova	R							12375	0.28	61
Molette	W	30	0.00	691	29	0.00	834	18	0.00	878
Molinara	R	1637	0.03	226	717	0.02	320	609	0.01	326
Mollard	R	23	0.00	719	23	0.00	876	23	0.00	844
Monarch	R							10	0.00	970
Monastrell	R	76304	1.56	14	69742	1.51	16	51930	1.16	18
Monbadon	W	657	0.01	337	498	0.01	378			
Mondet	R				0	0.00	1314	0	0.00	1395
Mondeuse Blanche	W	22	0.00	727	6	0.00	1070	7	0.00	1022
Mondeuse Noire	R	1404	0.03	246	303	0.01	455	287	0.01	433
Monemvassia	W	418	0.01	402	481	0.01	381	81	0.00	619
Monerac	R	2	0.00	919	3	0.00	1131	3	0.00	1134
Monica Nera	R	2835	0.06	169	1404	0.03	222	1203	0.03	235
Monstruosa	W	2	0.00	893	1	0.00	1245			
Montepulciano	R	28728	0.59	36	34956	0.76	28	32935	0.73	28
Montils	W	131	0.00	522	164	0.00	546	165	0.00	511
Montonico Bianco	W	656	0.01	338	734	0.02	315	567	0.01	338
Montreal Blues	R							1	0.00	1343
Montu	W	1091	0.02	282						
Monvedro	R	14	0.00	774	6	0.00	1072	4	0.00	1102
Moore's Diamond	W	19	0.00	743	42	0.00	775	24	0.00	835
Moradella	R				6	0.00	1069	1	0.00	1248
Morava	W							34	0.00	758
Moravia Agria	R	1092	0.02	281	550	0.01	359	222	0.00	468
Morenillo	R	39	0.00	668						
Morio-Muskat	W	1188	0.02	272	526	0.01	362	440	0.01	375
Moristel	R				147	0.00	561	247	0.01	451
Mornen Noir	R	1	0.00	944						
Morone	R	22	0.00	726	13	0.00	958	7	0.00	1009
Morrastel Bouschet	R	1	0.00	947				4	0.00	1106
Moscadet	W				4	0.00	1113	3	0.00	1129
Moscargo	R				0	0.00	1348	0	0.00	1463
Moscatel Lilaz	W				0	0.00	1385	0	0.00	1550
Moscatello Selvatico	W	105	0.00	553	35	0.00	800	5	0.00	1074
Moscato di Scanzo	R	73	0.00	600	53	0.00	738	10	0.00	975
Moscato di Terracina	W	229	0.00	473	138	0.00	572	22	0.00	847
Moscato Embrapa	W				862	0.02	295	683	0.02	306
Moscato Giallo	W	542	0.01	374	1467	0.03	215	1634	0.04	200
Moscato Nazareno	W				68	0.00	695	40	0.00	725

Table 5 (cont.): Global winegrape area, share of global area, and global ranks of each prime variety (alphabetic), 2000, 2010 and 2016 (ha and %)

Prime vairiety	*Col*	2000 *Global area (ha)*	2000 *Global share %*	2000 *Global rank*	2010 *Global area (ha)*	2010 *Global share %*	2010 *Global rank*	2016 *Global area (ha)*	2016 *Global share %*	2016 *Global rank*
Moscato Rosa del Trentino	R	93	0.00	569	81	0.00	659	39	0.00	733
Moschofilero	G	718	0.01	329	1111	0.02	257	1088	0.02	249
Moschomavro	R	2295	0.05	192	1428	0.03	219	113	0.00	565
Mostosa	W	95	0.00	566	24	0.00	872	16	0.00	895
Mourisco de Semente	R				60	0.00	714	61	0.00	660
Mourisco de Trevoes	R				2	0.00	1183	2	0.00	1192
Mourvaison	R	9	0.00	812	3	0.00	1148	3	0.00	1152
Mouyssagues	R	0	0.00	999	0	0.00	1363	0	0.00	1511
MRAC 1087	R							0	0.00	1452
MRAC 1099	R							0	0.00	1438
MRAC 1626	R							0	0.00	1543
MRAC 1817	R							0	0.00	1476
MRAC 40	R							0	0.00	1544
Mskhali	W	1093	0.02	280	1093	0.02	261			
Mtsvane Kakhuri	W	249	0.01	465	319	0.01	443	319	0.01	411
Müller-Thurgau	W	33587	0.69	30	22917	0.50	39	19501	0.43	41
Musann Blanc	W							5	0.00	1072
Muscadelle	W	2207	0.05	199	1637	0.04	204	1509	0.03	212
Muscardin	R	19	0.00	744	17	0.00	914	17	0.00	884
Muscaris	W							4	0.00	1092
Muscat	W							744	0.02	290
Muscat Bailey A	R	1372	0.03	249	1422	0.03	220	1821	0.04	186
Muscat Blanc à Petits Grains	W	29979	0.61	34	31259	0.68	35	33739	0.75	26
Muscat Blanc à Petits Grains (G)	W	10442	0.21	76	8761	0.19	82	8258	0.18	85
Muscat Blanc à Petits Grains (R)	W	1154	0.02	276	1459	0.03	218	1438	0.03	217
Muscat Bleu	R							3	0.00	1155
Muscat de Bugeac	R							2	0.00	1224
Muscat Fleur d'Oranger	W	36	0.00	675	91	0.00	637	299	0.01	421
Muscat of Alexandria	W	29590	0.61	35	27648	0.60	37	34805	0.78	25
Muscat of Alexandria (R)	W	11	0.00	794	6	0.00	1067	3	0.00	1150
Muscat of Hamburg	R	7068	0.14	99	8140	0.18	85	7680	0.17	90
Muscat Ottonel	W	12259	0.25	66	10340	0.22	70	12464	0.28	60
Muscat Swenson	W				24	0.00	861	37	0.00	746
Muscat Timpuriu de Bucuresti	W							15	0.00	908
Muscat Yantarnyi	W							683	0.02	304
Muscatin	W							0	0.00	1466
Muskat de Yaloven	W	20	0.00	735	32	0.00	815	16	0.00	894
Muskat Moravsky	W				514	0.01	368			
Muskat Zhemchuzhnyi	W							1	0.00	1239
Mustoasă de Măderat	W				255	0.01	483	282	0.01	435
Naia	W				0	0.00	1296	0	0.00	1411
Naparo	R	0	0.00	977	1	0.00	1250	0	0.00	1556
Narince	W	537	0.01	378	769	0.02	308	787	0.02	280
Nascetta	W				21	0.00	888	17	0.00	882
Nasco	W	166	0.00	505	141	0.00	568	91	0.00	596
Nebbiera	R	20	0.00	738	12	0.00	970	9	0.00	985
Nebbiolo	R	5264	0.11	118	6125	0.13	104	7997	0.18	87
Negoska	R	96	0.00	564	143	0.00	567	17	0.00	883

Table 5 (cont.): Global winegrape area, share of global area, and global ranks of each prime variety (alphabetic), 2000, 2010 and 2016 (ha and %)

		2000			2010			2016		
Prime vairiety	*Col*	*Global area (ha)*	*Global share %*	*Global rank*	*Global area (ha)*	*Global share %*	*Global rank*	*Global area (ha)*	*Global share %*	*Global rank*
Negramoll	R	3557	0.07	144	3195	0.07	143	3013	0.07	145
Négrette	R	1319	0.03	255	1202	0.03	246	1112	0.02	247
Negretto	R	280	0.01	449	75	0.00	674	35	0.00	749
Negroamaro	R	16619	0.34	50	11492	0.25	64	11449	0.26	65
Negru Aromat	R				1	0.00	1275	1	0.00	1338
Negru de Drăgăşani	R				16	0.00	929	18	0.00	874
Negru de Yaloven	R	15	0.00	766	15	0.00	936	141	0.00	530
Nehelescol	W	172	0.00	497	25	0.00	860	6	0.00	1041
Nektár	W				21	0.00	887	22	0.00	846
Ner d'Ala	R	8	0.00	823	30	0.00	826	10	0.00	965
Nerello Cappuccio	R	1501	0.03	235	508	0.01	372	125	0.00	543
Nerello Mascalese	R	4167	0.09	127	2883	0.06	154	1805	0.04	187
Neretta Cuneese	R	374	0.01	413	132	0.00	583	119	0.00	558
Neretto di Bairo	R	53	0.00	642	34	0.00	806	19	0.00	868
Nero Buono di Cori	R	114	0.00	542	135	0.00	577	58	0.00	668
Nero d'Avola	R	11323	0.23	69	16649	0.36	48	14281	0.32	53
Nero di Troia	R	1765	0.04	215	2572	0.06	165	2512	0.06	162
Neronet	R				72	0.00	680	6	0.00	1044
Neuburger	W	1434	0.03	241	1030	0.02	271	578	0.01	334
Nevoeira	R				0	0.00	1307	0	0.00	1383
New York Muscat	R				5	0.00	1087	12	0.00	930
New York Muscat and VG4111	R							2	0.00	1184
Neyret	R	76	0.00	593	41	0.00	776	12	0.00	929
Niagara	W	15343	0.31	54	4670	0.10	116	3264	0.07	141
Niagara Red	R							469	0.01	366
Nieddera	R	58	0.00	629	107	0.00	608	91	0.00	597
Nigra	R	7	0.00	828	3	0.00	1155	1	0.00	1276
Nincusa	R				17	0.00	920			
Noah	W	260	0.01	457	563	0.01	354	200	0.00	483
Nobling	W	102	0.00	559	1	0.00	1246	52	0.00	686
Nocera	R	27	0.00	700	15	0.00	938	5	0.00	1068
Noir Fleurien	R	0	0.00	991	0	0.00	1342	0	0.00	1507
Noiret	R				33	0.00	812	25	0.00	826
Noria	W				1	0.00	1209			
Norton	R	0	0.00	979	329	0.01	436	328	0.01	406
Nosiola	W	191	0.00	485	79	0.00	665	65	0.00	656
Nosztori Rizling	W				1	0.00	1276	0	0.00	1444
Notardomenico	R	13	0.00	780	10	0.00	1001	9	0.00	977
Noual	W	1	0.00	962						
Nouvelle	W	2	0.00	909	422	0.01	396	428	0.01	379
Novac	R				73	0.00	676	74	0.00	637
Nuragus	W	3186	0.07	155	1345	0.03	229	1008	0.02	255
Oberlin	R	0	0.00	990	64	0.00	705	26	0.00	819
Oberlin White	W	1	0.00	943						
Odessky Cherny	R	1694	0.03	219	2686	0.06	161	2508	0.06	163
Odola	R	1	0.00	925	0	0.00	1318	0	0.00	1405
Odysseus	W				4	0.00	1098	25	0.00	822
Oeillade Bousche	R	10	0.00	798	1	0.00	1272	1	0.00	1339
Oeillade Noire	R				18	0.00	907	18	0.00	877
Ofthalmo	R				141	0.00	569			
Ohanes	W	0	0.00	966	16	0.00	925	15	0.00	904
Ojaleshi	R	25	0.00	709	32	0.00	816	32	0.00	773

Table 5 (cont.): Global winegrape area, share of global area, and global ranks of each prime variety (alphabetic), 2000, 2010 and 2016 (ha and %)

		2000			2010			2016		
Prime vairiety	*Col*	*Global area (ha)*	*Global share %*	*Global rank*	*Global area (ha)*	*Global share %*	*Global rank*	*Global area (ha)*	*Global share %*	*Global rank*
Okanagan Riesling	W							1	0.00	1361
Öküzgözü	R	1033	0.02	290	1479	0.03	214	1601	0.04	205
Olivette Blanche	W				2	0.00	1187	2	0.00	1206
Olivette de Laconnex	W							0	0.00	1555
Olivette Noire	R	0	0.00	968	16	0.00	924	16	0.00	896
Ondenc	W	12	0.00	786	8	0.00	1027	1	0.00	1356
Onitskanskii Belyi	W	5	0.00	846	71	0.00	682	490	0.01	357
Ontario	W							0	0.00	1554
Optima	W	239	0.00	469	65	0.00	704	38	0.00	738
Ora	W				38	0.00	786	32	0.00	770
Oraniensteiner	W				3	0.00	1151	2	0.00	1177
Original	R							5	0.00	1079
Orion	W	8	0.00	821	13	0.00	956	1	0.00	1295
Orpheus	W				0	0.00	1316	2	0.00	1233
Orpicchio	W				1	0.00	1290	0	0.00	1537
Ortega	W	1054	0.02	288	667	0.01	332	532	0.01	346
Ortrugo	W	485	0.01	386	611	0.01	345	709	0.02	296
Osceola Muscat	W							7	0.00	1012
Oseleta	R				15	0.00	940	16	0.00	889
Osennii Ciornai	R							8	0.00	996
Osteiner	W	3	0.00	870	1	0.00	1237	1	0.00	1327
Otskhanuri Sapere	R	5	0.00	845	6	0.00	1053	6	0.00	1029
Padeiro	R				86	0.00	647	88	0.00	604
Palas	R							7	0.00	1015
Palatina	W				6	0.00	1065	3	0.00	1123
Palava	G				230	0.00	500			
Pallagrello Bianco	W				55	0.00	730	6	0.00	1052
Pallagrello Nero	R				169	0.00	541	107	0.00	575
Palomino Fino	W	30513	0.62	33	22693	0.49	40	23190	0.52	36
Palot	W	4	0.00	863						
Pamid	R	22718	0.46	44	9827	0.21	74	9961	0.22	71
Pampanaro	W				5	0.00	1095	1	0.00	1322
Pampanuto	W	277	0.01	450	356	0.01	423	33	0.00	768
Pamyati Negrulya	R							123	0.00	549
Pannon Frankos	R				16	0.00	923	12	0.00	932
Panonia	W							10	0.00	962
Panse Valenciano	W	649	0.01	340	1	0.00	1248			
Paolina	W				1	0.00	1206	0	0.00	1474
Papazkarası	R	122	0.00	531	175	0.00	537	204	0.00	480
Pardillo	W	7272	0.15	93	4364	0.09	122	3283	0.07	137
Parellada	W	11188	0.23	70	8847	0.19	81	7137	0.16	94
Parraleta	R	167	0.00	503	348	0.01	427	212	0.00	476
Parreira Matias	R				1	0.00	1269	1	0.00	1333
Pascal Blanc	W	0	0.00	987	0	0.00	1378	0	0.00	1531
Pascale	R	1573	0.03	231	375	0.01	411	289	0.01	430
Passau	R	12	0.00	787	5	0.00	1076	3	0.00	1111
Passerina	W	715	0.01	330	894	0.02	287	933	0.02	262
Patagonia	W	29	0.00	692	40	0.00	780			
Patorra	R				10	0.00	997	10	0.00	964
Pátria	W				3	0.00	1139	5	0.00	1081
Patrizia Rosa	G	10	0.00	795	153	0.00	556			
Pavana	R	69	0.00	611	32	0.00	817	20	0.00	860
Pe Comprido	W				1	0.00	1229	1	0.00	1351

Table 5 (cont.): Global winegrape area, share of global area, and global ranks of each prime variety (alphabetic), 2000, 2010 and 2016 (ha and %)

		2000			2010			2016		
Prime vairiety	*Col*	*Global area (ha)*	*Global share %*	*Global rank*	*Global area (ha)*	*Global share %*	*Global rank*	*Global area (ha)*	*Global share %*	*Global rank*
Pecorello	W	16	0.00	756	34	0.00	807	9	0.00	978
Pecorino	W	166	0.00	504	1228	0.03	241	1742	0.04	192
Pecsi Szagos	W				1	0.00	1230	1	0.00	1320
Pedral	R	179	0.00	491	151	0.00	558	80	0.00	623
Pedro Giménez	W	14862	0.30	56	13502	0.29	57	15576	0.35	51
Pedro Ximénez	W	17272	0.35	48	9235	0.20	78	8810	0.20	80
Pelaverga	R	28	0.00	697	55	0.00	728	46	0.00	712
Pelaverga Piccolo	R	24	0.00	716	6	0.00	1068	6	0.00	1049
Peloursin	R	0	0.00	963						
Pelso	W				1	0.00	1273	1	0.00	1337
Pepella	W				3	0.00	1140	0	0.00	1441
Perdea	W	13	0.00	778	2	0.00	1160	2	0.00	1191
Perera	W	25	0.00	711	4	0.00	1107	2	0.00	1199
Perigo	W				4	0.00	1105	4	0.00	1089
Perla dei Vivi	R				1	0.00	1266	1	0.00	1350
Perlaut	W				1	0.00	1226	1	0.00	1278
Perle	G	121	0.00	533	34	0.00	804	18	0.00	873
Perlette	W	14	0.00	771	1	0.00	1214	2	0.00	1174
Perlita	W	1	0.00	931	1	0.00	1249			
Perola	W	68	0.00	615						
Perricone	R	580	0.01	363	228	0.00	501	80	0.00	624
Perruno	W	2831	0.06	170	1509	0.03	211	745	0.02	289
Persan	R	3	0.00	872	12	0.00	971	12	0.00	939
Pervenets Magaracha	W	2837	0.06	168	2881	0.06	155	2755	0.06	153
Pervomaisky	R	64	0.00	619	64	0.00	706			
Petit Bouschet	R	1	0.00	959	15	0.00	931	120	0.00	554
Petit Courbu	W	75	0.00	596	102	0.00	618	1	0.00	1271
Petit Manseng	W	613	0.01	352	1109	0.02	258	1299	0.03	224
Petit Meslier	W	3	0.00	866	4	0.00	1104	3	0.00	1117
Petit Rouge	R	100	0.00	560	84	0.00	652	68	0.00	650
Petit Verdot	R	1640	0.03	225	7195	0.16	94	8124	0.18	86
Petite Amie	W							0	0.00	1516
Petite Milo	G							6	0.00	1050
Petite Pearl	R							11	0.00	957
Pexem	R				3	0.00	1122	3	0.00	1144
Phoenix	W	24	0.00	715	67	0.00	696	46	0.00	711
Picapoll Blanco	W	34	0.00	680	37	0.00	790	40	0.00	724
Picardan	W				1	0.00	1263	1	0.00	1323
Piccola Nera	G	17	0.00	750	17	0.00	916	6	0.00	1056
Picolit	W	93	0.00	570	128	0.00	588	121	0.00	553
Piculit Neri	R	126	0.00	529	22	0.00	885	8	0.00	994
Piedirosso	R	896	0.02	306	699	0.02	323	593	0.01	330
Pignola Valtellinese	R	70	0.00	607	49	0.00	758	28	0.00	798
Pignoletto	W	6009	0.12	110	1707	0.04	201	1174	0.03	240
Pignolo	R	18	0.00	748	93	0.00	633	50	0.00	693
Pineau d'Aunis	R	430	0.01	399	437	0.01	394	413	0.01	384
Pinella	W	66	0.00	617	72	0.00	681	128	0.00	539
Pinorico	R							0	0.00	1400
Pinot Blanc	W	16983	0.35	49	14812	0.32	54	13779	0.31	55
Pinot Gris	G	18893	0.39	46	43773	0.95	21	48570	1.08	20
Pinot Meunier	R	13131	0.27	63	13566	0.29	56	14695	0.33	52
Pinot Noir	R	68810	1.41	16	98623	2.14	12	105480	2.35	12
Pinot Noir Précoce	R	85	0.00	578	273	0.01	468	251	0.01	449

Table 5 (cont.): Global winegrape area, share of global area, and global ranks of each prime variety (alphabetic), 2000, 2010 and 2016 (ha and %)

		2000			2010			2016		
Prime vairiety	*Col*	*Global area (ha)*	*Global share %*	*Global rank*	*Global area (ha)*	*Global share %*	*Global rank*	*Global area (ha)*	*Global share %*	*Global rank*
Pinotage	R	6574	0.13	104	6404	0.14	98	7132	0.16	95
Pinotin	R							0	0.00	1456
Pionnier	R							1	0.00	1245
Piquepoul Blanc	W	975	0.02	296	1492	0.03	213	1565	0.03	208
Piquepoul Bousch	R	0	0.00	965						
Piquepoul Gris	G	9	0.00	806	2	0.00	1196	2	0.00	1231
Piquepoul Noir	R	103	0.00	556	70	0.00	688	69	0.00	647
Piroso	R							4	0.00	1093
Plant Droit	R	40	0.00	666	19	0.00	900	19	0.00	867
Planta Mula	R	1	0.00	932	24	0.00	868	11	0.00	958
Planta Nova	W	2029	0.04	206	1395	0.03	223	888	0.02	271
Plantet	R	209	0.00	481	1060	0.02	269	420	0.01	381
Plassa	R	41	0.00	661	91	0.00	635	86	0.00	608
Plavac Mali	R	6539	0.13	106	1569	0.03	209	1714	0.04	193
Plavaie	W				209	0.00	513	152	0.00	520
Plavec Žuti	W				13	0.00	957	82	0.00	615
Plavina	R				643	0.01	337	683	0.02	305
Podarok Magaracha	W	148	0.00	515	504	0.01	374	292	0.01	427
Podarok Zaporozju	W							31	0.00	777
Pollera Nera	R	19	0.00	741	54	0.00	731	32	0.00	771
Pölöskei Muskotály	W				103	0.00	616	207	0.00	479
Portan	R	368	0.01	415	264	0.01	477	256	0.01	445
Portland	W	9	0.00	813	12	0.00	963	39	0.00	729
Pošip Bijeli	W	6539	0.13	105	253	0.01	485			
Pougnet	R	0	0.00	973						
Poulsard	R	295	0.01	442	307	0.01	452	90	0.00	600
Poulsard Blanc	W	14	0.00	775	0	0.00	1379	0	0.00	1541
Pozsonyi Fehér	W				10	0.00	990	10	0.00	972
Praca	W				166	0.00	544	169	0.00	508
Prairie Star	W				21	0.00	889	23	0.00	839
Prensal	W	114	0.00	543	105	0.00	613	129	0.00	537
Preto Cardana	R				5	0.00	1091	5	0.00	1083
Preto Martinho	R	428	0.01	400	163	0.00	548	163	0.00	515
Prezentabil	W							215	0.00	472
Prie	W	36	0.00	677	33	0.00	810	24	0.00	836
Prieto Picudo	R	3256	0.07	152	4587	0.10	118	4293	0.10	129
Prima	R				84	0.00	653	81	0.00	621
Primavera	R				40	0.00	779	39	0.00	730
Primetta	G	17	0.00	749	24	0.00	874	14	0.00	911
Prinzipal	W							2	0.00	1188
Prior	R							14	0.00	916
Prodest	R	2	0.00	889						
Prokupac	R	15180	0.31	55	15180	0.33	53	1361	0.03	220
Promissao	W				6	0.00	1057	3	0.00	1121
Prosecco	W	7507	0.15	90	18437	0.40	45	20109	0.45	39
Prosecco Lungo	W				1367	0.03	227	1450	0.03	214
Provareau	R	2	0.00	908						
Prunelard	R	2	0.00	895	20	0.00	893	19	0.00	862
Prunesta	R	92	0.00	571	36	0.00	795	31	0.00	778
Pugnitello	R				28	0.00	836	15	0.00	906
Purcsin	R							0	0.00	1406
Putzscheere	W	0	0.00	1001						
Quagliano	R	8	0.00	816	9	0.00	1010	9	0.00	979

Table 5 (cont.): Global winegrape area, share of global area, and global ranks of each prime variety (alphabetic), 2000, 2010 and 2016 (ha and %)

		2000			2010			2016		
Prime vairiety	*Col*	*Global area (ha)*	*Global share %*	*Global rank*	*Global area (ha)*	*Global share %*	*Global rank*	*Global area (ha)*	*Global share %*	*Global rank*
Quebranta	R	230	0.00	472	345	0.01	430	330	0.01	403
Quiebratinajas Tinto	R	6	0.00	837	5	0.00	1083	9	0.00	984
Rabigato	W	1133	0.02	278	1273	0.03	237	1969	0.04	178
Rabigato Moreno	W				1	0.00	1256	1	0.00	1304
Rabo de Anho	R				99	0.00	623	86	0.00	607
Rabo de Lobo	R				3	0.00	1126	3	0.00	1124
Rabo de Ovelha	W	2330	0.05	189	908	0.02	285	563	0.01	339
Raboso Piave	R	1334	0.03	252	776	0.02	307	665	0.01	309
Raboso Veronese	R	307	0.01	435	277	0.01	465	295	0.01	425
Radisson	R							8	0.00	1004
Raffiat de Moncade	W	16	0.00	755	7	0.00	1048	6	0.00	1030
Raisaine	W	1	0.00	934						
Rambella	W				6	0.00	1059	6	0.00	1031
Ramisco	R	72	0.00	602	34	0.00	808	33	0.00	767
Ranfol	W				134	0.00	581			
Ranna Melnishka Loza	R				249	0.01	488			
Rathay	R				9	0.00	1019	32	0.00	775
Räuschling	W	23	0.00	720	23	0.00	881	23	0.00	840
Ravat	R	1	0.00	945	1	0.00	1268	1	0.00	1329
Ravat Blanc	W	2	0.00	898	7	0.00	1049	6	0.00	1040
Rayada Melonera	R				1	0.00	1220	1	0.00	1300
Rayon d'Or	W	1	0.00	946	6	0.00	1066	6	0.00	1051
Reberger	R							2	0.00	1187
Rebo	R	37	0.00	673	125	0.00	589	92	0.00	592
Recantina	R				9	0.00	1006	4	0.00	1100
Red Globe	R	2113	0.04	204	242	0.01	492	242	0.01	453
Red Millennium	R							2	0.00	1226
Refosco	R							1341	0.03	221
Refosco dal Peduncolo Rosso	R	711	0.01	331	1082	0.02	263	1272	0.03	225
Refosco di Faedis	R	256	0.01	461	217	0.00	505	185	0.00	495
Refrén	W				0	0.00	1303	0	0.00	1381
Regent	R	340	0.01	424	2187	0.05	177	1974	0.04	177
Regner	W	150	0.00	512	46	0.00	763	21	0.00	852
Reichensteiner	W	319	0.01	431	247	0.01	489	120	0.00	557
Reliance	R				4	0.00	1117	4	0.00	1103
Réselle	W							1	0.00	1279
Retagliado Bianco	W	26	0.00	707	28	0.00	837	11	0.00	951
Reze	W	1	0.00	949						
Ribol	R				147	0.00	562	141	0.00	531
Ribolla Gialla	W	1406	0.03	245	1178	0.03	252	959	0.02	260
Rieslaner	W	70	0.00	606	84	0.00	650	73	0.00	638
Rieslina	W	219	0.00	477	174	0.00	539	103	0.00	579
Riesling	W	43316	0.89	25	50014	1.08	20	59805	1.33	14
Riesling Forte	W							2	0.00	1162
Riesus	W				115	0.00	594	115	0.00	561
Rio Grande	W	4	0.00	861	1	0.00	1288	0	0.00	1363
Ripolo	W				1	0.00	1224	1	0.00	1273
Riton	W	2	0.00	906	257	0.01	480	568	0.01	337
Rkatsiteli	W	67354	1.38	17	58641	1.27	19	51374	1.15	19
Roal	R				1	0.00	1222	1	0.00	1317
Robola	W	359	0.01	417	471	0.01	384	152	0.00	519
Roditis	G	299	0.01	437	4668	0.10	117	8463	0.19	84

Table 5 (cont.): Global winegrape area, share of global area, and global ranks of each prime variety (alphabetic), 2000, 2010 and 2016 (ha and %)

Prime vairiety	*Col*	2000 *Global area (ha)*	2000 *Global share %*	2000 *Global rank*	2010 *Global area (ha)*	2010 *Global share %*	2010 *Global rank*	2016 *Global area (ha)*	2016 *Global share %*	2016 *Global rank*
Roditis (R)	G	6945	0.14	100	3826	0.08	131	828	0.02	277
Rojal Tinta	R	2845	0.06	166	1801	0.04	197	736	0.02	292
Rollo	W	117	0.00	540	51	0.00	748	15	0.00	901
Rome	R	2	0.00	914	297	0.01	456	172	0.00	507
Romeiko	R	382	0.01	410	1597	0.03	207	1131	0.03	245
Romorantin	W	81	0.00	583	72	0.00	679	69	0.00	646
Romulus	W							1	0.00	1285
Rondinella	R	2797	0.06	171	2480	0.05	168	2684	0.06	158
Rondo	R				40	0.00	777	51	0.00	692
Roobernet	R	78	0.00	590	139	0.00	571	269	0.01	441
Rosa Arica	R	1	0.00	952	1	0.00	1238	1	0.00	1291
Rosaki	W							2	0.00	1180
Rosciola	G				2	0.00	1162	1	0.00	1280
Rose Ciotat	W							2	0.00	1196
Rose du Var	G	129	0.00	525	56	0.00	723	54	0.00	675
Rosina	W				1	0.00	1228	1	0.00	1266
Rösler	R				160	0.00	550	217	0.00	471
Rossara Trentina	R	29	0.00	694	8	0.00	1028	6	0.00	1035
Rossese	R	232	0.00	471	312	0.01	447	164	0.00	514
Rossese Bianco	W				7	0.00	1033	5	0.00	1065
Rossignola	R	295	0.01	443	188	0.00	529	49	0.00	703
Rossola Nera	R	102	0.00	558	86	0.00	645	29	0.00	793
Rotberger	R	26	0.00	705	17	0.00	915	11	0.00	947
Roter Milan	R							0	0.00	1455
Roter Veltliner	G	258	0.01	459	199	0.00	521	198	0.00	486
Rotgipfler	W	118	0.00	537	105	0.00	612	123	0.00	548
Roublot	W	0	0.00	967						
Rouge de Fully	R							1	0.00	1341
Rouge du Pays	R							136	0.00	535
Rougeon	R	36	0.00	674	42	0.00	774	21	0.00	849
Roussanne	W	874	0.02	311	1851	0.04	194	2137	0.05	173
Roussette d'Ayze	W	3	0.00	879	1	0.00	1215	1	0.00	1319
Roussin	R	3	0.00	878	3	0.00	1132	2	0.00	1230
Roviello Bianco	W				2	0.00	1200	1	0.00	1326
Roxo de Vila Flor	R				0	0.00	1336	0	0.00	1447
Roxo Rei	G				0	0.00	1310	0	0.00	1390
Royal de Alloza	R				29	0.00	833	6	0.00	1055
Royalty	R	338	0.01	426	97	0.00	627	93	0.00	590
Roz de Minis	G				6	0.00	1055	7	0.00	1023
Rozala Bianca	W				2	0.00	1199	2	0.00	1182
Rózsakő	W				19	0.00	903	17	0.00	879
Rúbea	R				81	0.00	658	181	0.00	499
Rubienne	R							1	0.00	1312
Rubilande	G	3	0.00	881	8	0.00	1029	7	0.00	1021
Rubin Golodrigi	R				82	0.00	656	82	0.00	618
Rubin Tairovsky	R	2	0.00	902	2	0.00	1174	5	0.00	1062
Rubinet	R							15	0.00	903
Rubinovy Magaracha	R	0	0.00	993	0	0.00	1346	0	0.00	1460
Rubintos	R				18	0.00	908	13	0.00	923
Rubired	R	4153	0.08	128	4556	0.10	120	4916	0.11	115
Ruby	R	6	0.00	833	9	0.00	1008	9	0.00	976
Ruby Cabernet	R	7420	0.15	92	5729	0.12	106	5309	0.12	110
Ruby Seedless	R	25	0.00	712						

Table 5 (cont.): Global winegrape area, share of global area, and global ranks of each prime variety (alphabetic), 2000, 2010 and 2016 (ha and %)

		2000			2010			2016		
Prime vairiety	*Col*	*Global area (ha)*	*Global share %*	*Global rank*	*Global area (ha)*	*Global share %*	*Global rank*	*Global area (ha)*	*Global share %*	*Global rank*
Ruche	R	46	0.00	651	100	0.00	622	100	0.00	583
Rufete	R	3397	0.07	146	4833	0.10	113	1859	0.04	183
Ruggine	W				1	0.00	1204	1	0.00	1310
Ryugan	W							27	0.00	803
Sabrevois	R				10	0.00	996	25	0.00	820
Sacy	W	63	0.00	620	10	0.00	999	8	0.00	992
Sagrantino	R	351	0.01	420	995	0.02	273	1026	0.02	254
Saint Jeannet	W	68	0.00	613	56	0.00	724	43	0.00	720
Saint Macaire	R	1	0.00	960				6	0.00	1048
Saint-Cliche	W							0	0.00	1483
Saint-Pierre Dore	W	1	0.00	961	0	0.00	1304	0	0.00	1428
Salvador	R	572	0.01	365	394	0.01	405	351	0.01	399
Samarrinho	W				1	0.00	1216	1	0.00	1311
San Giuseppe Nero	R	348	0.01	422	192	0.00	526	82	0.00	616
San Lunardo	W	22	0.00	728	10	0.00	1002	4	0.00	1107
San Martino	R	44	0.00	653	21	0.00	891	6	0.00	1036
San Michele	R	120	0.00	534	57	0.00	722	37	0.00	744
Sanforte	R				1	0.00	1289	0	0.00	1473
Sangiovese	R	68877	1.41	15	78030	1.69	14	73464	1.64	13
Sankt Laurent	R	2555	0.05	181	3664	0.08	133	3272	0.07	140
Santa Maria	W	15	0.00	768	3	0.00	1142	2	0.00	1219
Santarena	R				739	0.02	313	724	0.02	294
Santoal	W				9	0.00	1011	4	0.00	1090
Sao Mamede	W				1	0.00	1278	1	0.00	1357
Saperavi	R	6707	0.14	102	8126	0.18	86	6478	0.14	99
Saperavi Severny	R	25	0.00	710	350	0.01	424	325	0.01	408
Saphira	W							8	0.00	990
Şarbă	W				265	0.01	475	266	0.01	442
Sárfehér	W							2	0.00	1202
Satin Noir	R							0	0.00	1514
Sauvignac	W							2	0.00	1179
Sauvignon Blanc	W	65190	1.33	18	111552	2.42	9	124700	2.78	10
Sauvignon Blanc (G)	W	76	0.00	595	698	0.02	324	1076	0.02	250
Sauvignonasse	W	5452	0.11	116	4563	0.10	119	3861	0.09	133
Savagnin Blanc	W	441	0.01	395	1950	0.04	187	2267	0.05	169
Savagnin Rose	G	883	0.02	309	884	0.02	292	48	0.00	707
Savatiano	W	12747	0.26	65	9920	0.21	73	10268	0.23	69
Scheurebe	W	3655	0.07	142	2039	0.04	184	1626	0.04	201
Schiava	R	1231	0.03	263	517	0.01	367	236	0.01	457
Schiava Gentile	R	1158	0.02	274	694	0.02	325	165	0.00	513
Schiava Grigia	R	79	0.00	588	66	0.00	698	4	0.00	1104
Schiava Grossa	R	3789	0.08	138	3011	0.07	150	2256	0.05	170
Schiava Lombarda	R				0	0.00	1360	701	0.02	300
Schioppettino	R	93	0.00	567	154	0.00	555	87	0.00	606
Schönburger	G	39	0.00	667	68	0.00	692	35	0.00	748
Sciaglin	W	4	0.00	862	6	0.00	1054	3	0.00	1113
Sciascinoso	R	253	0.01	462	94	0.00	632	59	0.00	664
Scimiscia	W				5	0.00	1075	2	0.00	1225
Scuppernong	W							27	0.00	805
Seara Nova	W	1213	0.02	267	681	0.01	330	471	0.01	364
Segalin	R	54	0.00	638	65	0.00	703	61	0.00	661
Seibel	R	1991	0.04	208	592	0.01	346	482	0.01	360
Seibel White	W	15	0.00	767	0	0.00	1359	0	0.00	1503

Table 5 (cont.): Global winegrape area, share of global area, and global ranks of each prime variety (alphabetic), 2000, 2010 and 2016 (ha and %)

		2000			2010			2016		
Prime vairiety	*Col*	*Global area (ha)*	*Global share %*	*Global rank*	*Global area (ha)*	*Global share %*	*Global rank*	*Global area (ha)*	*Global share %*	*Global rank*
Seinoir	R	15	0.00	769	87	0.00	642	50	0.00	699
Select	W	8	0.00	825	7	0.00	1036	7	0.00	1010
Semebat	R	2	0.00	892	2	0.00	1193	1	0.00	1241
Semidano	W	48	0.00	647	36	0.00	792	34	0.00	762
Sémillon	W	26239	0.54	40	22157	0.48	41	18693	0.42	44
Sennen	R				10	0.00	993	2	0.00	1171
Septimer	G							1	0.00	1298
Sercial	W	306	0.01	436	106	0.00	610	85	0.00	610
Sercialinho	W				9	0.00	1020	4	0.00	1099
Serna	G	19	0.00	745	36	0.00	796			
Servanin	R	0	0.00	978						
Servant	W	530	0.01	379	183	0.00	531	138	0.00	534
Sevilhao	R				14	0.00	947	14	0.00	915
Seyval Blanc	W	389	0.01	408	569	0.01	353	2699	0.06	155
Seyval Noir	R							76	0.00	635
Seyve Villard 23-512	R				29	0.00	835			
Sgavetta	R	61	0.00	623	47	0.00	760	26	0.00	812
Shalistin	W							0	0.00	1557
Sheridan	R	500	0.01	384	500	0.01	376	500	0.01	355
Shiroka Melnishka	R	3804	0.08	137	1580	0.03	208	1205	0.03	234
Sicilien	W				5	0.00	1090	5	0.00	1082
Siegerrebe	G	167	0.00	502	131	0.00	584	102	0.00	582
Siegfriedrebe	W							2	0.00	1210
Silcher	W	7	0.00	829				1	0.00	1290
Silvaner	W	11047	0.23	71	7395	0.16	93	6072	0.14	102
Silvaner (R)	W	2	0.00	905	38	0.00	785	25	0.00	825
Siramé	R							0	0.00	1394
Síria	W	2791	0.06	172	7898	0.17	89	7037	0.16	96
Sirio	W	23	0.00	722	14	0.00	948	5	0.00	1061
Sirius	W	2	0.00	907						
Škrlet	W				61	0.00	711			
Slankamenka	W				53	0.00	737	23	0.00	841
Solaris	W				81	0.00	662	118	0.00	559
Somerset Seedless	G							1	0.00	1246
Somszoekoe Kék	R							0	0.00	1520
Soperga	R	32	0.00	685	22	0.00	883	17	0.00	885
Soreli	W							0	0.00	1499
Souvignier Gris	G							3	0.00	1135
Sovereign Coronation	R							1	0.00	1307
Sovereign Opal	W				3	0.00	1145	3	0.00	1133
Spergola	W				110	0.00	604	115	0.00	562
Sremska Zelenika	W				0	0.00	1306	0	0.00	1488
St Croix	R				25	0.00	859	45	0.00	716
St Pepin	W				19	0.00	898	37	0.00	742
St Vincent	R				23	0.00	879	20	0.00	856
Stanušina Crna	R							400	0.01	387
Staufer	W	4	0.00	856						
Stavroto	R	104	0.00	555	11	0.00	981	0	0.00	1433
Stepnyak	W				144	0.00	564	144	0.00	527
Steuben	R				39	0.00	782	35	0.00	754
Storgozia	R				295	0.01	457			
Sugrafive	W	118	0.00	538	2	0.00	1178	0	0.00	1536
Sukholimansky Bely	W	1631	0.03	227	2156	0.05	180	405	0.01	386

Table 5 (cont.): Global winegrape area, share of global area, and global ranks of each prime variety (alphabetic), 2000, 2010 and 2016 (ha and %)

		2000			2010			2016		
Prime vairiety	*Col*	*Global area (ha)*	*Global share %*	*Global rank*	*Global area (ha)*	*Global share %*	*Global rank*	*Global area (ha)*	*Global share %*	*Global rank*
Sulima	W				0	0.00	1369	0	0.00	1530
Sultaniye	W	12122	0.25	67	3413	0.07	142	5325	0.12	109
Summerland	R							0	0.00	1481
Sumoll	R	1401	0.03	247	83	0.00	655	16	0.00	893
Sun Muscat	R							23	0.00	838
Superior Seedless	W	6	0.00	836	9	0.00	1016	9	0.00	980
Suscan	R				5	0.00	1074			
Susumaniello	R	62	0.00	621	50	0.00	752	8	0.00	1000
Swenson Red	R				11	0.00	977	10	0.00	971
Swenson White	W							2	0.00	1165
Symphony	W	184	0.00	489	324	0.01	441	647	0.01	315
Syrah	R	102490	2.10	10	185117	4.01	6	181185	4.04	7
Táltos	W				1	0.00	1234	1	0.00	1353
Tamarez	W	585	0.01	360	343	0.01	433	298	0.01	423
Tamarugal	W							1	0.00	1328
Taminga	W	46	0.00	652				2	0.00	1212
Tannat	R	5595	0.11	115	5765	0.12	105	5611	0.13	105
Tardia de Caxias	G							0	0.00	1386
Tarrango	R	120	0.00	535	72	0.00	678	16	0.00	888
Tauberschwarz	R	8	0.00	822	11	0.00	980	16	0.00	890
Taurus	W							0	0.00	1408
Tavkveri	R	29	0.00	693	37	0.00	788	37	0.00	740
Taylor	W	0	0.00	989						
Tazzelenghe	R	68	0.00	614	55	0.00	727	45	0.00	713
Tedi's Best	R							0	0.00	1497
Teinturier	R	1	0.00	921	7	0.00	1032	9	0.00	981
Tempranillo	R	93370	1.91	12	232988	5.05	4	219379	4.89	3
Tempranillo (W)	R				5	0.00	1086	110	0.00	570
Teoulier Noir	R	1	0.00	955	0	0.00	1371	0	0.00	1517
Termarina Rossa	G				20	0.00	895	2	0.00	1167
Teroldego	R	682	0.01	334	839	0.02	302	772	0.02	288
Terrano	R	1461	0.03	237	1914	0.04	190	209	0.00	478
Terrantez	W	27	0.00	701	12	0.00	969	11	0.00	950
Terrantez do Pico	W				0	0.00	1311	0	0.00	1410
Terras 20	R	13	0.00	779	0	0.00	1349	0	0.00	1464
Terret	W	2703	0.06	175	1390	0.03	225	872	0.02	274
Terret Gris	W	262	0.01	454	78	0.00	668	76	0.00	634
Terret Noir	R	370	0.01	414	143	0.00	566	139	0.00	532
Therona	W	185	0.00	488	99	0.00	624	67	0.00	651
Thrapsathiri	W				31	0.00	822	27	0.00	807
Tibouren	R	457	0.01	392	443	0.01	393	432	0.01	378
Tihanyi Kék	R							0	0.00	1518
Tilki Kuyrugu	W	172	0.00	498	246	0.01	490			
Timorasso	W	19	0.00	747	129	0.00	585	123	0.00	550
Timpuriu de Cluj	W				1	0.00	1253	1	0.00	1282
Tinta Aguiar	R				75	0.00	675	77	0.00	632
Tinta Aurelio	R				0	0.00	1386	0	0.00	1553
Tinta Barroca	R	6052	0.12	109	6172	0.13	101	4926	0.11	114
Tinta Bragao	R				63	0.00	708	64	0.00	658
Tinta Carvalha	R	1920	0.04	211	1311	0.03	234	1113	0.02	246
Tinta Castañal	R							3	0.00	1125
Tinta da Barca	R				345	0.01	431	352	0.01	397
Tinta da Melra	R				1	0.00	1282	1	0.00	1347

Table 5 (cont.): Global winegrape area, share of global area, and global ranks of each prime variety (alphabetic), 2000, 2010 and 2016 (ha and %)

		2000			2010			2016		
Prime vairiety	*Col*	*Global area (ha)*	*Global share %*	*Global rank*	*Global area (ha)*	*Global share %*	*Global rank*	*Global area (ha)*	*Global share %*	*Global rank*
Tinta de Alcoa	R				24	0.00	869	24	0.00	829
Tinta de Cidadelhe	R				1	0.00	1271	1	0.00	1330
Tinta de Pegoes	R				195	0.00	524	191	0.00	493
Tinta do Rodo	R				3	0.00	1123	3	0.00	1115
Tinta Engomada	R				4	0.00	1108	4	0.00	1105
Tinta Francisca	R				53	0.00	739	55	0.00	670
Tinta Malandra	R				0	0.00	1384	0	0.00	1551
Tinta Martins	R				11	0.00	985	10	0.00	959
Tinta Mesquita	R				1	0.00	1221	1	0.00	1265
Tinta Penajoia	R				53	0.00	740	53	0.00	683
Tinta Pereira	R				1	0.00	1231	1	0.00	1286
Tinta Pomar	R				30	0.00	830	30	0.00	784
Tinta Roseira	R				4	0.00	1115	4	0.00	1108
Tinta Valdosa	R				1	0.00	1286	1	0.00	1354
Tinta Varejoa	R				1	0.00	1232	1	0.00	1281
Tintem	R				9	0.00	1013	9	0.00	982
Tintilia del Molise	R				111	0.00	601	66	0.00	653
Tinto Cão	R	556	0.01	370	369	0.01	414	372	0.01	391
Tinto de Zafra	R	4	0.00	853	2	0.00	1180			
Tinto Jeroma	R	1	0.00	924						
Tinto Velasco	R	7998	0.16	87	7829	0.17	90	5369	0.12	108
Torbato	W	168	0.00	500	52	0.00	743	46	0.00	709
Torrontes Mendocino	W	780	0.02	319	661	0.01	334	653	0.01	313
Torrontés Riojano	W	8197	0.17	86	8937	0.19	79	8859	0.20	79
Torrontés Sanjuanino	W	3170	0.06	157	2531	0.05	167	3656	0.08	134
Tortosi	W	930	0.02	303	503	0.01	375	325	0.01	407
Touriga Femea	R				15	0.00	939	15	0.00	905
Touriga Franca	R	6674	0.14	103	11590	0.25	63	14224	0.32	54
Touriga Nacional	R	4263	0.09	126	10446	0.23	69	11722	0.26	63
Trajadura	W	2416	0.05	186	2169	0.05	179	2492	0.06	164
Traminette	W	5	0.00	849	240	0.01	494	239	0.01	455
Trbljan	W				231	0.01	498			
Trebbianina	W				128	0.00	587	30	0.00	785
Trebbiano d'Abruzzo	W	8435	0.17	82	5091	0.11	110	2630	0.06	160
Trebbiano Giallo	W	3984	0.08	133	10664	0.23	67	2275	0.05	168
Trebbiano Modenese	W	583	0.01	361	363	0.01	417	287	0.01	432
Trebbiano Romagnolo	W	19492	0.40	45	15893	0.34	51	19059	0.43	43
Trebbiano Spoletino	W	242	0.00	468	200	0.00	519	121	0.00	552
Trebbiano Toscano	W	137201	2.81	8	111290	2.41	10	120343	2.68	11
Trepat	R	1763	0.04	216	1358	0.03	228	1199	0.03	236
Tressot	R	0	0.00	982	0	0.00	1312	0	0.00	1392
Trevisana Nera	R	33	0.00	684	15	0.00	930	11	0.00	942
Tribidrag	R	26922	0.55	38	32755	0.71	32	33649	0.75	27
Trilla	W				1	0.00	1227	1	0.00	1275
Trincadeira	R	7265	0.15	95	9270	0.20	77	10510	0.23	68
Trincadeira das Pratas	W	216	0.00	480	239	0.01	496	124	0.00	545
Trincadeiro Branco	W				59	0.00	717	49	0.00	704
Triomphe	R				15	0.00	935	3	0.00	1148
Triplett Blanc	W				244	0.01	491	412	0.01	385
Triunfo	R				2	0.00	1184	2	0.00	1194
Trnjak	R				15	0.00	932			
Trobat	R	83	0.00	582	1	0.00	1239	3	0.00	1131
Tronto	R				2	0.00	1194	1	0.00	1284

Table 5 (cont.): Global winegrape area, share of global area, and global ranks of each prime variety (alphabetic), 2000, 2010 and 2016 (ha and %)

		2000			2010			2016		
Prime vairiety	*Col*	*Global area (ha)*	*Global share %*	*Global rank*	*Global area (ha)*	*Global share %*	*Global rank*	*Global area (ha)*	*Global share %*	*Global rank*
Trousseau	R	2223	0.05	196	3450	0.07	141	1263	0.03	227
Tsimlyansky Cherny	R				451	0.01	391	451	0.01	371
Tsitska	W	2839	0.06	167	3642	0.08	134	3642	0.08	135
Tsolikouri	W	6161	0.13	108	7903	0.17	88	7903	0.18	88
Tsulukidzis Tetra	W	152	0.00	510	195	0.00	523	195	0.00	491
Tsvetochny	W				169	0.00	542	169	0.00	509
Turán	R				177	0.00	535	183	0.00	498
Turchetta	R				3	0.00	1134	2	0.00	1163
Tyrian	R							38	0.00	736
Ucelut	W	10	0.00	796	10	0.00	1000	10	0.00	973
UD 31103	R							0	0.00	1490
Unirea	W				1	0.00	1285	1	0.00	1345
Úrréti	W				0	0.00	1357	0	0.00	1397
Usakhelouri	R	8	0.00	819	10	0.00	991	10	0.00	961
Uva Cao	W	33	0.00	683	1	0.00	1262	0	0.00	1385
Uva del Fantini	R				0	0.00	1350	0	0.00	1549
Uva del Tunde	R				2	0.00	1198	2	0.00	1228
Uva Longanesi	R				512	0.01	369	539	0.01	343
Uva Rara	R	570	0.01	366	460	0.01	389	197	0.00	488
Uva Tosca	R	84	0.00	580	71	0.00	684	29	0.00	789
Uvalino	R				1	0.00	1211	1	0.00	1249
Valais Noir	R	4	0.00	854						
Valbom	R				166	0.00	545	162	0.00	516
Valdiguié	R	79	0.00	587	272	0.01	469	126	0.00	540
Valenci Tinto	R	5	0.00	850	27	0.00	845	26	0.00	810
Valentino Nero	R	56	0.00	632	21	0.00	892	20	0.00	855
Valerien	W	2	0.00	900	24	0.00	862	23	0.00	842
Valiant	R				11	0.00	982	11	0.00	952
Valveirinha	W				0	0.00	1319	0	0.00	1399
Valvin Muscat	W				6	0.00	1060			
Vandal-Cliche	W							14	0.00	913
Varousset	R	12	0.00	788	5	0.00	1093	5	0.00	1085
Vasilaki	W	3	0.00	882	4	0.00	1111	4	0.00	1096
VB 32-7	W							3	0.00	1120
VB 91-26-25	R							1	0.00	1321
VB 91-26-26	R							0	0.00	1396
VB 91-26-27	R							0	0.00	1421
VB Cal 1-14	R							0	0.00	1512
VB Cal 1-29	R							0	0.00	1546
VB Cal 1-33	R							0	0.00	1547
VB Cal 6-04 N5	R							0	0.00	1380
VB Jura 25	R							0	0.00	1515
Vega	W	27	0.00	699	35	0.00	801	27	0.00	806
Ventura	W	39	0.00	669	24	0.00	864	10	0.00	963
Vênus	R				1	0.00	1223	0	0.00	1425
Verdea	W	107	0.00	550	83	0.00	654	39	0.00	732
Verdeca	W	2208	0.05	198	796	0.02	306	913	0.02	265
Verdejo	W	4453	0.09	121	16578	0.36	49	17931	0.40	45
Verdelet	W				1	0.00	1280	40	0.00	726
Verdelho	W	1643	0.03	224	2009	0.04	185	1516	0.03	211
Verdelho l'Anjou	W				0	0.00	1345	0	0.00	1465
Verdelho Tinto	R	1	0.00	957	28	0.00	839	29	0.00	792
Verdello	W	662	0.01	335	287	0.01	459	179	0.00	501

Table 5 (cont.): Global winegrape area, share of global area, and global ranks of each prime variety (alphabetic), 2000, 2010 and 2016 (ha and %)

Prime vairiety	*Col*	2000 *Global area (ha)*	2000 *Global share %*	2000 *Global rank*	2010 *Global area (ha)*	2010 *Global share %*	2010 *Global rank*	2016 *Global area (ha)*	2016 *Global share %*	2016 *Global rank*
Verdesse	W	8	0.00	818	10	0.00	995	3	0.00	1136
Verdial	W				1	0.00	1291	0	0.00	1427
Verdial Tinto	R				3	0.00	1153	3	0.00	1159
Verdicchio Bianco	W	5043	0.10	119	3532	0.08	138	4682	0.10	121
Verdil	W	131	0.00	524	57	0.00	721	50	0.00	696
Verdiso	W	71	0.00	603	68	0.00	694	52	0.00	687
Verdoncho	W	3092	0.06	162	2124	0.05	181			
Verduschia	W	15	0.00	770	11	0.00	975	9	0.00	986
Verduzzo Friulano	W	1598	0.03	229	812	0.02	304	690	0.02	303
Verduzzo Trevigiano	W	1657	0.03	223	708	0.02	322	531	0.01	347
Vermentino	W	5838	0.12	113	8874	0.19	80	11483	0.26	64
Vermentino Nero	R	143	0.00	519	210	0.00	512	124	0.00	544
Vernaccia di Oristano	W	565	0.01	369	272	0.01	470	246	0.01	452
Vernaccia di San Gimignano	W	854	0.02	312	522	0.01	363	884	0.02	273
Vertes Csillaga	W				21	0.00	886	11	0.00	941
Vertzami	R	491	0.01	385	335	0.01	435	60	0.00	663
Veruccese	R				0	0.00	1309	0	0.00	1436
Vespaiola	W	105	0.00	554	94	0.00	630	90	0.00	598
Vespolina	R	103	0.00	557	134	0.00	579	88	0.00	605
Victoria	W	145	0.00	517	52	0.00	741	620	0.01	324
Vidadillo	R							38	0.00	737
Vidal	W	611	0.01	353	1644	0.04	203	1936	0.04	181
Vidal Noir	R	0	0.00	970						
Vidvizhenets	W				271	0.01	471	271	0.01	439
Vien de Nus	R	25	0.00	708	13	0.00	961	9	0.00	983
Vignoles	W	68	0.00	612	254	0.01	484	241	0.01	454
Vijariego	W	510	0.01	382	285	0.01	461	369	0.01	392
Viktor	W				0	0.00	1339			
Viktória Gyöngye	W				190	0.00	527	198	0.00	485
Vilana	W	506	0.01	383	579	0.01	352	650	0.01	314
Vilana (R)	W	60	0.00	626						
Villard Blanc	W	746	0.02	326	654	0.01	335	743	0.02	291
Villard Noir	R	601	0.01	356	1273	0.03	236	777	0.02	287
Vincent	R				11	0.00	976	8	0.00	998
Vineland 53035	W							4	0.00	1101
Vineti	W	0	0.00	971						
Vinhao	R	5937	0.12	112	3160	0.07	144	4468	0.10	124
Viognier	W	3160	0.06	158	11785	0.26	62	16063	0.36	49
Violeta	R				98	0.00	625	636	0.01	317
Viorika	W	40	0.00	664	347	0.01	428	558	0.01	341
Viosinho	W	17	0.00	753	225	0.00	504	916	0.02	264
Vital	W	2246	0.05	195	1182	0.03	250	659	0.01	311
Vitovska	W	42	0.00	658	50	0.00	751	51	0.00	691
Vitovska Grganja	W							33	0.00	763
Voskeat	W	809	0.02	317	809	0.02	305			
Vostorg	W							17	0.00	881
Vranac	R				149	0.00	559	9503	0.21	77
Vugava	W				36	0.00	798			
Vuillermin	R	0	0.00	997	4	0.00	1101	3	0.00	1141
Vulcanus	W				5	0.00	1094	5	0.00	1059
Weldra	W	54	0.00	636	14	0.00	951	2	0.00	1190
Würzer	W	108	0.00	548	70	0.00	687	54	0.00	681

Table 5 (cont.): Global winegrape area, share of global area, and global ranks of each prime variety (alphabetic), 2000, 2010 and 2016 (ha and %)

		2000			2010			2016		
Prime vairiety	*Col*	*Global area (ha)*	*Global share %*	*Global rank*	*Global area (ha)*	*Global share %*	*Global rank*	*Global area (ha)*	*Global share %*	*Global rank*
Xara	R				0	0.00	1375	0	0.00	1524
Xarello	W	10299	0.21	77	8394	0.18	83	8534	0.19	83
Xinomavro	R	1816	0.04	213	1971	0.04	186	2135	0.05	175
Xinomavro (W)	R	3	0.00	877						
Xynisteri	W	2742	0.06	174	2092	0.05	183	1946	0.04	180
Yalovenskii Ustoichivyi	W							129	0.00	538
Yama Sauvignon	R							24	0.00	834
Yamabudo	R							35	0.00	750
Yamasachi	R							20	0.00	858
Yan 73	R							4800	0.11	117
Yaqui	R	22	0.00	724	2	0.00	1179	1	0.00	1289
Yubilei Zhuravlya	R							1	0.00	1262
Zalagyöngye	W	4330	0.09	125	1948	0.04	188	1259	0.03	228
Zalema	W	5969	0.12	111	4097	0.09	126	4015	0.09	131
Žametovka	R				914	0.02	284	822	0.02	278
Zefír	W				49	0.00	755	15	0.00	907
Zelen	W							75	0.00	636
Zengő	W				264	0.01	476	226	0.01	466
Zenit	W	405	0.01	405	580	0.01	351	660	0.01	310
Zéta	W				118	0.00	590	118	0.00	560
Zeusz	W				28	0.00	843	27	0.00	801
Zghihară de Huşi	W				87	0.00	643	54	0.00	680
Zierfandler	G	98	0.00	562	117	0.00	592	105	0.00	577
Žilavka	W							185	0.00	494
Žlahtina	W				135	0.00	578			
Zlatarica Vrgorska	W				19	0.00	901			
Župljanka	W				4	0.00	1103	505	0.01	353
Zweigelt	R	7267	0.15	94	10029	0.22	71	9068	0.20	78
other grey varieties	G							8639	0.19	
other red varieties	R	195563	4.00		98852	2.14		188270	4.20	
other white varieties	W	333660	6.83		127923	2.77		149126	3.33	
TOTAL, grey varieties*		66677	1.36		97702	2.12		109891	2.45	
TOTAL, red varieties*		2413939	49.39		2585934	56.02		2527059	56.37	
TOTAL, white varieties*		2407014	49.25		1932125	41.86		1846178	41.18	
TOTAL, all varieties*		**4887629**	**100.00**		**4615761**	**100.00**		**4483128**	**100.00**	

* Number of primes in Column B

Table 6: Global ranking of prime winegrape varieties by area, and changes, 2000, 2010 and 2016 (ha and %)

		2000			2010			2016			2010-2000	2016-2010	2016-2000	2010-2000	2016-2010	2016-2000
Prime vairiety	*Col*	*Global area (ha)*	*Global share %*	*Global rank*	*Global area (ha)*	*Global share %*	*Global rank*	*Global area (ha)*	*Global share %*	*Global rank*	*Change (ha)*	*Change (ha)*	*Change (ha)*	*Changes %*	*Changes %*	*Changes %*
Cabernet Sauvignon	R	223074	4.56	3	290083	6.28	1	310671	6.93	1	67009	20588	87597	30	7	39
Merlot	R	213368	4.37	5	267888	5.80	2	266440	5.94	2	54520	-1448	53072	26	-1	25
Tempranillo	R	93370	1.91	12	232988	5.05	4	219379	4.89	3	139618	-13609	126009	150	-6	135
Airén	W	387978	7.94	1	252364	5.47	3	203801	4.55	4	-135614	-48563	-184177	-35	-19	-47
Chardonnay	W	145543	2.98	7	199743	4.33	5	201649	4.50	5	54201	1905	56106	37	1	39
Syrah	R	102490	2.10	10	185117	4.01	6	181185	4.04	7	82627	-3932	78695	81	-2	77
Garnacha Tinta	R	216349	4.43	4	181553	3.93	7	150096	3.35	8	-34796	-31457	-66253	-16	-17	-31
Sauvignon Blanc	W	65190	1.33	18	111552	2.42	9	124700	2.78	10	46362	13148	59510	71	12	91
Trebbiano Toscano	W	137201	2.81	8	111290	2.41	10	120343	2.68	11	-25911	9053	-16858	-19	8	-12
Pinot Noir	R	68810	1.41	16	98623	2.14	12	105480	2.35	12	29812	6857	36670	43	7	53
Sangiovese	R	68877	1.41	15	78030	1.69	14	73464	1.64	13	9153	-4566	4587	13	-6	7
Riesling	W	43316	0.89	25	50014	1.08	20	59805	1.33	14	6698	9791	16489	15	20	38
Bobal	R	100128	2.05	11	80120	1.74	13	59189	1.32	15	-20008	-20931	-40940	-20	-26	-41
Cabernet Franc	R	51974	1.06	20	61295	1.33	17	56052	1.25	16	9320	-5243	4078	18	-9	8
Côt	R	26285	0.54	39	38158	0.83	25	52233	1.17	17	11873	14075	25948	45	37	99
Monastrell	R	76304	1.56	14	69742	1.51	16	51930	1.16	18	-6563	-17812	-24374	-9	-26	-32
Rkatsiteli	W	67354	1.38	17	58641	1.27	19	51374	1.15	19	-8713	-7267	-15980	-13	-12	-24
Pinot Gris	G	18893	0.39	46	43773	0.95	21	48570	1.08	20	24880	4797	29677	132	11	157
Mazuelo	R	127692	2.61	9	75716	1.64	15	47312	1.06	21	-51976	-28404	-80380	-41	-38	-63
Macabeo	W	48128	0.98	23	40864	0.89	22	38625	0.86	22	-7264	-2240	-9504	-15	-5	-20
Cayetana Blanca	W	55776	1.14	19	39781	0.86	23	36401	0.81	23	-15994	-3380	-19374	-29	-8	-35
Alicante Henri Bouschet	R	37157	0.76	28	38462	0.83	24	36031	0.80	24	1306	-2431	-1125	4	-6	-3
Muscat of Alexandria	W	29590	0.61	35	27648	0.60	37	34805	0.78	25	-1942	7157	5215	-7	26	18
Muscat Blanc à Petits Grains	W	29979	0.61	34	31259	0.68	35	33739	0.75	26	1281	2480	3761	4	8	13
Tribidrag	R	26922	0.55	38	32755	0.71	32	33649	0.75	27	5833	895	6728	22	3	25
Montepulciano	R	28728	0.59	36	34956	0.76	28	32935	0.73	28	6228	-2021	4207	22	-6	15
Chenin Blanc	W	45761	0.94	24	35703	0.77	27	32221	0.72	29	-10058	-3482	-13540	-22	-10	-30
Colombard	W	38632	0.79	26	32944	0.71	31	29996	0.67	30	-5687	-2948	-8636	-15	-9	-22
Cereza	G	31113	0.64	32	29934	0.65	36	28887	0.64	31	-1179	-1048	-2227	-4	-3	-7
Catarratto Bianco	W	50711	1.04	21	34863	0.76	29	28613	0.64	32	-15848	-6249	-22098	-31	-18	-44
Aligoté	W	35668	0.73	29	36120	0.78	26	26929	0.60	33	452	-9191	-8739	1	-25	-24
Gamay Noir	R	37798	0.77	27	31927	0.69	34	26221	0.58	34	-5871	-5706	-11577	-16	-18	-31
Graševina	W	92306	1.89	13	61200	1.33	18	24384	0.54	35	-31106	-36816	-67922	-34	-60	-74
Palomino Fino	W	30513	0.62	33	22693	0.49	40	23190	0.52	36	-7820	497	-7323	-26	2	-24
Cinsaut	R	48428	0.99	22	34751	0.75	30	22926	0.51	37	-13677	-11825	-25502	-28	-34	-53
Carmenère	R	5711	0.12	114	11366	0.25	65	22486	0.50	38	5655	11120	16775	99	98	294
Prosecco	W	7507	0.15	90	18437	0.40	45	20109	0.45	39	10930	1672	12601	146	9	168

Table 6 (cont.): Global ranking of prime winegrape varieties by area, and changes, 2000, 2010 and 2016 (ha and %)

Table 6 (cont.): Global ranking of prime winegrape varieties by area, and changes, 2000, 2010 and 2016 (ha and %)

		2000			2010			2016			2010-2000	2016-2010	2016-2000	2010-2000	2016-2010	2016-2000
Prime vairiety	*Col*	*Global area (ha)*	*Global share %*	*Global rank*	*Global area (ha)*	*Global share %*	*Global rank*	*Global area (ha)*	*Global share %*	*Global rank*	*Change (ha)*	*Change (ha)*	*Change (ha)*	*Changes %*	*Changes %*	*Changes %*
Douce Noire	R	18323	0.37	47	19630	0.43	43	19733	0.44	40	1307	102	1409	7	1	8
Müller-Thurgau	W	33587	0.69	30	22917	0.50	39	19501	0.43	41	-10670	-3415	-14085	-32	-15	-42
Grüner Veltliner	W	23604	0.48	43	18834	0.41	44	19118	0.43	42	-4770	284	-4486	-20	2	-19
Trebbiano Romagnolo	W	19492	0.40	45	15893	0.34	51	19059	0.43	43	-3598	3166	-432	-18	20	-2
Sémillon	W	26239	0.54	40	22157	0.48	41	18693	0.42	44	-4081	-3464	-7546	-16	-16	-29
Verdejo	W	4453	0.09	121	16578	0.36	49	17931	0.40	45	12126	1353	13479	272	8	303
Barbera	R	33041	0.68	31	24366	0.53	38	17824	0.40	46	-8676	-6542	-15217	-26	-27	-46
Isabella	R	27450	0.56	37	32494	0.70	33	17813	0.40	47	5044	-14681	-9637	18	-45	-35
Blaufränkisch	R	13997	0.29	59	17890	0.39	46	17180	0.38	48	3893	-710	3184	28	-4	23
Viognier	W	3160	0.06	158	11785	0.26	62	16063	0.36	49	8625	4278	12903	273	36	408
Criolla Grande	R	24264	0.50	41	20745	0.45	42	15596	0.35	50	-3519	-5149	-8668	-15	-25	-36
Pedro Giménez	W	14862	0.30	56	13502	0.29	57	15576	0.35	51	-1360	2074	714	-9	15	5
Pinot Meunier	R	13131	0.27	63	13566	0.29	56	14695	0.33	52	435	1129	1564	3	8	12
Nero d'Avola	R	11323	0.23	69	16649	0.36	48	14281	0.32	53	5326	-2368	2958	47	-14	26
Touriga Franca	R	6674	0.14	103	11590	0.25	63	14224	0.32	54	4916	2634	7550	74	23	113
Pinot Blanc	W	16983	0.35	49	14812	0.32	54	13779	0.31	55	-2171	-1033	-3204	-13	-7	-19
Fetească Albă	W	23828	0.49	42	17469	0.38	47	13382	0.30	56	-6359	-4087	-10446	-27	-23	-44
Fetească Regală	W	2578	0.05	180	13136	0.28	58	12991	0.29	57	10558	-145	10413	410	-1	404
Gewürztraminer	W	10670	0.22	75	14355	0.31	55	12823	0.29	58	3685	-1532	2152	35	-11	20
Castelão	R	14424	0.30	58	11088	0.24	66	12580	0.28	59	-3336	1492	-1844	-23	13	-13
Muscat Ottonel	W	12259	0.25	66	10340	0.22	70	12464	0.28	60	-1919	2124	205	-16	21	2
Moldova	R							12375	0.28	61						
Fernão Pires	W	14545	0.30	57	9609	0.21	75	12211	0.27	62	-4936	2602	-2334	-34	27	-16
Touriga Nacional	R	4263	0.09	126	10446	0.23	69	11722	0.26	63	6183	1276	7459	145	12	175
Vermentino	W	5838	0.12	113	8874	0.19	80	11483	0.26	64	3036	2609	5645	52	29	97
Negroamaro	R	16619	0.34	50	11492	0.25	64	11449	0.26	65	-5127	-43	-5170	-31	0	-31
Mencía	R	13138	0.27	62	10658	0.23	68	11052	0.25	66	-2479	394	-2085	-19	4	-16
Concord	R	11816	0.24	68	12238	0.27	61	10544	0.24	67	422	-1694	-1273	4	-14	-11
Trincadeira	R	7265	0.15	95	9270	0.20	77	10510	0.23	68	2005	1240	3245	28	13	45
Savatiano	W	12747	0.26	65	9920	0.21	73	10268	0.23	69	-2827	348	-2479	-22	4	-19
Listán Prieto	R	16589	0.34	51	4985	0.11	112	10267	0.23	70	-11604	5282	-6322	-70	106	-38
Pamid	R	22718	0.46	44	9827	0.21	74	9961	0.22	71	-12891	135	-12757	-57	1	-56
Bianca	W	2180	0.04	201	6462	0.14	97	9766	0.22	72	4282	3304	7586	196	51	348
Aglianico	R	9346	0.19	78	9995	0.22	72	9734	0.22	73	649	-260	388	7	-3	4
Dimyat	W	7740	0.16	88	2401	0.05	169	9696	0.22	74	-5339	7295	1956	-69	304	25
Malvasia Bianca di Candia	W	12889	0.26	64	9351	0.20	76	9685	0.22	75	-3538	334	-3204	-27	4	-25
Melon	W	13253	0.27	61	12306	0.27	60	9551	0.21	76	-948	-2755	-3703	-7	-22	-28
Vranac	R				149	0.00	559	9503	0.21	77		9354			6297	

Table 6 (cont.): Global ranking of prime winegrape varieties by area, and changes, 2000, 2010 and 2016 (ha and %)

		2000			2010			2016			2010-2000	2016-2010	2016-2000	2010-2000	2016-2010	2016-2000
Prime vairiety	*Col*	*Global area (ha)*	*Global share %*	*Global rank*	*Global area (ha)*	*Global share %*	*Global rank*	*Global area (ha)*	*Global share %*	*Global rank*	*Change (ha)*	*Change (ha)*	*Change (ha)*	*Changes %*	*Changes %*	*Changes %*
Zweigelt	R	7267	0.15	94	10029	0.22	71	9068	0.20	78	2761	-961	1801	38	-10	25
Torrontés Riojano	W	8197	0.17	86	8937	0.19	79	8859	0.20	79	740	-78	662	9	-1	8
Pedro Ximénez	W	17272	0.35	48	9235	0.20	78	8810	0.20	80	-8037	-425	-8461	-47	-5	-49
Garganega	W	16553	0.34	53	15397	0.33	52	8554	0.19	82	-1157	-6843	-8000	-7	-44	-48
Xarello	W	10299	0.21	77	8394	0.18	83	8534	0.19	83	-1905	140	-1765	-18	2	-17
Roditis	G	299	0.01	437	4668	0.10	117	8463	0.19	84	4369	3795	8164	1461	81	2730
Muscat Blanc à Petits Grains (G)	W	10442	0.21	76	8761	0.19	82	8258	0.18	85	-1681	-503	-2185	-16	-6	-21
Petit Verdot	R	1640	0.03	225	7195	0.16	94	8124	0.18	86	5555	929	6484	339	13	395
Nebbiolo	R	5264	0.11	118	6125	0.13	104	7997	0.18	87	860	1872	2732	16	31	52
Tsolikouri	W	6161	0.13	108	7903	0.17	88	7903	0.18	88	1742	0	1742	28	0	28
Dornfelder	R	3766	0.08	139	8182	0.18	84	7871	0.18	89	4416	-311	4105	117	-4	109
Muscat of Hamburg	R	7068	0.14	99	8140	0.18	85	7680	0.17	90	1073	-460	613	15	-6	9
Garnacha Blanca	W	10821	0.22	74	7398	0.16	92	7409	0.17	91	-3423	11	-3412	-32	0	-32
Grillo	W	1803	0.04	214	6295	0.14	100	7383	0.16	92	4491	1088	5579	249	17	309
Chasselas	W	13318	0.27	60	13119	0.28	59	7377	0.16	93	-199	-5742	-5941	-1	-44	-45
Parellada	W	11188	0.23	70	8847	0.19	81	7137	0.16	94	-2341	-1710	-4051	-21	-19	-36
Pinotage	R	6574	0.13	104	6404	0.14	98	7132	0.16	95	-170	728	558	-3	11	8
Síria	W	2791	0.06	172	7898	0.17	89	7037	0.16	96	5107	-861	4246	183	-11	152
Baga	R	6730	0.14	101	4108	0.09	125	6750	0.15	97	-2621	2642	21	-39	64	0
Blauer Portugieser	R	9156	0.19	81	8027	0.17	87	6590	0.15	98	-1129	-1437	-2566	-12	-18	-28
Saperavi	R	6707	0.14	102	8126	0.18	86	6478	0.14	99	1418	-1647	-229	21	-20	-3
Corvina Veronese	R	4800	0.10	120	7496	0.16	91	6240	0.14	100	2696	-1255	1440	56	-17	30
Lambrusco Salamino	R	4147	0.08	129	5003	0.11	111	6228	0.14	101	856	1225	2081	21	24	50
Silvaner	W	11047	0.23	71	7395	0.16	93	6072	0.14	102	-3652	-1323	-4975	-33	-18	-45
Bonarda Piemontese	R	23	0.00	718	6	0.00	1064	5926	0.13	103	-17	5920	5903	-75	99336	25183
Lambrusco Maestri	R	1513	0.03	233	2312	0.05	175	5657	0.13	104	799	3346	4145	53	145	274
Tannat	R	5595	0.11	115	5765	0.12	105	5611	0.13	105	170	-154	16	3	-3	0
Alvarinho	W							5545	0.12	106						
Arinto de Bucelas	W	3966	0.08	134	4482	0.10	121	5409	0.12	107	516	927	1443	13	21	36
Tinto Velasco	R	7998	0.16	87	7829	0.17	90	5369	0.12	108	-169	-2460	-2629	-2	-31	-33
Sultaniye	W	12122	0.25	67	3413	0.07	142	5325	0.12	109	-8709	1912	-6797	-72	56	-56
Ruby Cabernet	R	7420	0.15	92	5729	0.12	106	5309	0.12	110	-1691	-420	-2112	-23	-7	-28
Caladoc	R	1427	0.03	242	3675	0.08	132	5258	0.12	111	2247	1583	3830	157	43	268
Italia	W	1076	0.02	286	1463	0.03	216	5188	0.12	112	387	3724	4111	36	255	382
Chelva	W	10877	0.22	73	6168	0.13	102	5029	0.11	113	-4709	-1139	-5848	-43	-18	-54
Tinta Barroca	R	6052	0.12	109	6172	0.13	101	4926	0.11	114	120	-1246	-1126	2	-20	-19
Rubired	R	4153	0.08	128	4556	0.10	120	4916	0.11	115	403	360	763	10	8	18
Durif	R	1197	0.02	269	3557	0.08	137	4807	0.11	116	2360	1249	3609	197	35	301

Table 6 (cont.): Global ranking of prime winegrape varieties by area, and changes, 2000, 2010 and 2016 (ha and %)

		2000			2010			2016			2010-2000	2016-2010	2016-2000	2010-2000	2016-2010	2016-2000
Prime vairiety	*Col*	*Global area (ha)*	*Global share %*	*Global rank*	*Global area (ha)*	*Global share %*	*Global rank*	*Global area (ha)*	*Global share %*	*Global rank*	*Change (ha)*	*Change (ha)*	*Change (ha)*	*Changes %*	*Changes %*	*Changes %*
Yan 73	R							4800	0.11	117						
Inzolia	W	9259	0.19	79	6133	0.13	103	4740	0.11	118	-3127	-1392	-4519	-34	-23	-49
Loureiro	W	4392	0.09	123	4054	0.09	128	4696	0.10	119	-339	643	304	-8	16	7
Marufo	R	6339	0.13	107	6579	0.14	95	4683	0.10	120	240	-1895	-1656	4	-29	-26
Verdicchio Bianco	W	5043	0.10	119	3532	0.08	138	4682	0.10	121	-1511	1150	-361	-30	33	-7
Gaglioppo	R	3592	0.07	143	4214	0.09	123	4626	0.10	122	622	412	1034	17	10	29
Dolcetto	R	7197	0.15	96	6333	0.14	99	4545	0.10	123	-864	-1788	-2652	-12	-28	-37
Vinhao	R	5937	0.12	112	3160	0.07	144	4468	0.10	124	-2777	1308	-1469	-47	41	-25
Furmint	W	3481	0.07	145	5276	0.11	109	4435	0.10	125	1794	-841	953	52	-16	27
Alarije	W	1686	0.03	220	1726	0.04	199	4407	0.10	126	40	2681	2721	2	155	161
Misket Cherven	G				4159	0.09	124	4349	0.10	127		190			5	
Cserszegi Fűszeres	G	2185	0.04	200	3609	0.08	135	4299	0.10	128	1423	691	2114	65	19	97
Prieto Picudo	R	3256	0.07	152	4587	0.10	118	4293	0.10	129	1331	-294	1037	41	-6	32
Aspiran Bouschet	R	433	0.01	398	2245	0.05	176	4088	0.09	130	1812	1843	3655	419	82	845
Zalema	W	5969	0.12	111	4097	0.09	126	4015	0.09	131	-1872	-82	-1954	-31	-2	-33
Marselan	R	176	0.00	494	2731	0.06	160	3941	0.09	132	2555	1210	3765	1455	44	2143
Sauvignonasse	W	5452	0.11	116	4563	0.10	119	3861	0.09	133	-889	-702	-1591	-16	-15	-29
Torrontés Sanjuanino	W	3170	0.06	157	2531	0.05	167	3656	0.08	134	-639	1125	487	-20	44	15
Tsitska	W	2839	0.06	167	3642	0.08	134	3642	0.08	135	803	0	803	28	0	28
Falanghina Flegrea	W							3634	0.08	136						
Pardillo	W	7272	0.15	93	4364	0.09	122	3283	0.07	137	-2908	-1081	-3989	-40	-25	-55
Malvasia Fina	W	7102	0.15	98	3501	0.08	140	3282	0.07	138	-3601	-219	-3820	-51	-6	-54
Agiorgitiko	R	2320	0.05	190	2905	0.06	153	3272	0.07	139	586	367	953	25	13	41
Sankt Laurent	R	2555	0.05	181	3664	0.08	133	3272	0.07	140	1110	-392	717	43	-11	28
Niagara	W	15343	0.31	54	4670	0.10	116	3264	0.07	141	-10673	-1405	-12078	-70	-30	-79
Fetească Neagră	R	1214	0.02	266	1719	0.04	200	3248	0.07	142	505	1529	2034	42	89	168
Mavro	R	10969	0.22	72	3575	0.08	136	3187	0.07	143	-7394	-388	-7782	-67	-11	-71
Gros Manseng	W	2160	0.04	202	2960	0.06	151	3069	0.07	144	800	109	909	37	4	42
Negramoll	R	3557	0.07	144	3195	0.07	143	3013	0.07	145	-362	-182	-544	-10	-6	-15
Graciano	R	1960	0.04	210	3123	0.07	145	2910	0.06	146	1163	-213	950	59	-7	48
Alvarelhão	R	5272	0.11	117	5701	0.12	107	2910	0.06	147	429	-2791	-2362	8	-49	-45
Kerner	W	7129	0.15	97	4093	0.09	127	2891	0.06	148	-3036	-1202	-4238	-43	-29	-59
Auxerrois	W	2302	0.05	191	2785	0.06	156	2853	0.06	149	483	68	551	21	2	24
Listan Negro	R	3291	0.07	151	2666	0.06	162	2847	0.06	150	-625	181	-444	-19	7	-13
Malvazija Istarska	W	7559	0.15	89	2740	0.06	158	2788	0.06	151	-4818	48	-4770	-64	2	-63
Kyoho (4N)	R	4003	0.08	132	4003	0.09	129	2762	0.06	152	0	-1241	-1241	0	-31	-31
Pervenets Magaracha	W	2837	0.06	168	2881	0.06	155	2755	0.06	153	44	-125	-81	2	-4	-3
Ancellotta	R	4405	0.09	122	4681	0.10	115	2739	0.06	154	277	-1942	-1665	6	-41	-38

Table 6 (cont.): Global ranking of prime winegrape varieties by area, and changes, 2000, 2010 and 2016 (ha and %)

		2000			2010			2016			2010-2000	2016-2010	2016-2000	2010-2000	2016-2010	2016-2000
Prime vairiety	*Col*	*Global area (ha)*	*Global share %*	*Global rank*	*Global area (ha)*	*Global share %*	*Global rank*	*Global area (ha)*	*Global share %*	*Global rank*	*Change (ha)*	*Change (ha)*	*Change (ha)*	*Changes %*	*Changes %*	*Changes %*
Seyval Blanc	W	389	0.01	408	569	0.01	353	2699	0.06	155	181	2130	2310	46	374	594
Băbească Neagră	R	3722	0.08	141	3122	0.07	146	2696	0.06	156	-600	-426	-1026	-16	-14	-28
Croatina	R	3116	0.06	160	5700	0.12	108	2695	0.06	157	2584	-3005	-421	83	-53	-14
Rondinella	R	2797	0.06	171	2480	0.05	168	2684	0.06	158	-317	204	-113	-11	8	-4
Liatiko	R	2476	0.05	184	1211	0.03	243	2633	0.06	159	-1265	1422	157	-51	117	6
Trebbiano d'Abruzzo	W	8435	0.17	82	5091	0.11	110	2630	0.06	160	-3345	-2461	-5805	-40	-48	-69
Beba	W	8250	0.17	84	6524	0.14	96	2556	0.06	161	-1726	-3968	-5694	-21	-61	-69
Nero di Troia	R	1765	0.04	215	2572	0.06	165	2512	0.06	162	807	-60	747	46	-2	42
Odessky Cherny	R	1694	0.03	219	2686	0.06	161	2508	0.06	163	991	-178	814	59	-7	48
Trajadura	W	2416	0.05	186	2169	0.05	179	2492	0.06	164	-246	323	77	-10	15	3
Clairette	W	4359	0.09	124	3057	0.07	147	2420	0.05	165	-1303	-636	-1939	-30	-21	-44
Cortese	W	3113	0.06	161	2953	0.06	152	2405	0.05	166	-161	-548	-708	-5	-19	-23
Merseguera	W	7460	0.15	91	3946	0.09	130	2373	0.05	167	-3514	-1573	-5087	-47	-40	-68
Trebbiano Giallo	W	3984	0.08	133	10664	0.23	67	2275	0.05	168	6680	-8389	-1709	168	-79	-43
Savagnin Blanc	W	441	0.01	395	1950	0.04	187	2267	0.05	169	1509	317	1826	342	16	414
Schiava Grossa	R	3789	0.08	138	3011	0.07	150	2256	0.05	170	-778	-755	-1533	-21	-25	-40
Fiano	W	758	0.02	323	1377	0.03	226	2187	0.05	171	619	810	1429	82	59	188
Malvasia	W	61	0.00	624	45	0.00	766	2184	0.05	172	-15	2139	2124	-26	4749	3509
Roussanne	W	874	0.02	311	1851	0.04	194	2137	0.05	173	978	286	1264	112	15	145
Couderc Noir	R	614	0.01	351	3517	0.08	139	2136	0.05	174	2903	-1381	1522	473	-39	248
Xinomavro	R	1816	0.04	213	1971	0.04	186	2135	0.05	175	155	164	319	9	8	18
Greco Bianco	W	660	0.01	336	1604	0.03	206	2050	0.05	176	944	446	1390	143	28	211
Regent	R	340	0.01	424	2187	0.05	177	1974	0.04	177	1847	-213	1634	542	-10	480
Rabigato	W	1133	0.02	278	1273	0.03	237	1969	0.04	178	139	696	836	12	55	74
Grolleau Noir	R	3006	0.06	163	2759	0.06	157	1949	0.04	179	-247	-810	-1057	-8	-29	-35
Xynisteri	W	2742	0.06	174	2092	0.05	183	1946	0.04	180	-650	-146	-796	-24	-7	-29
Vidal	W	611	0.01	353	1644	0.04	203	1936	0.04	181	1033	293	1326	169	18	217
Malvasia Preta	R	2210	0.05	197	1903	0.04	191	1933	0.04	182	-307	30	-277	-14	2	-13
Rufete	R	3397	0.07	146	4833	0.10	113	1859	0.04	183	1436	-2975	-1539	42	-62	-45
Marsanne	W	1512	0.03	234	1763	0.04	198	1838	0.04	184	251	75	326	17	4	22
Grechetto di Orvieto	W	1177	0.02	273	1501	0.03	212	1824	0.04	185	324	322	647	28	21	55
Muscat Bailey A	R	1372	0.03	249	1422	0.03	220	1821	0.04	186	50	399	450	4	28	33
Nerello Mascalese	R	4167	0.09	127	2883	0.06	154	1805	0.04	187	-1284	-1078	-2362	-31	-37	-57
Irsai Olivér	W				1414	0.03	221	1790	0.04	188		376			27	
Assyrtiko	W	1106	0.02	279	902	0.02	286	1770	0.04	189	-204	868	664	-18	96	60
Antao Vaz	W	376	0.01	412	1252	0.03	239	1768	0.04	190	876	516	1392	233	41	370
Bacchus	W	3374	0.07	147	2113	0.05	182	1759	0.04	191	-1261	-355	-1616	-37	-17	-48
Pecorino	W	166	0.00	504	1228	0.03	241	1742	0.04	192	1062	514	1576	639	42	949

Table 6 (cont.): Global ranking of prime winegrape varieties by area, and changes, 2000, 2010 and 2016 (ha and %)

		2000			2010			2016			2010-2000	2016-2010	2016-2000	2010-2000	2016-2010	2016-2000
Prime vairiety	*Col*	*Global area (ha)*	*Global share %*	*Global rank*	*Global area (ha)*	*Global share %*	*Global rank*	*Global area (ha)*	*Global share %*	*Global rank*	*Change (ha)*	*Change (ha)*	*Change (ha)*	*Changes %*	*Changes %*	*Changes %*
Plavac Mali	R	6539	0.13	106	1569	0.03	209	1714	0.04	193	-4970	145	-4825	-76	9	-74
Brachetto del Piemonte	R	1534	0.03	232	1460	0.03	217	1694	0.04	194	-74	234	160	-5	16	10
Fer	R	1626	0.03	228	1854	0.04	193	1686	0.04	195	229	-168	60	14	-9	4
Aletta	W				723	0.02	318	1676	0.04	196		953			132	
Cococciola	W	887	0.02	308	983	0.02	275	1671	0.04	197	96	688	784	11	70	88
Cardinal	R	3870	0.08	136	536	0.01	361	1660	0.04	198	-3334	1124	-2210	-86	210	-57
Mavrouda	R	349	0.01	421	520	0.01	364	1658	0.04	199	171	1138	1309	49	219	375
Moscato Giallo	W	542	0.01	374	1467	0.03	215	1634	0.04	200	926	166	1092	171	11	202
Scheurebe	W	3655	0.07	142	2039	0.04	184	1626	0.04	201	-1617	-413	-2030	-44	-20	-56
Kadarka	R	2630	0.05	179	1181	0.03	251	1625	0.04	202	-1450	444	-1005	-55	38	-38
Damaschino	W	3187	0.07	154	2171	0.05	178	1622	0.04	203	-1016	-549	-1565	-32	-25	-49
Hárslevelű	W	1296	0.03	257	1856	0.04	192	1618	0.04	204	560	-237	322	43	-13	25
Öküzgözü	R	1033	0.02	290	1479	0.03	214	1601	0.04	205	446	123	569	43	8	55
Beibinghong	R							1600	0.04	206						
Folle Blanche	W	2648	0.05	178	1803	0.04	196	1574	0.04	207	-844	-229	-1073	-32	-13	-41
Piquepoul Blanc	W	975	0.02	296	1492	0.03	213	1565	0.03	208	516	73	589	53	5	60
Juan García	R	2077	0.04	205	1707	0.04	202	1545	0.03	209	-370	-162	-532	-18	-10	-26
Mauzac Blanc	W	3310	0.07	149	1933	0.04	189	1526	0.03	210	-1378	-407	-1784	-42	-21	-54
Verdelho	W	1643	0.03	224	2009	0.04	185	1516	0.03	211	365	-493	-127	22	-25	-8
Muscadelle	W	2207	0.05	199	1637	0.04	204	1509	0.03	212	-569	-129	-698	-26	-8	-32
Garnacha Roja (Gris)	G	2761	0.06	173	2366	0.05	173	1462	0.03	213	-396	-903	-1299	-14	-38	-47
Prosecco Lungo	W				1367	0.03	227	1450	0.03	214		83			6	
Azal	W	3302	0.07	150	1072	0.02	267	1443	0.03	215	-2230	371	-1860	-68	35	-56
Jacquez	R	226	0.00	474	2368	0.05	172	1443	0.03	216	2142	-925	1217	949	-39	539
Muscat Blanc à Petits Grains (R)	W	1154	0.02	276	1459	0.03	218	1438	0.03	217	305	-22	284	26	-1	25
Boğazkere	R	773	0.02	321	1106	0.02	259	1436	0.03	218	334	330	664	43	30	86
Godello	W	1489	0.03	236	1332	0.03	231	1406	0.03	219	-156	73	-83	-11	6	-6
Prokupac	R	15180	0.31	55	15180	0.33	53	1361	0.03	220	0	-13819	-13819	0	-91	-91
Refosco	R							1341	0.03	221						
Kotsifali	R	1148	0.02	277	2330	0.05	174	1338	0.03	222	1183	-992	190	103	-43	17
Biancame	W	1330	0.03	253	2599	0.06	164	1336	0.03	223	1269	-1263	6	95	-49	0
Petit Manseng	W	613	0.01	352	1109	0.02	258	1299	0.03	224	496	191	687	81	17	112
Refosco dal Peduncolo Rosso	R	711	0.01	331	1082	0.02	263	1272	0.03	225	372	190	561	52	18	79
Malvasia Nera di Brindisi	R	3174	0.06	156	1314	0.03	233	1264	0.03	226	-1859	-51	-1910	-59	-4	-60
Trousseau	R	2223	0.05	196	3450	0.07	141	1263	0.03	227	1227	-2187	-960	55	-63	-43
Zalagyöngye	W	4330	0.09	125	1948	0.04	188	1259	0.03	228	-2382	-689	-3072	-55	-35	-71
Malvasia Bianca Lunga	W	3937	0.08	135	2544	0.06	166	1247	0.03	229	-1393	-1296	-2690	-35	-51	-68
Chinuri	W	955	0.02	298	1225	0.03	242	1225	0.03	230	270	0	270	28	0	28

Table 6 (cont.): Global ranking of prime winegrape varieties by area, and changes, 2000, 2010 and 2016 (ha and %)

		2000			2010			2016			2010-2000	2016-2010	2016-2000	2010-2000	2016-2010	2016-2000
Prime vairiety	*Col*	*Global area (ha)*	*Global share %*	*Global rank*	*Global area (ha)*	*Global share %*	*Global rank*	*Global area (ha)*	*Global share %*	*Global rank*	*Change (ha)*	*Change (ha)*	*Change (ha)*	*Changes %*	*Changes %*	*Changes %*
Blauburger	R	1002	0.02	292	1339	0.03	230	1223	0.03	231	337	-116	221	34	-9	22
Alfrocheiro	R	523	0.01	380	1188	0.03	249	1216	0.03	232	665	28	693	127	2	133
Malvasia di Candia Aromatica	W	1754	0.04	217	927	0.02	280	1208	0.03	233	-827	282	-546	-47	30	-31
Shiroka Melnishka	R	3804	0.08	137	1580	0.03	208	1205	0.03	234	-2224	-375	-2599	-58	-24	-68
Monica Nera	R	2835	0.06	169	1404	0.03	222	1203	0.03	235	-1431	-201	-1632	-50	-14	-58
Trepat	R	1763	0.04	216	1358	0.03	228	1199	0.03	236	-405	-159	-565	-23	-12	-32
Mavrud	R	647	0.01	341	1296	0.03	235	1193	0.03	237	649	-103	546	100	-8	84
Aramon Noir	R	9157	0.19	80	2601	0.06	163	1181	0.03	238	-6556	-1419	-7976	-72	-55	-87
Arneis	W	738	0.02	327	1122	0.02	256	1179	0.03	239	385	56	441	52	5	60
Pignoletto	W	6009	0.12	110	1707	0.04	201	1174	0.03	240	-4302	-533	-4835	-72	-31	-80
Albillo Mayor	W	5	0.00	848	1319	0.03	232	1152	0.03	241	1314	-167	1148	26886	-13	23475
Bombino Bianco	W	2903	0.06	165	1239	0.03	240	1147	0.03	242	-1664	-92	-1756	-57	-7	-60
Kodryanka	R							1143	0.03	243						
Corvinone	R	88	0.00	573	930	0.02	279	1140	0.03	244	842	211	1052	952	23	1191
Romeiko	R	382	0.01	410	1597	0.03	207	1131	0.03	245	1215	-466	749	318	-29	196
Tinta Carvalha	R	1920	0.04	211	1311	0.03	234	1113	0.02	246	-609	-198	-807	-32	-15	-42
Négrette	R	1319	0.03	255	1202	0.03	246	1112	0.02	247	-117	-90	-208	-9	-8	-16
Diagalves	W	1088	0.02	283	1156	0.03	254	1090	0.02	248	68	-66	2	6	-6	0
Moschofilero	G	718	0.01	329	1111	0.02	257	1088	0.02	249	394	-24	370	55	-2	52
Sauvignon Blanc (G)	W	76	0.00	595	698	0.02	324	1076	0.02	250	622	378	1000	822	54	1323
Bical	W	912	0.02	305	924	0.02	281	1076	0.02	251	12	152	164	1	16	18
Lambrusco Marani	R	2280	0.05	193	1394	0.03	224	1074	0.02	252	-886	-319	-1206	-39	-23	-53
Canaiolo Nero	R	2418	0.05	185	1068	0.02	268	1033	0.02	253	-1349	-36	-1385	-56	-3	-57
Sagrantino	R	351	0.01	420	995	0.02	273	1026	0.02	254	644	31	675	183	3	192
Nuragus	W	3186	0.07	155	1345	0.03	229	1008	0.02	255	-1841	-337	-2178	-58	-25	-68
Longyan	R							1000	0.02	256						
Kunleány	W	1376	0.03	248	1211	0.03	244	974	0.02	257	-165	-236	-402	-12	-20	-29
Elbling	W	1208	0.02	268	935	0.02	277	972	0.02	258	-273	36	-236	-23	4	-20
Chambourcin	R	257	0.01	460	1097	0.02	260	968	0.02	259	840	-129	710	326	-12	276
Ribolla Gialla	W	1406	0.03	245	1178	0.03	252	959	0.02	260	-229	-219	-447	-16	-19	-32
Lambrusco Grasparossa	R	1720	0.04	218	2734	0.06	159	954	0.02	261	1014	-1780	-765	59	-65	-45
Passerina	W	715	0.01	330	894	0.02	287	933	0.02	262	179	38	217	25	4	30
Mandilaria	R	845	0.02	315	885	0.02	291	932	0.02	263	39	47	87	5	5	10
Viosinho	W	17	0.00	753	225	0.00	504	916	0.02	264	209	691	899	1251	307	5392
Verdeca	W	2208	0.05	198	796	0.02	306	913	0.02	265	-1413	117	-1296	-64	15	-59
Mammolo	R	777	0.02	320	841	0.02	301	911	0.02	266	65	70	135	8	8	17
Grignolino	R	1353	0.03	250	915	0.02	283	911	0.02	267	-438	-4	-441	-32	0	-33
Garnacha Peluda	R	2024	0.04	207	1206	0.03	245	898	0.02	268	-818	-308	-1126	-40	-26	-56

Table 6 (cont.): Global ranking of prime winegrape varieties by area, and changes, 2000, 2010 and 2016 (ha and %)

		2000			2010			2016			2010-2000	2016-2010	2016-2000	2010-2000	2016-2010	2016-2000
Prime vairiety	*Col*	*Global area (ha)*	*Global share %*	*Global rank*	*Global area (ha)*	*Global share %*	*Global rank*	*Global area (ha)*	*Global share %*	*Global rank*	*Change (ha)*	*Change (ha)*	*Change (ha)*	*Changes %*	*Changes %*	*Changes %*
Ciliegiolo	R	2527	0.05	182	1830	0.04	195	897	0.02	269	-697	-933	-1630	-28	-51	-65
Levokumskij	R				890	0.02	290	890	0.02	270		0			0	
Planta Nova	W	2029	0.04	206	1395	0.03	223	888	0.02	271	-634	-507	-1141	-31	-36	-56
Magaracha Rannii	R							884	0.02	272						
Vernaccia di San Gimignano	W	854	0.02	312	522	0.01	363	884	0.02	273	-332	362	30	-39	69	4
Terret	W	2703	0.06	175	1390	0.03	225	872	0.02	274	-1313	-518	-1831	-49	-37	-68
Bombino Nero	R	1156	0.02	275	1201	0.03	247	865	0.02	275	45	-336	-290	4	-28	-25
Lambrusco di Sorbara	R	1409	0.03	244	1606	0.03	205	858	0.02	276	196	-748	-552	14	-47	-39
Roditis (R)	G	6945	0.14	100	3826	0.08	131	828	0.02	277	-3119	-2998	-6117	-45	-78	-88
Žametovka	R				914	0.02	284	822	0.02	278		-92			-10	
Çalkarası	R	436	0.01	396	625	0.01	343	806	0.02	279	189	181	369	43	29	85
Narince	W	537	0.01	378	769	0.02	308	787	0.02	280	232	18	250	43	2	47
Marzemino	R	994	0.02	293	1091	0.02	262	785	0.02	281	97	-305	-209	10	-28	-21
Duras	R	972	0.02	297	892	0.02	288	785	0.02	282	-79	-107	-187	-8	-12	-19
Gibi	W	1227	0.03	264	1074	0.02	266	785	0.02	283	-153	-289	-442	-12	-27	-36
Királyleányka	W				855	0.02	299	784	0.02	284		-72			-8	
Albana	W	2487	0.05	183	1523	0.03	210	782	0.02	285	-964	-741	-1705	-39	-49	-69
Alb de Suruceni	W							780	0.02	286						
Villard Noir	R	601	0.01	356	1273	0.03	236	777	0.02	287	672	-496	176	112	-39	29
Teroldego	R	682	0.01	334	839	0.02	302	772	0.02	288	157	-67	90	23	-8	13
Perruno	W	2831	0.06	170	1509	0.03	211	745	0.02	289	-1322	-764	-2086	-47	-51	-74
Muscat	W							744	0.02	290						
Villard Blanc	W	746	0.02	326	654	0.01	335	743	0.02	291	-92	89	-4	-12	14	0
Rojal Tinta	R	2845	0.06	166	1801	0.04	197	736	0.02	292	-1044	-1065	-2109	-37	-59	-74
Baco Noir	R	397	0.01	407	475	0.01	383	735	0.02	293	77	261	338	19	55	85
Santarena	R				739	0.02	313	724	0.02	294		-15			-2	
Leányka	W				838	0.02	303	719	0.02	295		-119			-14	
Ortrugo	W	485	0.01	386	611	0.01	345	709	0.02	296	126	98	224	26	16	46
Grand Noir	R	949	0.02	299	955	0.02	276	707	0.02	297	6	-249	-242	1	-26	-26
Kalecik Karası	R	601	0.01	355	861	0.02	297	704	0.02	298	260	-156	103	43	-18	17
Dimrit	R	602	0.01	354	863	0.02	293	704	0.02	299	260	-158	102	43	-18	17
Schiava Lombarda	R				0	0.00	1360	701	0.02	300		701			700955	
Avesso	W	636	0.01	344	685	0.01	328	699	0.02	301	50	14	63	8	2	10
Koshu	G	118	0.00	539	168	0.00	543	690	0.02	302	51	522	573	43	310	487
Verduzzo Friulano	W	1598	0.03	229	812	0.02	304	690	0.02	303	-785	-123	-908	-49	-15	-57
Muscat Yantarnyi	W							683	0.02	304						
Plavina	R				643	0.01	337	683	0.02	305		40			6	
Moscato Embrapa	W				862	0.02	295	683	0.02	306		-179			-21	

Table 6 (cont.): Global ranking of prime winegrape varieties by area, and changes, 2000, 2010 and 2016 (ha and %)

		2000			2010			2016			2010-2000	2016-2010	2016-2000	2010-2000	2016-2010	2016-2000
Prime vairiety	*Col*	*Global area (ha)*	*Global share %*	*Global rank*	*Global area (ha)*	*Global share %*	*Global rank*	*Global area (ha)*	*Global share %*	*Global rank*	*Change (ha)*	*Change (ha)*	*Change (ha)*	*Changes %*	*Changes %*	*Changes %*
Malvasia del Lazio	W	2366	0.05	188	590	0.01	347	680	0.02	307	-1775	89	-1686	-75	15	-71
Magliocco Canino	R	579	0.01	364	539	0.01	360	679	0.02	308	-40	140	100	-7	26	17
Raboso Piave	R	1334	0.03	252	776	0.02	307	665	0.01	309	-557	-111	-669	-42	-14	-50
Zenit	W	405	0.01	405	580	0.01	351	660	0.01	310	175	80	256	43	14	63
Vital	W	2246	0.05	195	1182	0.03	250	659	0.01	311	-1065	-523	-1587	-47	-44	-71
Kövidinka	G	1214	0.02	265	1076	0.02	264	658	0.01	312	-138	-418	-556	-11	-39	-46
Torrontes Mendocino	W	780	0.02	319	661	0.01	334	653	0.01	313	-119	-8	-127	-15	-1	-16
Vilana	W	506	0.01	383	579	0.01	352	650	0.01	314	73	71	144	14	12	28
Symphony	W	184	0.00	489	324	0.01	441	647	0.01	315	140	323	463	76	100	251
Bayanshira	W	451	0.01	394	645	0.01	336	645	0.01	316	195	0	195	43	0	43
Violeta	R				98	0.00	625	636	0.01	317		538			551	
Ezerjó	W	3157	0.06	159	1074	0.02	265	636	0.01	318	-2083	-438	-2521	-66	-41	-80
Alphonse Lavallée	R	15	0.00	764	862	0.02	294	634	0.01	319	847	-228	619	5600	-26	4092
Grasă de Cotnari	W	850	0.02	313	685	0.01	327	632	0.01	320	-164	-53	-218	-19	-8	-26
Catawba	R	635	0.01	346	633	0.01	339	626	0.01	321	-2	-8	-10	0	-1	-2
Hondarribi Zuri	W							624	0.01	322						
Jacquere	W	1086	0.02	284	1014	0.02	272	621	0.01	323	-73	-393	-465	-7	-39	-43
Victoria	W	145	0.00	517	52	0.00	741	620	0.01	324	-93	568	475	-64	1084	326
Béquignol Noir	R	1083	0.02	285	891	0.02	289	616	0.01	325	-192	-275	-467	-18	-31	-43
Molinara	R	1637	0.03	226	717	0.02	320	609	0.01	326	-920	-109	-1028	-56	-15	-63
Jurançon Noir	R	1294	0.03	258	663	0.01	333	605	0.01	327	-631	-58	-689	-49	-9	-53
Len de l'El	W	734	0.02	328	629	0.01	341	603	0.01	328	-104	-26	-131	-14	-4	-18
Albillo Real	W	3368	0.07	148	861	0.02	296	601	0.01	329	-2507	-260	-2768	-74	-30	-82
Piedirosso	R	896	0.02	306	699	0.02	323	593	0.01	330	-198	-106	-303	-22	-15	-34
Arany Sárfehér	W	2914	0.06	164	1133	0.02	255	586	0.01	331	-1781	-547	-2328	-61	-48	-80
Gouveio Real	W				582	0.01	350	581	0.01	332		-1			0	
Frappato	R	784	0.02	318	752	0.02	310	580	0.01	333	-31	-173	-204	-4	-23	-26
Neuburger	W	1434	0.03	241	1030	0.02	271	578	0.01	334	-404	-452	-855	-28	-44	-60
Athiri	W	1350	0.03	251	748	0.02	312	577	0.01	335	-603	-170	-773	-45	-23	-57
Cora	R							570	0.01	336						
Riton	W	2	0.00	906	257	0.01	480	568	0.01	337	255	311	566	12750	121	28308
Montonico Bianco	W	656	0.01	338	734	0.02	315	567	0.01	338	78	-167	-89	12	-23	-14
Rabo de Ovelha	W	2330	0.05	189	908	0.02	285	563	0.01	339	-1422	-345	-1767	-61	-38	-76
Krasnostop Zolotovsky	R				562	0.01	355	562	0.01	340		0			0	
Viorika	W	40	0.00	664	347	0.01	428	558	0.01	341	307	211	518	768	61	1294
Chasan	W	914	0.02	304	749	0.02	311	549	0.01	342	-164	-200	-364	-18	-27	-40
Uva Longanesi	R				512	0.01	369	539	0.01	343		26			5	
Aubun	R	1411	0.03	243	553	0.01	358	537	0.01	344	-858	-16	-874	-61	-3	-62

Table 6 (cont.): Global ranking of prime winegrape varieties by area, and changes, 2000, 2010 and 2016 (ha and %)

		2000			2010			2016			2010-2000	2016-2010	2016-2000	2010-2000	2016-2010	2016-2000
Prime vairiety	*Col*	*Global area (ha)*	*Global share %*	*Global rank*	*Global area (ha)*	*Global share %*	*Global rank*	*Global area (ha)*	*Global share %*	*Global rank*	*Change (ha)*	*Change (ha)*	*Change (ha)*	*Changes %*	*Changes %*	*Changes %*
Forcallat Tinta	R	2690	0.06	176	1163	0.03	253	535	0.01	345	-1527	-628	-2154	-57	-54	-80
Ortega	W	1054	0.02	288	667	0.01	332	532	0.01	346	-387	-135	-522	-37	-20	-50
Verduzzo Trevigiano	W	1657	0.03	223	708	0.02	322	531	0.01	347	-949	-177	-1126	-57	-25	-68
Baco Blanc	W	2137	0.04	203	739	0.02	314	528	0.01	348	-1398	-211	-1609	-65	-29	-75
Freisa	R	1450	0.03	240	1054	0.02	270	519	0.01	349	-396	-535	-931	-27	-51	-64
Borraçal	R	2654	0.05	177	683	0.01	329	512	0.01	350	-1972	-170	-2142	-74	-25	-81
Carrega Branco	W				507	0.01	373	512	0.01	351		5			1	
Cornifesto	R	259	0.01	458	499	0.01	377	509	0.01	352	239	10	250	92	2	96
Župljanka	W				4	0.00	1103	505	0.01	353		501			12167	
Bourboulenc	W	772	0.02	322	585	0.01	349	501	0.01	354	-187	-84	-271	-24	-14	-35
Sheridan	R	500	0.01	384	500	0.01	376	500	0.01	355	0	0	0	0	0	0
Lorena	W				519	0.01	366	500	0.01	356		-19			-4	
Onitskanskii Belyi	W	5	0.00	846	71	0.00	682	490	0.01	357	66	419	485	1320	590	9695
Arinarnoa	R	150	0.00	513	189	0.00	528	486	0.01	358	39	297	336	26	158	224
Dunavski Lazur	W				483	0.01	380	483	0.01	359		0			0	
Seibel	R	1991	0.04	208	592	0.01	346	482	0.01	360	-1399	-110	-1508	-70	-19	-76
Durella	W	599	0.01	357	470	0.01	385	480	0.01	361	-130	11	-119	-22	2	-20
Acolon	R				490	0.01	379	477	0.01	362		-13			-3	
Couderc 13	W							474	0.01	363						
Seara Nova	W	1213	0.02	267	681	0.01	330	471	0.01	364	-532	-210	-742	-44	-31	-61
Fortana	R	1252	0.03	262	642	0.01	338	469	0.01	365	-610	-173	-783	-49	-27	-63
Niagara Red	R							469	0.01	366						
Listain de Huelva	W	596	0.01	359	350	0.01	425	466	0.01	367	-246	116	-130	-41	33	-22
Huxelrebe	W	1289	0.03	259	630	0.01	340	466	0.01	368	-659	-164	-823	-51	-26	-64
Codega de Larinho	W	4058	0.08	131	629	0.01	342	455	0.01	369	-3429	-174	-3603	-85	-28	-89
Chenanson	R	636	0.01	345	466	0.01	388	452	0.01	370	-170	-13	-184	-27	-3	-29
Tsimlyansky Cherny	R				451	0.01	391	451	0.01	371		0			0	
Crimposie	W				453	0.01	390	450	0.01	372		-3			-1	
Cesanese	R	1024	0.02	291	679	0.01	331	446	0.01	373	-345	-233	-578	-34	-34	-56
Gamaret	R	71	0.00	604	405	0.01	403	441	0.01	374	335	36	370	471	9	522
Morio-Muskat	W	1188	0.02	272	526	0.01	362	440	0.01	375	-662	-86	-748	-56	-16	-63
Greco Nero	R	3229	0.07	153	1256	0.03	238	437	0.01	376	-1973	-819	-2792	-61	-65	-86
Blauer Wildbacher	R	472	0.01	388	368	0.01	415	437	0.01	377	-104	69	-36	-22	19	-8
Tibouren	R	457	0.01	392	443	0.01	393	432	0.01	378	-14	-11	-25	-3	-2	-5
Nouvelle	W	2	0.00	909	422	0.01	396	428	0.01	379	420	6	426	22241	1	22553
Delaware	G	234	0.00	470	227	0.00	503	421	0.01	380	-7	194	187	-3	85	80
Plantet	R	209	0.00	481	1060	0.02	269	420	0.01	381	851	-639	211	407	-60	101
Counoise	R	638	0.01	343	408	0.01	400	418	0.01	382	-230	10	-220	-36	3	-34

Table 6 (cont.): Global ranking of prime winegrape varieties by area, and changes, 2000, 2010 and 2016 (ha and %)

		2000			2010			2016			2010-2000	2016-2010	2016-2000	2010-2000	2016-2010	2016-2000
Prime vairiety	*Col*	*Global area (ha)*	*Global share %*	*Global rank*	*Global area (ha)*	*Global share %*	*Global rank*	*Global area (ha)*	*Global share %*	*Global rank*	*Change (ha)*	*Change (ha)*	*Change (ha)*	*Changes %*	*Changes %*	*Changes %*
Galbenă de Odobești	W	546	0.01	372	385	0.01	407	417	0.01	383	-161	32	-129	-30	8	-24
Pineau d'Aunis	R	430	0.01	399	437	0.01	394	413	0.01	384	7	-24	-16	2	-5	-4
Triplett Blanc	W				244	0.01	491	412	0.01	385		168			69	
Sukholimansky Bely	W	1631	0.03	227	2156	0.05	180	405	0.01	386	526	-1751	-1226	32	-81	-75
Stanušina Crna	R							400	0.01	387						
Frühroter Veltliner	R	632	0.01	347	856	0.02	298	388	0.01	388	224	-468	-244	35	-55	-39
Kuldzhinskii	G	269	0.01	452	385	0.01	406	385	0.01	389	116	0	116	43	0	43
Domina	R	187	0.00	486	407	0.01	401	375	0.01	390	220	-32	188	117	-8	100
Tinto Cão	R	556	0.01	370	369	0.01	414	372	0.01	391	-188	3	-184	-34	1	-33
Vijariego	W	510	0.01	382	285	0.01	461	369	0.01	392	-225	84	-141	-44	29	-28
Frâncușă	W				621	0.01	344	365	0.01	393		-256			-41	
Espadeiro	R	1682	0.03	221	469	0.01	386	357	0.01	394	-1213	-111	-1324	-72	-24	-79
Cabinda	R				362	0.01	418	355	0.01	395		-7			-2	
Mazuelo (W)	R	1035	0.02	289	3016	0.07	149	355	0.01	396	1981	-2661	-680	192	-88	-66
Tinta da Barca	R				345	0.01	431	352	0.01	397		7			2	
Lairen	W	298	0.01	440	214	0.00	507	351	0.01	398	-84	137	53	-28	64	18
Salvador	R	572	0.01	365	394	0.01	405	351	0.01	399	-178	-43	-221	-31	-11	-39
Busuioacă de Bohotin	G				268	0.01	472	343	0.01	400		75			28	
Manzoni Bianco	W	8290	0.17	83	382	0.01	408	339	0.01	401	-7907	-43	-7951	-95	-11	-96
Faberrebe	W	1586	0.03	230	554	0.01	357	331	0.01	402	-1032	-223	-1255	-65	-40	-79
Quebranta	R	230	0.00	472	345	0.01	430	330	0.01	403	115	-15	100	50	-4	43
Generosa	W	9	0.00	808	107	0.00	609	328	0.01	404	97	221	319	1053	208	3450
Carmem	R							328	0.01	405						
Norton	R	0	0.00	979	329	0.01	436	328	0.01	406	329	-2	328	131697	-1	131025
Tortosi	W	930	0.02	303	503	0.01	375	325	0.01	407	-427	-178	-605	-46	-35	-65
Saperavi Severny	R	25	0.00	710	350	0.01	424	325	0.01	408	325	-25	300	1300	-7	1200
Mavrodafni	R	537	0.01	377	345	0.01	432	324	0.01	409	-192	-20	-213	-36	-6	-40
Falanghina	W	1658	0.03	222	3037	0.07	148	323	0.01	410	1379	-2715	-1336	83	-89	-81
Mtsvane Kakhuri	W	249	0.01	465	319	0.01	443	319	0.01	411	70	0	70	28	0	28
Crouchen	W	2259	0.05	194	725	0.02	316	319	0.01	412	-1534	-406	-1940	-68	-56	-86
Erbaluce	W	329	0.01	429	319	0.01	444	316	0.01	413	-9	-4	-13	-3	-1	-4
Cabernet Mitos	R				322	0.01	442	312	0.01	414		-10			-3	
Iordan	W				315	0.01	446	311	0.01	415		-4			-1	
Manto Negro	R	470	0.01	390	273	0.01	467	311	0.01	416	-197	38	-159	-42	14	-34
Abouriou	R	419	0.01	401	329	0.01	437	310	0.01	417	-90	-19	-109	-22	-6	-26
Citronny Magarach	W				307	0.01	453	307	0.01	418		0			0	
Camaralet de Lasseube	W	691	0.01	333	520	0.01	365	306	0.01	419	-171	-214	-385	-25	-41	-56
Arkadia	W							303	0.01	420						

Table 6 (cont.): Global ranking of prime winegrape varieties by area, and changes, 2000, 2010 and 2016 (ha and %)

		2000			2010			2016			2010-2000	2016-2010	2016-2000	2010-2000	2016-2010	2016-2000
Prime vairiety	*Col*	*Global area (ha)*	*Global share %*	*Global rank*	*Global area (ha)*	*Global share %*	*Global rank*	*Global area (ha)*	*Global share %*	*Global rank*	*Change (ha)*	*Change (ha)*	*Change (ha)*	*Changes %*	*Changes %*	*Changes %*
Muscat Fleur d'Oranger	W	36	0.00	675	91	0.00	637	299	0.01	421	55	208	263	150	228	721
Agronomica	R	19	0.00	742	327	0.01	439	299	0.01	422	308	-28	280	1612	-9	1466
Tamarez	W	585	0.01	360	343	0.01	433	298	0.01	423	-243	-45	-287	-41	-13	-49
Băbească Neagră (G)	R				328	0.01	438	297	0.01	424		-31			-9	
Raboso Veronese	R	307	0.01	435	277	0.01	465	295	0.01	425	-30	19	-11	-10	7	-4
Ezerfürtű	W	405	0.01	404	406	0.01	402	295	0.01	426	2	-112	-110	0	-27	-27
Podarok Magaracha	W	148	0.00	515	504	0.01	374	292	0.01	427	356	-212	144	240	-42	97
Brandam	W				312	0.01	448	291	0.01	428		-20			-7	
Dunkelfelder	R	280	0.01	448	356	0.01	422	291	0.01	429	76	-65	11	27	-18	4
Pascale	R	1573	0.03	231	375	0.01	411	289	0.01	430	-1198	-86	-1284	-76	-23	-82
Goruli Mtsvane	W	224	0.00	475	287	0.01	458	287	0.01	431	63	0	63	28	0	28
Trebbiano Modenese	W	583	0.01	361	363	0.01	417	287	0.01	432	-221	-75	-296	-38	-21	-51
Mondeuse Noire	R	1404	0.03	246	303	0.01	455	287	0.01	433	-1101	-16	-1117	-78	-5	-80
Egiodola	R	315	0.01	433	349	0.01	426	285	0.01	434	34	-64	-30	11	-18	-9
Mustoasă de Măderat	W				255	0.01	483	282	0.01	435		27			11	
Aleksandrouli	R	219	0.00	476	281	0.01	463	281	0.01	436	62	0	62	28	0	28
Argaman	R	202	0.00	483	202	0.00	518	275	0.01	437	0	73	73	0	36	36
Cabernet Dorsa	R	43	0.00	655	252	0.01	486	272	0.01	438	209	20	229	485	8	532
Vidvizhenets	W				271	0.01	471	271	0.01	439		0			0	
Casculho	R				267	0.01	474	269	0.01	440		2			1	
Roobernet	R	78	0.00	590	139	0.00	571	269	0.01	441	61	130	191	79	93	247
Şarbă	W				265	0.01	475	266	0.01	442		1			1	
Cerceal Branco	W	597	0.01	358	379	0.01	410	261	0.01	443	-218	-118	-337	-37	-31	-56
Maticha	W	354	0.01	419	311	0.01	449	257	0.01	444	-43	-54	-97	-12	-17	-27
Portan	R	368	0.01	415	264	0.01	477	256	0.01	445	-103	-9	-112	-28	-3	-30
Aurore	W	299	0.01	439	268	0.01	473	255	0.01	446	-31	-13	-44	-10	-5	-15
Gamay Teinturier de Bouze	R	318	0.01	432	278	0.01	464	255	0.01	447	-40	-23	-64	-13	-8	-20
Lacrima di Morro d'Alba	R	652	0.01	339	421	0.01	397	252	0.01	448	-231	-169	-400	-35	-40	-61
Pinot Noir Précoce	R	85	0.00	578	273	0.01	468	251	0.01	449	188	-22	166	220	-8	194
Lagrein	R	471	0.01	389	718	0.02	319	251	0.01	450	247	-467	-220	52	-65	-47
Moristel	R				147	0.00	561	247	0.01	451		100			68	
Vernaccia di Oristano	W	565	0.01	369	272	0.01	470	246	0.01	452	-293	-26	-320	-52	-10	-57
Red Globe	R	2113	0.04	204	242	0.01	492	242	0.01	453	-1871	0	-1871	-89	0	-89
Vignoles	W	68	0.00	612	254	0.01	484	241	0.01	454	186	-13	172	272	-5	252
Traminette	W	5	0.00	849	240	0.01	494	239	0.01	455	235	0	235	4839	0	4832
Campbell Early	R	43	0.00	656	61	0.00	713	238	0.01	456	18	177	195	43	290	459
Schiava	R	1231	0.03	263	517	0.01	367	236	0.01	457	-714	-281	-995	-58	-54	-81
Elvira	W	344	0.01	423	263	0.01	478	231	0.01	458	-81	-32	-113	-23	-12	-33

Table 6 (cont.): Global ranking of prime winegrape varieties by area, and changes, 2000, 2010 and 2016 (ha and %)

		2000			2010			2016			2010-2000	2016-2010	2016-2000	2010-2000	2016-2010	2016-2000
Prime vairiety	*Col*	*Global area (ha)*	*Global share %*	*Global rank*	*Global area (ha)*	*Global share %*	*Global rank*	*Global area (ha)*	*Global share %*	*Global rank*	*Change (ha)*	*Change (ha)*	*Change (ha)*	*Changes %*	*Changes %*	*Changes %*
Fiesta	W	161	0.00	508	230	0.00	499	230	0.01	459	69	0	69	43	0	43
Garanoir	R	76	0.00	594	216	0.00	506	229	0.01	460	140	13	153	186	6	203
Maréchal Foch	R	173	0.00	496	356	0.01	421	229	0.01	461	183	-127	56	106	-36	32
Kodrinskii	R	5	0.00	843	5	0.00	1084	229	0.01	462	0	224	224	0	4475	4475
Doina	R				227	0.00	502	227	0.01	463		0			0	
Altesse	W	294	0.01	444	359	0.01	419	227	0.01	464	65	-132	-67	22	-37	-23
Lacrima Christi	R				85	0.00	649	226	0.01	465		141			165	
Zengő	W				264	0.01	476	226	0.01	466		-39			-15	
Bouvier	W	365	0.01	416	250	0.01	487	224	0.00	467	-115	-26	-141	-32	-10	-39
Moravia Agria	R	1092	0.02	281	550	0.01	359	222	0.00	468	-542	-328	-871	-50	-60	-80
Malbo Gentile	R	106	0.00	552	211	0.00	510	219	0.00	469	105	8	113	99	4	106
Cayuga White	W	108	0.00	549	212	0.00	509	217	0.00	470	104	6	110	97	3	102
Rösler	R				160	0.00	550	217	0.00	471		56			35	
Prezentabil	W							215	0.00	472						
Abrusco	R	399	0.01	406	423	0.01	395	215	0.00	473	24	-208	-184	6	-49	-46
Frontenac	R				135	0.00	576	212	0.00	474		77			57	
Fokiano	R	162	0.00	507	262	0.01	479	212	0.00	475	100	-51	49	62	-19	30
Parraleta	R	167	0.00	503	348	0.01	427	212	0.00	476	181	-136	45	109	-39	27
Afus Ali	W	1837	0.04	212	381	0.01	409	211	0.00	477	-1456	-170	-1626	-79	-45	-88
Terrano	R	1461	0.03	237	1914	0.04	190	209	0.00	478	453	-1705	-1252	31	-89	-86
Pölöskei Muskotály	W				103	0.00	616	207	0.00	479		104			101	
Papazkarası	R	122	0.00	531	175	0.00	537	204	0.00	480	53	29	82	43	17	67
Dona Branca	W	296	0.01	441	276	0.01	466	204	0.00	481	-20	-72	-92	-7	-26	-31
Danlas	W				255	0.01	482	203	0.00	482		-52			-20	
Noah	W	260	0.01	457	563	0.01	354	200	0.00	483	303	-363	-60	117	-64	-23
Kraljevina	W				447	0.01	392	199	0.00	484		-248			-55	
Viktória Gyöngye	W				190	0.00	527	198	0.00	485		7			4	
Roter Veltliner	G	258	0.01	459	199	0.00	521	198	0.00	486	-59	-1	-60	-23	-1	-23
Menu Pineau	W	380	0.01	411	205	0.00	516	197	0.00	487	-174	-8	-182	-46	-4	-48
Uva Rara	R	570	0.01	366	460	0.01	389	197	0.00	488	-111	-263	-374	-19	-57	-66
Concord Clone 30	R							196	0.00	489						
Juhfark	W				186	0.00	530	195	0.00	490		9			5	
Tsulukidzis Tetra	W	152	0.00	510	195	0.00	523	195	0.00	491	43	0	43	28	0	28
Arvine	W	61	0.00	622	172	0.00	540	192	0.00	492	111	20	131	183	11	215
Tinta de Pegoes	R				195	0.00	524	191	0.00	493		-4			-2	
Žilavka	W							185	0.00	494						
Refosco di Faedis	R	256	0.01	461	217	0.00	505	185	0.00	495	-38	-33	-71	-15	-15	-28
Fintendo	R	144	0.00	518	118	0.00	591	185	0.00	496	-26	66	40	-18	56	28

Table 6 (cont.): Global ranking of prime winegrape varieties by area, and changes, 2000, 2010 and 2016 (ha and %)

		2000			2010			2016			2010-2000	2016-2010	2016-2000	2010-2000	2016-2010	2016-2000
Prime vairiety	*Col*	*Global area (ha)*	*Global share %*	*Global rank*	*Global area (ha)*	*Global share %*	*Global rank*	*Global area (ha)*	*Global share %*	*Global rank*	*Change (ha)*	*Change (ha)*	*Change (ha)*	*Changes %*	*Changes %*	*Changes %*
Bellone	W	1315	0.03	256	511	0.01	370	184	0.00	497	-804	-327	-1131	-61	-64	-86
Turán	R				177	0.00	535	183	0.00	498		6			3	
Rúbea	R				81	0.00	658	181	0.00	499		100			123	
Bastardo Magarachsky	R	1969	0.04	209	2370	0.05	171	180	0.00	500	401	-2190	-1789	20	-92	-91
Verdello	W	662	0.01	335	287	0.01	459	179	0.00	501	-374	-108	-482	-57	-38	-73
Enantio	R	1062	0.02	287	724	0.02	317	178	0.00	502	-338	-547	-884	-32	-75	-83
Emerald Riesling	W	937	0.02	301	508	0.01	371	177	0.00	503	-428	-331	-759	-46	-65	-81
Limnio	R	95	0.00	565	372	0.01	413	176	0.00	504	277	-196	81	292	-53	85
Csaba Gyöngye	W				89	0.00	638	175	0.00	505		86			96	
Crystal	W	1	0.00	926	175	0.00	538	175	0.00	506	173	0	173	12393	0	12393
Rome	R	2	0.00	914	297	0.01	456	172	0.00	507	295	-125	171	18697	-42	10795
Praca	W				166	0.00	544	169	0.00	508		3			2	
Tsvetochny	W				169	0.00	542	169	0.00	509		0			0	
Marquette	R				88	0.00	640	166	0.00	510		79			90	
Montils	W	131	0.00	522	164	0.00	546	165	0.00	511	33	1	34	25	1	26
Aleatico	R	458	0.01	391	346	0.01	429	165	0.00	512	-112	-181	-293	-24	-52	-64
Schiava Gentile	R	1158	0.02	274	694	0.02	325	165	0.00	513	-465	-529	-994	-40	-76	-86
Rossese	R	232	0.00	471	312	0.01	447	164	0.00	514	80	-148	-68	34	-47	-29
Preto Martinho	R	428	0.01	400	163	0.00	548	163	0.00	515	-265	-1	-265	-62	-1	-62
Valbom	R				166	0.00	545	162	0.00	516		-3			-2	
Folgasao	W	409	0.01	403	182	0.00	533	162	0.00	517	-227	-19	-246	-56	-11	-60
Avarengo	R	1453	0.03	238	987	0.02	274	153	0.00	518	-466	-834	-1300	-32	-84	-89
Robola	W	359	0.01	417	471	0.01	384	152	0.00	519	111	-319	-207	31	-68	-58
Plavaie	W				209	0.00	513	152	0.00	520		-57			-27	
Bonamico	R	336	0.01	427	233	0.01	497	149	0.00	521	-103	-84	-187	-31	-36	-56
Cornalin	R	93	0.00	568	256	0.01	481	147	0.00	522	163	-109	54	175	-43	58
Amur	R				146	0.00	563	146	0.00	523		0			0	
Lakhegyi Mézes	W	567	0.01	367	306	0.01	454	145	0.00	524	-261	-160	-421	-46	-52	-74
Castelino	R				147	0.00	560	144	0.00	525		-3			-2	
Giro Nero	R	537	0.01	376	200	0.00	520	144	0.00	526	-338	-56	-393	-63	-28	-73
Stepnyak	W				144	0.00	564	144	0.00	527		0			0	
Black Queen	R	566	0.01	368	713	0.02	321	143	0.00	528	147	-570	-423	26	-80	-75
Gamay Teinturier de Chaudenay	R	267	0.01	453	157	0.00	553	142	0.00	529	-109	-15	-125	-41	-10	-47
Negru de Yaloven	R	15	0.00	766	15	0.00	936	141	0.00	530	0	126	126	0	842	842
Ribol	R				147	0.00	562	141	0.00	531		-6			-4	
Terret Noir	R	370	0.01	414	143	0.00	566	139	0.00	532	-226	-5	-231	-61	-3	-62
Callet	R	151	0.00	511	154	0.00	554	138	0.00	533	3	-16	-12	2	-10	-8
Servant	W	530	0.01	379	183	0.00	531	138	0.00	534	-348	-45	-392	-66	-24	-74

Table 6 (cont.): Global ranking of prime winegrape varieties by area, and changes, 2000, 2010 and 2016 (ha and %)

		2000			2010			2016			2010-2000	2016-2010	2016-2000	2010-2000	2016-2010	2016-2000
Prime vairiety	*Col*	*Global area (ha)*	*Global share %*	*Global rank*	*Global area (ha)*	*Global share %*	*Global rank*	*Global area (ha)*	*Global share %*	*Global rank*	*Change (ha)*	*Change (ha)*	*Change (ha)*	*Changes %*	*Changes %*	*Changes %*
Rouge du Pays	R							136	0.00	535						
Encruzado	W	291	0.01	445	282	0.01	462	132	0.00	536	-9	-150	-158	-3	-53	-54
Prensal	W	114	0.00	543	105	0.00	613	129	0.00	537	-9	24	15	-8	23	13
Yalovenskii Ustoichivyi	W							129	0.00	538						
Pinella	W	66	0.00	617	72	0.00	681	128	0.00	539	5	56	61	8	78	93
Valdiguié	R	79	0.00	587	272	0.01	469	126	0.00	540	193	-146	47	244	-54	60
Millot-Foch	R							126	0.00	541						
Malagousia	W	23	0.00	721	182	0.00	532	126	0.00	542	160	-57	103	700	-31	451
Nerello Cappuccio	R	1501	0.03	235	508	0.01	372	125	0.00	543	-993	-383	-1376	-66	-75	-92
Vermentino Nero	R	143	0.00	519	210	0.00	512	124	0.00	544	67	-86	-19	46	-41	-13
Trincadeira das Pratas	W	216	0.00	480	239	0.01	496	124	0.00	545	23	-115	-92	11	-48	-43
Medina	R				159	0.00	551	124	0.00	546		-36			-22	
Carnelian	R	625	0.01	348	316	0.01	445	123	0.00	547	-308	-193	-501	-49	-61	-80
Rotgipfler	W	118	0.00	537	105	0.00	612	123	0.00	548	-13	18	5	-11	17	4
Pamyati Negrulya	R							123	0.00	549						
Timorasso	W	19	0.00	747	129	0.00	585	123	0.00	550	111	-7	104	597	-5	561
Diolinoir	R	31	0.00	687	114	0.00	596	122	0.00	551	83	8	91	266	7	291
Trebbiano Spoletino	W	242	0.00	468	200	0.00	519	121	0.00	552	-42	-79	-120	-17	-39	-50
Picolit	W	93	0.00	570	128	0.00	588	121	0.00	553	35	-7	28	37	-5	30
Petit Bouschet	R	1	0.00	959	15	0.00	931	120	0.00	554	15	105	120	2698	683	21802
Canada Muscat	W	49	0.00	645				120	0.00	555			72			147
Asprouda	W	433	0.01	397	113	0.00	598	120	0.00	556	-320	7	-313	-74	6	-72
Reichensteiner	W	319	0.01	431	247	0.01	489	120	0.00	557	-72	-127	-199	-23	-52	-63
Neretta Cuneese	R	374	0.01	413	132	0.00	583	119	0.00	558	-242	-12	-254	-65	-9	-68
Solaris	W				81	0.00	662	118	0.00	559		38			47	
Zéta	W				118	0.00	590	118	0.00	560		-1			-1	
Riesus	W				115	0.00	594	115	0.00	561		0			0	
Spergola	W				110	0.00	604	115	0.00	562		5			4	
Barbera Bianca	W	251	0.01	464	181	0.00	534	114	0.00	563	-70	-67	-137	-28	-37	-55
Coarnă Neagră	R							114	0.00	564						
Moschomavro	R	2295	0.05	192	1428	0.03	219	113	0.00	565	-867	-1315	-2182	-38	-92	-95
Malvasia di Lipari	W	516	0.01	381	310	0.01	450	113	0.00	566	-206	-198	-403	-40	-64	-78
Herbemont	R	1453	0.03	239	764	0.02	309	112	0.00	567	-688	-652	-1341	-47	-85	-92
Heroldrebe	R	199	0.00	484	134	0.00	580	112	0.00	568	-65	-22	-87	-33	-16	-44
Johanniter	W				86	0.00	646	111	0.00	569		25			29	
Tempranillo (W)	R				5	0.00	1086	110	0.00	570		105			2103	
Malvasia Branca de Sao Jorge	W	55	0.00	634	110	0.00	603	110	0.00	571	56	0	55	102	0	101
Maiskii Chernyi	R	77	0.00	591	110	0.00	605	110	0.00	572	33	0	33	43	0	43

Table 6 (cont.): Global ranking of prime winegrape varieties by area, and changes, 2000, 2010 and 2016 (ha and %)

		2000			2010			2016			2010-2000	2016-2010	2016-2000	2010-2000	2016-2010	2016-2000
Prime vairiety	*Col*	*Global area (ha)*	*Global share %*	*Global rank*	*Global area (ha)*	*Global share %*	*Global rank*	*Global area (ha)*	*Global share %*	*Global rank*	*Change (ha)*	*Change (ha)*	*Change (ha)*	*Changes %*	*Changes %*	*Changes %*
Bíborkadarka	R	202	0.00	482	136	0.00	575	109	0.00	573	-67	-26	-93	-33	-19	-46
Coloraillo	R	614	0.01	350	374	0.01	412	109	0.00	574	-240	-265	-505	-39	-71	-82
Pallagrello Nero	R				169	0.00	541	107	0.00	575		-63			-37	
Korinthiaki	R	834	0.02	316	54	0.00	732	106	0.00	576	-780	52	-728	-93	96	-87
Zierfandler	G	98	0.00	562	117	0.00	592	105	0.00	577	19	-12	7	19	-10	7
Lambrusco Oliva	R				112	0.00	600	104	0.00	578		-7			-7	
Rieslina	W	219	0.00	477	174	0.00	539	103	0.00	579	-45	-71	-116	-20	-41	-53
Complexa	R	6	0.00	834	103	0.00	617	103	0.00	580	96	0	96	1506	0	1506
De Chaunac	R	186	0.00	487	91	0.00	636	102	0.00	581	-95	11	-84	-51	12	-45
Siegerrebe	G	167	0.00	502	131	0.00	584	102	0.00	582	-36	-29	-65	-22	-22	-39
Ruche	R	46	0.00	651	100	0.00	622	100	0.00	583	54	0	54	118	0	118
Malvasia di Casorzo	R	98	0.00	563	107	0.00	607	99	0.00	584	10	-9	1	10	-8	1
Lival	R				101	0.00	621	99	0.00	585		-3			-3	
Goldburger	W	309	0.01	434	140	0.00	570	98	0.00	586	-168	-42	-211	-55	-30	-68
Albarola	W	4090	0.08	130	197	0.00	522	95	0.00	587	-3893	-101	-3995	-95	-52	-98
La Crescent	W				77	0.00	672	94	0.00	588		16			21	
Amaral	R	582	0.01	362	92	0.00	634	93	0.00	589	-490	1	-489	-84	1	-84
Royalty	R	338	0.01	426	97	0.00	627	93	0.00	590	-241	-4	-245	-71	-4	-72
Bouchales	R	108	0.00	547	95	0.00	628	93	0.00	591	-13	-3	-16	-12	-3	-14
Rebo	R	37	0.00	673	125	0.00	589	92	0.00	592	88	-32	55	238	-26	150
Casavecchia	R				136	0.00	574	92	0.00	593		-44			-32	
Frontenac (G)	R				59	0.00	720	92	0.00	594		33			55	
Jubileum 75	R				194	0.00	525	91	0.00	595		-103			-53	
Nasco	W	166	0.00	505	141	0.00	568	91	0.00	596	-25	-50	-75	-15	-36	-45
Nieddera	R	58	0.00	629	107	0.00	608	91	0.00	597	49	-16	33	83	-15	56
Vespaiola	W	105	0.00	554	94	0.00	630	90	0.00	598	-10	-4	-14	-10	-4	-13
Chardonel	W				144	0.00	565	90	0.00	599		-54			-37	
Poulsard	R	295	0.01	442	307	0.01	452	90	0.00	600	12	-217	-205	4	-71	-69
Chasselas (R)	W	11	0.00	791	95	0.00	629	90	0.00	601	84	-5	79	756	-5	709
Adakarası	R	48	0.00	648	69	0.00	691	89	0.00	602	21	20	41	43	29	85
Emir	W	480	0.01	387	688	0.01	326	89	0.00	603	207	-599	-392	43	-87	-82
Padeiro	R				86	0.00	647	88	0.00	604		2			2	
Vespolina	R	103	0.00	557	134	0.00	579	88	0.00	605	31	-47	-16	30	-35	-15
Schioppettino	R	93	0.00	567	154	0.00	555	87	0.00	606	61	-67	-6	65	-43	-6
Rabo de Anho	R				99	0.00	623	86	0.00	607		-14			-14	
Plassa	R	41	0.00	661	91	0.00	635	86	0.00	608	50	-6	45	124	-6	110
Léon Millot	R	17	0.00	751	102	0.00	619	85	0.00	609	85	-16	69	507	-16	409
Sercial	W	306	0.01	436	106	0.00	610	85	0.00	610	-200	-21	-220	-65	-19	-72

Table 6 (cont.): Global ranking of prime winegrape varieties by area, and changes, 2000, 2010 and 2016 (ha and %)

		2000			2010			2016			2010-2000	2016-2010	2016-2000	2010-2000	2016-2010	2016-2000
Prime vairiety	*Col*	*Global area (ha)*	*Global share %*	*Global rank*	*Global area (ha)*	*Global share %*	*Global rank*	*Global area (ha)*	*Global share %*	*Global rank*	*Change (ha)*	*Change (ha)*	*Change (ha)*	*Changes %*	*Changes %*	*Changes %*
Canari Noir	R	218	0.00	479	163	0.00	549	84	0.00	611	-55	-79	-135	-25	-49	-62
Caddiu	R	978	0.02	295	309	0.01	451	83	0.00	612	-669	-225	-894	-68	-73	-91
Baroque	W	169	0.00	499	94	0.00	631	83	0.00	613	-75	-11	-85	-44	-11	-51
Ehrenfelser	W	289	0.01	446	113	0.00	599	82	0.00	614	-176	-30	-207	-61	-27	-72
Plavec Žuti	W				13	0.00	957	82	0.00	615		69			533	
San Giuseppe Nero	R	348	0.01	422	192	0.00	526	82	0.00	616	-156	-109	-266	-45	-57	-76
Malvasia di Schierano	R	181	0.00	490	89	0.00	639	82	0.00	617	-92	-7	-98	-51	-8	-55
Rubin Golodrigi	R				82	0.00	656	82	0.00	618		0			0	
Monemvassia	W	418	0.01	402	481	0.01	381	81	0.00	619	62	-399	-337	15	-83	-81
Blanc du Bois	W				28	0.00	838	81	0.00	620		53			187	
Prima	R				84	0.00	653	81	0.00	621		-2			-3	
Drupeggio	W	617	0.01	349	286	0.01	460	81	0.00	622	-331	-205	-536	-54	-72	-87
Pedral	R	179	0.00	491	151	0.00	558	80	0.00	623	-28	-71	-99	-15	-47	-55
Perricone	R	580	0.01	363	228	0.00	501	80	0.00	624	-351	-149	-500	-61	-65	-86
Luisa Blanca	W							80	0.00	625						
Labrusco	R				81	0.00	660	79	0.00	626		-2			-2	
Gamay Teinturier Freaux	R	132	0.00	521	104	0.00	614	79	0.00	627	-28	-25	-53	-21	-24	-40
Groppello Gentile	R	219	0.00	478	326	0.01	440	78	0.00	628	108	-248	-141	49	-76	-64
Dekabrskii	R				78	0.00	669	78	0.00	629		0			0	
Caíño Blanco	W	69	0.00	609	128	0.00	586	77	0.00	630	59	-51	8	85	-40	12
Coda di Volpe Bianca	W	980	0.02	294	586	0.01	348	77	0.00	631	-393	-510	-903	-40	-87	-92
Tinta Aguiar	R				75	0.00	675	77	0.00	632		1			2	
Agua Santa	R				78	0.00	671	76	0.00	633		-1			-2	
Terret Gris	W	262	0.01	454	78	0.00	668	76	0.00	634	-184	-3	-186	-70	-3	-71
Seyval Noir	R							76	0.00	635						
Zelen	W							75	0.00	636						
Novac	R				73	0.00	676	74	0.00	637		1			1	
Rieslaner	W	70	0.00	606	84	0.00	650	73	0.00	638	14	-11	3	20	-13	4
Incrocio Manzoni 2.15	R	166	0.00	506	86	0.00	648	72	0.00	639	-80	-14	-94	-48	-16	-57
Chatus	R	15	0.00	765	79	0.00	667	71	0.00	640	64	-8	56	420	-10	369
Königin der Weingärten	W	750	0.02	325	61	0.00	712	70	0.00	641	-689	9	-680	-92	15	-91
Cienna	R							70	0.00	642						
Dorinto	W				115	0.00	595	70	0.00	643		-45			-39	
Barbera Sarda	R	326	0.01	430	84	0.00	651	70	0.00	644	-242	-14	-256	-74	-17	-79
Albarossa	R	5	0.00	851	80	0.00	664	70	0.00	645	75	-10	65	1582	-13	1362
Romorantin	W	81	0.00	583	72	0.00	679	69	0.00	646	-8	-3	-11	-10	-4	-14
Piquepoul Noir	R	103	0.00	556	70	0.00	688	69	0.00	647	-34	-1	-34	-32	-1	-33
Foglia Tonda	R	40	0.00	663	101	0.00	620	68	0.00	648	61	-33	28	153	-33	71

Table 6 (cont.): Global ranking of prime winegrape varieties by area, and changes, 2000, 2010 and 2016 (ha and %)

		2000			2010			2016			2010-2000	2016-2010	2016-2000	2010-2000	2016-2010	2016-2000
Prime vairiety	*Col*	*Global area (ha)*	*Global share %*	*Global rank*	*Global area (ha)*	*Global share %*	*Global rank*	*Global area (ha)*	*Global share %*	*Global rank*	*Change (ha)*	*Change (ha)*	*Change (ha)*	*Changes %*	*Changes %*	*Changes %*
Dakapo	R				51	0.00	747	68	0.00	649		18			35	
Petit Rouge	R	100	0.00	560	84	0.00	652	68	0.00	650	-16	-16	-32	-16	-19	-32
Therona	W	185	0.00	488	99	0.00	624	67	0.00	651	-86	-32	-118	-47	-32	-64
Aromat de Iasi	W				62	0.00	710	66	0.00	652		5			7	
Tintilia del Molise	R				111	0.00	601	66	0.00	653		-45			-41	
Dostoinyi	R				65	0.00	702	65	0.00	654		0			0	
L´Acadie Blanc	W							65	0.00	655						
Nosiola	W	191	0.00	485	79	0.00	665	65	0.00	656	-112	-15	-127	-59	-18	-66
Donzelinho Branco	W	59	0.00	628	65	0.00	701	64	0.00	657	6	-1	6	11	-1	10
Tinta Bragao	R				63	0.00	708	64	0.00	658		1			2	
Cabernet Cubin	R				60	0.00	715	62	0.00	659		2			4	
Mourisco de Semente	R				60	0.00	714	61	0.00	660		1			1	
Segalin	R	54	0.00	638	65	0.00	703	61	0.00	661	11	-4	7	20	-6	13
Menoir	R				65	0.00	699	61	0.00	662		-5			-7	
Vertzami	R	491	0.01	385	335	0.01	435	60	0.00	663	-156	-275	-431	-32	-82	-88
Sciascinoso	R	253	0.01	462	94	0.00	632	59	0.00	664	-159	-34	-193	-63	-37	-77
Aladasturi	R	46	0.00	650	59	0.00	718	59	0.00	665	13	0	13	28	0	28
Lambrusco Viadanese	R	277	0.01	451	240	0.01	493	59	0.00	666	-37	-182	-218	-13	-76	-79
Lambrusca di Alessandria	R	888	0.02	307	137	0.00	573	58	0.00	667	-751	-79	-830	-85	-57	-93
Nero Buono di Cori	R	114	0.00	542	135	0.00	577	58	0.00	668	21	-77	-56	19	-57	-49
Mazzese	R	80	0.00	585	76	0.00	673	57	0.00	669	-4	-19	-23	-5	-25	-28
Tinta Francisca	R				53	0.00	739	55	0.00	670		2			5	
Flame Seedless	R	548	0.01	371	55	0.00	726	55	0.00	671	-493	0	-493	-90	0	-90
Kishmish Luchistyi	G							55	0.00	672						
Marzemina Bianca	W	78	0.00	589	54	0.00	733	55	0.00	673	-24	0	-24	-31	1	-30
Lambrusco	R	42	0.00	659	45	0.00	765	54	0.00	674	4	9	13	9	20	31
Rose du Var	G	129	0.00	525	56	0.00	723	54	0.00	675	-73	-1	-75	-57	-2	-58
Bukettraube	W	280	0.01	447	71	0.00	685	54	0.00	676	-210	-16	-226	-75	-23	-81
Arriloba	W	59	0.00	627	55	0.00	729	54	0.00	677	-4	0	-5	-7	-1	-8
Malaga Blanc	W	11	0.00	789	16	0.00	922	54	0.00	678	5	38	43	43	234	378
Grossa	R				73	0.00	677	54	0.00	679		-19			-26	
Zghihară de Huşi	W				87	0.00	643	54	0.00	680		-33			-38	
Würzer	W	108	0.00	548	70	0.00	687	54	0.00	681	-38	-16	-54	-35	-23	-50
Garrido Fino	W	174	0.00	495	59	0.00	719	54	0.00	682	-115	-5	-120	-66	-9	-69
Tinta Penajoia	R				53	0.00	740	53	0.00	683		0			1	
Hrvatica	R	245	0.01	466	116	0.00	593	53	0.00	684	-129	-63	-191	-53	-54	-78
Concieira	R				52	0.00	742	53	0.00	685		1			2	
Nobling	W	102	0.00	559	1	0.00	1246	52	0.00	686	-101	51	-50	-99	5147	-49

Table 6 (cont.): Global ranking of prime winegrape varieties by area, and changes, 2000, 2010 and 2016 (ha and %)

		2000			2010			2016			2010-2000	2016-2010	2016-2000	2010-2000	2016-2010	2016-2000
Prime vairiety	*Col*	*Global area (ha)*	*Global share %*	*Global rank*	*Global area (ha)*	*Global share %*	*Global rank*	*Global area (ha)*	*Global share %*	*Global rank*	*Change (ha)*	*Change (ha)*	*Change (ha)*	*Changes %*	*Changes %*	*Changes %*
Verdiso	W	71	0.00	603	68	0.00	694	52	0.00	687	-4	-15	-19	-5	-23	-27
Fonte Cal	W	355	0.01	418	111	0.00	602	52	0.00	688	-244	-59	-303	-69	-53	-85
Magliocco Dolce	R	243	0.00	467	87	0.00	644	51	0.00	689	-157	-35	-192	-64	-41	-79
Folignan	W				51	0.00	746	51	0.00	690		0			1	
Vitovska	W	42	0.00	658	50	0.00	751	51	0.00	691	8	1	9	19	2	21
Rondo	R				40	0.00	777	51	0.00	692		10			26	
Pignolo	R	18	0.00	748	93	0.00	633	50	0.00	693	75	-43	32	407	-46	172
Golden Muscat	W	1190	0.02	270	1191	0.03	248	50	0.00	694	1	-1141	-1140	0	-96	-96
Fioletovy Ranny	R				50	0.00	750	50	0.00	695		0			0	
Verdil	W	131	0.00	524	57	0.00	721	50	0.00	696	-74	-7	-81	-56	-13	-62
Bosco	W	88	0.00	574	82	0.00	657	50	0.00	697	-6	-32	-38	-6	-39	-43
Göcseji Zamatos	W				55	0.00	725	50	0.00	698		-6			-10	
Seinoir	R	15	0.00	769	87	0.00	642	50	0.00	699	72	-37	35	494	-43	238
Kéknyelű	W				43	0.00	773	50	0.00	700		7			16	
Mátrai Muskotály	W				67	0.00	697	49	0.00	701		-17			-26	
Bailey	R				34	0.00	803	49	0.00	702		15			42	
Rossignola	R	295	0.01	443	188	0.00	529	49	0.00	703	-107	-139	-246	-36	-74	-83
Trincadeiro Branco	W				59	0.00	717	49	0.00	704		-11			-18	
Madeleine × Angevine 7672	W				52	0.00	745	48	0.00	705		-3			-6	
Albarín Blanco	W				23	0.00	878	48	0.00	706		25			109	
Savagnin Rose	G	883	0.02	309	884	0.02	292	48	0.00	707	1	-837	-836	0	-95	-95
Italica	W	178	0.00	492	367	0.01	416	47	0.00	708	189	-320	-131	106	-87	-73
Torbato	W	168	0.00	500	52	0.00	743	46	0.00	709	-116	-6	-122	-69	-11	-73
Krakhuna	W	36	0.00	676	46	0.00	762	46	0.00	710	10	0	10	28	0	28
Phoenix	W	24	0.00	715	67	0.00	696	46	0.00	711	43	-21	22	179	-31	92
Pelaverga	R	28	0.00	697	55	0.00	728	46	0.00	712	27	-9	19	99	-16	67
Tazzelenghe	R	68	0.00	614	55	0.00	727	45	0.00	713	-13	-10	-22	-19	-17	-33
Duna Gyöngye	R				63	0.00	707	45	0.00	714		-18			-28	
Caloria	R	129	0.00	526	108	0.00	606	45	0.00	715	-21	-63	-83	-16	-58	-65
St Croix	R				25	0.00	859	45	0.00	716		20			79	
Merlot Blanc	W	176	0.00	493	46	0.00	761	44	0.00	717	-129	-2	-132	-74	-5	-75
Blanca Ovoide	W	107	0.00	551	40	0.00	778	44	0.00	718	-67	4	-63	-63	9	-59
Karát	W				50	0.00	749	44	0.00	719		-7			-14	
Saint Jeannet	W	68	0.00	613	56	0.00	724	43	0.00	720	-12	-13	-25	-18	-23	-37
Merille	R	131	0.00	523	44	0.00	768	42	0.00	721	-87	-2	-89	-66	-5	-68
Amigne	W	21	0.00	732	43	0.00	771	42	0.00	722	22	-2	21	107	-4	98
Lyana	W							41	0.00	723						
Picapoll Blanco	W	34	0.00	680	37	0.00	790	40	0.00	724	3	3	6	8	8	17

Table 6 (cont.): Global ranking of prime winegrape varieties by area, and changes, 2000, 2010 and 2016 (ha and %)

		2000			2010			2016			2010-2000	2016-2010	2016-2000	2010-2000	2016-2010	2016-2000
Prime vairiety	*Col*	*Global area (ha)*	*Global share %*	*Global rank*	*Global area (ha)*	*Global share %*	*Global rank*	*Global area (ha)*	*Global share %*	*Global rank*	*Change (ha)*	*Change (ha)*	*Change (ha)*	*Changes %*	*Changes %*	*Changes %*
Moscato Nazareno	W				68	0.00	695	40	0.00	725		-28			-41	
Verdelet	W				1	0.00	1280	40	0.00	726		39			6459	
Malvasia Nera di Basilicata	R	754	0.02	324	114	0.00	597	39	0.00	727	-640	-75	-715	-85	-66	-95
Bianco d'Alessano	W	941	0.02	300	419	0.01	398	39	0.00	728	-522	-380	-902	-55	-91	-96
Portland	W	9	0.00	813	12	0.00	963	39	0.00	729	4	27	30	43	215	350
Primavera	R				40	0.00	779	39	0.00	730		-1			-2	
Maceratino	W	122	0.00	532	177	0.00	536	39	0.00	731	55	-138	-83	45	-78	-68
Verdea	W	107	0.00	550	83	0.00	654	39	0.00	732	-24	-45	-69	-22	-54	-64
Moscato Rosa del Trentino	R	93	0.00	569	81	0.00	659	39	0.00	733	-12	-42	-55	-13	-52	-59
Caberinta	R	85	0.00	577	69	0.00	690	38	0.00	734	-17	-30	-47	-19	-44	-55
Cabernet Cortis	R							38	0.00	735						
Tyrian	R							38	0.00	736						
Vidadillo	R							38	0.00	737						
Optima	W	239	0.00	469	65	0.00	704	38	0.00	738	-174	-27	-201	-73	-42	-84
Chancellor	R	27	0.00	702	49	0.00	753	38	0.00	739	23	-12	11	85	-24	41
Tavkveri	R	29	0.00	693	37	0.00	788	37	0.00	740	8	0	8	28	0	28
Landal	R	16	0.00	760	43	0.00	772	37	0.00	741	27	-6	21	172	-13	135
St Pepin	W				19	0.00	898	37	0.00	742		18			91	
Golubok	R	50	0.00	644	87	0.00	641	37	0.00	743	37	-50	-13	74	-57	-26
San Michele	R	120	0.00	534	57	0.00	722	37	0.00	744	-63	-20	-83	-53	-35	-69
Lumassina	W	111	0.00	546	98	0.00	626	37	0.00	745	-13	-61	-74	-12	-62	-67
Muscat Swenson	W				24	0.00	861	37	0.00	746		12			50	
Doral	W	3	0.00	876	27	0.00	846	35	0.00	747	24	8	33	872	32	1180
Schönburger	G	39	0.00	667	68	0.00	692	35	0.00	748	29	-33	-4	75	-48	-10
Negretto	R	280	0.01	449	75	0.00	674	35	0.00	749	-204	-40	-245	-73	-53	-87
Yamabudo	R							35	0.00	750						
Lora	W							35	0.00	751						
Grechetto Rosso	R	111	0.00	545	49	0.00	756	35	0.00	752	-62	-14	-76	-56	-29	-69
Carricante	W	252	0.01	463	205	0.00	517	35	0.00	753	-47	-170	-217	-19	-83	-86
Steuben	R				39	0.00	782	35	0.00	754		-4			-11	
Galotta	R				13	0.00	954	35	0.00	755		21			163	
Cabernet Dorio	R				36	0.00	797	34	0.00	756		-2			-5	
Branco Gouvaes	W				36	0.00	793	34	0.00	757		-2			-6	
Morava	W							34	0.00	758						
Jampal	W	127	0.00	527	71	0.00	683	34	0.00	759	-57	-37	-94	-44	-52	-73
Donzelinho Tinto	R				33	0.00	811	34	0.00	760		0			1	
Biancone di Portoferraio	W	67	0.00	616	78	0.00	670	34	0.00	761	11	-44	-33	16	-57	-50
Semidano	W	48	0.00	647	36	0.00	792	34	0.00	762	-12	-3	-15	-25	-8	-30

Table 6 (cont.): Global ranking of prime winegrape varieties by area, and changes, 2000, 2010 and 2016 (ha and %)

		2000			2010			2016			2010-2000	2016-2010	2016-2000	2010-2000	2016-2010	2016-2000
Prime vairiety	*Col*	*Global area (ha)*	*Global share %*	*Global rank*	*Global area (ha)*	*Global share %*	*Global rank*	*Global area (ha)*	*Global share %*	*Global rank*	*Change (ha)*	*Change (ha)*	*Change (ha)*	*Changes %*	*Changes %*	*Changes %*
Vitovska Grganja	W							33	0.00	763						
Centurian	R	134	0.00	520	34	0.00	805	33	0.00	764	-100	-1	-101	-75	-2	-75
Chenel	W	339	0.01	425	79	0.00	666	33	0.00	765	-260	-46	-306	-77	-58	-90
Caracol	W	14	0.00	773	33	0.00	814	33	0.00	766	19	0	19	140	0	140
Ramisco	R	72	0.00	602	34	0.00	808	33	0.00	767	-38	-1	-39	-53	-3	-54
Pampanuto	W	277	0.01	450	356	0.01	423	33	0.00	768	78	-323	-245	28	-91	-88
Courbu Blanc	W	47	0.00	649	43	0.00	770	32	0.00	769	-4	-11	-15	-9	-25	-32
Ora	W				38	0.00	786	32	0.00	770		-5			-14	
Pollera Nera	R	19	0.00	741	54	0.00	731	32	0.00	771	35	-22	13	183	-41	68
Kuban	R							32	0.00	772						
Ojaleshi	R	25	0.00	709	32	0.00	816	32	0.00	773	7	0	7	28	0	28
Malvasia Bianca di Basilicata	W	875	0.02	310	210	0.00	511	32	0.00	774	-665	-178	-843	-76	-85	-96
Rathay	R				9	0.00	1019	32	0.00	775		23			265	
Ferral	R	48	0.00	646	30	0.00	823	31	0.00	776	-18	1	-17	-37	3	-35
Podarok Zaporozju	W							31	0.00	777						
Prunesta	R	92	0.00	571	36	0.00	795	31	0.00	778	-56	-5	-62	-61	-15	-67
Magna	R							30	0.00	779						
Kreaca	W				29	0.00	831	30	0.00	780		1			3	
Aglianicone	R	148	0.00	514	62	0.00	709	30	0.00	781	-86	-32	-118	-58	-52	-80
Kabar	W				18	0.00	909	30	0.00	782		12			67	
Barkhatnyi	W				30	0.00	827	30	0.00	783		0			0	
Tinta Pomar	R				30	0.00	830	30	0.00	784		0			1	
Trebbianina	W				128	0.00	587	30	0.00	785		-98			-77	
Arnsburger	W	3	0.00	871	30	0.00	824	29	0.00	786	27	-1	26	913	-3	880
Barbaroux	G	79	0.00	586	30	0.00	828	29	0.00	787	-49	-1	-50	-62	-2	-63
Gargiulo 2539	R	69	0.00	610	49	0.00	757	29	0.00	788	-20	-20	-40	-29	-40	-58
Uva Tosca	R	84	0.00	580	71	0.00	684	29	0.00	789	-13	-42	-54	-15	-59	-65
Humagne	W	9	0.00	811	30	0.00	825	29	0.00	790	21	-1	20	241	-4	227
Manseng Noir	R	10	0.00	797	32	0.00	819	29	0.00	791	22	-3	19	211	-9	184
Verdelho Tinto	R	1	0.00	957	28	0.00	839	29	0.00	792	28	0	28	4673	2	4752
Rossola Nera	R	102	0.00	558	86	0.00	645	29	0.00	793	-16	-58	-74	-16	-67	-72
Fredonia	R	86	0.00	575	37	0.00	791	28	0.00	794	-49	-8	-57	-57	-23	-67
Kakotrygis	W				103	0.00	615	28	0.00	795		-75			-73	
Kanzler	W	53	0.00	640	33	0.00	813	28	0.00	796	-20	-5	-25	-38	-15	-47
Madrasa	R	20	0.00	739	28	0.00	840	28	0.00	797	8	0	8	43	0	43
Pignola Valtellinese	R	70	0.00	607	49	0.00	758	28	0.00	798	-21	-21	-42	-31	-43	-60
Kishmish Moldavskii	R							28	0.00	799						
Albaranzeuli Nero	R	40	0.00	665	49	0.00	754	28	0.00	800	10	-22	-12	24	-44	-31

Table 6 (cont.): Global ranking of prime winegrape varieties by area, and changes, 2000, 2010 and 2016 (ha and %)

		2000			2010			2016			2010-2000	2016-2010	2016-2000	2010-2000	2016-2010	2016-2000
Prime vairiety	*Col*	*Global area (ha)*	*Global share %*	*Global rank*	*Global area (ha)*	*Global share %*	*Global rank*	*Global area (ha)*	*Global share %*	*Global rank*	*Change (ha)*	*Change (ha)*	*Change (ha)*	*Changes %*	*Changes %*	*Changes %*
Zeusz	W				28	0.00	843	27	0.00	801		0			0	
Cabernet Jura	R				19	0.00	899	27	0.00	802		8			43	
Ryugan	W							27	0.00	803						
Alcañon	W	54	0.00	637	60	0.00	716	27	0.00	804	6	-33	-27	11	-55	-50
Scuppernong	W							27	0.00	805						
Vega	W	27	0.00	699	35	0.00	801	27	0.00	806	7	-8	0	27	-22	-1
Thrapsathiri	W				31	0.00	822	27	0.00	807		-4			-14	
Admirable de Courtiller	W				28	0.00	841	27	0.00	808		-1			-4	
Barcelo	W	34	0.00	682	23	0.00	880	26	0.00	809	-11	4	-7	-32	16	-22
Valenci Tinto	R	5	0.00	850	27	0.00	845	26	0.00	810	22	-1	22	461	-2	448
Frontenac (W)	R							26	0.00	811						
Sgavetta	R	61	0.00	623	47	0.00	760	26	0.00	812	-14	-21	-34	-23	-44	-57
Codană	R				24	0.00	866	26	0.00	813		2			8	
La Crosse	W				25	0.00	857	26	0.00	814		1			3	
Chkhaveri	G	20	0.00	736	26	0.00	852	26	0.00	815	6	0	6	28	0	28
Kisi	W	20	0.00	737	26	0.00	853	26	0.00	816	6	0	6	28	0	28
Calitor Noir	R	85	0.00	579	26	0.00	849	26	0.00	817	-59	-1	-59	-69	-3	-70
Columna	W				24	0.00	870	26	0.00	818		2			7	
Oberlin	R	0	0.00	990	64	0.00	705	26	0.00	819	64	-39	25	45756	-60	18175
Sabrevois	R				10	0.00	996	25	0.00	820		16			157	
Lambrusco Montericco	R	262	0.01	455	70	0.00	686	25	0.00	821	-191	-45	-236	-73	-64	-90
Odysseus	W				4	0.00	1098	25	0.00	822		21			468	
Fumin	R	73	0.00	598	31	0.00	821	25	0.00	823	-42	-6	-48	-57	-20	-66
Centesimino	R				24	0.00	871	25	0.00	824		1			5	
Silvaner (R)	W	2	0.00	905	38	0.00	785	25	0.00	825	36	-13	23	1800	-34	1150
Noiret	R				33	0.00	812	25	0.00	826		-8			-25	
Csillám	W				20	0.00	897	25	0.00	827		5			26	
Groppello di Mocasina	R	120	0.00	536	81	0.00	661	24	0.00	828	-39	-57	-95	-32	-70	-80
Tinta de Alcoa	R				24	0.00	869	24	0.00	829		0			1	
Adirondac	R							24	0.00	830						
Malvasia Fina Roxa	G				25	0.00	858	24	0.00	831		-1			-4	
Alb Aromat	W				24	0.00	863	24	0.00	832		0			-1	
Goldriesling	W	10	0.00	800	21	0.00	890	24	0.00	833	11	3	14	109	14	139
Yama Sauvignon	R							24	0.00	834						
Moore's Diamond	W	19	0.00	743	42	0.00	775	24	0.00	835	23	-18	5	121	-43	26
Prie	W	36	0.00	677	33	0.00	810	24	0.00	836	-2	-9	-12	-7	-28	-33
Kármin	R				36	0.00	794	24	0.00	837		-12			-34	
Sun Muscat	R							23	0.00	838						

Table 6 (cont.): Global ranking of prime winegrape varieties by area, and changes, 2000, 2010 and 2016 (ha and %)

		2000			2010			2016			2010-2000	2016-2010	2016-2000	2010-2000	2016-2010	2016-2000
Prime vairiety	*Col*	*Global area (ha)*	*Global share %*	*Global rank*	*Global area (ha)*	*Global share %*	*Global rank*	*Global area (ha)*	*Global share %*	*Global rank*	*Change (ha)*	*Change (ha)*	*Change (ha)*	*Changes %*	*Changes %*	*Changes %*
Prairie Star	W				21	0.00	889	23	0.00	839		2			10	
Räuschling	W	23	0.00	720	23	0.00	881	23	0.00	840	0	0	0	-2	1	-1
Slankamenka	W				53	0.00	737	23	0.00	841		-30			-57	
Valerien	W	2	0.00	900	24	0.00	862	23	0.00	842	22	-2	21	1108	-7	1023
Biancolella	W	385	0.01	409	164	0.00	547	23	0.00	843	-222	-141	-363	-58	-86	-94
Mollard	R	23	0.00	719	23	0.00	876	23	0.00	844	0	-1	-1	-1	-2	-3
Manzoni Rosa	G				29	0.00	832	23	0.00	845		-7			-23	
Nektár	W				21	0.00	887	22	0.00	846		1			5	
Moscato di Terracina	W	229	0.00	473	138	0.00	572	22	0.00	847	-91	-116	-207	-40	-84	-90
Cascade	R							22	0.00	848						
Rougeon	R	36	0.00	674	42	0.00	774	21	0.00	849	6	-21	-15	17	-50	-41
Lucie Kuhlmann	R							21	0.00	850						
Greco	W	1325	0.03	254	158	0.00	552	21	0.00	851	-1167	-137	-1304	-88	-87	-98
Regner	W	150	0.00	512	46	0.00	763	21	0.00	852	-104	-25	-129	-69	-54	-86
Brianna	W				12	0.00	964	21	0.00	853		8			69	
Hibernal	W							20	0.00	854						
Valentino Nero	R	56	0.00	632	21	0.00	892	20	0.00	855	-36	0	-36	-64	-1	-64
St Vincent	R				23	0.00	879	20	0.00	856		-3			-12	
Marmajuelo	W	37	0.00	672	24	0.00	867	20	0.00	857	-13	-4	-17	-36	-17	-46
Yamasachi	R							20	0.00	858						
Goethe	R							20	0.00	859						
Pavana	R	69	0.00	611	32	0.00	817	20	0.00	860	-37	-12	-49	-53	-38	-71
Fuella Nera	R	4	0.00	857	20	0.00	894	20	0.00	861	16	0	16	416	-2	403
Prunelard	R	2	0.00	895	20	0.00	893	19	0.00	862	18	-1	17	834	-6	781
Manzoni Moscato	R				20	0.00	896	19	0.00	863		-1			-3	
Minella Bianca	W	73	0.00	599	65	0.00	700	19	0.00	864	-8	-46	-54	-11	-71	-74
Mechta	R							19	0.00	865						
Goustolidi	W	112	0.00	544	68	0.00	693	19	0.00	866	-44	-49	-94	-40	-73	-83
Plant Droit	R	40	0.00	666	19	0.00	900	19	0.00	867	-21	-1	-21	-52	-3	-53
Neretto di Bairo	R	53	0.00	642	34	0.00	806	19	0.00	868	-19	-15	-34	-35	-45	-65
Abbuoto	R	696	0.01	332	37	0.00	789	18	0.00	869	-659	-19	-678	-95	-50	-97
Avana	R	53	0.00	641	28	0.00	842	18	0.00	870	-25	-9	-34	-48	-33	-65
Gualarido	W							18	0.00	871						
Castelão Branco	W				37	0.00	787	18	0.00	872		-19			-52	
Perle	G	121	0.00	533	34	0.00	804	18	0.00	873	-87	-16	-103	-72	-47	-85
Negru de Drăgăşani	R				16	0.00	929	18	0.00	874		2			16	
Crâmpoşie Selecţionată	W				409	0.01	399	18	0.00	875		-391			-96	
Folgasao Roxo	R				18	0.00	904	18	0.00	876		0			-2	

Table 6 (cont.): Global ranking of prime winegrape varieties by area, and changes, 2000, 2010 and 2016 (ha and %)

		2000			2010			2016			2010-2000	2016-2010	2016-2000	2010-2000	2016-2010	2016-2000
Prime vairiety	*Col*	*Global area (ha)*	*Global share %*	*Global rank*	*Global area (ha)*	*Global share %*	*Global rank*	*Global area (ha)*	*Global share %*	*Global rank*	*Change (ha)*	*Change (ha)*	*Change (ha)*	*Changes %*	*Changes %*	*Changes %*
Oeillade Noire	R				18	0.00	907	18	0.00	877		-1			-3	
Molette	W	30	0.00	691	29	0.00	834	18	0.00	878	-1	-11	-12	-3	-39	-41
Rózsakő	W				19	0.00	903	17	0.00	879		-1			-6	
Carrega Tinto	R				17	0.00	917	17	0.00	880		0			2	
Vostorg	W							17	0.00	881						
Nascetta	W				21	0.00	888	17	0.00	882		-4			-19	
Negoska	R	96	0.00	564	143	0.00	567	17	0.00	883	46	-126	-79	48	-88	-82
Muscardin	R	19	0.00	744	17	0.00	914	17	0.00	884	-2	0	-2	-9	-3	-11
Soperga	R	32	0.00	685	22	0.00	883	17	0.00	885	-9	-6	-15	-29	-26	-48
Colombana Nera	R	126	0.00	528	38	0.00	784	16	0.00	886	-88	-22	-110	-70	-57	-87
Bakator Roz	R				16	0.00	928	16	0.00	887		1			4	
Tarrango	R	120	0.00	535	72	0.00	678	16	0.00	888	-48	-56	-104	-40	-78	-87
Oseleta	R				15	0.00	940	16	0.00	889		1			9	
Tauberschwarz	R	8	0.00	822	11	0.00	980	16	0.00	890	3	5	8	38	45	100
Edelweiss	W				32	0.00	820	16	0.00	891		-16			-49	
Guardavalle	W	168	0.00	501	33	0.00	809	16	0.00	892	-134	-18	-152	-80	-53	-91
Sumoll	R	1401	0.03	247	83	0.00	655	16	0.00	893	-1318	-67	-1385	-94	-81	-99
Muskat de Yaloven	W	20	0.00	735	32	0.00	815	16	0.00	894	12	-17	-4	61	-51	-22
Mostosa	W	95	0.00	566	24	0.00	872	16	0.00	895	-71	-8	-79	-75	-34	-84
Olivette Noire	R	0	0.00	968	16	0.00	924	16	0.00	896	16	0	15	3523	-2	3433
Gyöngyrizling	W				23	0.00	875	16	0.00	897		-8			-34	
Gringet	W	74	0.00	597	25	0.00	856	15	0.00	898	-49	-10	-59	-66	-39	-79
Blasius	W				14	0.00	946	15	0.00	899		1			7	
Hondarribi Beltza	R	11	0.00	793	53	0.00	736	15	0.00	900	43	-38	4	392	-72	40
Rollo	W	117	0.00	540	51	0.00	748	15	0.00	901	-67	-35	-102	-57	-70	-87
Juwel	W	42	0.00	657	22	0.00	884	15	0.00	902	-20	-7	-27	-48	-32	-64
Rubinet	R							15	0.00	903						
Ohanes	W	0	0.00	966	16	0.00	925	15	0.00	904	15	-1	14	3441	-7	3193
Touriga Femea	R				15	0.00	939	15	0.00	905		0			0	
Pugnitello	R				28	0.00	836	15	0.00	906		-14			-48	
Zefir	W				49	0.00	755	15	0.00	907		-35			-70	
Muscat Timpuriu de Bucuresti	W							15	0.00	908						
Cesar	R	8	0.00	815	10	0.00	989	15	0.00	909	2	4	6	25	40	76
Fogoneu	R	36	0.00	678	35	0.00	799	15	0.00	910	-1	-20	-21	-2	-58	-59
Primetta	G	17	0.00	749	24	0.00	874	14	0.00	911	6	-9	-3	37	-39	-16
Aramon Noir (W)	R	43	0.00	654	15	0.00	937	14	0.00	912	-28	0	-29	-65	-3	-67
Vandal-Cliche	W							14	0.00	913						
Bastardo Branco	W				15	0.00	942	14	0.00	914		0			-2	

Table 6 (cont.): Global ranking of prime winegrape varieties by area, and changes, 2000, 2010 and 2016 (ha and %)

		2000			2010			2016			2010-2000	2016-2010	2016-2000	2010-2000	2016-2010	2016-2000
Prime vairiety	*Col*	*Global area (ha)*	*Global share %*	*Global rank*	*Global area (ha)*	*Global share %*	*Global rank*	*Global area (ha)*	*Global share %*	*Global rank*	*Change (ha)*	*Change (ha)*	*Change (ha)*	*Changes %*	*Changes %*	*Changes %*
Sevilhao	R				14	0.00	947	14	0.00	915		0			2	
Prior	R							14	0.00	916						
Debina	W	455	0.01	393	239	0.01	495	14	0.00	917	-216	-225	-441	-48	-94	-97
Helfensteiner	R	26	0.00	706	19	0.00	902	14	0.00	918	-7	-5	-12	-27	-26	-46
Malvasia Nera Lunga	R				38	0.00	783	14	0.00	919		-24			-64	
Geisenheim 318-57	W				106	0.00	611	14	0.00	920		-92			-87	
Floricica	W							14	0.00	921						
Casetta	R				12	0.00	962	14	0.00	922		1			9	
Rubintos	R				18	0.00	908	13	0.00	923		-5			-26	
GR 7	R				32	0.00	818	13	0.00	924		-19			-58	
Maturana Blanca	W				18	0.00	912	13	0.00	925		-5			-26	
Maiolica	R	70	0.00	605	26	0.00	851	13	0.00	926	-44	-13	-58	-63	-51	-82
Arvesiniadu	W	147	0.00	516	30	0.00	829	13	0.00	927	-117	-17	-134	-80	-58	-91
Meslier Saint-Francois	W	55	0.00	635	15	0.00	933	13	0.00	928	-40	-3	-42	-72	-17	-77
Neyret	R	76	0.00	593	41	0.00	776	12	0.00	929	-35	-28	-63	-46	-70	-84
New York Muscat	R				5	0.00	1087	12	0.00	930		7			152	
Corbina Vicentina	R				12	0.00	965	12	0.00	931		0			0	
Pannon Frankos	R				16	0.00	923	12	0.00	932		-4			-24	
Incrocio Terzi 1	R	65	0.00	618	44	0.00	769	12	0.00	933	-22	-31	-53	-34	-72	-82
Deckrot	R	30	0.00	690	2	0.00	1172	12	0.00	934	-28	10	-18	-93	501	-60
Albalonga	W	57	0.00	631	15	0.00	934	12	0.00	935	-42	-3	-45	-74	-20	-79
Braquet Noir	R	8	0.00	817	12	0.00	968	12	0.00	936	4	0	4	48	-2	45
Groppello di Revo	R				12	0.00	967	12	0.00	937		0			-3	
Flora	G	6	0.00	838	8	0.00	1030	12	0.00	938	2	4	6	28	52	95
Persan	R	3	0.00	872	12	0.00	971	12	0.00	939	9	0	9	317	-3	303
Bianchetta Trevigiana	W	53	0.00	639	13	0.00	952	12	0.00	940	-40	-2	-42	-75	-14	-78
Vertes Csillaga	W				21	0.00	886	11	0.00	941		-10			-46	
Trevisana Nera	R	33	0.00	684	15	0.00	930	11	0.00	942	-18	-4	-22	-54	-26	-66
Florental	R	1	0.00	956	26	0.00	854	11	0.00	943	25	-14	11	4157	-56	1792
Bakator Belyi	W				11	0.00	987	11	0.00	944		1			7	
Carminoir	R				10	0.00	992	11	0.00	945		1			11	
Dunav	R							11	0.00	946						
Rotberger	R	26	0.00	705	17	0.00	915	11	0.00	947	-9	-6	-15	-34	-35	-57
Corot Noir	R				27	0.00	848	11	0.00	948		-15			-58	
Dona Joaquina	W				24	0.00	873	11	0.00	949		-13			-53	
Terrantez	W	27	0.00	701	12	0.00	969	11	0.00	950	-15	-1	-16	-55	-9	-59
Retagliado Bianco	W	26	0.00	707	28	0.00	837	11	0.00	951	3	-17	-15	10	-61	-57
Valiant	R				11	0.00	982	11	0.00	952		0			1	

Table 6 (cont.): Global ranking of prime winegrape varieties by area, and changes, 2000, 2010 and 2016 (ha and %)

		2000			2010			2016			2010-2000	2016-2010	2016-2000	2010-2000	2016-2010	2016-2000
Prime vairiety	*Col*	*Global area (ha)*	*Global share %*	*Global rank*	*Global area (ha)*	*Global share %*	*Global rank*	*Global area (ha)*	*Global share %*	*Global rank*	*Change (ha)*	*Change (ha)*	*Change (ha)*	*Changes %*	*Changes %*	*Changes %*
Manteudo Preto	R				16	0.00	921	11	0.00	953		-5			-33	
Brun Argente	R	14	0.00	772	11	0.00	978	11	0.00	954	-3	0	-3	-19	-3	-21
Cabernet Carbon	R							11	0.00	955						
Bondola	R	17	0.00	752	13	0.00	959	11	0.00	956	-4	-2	-6	-24	-14	-35
Petite Pearl	R							11	0.00	957						
Planta Mula	R	1	0.00	932	24	0.00	868	11	0.00	958	23	-13	10	1934	-55	806
Tinta Martins	R				11	0.00	985	10	0.00	959		0			-2	
Gargiulo 45803	W	4	0.00	860	8	0.00	1023	10	0.00	960	4	2	6	111	27	168
Usakhelouri	R	8	0.00	819	10	0.00	991	10	0.00	961	2	0	2	28	0	28
Panonia	W							10	0.00	962						
Ventura	W	39	0.00	669	24	0.00	864	10	0.00	963	-15	-14	-29	-38	-58	-74
Patorra	R				10	0.00	997	10	0.00	964		0			2	
Ner d'Ala	R	8	0.00	823	30	0.00	826	10	0.00	965	22	-20	2	288	-67	29
Elbling (R)	W	4	0.00	855	9	0.00	1014	10	0.00	966	5	1	6	125	11	150
Divico	R							10	0.00	967						
Charmont	W	7	0.00	831	10	0.00	1003	10	0.00	968	3	0	3	39	4	45
Mara	R							10	0.00	969						
Monarch	R							10	0.00	970						
Swenson Red	R				11	0.00	977	10	0.00	971		-2			-15	
Pozsonyi Fehér	W				10	0.00	990	10	0.00	972		-1			-8	
Ucelut	W	10	0.00	796	10	0.00	1000	10	0.00	973	-1	0	-1	-7	-1	-8
Garonnet	R	7	0.00	832	14	0.00	945	10	0.00	974	8	-5	3	116	-33	44
Moscato di Scanzo	R	73	0.00	600	53	0.00	738	10	0.00	975	-19	-44	-63	-27	-82	-87
Ruby	R	6	0.00	833	9	0.00	1008	9	0.00	976	3	0	3	43	0	43
Notardomenico	R	13	0.00	780	10	0.00	1001	9	0.00	977	-3	0	-4	-26	-5	-30
Pecorello	W	16	0.00	756	34	0.00	807	9	0.00	978	17	-25	-7	106	-73	-44
Quagliano	R	8	0.00	816	9	0.00	1010	9	0.00	979	1	0	1	11	-1	9
Superior Seedless	W	6	0.00	836	9	0.00	1016	9	0.00	980	3	0	3	43	0	43
Teinturier	R	1	0.00	921	7	0.00	1032	9	0.00	981	6	2	8	401	20	503
Tintem	R				9	0.00	1013	9	0.00	982		0			-2	
Vien de Nus	R	25	0.00	708	13	0.00	961	9	0.00	983	-13	-4	-17	-50	-31	-66
Quiebratinajas Tinto	R	6	0.00	837	5	0.00	1083	9	0.00	984	-1	4	2	-20	74	39
Nebbiera	R	20	0.00	738	12	0.00	970	9	0.00	985	-7	-3	-11	-38	-29	-56
Verduschia	W	15	0.00	770	11	0.00	975	9	0.00	986	-3	-3	-6	-22	-25	-41
Arrufiac	W	126	0.00	530	80	0.00	663	9	0.00	987	-46	-71	-117	-36	-89	-93
Alionza	W	41	0.00	662	11	0.00	986	9	0.00	988	-30	-2	-32	-74	-19	-79
Barsaglina	R	20	0.00	734	17	0.00	918	9	0.00	989	-3	-9	-12	-16	-50	-58
Saphira	W							8	0.00	990						

Table 6 (cont.): Global ranking of prime winegrape varieties by area, and changes, 2000, 2010 and 2016 (ha and %)

		2000			2010			2016			2010-2000	2016-2010	2016-2000	2010-2000	2016-2010	2016-2000
Prime vairiety	*Col*	*Global area (ha)*	*Global share %*	*Global rank*	*Global area (ha)*	*Global share %*	*Global rank*	*Global area (ha)*	*Global share %*	*Global rank*	*Change (ha)*	*Change (ha)*	*Change (ha)*	*Changes %*	*Changes %*	*Changes %*
Forastera	W	543	0.01	373	208	0.00	514	8	0.00	991	-334	-200	-534	-62	-96	-98
Sacy	W	63	0.00	620	10	0.00	999	8	0.00	992	-54	-1	-55	-85	-14	-87
Frumoasa Alba	W							8	0.00	993						
Piculit Neri	R	126	0.00	529	22	0.00	885	8	0.00	994	-104	-14	-118	-83	-62	-94
Blattner Reds	R				39	0.00	781	8	0.00	995		-31			-79	
Osennii Ciornai	R							8	0.00	996						
Cornarea	R	22	0.00	731	13	0.00	953	8	0.00	997	-8	-5	-14	-39	-39	-63
Vincent	R				11	0.00	976	8	0.00	998		-3			-29	
Batoca	W	80	0.00	584	11	0.00	974	8	0.00	999	-69	-3	-72	-86	-30	-90
Susumaniello	R	62	0.00	621	50	0.00	752	8	0.00	1000	-12	-42	-54	-20	-84	-87
Ehrenbreitsteiner	W	13	0.00	782	10	0.00	994	8	0.00	1001	-3	-2	-5	-23	-20	-38
Colobel	R	3	0.00	874	9	0.00	1007	8	0.00	1002	6	-1	5	228	-15	178
Merlese	R				14	0.00	943	8	0.00	1003		-7			-47	
Radisson	R							8	0.00	1004						
Fino de Ribera del Fresno	W	332	0.01	428	45	0.00	767	8	0.00	1005	-287	-37	-325	-86	-83	-98
Crimson Seedless	R	1	0.00	927	8	0.00	1031	8	0.00	1006	6	0	6	440	0	440
Miorita	W				7	0.00	1041	8	0.00	1007		1			7	
Bonda	R	3	0.00	873	7	0.00	1047	7	0.00	1008	4	1	5	139	9	159
Morone	R	22	0.00	726	13	0.00	958	7	0.00	1009	-9	-6	-15	-41	-43	-67
Select	W	8	0.00	825	7	0.00	1036	7	0.00	1010	0	0	0	-5	1	-4
Bácska	G							7	0.00	1011						
Osceola Muscat	W							7	0.00	1012						
Galego Dourado	W	51	0.00	643	16	0.00	927	7	0.00	1013	-36	-9	-44	-69	-55	-86
Hegel	R	10	0.00	803	9	0.00	1015	7	0.00	1014	-1	-2	-3	-10	-22	-30
Palas	R							7	0.00	1015						
Blattner Whites	W				25	0.00	855	7	0.00	1016		-18			-73	
Criolla Mediana	R	1	0.00	920	3	0.00	1120	7	0.00	1017	2	3	5	134	98	363
Catanese Nero	R	76	0.00	592	15	0.00	941	7	0.00	1018	-61	-8	-69	-81	-53	-91
Catalanesca	W				54	0.00	735	7	0.00	1019		-47			-87	
Castela	R				8	0.00	1026	7	0.00	1020		-1			-15	
Rubilande	G	3	0.00	881	8	0.00	1029	7	0.00	1021	5	-1	4	195	-12	158
Mondeuse Blanche	W	22	0.00	727	6	0.00	1070	7	0.00	1022	-16	1	-15	-74	18	-69
Roz de Minis	G				6	0.00	1055	7	0.00	1023		0			7	
Madeleines	W				7	0.00	1044	7	0.00	1024		0			-3	
Kolor	R				2	0.00	1173	7	0.00	1025		5			234	
Jubiläumsrebe	G	30	0.00	689	14	0.00	950	7	0.00	1026	-17	-7	-24	-55	-51	-78
Forsellina	R	9	0.00	807	7	0.00	1037	7	0.00	1027	-2	-1	-3	-23	-8	-30
Khikhvi	W	5	0.00	844	6	0.00	1052	6	0.00	1028	1	0	1	28	0	28

Table 6 (cont.): Global ranking of prime winegrape varieties by area, and changes, 2000, 2010 and 2016 (ha and %)

		2000			2010			2016			2010-2000	2016-2010	2016-2000	2010-2000	2016-2010	2016-2000
Prime vairiety	*Col*	*Global area (ha)*	*Global share %*	*Global rank*	*Global area (ha)*	*Global share %*	*Global rank*	*Global area (ha)*	*Global share %*	*Global rank*	*Change (ha)*	*Change (ha)*	*Change (ha)*	*Changes %*	*Changes %*	*Changes %*
Otskhanuri Sapere	R	5	0.00	845	6	0.00	1053	6	0.00	1029	1	0	1	28	0	28
Raffiat de Moncade	W	16	0.00	755	7	0.00	1048	6	0.00	1030	-10	0	-10	-60	-3	-61
Rambella	W				6	0.00	1059	6	0.00	1031		0			4	
Helios	W							6	0.00	1032						
Cabernet Blanc	W							6	0.00	1033						
Bronner	W				9	0.00	1017	6	0.00	1034		-3			-31	
Rossara Trentina	R	29	0.00	694	8	0.00	1028	6	0.00	1035	-21	-2	-23	-73	-21	-79
San Martino	R	44	0.00	653	21	0.00	891	6	0.00	1036	-24	-14	-38	-53	-70	-86
Cabernet Carol	R							6	0.00	1037						
Doux d'Henry	R	26	0.00	704	9	0.00	1005	6	0.00	1038	-17	-3	-20	-64	-35	-77
Cove	W	56	0.00	633	6	0.00	1051	6	0.00	1039	-49	0	-50	-88	-6	-89
Ravat Blanc	W	2	0.00	898	7	0.00	1049	6	0.00	1040	4	0	4	217	-7	195
Nehelescol	W	172	0.00	497	25	0.00	860	6	0.00	1041	-147	-19	-166	-86	-75	-96
Freisamer	W	17	0.00	754	8	0.00	1022	6	0.00	1042	-8	-2	-11	-49	-28	-64
Dolciame	W	6	0.00	835	11	0.00	988	6	0.00	1043	4	-5	0	65	-43	-6
Neronet	R				72	0.00	680	6	0.00	1044		-66			-92	
Budai Zöld	W				6	0.00	1056	6	0.00	1045		0			-3	
Clarin	W	11	0.00	792	6	0.00	1063	6	0.00	1046	-5	0	-5	-45	-3	-47
Mayolet	R	4	0.00	858	7	0.00	1034	6	0.00	1047	3	-2	2	89	-21	49
Saint Macaire	R	1	0.00	960				6	0.00	1048			5			999
Pelaverga Piccolo	R	24	0.00	716	6	0.00	1068	6	0.00	1049	-18	0	-18	-76	-2	-76
Petite Milo	G							6	0.00	1050						
Rayon d'Or	W	1	0.00	946	6	0.00	1066	6	0.00	1051	5	0	5	631	-4	600
Pallagrello Bianco	W				55	0.00	730	6	0.00	1052		-49			-90	
Lagarino Bianco	W				23	0.00	882	6	0.00	1053		-17			-75	
Madeleine Sylvaner	W				7	0.00	1035	6	0.00	1054		-2			-23	
Royal de Alloza	R				29	0.00	833	6	0.00	1055		-23			-81	
Piccola Nera	G	17	0.00	750	17	0.00	916	6	0.00	1056	0	-12	-12	0	-68	-68
Carica l'Asino	W	299	0.01	438	17	0.00	913	5	0.00	1057	-282	-12	-294	-94	-69	-98
Biancu Gentile	W	1	0.00	940	9	0.00	1018	5	0.00	1058	8	-3	4	781	-39	441
Vulcanus	W				5	0.00	1094	5	0.00	1059		1			16	
Aranel	W	22	0.00	729	5	0.00	1080	5	0.00	1060	-17	0	-16	-77	7	-75
Sirio	W	23	0.00	722	14	0.00	948	5	0.00	1061	-9	-8	-17	-39	-62	-77
Rubin Tairovsky	R	2	0.00	902	2	0.00	1174	5	0.00	1062	0	3	3	0	161	161
Louise Swenson	W				3	0.00	1138	5	0.00	1063		2			76	
Gorgollasa	R							5	0.00	1064						
Rossese Bianco	W				7	0.00	1033	5	0.00	1065		-2			-31	
Caverdella	W							5	0.00	1066						

Table 6 (cont.): Global ranking of prime winegrape varieties by area, and changes, 2000, 2010 and 2016 (ha and %)

		2000			2010			2016			2010-2000	2016-2010	2016-2000	2010-2000	2016-2010	2016-2000
Prime vairiety	*Col*	*Global area (ha)*	*Global share %*	*Global rank*	*Global area (ha)*	*Global share %*	*Global rank*	*Global area (ha)*	*Global share %*	*Global rank*	*Change (ha)*	*Change (ha)*	*Change (ha)*	*Changes %*	*Changes %*	*Changes %*
Asirtiko Red	R	22	0.00	730	5	0.00	1088	5	0.00	1067	-17	0	-17	-78	6	-76
Nocera	R	27	0.00	700	15	0.00	938	5	0.00	1068	-12	-10	-22	-45	-66	-81
Marquinhas	W				11	0.00	983	5	0.00	1069		-6			-53	
Dindarella	R	9	0.00	814	7	0.00	1045	5	0.00	1070	-2	-2	-4	-20	-27	-41
Ar99	G	3	0.00	864	5	0.00	1085	5	0.00	1071	2	0	2	43	0	43
Musann Blanc	W							5	0.00	1072						
Andre	W				477	0.01	382	5	0.00	1073		-472			-99	
Moscatello Selvatico	W	105	0.00	553	35	0.00	800	5	0.00	1074	-71	-30	-100	-67	-86	-95
Invernenga	W	32	0.00	686	7	0.00	1050	5	0.00	1075	-25	-2	-27	-79	-24	-84
Adalmiina	W							5	0.00	1076						
Goldtraminer	W				9	0.00	1012	5	0.00	1077		-4			-47	
Completer	W	2	0.00	899	3	0.00	1137	5	0.00	1078	1	2	3	46	63	138
Original	R							5	0.00	1079						
Lariao	W				4	0.00	1114	5	0.00	1080		1			31	
Pátria	W				3	0.00	1139	5	0.00	1081		2			64	
Sicilien	W				5	0.00	1090	5	0.00	1082		0			-2	
Preto Cardana	R				5	0.00	1091	5	0.00	1083		0			-2	
Isa	W				9	0.00	1004	5	0.00	1084		-5			-51	
Varousset	R	12	0.00	788	5	0.00	1093	5	0.00	1085	-7	0	-7	-59	-3	-61
Etraire de l'Adui	R	8	0.00	826	5	0.00	1089	5	0.00	1086	-3	0	-3	-36	-5	-40
Krassato	R	38	0.00	670	52	0.00	744	5	0.00	1087	13	-47	-34	35	-91	-88
Castiglione	R	83	0.00	581	18	0.00	906	4	0.00	1088	-65	-14	-79	-78	-76	-95
Perigo	W				4	0.00	1105	4	0.00	1089		0			9	
Santoal	W				9	0.00	1011	4	0.00	1090		-5			-53	
Cividin	W				4	0.00	1102	4	0.00	1091		0			2	
Muscaris	W							4	0.00	1092						
Piroso	R							4	0.00	1093						
Bokay	W							4	0.00	1094						
Malvasia Parda	W				4	0.00	1099	4	0.00	1095		0			-6	
Vasilaki	W	3	0.00	882	4	0.00	1111	4	0.00	1096	1	0	2	43	11	58
Karalahna	R	3	0.00	883	4	0.00	1110	4	0.00	1097	1	0	2	43	11	58
Karasakiz	R	3	0.00	884	4	0.00	1109	4	0.00	1098	1	0	2	43	11	58
Sercialinho	W				9	0.00	1020	4	0.00	1099		-5			-53	
Recantina	R				9	0.00	1006	4	0.00	1100		-5			-57	
Vineland 53035	W							4	0.00	1101						
Monvedro	R	14	0.00	774	6	0.00	1072	4	0.00	1102	-8	-2	-10	-59	-29	-71
Reliance	R				4	0.00	1117	4	0.00	1103		0			9	
Schiava Grigia	R	79	0.00	588	66	0.00	698	4	0.00	1104	-13	-62	-75	-16	-94	-95

Table 6 (cont.): Global ranking of prime winegrape varieties by area, and changes, 2000, 2010 and 2016 (ha and %)

		2000			2010			2016			2010-2000	2016-2010	2016-2000	2010-2000	2016-2010	2016-2000
Prime vairiety	*Col*	*Global area (ha)*	*Global share %*	*Global rank*	*Global area (ha)*	*Global share %*	*Global rank*	*Global area (ha)*	*Global share %*	*Global rank*	*Change (ha)*	*Change (ha)*	*Change (ha)*	*Changes %*	*Changes %*	*Changes %*
Tinta Engomada	R				4	0.00	1108	4	0.00	1105		0			2	
Morrastel Bouschet	R	1	0.00	947				4	0.00	1106			3			403
San Lunardo	W	22	0.00	728	10	0.00	1002	4	0.00	1107	-12	-6	-18	-56	-61	-83
Tinta Roseira	R				4	0.00	1115	4	0.00	1108		0			2	
Bracciola Nera	R	89	0.00	572	26	0.00	850	4	0.00	1109	-63	-22	-85	-71	-86	-96
Incrocio Bruni 54	W	12	0.00	784	12	0.00	966	3	0.00	1110	0	-9	-9	-1	-72	-72
Passau	R	12	0.00	787	5	0.00	1076	3	0.00	1111	-7	-2	-9	-57	-34	-71
Ervi	R	5	0.00	840	4	0.00	1100	3	0.00	1112	-1	-1	-2	-21	-20	-36
Sciaglin	W	4	0.00	862	6	0.00	1054	3	0.00	1113	3	-3	0	73	-46	-7
Ekigaina	R	5	0.00	841	4	0.00	1118	3	0.00	1114	-2	0	-2	-30	-5	-33
Tinta do Rodo	R				3	0.00	1123	3	0.00	1115		0			-2	
Gouget Noir	R	1	0.00	948	10	0.00	998	3	0.00	1116	9	-6	3	1265	-66	367
Petit Meslier	W	3	0.00	866	4	0.00	1104	3	0.00	1117	1	-1	0	19	-19	-4
Lomanto	R							3	0.00	1118						
Amurg	R				3	0.00	1133	3	0.00	1119		0			7	
VB 32-7	W							3	0.00	1120						
Promissao	W				6	0.00	1057	3	0.00	1121		-3			-48	
Malvarisco	R				3	0.00	1129	3	0.00	1122		0			1	
Palatina	W				6	0.00	1065	3	0.00	1123		-3			-46	
Rabo de Lobo	R				3	0.00	1126	3	0.00	1124		0			-2	
Tinta Castañal	R							3	0.00	1125						
Exalta	W				4	0.00	1116	3	0.00	1126		0			-13	
Ganson	R	28	0.00	695	3	0.00	1127	3	0.00	1127	-25	0	-25	-89	-3	-89
Landot Noir	R	0	0.00	976	3	0.00	1128	3	0.00	1128	3	0	3	901	-3	868
Moscadet	W				4	0.00	1113	3	0.00	1129		-1			-17	
Haiduc	R				3	0.00	1143	3	0.00	1130		0			7	
Trobat	R	83	0.00	582	1	0.00	1239	3	0.00	1131	-82	2	-79	-99	206	-96
Bussanello	W	8	0.00	824	12	0.00	973	3	0.00	1132	4	-9	-5	55	-74	-61
Sovereign Opal	W				3	0.00	1145	3	0.00	1133		0			7	
Monerac	R	2	0.00	919	3	0.00	1131	3	0.00	1134	2	0	2	108	-3	101
Souvignier Gris	G							3	0.00	1135						
Verdesse	W	8	0.00	818	10	0.00	995	3	0.00	1136	2	-7	-5	23	-70	-63
Madeleine Royale	W				3	0.00	1125	3	0.00	1137		0			-12	
Fogarina	R				5	0.00	1078	3	0.00	1138		-2			-42	
Castel	R	0	0.00	992	2	0.00	1169	3	0.00	1139	2	1	3	1695	37	2355
Malvasia di Sardegna Rosada	G				3	0.00	1141	3	0.00	1140		0			2	
Vuillermin	R	0	0.00	997	4	0.00	1101	3	0.00	1141	4	-1	3	4556	-32	3079
Favorit	W							3	0.00	1142						

Table 6 (cont.): Global ranking of prime winegrape varieties by area, and changes, 2000, 2010 and 2016 (ha and %)

		2000			2010			2016			2010-2000	2016-2010	2016-2000	2010-2000	2016-2010	2016-2000
Prime vairiety	Col	Global area (ha)	Global share %	Global rank	Global area (ha)	Global share %	Global rank	Global area (ha)	Global share %	Global rank	Change (ha)	Change (ha)	Change (ha)	Changes %	Changes %	Changes %
Batuta Neagra	R				3	0.00	1150	3	0.00	1143		0			7	
Pexem	R				3	0.00	1122	3	0.00	1144		-1			-18	
Aromella	W							3	0.00	1145						
Canela	G	0	0.00	984	1	0.00	1213	3	0.00	1146	1	1	3	601	114	1401
Forgiarin	R	2	0.00	912	4	0.00	1106	3	0.00	1147	2	-1	1	140	-30	67
Triomphe	R				15	0.00	935	3	0.00	1148		-12			-82	
Gramon	R	14	0.00	776	3	0.00	1144	3	0.00	1149	-11	0	-11	-79	-3	-79
Muscat of Alexandria (R)	W	11	0.00	794	6	0.00	1067	3	0.00	1150	-5	-3	-8	-45	-52	-74
Çavuş	W							3	0.00	1151						
Mourvaison	R	9	0.00	812	3	0.00	1148	3	0.00	1152	-6	0	-6	-68	-2	-69
Guzun	W							3	0.00	1153						
Lambrusco Barghi	R				18	0.00	905	3	0.00	1154		-16			-85	
Muscat Bleu	R							3	0.00	1155						
Flavis	W	12	0.00	785	3	0.00	1130	3	0.00	1156	-9	0	-9	-74	-15	-78
Gargiulo 4113	R	7	0.00	827	6	0.00	1058	3	0.00	1157	-1	-4	-5	-16	-57	-64
Cabaret Noir	R							3	0.00	1158						
Verdial Tinto	R				3	0.00	1153	3	0.00	1159		0			-2	
Lauzet	W	1	0.00	938	3	0.00	1152	3	0.00	1160	2	0	1	151	-3	144
Barreto de Semente	R				3	0.00	1158	3	0.00	1161		0			1	
Riesling Forte	W							2	0.00	1162						
Turchetta	R				3	0.00	1134	2	0.00	1163		-1			-19	
Albaranzeuli Bianco	W	72	0.00	601	7	0.00	1046	2	0.00	1164	-65	-4	-70	-91	-65	-97
Swenson White	W							2	0.00	1165						
Bakator Kék	R				3	0.00	1154	2	0.00	1166		0			-7	
Termarina Rossa	G				20	0.00	895	2	0.00	1167		-18			-88	
Fubiano	W	2	0.00	891	9	0.00	1009	2	0.00	1168	7	-7	0	294	-74	2
Grapariol	W				2	0.00	1168	2	0.00	1169		0			7	
Aledo	W				7	0.00	1043	2	0.00	1170		-5			-68	
Sennen	R				10	0.00	993	2	0.00	1171		-8			-78	
Kocsis Irma	W				11	0.00	984	2	0.00	1172		-8			-79	
Diamond Muscat	W							2	0.00	1173						
Perlette	W	14	0.00	771	1	0.00	1214	2	0.00	1174	-13	1	-12	-91	75	-84
Bequignol Gris	W				3	0.00	1149	2	0.00	1175		-1			-20	
Castelo Branco	W				5	0.00	1097	2	0.00	1176		-2			-53	
Oraniensteiner	W				3	0.00	1151	2	0.00	1177		-1			-20	
Colomino	W	16	0.00	758	5	0.00	1096	2	0.00	1178	-11	-2	-14	-71	-54	-87
Sauvignac	W							2	0.00	1179						
Rosaki	W							2	0.00	1180						

Table 6 (cont.): Global ranking of prime winegrape varieties by area, and changes, 2000, 2010 and 2016 (ha and %)

		2000			2010			2016			2010-2000	2016-2010	2016-2000	2010-2000	2016-2010	2016-2000
Prime vairiety	*Col*	*Global area (ha)*	*Global share %*	*Global rank*	*Global area (ha)*	*Global share %*	*Global rank*	*Global area (ha)*	*Global share %*	*Global rank*	*Change (ha)*	*Change (ha)*	*Change (ha)*	*Changes %*	*Changes %*	*Changes %*
Albanello	W	117	0.00	541	18	0.00	911	2	0.00	1181	-99	-16	-115	-85	-88	-98
Rozala Bianca	W				2	0.00	1199	2	0.00	1182		1			36	
Fertilia	R	13	0.00	777	3	0.00	1124	2	0.00	1183	-10	-1	-11	-75	-40	-85
New York Muscat and VG4111	R							2	0.00	1184						
Boal Vencedor	W				2	0.00	1161	2	0.00	1185		0			-16	
Hölder	W	13	0.00	781	5	0.00	1081	2	0.00	1186	-8	-3	-11	-62	-60	-85
Reberger	R							2	0.00	1187						
Prinzipal	W							2	0.00	1188						
Mariensteiner	W	9	0.00	810				2	0.00	1189			-7			-78
Weldra	W	54	0.00	636	14	0.00	951	2	0.00	1190	-41	-12	-53	-75	-85	-96
Perdea	W	13	0.00	778	2	0.00	1160	2	0.00	1191	-11	0	-11	-82	-18	-85
Mourisco de Trevoes	R				2	0.00	1183	2	0.00	1192		0			1	
Danuta	W				2	0.00	1171	2	0.00	1193		0			-3	
Triunfo	R				2	0.00	1184	2	0.00	1194		0			0	
Jurançon Blanc	W	24	0.00	717	7	0.00	1040	2	0.00	1195	-17	-5	-22	-70	-72	-92
Rose Ciotat	W							2	0.00	1196						
Cabral	R				2	0.00	1186	2	0.00	1197		0			1	
Jázmin	W							2	0.00	1198						
Perera	W	25	0.00	711	4	0.00	1107	2	0.00	1199	-21	-2	-23	-85	-50	-92
Campanario	R				2	0.00	1185	2	0.00	1200		0			-2	
Lado	W	1	0.00	953	1	0.00	1241	2	0.00	1201	0	1	1	47	88	176
Sárfehér	W							2	0.00	1202						
Branco Escola	W				2	0.00	1189	2	0.00	1203		0			2	
Espirit	W							2	0.00	1204						
Kay Gray	W				1	0.00	1217	2	0.00	1205		1			43	
Olivette Blanche	W				2	0.00	1187	2	0.00	1206		0			-2	
Cabernet Foch	R							2	0.00	1207						
Alvar Roxo	G				2	0.00	1192	2	0.00	1208		0			3	
Baratuciat	W				2	0.00	1191	2	0.00	1209		0			3	
Siegfriedrebe	W							2	0.00	1210						
Cordenossa	R				5	0.00	1077	2	0.00	1211		-3			-66	
Taminga	W	46	0.00	652				2	0.00	1212			-44			-96
Impigno	W	60	0.00	625	7	0.00	1038	2	0.00	1213	-53	-5	-58	-88	-76	-97
David Macgregor 8521-1	R							2	0.00	1214						
Eona	W							2	0.00	1215						
Blattner Cal 1-28	R							2	0.00	1216						
Alutus	R				2	0.00	1197	2	0.00	1217		0			7	
Mars	R				2	0.00	1182	2	0.00	1218		0			-15	

Table 6 (cont.): Global ranking of prime winegrape varieties by area, and changes, 2000, 2010 and 2016 (ha and %)

		2000			2010			2016			2010-2000	2016-2010	2016-2000	2010-2000	2016-2010	2016-2000
Prime vairiety	*Col*	*Global area (ha)*	*Global share %*	*Global rank*	*Global area (ha)*	*Global share %*	*Global rank*	*Global area (ha)*	*Global share %*	*Global rank*	*Change (ha)*	*Change (ha)*	*Change (ha)*	*Changes %*	*Changes %*	*Changes %*
Santa Maria	W	15	0.00	768	3	0.00	1142	2	0.00	1219	-12	-1	-13	-80	-43	-89
Margot	R							2	0.00	1220						
Folha de Figueira	W				3	0.00	1121	2	0.00	1221		-2			-53	
Cellerina	R				2	0.00	1195	2	0.00	1222		0			0	
Cabertin	R							2	0.00	1223						
Muscat de Bugeac	R							2	0.00	1224						
Scimiscia	W				5	0.00	1075	2	0.00	1225		-4			-70	
Red Millennium	R							2	0.00	1226						
Babosa de Madere	W				2	0.00	1170	2	0.00	1227		0			-24	
Uva del Tunde	R				2	0.00	1198	2	0.00	1228		0			-1	
Francavidda	W	86	0.00	576	13	0.00	960	2	0.00	1229	-73	-11	-84	-85	-88	-98
Roussin	R	3	0.00	878	3	0.00	1132	2	0.00	1230	0	-2	-1	11	-49	-43
Piquepoul Gris	G	9	0.00	806	2	0.00	1196	2	0.00	1231	-8	0	-8	-83	-3	-84
Csókaszőlő	R				2	0.00	1166	2	0.00	1232		-1			-31	
Orpheus	W				0	0.00	1316	2	0.00	1233		1			376	
Lafnetscha	W	1	0.00	928				2	0.00	1234			0			13
Erbamat	W				24	0.00	865	2	0.00	1235		-23			-94	
Chelois	R				1	0.00	1202	2	0.00	1236		0			3	
Cinsaut (G)	R							2	0.00	1237						
Melara	W	13	0.00	783	3	0.00	1147	1	0.00	1238	-10	-1	-11	-78	-47	-88
Muskat Zhemchuzhnyi	W							1	0.00	1239						
Cavrara	R				23	0.00	877	1	0.00	1240		-22			-94	
Semebat	R	2	0.00	892	2	0.00	1193	1	0.00	1241	-1	0	-1	-25	-15	-37
Bric	R	21	0.00	733	2	0.00	1165	1	0.00	1242	-19	-1	-19	-89	-37	-93
Carignan Bouschet	R	16	0.00	762	1	0.00	1203	1	0.00	1243	-14	0	-14	-91	-3	-91
Cabernet Cantor	R							1	0.00	1244						
Pionnier	R							1	0.00	1245						
Somerset Seedless	G							1	0.00	1246						
Apiren Alb	W							1	0.00	1247						
Moradella	R				6	0.00	1069	1	0.00	1248		-4			-76	
Uvalino	R				1	0.00	1211	1	0.00	1249		0			0	
Codivarta	W	3	0.00	886	2	0.00	1167	1	0.00	1250	0	-1	-1	-15	-39	-48
Champanel	R							1	0.00	1251						
Blattner Cal 1-36	R							1	0.00	1252						
Estreito Macio	W				3	0.00	1146	1	0.00	1253		-2			-53	
Courbu Noir	R	2	0.00	915	1	0.00	1207	1	0.00	1254	0	0	0	-12	-3	-15
Cianorie	R				2	0.00	1159	1	0.00	1255		-1			-46	
Arbane	W	1	0.00	930	2	0.00	1190	1	0.00	1256	0	0	0	29	-24	-2

Table 6 (cont.): Global ranking of prime winegrape varieties by area, and changes, 2000, 2010 and 2016 (ha and %)

		2000			2010			2016			2010-2000	2016-2010	2016-2000	2010-2000	2016-2010	2016-2000
Prime vairiety	*Col*	*Global area (ha)*	*Global share %*	*Global rank*	*Global area (ha)*	*Global share %*	*Global rank*	*Global area (ha)*	*Global share %*	*Global rank*	*Change (ha)*	*Change (ha)*	*Change (ha)*	*Changes %*	*Changes %*	*Changes %*
Liliorila	W	3	0.00	885	4	0.00	1119	1	0.00	1257	1	-2	-1	36	-63	-50
Hasansky Sladky	R							1	0.00	1258						
Milgranet	R	2	0.00	894	1	0.00	1208	1	0.00	1259	-1	0	-1	-38	-6	-42
Cabernet Dore	W				1	0.00	1219	1	0.00	1260		0			4	
Crimson Cabernet	R				1	0.00	1218	1	0.00	1261		0			4	
Yubilei Zhuravlya	R							1	0.00	1262						
Centennial Seedless	W	2	0.00	888				1	0.00	1263			-1			-51
Marquis	W							1	0.00	1264						
Tinta Mesquita	R				1	0.00	1221	1	0.00	1265		0			2	
Rosina	W				1	0.00	1228	1	0.00	1266		0			7	
Grangeal	R				1	0.00	1225	1	0.00	1267		0			2	
Graisse	W	22	0.00	725	14	0.00	949	1	0.00	1268	-9	-12	-21	-39	-92	-95
Marfal	W	35	0.00	679	2	0.00	1177	1	0.00	1269	-33	-1	-33	-94	-43	-97
Gouais Blanc	W	1	0.00	933				1	0.00	1270			0			1
Petit Courbu	W	75	0.00	596	102	0.00	618	1	0.00	1271	28	-101	-74	37	-99	-98
Castellana Blanca	W							1	0.00	1272						
Ripolo	W				1	0.00	1224	1	0.00	1273		0			-3	
Baron	R							1	0.00	1274						
Trilla	W				1	0.00	1227	1	0.00	1275		0			0	
Nigra	R	7	0.00	828	3	0.00	1155	1	0.00	1276	-5	-1	-6	-65	-56	-85
Batily	R	38	0.00	671	54	0.00	734	1	0.00	1277	16	-53	-37	41	-98	-97
Perlaut	W				1	0.00	1226	1	0.00	1278		0			-3	
Réselle	W							1	0.00	1279						
Rosciola	G				2	0.00	1162	1	0.00	1280		-1			-54	
Tinta Varejoa	R				1	0.00	1232	1	0.00	1281		0			2	
Timpuriu de Cluj	W				1	0.00	1253	1	0.00	1282		0			7	
Geilweilerhof Ga- 48- 12	W							1	0.00	1283						
Tronto	R				2	0.00	1194	1	0.00	1284		-1			-36	
Romulus	W							1	0.00	1285						
Tinta Pereira	R				1	0.00	1231	1	0.00	1286		0			-2	
Donaris	W				1	0.00	1254	1	0.00	1287		0			7	
Birstaler Muskat	W							1	0.00	1288						
Yaqui	R	22	0.00	724	2	0.00	1179	1	0.00	1289	-20	-1	-21	-91	-49	-95
Silcher	W	7	0.00	829				1	0.00	1290			-6			-86
Rosa Arica	R	1	0.00	952	1	0.00	1238	1	0.00	1291	0	0	0	43	0	43
Borsmenta	W							1	0.00	1292						
Merzling	W	5	0.00	842				1	0.00	1293			-4			-80
Imperial Seedless	G	1	0.00	951	1	0.00	1243	1	0.00	1294	0	0	0	43	0	43

Table 6 (cont.): Global ranking of prime winegrape varieties by area, and changes, 2000, 2010 and 2016 (ha and %)

		2000			2010			2016			2010-2000	2016-2010	2016-2000	2010-2000	2016-2010	2016-2000
Prime vairiety	*Col*	*Global area (ha)*	*Global share %*	*Global rank*	*Global area (ha)*	*Global share %*	*Global rank*	*Global area (ha)*	*Global share %*	*Global rank*	*Change (ha)*	*Change (ha)*	*Change (ha)*	*Changes %*	*Changes %*	*Changes %*
Orion	W	8	0.00	821	13	0.00	956	1	0.00	1295	5	-12	-7	63	-92	-88
Fontanara	W	2	0.00	904				1	0.00	1296			-1			-50
Accent	R							1	0.00	1297						
Septimer	G							1	0.00	1298						
Ar110	G	1	0.00	950	1	0.00	1244	1	0.00	1299	0	0	0	43	0	43
Rayada Melonera	R				1	0.00	1220	1	0.00	1300		0			-19	
Gascon	R	0	0.00	1003	1	0.00	1252	1	0.00	1301	1	0	1	4897	-4	4703
Korona	W				1	0.00	1284	1	0.00	1302		0			59	
Cabernet Soyhières	R							1	0.00	1303						
Rabigato Moreno	W				1	0.00	1256	1	0.00	1304		0			2	
Calrao	R				1	0.00	1233	1	0.00	1305		0			-13	
Arcas	R				1	0.00	1260	1	0.00	1306		0			7	
Sovereign Coronation	R							1	0.00	1307						
Dona Zillá	R							1	0.00	1308						
Joannes Seyve	R	3	0.00	887	1	0.00	1255	1	0.00	1309	-2	0	-2	-65	-3	-66
Ruggine	W				1	0.00	1204	1	0.00	1310		-1			-41	
Samarrinho	W				1	0.00	1216	1	0.00	1311		0			-33	
Rubienne	R							1	0.00	1312						
Ederena	R	0	0.00	964	1	0.00	1261	1	0.00	1313	0	0	0	73	0	72
Incrocio Bianco Fedit 51	W	11	0.00	790	5	0.00	1092	1	0.00	1314	-6	-4	-10	-58	-83	-93
Dedo de Dama	W				1	0.00	1259	1	0.00	1315		0			-3	
Fenile	W				5	0.00	1073	1	0.00	1316		-5			-85	
Roal	R				1	0.00	1222	1	0.00	1317		0			-32	
Coracao de Galo	R				1	0.00	1257	1	0.00	1318		0			-12	
Roussette d'Ayze	W	3	0.00	879	1	0.00	1215	1	0.00	1319	-1	0	-2	-52	-39	-71
Pecsi Szagos	W				1	0.00	1230	1	0.00	1320		0			-27	
VB 91-26-25	R							1	0.00	1321						
Pampanaro	W				5	0.00	1095	1	0.00	1322		-4			-84	
Picardan	W				1	0.00	1263	1	0.00	1323		0			-6	
Bouillet	R	2	0.00	896	1	0.00	1265	1	0.00	1324	-1	0	-1	-65	-3	-66
Ignea	G				1	0.00	1267	1	0.00	1325		0			-2	
Roviello Bianco	W				2	0.00	1200	1	0.00	1326		-1			-53	
Osteiner	W	3	0.00	870	1	0.00	1237	1	0.00	1327	-2	0	-2	-67	-30	-77
Tamarugal	W							1	0.00	1328						
Ravat	R	1	0.00	945	1	0.00	1268	1	0.00	1329	0	0	0	-18	-3	-21
Tinta de Cidadelhe	R				1	0.00	1271	1	0.00	1330		0			2	
Goncalo Pires	R				1	0.00	1270	1	0.00	1331		0			1	
Michele Palieri	R	24	0.00	714	1	0.00	1240	1	0.00	1332	-23	0	-24	-96	-31	-97

Table 6 (cont.): Global ranking of prime winegrape varieties by area, and changes, 2000, 2010 and 2016 (ha and %)

		2000			2010			2016			2010-2000	2016-2010	2016-2000	2010-2000	2016-2010	2016-2000
Prime vairiety	*Col*	*Global area (ha)*	*Global share %*	*Global rank*	*Global area (ha)*	*Global share %*	*Global rank*	*Global area (ha)*	*Global share %*	*Global rank*	*Change (ha)*	*Change (ha)*	*Change (ha)*	*Changes %*	*Changes %*	*Changes %*
Parreira Matias	R				1	0.00	1269	1	0.00	1333		0			-2	
Kosmopolita	G							1	0.00	1334						
Aubin Blanc	W	2	0.00	913	1	0.00	1258	1	0.00	1335	-1	0	-1	-46	-22	-57
Augster Weiss	W				1	0.00	1277	1	0.00	1336		0			10	
Pelso	W				1	0.00	1273	1	0.00	1337		0			1	
Negru Aromat	R				1	0.00	1275	1	0.00	1338		0			7	
Oeillade Bousche	R	10	0.00	798	1	0.00	1272	1	0.00	1339	-9	0	-10	-93	-3	-94
Claverie	W	3	0.00	865	1	0.00	1264	1	0.00	1340	-3	0	-3	-78	-15	-81
Rouge de Fully	R							1	0.00	1341						
Kentville White 94-2	W							1	0.00	1342						
Montreal Blues	R							1	0.00	1343						
Black Prince	R							1	0.00	1344						
Unirea	W				1	0.00	1285	1	0.00	1345		0			7	
Mandon	R	261	0.01	456				1	0.00	1346			-260			-100
Tinta da Melra	R				1	0.00	1282	1	0.00	1347		0			1	
Mézes Fehér	W				2	0.00	1163	1	0.00	1348		-2			-75	
Andor	W							1	0.00	1349						
Perla dei Vivi	R				1	0.00	1266	1	0.00	1350		0			-22	
Pe Comprido	W				1	0.00	1229	1	0.00	1351		0			-46	
BX 81-83	R							1	0.00	1352						
Táltos	W				1	0.00	1234	1	0.00	1353		0			-48	
Tinta Valdosa	R				1	0.00	1286	1	0.00	1354		0			-1	
Mindelo	R				1	0.00	1287	1	0.00	1355		0			0	
Ondenc	W	12	0.00	786	8	0.00	1027	1	0.00	1356	-4	-7	-12	-35	-93	-96
Sao Mamede	W				1	0.00	1278	1	0.00	1357		0			-16	
Cornichon Blanc	W	4	0.00	859				1	0.00	1358			-3			-87
Ginestra	W				4	0.00	1112	1	0.00	1359		-3			-87	
Himrod	W				0	0.00	1293	1	0.00	1360		0			4	
Okanagan Riesling	W							1	0.00	1361						
Capolongo	W				5	0.00	1079	0	0.00	1362		-5			-90	
Rio Grande	W	4	0.00	861	1	0.00	1288	0	0.00	1363	-3	0	-3	-86	-6	-87
Assaraky	W				1	0.00	1235	0	0.00	1364		-1			-53	
Meszi Kadarka	R							0	0.00	1365						
Gamba Rossa	R				1	0.00	1279	0	0.00	1366		0			-22	
Csomorika	W				0	0.00	1295	0	0.00	1367		0			0	
Balada	R				0	0.00	1297	0	0.00	1368		0			7	
Blattner Cal 1-20	R							0	0.00	1369						
Lecinaro	R				1	0.00	1274	0	0.00	1370		0			-33	

Table 6 (cont.): Global ranking of prime winegrape varieties by area, and changes, 2000, 2010 and 2016 (ha and %)

		2000			2010			2016			2010-2000	2016-2010	2016-2000	2010-2000	2016-2010	2016-2000
Prime vairiety	*Col*	*Global area (ha)*	*Global share %*	*Global rank*	*Global area (ha)*	*Global share %*	*Global rank*	*Global area (ha)*	*Global share %*	*Global rank*	*Change (ha)*	*Change (ha)*	*Change (ha)*	*Changes %*	*Changes %*	*Changes %*
Breidecker	W	28	0.00	696	7	0.00	1042	0	0.00	1371	-21	-7	-28	-75	-94	-98
Malingre Precoce	W	3	0.00	880	0	0.00	1294	0	0.00	1372	-2	0	-2	-82	-7	-83
Canadice	R				0	0.00	1326	0	0.00	1373		0			79	
Cabernet Diane	R				0	0.00	1301	0	0.00	1374		0			4	
Caramela	W				0	0.00	1292	0	0.00	1375		0			-15	
Apiren Roz	R							0	0.00	1376						
Michurinets	R				0	0.00	1300	0	0.00	1377		0			0	
Dattier de St. Vallier	W				0	0.00	1298	0	0.00	1378		0			-3	
Kunbarát	W				9	0.00	1021	0	0.00	1379		-8			-95	
VB Cal 6-04 N5	R							0	0.00	1380						
Refrén	W				0	0.00	1303	0	0.00	1381		0			0	
Kalina	R							0	0.00	1382						
Nevoeira	R				0	0.00	1307	0	0.00	1383		0			2	
Maiolina	R				1	0.00	1201	0	0.00	1384		-1			-74	
Uva Cao	W	33	0.00	683	1	0.00	1262	0	0.00	1385	-33	0	-33	-98	-52	-99
Tardia de Caxias	G							0	0.00	1386						
Mamaia	R				0	0.00	1313	0	0.00	1387		0			7	
Espadeiro Mole	R				0	0.00	1308	0	0.00	1388		0			-2	
Dodrelyabi	R							0	0.00	1389						
Roxo Rei	G				0	0.00	1310	0	0.00	1390		0			-1	
Blattner Cal 1-22	R							0	0.00	1391						
Tressot	R	0	0.00	982	0	0.00	1312	0	0.00	1392	0	0	0	62	2	64
Gouveio Preto	R				0	0.00	1315	0	0.00	1393		0			2	
Siramé	R							0	0.00	1394						
Mondet	R				0	0.00	1314	0	0.00	1395		0			1	
VB 91-26-26	R							0	0.00	1396						
Úrréti	W				0	0.00	1357	0	0.00	1397		0			197	
Minnesota Muscat	W							0	0.00	1398						
Valveirinha	W				0	0.00	1319	0	0.00	1399		0			2	
Pinorico	R							0	0.00	1400						
Abondant	W	0	0.00	974	0	0.00	1317	0	0.00	1401	0	0	0	-12	-1	-13
Jádorvány	W							0	0.00	1402						
Einset Seedless	G							0	0.00	1403						
Calmeria	W							0	0.00	1404						
Odola	R	1	0.00	925	0	0.00	1318	0	0.00	1405	-1	0	-1	-78	-3	-79
Purcsin	R							0	0.00	1406						
Ahmeur Bou Ahmeur	R	3	0.00	867	1	0.00	1247	0	0.00	1407	-2	-1	-3	-69	-70	-91
Taurus	W							0	0.00	1408						

Table 6 (cont.): Global ranking of prime winegrape varieties by area, and changes, 2000, 2010 and 2016 (ha and %)

		2000			2010			2016			2010-2000	2016-2010	2016-2000	2010-2000	2016-2010	2016-2000
Prime vairiety	*Col*	*Global area (ha)*	*Global share %*	*Global rank*	*Global area (ha)*	*Global share %*	*Global rank*	*Global area (ha)*	*Global share %*	*Global rank*	*Change (ha)*	*Change (ha)*	*Change (ha)*	*Changes %*	*Changes %*	*Changes %*
Augustovski	W							0	0.00	1409						
Terrantez do Pico	W				0	0.00	1311	0	0.00	1410		0			-13	
Naia	W				0	0.00	1296	0	0.00	1411		0			-37	
Magyarfrankos	R				0	0.00	1302	0	0.00	1412		0			-28	
Lameiro	W				0	0.00	1305	0	0.00	1413		0			-26	
Gesztus	W				0	0.00	1320	0	0.00	1414		0			0	
Kentville White 94-1	W							0	0.00	1415						
Excelsior	W							0	0.00	1416						
Leira	W				1	0.00	1281	0	0.00	1417		0			-53	
Boal Barreiro	W				1	0.00	1283	0	0.00	1418		0			-53	
Beogradska Crna	W							0	0.00	1419						
Danam	W				0	0.00	1322	0	0.00	1420		0			-2	
VB 91-26-27	R							0	0.00	1421						
Granho	W				0	0.00	1323	0	0.00	1422		0			2	
Crovassa	R	2	0.00	918	0	0.00	1325	0	0.00	1423	-1	0	-1	-84	6	-83
Clinton	R							0	0.00	1424						
Vênus	R				1	0.00	1223	0	0.00	1425		-1			-78	
Colmar Precoce Noir	R							0	0.00	1426						
Verdial	W				1	0.00	1291	0	0.00	1427		0			-53	
Saint-Pierre Dore	W	1	0.00	961	0	0.00	1304	0	0.00	1428	0	0	0	-22	-40	-53
Himbertscha	W	0	0.00	986				0	0.00	1429			0			35
Hamvas	W							0	0.00	1430						
Lourela	R				0	0.00	1328	0	0.00	1431		0			-2	
Astra	W				0	0.00	1332	0	0.00	1432		0			7	
Stavroto	R	104	0.00	555	11	0.00	981	0	0.00	1433	-93	-11	-104	-89	-98	-100
Malvia	W				0	0.00	1331	0	0.00	1434		0			2	
Eyholzer Rote	R							0	0.00	1435						
Veruccese	R				0	0.00	1309	0	0.00	1436		0			-42	
Mergeritar	W							0	0.00	1437						
MRAC 1099	R							0	0.00	1438						
KW 96-2	W							0	0.00	1439						
Kat.E.Lin	R							0	0.00	1440						
Pepella	W				3	0.00	1140	0	0.00	1441		-3			-93	
Cinsaut (W)	R	41	0.00	660	7	0.00	1039	0	0.00	1442	-34	-7	-41	-83	-97	-100
Branco Sr. Joao	W				0	0.00	1299	0	0.00	1443		0			-52	
Nosztori Rizling	W				1	0.00	1276	0	0.00	1444		0			-69	
Kozmopoliten	W							0	0.00	1445						
Merlot Khorus	R							0	0.00	1446						

Table 6 (cont.): Global ranking of prime winegrape varieties by area, and changes, 2000, 2010 and 2016 (ha and %)

		2000			2010			2016			2010-2000	2016-2010	2016-2000	2010-2000	2016-2010	2016-2000
Prime vairiety	*Col*	*Global area (ha)*	*Global share %*	*Global rank*	*Global area (ha)*	*Global share %*	*Global rank*	*Global area (ha)*	*Global share %*	*Global rank*	*Change (ha)*	*Change (ha)*	*Change (ha)*	*Changes %*	*Changes %*	*Changes %*
Roxo de Vila Flor	R				0	0.00	1336	0	0.00	1447		0			-2	
Chardoris	W							0	0.00	1448						
Heuréka	W							0	0.00	1449						
Delhro	R				0	0.00	1334	0	0.00	1450		0			-3	
Docal	R				0	0.00	1341	0	0.00	1451		0			1	
MRAC 1087	R							0	0.00	1452						
Blanc Dame	W	0	0.00	980	0	0.00	1338	0	0.00	1453	0	0	0	-16	-3	-18
Aramont	R	1	0.00	941				0	0.00	1454			-1			-81
Roter Milan	R							0	0.00	1455						
Pinotin	R							0	0.00	1456						
Alvar Branco	W				0	0.00	1329	0	0.00	1457		0			-26	
Divona	W							0	0.00	1458						
Cidreiro	R				0	0.00	1344	0	0.00	1459		0			-2	
Rubinovy Magaracha	R	0	0.00	993	0	0.00	1346	0	0.00	1460	0	0	0	43	0	43
Knipperlé	W	2	0.00	901	0	0.00	1352	0	0.00	1461	-2	0	-2	-93	4	-93
Mireille	W				0	0.00	1347	0	0.00	1462		0			-2	
Moscargo	R				0	0.00	1348	0	0.00	1463		0			-2	
Terras 20	R	13	0.00	779	0	0.00	1349	0	0.00	1464	-13	0	-13	-99	-3	-99
Verdelho l'Anjou	W				0	0.00	1345	0	0.00	1465		0			-7	
Muscatin	W							0	0.00	1466						
Malvasia Romana	W				0	0.00	1340	0	0.00	1467		0			-26	
Donzelinho Roxo	R				0	0.00	1353	0	0.00	1468		0			1	
Luzidio	W				0	0.00	1321	0	0.00	1469		0			-53	
Lusitano	R				0	0.00	1354	0	0.00	1470		0			-2	
Branco Valente	W				0	0.00	1330	0	0.00	1471		0			-40	
Jaoumet	W				0	0.00	1355	0	0.00	1472		0			-3	
Sanforte	R				1	0.00	1289	0	0.00	1473		0			-77	
Paolina	W				1	0.00	1206	0	0.00	1474		-1			-91	
Cabernet Sanzey	R							0	0.00	1475						
MRAC 1817	R							0	0.00	1476						
Deliciosa	R				0	0.00	1356	0	0.00	1477		0			-2	
Kurucvér	R				0	0.00	1343	0	0.00	1478		0			-32	
Cabernet x Maréchal Foch	R							0	0.00	1479						
Carla	R				0	0.00	1358	0	0.00	1480		0			-3	
Summerland	R							0	0.00	1481						
Elmer Swenson 10- 18- 30	W							0	0.00	1482						
Saint-Cliche	W							0	0.00	1483						
Delisle	W							0	0.00	1484						

Table 6 (cont.): Global ranking of prime winegrape varieties by area, and changes, 2000, 2010 and 2016 (ha and %)

		2000			2010			2016			2010-2000	2016-2010	2016-2000	2010-2000	2016-2010	2016-2000
Prime vairiety	*Col*	*Global area (ha)*	*Global share %*	*Global rank*	*Global area (ha)*	*Global share %*	*Global rank*	*Global area (ha)*	*Global share %*	*Global rank*	*Change (ha)*	*Change (ha)*	*Change (ha)*	*Changes %*	*Changes %*	*Changes %*
ES 10-18-14	W							0	0.00	1485						
Golia	W				0	0.00	1361	0	0.00	1486		0			7	
Augster Blau	R							0	0.00	1487						
Sremska Zelenika	W				0	0.00	1306	0	0.00	1488		0			-76	
Carbernet Volos	R							0	0.00	1489						
UD 31103	R							0	0.00	1490						
DU 31120	R							0	0.00	1491						
Cabernet Early	R							0	0.00	1492						
Gosen	R				1	0.00	1212	0	0.00	1493		-1			-93	
Castalia	W				0	0.00	1333	0	0.00	1494		0			-53	
Almafra	W				0	0.00	1335	0	0.00	1495		0			-53	
Blattner Cal 1-15	R							0	0.00	1496						
Tedi's Best	R							0	0.00	1497						
Fleurtai	W							0	0.00	1498						
Soreli	W							0	0.00	1499						
Esganinho	W				0	0.00	1351	0	0.00	1500		0			-53	
Bousquet Precoce	W	16	0.00	759	6	0.00	1071	0	0.00	1501	-10	-6	-16	-64	-99	-100
Grassen	R	1	0.00	942	0	0.00	1366	0	0.00	1502	-1	0	-1	-93	-2	-93
Seibel White	W	15	0.00	767	0	0.00	1359	0	0.00	1503	-15	0	-15	-99	-39	-100
Beclan	R	0	0.00	985	0	0.00	1324	0	0.00	1504	0	0	0	39	-74	-64
Jardovany Fekete	R							0	0.00	1505						
Melhorio	R				0	0.00	1367	0	0.00	1506		0			2	
Noir Fleurien	R	0	0.00	991	0	0.00	1342	0	0.00	1507	0	0	0	25	-66	-57
Blattner Cal 1-31	R							0	0.00	1508						
Canner Seedless	W				0	0.00	1368	0	0.00	1509		0			-3	
Cruciulita	W				0	0.00	1372	0	0.00	1510		0			7	
Mouyssagues	R	0	0.00	999	0	0.00	1363	0	0.00	1511	0	0	0	54	-31	7
VB Cal 1-14	R							0	0.00	1512						
Laska	R	1	0.00	935				0	0.00	1513			-1			-95
Satin Noir	R							0	0.00	1514						
VB Jura 25	R							0	0.00	1515						
Petite Amie	W							0	0.00	1516						
Teoulier Noir	R	1	0.00	955	0	0.00	1371	0	0.00	1517	-1	0	-1	-92	-2	-92
Tihanyi Kék	R							0	0.00	1518						
Agasfark	R							0	0.00	1519						
Somszoekoe Kék	R							0	0.00	1520						
Hajnalka	W				0	0.00	1373	0	0.00	1521		0			0	
Burdin	W	0	0.00	988	0	0.00	1365	0	0.00	1522	0	0	0	-52	-38	-70

Table 6 (cont.): Global ranking of prime winegrape varieties by area, and changes, 2000, 2010 and 2016 (ha and %)

		2000			2010			2016			2010-2000	2016-2010	2016-2000	2010-2000	2016-2010	2016-2000
Prime vairiety	*Col*	*Global area (ha)*	*Global share %*	*Global rank*	*Global area (ha)*	*Global share %*	*Global rank*	*Global area (ha)*	*Global share %*	*Global rank*	*Change (ha)*	*Change (ha)*	*Change (ha)*	*Changes %*	*Changes %*	*Changes %*
Clara	W							0	0.00	1523						
Xara	R				0	0.00	1375	0	0.00	1524		0			2	
Grachen	W	3	0.00	875	2	0.00	1164	0	0.00	1525	0	-2	-3	-17	-98	-99
Imperial Napoleon	R				12	0.00	972	0	0.00	1526		-12			-100	
IRAC 1933	R							0	0.00	1527						
Chasselas Sabor	W				0	0.00	1370	0	0.00	1528		0			-31	
Jeroma	R							0	0.00	1529						
Sulima	W				0	0.00	1369	0	0.00	1530		0			-45	
Pascal Blanc	W	0	0.00	987	0	0.00	1378	0	0.00	1531	0	0	0	-81	-2	-81
Magliasina	R							0	0.00	1532						
Gaillard	R	0	0.00	994	0	0.00	1362	0	0.00	1533	0	0	0	-20	-69	-75
Enfarine Noir	R				0	0.00	1376	0	0.00	1534		0			-37	
Buffalo	R							0	0.00	1535						
Sugrafive	W	118	0.00	538	2	0.00	1178	0	0.00	1536	-116	-2	-118	-98	-99	-100
Orpicchio	W				1	0.00	1290	0	0.00	1537		0			-96	
Esganacao Preto	R				0	0.00	1380	0	0.00	1538		0			2	
Alvarelhao Ceitao	R				0	0.00	1381	0	0.00	1539		0			-2	
De Cilindro	W	15	0.00	763				0	0.00	1540			-15			-100
Poulsard Blanc	W	14	0.00	775	0	0.00	1379	0	0.00	1541	-14	0	-14	-100	-27	-100
Arrouya	R				0	0.00	1382	0	0.00	1542		0			-3	
MRAC 1626	R							0	0.00	1543						
MRAC 40	R							0	0.00	1544						
Castets	R	0	0.00	998	0	0.00	1377	0	0.00	1545	0	0	0	-36	-66	-78
VB Cal 1-29	R							0	0.00	1546						
VB Cal 1-33	R							0	0.00	1547						
Gänsfüsser	R				0	0.00	1374	0	0.00	1548		0			-74	
Uva del Fantini	R				0	0.00	1350	0	0.00	1549		0			-93	
Moscatel Lilaz	W				0	0.00	1385	0	0.00	1550		0			2	
Tinta Malandra	R				0	0.00	1384	0	0.00	1551		0			2	
Fepiro	R				0	0.00	1383	0	0.00	1552		0			-2	
Tinta Aurelio	R				0	0.00	1386	0	0.00	1553		0			-2	
Ontario	W							0	0.00	1554						
Olivette de Laconnex	W							0	0.00	1555						
Naparo	R	0	0.00	977	1	0.00	1250	0	0.00	1556	1	-1	0	223	-100	-99
Shalistin	W							0	0.00	1557						
Esther	R							0	0.00	1558						
Lilla	W							0	0.00	1559						
Doukkali	R	16557	0.34	52	16557	0.36	50				0			0		

Table 6 (cont.): Global ranking of prime winegrape varieties by area, and changes, 2000, 2010 and 2016 (ha and %)

		2000			2010			2016			2010-2000	2016-2010	2016-2000	2010-2000	2016-2010	2016-2000
Prime vairiety	*Col*	*Global area (ha)*	*Global share %*	*Global rank*	*Global area (ha)*	*Global share %*	*Global rank*	*Global area (ha)*	*Global share %*	*Global rank*	*Change (ha)*	*Change (ha)*	*Change (ha)*	*Changes %*	*Changes %*	*Changes %*
Calagrano	W	8229	0.17	85	4794	0.10	114				-3435			-42		
Abbou	R	2375	0.05	187	2375	0.05	170				0			0		
Verdoncho	W	3092	0.06	162	2124	0.05	181				-968			-31		
Mskhali	W	1093	0.02	280	1093	0.02	261				0			0		
Garandmak	W	931	0.02	302	931	0.02	278				0			0		
Kokur Bely	W	641	0.01	342	918	0.02	282				277			43		
Kangun	W	850	0.02	314	850	0.02	300				0			0		
Voskeat	W	809	0.02	317	809	0.02	305				0			0		
Malvasia Moscata	W				554	0.01	356									
Muskat Moravsky	W				514	0.01	368									
Monbadon	W	657	0.01	337	498	0.01	378				-159			-24		
Miguel del Arco	R	1267	0.03	260	468	0.01	387				-799			-63		
Debit	W				403	0.01	404									
Babić	R	1189	0.02	271	359	0.01	420				-830			-70		
Misket Varnenski	W				336	0.01	434									
Storgozia	R				295	0.01	457									
Pošip Bijeli	W	6539	0.13	105	253	0.01	485				-6287			-96		
Ranna Melnishka Loza	R				249	0.01	488									
Tilki Kuyrugu	W	172	0.00	498	246	0.01	490				74			43		
Trbljan	W				231	0.01	498									
Palava	G				230	0.00	500									
Cabernet Moravia	R				212	0.00	508									
Kujundžuša	W				206	0.00	515									
Patrizia Rosa	G	10	0.00	795	153	0.00	556				143			1366		
Maratheftiko	R				152	0.00	557									
Ofthalmo	R				141	0.00	569									
Žlahtina	W				135	0.00	578									
Ranfol	W				134	0.00	581									
Devin	W				133	0.00	582									
Aurelius	W				70	0.00	689									
Pervomaisky	R	64	0.00	619	64	0.00	706				0			0		
Škrlet	W				61	0.00	711									
Bogdanuša	W				48	0.00	759									
Dunaj	R				46	0.00	764									
Patagonia	W	29	0.00	692	40	0.00	780				10			36		
Serna	G	19	0.00	745	36	0.00	796				17			94		
Vugava	W				36	0.00	798									
Cetinka	W				35	0.00	802									

Table 6 (cont.): Global ranking of prime winegrape varieties by area, and changes, 2000, 2010 and 2016 (ha and %)

		2000			2010			2016			2010-2000	2016-2010	2016-2000	2010-2000	2016-2010	2016-2000
Prime vairiety	*Col*	*Global area (ha)*	*Global share %*	*Global rank*	*Global area (ha)*	*Global share %*	*Global rank*	*Global area (ha)*	*Global share %*	*Global rank*	*Change (ha)*	*Change (ha)*	*Change (ha)*	*Changes %*	*Changes %*	*Changes %*
Seyve Villard 23-512	R				29	0.00	835									
Ensanyo Tintas	R				27	0.00	844									
Ekim Kara	R	19	0.00	746	27	0.00	847				8			43		
Zlatarica Vrgorska	W				19	0.00	901									
Babica	R				18	0.00	910									
GM 322	R				17	0.00	919									
Nincusa	R				17	0.00	920									
Ives	R	23	0.00	723	16	0.00	926				-7			-30		
Trnjak	R				15	0.00	932									
Lasina	R				14	0.00	944									
Cinsaut Seedless	R	9	0.00	809	13	0.00	955				4			44		
Gegić	W				11	0.00	979									
Canelon	R	8	0.00	820	8	0.00	1024				0			0		
Grand Manchen	W				8	0.00	1025									
Valvin Muscat	W				6	0.00	1060									
Agni	R				6	0.00	1061									
Laurot	R				6	0.00	1062									
Suscan	R				5	0.00	1074									
Ilichevskii Rannii	R	5	0.00	847	5	0.00	1082				0			0		
Droujba	W	3	0.00	869	3	0.00	1135				0			0		
Ariana	R				3	0.00	1136									
Chaouch Blanc	W	2	0.00	911	3	0.00	1156				1			43		
Gargiulo 26879	W	4	0.00	852	3	0.00	1157				-2			-40		
Alb de Ialoveni	W	2	0.00	903	2	0.00	1175				0			0		
Gateta	R	7	0.00	830	2	0.00	1176				-5			-71		
Tinto de Zafra	R	4	0.00	853	2	0.00	1180				-2			-51		
Gutenborner	W				2	0.00	1181									
Dišeća Ranina	W				2	0.00	1188									
Joubertin	R	2	0.00	916	1	0.00	1205				0			-6		
Noria	W				1	0.00	1209									
Milia	R				1	0.00	1210									
Gargiulo 14260	R	1	0.00	939	1	0.00	1236				0			0		
Dominga	W	0	0.00	981	1	0.00	1242				1			376		
Monstruosa	W	2	0.00	893	1	0.00	1245				-1			-55		
Panse Valenciano	W	649	0.01	340	1	0.00	1248				-648			-100		
Perlita	W	1	0.00	931	1	0.00	1249				0			-20		
Ensayo Blancas	W				1	0.00	1251									
Juliana	W				0	0.00	1327									

Table 6 (cont.): Global ranking of prime winegrape varieties by area, and changes, 2000, 2010 and 2016 (ha and %)

		2000			2010			2016			2010-2000	2016-2010	2016-2000	2010-2000	2016-2010	2016-2000
Prime vairiety	*Col*	*Global area (ha)*	*Global share %*	*Global rank*	*Global area (ha)*	*Global share %*	*Global rank*	*Global area (ha)*	*Global share %*	*Global rank*	*Change (ha)*	*Change (ha)*	*Change (ha)*	*Changes %*	*Changes %*	*Changes %*
K.35	W				0	0.00	1337									
Viktor	W				0	0.00	1339									
Kozma CS. 2	R				0	0.00	1364									
Misket	W	3764	0.08	140												
Agadai	W	1265	0.03	261												
Montu	W	1091	0.02	282												
Bonarda Grande	R	538	0.01	375												
Canorroyo	W	157	0.00	509												
Dalkauer	W	100	0.00	561												
Liatiko (W)	R	70	0.00	608												
Perola	W	68	0.00	615												
Vilana (R)	W	60	0.00	626												
Fokiano (W)	R	57	0.00	630												
Morenillo	R	39	0.00	668												
Cabernet Malbec	R	34	0.00	681												
Cartouche	W	31	0.00	688												
Limnio (W)	R	27	0.00	698												
Batili	W	27	0.00	703												
Ruby Seedless	R	25	0.00	712												
Mazuelo (G)	R	25	0.00	713												
Dawn Seedless	W	19	0.00	740												
Barbarossa	G	16	0.00	757												
Emerald seedless	W	16	0.00	761												
Blush Seedless	R	10	0.00	799												
Listrao Roxo	R	10	0.00	801												
Aramon Bouschet	R	10	0.00	802												
Mauzac Noir	R	10	0.00	804												
Carmine	R	10	0.00	805												
Gargiulo 26189	R	6	0.00	839												
Valais Noir	R	4	0.00	854												
Staufer	W	4	0.00	856												
Palot	W	4	0.00	863												
Mavro Messenikola	R	3	0.00	868												
Xinomavro (W)	R	3	0.00	877												
Prodest	R	2	0.00	889												
Brocada	W	2	0.00	890												
Mecle de Bourgoin	R	2	0.00	897												
Sirius	W	2	0.00	907												

Table 6 (cont.): Global ranking of prime winegrape varieties by area, and changes, 2000, 2010 and 2016 (ha and %)

		2000			2010			2016			2010-2000	2016-2010	2016-2000	2010-2000	2016-2010	2016-2000
Prime vairiety	*Col*	*Global area (ha)*	*Global share %*	*Global rank*	*Global area (ha)*	*Global share %*	*Global rank*	*Global area (ha)*	*Global share %*	*Global rank*	*Change (ha)*	*Change (ha)*	*Change (ha)*	*Changes %*	*Changes %*	*Changes %*
Provareau	R	2	0.00	908												
Leopoldo III	R	2	0.00	910												
Bellandais	R	2	0.00	917												
Chenivesse	R	1	0.00	922												
Baleille	W	1	0.00	923												
Tinto Jeroma	R	1	0.00	924												
Boiziau	R	1	0.00	929												
Raisaine	W	1	0.00	934												
Malegue	W	1	0.00	936												
Lacoste	R	1	0.00	937												
Oberlin White	W	1	0.00	943												
Mornen Noir	R	1	0.00	944												
Reze	W	1	0.00	949												
Dunze	R	1	0.00	954												
Blanqueiro	W	1	0.00	958												
Noual	W	1	0.00	962												
Peloursin	R	0	0.00	963												
Piquepoul Bousch	R	0	0.00	965												
Roublot	W	0	0.00	967												
Maliverne	W	0	0.00	969												
Vidal Noir	R	0	0.00	970												
Vineti	W	0	0.00	971												
Bariadorgia	W	0	0.00	972												
Pougnet	R	0	0.00	973												
Castonotal	W	0	0.00	975												
Servanin	R	0	0.00	978												
Fontan	R	0	0.00	983												
Taylor	W	0	0.00	989												
Guillemot	W	0	0.00	995												
Cep Rouge	R	0	0.00	996												
Feunate	R	0	0.00	1000												
Putzscheere	W	0	0.00	1001												
Jurie	W	0	0.00	1002												
other varieties		529222	10.83		226775	4.91		346035	7.72		-302447	119260	-183187	-57	53	-35
Total		**4887629**	**100**		**4615761**	**100**		**4483128**	**100**		**-271868**	**-132633**	**-404501**	**-6**	**-3**	**-8**

Table 7: National and global winegrape bearing area of top 50 varieties, 1990 (hectares and % of world)

Colour	*Country of planting→*	*Argentina*	*Australia*	*Austria* [ab]	*Bulgaria*	*France* [a]	*Georgia*	*Germany* [a]	*Italy*	*United States* [ad]	*Other countries*	*WORLD*	%
W	Airén										476396	**476396**	8.96
R	Garnacha Tinta		2129			86700			12362	5297	198377	**304602**	5.73
W	Rkatsiteli				18700		86802				175067	**280569**	5.28
W	Sultaniye		2909								268919	**271828**	5.11
W	Trebbiano Toscano	2229	1254		2400	103000			61600	239	36720	**207442**	3.90
R	Mazuelo					116100			3000	4136	79633	**202869**	3.82
R	Merlot	1160	318		15300	60000	66		31900	3314	42695	**154752**	2.91
R	Cabernet Sauvignon	2347	3684		18700	36500	235		3200	13831	49182	**127678**	2.40
R	Monastrell		620			5600				125	101868	**108213**	2.04
R	Bobal										106149	**106149**	2.00
R	Sangiovese					1400			87400	94	10052	**98946**	1.86
W	Catarratto Bianco								74800		5328	**80128**	1.51
W	Chardonnay	908	3123		4000	19900	43		6200	22909	12200	**69282**	1.30
R	Criolla Grande										68513	**68513**	1.29
R	Barbera	958							47200	4145	15684	**67987**	1.28
W	Cayetana Blanca		863								65276	**66139**	1.24
W	Muscat of Alexandria		3193			3200			1400		56430	**64224**	1.21
R	Cinsaut					48200				36	14936	**63171**	1.19
W	Chenin Blanc	4031	508			9100				12493	33842	**59974**	1.13
W	Aligoté				4000	1300	292				48838	**54430**	1.02
W	Riesling	293	3615	1250		2900		22013	400	1690	20003	**52164**	0.98
W	Palomino Fino								4000	502	46043	**50545**	0.95
W	Pedro Ximénez										47915	**47915**	0.90
R	Tempranillo	5659								216	41555	**47429**	0.89
W	Sauvignon Blanc	278	732			12000			3000	5372	23295	**44677**	0.84
W	Macabeo										43504	**43504**	0.82
G	Cereza										42937	**42937**	0.81
W	Malvasia Bianca di Candia								23100		19554	**42654**	0.80
R	Tribidrag								17300	13909	10474	**41683**	0.78
R	Pinot Noir	232	801	300		22000	878	6449	3500	3877	3501	**41539**	0.78
R	Negroamaro								31400		8664	**40064**	0.75
R	Cabernet Franc	76	212			30300				675	8356	**39619**	0.75
W	Müller-Thurgau			5150				24600	1000		5631	**36381**	0.68
W	Colombard		618			4900				22597	8023	**36138**	0.68
R	Syrah	687	4711			27000			100	167	2421	**35086**	0.66
R	Gamay Noir					33600				1135	270	**35005**	0.66
R	Montepulciano								31000		1982	**32982**	0.62
W	Sémillon	1255	2526			17600				833	10306	**32520**	0.61
R	Concord										30513	**30513**	0.57
R	Korinthiaki		18								28108	**28126**	0.53
W	Trebbiano Romagnolo								21300			**21300**	0.40
R	Isabella										21003	**21003**	0.40
W	Grüner Veltliner			20600							160	**20760**	0.39
W	Pedro Giménez	20647										**20647**	0.39
W	Savatiano										20396	**20396**	0.38
R	Parmak Cerven										20275	**20275**	0.38
R	Flame Seedless		13								20181	**20194**	0.38
R	Alicante Henri Bouschet					15800				768	3019	**19587**	0.37
W	Graševina			4900	4000				2000	118	8367	**19384**	0.36
W	Muscat Blanc à Petits Grains		542			4600			13400	509		**19052**	0.36
R	other red varieties	82401	1420	9550	27011	168808	3765	11936	216392	15748	149010	**686042**	12.90
W	other white varieties	78985	3530	1750	19689	86492	23458	38779	185108	12152	242943	692885	13.03
	Total	**202146**	**37339**	**43500**	**113800**	**917000**	**115539**	**103777**	**881800**	**146886**	**2754515**	**5316301**	**100**
	% of world	3.80	0.70	0.82	2.14	17.25	2.17	1.95	16.59	2.76	51.81	100.0	

Table 8: Shares of red, white and grey winegrapes in national winegrape area, 1990, 2000, 2010 and 2016 (%)

	1990			*2000*			*2010*			*2016*		
Country	*Red*	*White*	*Grey*	*Red*	*White*	*Grey*	*Red*	*White*	*Grey*	*Red*	*White*	*Grey*
Algeria				100			100			96	4	
Argentina	46	54		53	31	16	59	27	14	63	23	14
Armenia					100			100		30	70	
Australia	37	63		59	41		61	37	2	64	34	3
Austria	23	77		27	72	1	35	63	1	34	65	1
Brazil				60	40		83	17	0	84	16	0
Bulgaria	54	45	2	61	39		63	29	7	59	33	8
Cambodia										100		
Canada				56	42	2	45	49	6	42	53	5
Chile				76	24	0	73	27	0	69	30	0
China							96	4	0	86	14	
Croatia				25	75		34	65	1	38	62	
Cyprus				80	20		66	34		62	38	
Czechia				30	70		34	60	6	32	62	6
Ethiopia							66	34		66	34	
France	71	29	0	69	30	1	67	32	1	66	34	1
Georgia	4	96	0	12	88	0	12	88	0	12	88	0
Germany	18	79	3	24	72	4	36	59	5	36	59	5
Greece				37	47	16	44	37	19	42	37	21
Hungary				24	71	5	30	61	9	30	60	10
India										26	74	
Israel				61	39		61	39		83	17	
Italy	55	45	0	52	47	1	57	40	3	54	43	3
Japan							60	35	6	46	30	24
Kazakhstan							14	81	6	14	81	6
Korea, Rep.				98		2	98		2	98		2
Lebanon										55	45	
Luxembourg				5	84	11	8	81	11	10	75	15
Mexico							62	38		65	35	
Moldova				40	58	2	40	58	2	51	47	2
Morocco				82	18		82	18		50	50	
Myanmar							59	41		56	44	
New Zealand				28	70	1	24	71	5	22	71	7
North Macedonia										60	40	
Norway										34	66	
Peru							60	40	0	60	40	0
Portugal				58	42		67	33	0	65	31	4
Romania				28	71	1	31	68	1	34	65	1
Russia				19	81		38	61	0	37	63	0
Serbia				31	69		31	69		54	45	1
Slovakia				17	83		29	69	2	40	60	
Slovenia				26	74		32	65	3	31	66	3
South Africa				36	64	0	44	56	0	44	55	0
Spain				39	61	0	54	46	0	53	47	0
Switzerland				53	46	1	58	41	1	58	41	2
Taiwan				54	46		54	46		63	37	
Thailand							66	30	4	53	44	2
Tunisia				100			100			50	50	
Turkey							67	33		68	32	
Ukraine							31	68	1	36	64	
United Kingdom				23	77		34	61	5	47	50	3
United States	46	54		55	45	1	63	35	2	64	33	3
Uruguay				96	4		80	19	0	81	19	0
Missing 9				54	45	1						
Other countries	38	61	2									
Old World subtotal				**48**	**51**	**1**	**55**	**44**	**2**	**53**	**45**	**2**
New World subtotal				**57**	**39**	**4**	**61**	**34**	**4**	**65**	**31**	**4**
World total	**46**	**53**	**1**	**49**	**49**	**1**	**56**	**42**	**2**	**56**	**41**	**2**

Table 9: Red winegrape area and share of all varieties, by country, 2000, 2010 and 2016, and change between 2000 and 2016 (ha and %)

	Area (ha) of red varieties			*Changes in red varieties 2000 to 2016*		*Red varieties' share (%) of national area*		
Country	*2000*	*2010*	*2016*	*Hectares*	*%*	*2000*	*2010*	*2016*
Algeria	30200	30200	8000	-22200	-74	100	100	96
Argentina	105606	125271	129381	23775	23	53	59	63
Armenia			4405					30
Australia	77372	92200	84141	6768	9	59	61	64
Austria	12977	16137	15306	2329	18	27	35	34
Brazil	31721	40874	27998	-3722	-12	60	83	84
Bulgaria	58273	35543	31196	-27077	-46	61	63	59
Cambodia			10					100
Canada	4740	4548	5272	531	11	56	45	42
Chile	86936	81017	101275	14339	16	76	73	69
China		28350	152353				96	86
Croatia	15084	7085	4405	-10679	-71	25	34	38
Cyprus	14625	5707	3187	-11438	-78	80	66	62
Czechia	3399	5556	4392	993	29	30	34	32
Ethiopia		111	111				66	66
France	596377	561527	534071	-62306	-10	69	67	66
Georgia	4563	5853	5853	1290	28	12	12	12
Germany	25180	36792	34175	8995	36	24	36	36
Greece	18901	23965	21491	2590	14	37	44	42
Hungary	20939	21090	19111	-1829	-9	24	30	30
India			700					26
Israel	2970	2970	4148	1177	40	61	61	83
Italy	328902	356033	326428	-2474	-1	52	57	54
Japan		2213	1769				60	46
Kazakhstan		955	944				14	14
Korea, Rep.	5300	5300	5300			98	98	98
Lebanon			2182					55
Luxembourg	67	101	126	59	89	5	8	10
Mexico		3409	3534				62	65
Moldova	35741	35741	42442	6701	19	40	40	51
Morocco	40699	40165	8830	-31870	-78	82	82	50
Myanmar		44	40				59	56
New Zealand	2825	7689	7851	5027	178	28	24	22
North Macedonia			14826					60
Norway			4					34
Peru		2292	2292				60	60
Portugal	119300	109783	118392	-908	-1	58	67	65
Romania	61576	52817	61869	292	0	28	31	34
Russia	10706	9795	18936	8230	77	19	38	37
Serbia	21390	21390	11970	-9420	-44	31	31	54
Slovakia	2649	3723	3063	414	16	17	29	40
Slovenia	6103	5282	4926	-1177	-19	26	32	31
South Africa	33512	44049	42561	9049	27	36	44	44
Spain	460529	558859	469385	8857	2	39	54	53
Switzerland	7913	8574	8515	602	8	53	58	58
Taiwan	1530	1530	94	-1436	-94	54	54	63
Thailand		99	111				66	53
Tunisia	16836	16836	1707	-15129	-90	100	100	50
Turkey		8677	9355				67	68
Ukraine		16149	9170				31	36
United Kingdom	200	405	867	667	333	23	34	47
United States	96153	143076	153116	56963	59	55	63	64
Uruguay	8555	6152	5476	-3079	-36	96	80	81
Missing 9	43589					54		
Old World subtotal	**1915900**	**1997305**	**1802804**	**-113096**	**-6**	**48**	**55**	**53**
New World subtotal	**454450**	**588629**	**724255**	**269805**	**59**	**57**	**61**	**65**
World total	**2413939**	**2585934**	**2527059**	**113120**	**5**	**49**	**56**	**56**

Table 10: White winegrape area and share of all varieties, by country, 2000, 2010 and 2016, and change between 2000 and 2016 (ha and %)

	Area (ha) of white varieties			*Changes in white varieties 2000 to 2016*		*White varieties' share (%) of national area*		
Country	*2000*	*2010*	*2016*	*Hectares*	*%*	*2000*	*2010*	*2016*
Algeria			300					4
Argentina	60656	57591	47671	-12985	-21	31	27	23
Armenia	11206	11206	10300	-906	-8	100	100	70
Australia	53230	56292	44641	-8589	-16	41	37	34
Austria	34840	28890	29623	-5217	-15	72	63	65
Brazil	21119	8523	5199	-15921	-75	40	17	16
Bulgaria	37724	16431	17429	-20295	-54	39	29	33
Cambodia								
Canada	3548	4971	6637	3089	87	42	49	53
Chile	27028	30408	44160	17133	63	24	27	30
China		1193	25647				4	14
Croatia	44364	13394	7341	-37023	-83	75	65	62
Cyprus	3656	2901	1946	-1710	-47	20	34	38
Czechia	7932	9782	8382	450	6	70	60	62
Ethiopia		58	58				34	34
France	263593	269576	276551	12958	5	30	32	34
Georgia	32836	42122	42121	9285	28	88	88	88
Germany	75241	59837	55328	-19913	-26	72	59	59
Greece	24052	20174	18860	-5191	-22	47	37	37
Hungary	61657	42335	38187	-23470	-38	71	61	60
India			2000					74
Israel	1881	1881	852	-1029	-55	39	39	17
Italy	301118	252293	259257	-41861	-14	47	40	43
Japan		1282	1155				35	30
Kazakhstan		5598	5609				81	81
Korea, Rep.								
Lebanon			1818					45
Luxembourg	1126	1057	978	-148	-13	84	81	75
Mexico		2056	1931				38	35
Moldova	52061	52061	38896	-13165	-25	58	58	47
Morocco	8901	8835	8760	-140	-2	18	18	50
Myanmar		31	31				41	44
New Zealand	6984	22772	25187	18203	261	70	71	71
North Macedonia			9951					40
Norway			8					66
Peru		1532	1532				40	40
Portugal	85703	53704	57011	-28691	-33	42	33	31
Romania	158209	115899	118982	-39227	-25	71	68	65
Russia	45626	15755	31780	-13846	-30	81	61	63
Serbia	47609	47609	9932	-37677	-79	69	69	45
Slovakia	12932	8671	4685	-8246	-64	83	69	60
Slovenia	17369	10571	10555	-6814	-39	74	65	66
South Africa	60040	56704	52834	-7205	-12	64	56	55
Spain	721155	469322	412658	-308497	-43	61	46	47
Switzerland	6979	6029	6045	-934	-13	46	41	41
Taiwan	1303	1303	55	-1248	-96	46	46	37
Thailand		44	91				30	44
Tunisia			1693					50
Turkey		4179	4349				33	32
Ukraine		35458	15996				68	64
United Kingdom	673	734	919	246	37	77	61	50
United States	78616	79566	78987	370	0	45	35	33
Uruguay	325	1493	1258	933	287	4	19	19
Missing 9	35720					45		
Old World subtotal	**2057771**	**1605571**	**1506176**	**-551595**	**-27**	**51**	**44**	**45**
New World subtotal	**313522**	**326554**	**340001**	**26479**	**8**	**39**	**34**	**31**
World total	**2407014**	**1932125**	**1846178**	**-560836**	**-23**	**49**	**42**	**41**

Table 11: Grey winegrape area and share of all varieties, by country, 2000, 2010 and 2016, and change between 2000 and 2016 (ha and %)

	Area (ha) of grey varieties			*Changes in grey varieties 2000 to 2016*		*Grey varieties' share (%) of national area*		
Country	*2000*	*2010*	*2016*	*Hectares*	*%*	*2000*	*2010*	*2016*
Algeria								
Argentina	31156	30510	29290	-1865	-6	16	14	14
Armenia								
Australia		3296	3653				2	3
Austria	679	507	510	-169	-25	1	1	1
Brazil		15	8				0	0
Bulgaria		4159	4349				7	8
Cambodia								
Canada	210	578	692	481	229	2	6	5
Chile	2	100	437	435	21767	0	0	0
China		2					0	
Croatia		275					1	
Cyprus								
Czechia		904	826				6	6
Ethiopia								
France	4876	4450	4260	-615	-13	1	1	1
Georgia	20	26	26	6	28	0	0	0
Germany	3812	5506	4998	1186	31	4	5	5
Greece	7962	10251	10493	2531	32	16	19	21
Hungary	4290	6290	6583	2293	53	5	9	10
India								
Israel								
Italy	6643	17374	18867	12224	184	1	3	3
Japan		220	945				6	24
Kazakhstan		385	385				6	6
Korea, Rep.	100	100	100			2	2	2
Lebanon								
Luxembourg	155	146	196	41	26	11	11	15
Mexico								
Moldova	2042	2042	1263	-779	-38	2	2	2
Morocco								
Myanmar								
New Zealand	133	1503	2425	2292	1723	1	5	7
North Macedonia								
Norway								
Peru		7	7				0	0
Portugal		35	7245				0	4
Romania	2388	1576	1911	-476	-20	1	1	1
Russia		78	78				0	0
Serbia			112					1
Slovakia		242					2	
Slovenia		501	508				3	3
South Africa	104	264	380	276	266	0	0	0
Spain	122	77	1514	1392	1140	0	0	0
Switzerland	149	216	233	84	56	1	1	2
Taiwan								
Thailand		6	5				4	2
Tunisia								
Turkey								
Ukraine		685					1	
United Kingdom		59	53				5	3
United States	923	5307	7529	6605	715	1	2	3
Uruguay		12	9				0	0
Missing 9	912					1		
Old World subtotal	**33137**	**55725**	**64358**	**31221**	**94**	**1**	**2**	**2**
New World subtotal	**32628**	**41977**	**45533**	**12905**	**40**	**4**	**4**	**4**
World total	**66677**	**97702**	**109891**	**43215**	**65**	**1**	**2**	**2**

Table 12: Number of prime varieties and their bearing area, and their global shares, by country of origin of prime, 2000, 2010 and 2016 (ha and %)

		No. of primes	*Global hectares of prime*			*Share (%) of number of primes*			*% of global bearing area*		
			2000	*2010*	*2016*	*2000*	*2010*	*2016*	*2000*	*2010*	*2016*
OW	France	280	1359212	1720234	1744155	24.1	17.2	15.9	31.2	39.2	42.2
OW	Spain	117	1328846	1142234	957371	9.6	7.0	5.8	30.5	26.0	23.1
OW	Italy	347	616536	545383	516172	26.9	23.8	21.8	14.1	12.4	12.5
OW	Portugal	202	135318	131816	141887	6.5	14.3	12.8	3.1	3.0	3.4
OW	Greece	47	111004	107100	119501	4.4	2.8	2.6	2.5	2.4	2.9
OW	Germany	90	115274	112929	114391	5.3	4.4	5.5	2.6	2.6	2.8
NW	Argentina	19	82907	76763	73460	1.9	1.3	0.8	1.9	1.7	1.8
OW	Georgia	19	85054	80868	71954	1.8	1.3	1.2	2.0	1.8	1.7
OW	Croatia	27	144414	107048	66247	0.8	1.9	0.6	3.3	2.4	1.6
OW	Austria	20	71983	70215	64732	1.7	1.3	1.3	1.7	1.6	1.6
NW	United States	116	80677	70539	51703	4.5	5.2	6.8	1.9	1.6	1.2
OW	Hungary	98	38416	48978	48515	1.7	5.7	6.0	0.9	1.1	1.2
OW	Moldova	40	25208	20992	35704	1.0	1.1	2.4	0.6	0.5	0.9
OW	Bulgaria	11	34913	20628	26899	0.5	0.7	0.4	0.8	0.5	0.7
OW	Romania	36	7696	20277	19103	0.4	2.5	2.3	0.2	0.5	0.5
OW	Turkey	16	16764	10060	11763	1.4	1.0	1.0	0.4	0.2	0.3
OW	Montenegro	1	na	149	9503		0.1	0.1	na	0.0	0.2
OW	Switzerland	70	13629	14355	8992	1.6	1.2	4.4	0.3	0.3	0.2
NW	South Africa	8	7251	7164	7933	0.8	0.6	0.5	0.2	0.2	0.2
NW	United Kingdom	1	7068	8140	7680	0.1	0.1	0.1	0.2	0.2	0.2
OW	Ukraine	21	12760	12141	7632	1.2	1.0	1.1	0.3	0.3	0.2
NW	China	4	na	175	7575	0.1	0.1	0.3	na	0.0	0.2
NW	Japan	11	6059	6306	5529	0.4	0.3	0.7	0.1	0.1	0.1
OW	Cyprus	4	13711	5960	5133	0.2	0.3	0.1	0.3	0.1	0.1
OW	Armenia	5	3854	3928	na	0.5	0.4		0.1	0.1	na
OW	Russia	17	1435	3045	3639	0.3	0.9	1.0	0.0	0.1	0.1
NW	Brazil	15	10	1714	3596	0.1	0.4	0.8	0.0	0.0	0.1
OW	Serbia	9	15180	15267	1972	0.1	0.3	0.6	0.3	0.3	0.0
OW	Slovenia	5	365	1297	1154	0.1	0.2	0.3	0.0	0.0	0.0
OW	Azerbaijan	2	470	673	673	0.2	0.1	0.1	0.0	0.0	0.0
OW	North Macedonia	1	na	na	400			0.1	na	na	0.0
OW	Kazakhstan	1	269	385	385	0.1	0.1	0.1	0.0	0.0	0.0
NW	Peru	5	236	353	338	0.5	0.4	0.3	0.0	0.0	0.0
OW	Israel	2	374	227	281	0.2	0.1	0.1	0.0	0.0	0.0
OW	Morocco	3	19286	19243	257	0.3	0.2	0.1	0.4	0.4	0.0
OW	Lebanon	1	1837	381	211	0.1	0.1	0.1	0.0	0.0	0.0
OW	Bosnia and Herzegovina	1	na	na	185			0.1	na	na	0.0
NW	Canada	15	39	55	182	0.1	0.3	0.9	0.0	0.0	0.0
OW	Slovakia	4	na	182	na		0.3		na	0.0	na
NW	Australia	9	166	72	156	0.2	0.1	0.6	0.0	0.0	0.0
OW	Uzbekistan	1	64	64	na	0.1	0.1		0.0	0.0	na
NW	Thailand	1	11	16	54	0.1	0.1	0.1	0.0	0.0	0.0
NW	Chile	2	107	40	44	0.1	0.1	0.1	0.0	0.0	0.0
OW	Czechia	10	na	1589	26		0.6	0.2	na	0.0	0.0
NW	Taiwan	1	na	na	5			0.1	na	na	0.0
OW	Belgium	1	2	na	na	0.1	0.1	0.1	0.0	na	na
OW	Algeria	1	3	1	0	0.1			0.0	0.0	0.0
	Old World subtotal	**1507**	**4173811**	**4217585**	**3978838**	**91.1**	**91.2**	**87.9**	**95.8**	**96.1**	**96.2**
	New World subtotal	**209**	**184531**	**171337**	**158255**	**8.9**	**8.8**	**12.1**	**4.2**	**3.9**	**3.8**
	World total*	**1716**	**4358407**	**4388986**	**4137093**	**100**	**100**	**100**	**100**	**100**	**100**

* na means not available, because areas are hidden in the "other varieties" residual category for some countries. Hence the world total areas are less than in Table 2 etc.

Table 13: Prime varieties' colour, synonyms, country of origin, and global area, share and rank, 2010 and 2016 (ha and %)

				2010			2016		
Prime variety	*Col.*	*Synonyms*	*Country of origin*	*Area (hectares)*	*Share (%)*	*Rank*	*Area (hectares)*	*Share (%)*	*Rank*
Abbou	R	Abbo	Morocco	2375	0.051	168			
Abbuoto	R		Italy	37	0.001	787	18	0.000	866
Abondant	W		France	0	0.000	1315	0	0.000	1398
Abouriou	R		France	329	0.007	435	310	0.007	414
Abrusco	R	Colorino	Italy	423	0.009	393	215	0.005	470
Accent	R		Germany				1	0.000	1292
Acolon	R		Germany	490	0.011	377	477	0.011	359
Adakarası	R	Adakarasi	Turkey	69	0.001	689	89	0.002	599
Adalmiina	W	Aldamiina	United States				5	0.000	1073
Adirondac	R		United States				24	0.001	827
Admirable de Courtiller	W		France	28	0.001	839	27	0.001	805
Afus Ali	W	Karaburnu; Afuz-Ali; Roseti; Regina; Waltham Cross	Lebanon	381	0.008	407	211	0.005	474
Agadai	W		Russia						
Agasfark	R	Tarcali Kék	Hungary				0	0.00	1517
Agiorgitiko	R		Greece	2905	0.063	151	3272	0.073	136
Aglianico	R	Aglianico del Vulture; Aglianco	Italy	9995	0.217	70	9734	0.217	71
Aglianicone	R		Italy	62	0.001	707	30	0.001	778
Agni	R		Czechia	6	0.000	1060			
Agronomica	R		Portugal	327	0.007	437	299	0.007	419
Agua Santa	R		Portugal	78	0.002	669	76	0.002	630
Ahmeur Bou Ahmeur	R	Argelina	Algeria	1	0.000	1249	0	0.000	1405
Airén	W	Airen; Burra Blanca; Forcallat Blanca	Spain	252364	5.467	3	203801	4.546	4
Aladasturi	R		Georgia	59	0.001	716	59	0.001	662
Alarije	W	Malfar	Spain	1726	0.037	197	4407	0.098	123
Alb Aromat	W		Romania	24	0.001	861	24	0.001	829
Alb de Ialoveni	W		Moldova	2	0.000	1179			
Alb de Suruceni	W		Moldova				780	0.017	283
Albalonga	W		Germany	15	0.000	934	12	0.000	932
Albana	W		Italy	1523	0.033	208	782	0.017	282
Albanello	W		Italy	18	0.000	909	2	0.000	1178
Albaranzeuli Bianco	W		Italy	7	0.000	1044	2	0.000	1161
Albaranzeuli Nero	R		Italy	49	0.001	752	28	0.001	797
Albarín Blanco	W	Albarin Blanco; Branco Lexítimo- B	Spain	23	0.000	876	48	0.001	703
Albarola	W	Bianchetta Genovese	Italy	197	0.004	520	95	0.002	584
Albarossa	R		Italy	80	0.002	662	70	0.002	642
Albillo Mayor	W	Albilla	Spain	1319	0.029	230	1152	0.026	238
Albillo Real	W	Albillo	Spain	861	0.019	294	601	0.013	326
Alcañon	W	Alcanon	Spain	60	0.001	714	27	0.001	801
Aleatico	R		Italy	346	0.007	427	165	0.004	509
Aledo	W		Spain	7	0.000	1041	2	0.000	1167
Aleksandrouli	R	Alexandrouli	Georgia	281	0.006	461	281	0.006	433
Aletta	W		Hungary	723	0.016	316	1676	0.037	193
Alfrocheiro	R	Bastarda Negra; Bastardo Negro; Bruñal; Tinta Bastardinha	Portugal	1188	0.026	247	1216	0.027	229
Alicante Henri Bouschet	R	Alicante; Alicante Bouchet; Alicante Bouschet; Alicante H. Bouschet/Tintorera; Alicante H Bouschet Tintor; Alikante Mouse; Garnacha Tintorera; Tintoreras	France	38462	0.833	22	36031	0.804	22
Aligoté	W	Aligote; Mukhranuli	France	36120	0.783	24	26929	0.601	31
Alionza	W		Italy	11	0.000	984	9	0.000	985
Almafra	W	Almafre	Portugal	0	0.000	1334	0	0.000	1492
Alphonse Lavallée	R	Alfonso Lavalle; Alphonse Lavallee	France	862	0.019	292	634	0.014	316
Altesse	W		France	359	0.008	417	227	0.005	461
Alutus	R		Romania	2	0.000	1195	2	0.000	1214
Alvar Branco	W	Alvar	Portugal	0	0.000	1327	0	0.000	1454
Alvar Roxo	G		Portugal	2	0.000	1190	2	0.000	1205
Alvarelhão	R	Alvarelhao; Alvarinho; Brancellao; Pilongo	Portugal	5701	0.124	105	2910	0.065	144
Alvarelhao Ceitao	R		Portugal	0	0.000	1379	0	0.000	1536
Alvarinho	W	Albarifio; Albarino; Albariño;	Portugal				5545	0.124	103
Amaral	R		Portugal	92	0.002	632	93	0.002	586
Amigne	W		Switzerland	43	0.001	769	42	0.001	719
Amur	R		Moldova	146	0.003	561	146	0.003	520
Amurg	R		Romania	3	0.000	1131	3	0.000	1116
Ancellotta	R	Ancellota; Ancelota	Italy	4681	0.101	113	2739	0.061	151
Andor	W		Hungary				1	0.000	1346
Andre	W		Italy	477	0.010	380	5	0.000	1069
Antao Vaz	W		Portugal	1252	0.027	237	1768	0.039	187
Apiren Alb	W		Moldova				1	0.000	1244
Apiren Roz	R		Moldova				0	0.000	1373
Ar110	G		Peru	1	0.000	1248	1	0.000	1296
Ar99	G		Peru	5	0.000	1084	5	0.000	1070

Table 13 (cont.): Prime varieties' colour, synonyms, country of origin, and global area, share and rank, 2010 and 2016 (ha and %)

Prime variety	*Col.*	*Synonyms*	*Country of origin*	2010 *Area (hectares)*	2010 *Share (%)*	2010 *Rank*	2016 *Area (hectares)*	2016 *Share (%)*	2016 *Rank*
Aramon Bouschet	R		France						
Aramon Noir	R	Amor-Nao-Me-Deixes; Aramon;	France	2601	0.056	161	1181	0.026	235
Aramon Noir (W)	R	Aromon Blanc; Aramon Blanc	France	15	0.000	935	14	0.000	909
Aramont	R		France				0	0.000	1451
Aranel	W		France	5	0.000	1078	5	0.000	1057
Arany Sárfehér	W	Arany Sarfeher; Izsaki Sarfeher	Hungary	1133	0.025	253	586	0.013	328
Arbane	W		France	2	0.000	1188	1	0.000	1253
Arcas	R		Romania	1	0.000	1258	1	0.000	1303
Argaman	R		Israel	202	0.004	516	275	0.006	434
Ariana	R		Czechia	3	0.000	1134			
Arinarnoa	R	Merlot, Petit Verdot	France	189	0.004	526	486	0.011	355
Arinto de Bucelas	W	Arinto; Arinto Roxo	Portugal	4482	0.097	119	5409	0.121	104
Arkadia	W	Arcadia; Nastea	Ukraine				303	0.01	417
Arneis	W		Italy	1122	0.024	254	1179	0.026	236
Arnsburger	W		Germany	30	0.001	822	29	0.001	783
Aromat de Iasi	W		Romania	62	0.001	708	66	0.001	649
Aromella	W	Aromella (NY 76)	United States				3	0.000	1142
Arriloba	W		France	55	0.001	727	54	0.001	674
Arrouya	R		France	0	0.000	1380	0	0.000	1539
Arrufiac	W		France	80	0.002	661	9	0.000	984
Arvesiniadu	W		Italy	30	0.001	827	13	0.000	924
Arvine	W	Arvine (Petite); Petite Arvine	Switzerland	172	0.004	538	192	0.004	489
Asirtiko Red	R		Greece	5	0.000	1086	5	0.000	1064
Aspiran Bouschet	R	Aspirant Bouschet	France	2245	0.049	174	4088	0.091	127
Asprouda	W	Asproudi	Greece	113	0.002	596	120	0.003	553
Assaraky	W	Assaraka	Portugal	1	0.000	1233	0	0.000	1361
Assyrtiko	W	Asirtiko	Greece	902	0.020	284	1770	0.039	186
Astra	W		Romania	0	0.000	1330	0	0.000	1429
Athiri	W	Athiri Aspro	Greece	748	0.016	310	577	0.013	332
Aubin Blanc	W		France	1	0.000	1256	1	0.000	1332
Aubun	R	Corvo	France	553	0.012	356	537	0.012	341
Augster Blau	R	Kék Bajor; Fekete Fájú Bajor; Gohér	Austria				0	0.000	1484
Augster Weiss	W	Gohér; Goher	Hungary	1	0.000	1275	1	0.000	1333
Augustovski	W		Moldova				0	0.000	1406
Aurelius	W		Czechia	70	0.002	687			
Aurore	W	Aurora	France	268	0.006	471	255	0.006	443
Auxerrois	W	Aucerot	France	2785	0.060	154	2853	0.064	146
Avana	R		Italy	28	0.001	840	18	0.000	867
Avarengo	R		Italy	987	0.021	272	153	0.003	515
Avesso	W		Portugal	685	0.015	326	699	0.016	298
Azal	W	Azal Branco	Portugal	1072	0.023	265	1443	0.032	212
Băbească Neagră	R	Babeasca Neagra; Rara Neagra	Romania	3122	0.068	144	2696	0.060	153
Băbească Neagră (G)	R	Babeasca Gri; Băbească Gris; Babeasca Gris	Romania	328	0.007	436	297	0.007	421
Babić	R	Babic	Croatia	359	0.008	418			
Babica	R		Croatia	18	0.000	908			
Babosa de Madere	W	Babosa	Portugal	2	0.000	1168	2	0.000	1224
Bacchus	W		Germany	2113	0.046	180	1759	0.039	188
Baco Blanc	W		France	739	0.016	312	528	0.012	345
Baco Noir	R	Baco; Baco Noir; Baco Noir, Chambourcin, etc.	France	475	0.010	381	735	0.016	290
Bácska	G	Baska	Serbia				7	0.000	1008
Baga	R	Carrasquenho; Carrega Burros	Portugal	4108	0.089	123	6750	0.151	94
Bailey	R		United States	34	0.001	801	49	0.001	699
Bakator Belyi	W	Ardeleanca	Russia	11	0.000	985	11	0.000	941
Bakator Kék	R	Kek Bakator; Bakator Kek	Hungary	3	0.000	1152	2	0.000	1163
Bakator Roz	R	Piros Bakator	Hungary	16	0.000	926	16	0.000	884
Balada	R		Romania	0	0.000	1295	0	0.000	1365
Baleille	W		France						
Baratuciat	W		Italy	2	0.000	1189	2	0.000	1206
Barbarossa	G		Italy						
Barbaroux	G		France	30	0.001	826	29	0.001	784
Barbera	R	Barbera Nera	Italy	24366	0.528	36	17824	0.398	44
Barbera Bianca	W		Italy	181	0.004	532	114	0.003	560
Barbera Sarda	R		Italy	84	0.002	649	70	0.002	641
Barcelo	W		Portugal	23	0.000	878	26	0.001	806
Bariadorgia	W		Italy						
Barkhatnyi	W	Barkhatny	Russia	30	0.00	825	30	0.00	780
Baron	R		Germany				1	0.00	1271
Baroque	W		France	94	0.00	629	83	0.00	610
Barreto de Semente	R	Barreto	Portugal	3	0.00	1156	3	0.00	1158
Barsaglina	R		Italy	17	0.00	916	9	0.00	986
Bastardo Branco	W		Portugal	15	0.00	940	14	0.00	911

Table 13 (cont.): Prime varieties' colour, synonyms, country of origin, and global area, share and rank, 2010 and 2016 (ha and %)

Prime variety	Col.	Synonyms	Country of origin	2010 Area (hectares)	2010 Share (%)	2010 Rank	2016 Area (hectares)	2016 Share (%)	2016 Rank
Bastardo Magarachsky	R	Bastardo Magaraceskii; Bastardo Magarachskiy; Bastardo Magareceski	Ukraine	2370	0.05	169	180	0.00	497
Batili	W		Greece						
Batily	R		Greece	54	0.00	732	1	0.00	1274
Batoca	W		Portugal	11	0.00	972	8	0.00	996
Batuta Neagra	R		Moldova	3	0.00	1148	3	0.00	1140
Bayanshira	W	Bayan-Shirey	Azerbaijan	645	0.01	334	645	0.01	313
Beba		Breval; Iso; Teta de Vaca; Teneron; Valency	Spain	6524	0	94	2556	0.06	158
Beclan	R		France	0	0.00	1322	0	0.00	1501
Beibinghong	R		China				1600	0.04	203
Bellandais	R		France						
Bellone	W		Italy	511	0.01	368	184	0.00	494
Beogradska Crna	W		Serbia				0	0.00	1416
Bequignol Gris	W	Pintes	France	3	0.00	1147	2	0.00	1172
Béquignol Noir	R	Bequignol; Bequignol Noir	France	891	0.02	287	616	0.01	322
Bianca	W		Hungary	6462	0.14	95	9766	0.22	70
Biancame	W		Italy	2599	0.06	162	1336	0.03	220
Bianchetta Trevigiana	W		Italy	13	0.00	950	12	0.00	937
Bianco d'Alessano	W		Italy	419	0.01	396	39	0.00	725
Biancolella	W		Italy	164	0.00	545	23	0.00	840
Biancone di Portoferraio	W	Biancone	Italy	78	0.00	668	34	0.00	758
Biancu Gentile	W		France	9	0.00	1016	5	0.00	1055
Bíborkadarka	R	Bibor Kadarka; Biborkadarka; Bíbor kadarka	Hungary	136	0.00	573	109	0.00	570
Bical	W		Portugal	924	0.02	279	1076	0.02	248
Birstaler Muskat	W		Switzerland				1	0.00	1285
Black Prince	R	San Francisco	United States				1	0.00	1341
Black Queen	R	Black Queen/Pokdum	Japan	713	0.02	319	143	0.00	525
Blanc Dame	W		France	0	0.00	1336	0	0.00	1450
Blanc du Bois	W		United States	28	0.00	836	81	0.00	617
Blanca Ovoide	W		Chile	40	0.00	776	44	0.00	715
Blanqueiro	W		France						
Blasius	W		Romania	14	0.00	944	15	0.00	896
Blattner Cal 1-15	R	VB Cal 1-15	Switzerland				0	0.00	1493
Blattner Cal 1-20	R	VB Cal 1-20	Switzerland				0	0.00	1366
Blattner Cal 1-22	R	VB Cal 1-22	Switzerland				0	0.00	1388
Blattner Cal 1-28	R	VB Cal 1-28	Switzerland				2	0.00	1213
Blattner Cal 1-31	R	VB Cal 1-31	Switzerland				0	0.00	1505
Blattner Cal 1-36	R	VB Cal 1-36	Switzerland				1	0.00	1249
Blattner Reds	R		Switzerland	39	0.00	779	8	0.00	992
Blattner Whites	W		Switzerland	25	0.00	853	7	0.00	1013
Blauburger	R	Blauberger, Experimental Red Vinifera, Petite Sirah, Pinot Meunier, Sangiovese;	Austria	1339	0.03	228	1223	0.03	228
Blauer Portugieser	R	Portugieser Blau; Modry Portugal; Oporto; Portoghese; Portugais Bleu; Portugalka; Portugieser, Blauer; Portugizac; Portugues Azul; Portuguese, blue; Ranina; Kékoportó	Austria	8027	0.17	85	6590	0.15	95
Blauer Wildbacher	R	Wildbacher	Austria	368	0.01	413	437	0.01	374
Blaufränkisch	R	Blaufrankisch; Blaufränkisch/Lemberger; Borgona; Borgonja; Burgund Mare; Franconia; Frankovka; Frankovka Modra; Lemberger; Limberger, Blauer; Modra Frankinja; Kékfrankos	Austria	17890	0.39	44	17180	0.38	46
Blush Seedless	R		United States						
Boal Barreiro	W		Portugal	1	0.00	1281	0	0.00	1415
Boal Vencedor	W	Vencedor	Portugal	2	0.00	1159	2	0.00	1182
Bobal	R	Bobal, Provechon	Spain	80120	1.74	11	59189	1.32	13
Boğazkere	R	Bogazkere	Turkey	1106	0.02	257	1436	0.03	215
Bogdanuša	W	Bogdanusa	Croatia	48	0.00	757			
Boiziau	R		France						
Bokay	W		United States				4	0.00	1091
Bombino Bianco	W		Italy	1239	0.03	238	1147	0.03	239
Bombino Nero	R		Italy	1201	0.03	245	865	0.02	272
Bonamico	R	Buonamico	Italy	233	0.01	495	149	0.00	518
Bonarda Grande	R		Italy						
Bonarda Piemontese	R	Durasa	Italy	6	0.00	1062	5926	0.13	100
Bonda	R		Italy	7	0.00	1045	7	0.00	1005
Bondola	R		Switzerland	13	0.00	957	11	0.00	953
Borraçal	R	Borracal; Caino Tinto; Caíño Tinto	Portugal	683	0.01	327	512	0.01	347
Borsmenta	W		Hungary				1	0.00	1291

Table 13 (cont.): Prime varieties' colour, synonyms, country of origin, and global area, share and rank, 2010 and 2016 (ha and %)

				2010			2016		
Prime variety	*Col.*	*Synonyms*	*Country of origin*	*Area (hectares)*	*Share (%)*	*Rank*	*Area (hectares)*	*Share (%)*	*Rank*
Bosco	W		Italy	82	0.00	655	50	0.00	694
Bouchales	R		France	95	0.00	626	93	0.00	588
Bouillet	R		France	1	0.00	1263	1	0.00	1321
Bourboulenc	W		France	585	0.01	347	501	0.01	351
Bousquet Precoce	W		France	6	0.00	1069	0	0.00	1498
Bouvier	W	Bouvierovo Hrozno	Slovenia	250	0.01	485	224	0.00	464
Bracciola Nera	R		Italy	26	0.00	849	4	0.00	1106
Brachetto del Piemonte	R	Brachetto	Italy	1460	0.03	215	1694	0.04	191
Branco Escola	W		Portugal	2	0.00	1187	2	0.00	1200
Branco Gouvaes	W	Touriga Branca	Portugal	36	0.00	791	34	0.00	754
Branco Sr. Joao	W	Branco Joao	Portugal	0	0.00	1297	0	0.00	1440
Branco Valente	W	Valente	Portugal	0	0.00	1329	0	0.00	1468
Brandam	W		Portugal	312	0.01	446	291	0.01	425
Braquet Noir	R		France	12	0.00	966	12	0.00	933
Breidecker	W		Germany	7	0.00	1040	0	0.00	1368
Brianna	W		United States	12	0.00	962	21	0.00	850
Bric	R		Italy	2	0.00	1163	1	0.00	1239
Brocada	W		Spain						
Bronner	W		Germany	9	0.00	1015	6	0.00	1031
Brun Argente	R		France	11	0.00	976	11	0.00	951
Budai Zöld	W	Budai	Hungary	6	0.00	1054	6	0.00	1042
Buffalo	R		United States				0	0.00	1532
Bukettraube	W		Germany	71	0.00	683	54	0.00	673
Burdin	W		France	0	0.00	1363	0	0.00	1519
Bussanello	W		Italy	12	0.00	971	3	0.00	1129
Busuioacă de Bohotin	G	Busuioaca De Bohotin	Romania	268	0.01	470	343	0.01	397
BX 81-83	R		Switzerland				1	0.00	1349
Cabaret Noir	R	Cabernet Noir (VB 91-26-04)	Switzerland				3	0.00	1155
Caberinta	R		Argentina	69	0.00	688	38	0.00	731
Cabernet Blanc	W	Cabernet B; Cabernet Blanc (VB 91-26-01); Cabernet VB; CabVB	Switzerland				6	0.00	1030
Cabernet Cantor	R		Germany				1	0.00	1241
Cabernet Carbon	R		Germany				11	0.00	952
Cabernet Carol	R		Germany				6	0.00	1034
Cabernet Cortis	R		Germany				38	0.00	732
Cabernet Cubin	R		Germany	60	0.00	713	62	0.00	656
Cabernet Diane	R		United States	0	0.00	1299	0	0.00	1371
Cabernet Dore	W		United States	1	0.00	1217	1	0.00	1258
Cabernet Dorio	R		Germany	36	0.00	795	34	0.00	753
Cabernet Dorsa	R		Germany	252	0.01	484	272	0.01	435
Cabernet Early	R		Switzerland				0	0.00	1489
Cabernet Foch	R		Switzerland				2	0.00	1204
Cabernet Franc	R	Bordo; Cabernet Franc - Cabernet Franco;	France	61295	1.33	15	56052	1.25	14
Cabernet Jura	R		Switzerland	19	0.00	897	27	0.00	799
Cabernet Malbec	R		France						
Cabernet Mitos	R		Germany	322	0.01	440	312	0.01	411
Cabernet Moravia	R		Czechia	212	0.00	506			
Cabernet Sanzey	R		Australia				0	0.00	1472
Cabernet Sauvignon	R	Burdeos; Cabernet S.; Cabernet Sauvignon - Cabernet; Cabernet-Sauvignon	France	290083	6.28	1	310671	6.93	1
Cabernet Soyhières	R		Switzerland				1	0.00	1300
Cabernet x Maréchal Foch	R		Switzerland				0	0.00	1476
Cabertin	R	Cabertin (VB 91-26-18); VB 91-26-17	Germany				2	0.00	1220
Cabinda	R		Portugal	362	0.01	416	355	0.01	392
Cabral	R	Malvasia Cabral	Portugal	2	0.00	1184	2	0.00	1194
Caddiu	R		Italy	309	0.01	449	83	0.00	609
Caíño Blanco	W	Cainho; Cainho de Moreira; Caino Blanco;	Portugal	128	0.00	584	77	0.00	627
Caladoc	R		France	3675	0.08	130	5258	0.12	108
Calagrano	W		Spain	4794	0.10	112			
Calitor Noir	R		France	26	0.00	847	26	0.00	814
Çalkarası	R	Calkarasi; Çalkarasi	Turkey	625	0.01	341	806	0.02	276
Callet	R		Spain	154	0.00	552	138	0.00	530
Calmeria	W		United States				0	0.00	1401
Caloria	R		Italy	108	0.00	604	45	0.00	712
Calrao	R		Portugal	1	0.00	1231	1	0.00	1302
Camaralet de Lasseube	W	Camarate	France	520	0.01	363	306	0.01	416
Campanario	R		Portugal	2	0.00	1183	2	0.00	1197
Campbell Early	R		United States	61	0.00	711	238	0.01	453
Canada Muscat	W		United States				120	0.00	552
Canadice	R		United States	0	0.00	1324	0	0.00	1370

Table 13 (cont.): Prime varieties' colour, synonyms, country of origin, and global area, share and rank, 2010 and 2016 (ha and %)

				2010			2016		
Prime variety	*Col.*	*Synonyms*	*Country of origin*	*Area (hectares)*	*Share (%)*	*Rank*	*Area (hectares)*	*Share (%)*	*Rank*
Canaiolo Nero	R	Canaiolo; Canaiolo Rosa	Italy	1068	0.02	266	1033	0.02	250
Canari Noir	R	Canari; Folle Noire (Vidiella); Folle Noire	France	163	0.00	547	84	0.00	608
Canela	G		Argentina	1	0.00	1211	3	0.00	1143
Canelon	R		Argentina	8	0.00	1023			
Canner Seedless	W		United States	0	0.00	1366	0	0.00	1506
Canorroyo	W		Spain						
Capolongo	W		Italy	5	0.00	1077	0	0.00	1359
Caracol	W		Portugal	33	0.00	812	33	0.00	763
Caramela	W		Portugal	0	0.00	1290	0	0.00	1372
Carbernet Volos	R	UD-32.078	Italy				0	0.00	1488
Cardinal	R	Cardenal; Kardinal Crveni	United States	536	0.01	359	1660	0.04	195
Carica l'Asino	W		Italy	17	0.00	911	5	0.00	1054
Carignan Bouschet	R		France	1	0.00	1201	1	0.00	1240
Carla	R		France	0	0.00	1356	0	0.00	1477
Carmem	R	BRS Carmen	Brazil				328	0.01	402
Carmenère	R	Cabernet Gernischt; Carmenere; Carmenère - Grande Vidure; Carmenere Crni	France	11366	0.25	63	22486	0.50	36
Carmine	R		United States						
Carminoir	R	Carmi Noir	Switzerland	10	0.00	990	11	0.00	942
Carnelian	R		United States	316	0.01	443	123	0.00	544
Carrega Branco	W		Portugal	507	0.01	371	512	0.01	348
Carrega Tinto	R		Portugal	17	0.00	915	17	0.00	877
Carricante	W		Italy	205	0.00	515	35	0.00	750
Cartouche	W		Spain						
Casavecchia	R		Italy	136	0.00	572	92	0.00	590
Cascade	R		France				22	0.00	845
Casculho	R		Portugal	267	0.01	472	269	0.01	437
Casetta	R		Italy	12	0.00	960	14	0.00	919
Castalia	W		Portugal	0	0.00	1333	0	0.00	1491
Castel	R		France	2	0.00	1167	3	0.00	1136
Castela	R		Portugal	8	0.00	1024	7	0.00	1017
Castelão	R	Castelao; Casteloa; Periquita; Piriquita	Portugal	11088	0.24	64	12580	0.28	57
Castelão Branco	W	Castelao Branco	Portugal	37	0.00	785	18	0.00	869
Castelino	R		Portugal	147	0.00	558	144	0.00	522
Castellana Blanca	W		Spain				1	0.00	1269
Castelo Branco	W		Portugal	5	0.00	1095	2	0.00	1173
Castets	R		France	0	0.00	1375	0	0.00	1542
Castiglione	R		Italy	18	0.00	904	4	0.00	1085
Castonotal	W		Spain						
Catalanesca	W		Italy	54	0.00	733	7	0.00	1016
Catanese Nero	R		Italy	15	0.00	939	7	0.00	1015
Catarratto Bianco	W	Catarratto; Catarratto Bianco Comune; Catarratto Bianco Lucido	Italy	34863	0.76	27	28613	0.64	30
Catawba	R	Katawaba	United States	633	0.01	337	626	0.01	318
Caverdella	W		Australia				5	0.00	1063
Cavrara	R		Italy	23	0.00	875	1	0.00	1237
Çavuş	W		Turkey				3	0.00	1148
Cayetana Blanca	W	Baladi Verdejo; Blanca Cayetana; Mourisco Branco; Pardina; Robal; Sarigo; Doradilla	Spain	39781	0.86	21	36401	0.81	21
Cayuga White	W	Cayuga	United States	212	0.00	507	217	0.00	467
Cellerina	R	Slarina	Italy	2	0.00	1193	2	0.00	1219
Centennial Seedless	W	Centenial Seedless	United States				1	0.00	1260
Centesimino	R		Italy	24	0.00	869	25	0.00	821
Centurian	R		United States	34	0.00	803	33	0.00	761
Cep Rouge	R		France						
Cerceal Branco	W	Cercial	Portugal	379	0.01	408	261	0.01	440
Cereza	G		Argentina	29934	0.65	34	28887	0.64	29
Cesanese	R	Cesanese Comune; Cesanese d'Affile	Italy	679	0.01	329	446	0.01	370
Cesar	R	Cesar N	France	10	0.00	987	15	0.00	906
Cetinka	W		Croatia	35	0.00	800			
Chambourcin	R		France	1097	0.02	258	968	0.02	256
Champanel	R		United States				1	0.00	1248
Chancellor	R		France	49	0.00	751	38	0.00	736
Chaouch Blanc	W	Cavus	Turkey	3	0.00	1155			
Chardonel	W		United States	144	0.00	563	90	0.00	596
Chardonnay	W	Chardonnay - Pinot Chardonnay; Chardonnay Blanc; Chardonnay Musque and Chardonnay	France	199743	4.33	5	201649	4.50	5
Chardoris	W		Switzerland				0	0.00	1445
Charmont	W		Switzerland	10	0.00	1001	10	0.00	965
Chasan	W		France	749	0.02	309	549	0.01	339

Table 13 (cont.): Prime varieties' colour, synonyms, country of origin, and global area, share and rank, 2010 and 2016 (ha and %)

				2010			2016		
Prime variety	*Col.*	*Synonyms*	*Country of origin*	*Area (hectares)*	*Share (%)*	*Rank*	*Area (hectares)*	*Share (%)*	*Rank*
Chasselas	W	Chassela Blanc, Rose; Chaselas Dorada; Chaslas; Chasselas alba; Chasselas Dorato; Chasselas Doré; Chasselas musque; Chasselas/Gutedel; Chasselas Blanc; Gutedel Weisser; Gutedel, Weißer; Plemenka Bijela	Switzerland	13119	0.28	57	7377	0.16	90
Chasselas (R)	W	Chasselas rose; Chasselas Rouge; Chasselas Roxo	Switzerland	95	0.00	627	90	0.00	598
Chasselas Sabor	W		Portugal	0	0.00	1370	0	0.00	1525
Chatus	R		France	79	0.00	665	71	0.00	637
Chelois	R		France	1	0.00	1200	2	0.00	1233
Chelva	W	Forastera Blanca	Spain	6168	0.13	100	5029	0.11	110
Chenanson	R		France	466	0.01	386	452	0.01	367
Chenel	W		South Africa	79	0.00	664	33	0.00	762
Chenin Blanc	W	Chenin; Chenin Blanc - Chenin	France	35703	0.77	25	32221	0.72	27
Chenivesse	R		France						
Chinuri	W		Georgia	1225	0.03	240	1225	0.03	227
Chkhaveri	G		Georgia	26	0.00	851	26	0.00	813
Cianorie	R	Cjanorie	Italy	2	0.00	1157	1	0.00	1252
Cidreiro	R		Portugal	0	0.00	1342	0	0.00	1456
Cienna	R		Australia				70	0.00	639
Ciliegiolo	R		Italy	1830	0.04	193	897	0.02	266
Cinsaut	R	Bastardillo; Cargadora; Cinsault; Cinsaut Seedles; Ottavianello; Sao Saul; Senso	France	34751	0.75	28	22926	0.51	35
Cinsaut (G)	R	Cinsaut Gris	France				2	0.00	1234
Cinsaut (W)	R	Cinsaut Blanc	France	7	0.00	1037	0	0.00	1439
Cinsaut Seedless	R		Spain	13	0.00	954			
Citronny Magarach	W	Citronny of Magarach	Ukraine	307	0.01	451	307	0.01	415
Cividin	W		Italy	4	0.00	1100	4	0.00	1088
Clairette	W	Clairet; Clairette Blanche;	France	3057	0.07	145	2420	0.05	162
Clara	W	BRS Clara	Brazil				0	0.00	1520
Clarin	W		France	6	0.00	1061	6	0.00	1043
Claverie	W		France	1	0.00	1262	1	0.00	1337
Clinton	R		United States				0	0.00	1421
Coarnă Neagră	R	Coarna Neagra	Moldova				114	0.00	561
Cococciola	W		Italy	983	0.02	273	1671	0.04	194
Coda di Volpe Bianca	W	Guarnaccia	Italy	586	0.01	346	77	0.00	628
Codană	R	Codana	Romania	24	0.00	864	26	0.00	810
Codega de Larinho	W	Codega do Larinho	Portugal	629	0.01	340	455	0.01	366
Codivarta	W		France	2	0.00	1165	1	0.00	1247
Colmar Precoce Noir	R	Prococe de Colmar	France				0	0.00	1423
Colobel	R		France	9	0.00	1005	8	0.00	999
Colombana Nera	R		Italy	38	0.00	782	16	0.00	883
Colombard	W	Colombar; French Colombard	France	32944	0.71	29	29996	0.67	28
Colomino	W		South Africa	5	0.00	1094	2	0.00	1175
Coloraillo	R		Spain	374	0.01	410	109	0.00	571
Columna	W		Romania	24	0.00	868	26	0.00	815
Completer	W		Switzerland	3	0.00	1135	5	0.00	1075
Complexa	R		Portugal	103	0.00	615	103	0.00	577
Concieira	R		Portugal	52	0.00	740	53	0.00	682
Concord	R		United States	12238	0.27	59	10544	0.24	65
Concord Clone 30	R		Brazil				196	0.00	486
Cora	R	BRS Cora	Brazil				570	0.01	333
Coracao de Galo	R		Portugal	1	0.00	1255	1	0.00	1315
Corbina Vicentina	R	Corbina	Italy	12	0.00	963	12	0.00	928
Cordenossa	R		Italy	5	0.00	1075	2	0.00	1208
Cornalin	R	Humagne Rouge	Italy	256	0.01	479	147	0.00	519
Cornarea	R		Italy	13	0.00	951	8	0.00	994
Cornichon Blanc	W	De Cuerno; Decuerno	Italy				1	0.00	1355
Cornifesto	R		Portugal	499	0.01	375	509	0.01	349
Corot Noir	R		United States	27	0.00	846	11	0.00	945
Cortese	W		Italy	2953	0.06	150	2405	0.05	163
Corvina Veronese	R	Corvina	Italy	7496	0.16	89	6240	0.14	97
Corvinone	R		Italy	930	0.02	277	1140	0.03	241
Côt	R	Cot; Cot Rouge; Malbec; Malbek; Malbeck; Malbech	France	38158	0.83	23	52233	1.17	15
Couderc 13	W		France				474	0.01	360
Couderc Noir	R	Couderc	France	3517	0.08	137	2136	0.05	171
Counoise	R		France	408	0.01	398	418	0.01	379
Courbu Blanc	W		France	43	0.00	768	32	0.00	766
Courbu Noir	R		France	1	0.00	1205	1	0.00	1251
Cove	W		Italy	6	0.00	1049	6	0.00	1036
Crâmpoşie Selecţionată	W	Cramposie Selectionata	Romania	409	0.01	397	18	0.00	870

Table 13 (cont.): Prime varieties' colour, synonyms, country of origin, and global area, share and rank, 2010 and 2016 (ha and %)

				2010			2016		
Prime variety	*Col.*	*Synonyms*	*Country of origin*	*Area (hectares)*	*Share (%)*	*Rank*	*Area (hectares)*	*Share (%)*	*Rank*
Crimposie	W	Cramposie; Crâmpoşie	Romania	453	0.01	388	450	0.01	369
Crimson Cabernet	R		United States	1	0.00	1216	1	0.00	1257
Crimson Seedless	R		United States	8	0.00	1029	8	0.00	1003
Criolla Grande	R		Argentina	20745	0.45	40	15596	0.35	48
Criolla Mediana	R		Argentina	3	0.00	1118	7	0.00	1014
Croatina	R		Croatia	5700	0.12	106	2695	0.06	154
Crouchen	W		France	725	0.02	314	319	0.01	409
Crovassa	R		Italy	0	0.00	1323	0	0.00	1420
Cruciulita	W	Alb Rominesc	Romania	0	0.00	1369	0	0.00	1507
Crystal	W	Cristal	China	175	0.00	536	175	0.00	503
Csaba Gyöngye	W	Csabagyongye; Jemciug csaba; Perle of Csaba; Csaba Gyongye; Csabagyöngye	Hungary	89	0.00	636	175	0.00	502
Cserszegi Fűszeres	G	Cserszegi Fuszeres	Hungary	3609	0.08	133	4299	0.10	125
Csillám	W	Csillam	Hungary	20	0.00	895	25	0.00	824
Csókaszőlő	R	Csokaszolo	Hungary	2	0.00	1164	2	0.00	1229
Csomorika	W		Hungary	0	0.00	1293	0	0.00	1364
Dakapo	R		Germany	51	0.00	745	68	0.00	646
Dalkauer	W		Germany						
Damaschino	W	Alicante Branco; Planta Fina; Valenci	Italy	2171	0.05	176	1622	0.04	200
Danam	W		France	0	0.00	1320	0	0.00	1417
Danlas	W		France	255	0.01	480	203	0.00	479
Danuta	W		France	2	0.00	1169	2	0.00	1190
Dattier de St. Vallier	W		France	0	0.00	1296	0	0.00	1375
David Macgregor 8521-1	R	DM 8521-1	United States				2	0.00	1212
Dawn Seedless	W		United States						
De Chaunac	R	De Chaunac and Rosette	France	91	0.00	634	102	0.00	578
De Cilindro	W		Spain				0	0.00	1537
Debina	W	Dembina	Greece	239	0.01	493	14	0.00	914
Debit	W		Croatia	403	0.01	402			
Deckrot	R		Germany	2	0.00	1178	12	0.00	931
Dedo de Dama	W	Almenhaca; Alminhaca	Portugal	1	0.000	1257	1	0.000	1312
Dekabrskii	R	Dekabrsky	Moldova	78	0.00	667	78	0.00	626
Delaware	G		United States	227	0.00	501	421	0.01	377
Delhro	R		France	0	0.00	1332	0	0.00	1447
Deliciosa	R		Portugal	0	0.00	1354	0	0.00	1474
Delisle	W		United States				0	0.00	1482
Devin	W		Slovakia	133	0.00	580			
Diagalves	W		Portugal	1156	0.03	252	1090	0.02	245
Diamond Muscat	W		United States				2	0.00	1170
Dimrit	R		Turkey	863	0.02	291	704	0.02	296
Dimyat	W	Smederevka; Zoumiatiko	Bulgaria	2401	0.05	167	9696	0.22	72
Dindarella	R		Italy	7	0.00	1043	5	0.00	1067
Diolinoir	R		Switzerland	114	0.00	594	122	0.00	548
Dišeća Ranina	W	Diseca Ranina	Croatia	2	0.00	1186			
Divico	R	Divico (IRAC 2091)	Switzerland				10	0.00	964
Divona	W	IRAC 2060	Switzerland				0	0.00	1455
Docal	R		Portugal	0	0.00	1339	0	0.00	1448
Dodrelyabi	R	Gros Colman	Georgia				0	0.00	1386
Doina	R		Moldova	227	0.00	500	227	0.01	460
Dolcetto	R		Italy	6333	0.14	97	4545	0.10	120
Dolciame	W		Italy	11	0.00	986	6	0.00	1040
Domina	R		Germany	407	0.01	399	375	0.01	387
Dominga	W		Spain	1	0.00	1247			
Dona Branca	W		Portugal	276	0.01	464	204	0.00	478
Dona Joaquina	W		Portugal	24	0.00	871	11	0.00	946
Dona Zillá	R		Brazil				1	0.00	1305
Donaris	W		Romania	1	0.00	1252	1	0.00	1284
Donzelinho Branco	W		Portugal	65	0.00	699	64	0.00	654
Donzelinho Roxo	R		Portugal	0	0.00	1352	0	0.00	1465
Donzelinho Tinto	R		Portugal	33	0.00	809	34	0.00	757
Doral	W		Switzerland	27	0.00	844	35	0.00	744
Dorinto	W	Arinto do Interior	Portugal	115	0.00	593	70	0.00	640
Dornfelder	R	Dom Felder	Germany	8182	0.18	82	7871	0.18	86
Dostoinyi	R		Russia	65	0.00	700	65	0.00	651
Douce Noire	R	Corbeau; Serbina; Bonarda; Charbono; Turca	France	19630	0.43	41	19733	0.44	38
Doukkali	R		Morocco	16557	0.36	48			
Doux d'Henry	R		Italy	9	0.00	1003	6	0.00	1035
Droujba	W	Drujba	Bulgaria	3	0.00	1133			
Drupeggio	W	Canaiolo Bianco	Italy	286	0.01	458	81	0.00	619
DU 31120	R	UD-31.120	Italy				0	0.00	1487
Duna Gyöngye	R		Hungary	63	0.00	705	45	0.00	711
Dunaj	R		Slovakia	46	0.00	762			

Table 13 (cont.): Prime varieties' colour, synonyms, country of origin, and global area, share and rank, 2010 and 2016 (ha and %)

				2010			2016		
				Area	Share		Area	Share	
Prime variety	*Col.*	*Synonyms*	*Country of origin*	*(hectares)*	*(%)*	*Rank*	*(hectares)*	*(%)*	*Rank*
Dunav	R		Bulgaria				11	0.00	943
Dunavski Lazur	W		Bulgaria	483	0.01	378	483	0.01	356
Dunkelfelder	R		Germany	356	0.01	420	291	0.01	426
Dunze	R		Switzerland						
Duras	R		France	892	0.02	286	785	0.02	279
Durella	W		Italy	470	0.01	383	480	0.01	358
Durif	R	Petit Syrah; Petite Sirah; Petite Syrah	France	3557	0.08	135	4807	0.11	113
Edelweiss	W		United States	32	0.00	818	16	0.00	888
Ederena	R		France	1	0.00	1259	1	0.00	1310
Egiodola	R		France	349	0.01	424	285	0.01	431
Ehrenbreitsteiner	W		Germany	10	0.00	992	8	0.00	998
Ehrenfelser	W		Germany	113	0.00	597	82	0.00	611
Einset Seedless	G	Einset	United States				0	0.00	1400
Ekigaina	R		France	4	0.00	1116	3	0.00	1111
Ekim Kara	R		Ukraine	27	0.00	845			
Elbling	W	Burger; Elbling Weisser; Elbling, Weißer	Germany	935	0.02	275	972	0.02	255
Elbling (R)	W	Elbling Rot; Elbling, Roter	Germany	9	0.00	1014	10	0.00	963
Elmer Swenson 10- 18-30	W	ES 10-18-30	United States				0	0.00	1481
Elvira	W		United States	263	0.01	476	231	0.01	455
Emerald Riesling	W		United States	508	0.01	369	177	0.00	500
Emerald seedless	W		United States						
Emir	W		Turkey	688	0.01	324	89	0.00	600
Enantio	R	Lambrusco a Foglia Frastagliata	Italy	724	0.02	315	178	0.00	499
Encruzado	W		Portugal	282	0.01	460	132	0.00	533
Enfarine Noir	R		France	0	0.00	1374	0	0.00	1531
Ensanyo Tintas	R		Spain	27	0.00	843			
Ensayo Blancas	W		Spain	1	0.00	1246			
Eona	W		United States				2	0.00	1211
Erbaluce	W		Italy	319	0.01	442	316	0.01	410
Erbamat	W	Verdealbara	Italy	24	0.00	863	2	0.00	1232
Ervi	R		Italy	4	0.00	1098	3	0.00	1109
ES 10-18-14	W		United States				0	0.00	1480
Esganacao Preto	R	Esgana Cao Tinto	Portugal	0	0.00	1378	0	0.00	1535
Esganinho	W		Portugal	0	0.00	1349	0	0.00	1497
Espadeiro	R		Portugal	469	0.01	384	357	0.01	391
Espadeiro Mole	R		Portugal	0	0.00	1307	0	0.00	1385
Espirit	W		United States				2	0.00	1201
Esther	R	Königliche Esther	Hungary				0	0.00	1555
Estreito Macio	W		Portugal	3	0.00	1144	1	0.00	1250
Etraire de l'Adui	R		France	5	0.00	1087	5	0.00	1083
Exalta	W		France	4	0.00	1114	3	0.00	1123
Excelsior	W		France				0	0.00	1413
Eyholzer Rote	R	Roter Eyholzer	Switzerland				0	0.00	1432
Ezerfürtü	W	Ezerfurtu	Hungary	406	0.01	400	295	0.01	423
Ezerjó	W	Ezerjo	Hungary	1074	0.02	263	636	0.01	315
Faberrebe	W	Faber	Germany	554	0.01	355	331	0.01	399
Falanghina	W		Italy	3037	0.07	146	323	0.01	407
Falanghina Flegrea	W		Italy				3634	0.08	133
Favorit	W		Hungary				3	0.00	1139
Fenile	W		Italy	5	0.00	1071	1	0.00	1313
Fepiro	R		Portugal	0	0.00	1384	0	0.00	1550
Fer	R		France	1854	0.04	191	1686	0.04	192
Fernão Pires	W	Fernao Pires; Molinha; Fernao Pires Rosado	Portugal	9609	0.21	73	12211	0.27	60
Ferral	R		Spain	30	0.00	821	31	0.00	773
Fertilia	R		Italy	3	0.00	1122	2	0.00	1180
Fetească Albă	W	Dievcie Hrozno; Feteasca Alba;	Moldova	17469	0.38	45	13382	0.30	54
Fetească Neagră	R	Feteasca Neagra; Feteasca neagră; Fekete leányka	Moldova	1719	0.04	198	3248	0.07	139
Fetească Regală	W	Feteasca Regala; Feteasca regală	Romania	13136	0.28	56	12991	0.29	55
Feunate	R		France						
Fiano	W		Italy	1377	0.03	224	2187	0.05	168
Fiesta	W		United States	230	0.00	497	230	0.01	456
Fino de Ribera del Fresno	W		Spain	45	0.00	765	8	0.00	1002
Fintendo	R		Spain	118	0.00	589	185	0.00	493
Fioletovy Ranny	R	Fioletovy Ranny	Russia	50	0.00	748	50	0.00	692
Flame Seedless	R		United States	55	0.00	724	55	0.00	668
Flavis	W		Italy	3	0.00	1128	3	0.00	1153
Fleurtai	W	Fleurtai (UD-34.111)	Italy				0	0.00	1496
Flora	G		United States	8	0.00	1028	12	0.00	935
Florental	R	Bourdin (S)	France	26	0.00	852	11	0.00	940

Table 13 (cont.): Prime varieties' colour, synonyms, country of origin, and global area, share and rank, 2010 and 2016 (ha and %)

				2010			2016		
				Area	Share		Area	Share	
Prime variety	*Col.*	*Synonyms*	*Country of origin*	*(hectares)*	*(%)*	*Rank*	*(hectares)*	*(%)*	*Rank*
Floricica	W		Moldova				14	0.00	918
Fogarina	R		Italy	5	0.00	1076	3	0.00	1135
Foglia Tonda	R		Italy	101	0.00	618	68	0.00	645
Fogoneu	R		Spain	35	0.00	797	15	0.00	907
Fokiano	R		Greece	262	0.01	477	212	0.00	472
Fokiano (W)	R	Fokiano (White)	Greece						
Folgasao	W		Portugal	182	0.00	531	162	0.00	514
Folgasao Roxo	R		Portugal	18	0.00	902	18	0.00	873
Folha de Figueira	W		Portugal	3	0.00	1120	2	0.00	1218
Folignan	W		France	51	0.00	744	51	0.00	687
Folle Blanche	W		France	1803	0.04	194	1574	0.04	204
Fontan	R		France						
Fontanara	W		Germany				1	0.00	1290
Fonte Cal	W		Portugal	111	0.00	600	52	0.00	685
Forastera	W		Italy	208	0.00	512	8	0.00	988
Forcallat Tinta	R		Spain	1163	0.03	251	535	0.01	342
Forgiarin	R		Italy	4	0.00	1104	3	0.00	1144
Forsellina	R		Italy	7	0.00	1035	7	0.00	1024
Fortana	R	Canina Nera	Italy	642	0.01	336	469	0.01	362
Francavidda	W		Italy	13	0.00	958	2	0.00	1226
Frâncuşă	W	Francusa	Romania	621	0.01	342	365	0.01	390
Frappato	R	Frappato Di Vittoria	Italy	752	0.02	308	580	0.01	330
Fredonia	R		United States	37	0.00	789	28	0.00	791
Freisa	R		Italy	1054	0.02	268	519	0.01	346
Freisamer	W	Freiburger	Germany	8	0.00	1020	6	0.00	1039
Frontenac	R	Frontenac Noir; Frontenac Rouge	United States	135	0.00	574	212	0.00	471
Frontenac (G)	R	Frontenac Gris; Frontenac Grus	United States	59	0.00	718	92	0.00	591
Frontenac (W)	R	Frontenac Blanc	United States				26	0.00	808
Frühroter Veltliner	R	Fruhroter Veltliner; Korai Piros Veltelini; Malvasier; Malvasier, Früher Roter; Veltlinske Cervene Rane; Veltlinske Cervene Skore; Korai Piros Veltelíni	Austria	856	0.02	296	388	0.01	385
Frumoasa Alba	W		Moldova				8	0.00	990
Fubiano	W		Italy	9	0.00	1007	2	0.00	1165
Fuella Nera	R		France	20	0.00	892	20	0.00	858
Fumin	R		Italy	31	0.00	819	25	0.00	820
Furmint	W	Malvasia Verde; Moslavac; Sipelj; Sipon; Šipon	Hungary	5276	0.11	107	4435	0.10	122
Gaglioppo	R		Italy	4214	0.09	121	4626	0.10	119
Gaillard	R		France	0	0.00	1360	0	0.00	1530
Galbenă de Odobeşti	W	Galbena De Odobesti; Galbenade Odobesti	Romania	385	0.01	405	417	0.01	380
Galego Dourado	W	Galego; Pedro Luis; Dourado	Portugal	16	0.00	925	7	0.00	1010
Galotta	R	Gallota	Switzerland	13	0.00	952	35	0.00	752
Gamaret	R		Switzerland	405	0.01	401	441	0.01	371
Gamay Noir	R	Gamay; Gamay Beaujolais; Gamay cl 565; Gamay St Romain; Plant Robert, Gamay and Zweigeltrebe	France	31927	0.69	32	26221	0.58	32
Gamay Teinturier de Bouze	R	Gamay De Teinturier Bouze; Gammay de Bouze	France	278	0.01	462	255	0.01	444
Gamay Teinturier de Chaudenay	R	Gamay de Chaudenay	France	157	0.00	551	142	0.00	526
Gamay Teinturier Freaux	R	Gammay Freaux	France	104	0.00	612	79	0.00	624
Gamba Rossa	R		Italy	1	0.00	1277	0	0.00	1363
Gänsfüsser	R	Gansfusser	Germany	0	0.00	1372	0	0.00	1545
Ganson	R		France	3	0.00	1125	3	0.00	1124
Garandmak	W		Armenia	931	0.02	276			
Garanoir	R	Granoir	Switzerland	216	0.00	504	229	0.01	457
Garganega	W	Grecanico Dorato	Italy	15397	0.33	50	8554	0.19	79
Gargiulo 14260	R	C.G. 14260 (Inta)	Argentina	1	0.00	1245			
Gargiulo 2539	R	C.G.2539 (Inta)	Argentina	49	0.00	755	29	0.00	785
Gargiulo 26189	R	C.G. 26189 (Inta)	Argentina						
Gargiulo 26879	W	C G 26879 (Inta)	Argentina	3	0.00	1154			
Gargiulo 4113	R	C.G.4113 (Inta)	Argentina	6	0.00	1056	3	0.00	1154
Gargiulo 45803	W	C G 45803 (Inta)	Argentina	8	0.00	1021	10	0.00	957
Garnacha Blanca	W	Garnachablanca; Grenache Blanc	Spain	7398	0.16	90	7409	0.17	88
Garnacha Peluda	R	Lledoner Pelut	Spain	1206	0.03	243	898	0.02	265
Garnacha Roja (Gris)	G	Garnacha Roja; Garnacha Rose; Grenache Gris; Grenache Rose	Spain	2366	0.05	171	1462	0.03	210

Table 13 (cont.): Prime varieties' colour, synonyms, country of origin, and global area, share and rank, 2010 and 2016 (ha and %)

				2010			2016		
Prime variety	*Col.*	*Synonyms*	*Country of origin*	*Area (hectares)*	*Share (%)*	*Rank*	*Area (hectares)*	*Share (%)*	*Rank*
Garnacha Tinta	R	Cannonau; Garnacha; Garnacha Tinta; Grenache Noir; Grenache; Tinto Basto; Tocai Rosso; Vernaccia Nera; Vernaccia Nera Grossa; Vernaccina Nero	Spain	181553	3.93	7	150096	3.35	7
Garonnet	R		France	14	0.00	943	10	0.00	971
Garrido Fino	W		Spain	59	0.00	717	54	0.00	679
Gascon	R		France	1	0.00	1250	1	0.00	1298
Gateta	R		Spain	2	0.00	1177			
Gegić	W	Gegic	Croatia	11	0.00	977			
Geilweilerhof Ga- 48- 12	W	GF 48-12	Germany				1	0.00	1280
Geisenheim 318-57	W	Geisenheim 813-57; Geisenheim	Germany	106	0.00	609	14	0.00	917
Generosa	W		Portugal	107	0.00	607	328	0.01	401
Gesztus	W		Hungary	0	0.00	1318	0	0.00	1411
Gewürztraminer	W	Gewurztraminer; Tramin Cerveny; Traminac Crveni; Traminer; Traminer Aromatico; Traminer Rose; Traminer rosse; Traminer Rot; Traminer Roz; Traminer, Roter; Tramini	Germany	14355	0.31	53	12823	0.29	56
Gibi	W		Spain	1074	0.02	264	785	0.02	280
Ginestra	W		Italy	4	0.00	1110	1	0.00	1356
Giro Nero	R	Giro	Italy	200	0.00	518	144	0.00	523
GM 322	R		Canada	17	0.00	917			
Göcseji Zamatos	W	Gecsei Zamatos; Gocseji Zamatos	Hungary	55	0.00	723	50	0.00	695
Godello	W	Agudello; Godelho; Gouveio; Gouveio Roxo	Spain	1332	0.03	229	1406	0.03	216
Goethe	R		United States				20	0.00	856
Goldburger	W		Austria	140	0.00	568	98	0.00	583
Golden Muscat	W		United States	1191	0.03	246	50	0.00	691
Goldriesling	W		France	21	0.00	888	24	0.00	830
Goldtraminer	W		Italy	9	0.00	1010	5	0.00	1074
Golia	W		Romania	0	0.00	1359	0	0.00	1483
Golubok	R	Goluboc	Ukraine	87	0.00	639	37	0.00	740
Goncalo Pires	R		Portugal	1	0.00	1269	1	0.00	1328
Gorgollasa	R		Spain				5	0.00	1061
Goruli Mtsvane	W		Georgia	287	0.01	456	287	0.01	428
Gosen	R		Italy	1	0.00	1210	0	0.00	1490
Gouais Blanc	W	Gouais/Gwäss	France				1	0.00	1267
Gouget Noir	R		France	10	0.00	996	3	0.00	1113
Goustolidi	W		Greece	68	0.00	691	19	0.00	863
Gouveio Preto	R		Portugal	0	0.00	1313	0	0.00	1390
Gouveio Real	W	Gouveio Estimado	Portugal	582	0.01	348	581	0.01	329
GR 7	R		United States	32	0.00	816	13	0.00	921
Grachen	W		South Africa	2	0.00	1162	0	0.00	1522
Graciano	R	Bovale; Cagnulari; Graciana; Tinta Fontes; Tinta Miuda; Tintilla de Rota	Spain	3123	0.07	143	2910	0.06	143
Graisse	W		France	14	0.00	947	1	0.00	1265
Gramon	R		France	3	0.00	1142	3	0.00	1146
Grand Manchen	W		Spain	8	0.00	1022			
Grand Noir	R	Gran Negro; Grand Noir de la Calmette	France	955	0.02	274	707	0.02	294
Grangeal	R		Portugal	1	0.00	1223	1	0.00	1264
Granho	W		Portugal	0	0.00	1321	0	0.00	1419
Grapariol	W		Italy	2	0.00	1166	2	0.00	1166
Grasă de Cotnari	W	Grasa de Cotnari; Grasade Cotnari; Koverszolo; Kövérszőlő	Romania	685	0.01	325	632	0.01	317
Graševina	W	Borba; Grasevina; Grasvena; Welschriesling; Italian Riesling; Laki Rizling; Laški Rizling; Riesling Italian; Riesling Italico; Rizling Vlasky; Ryzlink Vlaky; Olasz Rizling	Croatia	61200	1.33	16	24384	0.54	33
Grassen	R		France	0	0.00	1364	0	0.00	1499
Grechetto di Orvieto	W	Grechetto Bianco; Grechetto	Italy	1501	0.03	210	1824	0.04	182
Grechetto Rosso	R		Italy	49	0.00	754	35	0.00	749
Greco	W	Asprinio; Greco di Tufo	Italy	158	0.00	550	21	0.00	848
Greco Bianco	W		Italy	1604	0.03	204	2050	0.05	173
Greco Nero	R		Italy	1256	0.03	236	437	0.01	373
Grignolino	R		Italy	915	0.02	281	911	0.02	264
Grillo	W		Italy	6295	0.14	98	7383	0.16	89
Gringet	W		France	25	0.00	854	15	0.00	895
Grolleau Noir	R	Grolleau Gris; Grolleau	France	2759	0.06	155	1949	0.04	176
Groppello di Mocasina	R	Groppello di Santo Stefano	Italy	81	0.00	659	24	0.00	825
Groppello di Revo	R		Italy	12	0.00	965	12	0.00	934
Groppello Gentile	R		Italy	326	0.01	438	78	0.00	625

Table 13 (cont.): Prime varieties' colour, synonyms, country of origin, and global area, share and rank, 2010 and 2016 (ha and %)

				2010			2016		
Prime variety	*Col.*	*Synonyms*	*Country of origin*	*Area (hectares)*	*Share (%)*	*Rank*	*Area (hectares)*	*Share (%)*	*Rank*
Gros Manseng	W	Manseng Gros Blanc; Gross Manseng	France	2960	0.06	149	3069	0.07	141
Grossa	R		Portugal	73	0.00	675	54	0.00	676
Grüner Veltliner	W	Falkensteiner; Veltliner Grun; Gruner Veltliner; Veltliner, Grüner; Veltlinske Zelene; Zöld Veltelini	Austria	18834	0.41	42	19118	0.43	40
Gualarido	W		Spain				18	0.00	868
Guardavalle	W		Italy	33	0.00	807	16	0.00	889
Guillemot	W		France						
Gutenborner	W		Germany	2	0.00	1176			
Guzun	W		Moldova				3	0.00	1150
Gyöngyrizling	W	Gyongyrizling	Hungary	23	0.00	873	16	0.00	894
Haiduc	R		Romania	3	0.00	1141	3	0.00	1127
Hajnalka	W	Hajnal	Hungary	0	0.00	1371	0	0.00	1518
Hamvas	W	Barátcsuha	Hungary				0	0.00	1427
Hárslevelű	W	Feuille de Tilleul; Harslevelu; Lipovina	Hungary	1856	0.04	190	1618	0.04	201
Hasansky Sladky	R	Baltica	Ukraine				1	0.00	1255
Hegel	R		Germany	9	0.00	1013	7	0.00	1012
Helfensteiner	R		Germany	19	0.00	900	14	0.00	915
Helios	W		Germany				6	0.00	1029
Herbemont	R		United States	764	0.02	307	112	0.00	564
Heroldrebe	R		Germany	134	0.00	578	112	0.00	565
Heuréka	W		Hungary				0	0.00	1446
Hibernal	W		Germany				20	0.00	851
Himbertscha	W		Switzerland				0	0.00	1426
Himrod	W		United States	0	0.00	1291	1	0.00	1357
Hölder	W	Holder	Germany	5	0.00	1083	2	0.00	1186
Hondarribi Beltza	R	Beltza; Hondarrabi Beltza; Ondarrabi Beltza	Spain	53	0.00	734	15	0.00	897
Hondarribi Zuri	W		France				624	0.01	319
Hrvatica	R	Negrara	Croatia	116	0.00	591	53	0.00	681
Humagne	W	Humagne Blanc	Switzerland	30	0.00	823	29	0.00	787
Huxelrebe	W		Germany	630	0.01	338	466	0.01	365
Ignea	G		Italy	1	0.00	1265	1	0.00	1322
Ilichevskii Rannii	R	Iliciovski Ciornai Rannii	Ukraine	5	0.00	1082			
Imperial Napoleon	R	Don Mariano	Spain	12	0.00	970	0	0.00	1523
Imperial Seedless	G		Peru	1	0.00	1244	1	0.00	1295
Impigno	W		Italy	7	0.00	1036	2	0.00	1210
Incrocio Bianco Fedit 51	W		Italy	5	0.00	1090	1	0.00	1311
Incrocio Bruni 54	W		Italy	12	0.00	964	3	0.00	1107
Incrocio Manzoni 2.15	R		Italy	86	0.00	646	72	0.00	636
Incrocio Terzi 1	R		Italy	44	0.00	767	12	0.00	930
Invernenga	W		Italy	7	0.00	1048	5	0.00	1072
Inzolia	W	Ansonica	Italy	6133	0.13	101	4740	0.11	115
Iordan	W	Iordană	Romania	315	0.01	444	311	0.01	412
IRAC 1933	R		Switzerland				0	0.00	1524
Irsai Olivér	W	Irsai Oliver; Irşai Oliver; Irsay Olivier; Muscat Oliver	Hungary	1414	0.03	219	1790	0.04	185
Isa	W		France	9	0.00	1002	5	0.00	1081
Isabella	R	Frutilla; Isabel; Isabel Precoce; Lidia	United States	32494	0.70	31	17813	0.40	45
Italia	W	Italia (Pirovano 65) (PE); Italian, Portugese whites	Italy	1463	0.03	214	5188	0.12	109
Italica	W		Italy	367	0.01	414	47	0.00	705
Ives	R		United States	16	0.00	924			
Jacquere	W		France	1014	0.02	270	621	0.01	320
Jacquez	R	Black Spanish; Lenoir	United States	2368	0.05	170	1443	0.03	213
Jádorvány	W		Hungary				0	0.00	1399
Jampal	W	Pinheira Branca	Portugal	71	0.00	681	34	0.00	756
Jaoumet	W		France	0	0.00	1353	0	0.00	1469
Jardovany Fekete	R	Fekete Jádorvány	Hungary				0	0.00	1502
Jázmin	W	Jázmin (8/1)	Hungary				2	0.00	1195
Jeroma	R		Spain				0	0.00	1526
Joannes Seyve	R		France	1	0.00	1253	1	0.00	1306
Johanniter	W		Germany	86	0.00	645	111	0.00	566
Joubertin	R		France	1	0.00	1203			
Juan García	R	Gorda; Juan Garcia; Mouraton; Tinta Gorda	Spain	1707	0.04	200	1545	0.03	206
Jubiläumsrebe	G	Jubilaumsrebe; Jubileumsrebe	Austria	14	0.00	948	7	0.00	1023
Jubileum 75	R		Hungary	194	0.00	523	91	0.00	592
Juhfark	W		Hungary	186	0.00	528	195	0.00	487
Juliana	W		Brazil	0	0.00	1325			
Jurançon Blanc	W	Jurancon Blanc	France	7	0.00	1038	2	0.00	1192
Jurançon Noir	R	Jurancon Noir	France	663	0.01	331	605	0.01	324
Jurie	W		France						

Table 13 (cont.): Prime varieties' colour, synonyms, country of origin, and global area, share and rank, 2010 and 2016 (ha and %)

Prime variety	Col.	Synonyms	Country of origin	2010 Area (hectares)	2010 Share (%)	2010 Rank	2016 Area (hectares)	2016 Share (%)	2016 Rank
Juwel	W		Germany	22	0.00	882	15	0.00	900
K.35	W		Hungary	0	0.00	1335			
Kabar	W		Hungary	18	0.00	907	30	0.00	779
Kadarka	R	Cadarcă; Gamza	Hungary	1181	0.03	249	1625	0.04	199
Kakotrygis	W		Greece	103	0.00	613	28	0.00	792
Kalecik Karası	R	Kalecik Karasi	Turkey	861	0.02	295	704	0.02	295
Kalina	R		Switzerland				0	0.00	1379
Kangun	W		Armenia	850	0.02	298			
Kanzler	W		Germany	33	0.00	811	28	0.00	793
Karalahna	R		Turkey	4	0.00	1109	4	0.00	1095
Karasakiz	R	Kuntra	Turkey	4	0.00	1108	4	0.00	1094
Karát	W	Karat	Hungary	50	0.00	747	44	0.00	716
Kármin	R	Karmin	Hungary	36	0.00	792	24	0.00	834
Kat.E.Lin	R	Katelin	Canada				0	0.00	1437
Kay Gray	W	Key Gray	United States	1	0.00	1215	2	0.00	1202
Kéknyelű	W	Keknyelu	Hungary	43	0.00	771	50	0.00	697
Kentville White 94-1	W	KW 94-1	Canada				0	0.00	1412
Kentville White 94-2	W	KW 94-2	Canada				1	0.00	1339
Kerner	W	Kerner Bijeli; Kernling; Riesling Trollinger	Germany	4093	0.09	125	2891	0.06	145
Khikhvi	W		Georgia	6	0.00	1051	6	0.00	1026
Királyleányka	W	Kiralyleanyka	Hungary	855	0.02	297	784	0.02	281
Kishmish Luchistyi	G	Kişmiş lucistăi	Moldova				55	0.00	669
Kishmish Moldavskii	R	Kişmiş moldovenesc	Moldova				28	0.00	796
Kisi	W		Georgia	26	0.00	850	26	0.00	812
Knipperlé	W	Knipperle	France	0	0.00	1350	0	0.00	1458
Kocsis Irma	W	Koksis Irma	Hungary	11	0.00	982	2	0.00	1169
Kodrinskii	R	Codrinski	Moldova	5	0.00	1081	229	0.01	459
Kodryanka	R	Black Magic; BlackMagic; Codreanca	Moldova				1143	0.03	240
Kokur Bely	W	Cokur White	Ukraine	918	0.02	280			
Kolor	R		Germany	2	0.00	1175	7	0.00	1022
Königin der Weingärten	W	Koenigin Der Weingaerten; Regina dei Vigneti; Regina viilor	Hungary	61	0.00	710	70	0.00	638
Korinthiaki	R	Corinto; Corinto Nero	Greece	54	0.00	730	106	0.00	573
Korona	W		Hungary	1	0.00	1282	1	0.00	1299
Koshu	G		Japan	168	0.00	541	690	0.02	299
Kosmopolita	G		Serbia				1	0.00	1331
Kotsifali	R		Greece	2330	0.05	172	1338	0.03	219
Kövidinka	G	Koevidinka; Kovidinka; Ruzica Crvena; Ružicacrvena; Vörös Dinka	Hungary	1076	0.02	262	658	0.01	309
Kozma CS. 2	R	Cs.2	Hungary	0	0.00	1362			
Kozmopoliten	W		Hungary				0	0.00	1442
Krakhuna	W	Krakhuna Tetri	Georgia	46	0.00	760	46	0.00	707
Kraljevina	W		Croatia	447	0.01	390	199	0.00	481
Krasnostop Zolotovsky	R	Krasnostop Anapsky	Russia	562	0.01	353	562	0.01	337
Krassato	R		Greece	52	0.00	742	5	0.00	1084
Kreaca	W	Banati Rizling; Bánáti Rizling	Serbia	29	0.00	829	30	0.00	777
Kuban	R	Kubani	Russia				32	0.00	769
Kujundžuša	W	Kujundzusa;	Croatia	206	0.00	513			
Kuldzhinskii	G	Kuldzhinskiy	Kazakhstan	385	0.01	404	385	0.01	386
Kunbarát	W	Kunbarat	Hungary	9	0.00	1019	0	0.00	1376
Kunleány	W	Kunleany	Hungary	1211	0.03	242	974	0.02	254
Kurucvér	R	Kurucver	Hungary	0	0.00	1341	0	0.00	1475
KW 96-2	W		Canada				0	0.00	1436
Kyoho (4N)	R	Kyoho; Kyoko	Japan	4003	0.09	127	2762	0.06	149
L´Acadie Blanc	W	Acadie blanc; L´Acadie	Canada				65	0.00	652
La Crescent	W		United States	77	0.00	670	94	0.00	585
La Crosse	W		United States	25	0.00	855	26	0.00	811
Labrusco	R		Portugal	81	0.00	658	79	0.00	623
Lacoste	R		France						
Lacrima Christi	R		Italy	85	0.00	647	226	0.01	462
Lacrima di Morro d'Alba	R	Lacrima	Italy	421	0.01	395	252	0.01	445
Lado	W		Spain	1	0.00	1243	2	0.00	1198
Lafnetscha	W		Switzerland				2	0.00	1231
Lagarino Bianco	W		Italy	23	0.00	880	6	0.00	1050
Lagrein	R		Italy	718	0.02	317	251	0.01	447
Lairen	W	Malvar	Spain	214	0.00	505	351	0.01	395
Lakhegyi Mézes	W	Lakhegyi Mezes	Hungary	306	0.01	452	145	0.00	521
Lambrusca di Alessandria	R		Italy	137	0.00	571	58	0.00	664
Lambrusco	R		Italy	45	0.00	763	54	0.00	671
Lambrusco Barghi	R		Italy	18	0.00	903	3	0.00	1151
Lambrusco di Sorbara	R		Italy	1606	0.03	203	858	0.02	273
Lambrusco Grasparossa	R		Italy	2734	0.06	157	954	0.02	258

Table 13 (cont.): Prime varieties' colour, synonyms, country of origin, and global area, share and rank, 2010 and 2016 (ha and %)

				2010			2016		
				Area	Share		Area	Share	
Prime variety	*Col.*	*Synonyms*	*Country of origin*	*(hectares)*	*(%)*	*Rank*	*(hectares)*	*(%)*	*Rank*
Lambrusco Maestri	R		Italy	2312	0.05	173	5657	0.13	101
Lambrusco Marani	R		Italy	1394	0.03	222	1074	0.02	249
Lambrusco Montericco	R		Italy	70	0.00	684	25	0.00	818
Lambrusco Oliva	R		Italy	112	0.00	598	104	0.00	575
Lambrusco Salamino	R		Italy	5003	0.11	109	6228	0.14	98
Lambrusco Viadanese	R		Italy	240	0.01	491	59	0.00	663
Lameiro	W		Portugal	0	0.00	1303	0	0.00	1410
Landal	R		France	43	0.00	770	37	0.00	738
Landot Noir	R	Landot	France	3	0.00	1126	3	0.00	1125
Lariao	W		Portugal	4	0.00	1112	5	0.00	1077
Lasina	R		Croatia	14	0.00	942			
Laska	R		Austria				0	0.00	1510
Laurot	R		Czechia	6	0.00	1059			
Lauzet	W		France	3	0.00	1150	3	0.00	1157
Leányka	W	Leanyka	Hungary	838	0.02	301	719	0.02	292
Lecinaro	R		Italy	1	0.00	1272	0	0.00	1367
Leira	W		Portugal	1	0.00	1279	0	0.00	1414
Len de l'El	W		France	629	0.01	339	603	0.01	325
Léon Millot	R	Leon Millet; Leon Millot	France	102	0.00	617	85	0.00	606
Leopoldo III	R		Belgium						
Levokumskij	R		Russia	890	0.02	288	890	0.02	267
Liatiko	R		Greece	1211	0.03	241	2633	0.06	156
Liatiko (W)	R	Liatiko(White)	Greece						
Liliorila	W		France	4	0.00	1117	1	0.00	1254
Lilla	W		Hungary				0	0.00	1556
Limnio	R		Greece	372	0.01	411	176	0.00	501
Limnio (W)	R	Limnio(white)	Greece						
Listain de Huelva	W	Manteudo	Spain	350	0.01	423	466	0.01	364
Listan Negro	R	Almuneco; Negra Commun	Spain	2666	0.06	160	2847	0.06	147
Listán Prieto	R	California; Criolla Chica; Mission; Moscatel Negra; Moscatel Negro; Pais; Pais - Mission, Criolla	Spain	4985	0.11	110	10267	0.23	68
Listrao Roxo	R		Portugal						
Lival	R		France	101	0.00	619	99	0.00	582
Lomanto	R		United States				3	0.00	1115
Longyan	R	Long Yan	China				1000	0.02	253
Lora	W		Ukraine				35	0.00	748
Lorena	W	BRS Lorena	Brazil	519	0.01	364	500	0.01	353
Louise Swenson	W		United States	3	0.00	1136	5	0.00	1060
Loureiro	W	Loureiro Blanca; Loureiro Blanco	Portugal	4054	0.09	126	4696	0.10	116
Lourela	R		Portugal	0	0.00	1326	0	0.00	1428
Lucie Kuhlmann	R		France				21	0.00	847
Luisa Blanca	W		Spain				80	0.00	622
Lumassina	W		Italy	98	0.00	624	37	0.00	742
Lusitano	R		Portugal	0	0.00	1351	0	0.00	1467
Luzidio	W		Portugal	0	0.00	1319	0	0.00	1466
Lyana	W	Leana	Moldova				41	0.00	720
Macabeo	W	Macabeu; Maccabeo; Viura	Spain	40864	0.89	20	38625	0.86	20
Maceratino	W		Italy	177	0.00	534	39	0.00	728
Madeleine × Angevine 7672	W	Madeleine Angevine	Germany	52	0.00	743	48	0.00	702
Madeleine Royale	W	Branco Especial	France	3	0	1123	3	0	1134
Madeleine Sylvaner	W		Portugal	7	0.00	1033	6	0.00	1051
Madeleines	W		France	7	0.00	1042	7	0.00	1021
Madrasa	R	Matrasa	Azerbaijan	28	0.00	838	28	0.00	794
Magaracha Rannii	R	Rannii Magaracea	Ukraine				884	0.02	269
Magliasina	R	Magliasino	Italy				0	0.00	1529
Magliocco Canino	R		Italy	539	0.01	358	679	0.02	305
Magliocco Dolce	R	Marsigliana Nera	Italy	87	0.00	642	51	0.00	686
Magna	R	BRS Magna	Brazil				30	0.00	776
Magyarfrankos	R	Magyar Frankos	Hungary	0	0.00	1300	0	0.00	1409
Maiolica	R		Italy	26	0.00	848	13	0.00	923
Maiolina	R		Italy	1	0.00	1199	0	0.00	1381
Maiskii Chernyi	R	Mayskiy	Moldova	110	0.00	603	110	0.00	569
Malaga Blanc	W		Thailand	16	0.00	920	54	0.00	675
Malagousia	W	Malagouzia	Greece	182	0.00	530	126	0.00	539
Malbo Gentile	R		Italy	211	0.00	508	219	0.00	466
Malegue	W		France						
Malingre Precoce	W	Precoce De Malingre	France	0	0.00	1292	0	0.00	1369
Maliverne	W		France						
Malvarisco	R		Portugal	3	0.00	1127	3	0.00	1119
Malvasia	W		Greece	45	0.00	764	2184	0.05	169
Malvasia Bianca di Basilicata	W		Italy	210	0.00	509	32	0.00	771

Table 13 (cont.): Prime varieties' colour, synonyms, country of origin, and global area, share and rank, 2010 and 2016 (ha and %)

				2010			2016		
Prime variety	*Col.*	*Synonyms*	*Country of origin*	*Area (hectares)*	*Share (%)*	*Rank*	*Area (hectares)*	*Share (%)*	*Rank*
Malvasia Bianca di Candia	W	Malvasia Amarela; Malvasia Bianca; Malvasia di Candia	Italy	9351	0.20	74	9685	0.22	73
Malvasia Bianca Lunga	W	Malvasia Chianti; Marastina; Maraština	Italy	2544	0.06	164	1247	0.03	226
Malvasia Branca de Sao Jorge	W		Portugal	110	0.00	601	110	0.00	568
Malvasia del Lazio	W		Italy	590	0.01	345	680	0.02	304
Malvasia di Candia Aromatica	W	Malvasia Aromatica; Malvazija Aromatična	Italy	927	0.02	278	1208	0.03	230
Malvasia di Casorzo	R		Italy	107	0.00	605	99	0.00	581
Malvasia di Lipari	W	Malvasia Candida; Malvasia di Sardegna; Malvasia Dubrovacka Bijela; Malvasia Sitges; MalvasiaDubrovačka	Italy	310	0.01	448	113	0.00	563
Malvasia di Sardegna Rosada	G	Malvasia Candida Roxa	Spain	3	0.00	1140	3	0.00	1137
Malvasia di Schierano	R		Italy	89	0.00	637	82	0.00	614
Malvasia Fina	W	Ratinho	Portugal	3501	0.08	138	3282	0.07	135
Malvasia Fina Roxa	G		Portugal	25	0.00	856	24	0.00	828
Malvasia Moscata	W	Malvasia Bianca di Piemonte	Italy	554	0.01	354			
Malvasia Nera di Basilicata	R		Italy	114	0.00	595	39	0.00	724
Malvasia Nera di Brindisi	R	Malvasia Negra; Malvasia Nera; Malvasia Nera di Lecce	Italy	1314	0.03	231	1264	0.03	223
Malvasia Nera Lunga	R		Italy	38	0.00	781	14	0.00	916
Malvasia Parda	W	Farinheira	Portugal	4	0	1097	4	0	1092
Malvasia Preta	R	Moreto; Pinheira Roxa	Portugal	1903	0.04	189	1933	0.04	179
Malvasia Romana	W		Portugal	0	0.00	1338	0	0.00	1464
Malvazija Istarska	W	Malvasia Istriana; Malvazija	Croatia	2740	0.06	156	2788	0.06	148
Malvia	W		Portugal	0	0.00	1328	0	0.00	1431
Mamaia	R		Romania	0	0.00	1311	0	0.00	1384
Mammolo	R	Sciaccarello	Italy	841	0.02	299	911	0.02	263
Mandilaria	R		Greece	885	0.02	289	932	0.02	260
Mandon	R		Spain				1	0.00	1343
Manseng Noir	R	Ferrol; Ferron	France	32	0.00	817	29	0.00	788
Manteudo Preto	R		Portugal	16	0.00	919	11	0.00	950
Manto Negro	R		Spain	273	0.01	466	311	0.01	413
Manzoni Bianco	W		Italy	382	0.01	406	339	0.01	398
Manzoni Moscato	R		Italy	20	0.00	894	19	0.00	860
Manzoni Rosa	G		Italy	29	0.00	830	23	0.00	842
Mara	R	Mara (RAC 3022, C41)	Switzerland				10	0.00	966
Maratheftiko	R		Cyprus	152	0.00	555			
Maréchal Foch	R	Foch; Foch (Marechal); Marechal Foch; Marechel Foch	France	356	0.01	419	229	0.01	458
Marfal	W	Huerta del Rey	Spain	2	0.00	1174	1	0.00	1266
Margot	R	BRS Margot	Brazil				2	0.00	1217
Mariensteiner	W		Germany				2	0.00	1185
Marmajuelo	W	Bermejuela; Vermejuela	Spain	24	0.00	866	20	0.00	854
Marquette	R	Marquette	United States	88	0.00	638	166	0.00	507
Marquinhas	W		Portugal	11	0.00	981	5	0.00	1066
Marquis	W		United States				1	0.00	1261
Mars	R		United States	2	0.00	1180	2	0.00	1215
Marsanne	W	Marsanne Blanche; Marsanne Blanche/Ermitage	France	1763	0.04	196	1838	0.04	181
Marselan	R	Marcelan	France	2731	0.06	158	3941	0.09	129
Marufo	R	Moravia Dulce; Mourisco; Mourisco Roxo	Portugal	6579	0.14	93	4683	0.10	117
Marzemina Bianca	W		Italy	54	0.00	731	55	0.00	670
Marzemino	R	Marzemina; Marzemina Grossa	Italy	1091	0.02	260	785	0.02	278
Maticha	W		Morocco	311	0.01	447	257	0.01	441
Mátrai Muskotály	W	Matrai Muskotaly	Hungary	67	0.00	695	49	0.00	698
Maturana Blanca	W	Maturano Bianco	Spain	18	0.00	910	13	0.00	922
Mauzac Blanc	W	Mauzac	France	1933	0.04	187	1526	0.03	207
Mauzac Noir	R		France						
Mavro	R		Cyprus	3575	0.08	134	3187	0.07	140
Mavro Messenikola	R		Greece						
Mavrodafni	R		Greece	345	0.01	430	324	0.01	406
Mavrouda	R	Mavroudi	Greece	520	0.01	362	1658	0.04	196
Mavrud	R		Bulgaria	1296	0.03	233	1193	0.03	234
Mayolet	R		Italy	7	0.00	1032	6	0.00	1044
Mazuelo	R	Bovale Grande; Carignan; Carignan - Carignane, Cariñena; Carignane; Carignan Noir Mazuela; Carignano; Mazuela; Carignan Noir; Karinian	Spain	75716	1.64	13	47312	1.06	19

Table 13 (cont.): Prime varieties' colour, synonyms, country of origin, and global area, share and rank, 2010 and 2016 (ha and %)

Prime variety	*Col.*	*Synonyms*	*Country of origin*	2010 *Area (hectares)*	2010 *Share (%)*	2010 *Rank*	2016 *Area (hectares)*	2016 *Share (%)*	2016 *Rank*
Mazuelo (G)	R	Carignan Gris	Spain						
Mazuelo (W)	R	Carignan Blanc; Carinena Blanco	Spain	3016	0.07	147	355	0.01	393
Mazzese	R		Italy	76	0.00	671	57	0.00	666
Mechta	R	Mecita	Ukraine				19	0.00	862
Mecle de Bourgoin	R		France						
Medina	R		Hungary	159	0.00	549	124	0.00	543
Melara	W		Italy	3	0.00	1145	1	0.00	1235
Melhorio	R		Portugal	0	0.00	1365	0	0.00	1503
Melon	W	Melón; Muscadet	France	12306	0.27	58	9551	0.21	74
Mencía	R	Jaen; Jaen Tinto; Loureiro Tinto; Mencia	Spain	10658	0.23	66	11052	0.25	64
Menoir	R	Menoire	Hungary	65	0.00	697	61	0.00	659
Menu Pineau	W	Orbois	France	205	0.00	514	197	0.00	484
Mergeritar	W	Mărgăritar	Russia				0	0.00	1434
Merille	R		France	44	0.00	766	42	0.00	718
Merlese	R		Italy	14	0.00	941	8	0.00	1000
Merlot	R	Merlot Noir	France	267888	5.80	2	266440	5.94	2
Merlot Blanc	W		France	46	0.00	759	44	0.00	714
Merlot Khorus	R	UD-31.125	Italy				0	0.00	1443
Merseguera	W		Spain	3946	0.09	128	2373	0.05	164
Merzling	W		Germany				1	0.00	1289
Meslier Saint-Francois	W	Meslier Saint Francois	France	15	0.00	931	13	0.00	925
Meszi Kadarka	R	Mészikadar	Hungary				0	0.00	1362
Mézes Fehér	W	Mezes; Mezes Feher; Mézes	Hungary	2	0.00	1161	1	0.00	1345
Michele Palieri	R	Michele Parlieri; Michele Palieri; Michelle Parlieri	Italy	1	0.00	1242	1	0.00	1329
Michurinets	R	Michurinetz	Russia	0	0.00	1298	0	0.00	1374
Miguel del Arco	R	Miguel De Arco;;Vidau	Spain	468	0.01	385			
Milgranet	R		France	1	0.00	1206	1	0.00	1256
Milia	R		Slovakia	1	0.00	1208			
Millot-Foch	R	Léon Millot x Maréchal Foch; VB 85-1	Switzerland				126	0.00	538
Mindelo	R		Portugal	1	0.00	1285	1	0.00	1352
Minella Bianca	W		Italy	65	0.00	698	19	0.00	861
Minnesota Muscat	W		United States				0	0.00	1395
Miorita	W		Romania	7	0.00	1039	8	0.00	1004
Mireille	W		France	0	0.00	1345	0	0.00	1459
Misket	W		Ukraine						
Misket Cherven	G		Bulgaria	4159	0.09	122	4349	0.10	124
Misket Varnenski	W		Bulgaria	336	0.01	432			
Moldova	R		Moldova				12375	0.28	59
Molette	W		France	29	0.00	832	18	0.00	875
Molinara	R	Molinera	Italy	717	0.02	318	609	0.01	323
Mollard	R		France	23	0.00	874	23	0.00	841
Monarch	R		Germany				10	0.00	967
Monastrell	R	Garrut; Mataro; Mourvedre; Nonastrell; Mourvèdre; Monastrell; Mourvèdre	Spain	69742	1.51	14	51930	1.16	16
Monbadon	W		France	498	0.01	376			
Mondet	R		Portugal	0	0.00	1312	0	0.00	1392
Mondeuse Blanche	W		France	6	0.00	1068	7	0.00	1019
Mondeuse Noire	R	Mondeuse; Mondeuse Rouge	France	303	0.01	453	287	0.01	430
Monemvassia	W	Monemvasia	Greece	481	0.01	379	81	0.00	616
Monerac	R		France	3	0.00	1129	3	0.00	1131
Monica Nera	R		Italy	1404	0.03	220	1203	0.03	232
Monstruosa	W	Monstruosa de Monterrei	Spain	1	0.00	1241			
Montepulciano	R	Cordisco	Italy	34956	0.76	26	32935	0.73	26
Montils	W		France	164	0.00	544	165	0.00	508
Montonico Bianco	W		Italy	734	0.02	313	567	0.01	335
Montreal Blues	R	Montreal blue	United States				1	0.00	1340
Montu	W		Italy						
Monvedro	R		Portugal	6	0.00	1070	4	0.00	1099
Moore's Diamond	W	Moores Diamond	United States	42	0.00	773	24	0.00	832
Moradella	R		Italy	6	0.00	1067	1	0.00	1245
Morava	W		Serbia				34	0.00	755
Moravia Agria	R		Spain	550	0.01	357	222	0.00	465
Morenillo	R		Spain						
Morio-Muskat	W	Morio Muscat; Muscat Morio	Germany	526	0.01	360	440	0.01	372
Moristel	R	Juan Ibáñez; Miguel Del Arco	Spain	147	0.00	559	247	0.01	448
Mornen Noir	R		France						
Morone	R		Italy	13	0.00	956	7	0.00	1006
Morrastel Bouschet	R		France				4	0.00	1103
Moscadet	W		Portugal	4	0.00	1111	3	0.00	1126
Moscargo	R		Portugal	0	0.00	1348	0	0.00	1460

Table 13 (cont.): Prime varieties' colour, synonyms, country of origin, and global area, share and rank, 2010 and 2016 (ha and %)

				2010			2016		
Prime variety	*Col.*	*Synonyms*	*Country of origin*	*Area (hectares)*	*Share (%)*	*Rank*	*Area (hectares)*	*Share (%)*	*Rank*
Moscatel Lilaz	W	Lilas	Portugal	0	0.00	1383	0	0.00	1548
Moscatello Selvatico	W		Italy	35	0.00	798	5	0.00	1071
Moscato di Scanzo	R		Italy	53	0.00	736	10	0.00	972
Moscato di Terracina	W		Italy	138	0.00	570	22	0.00	844
Moscato Embrapa	W		Brazil	862	0.02	293	683	0.02	303
Moscato Giallo	W	Moscatel Amarilla; Moscatel Amarillo; Moscato Gaillo; Moscato Giallo/Moscatel del Trentino; Muskat Zuti	Italy	1467	0.03	213	1634	0.04	197
Moscato Nazareno	W		Portugal	68	0.00	693	40	0.00	722
Moscato Rosa del Trentino	R	Moscato Rosa; Moscato Rosado; Muskat Ruza Crni	Italy	81	0.00	657	39	0.00	730
Moschofilero	G	Fileri, Moschofilero Ii	Greece	1111	0.02	255	1088	0.02	246
Moschomavro	R	Moschato	Greece	1428	0.03	217	113	0.00	562
Mostosa	W		Italy	24	0.00	870	16	0.00	892
Mourisco de Semente	R		Portugal	60	0.00	712	61	0.00	657
Mourisco de Trevoes	R		Portugal	2	0.00	1181	2	0.00	1189
Mourvaison	R		France	3	0.00	1146	3	0.00	1149
Mouyssagues	R		France	0	0.00	1361	0	0.00	1508
MRAC 1087	R		Switzerland				0	0.00	1449
MRAC 1099	R		Switzerland				0	0.00	1435
MRAC 1626	R		Switzerland				0	0.00	1541
MRAC 1817	R		Switzerland				0	0.00	1473
MRAC 40	R		Switzerland				0	0.00	1540
Mskhali	W		Armenia	1093	0.02	259			
Mtsvane Kakhuri	W	Mtsvane	Georgia	319	0.01	441	319	0.01	408
Müller-Thurgau	W	Findling; Muller Thurgau; Muller Thurgau Weiss; Müller Thurgau Rivaner; Rivaner; Rizvanac; Rizlingszilváni	Germany	22917	0.50	37	19501	0.43	39
Musann Blanc	W		Taiwan				5	0.00	1068
Muscadelle	W	Muscadelle (Tokay)	France	1637	0.04	202	1509	0.03	209
Muscardin	R		France	17	0.00	912	17	0.00	881
Muscaris	W	Muscaris (FR 493-87)	Germany				4	0.00	1089
Muscat	W	Moscatello; Moscato; Muscat Varieties; Muscat/Muskateller	Greece				744	0.02	287
Muscat Bailey A	R	Muscat Bailey	Japan	1422	0.03	218	1821	0.04	183
Muscat Blanc à Petits Grains	W	Bornova Misketi; Moscatel; Moscatel De Frontignan; Muscat Frontignan; Muscat frontignian; Muscat de Frontignan; Moscatel Frontignan; Moscatel de Grano Menudo; Moscatel Galego Branco; Moscato Bianco; Moscato Bianco R2; Moscato Canelli; Muscat Canelli; Moschato; Muscadel; Moscatel Nunes; Moscatel Branco; Muscat A Petit Grains Blanc (Frontignac); Muscat a Petits Grains Blanc; Muscat A Petits Grains Blancs; Muscat Blanc; Muscat Blanc (du Valais); Muscat Blanc a Petit Grains; Muscat Petit Grain; Muscat Petits Grains; Muscat White; Muscat Blanc à Petits Grains; Muskat; Muskat Bijeli; Muskat Zlty; Muskateller; Muskateller Gelber; Muskateller, Gelber; MuškatŽuti; Rumeni Muskat; Rumeni Muškat; Tamaioasa Romaneasca; Tămâioasă românească; Tamaioasa Romaneasca; Tamiosa Romaneasca; Temjanika; Fekete Muskotály; Sárga Muskotály	Greece	31259	0.68	33	33739	0.75	24
Muscat Blanc à Petits Grains (G)	W	Moscatel Rosada; Moscatel Rosada (Pastilla); Moscatel Rosado	Greece	8761	0.19	80	8258	0.18	82
Muscat Blanc à Petits Grains (R)	W	Muscadel Red; Muscat A Petit Grains Rouge/Rose (Frontignac); Muscat A Petits Grains Rouge; Muscat Rose; Muscat Violet; Muskateller, Roter; Muscat A Petits Grains Noirs; Moscatel Galego Tinto; Muscat A Petits Grains Roses; Moscatel Galego Roxo; Muscadel (red)	Greece	1459	0.03	216	1438	0.03	214
Muscat Bleu	R		Switzerland				3	0.00	1152
Muscat de Bugeac	R		Moldova				2	0.00	1221

Table 13 (cont.): Prime varieties' colour, synonyms, country of origin, and global area, share and rank, 2010 and 2016 (ha and %)

				2010			2016		
				Area	Share		Area	Share	
Prime variety	*Col.*	*Synonyms*	*Country of origin*	*(hectares)*	*(%)*	*Rank*	*(hectares)*	*(%)*	*Rank*
Muscat Fleur d'Oranger	W	Muscat Croquant; Muscat Fleur d'Oranger; Orange Muscat	France	91	0.00	635	299	0.01	418
Muscat of Alexandria	W	Hanepoot; Malaga; Moscatel Blanco; Moscatel Blanca; Moscatel de Alejandr¡a; Moscatel de Alejandria; Moscatel De Alejandría; Moscatel De Alejandría - Blanca Italia; Moscatel de Malaga; Moscatel Graudo; Moscato de Alexandria; Muscat Amber; Muscat Alexandria; Muscat d'Alexandrie; Muscat Gordo Blanco; Zibibbo; Muscat Gordo Blanco	Greece	27648	0.60	35	34805	0.78	23
Muscat of Alexandria (R)	W	Red Hanepoot; Muscat Of Alexandria Red	Greece	6	0.00	1065	3	0.00	1147
Muscat of Hamburg	R	Hamburgi Muskotaly; Moscatel de Hamburg; Moscatel De Hamburgo; Moscatel Hamburg; Moscato d'Amburgo; Moscato Nero di Acqui; Muscat de Hamburg; Muscat Hambourg; Muscat Hamburg; Muscat de Hambourg; Muskat Hungarian; Muskattrollinger; Muskat-Trollinger; Roter Muskateller; Hamburgi Muskotály; Black Muscat	United Kingdom	8140	0.18	83	7680	0.17	87
Muscat Ottonel	W	Moscato Ottonel; Muscat Ottonel; Muškat Ottonel; Muskat, Ottonel; Muskat-Ottonel; Ottonel Muskotály	France	10340	0.22	68	12464	0.28	58
Muscat Swenson	W	Muscats	United States	24	0.00	859	37	0.00	743
Muscat Timpuriu de Bucuresti	W	Muscat timpuriu	Romania				15	0.00	905
Muscat Yantarnyi	W	Muscat iantarnîi	Moldova				683	0.02	301
Muscatin	W		Switzerland				0	0.00	1463
Muskat de Yaloven	W	Trigueira; Muscat de Ialoveni; Muscat de Yaloven; Malvasia Trigueira	Moldova	32	0	813	16	0	891
Muskat Moravsky	W		Czechia	514	0.01	366			
Muskat Zhemchuzhnyi	W	Muscat jemciujenîi	Ukraine				1	0.00	1236
Mustoasă de Măderat	W	Mustoasa de Maderat; Mustoasa de Măderat; Mustoasade Măderat	Romania	255	0.01	481	282	0.01	432
Naia	W		Portugal	0	0.00	1294	0	0.00	1408
Naparo	R		Spain	1	0.00	1240	0	0.00	1553
Narince	W		Turkey	769	0.02	306	787	0.02	277
Nascetta	W		Italy	21	0.00	886	17	0.00	879
Nasco	W		Italy	141	0.00	566	91	0.00	593
Nebbiera	R		Italy	12	0.00	968	9	0.00	982
Nebbiolo	R	Nebiolo; Nebiollo	Italy	6125	0.13	102	7997	0.18	84
Negoska	R	Negkoska	Greece	143	0.00	565	17	0.00	880
Negramoll	R	Molar; Mollar; Mulata; Negra Criolla; Negra Mole; Rabo de Ovelha Tinto; Saborinho; Tinta Madeira; Tinta Negra; Tinta Molle; Tinta Porto Santo	Spain	3195	0.07	141	3013	0.07	142
Négrette	R	Negrette; Pinot Saint George	France	1202	0.03	244	1112	0.02	244
Negretto	R		Italy	75	0.00	672	35	0.00	746
Negroamaro	R	Negro Amaro	Italy	11492	0.25	62	11449	0.26	63
Negru Aromat	R		Romania	1	0.00	1273	1	0.00	1335
Negru de Drăgăşani	R	Negru de Dragasani; Negru de Drăgăşani; Negrude Drăgăşani	Romania	16	0.00	927	18	0.00	872
Negru de Yaloven	R	Negru de Ialoveni	Moldova	15	0.00	933	141	0.00	527
Nehelescol	W	Kanaan	Israel	25	0.00	858	6	0.00	1038
Nektár	W	Nektar	Hungary	21	0.00	885	22	0.00	843
Ner d'Ala	R		Italy	30	0.00	824	10	0.00	962
Nerello Cappuccio	R		Italy	508	0.01	370	125	0.00	540
Nerello Mascalese	R		Italy	2883	0.06	152	1805	0.04	184
Neretta Cuneese	R		Italy	132	0.00	581	119	0.00	555
Neretto di Bairo	R		Italy	34	0.00	804	19	0.00	865
Nero Buono di Cori	R	Nero Buono	Italy	135	0.00	575	58	0.00	665
Nero d'Avola	R	Calabrese; Nero; Nero Davola	Italy	16649	0.36	46	14281	0.32	51
Nero di Troia	R	Uva Di Troia	Italy	2572	0.06	163	2512	0.06	159
Neronet	R		Czechia	72	0.00	678	6	0.00	1041
Neuburger	W	Neuburske	Austria	1030	0.02	269	578	0.01	331
Nevoeira	R		Portugal	0	0.00	1305	0	0.00	1380
New York Muscat	R	NY Muscat	United States	5	0.00	1085	12	0.00	927
New York Muscat and VG4111	R		United States				2	0.00	1181
Neyret	R		Italy	41	0.00	774	12	0.00	926

Table 13 (cont.): Prime varieties' colour, synonyms, country of origin, and global area, share and rank, 2010 and 2016 (ha and %)

				2010			2016		
Prime variety	*Col.*	*Synonyms*	*Country of origin*	*Area (hectares)*	*Share (%)*	*Rank*	*Area (hectares)*	*Share (%)*	*Rank*
Niagara	W	Niagara and Wiley White; Niagara Branca; Niagara White	United States	4670	0.10	114	3264	0.07	138
Niagara Red	R	Niagara Rosada	Brazil				469	0.01	363
Nieddera	R		Italy	107	0.00	606	91	0.00	594
Nigra	R		Italy	3	0.00	1153	1	0.00	1273
Nincusa	R		Croatia	17	0.00	918			
Noah	W	Ondarrabi Zuri	United States	563	0.01	352	200	0.00	480
Nobling	W		Germany	1	0.00	1239	52	0.00	683
Nocera	R		Italy	15	0.00	936	5	0.00	1065
Noir Fleurien	R		France	0	0.00	1340	0	0.00	1504
Noiret	R		United States	33	0.00	810	25	0.00	823
Noria	W		Slovakia	1	0.00	1207			
Norton	R	Cynthiana	United States	329	0.01	434	328	0.01	403
Nosiola	W		Italy	79	0.00	663	65	0.00	653
Nosztori Rizling	W		Hungary	1	0.00	1274	0	0.00	1441
Notardomenico	R		Italy	10	0.00	999	9	0.00	974
Noual	W		France						
Nouvelle	W		South Africa	422	0.01	394	428	0.01	376
Novac	R		Romania	73	0.00	674	74	0.00	634
Nuragus	W		Italy	1345	0.03	227	1008	0.02	252
Oberlin	R		France	64	0.00	703	26	0.00	816
Oberlin White	W		France						
Odessky Cherny	R	Alibernet; Odessa Black; Odessky Chernyi	Ukraine	2686	0.06	159	2508	0.06	160
Odola	R		France	0	0.00	1317	0	0.00	1402
Odysseus	W		Hungary	4	0.00	1096	25	0.00	819
Oeillade Bousche	R		France	1	0.00	1270	1	0.00	1336
Oeillade Noire	R		France	18	0.00	905	18	0.00	874
Ofthalmo	R		Cyprus	141	0.00	567			
Ohanes	W	Valensi; Valensi du Maroc	Spain	16	0	923	15	0	901
Ojaleshi	R		Georgia	32	0.00	814	32	0.00	770
Okanagan Riesling	W		United States				1	0.00	1358
Öküzgözü	R	Okuzgozu	Turkey	1479	0.03	212	1601	0.04	202
Olivette Blanche	W		France	2	0.00	1185	2	0.00	1203
Olivette de Laconnex	W		Switzerland				0	0.00	1552
Olivette Noire	R		Hungary	16	0.00	922	16	0.00	893
Ondenc	W		France	8	0.00	1025	1	0.00	1353
Onitskanskii Belyi	W	Alb de Onitcani; Onitcani	Moldova	71	0	680	490	0	354
Ontario	W		United States				0	0.00	1551
Optima	W	Optima 113	Germany	65	0.00	702	38	0.00	735
Ora	W		France	38	0.00	784	32	0.00	767
Oraniensteiner	W		Germany	3	0.00	1149	2	0.00	1174
Original	R		Ukraine				5	0.00	1076
Orion	W		Germany	13	0.00	953	1	0.00	1293
Orpheus	W		Hungary	0	0.00	1314	2	0.00	1230
Orpicchio	W		Italy	1	0.00	1289	0	0.00	1534
Ortega	W		Germany	667	0.01	330	532	0.01	343
Ortrugo	W		Italy	611	0.01	343	709	0.02	293
Osceola Muscat	W		United States				7	0.00	1009
Oseleta	R		Italy	15	0.00	938	16	0.00	886
Osennii Ciornai	R	Osennii ciornîi	Moldova				8	0.00	993
Osteiner	W		Germany	1	0.00	1238	1	0.00	1325
Otskhanuri Sapere	R	Sapere Otskhanuri	Georgia	6	0.00	1050	6	0.00	1025
Padeiro	R		Portugal	86	0.00	644	88	0.00	601
Palas	R		Germany				7	0.00	1011
Palatina	W		Hungary	6	0.00	1063	3	0.00	1120
Palava	G		Czechia	230	0.00	498			
Pallagrello Bianco	W		Italy	55	0.00	728	6	0.00	1049
Pallagrello Nero	R		Italy	169	0.00	539	107	0.00	572
Palomino Fino	W	Listan; Madera; Malvasia Rei; Palamino; Palomino; Palomino Superior	Spain	22693	0.49	38	23190	0.52	34
Palot	W		Spain						
Pamid	R	Pamiti; Piros Szlanka; Plovdina; Rosioara; Roşioară	Bulgaria	9827	0.21	72	9961	0.22	69
Pampanaro	W		Italy	5	0.00	1093	1	0.00	1319
Pampanuto	W		Italy	356	0.01	421	33	0.00	765
Pamyati Negrulya	R	Pameati Negrulea	Moldova				123	0.00	546
Pannon Frankos	R		Hungary	16	0.00	921	12	0.00	929
Panonia	W	SK 90-2/19 (Pannonija)	Serbia				10	0.00	959
Panse Valenciano	W	Panse Valenciana	Spain	1	0.00	1237			
Paolina	W		Italy	1	0.00	1204	0	0.00	1471
Papazkarası	R	Papazkarasi	Turkey	175	0.00	535	204	0.00	477
Pardillo	W	Marisancho	Spain	4364	0.09	120	3283	0.07	134

Table 13 (cont.): Prime varieties' colour, synonyms, country of origin, and global area, share and rank, 2010 and 2016 (ha and %)

				2010			2016		
				Area	Share		Area	Share	
Prime variety	*Col.*	*Synonyms*	*Country of origin*	*(hectares)*	*(%)*	*Rank*	*(hectares)*	*(%)*	*Rank*
Parellada	W		Spain	8847	0.19	79	7137	0.16	91
Parraleta	R	Bonvedro; Caricagiola; Pau Ferro; Tinta Caiada; Tinta Lameira	Spain	348	0.01	425	212	0.00	473
Parreira Matias	R		Portugal	1	0.00	1267	1	0.00	1330
Pascal Blanc	W		France	0	0.00	1376	0	0.00	1528
Pascale	R	Nieddu Mannu	Italy	375	0.01	409	289	0.01	427
Passau	R		Italy	5	0.00	1074	3	0.00	1108
Passerina	W		Italy	894	0.02	285	933	0.02	259
Patagonia	W		Argentina	40	0.00	778			
Patorra	R		Portugal	10	0.00	995	10	0.00	961
Pátria	W	Patria	Hungary	3	0.00	1137	5	0.00	1078
Patrizia Rosa	G	Patricia	Brazil	153	0.00	554			
Pavana	R		Italy	32	0.00	815	20	0.00	857
Pe Comprido	W		Portugal	1	0.00	1227	1	0.00	1348
Pecorello	W		Italy	34	0.00	805	9	0.00	975
Pecorino	W	Uvina	Italy	1228	0.03	239	1742	0.04	189
Pecsi Szagos	W	Zold Szagos; Zöld Szagos	Hungary	1	0.00	1228	1	0.00	1317
Pedral	R		Portugal	151	0.00	556	80	0.00	620
Pedro Giménez	W	Pedro Gimenez; Pedro Gimenez Rio Colorado; Pedro Gimenez Río Colorado; Pedro Jimenez; Pedro Jimenez	Argentina	13502	0.29	55	15576	0.35	49
Pedro Ximénez	W	Pedro; Pedro Ximenez; Perrum; Pedro Ximenes	Spain	9235	0.20	76	8810	0.20	78
Pelaverga	R		Italy	55	0.00	726	46	0.00	709
Pelaverga Piccolo	R		Italy	6	0.00	1066	6	0.00	1046
Peloursin	R		France						
Pelso	W		Hungary	1	0.00	1271	1	0.00	1334
Pepella	W		Italy	3	0.00	1138	0	0.00	1438
Perdea	W		France	2	0.00	1158	2	0.00	1188
Perera	W		Italy	4	0.00	1105	2	0.00	1196
Perigo	W		Portugal	4	0.00	1103	4	0.00	1086
Perla dei Vivi	R		Italy	1	0.00	1264	1	0.00	1347
Perlaut	W		France	1	0.00	1224	1	0.00	1275
Perle	G		Germany	34	0.00	802	18	0.00	871
Perlette	W		United States	1	0.00	1212	2	0.00	1171
Perlita	W		Spain	1	0.00	1236			
Perola	W		Portugal						
Perricone	R		Italy	228	0.00	499	80	0.00	621
Perruno	W	Royal	Spain	1509	0.03	209	745	0.02	286
Persan	R	Becuet	France	12	0.00	969	12	0.00	936
Pervenets Magaracha	W	Magaracha's Firstborn; Pervenec Magaracha; Perveneț Magaracia; Pervenets of Magarach	Ukraine	2881	0.06	153	2755	0.06	150
Pervomaisky	R	Pervomaiskii	Uzbekishtan	64	0.00	704			
Petit Bouschet	R	Tintinha	France	15	0.00	929	120	0.00	551
Petit Courbu	W	Petit Courbou Ondarrzurizerrat	France	102	0.00	616	1	0.00	1268
Petit Manseng	W	Manseng Petit Blanc	France	1109	0.02	256	1299	0.03	221
Petit Meslier	W		France	4	0.00	1102	3	0.00	1114
Petit Rouge	R		Italy	84	0.00	650	68	0.00	647
Petit Verdot	R	Verdot	France	7195	0.16	92	8124	0.18	83
Petite Amie	W		United States				0	0.00	1513
Petite Milo	G		Switzerland				6	0.00	1047
Petite Pearl	R	Petite Perle	United States				11	0.00	954
Pexem	R		Portugal	3	0.00	1119	3	0.00	1141
Phoenix	W		Germany	67	0.00	694	46	0.00	708
Picapoll Blanco	W		Spain	37	0.00	788	40	0.00	721
Picardan	W		France	1	0.00	1261	1	0.00	1320
Piccola Nera	G		Italy	17	0.00	914	6	0.00	1053
Picolit	W		Italy	128	0.00	586	121	0.00	550
Piculit Neri	R		Italy	22	0.00	883	8	0.00	991
Piedirosso	R		Italy	699	0.02	321	593	0.01	327
Pignola Valtellinese	R	Pignola	Italy	49	0.00	756	28	0.00	795
Pignoletto	W		Italy	1707	0.04	199	1174	0.03	237
Pignolo	R		Italy	93	0.00	631	50	0.00	690
Pineau d'Aunis	R		France	437	0.01	392	413	0.01	381
Pinella	W	Pinela	Italy	72	0.00	679	128	0.00	536
Pinorico	R		Switzerland				0	0.00	1397
Pinot Blanc	W	Beli Pinot; Burgunder; Burgunder, Weißer; Pinot Bianco; Pinot Bijeli; Pinot Blanc; Weisser Burgunder; Weissburgunder; Pinot Blanco; Pinot White; Rulandske Bile; Weißburgunder	France	14812	0.32	52	13779	0.31	53

Table 13 (cont.): Prime varieties' colour, synonyms, country of origin, and global area, share and rank, 2010 and 2016 (ha and %)

				2010			2016		
Prime variety	*Col.*	*Synonyms*	*Country of origin*	*Area (hectares)*	*Share (%)*	*Rank*	*Area (hectares)*	*Share (%)*	*Rank*
Pinot Gris	G	Grauer Burgunder; Pinot Grey; Pinot Grigio; Pinot Gris (Grigio); Pinot Gris/Grigio; Pinot Gris/Malvoisie/Grauburgunder; Pinot Sivi; Pinotsivi; Rulander; Ruländer (Burgunder, Grauer); Rulandske Sede; Sivi Pinot; Szürkebarát	France	43773	0.95	19	48570	1.08	18
Pinot Meunier	R	Meunier; Müllerrebe (Schwarzriesling); Pinot Menier; Pinot Menieur; Schwarzriesling	France	13566	0.29	54	14695	0.33	50
Pinot Noir	R	Pinot Noire; Blauer Burgunder; Gros Bec; Mario Feld; Modri Pinot; Pinot 386; Pinot Crni; Pinot Negro; Pinot Nero; Pinot Noir - Pinot Negro; Pinot Noir / Blauer burgunder; Pinot Noir, Blue (including velvet red); Pinot Noir/Blauburgunder; Pinot Noir; Pinot Noir Blauer Burgunder; Rulandske Modre; Servagnin; Spätburgunder, Blauer (einschl. Samtrot)	France	98623	2.14	10	105480	2.35	10
Pinot Noir Précoce	R	Pinot Noir Precoce; Fruhburgunder; Frühburgunder, Blauer; Pinot Noir Précoce/Frühburgunder	France	273	0.01	465	251	0.01	446
Pinotage	R		South Africa	6404	0.14	96	7132	0.16	92
Pinotin	R		Germany				0	0.00	1453
Pionnier	R		United States				1	0.00	1243
Piquepoul Blanc	W		France	1492	0.03	211	1565	0.03	205
Piquepoul Bousch	R		France						
Piquepoul Gris	G		France	2	0.00	1194	2	0.00	1228
Piquepoul Noir	R	Pical; Picapoll Negro	France	70	0.00	686	69	0.00	644
Piroso	R		Germany				4	0.00	1090
Plant Droit	R		France	19	0.00	898	19	0.00	864
Planta Mula	R		Spain	24	0.00	865	11	0.00	955
Planta Nova	W	Alvarelhao Branco	Spain	1395	0.03	221	888	0.02	268
Plantet	R		France	1060	0.02	267	420	0.01	378
Plassa	R		Italy	91	0.00	633	86	0.00	605
Plavac Mali	R	Plavac Mali Crni; Plavec Mali; Plavecmal	Croatia	1569	0.03	207	1714	0.04	190
Plavaie	W	Plăvaie; Plavay	Romania	209	0.00	511	152	0.00	517
Plavec Žuti	W	Plavec Zuti; Rumeni Plavec; Plavac Zuti	Croatia	13	0.00	955	82	0.00	612
Plavina	R	Plavina Crna; Plavinacrna	Croatia	643	0.01	335	683	0.02	302
Podarok Magaracha	W	Magaracha's Gift; Podarok of Magarach	Ukraine	504	0.01	372	292	0.01	424
Podarok Zaporozju	W	Novîi podaroc Zaporojiu	Ukraine				31	0.00	774
Pollera Nera	R		Italy	54	0.00	729	32	0.00	768
Pölöskei Muskotály	W	Poeloeske Muskotaly; Poloske; Poloskei Muskotaly; Muscat Poloskey; Pölöskei Muskotaly	Hungary	103	0.00	614	207	0.00	476
Portan	R		France	264	0.01	475	256	0.01	442
Portland	W		United States	12	0.00	961	39	0.00	726
Pošip Bijeli	W	Posip Bijeli	Croatia	253	0.01	483			
Pougnet	R		France						
Poulsard	R		France	307	0.01	450	90	0.00	597
Poulsard Blanc	W		France	0	0.00	1377	0	0.00	1538
Pozsonyi Fehér	W	Pozsonyi Feher; Pozsonyi	Hungary	10	0.00	988	10	0.00	969
Praca	W		Portugal	166	0.00	542	169	0.00	505
Prairie Star	W		United States	21	0.00	887	23	0.00	836
Prensal	W	Moll; Pensal Blanco	Spain	105	0.00	611	129	0.00	534
Preto Cardana	R		Portugal	5	0.00	1089	5	0.00	1080
Preto Martinho	R	Amostrinha	Portugal	163	0.00	546	163	0.00	512
Prezentabil	W		Moldova				215	0.00	469
Prie	W	Prie Blanc	Italy	33	0.00	808	24	0.00	833
Prieto Picudo	R		Spain	4587	0.10	116	4293	0.10	126
Prima	R		France	84	0.00	651	81	0.00	618
Primavera	R		Portugal	40	0.00	777	39	0.00	727
Primetta	G	Prie Rouge	Italy	24	0.00	872	14	0.00	908
Prinzipal	W		Germany				2	0.00	1184
Prior	R		Germany				14	0.00	913
Prodest	R		Italy						
Prokupac	R	Prokupec	Serbia	15180	0.33	51	1361	0.03	217
Promissao	W		Portugal	6	0.00	1055	3	0.00	1118
Prosecco	W		Italy	18437	0.40	43	20109	0.45	37
Prosecco Lungo	W		Italy	1367	0.03	225	1450	0.03	211

Table 13 (cont.): Prime varieties' colour, synonyms, country of origin, and global area, share and rank, 2010 and 2016 (ha and %)

				2010			2016		
Prime variety	*Col.*	*Synonyms*	*Country of origin*	*Area (hectares)*	*Share (%)*	*Rank*	*Area (hectares)*	*Share (%)*	*Rank*
Provareau	R		France						
Prunelard	R		France	20	0.00	891	19	0.00	859
Prunesta	R		Italy	36	0.00	793	31	0.00	775
Pugnitello	R		Italy	28	0.00	834	15	0.00	903
Purcsin	R		Hungary				0	0.00	1403
Putzscheere	W		Hungary						
Quagliano	R		Italy	9	0.00	1008	9	0.00	976
Quebranta	R		Peru	345	0.01	428	330	0.01	400
Quiebratinajas Tinto	R	Colgadero; Quiebratinajas	Spain	5	0.00	1080	9	0.00	981
Rabigato	W	Rabigato Franco	Portugal	1273	0.03	235	1969	0.04	175
Rabigato Moreno	W		Portugal	1	0.00	1254	1	0.00	1301
Rabo de Anho	R		Portugal	99	0.00	621	86	0.00	604
Rabo de Lobo	R		Portugal	3	0.00	1124	3	0.00	1121
Rabo de Ovelha	W		Portugal	908	0.02	283	563	0.01	336
Raboso Piave	R	Raboso	Italy	776	0.02	305	665	0.01	306
Raboso Veronese	R		Italy	277	0.01	463	295	0.01	422
Radisson	R		United States				8	0.00	1001
Raffiat de Moncade	W		France	7	0.00	1046	6	0.00	1027
Raisaine	W		France						
Rambella	W	Famoso	Italy	6	0.00	1057	6	0.00	1028
Ramisco	R	Ramisco Tinto	Portugal	34	0.00	806	33	0.00	764
Ranfol	W		Slovenia	134	0.00	579			
Ranna Melnishka Loza	R		Bulgaria	249	0.01	486			
Rathay	R		Austria	9	0.00	1017	32	0.00	772
Räuschling	W	Rauschling; Roter Räuschling	Germany	23	0.00	879	23	0.00	837
Ravat	R		France	1	0.00	1266	1	0.00	1326
Ravat Blanc	W		France	7	0.00	1047	6	0.00	1037
Rayada Melonera	R	Corropio	Spain	1	0.00	1218	1	0.00	1297
Rayon d'Or	W		France	6	0.00	1064	6	0.00	1048
Reberger	R		Germany				2	0.00	1183
Rebo	R		Italy	125	0.00	587	92	0.00	589
Recantina	R		Italy	9	0.00	1004	4	0.00	1097
Red Globe	R		United States	242	0.01	490	242	0.01	450
Red Millennium	R		Japan				2	0.00	1223
Refosco	R	Refosk; Refošk	Italy				1341	0.03	218
Refosco dal Peduncolo Rosso	R		Italy	1082	0.02	261	1272	0.03	222
Refosco di Faedis	R	Refosco Nostrano	Italy	217	0.00	503	185	0.00	492
Refrén	W	Refren	Hungary	0	0.00	1301	0	0.00	1378
Regent	R		Germany	2187	0.05	175	1974	0.04	174
Regner	W		Germany	46	0.00	761	21	0.00	849
Reichensteiner	W		Germany	247	0.01	487	120	0.00	554
Reliance	R		United States	4	0.00	1115	4	0.00	1100
Réselle	W		Switzerland				1	0.00	1276
Retagliado Bianco	W		Italy	28	0.00	835	11	0.00	948
Reze	W		Switzerland						
Ribol	R		France	147	0.00	560	141	0.00	528
Ribolla Gialla	W	Rebula	Italy	1178	0.03	250	959	0.02	257
Rieslaner	W		Germany	84	0.00	648	73	0.00	635
Rieslina	W	Rieslina (Inta C.G. 38049)	Argentina	174	0.00	537	103	0.00	576
Riesling	W	J. Riesling; Rajinski Rizling; Renski Rizling; Rèze; Rheine Riesling; Rheinriesling; Riesling de Rhin; Riesling Lion; Riesling Renan; Riesling Renano; Riesling Weisser; Riesling, Weißer; Riesling-Lion; Rizling Rynsky; Ryzlink Rynsky; White Riesling; Rajnai rizling	Germany	50014	1.08	18	59805	1.33	12
Riesling Forte	W		Japan				2	0.00	1159
Riesus	W		Ukraine	115	0.00	592	115	0.00	558
Rio Grande	W		Portugal	1	0.00	1287	0	0.00	1360
Ripolo	W		Italy	1	0.00	1222	1	0.00	1270
Riton	W		Moldova	257	0.01	478	568	0.01	334
Rkatsiteli	W	Rkaciteli; Rkatiteli; Rkaţiteli; Rkatziteli	Georgia	58641	1.27	17	51374	1.15	17
Roal	R	Rual	Portugal	1	0.00	1220	1	0.00	1314
Robola	W	Rombola	Greece	471	0.01	382	152	0.00	516
Roditis	G		Greece	4668	0.10	115	8463	0.19	81
Roditis (R)	G	Roditis (Red)	Greece	3826	0.08	129	828	0.02	274
Rojal Tinta	R		Spain	1801	0.04	195	736	0.02	289
Rollo	W	Livornese Bianca	Italy	51	0.00	746	15	0.00	898
Rome	R	Rome Tinto	Spain	297	0.01	454	172	0.00	504
Romeiko	R		Greece	1597	0.03	205	1131	0.03	242
Romorantin	W		France	72	0.00	677	69	0.00	643

Table 13 (cont.): Prime varieties' colour, synonyms, country of origin, and global area, share and rank, 2010 and 2016 (ha and %)

				2010			2016		
Prime variety	*Col.*	*Synonyms*	*Country of origin*	*Area (hectares)*	*Share (%)*	*Rank*	*Area (hectares)*	*Share (%)*	*Rank*
Romulus	W		United States				1	0.00	1282
Rondinella	R		Italy	2480	0.05	166	2684	0.06	155
Rondo	R		Germany	40	0.00	775	51	0.00	689
Roobernet	R		South Africa	139	0.00	569	269	0.01	438
Rosa Arica	R		Peru	1	0.00	1235	1	0.00	1294
Rosaki	W		Turkey				2	0.00	1177
Rosciola	G	Rosciola rose	Italy	2	0.00	1160	1	0.00	1277
Rose Ciotat	W		Japan				2	0.00	1193
Rose du Var	G		France	56	0.00	721	54	0.00	672
Rosina	W		Romania	1	0.00	1226	1	0.00	1263
Rösler	R	Roesler	Austria	160	0.00	548	217	0.00	468
Rossara Trentina	R	Rossara	Italy	8	0.00	1026	6	0.00	1032
Rossese	R		Italy	312	0.01	445	164	0.00	511
Rossese Bianco	W		Italy	7	0.00	1031	5	0.00	1062
Rossignola	R		Italy	188	0.00	527	49	0.00	700
Rossola Nera	R		Italy	86	0.00	643	29	0.00	790
Rotberger	R		Germany	17	0.00	913	11	0.00	944
Roter Milan	R		Switzerland				0	0.00	1452
Roter Veltliner	G	Piros Veltelini; Red Veltliner	Austria	199	0.00	519	198	0.00	483
Rotgipfler	W		Austria	105	0.00	610	123	0.00	545
Roublot	W		France						
Rouge de Fully	R	Durize	Switzerland				1	0.00	1338
Rouge du Pays	R	Cornalin/Landroter	Switzerland				136	0.00	532
Rougeon	R		France	42	0.00	772	21	0.00	846
Roussanne	W	Rousanne	France	1851	0.04	192	2137	0.05	170
Roussette d'Ayze	W		France	1	0.00	1213	1	0.00	1316
Roussin	R		Italy	3	0.00	1130	2	0.00	1227
Roviello Bianco	W		Italy	2	0.00	1198	1	0.00	1323
Roxo de Vila Flor	R	Roxo Flor	Portugal	0	0.00	1331	0	0.00	1444
Roxo Rei	G		Portugal	0	0.00	1308	0	0.00	1387
Royal de Alloza	R	Derechero	Spain	29	0.00	831	6	0.00	1052
Royalty	R		United States	97	0.00	625	93	0.00	587
Roz de Minis	G		Romania	6	0.00	1053	7	0.00	1020
Rozala Bianca	W	Rozalia; Rozália	Hungary	2	0.00	1197	2	0.00	1179
Rózsakő	W	Rozsakoe	Hungary	19	0.00	901	17	0.00	876
Rúbea	R		Brazil	81	0.00	656	181	0.00	496
Rubienne	R		Australia				1	0.00	1309
Rubilande	G		France	8	0.00	1027	7	0.00	1018
Rubin Golodrigi	R	Ruby of Golodryga	Ukraine	82	0.00	654	82	0.00	615
Rubin Tairovsky	R	Rubin Tairovski	Ukraine	2	0.00	1173	5	0.00	1059
Rubinet	R		Czechia				15	0.00	899
Rubinovy Magaracha	R	Ruby Magaracha	Ukraine	0	0.00	1344	0	0.00	1457
Rubintos	R		Hungary	18	0.00	906	13	0.00	920
Rubired	R	Tintoria	United States	4556	0.10	118	4916	0.11	112
Ruby	R		United States	9	0.00	1006	9	0.00	973
Ruby Cabernet	R	Rubi Cabernet	United States	5729	0.12	104	5309	0.12	107
Ruby Seedless	R		United States						
Ruche	R		Italy	100	0.00	620	100	0.00	580
Rufete	R	Castellana	Portugal	4833	0.10	111	1859	0.04	180
Ruggine	W		Italy	1	0.00	1202	1	0.00	1307
Ryugan	W		Japan				27	0.00	800
Sabrevois	R		United States	10	0.00	994	25	0.00	817
Sacy	W		France	10	0.00	997	8	0.00	989
Sagrantino	R		Italy	995	0.02	271	1026	0.02	251
Saint Jeannet	W	Saint Jeannett	Argentina	56	0.00	722	43	0.00	717
Saint Macaire	R		France				6	0.00	1045
Saint-Cliche	W		Canada				0	0.00	1479
Saint-Pierre Dore	W		France	0	0.00	1302	0	0.00	1425
Salvador	R		France	394	0.01	403	351	0.01	396
Samarrinho	W		Portugal	1	0.00	1214	1	0.00	1308
San Giuseppe Nero	R		Italy	192	0.00	524	82	0.00	613
San Lunardo	W		Italy	10	0.00	1000	4	0.00	1104
San Martino	R		Italy	21	0.00	889	6	0.00	1033
San Michele	R		Italy	57	0.00	720	37	0.00	741
Sanforte	R		Italy	1	0.00	1286	0	0.00	1470
Sangiovese	R	Nielluccio; Prugnolo Gentile; Sangiovese/ Nielluccio; Brunello	Italy	78030	1.69	12	73464	1.64	11
Sankt Laurent	R	Saint Laurent; Šentlovrenka; St Laurent; Svatovavrinecke; Szentlorinc; Szentlőrinc	Austria	3664	0.08	131	3272	0.07	137
Santa Maria	W		Italy	3	0.00	1139	2	0.00	1216
Santarena	R	Santareno	Portugal	739	0.02	311	724	0.02	291
Santoal	W		Portugal	9	0.00	1009	4	0.00	1087
Sao Mamede	W	S. Mamede	Portugal	1	0.00	1276	1	0.00	1354

Table 13 (cont.): Prime varieties' colour, synonyms, country of origin, and global area, share and rank, 2010 and 2016 (ha and %)

				2010			2016		
Prime variety	*Col.*	*Synonyms*	*Country of origin*	*Area (hectares)*	*Share (%)*	*Rank*	*Area (hectares)*	*Share (%)*	*Rank*
Saperavi	R		Georgia	8126	0.18	84	6478	0.14	96
Saperavi Severny	R	Saperavi Nothern; Saperavi Severnii	Russia	350	0.01	422	325	0.01	405
Saphira	W		Germany				8	0.00	987
Şarbă	W	Sarba	Romania	265	0.01	473	266	0.01	439
Sárfehér	W	Sarfeher	Hungary				2	0.00	1199
Satin Noir	R	VB 91-26-29	Switzerland				0	0.00	1512
Sauvignac	W	VB Cal 6-04	Switzerland				2	0.00	1176
Sauvignon Blanc	W	Sauvignon; Sauvignon B.; Sauvignon Blanca; Sauvignon Blanco; Sauvignon Musque	France	111552	2.42	8	124700	2.78	8
Sauvignon Blanc (G)	W	Sauvignon Gris; Sauvignon Rose	France	698	0.02	322	1076	0.02	247
Sauvignonasse	W	Friulano; Sauvignon Vert; Tocai; Tocai Friulano; Zeleni Sauvignon	France	4563	0.10	117	3861	0.09	130
Savagnin Blanc	W	Savagnin; Traminac Bijeli; Traminec; Traminec Bel; Traminer; Savagnin Blanc/Heida	France	1950	0.04	185	2267	0.05	166
Savagnin Rose	G	Piros Tramini	France	884	0.02	290	48	0.00	704
Savatiano	W	Savvatiano	Greece	9920	0.21	71	10268	0.23	67
Scheurebe	W		Germany	2039	0.04	182	1626	0.04	198
Schiava	R		Italy	517	0.01	365	236	0.01	454
Schiava Gentile	R		Italy	694	0.02	323	165	0.00	510
Schiava Grigia	R		Italy	66	0.00	696	4	0.00	1101
Schiava Grossa	R	Blauer Trollinger; Tschaggele; Vernatsch; Trollinger	Italy	3011	0.07	148	2256	0.05	167
Schiava Lombarda	R	Erbanno	Italy	0	0.00	1358	701	0.02	297
Schioppettino	R		Italy	154	0.00	553	87	0.00	603
Schönburger	G	Schonburger	Germany	68	0.00	690	35	0.00	745
Sciaglin	W		Italy	6	0.00	1052	3	0.00	1110
Sciascinoso	R	Olivella Nera	Italy	94	0.00	630	59	0.00	661
Scimiscia	W		Italy	5	0.00	1073	2	0.00	1222
Scuppernong	W	Muscadine	United States				27	0.00	802
Seara Nova	W		Portugal	681	0.01	328	471	0.01	361
Segalin	R		France	65	0.00	701	61	0.00	658
Seibel	R	Seibel 2 (seibeleto)	France	592	0.01	344	482	0.01	357
Seibel White	W		France	0	0.00	1357	0	0.00	1500
Seinoir	R		France	87	0.00	640	50	0.00	696
Select	W	Seleções	France	7	0.00	1034	7	0.00	1007
Semebat	R		France	2	0.00	1191	1	0.00	1238
Semidano	W		Italy	36	0.00	790	34	0.00	759
Sémillon	W	Semigion; Semilao; Semilion; Semillon	France	22157	0.48	39	18693	0.42	42
Sennen	R		Italy	10	0.00	991	2	0.00	1168
Septimer	G		Germany				1	0.00	1288
Sercial	W	Esganoso	Portugal	106	0.00	608	85	0.00	607
Sercialinho	W		Portugal	9	0.00	1018	4	0.00	1096
Serna	G		Argentina	36	0.00	794			
Servanin	R		France						
Servant	W	Raisin Blanc	France	183	0.00	529	138	0.00	531
Sevilhao	R		Portugal	14	0.00	945	14	0.00	912
Seyval Blanc	W	Seyval; Seyval (Blanc); Seyval Blanc and Vidal 256; Seyve Villard; Seyve Villard 5276	France	569	0.01	351	2699	0.06	152
Seyval Noir	R	Seyve Willard (Tinta)	Canada				76	0.00	632
Seyve Villard 23-512	R	S.V. 23-512	France	29	0.00	833			
Sgavetta	R		Italy	47	0.00	758	26	0.00	809
Shalistin	W		Australia				0	0.00	1554
Sheridan	R	Seridan	United States	500	0.01	374	500	0.01	352
Shiroka Melnishka	R	Melnik; Shiroka Melnishka Loza	Bulgaria	1580	0.03	206	1205	0.03	231
Sicilien	W		France	5	0.00	1088	5	0.00	1079
Siegerrebe	G		Germany	131	0.00	582	102	0.00	579
Siegfriedrebe	W	Siegfried	Germany				2	0.00	1207
Silcher	W		Germany				1	0.00	1287
Silvaner	W	Silvanac Zeleni; Silvaner Grun; Silvaner, Grüner; Silvanske Zelene; Sylvaner; Sylvaner Verde; Sylvaner/Rhin; Sylvanske Zelene; Zeleni Silvanec; Zold Szilvani; Zöld szilváni	Austria	7395	0.16	91	6072	0.14	99
Silvaner (R)	W	Blauer Silvaner; Silvaner, Blauer; Roter Silvaner	Austria	38	0.00	783	25	0.00	822
Siramé	R		Switzerland				0	0.00	1391

Table 13 (cont.): Prime varieties' colour, synonyms, country of origin, and global area, share and rank, 2010 and 2016 (ha and %)

				2010			2016		
Prime variety	*Col.*	*Synonyms*	*Country of origin*	*Area (hectares)*	*Share (%)*	*Rank*	*Area (hectares)*	*Share (%)*	*Rank*
Síria	W	Alvadurao; Ciguente; Crato Espanhol; Dona Blanca; Doña Blanca; Malvasia Branca; Roupeiro Branco; Sabro; Siria; Valenciana Blanca	Portugal	7898	0.17	87	7037	0.16	93
Sirio	W		Italy	14	0.00	946	5	0.00	1058
Sirius	W		Germany						
Škrlet	W	Skrlet	Croatia	61	0.00	709			
Slankamenka	W	Majarcă Albă	Serbia	53	0.00	735	23	0.00	838
Solaris	W	Solaris (FR 240-75)	Germany	81	0.00	660	118	0.00	556
Somerset Seedless	G	Somerset	United States				1	0.00	1242
Somszoekoe Kék	R	Hajnos Kék	Hungary				0	0.00	1516
Soperga	R		Italy	22	0.00	881	17	0.00	882
Soreli	W	Soreli (UD-31.113)	Italy				0	0.00	1495
Souvignier Gris	G	Souvignier Gris (FR 392-83)	Germany				3	0.00	1132
Sovereign Coronation	R	Sovereign, Coronation, etc.	Canada				1	0.00	1304
Sovereign Opal	W		Canada	3	0.00	1143	3	0.00	1130
Spergola	W		Italy	110	0.00	602	115	0.00	559
Sremska Zelenika	W	Szerémi Zöld; Szeremi Zold	Serbia	0	0.00	1304	0	0.00	1485
St Croix	R		United States	25	0.00	857	45	0.00	713
St Pepin	W		United States	19	0.00	896	37	0.00	739
St Vincent	R		United States	23	0.00	877	20	0.00	853
Stanušina Crna	R	Stanushina; Stanušina	North Macedonia				400	0.01	384
Staufer	W		Germany						
Stavroto	R		Greece	11	0.00	979	0	0.00	1430
Stepnyak	W	Stepniak	Russia	144	0.00	562	144	0.00	524
Steuben	R		United States	39	0.00	780	35	0.00	751
Storgozia	R		Bulgaria	295	0.01	455			
Sugrafive	W	Sugra Five;	United States	2	0.00	1172	0	0.00	1533
Sukholimansky Bely	W	Suholimanski belii; Suholimanskiy White; Suholimenschii Belii; Sukholimansky	Ukraine	2156	0.05	178	405	0.01	383
Sulima	W		France	0	0.00	1367	0	0.00	1527
Sultaniye	W	Soultanina; Sultana; Sultanina; Thompson Seedless	Turkey	3413	0.07	140	5325	0.12	106
Summerland	R		Canada				0	0.00	1478
Sumoll	R	Vijiriego Negro	Spain	83	0.00	653	16	0.00	890
Sun Muscat	R		Australia				23	0.00	835
Superior Seedless	W		United States	9	0.00	1012	9	0.00	977
Suscan	R	Brajda Crna	Croatia	5	0.00	1072			
Susumaniello	R		Italy	50	0.00	750	8	0.00	997
Swenson Red	R		United States	11	0.00	975	10	0.00	968
Swenson White	W		United States				2	0.00	1162
Symphony	W		United States	324	0.01	439	647	0.01	312
Syrah	R	Chirac; Shiraz; Sira (falsa); Sirah; Syrach	France	185117	4.01	6	181185	4.04	6
Táltos	W	Taltos	Hungary	1	0.00	1232	1	0.00	1350
Tamarez	W		Portugal	343	0.01	431	298	0.01	420
Tamarugal	W		Chile				1	0.00	1324
Taminga	W		Australia				2	0.00	1209
Tannat	R	Tannat (Harriague)	France	5765	0.12	103	5611	0.13	102
Tardia de Caxias	G		Brazil				0	0.00	1383
Tarrango	R		Australia	72	0.00	676	16	0.00	885
Tauberschwarz	R	Tauberschwarz (Hängling, Blauer)	Germany	11	0.00	978	16	0.00	887
Taurus	W		Hungary				0	0.00	1404
Tavkveri	R		Georgia	37	0.00	786	37	0.00	737
Taylor	W		United States						
Tazzelenghe	R		Italy	55	0.00	725	45	0.00	710
Tedi's Best	R		Switzerland				0	0.00	1494
Teinturier	R	Farbertraube; Färbertraube; Pontak	France	7	0.00	1030	9	0.00	978
Tempranillo	R	Aragonez; Tempranillo Tinto; Tinta Roriz; Tinto de Toro	Spain	232988	5.05	4	219379	4.89	3
Tempranillo (W)	R	Tempranillo Blanco	Spain	5	0.00	1079	110	0.00	567
Teoulier Noir	R		France	0	0.00	1368	0	0.00	1514
Termarina Rossa	G	Termarina	Italy	20	0.00	893	2	0.00	1164
Teroldego	R	Teroldego, Lagrein, Dolcetto	Italy	839	0.02	300	772	0.02	285
Terrano	R	Teran	Italy	1914	0.04	188	209	0.00	475
Terrantez	W		Portugal	12	0.00	967	11	0.00	947
Terrantez do Pico	W		Portugal	0	0.00	1309	0	0.00	1407
Terras 20	R		France	0	0.00	1347	0	0.00	1461
Terret	W	Terret Blanc	France	1390	0.03	223	872	0.02	271
Terret Gris	W		France	78	0.00	666	76	0.00	631
Terret Noir	R		France	143	0.00	564	139	0.00	529
Therona	W		South Africa	99	0.00	622	67	0.00	648
Thrapsathiri	W		Greece	31	0.00	820	27	0.00	804

Table 13 (cont.): Prime varieties' colour, synonyms, country of origin, and global area, share and rank, 2010 and 2016 (ha and %)

				2010			2016		
Prime variety	*Col.*	*Synonyms*	*Country of origin*	*Area (hectares)*	*Share (%)*	*Rank*	*Area (hectares)*	*Share (%)*	*Rank*
Tibouren	R		France	443	0.01	391	432	0.01	375
Tihanyi Kék	R		Hungary				0	0.00	1515
Tilki Kuyrugu	W	Telti Kyryk	Armenia	246	0.01	488			
Timorasso	W		Italy	129	0.00	583	123	0.00	547
Timpuriu de Cluj	W		Romania	1	0.00	1251	1	0.00	1279
Tinta Aguiar	R		Portugal	75	0.00	673	77	0.00	629
Tinta Aurelio	R	Tinto do Aurelio	Portugal	0	0.00	1382	0	0.00	1549
Tinta Barroca	R	Barroco	Portugal	6172	0.13	99	4926	0.11	111
Tinta Bragao	R	Bragao	Portugal	63	0.00	706	64	0.00	655
Tinta Carvalha	R		Portugal	1311	0.03	232	1113	0.02	243
Tinta Castañal	R	Castañal	Spain				3	0.00	1122
Tinta da Barca	R	Barca	Portugal	345	0.01	429	352	0.01	394
Tinta da Melra	R	Melra	Portugal	1	0.00	1280	1	0.00	1344
Tinta de Alcoa	R	Alcoa	Portugal	24	0.00	867	24	0.00	826
Tinta de Cidadelhe	R	Cidadelhe	Portugal	1	0.00	1268	1	0.00	1327
Tinta de Pegoes	R	Tinto Pegoes	Portugal	195	0.00	522	191	0.00	490
Tinta do Rodo	R	Rodo	Portugal	3	0.00	1121	3	0.00	1112
Tinta Engomada	R	Engomada	Portugal	4	0.00	1106	4	0.00	1102
Tinta Francisca	R		Portugal	53	0.00	737	55	0.00	667
Tinta Malandra	R	Malandra	Portugal	0	0.00	1381	0	0.00	1547
Tinta Martins	R		Portugal	11	0.00	983	10	0.00	956
Tinta Mesquita	R		Portugal	1	0.00	1219	1	0.00	1262
Tinta Penajoia	R		Portugal	53	0.00	738	53	0.00	680
Tinta Pereira	R		Portugal	1	0.00	1229	1	0.00	1283
Tinta Pomar	R		Portugal	30	0.00	828	30	0.00	781
Tinta Roseira	R	Roseira	Portugal	4	0.00	1113	4	0.00	1105
Tinta Valdosa	R	Valdosa	Portugal	1	0.00	1284	1	0.00	1351
Tinta Varejoa	R	Varejoa	Portugal	1	0.00	1230	1	0.00	1278
Tintem	R		Portugal	9	0.00	1011	9	0.00	979
Tintilia del Molise	R	Tintilia	Italy	111	0.00	599	66	0.00	650
Tinto Cão	R	Tinto Cao; Tinta Cao	Portugal	369	0.01	412	372	0.01	388
Tinto de Zafra	R		Spain	2	0.00	1171			
Tinto Jeroma	R		Spain						
Tinto Velasco	R	Tinto de la Pampa Blanca; Tinto Velasco, Frasco	Spain	7829	0.17	88	5369	0.12	105
Torbato	W		Italy	52	0.00	741	46	0.00	706
Torrontes Mendocino	W	Torrontés Mendocino	Argentina	661	0.01	332	653	0.01	310
Torrontés Riojano	W	Torontel; Torrontes Riojano	Argentina	8937	0.19	77	8859	0.20	77
Torrontés Sanjuanino	W	Moscatel Austria; Moscatel De Austria; Torrontes Sanjuanino	Argentina	2531	0.05	165	3656	0.08	131
Tortosi	W	Tortosina	Spain	503	0.01	373	325	0.01	404
Touriga Femea	R		Portugal	15	0.00	937	15	0.00	902
Touriga Franca	R	Touriga Francesa	Portugal	11590	0.25	61	14224	0.32	52
Touriga Nacional	R	Touriga; Touriga Nacional N; Touriga National	Portugal	10446	0.23	67	11722	0.26	61
Trajadura	W	Treixadura	Portugal	2169	0.05	177	2492	0.06	161
Traminette	W		United States	240	0.01	492	239	0.01	452
Trbljan	W	Trbljan Bijeli	Croatia	231	0.01	496			
Trebbianina	W		Italy	128	0.00	585	30	0.00	782
Trebbiano d'Abruzzo	W	Trebbiano Abruzzese	Italy	5091	0.11	108	2630	0.06	157
Trebbiano Giallo	W		Italy	10664	0.23	65	2275	0.05	165
Trebbiano Modenese	W		Italy	363	0.01	415	287	0.01	429
Trebbiano Romagnolo	W		Italy	15893	0.34	49	19059	0.43	41
Trebbiano Spoletino	W		Italy	200	0.00	517	121	0.00	549
Trebbiano Toscano	W	St. Emilion; Talia; Trebbiano; Trebbiano Toscana; Trebbiano Toscano / Ugni blanc; Trebbiano Toscano Ugni blanc; Ugni blanc; Ugni Blanc (Trebbiano Toscana); Ugni	Italy	111290	2.41	9	120343	2.68	9
Trepat	R	Bonicaire; Parrel Verdal	Spain	1358	0.03	226	1199	0.03	233
Tressot	R		France	0	0.00	1310	0	0.00	1389
Trevisana Nera	R		Italy	15	0.00	928	11	0.00	939
Tribidrag	R	Crljenak Kastelanski; Kratoshija; Kratošija; Primitivo; Zinfandel	Croatia	32755	0.71	30	33649	0.75	25
Trilla	W		Hungary	1	0.00	1225	1	0.00	1272
Trincadeira	R	Tinta Amarela; Tinta Amarella; Trincadeira Preta; Trincdeira	Portugal	9270	0.20	75	10510	0.23	66
Trincadeira das Pratas	W		Portugal	239	0.01	494	124	0.00	542
Trincadeiro Branco	W	Trincadeira Branca	Portugal	59	0.00	715	49	0.00	701
Triomphe	R	Triomphe d´Alsace	France	15	0.00	932	3	0.00	1145
Triplett Blanc	W		United States	244	0.01	489	412	0.01	382
Triunfo	R		Portugal	2	0.00	1182	2	0.00	1191
Trnjak	R		Croatia	15	0.00	930			
Trobat	R	Panse Negro	Spain	1	0.00	1234	3	0.00	1128

Table 13 (cont.): Prime varieties' colour, synonyms, country of origin, and global area, share and rank, 2010 and 2016 (ha and %)

				2010			2016		
Prime variety	*Col.*	*Synonyms*	*Country of origin*	*Area (hectares)*	*Share (%)*	*Rank*	*Area (hectares)*	*Share (%)*	*Rank*
Tronto	R		Italy	2	0.00	1192	1	0.00	1281
Trousseau	R	Bastarda; Bastardo; Bastardo Roxo;Bastardo Tinto; Maturana Tinta; Merenzao; Tinta; Tinta Lisboa; Verdejo Negro; Trousseau Gris; Trousseau Noir; Bastardo Do Castello	France	3450	0.07	139	1263	0.03	224
Tsimlyansky Cherny	R	Tsymlansky Black	Russia	451	0.01	389	451	0.01	368
Tsitska	W		Georgia	3642	0.08	132	3642	0.08	132
Tsolikouri	W		Georgia	7903	0.17	86	7903	0.18	85
Tsulukidzis Tetra	W		Georgia	195	0.00	521	195	0.00	488
Tsvetochny	W	Tsvetochny (Floral); Tsvetochny (Flowery)	Russia	169	0.00	540	169	0.00	506
Turán	R	Agria; Turan	Hungary	177	0.00	533	183	0.00	495
Turchetta	R		Italy	3	0.00	1132	2	0.00	1160
Tyrian	R		Australia				38	0.00	733
Ucelut	W		Italy	10	0.00	998	10	0.00	970
UD 31103	R	UD-31.103	Italy				0	0.00	1486
Unirea	W		Romania	1	0.00	1283	1	0.00	1342
Úrréti	W	Urreti	Hungary	0	0.00	1355	0	0.00	1394
Usakhelouri	R		Georgia	10	0.00	989	10	0.00	958
Uva Cao	W		Portugal	1	0.00	1260	0	0.00	1382
Uva del Fantini	R		Italy	0	0.00	1346	0	0.00	1546
Uva del Tunde	R		Italy	2	0.00	1196	2	0.00	1225
Uva Longanesi	R		Italy	512	0.01	367	539	0.01	340
Uva Rara	R		Italy	460	0.01	387	197	0.00	485
Uva Tosca	R		Italy	71	0.00	682	29	0.00	786
Uvalino	R		Italy	1	0.00	1209	1	0.00	1246
Valais Noir	R		France						
Valbom	R		Portugal	166	0.00	543	162	0.00	513
Valdiguié	R	Napa Gamay; Valdiguie	France	272	0.01	467	126	0.00	537
Valenci Tinto	R		Spain	27	0.00	842	26	0.00	807
Valentino Nero	R		Italy	21	0.00	890	20	0.00	852
Valerien	W		France	24	0.00	860	23	0.00	839
Valiant	R		United States	11	0.00	980	11	0.00	949
Valveirinha	W	Valveirinho	Portugal	0	0.00	1316	0	0.00	1396
Valvin Muscat	W		United States	6	0.00	1058			
Vandal-Cliche	W		Canada				14	0.00	910
Varousset	R		France	5	0.00	1091	5	0.00	1082
Vasilaki	W		Turkey	4	0.00	1107	4	0.00	1093
VB 32-7	W	Sauvignon Soyhières (VB 32-7); Blattner Vb 32-7	Switzerland				3	0.00	1117
VB 91-26-25	R		Switzerland				1	0.00	1318
VB 91-26-26	R		Switzerland				0	0.00	1393
VB 91-26-27	R		Switzerland				0	0.00	1418
VB Cal 1-14	R		Switzerland				0	0.00	1509
VB Cal 1-29	R		Switzerland				0	0.00	1544
VB Cal 1-33	R		Switzerland				0	0.00	1543
VB Cal 6-04 N5	R		Switzerland				0	0.00	1377
VB Jura 25	R		Switzerland				0	0.00	1511
Vega	W		Italy	35	0.00	799	27	0.00	803
Ventura	W		Canada	24	0.00	862	10	0.00	960
Vênus	R	Venus	United States	1	0.00	1221	0	0.00	1422
Verdea	W		Italy	83	0.00	652	39	0.00	729
Verdeca	W	Lagorthi	Italy	796	0.02	304	913	0.02	262
Verdejo	W	Verdejo Blanco	Spain	16578	0.36	47	17931	0.40	43
Verdelet	W		France	1	0.00	1278	40	0.00	723
Verdelho	W		Portugal	2009	0.04	183	1516	0.03	208
Verdelho l'Anjou	W		France	0	0.00	1343	0	0.00	1462
Verdelho Tinto	R		Portugal	28	0.00	837	29	0.00	789
Verdello	W		Italy	287	0.01	457	179	0.00	498
Verdesse	W		France	10	0.00	993	3	0.00	1133
Verdial	W	Verdial Branco	Portugal	1	0.00	1288	0	0.00	1424
Verdial Tinto	R		Portugal	3	0.00	1151	3	0.00	1156
Verdicchio Bianco	W	Boschera; Peverella; Trebbiano di Soave; Verdicchio	Italy	3532	0.08	136	4682	0.10	118
Verdil	W		Spain	57	0.00	719	50	0.00	693
Verdiso	W	Pedevenda	Italy	68	0.00	692	52	0.00	684
Verdoncho	W		Spain	2124	0.05	179			
Verduschia	W		Italy	11	0.00	973	9	0.00	983
Verduzzo Friulano	W	Verduzzo	Italy	812	0.02	302	690	0.02	300
Verduzzo Trevigiano	W		Italy	708	0.02	320	531	0.01	344
Vermentino	W	Favorita; Favorita Diaz; Pigato; Vermentino B	Italy	8874	0.19	78	11483	0.26	62
Vermentino Nero	R	Vermintino Nero	Italy	210	0.00	510	124	0.00	541

Table 13 (cont.): Prime varieties' colour, synonyms, country of origin, and global area, share and rank, 2010 and 2016 (ha and %)

Prime variety	Col.	Synonyms	Country of origin	2010 Area (hectares)	2010 Share (%)	2010 Rank	2016 Area (hectares)	2016 Share (%)	2016 Rank
Vernaccia di Oristano	W	Vernaccia	Italy	272	0.01	468	246	0.01	449
Vernaccia di San Gimignano	W	Bervedino	Italy	522	0.01	361	884	0.02	270
Vertes Csillaga	W	Vértes Csillaga	Hungary	21	0.00	884	11	0.00	938
Vertzami	R	Lefkada	Greece	335	0.01	433	60	0.00	660
Veruccese	R		Italy	0	0.00	1306	0	0.00	1433
Vespaiola	W		Italy	94	0.00	628	90	0.00	595
Vespolina	R		Italy	134	0.00	577	88	0.00	602
Victoria	W	Viktorija	Russia	52	0.00	739	620	0.01	321
Vidadillo	R		Spain				38	0.00	734
Vidal	W	Vid al Blanc; Vidal 256; Vidal Blanc	France	1644	0.04	201	1936	0.04	178
Vidal Noir	R	Vidal Red	France						
Vidvizhenets	W		Russia	271	0.01	469	271	0.01	436
Vien de Nus	R		Italy	13	0.00	959	9	0.00	980
Vignoles	W		France	254	0.01	482	241	0.01	451
Vijariego	W	Diego; Vijiriego	Spain	285	0.01	459	369	0.01	389
Viktor	W		Hungary	0	0.00	1337			
Viktória Gyöngye	W	Viktoria Gyongye	Hungary	190	0.00	525	198	0.00	482
Vilana	W		Greece	579	0.01	350	650	0.01	311
Vilana (R)	W		Greece						
Villard Blanc	W	Seyve Villard 12375	France	654	0.01	333	743	0.02	288
Villard Noir	R		France	1273	0.03	234	777	0.02	284
Vincent	R		Canada	11	0.00	974	8	0.00	995
Vineland 53035	W	V 53	Canada				4	0.00	1098
Vineti	W		Spain						
Vinhao	R	Sousao; Souson; Souzao	Portugal	3160	0.07	142	4468	0.10	121
Viognier	W	Viogner	France	11785	0.26	60	16063	0.36	47
Violeta	R	BRS Violeta	Brazil	98	0.00	623	636	0.01	314
Viorika	W	Viorica	Moldova	347	0.01	426	558	0.01	338
Viosinho	W		Portugal	225	0.00	502	916	0.02	261
Vital	W		Portugal	1182	0.03	248	659	0.01	308
Vitovska	W	Vitouska	Italy	50	0.00	749	51	0.00	688
Vitovska Grganja	W		Slovenia				33	0.00	760
Voskeat	W	Voskehat	Armenia	809	0.02	303			
Vostorg	W		Russia				17	0.00	878
Vranac	R	Vranec	Montenegro	149	0.00	557	9503	0.21	75
Vugava	W		Croatia	36	0.00	796			
Vuillermin	R		Italy	4	0.00	1099	3	0.00	1138
Vulcanus	W		Hungary	5	0.00	1092	5	0.00	1056
Weldra	W		South Africa	14	0.00	949	2	0.00	1187
Würzer	W	Wurzer	Germany	70	0.00	685	54	0.00	678
Xara	R		Portugal	0	0.00	1373	0	0.00	1521
Xarello	W	Xarello Blanco	Spain	8394	0.18	81	8534	0.19	80
Xinomavro	R	Xynomavro	Greece	1971	0.04	184	2135	0.05	172
Xinomavro (W)	R	Xinomavro (White)	Greece						
Xynisteri	W		Cyprus	2092	0.05	181	1946	0.04	177
Yalovenskii Ustoichivyi	W	Ialovenschi ustoicivîi	Moldova				129	0.00	535
Yama Sauvignon	R		Japan				24	0.00	831
Yamabudo	R		Japan				35	0.00	747
Yamasachi	R		Japan				20	0.00	855
Yan 73	R		China				4800	0.11	114
Yaqui	R		Spain	2	0.00	1170	1	0.00	1286
Yubilei Zhuravlya	R	Iubilei Juravelea	Moldova				1	0.00	1259
Zalagyöngye	W	Perle von Zala; Zala Dende; Zala Gyöngye; Zala Gyoengye; Zalagyongye	Hungary	1948	0.04	186	1259	0.03	225
Zalema	W		Spain	4097	0.09	124	4015	0.09	128
Žametovka	R	Kavcina Crna; Zametovka	Slovenia	914	0.02	282	822	0.02	275
Zefír	W	Zefir	Hungary	49	0.00	753	15	0.00	904
Zelen	W		Slovenia				75	0.00	633
Zengő	W	Zengo	Hungary	264	0.01	474	226	0.01	463
Zenit	W	Zenith	Hungary	580	0.01	349	660	0.01	307
Zéta	W	Zeta	Hungary	118	0.00	588	118	0.00	557
Zeusz	W	Zeus	Hungary	28	0.00	841	27	0.00	798
Zghihară de Huşi	W	Zghihara de Husi	Romania	87	0.00	641	54	0.00	677
Zierfandler	G	Cirfandli	Austria	117	0.00	590	105	0.00	574
Žilavka	W	Zilavka	Bosnia and Herzegovina				185	0.00	491
Žlahtina	W	Zlahtina	Croatia	135	0.00	576			
Zlatarica Vrgorska	W		Croatia	19	0.00	899			
Župljanka	W	Zupjanka; Zupljanka	Serbia	4	0.00	1101	505	0.01	350
Zweigelt	R	Zweigeltrebe Blau; Zweigelt, Blauer; Zweigeltrebe	Austria	10029	0.22	69	9068	0.20	76

Table 14: Prime varieties' global area and share of all varieties globally, by prime's country of origin, 2000, 2010 and 2016 (ha and %)

	2000		2010		2016	
	Hectares	*Share (%)*	*Hectares*	*Share (%)*	*Hectares*	*Share (%)*
France						
Abondant	0	0.00	0	0.00	0	0.00
Abouriou	419	0.01	329	0.01	310	0.01
Admirable de Courtiller	0	0.00	28	0.00	27	0.00
Alicante Henri Bouschet	37157	0.80	38462	0.83	36031	0.80
Aligoté	35668	0.77	36120	0.78	26929	0.60
Alphonse Lavallée	15	0.00	862	0.02	634	0.01
Altesse	294	0.01	359	0.01	227	0.01
Aramon Bouschet	10	0.00	0	0.00	0	0.00
Aramon Noir	9157	0.20	2601	0.06	1181	0.03
Aramon Noir (W)	43	0.00	15	0.00	14	0.00
Aramont	1	0.00	0	0.00	0	0.00
Aranel	22	0.00	5	0.00	5	0.00
Arbane	1	0.00	2	0.00	1	0.00
Arinarnoa	150	0.00	189	0.00	486	0.01
Arriloba	59	0.00	55	0.00	54	0.00
Arrouya	0	0.00	0	0.00	0	0.00
Arrufiac	126	0.00	80	0.00	9	0.00
Aspiran Bouschet	433	0.01	2245	0.05	4088	0.09
Aubin Blanc	2	0.00	1	0.00	1	0.00
Aubun	1411	0.03	553	0.01	537	0.01
Aurore	299	0.01	268	0.01	255	0.01
Auxerrois	2302	0.05	2785	0.06	2853	0.06
Baco Blanc	2137	0.05	739	0.02	528	0.01
Baco Noir	397	0.01	475	0.01	735	0.02
Baleille	1	0.00	0	0.00	0	0.00
Barbaroux	79	0.00	30	0.00	29	0.00
Baroque	169	0.00	94	0.00	83	0.00
Beclan	0	0.00	0	0.00	0	0.00
Bellandais	2	0.00	0	0.00	0	0.00
Béquignol Noir	1083	0.02	891	0.02	616	0.01
Biancu Gentile	1	0.00	9	0.00	5	0.00
Blanc Dame	0	0.00	0	0.00	0	0.00
Blanqueiro	1	0.00	0	0.00	0	0.00
Boiziau	1	0.00	0	0.00	0	0.00
Bouchales	108	0.00	95	0.00	93	0.00
Bouillet	2	0.00	1	0.00	1	0.00
Bourboulenc	772	0.02	585	0.01	501	0.01
Bousquet Precoce	16	0.00	6	0.00	0	0.00
Braquet Noir	8	0.00	12	0.00	12	0.00
Brun Argente	14	0.00	11	0.00	11	0.00
Burdin	0	0.00	0	0.00	0	0.00

Table 14 (cont.): Prime varieties' global area and share of all varieties globally, by prime's country of origin, 2000, 2010 and 2016

	2000		2010		2016	
	Hectares	*Share (%)*	*Hectares*	*Share (%)*	*Hectares*	*Share (%)*
France (continued)						
Cabernet Franc	51974	1.13	61295	1.33	56052	1.25
Cabernet Malbec	34	0.00	0	0.00	0	0.00
Cabernet Sauvignon	223074	4.83	290083	6.28	310671	6.93
Caladoc	1427	0.03	3675	0.08	5258	0.12
Calitor Noir	85	0.00	26	0.00	26	0.00
Camaralet de Lasseube	691	0.01	520	0.01	306	0.01
Canari Noir	218	0.00	163	0.00	84	0.00
Carignan Bouschet	16	0.00	1	0.00	1	0.00
Carla	0	0.00	0	0.00	0	0.00
Carmenère	5711	0.12	11366	0.25	22486	0.50
Cascade	0	0.00	0	0.00	22	0.00
Castel	0	0.00	2	0.00	3	0.00
Castets	0	0.00	0	0.00	0	0.00
Cep Rouge	0	0.00	0	0.00	0	0.00
Cesar	8	0.00	10	0.00	15	0.00
Chambourcin	257	0.01	1097	0.02	968	0.02
Chancellor	27	0.00	49	0.00	38	0.00
Chardonnay	145543	3.15	199743	4.33	201649	4.50
Chasan	914	0.02	749	0.02	549	0.01
Chatus	15	0.00	79	0.00	71	0.00
Chelois	0	0.00	1	0.00	2	0.00
Chenanson	636	0.01	466	0.01	452	0.01
Chenin Blanc	45761	0.99	35703	0.77	32221	0.72
Chenivesse	1	0.00	0	0.00	0	0.00
Cinsaut	48428	1.05	34751	0.75	22926	0.51
Cinsaut (G)	0	0.00	0	0.00	2	0.00
Cinsaut (W)	41	0.00	7	0.00	0	0.00
Clairette	4359	0.09	3057	0.07	2420	0.05
Clarin	11	0.00	6	0.00	6	0.00
Claverie	3	0.00	1	0.00	1	0.00
Codivarta	3	0.00	2	0.00	1	0.00
Colmar Precoce Noir	0	0.00	0	0.00	0	0.00
Colobel	3	0.00	9	0.00	8	0.00
Colombard	38632	0.84	32944	0.71	29996	0.67
Côt	26285	0.57	38158	0.83	52233	1.17
Couderc 13	0	0.00	0	0.00	474	0.01
Couderc Noir	614	0.01	3517	0.08	2136	0.05
Counoise	638	0.01	408	0.01	418	0.01
Courbu Blanc	47	0.00	43	0.00	32	0.00
Courbu Noir	2	0.00	1	0.00	1	0.00
Crouchen	2259	0.05	725	0.02	319	0.01
Danam	0	0.00	0	0.00	0	0.00

Table 14 (cont.): Prime varieties' global area and share of all varieties globally, by prime's country of origin, 2000, 2010 and 2016

	2000		2010		2016	
	Hectares	*Share (%)*	*Hectares*	*Share (%)*	*Hectares*	*Share (%)*
France (continued)						
Danlas	0	0.00	255	0.01	203	0.00
Danuta	0	0.00	2	0.00	2	0.00
Dattier de St. Vallier	0	0.00	0	0.00	0	0.00
De Chaunac	186	0.00	91	0.00	102	0.00
Delhro	0	0.00	0	0.00	0	0.00
Douce Noire	18323	0.40	19630	0.43	19733	0.44
Duras	972	0.02	892	0.02	785	0.02
Durif	1197	0.03	3557	0.08	4807	0.11
Ederena	0	0.00	1	0.00	1	0.00
Egiodola	315	0.01	349	0.01	285	0.01
Ekigaina	5	0.00	4	0.00	3	0.00
Enfarine Noir	0	0.00	0	0.00	0	0.00
Etraire de l'Adui	8	0.00	5	0.00	5	0.00
Exalta	0	0.00	4	0.00	3	0.00
Excelsior	0	0.00	0	0.00	0	0.00
Fer	1626	0.04	1854	0.04	1686	0.04
Feunate	0	0.00	0	0.00	0	0.00
Florental	1	0.00	26	0.00	11	0.00
Folignan	0	0.00	51	0.00	51	0.00
Folle Blanche	2648	0.06	1803	0.04	1574	0.04
Fontan	0	0.00	0	0.00	0	0.00
Fuella Nera	4	0.00	20	0.00	20	0.00
Gaillard	0	0.00	0	0.00	0	0.00
Gamay Noir	37798	0.82	31927	0.69	26221	0.58
Gamay Teinturier de Bouze	318	0.01	278	0.01	255	0.01
Gamay Teinturier de Chaudenay	267	0.01	157	0.00	142	0.00
Gamay Teinturier Freaux	132	0.00	104	0.00	79	0.00
Ganson	28	0.00	3	0.00	3	0.00
Garonnet	7	0.00	14	0.00	10	0.00
Gascon	0	0.00	1	0.00	1	0.00
Goldriesling	10	0.00	21	0.00	24	0.00
Gouais Blanc	1	0.00	0	0.00	1	0.00
Gouget Noir	1	0.00	10	0.00	3	0.00
Graisse	22	0.00	14	0.00	1	0.00
Gramon	14	0.00	3	0.00	3	0.00
Grand Noir	949	0.02	955	0.02	707	0.02
Grassen	1	0.00	0	0.00	0	0.00
Gringet	74	0.00	25	0.00	15	0.00
Grolleau Noir	3006	0.07	2759	0.06	1949	0.04
Gros Manseng	2160	0.05	2960	0.06	3069	0.07
Guillemot	0	0.00	0	0.00	0	0.00
Hondarribi Zuri	0	0.00	0	0.00	624	0.01
Isa	0	0.00	9	0.00	5	0.00

Table 14 (cont.): Prime varieties' global area and share of all varieties globally, by prime's country of origin, 2000, 2010 and 2016

	2000		2010		2016	
	Hectares	*Share (%)*	*Hectares*	*Share (%)*	*Hectares*	*Share (%)*
France (continued)						
Jacquere	1086	0.02	1014	0.02	621	0.01
Jaoumet	0	0.00	0	0.00	0	0.00
Joannes Seyve	3	0.00	1	0.00	1	0.00
Joubertin	2	0.00	1	0.00	0	0.00
Jurançon Blanc	24	0.00	7	0.00	2	0.00
Jurançon Noir	1294	0.03	663	0.01	605	0.01
Jurie	0	0.00	0	0.00	0	0.00
Knipperlé	2	0.00	0	0.00	0	0.00
Lacoste	1	0.00	0	0.00	0	0.00
Landal	16	0.00	43	0.00	37	0.00
Landot Noir	0	0.00	3	0.00	3	0.00
Lauzet	1	0.00	3	0.00	3	0.00
Len de l'El	734	0.02	629	0.01	603	0.01
Léon Millot	17	0.00	102	0.00	85	0.00
Liliorila	3	0.00	4	0.00	1	0.00
Lival	0	0.00	101	0.00	99	0.00
Lucie Kuhlmann	0	0.00	0	0.00	21	0.00
Madeleine Royale	0	0.00	0	0.00	0	0.00
Madeleines	0	0.00	7	0.00	7	0.00
Malegue	1	0.00	0	0.00	0	0.00
Malingre Precoce	3	0.00	0	0.00	0	0.00
Maliverne	0	0.00	0	0.00	0	0.00
Manseng Noir	10	0.00	32	0.00	29	0.00
Maréchal Foch	173	0.00	356	0.01	229	0.01
Marsanne	1512	0.03	1763	0.04	1838	0.04
Marselan	176	0.00	2731	0.06	3941	0.09
Mauzac Blanc	3310	0.07	1933	0.04	1526	0.03
Mauzac Noir	10	0.00	0	0.00	0	0.00
Mecle de Bourgoin	2	0.00	0	0.00	0	0.00
Melon	13253	0.29	12306	0.27	9551	0.21
Menu Pineau	380	0.01	205	0.00	197	0.00
Merille	131	0.00	44	0.00	42	0.00
Merlot	213368	4.62	267888	5.80	266440	5.94
Merlot Blanc	176	0.00	46	0.00	44	0.00
Meslier Saint-Francois	55	0.00	15	0.00	13	0.00
Milgranet	2	0.00	1	0.00	1	0.00
Mireille	0	0.00	0	0.00	0	0.00
Molette	30	0.00	29	0.00	18	0.00
Mollard	23	0.00	23	0.00	23	0.00
Monbadon	657	0.01	498	0.01	0	0.00
Mondeuse Blanche	22	0.00	6	0.00	7	0.00
Mondeuse Noire	1404	0.03	303	0.01	287	0.01

Table 14 (cont.): Prime varieties' global area and share of all varieties globally, by prime's country of origin, 2000, 2010 and 2016

	2000		2010		2016	
	Hectares	*Share (%)*	Hectares	*Share (%)*	Hectares	*Share (%)*
France (continued)						
Monerac	2	0.00	3	0.00	3	0.00
Montils	131	0.00	164	0.00	165	0.00
Mornen Noir	1	0.00	0	0.00	0	0.00
Morrastel Bouschet	1	0.00	0	0.00	4	0.00
Mourvaison	9	0.00	3	0.00	3	0.00
Mouyssagues	0	0.00	0	0.00	0	0.00
Muscadelle	2207	0.05	1637	0.04	1509	0.03
Muscardin	19	0.00	17	0.00	17	0.00
Muscat Fleur d'Oranger	36	0.00	91	0.00	299	0.01
Muscat Ottonel	12259	0.27	10340	0.22	12464	0.28
Négrette	1319	0.03	1202	0.03	1112	0.02
Noir Fleurien	0	0.00	0	0.00	0	0.00
Noual	1	0.00	0	0.00	0	0.00
Oberlin	0	0.00	64	0.00	26	0.00
Oberlin White	1	0.00	0	0.00	0	0.00
Odola	1	0.00	0	0.00	0	0.00
Oeillade Bousche	10	0.00	1	0.00	1	0.00
Oeillade Noire	0	0.00	18	0.00	18	0.00
Olivette Blanche	0	0.00	2	0.00	2	0.00
Olivette Noire	0	0.00	16	0.00	16	0.00
Ondenc	12	0.00	8	0.00	1	0.00
Ora	0	0.00	38	0.00	32	0.00
Pascal Blanc	0	0.00	0	0.00	0	0.00
Peloursin	0	0.00	0	0.00	0	0.00
Perdea	13	0.00	2	0.00	2	0.00
Perlaut	0	0.00	1	0.00	1	0.00
Persan	3	0.00	12	0.00	12	0.00
Petit Bouschet	1	0.00	15	0.00	120	0.00
Petit Courbu	75	0.00	102	0.00	1	0.00
Petit Manseng	613	0.01	1109	0.02	1299	0.03
Petit Meslier	3	0.00	4	0.00	3	0.00
Petit Verdot	1640	0.04	7195	0.16	8124	0.18
Picardan	0	0.00	1	0.00	1	0.00
Pineau d'Aunis	430	0.01	437	0.01	413	0.01
Pinot Blanc	16983	0.37	14812	0.32	13779	0.31
Pinot Gris	18893	0.41	43773	0.95	48570	1.08
Pinot Meunier	13131	0.28	13566	0.29	14695	0.33
Pinot Noir	68810	1.49	98623	2.14	105480	2.35
Pinot Noir Précoce	85	0.00	273	0.01	251	0.01
Piquepoul Blanc	975	0.02	1492	0.03	1565	0.03
Piquepoul Bousch	0	0.00	0	0.00	0	0.00
Piquepoul Gris	9	0.00	2	0.00	2	0.00

Table 14 (cont.): Prime varieties' global area and share of all varieties globally, by prime's country of origin, 2000, 2010 and 2016

	2000		2010		2016	
	Hectares	*Share (%)*	*Hectares*	*Share (%)*	*Hectares*	*Share (%)*
France (continued)						
Piquepoul Noir	103	0.00	70	0.00	69	0.00
Plant Droit	40	0.00	19	0.00	19	0.00
Plantet	209	0.00	1060	0.02	420	0.01
Portan	368	0.01	264	0.01	256	0.01
Pougnet	0	0.00	0	0.00	0	0.00
Poulsard	295	0.01	307	0.01	90	0.00
Poulsard Blanc	14	0.00	0	0.00	0	0.00
Prima	0	0.00	84	0.00	81	0.00
Provareau	2	0.00	0	0.00	0	0.00
Prunelard	2	0.00	20	0.00	19	0.00
Raffiat de Moncade	16	0.00	7	0.00	6	0.00
Raisaine	1	0.00	0	0.00	0	0.00
Ravat	1	0.00	1	0.00	1	0.00
Ravat Blanc	2	0.00	7	0.00	6	0.00
Rayon d'Or	1	0.00	6	0.00	6	0.00
Ribol	0	0.00	147	0.00	141	0.00
Romorantin	81	0.00	72	0.00	69	0.00
Rose du Var	129	0.00	56	0.00	54	0.00
Roublot	0	0.00	0	0.00	0	0.00
Rougeon	36	0.00	42	0.00	21	0.00
Roussanne	874	0.02	1851	0.04	2137	0.05
Roussette d'Ayze	3	0.00	1	0.00	1	0.00
Rubilande	3	0.00	8	0.00	7	0.00
Sacy	63	0.00	10	0.00	8	0.00
Saint Macaire	1	0.00	0	0.00	6	0.00
Saint-Pierre Dore	1	0.00	0	0.00	0	0.00
Salvador	572	0.01	394	0.01	351	0.01
Sauvignon Blanc	65190	1.41	111552	2.42	124700	2.78
Sauvignon Blanc (G)	76	0.00	698	0.02	1076	0.02
Sauvignonasse	5452	0.12	4563	0.10	3861	0.09
Savagnin Blanc	441	0.01	1950	0.04	2267	0.05
Savagnin Rose	883	0.02	884	0.02	48	0.00
Segalin	54	0.00	65	0.00	61	0.00
Seibel	1991	0.04	592	0.01	482	0.01
Seibel White	15	0.00	0	0.00	0	0.00
Seinoir	15	0.00	87	0.00	50	0.00
Select	8	0.00	7	0.00	7	0.00
Semebat	2	0.00	2	0.00	1	0.00
Sémillon	26239	0.57	22157	0.48	18693	0.42
Servanin	0	0.00	0	0.00	0	0.00
Servant	530	0.01	183	0.00	138	0.00

Table 14 (cont.): Prime varieties' global area and share of all varieties globally, by prime's country of origin, 2000, 2010 and 2016

	2000		2010		2016	
	Hectares	*Share (%)*	*Hectares*	*Share (%)*	*Hectares*	*Share (%)*
France (continued)						
Seyval Blanc	389	0.01	569	0.01	2699	0.06
Seyve Villard 23-512	0	0.00	29	0.00	0	0.00
Sicilien	0	0.00	5	0.00	5	0.00
Sulima	0	0.00	0	0.00	0	0.00
Syrah	102490	2.22	185117	4.01	181185	4.04
Tannat	5595	0.12	5765	0.12	5611	0.13
Teinturier	1	0.00	7	0.00	9	0.00
Teoulier Noir	1	0.00	0	0.00	0	0.00
Terras 20	13	0.00	0	0.00	0	0.00
Terret	2703	0.06	1390	0.03	872	0.02
Terret Gris	262	0.01	78	0.00	76	0.00
Terret Noir	370	0.01	143	0.00	139	0.00
Tibouren	457	0.01	443	0.01	432	0.01
Tressot	0	0.00	0	0.00	0	0.00
Triomphe	0	0.00	15	0.00	3	0.00
Trousseau	2223	0.05	3450	0.07	1263	0.03
Valais Noir	4	0.00	0	0.00	0	0.00
Valdiguié	79	0.00	272	0.01	126	0.00
Valerien	2	0.00	24	0.00	23	0.00
Varousset	12	0.00	5	0.00	5	0.00
Verdelet	0	0.00	1	0.00	40	0.00
Verdelho l'Anjou	0	0.00	0	0.00	0	0.00
Verdesse	8	0.00	10	0.00	3	0.00
Vidal	611	0.01	1644	0.04	1936	0.04
Vidal Noir	0	0.00	0	0.00	0	0.00
Vignoles	68	0.00	254	0.01	241	0.01
Villard Blanc	746	0.02	654	0.01	743	0.02
Villard Noir	601	0.01	1273	0.03	777	0.02
Viognier	3160	0.07	11785	0.26	16063	0.36
Total (280 varieties)	**1361768**	**29.46**	**1723075**	**37.29**	**1746654**	**38.92**
Spain						
Airén	387978	8.41	252364	5.47	203801	4.55
Alcañon	54	0.00	60	0.00	27	0.00
Aledo	0	0.00	7	0.00	2	0.00
Beba	4762	0.10	3036	0.07	2556	0.06
Bobal	100128	2.17	80120	1.74	59189	1.32
Brocada	2	0.00	0	0.00	0	0.00
Calagrano	8229	0.18	4794	0.10	0	0.00
Callet	151	0.00	154	0.00	138	0.00
Canela	0	0.00	1	0.00	3	0.00
Canorroyo	157	0.00	0	0.00	0	0.00

Table 14 (cont.): Prime varieties' global area and share of all varieties globally, by prime's country of origin, 2000, 2010 and 2016

	2000		2010		2016	
	Hectares	*Share (%)*	*Hectares*	*Share (%)*	*Hectares*	*Share (%)*
Spain (continued)						
Cartouche	31	0.00	0	0.00	0	0.00
Castellana Blanca	0	0.00	0	0.00	1	0.00
Castonotal	0	0.00	0	0.00	0	0.00
Cayetana Blanca	55776	1.21	39781	0.86	36401	0.81
Chelva	10877	0.24	6168	0.13	5029	0.11
Cinsaut Seedless	9	0.00	13	0.00	0	0.00
Coloraillo	614	0.01	374	0.01	109	0.00
Corropio	0	0.00	1	0.00	1	0.00
De Cilindro	15	0.00	0	0.00	0	0.00
Decuerno	4	0.00	0	0.00	0	0.00
Dominga	0	0.00	1	0.00	0	0.00
Ensanyo Tintas	0	0.00	27	0.00	0	0.00
Ensayo Blancas	0	0.00	1	0.00	0	0.00
Ferral	48	0.00	30	0.00	31	0.00
Fino de Ribera del Fresno	332	0.01	45	0.00	8	0.00
Fintendo	144	0.00	118	0.00	185	0.00
Fogoneu	36	0.00	35	0.00	15	0.00
Forcallat Tinta	2690	0.06	1163	0.03	535	0.01
Garnacha Blanca	10821	0.23	7398	0.16	7409	0.17
Garnacha Peluda	2024	0.04	1206	0.03	898	0.02
Garnacha Roja (Gris)	2761	0.06	2366	0.05	1462	0.03
Garnacha Tinta	216349	4.69	181553	3.93	150096	3.35
Garrido Fino	174	0.00	59	0.00	54	0.00
Gateta	7	0.00	2	0.00	0	0.00
Gibi	1227	0.03	1074	0.02	785	0.02
Godello	1489	0.03	1332	0.03	1406	0.03
Gorgollasa	0	0.00	0	0.00	5	0.00
Graciano	1960	0.04	3123	0.07	2910	0.06
Grand Manchen	0	0.00	8	0.00	0	0.00
Gualarido	0	0.00	0	0.00	18	0.00
Hondarribi Beltza	11	0.00	53	0.00	15	0.00
Huerta del Rey	35	0.00	2	0.00	0	0.00
Imperial Napoleon	0	0.00	12	0.00	0	0.00
Jeroma	0	0.00	0	0.00	0	0.00
Juan García	2077	0.04	1707	0.04	1545	0.03
Lado	1	0.00	1	0.00	2	0.00
Lairen	298	0.01	214	0.00	351	0.01
Listain de Huelva	596	0.01	350	0.01	466	0.01
Listan Negro	3291	0.07	2666	0.06	2847	0.06
Listan Prieto	16232	0.35	4659	0.10	10267	0.23
Luisa Blanca	0	0.00	0	0.00	80	0.00
Macabeo	48128	1.04	40864	0.89	38625	0.86
Malvasia di Sardegna Rosada	0	0.00	3	0.00	3	0.00

Table 14 (cont.): Prime varieties' global area and share of all varieties globally, by prime's country of origin, 2000, 2010 and 2016

	2000		2010		2016	
	Hectares	*Share (%)*	*Hectares*	*Share (%)*	*Hectares*	*Share (%)*
Spain (continued)						
Mandon	261	0.01	0	0.00	1	0.00
Manto Negro	470	0.01	273	0.01	311	0.01
Marfal	0	0.00	0	0.00	1	0.00
Marmajuelo	37	0.00	24	0.00	20	0.00
Maturana Blanca	0	0.00	18	0.00	13	0.00
Mazuelo	127692	2.77	75716	1.64	47312	1.06
Mazuelo (G)	25	0.00	0	0.00	0	0.00
Mazuelo (W)	1035	0.02	3016	0.07	355	0.01
Mencía	13138	0.28	10658	0.23	11052	0.25
Merseguera	7460	0.16	3946	0.09	2373	0.05
Miguel del Arco	1267	0.03	468	0.01	0	0.00
Monastrell	76304	1.65	69742	1.51	51930	1.16
Monstruosa	2	0.00	1	0.00	0	0.00
Moravia Agria	1092	0.02	550	0.01	222	0.00
Morenillo	39	0.00	0	0.00	0	0.00
Moristel	0	0.00	147	0.00	247	0.01
Naparo	0	0.00	1	0.00	0	0.00
Negramoll	3557	0.08	3193	0.07	3011	0.07
Ohanes	0	0.00	1	0.00	0	0.00
Ondarrabi Zuri	189	0.00	492	0.01	0	0.00
Palomino Fino	30513	0.66	22693	0.49	23190	0.52
Palot	4	0.00	0	0.00	0	0.00
Panse Valenciano	649	0.01	1	0.00	0	0.00
Pardillo	7272	0.16	4364	0.09	3283	0.07
Parellada	11188	0.24	8847	0.19	7137	0.16
Parraleta	167	0.00	348	0.01	212	0.00
Pedro Ximénez	17272	0.37	9235	0.20	8810	0.20
Perlita	1	0.00	1	0.00	0	0.00
Perruno	2831	0.06	1509	0.03	745	0.02
Picapoll Blanco	34	0.00	37	0.00	40	0.00
Planta Mula	1	0.00	24	0.00	11	0.00
Planta Nova	2029	0.04	1395	0.03	888	0.02
Prensal	114	0.00	105	0.00	129	0.00
Prieto Picudo	3256	0.07	4587	0.10	4293	0.10
Quiebratinajas Tinto	6	0.00	5	0.00	9	0.00
Rojal Tinta	2845	0.06	1801	0.04	736	0.02
Rome	2	0.00	297	0.01	172	0.00
Royal de Alloza	0	0.00	29	0.00	6	0.00
Sumoll	1401	0.03	83	0.00	16	0.00
Tempranillo	93370	2.02	232988	5.05	219379	4.89
Tempranillo (W)	0	0.00	5	0.00	110	0.00
Teneron	3488	0.08	3488	0.08	0	0.00
Tinta Castañal	0	0.00	0	0.00	3	0.00

Table 14 (cont.): Prime varieties' global area and share of all varieties globally, by prime's country of origin, 2000, 2010 and 2016

	2000		2010		2016	
	Hectares	*Share (%)*	*Hectares*	*Share (%)*	*Hectares*	*Share (%)*
Spain (continued)						
Tinto de Zafra	4	0.00	2	0.00	0	0.00
Tinto Jeroma	1	0.00	0	0.00	0	0.00
Tinto Velasco	7998	0.17	7829	0.17	5369	0.12
Tortosina	930	0.02	503	0.01	325	0.01
Trepat	1763	0.04	1358	0.03	1199	0.03
Trobat	83	0.00	1	0.00	3	0.00
Valenci Tinto	5	0.00	27	0.00	26	0.00
Valensi du Maroc	0	0.00	15	0.00	15	0.00
Verdejo	4453	0.10	16578	0.36	17931	0.40
Verdil	131	0.00	57	0.00	50	0.00
Verdoncho	3092	0.07	2124	0.05	0	0.00
Vidadillo	0	0.00	0	0.00	38	0.00
Vijariego	510	0.01	285	0.01	369	0.01
Vineti	0	0.00	0	0.00	0	0.00
Xarello	10299	0.22	8394	0.18	8534	0.19
Yaqui	22	0.00	2	0.00	1	0.00
Zalema	5969	0.13	4097	0.09	4015	0.09
Total (117 varieties)	**1323786**	**28.68**	**1138305**	**24.66**	**951163**	**21.22**
Italy						
Abbuoto	696	0.02	37	0.00	18	0.00
Abrusco	399	0.01	423	0.01	215	0.00
Aglianico	9346	0.20	9995	0.22	9734	0.22
Aglianicone	148	0.00	62	0.00	30	0.00
Albana	2487	0.05	1523	0.03	782	0.02
Albanello	117	0.00	18	0.00	2	0.00
Albaranzeuli Bianco	72	0.00	7	0.00	2	0.00
Albaranzeuli Nero	40	0.00	49	0.00	28	0.00
Albarola	4090	0.09	197	0.00	95	0.00
Albarossa	5	0.00	80	0.00	70	0.00
Aleatico	458	0.01	346	0.01	165	0.00
Alionza	41	0.00	11	0.00	9	0.00
Ancellotta	4405	0.10	4681	0.10	2739	0.06
Arneis	738	0.02	1122	0.02	1179	0.03
Arvesiniadu	147	0.00	30	0.00	13	0.00
Avana	53	0.00	28	0.00	18	0.00
Avarengo	1453	0.03	987	0.02	153	0.00
Baratuciat	0	0.00	2	0.00	2	0.00
Barbarossa	16	0.00	0	0.00	0	0.00
Barbera	33041	0.72	24366	0.53	17824	0.40
Barbera Bianca	251	0.01	181	0.00	114	0.00

Table 14 (cont.): Prime varieties' global area and share of all varieties globally, by prime's country of origin, 2000, 2010 and 2016

	2000		2010		2016	
	Hectares	*Share (%)*	*Hectares*	*Share (%)*	*Hectares*	*Share (%)*
Italy (continued)						
Barbera Sarda	326	0.01	84	0.00	70	0.00
Bariadorgia	0	0.00	0	0.00	0	0.00
Barsaglina	20	0.00	17	0.00	9	0.00
Bellone	1315	0.03	511	0.01	184	0.00
Biancame	1330	0.03	2599	0.06	1336	0.03
Bianchetta Trevigiana	53	0.00	13	0.00	12	0.00
Bianco d'Alessano	941	0.02	419	0.01	39	0.00
Biancolella	385	0.01	164	0.00	23	0.00
Biancone di Portoferraio	67	0.00	78	0.00	34	0.00
Bombino Bianco	2903	0.06	1239	0.03	1147	0.03
Bombino Nero	1156	0.03	1201	0.03	865	0.02
Bonamico	336	0.01	233	0.01	149	0.00
Bonarda Grande	538	0.01	0	0.00	0	0.00
Bonarda Piemontese	23	0.00	6	0.00	5926	0.13
Bonda	3	0.00	7	0.00	7	0.00
Bosco	88	0.00	82	0.00	50	0.00
Bracciola Nera	89	0.00	26	0.00	4	0.00
Brachetto del Piemonte	1534	0.03	1460	0.03	1694	0.04
Bric	21	0.00	2	0.00	1	0.00
Bussanello	8	0.00	12	0.00	3	0.00
Caddiu	978	0.02	309	0.01	83	0.00
Caloria	129	0.00	108	0.00	45	0.00
Canaiolo Nero	2418	0.05	1068	0.02	1033	0.02
Capolongo	0	0.00	5	0.00	0	0.00
Carbernet Volos	0	0.00	0	0.00	0	0.00
Carica l'Asino	299	0.01	17	0.00	5	0.00
Carricante	252	0.01	205	0.00	35	0.00
Casavecchia	0	0.00	136	0.00	92	0.00
Casetta	0	0.00	12	0.00	14	0.00
Castiglione	83	0.00	18	0.00	4	0.00
Catalanesca	0	0.00	54	0.00	7	0.00
Catanese Nero	76	0.00	15	0.00	7	0.00
Catarratto Bianco	50711	1.10	34863	0.76	28613	0.64
Cavrara	0	0.00	23	0.00	1	0.00
Cellerina	0	0.00	2	0.00	2	0.00
Centesimino	0	0.00	24	0.00	25	0.00
Cesanese	1024	0.02	679	0.01	446	0.01
Cianorie	0	0.00	2	0.00	1	0.00
Ciliegiolo	2527	0.05	1830	0.04	897	0.02
Cividin	0	0.00	4	0.00	4	0.00
Cococciola	887	0.02	983	0.02	1671	0.04
Coda di Volpe Bianca	980	0.02	586	0.01	77	0.00

Table 14 (cont.): Prime varieties' global area and share of all varieties globally, by prime's country of origin, 2000, 2010 and 2016

	2000		2010		2016	
	Hectares	*Share (%)*	*Hectares*	*Share (%)*	*Hectares*	*Share (%)*
Italy (continued)						
Colombana Nera	126	0.00	38	0.00	16	0.00
Corbina Vicentina	0	0.00	12	0.00	12	0.00
Cordenossa	0	0.00	5	0.00	2	0.00
Cornalin	93	0.00	256	0.01	147	0.00
Cornarea	22	0.00	13	0.00	8	0.00
Cornichon Blanc	0	0.00	0	0.00	1	0.00
Cortese	3113	0.07	2953	0.06	2405	0.05
Corvina Veronese	4800	0.10	7496	0.16	6240	0.14
Corvinone	88	0.00	930	0.02	1140	0.03
Cove	56	0.00	6	0.00	6	0.00
Crovassa	2	0.00	0	0.00	0	0.00
Damaschino	3187	0.07	2171	0.05	1622	0.04
Dindarella	9	0.00	7	0.00	5	0.00
Dolcetto	7197	0.16	6333	0.14	4545	0.10
Dolciame	6	0.00	11	0.00	6	0.00
Doux d'Henry	26	0.00	9	0.00	6	0.00
Drupeggio	617	0.01	286	0.01	81	0.00
DU 31120	0	0.00	0	0.00	0	0.00
Durella	599	0.01	470	0.01	480	0.01
Enantio	1062	0.02	724	0.02	178	0.00
Erbaluce	329	0.01	319	0.01	316	0.01
Erbamat	0	0.00	24	0.00	2	0.00
Ervi	5	0.00	4	0.00	3	0.00
Falanghina	1658	0.04	3037	0.07	323	0.01
Falanghina Flegrea	0	0.00	0	0.00	3634	0.08
Famoso	0	0.00	6	0.00	6	0.00
Fenile	0	0.00	5	0.00	1	0.00
Fertilia	13	0.00	3	0.00	2	0.00
Fiano	758	0.02	1377	0.03	2187	0.05
Flavis	12	0.00	3	0.00	3	0.00
Fleurtai	0	0.00	0	0.00	0	0.00
Fogarina	0	0.00	5	0.00	3	0.00
Foglia Tonda	40	0.00	101	0.00	68	0.00
Forastera	543	0.01	208	0.00	8	0.00
Forgiarin	2	0.00	4	0.00	3	0.00
Forsellina	9	0.00	7	0.00	7	0.00
Fortana	1252	0.03	642	0.01	469	0.01
Francavidda	86	0.00	13	0.00	2	0.00
Frappato	784	0.02	752	0.02	580	0.01
Freisa	1450	0.03	1054	0.02	519	0.01
Fubiano	2	0.00	9	0.00	2	0.00
Fumin	73	0.00	31	0.00	25	0.00
Gaglioppo	3592	0.08	4214	0.09	4626	0.10

Table 14 (cont.): Prime varieties' global area and share of all varieties globally, by prime's country of origin, 2000, 2010 and 2016

	2000		2010		2016	
	Hectares	*Share (%)*	*Hectares*	*Share (%)*	*Hectares*	*Share (%)*
Italy (continued)						
Gamba Rossa	0	0.00	1	0.00	0	0.00
Garganega	16553	0.36	15397	0.33	8554	0.19
Ginestra	0	0.00	4	0.00	1	0.00
Giro Nero	537	0.01	200	0.00	144	0.00
Goldtraminer	0	0.00	9	0.00	5	0.00
Gosen	0	0.00	1	0.00	0	0.00
Grapariol	0	0.00	2	0.00	2	0.00
Grechetto di Orvieto	1177	0.03	1501	0.03	1824	0.04
Grechetto Rosso	111	0.00	49	0.00	35	0.00
Greco	1325	0.03	158	0.00	21	0.00
Greco Bianco	660	0.01	1604	0.03	2050	0.05
Greco Nero	3229	0.07	1256	0.03	437	0.01
Grignolino	1353	0.03	915	0.02	911	0.02
Grillo	1803	0.04	6295	0.14	7383	0.16
Groppello di Mocasina	120	0.00	81	0.00	24	0.00
Groppello di Revo	0	0.00	12	0.00	12	0.00
Groppello Gentile	219	0.00	326	0.01	78	0.00
Guardavalle	168	0.00	33	0.00	16	0.00
Ignea	0	0.00	1	0.00	1	0.00
Impigno	60	0.00	7	0.00	2	0.00
Incrocio Bianco Fedit 51	11	0.00	5	0.00	1	0.00
Incrocio Bruni 54	12	0.00	12	0.00	3	0.00
Incrocio Manzoni 2.15	166	0.00	86	0.00	72	0.00
Incrocio Terzi 1	65	0.00	44	0.00	12	0.00
Invernenga	32	0.00	7	0.00	5	0.00
Inzolia	9259	0.20	6133	0.13	4740	0.11
Italia	1076	0.02	1463	0.03	5188	0.12
Italica	178	0.00	367	0.01	47	0.00
Lacrima Christi	0	0.00	85	0.00	226	0.01
Lacrima di Morro d'Alba	652	0.01	421	0.01	252	0.01
Lagarino Bianco	0	0.00	23	0.00	6	0.00
Lagrein	471	0.01	718	0.02	251	0.01
Lambrusca di Alessandria	888	0.02	137	0.00	58	0.00
Lambrusco	42	0.00	45	0.00	54	0.00
Lambrusco Barghi	0	0.00	18	0.00	3	0.00
Lambrusco di Sorbara	1409	0.03	1606	0.03	858	0.02
Lambrusco Grasparossa	1720	0.04	2734	0.06	954	0.02
Lambrusco Maestri	1513	0.03	2312	0.05	5657	0.13
Lambrusco Marani	2280	0.05	1394	0.03	1074	0.02
Lambrusco Montericco	262	0.01	70	0.00	25	0.00
Lambrusco Oliva	0	0.00	112	0.00	104	0.00
Lambrusco Salamino	4147	0.09	5003	0.11	6228	0.14
Lambrusco Viadanese	277	0.01	240	0.01	59	0.00

Table 14 (cont.): Prime varieties' global area and share of all varieties globally, by prime's country of origin, 2000, 2010 and 2016

	2000		2010		2016	
	Hectares	*Share (%)*	*Hectares*	*Share (%)*	*Hectares*	*Share (%)*
Italy (continued)						
Lecinaro	0	0.00	1	0.00	0	0.00
Lumassina	111	0.00	98	0.00	37	0.00
Maceratino	122	0.00	177	0.00	39	0.00
Magliasina	0	0.00	0	0.00	0	0.00
Magliocco Canino	579	0.01	539	0.01	679	0.02
Magliocco Dolce	243	0.01	87	0.00	51	0.00
Maiolica	70	0.00	26	0.00	13	0.00
Maiolina	0	0.00	1	0.00	0	0.00
Malbo Gentile	106	0.00	211	0.00	219	0.00
Malvasia Bianca di Basilicata	875	0.02	210	0.00	32	0.00
Malvasia Bianca di Candia	12889	0.28	9351	0.20	9685	0.22
Malvasia Bianca Lunga	3937	0.09	2544	0.06	1247	0.03
Malvasia del Lazio	2366	0.05	590	0.01	680	0.02
Malvasia di Candia Aromatica	1754	0.04	927	0.02	1208	0.03
Malvasia di Casorzo	98	0.00	107	0.00	99	0.00
Malvasia di Lipari	516	0.01	310	0.01	113	0.00
Malvasia di Schierano	181	0.00	89	0.00	82	0.00
Malvasia Moscata	0	0.00	554	0.01	0	0.00
Malvasia Nera di Basilicata	754	0.02	114	0.00	39	0.00
Malvasia Nera di Brindisi	3174	0.07	1314	0.03	1264	0.03
Malvasia Nera Lunga	0	0.00	38	0.00	14	0.00
Mammolo	777	0.02	841	0.02	911	0.02
Manzoni Bianco	8290	0.18	382	0.01	339	0.01
Manzoni Moscato	0	0.00	20	0.00	19	0.00
Manzoni Rosa	0	0.00	29	0.00	23	0.00
Marzemina Bianca	78	0.00	54	0.00	55	0.00
Marzemino	994	0.02	1091	0.02	785	0.02
Mayolet	4	0.00	7	0.00	6	0.00
Mazzese	80	0.00	76	0.00	57	0.00
Melara	13	0.00	3	0.00	1	0.00
Merlese	0	0.00	14	0.00	8	0.00
Merlot Khorus	0	0.00	0	0.00	0	0.00
Michele Palieri	24	0.00	1	0.00	1	0.00
Minella Bianca	73	0.00	65	0.00	19	0.00
Molinara	1637	0.04	717	0.02	609	0.01
Monica Nera	2835	0.06	1404	0.03	1203	0.03
Montepulciano	28728	0.62	34956	0.76	32935	0.73
Montonico Bianco	656	0.01	734	0.02	567	0.01
Montu	1091	0.02	0	0.00	0	0.00
Moradella	0	0.00	6	0.00	1	0.00
Morone	22	0.00	13	0.00	7	0.00
Moscatello Selvatico	105	0.00	35	0.00	5	0.00
Moscato di Scanzo	73	0.00	53	0.00	10	0.00

Table 14 (cont.): Prime varieties' global area and share of all varieties globally, by prime's country of origin, 2000, 2010 and 2016

	2000		2010		2016	
	Hectares	*Share (%)*	*Hectares*	*Share (%)*	*Hectares*	*Share (%)*
Italy (continued)						
Moscato di Terracina	229	0.00	138	0.00	22	0.00
Moscato Giallo	542	0.01	1467	0.03	1634	0.04
Moscato Rosa del Trentino	93	0.00	81	0.00	39	0.00
Mostosa	95	0.00	24	0.00	16	0.00
Nascetta	0	0.00	21	0.00	17	0.00
Nasco	166	0.00	141	0.00	91	0.00
Nebbiera	20	0.00	12	0.00	9	0.00
Nebbiolo	5264	0.11	6125	0.13	7997	0.18
Negretto	280	0.01	75	0.00	35	0.00
Negroamaro	16619	0.36	11492	0.25	11449	0.26
Ner d'Ala	8	0.00	30	0.00	10	0.00
Nerello Cappuccio	1501	0.03	508	0.01	125	0.00
Nerello Mascalese	4167	0.09	2883	0.06	1805	0.04
Neretta Cuneese	374	0.01	132	0.00	119	0.00
Neretto di Bairo	53	0.00	34	0.00	19	0.00
Nero Buono di Cori	114	0.00	135	0.00	58	0.00
Nero di Troia	1765	0.04	2572	0.06	2512	0.06
Neyret	76	0.00	41	0.00	12	0.00
Nieddera	58	0.00	107	0.00	91	0.00
Nigra	7	0.00	3	0.00	1	0.00
Nocera	27	0.00	15	0.00	5	0.00
Nosiola	191	0.00	79	0.00	65	0.00
Notardomenico	13	0.00	10	0.00	9	0.00
Nuragus	3186	0.07	1345	0.03	1008	0.02
Orpicchio	0	0.00	1	0.00	0	0.00
Ortrugo	485	0.01	611	0.01	709	0.02
Oseleta	0	0.00	15	0.00	16	0.00
Pallagrello Bianco	0	0.00	55	0.00	6	0.00
Pallagrello Nero	0	0.00	169	0.00	107	0.00
Pampanaro	0	0.00	5	0.00	1	0.00
Pampanuto	277	0.01	356	0.01	33	0.00
Paolina	0	0.00	1	0.00	0	0.00
Pascale	1573	0.03	375	0.01	289	0.01
Passau	12	0.00	5	0.00	3	0.00
Passerina	715	0.02	894	0.02	933	0.02
Pavana	69	0.00	32	0.00	20	0.00
Pecorello	16	0.00	34	0.00	9	0.00
Pecorino	166	0.00	1228	0.03	1742	0.04
Pelaverga	28	0.00	55	0.00	46	0.00
Pelaverga Piccolo	24	0.00	6	0.00	6	0.00
Pepella	0	0.00	3	0.00	0	0.00
Perera	25	0.00	4	0.00	2	0.00
Perla dei Vivi	0	0.00	1	0.00	1	0.00

Table 14 (cont.): Prime varieties' global area and share of all varieties globally, by prime's country of origin, 2000, 2010 and 2016

	2000		2010		2016	
	Hectares	*Share (%)*	*Hectares*	*Share (%)*	*Hectares*	*Share (%)*
Italy (continued)						
Perricone	580	0.01	228	0.00	80	0.00
Petit Rouge	100	0.00	84	0.00	68	0.00
Piccola Nera	17	0.00	17	0.00	6	0.00
Picolit	93	0.00	128	0.00	121	0.00
Piculit Neri	126	0.00	22	0.00	8	0.00
Piedirosso	896	0.02	699	0.02	593	0.01
Pignola Valtellinese	70	0.00	49	0.00	28	0.00
Pignoletto	6009	0.13	1707	0.04	1174	0.03
Pignolo	18	0.00	93	0.00	50	0.00
Pinella	66	0.00	72	0.00	128	0.00
Plassa	41	0.00	91	0.00	86	0.00
Pollera Nera	19	0.00	54	0.00	32	0.00
Prie	36	0.00	33	0.00	24	0.00
Primetta	17	0.00	24	0.00	14	0.00
Prodest	2	0.00	0	0.00	0	0.00
Prosecco	7507	0.16	18437	0.40	20109	0.45
Prosecco Lungo	0	0.00	1367	0.03	1450	0.03
Prunesta	92	0.00	36	0.00	31	0.00
Pugnitello	0	0.00	28	0.00	15	0.00
Quagliano	8	0.00	9	0.00	9	0.00
Raboso Piave	1334	0.03	776	0.02	665	0.01
Raboso Veronese	307	0.01	277	0.01	295	0.01
Rebo	37	0.00	125	0.00	92	0.00
Recantina	0	0.00	9	0.00	4	0.00
Refosco	0	0.00	0	0.00	1341	0.03
Refosco dal Peduncolo Rosso	711	0.02	1082	0.02	1272	0.03
Refosco di Faedis	256	0.01	217	0.00	185	0.00
Retagliado Bianco	26	0.00	28	0.00	11	0.00
Ribolla Gialla	1406	0.03	1178	0.03	959	0.02
Ripolo	0	0.00	1	0.00	1	0.00
Rollo	117	0.00	51	0.00	15	0.00
Rondinella	2797	0.06	2480	0.05	2684	0.06
Rosciola	0	0.00	2	0.00	1	0.00
Rossara Trentina	29	0.00	8	0.00	6	0.00
Rossese	232	0.01	312	0.01	164	0.00
Rossese Bianco	0	0.00	7	0.00	5	0.00
Rossignola	295	0.01	188	0.00	49	0.00
Rossola Nera	102	0.00	86	0.00	29	0.00
Roussin	3	0.00	3	0.00	2	0.00
Roviello Bianco	0	0.00	2	0.00	1	0.00
Ruche	46	0.00	100	0.00	100	0.00
Ruggine	0	0.00	1	0.00	1	0.00
Sagrantino	351	0.01	995	0.02	1026	0.02

Table 14 (cont.): Prime varieties' global area and share of all varieties globally, by prime's country of origin, 2000, 2010 and 2016

	2000		2010		2016	
	Hectares	*Share (%)*	*Hectares*	*Share (%)*	*Hectares*	*Share (%)*
Italy (continued)						
San Giuseppe Nero	348	0.01	192	0.00	82	0.00
San Lunardo	22	0.00	10	0.00	4	0.00
San Martino	44	0.00	21	0.00	6	0.00
San Michele	120	0.00	57	0.00	37	0.00
Sanforte	0	0.00	1	0.00	0	0.00
Sangiovese	68877	1.49	78030	1.69	73464	1.64
Santa Maria	15	0.00	3	0.00	2	0.00
Schiava	1231	0.03	517	0.01	236	0.01
Schiava Gentile	1158	0.03	694	0.02	165	0.00
Schiava Grigia	79	0.00	66	0.00	4	0.00
Schiava Grossa	3789	0.08	3011	0.07	2256	0.05
Schiava Lombarda	0	0.00	0	0.00	701	0.02
Schioppettino	93	0.00	154	0.00	87	0.00
Sciaglin	4	0.00	6	0.00	3	0.00
Sciascinoso	253	0.01	94	0.00	59	0.00
Scimiscia	0	0.00	5	0.00	2	0.00
Semidano	48	0.00	36	0.00	34	0.00
Sennen	0	0.00	10	0.00	2	0.00
Sgavetta	61	0.00	47	0.00	26	0.00
Sirio	23	0.00	14	0.00	5	0.00
Soperga	32	0.00	22	0.00	17	0.00
Soreli	0	0.00	0	0.00	0	0.00
Spergola	0	0.00	110	0.00	115	0.00
Susumaniello	62	0.00	50	0.00	8	0.00
Tazzelenghe	68	0.00	55	0.00	45	0.00
Termarina Rossa	0	0.00	20	0.00	2	0.00
Teroldego	682	0.01	839	0.02	772	0.02
Terrano	1461	0.03	1914	0.04	209	0.00
Timorasso	19	0.00	129	0.00	123	0.00
Tintilia del Molise	0	0.00	111	0.00	66	0.00
Torbato	168	0.00	52	0.00	46	0.00
Trebbianina	0	0.00	128	0.00	30	0.00
Trebbiano d'Abruzzo	8435	0.18	5091	0.11	2630	0.06
Trebbiano Giallo	3984	0.09	10664	0.23	2275	0.05
Trebbiano Modenese	583	0.01	363	0.01	287	0.01
Trebbiano Romagnolo	19492	0.42	15893	0.34	19059	0.43
Trebbiano Spoletino	242	0.01	200	0.00	121	0.00
Trebbiano Toscano	137201	2.97	111290	2.41	120343	2.68
Trevisana Nera	33	0.00	15	0.00	11	0.00
Tronto	0	0.00	2	0.00	1	0.00
Turchetta	0	0.00	3	0.00	2	0.00
Ucelut	10	0.00	10	0.00	10	0.00
UD 31103	0	0.00	0	0.00	0	0.00

Table 14 (cont.): Prime varieties' global area and share of all varieties globally, by prime's country of origin, 2000, 2010 and 2016

	2000		2010		2016	
	Hectares	*Share (%)*	*Hectares*	*Share (%)*	*Hectares*	*Share (%)*
Italy (continued)						
Uva del Fantini	0	0.00	0	0.00	0	0.00
Uva del Tunde	0	0.00	2	0.00	2	0.00
Uva Longanesi	0	0.00	512	0.01	539	0.01
Uva Rara	570	0.01	460	0.01	197	0.00
Uva Tosca	84	0.00	71	0.00	29	0.00
Uvalino	0	0.00	1	0.00	1	0.00
Valentino Nero	56	0.00	21	0.00	20	0.00
Vega	27	0.00	35	0.00	27	0.00
Verdea	107	0.00	83	0.00	39	0.00
Verdeca	2208	0.05	796	0.02	913	0.02
Verdello	662	0.01	287	0.01	179	0.00
Verdicchio Bianco	5043	0.11	3532	0.08	4682	0.10
Verdiso	71	0.00	68	0.00	52	0.00
Verduschia	15	0.00	11	0.00	9	0.00
Verduzzo Friulano	1598	0.03	812	0.02	690	0.02
Verduzzo Trevigiano	1657	0.04	708	0.02	531	0.01
Vermentino	5838	0.13	8874	0.19	11483	0.26
Vermentino Nero	143	0.00	210	0.00	124	0.00
Vernaccia di Oristano	565	0.01	272	0.01	246	0.01
Vernaccia di San Gimignano	854	0.02	522	0.01	884	0.02
Veruccese	0	0.00	0	0.00	0	0.00
Vespaiola	105	0.00	94	0.00	90	0.00
Vespolina	103	0.00	134	0.00	88	0.00
Vien de Nus	25	0.00	13	0.00	9	0.00
Vitovska	42	0.00	50	0.00	51	0.00
Vuillermin	0	0.00	4	0.00	3	0.00
Total (347 varieties)	**616536**	**13.36**	**545383**	**11.82**	**516172**	**11.51**
Portugal						
Agronomica	19	0.00	327	0.01	299	0.01
Agua Santa	0	0.00	78	0.00	76	0.00
Alfrocheiro	523	0.01	1188	0.03	1216	0.03
Almafre	0	0.00	0	0.00	0	0.00
Alminhaca	0	0.00	1	0.00	1	0.00
Alvar Branco	0	0.00	0	0.00	0	0.00
Alvar Roxo	0	0.00	2	0.00	2	0.00
Alvarelhão	5272	0.11	5701	0.12	2910	0.06
Alvarelhao Ceitao	0	0.00	0	0.00	0	0.00
Alvarinho	0	0.00	0	0.00	5545	0.12
Amaral	582	0.01	92	0.00	93	0.00
Antao Vaz	376	0.01	1252	0.03	1768	0.04
Arinto de Bucelas	3966	0.09	4482	0.10	5409	0.12

Table 14 (cont.): Prime varieties' global area and share of all varieties globally, by prime's country of origin, 2000, 2010 and 2016

	2000		2010		2016	
	Hectares	*Share (%)*	*Hectares*	*Share (%)*	*Hectares*	*Share (%)*
Portugal (continued)						
Arjuncao	0	0.00	1	0.00	1	0.00
Assaraky	0	0.00	1	0.00	0	0.00
Avesso	636	0.01	685	0.01	699	0.02
Azal	3302	0.07	1072	0.02	1443	0.03
Babosa de Madere	0	0.00	2	0.00	2	0.00
Baga	6730	0.15	4108	0.09	6750	0.15
Barcelo	34	0.00	23	0.00	26	0.00
Barreto de Semente	0	0.00	3	0.00	3	0.00
Bastardo Branco	0	0.00	15	0.00	14	0.00
Batoca	80	0.00	11	0.00	8	0.00
Bical	912	0.02	924	0.02	1076	0.02
Boal Barreiro	0	0.00	1	0.00	0	0.00
Boal Vencedor	0	0.00	2	0.00	2	0.00
Borraçal	2654	0.06	683	0.01	512	0.01
Branco Especial	0	0.00	3	0.00	3	0.00
Branco Sr. Joao	0	0.00	0	0.00	0	0.00
Branco Valente	0	0.00	0	0.00	0	0.00
Brandam	0	0.00	312	0.01	291	0.01
Cabinda	0	0.00	362	0.01	355	0.01
Cabral	0	0.00	2	0.00	2	0.00
Caíño Blanco	69	0.00	128	0.00	77	0.00
Calrao	0	0.00	1	0.00	1	0.00
Campanario	0	0.00	2	0.00	2	0.00
Caracol	14	0.00	33	0.00	33	0.00
Caramela	0	0.00	0	0.00	0	0.00
Carrega Branco	0	0.00	507	0.01	512	0.01
Carrega Tinto	0	0.00	17	0.00	17	0.00
Casculho	0	0.00	267	0.01	269	0.01
Castalia	0	0.00	0	0.00	0	0.00
Castela	0	0.00	8	0.00	7	0.00
Castelão	14424	0.31	11088	0.24	12580	0.28
Castelão Branco	0	0.00	37	0.00	18	0.00
Castelino	0	0.00	147	0.00	144	0.00
Castelo Branco	0	0.00	5	0.00	2	0.00
Cerceal Branco	597	0.01	379	0.01	261	0.01
Chasselas Sabor	0	0.00	0	0.00	0	0.00
Cidreiro	0	0.00	0	0.00	0	0.00
Codega de Larinho	4058	0.09	629	0.01	455	0.01
Complexa	6	0.00	103	0.00	103	0.00
Concieira	0	0.00	52	0.00	53	0.00
Coracao de Galo	0	0.00	1	0.00	1	0.00
Cornifesto	259	0.01	499	0.01	509	0.01
Deliciosa	0	0.00	0	0.00	0	0.00

Table 14 (cont.): Prime varieties' global area and share of all varieties globally, by prime's country of origin, 2000, 2010 and 2016

	2000		2010		2016	
	Hectares	*Share (%)*	*Hectares*	*Share (%)*	*Hectares*	*Share (%)*
Portugal (continued)						
Diagalves	1088	0.02	1156	0.03	1090	0.02
Docal	0	0.00	0	0.00	0	0.00
Dona Branca	296	0.01	276	0.01	204	0.00
Dona Joaquina	0	0.00	24	0.00	11	0.00
Donzelinho Branco	59	0.00	65	0.00	64	0.00
Donzelinho Roxo	0	0.00	0	0.00	0	0.00
Donzelinho Tinto	0	0.00	33	0.00	34	0.00
Dorinto	0	0.00	115	0.00	70	0.00
Encruzado	291	0.01	282	0.01	132	0.00
Esganacao Preto	0	0.00	0	0.00	0	0.00
Esganinho	0	0.00	0	0.00	0	0.00
Espadeiro	1682	0.04	469	0.01	357	0.01
Espadeiro Mole	0	0.00	0	0.00	0	0.00
Estreito Macio	0	0.00	3	0.00	1	0.00
Farinheira	0	0.00	3	0.00	3	0.00
Fepiro	0	0.00	0	0.00	0	0.00
Fernão Pires	14545	0.32	9609	0.21	12211	0.27
Folgasao	409	0.01	182	0.00	162	0.00
Folgasao Roxo	0	0.00	18	0.00	18	0.00
Folha de Figueira	0	0.00	3	0.00	2	0.00
Fonte Cal	355	0.01	111	0.00	52	0.00
Galego Dourado	51	0.00	16	0.00	7	0.00
Generosa	9	0.00	107	0.00	328	0.01
Goncalo Pires	0	0.00	1	0.00	1	0.00
Gouveio Preto	0	0.00	0	0.00	0	0.00
Gouveio Real	0	0.00	582	0.01	581	0.01
Grangeal	0	0.00	1	0.00	1	0.00
Granho	0	0.00	0	0.00	0	0.00
Grossa	0	0.00	73	0.00	54	0.00
Jampal	127	0.00	71	0.00	34	0.00
Labrusco	0	0.00	81	0.00	79	0.00
Lameiro	0	0.00	0	0.00	0	0.00
Lariao	0	0.00	4	0.00	5	0.00
Leira	0	0.00	1	0.00	0	0.00
Listrao Roxo	10	0.00	0	0.00	0	0.00
Loureiro	4392	0.10	4054	0.09	4696	0.10
Lourela	0	0.00	0	0.00	0	0.00
Lusitano	0	0.00	0	0.00	0	0.00
Luzidio	0	0.00	0	0.00	0	0.00
Madeleine Sylvaner	0	0.00	7	0.00	6	0.00
Malvarisco	0	0.00	3	0.00	3	0.00
Malvasia Branca de Sao Jorge	55	0.00	110	0.00	110	0.00

Table 14 (cont.): Prime varieties' global area and share of all varieties globally, by prime's country of origin, 2000, 2010 and 2016

	2000		2010		2016	
	Hectares	*Share (%)*	*Hectares*	*Share (%)*	*Hectares*	*Share (%)*
Portugal (continued)						
Malvasia Fina	7102	0.15	3501	0.08	3282	0.07
Malvasia Fina Roxa	0	0.00	25	0.00	24	0.00
Malvasia Parda	0	0.00	1	0.00	1	0.00
Malvasia Preta	2210	0.05	1903	0.04	1933	0.04
Malvasia Romana	0	0.00	0	0.00	0	0.00
Malvasia Trigueira	0	0.00	12	0.00	12	0.00
Malvia	0	0.00	0	0.00	0	0.00
Manteudo Preto	0	0.00	16	0.00	11	0.00
Marquinhas	0	0.00	11	0.00	5	0.00
Marufo	6339	0.14	6579	0.14	4683	0.10
Melhorio	0	0.00	0	0.00	0	0.00
Mindelo	0	0.00	1	0.00	1	0.00
Mondet	0	0.00	0	0.00	0	0.00
Monvedro	14	0.00	6	0.00	4	0.00
Moscadet	0	0.00	4	0.00	3	0.00
Moscargo	0	0.00	0	0.00	0	0.00
Moscatel Lilaz	0	0.00	0	0.00	0	0.00
Moscato Nazareno	0	0.00	68	0.00	40	0.00
Mourisco de Semente	0	0.00	60	0.00	61	0.00
Mourisco de Trevoes	0	0.00	2	0.00	2	0.00
Naia	0	0.00	0	0.00	0	0.00
Nevoeira	0	0.00	0	0.00	0	0.00
Padeiro	0	0.00	86	0.00	88	0.00
Parreira Matias	0	0.00	1	0.00	1	0.00
Patorra	0	0.00	10	0.00	10	0.00
Pe Comprido	0	0.00	1	0.00	1	0.00
Pedral	179	0.00	151	0.00	80	0.00
Perigo	0	0.00	4	0.00	4	0.00
Perola	68	0.00	0	0.00	0	0.00
Pexem	0	0.00	3	0.00	2	0.00
Pintosa	0	0.00	2	0.00	2	0.00
Praca	0	0.00	166	0.00	169	0.00
Preto Cardana	0	0.00	5	0.00	5	0.00
Preto Martinho	428	0.01	163	0.00	163	0.00
Primavera	0	0.00	40	0.00	39	0.00
Promissao	0	0.00	6	0.00	3	0.00
Rabigato	1133	0.02	1273	0.03	1969	0.04
Rabigato Moreno	0	0.00	1	0.00	1	0.00
Rabo de Anho	0	0.00	99	0.00	86	0.00
Rabo de Lobo	0	0.00	3	0.00	3	0.00
Rabo de Ovelha	2330	0.05	908	0.02	563	0.01
Ramisco	72	0.00	34	0.00	33	0.00

Table 14 (cont.): Prime varieties' global area and share of all varieties globally, by prime's country of origin, 2000, 2010 and 2016

	2000		2010		2016	
	Hectares	*Share (%)*	*Hectares*	*Share (%)*	*Hectares*	*Share (%)*
Portugal (continued)						
Rio Grande	4	0.00	1	0.00	0	0.00
Roxo de Vila Flor	0	0.00	0	0.00	0	0.00
Roxo Rei	0	0.00	0	0.00	0	0.00
Rual	0	0.00	1	0.00	1	0.00
Rufete	3397	0.07	4833	0.10	1859	0.04
Samarrinho	0	0.00	1	0.00	1	0.00
Santarena	0	0.00	739	0.02	724	0.02
Santoal	0	0.00	9	0.00	4	0.00
Sao Mamede	0	0.00	1	0.00	1	0.00
Seara Nova	1213	0.03	681	0.01	471	0.01
Sercial	306	0.01	106	0.00	85	0.00
Sercialinho	0	0.00	9	0.00	4	0.00
Sevilhao	0	0.00	14	0.00	14	0.00
Síria	2791	0.06	7898	0.17	7037	0.16
Tamarez	585	0.01	343	0.01	298	0.01
Terrantez	27	0.00	12	0.00	11	0.00
Terrantez do Pico	0	0.00	0	0.00	0	0.00
Tinta Aguiar	0	0.00	75	0.00	77	0.00
Tinta Barroca	6052	0.13	6172	0.13	4926	0.11
Tinta Bragao	0	0.00	63	0.00	64	0.00
Tinta Carvalha	1920	0.04	1311	0.03	1113	0.02
Tinta da Barca	0	0.00	345	0.01	352	0.01
Tinta da Melra	0	0.00	1	0.00	1	0.00
Tinta de Alcoa	0	0.00	24	0.00	24	0.00
Tinta de Cidadelhe	0	0.00	1	0.00	1	0.00
Tinta de Pegoes	0	0.00	195	0.00	191	0.00
Tinta de Porto Santo	0	0.00	2	0.00	2	0.00
Tinta do Rodo	0	0.00	3	0.00	3	0.00
Tinta Engomada	0	0.00	4	0.00	4	0.00
Tinta Francisca	0	0.00	53	0.00	55	0.00
Tinta Malandra	0	0.00	0	0.00	0	0.00
Tinta Martins	0	0.00	11	0.00	10	0.00
Tinta Mesquita	0	0.00	1	0.00	1	0.00
Tinta Penajoia	0	0.00	53	0.00	53	0.00
Tinta Pereira	0	0.00	1	0.00	1	0.00
Tinta Pomar	0	0.00	30	0.00	30	0.00
Tinta Roseira	0	0.00	4	0.00	4	0.00
Tinta Valdosa	0	0.00	1	0.00	1	0.00
Tinta Varejoa	0	0.00	1	0.00	1	0.00
Tintem	0	0.00	9	0.00	9	0.00
Tinto Cão	556	0.01	369	0.01	372	0.01
Tinto do Aurelio	0	0.00	0	0.00	0	0.00

Table 14 (cont.): Prime varieties' global area and share of all varieties globally, by prime's country of origin, 2000, 2010 and 2016

	2000		2010		2016	
	Hectares	*Share (%)*	*Hectares*	*Share (%)*	*Hectares*	*Share (%)*
Portugal (continued)						
Touriga Branca	0	0.00	36	0.00	34	0.00
Touriga Femea	0	0.00	15	0.00	15	0.00
Touriga Franca	6674	0.14	11590	0.25	14224	0.32
Touriga Nacional	4263	0.09	10446	0.23	11722	0.26
Trajadura	2416	0.05	2169	0.05	2492	0.06
Trincadeira	7265	0.16	9270	0.20	10510	0.23
Trincadeira das Pratas	216	0.00	239	0.01	124	0.00
Trincadeiro Branco	0	0.00	59	0.00	49	0.00
Triunfo	0	0.00	2	0.00	2	0.00
Uva Cao	33	0.00	1	0.00	0	0.00
Valbom	0	0.00	166	0.00	162	0.00
Valveirinha	0	0.00	0	0.00	0	0.00
Verdelho	1643	0.04	2009	0.04	1516	0.03
Verdelho Tinto	1	0.00	28	0.00	29	0.00
Verdial	0	0.00	1	0.00	0	0.00
Verdial Tinto	0	0.00	3	0.00	3	0.00
Vinhao	5937	0.13	3160	0.07	4468	0.10
Viosinho	17	0.00	225	0.00	916	0.02
Vital	2246	0.05	1182	0.03	659	0.01
Xara	0	0.00	0	0.00	0	0.00
Total (202 varieties)	**135318**	**2.93**	**131816**	**2.86**	**141887**	**3.16**
Greece						
Agiorgitiko	2320	0.05	2905	0.06	3272	0.07
Asirtiko Red	22	0.00	5	0.00	5	0.00
Asprouda	433	0.01	113	0.00	120	0.00
Assyrtiko	1106	0.02	902	0.02	1770	0.04
Athiri	1350	0.03	748	0.02	577	0.01
Batili	27	0.00	0	0.00	0	0.00
Batily	38	0.00	54	0.00	1	0.00
Debina	455	0.01	239	0.01	14	0.00
Fokiano	162	0.00	262	0.01	212	0.00
Fokiano (W)	57	0.00	0	0.00	0	0.00
Goustolidi	112	0.00	68	0.00	19	0.00
Kakotrygis	0	0.00	103	0.00	28	0.00
Korinthiaki	834	0.02	54	0.00	106	0.00
Kotsifali	1148	0.02	2330	0.05	1338	0.03
Krassato	38	0.00	52	0.00	5	0.00
Liatiko	2476	0.05	1211	0.03	2633	0.06
Liatiko (W)	70	0.00	0	0.00	0	0.00
Limnio	95	0.00	372	0.01	176	0.00

Table 14 (cont.): Prime varieties' global area and share of all varieties globally, by prime's country of origin, 2000, 2010 and 2016

	2000		2010		2016	
	Hectares	*Share (%)*	*Hectares*	*Share (%)*	*Hectares*	*Share (%)*
Greece (continued)						
Limnio (W)	27	0.00	0	0.00	0	0.00
Malagousia	23	0.00	182	0.00	126	0.00
Malvasia	61	0.00	45	0.00	2184	0.05
Mandilaria	845	0.02	885	0.02	932	0.02
Mavro Messenikola	3	0.00	0	0.00	0	0.00
Mavrodafni	537	0.01	345	0.01	324	0.01
Mavrouda	349	0.01	520	0.01	1658	0.04
Monemvassia	418	0.01	481	0.01	81	0.00
Moschofilero	718	0.02	1111	0.02	1088	0.02
Moschomavro	2295	0.05	1428	0.03	113	0.00
Muscat	0	0.00	0	0.00	744	0.02
Muscat Blanc à Petits Grains	29979	0.65	31259	0.68	33739	0.75
Muscat Blanc à Petits Grains (G)	10442	0.23	8761	0.19	8258	0.18
Muscat Blanc à Petits Grains (R)	1154	0.03	1459	0.03	1438	0.03
Muscat of Alexandria	29590	0.64	27648	0.60	34805	0.78
Muscat of Alexandria (R)	11	0.00	6	0.00	3	0.00
Negoska	96	0.00	143	0.00	17	0.00
Robola	359	0.01	471	0.01	152	0.00
Roditis	299	0.01	4668	0.10	8463	0.19
Roditis (R)	6945	0.15	3826	0.08	828	0.02
Romeiko	382	0.01	1597	0.03	1131	0.03
Savatiano	12747	0.28	9920	0.21	10268	0.23
Stavroto	104	0.00	11	0.00	0	0.00
Thrapsathiri	0	0.00	31	0.00	27	0.00
Vertzami	491	0.01	335	0.01	60	0.00
Vilana	506	0.01	579	0.01	650	0.01
Vilana (R)	60	0.00	0	0.00	0	0.00
Xinomavro	1816	0.04	1971	0.04	2135	0.05
Xinomavro (W)	3	0.00	0	0.00	0	0.00
Total (47 varieties)	**111004**	**2.40**	**107100**	**2.32**	**119501**	**2.67**
Germany						
Accent	0	0.00	0	0.00	1	0.00
Acolon	0	0.00	490	0.01	477	0.01
Albalonga	57	0.00	15	0.00	12	0.00
Arnsburger	3	0.00	30	0.00	29	0.00
Bacchus	3374	0.07	2113	0.05	1759	0.04
Baron	0	0.00	0	0.00	1	0.00
Breidecker	28	0.00	7	0.00	0	0.00
Bronner	0	0.00	9	0.00	6	0.00
Bukettraube	280	0.01	71	0.00	54	0.00

Table 14 (cont.): Prime varieties' global area and share of all varieties globally, by prime's country of origin, 2000, 2010 and 2016

	2000		2010		2016	
	Hectares	*Share (%)*	*Hectares*	*Share (%)*	*Hectares*	*Share (%)*
Germany (continued)						
Cabernet Cantor	0	0.00	0	0.00	1	0.00
Cabernet Carbon	0	0.00	0	0.00	11	0.00
Cabernet Carol	0	0.00	0	0.00	6	0.00
Cabernet Cortis	0	0.00	0	0.00	38	0.00
Cabernet Cubin	0	0.00	60	0.00	62	0.00
Cabernet Dorio	0	0.00	36	0.00	34	0.00
Cabernet Dorsa	43	0.00	252	0.01	272	0.01
Cabernet Mitos	0	0.00	322	0.01	312	0.01
Cabertin	0	0.00	0	0.00	2	0.00
Dakapo	0	0.00	51	0.00	68	0.00
Dalkauer	100	0.00	0	0.00	0	0.00
Deckrot	30	0.00	2	0.00	12	0.00
Domina	187	0.00	407	0.01	375	0.01
Dornfelder	3766	0.08	8182	0.18	7871	0.18
Dunkelfelder	280	0.01	356	0.01	291	0.01
Ehrenbreitsteiner	13	0.00	10	0.00	8	0.00
Ehrenfelser	289	0.01	113	0.00	82	0.00
Elbling	1208	0.03	935	0.02	972	0.02
Elbling (R)	4	0.00	9	0.00	10	0.00
Faberrebe	1586	0.03	554	0.01	331	0.01
Fontanara	2	0.00	0	0.00	1	0.00
Freisamer	17	0.00	8	0.00	6	0.00
Gänsfüsser	0	0.00	0	0.00	0	0.00
Geilweilerhof Ga- 48- 12	0	0.00	0	0.00	1	0.00
Geisenheim 318-57	0	0.00	106	0.00	14	0.00
Gewürztraminer	10670	0.23	14355	0.31	12823	0.29
Gutenborner	0	0.00	2	0.00	0	0.00
Hegel	10	0.00	9	0.00	7	0.00
Helfensteiner	26	0.00	19	0.00	14	0.00
Helios	0	0.00	0	0.00	6	0.00
Heroldrebe	199	0.00	134	0.00	112	0.00
Hibernal	0	0.00	0	0.00	20	0.00
Hölder	13	0.00	5	0.00	2	0.00
Huxelrebe	1289	0.03	630	0.01	466	0.01
Johanniter	0	0.00	86	0.00	111	0.00
Juwel	42	0.00	22	0.00	15	0.00
Kanzler	53	0.00	33	0.00	28	0.00
Kerner	7129	0.15	4093	0.09	2891	0.06
Kolor	0	0.00	2	0.00	7	0.00
Madeleine × Angevine 7672	0	0.00	52	0.00	48	0.00
Mariensteiner	9	0.00	0	0.00	2	0.00
Merzling	5	0.00	0	0.00	1	0.00

Table 14 (cont.): Prime varieties' global area and share of all varieties globally, by prime's country of origin, 2000, 2010 and 2016

	2000		2010		2016	
	Hectares	*Share (%)*	*Hectares*	*Share (%)*	*Hectares*	*Share (%)*
Germany (continued)						
Monarch	0	0.00	0	0.00	10	0.00
Morio-Muskat	1188	0.03	526	0.01	440	0.01
Müller-Thurgau	33587	0.73	22917	0.50	19501	0.43
Muscaris	0	0.00	0	0.00	4	0.00
Nobling	102	0.00	1	0.00	52	0.00
Optima	239	0.01	65	0.00	38	0.00
Oraniensteiner	0	0.00	3	0.00	2	0.00
Orion	8	0.00	13	0.00	1	0.00
Ortega	1054	0.02	667	0.01	532	0.01
Osteiner	3	0.00	1	0.00	1	0.00
Palas	0	0.00	0	0.00	7	0.00
Perle	121	0.00	34	0.00	18	0.00
Phoenix	24	0.00	67	0.00	46	0.00
Pinotin	0	0.00	0	0.00	0	0.00
Piroso	0	0.00	0	0.00	4	0.00
Prinzipal	0	0.00	0	0.00	2	0.00
Prior	0	0.00	0	0.00	14	0.00
Räuschling	23	0.00	23	0.00	23	0.00
Reberger	0	0.00	0	0.00	2	0.00
Regent	340	0.01	2187	0.05	1974	0.04
Regner	150	0.00	46	0.00	21	0.00
Reichensteiner	319	0.01	247	0.01	120	0.00
Rieslaner	70	0.00	84	0.00	73	0.00
Riesling	43316	0.94	50014	1.08	59805	1.33
Rondo	0	0.00	40	0.00	51	0.00
Rotberger	26	0.00	17	0.00	11	0.00
Saphira	0	0.00	0	0.00	8	0.00
Scheurebe	3655	0.08	2039	0.04	1626	0.04
Schönburger	39	0.00	68	0.00	35	0.00
Septimer	0	0.00	0	0.00	1	0.00
Siegerrebe	167	0.00	131	0.00	102	0.00
Siegfriedrebe	0	0.00	0	0.00	2	0.00
Silcher	7	0.00	0	0.00	1	0.00
Sirius	2	0.00	0	0.00	0	0.00
Solaris	0	0.00	81	0.00	118	0.00
Souvignier Gris	0	0.00	0	0.00	3	0.00
Staufer	4	0.00	0	0.00	0	0.00
Tauberschwarz	8	0.00	11	0.00	16	0.00
Würzer	108	0.00	70	0.00	54	0.00
Total (90 varieties)	**115274**	**2.50**	**112929**	**2.45**	**114391**	**2.55**

Table 14 (cont.): Prime varieties' global area and share of all varieties globally, by prime's country of origin, 2000, 2010 and 2016

	2000		2010		2016	
	Hectares	*Share (%)*	*Hectares*	*Share (%)*	*Hectares*	*Share (%)*
Argentina						
Caberinta	85	0.00	69	0.00	38	0.00
Canelon	8	0.00	8	0.00	0	0.00
Cereza	31113	0.67	29934	0.65	28887	0.64
Criolla Grande	24264	0.53	20745	0.45	15596	0.35
Criolla Mediana	1	0.00	3	0.00	7	0.00
Gargiulo 14260	1	0.00	1	0.00	0	0.00
Gargiulo 2539	69	0.00	49	0.00	29	0.00
Gargiulo 26189	6	0.00	0	0.00	0	0.00
Gargiulo 26879	4	0.00	3	0.00	0	0.00
Gargiulo 4113	7	0.00	6	0.00	3	0.00
Gargiulo 45803	4	0.00	8	0.00	10	0.00
Patagonia	29	0.00	40	0.00	0	0.00
Pedro Giménez	14862	0.32	13502	0.29	15576	0.35
Rieslina	219	0.00	174	0.00	103	0.00
Saint Jeannet	68	0.00	56	0.00	43	0.00
Serna	19	0.00	36	0.00	0	0.00
Torrontes Mendocino	780	0.02	661	0.01	653	0.01
Torrontés Riojano	8197	0.18	8937	0.19	8859	0.20
Torrontés Sanjuanino	3170	0.07	2531	0.05	3656	0.08
Total (19 varieties)	**82907**	**1.80**	**76763**	**1.66**	**73460**	**1.64**
Georgia						
Aladasturi	46	0.00	59	0.00	59	0.00
Aleksandrouli	219	0.00	281	0.01	281	0.01
Chinuri	955	0.02	1225	0.03	1225	0.03
Chkhaveri	20	0.00	26	0.00	26	0.00
Dodrelyabi	0	0.00	0	0.00	0	0.00
Goruli Mtsvane	224	0.00	287	0.01	287	0.01
Khikhvi	5	0.00	6	0.00	6	0.00
Kisi	20	0.00	26	0.00	26	0.00
Krakhuna	36	0.00	46	0.00	46	0.00
Mtsvane Kakhuri	249	0.01	319	0.01	319	0.01
Ojaleshi	25	0.00	32	0.00	32	0.00
Otskhanuri Sapere	5	0.00	6	0.00	6	0.00
Rkatsiteli	67354	1.46	58641	1.27	51374	1.15
Saperavi	6707	0.15	8126	0.18	6478	0.14
Tavkveri	29	0.00	37	0.00	37	0.00
Tsitska	2839	0.06	3642	0.08	3642	0.08
Tsolikouri	6161	0.13	7903	0.17	7903	0.18
Tsulukidzis Tetra	152	0.00	195	0.00	195	0.00
Usakhelouri	8	0.00	10	0.00	10	0.00
Total (19 varieties)	**85054**	**1.84**	**80868**	**1.75**	**71954**	**1.61**

Table 14 (cont.): Prime varieties' global area and share of all varieties globally, by prime's country of origin, 2000, 2010 and 2016

	2000		2010		2016	
	Hectares	*Share (%)*	*Hectares*	*Share (%)*	*Hectares*	*Share (%)*
Croatia						
Babić	1189	0.03	359	0.01	0	0.00
Babica	0	0.00	18	0.00	0	0.00
Bogdanuša	0	0.00	48	0.00	0	0.00
Cetinka	0	0.00	35	0.00	0	0.00
Croatina	3116	0.07	5700	0.12	2695	0.06
Debit	0	0.00	403	0.01	0	0.00
Dišeća Ranina	0	0.00	2	0.00	0	0.00
Gegić	0	0.00	11	0.00	0	0.00
Graševina	92306	2.00	61200	1.33	24384	0.54
Hrvatica	245	0.01	116	0.00	53	0.00
Kraljevina	0	0.00	447	0.01	199	0.00
Kujundžuša	0	0.00	206	0.00	0	0.00
Lasina	0	0.00	14	0.00	0	0.00
Malvazija Istarska	7559	0.16	2740	0.06	2788	0.06
Nincusa	0	0.00	17	0.00	0	0.00
Plavac Mali	6539	0.14	1569	0.03	1714	0.04
Plavec Žuti	0	0.00	13	0.00	82	0.00
Plavina	0	0.00	643	0.01	683	0.02
Pošip Bijeli	6539	0.14	253	0.01	0	0.00
Škrlet	0	0.00	61	0.00	0	0.00
Suscan	0	0.00	5	0.00	0	0.00
Trbljan	0	0.00	231	0.01	0	0.00
Tribidrag	26922	0.58	32755	0.71	33649	0.75
Trnjak	0	0.00	15	0.00	0	0.00
Vugava	0	0.00	36	0.00	0	0.00
Žlahtina	0	0.00	135	0.00	0	0.00
Zlatarica Vrgorska	0	0.00	19	0.00	0	0.00
Total (27 varieties)	**144414**	**3.13**	**107048**	**2.32**	**66247**	**1.48**
Austria						
Augster Blau	0	0.00	0	0.00	0	0.00
Blauburger	1002	0.02	1339	0.03	1223	0.03
Blauer Portugieser	9156	0.20	8027	0.17	6590	0.15
Blauer Wildbacher	472	0.01	368	0.01	437	0.01
Blaufränkisch	13997	0.30	17890	0.39	17180	0.38
Frühroter Veltliner	632	0.01	856	0.02	388	0.01
Goldburger	309	0.01	140	0.00	98	0.00
Grüner Veltliner	23604	0.51	18834	0.41	19118	0.43
Jubiläumsrebe	30	0.00	14	0.00	7	0.00
Laska	1	0.00	0	0.00	0	0.00
Neuburger	1434	0.03	1030	0.02	578	0.01

Table 14 (cont.): Prime varieties' global area and share of all varieties globally, by prime's country of origin, 2000, 2010 and 2016

	2000		2010		2016	
	Hectares	*Share (%)*	*Hectares*	*Share (%)*	*Hectares*	*Share (%)*
Austria (continued)						
Rathay	0	0.00	9	0.00	32	0.00
Rösler	0	0.00	160	0.00	217	0.00
Roter Veltliner	258	0.01	199	0.00	198	0.00
Rotgipfler	118	0.00	105	0.00	123	0.00
Sankt Laurent	2555	0.06	3664	0.08	3272	0.07
Silvaner	11047	0.24	7395	0.16	6072	0.14
Silvaner (R)	2	0.00	38	0.00	25	0.00
Zierfandler	98	0.00	117	0.00	105	0.00
Zweigelt	7267	0.16	10029	0.22	9068	0.20
Total (20 varieties)	**71983**	**1.56**	**70215**	**1.52**	**64732**	**1.44**
United States						
Adalmiina	0	0.00	0	0.00	5	0.00
Adirondac	0	0.00	0	0.00	24	0.00
Aromella	0	0.00	0	0.00	3	0.00
Bailey	0	0.00	34	0.00	49	0.00
Black Prince	0	0.00	0	0.00	1	0.00
Blanc du Bois	0	0.00	28	0.00	81	0.00
Blush Seedless	10	0.00	0	0.00	0	0.00
Bokay	0	0.00	0	0.00	4	0.00
Brianna	0	0.00	12	0.00	21	0.00
Buffalo	0	0.00	0	0.00	0	0.00
Cabernet Diane	0	0.00	0	0.00	0	0.00
Cabernet Dore	0	0.00	1	0.00	1	0.00
California	357	0.01	326	0.01	0	0.00
Calmeria	0	0.00	0	0.00	0	0.00
Campbell Early	43	0.00	61	0.00	238	0.01
Canada Muscat	49	0.00	0	0.00	120	0.00
Canadice	0	0.00	0	0.00	0	0.00
Canner Seedless	0	0.00	0	0.00	0	0.00
Cardinal	3870	0.08	536	0.01	1660	0.04
Carmine	10	0.00	0	0.00	0	0.00
Carnelian	625	0.01	316	0.01	123	0.00
Catawba	635	0.01	633	0.01	626	0.01
Cayuga White	108	0.00	212	0.00	217	0.00
Centennial Seedless	2	0.00	0	0.00	1	0.00
Centurian	134	0.00	34	0.00	33	0.00
Champanel	0	0.00	0	0.00	1	0.00
Chardonel	0	0.00	144	0.00	90	0.00
Clinton	0	0.00	0	0.00	0	0.00
Concord	11816	0.26	12238	0.27	10544	0.24

Table 14 (cont.): Prime varieties' global area and share of all varieties globally, by prime's country of origin, 2000, 2010 and 2016

	2000		2010		2016	
	Hectares	*Share (%)*	Hectares	*Share (%)*	Hectares	*Share (%)*
USA (continued)						
Corot Noir	0	0.00	27	0.00	11	0.00
Crimson Cabernet	0	0.00	1	0.00	1	0.00
Crimson Seedless	1	0.00	8	0.00	8	0.00
David Macgregor 8521-1	0	0.00	0	0.00	2	0.00
Dawn Seedless	19	0.00	0	0.00	0	0.00
Delaware	234	0.01	227	0.00	421	0.01
Delisle	0	0.00	0	0.00	0	0.00
Diamond Muscat	0	0.00	0	0.00	2	0.00
Edelweiss	0	0.00	32	0.00	16	0.00
Einset Seedless	0	0.00	0	0.00	0	0.00
Elmer Swenson 10- 18- 30	0	0.00	0	0.00	0	0.00
Elvira	344	0.01	263	0.01	231	0.01
Emerald Riesling	937	0.02	508	0.01	177	0.00
Emerald seedless	16	0.00	0	0.00	0	0.00
Eona	0	0.00	0	0.00	2	0.00
ES 10-18-14	0	0.00	0	0.00	0	0.00
Espirit	0	0.00	0	0.00	2	0.00
Fiesta	161	0.00	230	0.00	230	0.01
Flame Seedless	548	0.01	55	0.00	55	0.00
Flora	6	0.00	8	0.00	12	0.00
Fredonia	86	0.00	37	0.00	28	0.00
Frontenac	0	0.00	135	0.00	212	0.00
Frontenac (G)	0	0.00	59	0.00	92	0.00
Frontenac (W)	0	0.00	0	0.00	26	0.00
Goethe	0	0.00	0	0.00	20	0.00
Golden Muscat	1190	0.03	1191	0.03	50	0.00
GR 7	0	0.00	32	0.00	13	0.00
Herbemont	1453	0.03	764	0.02	112	0.00
Himrod	0	0.00	0	0.00	1	0.00
Isabella	27450	0.59	32494	0.70	17813	0.40
Ives	23	0.00	16	0.00	0	0.00
Jacquez	226	0.00	2368	0.05	1443	0.03
Kay Gray	0	0.00	1	0.00	2	0.00
La Crescent	0	0.00	77	0.00	94	0.00
La Crosse	0	0.00	25	0.00	26	0.00
Lomanto	0	0.00	0	0.00	3	0.00
Louise Swenson	0	0.00	3	0.00	5	0.00
Marquette	0	0.00	88	0.00	166	0.00
Marquis	0	0.00	0	0.00	1	0.00
Mars	0	0.00	2	0.00	2	0.00
Minnesota Muscat	0	0.00	0	0.00	0	0.00
Montreal Blues	0	0.00	0	0.00	1	0.00

Table 14 (cont.): Prime varieties' global area and share of all varieties globally, by prime's country of origin, 2000, 2010 and 2016

	2000		2010		2016	
	Hectares	*Share (%)*	*Hectares*	*Share (%)*	*Hectares*	*Share (%)*
USA (continued)						
Moore's Diamond	19	0.00	42	0.00	24	0.00
Muscat Swenson	0	0.00	24	0.00	37	0.00
New York Muscat	0	0.00	5	0.00	12	0.00
New York Muscat and VG4111	0	0.00	0	0.00	2	0.00
Niagara	15343	0.33	4670	0.10	3264	0.07
Noah	71	0.00	71	0.00	200	0.00
Noiret	0	0.00	33	0.00	25	0.00
Norton	0	0.00	329	0.01	328	0.01
Okanagan Riesling	0	0.00	0	0.00	1	0.00
Ontario	0	0.00	0	0.00	0	0.00
Osceola Muscat	0	0.00	0	0.00	7	0.00
Perlette	14	0.00	1	0.00	2	0.00
Petite Amie	0	0.00	0	0.00	0	0.00
Petite Pearl	0	0.00	0	0.00	11	0.00
Pionnier	0	0.00	0	0.00	1	0.00
Portland	9	0.00	12	0.00	39	0.00
Prairie Star	0	0.00	21	0.00	23	0.00
Radisson	0	0.00	0	0.00	8	0.00
Red Globe	2113	0.05	242	0.01	242	0.01
Reliance	0	0.00	4	0.00	4	0.00
Romulus	0	0.00	0	0.00	1	0.00
Royalty	338	0.01	97	0.00	93	0.00
Rubired	4153	0.09	4556	0.10	4916	0.11
Ruby	6	0.00	9	0.00	9	0.00
Ruby Cabernet	7420	0.16	5729	0.12	5309	0.12
Ruby Seedless	25	0.00	0	0.00	0	0.00
Sabrevois	0	0.00	10	0.00	25	0.00
Scuppernong	0	0.00	0	0.00	27	0.00
Sheridan	500	0.01	500	0.01	500	0.01
Somerset Seedless	0	0.00	0	0.00	1	0.00
St Croix	0	0.00	25	0.00	45	0.00
St Pepin	0	0.00	19	0.00	37	0.00
St Vincent	0	0.00	23	0.00	20	0.00
Steuben	0	0.00	39	0.00	35	0.00
Sugrafive	118	0.00	2	0.00	0	0.00
Superior Seedless	6	0.00	9	0.00	9	0.00
Swenson Red	0	0.00	11	0.00	10	0.00
Swenson White	0	0.00	0	0.00	2	0.00
Symphony	184	0.00	324	0.01	647	0.01
Taylor	0	0.00	0	0.00	0	0.00
Traminette	5	0.00	240	0.01	239	0.01
Triplett Blanc	0	0.00	244	0.01	412	0.01
Valiant	0	0.00	11	0.00	11	0.00
Valvin Muscat	0	0.00	6	0.00	0	0.00
Vênus	0	0.00	1	0.00	0	0.00
Total (116 varieties)	**80677**	**1.75**	**70539**	**1.53**	**51703**	**1.15**

Table 14 (cont.): Prime varieties' global area and share of all varieties globally, by prime's country of origin, 2000, 2010 and 2016

	2000		2010		2016	
	Hectares	*Share (%)*	*Hectares*	*Share (%)*	*Hectares*	*Share (%)*
Hungary						
Aletta	0	0.00	723	0.02	1676	0.04
Andor	0	0.00	0	0.00	1	0.00
Arany Sárfehér	2914	0.06	1133	0.02	586	0.01
Augster Weiss	0	0.00	1	0.00	1	0.00
Bakator Kék	0	0.00	3	0.00	2	0.00
Bakator Roz	0	0.00	16	0.00	16	0.00
Bianca	2180	0.05	6462	0.14	9766	0.22
Bíborkadarka	202	0.00	136	0.00	109	0.00
Borsmenta	0	0.00	0	0.00	1	0.00
Budai Zöld	0	0.00	6	0.00	6	0.00
Csaba Gyöngye	0	0.00	89	0.00	175	0.00
Cserszegi Fűszeres	2185	0.05	3609	0.08	4299	0.10
Csillám	0	0.00	20	0.00	25	0.00
Csókaszőlő	0	0.00	2	0.00	2	0.00
Csomorika	0	0.00	0	0.00	0	0.00
Duna Gyöngye	0	0.00	63	0.00	45	0.00
Esther	0	0.00	0	0.00	0	0.00
Ezerfürtű	405	0.01	406	0.01	295	0.01
Ezerjó	3157	0.07	1074	0.02	636	0.01
Favorit	0	0.00	0	0.00	3	0.00
Fekete Jádorvány	0	0.00	0	0.00	0	0.00
Furmint	3481	0.08	5276	0.11	4435	0.10
Gesztus	0	0.00	0	0.00	0	0.00
Göcseji Zamatos	0	0.00	55	0.00	50	0.00
Gyöngyrizling	0	0.00	23	0.00	16	0.00
Hajnalka	0	0.00	0	0.00	0	0.00
Hamvas	0	0.00	0	0.00	0	0.00
Hárslevelű	1296	0.03	1856	0.04	1618	0.04
Heuréka	0	0.00	0	0.00	0	0.00
Irsai Olivér	0	0.00	1414	0.03	1790	0.04
Jádorvány	0	0.00	0	0.00	0	0.00
Jázmin	0	0.00	0	0.00	2	0.00
Jubileum 75	0	0.00	194	0.00	91	0.00
Juhfark	0	0.00	186	0.00	195	0.00
K.35	0	0.00	0	0.00	0	0.00
Kabar	0	0.00	18	0.00	30	0.00
Kadarka	2630	0.06	1181	0.03	1625	0.04
Karát	0	0.00	50	0.00	44	0.00
Kármin	0	0.00	36	0.00	24	0.00

Table 14 (cont.): Prime varieties' global area and share of all varieties globally, by prime's country of origin, 2000, 2010 and 2016

	2000		2010		2016	
	Hectares	*Share (%)*	*Hectares*	*Share (%)*	*Hectares*	*Share (%)*
Hungary (continued)						
Kéknyelű	0	0.00	43	0.00	50	0.00
Királyleányka	0	0.00	855	0.02	784	0.02
Kocsis Irma	0	0.00	11	0.00	2	0.00
Königin der Weingärten	750	0.02	61	0.00	70	0.00
Korona	0	0.00	1	0.00	1	0.00
Kövidinka	1214	0.03	1076	0.02	658	0.01
Kozma CS. 2	0	0.00	0	0.00	0	0.00
Kozmopoliten	0	0.00	0	0.00	0	0.00
Kunbarát	0	0.00	9	0.00	0	0.00
Kunleány	1376	0.03	1211	0.03	974	0.02
Kurucvér	0	0.00	0	0.00	0	0.00
Lakhegyi Mézes	567	0.01	306	0.01	145	0.00
Leányka	0	0.00	838	0.02	719	0.02
Lilla	0	0.00	0	0.00	0	0.00
Magyarfrankos	0	0.00	0	0.00	0	0.00
Mátrai Muskotály	0	0.00	67	0.00	49	0.00
Medina	0	0.00	159	0.00	124	0.00
Menoir	0	0.00	65	0.00	61	0.00
Meszi Kadarka	0	0.00	0	0.00	0	0.00
Mézes Fehér	0	0.00	2	0.00	1	0.00
Nektár	0	0.00	21	0.00	22	0.00
Nero d'Avola	11323	0.25	16649	0.36	14281	0.32
Nosztori Rizling	0	0.00	1	0.00	0	0.00
Odysseus	0	0.00	4	0.00	25	0.00
Orpheus	0	0.00	0	0.00	2	0.00
Palatina	0	0.00	6	0.00	3	0.00
Pannon Frankos	0	0.00	16	0.00	12	0.00
Pátria	0	0.00	3	0.00	5	0.00
Pecsi Szagos	0	0.00	1	0.00	1	0.00
Pelso	0	0.00	1	0.00	1	0.00
Pintes	0	0.00	3	0.00	2	0.00
Pölöskei Muskotály	0	0.00	103	0.00	207	0.00
Pozsonyi Fehér	0	0.00	10	0.00	10	0.00
Purcsin	0	0.00	0	0.00	0	0.00
Putzscheere	0	0.00	0	0.00	0	0.00
Refrén	0	0.00	0	0.00	0	0.00
Rozala Bianca	0	0.00	2	0.00	2	0.00
Rózsakő	0	0.00	19	0.00	17	0.00
Rubintos	0	0.00	18	0.00	13	0.00
Sárfehér	0	0.00	0	0.00	2	0.00
Somszoekoe Kék	0	0.00	0	0.00	0	0.00
Sremska Zelenika	0	0.00	0	0.00	0	0.00

Table 14 (cont.): Prime varieties' global area and share of all varieties globally, by prime's country of origin, 2000, 2010 and 2016

	2000		2010		2016	
	Hectares	*Share (%)*	*Hectares*	*Share (%)*	*Hectares*	*Share (%)*
Hungary (continued)						
Táltos	0	0.00	1	0.00	1	0.00
Tarcali Kék	0	0.00	0	0.00	0	0.00
Taurus	0	0.00	0	0.00	0	0.00
Tihanyi Kék	0	0.00	0	0.00	0	0.00
Trilla	0	0.00	1	0.00	1	0.00
Turán	0	0.00	177	0.00	183	0.00
Úrréti	0	0.00	0	0.00	0	0.00
Vertes Csillaga	0	0.00	21	0.00	11	0.00
Viktor	0	0.00	0	0.00	0	0.00
Viktória Gyöngye	0	0.00	190	0.00	198	0.00
Vulcanus	0	0.00	5	0.00	5	0.00
Zalagyöngye	4330	0.09	1948	0.04	1259	0.03
Zefír	0	0.00	49	0.00	15	0.00
Zengő	0	0.00	264	0.01	226	0.01
Zenit	405	0.01	580	0.01	660	0.01
Zéta	0	0.00	118	0.00	118	0.00
Zeusz	0	0.00	28	0.00	27	0.00
Total (98 varieties)	**38416**	**0.83**	**48978**	**1.06**	**48515**	**1.08**
Moldova						
Alb de Ialoveni	2	0.00	2	0.00	0	0.00
Alb de Oniţcni	0	0.00	0	0.00	424	0.01
Alb de Suruceni	0	0.00	0	0.00	780	0.02
Amur	0	0.00	146	0.00	146	0.00
Apiren Alb	0	0.00	0	0.00	1	0.00
Apiren Roz	0	0.00	0	0.00	0	0.00
Augustovski	0	0.00	0	0.00	0	0.00
Batuta Neagra	0	0.00	3	0.00	3	0.00
Coarnă Neagră	0	0.00	0	0.00	114	0.00
Dekabrskii	0	0.00	78	0.00	78	0.00
Doina	0	0.00	227	0.00	227	0.01
Fetească Albă	23828	0.52	17469	0.38	13382	0.30
Fetească Neagră	1214	0.03	1719	0.04	3248	0.07
Floricica	0	0.00	0	0.00	14	0.00
Frumoasa Alba	0	0.00	0	0.00	8	0.00
Guzun	0	0.00	0	0.00	3	0.00
Iordan	0	0.00	315	0.01	311	0.01
Kishmish Luchistyi	0	0.00	0	0.00	55	0.00
Kishmish Moldavskii	0	0.00	0	0.00	28	0.00
Kodrinskii	5	0.00	5	0.00	229	0.01
Kodryanka	0	0.00	0	0.00	1143	0.03

Table 14 (cont.): Prime varieties' global area and share of all varieties globally, by prime's country of origin, 2000, 2010 and 2016

	2000		2010		2016	
	Hectares	*Share (%)*	*Hectares*	*Share (%)*	*Hectares*	*Share (%)*
Moldova (continued)						
Lyana	0	0.00	0	0.00	41	0.00
Maiskii Chernyi	77	0.00	110	0.00	110	0.00
Moldova	0	0.00	0	0.00	12375	0.28
Muscat de Bugeac	0	0.00	0	0.00	2	0.00
Muscat de Yaloven	0	0.00	0	0.00	3	0.00
Muscat Yantarnyi	0	0.00	0	0.00	683	0.02
Muskat de Yaloven	20	0.00	20	0.00	0	0.00
Nastea	0	0.00	0	0.00	303	0.01
Negru de Yaloven	15	0.00	15	0.00	141	0.00
Novîi podaroc Zaporojiu	0	0.00	0	0.00	31	0.00
Onitskanskii Belyi	5	0.00	71	0.00	66	0.00
Osennii Ciornai	0	0.00	0	0.00	8	0.00
Pamyati Negrulya	0	0.00	0	0.00	123	0.00
Plavay	0	0.00	209	0.00	152	0.00
Prezentabil	0	0.00	0	0.00	215	0.00
Riton	2	0.00	257	0.01	568	0.01
Viorika	40	0.00	347	0.01	558	0.01
Yalovenskii Ustoichivyi	0	0.00	0	0.00	129	0.00
Yubilei Zhuravlya	0	0.00	0	0.00	1	0.00
Total (40 varieties)	**25208**	**0.55**	**20992**	**0.45**	**35704**	**0.80**
Bulgaria						
Dimyat	7740	0.17	2401	0.05	9696	0.22
Droujba	3	0.00	3	0.00	0	0.00
Dunav	0	0.00	0	0.00	11	0.00
Dunavski Lazur	0	0.00	483	0.01	483	0.01
Mavrud	647	0.01	1296	0.03	1193	0.03
Misket Cherven	0	0.00	4159	0.09	4349	0.10
Misket Varnenski	0	0.00	336	0.01	0	0.00
Pamid	22718	0.49	9827	0.21	9961	0.22
Ranna Melnishka Loza	0	0.00	249	0.01	0	0.00
Shiroka Melnishka	3804	0.08	1580	0.03	1205	0.03
Storgozia	0	0.00	295	0.01	0	0.00
Total (11 varieties)	**34913**	**0.76**	**20628**	**0.45**	**26899**	**0.60**

Table 14 (cont.): Prime varieties' global area and share of all varieties globally, by prime's country of origin, 2000, 2010 and 2016

	2000		2010		2016	
	Hectares	*Share (%)*	*Hectares*	*Share (%)*	*Hectares*	*Share (%)*
Romania						
Alb Aromat	0	0.00	24	0.00	24	0.00
Alutus	0	0.00	2	0.00	2	0.00
Amurg	0	0.00	3	0.00	3	0.00
Arcas	0	0.00	1	0.00	1	0.00
Aromat de Iasi	0	0.00	62	0.00	66	0.00
Astra	0	0.00	0	0.00	0	0.00
Băbească Neagră	3722	0.08	3122	0.07	2696	0.06
Băbească Neagră (G)	0	0.00	328	0.01	297	0.01
Balada	0	0.00	0	0.00	0	0.00
Blasius	0	0.00	14	0.00	15	0.00
Busuioacă de Bohotin	0	0.00	268	0.01	343	0.01
Codană	0	0.00	24	0.00	26	0.00
Columna	0	0.00	24	0.00	26	0.00
Crâmpoşie Selecţionată	0	0.00	409	0.01	18	0.00
Crimposie	0	0.00	453	0.01	450	0.01
Cruciulita	0	0.00	0	0.00	0	0.00
Donaris	0	0.00	1	0.00	1	0.00
Fetească Regală	2578	0.06	13136	0.28	12991	0.29
Frâncuşă	0	0.00	621	0.01	365	0.01
Galbenă de Odobeşti	546	0.01	385	0.01	417	0.01
Golia	0	0.00	0	0.00	0	0.00
Grasă de Cotnari	850	0.02	685	0.01	632	0.01
Haiduc	0	0.00	3	0.00	3	0.00
Mamaia	0	0.00	0	0.00	0	0.00
Miorita	0	0.00	7	0.00	8	0.00
Muscat Timpuriu de Bucuresti	0	0.00	0	0.00	15	0.00
Mustoasă de Măderat	0	0.00	255	0.01	282	0.01
Negru Aromat	0	0.00	1	0.00	1	0.00
Negru de Drăgăşani	0	0.00	16	0.00	18	0.00
Novac	0	0.00	73	0.00	74	0.00
Rosina	0	0.00	1	0.00	1	0.00
Roz de Minis	0	0.00	6	0.00	7	0.00
Şarbă	0	0.00	265	0.01	266	0.01
Timpuriu de Cluj	0	0.00	1	0.00	1	0.00
Unirea	0	0.00	1	0.00	1	0.00
Zghihară de Huşi	0	0.00	87	0.00	54	0.00
Total (36 varieties)	**7696**	**0.17**	**20277**	**0.44**	**19103**	**0.43**

Table 14 (cont.): Prime varieties' global area and share of all varieties globally, by prime's country of origin, 2000, 2010 and 2016

	2000		2010		2016	
	Hectares	*Share (%)*	*Hectares*	*Share (%)*	*Hectares*	*Share (%)*
Turkey						
Adakarası	48	0.00	69	0.00	89	0.00
Boğazkere	773	0.02	1106	0.02	1436	0.03
Çalkarası	436	0.01	625	0.01	806	0.02
Çavuş	0	0.00	0	0.00	3	0.00
Chaouch Blanc	2	0.00	3	0.00	0	0.00
Dimrit	602	0.01	863	0.02	704	0.02
Emir	480	0.01	688	0.01	89	0.00
Kalecik Karası	601	0.01	861	0.02	704	0.02
Karalahna	3	0.00	4	0.00	4	0.00
Karasakiz	3	0.00	4	0.00	4	0.00
Narince	537	0.01	769	0.02	787	0.02
Öküzgözü	1033	0.02	1479	0.03	1601	0.04
Papazkarası	122	0.00	175	0.00	204	0.00
Rosaki	0	0.00	0	0.00	2	0.00
Sultaniye	12122	0.26	3413	0.07	5325	0.12
Vasilaki	3	0.00	4	0.00	4	0.00
Vranac	0	0.00	149	0.00	9503	0.21
Total (16 varieties)	**16764**	**0.36**	**10208**	**0.22**	**21266**	**0.47**
Montenegro						
Vranac	0	0.00	149	0.00	9503	0.21
Total (1 variety)	**0**	**0.00**	**149**	**0.00**	**9503**	**0.21**
Switzerland						
Arvine	61	0.00	172	0.00	192	0.00
Birstaler Muskat	0	0.00	0	0.00	1	0.00
Blattner Cal 1-15	0	0.00	0	0.00	0	0.00
Blattner Cal 1-20	0	0.00	0	0.00	0	0.00
Blattner Cal 1-22	0	0.00	0	0.00	0	0.00
Blattner Cal 1-28	0	0.00	0	0.00	2	0.00
Blattner Cal 1-36	0	0.00	0	0.00	1	0.00
Blattner Reds	0	0.00	39	0.00	8	0.00
Blattner Whites	0	0.00	25	0.00	7	0.00
Bondola	17	0.00	13	0.00	11	0.00
BX 81-83	0	0.00	0	0.00	1	0.00
Cabaret Noir	0	0.00	0	0.00	3	0.00
Cabernet Blanc	0	0.00	0	0.00	6	0.00
Cabernet Early	0	0.00	0	0.00	0	0.00
Cabernet Foch	0	0.00	0	0.00	2	0.00

Table 14 (cont.): Prime varieties' global area and share of all varieties globally, by prime's country of origin, 2000, 2010 and 2016

	2000		2010		2016	
	Hectares	*Share (%)*	*Hectares*	*Share (%)*	*Hectares*	*Share (%)*
Switzerland (continued)						
Cabernet Jura	0	0.00	19	0.00	27	0.00
Cabernet Soyhières	0	0.00	0	0.00	1	0.00
Cabernet x Maréchal Foch	0	0.00	0	0.00	0	0.00
Carminoir	0	0.00	10	0.00	11	0.00
Chardoris	0	0.00	0	0.00	0	0.00
Charmont	7	0.00	10	0.00	10	0.00
Chasselas	13318	0.29	13119	0.28	7377	0.16
Chasselas (R)	11	0.00	95	0.00	90	0.00
Completer	2	0.00	3	0.00	5	0.00
Diolinoir	31	0.00	114	0.00	122	0.00
Divico	0	0.00	0	0.00	10	0.00
Divona	0	0.00	0	0.00	0	0.00
Doral	3	0.00	27	0.00	35	0.00
Dunze	1	0.00	0	0.00	0	0.00
Eyholzer Rote	0	0.00	0	0.00	0	0.00
Galotta	0	0.00	13	0.00	35	0.00
Gamaret	71	0.00	405	0.01	441	0.01
Garanoir	76	0.00	216	0.00	229	0.01
Himbertscha	0	0.00	0	0.00	0	0.00
Humagne	9	0.00	30	0.00	29	0.00
IRAC 1933	0	0.00	0	0.00	0	0.00
Kalina	0	0.00	0	0.00	0	0.00
Lafnetscha	1	0.00	0	0.00	2	0.00
Mara	0	0.00	0	0.00	10	0.00
Millot-Foch	0	0.00	0	0.00	126	0.00
MRAC 1087	0	0.00	0	0.00	0	0.00
MRAC 1099	0	0.00	0	0.00	0	0.00
MRAC 1626	0	0.00	0	0.00	0	0.00
MRAC 1817	0	0.00	0	0.00	0	0.00
MRAC 40	0	0.00	0	0.00	0	0.00
Muscat Bleu	0	0.00	0	0.00	3	0.00
Muscatin	0	0.00	0	0.00	0	0.00
Olivette de Laconnex	0	0.00	0	0.00	0	0.00
Petite Milo	0	0.00	0	0.00	6	0.00
Pinorico	0	0.00	0	0.00	0	0.00
Réselle	0	0.00	0	0.00	1	0.00
Reze	1	0.00	0	0.00	0	0.00
Roter Milan	0	0.00	0	0.00	0	0.00
Rouge de Fully	0	0.00	0	0.00	1	0.00
Rouge du Pays	0	0.00	0	0.00	136	0.00
Satin Noir	0	0.00	0	0.00	0	0.00

Table 14 (cont.): Prime varieties' global area and share of all varieties globally, by prime's country of origin, 2000, 2010 and 2016

	2000		2010		2016	
	Hectares	*Share (%)*	*Hectares*	*Share (%)*	*Hectares*	*Share (%)*
Switzerland (continued)						
Sauvignac	0	0.00	0	0.00	2	0.00
Siramé	0	0.00	0	0.00	0	0.00
Tedi's Best	0	0.00	0	0.00	0	0.00
VB 32-7	0	0.00	0	0.00	3	0.00
VB 91-26-25	0	0.00	0	0.00	1	0.00
VB 91-26-26	0	0.00	0	0.00	0	0.00
VB 91-26-27	0	0.00	0	0.00	0	0.00
VB Cal 1-14	0	0.00	0	0.00	0	0.00
VB Cal 1-29	0	0.00	0	0.00	0	0.00
VB Cal 1-31	0	0.00	0	0.00	0	0.00
VB Cal 1-33	0	0.00	0	0.00	0	0.00
VB Cal 6-04 N5	0	0.00	0	0.00	0	0.00
VB Jura 25	0	0.00	0	0.00	0	0.00
Total (70 varieties)	**13608**	**0.29**	**14312**	**0.31**	**8950**	**0.20**
South Africa						
Chenel	339	0.01	79	0.00	33	0.00
Colomino	16	0.00	5	0.00	2	0.00
Grachen	3	0.00	2	0.00	0	0.00
Nouvelle	2	0.00	422	0.01	428	0.01
Pinotage	6574	0.14	6404	0.14	7132	0.16
Roobernet	78	0.00	139	0.00	269	0.01
Therona	185	0.00	99	0.00	67	0.00
Weldra	54	0.00	14	0.00	2	0.00
Total (8 varieties)	**7251**	**0.16**	**7164**	**0.16**	**7933**	**0.18**
United Kingdom						
Muscat of Hamburg	7068	0.15	8140	0.18	7680	0.17
Total (1 variety)	**7068**	**0.15**	**8140**	**0.18**	**7680**	**0.17**
Ukraine						
Bastardo Magarachsky	1969	0.04	2370	0.05	180	0.00
Citronny Magarach	0	0.00	307	0.01	307	0.01
Ekim Kara	19	0.00	27	0.00	0	0.00
Golubok	50	0.00	87	0.00	37	0.00
Hasansky Sladky	0	0.00	0	0.00	1	0.00
Ilichevskii Rannii	5	0.00	5	0.00	0	0.00
Kokur Bely	641	0.01	918	0.02	0	0.00

Table 14 (cont.): Prime varieties' global area and share of all varieties globally, by prime's country of origin, 2000, 2010 and 2016

	2000		2010		2016	
	Hectares	*Share (%)*	*Hectares*	*Share (%)*	*Hectares*	*Share (%)*
Ukraine (continued)						
Lora	0	0.00	0	0.00	35	0.00
Magaracha Rannii	0	0.00	0	0.00	884	0.02
Mechta	0	0.00	0	0.00	19	0.00
Misket	3764	0.08	0	0.00	0	0.00
Muskat Zhemchuzhnyi	0	0.00	0	0.00	1	0.00
Odessky Cherny	1694	0.04	2686	0.06	2508	0.06
Original	0	0.00	0	0.00	5	0.00
Pervenets Magaracha	2837	0.06	2881	0.06	2755	0.06
Podarok Magaracha	148	0.00	504	0.01	292	0.01
Riesus	0	0.00	115	0.00	115	0.00
Rubin Golodrigi	0	0.00	82	0.00	82	0.00
Rubin Tairovsky	2	0.00	2	0.00	5	0.00
Rubinovy Magaracha	0	0.00	0	0.00	0	0.00
Sukholimansky Bely	1631	0.04	2156	0.05	405	0.01
Total (21 varieties)	**12760**	**0.28**	**12141**	**0.26**	**7632**	**0.17**
China						
Beibinghong	0	0.00	0	0.00	1600	0.04
Crystal	1	0.00	175	0.00	175	0.00
Longyan	0	0.00	0	0.00	1000	0.02
Yan 73	0	0.00	0	0.00	4800	0.11
Total (4 varieties)	**1**	**0.00**	**175**	**0.00**	**7575**	**0.17**
Japan						
Black Queen	566	0.01	713	0.02	143	0.00
Koshu	118	0.00	168	0.00	690	0.02
Kyoho (4N)	4003	0.09	4003	0.09	2762	0.06
Muscat Bailey A	1372	0.03	1422	0.03	1821	0.04
Red Millennium	0	0.00	0	0.00	2	0.00
Riesling Forte	0	0.00	0	0.00	2	0.00
Rose Ciotat	0	0.00	0	0.00	2	0.00
Ryugan	0	0.00	0	0.00	27	0.00
Yama Sauvignon	0	0.00	0	0.00	24	0.00
Yamabudo	0	0.00	0	0.00	35	0.00
Yamasachi	0	0.00	0	0.00	20	0.00
Total (11 varieties)	**6059**	**0.13**	**6306**	**0.14**	**5529**	**0.12**

Table 14 (cont.): Prime varieties' global area and share of all varieties globally, by prime's country of origin, 2000, 2010 and 2016

	2000		2010		2016	
	Hectares	*Share (%)*	*Hectares*	*Share (%)*	*Hectares*	*Share (%)*
Cyprus						
Maratheftiko	0	0.00	152	0.00	0	0.00
Mavro	10969	0.24	3575	0.08	3187	0.07
Ofthalmo	0	0.00	141	0.00	0	0.00
Xynisteri	2742	0.06	2092	0.05	1946	0.04
Total (4 varieties)	**13711**	**0.30**	**5960**	**0.13**	**5133**	**0.11**
Russia						
Agadai	1265	0.03	0	0.00	0	0.00
Bakator Belyi	0	0.00	11	0.00	11	0.00
Barkhatnyi	0	0.00	30	0.00	30	0.00
Dostoinyi	0	0.00	65	0.00	65	0.00
Fioletovy Ranny	0	0.00	50	0.00	50	0.00
Krasnostop Zolotovsky	0	0.00	562	0.01	562	0.01
Kuban	0	0.00	0	0.00	32	0.00
Levokumskij	0	0.00	890	0.02	890	0.02
Mergeritar	0	0.00	0	0.00	0	0.00
Michurinets	0	0.00	0	0.00	0	0.00
Saperavi Severny	25	0.00	350	0.01	325	0.01
Stepnyak	0	0.00	144	0.00	144	0.00
Tsimlyansky Cherny	0	0.00	451	0.01	451	0.01
Tsvetochny	0	0.00	169	0.00	169	0.00
Victoria	145	0.00	52	0.00	620	0.01
Vidvizhenets	0	0.00	271	0.01	271	0.01
Vostorg	0	0.00	0	0.00	17	0.00
Total (17 varieties)	**1435**	**0.03**	**3045**	**0.07**	**3639**	**0.08**
Brazil						
Carmem	0	0.00	0	0.00	328	0.01
Clara	0	0.00	0	0.00	0	0.00
Concord Clone 30	0	0.00	0	0.00	196	0.00
Cora	0	0.00	0	0.00	570	0.01
Dona Zillá	0	0.00	0	0.00	1	0.00
Juliana	0	0.00	0	0.00	0	0.00
Lorena	0	0.00	519	0.01	500	0.01
Magna	0	0.00	0	0.00	30	0.00
Margot	0	0.00	0	0.00	2	0.00
Moscato Embrapa	0	0.00	862	0.02	683	0.02
Niagara Red	0	0.00	0	0.00	469	0.01
Patrizia Rosa	10	0.00	153	0.00	0	0.00
Rúbea	0	0.00	81	0.00	181	0.00
Tardia de Caxias	0	0.00	0	0.00	0	0.00
Violeta	0	0.00	98	0.00	636	0.01
Total (15 varieties)	**10**	**0.00**	**1714**	**0.04**	**3596**	**0.08**

Table 14 (cont.): Prime varieties' global area and share of all varieties globally, by prime's country of origin, 2000, 2010 and 2016

	2000		2010		2016	
	Hectares	*Share (%)*	*Hectares*	*Share (%)*	*Hectares*	*Share (%)*
Serbia						
Bácska	0	0.00	0	0.00	7	0.00
Beogradska Crna	0	0.00	0	0.00	0	0.00
Kosmopolita	0	0.00	0	0.00	1	0.00
Kreaca	0	0.00	29	0.00	30	0.00
Morava	0	0.00	0	0.00	34	0.00
Panonia	0	0.00	0	0.00	10	0.00
Prokupac	15180	0.33	15180	0.33	1361	0.03
Slankamenka	0	0.00	53	0.00	23	0.00
Župljanka	0	0.00	4	0.00	505	0.01
Total (9 varieties)	**30360**	**0.66**	**30533**	**0.66**	**3944**	**0.09**
Slovenia						
Bouvier	365	0.01	250	0.01	224	0.00
Ranfol	0	0.00	134	0.00	0	0.00
Vitovska Grganja	0	0.00	0	0.00	33	0.00
Žametovka	0	0.00	914	0.02	822	0.02
Zelen	**365**	**0.01**	**1297**	**0.03**	**1079**	**0.02**
Total (5 varieties)	**365**	**0.01**	**1297**	**0.03**	**1079**	**0.02**
Azerbaijan						
Bayanshira	451	0.01	645	0.01	645	0.01
Madrasa	20	0.00	28	0.00	28	0.00
Total (2 varieties)	**470**	**0.01**	**673**	**0.01**	**673**	**0.02**
North Macedonia						
Stanušina Crna	0	0.00	0	0.00	400	0.01
Total (1 variety)	**0**	**0.00**	**0**	**0.00**	**400**	**0.01**
Kazakhstan						
Kuldzhinskii	269	0.01	385	0.01	385	0.01
Total (1 variety)	**269**	**0.01**	**385**	**0.01**	**385**	**0.01**

Table 14 (cont.): Prime varieties' global area and share of all varieties globally, by prime's country of origin, 2000, 2010 and 2016

	2000		2010		2016	
	Hectares	*Share (%)*	*Hectares*	*Share (%)*	*Hectares*	*Share (%)*
Peru						
Ar110	1	0.00	1	0.00	1	0.00
Ar99	3	0.00	5	0.00	5	0.00
Imperial Seedless	1	0.00	1	0.00	1	0.00
Quebranta	230	0.00	345	0.01	330	0.01
Rosa Arica	1	0.00	1	0.00	1	0.00
Total (5 varieties)	**236**	**0.01**	**353**	**0.01**	**338**	**0.01**
Israel						
Argaman	202	0.00	202	0.00	275	0.01
Nehelescol	172	0.00	25	0.00	6	0.00
Total (2 varieties)	**374**	**0.01**	**227**	**0.00**	**281**	**0.01**
Morocco						
Abbou	2375	0.05	2375	0.05	0	0.00
Doukkali	16557	0.36	16557	0.36	0	0.00
Maticha	354	0.01	311	0.01	257	0.01
Total (3 varieties)	**19286**	**0.42**	**19243**	**0.42**	**257**	**0.01**
Lebanon						
Afus Ali	1837	0.04	381	0.01	211	0.00
Total (1 variety)	**1837**	**0.04**	**381**	**0.01**	**211**	**0.00**
Bosnia and Herzegovina						
Žilavka	0	0.00	0	0.00	185	0.00
Total (1 variety)	**0**	**0.00**	**0**	**0.00**	**185**	**0.00**
Canada						
GM 322	0	0.00	17	0.00	0	0.00
Kat.E.Lin	0	0.00	0	0.00	0	0.00
Kentville White 94-1	0	0.00	0	0.00	0	0.00
Kentville White 94-2	0	0.00	0	0.00	1	0.00
KW 96-2	0	0.00	0	0.00	0	0.00
L´Acadie Blanc	0	0.00	0	0.00	65	0.00
Saint-Cliche	0	0.00	0	0.00	0	0.00
Seyval Noir	0	0.00	0	0.00	76	0.00
Sovereign Coronation	0	0.00	0	0.00	1	0.00

Table 14 (cont.): Prime varieties' global area and share of all varieties globally, by prime's country of origin, 2000, 2010 and 2016

	2000		2010		2016	
	Hectares	*Share (%)*	*Hectares*	*Share (%)*	*Hectares*	*Share (%)*
Canada (continued)						
Sovereign Opal	0	0.00	3	0.00	3	0.00
Summerland	0	0.00	0	0.00	0	0.00
Vandal-Cliche	0	0.00	0	0.00	14	0.00
Ventura	39	0.00	24	0.00	10	0.00
Vincent	0	0.00	11	0.00	8	0.00
Vineland 53035	0	0.00	0	0.00	4	0.00
Total (15 varieties)	**39**	**0.00**	**55**	**0.00**	**182**	**0.00**
Australia						
Cabernet Sanzey	0	0.00	0	0.00	0	0.00
Caverdella	0	0.00	0	0.00	5	0.00
Cienna	0	0.00	0	0.00	70	0.00
Rubienne	0	0.00	0	0.00	1	0.00
Shalistin	0	0.00	0	0.00	0	0.00
Sun Muscat	0	0.00	0	0.00	23	0.00
Taminga	46	0.00	0	0.00	2	0.00
Tarrango	120	0.00	72	0.00	16	0.00
Tyrian	0	0.00	0	0.00	38	0.00
Total (9 varieties)	**166**	**0.00**	**72**	**0.00**	**156**	**0.00**
Thailand						
Malaga Blanc	11	0.00	16	0.00	54	0.00
Total (1 variety)	**11**	**0.00**	**16**	**0.00**	**54**	**0.00**
Chile						
Blanca Ovoide	107	0.00	40	0.00	44	0.00
Tamarugal	0	0.00	0	0.00	1	0.00
Total (2 varieties)	**107**	**0.00**	**40**	**0.00**	**44**	**0.00**

Table 14 (cont.): Prime varieties' global area and share of all varieties globally, by prime's country of origin, 2000, 2010 and 2016

	2000		2010		2016	
	Hectares	*Share (%)*	*Hectares*	*Share (%)*	*Hectares*	*Share (%)*
Czechia						
Agni	0	0.00	6	0.00	0	0.00
Andre	0	0.00	477	0.01	5	0.00
Ariana	0	0.00	3	0.00	0	0.00
Aurelius	0	0.00	70	0.00	0	0.00
Cabernet Moravia	0	0.00	212	0.00	0	0.00
Laurot	0	0.00	6	0.00	0	0.00
Muskat Moravsky	0	0.00	514	0.01	0	0.00
Neronet	0	0.00	72	0.00	6	0.00
Palava	0	0.00	230	0.00	0	0.00
Rubinet	0	0.00	0	0.00	15	0.00
Total (10 varieties)	**0**	**0.00**	**1589**	**0.03**	**26**	**0.00**
Taiwan						
Musann Blanc	0	0.00	0	0.00	5	0.00
Total (1 variety)	**0**	**0.00**	**0**	**0.00**	**5**	**0.00**
Algeria						
Ahmeur Bou Ahmeur	3	0.00	1	0.00	0	0.00
Total (1 variety)	**3**	**0.00**	**1**	**0.00**	**0**	**0.00**
Uzbekishtan						
Pervomaisky	64	0.00	64	0.00	0	0.00
Total (1 variety)	**64**	**0.00**	**64**	**0.00**	**0**	**0.00**
Armenia						
Garandmak	931	0.02	931	0.02	0	0.00
Victoria	145	0.00	52	0.00	620	0.01
Vidvizhenets	0	0.00	271	0.01	271	0.01
Vostorg	0	0.00	0	0.00	17	0.00
Carmem	0	0.00	0	0.00	328	0.01
Total (5 varieties)	**1076**	**0.02**	**1254**	**0.03**	**1236**	**0.03**

Table 14 (cont.): Prime varieties' global area and share of all varieties globally, by prime's country of origin, 2000, 2010 and 2016

	2000		2010		2016	
	Hectares	*Share (%)*	*Hectares*	*Share (%)*	*Hectares*	*Share (%)*
Belgium						
Leopoldo III	2	0.00	0	0.00	0	0.00
Total (1 variety)	**2**	**0.00**	**0**	**0.00**	**0**	**0.00**
Slovakia						
Devin	0	0.00	133	0.00	0	0.00
Dunaj	0	0.00	46	0.00	0	0.00
Milia	0	0.00	1	0.00	0	0.00
Noria	0	0.00	1	0.00	0	0.00
Total (4 varieties)	**0**	**0.00**	**182**	**0.00**	**0**	**0.00**

Table 15: Synonymns and their prime variety and colour

Synonym	Prime	Colour (of prime)
Abbo	Abbou	R
Acadie blanc	L´Acadie Blanc	W
Adakarasi	Adakarası	R
Afuz-Ali	Afus Ali	W
Aglianco	Aglianico	R
Aglianico del Vulture	Aglianico	R
Agria	Turán	R
Agudello	Godello	W
Airen	Airén	W
Alb de Onitcani	Onitskanskii Belyi	W
Alb Rominesc	Cruciulita	W
Albarifio	Alvarinho	W
Albarin Blanco	Albarín Blanco	W
Albarino	Alvarinho	W
Albariño	Alvarinho	W
Albilla	Albillo Mayor	W
Albillo	Albillo Real	W
Albilo Krimski	Albillo Krimskii	W
Alcanon	Alcañon	W
Alcoa	Tinta de Alcoa	R
Aldamiina	Adalmiina	W
Alexandrouli	Aleksandrouli	R
Alfonso Lavalle	Alphonse Lavallée	R
Alibernet	Odessky Cherny	R
Alicante	Alicante Henri Bouschet	R
Alicante Bouchet	Alicante Henri Bouschet	R
Alicante Bouschet	Alicante Henri Bouschet	R
Alicante Branco	Damaschino	W
Alicante H Bouschet Tintor	Alicante Henri Bouschet	R
Alicante H. Bouschet/Tintorera	Alicante Henri Bouschet	R
Aligote	Aligoté	W
Alikante Mouse	Alicante Henri Bouschet	R
Almafre	Almafra	W
Alminhaca	Almenhaca	W
Almuneco	Listan Negro	R
Alphonse Lavallee	Alphonse Lavallée	R
Alvadurao	Síria	W
Alvar	Alvar Branco	W
Alvarelhao	Alvarelhão	R
Alvarelhao Branco	Planta Nova	W
Alvarinho	Alvarelhão	R
Amor-Nao-Me-Deixes	Aramon Noir	R
Amostrinha	Preto Martinho	R
Ancellota	Ancellotta	R
Ancelota	Ancellotta	R
Ansonica	Inzolia	W
Aragonez	Tempranillo	R
Aramon	Aramon Noir	R
Aramon Blanc	Aramon Noir (W)	R
Arany Sarfeher	Arany Sárfehér	W
Arcadia	Nastea	W
Ardeleanca	Bakator Belyi	W
Argelina	Ahmeur Bou Ahmeur	R
Arinto	Arinto de Bucelas	W
Arinto do Interior	Dorinto	W
Arinto Roxo	Arinto de Bucelas	W
Aromella (NY 76)	Aromella	W
Aromon Blanc	Aramon Noir (W)	R
Arvine (Petite)	Arvine	W
Asirtiko	Assyrtiko	W
Aspirant Bouschet	Aspiran Bouschet	R
Asprinio	Greco	W
Asproudi	Asprouda	W
Assaraka	Assaraky	W
Athiri Aspro	Athiri	W
Aucerot	Auxerrois	W
Aurora	Aurore	W
Azal Branco	Azal	W
Babeasca Gri	Băbească Neagră (G)	R
Babeasca Gris	Băbească Neagră (G)	R
Băbească Gris	Băbească Neagră (G)	R
Babeasca Negra	Băbească Neagră	R
Babic	Babić	R
Babosa	Babosa de Madere	W
Baco	Baco Noir	R
Baco Noir	Baco Noir	R
Baco Noir, Chambourcin, etc.	Baco Noir	R
Bakator Kek	Bakator Kék	R
Baladi Verdejo	Cayetana Blanca	W
Baltica	Hasansky Sladky	R
Banati Rizling	Kreaca	W
Bánáti Rizling	Kreaca	W
Barátcsuha	Hamvas	W
Barbera Nera	Barbera	R
Barca	Tinta da Barca	R
Barkhatny	Barkhatnyi	W
Barreto	Barreto de Semente	R
Barroco	Tinta Barroca	R
Baska	Bácska	G
Baska	Bácska	G
Bastarda	Trousseau	R
Bastarda Negra	Alfrocheiro	R
Bastardillo	Cinsaut	R
Bastardo	Trousseau	R
Bastardo Do Castello	Trousseau	R
Bastardo Magaraceskii	Bastardo Magarachsky	R
Bastardo Magarachskiy	Bastardo Magarachsky	R
Bastardo Magareceski	Bastardo Magarachsky	R
Bastardo Negro	Alfrocheiro	R
Bastardo Roxo	Trousseau	R
Bastardo Tinto	Trousseau	R
Bayan-Shirey	Bayanshira	W
Becuet	Persan	R
Beli Pinot	Pinot Blanc	W
Beltza	Hondarribi Beltza	R
Bequignol	Béquignol Noir	R
Bequignol Noir	Béquignol Noir	R
Bermejuela	Marmajuelo	W
Bervedino	Vernaccia di San Gimignano	W
Bianchetta Genovese	Albarola	W
Biancone	Biancone di Portoferraio	W
Bibor Kadarka	Bíborkadarka	R
Bíbor kadarka	Bíborkadarka	R
Biborkadarka	Bíborkadarka	R
Black Magic	Kodryanka	R
Black Muscat	Muscat of Hamburg	R
Black Queen/Pokdum	Black Queen	R
Black Spanish	Jacquez	R
BlackMagic	Kodryanka	R
Blanca Cayetana	Cayetana Blanca	W
Blattner Vb 32-7	VB 32-7	W
Blauberger, Experimental Red Vinifera, Petite Sirah, Pinot Meunier, Sangiovese	Blauburger	R
Blauer Burgunder	Pinot Noir	R
Blauer Silvaner	Silvaner (R)	W
Blauer Trollinger	Schiava Grossa	R
Blaufrankisch	Blaufränkisch	R
Blaufränkisch/Lemberger	Blaufränkisch	R
Boal	Malvasia Fina	W
Boal Branco	Malvasia Fina	W
Boal Espinho	Malvasia Fina	W
Bobal, Provechon	Bobal	R
Bogazkere	Boğazkere	R
Bogdanusa	Bogdanuša	W
Bonarda	Douce Noire	R
Bonicaire	Trepat	R
Bonvedro	Parraleta	R
Borba	Graševina	W
Bordo	Cabernet Franc	R
Borgona	Blaufränkisch	R
Borgonja	Blaufränkisch	R
Bornova Misketi	Muscat Blanc à Petits Grains	W
Borracal	Borraçal	R
Boschera	Verdicchio Bianco	W
Bourdin (S)	Florental	R
Bouvierovo Hrozno	Bouvier	W
Bovale	Graciano	R
Bovale Grande	Mazuelo	R
Brachetto	Brachetto del Piemonte	R
Bragao	Tinta Bragao	R
Brajda Crna	Suscan	R
Brancellao	Alvarelhão	R
Branco Joao	Branco Sr. Joao	W
Branco Lexítimo- B	Albarín Blanco	W
Breval	Beba	W
BRS Carmen	Carmem	R
BRS Clara	Clara	W
BRS Cora	Cora	R
BRS Lorena	Lorena	W
BRS Magna	Magna	R
BRS Margot	Margot	R
BRS Violeta	Violeta	R
Bruñal	Alfrocheiro	R
Brunello	Sangiovese	R
Budai	Budai Zöld	W
Buera	Bua Kurdzeni	W
Buonamico	Bonamico	R

Table 15 (cont.): Synonymns and their prime variety and colour

Synonym	Prime	Colour (of prime)	Synonym	Prime	Colour (of prime)
Burdeos	Cabernet Sauvignon	R	Chasselas Dorato	Chasselas	W
Burger	Elbling	W	Chasselas Doré	Chasselas	W
Burgund Mare	Blaufränkisch	R	Chasselas musque	Chasselas	W
Burgunder	Pinot Blanc	W	Chasselas rose	Chasselas (R)	W
Burgunder, Weißer	Pinot Blanc	W	Chasselas Rouge	Chasselas (R)	W
Burra Blanca	Airén	W	Chasselas Roxo	Chasselas (R)	W
Busuioaca De Bohotin	Busuioacă de Bohotin	G	Chasselas/Gutedel	Chasselas	W
C.G. 14260 (Inta)	Gargiulo 14260	R	Chenin	Chenin Blanc	W
C.G.2539 (Inta)	Gargiulo 2539	R	Chenin Blanc - Chenin	Chenin Blanc	W
C.G. 26189 (Inta)	Gargiulo 26189	R	Chirac	Syrah	R
C G 26879 (Inta)	Gargiulo 26879	W	Cidadelhe	Tinta de Cidadelhe	R
C.G.4113 (Inta)	Gargiulo 4113	R	Ciguente	Síria	W
C G 45803 (Inta)	Gargiulo 45803	W	Cinsault	Cinsaut	R
Cabernet B	Cabernet Blanc	W	Cinsaut Blanc	Cinsaut (W)	R
Cabernet Blanc (VB 91-26-01)	Cabernet Blanc	W	Cinsaut Gris	Cinsaut (G)	R
Cabernet Franc - Cabernet Franco	Cabernet Franc	R	Cinsaut Seedles	Cinsaut	R
Cabernet Gernischt	Carmenère	R	Cirfandli	Zierfandler	G
Cabernet Noir (VB 91-26-04)	Cabaret Noir	R	Citronny of Magarach	Citronny Magarach	W
Cabernet S.	Cabernet Sauvignon	R	Cjanorie	Cianorie	R
Cabernet Sauvignon - Cabernet	Cabernet Sauvignon	R	Clairet	Clairette	W
Cabernet VB	Cabernet Blanc	W	Clairette Blanche	Clairette	W
Cabernet-Sauvignon	Cabernet Sauvignon	R	Coarna Neagra	Coarnă Neagră	R
Cabertin (VB 91-26-18)	Cabertin	R	Codana	Codană	R
CabVB	Cabernet Blanc	W	Codega do Larinho	Codega de Larinho	W
Cadarcă	Kadarka	R	Codreanca	Kodryanka	R
Cagnulari	Graciano	R	Codrinski	Kodrinskii	R
Cainho	Caíño Blanco	W	Cokur White	Kokur Bely	W
Cainho de Moreira	Caíño Blanco	W	Colgadero	Quiebratinajas Tinto	R
Caino Blanco	Caíño Blanco	W	Colombar	Colombard	W
Caino Tinto	Borraçal	R	Colorino	Abrusco	R
Caíño Tinto	Borraçal	R	Corbina	Corbina Vicentina	R
Calabrese	Nero d'Avola	R	Cordisco	Montepulciano	R
Çalkarasi	Çalkarası	R	Corinto	Korinthiaki	R
Calkarasi	Çalkarası	R	Corinto Nero	Korinthiaki	R
Camarate	Camaralet de Lasseube	W	Cornalin/Landroter	Rouge du Pays	R
Canaiolo	Canaiolo Nero	R	Corvina	Corvina Veronese	R
Canaiolo Bianco	Drupeggio	W	Corvo	Aubun	R
Canaiolo Rosa	Canaiolo Nero	R	Cot	Côt	R
Canari	Canari Noir	R	Cot Rouge	Côt	R
Canina Nera	Fortana	R	Couderc	Couderc Noir	R
Cannon Hall Muscat (4N)	Muscat Cannon Hall (4N)	W	Cramposie	Crimposie	W
Cannonau	Garnacha Tinta	R	Crâmpoşie	Crimposie	W
Cardenal	Cardinal	R	Cramposie Selectionata	Crâmpoşie Selecţionată	W
Cargadora	Cinsaut	R	Crato Espanhol	Síria	W
Caricagiola	Parraleta	R	Criolla Chica	Listán Prieto	R
Carignan	Mazuelo	R	Cristal	Crystal	W
Carignan - Carignane, Cariñena	Mazuelo	R	Crljenak Kastelanski	Tribidrag	R
Carignan Blanc	Mazuelo (W)	R	Cs.2	Kozma CS. 2	R
Carignan Gris	Mazuelo (G)	R	Csaba Gyongye	Csaba Gyöngye	W
Carignan Noir	Mazuelo	R	Csabagyongye	Csaba Gyöngye	W
Carignan Noir Mazuela	Mazuelo	R	Csabagyöngye	Csaba Gyöngye	W
Carignane	Mazuelo	R	Cserszegi Fuszeres	Cserszegi Fűszeres	G
Carignano	Mazuelo	R	Csillam	Csillám	W
Carinena Blanco	Mazuelo (W)	R	Csokaszolo	Csókaszőlő	R
Carmenere	Carmenère	R	Cynthiana	Norton	R
Carmenère - Grande Vidure	Carmenère	R	De Chaunac and Rosette	De Chaunac	R
Carmenere Crni	Carmenère	R	De Cuerno	Cornichon Blanc	W
Carmi Noir	Carminoir	R	Dekabrsky	Dekabrskii	R
Carrasquenho	Baga	R	Dembina	Debina	W
Carrega Burros	Baga	R	Derechero	Royal de Alloza	R
Castañal	Tinta Castañal	R	Diego	Vijariego	W
Castelao	Castelão	R	Dievcie Hrozno	Fetească Albă	W
Castelao Branco	Castelão Branco	W	Diseca Ranina	Dišeća Ranina	W
Castellana	Rufete	R	Divico (IRAC 2091)	Divico	R
Casteloa	Castelão	R	DM 8521-1	David Macgregor 8521-1	R
Catarratto	Catarratto Bianco	W	Dom Felder	Dornfelder	R
Catarratto Bianco Comune	Catarratto Bianco	W	Don Mariano	Imperial Napoleon	R
Catarratto Bianco Lucido	Catarratto Bianco	W	Dona Blanca	Síria	W
Cavus	Chaouch Blanc	W	Doña Blanca	Síria	W
Cayuga	Cayuga White	W	Doradilla	Cayetana Blanca	W
Centenial Seedless	Centennial Seedless	W	Dourado	Galego Dourado	W
Cercial	Cerceal Branco	W	Drujba	Droujba	W
Cesanese Comune	Cesanese	R	Durasa	Bonarda Piemontese	R
Cesanese d'Affile	Cesanese	R	Durize	Rouge de Fully	R
Cesar N	Cesar	R	Einset	Einset Seedless	R
Charbono	Douce Noire	R	Elbling Rot	Elbling (R)	W
Chardonnay - Pinot Chardonnay	Chardonnay	W	Elbling Weisser	Elbling	W
Chardonnay Blanc	Chardonnay	W	Elbling, Roter	Elbling (R)	W
Chardonnay Musque and Chardonnay	Chardonnay	W	Elbling, Weißer	Elbling	W
Chaselas Dorada	Chasselas	W	Engomada	Tinta Engomada	R
Chaslas	Chasselas	W	Erbanno	Schiava Lombarda	R
Chassela Blanc, Rose	Chasselas	W	ES 10-18-30	Elmer Swenson 10- 18- 30	W
Chasselas alba	Chasselas	W	Esgana Cao Tinto	Esganacao Preto	R
Chasselas Blanc	Chasselas	W	Esganoso	Sercial	W

Table 15 (cont.): Synonymns and their prime variety and colour

Synonym	Prime	Colour (of prime)	Synonym	Prime	Colour (of prime)
Ezerfurtu	Ezerfürtü	W	Gouveio Estimado	Gouveio Real	W
Ezerjo	Ezerjó	W	Gouveio Roxo	Godello	W
Faber	Faberrebe	W	Graciana	Graciano	R
Falkensteiner	Grüner Veltliner	W	Gran Negro	Grand Noir	R
Farbertraube	Teinturier	R	Grand Noir de la Calmette	Grand Noir	R
Färbertraube	Teinturier	R	Granoir	Garanoir	R
Favorita	Vermentino	W	Grasa de Cotnari	Grasă de Cotnari	W
Favorita Diaz	Vermentino	W	Grasade Cotnari	Grasă de Cotnari	W
Fekete Fájú Bajor	Augster Blau	R	Grasevina	Graševina	W
Fekete leányka	Fetească Neagră	R	Grasvena	Graševina	W
Fekete Muskotály	Muscat Blanc à Petits Grains	W	Grauer Burgunder	Pinot Gris	G
Fernao Pires	Fernão Pires	W	Grecanico Dorato	Garganega	W
Fernao Pires Rosado	Fernão Pires	W	Grechetto	Grechetto di Orvieto	W
Ferrol	Manseng Noir	R	Grechetto Bianco	Grechetto di Orvieto	W
Ferron	Manseng Noir	R	Greco di Tufo	Greco	W
Feteasca Alba	Fetească Albă	W	Grenache	Garnacha Tinta	R
Feteasca Neagra	Fetească Neagră	R	Grenache Blanc	Garnacha Blanca	W
Feteasca neagră	Fetească Neagră	R	Grenache Gris	Garnacha Roja (Gris)	G
Feteasca Regala	Fetească Regală	W	Grenache Noir	Garnacha Tinta	R
Feteasca regală	Fetească Regală	W	Grenache Rose	Garnacha Roja (Gris)	G
Feuille de Tilleul	Hárslevelű	W	Grolleau	Grolleau Noir	R
Fileri	Moschofilero	G	Grolleau Gris	Grolleau Noir	R
Findling	Müller-Thurgau	W	Groppello di Santo Stefano	Groppello di Mocasina	R
Fioletovy Ranny	Fioletovy Ranny	R	Gros Bec	Pinot Noir	R
Fleurtai (UD-34.111)	Fleurtai	W	Gros Colman	Dodrelyabi	R
Foch	Maréchal Foch	R	Gross Manseng	Gros Manseng	W
Foch (Marechal)	Maréchal Foch	R	Gruner Veltliner	Grüner Veltliner	W
Fokiano (White)	Fokiano (W)	R	Gual	Malvasia Fina	W
Folle Noire	Canari Noir	R	Guarnaccia	Coda di Volpe Bianca	W
Folle Noire (Vidiella)	Canari Noir	R	Gutedel Weisser	Chasselas	W
Forastera Blanca	Chelva	W	Gutedel, Weißer	Chasselas	W
Forcallat Blanca	Airén	W	Gyongyrizling	Gyöngyrizling	W
Franconia	Blaufränkisch	R	Hajnal	Hajnalka	W
Francusa	Francuşă	W	Hajnos Kék	Somszoekoe Kék	R
Frankovka	Blaufränkisch	R	Hamburgi Muskotaly	Muscat of Hamburg	R
Frankovka Modra	Blaufränkisch	R	Hamburgi Muskotály	Muscat of Hamburg	R
Frappato Di Vittoria	Frappato	R	Hanepoot	Muscat of Alexandria	W
Freiburger	Freisamer	W	Harslevelu	Hárslevelű	W
French Colombard	Colombard	W	Holder	Hölder	W
Friulano	Sauvignonasse	W	Hondarrabi Beltza	Hondarribi Beltza	R
Frontenac Blanc	Frontenac (W)	R	Humagne Blanc	Humagne	W
Frontenac Gris	Frontenac (G)	R	Humagne Rouge	Cornalin	R
Frontenac Grus	Frontenac (G)	R	Ialovenschi ustoicivii	Yalovenskii Ustoichivyi	W
Frontenac Noir	Frontenac	R	Iliciovski Ciornai Rannii	Ilichevskii Rannii	R
Frontenac Rouge	Frontenac	R	Iordană	Iordan	W
Fruhburgunder	Pinot Noir Précoce	R	IRAC 2060	Divona	W
Frühburgunder, Blauer	Pinot Noir Précoce	R	Irsai Oliver	Irsai Olivér	W
Fruhroter Veltliner	Frühroter Veltliner	R	Irşai Oliver	Irsai Olivér	W
Frutilla	Isabella	R	Irsay Olivier	Irsai Olivér	W
Galbena De Odobesti	Galbenă de Odobeşti	W	Isabel	Isabella	R
Galbenade Odobesti	Galbenă de Odobeşti	W	Isabel Precoce	Isabella	R
Galego	Galego Dourado	W	Iso	Beba	W
Gallota	Galotta	R	Italia (Pirovano 65) (PE)	Italia	W
Gamay	Gamay Noir	R	Italian Riesling	Graševina	W
Gamay Beaujolais	Gamay Noir	R	Italian, Portugese whites	Italia	W
Gamay cl 565	Gamay Noir	R	Iubilei Juravelea	Yubilei Zhuravlya	R
Gamay de Chaudenay	Gamay Teinturier de Chaudenay	R	Izsaki Sarfeher	Arany Sárfehér	W
Gamay De Teinturier Bouze	Gamay Teinturier de Bouze	R	J. Riesling	Riesling	W
Gamay St Romain	Gamay Noir	R	Jaen	Mencía	R
Gammay de Bouze	Gamay Teinturier de Bouze	R	Jaen Tinto	Mencía	R
Gammay Freaux	Gamay Teinturier Freaux	R	Jázmin (8/1)	Jázmin	W
Gamza	Kadarka	R	Jemciug csaba	Csaba Gyöngye	W
Ganjuri	Tebrizi	W	Juan Garcia	Juan García	R
Gansfusser	Gänsfüsser	R	Juan Ibáñez	Moristel	R
Garnacha	Garnacha Tinta	R	Jubilaumsrebe	Jubiläumsrebe	G
Garnacha Roja	Garnacha Roja (Gris)	G	Jubileumsrebe	Jubiläumsrebe	G
Garnacha Rose	Garnacha Roja (Gris)	G	Jurancon Blanc	Jurançon Blanc	W
Garnacha Tinta	Garnacha Tinta	R	Jurancon Noir	Jurançon Noir	R
Garnacha Tintorera	Alicante Henri Bouschet	R	Kalecik Karasi	Kalecik Karası	R
Garnachablanca	Garnacha Blanca	W	Kanaan	Nehelescol	W
Garrut	Monastrell	R	Karaburnu	Afus Ali	W
Gecsei Zamatos	Göcseji Zamatos	W	Karat	Karát	W
Gegic	Gegić	W	Kardinal Crveni	Cardinal	R
Geisenheim	Geisenheim 318-57	W	Karinian	Mazuelo	R
Geisenheim 813-57	Geisenheim 318-57	W	Karmin	Kármin	R
Gewurztraminer	Gewürztraminer	W	Katawaba	Catawba	R
GF 48-12	Geilweilerhof Ga- 48- 12	W	Katelin	Kat.E.Lin	R
Giro	Giro Nero	R	Kavcina Crna	Žametovka	R
Gocseji Zamatos	Göcseji Zamatos	W	Kék Bajor	Augster Blau	R
Godelho	Godello	W	Kek Bakator	Bakator Kék	R
Goher	Augster Weiss	W	Kékfrankos	Blaufränkisch	R
Gohér	Augster Blau	R	Keknyelu	Kéknyelű	W
Gohér	Augster Weiss	W	Kékoportó	Blauer Portugieser	R
Goluboc	Golubok	R	Kerner Bijeli	Kerner	W
Gorda	Juan García	R	Kernling	Kerner	W
Gouais/Gwäss	Gouais Blanc	W	Key Gray	Kay Gray	W
Gouveio	Godello	W	Kiralyleanyka	Királyleányka	W

Table 15 (cont.): Synonymns and their prime variety and colour

Synonym	Prime	Colour (of prime)
Kirovabadskii Stolovyi (Tabriz, Ganjuri)	Tebrizi	W
Kişmiş lucistăi	Kishmish Luchistyi	R
Kişmiş moldovenesc	Kishmish Moldavskii	R
Knipperle	Knipperlé	W
Koenigin Der Weingaerten	Königin der Weingärten	W
Koevidinka	Kövidinka	G
Koksis Irma	Kocsis Irma	W
Königliche Esther	Esther	R
Korai Piros Veltelini	Frühroter Veltliner	R
Korai Piros Veltelíni	Frühroter Veltliner	R
Koverszolo	Grasă de Cotnari	W
Kövérszőlő	Grasă de Cotnari	W
Kovidinka	Kövidinka	G
Krakhuna Tetri	Krakhuna	W
Krasnostop Anapsky	Krasnostop Zolotovsky	R
Kratoshija	Tribidrag	R
Kratošija	Tribidrag	R
Kubani	Kuban	R
Kujundzusa	Kujundžuša	W
Kuldzhinskiy	Kuldzhinskii	R
Kunbarat	Kunbarát	W
Kundza Tetri	Kundza	W
Kunleany	Kunleány	W
Kuntra	Karasakiz	R
Kurucver	Kurucvér	R
KW 94-1	Kentville White 94-1	W
KW 94-2	Kentville White 94-2	W
Kyoho	Kyoho (4N)	R
Kyoko	Kyoho (4N)	R
L´Acadie	L´Acadie Blanc	W
Lacrima	Lacrima di Morro d'Alba	R
Lagorthi	Verdeca	W
Lakhegyi Mezes	Lakhegyi Mézes	W
Laki Rizling	Graševina	W
Lambrusco a Foglia Frastagliata	Enantio	R
Landot	Landot Noir	R
Laški Rizling	Graševina	W
Leana	Lyana	W
Leanyka	Leányka	W
Lefkada	Vertzami	R
Lemberger	Blaufränkisch	R
Lenoir	Jacquez	R
Leon Millet	Léon Millot	R
Leon Millot	Léon Millot	R
Léon Millot x Maréchal Foch	Millot-Foch	R
Liatiko(White)	Liatiko (W)	R
Lidia	Isabella	R
Lilas	Moscatel Lilaz	W
Limberger, Blauer	Blaufränkisch	R
Limnio(white)	Limnio (W)	R
Lipovina	Hárslevelü	W
Listan	Palomino Fino	W
Livornese Bianca	Rollo	W
Lledoner Pelut	Garnacha Peluda	R
Long Yan	Longyan	R
Loureiro Blanca	Loureiro	W
Loureiro Blanco	Loureiro	W
Loureiro Tinto	Mencía	R
Macabeu	Macabeo	W
Maccabeo	Macabeo	W
Machanauri Sapere	Machanauri	R
Madeleine Angevine	Madeleine × Angevine 7672	W
Madera	Palomino Fino	W
Magaracha's Firstborn	Pervenets Magaracha	W
Magaracha's Gift	Podarok Magaracha	W
Magliasino	Magliasina	R
Magyar Frankos	Magyarfrankos	R
Majarcă Albă	Slankamenka	W
Malaga	Muscat of Alexandria	W
Malagouzia	Malagousia	W
Malandra	Tinta Malandra	R
Malbec	Côt	R
Malbech	Côt	R
Malbeck	Côt	R
Malbek	Côt	R
Malfar	Alarije	W
Malvar	Lairen	W
Malvasia Amarela	Malvasia Bianca di Candia	W
Malvasia Aromatica	Malvasia di Candia Aromatica	W
Malvasia Bianca	Malvasia Bianca di Candia	W
Malvasia Bianca di Piemonte	Malvasia Moscata	W
Malvasia Branca	Síria	W
Malvasia Cabral	Cabral	R
Malvasia Candida	Malvasia di Lipari	W
Malvasia Candida Roxa	Malvasia di Sardegna Rosada	G
Malvasia Chianti	Malvasia Bianca Lunga	W
Malvasia di Candia	Malvasia Bianca di Candia	W
Malvasia di Sardegna	Malvasia di Lipari	W
Malvasia Dubrovacka Bijela	Malvasia di Lipari	W
Malvasia Istriana	Malvazija Istarska	W
Malvasia Negra	Malvasia Nera di Brindisi	R
Malvasia Nera	Malvasia Nera di Brindisi	R
Malvasia Nera di Lecce	Malvasia Nera di Brindisi	R
Malvasia Rei	Palomino Fino	W
Malvasia Sitges	Malvasia di Lipari	W
Malvasia Verde	Furmint	W
MalvasiaDubrovačka	Malvasia di Lipari	W
Malvasier	Frühroter Veltliner	R
Malvasier, Früher Roter	Frühroter Veltliner	R
Malvazija	Malvazija Istarska	W
Malvazija Aromatična	Malvasia di Candia Aromatica	W
Manseng Gros Blanc	Gros Manseng	W
Manseng Petit Blanc	Petit Manseng	W
Manteudo	Listain de Huelva	W
Mara (RAC 3022, C41)	Mara	R
Marastina	Malvasia Bianca Lunga	W
Maraština	Malvasia Bianca Lunga	W
Marcelan	Marselan	R
Marechal Foch	Maréchal Foch	R
Marechel Foch	Maréchal Foch	R
Mărgăritar	Mergeritar	W
Mario Feld	Pinot Noir	R
Marisancho	Pardillo	W
Marquette	Marquette	R
Marsanne Blanche	Marsanne	W
Marsanne Blanche/Ermitage	Marsanne	W
Marsigliana Nera	Magliocco Dolce	R
Marzemina	Marzemino	R
Marzemina Grossa	Marzemino	R
Mataro	Monastrell	R
Matrai Muskotaly	Mátrai Muskotály	W
Matrasa	Madrasa	R
Maturana Tinta	Trousseau	R
Maturano Bianco	Maturana Blanca	W
Mauzac	Mauzac Blanc	W
Mavroudi	Mavrouda	R
Mayskiy	Maiskii Chernyi	R
Mazuela	Mazuelo	R
Mecita	Mechta	R
Melnik	Shiroka Melnishka	R
Melón	Melon	W
Melra	Tinta da Melra	R
Mencia	Mencía	R
Menoire	Menoir	R
Merenzao	Trousseau	R
Merlot Noir	Merlot	R
Merlot, Petit Verdot	Arinarnoa	R
Meslier Saint Francois	Meslier Saint-Francois	W
Mészikadar	Meszi Kadarka	R
Meunier	Pinot Meunier	R
Mezes	Mézes Fehér	W
Mézes	Mézes Fehér	W
Mezes Feher	Mézes Fehér	W
Michele Palieri	Michele Palieri	R
Michele Parlieri	Michele Palieri	R
Michelle Parlieri	Michele Palieri	R
Michurinetz	Michurinets	R
Miguel De Arco	Miguel del Arco	R
Miguel Del Arco	Moristel	R
Mission	Listan Prieto	R
Modra Frankinja	Blaufränkisch	R
Modri Pinot	Pinot Noir	R
Modry Portugal	Blauer Portugieser	R
Molar	Negramoll	R
Molinera	Molinara	R
Molinha	Fernão Pires	W
Moll	Prensal	W
Mollar	Negramoll	R
Monastrell	Monastrell	R
Mondeuse	Mondeuse Noire	R
Mondeuse Rouge	Mondeuse Noire	R
Monemvasia	Monemvassia	W
Monstruosa de Monterrei	Monstruosa	W
Montreal blue	Montreal Blues	R
Moores Diamond	Moore's Diamond	W
Moravia Dulce	Marufo	R
Moreto	Malvasia Preta	R
Morio Muscat	Morio-Muskat	W

Table 15 (cont.): Synonymns and their prime variety and colour

Synonym	Prime	Colour (of prime)	Synonym	Prime	Colour (of prime)
Moscatel	Muscat Blanc à Petits Grains	W	Muscaris (FR 493-87)	Muscaris	W
Moscatel Amarilla	Moscato Giallo	W	Muscat A Petit Grains Blanc (Frontignac)	Muscat Blanc à Petits Grains	W
Moscatel Amarillo	Moscato Giallo	W	Muscat A Petit Grains Rouge/Rose (Frontignac)	Muscat Blanc à Petits Grains (R)	W
Moscatel Austria	Torrontés Sanjuanino	W	Muscat a Petits Grains Blanc	Muscat Blanc à Petits Grains	W
Moscatel Blanca	Muscat of Alexandria	W	Muscat A Petits Grains Blancs	Muscat Blanc à Petits Grains	W
Moscatel Blanco	Muscat of Alexandria	W	Muscat A Petits Grains Noirs	Muscat Blanc à Petits Grains (R)	W
Moscatel Branco	Muscat Blanc à Petits Grains	W	Muscat A Petits Grains Roses	Muscat Blanc à Petits Grains (R)	W
Moscatel de Alejandr¡a	Muscat of Alexandria	W	Muscat A Petits Grains Rouge	Muscat Blanc à Petits Grains (R)	W
Moscatel de Alejandria	Muscat of Alexandria	W	Muscat Alexandria	Muscat of Alexandria	W
Moscatel De Alejandría	Muscat of Alexandria	W	Muscat Amber	Muscat of Alexandria	W
Moscatel De Alejandría - Blanca Italia	Muscat of Alexandria	W	Muscat Bailey	Muscat Bailey A	R
Moscatel De Austria	Torrontés Sanjuanino	W	Muscat Blanc	Muscat Blanc à Petits Grains	W
Moscatel De Frontignan	Muscat Blanc à Petits Grains	W	Muscat Blanc (du Valais)	Muscat Blanc à Petits Grains	W
Moscatel de Grano Menudo	Muscat Blanc à Petits Grains	W	Muscat Blanc a Petit Grains	Muscat Blanc à Petits Grains	W
Moscatel de Hamburg	Muscat of Hamburg	R	Muscat Blanc à Petits Grains	Muscat Blanc à Petits Grains	W
Moscatel De Hamburgo	Muscat of Hamburg	R	Muscat Canelli	Muscat Blanc à Petits Grains	W
Moscatel de Malaga	Muscat of Alexandria	W	Muscat Croquant	Muscat Fleur d'Oranger	W
Moscatel Frontignan	Muscat Blanc à Petits Grains	W	Muscat d'Alexandrie	Muscat of Alexandria	W
Moscatel Galego Branco	Muscat Blanc à Petits Grains	W	Muscat de Frontignan	Muscat Blanc à Petits Grains	W
Moscatel Galego Roxo	Muscat Blanc à Petits Grains (R)	W	Muscat de Hambourg	Muscat of Hamburg	R
Moscatel Galego Tinto	Muscat Blanc à Petits Grains (R)	W	Muscat de Hamburg	Muscat of Hamburg	R
Moscatel Graudo	Muscat of Alexandria	W	Muscat de ialoveni	Muscat de Yaloven	W
Moscatel Hamburg	Muscat of Hamburg	R	Muscat De Ialoveni	Muskat de Yaloven	W
Moscatel Negra	Listan Prieto	R	Muscat Fleur d'Oranger	Muscat Fleur d'Oranger	W
Moscatel Negro	Listan Prieto	R	Muscat Frontignan	Muscat Blanc à Petits Grains	W
Moscatel Nunes	Muscat Blanc à Petits Grains	W	Muscat frontignian	Muscat Blanc à Petits Grains	W
Moscatel Rosada	Muscat Blanc à Petits Grains (G)	W	Muscat Gordo Blanco	Muscat of Alexandria	W
Moscatel Rosada (Pastilla)	Muscat Blanc à Petits Grains (G)	W	Muscat Gordo Blanco	Muscat of Alexandria	W
Moscatel Rosado	Muscat Blanc à Petits Grains (G)	W	Muscat Hambourg	Muscat of Hamburg	R
Moscatello	Muscat	W	Muscat Hamburg	Muscat of Hamburg	R
Moscato	Muscat	W	Muscat iantarnii	Muscat Yantarnyi	W
Moscato Bianco	Muscat Blanc à Petits Grains	W	Muscat jemciujenii	Muskat Zhemchuzhnyi	W
Moscato Bianco R2	Muscat Blanc à Petits Grains	W	Muscat Morio	Morio-Muskat	W
Moscato Canelli	Muscat Blanc à Petits Grains	W	Muscat Of Alexandria Red	Muscat of Alexandria (R)	W
Moscato d'Amburgo	Muscat of Hamburg	R	Muscat Oliver	Irsai Olivér	W
Moscato de Alexandria	Muscat of Alexandria	W	Muscat Ottonel	Muscat Ottonel	W
Moscato Gaillo	Moscato Giallo	W	Muscat Petit Grain	Muscat Blanc à Petits Grains	W
Moscato Giallo/Moscatel del Trentino	Moscato Giallo	W	Muscat Petits Grains	Muscat Blanc à Petits Grains	W
Moscato Nero di Acqui	Muscat of Hamburg	R	Muscat Poloskey	Pölöskei Muskotály	W
Moscato Ottonel	Muscat Ottonel	W	Muscat Poloskey	Pölöskei Muskotály	W
Moscato Rosa	Moscato Rosa del Trentino	R	Muscat Rose	Muscat Blanc à Petits Grains (R)	W
Moscato Rosado	Moscato Rosa del Trentino	R	Muscat timpuriu	Muscat Timpuriu de Bucuresti	W
Moschato	Moschomavro	R	Muscat Varieties	Muscat	W
Moschato	Muscat Blanc à Petits Grains	W	Muscat Violet	Muscat Blanc à Petits Grains (R)	W
Moschofilero Ii	Moschofilero	G	Muscat White	Muscat Blanc à Petits Grains	W
Moslavac	Furmint	W	Muscat/Muskateller	Muscat	W
Mouraton	Juan García	R	Muscat White	Muscat Blanc à Petits Grains	W
Mourisco	Marufo	R	Muscat/Muskateller	Muscat	W
Mourisco Branco	Cayetana Blanca	W	Muscats	Muscat Swenson	W
Mourisco Roxo	Marufo	R	Muskat	Muscat Blanc à Petits Grains	W
Mourvedre	Monastrell	R	Muskat Bijeli	Muscat Blanc à Petits Grains	W
Mourvèdre	Monastrell	R	Muskat Hungarian	Muscat of Hamburg	R
Mourvèdre	Monastrell	R	Muškat Ottonel	Muscat Ottonel	W
Mtsvane	Mtsvane Kakhuri	W	Muskat Ruza Crni	Moscato Rosa del Trentino	R
Mukhranuli	Aligoté	W	Muskat Zlty	Muscat Blanc à Petits Grains	W
Mulata	Negramoll	R	Muskat Zuti	Moscato Giallo	W
Muller Thurgau	Müller-Thurgau	W	Muskat, Ottonel	Muscat Ottonel	W
Müller Thurgau Rivaner	Müller-Thurgau	W	Muskateller	Muscat Blanc à Petits Grains	W
Muller Thurgau Weiss	Müller-Thurgau	W	Muskateller Gelber	Muscat Blanc à Petits Grains	W
Müllerrebe (Schwarzriesling)	Pinot Meunier	R	Muskateller, Gelber	Muscat Blanc à Petits Grains	W
Muscadel	Muscat Blanc à Petits Grains	W	Muskateller, Roter	Muscat Blanc à Petits Grains (R)	W
Muscadel (red)	Muscat Blanc à Petits Grains (R)	W	Muskat-Ottonel	Muscat Ottonel	W
Muscadel Red	Muscat Blanc à Petits Grains (R)	W	Muskattrollinger	Muscat of Hamburg	R
Muscadelle (Tokay)	Muscadelle	W	Muskat-Trollinger	Muscat of Hamburg	R
Muscadet	Melon	W	MuškatŽuti	Muscat Blanc à Petits Grains	W
Muscadine	Scuppernong	W			

Table 15 (cont.): Synonymns and their prime variety and colour

Synonym	Prime	Colour (of prime)	Synonym	Prime	Colour (of prime)
Mustoasa de Maderat	Mustoasă de Măderat	W	Pical	Piquepoul Noir	R
Mustoasa dc Măderat	Mustoasă de Măderat	W	Picapoll Negro	Piquepoul Noir	R
Mustoasade Măderat	Mustoasă de Măderat	W	Pigato	Vermentino	W
Napa Gamay	Valdiguié	R	Pignola	Pignola Valtellinese	R
Nebiollo	Nebbiolo	R	Pilongo	Alvarelhão	R
Nebiolo	Nebbiolo	R	Pinela	Pinella	W
Negkoska	Negoska	R	Pinheira Branca	Jampal	W
Negra Commun	Listan Negro	R	Pinheira Roxa	Malvasia Preta	R
Negra Criolla	Negramoll	R	Pinot 386	Pinot Noir	R
Negra Mole	Negramoll	R	Pinot Bianco	Pinot Blanc	W
Negrara	Hrvatica	R	Pinot Bijeli	Pinot Blanc	W
Negrette	Négrette	R	Pinot Blanc	Pinot Blanc	W
Negro Amaro	Negroamaro	R	Pinot Blanco	Pinot Blanc	W
Negru de Dragasani	Negru de Drăgăşani	R	Pinot Crni	Pinot Noir	R
Negru de Drăgăşani	Negru de Drăgăşani	R	Pinot Grey	Pinot Gris	G
Negru de Ialoveni	Negru de Yaloven	R	Pinot Grigio	Pinot Gris	G
Negrude Drăgăşani	Negru de Drăgăşani	R	Pinot Gris (Grigio)	Pinot Gris	G
Nektar	Nektár	W	Pinot Gris/Grigio	Pinot Gris	G
Nero	Nero d'Avola	R	Pinot Gris/Malvoisie/Grauburgunder	Pinot Gris	G
Nero Buono	Nero Buono di Cori	R	Pinot Menier	Pinot Meunier	R
Nero Davola	Nero d'Avola	R	Pinot Menieur	Pinot Meunier	R
Neuburske	Neuburger	W	Pinot Negro	Pinot Noir	R
Niagara and Wiley White	Niagara	W	Pinot Nero	Pinot Noir	R
Niagara Branca	Niagara	W	Pinot Noir	Pinot Noir	R
Niagara Rosada	Niagara Red	R	Pinot Noir - Pinot Negro	Pinot Noir	R
Niagara White	Niagara	W	Pinot Noir / Blauer burgunder	Pinot Noir	R
Nieddu Mannu	Pascale	R	Pinot Noir Blauer Burgunder	Pinot Noir	R
Nielluccio	Sangiovese	R	Pinot Noir Precoce	Pinot Noir Précoce	R
Nonastrell	Monastrell	R	Pinot Noir Précoce/Frühburgunder	Pinot Noir Précoce	R
NY Muscat	New York Muscat	R	Pinot Noir, Blue (including velvet red)	Pinot Noir	R
Odessa Black	Odessky Cherny	R	Pinot Noir/Blauburgunder	Pinot Noir	R
Odessky Chernyi	Odessky Cherny	R	Pinot Noire	Pinot Noir	R
Okuzgozu	Öküzgözü	R	Pinot Saint George	Négrette	R
Olasz Rizling	Graševina	W	Pinot Sivi	Pinot Gris	G
Olivella Nera	Sciascinoso	R	Pinot White	Pinot Blanc	W
Ondarrabi Beltza	Hondarribi Beltza	R	Pinotsivi	Pinot Gris	G
Onitcani	Onitskanskii Belyi	W	Piriquita	Castelão	R
Oporto	Blauer Portugieser	R	Piros Bakator	Bakator Roz	R
Optima 113	Optima	W	Piros Szlanka	Pamid	R
Orange Muscat	Muscat Fleur d'Oranger	W	Piros Tramini	Savagnin Rose	G
Orbois	Menu Pineau	W	Piros Veltelini	Roter Veltliner	G
Osennii ciornii	Osennii Ciornai	R	Plant Robert, Gamay and Zweigeltrebe	Gamay Noir	R
Ottavianello	Cinsaut	R	Planta Fina	Damaschino	W
Ottonel Muskotály	Muscat Ottonel	W	Plavac Mali Crni	Plavac Mali	R
Pais	Listan Prieto	R	Plavac Zuti	Plavec Žuti	W
Pais - Mission, Criolla	Listan Prieto	R	Plăvaie	Plavay	W
Palamino	Palomino Fino	W	Plavec Mali	Plavac Mali	R
Palomino	Palomino Fino	W	Plavec Zuti	Plavec Žuti	W
Palomino Superior	Palomino Fino	W	Plavecmal	Plavac Mali	R
Pameati Negrulea	Pamyati Negrulya	R	Plavina Crna	Plavina	R
Pamiti	Pamid	R	Plavinacrna	Plavina	R
Panse Negro	Trobat	R	Plemenka Bijela	Chasselas	W
Panse Valenciana	Panse Valenciano	W	Plovdina	Pamid	R
Papazkarasi	Papazkarası	R	Podarok of Magarach	Podarok Magaracha	W
Pardina	Cayetana Blanca	W	Poeloeske Muskotaly	Pölöskei Muskotály	W
Parrel Verdal	Trepat	R	Poeloeske Muskotaly	Pölöskei Muskotály	W
Patria	Pátria	W	Poloske	Pölöskei Muskotály	W
Patricia	Patrizia Rosa	R	Poloske	Pölöskei Muskotály	W
Pau Ferro	Parraleta	R	Poloskei Muskotaly	Pölöskei Muskotály	W
Pedevenda	Verdiso	W	Poloskei Muskotaly	Pölöskei Muskotály	W
Pedro	Pedro Ximénez	W	Pölöskei Muskotaly	Pölöskei Muskotály	W
Pedro Gimenez	Pedro Giménez	W	Pölöskei Muskotaly	Pölöskei Muskotály	W
Pedro Gimenez Rio Colorado	Pedro Giménez	W	Pontak	Teinturier	R
Pedro Gimenez Río Colorado	Pedro Giménez	W	Portoghese	Blauer Portugieser	R
Pedro Jimenez	Pedro Giménez	W	Portugais Bleu	Blauer Portugieser	R
Pedro Jimenez	Pedro Giménez	W	Portugalka	Blauer Portugieser	R
Pedro Luis	Galego Dourado	W	Portugieser Blau	Blauer Portugieser	R
Pedro Ximenes	Pedro Ximénez	W	Portugieser, Blauer	Blauer Portugieser	R
Pedro Ximenez	Pedro Ximénez	W	Portugizac	Blauer Portugieser	R
Pensal Blanco	Prensal	W	Portugues Azul	Blauer Portugieser	R
Periquita	Castelão	R	Portuguese, blue	Blauer Portugieser	R
Perle of Csaba	Csaba Gyöngye	W	Posip Bijeli	Pošip Bijeli	W
Perle von Zala	Zalagyöngye	W	Pozsonyi	Pozsonyi Fehér	W
Perrum	Pedro Ximénez	W	Pozsonyi Feher	Pozsonyi Fehér	W
Pervenec Magaracha	Pervenets Magaracha	W	Precoce De Malingre	Malingre Precoce	W
Perveneţ Magaracia	Pervenets Magaracha	W	Prie Blanc	Prie	W
Pervenets of Magarach	Pervenets Magaracha	W	Prie Rouge	Primetta	G
Pervomaiskii	Pervomaisky	R	Primitivo	Tribidrag	R
Petit Courbou Ondarrzurizerrat	Petit Courbu	W	Prococe de Colmar	Colmar Precoce Noir	R
Petit Syrah	Durif	R	Prokupec	Prokupac	R
Petite Arvine	Arvine	W	Prugnolo Gentile	Sangiovese	R
Petite Perle	Petite Pearl	R	Quiebratinajas	Quiebratinajas Tinto	R
Petite Sirah	Durif	R	Rabigato Franco	Rabigato	W
Petite Syrah	Durif	R	Rabo de Ovelha Tinto	Negramoll	R
Peverella	Verdicchio Bianco	W	Raboso	Raboso Piave	R

Table 15 (cont.): Synonymns and their prime variety and colour

Synonym	Prime	Colour (of prime)
Raisin Blanc	Servant	W
Rajinski Rizling	Riesling	W
Rajnai rizling	Riesling	W
Ramisco Tinto	Ramisco	R
Ranina	Blauer Portugieser	R
Rannii Magaracea	Magaracha Rannii	R
Rara Neagra	Băbească Neagră	R
Ratinho	Malvasia Fina	W
Rauschling	Räuschling	W
Rebula	Ribolla Gialla	W
Red Hanepoot	Muscat of Alexandria (R)	W
Red Veltliner	Roter Veltliner	G
Refosco Nostrano	Refosco di Faedis	R
Refosk	Refosco	R
Refošk	Refosco	R
Refren	Refrén	W
Regina	Afus Ali	W
Regina dei Vigneti	Königin der Weingärten	W
Regina viilor	Königin der Weingärten	W
Renski Rizling	Riesling	W
Rèze	Riesling	W
Rheine Riesling	Riesling	W
Rheinriesling	Riesling	W
Rieslina (Inta C.G. 38049)	Rieslina	W
Riesling de Rhin	Riesling	W
Riesling Italian	Graševina	W
Riesling Italico	Graševina	W
Riesling Lion	Riesling	W
Riesling Renan	Riesling	W
Riesling Renano	Riesling	W
Riesling Trollinger	Kerner	W
Riesling Weisser	Riesling	W
Riesling, Weißer	Riesling	W
Riesling-Lion	Riesling	W
Rivaner	Müller-Thurgau	W
Rizling Rynsky	Riesling	W
Rizling Vlasky	Graševina	W
Rizlingszilváni	Müller-Thurgau	W
Rizvanac	Müller-Thurgau	W
Rkaciteli	Rkatsiteli	W
Rkatiteli	Rkatsiteli	W
Rkaţiteli	Rkatsiteli	W
Rkatziteli	Rkatsiteli	W
Roal	Rual	R
Robal	Cayetana Blanca	W
Roditis (Red)	Roditis (R)	G
Rodo	Tinta do Rodo	R
Roesler	Rösler	R
Rombola	Robola	W
Rome Tinto	Rome	R
Rosciola rose	Rosciola	G
Roseira	Tinta Roseira	R
Roseti	Afus Ali	W
Rosioara	Pamid	R
Roșioară	Pamid	R
Rossara	Rossara Trentina	R
Roter Eyholzer	Eyholzer Rote	R
Roter Muskateller	Muscat of Hamburg	R
Roter Räuschling	Räuschling	W
Roter Silvaner	Silvaner (R)	W
Roupeiro Branco	Síria	W
Rousanne	Roussanne	W
Roxo Flor	Roxo de Vila Flor	R
Royal	Perruno	W
Rozalia	Rozala Bianca	W
Rozália	Rozala Bianca	W
Rozsakoe	Rózsakő	W
Rubi Cabernet	Ruby Cabernet	R
Rubin Tairovski	Rubin Tairovsky	R
Ruby Magaracha	Rubinovy Magaracha	R
Ruby of Golodryga	Rubin Golodrigi	R
Rulander	Pinot Gris	G
Ruländer (Burgunder, Grauer)	Pinot Gris	G
Rulandske Bile	Pinot Blanc	W
Rulandske Modre	Pinot Noir	R
Rulandske Sede	Pinot Gris	G
Rumeni Muskat	Muscat Blanc à Petits Grains	W
Rumeni Muškat	Muscat Blanc à Petits Grains	W
Rumeni Plavec	Plavec Žuti	W
Ruzica Crvena	Kövidinka	G
Ružicacrvena	Kövidinka	G
Ryzlink Rynsky	Riesling	W
Ryzlink Vlaky	Graševina	W
S. Mamede	Sao Mamede	W
S.V. 23-512	Seyve Villard 23-512	R
Saborinho	Negramoll	R

Synonym	Prime	Colour (of prime)
Sabro	Síria	W
Saint Jeannett	Saint Jeannet	W
Saint Laurent	Sankt Laurent	R
San Francisco	Black Prince	R
Sangiovese/ Nielluccio	Sangiovese	R
Santareno	Santarena	R
Sao Saul	Cinsaut	R
Saperavi Nothern	Saperavi Severny	R
Saperavi Severnii	Saperavi Severny	R
Sapere Otskhanuri	Otskhanuri Sapere	R
Sarba	Şarbă	W
Sarfeher	Sárfehér	W
Sárga Muskotály	Muscat Blanc à Petits Grains	W
Sarigo	Cayetana Blanca	W
Sauvignon	Sauvignon Blanc	W
Sauvignon B.	Sauvignon Blanc	W
Sauvignon Blanca	Sauvignon Blanc	W
Sauvignon Blanco	Sauvignon Blanc	W
Sauvignon Gris	Sauvignon Blanc (G)	W
Sauvignon Gris	Sauvignon Blanc (G)	W
Sauvignon Musque	Sauvignon Blanc	W
Sauvignon Rose	Sauvignon Blanc (G)	W
Sauvignon Rose	Sauvignon Blanc (G)	W
Sauvignon Soyhières (VB 32-7)	VB 32-7	W
Sauvignon Vert	Sauvignonasse	W
Savagnin	Savagnin Blanc	W
Savagnin Blanc/Heida	Savagnin Blanc	W
Savvatiano	Savatiano	W
Schonburger	Schönburger	G
Schwarzriesling	Pinot Meunier	R
Sciaccarello	Mammolo	R
Seibel 2 (seibeleto)	Seibel	R
Seleções	Select	W
Semigion	Sémillon	W
Semilao	Sémillon	W
Semilion	Sémillon	W
Semillon	Sémillon	W
Senso	Cinsaut	R
Šentlovrenka	Sankt Laurent	R
Serbina	Douce Noire	R
Seridan	Sheridan	R
Servagnin	Pinot Noir	R
Seyval	Seyval Blanc	W
Seyval (Blanc)	Seyval Blanc	W
Seyval Blanc and Vidal 256	Seyval Blanc	W
Seyve Villard	Seyval Blanc	W
Seyve Villard 12375	Villard Blanc	W
Seyve Villard 5276	Seyval Blanc	W
Seyve Willard (Tinta)	Seyval Noir	R
Shiraz	Syrah	R
Shiroka Melnishka Loza	Shiroka Melnishka	R
Siegfried	Siegfriedrebe	W
Silvanac Zeleni	Silvaner	W
Silvaner Grun	Silvaner	W
Silvaner, Blauer	Silvaner (R)	W
Silvaner, Grüner	Silvaner	W
Silvanske Zelene	Silvaner	W
Sipelj	Furmint	W
Sipon	Furmint	W
Šipon	Furmint	W
Sira (falsa)	Syrah	R
Sirah	Syrah	R
Siria	Síria	W
Sivi Pinot	Pinot Gris	G
SK 90-2/19 (Pannonija)	Panonia	W
Skrlet	Škrlet	W
Slarina	Cellerina	R
Smederevka	Dimyat	W
Solaris (FR 240-75)	Solaris	W
Somerset	Somerset Seedless	R
Soreli (UD-31.113)	Soreli	W
Soultanina	Sultaniye	W
Sousao	Vinhao	R
Souson	Vinhao	R
Souvignier Gris (FR 392-83)	Souvignier Gris	G
Souzao	Vinhao	R
Sovereign, Coronation, etc.	Sovereign Coronation	R
Spätburgunder, Blauer (einschl. Samtrot)	Pinot Noir	R
St Laurent	Sankt Laurent	R
St. Emilion	Trebbiano Toscano	W
Stanushina	Stanušina Crna	R
Stanušina	Stanušina Crna	R
Stepniak	Stepnyak	W
Sugra Five	Sugrafive	W
Suholimanski belii	Sukholimansky Bely	W
Suholimanskiy White	Sukholimansky Bely	W

Table 15 (cont.): Synonymns and their prime variety and colour

Synonym	Prime	Colour (of prime)
Suholimenschii Belii	Sukholimansky Bely	W
Sukholimansky	Sukholimansky Bely	W
Sultana	Sultaniye	W
Sultanina	Sultaniye	W
Svatovavrinecke	Sankt Laurent	R
Sylvaner	Silvaner	W
Sylvaner Verde	Silvaner	W
Sylvaner/Rhin	Silvaner	W
Sylvanske Zelene	Silvaner	W
Syrach	Syrah	R
Szentlorinc	Sankt Laurent	R
Szentlőrinc	Sankt Laurent	R
Szeremi Zold	Sremska Zelenika	W
Szerémi Zöld	Sremska Zelenika	W
Szürkebarát	Pinot Gris	G
Tabriz	Tebrizi	W
Talia	Trebbiano Toscano	W
Taltos	Táltos	W
Tamaioasa Romaneasca	Muscat Blanc à Petits Grains	W
Tamaioasa Romaneasca	Muscat Blanc à Petits Grains	W
Tămâioasă românească	Muscat Blanc à Petits Grains	W
Tamiosa Romaneasca	Muscat Blanc à Petits Grains	W
Tannat (Harriague)	Tannat	R
Tauberschwarz (Hängling, Blauer)	Tauberschwarz	R
Temjanika	Muscat Blanc à Petits Grains	W
Tempranillo Blanco	Tempranillo (W)	R
Tempranillo Tinto	Tempranillo	R
Teran	Terrano	R
Termarina	Termarina Rossa	G
Terret Blanc	Terret	W
Teta de Vaca	Beba	W
Thompson Seedless	Sultaniye	W
Tinta	Trousseau	R
Tinta Amarela	Trincadeira	R
Tinta Amarella	Trincadeira	R
Tinta Aurelio	Tinto do Aurelio	R
Tinta Bastardinha	Alfrocheiro	R
Tinta Caiada	Parraleta	R
Tinta Cao	Tinto Cão	R
Tinta Fontes	Graciano	R
Tinta Gorda	Juan García	R
Tinta Lameira	Parraleta	R
Tinta Lisboa	Trousseau	R
Tinta Madeira	Negramoll	R
Tinta Miuda	Graciano	R
Tinta Molle	Negramoll	R
Tinta Negra	Negramoll	R
Tinta Porto Santo	Tinta de Porto Santo	R
Tinta Roriz	Tempranillo	R
Tintilia	Tintilia del Molise	R
Tintilla de Rota	Graciano	R
Tintinha	Petit Bouschet	R
Tinto Basto	Garnacha Tinta	R
Tinto Cao	Tinto Cão	R
Tinto de la Pampa Blanca	Tinto Velasco	R
Tinto de Toro	Tempranillo	R
Tinto Pegoes	Tinta de Pegoes	R
Tinto Velasco, Frasco	Tinto Velasco	R
Tintoreras	Alicante Henri Bouschet	R
Tintoria	Rubired	R
Tocai	Sauvignonasse	W
Tocai Friulano	Sauvignonasse	W
Tocai Rosso	Garnacha Tinta	R
Torontel	Torrontés Riojano	W
Torrontes	Malvasia Fina	W
Torrontés Mendocino	Torrontes Mendocino	W
Torrontes Riojano	Torrontés Riojano	W
Torrontes Sanjuanino	Torrontés Sanjuanino	W
Touriga	Touriga Nacional	R
Touriga Francesa	Touriga Franca	R
Touriga Nacional N	Touriga Nacional	R
Touriga National	Touriga Nacional	R
Tramin Cerveny	Gewürztraminer	W
Traminac Bijeli	Savagnin Blanc	W
Traminac Crveni	Gewürztraminer	W
Traminec	Savagnin Blanc	W
Traminec Bel	Savagnin Blanc	W
Traminer	Gewürztraminer	W
Traminer	Savagnin Blanc	W
Traminer Aromatico	Gewürztraminer	W
Traminer Rose	Gewürztraminer	W
Traminer rosse	Gewürztraminer	W
Traminer Rot	Gewürztraminer	W
Traminer Roz	Gewürztraminer	W
Traminer, Roter	Gewürztraminer	W
Tramini	Gewürztraminer	W
Trbljan Bijeli	Trbljan	W
Trebbiano	Trebbiano Toscano	W
Trebbiano Abruzzese	Trebbiano d'Abruzzo	W
Trebbiano di Soave	Verdicchio Bianco	W
Trebbiano Toscana	Trebbiano Toscano	W
Trebbiano Toscano / Ugni blanc	Trebbiano Toscano	W
Trebbiano Toscano Ugni blanc	Trebbiano Toscano	W
Treixadura	Trajadura	W
Trigueira	Malvasia Trigueira	R
Trincadeira Branca	Trincadeiro Branco	W
Trincadeira Preta	Trincadeira	R
Trincdeira	Trincadeira	R
Triomphe d´Alsace	Triomphe	R
Trollinger	Schiava Grossa	R
Trousseau Gris	Trousseau	R
Trousseau Noir	Trousseau	R
Tschaggele	Schiava Grossa	R
Tskhvediani Tetra	Tskhvedianis Tetra	W
Tsvetochny (Floral)	Tsvetochny	W
Tsvetochny (Flowery)	Tsvetochny	W
Tsymlansky Black	Tsimlyansky Cherny	R
Turan	Turán	R
Turca	Douce Noire	R
UD-31.103	UD 31103	R
UD-31.120	DU 31120	R
UD-31.125	Merlot Khorus	R
UD-32.078	Carbernet Volos	R
Ugni	Trebbiano Toscano	W
Ugni blanc	Trebbiano Toscano	W
Ugni Blanc (Trebbiano Toscana)	Trebbiano Toscano	W
Urreti	Urréti	W
Uva Di Troia	Nero di Troia	R
Uvina	Pecorino	W
V 53	Vineland 53035	W
Valdiguie	Valdiguié	R
Valdosa	Tinta Valdosa	R
Valenci	Damaschino	W
Valenciana Blanca	Síria	W
Valency	Teneron	R
Valensi	Valensi du Maroc	W
Valente	Branco Valente	W
Valveirinho	Valveirinha	W
Varejoa	Tinta Varejoa	R
VB 85-1	Millot-Foch	R
VB 91-26-17	Cabertin	R
VB 91-26-29	Satin Noir	R
VB Cal 1-15	Blattner Cal 1-15	R
VB Cal 1-20	Blattner Cal 1-20	R
VB Cal 1-22	Blattner Cal 1-22	R
VB Cal 1-28	Blattner Cal 1-28	R
VB Cal 1-36	Blattner Cal 1-36	R
VB Cal 6-04	Sauvignac	W
Veltliner Grun	Grüner Veltliner	W
Veltliner, Grüner	Grüner Veltliner	W
Veltlinske Cervene Rane	Frühroter Veltliner	R
Veltlinske Cervene Skore	Frühroter Veltliner	R
Veltlinske Zelene	Grüner Veltliner	W
Vencedor	Boal Vencedor	W
Venus	Vênus	R
Verdealbara	Erbamat	W
Verdejo Blanco	Verdejo	W
Verdejo Negro	Trousseau	R
Verdial Branco	Verdial	W
Verdicchio	Verdicchio Bianco	W
Verdot	Petit Verdot	R
Verduzzo	Verduzzo Friulano	W
Vermejuela	Marmajuelo	W
Vermentino B	Vermentino	W
Vermintino Nero	Vermentino Nero	R
Vernaccia	Vernaccia di Oristano	W
Vernaccia Nera	Garnacha Tinta	R
Vernaccia Nera Grossa	Garnacha Tinta	R
Vernaccina Nero	Garnacha Tinta	R
Vernatsch	Schiava Grossa	R
Vértes Csillaga	Vertes Csillaga	W
Vid al Blanc	Vidal	W
Vidal 256	Vidal	W
Vidal Blanc	Vidal	W

Table 15 (cont.): Synonymns and their prime variety and colour

Synonym	Prime	Colour (of prime)
Vidal Red	Vidal Noir	R
Vidau	Miguel del Arco	R
Vijiriego	Vijariego	W
Vijiriego Negro	Sumoll	R
Viktoria Gyongye	Viktória Gyöngye	W
Viktorija	Victoria	W
Viogner	Viognier	W
Viorica	Viorika	W
Vitouska	Vitovska	W
Viura	Macabeo	W
Vörös Dinka	Kövidinka	G
Voskehat	Voskeat	W
Vranec	Vranac	R
Waltham Cross	Afus Ali	W
Weissburgunder	Pinot Blanc	W
Weißburgunder	Pinot Blanc	W
Weisser Burgunder	Pinot Blanc	W
Welschriesling	Graševina	W
White Riesling	Riesling	W
Wildbacher	Blauer Wildbacher	R
Wurzer	Würzer	W
Xarello Blanco	Xarello	W
Xinomavro (White)	Xinomavro (W)	R
Xynomavro	Xinomavro	R
Zala Dende	Zalagyöngye	W
Zala Gyoengye	Zalagyöngye	W
Zala Gyöngye	Zalagyöngye	W
Zalagyongye	Zalagyöngye	W
Zametovka	Žametovka	R
Zefir	Zefír	W
Zeleni Sauvignon	Sauvignonasse	W
Zeleni Silvanec	Silvaner	W
Zengo	Zengő	W
Zenith	Zenit	W
Zeta	Zéta	W
Zeus	Zeusz	W
Zghihara de Husi	Zghihară de Huşi	W
Zibibbo	Muscat of Alexandria	W
Zilavka	Žilavka	W
Zinfandel	Tribidrag	R
Zlahtina	Žlahtina	W
Zold Szagos	Pecsi Szagos	W
Zöld Szagos	Pecsi Szagos	W
Zold Szilvani	Silvaner	W
Zöld szilváni	Silvaner	W
Zöld Veltelíni	Grüner Veltliner	W
Zoumiatiko	Dimyat	W
Zupjanka	Župljanka	W
Zupljanka	Župljanka	W
Zweigelt, Blauer	Zweigelt	R
Zweigeltrebe	Zweigelt	R

Table 16: Shares of national winegrape area by varietal country of origin, 2000 (%)

Country of planting → Country of origin↓	*Algeria*	*Argentina*	*Armenia*	*Australia*	*Austria*	*Brazil*	*Bulgaria*	*Canada*	*Chile*
Algeria									
Argentina		42.0							0.1
Armenia			32.9						
Australia				0.1					
Austria		0.0			64.3				
Azerbaijan									
Belgium									
Brazil		0.0							
Bulgaria							36.1		
Canada									
Chile									0.1
China									
Croatia		0.0			8.9	1.7	3.8		0.1
Cyprus									
France	55.0	39.4		74.3	9.9	15.4	30.4	60.7	68.9
Georgia			22.0				9.8		
Germany		0.1		2.8	12.0	0.3	1.7	8.5	0.4
Greece		8.2	4.7	3.0	0.3	1.5			
Hungary		0.0			0.0		2.3		
Israel									
Italy		4.6		0.9		1.3	1.9		0.1
Japan									
Kazakhstan									
Lebanon				0.3					
Moldova									
Morocco		0.2							
Peru									
Portugal		0.0		1.0					3.8
Romania									
Russia		0.1							
Serbia									
Slovenia					0.8				
South Africa									
Spain	45.0	3.6		2.8					16.0
Switzerland									0.4
Thailand									
Turkey				7.9					
Ukraine							3.9		
United Kingdom		0.0					0.5		
United States		1.8		1.9		60.3	3.2	18.6	
Uzbekistan									
Unknown origin & other varieties		0.0	40.4	5.0	3.8	19.5	6.5	12.2	10.3
Old World subtotal	**100.0**	**56.2**	**59.6**	**93.0**	**96.2**	**20.2**	**89.9**	**69.2**	**89.6**
New World subtotal	**0.0**	**43.8**	**0.0**	**2.0**	**0.0**	**60.3**	**3.6**	**18.6**	**0.2**
World total	**100**	**100**	**100**	**100**	**100**	**100**	**100**	**100**	**100**

Table 16 (cont.) Shares of national winegrape area by varietal country of origin, 2000 (%)

Country of planting → Country of origin↓	*Croatia*	*Cyprus*	*Czechia*	*France*	*Georgia*	*Germany*	*Greece*	*Hungary*	*Israel*
Algeria									
Argentina									
Armenia									
Australia									
Austria	2.0		40.0	0.3		12.6		15.4	
Azerbaijan									
Belgium									
Brazil									
Bulgaria							0.2	0.1	
Canada									
Chile									
China									
Croatia	63.0		11.0	0.0				7.7	
Cyprus		75.0							
France	2.0		5.0	61.3	1.3	16.8	2.5	9.1	46.3
Georgia					92.0				
Germany			21.0	0.7		66.2		5.6	
Greece				1.2		0.1	80.9	1.8	4.2
Hungary				0.0				23.6	
Israel									4.2
Italy				11.0		2.4	1.5		
Japan									
Kazakhstan									
Lebanon									
Moldova								1.1	
Morocco									
Peru									
Portugal									
Romania								0.7	
Russia									
Serbia									
Slovenia									
South Africa									
Spain				25.1			0.1		20.0
Switzerland				0.1		1.1		2.2	
Thailand									
Turkey									
Ukraine									
United Kingdom				0.1					
United States				0.0					7.1
Uzbekishtan									
Unknown origin & other varieties	33.0	25.0	23.0	0.2	6.7	0.8	14.9	32.7	18.3
Old World subtotal	**67.0**	**75.0**	**77.0**	**99.7**	**93.3**	**99.2**	**85.1**	**67.3**	**74.7**
New World subtotal	**0.0**	**0.0**	**0.0**	**0.1**	**0.0**	**0.0**	**0.0**	**0.0**	**7.1**
World total	**100**	**100**	**100**	**100**	**100**	**100**	**100**	**100**	**100**

Table 16 (cont.) Shares of national winegrape area by varietal country of origin, 2000 (%)

Country of planting →	*Italy*	*Korea, Rep.*	*Luxembourg*	*Moldova*	*Morocco*	*New Zealand*	*Portugal*	*Romania*	*Russia*
Country of origin↓									
Algeria									
Argentina									
Armenia									
Australia									
Austria	0.1		0.1	0.1		0.2			
Azerbaijan									
Belgium									
Brazil									
Bulgaria				0.0					
Canada									
Chile									
China									
Croatia	2.1							6.8	
Cyprus									
France	11.9		39.8	61.6	10.2	84.1	1.7	18.0	14.3
Georgia				13.6				0.2	25.0
Germany	0.3		60.1	4.7		11.5		0.2	2.4
Greece	2.2			0.2	7.4		0.9	0.5	
Hungary	1.8			0.0					7.0
Israel									
Italy	77.0						0.9		
Japan		74.1							
Kazakhstan									
Lebanon	0.2								
Moldova				4.9				8.7	
Morocco					38.2				
Peru									
Portugal							56.8		
Romania				0.1				3.0	
Russia				0.0					2.2
Serbia									
Slovenia									
South Africa						0.7			
Spain	1.5				12.1	0.2	6.1		
Switzerland	0.0					0.3			
Thailand									
Turkey									
Ukraine				1.9					4.2
United Kingdom	0.0								
United States		11.1		12.8		0.1			
Uzbekishtan				0.1					
Unknown origin & other varieties	2.7	14.8			32.2	3.0	33.5	62.6	44.8
Old World subtotal	**97.3**	**0.0**	**100.0**	**87.2**	**67.8**	**96.2**	**66.5**	**37.4**	**55.2**
New World subtotal	**0.0**	**85.2**	**0.0**	**12.8**	**0.0**	**0.8**	**0.0**	**0.0**	**0.0**
World total	**100**	**100**	**100**	**100**	**100**	**100**	**100**	**100**	**100**

Table 16 (cont.) Shares of national winegrape area by varietal country of origin, 2000 (%)

Country of planting →	*Serbia*	*Slovakia*	*Slovenia*	*South Africa*	*Spain*	*Switzerland*	*Taiwan*	*Tunisia*	*UK*
Country of origin↓									
Algeria					0.0				
Argentina									
Armenia									
Australia									
Austria		33.0			0.0	1.4			12.0
Azerbaijan									
Belgium					0.0				
Brazil									
Bulgaria									
Canada									
Chile									
China									
Croatia	48.0	25.0	15.2	0.0	0.2			2.0	
Cyprus									
France		9.0	16.9	79.5	2.4	53.9		14.0	24.2
Georgia									
Germany		12.0		1.1	0.0	5.1			37.8
Greece				5.5	0.5	0.3			
Hungary				0.3					
Israel				0.2					
Italy			10.2	0.2	0.1	0.6	4.0	5.0	
Japan							54.0		
Kazakhstan									
Lebanon					0.0				
Moldova		2.0							
Morocco									
Peru									
Portugal				0.9	1.0				
Romania		2.0							
Russia									
Serbia	22.0								
Slovenia									
South Africa				7.7					
Spain				1.9	86.2			59.0	
Switzerland	5.0				0.0	37.4			
Thailand									
Turkey					0.0				
Ukraine									
United Kingdom	4.0				0.0				
United States				2.6	0.0		42.0		
Uzbekishtan									
Unknown origin & other varieties	21.0	17.0	57.7	0.1	9.6	1.3		20.0	26.0
Old World subtotal	**75.0**	**83.0**	**42.3**	**89.6**	**90.4**	**98.7**	**4.0**	**80.0**	**74.0**
New World subtotal	**4.0**	**0.0**	**0.0**	**10.2**	**0.0**	**0.0**	**96.0**	**0.0**	**0.0**
World total	**100**	**100**	**100**	**100**	**100**	**100**	**100**	**100**	**100**

Table 16 (cont.) Shares of national winegrape area by varietal country of origin, 2000 (%)

Country of planting → / Country of origin↓	*USA*	*Uruguay*	*Missing 9*	***Old World***	***New World***	***World***
Algeria				**0.0**		**0.0**
Argentina			0.0		**10.4**	**1.7**
Armenia			0.2	**0.1**		**0.1**
Australia					**0.0**	**0.0**
Austria	0.0		0.5	**1.8**	**0.0**	**1.5**
Azerbaijan			0.6			**0.0**
Belgium				**0.0**		**0.0**
Brazil					**0.0**	**0.0**
Bulgaria				**0.9**		**0.7**
Canada	0.0				**0.0**	**0.0**
Chile					**0.0**	**0.0**
China			0.0			**0.0**
Croatia	10.6			**3.1**	**2.5**	**3.0**
Cyprus				**0.3**		**0.3**
France	65.9	54.9	55.2	**21.1**	**58.7**	**27.8**
Georgia			14.8	**1.8**		**1.7**
Germany	1.5		4.1	**2.5**	**1.3**	**2.4**
Greece	1.4		1.4	**2.0**	**3.6**	**2.3**
Hungary				**1.0**	**0.0**	**0.8**
Israel				**0.0**	**0.0**	**0.0**
Italy	3.8		1.1	**14.9**	**2.3**	**12.6**
Japan			0.7		**0.7**	**0.1**
Kazakhstan			0.3			**0.0**
Lebanon				**0.0**	**0.0**	**0.0**
Moldova			0.1	**0.6**		**0.5**
Morocco				**0.5**	**0.0**	**0.4**
Peru			0.3			**0.0**
Portugal	0.0			**3.2**	**0.8**	**2.8**
Romania				**0.2**		**0.2**
Russia				**0.0**	**0.0**	**0.0**
Serbia				**0.4**		**0.3**
Slovenia				**0.0**		**0.0**
South Africa					**0.9**	**0.1**
Spain	4.9		2.0	**32.1**	**4.9**	**27.2**
Switzerland				**0.3**	**0.1**	**0.3**
Thailand			0.0			**0.0**
Turkey			8.1	**0.0**	**1.3**	**0.3**
Ukraine			6.1	**0.2**		**0.3**
United Kingdom	0.0	32.5		**0.1**	**0.4**	**0.1**
United States	11.1		3.1	**0.4**	**7.9**	**1.7**
Uzbekishtan				**0.0**		**0.0**
Unknown origin & other varieties	0.6	12.6	1.3	**12.3**	**4.2**	**10.8**
Old World subtotal	**88.3**	**54.9**	**94.6**	**87.2**	**75.6**	**85.4**
New World subtotal	**11.1**	**32.5**	**4.0**	**0.5**	**20.2**	**3.8**
World total	**100**	**100**	**100**	**100**	**100**	**100**

Table 17: Shares of national winegrape area by varietal country of origin, 2010 (%)

Country of planting → Country of origin↓	*Algeria*	*Argentina*	*Armenia*	*Australia*	*Austria*	*Brazil*	*Bulgaria*	*Canada*	*Chile*
Algeria									
Argentina		35.7							0.5
Armenia			32.9						
Australia				0.0					
Austria		0.0			62.9	0.0		0.5	0.0
Azerbaijan									
Brazil		0.1				3.2			
Bulgaria							30.5		
Canada								0.2	
Chile									0.0
China									0.0
Croatia		0.0		0.1	7.6	0.4		0.1	0.1
Cyprus									
Czechia									
France	55.0	50.1		88.5	14.9	31.1	44.0	77.3	93.4
Georgia			22.0				5.6		
Germany		0.1		3.3	10.1	0.0	1.3	15.1	0.6
Greece		6.1	4.7	1.8	1.1	2.0			1.0
Hungary		0.0			0.0	0.0	1.0	0.0	
Israel									
Italy		3.7		0.8		1.5	1.3	0.1	0.2
Japan						0.0			
Kazakhstan									
Lebanon									
Moldova									
Montenegro									
Morocco		0.1							
Peru									
Portugal		0.0		1.0		0.2		0.1	0.0
Romania									
Russia								0.0	
Serbia									
Slovakia									
Slovenia					0.5				
South Africa						0.2		0.1	
Spain	45.0	4.0		1.9		0.0		0.1	4.0
Switzerland						0.0		0.7	0.1
Thailand									
Turkey				0.3					
Ukraine									
United Kingdom		0.0							
United States		0.2		0.6		57.1		4.4	
Uzbekistan									
Unknown origin & other varieties		0.0	40.4	1.6	2.9	4.2	16.4	1.4	0.1
Old World subtotal	**100.0**	**64.0**	**59.6**	**97.7**	**97.1**	**35.3**	**83.6**	**94.0**	**99.4**
New World subtotal	**0.0**	**35.9**	**0.0**	**0.7**	**0.0**	**60.4**	**0.0**	**4.7**	**0.5**
World total	**100**	**100**	**100**	**100**	**100**	**100**	**100**	**100**	**100**

Table 17 (cont.) Shares of national winegrape area by varietal country of origin, 2010 (%)

Country of planting → Country of origin↓	*China*	*Croatia*	*Cyprus*	*Czechia*	*Ethiopia*	*France*	*Georgia*	*Germany*	*Greece*
Algeria									
Argentina									
Armenia									
Australia									
Austria		4.2		37.3		0.2		11.7	0.0
Azerbaijan									
Brazil									
Bulgaria		0.0							0.1
Canada									
Chile									
China									
Croatia		52.2		7.1		0.0			
Cyprus			69.2						
Czechia				6.9				0.0	
France	98.3	15.8	11.7	25.1	31.9	66.5	1.3	25.8	8.1
Georgia		0.3					92.0		
Germany	1.5	4.7		21.4		0.8		58.0	0.0
Greece		0.3	2.5			1.3		0.2	73.4
Hungary		1.2		0.4		0.0			
Israel									
Italy		4.6			53.2	10.7		2.4	0.6
Japan									
Kazakhstan									
Lebanon						0.0			
Moldova									
Montenegro		0.7							
Morocco									
Peru									
Portugal									
Romania									
Russia		0.3							
Serbia		0.2							
Slovakia									
Slovenia		0.6							
South Africa									
Spain	0.0	0.8	8.6			19.0			1.3
Switzerland		0.1				0.3		1.1	
Thailand									
Turkey			4.3		2.5	0.0			
Ukraine				0.1					
United Kingdom		0.1				0.4		0.1	
United States	0.1	0.4			12.3	0.0			
Uzbekistan									
Unknown origin & other varieties		13.7	3.7	1.8		0.8	6.7	0.8	16.6
Old World subtotal	**99.9**	**85.8**	**96.3**	**98.2**	**87.7**	**98.8**	**93.3**	**99.1**	**83.4**
New World subtotal	**0.1**	**0.5**	**0.0**	**0.0**	**12.3**	**0.4**	**0.0**	**0.1**	**0.0**
World total	**100**	**100**	**100**	**100**	**100**	**100**	**100**	**100**	**100**

Table 17 (cont.) Shares of national winegrape area by varietal country of origin, 2010 (%)

Country of planting → Country of origin↓	*Hungary*	*Israel*	*Italy*	*Japan*	*Kazakhstan*	*Korea, Rep.*	*Luxembourg*	*Mexico*	*Moldova*
Algeria									
Argentina									
Armenia									
Australia									
Austria	19.1		0.1	6.2					0.1
Azerbaijan					9.7				
Brazil									
Bulgaria	0.1								0.0
Canada									
Chile									
China									
Croatia	6.7		3.2						
Cyprus									
Czechia									
France	21.8	46.3	17.9	52.5	7.7		46.6	39.7	61.6
Georgia					57.4				13.6
Germany	4.9		0.5	14.5	1.6		53.3		4.7
Greece	1.0	4.2	2.1		3.7			4.5	0.2
Hungary	37.5		2.7						0.0
Israel		4.2							
Italy	0.0		68.5					3.3	
Japan				19.9		74.1			
Kazakhstan					5.6				
Lebanon			0.1						
Moldova					1.6				4.9
Montenegro									
Morocco									
Peru									
Portugal	0.1								
Romania	0.1								0.1
Russia									0.0
Serbia	0.0								
Slovakia									
Slovenia	0.0								
South Africa									
Spain		20.0	1.4					16.9	
Switzerland	2.7		0.0						
Thailand									
Turkey								15.4	
Ukraine	0.0				0.0				1.9
United Kingdom	0.0		0.0						
United States		7.1		6.8	0.1	11.1		8.7	12.8
Uzbekistan									0.1
Unknown origin & other varieties	5.9	18.3	3.5		12.7	14.8	0.1	11.4	
Old World subtotal	**94.1**	**74.7**	**96.4**	**73.3**	**87.2**	**0.0**	**99.9**	**79.8**	**87.2**
New World subtotal	**0.0**	**7.1**	**0.0**	**26.7**	**0.1**	**85.2**	**0.0**	**8.7**	**12.8**
World total	**100**	**100**	**100**	**100**	**100**	**100**	**100**	**100**	**100**

Table 17 (cont.) Shares of national winegrape area by varietal country of origin, 2010 (%)

Country of planting → Country of origin↓	*Morocco*	*Myanmar*	*New Zealand*	*Peru*	*Portugal*	*Romania*	*Russia*
Algeria							
Argentina				0.2			
Armenia							
Australia							
Austria			0.0	7.6	0.0	0.5	
Azerbaijan							
Brazil							
Bulgaria						1.7	1.9
Canada							
Chile							
China				0.1			0.7
Croatia			0.0			4.4	
Cyprus							
Czechia							
France	10.3	80.2	91.9	2.2	9.7	19.9	42.0
Georgia						0.2	5.5
Germany			4.6		0.0	0.2	3.9
Greece	7.5	9.5		9.4	0.8	0.5	0.6
Hungary						0.1	14.8
Israel							
Italy			0.1	29.4	0.8	0.1	0.3
Japan							
Kazakhstan							
Lebanon							
Moldova						8.9	4.2
Montenegro							
Morocco	38.6						
Peru				8.8			
Portugal			0.0		72.2	0.0	
Romania						11.7	
Russia						0.0	11.5
Serbia						0.0	
Slovakia							
Slovenia							
South Africa			0.2				
Spain	12.2	10.3	0.1	32.9	15.5	0.0	
Switzerland			0.0		0.1		0.1
Thailand							
Turkey				0.4			
Ukraine							12.3
United Kingdom							0.7
United States			0.0	9.1			0.6
Uzbekistan							
Unknown origin & other varieties	31.4		3.1		0.8	51.7	0.9
Old World subtotal	**68.6**	**100.0**	**96.6**	**81.9**	**99.2**	**48.3**	**97.1**
New World subtotal	**0.0**	**0.0**	**0.2**	**18.1**	**0.0**	**0.0**	**2.0**
World total	**100**	**100**	**100**	**100**	**100**	**100**	**100**

Table 17 (cont.) Shares of national winegrape area by varietal country of origin, 2010 (%)

Country of planting → Country of origin↓	*Serbia*	*Slovakia*	*Slovenia*	*South Africa*	*Spain*	*Switzerland*	*Taiwan*	*Thailand*	*Tunisia*
Algeria					0.0				
Argentina									
Armenia									
Australia									
Austria		39.1	4.2			1.8			
Azerbaijan									
Brazil									
Bulgaria									
Canada									
Chile									
China									
Croatia	48.0	13.1	20.0	0.0	0.1				2.0
Cyprus									
Czechia		3.7							
France		14.8	31.5	85.2	9.2	55.8		66.2	14.0
Georgia									
Germany		14.3	4.1	0.4	0.0	5.1		1.3	
Greece		0.4	2.2	3.2	0.9	0.3		2.2	
Hungary		4.9	4.0	0.1		0.0			
Israel				0.0					
Italy			13.3	0.2	0.1	1.8	4.0	1.4	5.0
Japan							54.0	11.4	
Kazakhstan									
Lebanon					0.0				
Moldova		1.7	0.4						
Montenegro									
Morocco									
Peru					0.0				
Portugal				0.6	1.1				
Romania		1.8							
Russia									
Serbia	22.0								
Slovakia		1.4							
Slovenia		0.0	5.6						
South Africa				6.9					
Spain				1.0	86.8			6.7	59.0
Switzerland	5.0				0.0	33.9			
Thailand								10.9	
Turkey									
Ukraine		1.9							
United Kingdom	4.0				0.0				
United States				2.3	0.0		42.0		
Uzbekistan									
Unknown origin & other varieties	21.0	2.9	14.8	0.0	1.9	1.3			20.0
Old World subtotal	**75.0**	**97.1**	**85.2**	**90.7**	**98.1**	**98.7**	**4.0**	**77.7**	**80.0**
New World subtotal	**4.0**	**0.0**	**0.0**	**9.2**	**0.0**	**0.0**	**96.0**	**22.3**	**0.0**
World total	**100**	**100**	**100**	**100**	**100**	**100**	**100**	**100**	**100**

Table 17 (cont.) Shares of national winegrape area by varietal country of origin, 2010 (%)

Country of planting →	*Turkey*	*Ukraine*	*UK*	*USA*	*Uruguay*	***Old World***	***New World***	***World***
Country of origin↓								
Algeria						**0.0**		**0.0**
Argentina							**8.0**	**1.7**
Armenia		0.5				**0.1**		**0.1**
Australia							**0.0**	**0.0**
Austria			0.1	0.0		**1.9**	**0.1**	**1.5**
Azerbaijan						**0.0**		**0.0**
Brazil							**0.2**	**0.0**
Bulgaria						**0.6**		**0.4**
Canada				0.0			**0.0**	**0.0**
Chile							**0.0**	**0.0**
China						**0.0**	**0.0**	**0.0**
Croatia				8.7		**2.4**	**2.1**	**2.3**
Cyprus						**0.2**		**0.1**
Czechia						**0.0**		**0.0**
France	32.7	48.2	57.5	70.9	58.9	**28.3**	**71.5**	**37.3**
Georgia		25.0				**2.2**		**1.8**
Germany	0.0	7.0	42.3	2.6	0.5	**2.6**	**1.7**	**2.4**
Greece	0.9	1.3		0.9	0.1	**2.3**	**2.5**	**2.3**
Hungary						**1.3**	**0.0**	**1.1**
Israel						**0.0**	**0.0**	**0.0**
Italy	0.1			2.0	10.3	**14.4**	**1.8**	**11.8**
Japan							**0.7**	**0.1**
Kazakhstan						**0.0**		**0.0**
Lebanon						**0.0**		**0.0**
Moldova						**0.6**		**0.5**
Montenegro						**0.0**		**0.0**
Morocco						**0.5**	**0.0**	**0.4**
Peru						**0.0**	**0.0**	**0.0**
Portugal			0.1	0.1	0.1	**3.5**	**0.3**	**2.9**
Romania						**0.6**		**0.4**
Russia						**0.1**	**0.0**	**0.1**
Serbia						**0.4**		**0.3**
Slovakia						**0.0**		**0.0**
Slovenia						**0.0**		**0.0**
South Africa				0.0			**0.7**	**0.2**
Spain	1.0			2.4	0.1	**30.5**	**2.6**	**24.7**
Switzerland						**0.4**	**0.0**	**0.3**
Thailand							**0.0**	**0.0**
Turkey	65.3					**0.2**	**0.1**	**0.2**
Ukraine		13.5				**0.3**		**0.3**
United Kingdom				0.0	19.6	**0.2**	**0.2**	**0.2**
United States		4.6		8.9	3.7	**0.4**	**5.8**	**1.5**
Uzbekistan						**0.0**		**0.0**
Unknown origin & other varieties				3.4	6.8	**5.8**	**1.6**	**4.9**
Old World subtotal	**100.0**	**95.4**	**100.0**	**87.6**	**70.0**	**93.6**	**82.7**	**91.4**
New World subtotal	**0.0**	**4.6**	**0.0**	**9.0**	**23.2**	**0.6**	**15.7**	**3.7**
World total	**100**	**100**	**100**	**100**	**100**	**100**	**100**	**100**

Table 18: Shares of national winegrape area by varietal country of origin, 2016 (%)

Country of planting → Country of origin↓	*Algeria*	*Argentina*	*Armenia*	*Australia*	*Austria*	*Brazil*	*Bulgaria*	*Cambodia*	*Canada*
Algeria									
Argentina		32.3							
Australia				0.1					
Austria		0.0		0.0	62.4				0.4
Azerbaijan									
Bosnia and Herzegc									
Brazil						10.8			
Bulgaria							31.4		
Canada						0.2			0.8
Chile									
China									
Croatia		0.0		0.1	7.1	0.6			0.1
Cyprus									
Czechia									
France	36.1	56.1		89.6	16.7	34.6	53.7	70.0	76.0
Georgia				0.0			10.2		
Germany		0.0		2.5	9.1	0.0	1.1		14.1
Greece	2.4	4.5		2.5	1.8	2.1			0.1
Hungary		0.0		0.0	0.0	0.0	2.2		0.0
Israel									
Italy		3.1		1.1		2.2	1.4		0.1
Japan								30.0	
Kazakhstan									
Lebanon				0.0					
Moldova									
Montenegro									
Morocco		0.1							
North Macedonia									
Peru									
Portugal		0.0		0.8		0.1			0.0
Romania									
Russia									0.0
Serbia									
Slovenia					0.5				
South Africa				0.0		0.0			0.0
Spain	60.2	3.7		2.3		0.1			0.1
Switzerland									1.2
Taiwan									
Thailand									
Turkey	1.2			0.0					
Ukraine									0.0
United Kingdom				0.0					
United States		0.0		0.8		49.2			4.6
Unknown origin & other varieties		0.1	100.0	0.1	2.4				2.5
Old World subtota	**100.0**	**67.6**	**0.0**	**98.9**	**97.6**	**39.7**	**100.0**	**70.0**	**92.0**
New World subtot	**0.0**	**32.3**	**0.0**	**1.0**	**0.0**	**60.3**	**0.0**	**30.0**	**5.4**
World total	**100**	**100**	**100**	**100**	**100**	**100**	**100**	**100**	**100**

Table 18 (cont.): Shares of national winegrape area by varietal country of origin, 2016 (%)

Country of planting → / Country of origin↓	*Chile*	*China*	*Croatia*	*Cyprus*	*Czechia*	*Ethiopia*	*France*	*Georgia*	*Germany*
Algeria	0.0								
Argentina	4.7								
Australia									
Austria	0.0		4.4		38.5		0.1		10.9
Azerbaijan									
Bosnia and Herzegc									
Brazil									
Bulgaria									
Canada									
Chile	0.0								
China	0.0	4.2							
Croatia	0.0	1.7	71.6		8.2		0.0		
Cyprus				100.0					
Czechia									0.0
France	81.7	44.8	18.7		29.5	31.9	63.9	1.3	27.6
Georgia	0.0							92.0	
Germany	0.5	0.9	5.3		23.8		0.9		57.3
Greece	5.0	1.7					1.3		0.3
Hungary							0.0		0.0
Israel									
Italy	0.4	0.8				53.2	10.6		2.3
Japan									
Kazakhstan									
Lebanon							0.0		
Moldova									
Montenegro									
Morocco									
North Macedonia									
Peru									
Portugal	0.0								
Romania									
Russia									
Serbia									
Slovenia									
South Africa									
Spain	7.5	2.3					15.8		
Switzerland	0.1						0.1		1.1
Taiwan									
Thailand									
Turkey						2.5	0.0		
Ukraine									
United Kingdom							0.3		0.1
United States	0.0					12.3	0.0		
Unknown origin & other varieties		43.6					7.1	6.7	0.3
Old World subtota	**95.3**	**52.2**	**100.0**	**100.0**	**100.0**	**87.7**	**92.6**	**93.3**	**99.6**
New World subtot	**4.7**	**4.2**	**0.0**	**0.0**	**0.0**	**12.3**	**0.3**	**0.0**	**0.1**
World total	**100**	**100**	**100**	**100**	**100**	**100**	**100**	**100**	**100**

Table 18 (cont.): Shares of national winegrape area by varietal country of origin, 2016 (%)

Country of planting →	*Greece*	*Hungary*	*India*	*Israel*	*Italy*	*Japan*	*Kazakhstan*	*Korea, Rep.*
Country of origin↓								
Algeria								
Argentina								
Australia								
Austria	0.0	18.5			0.0	1.5		
Azerbaijan							9.7	
Bosnia and Herzeg								
Brazil								
Bulgaria	0.1	0.1						
Canada								
Chile								
China								
Croatia		6.2			3.0			
Cyprus								
Czechia								
France	11.5	22.1	48.1	63.8	17.9	12.2	7.7	
Georgia							57.4	
Germany	0.0	5.7			0.7	3.1	1.6	
Greece	79.8	1.2	3.7	4.4	2.4		3.7	
Hungary		41.9			2.3			
Israel				5.5				
Italy	0.4	0.0	11.1		66.4			
Japan						36.6		74.1
Kazakhstan							5.6	
Lebanon					0.0			
Moldova		0.0					1.6	
Montenegro		0.0						
Morocco								
North Macedonia								
Peru								
Portugal		0.5						
Romania		0.1						
Russia								
Serbia		0.0						
Slovenia		0.0						
South Africa								
Spain	0.3			20.9	1.3			
Switzerland		1.8			0.0			
Taiwan								
Thailand								
Turkey			37.0					
Ukraine		0.0					0.0	
United Kingdom	4.5	0.0			0.0			
United States				2.2		36.1	0.1	11.1
Unknown origin & other varieties	3.5	1.8		3.2	5.9	10.5	12.7	14.8
Old World subtot	**92.0**	**98.1**	**100.0**	**94.6**	**94.1**	**16.8**	**87.1**	**0.0**
New World subto	**4.5**	**0.0**	**0.0**	**2.2**	**0.0**	**72.7**	**0.1**	**85.2**
World total	**100**	**100**	**100**	**100**	**100**	**100**	**100**	**100**

Table 18 (cont.): Shares of national winegrape area by varietal country of origin, 2016 (%)

Country of planting →	*Lebanon*	*Luxembourg*	*Mexico*	*Moldova*	*Morocco*	*Myanmar*	*New Zealand*	*North Macedonia*
Country of origin↓								
Algeria								
Argentina								
Australia								
Austria		0.3		0.0			0.1	
Azerbaijan								
Bosnia and Herze								0.7
Brazil								
Bulgaria				0.0				27.2
Canada								
Chile								
China								
Croatia							0.0	5.3
Cyprus								
Czechia								
France	82.5	53.6	39.7	50.3	39.1	85.0	96.2	14.9
Georgia				5.4				1.9
Germany		45.0		3.4			3.0	3.6
Greece		0.1	4.5	0.1	11.9	10.0	0.1	1.6
Hungary				2.2				
Israel								
Italy			3.3	0.5	19.0		0.1	
Japan								
Kazakhstan								
Lebanon				0.1				
Moldova				22.6				
Montenegro								38.3
Morocco								
North Macedonia								1.6
Peru								
Portugal							0.1	
Romania				0.5				
Russia				0.7				
Serbia								2.8
Slovenia								
South Africa							0.1	
Spain			16.9		14.0	5.0	0.1	
Switzerland				0.4				
Taiwan								
Thailand								
Turkey			15.4		4.1			
Ukraine				2.4				
United Kingdom				0.3				1.4
United States			8.7	5.0	3.5		0.0	
Unknown origin & other varieties	17.5	1.0	11.4	6.2	8.4		0.1	0.5
Old World subto	**82.5**	**99.0**	**79.8**	**88.5**	**88.0**	**100.0**	**99.8**	**98.1**
New World subt	**0.0**	**0.0**	**8.7**	**5.3**	**3.5**	**0.0**	**0.1**	**1.4**
World total	**100**	**100**	**100**	**100**	**100**	**100**	**100**	**100**

Table 18 (cont.): Shares of national winegrape area by varietal country of origin, 2016 (%)

Country of planting →	*Norway*	*Peru*	*Portugal*	*Romania*	*Russia*	*Serbia*	*Slovakia*	*Slovenia*
Country of origin↓								
Algeria								
Argentina		0.2						
Australia								
Austria		7.6	0.0	0.7		3.3	45.9	5.5
Azerbaijan								
Bosnia and He								
Brazil								
Bulgaria				1.5	1.0	1.2		
Canada								
Chile								
China		0.1			0.3			
Croatia				0.8		9.3	5.9	19.6
Cyprus								
Czechia								
France		2.2	9.0	21.5	50.2	34.2	11.4	32.9
Georgia				0.2	14.2	0.3		
Germany	90.0		0.0	3.6	5.6	6.8	14.6	5.3
Greece		9.4	1.3	0.9	1.0	0.2		3.7
Hungary				0.1	7.5	0.1		3.4
Israel								
Italy		29.4	0.6	0.1	0.1			13.1
Japan								
Kazakhstan								
Lebanon						0.3		
Moldova				8.7	2.1			
Montenegro						0.0		
Morocco								
North Macedo								
Peru		8.8						
Portugal			70.6	0.0				
Romania				10.2				
Russia				0.0	5.8	0.3		
Serbia				0.0		5.5		
Slovenia								5.8
South Africa								
Spain		32.9	13.8	0.0				
Switzerland			0.0	0.1	0.0			
Taiwan								
Thailand								
Turkey		0.4						
Ukraine					8.7			
United Kingdo					0.4	2.8		
United States		9.1			2.7	0.0		
Unknown origin &	10.0		4.7	51.7	0.5	35.7	22.2	10.7
Old World su	**90.0**	**81.9**	**95.3**	**48.3**	**96.2**	**61.4**	**77.8**	**89.3**
New World s	**0.0**	**18.1**	**0.0**	**0.0**	**3.4**	**2.8**	**0.0**	**0.0**
World total	**100**	**100**	**100**	**100**	**100**	**100**	**100**	**100**

Table 18 (cont.): Shares of national winegrape area by varietal country of origin, 2016 (%)

Country of planting → Country of origin↓	*South Africa*	*Spain*	*Switzerland*	*Taiwan*	*Thailand*	*Tunisia*	*Turkey*	*Ukraine*
Algeria								
Argentina								
Australia								
Austria	0.0		1.9					
Azerbaijan								
Bosnia and He								
Brazil								
Bulgaria								
Canada								
Chile								
China								
Croatia	0.0	0.1	0.0					
Cyprus								
Czechia								
France	83.4	10.0	55.6		55.2	39.1	32.7	58.6
Georgia								22.9
Germany	0.4	0.1	5.4		1.4		0.0	7.4
Greece	3.2	1.4	0.3		1.5	11.9	0.9	1.3
Hungary	0.1	0.0	0.1					
Israel	0.0							
Italy	0.3	0.0	1.1		1.0	19.0	0.1	
Japan				63.2	8.7			
Kazakhstan								
Lebanon								
Moldova								
Montenegro								
Morocco								
North Macedo								
Peru								
Portugal	0.5	1.2	0.0		0.5			
Romania	0.0							
Russia								
Serbia	0.0							
Slovenia			0.0					
South Africa	8.2		0.0					
Spain	1.3	85.2	0.0		4.3	14.0	1.3	
Switzerland		0.0	35.0					
Taiwan				3.4				
Thailand					26.0			
Turkey		0.0				4.1	64.9	
Ukraine								5.0
United Kingdo		0.0	0.0					
United States	2.5	0.0	0.0	33.4		3.5		4.8
Unknown origin & other	0.0	2.0	0.7		1.3	8.4		
Old World su	**89.3**	**98.0**	**99.3**	**0.0**	**64.0**	**88.0**	**100.0**	**95.2**
New World su	**10.7**	**0.0**	**0.0**	**100.0**	**34.7**	**3.5**	**0.0**	**4.8**
World total	**100**	**100**	**100**	**100**	**100**	**100**	**100**	**100**

Table 18 (cont.): Shares of national winegrape area by varietal country of origin, 2016 (%)

Country of planting →	*UK*	*USA*	*Uruguay*	***Old World***	***New World***	***World***
Country of origin↓						
Algeria					**0.0**	**0.0**
Argentina					**6.6**	**1.6**
Australia					**0.0**	**0.0**
Austria		0.0		**1.9**	**0.0**	**1.4**
Azerbaijan				**0.0**		**0.0**
Bosnia and Herzegc				**0.0**		**0.0**
Brazil					**0.3**	**0.1**
Bulgaria				**0.8**		**0.6**
Canada		0.0			**0.0**	**0.0**
Chile					**0.0**	**0.0**
China				**0.0**	**0.7**	**0.2**
Croatia		7.7		**1.3**	**2.0**	**1.5**
Cyprus				**0.2**		**0.1**
Czechia				**0.0**		**0.0**
France	77.0	72.4	58.8	**29.4**	**67.7**	**38.9**
Georgia				**2.1**	**0.0**	**1.6**
Germany	18.2	2.6	0.4	**2.9**	**1.4**	**2.6**
Greece		1.3	0.5	**2.6**	**2.8**	**2.7**
Hungary				**1.4**	**0.0**	**1.1**
Israel				**0.0**	**0.0**	**0.0**
Italy		1.6	10.5	**14.8**	**1.6**	**11.5**
Japan					**0.5**	**0.1**
Kazakhstan				**0.0**		**0.0**
Lebanon				**0.0**	**0.0**	**0.0**
Moldova				**1.1**		**0.8**
Montenegro				**0.3**		**0.2**
Morocco					**0.0**	**0.0**
North Macedonia				**0.0**		**0.0**
Peru					**0.0**	**0.0**
Portugal		0.1	0.1	**4.1**	**0.2**	**3.2**
Romania				**0.6**	**0.0**	**0.4**
Russia				**0.1**	**0.0**	**0.1**
Serbia				**0.1**	**0.0**	**0.0**
Slovenia				**0.0**		**0.0**
South Africa		0.0		**0.0**	**0.7**	**0.2**
Spain		2.0	0.1	**27.4**	**3.1**	**21.4**
Switzerland				**0.3**	**0.0**	**0.2**
Taiwan					**0.0**	**0.0**
Thailand					**0.0**	**0.0**
Turkey				**0.3**	**0.2**	**0.3**
Ukraine		0.0		**0.2**	**0.0**	**0.2**
United Kingdom		0.0	18.8	**0.2**	**0.1**	**0.2**
United States		8.6	1.9	**0.2**	**4.0**	**1.2**
Unknown origin & other varieties	4.8	3.4	9.0	**7.6**	**8.0**	**7.7**
Old World subtota	**95.2**	**87.9**	**70.3**	**92.0**	**79.0**	**88.8**
New World subtot	**0.0**	**8.6**	**20.7**	**0.4**	**13.0**	**3.5**
World total	**100**	**100**	**100**	**100**	**100**	**100**

Table 19: NVII of national winegrapes by varietal country of origin, 2000

Country of planting → Country of origin↓	*Algeria*	*Argentina*	*Armenia*	*Australia*	*Austria*	*Brazil*	*Bulgaria*	*Canada*	*Chile*
Algeria									
Argentina		16.3							-0.4
Armenia			0.8						
Australia				0.0					
Austria		-0.6			6.2				
Azerbaijan									
Belgium									
Brazil		0.0							
Bulgaria							7.0		
Canada									
Chile									0.0
China									
Croatia		-1.2			0.6	-0.1	0.2		-0.7
Cyprus									
France	1.7	4.7		12.4	-1.8	-1.3	0.5	0.6	9.6
Georgia			0.5				1.6		
Germany		-0.9		0.1	1.0	-0.2	-0.1	0.1	-0.5
Greece		2.4	0.1	0.2	-0.2	-0.1			
Hungary					-0.1		0.3		
Israel									
Italy		-3.3		-3.2		-1.2	-2.2		-3.0
Japan									
Kazakhstan									
Lebanon				0.1					
Moldova									
Morocco		-0.1							
Peru									
Portugal		-1.1		-0.5					0.2
Romania									
Russia		0.0							
Serbia									
Slovenia					0.1				
South Africa									
Spain	1.1	-9.5		-6.5					-2.6
Switzerland									0.0
Thailand									
Turkey				2.0					
Ukraine							0.7		
United Kingdom		-0.1					0.1		
United States		0.1		0.1		6.3	0.3	0.3	
Uzbekistan									
Unknown origin & other varieties		-8.7	1.4	-3.1	-1.4	1.9	-1.7	0.0	-0.3
Old World subtota	0.9	-11.8	-0.6	2.0	1.1	-7.1	0.9	-0.3	1.0
New World subtot:		16.2		-0.5		6.1	0.0	0.3	-0.8
World total	0.0	0.0	0.0	0.0	0.0	0.0	0.0	0.0	0.0

Table 19 (cont.): NVII of national winegrapes by varietal country of origin, 2000

Country of planting →	*Croatia*	*Cyprus*	*Czechia*	*France*	*Georgia*	*Germany*	*Greece*	*Hungary*	*Israel*
Country of origin↓									
Algeria									
Argentina									
Armenia									
Australia									
Austria	0.1		0.9	-2.2		2.4		2.5	
Azerbaijan									
Belgium									
Brazil									
Bulgaria							-0.1	-0.1	
Canada									
Chile									
China									
Croatia	7.3		0.2	-5.2				0.8	
Cyprus		2.8							
France	-3.1		-0.5	59.3	-2.0	-2.3	-2.6	-3.3	0.2
Georgia					6.9				
Germany			0.4	-2.9		13.6		0.6	
Greece				-1.9		-0.5	8.2	-0.1	0.0
Hungary				-1.0				4.1	
Israel									0.0
Italy				-3.2		-2.2	-1.2		
Japan									
Kazakhstan									
Lebanon									
Moldova								0.1	
Morocco									
Peru									
Portugal									
Romania								0.1	
Russia									
Serbia									
Slovenia									
South Africa									
Spain				-3.7			-2.8		-0.1
Switzerland				-0.3		0.2		0.3	
Thailand									
Turkey									
Ukraine									
United Kingdom				-0.1					
United States				-2.9					0.1
Uzbekistan									
Unknown origin & other varieties	5.4	1.1	0.6	-37.5	-0.6	-4.3	0.8	7.8	0.1
Old World subtotal	-2.2	-0.4	-0.2	25.3	0.6	3.0	0.0	-3.2	-0.1
New World subtotal				-6.5					0.0
World total	0.0	0.0	0.0	0.0	0.0	0.0	0.0	0.0	0.0

Table 19 (cont.): NVII of national winegrapes by varietal country of origin, 2000

Country of planting →	*Italy*	*Korea, Rep.*	*Luxembourg*	*Moldova*	*Morocco*	*New Zealand*	*Portugal*	*Romania*	*Russia*	*Serbia*
Country of origin↓										
Algeria										
Argentina										
Armenia										
Australia										
Austria	-1.8		0.0	-0.3		0.0				
Azerbaijan										
Belgium										
Brazil										
Bulgaria				-0.1						
Canada										
Chile										
China										
Croatia	-1.1							1.7		6.4
Cyprus										
France	-20.7		0.0	6.2	-1.8	1.1	-11.0	-4.5	-1.6	
Georgia				2.2				-0.7	2.7	
Germany	-2.6		0.2	0.4		0.2		-1.0	0.0	
Greece	0.0			-0.4	0.5		-0.6	-0.8		
Hungary	-0.7			-0.1					0.7	
Israel										
Italy	85.9						-5.0			
Japan		0.8								
Kazakhstan										
Lebanon	0.3									
Moldova				0.8				3.7		
Morocco					3.8					
Peru										
Portugal							22.7			
Romania				0.0				1.3		
Russia				0.0					0.3	
Serbia										3.1
Slovenia										
South Africa						0.0				
Spain	-33.5				-1.5	-0.5	-8.8			
Switzerland	-0.3					0.0				0.7
Thailand										
Turkey										
Ukraine				0.3					0.5	
United Kingdom	-0.1									0.5
United States		0.1		2.0		0.0				
Uzbekistan				0.0						
Unknown origin & other varieties	-21.2	0.1			4.3	-0.3	19.1	47.1	7.8	2.9
Old World subtotal	15.5		0.0	0.3	-1.8	0.2	-7.9	-21.8	-3.5	-1.5
New World subtotal	-4.9	0.9		1.7		-0.1				0.0
World total	0.0	0.0	0.0	0.0	0.0	0.0	0.0	0.0	0.0	0.0

Table 19 (cont.): NVII of national winegrapes by varietal country of origin, 2000

Country of planting →	*Serbia*	*Slovakia*	*Slovenia*	*South Africa*	*Spain*	*Switzerland*	*Taiwan*	*Tunisia*	*UK*
Country of origin↓									
Algeria					0.0				
Argentina									
Armenia									
Australia									
Austria		1.0			-3.6	0.0			0.0
Azerbaijan									
Belgium					0.0				
Brazil									
Bulgaria									
Canada									
Chile									
China									
Croatia	6.4	0.7	0.6	-0.6	-6.8			0.0	
Cyprus									
France		-0.6	-0.5	9.9	-61.5	0.8		-0.5	0.0
Georgia									
Germany		0.3		-0.2	-5.7	0.1			0.1
Greece				0.6	-4.2	-0.1			
Hungary				0.0					
Israel				0.0					
Italy			-0.1	-2.4	-30.7	-0.4	-0.1	-0.3	
Japan							0.3		
Kazakhstan									
Lebanon					-0.1				
Moldova		0.0							
Morocco									
Peru									
Portugal				-0.4	-4.2				
Romania		0.1							
Russia									
Serbia	3.1								
Slovenia									
South Africa				1.4					
Spain				-4.8	142.5			1.1	
Switzerland	0.7				-0.7	1.1			
Thailand									
Turkey					-0.8				
Ukraine									
United Kingdom	0.5				-0.3				
United States				0.2	-3.9		0.2		
Uzbekistan									
Unknown origin & other varieties	2.9	0.4	4.5	-4.1	-6.2	-0.6		0.6	0.1
Old World subtotal	-1.5	-0.1	-2.1	0.8	12.1	0.4	-0.5	-0.2	0.0
New World subtota	0.0			1.2	-9.1		0.5		
World total	0.0	0.0	0.0	0.0	0.0	0.0	0.0	0.0	0.0

Table 19 (cont.): NVII of national winegrapes by varietal country of origin, 2000

Country of planting →	*USA*	*Uruguay*	*Missing 9*	***Old World***	***New World***	***World***
Country of origin↓						
Algeria				0.0		0.0
Argentina			-0.6		14.2	0.0
Armenia			0.0	0.1		0.0
Australia					0.0	0.0
Austria	-0.5		-0.3	2.5	-2.4	0.0
Azerbaijan			0.2			0.0
Belgium				0.0		0.0
Brazil					0.0	0.0
Bulgaria				1.3		0.0
Canada	0.0				0.0	0.0
Chile					0.0	0.0
China			0.0			0.0
Croatia	2.7			1.3	-0.8	0.0
Cyprus				0.5		0.0
France	13.7	0.5	9.0	-55.2	50.7	0.0
Georgia			4.3	0.7		0.0
Germany	-0.3		0.6	1.5	-1.8	0.0
Greece	-0.3		-0.3	-2.0	2.1	0.0
Hungary				0.9	-0.8	0.0
Israel				0.0	0.0	0.0
Italy	-3.2		-3.8	19.3	-17.3	0.0
Japan			0.2		0.9	0.0
Kazakhstan			0.1			0.0
Lebanon				0.0	0.0	0.0
Moldova			-0.1	0.9		0.0
Morocco				0.6	-0.6	0.0
Peru			0.1			0.0
Portugal	-1.0			3.6	-3.2	0.0
Romania				0.3		0.0
Russia				0.0	0.0	0.0
Serbia				0.6		0.0
Slovenia				0.0		0.0
South Africa					1.2	0.0
Spain	-8.0		-8.3	40.6	-36.4	0.0
Switzerland				0.4	-0.4	0.0
Thailand			0.0			0.0
Turkey			2.5	-2.8	1.5	0.0
Ukraine			1.9	-0.5		0.0
United Kingdom	0.0	0.6		-0.3	0.4	0.0
United States	3.4		0.5	-10.4	10.2	0.0
Uzbekistan						
Unknown origin & other varieties	-7.3	0.1	-6.2	24.9	-21.8	0.0
Old World subtotal	1.0	-0.6	3.0	14.5	-16.1	0.0
New World subtota	2.6	0.5	0.1	-27.0	27.0	0.0
World total	0.0	0.0	0.0	0.0	0.0	0.0

Table 20: NVII of national winegrapes by varietal country of origin, 2010

Country of planting → Country of origin↓	*Algeria*	*Argentina*	*Armenia*	*Australia*	*Austria*	*Brazil*	*Bulgaria*	*Canada*	*Chile*
Algeria									
Argentina		15.7							-0.3
Armenia			0.8						
Australia				0.0					
Austria		-0.7			6.1	-0.2		0.0	-0.4
Azerbaijan									
Brazil		0.0				0.3			
Bulgaria							3.6		
Canada								0.0	
Chile									0.0
China									0.0
Croatia		-1.1		-0.7	0.5	-0.2		0.0	-0.5
Cyprus									
Czechia									
France	1.2	5.9		16.8	-2.2	-0.7	0.8	0.9	13.6
Georgia			0.5				0.5		
Germany		-1.1		0.3	0.8	-0.3	-0.1	0.3	-0.4
Greece		1.7	0.1	-0.2	-0.1	0.0			-0.3
Hungary					-0.1	-0.1	0.0	0.0	
Israel									
Italy		-3.9		-3.7		-1.1	-1.3	-0.3	-2.9
Japan						0.0			
Kazakhstan									
Lebanon									
Moldova									
Montenegro									
Morocco		-0.1							
Peru									
Portugal		-1.3		-0.6		-0.3		-0.1	-0.7
Romania									
Russia								0.0	
Serbia									
Slovakia									
Slovenia					0.0				
South Africa						0.0		0.0	
Spain	1.3	-9.6		-7.5		-2.6		-0.5	-5.0
Switzerland						0.0		0.0	-0.1
Thailand									
Turkey				0.0					
Ukraine									
United Kingdom		-0.1							
United States		-0.6		-0.3		5.9		0.1	
Uzbekistan									
Unknown origin & other varieties		-4.5	1.7	-2.2	-0.4	-0.1	2.8	-0.2	-2.3
Old World subtotal	0.6	-12.6	-0.8	2.1	0.6	-6.0	-0.9	0.1	1.9
New World subtotal		14.9		-1.0		6.1		0.0	-0.8
World total	0.0	0.0	0.0	0.0	0.0	0.0	0.0	0.0	0.0

Table 20 (cont.): NVII of national winegrapes by varietal country of origin, 2010

Country of planting → Country of origin↓	*China*	*Croatia*	*Cyprus*	*Czechia*	*Ethiopia*	*France*	*Georgia*	*Germany*	*Greece*
Algeria									
Argentina									
Armenia									
Australia									
Austria		0.1		1.3		-2.5		2.3	-0.2
Azerbaijan									
Brazil									
Bulgaria		0.0							0.0
Canada									
Chile									
China									
Croatia		2.2		0.2		-4.2			
Cyprus			1.3						
Czechia				0.2					
France	3.9	-1.0	-0.5	-0.4	0.0	52.9	-3.7	-2.5	-3.4
Georgia		-0.1					9.4		
Germany	-0.1	0.1		0.7		-3.0		12.3	-0.3
Greece		-0.1	0.0			-1.9		-0.5	8.4
Hungary		0.0		0.0		-1.3			
Israel									
Italy		-0.3		-0.4	0.0	-2.6		-2.2	-1.4
Japan									
Kazakhstan									
Lebanon						0.0			
Moldova									
Montenegro		0.0							
Morocco									
Peru									
Portugal									
Romania									
Russia		0.0							
Serbia		0.0							
Slovakia									
Slovenia		0.0							
South Africa									
Spain	-1.6	-1.1	-0.3			-10.4			-2.8
Switzerland		0.0				0.0		0.2	
Thailand									
Turkey			0.1		0.0	-0.4			
Ukraine				0.0					
United Kingdom		0.0				0.4		0.0	
United States	-0.1	0.0			0.0	-2.7			
Uzbekistan									
Unknown origin & other varieties		0.8	0.0	-0.2		-14.8	0.4	-1.8	2.8
Old World subtotal	0.5	-0.3	0.1	0.2	0.0	13.4	0.2	1.7	-0.9
New World subtotal	-0.2	-0.1			0.0	-5.9		-0.8	
World total	0.0	0.0	0.0	0.0	0.0	0.0	0.0	0.0	0.0

Table 20 (cont.): NVII of national winegrapes by varietal country of origin, 2010

Country of planting →	*Hungary*	*Israel*	*Italy*	*Japan*	*ıkhstan*	*Korea, Rep.*	*:embourg*	*Mexico*	*Moldova*
Country of origin↓									
Algeria									
Argentina									
Armenia									
Australia									
Austria	2.6		-2.0	0.0					-0.3
Azerbaijan					0.1				
Brazil									
Bulgaria	0.0								-0.1
Canada									
Chile									
China									
Croatia	0.7		1.2						
Cyprus									
Czechia									
France	-2.3	0.1	-26.3	0.1	-0.4		0.0	0.0	4.7
Georgia					0.8				2.3
Germany	0.4		-2.6	0.1	0.0		0.1		0.4
Greece	-0.2	0.0	-0.3		0.0			0.0	-0.4
Hungary	5.5		-0.9						-0.1
Israel		0.0							
Italy	-1.8		79.9					-0.1	
Japan				0.2		0.9			
Kazakhstan					0.1				
Lebanon			0.1						
Moldova					0.0				0.9
Montenegro									
Morocco									
Peru									
Portugal	-0.4								
Romania	-0.1								-0.1
Russia									0.0
Serbia	0.0								
Slovakia									
Slovenia	0.0								
South Africa									
Spain		0.0	-31.6					-0.1	
Switzerland	0.4		-0.4						
Thailand									
Turkey								0.2	
Ukraine	0.0				0.0				0.3
United Kingdom	0.0		-0.2						
United States		0.1		0.0	0.0	0.1		0.1	2.2
Uzbekistan									0.0
Unknown origin & other varieties	0.3	0.3	-3.7		0.2	0.2	0.0	0.2	
Old World subtotal	0.4	-0.2	6.9	-0.1	-0.1		0.0	-0.1	-0.8
New World subtotal	-0.6	0.0	-5.0	0.2	-0.1	1.0		0.1	1.8
World total	0.0	0.0	0.0	0.0	0.0	0.0	0.0	0.0	0.0

Table 20 (cont.): NVII of national winegrapes by varietal country of origin, 2010

Country of planting → Country of origin↓	*Morocco*	*Myanmar*	*New Zealand*	*Peru*	*Portugal*	*Romania*	*Russia*
Algeria							
Argentina				0.0			
Armenia							
Australia							
Austria			-0.1	0.1	-0.5	-0.4	
Azerbaijan							
Brazil							
Bulgaria						0.5	0.1
Canada							
Chile							
China				0.0			0.0
Croatia			-0.2			0.8	
Cyprus							
Czechia							
France	-2.9	0.0	3.8	-0.3	-9.8	-6.4	0.3
Georgia						-0.6	0.2
Germany			0.1		-0.9	-0.8	0.1
Greece	0.5	0.0		0.1	-0.5	-0.7	-0.1
Hungary						-0.2	0.8
Israel							
Italy			-0.8	0.1	-4.0	-4.5	-0.7
Japan							
Kazakhstan							
Lebanon							
Moldova						3.0	0.2
Montenegro							
Morocco	4.1						
Peru				0.1			
Portugal			-0.2		24.6	-1.1	
Romania						4.3	
Russia						0.0	0.6
Serbia						-0.1	
Slovakia							
Slovenia							
South Africa			0.0				
Spain	-1.3	0.0	-1.7	0.1	-3.3	-9.1	
Switzerland			0.0		-0.1		0.0
Thailand							
Turkey				0.0			
Ukraine							0.7
United Kingdom							0.0
United States			-0.1	0.1			-0.1
Uzbekistan							
Unknown origin & other varieties	5.6		-0.2		-2.9	34.5	-0.4
Old World subtotal	-2.4	0.0	0.4	-0.1	2.8	-15.9	0.3
New World subtotal			-0.2	0.1			-0.1
World total	0.0	0.0	0.0	0.0	0.0	0.0	0.0

Table 20 (cont.): NVII of national winegrapes by varietal country of origin, 2010

Country of planting →	*Serbia*	*Slovakia*	*Slovenia*	*South Africa*	*Spain*	*Switzerland*	*Taiwan*	*Thailand*
Country of origin↓								
Algeria					0.0			
Argentina								
Armenia								
Australia								
Austria		1.0	0.1			0.0		
Azerbaijan								
Brazil								
Bulgaria								
Canada								
Chile								
China								
Croatia	6.8	0.3	0.6	-0.5	-5.0			
Cyprus								
Czechia		0.1						
France		-0.6	-0.2	10.5	-62.6	0.6		0.0
Georgia								
Germany		0.3	0.1	-0.4	-5.4	0.1		0.0
Greece		-0.1	0.0	0.2	-3.1	-0.1		0.0
Hungary		0.1	0.1	-0.1		0.0		
Israel				0.0				
Italy		-0.3	0.0	-2.6	-27.0	-0.3	-0.1	0.0
Japan							0.3	0.0
Kazakhstan								
Lebanon					0.0			
Moldova		0.0						
Montenegro								
Morocco								
Peru					0.0			
Portugal				-0.5	-3.9			
Romania		0.0	0.0					
Russia								
Serbia	3.2							
Slovakia		0.0						
Slovenia		0.0	0.2					
South Africa				1.5				
Spain				-5.2	138.1			0.0
Switzerland	0.7				-0.7	1.1		
Thailand								0.0
Turkey								
Ukraine		0.0						
United Kingdom	0.6				-0.4			
United States				0.2	-3.3		0.2	
Uzbekistan								
Unknown origin & other varieties	4.8	-0.1	0.7	-2.1	-13.6	-0.2		
Old World subtotal	-2.4	0.2	-0.2	-0.1	15.0	0.2	-0.5	0.0
New World subtotal	0.0			1.2	-8.2		0.6	0.0
World total	0.0	0.0	0.0	0.0	0.0	0.0	0.0	0.0

Table 20 (cont.): NVII of national winegrapes by varietal country of origin, 2010

Country of planting → / Country of origin↓	*Tunisia*	*Turkey*	*Ukraine*	*UK*	*USA*	*Uruguay*	***Old World***	***New World***	***World***
Algeria							0.0		0.0
Argentina								13.2	0.0
Armenia			0.0				0.2		0.0
Australia								0.0	0.0
Austria				0.0	-0.7		3.0	-3.0	0.0
Azerbaijan							0.0		0.0
Brazil								0.3	0.0
Bulgaria							0.9		0.0
Canada					0.0			0.0	0.0
Chile								0.0	0.0
China							0.0	0.0	0.0
Croatia	0.0				3.2		0.4	-0.4	0.0
Cyprus							0.3		0.0
Czechia							0.0		0.0
France	-0.8	-0.1	1.2	0.1	16.6	0.4	-71.0	71.0	0.0
Georgia			2.6				3.6		0.0
Germany		-0.1	0.5	0.1	0.1	0.0	1.5	-1.5	0.0
Greece		0.0	-0.1		-0.7	0.0	-0.3	0.3	0.0
Hungary							1.4	-1.4	0.0
Israel							0.0	0.0	0.0
Italy	-0.3	-0.3			-5.0	0.0	21.6	-21.6	0.0
Japan								1.1	0.0
Kazakhstan							0.0		0.0
Lebanon							0.0		0.0
Moldova							0.9		0.0
Montenegro							0.0		0.0
Morocco							0.8	-0.8	0.0
Peru							-0.1	0.1	0.0
Portugal				0.0	-1.4	0.0	5.4	-5.4	0.0
Romania							0.9		0.0
Russia							0.1	-0.1	0.0
Serbia							0.7		0.0
Slovakia							0.0		0.0
Slovenia							0.1		0.0
South Africa					-0.1			1.2	0.0
Spain	1.2	-0.7			-11.0	-0.4	46.0	-46.0	0.0
Switzerland							0.6	-0.6	0.0
Thailand								0.0	0.0
Turkey		1.8					0.2	-0.2	0.0
Ukraine			1.5				0.5		0.0
United Kingdc					-0.1	0.3	0.0	0.0	0.0
United States			0.3		3.6	0.0	-8.9	8.9	0.0
Uzbekistan							0.0		0.0
Unknown orig	1.1				-1.5	0.1	13.7	-13.7	0.0
Old World su	-0.4		0.5	0.0	-1.8	-0.4	17.9	-17.9	0.0
New World si			0.1		2.6	0.3	-24.8	24.8	0.0
World total	0.0		0.0	0.0	0.0	0.0	0.0	0.0	0.0

Table 21: NVII of national winegrapes by varietal country of origin, 2016

Country of planting → Country of origin↓	*Algeria*	*Argentina*	*Armenia*	*Australia*	*Austria*	*Brazil*	*Bulgaria*	*Cambodia*	*Canada*
Algeria									
Argentina		14.1							
Australia				0.0					
Austria		-0.7		-0.4	6.2				0.0
Azerbaijan									
Bosnia and Herzegovina									
Brazil						0.8			
Bulgaria							3.6		
Canada						0.0			0.0
Chile									
China									
Croatia		-0.7		-0.4	0.6	-0.1			0.0
Cyprus									
Czechia									
France	-0.1	7.9		15.0	-2.2	-0.3	1.7	0.0	1.0
Georgia				-0.5			1.0		
Germany		-1.2		0.0	0.7	-0.2	-0.2		0.3
Greece	0.0	0.9		-0.1	-0.1	0.0			-0.1
Hungary				-0.2	-0.1	-0.1	0.2		0.0
Israel									
Italy		-4.0		-3.2		-0.7	-1.2		-0.3
Japan								0.0	
Kazakhstan									
Lebanon				0.0					
Moldova									
Montenegro									
Morocco		0.1							
North Macedonia									
Peru									
Portugal		-1.4		-0.7		-0.2			-0.1
Romania									
Russia									0.0
Serbia									
Slovenia					0.0				
South Africa				-0.1		0.0			0.0
Spain	0.7	-8.1		-5.6		-1.6			-0.6
Switzerland									0.0
Taiwan									
Thailand									
Turkey	0.0			-0.1					
Ukraine									0.0
United Kingdom				0.0					
United States		-0.5		-0.1		3.6			0.1
Unknown origin & other varieties		-7.1	6.1	-4.5	-1.1				-0.3
Old World subtotal	0.2	-9.7		3.0	0.9	-3.6	1.3	0.0	0.1
New World subtotal		13.2		-0.8		4.2		0.0	0.1
World total	0.0	0.0	0.0	0.0	0.0	0.0	0.0	0.0	0.0

Table 21 (cont.): NVII of national winegrapes by varietal country of origin, 2016

Country of planting → Country of origin↓	*Chile*	*China*	*Croatia*	*Cyprus*	*Czechia*	*Ethiopia*	*France*	*Georgia*	*Germany*
Algeria	0.0								
Argentina	1.0								
Australia									
Austria	-0.5		0.1		1.1		-2.4		2.0
Azerbaijan									
Bosnia and Herzegovina									
Brazil									
Bulgaria									
Canada									
Chile	0.0								
China	-0.1	1.6							
Croatia	-0.5	0.1	1.8		0.2		-2.7		
Cyprus				1.1					
Czechia									0.0
France	13.9	2.4	-0.5		-0.3	0.0	45.4	-4.0	-2.4
Georgia	-0.5							9.7	
Germany	-0.7	-0.7	0.1		0.6		-3.0		11.5
Greece	0.7	-0.4					-2.6		-0.5
Hungary							-1.4		-0.2
Israel									
Italy	-3.7	-4.4				0.0	-2.3		-2.0
Japan									
Kazakhstan									
Lebanon							0.0		
Moldova									
Montenegro									
Morocco									
North Macedonia									
Peru									
Portugal	-1.0								
Romania									
Russia									
Serbia									
Slovenia									
South Africa									
Spain	-4.5	-7.6					-10.1		
Switzerland	0.0						-0.2		0.2
Taiwan									
Thailand									
Turkey						0.0	-0.5		
Ukraine									
United Kingdom							0.2		0.0
United States	-0.4					0.0	-2.1		
Unknown origin & other varieties		28.5					-2.2	-0.2	-3.1
Old World subtotal	2.1	-14.5	0.3	0.1	0.3	0.0	7.0	0.5	2.3
New World subtotal	0.4	0.2				0.0	-5.9		-0.7
World total	0.0	0.0	0.0	0.0	0.0	0.0	0.0	0.0	0.0

Table 21 (cont.): NVII of national winegrapes by varietal country of origin, 2016

Country of planting → Country of origin↓	*Greece*	*Hungary*	*India*	*Israel*	*Italy*	*Japan*	*Kazakhstan*	*Korea, Rep.*
Algeria								
Argentina								
Australia								
Austria	-0.2	2.4			-1.9	0.0		
Azerbaijan							0.1	
Bosnia and Herzegovina								
Brazil								
Bulgaria	-0.1	-0.1						
Canada								
Chile								
China								
Croatia		0.7			2.1			
Cyprus								
Czechia								
France	-3.1	-2.4	0.1	0.3	-28.4	-0.2	-0.5	
Georgia							0.9	
Germany	-0.3	0.5			-2.5	0.0	0.0	
Greece	8.7	-0.2	0.0	0.0	-0.3		0.0	
Hungary		5.8			-1.0			
Israel				0.1				
Italy	-1.3	-1.7	0.0		76.8			
Japan						0.3		0.9
Kazakhstan							0.1	
Lebanon					0.0			
Moldova		-0.1					0.0	
Montenegro		0.0						
Morocco								
North Macedonia								
Peru								
Portugal		-0.4						
Romania		-0.1						
Russia								
Serbia		0.0						
Slovenia		0.0						
South Africa								
Spain	-2.4			0.0	-27.1			
Switzerland		0.2			-0.3			
Taiwan								
Thailand								
Turkey			0.2					
Ukraine		0.0					0.0	
United Kingdom	0.5	0.0			-0.2			
United States				0.0		0.3	0.0	0.1
Unknown origin & other varieties	-1.0	-1.7		-0.1	-4.8	0.0	0.2	0.2
Old World subtotal	0.4	1.3	0.1	0.1	7.1	-0.6	0.0	
New World subtotal	0.1	-0.5		0.0	-4.8	0.6	-0.1	1.0
World total	0.0	0.0	0.0	0.0	0.0	0.0	0.0	0.0

Table 21 (cont.): NVII of national winegrapes by varietal country of origin, 2016

Country of planting →	*Lebanon*	*Luxembourg*	*Mexico*	*Moldova*	*Morocco*	*Myanmar*	*New Zealand*	*North Macedonia*
Country of origin↓								
Algeria								
Argentina								
Australia								
Austria		0.0		-0.3			-0.1	
Azerbaijan								
Bosnia and Herzegovina								0.0
Brazil								
Bulgaria				-0.1				1.5
Canada								
Chile								
China								
Croatia							-0.1	0.2
Cyprus								
Czechia								
France	0.4	0.0	0.0	2.1	0.0	0.0	4.5	-1.3
Georgia				0.7				0.0
Germany		0.1		0.2			0.0	0.1
Greece		0.0	0.0	-0.5	0.4	0.0	-0.2	-0.1
Hungary				0.3				
Israel								
Italy			-0.1	-2.1	0.3		-0.9	
Japan								
Kazakhstan								
Lebanon				0.0				
Moldova				3.9				
Montenegro								2.1
Morocco								
North Macedonia								0.1
Peru								
Portugal							-0.2	
Romania				0.0				
Russia				0.1				
Serbia								0.2
Slovenia								
South Africa							0.0	
Spain			-0.1		-0.3	0.0	-1.7	
Switzerland				0.0				
Taiwan								
Thailand								
Turkey			0.2		0.2			
Ukraine				0.5				
United Kingdom				0.0				0.1
United States			0.1	0.7	0.1		-0.1	
Unknown origin & other varieties	0.2	0.0	0.1	-0.6	0.1		-1.2	-0.8
Old World subtotal	-0.1	0.0	-0.1	-0.1	0.0	0.0	0.9	0.5
New World subtotal			0.1	0.3	0.0		-0.3	-0.1
World total	0.0	0.0	0.0	0.0	0.0	0.0	0.0	0.0

Table 21 (cont.): NVII of national winegrapes by varietal country of origin, 2016

Country of planting → / Country of origin↓	*Norway*	*Peru*	*Portugal*	*Romania*	*Russia*	*Serbia*	*Slovakia*	*Slovenia*
Algeria								
Argentina		0.0						
Australia								
Austria		0.1	-0.6	-0.3		0.1	0.8	0.1
Azerbaijan								
Bosnia and Herzegovina								
Brazil								
Bulgaria				0.4	0.0	0.0		
Canada								
Chile								
China		0.0			0.0			
Croatia				-0.3		0.4	0.1	0.6
Cyprus								
Czechia								
France		-0.3	-12.2	-7.1	1.3	-0.2	-0.5	-0.2
Georgia				-0.6	1.4	-0.1		
Germany	0.0		-1.0	0.4	0.3	0.2	0.2	0.1
Greece		0.1	-0.5	-0.7	-0.2	-0.1		0.0
Hungary				-0.3	0.8	0.0		0.1
Israel								
Italy		0.1	-4.6	-4.8	-1.3			0.0
Japan								
Kazakhstan								
Lebanon						0.0		
Moldova				3.1	0.2			
Montenegro						0.0		
Morocco								
North Macedonia								
Peru		0.1						
Portugal			27.5	-1.3				
Romania				4.1				
Russia				0.0	0.7	0.0		
Serbia				0.0		0.3		
Slovenia								0.2
South Africa								
Spain		0.1	-3.1	-8.7				
Switzerland			-0.1	-0.1	0.0			
Taiwan								
Thailand								
Turkey		0.0						
Ukraine					1.0			
United Kingdom					0.0	0.1		
United States		0.1			0.2	-0.1		
Unknown origin & other varieties	0.0		-2.5	35.8	-1.6	2.8	0.5	0.2
Old World subtotal	0.0	-0.1	2.7	-16.5	0.8	-1.3	-0.2	0.0
New World subtotal		0.1			0.0	0.0		
World total	0.0	0.0	0.0	0.0	0.0	0.0	0.0	0.0

Table 21 (cont.): NVII of national winegrapes by varietal country of origin, 2016

Country of planting → Country of origin↓	*South Africa*	*Spain*	*Switzerland*	*Taiwan*	*Thailand*	*Tunisia*	*Turkey*	*Ukraine*
Algeria								
Argentina								
Australia								
Austria	-0.3		0.0					
Azerbaijan								
Bosnia and Herzegovina								
Brazil								
Bulgaria								
Canada								
Chile								
China								
Croatia	-0.3	-2.7	0.0					
Cyprus								
Czechia								
France	9.5	-56.9	0.5		0.0	0.0	-0.2	1.1
Georgia								1.2
Germany	-0.5	-4.9	0.1		0.0		-0.1	0.3
Greece	0.1	-2.5	-0.1		0.0	0.1	-0.1	-0.1
Hungary	-0.1		0.0					
Israel	0.0							
Italy	-2.5	-23.2	-0.4		0.0	0.1	-0.4	
Japan				0.0	0.0			
Kazakhstan								
Lebanon								
Moldova								
Montenegro								
Morocco								
North Macedonia								
Peru								
Portugal	-0.6	-3.9	-0.1		0.0			
Romania	-0.1							
Russia								
Serbia	0.0							
Slovenia			0.0					
South Africa	1.7		0.0					
Spain	-4.3	125.8	-0.7		0.0	-0.1	-0.6	
Switzerland		-0.4	1.1					
Taiwan				0.0				
Thailand					0.0			
Turkey		-0.5				0.0	2.0	
Ukraine								0.3
United Kingdom		-0.3	0.0					
United States	0.3	-2.3	0.0	0.0		0.0		0.2
Unknown origin & other varieties	-3.3	-22.6	-0.5		0.0	0.0		
Old World subtotal	0.1	18.2	0.3		0.0	0.0	0.3	0.4
New World subtotal	1.5	-6.9	-0.1	0.0	0.0	0.0		0.1
World total	0.0	0.0	0.0	0.0	0.0	0.0	0.0	0.0

Table 21 (cont.): NVII of national winegrapes by varietal country of origin, 2016

Country of planting →	*UK*	*USA*	*Uruguay*	***Old World***	***New World***	***World***
Country of origin↓						
Algeria					0.0	0.0
Argentina					12.3	0.0
Australia					0.0	0.0
Austria		-0.8		3.5	-3.5	0.0
Azerbaijan				0.0		0.0
Bosnia and Herzegovina				0.0		0.0
Brazil					0.6	0.0
Bulgaria				1.5		0.0
Canada		0.0			0.0	0.0
Chile					0.0	0.0
China				-1.2	1.2	0.0
Croatia		3.3		-1.2	1.2	0.0
Cyprus				0.3		0.0
Czechia				0.0		0.0
France	0.2	17.9	0.3	-71.4	71.4	0.0
Georgia				4.0	-4.0	0.0
Germany	0.1	0.0	0.0	2.8	-2.8	0.0
Greece		-0.7	0.0	-0.2	0.2	0.0
Hungary				1.9	-1.9	0.0
Israel				0.0	0.0	0.0
Italy		-5.5	0.0	25.4	-25.4	0.0
Japan					0.9	0.0
Kazakhstan				0.0		0.0
Lebanon				0.0	0.0	0.0
Moldova				1.9		0.0
Montenegro				0.5		0.0
Morocco					0.0	0.0
North Macedonia				0.0		0.0
Peru					0.1	0.0
Portugal		-1.6	0.0	7.4	-7.4	0.0
Romania				1.1	-1.1	0.0
Russia				0.2	-0.2	0.0
Serbia				0.1	-0.1	0.0
Slovenia				0.1		0.0
South Africa		-0.1		-1.3	1.3	0.0
Spain		-10.3	-0.3	45.3	-45.3	0.0
Switzerland				0.4	-0.4	0.0
Taiwan					0.0	0.0
Thailand					0.0	0.0
Turkey				0.2	-0.2	0.0
Ukraine		-0.1		0.4	-0.4	0.0
United Kingdom		-0.1	0.3	0.1	-0.1	0.0
United States		4.0	0.0	-6.9	6.9	0.0
Unknown origin & other varieties	0.0	-4.6	0.0	-1.5	1.5	0.0
Old World subtotal	0.0	-0.4	-0.3	24.1	-24.1	0.0
New World subtotal		2.7	0.3	-23.4	23.4	0.0
World total	0.0	0.0	0.0	0.0	0.0	0.0

Table 22: National shares of global winegrape area by varietal country of origin, 2000 (%)

Country of planting →	*Algeria*	*Argentina*	*Armenia*	*Australia*	*Austria*	*Brazil*	*Bulgaria*	*Canada*	*Chile*
Country of origin↓									
Algeria									
Argentina		99.9							0.1
Armenia			95.5						
Australia				100.0					
Austria		0.0			43.3				
Azerbaijan									
Belgium									
Brazil		100.0							
Bulgaria							99.3		
Canada									
Chile									100.0
China									
Croatia		0.0			3.0	0.6	2.5		0.1
Cyprus									
France	1.2	5.7		7.1	0.4	0.6	2.1	0.4	5.8
Georgia			2.9				11.1		
Germany		0.1		3.2	5.1	0.1	1.4	0.6	0.4
Greece		14.6	0.5	3.5	0.1	0.7			
Hungary					0.0		8.1		
Israel									
Italy		1.4		0.2		0.1	0.3		0.0
Japan									
Kazakhstan									
Lebanon				17.8					
Moldova									
Morocco		1.8							
Peru									
Portugal		0.0		1.0					3.2
Romania									
Russia		10.1							
Serbia									
Slovenia					100.0				
South Africa									
Spain	1.0	0.5		0.3					1.4
Switzerland									3.0
Thailand									
Turkey				61.4					
Ukraine							29.5		
United Kingdom		0.0					6.3		
United States		4.5		3.1		39.4	3.8	2.0	
Uzbekistan									
Unknown origin & other varieties		0.0	0.9	1.2	0.3	2.0	1.2	0.2	2.2
Old World subtotal	**0.7**	**2.7**	**0.2**	**2.9**	**1.1**	**0.3**	**2.1**	**0.1**	**2.4**
New World subtotal		**46.8**		**1.4**		**17.2**	**1.9**	**0.9**	**0.1**
World total	**0.6**	**4.0**	**0.2**	**2.7**	**1.0**	**1.1**	**2.0**	**0.2**	**2.3**

Table 22 (cont.): National shares of global winegrape area by varietal country of origin, 2000 (%)

Country of planting → Country of origin↓	*Croatia*	*Cyprus*	*Czechia*	*France*	*Georgia*	*Germany*	*Greece*	*Hungary*	*Israel*
Algeria									
Argentina									
Armenia									
Australia									
Austria	1.7		6.3	3.0		18.2		18.6	
Azerbaijan									
Belgium									
Brazil									
Bulgaria							0.3	0.3	
Canada									
Chile									
China									
Croatia	25.9		0.9	0.0				4.6	
Cyprus		100.0							
France	0.1		0.0	39.0	0.0	1.3	0.1	0.6	0.2
Georgia					40.5				
Germany			2.1	5.4		59.9		4.2	
Greece				9.2		0.1	37.1	1.4	0.2
Hungary				0.0				75.6	
Israel									54.1
Italy				15.2		0.4	0.1		
Japan									
Kazakhstan									
Lebanon									
Moldova								3.9	
Morocco									
Peru									
Portugal									
Romania								7.4	
Russia									
Serbia									
Slovenia									
South Africa									
Spain				16.3			0.0		0.1
Switzerland				7.0		8.8		14.0	
Thailand									
Turkey									
Ukraine									
United Kingdom				10.6					
United States				0.0					0.4
Uzbekishtan									
Unknown origin & other varieties	3.7	0.9	0.5	0.4	0.5	0.2	1.4	5.4	0.2
Old World subtotal	**1.0**	**0.3**	**0.2**	**20.7**	**0.8**	**2.5**	**1.0**	**1.4**	**0.1**
New World subtotal				**0.4**					**0.2**
World total	**1.2**	**0.4**	**0.2**	**17.7**	**0.8**	**2.1**	**1.0**	**1.8**	**0.1**

Table 22 (cont.): National shares of global winegrape area by varietal country of origin, 2000 (%)

Country of planting →	*Italy*	*Korea, Rep.*	*Luxembourg*	*Moldova*	*Morocco*	*New Zealand*	*Portugal*	*Romania*	*Russia*
Country of origin↓									
Algeria									
Argentina									
Armenia									
Australia									
Austria	0.6		0.0	0.1		0.0			
Azerbaijan									
Belgium									
Brazil									
Bulgaria				0.0					
Canada									
Chile									
China									
Croatia	9.4							10.4	
Cyprus									
France	5.6		0.0	4.1	0.4	0.6	0.3	2.9	0.6
Georgia				14.4				0.6	16.6
Germany	1.9		0.7	3.7		1.0		0.4	1.2
Greece	12.8			0.2	3.3		1.7	0.9	
Hungary	0.7			0.1					14.6
Israel									
Italy	79.9						0.3		
Japan		66.0							
Kazakhstan									
Lebanon	82.2								
Moldova				17.5				77.1	
Morocco					98.2				
Peru									
Portugal							86.0		
Romania				1.0				87.6	
Russia				1.7					88.1
Serbia									
Slovenia									
South Africa						1.0			
Spain	0.7				0.5	0.0	0.9		
Switzerland	0.5					0.2			
Thailand									
Turkey									
Ukraine				13.3					18.7
United Kingdom	2.8								
United States		0.7		14.2		0.0			
Uzbekishtan				100.0					
Unknown origin & other varieties	3.2	0.2			3.0	0.1	13.0	26.3	4.8
Old World subtotal	**14.8**		**0.0**	**1.9**	**0.8**	**0.2**	**3.3**	**2.0**	**0.7**
New World subtotal	**0.1**	**2.5**		**6.2**		**0.0**			
World total	**13.0**	**0.1**	**0.0**	**1.8**	**1.0**	**0.2**	**4.2**	**4.5**	**1.2**

Table 22 (cont.): National shares of global winegrape area by varietal country of origin, 2000 (%)

Country of planting →	*Serbia*	*Slovakia*	*Slovenia*	*South Africa*	*Spain*	*Switzerland*	*Taiwan*	*Tunisia*	*UK*
Country of origin↓									
Algeria					100.0				
Argentina									
Armenia									
Australia									
Austria		7.1			0.0	0.3			0.1
Azerbaijan									
Belgium					100.0				
Brazil									
Bulgaria									
Canada									
Chile									
China									
Croatia	22.9	2.7	2.5	0.0	1.3			0.2	
Cyprus									
France		0.1	0.3	5.5	2.1	0.6		0.2	0.0
Georgia									
Germany		1.6		0.9	0.1	0.7			0.3
Greece				4.7	5.7	0.0			
Hungary				1.0					
Israel				45.9					
Italy			0.4	0.0	0.3	0.0	0.0	0.1	
Japan							25.2		
Kazakhstan									
Lebanon					0.0				
Moldova		1.2							
Morocco									
Peru									
Portugal				0.6	9.1				
Romania		4.0							
Russia									
Serbia	100.0								
Slovenia									
South Africa				99.0					
Spain				0.1	76.6			0.7	
Switzerland	25.3				0.0	41.3			
Thailand									
Turkey					0.0				
Ukraine									
United Kingdom	39.1				0.1				
United States				3.0	0.5		1.5		
Uzbekishtan									
Unknown origin & other varieties	2.7	0.5	2.6	0.0	21.3	0.0		0.6	0.0
Old World subtotal	**1.2**	**0.3**	**0.2**	**2.0**	**25.6**	**0.4**	**0.0**	**0.3**	**0.0**
New World subtotal	**1.5**			**5.2**	**0.2**		**1.5**		
World total	**1.4**	**0.3**	**0.5**	**1.9**	**24.2**	**0.3**	**0.1**	**0.3**	**0.0**

Table 22 (cont.): National shares of global winegrape area by varietal country of origin, 2000 (%)

Country of planting →	*USA*	*Uruguay*	*Missing 9*	***Old World***	***New World***	***World***
Country of origin↓						
Algeria				**100.0**	**0.0**	**100**
Argentina			0.0	**0.0**	**100.0**	**100**
Armenia			4.5	**95.5**	**0.0**	**100**
Australia				**0.0**	**100.0**	**100**
Austria	0.1		0.5	**99.3**	**0.2**	**100**
Azerbaijan			100.0	**0.0**	**0.0**	**100**
Belgium				**100.0**	**0.0**	**100**
Brazil				**0.0**	**100.0**	**100**
Bulgaria				**100.0**	**0.0**	**100**
Canada	100.0			**0.0**	**100.0**	**100**
Chile				**0.0**	**100.0**	**100**
China			100.0	**0.0**	**0.0**	**100**
Croatia	12.9			**86.4**	**13.6**	**100**
Cyprus				**100.0**	**0.0**	**100**
France	8.5	0.4	3.3	**62.1**	**34.6**	**100**
Georgia			14.0	**86.0**	**0.0**	**100**
Germany	2.3		2.9	**88.3**	**8.9**	**100**
Greece	2.3		1.0	**73.2**	**25.7**	**100**
Hungary				**99.0**	**1.0**	**100**
Israel				**54.1**	**45.9**	**100**
Italy	1.1		0.1	**97.0**	**2.9**	**100**
Japan			8.7	**0.0**	**91.3**	**100**
Kazakhstan			100.0	**0.0**	**0.0**	**100**
Lebanon				**82.2**	**17.8**	**100**
Moldova			0.3	**99.7**	**0.0**	**100**
Morocco				**98.2**	**1.8**	**100**
Peru			100.0	**0.0**	**0.0**	**100**
Portugal	0.0			**95.2**	**4.8**	**100**
Romania				**100.0**	**0.0**	**100**
Russia				**89.9**	**10.1**	**100**
Serbia				**100.0**	**0.0**	**100**
Slovenia				**100.0**	**0.0**	**100**
South Africa				**0.0**	**100.0**	**100**
Spain	0.7		0.1	**96.9**	**3.0**	**100**
Switzerland				**96.9**	**3.1**	**100**
Thailand			100.0	**0.0**	**0.0**	**100**
Turkey			38.5	**0.0**	**61.4**	**100**
Ukraine			38.5	**61.5**	**0.0**	**100**
United Kingdom	0.2	40.8		**58.9**	**41.1**	**100**
United States	24.1		3.0	**18.8**	**78.1**	**100**
Uzbekishtan				**100.0**	**0.0**	**100**
Unknown origin & other varieties	0.2	0.2	0.2	**93.5**	**6.3**	**100**
Old World subtotal	**3.7**	**0.1**	**1.8**	**83.7**	**14.5**	**100**
New World subtotal	**10.6**	**1.6**	**1.8**	**10.5**	**87.8**	**100**
World total	**3.6**	**0.2**	**1.6**	**82**	**16**	**100**

Table 23: National shares of global winegrape area by varietal country of origin, 2010 (%)

Country of planting → Country of origin↓	*Algeria*	*Argentina*	*Armenia*	*Australia*	*Austria*	*Brazil*	*Bulgaria*	*Canada*	*Chile*
Algeria									
Argentina		99.2							0.7
Armenia			93.7						
Australia				100.0					
Austria		0.0			40.8	0.0		0.1	0.0
Azerbaijan									
Brazil		8.9				91.1			
Bulgaria							82.9		
Canada								35.8	
Chile									100.0
China									0.9
Croatia		0.0		0.1	3.2	0.2		0.0	0.1
Cyprus									
Czechia									
France	1.0	6.2		7.8	0.4	0.9	1.4	0.5	6.1
Georgia			3.1				3.9		
Germany		0.1		4.4	4.1	0.0	0.7	1.4	0.6
Greece		12.1	0.5	2.6	0.5	0.9			1.1
Hungary		0.0			0.0	0.0	1.1	0.0	
Israel									
Italy		1.4		0.2		0.1	0.1	0.0	0.0
Japan						0.3			
Kazakhstan									
Lebanon									
Moldova									
Montenegro									
Morocco		1.6							
Peru									
Portugal		0.0		1.2		0.1		0.0	0.0
Romania									
Russia								0.0	
Serbia									
Slovakia									
Slovenia					18.1				
South Africa						1.0		0.1	
Spain	1.2	0.7		0.3		0.0		0.0	0.4
Switzerland						0.0		0.5	0.6
Thailand									
Turkey				4.3					
Ukraine									
United Kingdom		0.0							
United States		0.5		1.4		40.0		0.6	
Uzbekistan									
Unknown origin & other varieties		0.0	2.0	1.1	0.6	0.9	4.1	0.1	0.0
Old World subtotal	**0.7**	**3.2**	**0.2**	**3.5**	**1.0**	**0.4**	**1.1**	**0.2**	**2.6**
New World subtotal		**44.6**		**0.6**		**17.4**		**0.3**	**0.4**
World total	**0.7**	**4.6**	**0.2**	**3.3**	**1.0**	**1.1**	**1.2**	**0.2**	**2.4**

Table 23 (cont.): National shares of global winegrape area by varietal country of origin, 2010 (%)

Country of planting → Country of origin↓	*China*	*Croatia*	*Cyprus*	*Czechia*	*Ethiopia*	*France*	*Georgia*	*Germany*	*Greece*
Algeria									
Argentina									
Armenia									
Australia									
Austria		1.2		8.6		2.0		17.0	0.0
Azerbaijan									
Brazil									
Bulgaria		0.0							0.2
Canada									
Chile									
China									
Croatia		10.1		1.1		0.0			
Cyprus			100.0						
Czechia				70.1				0.3	
France	1.7	0.2	0.1	0.2	0.0	32.3	0.0	1.5	0.3
Georgia		0.1					54.6		
Germany	0.4	0.9		3.1		5.9		52.4	0.0
Greece		0.1	0.2			10.0		0.2	37.3
Hungary		0.5		0.1		0.0			
Israel									
Italy		0.2			0.0	16.5		0.4	0.1
Japan									
Kazakhstan									
Lebanon						13.7			
Moldova									
Montenegro		100.0							
Morocco									
Peru									
Portugal									
Romania									
Russia		1.7							
Serbia		0.2							
Slovakia									
Slovenia		10.3							
South Africa									
Spain	0.0	0.0	0.1			13.9			0.1
Switzerland		0.1				17.1		7.9	
Thailand									
Turkey			3.7		0.0	0.0			
Ukraine				0.1					
United Kingdom		0.2				42.3		0.8	
United States	0.1	0.1			0.0	0.3			
Uzbekistan									
Unknown origin & other varieties		1.3	0.1	0.1		3.0	1.4	0.4	4.0
Old World subtotal	**0.7**	**0.4**	**0.2**	**0.4**	**0.0**	**19.6**	**1.1**	**2.4**	**1.1**
New World subtotal	**0.0**	**0.1**			**0.0**	**2.1**		**0.0**	
World total	**0.6**	**0.4**	**0.2**	**0.4**	**0.0**	**18.1**	**1.0**	**2.2**	**1.2**

Table 23 (cont.): National shares of global winegrape area by varietal country of origin, 2010 (%)

Country of planting → / Country of origin↓	*Hungary*	*Israel*	*Italy*	*Japan*	*Kazakhstan*	*Korea, Rep.2*	*Luxembourg*	*Mexico*	*Moldova*
Algeria									
Argentina									
Armenia									
Australia									
Austria	18.9		0.7	0.3					0.1
Azerbaijan					100.0				
Brazil									
Bulgaria	0.4								0.0
Canada									
Chile									
China									
Croatia	4.4		18.6						
Cyprus									
Czechia									
France	0.9	0.1	6.5	0.1	0.0		0.0	0.1	3.2
Georgia					4.9				15.1
Germany	3.0		2.9	0.5	0.1		0.6		3.8
Greece	0.7	0.2	12.2		0.2			0.2	0.2
Hungary	53.4		34.0						0.0
Israel		89.2							
Italy	0.0		78.6					0.0	
Japan				11.7		63.4			
Kazakhstan					100.0				
Lebanon			84.2						
Moldova					0.5				21.1
Montenegro									
Morocco									
Peru									
Portugal	0.1								
Romania	0.2								0.4
Russia									0.8
Serbia	0.0								
Slovakia									
Slovenia	0.7								
South Africa									
Spain		0.1	0.8					0.1	
Switzerland	13.2		0.6						
Thailand									
Turkey								8.4	
Ukraine	0.1				0.0				14.0
United Kingdom	0.1		0.9						
United States		0.5		0.4	0.0	0.9		0.7	16.3
Uzbekistan									100.0
Unknown origin & other varieties	1.8	0.4	9.8		0.4	0.4	0.0	0.3	
Old World subtotal	**1.6**	**0.1**	**14.3**	**0.1**	**0.1**		**0.0**	**0.1**	**1.9**
New World subtotal	**0.0**	**0.2**	**0.0**	**0.6**	**0.0**	**2.7**		**0.3**	**6.7**
World total	**1.5**	**0.1**	**13.6**	**0.1**	**0.2**	**0.1**	**0.0**	**0.1**	**1.9**

Table 23 (cont.): National shares of global winegrape area by varietal country of origin, 2010 (%)

Country of planting → Country of origin↓	*Morocco*	*Myanmar*	*New Zealand2*	*Peru*	*Portugal*	*Romania*	*Russia*
Algeria							
Argentina				0.0			
Armenia							
Australia							
Austria			0.0	0.4	0.1	1.3	
Azerbaijan							
Brazil							
Bulgaria						14.2	2.3
Canada							
Chile							
China				1.1			98.0
Croatia			0.0			7.0	
Cyprus							
Czechia							
France	0.3	0.0	1.7	0.0	0.9	2.0	0.6
Georgia						0.4	1.8
Germany			1.3		0.0	0.3	0.9
Greece	3.4	0.0		0.3	1.3	0.8	0.2
Hungary						0.3	7.7
Israel							
Italy			0.0	0.2	0.2	0.0	0.0
Japan							
Kazakhstan							
Lebanon							
Moldova						71.9	5.1
Montenegro							
Morocco	98.4						
Peru				95.8			
Portugal			0.0		89.6	0.0	
Romania						98.2	
Russia						0.3	97.1
Serbia						0.3	
Slovakia							
Slovenia							
South Africa			1.0				
Spain	0.5	0.0	0.0	0.1	2.2	0.0	
Switzerland			0.0		0.6		0.1
Thailand							
Turkey				0.1			
Ukraine							26.0
United Kingdom							2.2
United States			0.0	0.5			0.2
Uzbekistan							
Unknown origin & other varieties	6.8		0.4		0.6	38.8	0.1
Old World subtotal	**0.8**	**0.0**	**0.7**	**0.1**	**3.8**	**2.0**	**0.6**
New World subtotal			**0.0**	**0.4**			**0.3**
World total	**1.1**	**0.0**	**0.7**	**0.1**	**3.5**	**3.7**	**0.6**

Table 23 (cont.): National shares of global winegrape area by varietal country of origin, 2010 (%)

Country of planting → Country of origin↓	*Serbia*	*Slovakia*	*Slovenia*	*South Africa*	*Spain*	*Switzerland*	*Taiwan*	*Thailand*	*Tunisia*
Algeria					100.0				
Argentina									
Armenia									
Australia									
Austria		7.0	1.0			0.4			
Azerbaijan									
Brazil									
Bulgaria									
Canada									
Chile									
China									
Croatia	30.9	1.5	3.1	0.0	0.7				0.3
Cyprus									
Czechia		29.6							
France		0.1	0.3	5.0	5.5	0.5		0.0	0.1
Georgia									
Germany		1.6	0.6	0.4	0.4	0.7		0.0	
Greece		0.0	0.3	3.0	8.9	0.0		0.0	
Hungary		1.3	1.3	0.1		0.0			
Israel				10.8					
Italy			0.4	0.0	0.1	0.0	0.0	0.0	0.2
Japan							24.3	0.3	
Kazakhstan									
Lebanon					2.1				
Moldova		1.0	0.3						
Montenegro									
Morocco									
Peru					4.2				
Portugal				0.4	8.5				
Romania		1.1							
Russia									
Serbia	99.4								
Slovakia		100.0							
Slovenia		0.4	70.4						
South Africa				97.7					
Spain				0.1	78.1			0.0	0.9
Switzerland	24.0				0.1	35.0			
Thailand								100.0	
Turkey									
Ukraine		1.9							
United Kingdom	33.9				0.0				
United States				3.3	0.1		1.7		
Uzbekistan									
Unknown origin & other varieties	6.4	0.2	1.1	0.0	8.4	0.1			1.5
Old World subtotal	**1.2**	**0.3**	**0.3**	**2.2**	**23.9**	**0.3**	**0.0**	**0.0**	**0.3**
New World subtotal	**1.6**			**5.4**	**0.3**		**1.6**	**0.0**	
World total	**1.5**	**0.3**	**0.4**	**2.2**	**22.3**	**0.3**	**0.1**	**0.0**	**0.4**

Table 23 (cont.): National shares of global winegrape area by varietal country of origin, 2010 (%)

Country of planting →	*Turkey*	*Ukraine*	*UK*	*USA*	*Uruguay*	***Old World***	***New World***	***World***
Country of origin↓								
Algeria						**100.0**	**0.0**	**100**
Argentina						**0.0**	**100.0**	**100**
Armenia		6.3				**100.0**	**0.0**	**100**
Australia						**0.0**	**100.0**	**100**
Austria			0.0	0.0		**99.1**	**0.9**	**100**
Azerbaijan						**100.0**	**0.0**	**100**
Brazil						**0.0**	**100.0**	**100**
Bulgaria						**100.0**	**0.0**	**100**
Canada				64.2		**0.0**	**100.0**	**100**
Chile						**0.0**	**100.0**	**100**
China						**98.0**	**2.0**	**100**
Croatia				18.5		**81.0**	**19.0**	**100**
Cyprus						**100.0**	**0.0**	**100**
Czechia						**100.0**	**0.0**	**100**
France	0.2	1.5	0.0	9.4	0.3	**60.2**	**39.8**	**100**
Georgia		16.2				**100.0**	**0.0**	**100**
Germany	0.0	3.2	0.4	5.3	0.0	**85.2**	**14.8**	**100**
Greece	0.1	0.6		1.9	0.0	**77.8**	**22.2**	**100**
Hungary						**99.8**	**0.2**	**100**
Israel						**89.2**	**10.8**	**100**
Italy	0.0			0.8	0.1	**97.0**	**3.0**	**100**
Japan						**0.0**	**100.0**	**100**
Kazakhstan						**100.0**	**0.0**	**100**
Lebanon						**100.0**	**0.0**	**100**
Moldova						**100.0**	**0.0**	**100**
Montenegro						**100.0**	**0.0**	**100**
Morocco						**98.4**	**1.6**	**100**
Peru						**4.2**	**95.8**	**100**
Portugal			0.0	0.1	0.0	**98.1**	**1.9**	**100**
Romania						**100.0**	**0.0**	**100**
Russia						**100.0**	**0.0**	**100**
Serbia						**100.0**	**0.0**	**100**
Slovakia						**100.0**	**0.0**	**100**
Slovenia						**100.0**	**0.0**	**100**
South Africa				0.1		**0.0**	**100.0**	**100**
Spain	0.0			0.5	0.0	**97.8**	**2.2**	**100**
Switzerland						**98.9**	**1.1**	**100**
Thailand						**0.0**	**100.0**	**100**
Turkey	83.5					**87.2**	**12.8**	**100**
Ukraine		57.9				**100.0**	**0.0**	**100**
United Kingdom				1.1	18.4	**80.4**	**19.6**	**100**
United States		3.4		28.8	0.4	**21.4**	**78.6**	**100**
Uzbekistan						**100.0**	**0.0**	**100**
Unknown origin & other varieties				3.4	0.2	**93.2**	**6.8**	**100**
Old World subtotal	**0.3**	**1.2**	**0.0**	**4.7**	**0.1**	**81.2**	**18.8**	**100**
New World subtotal		**1.4**		**11.9**	**1.0**	**12.8**	**87.2**	**100**
World total	**0.3**	**1.1**	**0.0**	**4.9**	**0.2**	**79.3**	**20.7**	**100**

Table 24: National shares of global winegrape area by varietal country of origin, 2016 (%)

Country of planting → Country of origin↓	*Algeria*	*Argentina*	*Armenia*	*Australia*	*Austria*	*Brazil*	*Bulgaria*	*Cambodia*	*Canada*
Algeria									
Argentina		90.7							
Australia				100.0					
Austria		0.0		0.0	43.8				0.1
Azerbaijan									
Bosnia and Herzegovina									
Brazil						100.0			
Bulgaria							61.8		
Canada						34.7			55.4
Chile									
China									
Croatia		0.0		0.1	4.9	0.3			0.0
Cyprus									
Czechia									
France	0.2	6.6		6.8	0.4	0.7	1.6	0.0	0.5
Georgia				0.0			7.5		
Germany		0.1		2.9	3.6	0.0	0.5		1.6
Greece	0.2	7.8		2.7	0.7	0.6			0.0
Hungary				0.0	0.0	0.0	3.4		0.0
Israel									
Italy		1.2		0.3		0.1	0.1		0.0
Japan								0.1	
Kazakhstan									
Lebanon				0.1					
Moldova									
Montenegro									
Morocco		100.0							
North Macedonia									
Peru									
Portugal		0.0		0.8		0.0			0.0
Romania									
Russia									0.0
Serbia									
Slovenia					18.7				
South Africa				0.0		0.2			0.1
Spain	0.5	0.8		0.3		0.0			0.0
Switzerland									1.7
Taiwan									
Thailand									
Turkey	0.9			0.4					
Ukraine									0.0
United Kingdom				0.5					
United States		0.0		2.1		31.6			1.1
Unknown origin & other varieties		0.0	4.2	0.0	0.3				0.1
Old World subtotal	**0.2**	**3.5**		**3.3**	**1.1**	**0.3**	**1.3**	**0.0**	**0.3**
New World subtotal		**42.1**		**0.8**		**12.6**		**0.0**	**0.4**
World total	**0.2**	**4.6**	**0.3**	**3.0**	**1.0**	**0.7**	**1.2**	**0.0**	**0.3**

Table 24 (cont.): National shares of global winegrape area by varietal country of origin, 2016 (%)

Country of planting → Country of origin↓	*Chile*	*China*	*Croatia*	*Cyprus*	*Czechia*	*Ethiopia*	*France*	*Georgia*	*Germany*
Algeria	100.0								
Argentina	9.2								
Australia									
Austria	0.0		0.8		8.1		1.6		16.0
Azerbaijan									
Bosnia and Herzegovina									
Brazil									
Bulgaria									
Canada									
Chile	100.0								
China	0.0	97.7							
Croatia	0.1	4.5	12.7		1.7		0.0		
Cyprus				100.0					
Czechia									100.0
France	6.8	4.6	0.1		0.2	0.0	29.8	0.0	1.5
Georgia	0.0							61.4	
Germany	0.7	1.4	0.5		2.8		6.4		47.4
Greece	6.1	2.5					8.5		0.2
Hungary							0.0		0.0
Israel									
Italy	0.1	0.3				0.0	16.2		0.4
Japan									
Kazakhstan									
Lebanon							23.5		
Moldova									
Montenegro									
Morocco									
North Macedonia									
Peru									
Portugal	0.0								
Romania									
Russia									
Serbia									
Slovenia									
South Africa									
Spain	1.1	0.4					13.4		
Switzerland	2.2						7.0		11.6
Taiwan									
Thailand									
Turkey						0.0	0.0		
Ukraine									
United Kingdom							30.3		1.1
United States	0.0					0.0	0.4		
Unknown origin & other varieties		22.4					16.7	0.9	0.1
Old World subtotal	**3.5**	**2.3**	**0.3**	**0.1**	**0.3**	**0.0**	**19.0**	**1.1**	**2.4**
New World subtotal	**4.3**	**4.7**				**0.0**	**1.6**		**0.1**
World total	**3.3**	**4.0**	**0.3**	**0.1**	**0.3**	**0.0**	**18.2**	**1.1**	**2.1**

Table 24 (cont.): National shares of global winegrape area by varietal country of origin, 2016 (%)

Country of planting →	*Greece*	*Hungary*	*India*	*Israel*	*Italy*	*Japan*	*Kazakhstan*	*Korea, Rep.*
Country of origin↓								
Algeria								
Argentina								
Australia								
Austria	0.0	18.3			0.2	0.1		
Azerbaijan							100.0	
Bosnia and Herzegovina								
Brazil								
Bulgaria	0.1	0.1						
Canada								
Chile								
China								
Croatia		5.9			27.4			
Cyprus								
Czechia								
France	0.3	0.8	0.1	0.2	6.2	0.0	0.0	
Georgia							5.5	
Germany	0.0	3.2			3.6	0.1	0.1	
Greece	33.9	0.6	0.1	0.2	12.3		0.2	
Hungary		77.9			0.0			
Israel				97.8				
Italy	0.0	0.0	0.1		78.4			
Japan						25.6		72.3
Kazakhstan							100.0	
Lebanon					19.5			
Moldova		0.0					0.3	
Montenegro		0.0						
Morocco								
North Macedonia								
Peru								
Portugal		0.2						
Romania		0.2						
Russia								
Serbia		0.9						
Slovenia		0.6						
South Africa								
Spain	0.0			0.1	0.8			
Switzerland		12.9			0.5			
Taiwan								
Thailand								
Turkey			8.5					
Ukraine		0.1					0.0	
United Kingdom	29.8	0.2			0.3			
United States				0.2		2.7	0.0	1.2
Unknown origin & other varieties	0.5	0.3		0.0	10.4	0.1	0.3	0.2
Old World subtotal	**1.2**	**1.6**	**0.1**	**0.1**	**14.3**	**0.0**	**0.2**	
New World subtotal	**1.4**	**0.0**		**0.1**	**0.0**	**1.8**	**0.0**	**2.9**
World total	**1.1**	**1.4**	**0.1**	**0.1**	**13.5**	**0.1**	**0.2**	**0.1**

Table 24 (cont.): National shares of global winegrape area by varietal country of origin, 2016 (%)

Country of planting → Country of origin↓	*Lebanon*	*Luxembourg*	*Mexico*	*Moldova*	*Morocco*	*Myanmar*	*New Zealand*	*North Macedonia*
Algeria								
Argentina								
Australia								
Austria		0.0		0.0			0.1	
Azerbaijan								
Bosnia and Herzegovina								100.0
Brazil								
Bulgaria				0.0				25.1
Canada								
Chile								
China								
Croatia							0.0	2.0
Cyprus								
Czechia								
France	0.2	0.0	0.1	2.4	0.4	0.0	2.0	0.2
Georgia				6.2				0.6
Germany		0.5		2.5			0.9	0.8
Greece		0.0	0.2	0.0	1.8	0.0	0.0	0.3
Hungary				5.2				
Israel								
Italy			0.0	0.1	0.6		0.0	
Japan								
Kazakhstan								
Lebanon				22.5				
Moldova				52.5				
Montenegro								100.0
Morocco								
North Macedonia								100.0
Peru								
Portugal							0.0	
Romania				2.0				
Russia				16.9				
Serbia								35.2
Slovenia								
South Africa							0.5	
Spain			0.1		0.3	0.0	0.0	
Switzerland				3.7				
Taiwan								
Thailand								
Turkey			7.1		6.1			
Ukraine				28.9				
United Kingdom				3.3				4.6
United States			0.9	8.0	1.2		0.0	
Unknown origin & other varieties	0.2	0.0	0.2	1.5	0.4		0.0	0.0
Old World subtotal	**0.1**	**0.0**	**0.1**	**1.8**	**0.4**	**0.0**	**0.9**	**0.6**
New World subtotal			**0.3**	**2.8**	**0.4**		**0.0**	**0.2**
World total	**0.1**	**0.0**	**0.1**	**1.8**	**0.4**	**0.0**	**0.8**	**0.6**

Table 24 (cont.): National shares of global winegrape area by varietal country of origin, 2016 (%)

Country of planting → Country of origin↓	*Norway*	*Peru*	*Portugal*	*Romania*	*Russia*	*Serbia*	*Slovakia*	*Slovenia*
Algeria								
Argentina		0.0						
Australia								
Austria		0.4	0.1	1.9		1.1	5.5	1.4
Azerbaijan								
Bosnia and Herzegovina								
Brazil								
Bulgaria				10.1	1.8	1.0		
Canada								
Chile								
China		0.0			2.3			
Croatia				2.2		3.1	0.7	4.7
Cyprus								
Czechia								
France		0.0	0.9	2.3	1.5	0.4	0.1	0.3
Georgia				0.6	10.0	0.1		
Germany	0.0		0.0	5.8	2.5	1.3	1.0	0.7
Greece		0.3	2.1	1.3	0.4	0.0		0.5
Hungary				0.5	11.1	0.0		1.6
Israel								
Italy		0.2	0.2	0.0	0.0			0.4
Japan								
Kazakhstan								
Lebanon						34.4		
Moldova				44.1	3.1			
Montenegro						0.0		
Morocco								
North Macedonia								
Peru		100.0						
Portugal			90.8	0.0				
Romania				97.7				
Russia				0.3	81.3	1.5		
Serbia				2.6		61.3		
Slovenia								80.6
South Africa								
Spain		0.1	2.6	0.0				
Switzerland			0.9	1.4	0.2			
Taiwan								
Thailand								
Turkey		0.1						
Ukraine					55.2			
United Kingdom					2.3	8.1		
United States		0.7			2.6	0.0		
Unknown origin & other varieties	0.0		2.5	27.3	0.1	2.3	0.5	0.5
Old World subtotal		**0.1**	**4.4**	**2.2**	**1.2**	**0.3**	**0.2**	**0.4**
New World subtotal		**0.4**			**1.1**	**0.4**		
World total		**0.1**	**4.1**	**4.1**	**1.1**	**0.5**	**0.2**	**0.4**

Table 24 (cont.): National shares of global winegrape area by varietal country of origin, 2016 (%)

Country of planting → / Country of origin↓	*South Africa*	*Spain*	*Switzerland*	*Taiwan*	*Thailand*	*Tunisia*	*Turkey*	*Ukraine*
Algeria								
Argentina								
Australia								
Austria	0.0		0.4					
Azerbaijan								
Bosnia and Herzegovina								
Brazil								
Bulgaria								
Canada								
Chile								
China								
Croatia	0.0	1.6	0.0					
Cyprus								
Czechia								
France	4.6	5.1	0.5		0.0	0.1	0.3	0.8
Georgia								8.0
Germany	0.3	0.5	0.7		0.0		0.0	1.6
Greece	2.5	10.2	0.0		0.0	0.3	0.1	0.3
Hungary	0.2		0.0					
Israel	2.2							
Italy	0.1	0.1	0.0		0.0	0.1	0.0	
Japan				1.7	0.3			
Kazakhstan								
Lebanon								
Moldova								
Montenegro								
Morocco								
North Macedonia								
Peru								
Portugal	0.4	7.5	0.0		0.0			
Romania	0.1							
Russia								
Serbia	0.0							
Slovenia			0.0					
South Africa	99.0		0.0					
Spain	0.1	78.6	0.0		0.0	0.0	0.0	
Switzerland		0.3	57.6					
Taiwan				100.0				
Thailand					100.0			
Turkey		0.0				1.2	75.6	
Ukraine								15.7
United Kingdom		1.6	0.0					
United States	4.6	0.1	0.0	0.1		0.2		2.3
Unknown origin & other varieties	0.0	5.1	0.0		0.0	0.1		
Old World subtotal	**2.1**	**21.8**	**0.4**		**0.0**	**0.1**	**0.3**	**0.6**
New World subtotal	**6.5**	**0.1**	**0.0**	**0.1**	**0.0**	**0.1**		**0.8**
World total	**2.1**	**19.7**	**0.3**	**0.0**	**0.0**	**0.1**	**0.3**	**0.6**

Table 24 (cont.): National shares of global winegrape area by varietal country of origin, 2016 (%)

Country of planting → Country of origin↓	*UK*	*USA*	*Uruguay*	***Old World***	***New World***	***World***
Algeria				**0**	**100**	**100**
Argentina				**0**	**100**	**100**
Australia				**0**	**100**	**100**
Austria		0.1		**99**	**1**	**100**
Azerbaijan				**100**	**0**	**100**
Bosnia and Herzegovina				**100**	**0**	**100**
Brazil				**0**	**100**	**100**
Bulgaria				**100**	**0**	**100**
Canada		10.0		**0**	**100**	**100**
Chile				**0**	**100**	**100**
China				**2**	**98**	**100**
Croatia		28.0		**67**	**33**	**100**
Cyprus				**100**	**0**	**100**
Czechia				**100**	**0**	**100**
France	0.1	10.0	0.2	**57**	**43**	**100**
Georgia				**100**	**0**	**100**
Germany	0.3	5.5	0.0	**86**	**14**	**100**
Greece		2.7	0.0	**74**	**26**	**100**
Hungary				**100**	**0**	**100**
Israel				**98**	**2**	**100**
Italy		0.7	0.2	**97**	**3**	**100**
Japan				**0**	**100**	**100**
Kazakhstan				**100**	**0**	**100**
Lebanon				**100**	**0**	**100**
Moldova				**100**	**0**	**100**
Montenegro				**100**	**0**	**100**
Morocco				**0**	**100**	**100**
North Macedonia				**100**	**0**	**100**
Peru				**0**	**100**	**100**
Portugal		0.2	0.0	**99**	**1**	**100**
Romania				**100**	**0**	**100**
Russia				**100**	**0**	**100**
Serbia				**100**	**0**	**100**
Slovenia				**100**	**0**	**100**
South Africa		0.3		**0**	**100**	**100**
Spain		0.5	0.0	**96**	**4**	**100**
Switzerland				**96**	**4**	**100**
Taiwan				**0**	**100**	**100**
Thailand				**0**	**100**	**100**
Turkey				**84**	**16**	**100**
Ukraine		0.0		**100**	**0**	**100**
United Kingdom		1.5	16.5	**82**	**18**	**100**
United States		39.7	0.2	**15**	**85**	**100**
Unknown origin & other varieties	0.0	2.4	0.2	**74**	**26**	**100**
Old World subtotal	**0.0**	**5.3**	**0.1**	**78.0**	**22.0**	**100**
New World subtotal		**13.1**	**0.9**	**9.0**	**91.0**	**100**
World total	**0.0**	**5.3**	**0.2**	**75**	**25**	**100**

Table 25: Index of internationalization of prime varieties, by country of origin, 2000, 2010 and 2016

	Own-country varieties' share (%) of world winegrape area ***(A)***			*Own-country's share of world winegrape area* ***(B)***			*Index of internationalization of prime varieties* ***(A/B)***		
Country of origin ↓	2000	2010	2016	2000	2010	2016	2000	2010	2016
Croatia	2.95	2.32	1.48	1.20	0.45	0.26	2.46	5.21	5.69
United Kingdom	0.14	0.18	0.17	0.02	0.03	0.04	8.18	6.86	4.21
Greece	2.27	2.32	2.67	1.03	1.17	1.13	2.20	1.99	2.37
France	27.81	37.27	38.90	17.51	17.93	18.03	1.59	2.08	2.16
Georgia	1.74	1.75	1.61	0.76	1.03	1.06	2.30	1.70	1.51
Japan	0.12	0.14	0.12		0.08	0.09		1.71	1.44
Austria	1.47	1.52	1.44	0.98	0.98	1.01	1.50	1.56	1.44
Germany	2.36	2.45	2.55	2.11	2.19	2.09	1.12	1.12	1.22
Spain	27.19	24.75	21.35	23.92	22.06	19.55	1.14	1.12	1.09
Cyprus	0.28	0.13	0.11	0.37	0.18	0.11	0.76	0.70	1.01
Turkey	0.34	0.22	0.26	0.29	0.28	0.30	1.18	0.79	0.87
Italy	12.61	11.82	11.51	12.89	13.43	13.38	0.98	0.88	0.86
Portugal	2.77	2.86	3.16	4.15	3.51	4.04	0.67	0.81	0.78
Hungary	0.79	1.06	1.08	1.76	1.50	1.41	0.45	0.71	0.77
Switzerland	0.28	0.31	0.20	0.30	0.32	0.33	0.92	0.98	0.61
Bulgaria	0.71	0.45	0.60	1.94	1.20	1.17	0.37	0.37	0.51
Moldova	0.52	0.45	0.80	1.82	1.93	1.83	0.28	0.24	0.44
Argentina	1.70	1.66	1.64	4.00	4.58	4.57	0.42	0.36	0.36
Ukraine	0.26	0.26	0.17		1.12	0.56		0.23	0.31
Thailand	0.00	0.00	0.00	0.00	0.00	0.00		0.11	0.26
United States	1.65	1.53	1.15	3.56	4.89	5.30	0.46	0.31	0.22
Brazil	0.00	0.04	0.08	1.07	1.06	0.73	0.00	0.04	0.11
Romania	0.16	0.44	0.43	4.50	3.65	4.04	0.04	0.12	0.11
Serbia	0.31	0.33	0.04	1.40	1.48	0.49	0.22	0.22	0.09
Peru	0.00	0.01	0.01		0.08	0.08		0.09	0.09
South Africa	0.15	0.16	0.18	1.90	2.17	2.12	0.08	0.07	0.08
Slovenia	0.01	0.03	0.03	0.48	0.35	0.35	0.02	0.08	0.07
Russia	0.03	0.07	0.08	1.14	0.55	1.12	0.03	0.12	0.07
Israel	0.01	0.00	0.01	0.10	0.10	0.11	0.08	0.05	0.06
Kazakhstan	0.01	0.01	0.01		0.15	0.15		0.06	0.06
Lebanon	0.04	0.01	0.00		0.05	0.09		0.17	0.05
China	0.00	0.00	0.17		0.63	3.94		0.01	0.04
Taiwan			0.00	0.06	0.06	0.00			0.03
North Macedonia			0.01		0.31	0.55			0.02
Morocco	0.39	0.42	0.01	1.00	1.05	0.39	0.39	0.40	0.01
Canada	0.00	0.00	0.00	0.17	0.22	0.28	0.00	0.01	0.01
Czechia		0.03	0.00	0.23	0.35	0.30		0.10	0.00
Australia	0.00	0.00	0.00	2.64	3.26	2.93	0.00	0.00	0.00
Chile	0.00	0.00	0.00	2.31	2.39	3.23	0.00	0.00	0.00
Algeria	0.00	0.00	0.00	0.61	0.65	0.18	0.00	0.00	0.00
Armenia	0.08	0.09		0.23	0.24	0.33	0.35	0.35	
Slovakia		0.00		0.32	0.27	0.17		0.01	

Table 26: Index of internationalization of national varietal choice, by country of planting, 2000, 2010 and 2016

	% of country's area in rest of world's varieties ***(A)***			*% of rest of world's varieties in world winegrape area* ***(B)***			*Index of internationalization of national varietal choice* ***(A/B)***		
Country of planting↓	2000	2010	2016	2000	2010	2016	2000	2010	2016
Algeria	100.0	100.0	100.0	100.0	100.0	100.0	1.00	1.00	1.00
Argentina	58.0	64.3	67.7	98.3	98.3	98.4	0.59	0.65	0.69
Armenia	67.1	67.1	100.0	99.9	99.9	100.0	0.67	0.67	1.00
Australia	99.9	100.0	99.9	100.0	100.0	100.0	1.00	1.00	1.00
Austria	35.7	37.1	37.6	98.5	98.5	98.6	0.36	0.38	0.38
Azerbaijan	100.0	100.0	100.0	100.0	100.0	100.0	1.00	1.00	1.00
Brazil	100.0	96.8	89.2	100.0	100.0	99.9	1.00	0.97	0.89
Bulgaria	63.9	69.5	68.6	99.3	99.6	99.4	0.64	0.70	0.69
Canada	99.8	99.8	99.2	100.0	100.0	100.0	1.00	1.00	0.99
Chile	99.9	100.0	100.0	100.0	100.0	100.0	1.00	1.00	1.00
China	100.0		95.8						
Croatia	37.0	47.8	28.4	97.0	97.7	98.5	0.38	0.49	0.29
Cyprus	25.0	30.8	0.0	99.7	99.9	99.9	0.25	0.31	0.00
Czechia	100.0	93.1	100.0	100.0	100.0	100.0	1.00	0.93	1.00
France	38.7	33.5	36.1	72.2	62.7	61.1	0.54	0.53	0.59
Georgia	8.0	8.0	8.0	98.3	98.2	98.4	0.08	0.08	0.08
Germany	33.8	42.0	42.7	97.6	97.6	97.4	0.35	0.43	0.44
Greece	19.1	26.6	20.2	97.7	97.7	97.3	0.20	0.27	0.21
Hungary	76.4	62.5	58.1	99.2	98.9	98.9	0.77	0.63	0.59
Israel	95.8	95.8	94.5	100.0	100.0	100.0	0.96	0.96	0.95
Italy	23.0	31.5	33.6	87.4	88.2	88.5	0.26	0.36	0.38
Japan	100.0	80.1	63.4	99.9	99.9	99.9	1.00	0.80	0.64
Kazakhstan	100.0	94.4	94.4	100.0	100.0	100.0	1.00	0.94	0.94
Lebanon	100.0	100.0	100.0	100.0	100.0	100.0	1.00	1.00	1.00
Luxembourg	100.0	100.0	100.0	100.0	100.0	100.0	1.00	1.00	1.00
Mexico	100.0	100.0	100.0	100.0	100.0	100.0	1.00	1.00	1.00
Moldova	95.1	95.1	77.4	99.5	99.5	99.2	0.96	0.96	0.78
Montenegro	100.0	100.0	100.0	100.0	100.0	99.8	1.00	1.00	1.00
Morocco	61.8	61.4	100.0	99.6	99.6	100.0	0.62	0.62	1.00
Myanmar			100.0			100.0	1.00	1.00	1.00
New Zealand	100.0	100.0	100.0	100.0	100.0	100.0	1.00	1.00	1.00
North Macedonia			98.4			100.0			0.98
Peru	100.0	91.2	91.2	100.0	100.0	100.0	1.00	0.91	0.91
Portugal	43.2	27.8	29.4	97.2	97.1	96.8	0.44	0.29	0.30
Romania	97.0	88.3	89.8	99.8	99.6	99.6	0.97	0.89	0.90
Russia	97.8	88.5	94.2	100.0	99.9	99.9	0.98	0.89	0.94
South Africa	92.3	93.1	91.8	99.9	99.8	99.8	0.92	0.93	0.92
Serbia	78.0	78.0	94.5	99.7	99.7	100.0	0.78	0.78	0.95
Slovakia	100.0	98.6	100.0	100.0	100.0	100.0	1.00	0.99	1.00
Slovenia	100.0	94.4	94.2	100.0	100.0	100.0	1.00	0.94	0.94
Spain	13.8	13.2	14.8	72.8	75.3	78.6	0.19	0.18	0.19
Switzerland	62.6	66.1	65.0	99.7	99.7	99.8	0.63	0.66	0.65
Thailand	100.0	89.1	74.0	100.0	100.0	100.0	1.00	0.89	0.74
Turkey	100.0	34.7	35.1	99.7	99.8	99.7	1.00	0.35	0.35
United Kingdom	100.0	100.0	100.0	99.9	99.8	99.8	1.00	1.00	1.00
Ukraine	100.0	86.5	95.0	99.7	99.7	99.8	1.00	0.87	0.95
United States	88.9	91.1	91.4	98.3	98.5	98.8	0.90	0.92	0.93
Uzbekishtan	100.0	100.0	100.0	100.0	100.0	100.0	1.00	1.00	1.00

Table 27: Share of database's regions growing top 100 varieties, 2000, 2010 and 2016 (%)

	2000		2010		2016
Cabernet Sauvignon	40	Cabernet Sauvignon	42	Cabernet Sauvignon	31
Chardonnay	39	Chardonnay	42	Chardonnay	31
Merlot	37	Merlot	40	Merlot	30
Pinot Noir	31	Sauvignon Blanc	35	Sauvignon Blanc	27
Sauvignon Blanc	31	Pinot Noir	34	Syrah	27
Syrah	27	Syrah	33	Pinot Noir	25
Cabernet Franc	27	Cabernet Franc	28	Riesling	22
Riesling	26	Pinot Gris	26	Cabernet Franc	21
Muscat Blanc à Petits Grains	22	Riesling	25	Pinot Gris	20
Garnacha Tinta	20	Muscat Blanc à Petits Grains	22	Côt	19
Sangiovese	19	Côt	19	Viognier	16
Barbera	19	Viognier	18	Muscat Blanc à Petits Grains	14
Côt	18	Sangiovese	18	Tempranillo	14
Sémillon	18	Gewürztraminer	18	Gewürztraminer	13
Gewürztraminer	18	Garnacha Tinta	17	Sémillon	13
Trebbiano Toscano	18	Pinot Blanc	17	Petit Verdot	13
Muscat of Alexandria	17	Sémillon	17	Sangiovese	13
Alicante Henri Bouschet	16	Barbera	16	Garnacha Tinta	11
Pinot Gris	16	Muscat of Alexandria	16	Pinot Blanc	11
Pinot Blanc	16	Petit Verdot	15	Muscat of Alexandria	10
Chenin Blanc	15	Tempranillo	15	Barbera	10
Nebbiolo	13	Trebbiano Toscano	14	Tannat	9
Douce Noire	13	Alicante Henri Bouschet	13	Chenin Blanc	9
Mazuelo	12	Chenin Blanc	12	Monastrell	8
Tempranillo	12	Tribidrag	12	Müller-Thurgau	8
Tribidrag	11	Nebbiolo	11	Nebbiolo	8
Müller-Thurgau	11	Carmenère	9	Tribidrag	7
Viognier	11	Mazuelo	9	Alicante Henri Bouschet	7
Malvasia Bianca di Candia	10	Monastrell	9	Trebbiano Toscano	7
Monastrell	10	Müller-Thurgau	9	Savagnin Blanc	7
Manzoni Bianco	10	Moscato Giallo	8	Blaufränkisch	6
Sauvignonasse	10	Douce Noire	8	Muscat of Hamburg	6
Greco Nero	10	Graševina	8	Durif	6
Pignoletto	10	Chasselas	8	Mazuelo	6
Gamay Noir	9	Montepulciano	8	Graševina	6
Petit Verdot	9	Muscat of Hamburg	8	Douce Noire	5
Muscat of Hamburg	9	Gamay Noir	8	Verdelho	5
Palomino Fino	9	Malvasia Bianca di Candia	7	Grüner Veltliner	5
Montepulciano	9	Tannat	7	Zweigelt	5
Verdicchio Bianco	9	Blaufränkisch	7	Pinot Meunier	5
Malvasia Bianca Lunga	8	Sauvignonasse	7	Gamay Noir	5
Vermentino	8	Trebbiano Giallo	7	Marsanne	5
Malvasia Nera di Brindisi	8	Biancame	7	Torrontés Riojano	5
Graciano	8	Dolcetto	7	Vermentino	5
Moscato Giallo	8	Roussanne	7	Blauer Portugieser	5
Trebbiano Giallo	8	Malvasia Bianca Lunga	6	Carmenère	5
Aglianico	8	Vermentino	6	Chasselas	5
Albarola	8	Ciliegiolo	6	Montepulciano	5
Fortana	8	Malvasia Nera di Brindisi	6	Roussanne	4
Graševina	8	Savagnin Blanc	6	Touriga Nacional	4

Table 27 (cont.): Share of database's regions growing top 100 varieties, 2000 and 2016 (%)

2000		2010		2016	
Ciliegiolo	8	Ancellotta	6	Muscat Ottonel	4
Trebbiano Romagnolo	8	Verdicchio Bianco	6	Silvaner	4
Avarengo	7	Marsanne	6	Cereza	4
Malvasia di Lipari	7	Blauer Portugieser	6	Listan Prieto	4
Afus Ali	7	Palomino Fino	5	Palomino Fino	4
Damaschino	7	Garganega	5	Dolcetto	4
Malvasia di Candia Aromatic	7	Afus Ali	5	Muscat Blanc à Petits Grains	4
Aleatico	7	Graciano	5	Cinsaut	4
Ancellotta	7	Cinsaut	5	Sauvignonasse	4
Pinot Meunier	7	Verdelho	5	Graciano	3
Biancame	7	Colombard	5	Ancellotta	3
Colombard	7	Nero d'Avola	5	Colombard	3
Listan Prieto	7	Zweigelt	5	Moscato Giallo	3
Bombino Bianco	7	Bellone	5	Sankt Laurent	3
Chasselas	7	Torrontés Riojano	5	Pedro Giménez	3
Dolcetto	7	Durif	5	Ruby Cabernet	3
Garganega	7	Listan Prieto	5	Muscat Blanc à Petits Grains	3
Greco	7	Silvaner	5	Nero d'Avola	3
Greco Bianco	7	Trebbiano Romagnolo	4	Garnacha Blanca	3
Inzolia	7	Barbera Bianca	4	Regent	3
Malvasia del Lazio	7	Cereza	4	Dornfelder	3
San Giuseppe Nero	7	Fortana	4	Seyval Blanc	3
Cinsaut	6	Aglianico	4	Fiano	3
Korinthiaki	6	Malvasia di Candia Aromatica	4	Kerner	3
Albana	6	Muscat Blanc à Petits Grains (	4	Aglianico	3
Marsanne	6	Trebbiano d'Abruzzo	4	Arneis	3
Torrontés Riojano	6	Albarola	4	Criolla Grande	3
Barbera Bianca	6	Greco Bianco	4	Aspiran Bouschet	2
Bellone	6	Lambrusco Maestri	4	Blauburger	2
Blauer Portugieser	6	Muscat Ottonel	4	Irsai Olivér	2
Ruby Cabernet	6	Arneis	4	Malvasia Bianca di Candia	2
Uva Rara	6	Greco Nero	4	Prosecco	2
Cereza	6	Grüner Veltliner	4	Scheurebe	2
Nero d'Avola	6	Lambrusco Grasparossa	4	Chambourcin	2
Freisa	6	Lambrusco Salamino	4	Alvarinho	2
Lambrusco Grasparossa	6	Ruby Cabernet	4	Damaschino	2
Malvazija Istarska	6	Trebbianina	4	Torrontés Sanjuanino	2
Prosecco	6	Aleatico	4	Solaris	2
Silvaner	6	Prosecco	4	Furmint	2
Trebbiano d'Abruzzo	6	Lambrusca di Alessandria	4	Greco Nero	2
Canaiolo Nero	6	Marzemino	4	Vinhao	2
Abbuoto	6	Montonico Bianco	4	Bianca	2
Caddiu	6	San Giuseppe Nero	4	Gamaret	2
Marzemino	6	Albana	4	Maréchal Foch	2
Bonamico	6	Garnacha Blanca	4	Johanniter	2
Catarratto Bianco	6	Malvazija Istarska	4	Lambrusco Maestri	2
Lambrusco Maestri	6	Manzoni Bianco	4	Marselan	2
Trebbiano Modenese	6	Bombino Bianco	4	Teroldego	2
Drupeggio	6	Chambourcin	4	Aligoté	2
Neretta Cuneese	6	Croatina	4	Pinotage	2

Table 28: Shares of national top, top 3 and top 10 varieties, by winegrape area, 2000, 2010 and 2016 (%)

	Top variety			*Top 3*			Top 10		
Country	*2000*	*2010*	*2016*	*2000*	*2010*	*2016*	*2000*	*2010*	*2016*
Algeria	25	25	36	70	70	72	100	100	100
Argentina	16	14	20	37	37	43	77	78	80
Armenia	22	22	70	40	40	100	60	60	100
Australia	22	28	29	55	64	64	82	88	88
Austria	36	30	32	54	51	53	85	80	80
Brazil	27	37	35	60	61	62	76	86	82
Bulgaria	24	19	19	46	46	50	82	78	91
Cambodia			30			80			100
Canada	12	13	18	31	34	39	70	74	74
Chile	32	37	29	56	59	48	83	93	86
China		77	23		93	38		100	52
Croatia	27	23	38	50	38	66	67	60	100
Cyprus	60	42	62	75	71	100	75	91	100
Czechia	15	10	11	40	27	31	77	68	81
Ethiopia		53	53		93	93		100	100
France	12	14	13	34	35	33	72	72	67
Georgia	53	53	53	79	79	79	92	92	92
Germany	21	22	23	50	47	47	80	78	78
Greece	25	18	20	44	34	43	68	59	71
Hungary	8	11	11	20	24	26	42	50	54
India			37			74			100
Israel	20	20	20	46	46	53	82	82	86
Italy	10	11	11	24	23	23	47	45	47
Japan		22	18		51	46		95	78
Kazakhstan		51	51		67	67		86	86
KoreaRep	50	50	50	83	83	83	85	85	85
Lebanon			25			63			83
Luxembourg	34	28	24	60	57	54	100	100	99
Mexico		15	15		37	37		72	72
Missing9	25			47			70		
Moldova	18	18	15	43	43	34	90	90	71
Morocco	33	34	19	49	49	49	68	69	82
Myanmar		36	39		75	80		100	100
New Zealand	28	51	58	63	78	82	91	94	98
North Macedonia			38			70			90
Norway			60			90			90
Peru		33	33		68	68		98	98
Portugal	7	10	9	18	24	24	37	55	55
Romania	8	8	7	19	22	20	34	41	38
Russia	23	14	17	33	36	41	54	68	77
Serbia	48	48	10	75	75	28	79	79	57
Slovakia	25	17	21	56	41	46	82	73	78
Slovenia	15	14	12	27	30	28	42	65	63
South Africa	24	18	18	46	42	42	84	87	86
Spain	33	25	23	50	53	52	75	79	78
Switzerland	36	30	28	79	67	64	95	87	84
Taiwan	46	46	63	96	96	100	100	100	100
Thailand		44	36		67	71		98	96
Tunisia	45	45	19	62	62	49	80	80	82
Turkey		14	18		36	40		78	80
Ukraine		22	23		50	62		84	96
United Kingdom	13	20	30	36	49	70	74	80	91
United States	20	18	17	41	43	44	79	77	77
Uruguay	33	24	26	72	55	55	87	86	84
Old World subtotal	**10**	**7**	**6**	**19**	**19**	**18**	**37**	**40**	**40**
New World subtotal	**13**	**17**	**16**	**29**	**36**	**32**	**55**	**63**	**59**
World total	**8**	**6**	**7**	**17**	**18**	**18**	**36**	**42**	**42**

Table 29: Bearing areas of fungus-resistant grapevine varieties (PIWIs, or Pilzwiderstandsfähige Sorten, in German), by country of planting, 2016 (hectares)

colour	Variety	Australia	France	Germany	Hungary	Italy	NZ	Norway	S. Africa	Switz.	UK	USA	World
R	Allegro												**0.0**
W	Aromera												**0.0**
R	Baron			1.0						0.1			**1.1**
W	Bronner			4.0		1.7				0.5			**6.2**
R	Bundessortenamt												**0.0**
W	Cabernet Blanc									6.2			**6.2**
R	Cabernet Cantor									1.4			**1.4**
R	Cabernet Carbon			10.0						0.9			**10.9**
R	Cabernet Carol			6.0						0.1			**6.1**
R	Cabernet Cortis			34.0						4.2			**38.2**
R	Chambourcin	41.5	592.4				3.4		13.0	2.4		315.1	**967.8**
R	Divico									9.9			**9.9**
W	Felicia												**0.0**
W	Golden Muscat									49.6		0.4	**50.0**
W	Helios			6.0						0.2			**6.2**
W	Johanniter			92.0						18.6			**110.6**
R	Monarch			8.0						1.6			**9.6**
W	Muscaris									4.2			**4.2**
R	Muscat Bleu									2.7			**2.7**
	Pinot Cortis												**0.0**
R	Prior			13.0						1.2			**14.2**
R	Rathay												**0.0**
R	Reberger			2.0									**2.0**
R	Regent			1902.0	4.2	0.9				38.1	25.7	2.4	**1973.3**
R	Rondo			9.0				3.8		0.8	36.9		**50.5**
W	Sauvignac									2.1			**2.1**
W	Solaris			91.0	0.2			7.5		19.6			**118.3**
W	Sorila									0.1			**0.1**
W	Souvignier Gris									3.0			**3.0**
W	Villaris												**0.0**
R	Vinera												**0.0**
R	Vinorè												**0.0**
	Total	**41.5**	**592.4**	**2178.0**	**4.4**	**2.6**	**3.4**	**11.3**	**13.0**	**167.5**	**62.6**	**317.9**	**3394.6**

III. Winegrape areas for world's top varieties, by country

Table 30: National ranking of top 20 countries, 24 top red varieties, 2000, 2010 and 2016 (ha and %)

Cabernet Sauvignon	2000	bearing area (ha)	% of world	2010	bearing area (ha)	% of world	2016	bearing area (ha)	% of world
	1 France	53413	23.9	1 France	54434	18.8	1 France	46555	15.0
	2 Chile	35967	16.1	2 Chile	40728	14.0	2 Chile	42409	13.7
	3 Australia	24997	11.2	3 United States	34788	12.0	3 United States	40837	13.1
	4 United States	17573	7.9	4 Australia	25967	9.0	4 China	40300	13.0
	5 Argentina	13776	6.2	5 Spain	23237	8.0	5 Australia	23987	7.7
	6 Bulgaria	10441	4.7	6 China	22612	7.8	6 Spain	20139	6.5
	7 South Africa	8824	4.0	7 Argentina	17674	6.1	7 Argentina	15356	4.9
	8 Romania	8620	3.9	8 Italy	13724	4.7	8 Italy	14240	4.6
	9 Italy	7682	3.4	9 South Africa	12325	4.2	9 South Africa	10589	3.4
	10 Moldova	7590	3.4	10 Bulgaria	8436	2.9	10 Bulgaria	9327	3.0
	11 Spain	4519	2.0	11 Moldova	7590	2.6	11 Russia	8528	2.7
	12 Russia	1578	0.7	12 Ukraine	4869	1.7	12 Moldova	8169	2.6
	13 Algeria	1510	0.7	13 Romania	3718	1.3	13 Romania	5359	1.7
	14 Hungary	1052	0.5	14 Russia	3593	1.2	14 Ukraine	4935	1.6
	15 Greece	688	0.3	15 Hungary	2863	1.0	15 Hungary	2677	0.9
	16 Uruguay	675	0.3	16 Portugal	1671	0.6	16 Portugal	2346	0.8
	17 New Zealand	654	0.3	17 Greece	1550	0.5	17 Serbia	2111	0.7
	18 Israel	607	0.3	18 Algeria	1510	0.5	18 Greece	1929	0.6
	19 Brazil	587	0.3	19 Brazil	914	0.3	19 N. Macedonia	1020	0.3
	20 Canada	569	0.3	20 Mexico	756	0.3	20 Algeria	1000	0.3
	Others	21753	9.8	Others	7124	2.5	Others	8859	2.9
	Total	223074	100.0	Total	290083	100.0	Total	310671	100.0

Merlot	2000	bearing area (ha) (ha)	% of world	2010	bearing area (ha)	% of world	2016	bearing area (ha)	% of world
	1 France	101309	47.5	1 France	114675	42.8	1 France	108483	40.7
	2 Italy	21861	10.2	2 Italy	28042	10.5	2 Italy	24057	9.0
	3 United States	16875	7.9	3 United States	22729	8.5	3 United States	21251	8.0
	4 Chile	12825	6.0	4 Spain	15540	5.8	4 China	16700	6.3
	5 Bulgaria	11169	5.2	5 Romania	10988	4.1	5 Spain	12852	4.8
	6 Moldova	8123	3.8	6 Bulgaria	10573	3.9	6 Chile	12057	4.5
	7 Romania	7810	3.7	7 Chile	10041	3.7	7 Romania	11647	4.4
	8 Australia	7669	3.6	8 Australia	10028	3.7	8 Bulgaria	10050	3.8
	9 Argentina	6263	2.9	9 Moldova	8123	3.0	9 Australia	8415	3.2
	10 South Africa	4888	2.3	10 Argentina	6953	2.6	10 Moldova	7689	2.9
	11 Algeria	1510	0.7	11 South Africa	6497	2.4	11 Argentina	5632	2.1
	12 Slovenia	1197	0.6	12 China	3560	1.3	12 South Africa	5558	2.1
	13 Spain	1186	0.6	13 Ukraine	2820	1.1	13 Russia	2988	1.1
	14 Uruguay	1057	0.5	14 Hungary	1907	0.7	14 Serbia	1968	0.7
	15 Switzerland	848	0.4	15 Russia	1588	0.6	15 Hungary	1961	0.7
	16 Canada	674	0.3	16 Algeria	1510	0.6	16 Ukraine	1400	0.5
	17 New Zealand	657	0.3	17 New Zealand	1369	0.5	17 Greece	1393	0.5
	18 Israel	647	0.3	18 Greece	1248	0.5	18 N. Macedonia	1240	0.5
	19 Hungary	486	0.2	19 Switzerland	1028	0.4	19 New Zealand	1239	0.5
	20 Brazil	469	0.2	20 Canada	999	0.4	20 Switzerland	1124	0.4
	Others	5844	2.7	Others	7671	2.9	Others	8737	3.3
	Total	213368	100.0	Total	267888	100.0	Total	266440	100.0

Table 30 (cont.): National ranking of top 20 countries, 24 top red varieties, 2000, 2010 and 2016 (ha and %)

Tempranillo	**2000**	bearing area (ha)	% of world	**2010**	bearing area (ha)	% of world	**2016**	bearing area (ha)	% of world
1	Spain	79310	84.9	Spain	207677	89.1	Spain	193597	88.2
2	Portugal	7356	7.9	Portugal	16706	7.2	Portugal	17014	7.8
3	Argentina	4720	5.1	Argentina	6565	2.8	Argentina	6140	2.8
4	France	1549	1.7	France	680	0.3	Australia	681	0.3
5	United States	201	0.2	Australia	476	0.2	France	658	0.3
6	Australia	41	0.0	United States	414	0.2	United States	626	0.3
7	South Africa	14	0.0	Mexico	229	0.1	Mexico	229	0.1
8	Italy	7	0.0	Romania	70	0.0	Chile	127	0.1
9	Chile	1	0.0	Chile	48	0.0	South Africa	92	0.0
10				South Africa	38	0.0	Romania	67	0.0
11				Italy	23	0.0	Israel	55	0.0
12				Greece	19	0.0	Brazil	23	0.0
13				Brazil	16	0.0	Greece	22	0.0
14				Turkey	9	0.0	New Zealand	18	0.0
15				New Zealand	7	0.0	Italy	9	0.0
16				Canada	6	0.0	Canada	6	0.0
17				Thailand	4	0.0	Turkey	6	0.0
18				Myanmar	3	0.0	Thailand	4	0.0
19							Myanmar	4	0.0
20							Switzerland	0	0.0
	Others	171	0.2	Others			Others		
	Total	93370	100.0	Total	232988	100.0	Total	219379	100.0

Syrah	**2000**	bearing area (ha)	% of world	**2010**	bearing area (ha)	% of world	**2016**	bearing area (ha)	% of world
1	France	50676	49.4	France	67382	36.3	France	62211	34.3
2	Australia	29295	28.6	Australia	42675	23.0	Australia	38942	21.5
3	Argentina	8888	8.7	Spain	20000	10.8	Spain	19488	10.8
4	South Africa	5631	5.5	Argentina	13093	7.0	Argentina	12707	7.0
5	Chile	2040	2.0	South Africa	10136	5.5	South Africa	9946	5.5
6	Algeria	1510	1.5	United States	9197	4.9	United States	9083	5.0
7	United States	1509	1.5	Italy	6739	3.6	Chile	7994	4.4
8	Italy	1025	1.0	Chile	6027	3.2	Italy	7693	4.2
9	Tunisia	337	0.3	Portugal	3501	1.9	Portugal	4017	2.2
10	Spain	86	0.1	Algeria	1510	0.8	Turkey	1439	0.8
11	Uruguay	62	0.1	Turkey	1367	0.7	Greece	1042	0.6
12	New Zealand	60	0.1	Greece	641	0.3	China	1000	0.6
13	Switzerland	54	0.1	Romania	470	0.3	Algeria	1000	0.6
14	Greece	39	0.0	Tunisia	337	0.2	Bulgaria	804	0.4
15				New Zealand	293	0.2	Romania	504	0.3
16				Canada	274	0.1	India	500	0.3
17				Cyprus	244	0.1	New Zealand	436	0.2
18				China	223	0.1	Israel	385	0.2
19				Croatia	187	0.1	Morocco	347	0.2
20				Switzerland	181	0.1	Lebanon	300	0.2
	Others	1278	1.2	Others	1346	0.7	Others	1346	0.7
	Total	102490	100.0	Total	185822	100.0	Total	181185	100.0

Table 30 (cont.): National ranking of top 20 countries, 24 top red varieties, 2000, 2010 and 2016 (ha and %)

Garnacha Tinta	2000	bearing area (ha)	% of world	2010	bearing area (ha)	% of world	2016	bearing area (ha)	% of world
	1 Spain	98131	45.4	1 France	90991	50.1	1 France	78631	52.4
	2 France	95717	44.2	2 Spain	70202	38.7	2 Spain	54606	36.4
	3 Italy	6781	3.1	3 Italy	6372	3.5	3 Italy	5421	3.6
	4 Algeria	6040	2.8	4 Algeria	6040	3.3	4 China	4000	2.7
	5 United States	4519	2.1	5 United States	2666	1.5	5 United States	2213	1.5
	6 Australia	2139	1.0	6 Tunisia	2020	1.1	6 Algeria	2000	1.3
	7 Tunisia	2020	0.9	7 Australia	1748	1.0	7 Australia	1492	1.0
	8 Morocco	802	0.4	8 Morocco	802	0.4	8 Morocco	786	0.5
	9 South Africa	40	0.0	9 South Africa	187	0.1	9 South Africa	344	0.2
	10 Greece	22	0.0	10 Mexico	140	0.1	10 Chile	187	0.1
	11 Argentina	9	0.0	11 Croatia	103	0.1	11 Tunisia	152	0.1
	12			12 Portugal	84	0.0	12 Mexico	140	0.1
	13			13 Cyprus	84	0.0	13 Portugal	60	0.0
	14			14 Chile	37	0.0	14 Turkey	33	0.0
	15			15 Turkey	33	0.0	15 Argentina	22	0.0
	16			16 Argentina	22	0.0	16 Uruguay	4	0.0
	17			17 China	11	0.0	17 New Zealand	1	0.0
	18			18 Uruguay	5	0.0	18 Canada	1	0.0
	19			19 New Zealand	2	0.0	19 Switzerland	1	0.0
	20			20 Canada	2	0.0	20 Peru	1	0.0
	Others	129	0.1	Others	2	0.0	Others		
	Total	216349	100.0	Total	181553	100.0	Total	150096	100.0

Pinot Noir	2000	bearing area (ha)	% of world	2010	bearing area (ha)	% of world	2016	bearing area (ha)	% of world
	1 France	26526	38.5	1 France	30086	30.5	1 France	31602	30.0
	2 Germany	8643	12.6	2 United States	16776	17.0	2 United States	22998	21.8
	3 Moldova	6521	9.5	3 Germany	11724	11.9	3 Germany	11184	10.6
	4 United States	5343	7.8	4 Moldova	6521	6.6	4 New Zealand	5514	5.2
	5 Switzerland	4601	6.7	5 Italy	5046	5.1	5 Italy	5057	4.8
	6 Italy	3287	4.8	6 New Zealand	4776	4.8	6 Australia	4806	4.6
	7 Australia	3223	4.7	7 Australia	4690	4.8	7 Switzerland	4209	4.0
	8 Romania	1740	2.5	8 Switzerland	4402	4.5	8 Chile	4091	3.9
	9 Chile	1614	2.3	9 Chile	2884	2.9	9 Moldova	2366	2.2
	10 Algeria	1510	2.2	10 Argentina	1689	1.7	10 Romania	1930	1.8
	11 Argentina	1114	1.6	11 Algeria	1510	1.5	11 Argentina	1866	1.8
	12 New Zealand	1098	1.6	12 Hungary	1091	1.1	12 South Africa	1153	1.1
	13 Bulgaria	769	1.1	13 Romania	1089	1.1	13 Hungary	1092	1.0
	14 South Africa	487	0.7	14 Spain	1044	1.1	14 Spain	969	0.9
	15 Canada	457	0.7	15 South Africa	962	1.0	15 Russia	918	0.9
	16 Spain	417	0.6	16 Ukraine	767	0.8	16 Czechia	697	0.7
	17 Austria	409	0.6	17 Czechia	688	0.7	17 Canada	639	0.6
	18 Hungary	243	0.4	18 Austria	646	0.7	18 Serbia	633	0.6
	19 Luxembourg	66	0.1	19 Canada	640	0.6	19 Austria	614	0.6
	20			20 Russia	533	0.5	20 United Kingdom	546	0.5
	Others	741	1.1	Others	1058	1.1	Others	2597	2.5
	Total	68810	100.0	Total	98623	100.0	Total	105480	100.0

Table 30 (cont.): National ranking of top 20 countries, 24 top red varieties, 2000, 2010 and 2016 (ha and %)

Sangiovese	2000	bearing area (ha)	% of world	2010	bearing area (ha)	% of world	2016	bearing area (ha)	% of world
1	Italy	62761	91.1	Italy	71619	91.8	Italy	68428	93.1
2	Argentina	2490	3.6	Argentina	2231	2.9	Argentina	1837	2.5
3	France	1564	2.3	France	1511	1.9	France	1503	2.0
4	Tunisia	842	1.2	United States	852	1.1	United States	827	1.1
5	United States	682	1.0	Tunisia	842	1.1	Australia	430	0.6
6	Australia	372	0.5	Australia	589	0.8	Chile	152	0.2
7	Chile	123	0.2	Chile	100	0.1	Ethiopia	90	0.1
8	South Africa	35	0.1	Ethiopia	90	0.1	Romania	88	0.1
9				Romania	88	0.1	South Africa	70	0.1
10				South Africa	61	0.1	Turkey	18	0.0
11				Brazil	26	0.0	New Zealand	8	0.0
12				Turkey	9	0.0	Canada	4	0.0
13				New Zealand	6	0.0	Brazil	3	0.0
14				Canada	3	0.0	Spain	2	0.0
15				Thailand	2	0.0	Thailand	2	0.0
16				Hungary	1	0.0	Hungary	1	0.0
17							Switzerland	0	0.0
18									
19									
20									
	Others	8	0.0	Others			Others		
	Total	68877	100.0	Total	78030	100.0	Total	73464	100.0

Bobal	2000	bearing area (ha)	% of world	2010	bearing area (ha)	% of world	2016	bearing area (ha)	% of world
1	Spain	100128	100.0	Spain	80120	100.0	Spain	59189	100.0
2									
3									
4									
5									
6									
7									
8									
9									
10									
11									
12									
13									
14									
15									
16									
17									
18									
19									
20									
	Others			Others			Others		
	Total	100128	100.0	Total	80120	100.0	Total	59189	100.0

Table 30 (cont.): National ranking of top 20 countries, 24 top red varieties, 2000, 2010 and 2016 (ha and %)

Cabernet Franc	**2000**	bearing area (ha)	% of world	**2010**	bearing area (ha)	% of world	**2016**	bearing area (ha)	% of world
	1 France	36094	69.4	1 France	36302	59.2	1 France	32327	57.7
	2 Italy	6639	12.8	2 Brazil	8516	13.9	2 Brazil	6834	12.2
	3 Brazil	3784	7.3	3 Italy	6314	10.3	3 Italy	5590	10.0
	4 United States	1189	2.3	4 United States	2215	3.6	4 United States	2199	3.9
	5 Australia	744	1.4	5 Hungary	1352	2.2	5 Chile	1578	2.8
	6 Chile	689	1.3	6 Chile	1321	2.2	6 Hungary	1368	2.4
	7 Canada	567	1.1	7 South Africa	934	1.5	7 Argentina	929	1.7
	8 Hungary	526	1.0	8 Spain	849	1.4	8 South Africa	835	1.5
	9 South Africa	488	0.9	9 Canada	664	1.1	9 Canada	820	1.5
	10 Uruguay	364	0.7	10 Argentina	592	1.0	10 Moldova	756	1.3
	11 Argentina	252	0.5	11 Australia	591	1.0	11 Spain	680	1.2
	12 New Zealand	118	0.2	12 China	507	0.8	12 China	600	1.1
	13 Greece	40	0.1	13 Uruguay	334	0.5	13 Australia	328	0.6
	14 Spain	30	0.1	14 Cyprus	203	0.3	14 Uruguay	266	0.5
	15 Austria	27	0.1	15 New Zealand	163	0.3	15 Bulgaria	240	0.4
	16 Switzerland	16	0.0	16 Croatia	95	0.2	16 Israel	110	0.2
	17			17 Romania	73	0.1	17 New Zealand	109	0.2
	18			18 Kazakhstan	56	0.1	18 Serbia	79	0.1
	19			19 Austria	56	0.1	19 Romania	72	0.1
	20			20 Switzerland	54	0.1	20 Austria	64	0.1
	Others	409	0.8	Others	105	0.2	Others	268	0.5
	Total	51974	100.0	Total	61295	100.0	Total	56052	100.0

Côt	**2000**	bearing area (ha)	% of world	**2010**	bearing area (ha)	% of world	**2016**	bearing area (ha)	% of world
	1 Argentina	18230	69.4	1 Argentina	28543	74.8	1 Argentina	40401	77.3
	2 France	6129	23.3	2 France	6123	16.0	2 France	6100	11.7
	3 Chile	929	3.5	3 Chile	1264	3.3	3 Chile	2293	4.4
	4 Australia	429	1.6	4 United States	717	1.9	4 United States	1610	3.1
	5 Italy	251	1.0	5 South Africa	450	1.2	5 Australia	515	1.0
	6 United States	97	0.4	6 Australia	356	0.9	6 South Africa	452	0.9
	7 South Africa	76	0.3	7 Italy	260	0.7	7 Italy	178	0.3
	8 New Zealand	67	0.3	8 New Zealand	156	0.4	8 Moldova	162	0.3
	9 Moldova	39	0.1	9 Spain	93	0.2	9 New Zealand	129	0.2
	10 Spain	23	0.1	10 Uruguay	41	0.1	10 Spain	113	0.2
	11 Switzerland	0	0.0	11 Moldova	39	0.1	11 Israel	110	0.2
	12			12 Canada	39	0.1	12 Uruguay	43	0.1
	13			13 Brazil	37	0.1	13 Canada	41	0.1
	14			14 Turkey	13	0.0	14 Brazil	30	0.1
	15			15 Switzerland	10	0.0	15 Turkey	21	0.0
	16			16 Peru	10	0.0	16 Switzerland	15	0.0
	17			17 Romania	7	0.0	17 Peru	10	0.0
	18			18 Hungary	0	0.0	18 Romania	7	0.0
	19			19			19 Hungary	3	0.0
	20			20			20		
	Others	16	0.1	Others			Others		
	Total	26285	100.0	Total	38158	100.0	Total	52233	100.0

Table 30 (cont.): National ranking of top 20 countries, 24 top red varieties, 2000, 2010 and 2016 (ha and %)

Monastrell	**2000**	bearing area (ha)	% of world	**2010**	bearing area (ha)	% of world	**2016**	bearing area (ha)	% of world
	1 Spain	67160	88.0	1 Spain	58406	83.7	1 Spain	41303	79.5
	2 France	7634	10.0	2 France	9257	13.3	2 France	8754	16.9
	3 Australia	948	1.2	3 Australia	692	1.0	3 Australia	704	1.4
	4 Tunisia	337	0.4	4 United States	404	0.6	4 United States	515	1.0
	5 United States	187	0.2	5 South Africa	403	0.6	5 South Africa	473	0.9
	6 Chile	22	0.0	6 Tunisia	337	0.5	6 Chile	102	0.2
	7 South Africa	13	0.0	7 Cyprus	172	0.2	7 Israel	55	0.1
	8 Argentina	0	0.0	8 Chile	59	0.1	8 Argentina	12	0.0
	9			9 Turkey	4	0.0	9 Turkey	7	0.0
	10			10 Romania	4	0.0	10 Romania	3	0.0
	11			11 Canada	2	0.0	11 Canada	1	0.0
	12			12 Uruguay	1	0.0	12 Switzerland	0	0.0
	13			13 Argentina	1	0.0	13		
	14			14			14		
	15			15			15		
	16			16			16		
	17			17			17		
	18			18			18		
	19			19			19		
	20			20			20		
	Others	3	0.0	Others			Others		
	Total	76304	100.0	Total	69742	100.0	Total	51930	100.0

Mazuelo	**2000**	bearing area (ha)	% of world	**2010**	bearing area (ha)	% of world	**2016**	bearing area (ha)	% of world
	1 France	95745	75.0	1 France	47720	63.0	1 France	31760	67.1
	2 Spain	8103	6.3	2 Tunisia	7576	10.0	2 Spain	5461	11.5
	3 Tunisia	7576	5.9	3 Algeria	7550	10.0	3 Algeria	3000	6.3
	4 Algeria	7550	5.9	4 Spain	4749	6.3	4 Italy	1686	3.6
	5 United States	3088	2.4	5 Italy	2023	2.7	5 Morocco	1230	2.6
	6 Italy	1721	1.3	6 Morocco	1692	2.2	6 United States	1086	2.3
	7 Morocco	1692	1.3	7 United States	1441	1.9	7 Israel	935	2.0
	8 Israel	971	0.8	8 Israel	971	1.3	8 Chile	811	1.7
	9 Chile	641	0.5	9 Cyprus	481	0.6	9 Mexico	448	0.9
	10 Australia	90	0.1	10 Chile	477	0.6	10 Portugal	291	0.6
	11 South Africa	71	0.1	11 Mexico	448	0.6	11 Tunisia	238	0.5
	12 Argentina	57	0.0	12 Portugal	338	0.4	12 Turkey	130	0.3
	13 Greece	14	0.0	13 Turkey	84	0.1	13 South Africa	114	0.2
	14			14 South Africa	81	0.1	14 China	100	0.2
	15			15 Croatia	34	0.0	15 Argentina	13	0.0
	16			16 Argentina	30	0.0	16 Australia	8	0.0
	17			17 Greece	16	0.0	17 Greece	1	0.0
	18			18 Myanmar	4	0.0	18		
	19			19			19		
	20			20			20		
	Others	374	0.3	Others			Others		
	Total	127692	100.0	Total	75716	100.0	Total	47312	100.0

Table 30 (cont.): National ranking of top 20 countries, 24 top red varieties, 2000, 2010 and 2016 (ha and %)

Alicante Henri Bouschet	**2000**	bearing area (ha)	% of world	**2010**	bearing area (ha)	% of world	**2016**	bearing area (ha)	% of world
	1 Spain	18321	49.3	1 Spain	19551	50.8	1 Spain	19294	53.5
	2 France	8764	23.6	2 France	4322	11.2	2 Chile	6908	19.2
	3 Algeria	3020	8.1	3 Chile	4228	11.0	3 Portugal	4547	12.6
	4 Chile	2882	7.8	4 Portugal	3322	8.6	4 France	2607	7.2
	5 Morocco	1098	3.0	5 Algeria	3020	7.9	5 Morocco	919	2.6
	6 Tunisia	842	2.3	6 Morocco	1098	2.9	6 Turkey	532	1.5
	7 Portugal	675	1.8	7 Tunisia	842	2.2	7 United States	380	1.1
	8 United States	563	1.5	8 Italy	645	1.7	8 Italy	286	0.8
	9 Italy	510	1.4	9 Turkey	488	1.3	9 Tunisia	178	0.5
	10 Argentina	113	0.3	10 United States	431	1.1	10 Argentina	135	0.4
	11 Greece	21	0.1	11 Argentina	257	0.7	11 Brazil	101	0.3
	12 South Africa	8	0.0	12 Brazil	129	0.3	12 Greece	60	0.2
	13			13 Greece	56	0.1	13 Uruguay	24	0.1
	14			14 Uruguay	22	0.1	14 Romania	20	0.1
	15			15 Hungary	21	0.1	15 Australia	19	0.1
	16			16 Romania	20	0.1	16 Hungary	14	0.0
	17			17 South Africa	10	0.0	17 South Africa	7	0.0
	18			18 Myanmar	0	0.0	18 Canada	0	0.0
	19			19			19 Switzerland	0	0.0
	20			20			20		
	Others	341	0.9	Others			Others		
	Total	37157	100.0	Total	38462	100.0	Total	36031	100.0

Tribidrag	**2000**	bearing area (ha)	% of world	**2010**	bearing area (ha)	% of world	**2016**	bearing area (ha)	% of world
	1 United States	18630	69.2	1 United States	19857	60.6	1 United States	18551	55.1
	2 Italy	7828	29.1	2 Italy	12234	37.3	2 Italy	13896	41.3
	3 Tunisia	337	1.3	3 Tunisia	337	1.0	3 N. Macedonia	1000	3.0
	4 Chile	91	0.3	4 Australia	149	0.5	4 Australia	87	0.3
	5 South Africa	28	0.1	5 Croatia	65	0.2	5 Chile	66	0.2
	6 Argentina	6	0.0	6 Chile	58	0.2	6 South Africa	24	0.1
	7 France	1	0.0	7 South Africa	34	0.1	7 Romania	9	0.0
	8			8 Romania	8	0.0	8 Canada	8	0.0
	9			9 Canada	8	0.0	9 New Zealand	4	0.0
	10			10 New Zealand	4	0.0	10 France	1	0.0
	11			11 France	2	0.0	11 Spain	1	0.0
	12			12 Argentina	0	0.0	12 Argentina	1	0.0
	13			13 Brazil	0	0.0	13 Switzerland	1	0.0
	14			14			14		
	15			15			15		
	16			16			16		
	17			17			17		
	18			18			18		
	19			19			19		
	20			20			20		
	Others			Others			Others		
	Total	26922	100.0	Total	32755	100.0	Total	33649	100.0

Table 30 (cont.): National ranking of top 20 countries, 24 top red varieties, 2000, 2010 and 2016 (ha and %)

Montepulciano	**2000**	bearing area (ha)	% of world	**2010**	bearing area (ha)	% of world	**2016**	bearing area (ha)	% of world
1	Italy	28679	99.8	1 Italy	34824	99.6	1 Italy	32724	99.4
2	Argentina	49	0.2	2 Argentina	94	0.3	2 Argentina	82	0.2
3				3 United States	29	0.1	3 Australia	60	0.2
4				4 New Zealand	7	0.0	4 United States	58	0.2
5				5 Brazil	2	0.0	5 New Zealand	8	0.0
6				6			6 Chile	2	0.0
7				7			7 Brazil	1	0.0
8				8			8 Spain	0	0.0
9				9			9		
10				10			10		
11				11			11		
12				12			12		
13				13			13		
14				14			14		
15				15			15		
16				16			16		
17				17			17		
18				18			18		
19				19			19		
20				20			20		
	Others			Others			Others		
	Total	28728	100.0	Total	34956	100.0	Total	32935	100.0

Gamay Noir	**2000**	bearing area (ha)	% of world	**2010**	bearing area (ha)	% of world	**2016**	bearing area (ha)	% of world
1	France	34537	91.4	1 France	29698	93.0	1 France	24095	91.9
2	Switzerland	1977	5.2	2 Switzerland	1521	4.8	2 Switzerland	1349	5.1
3	United States	684	1.8	3 Canada	220	0.7	3 Canada	272	1.0
4	Canada	263	0.7	4 Turkey	206	0.6	4 Turkey	228	0.9
5	Italy	152	0.4	5 Italy	122	0.4	5 United States	123	0.5
6	South Africa	36	0.1	6 Luxembourg	98	0.3	6 Italy	64	0.2
7	Argentina	2	0.0	7 South Africa	19	0.1	7 Serbia	54	0.2
8	Spain	1	0.0	8 Brazil	16	0.0	8 South Africa	9	0.0
9	Luxembourg	1	0.0	9 New Zealand	12	0.0	9 New Zealand	7	0.0
10				10 Uruguay	9	0.0	10 Australia	6	0.0
11				11 Hungary	3	0.0	11 Brazil	5	0.0
12				12 Argentina	2	0.0	12 Slovenia	4	0.0
13				13 United Kingdom	1	0.0	13 Hungary	3	0.0
14				14 Chile	0	0.0	14 Uruguay	1	0.0
15				15 Portugal	0	0.0	15 Chile	0	0.0
16				16			16 Argentina	0	0.0
17				17			17 Portugal	0	0.0
18				18			18		
19				19			19		
20				20			20		
	Others	144	0.4	Others			Others		
	Total	37798	100.0	Total	31927	100.0	Total	26221	100.0

Table 30 (cont.): National ranking of top 20 countries, 24 top red varieties, 2000, 2010 and 2016 (ha and %)

Cinsaut	2000	bearing area (ha)	% of world	2010	bearing area (ha)	% of world	2016	bearing area (ha)	% of world
	1 France	31593	65.2	1 France	19505	56.1	1 France	15930	69.5
	2 Algeria	7550	15.6	2 Algeria	7550	21.7	2 Morocco	3239	14.1
	3 Morocco	3940	8.1	3 Morocco	3940	11.3	3 South Africa	1767	7.7
	4 South Africa	3533	7.3	4 South Africa	2052	5.9	4 Chile	848	3.7
	5 Tunisia	842	1.7	5 Tunisia	842	2.4	5 Tunisia	626	2.7
	6 Italy	274	0.6	6 Turkey	500	1.4	6 Turkey	430	1.9
	7 Chile	195	0.4	7 Chile	198	0.6	7 United States	45	0.2
	8 Greece	108	0.2	8 Italy	51	0.1	8 Portugal	12	0.1
	9 United States	33	0.1	9 United States	45	0.1	9 Australia	10	0.0
	10 Argentina	6	0.0	10 Greece	43	0.1	10 Spain	10	0.0
	11 Spain	3	0.0	11 Portugal	17	0.0	11 Greece	4	0.0
	12			12 Argentina	6	0.0	12 Italy	4	0.0
	13			13 China	3	0.0	13 Argentina	1	0.0
	14			14			14		
	15			15			15		
	16			16			16		
	17			17			17		
	18			18			18		
	19			19			19		
	20			20			20		
	Others	351	0.7	Others			Others		
	Total	48428	100.0	Total	34751	100.0	Total	22926	100.0

Carmenère	2000	bearing area (ha)	% of world	2010	bearing area (ha)	% of world	2016	bearing area (ha)	% of world
	1 Chile	4719	82.6	1 Chile	8827	77.7	1 China	11200	49.8
	2 Italy	43	0.7	2 China	1353	11.9	2 Chile	10503	46.7
	3 France	5	0.1	3 Italy	1074	9.4	3 Italy	635	2.8
	4			4 Argentina	33	0.3	4 Argentina	59	0.3
	5			5 France	29	0.3	5 France	28	0.1
	6			6 United States	22	0.2	6 United States	24	0.1
	7			7 Croatia	19	0.2	7 Australia	16	0.1
	8			8 Brazil	7	0.1	8 Brazil	10	0.0
	9			9 Canada	2	0.0	9 South Africa	8	0.0
	10			10 Hungary	0	0.0	10 Canada	3	0.0
	11			11			11 Hungary	0	0.0
	12			12			12 Switzerland	0	0.0
	13			13			13 Spain	0	0.0
	14			14			14		
	15			15			15		
	16			16			16		
	17			17			17		
	18			18			18		
	19			19			19		
	20			20			20		
	Others	945	16.5	Others			Others		
	Total	5711	2.7	Total	11366	100.0	Total	22486	100.0

Table 30 (cont.): National ranking of top 20 countries, 24 top red varieties, 2000, 2010 and 2016 (ha and %)

Douce Noire	**2000**	bearing area (ha)	% of world		**2010**	bearing area (ha)	% of world		**2016**	bearing area (ha)	% of world
1	Argentina	15659	85.5	1	Argentina	18775	95.6	1	Argentina	19072	96.6
2	Italy	2642	14.4	2	Italy	821	4.2	2	Italy	630	3.2
3	United States	21	0.1	3	United States	34	0.2	3	United States	31	0.2
4	France	1	0.0	4	France	0	0.0	4	France	0	0.0
5				5	Brazil	0	0.0	5			
6				6				6			
7				7				7			
8				8				8			
9				9				9			
10				10				10			
11				11				11			
12				12				12			
13				13				13			
14				14				14			
15				15				15			
16				16				16			
17				17				17			
18				18				18			
19				19				19			
20				20				20			
	Others				Others				Others		
	Total	18323	100.0		Total	19630	100.0		Total	19733	100.0

Barbera	**2000**	bearing area (ha)	% of world		**2010**	bearing area (ha)	% of world		**2016**	bearing area (ha)	% of world
1	Italy	27175	82.2	1	Italy	20524	84.2	1	Italy	15006	84.2
2	United States	4693	14.2	2	United States	2798	11.5	2	United States	2131	12.0
3	Argentina	1055	3.2	3	Argentina	733	3.0	3	Argentina	444	2.5
4	Australia	103	0.3	4	Slovenia	134	0.5	4	Australia	102	0.6
5	South Africa	15	0.0	5	Australia	116	0.5	5	Slovenia	98	0.5
6				6	South Africa	51	0.2	6	South Africa	35	0.2
7				7	Brazil	5	0.0	7	Chile	5	0.0
8				8	Chile	4	0.0	8	Brazil	2	0.0
9				9	Canada	1	0.0	9	Canada	1	0.0
10				10	Romania	0	0.0	10	Switzerland	0	0.0
11				11				11	Romania	0	0.0
12				12				12	Spain	0	0.0
13				13				13			
14				14				14			
15				15				15			
16				16				16			
17				17				17			
18				18				18			
19				19				19			
20				20				20			
	Others				Others				Others		
	Total	33041	100.0		Total	24366	100.0		Total	17824	100.0

Table 30 (cont.): National ranking of top 20 countries, 24 top red varieties, 2000, 2010 and 2016 (ha and %)

Isabella	**2000**	bearing area (ha)	% of world		**2010**	bearing area (ha)	% of world		**2016**	bearing area (ha)	% of world
1	Brazil	14285	52.0	1	Brazil	18279	56.3	1	Brazil	11664	65.5
2	Moldova	11401	41.5	2	Moldova	11401	35.1	2	Moldova	3468	19.5
3	Argentina	74	0.3	3	Ukraine	2396	7.4	3	Russia	1362	7.6
4	United States	16	0.1	4	Uruguay	256	0.8	4	Ukraine	1200	6.7
5				5	Russia	162	0.5	5	Uruguay	102	0.6
6				6				6	Australia	15	0.1
7				7				7	Switzerland	1	0.0
8				8				8			
9				9				9			
10				10				10			
11				11				11			
12				12				12			
13				13				13			
14				14				14			
15				15				15			
16				16				16			
17				17				17			
18				18				18			
19				19				19			
20				20				20			
	Others	1673	6.1		Others				Others		
	Total	27450	100.0		Total	32494	100.0		Total	17813	100.0

Blaufränkisch	**2000**	bearing area (ha)	% of world		**2010**	bearing area (ha)	% of world		**2016**	bearing area (ha)	% of world
1	Hungary	6920	49.4	1	Hungary	7998	44.7	1	Hungary	7260	42.3
2	Austria	2641	18.9	2	Austria	3228	18.0	2	Austria	2808	16.3
3	Croatia	1189	8.5	3	Germany	1749	9.8	3	Germany	1737	10.1
4	Germany	1118	8.0	4	Slovakia	1378	7.7	4	Slovakia	1216	7.1
5	Slovakia	1091	7.8	5	Czechia	1160	6.5	5	Czechia	1143	6.7
6	Czechia	680	4.9	6	Romania	760	4.2	6	Romania	729	4.2
7	Italy	111	0.8	7	Slovenia	680	3.8	7	Serbia	727	4.2
8	United States	45	0.3	8	Croatia	558	3.1	8	Slovenia	709	4.1
9				9	Peru	290	1.6	9	Croatia	521	3.0
10				10	Italy	59	0.3	10	Peru	290	1.7
11				11	United States	22	0.1	11	Italy	28	0.2
12				12	Canada	4	0.0	12	Canada	5	0.0
13				13	Switzerland	3	0.0	13	Switzerland	4	0.0
14				14				14	United States	3	0.0
15				15				15	Australia	1	0.0
16				16				16			
17				17				17			
18				18				18			
19				19				19			
20				20				20			
	Others	203	1.4		Others				Others		
	Total	13997	100.0		Total	17890	100.0		Total	17180	100.0

Table 30 (cont.): National ranking of top 20 countries, 24 top red varieties, 2000, 2010 and 2016 (ha and %)

Criolla Grande	2000	bearing area (ha)	% of world		2010	bearing area (ha)	% of world		2016	bearing area (ha)	% of world
1	Argentina	24264	100.0	1	Argentina	20745	100.0	1	Argentina	15596	100.0
2				2				2			
3				3				3			
4				4				4			
5				5				5			
6				6				6			
7				7				7			
8				8				8			
9				9				9			
10				10				10			
11				11				11			
12				12				12			
13				13				13			
14				14				14			
15				15				15			
16				16				16			
17				17				17			
18				18				18			
19				19				19			
20				20				20			
	Others				Others				Others		
	Total	24264	100.0		Total	20745	100.0		Total	15596	100.0

Pinot Meunier	2000	bearing area (ha)	% of world		2010	bearing area (ha)	% of world		2016	bearing area (ha)	% of world
1	France	10621	80.9	1	France	11087	81.7	1	France	12130	82.5
2	Germany	2289	17.4	2	Germany	2301	17.0	2	Germany	2002	13.6
3	Australia	107	0.8	3	United States	66	0.5	3	United Kingdom	202	1.4
4	United States	80	0.6	4	United Kingdom	50	0.4	4	Moldova	138	0.9
5	Argentina	11	0.1	5	New Zealand	19	0.1	5	Australia	82	0.6
6	Italy	10	0.1	6	Italy	14	0.1	6	United States	76	0.5
7	Spain	7	0.1	7	South Africa	13	0.1	7	New Zealand	21	0.1
8	South Africa	6	0.0	8	Argentina	11	0.1	8	South Africa	14	0.1
9				9	Canada	5	0.0	9	Argentina	11	0.1
10				10	Spain	1	0.0	10	Canada	9	0.1
11				11				11	Italy	5	0.0
12				12				12	Spain	2	0.0
13				13				13	Chile	2	0.0
14				14				14	Switzerland	0	0.0
15				15				15			
16				16				16			
17				17				17			
18				18				18			
19				19				19			
20				20				20			
	Others				Others				Others		
	Total	13131	100.0		Total	13566	100.0		Total	14695	100.0

Table 31: National ranking of top 20 countries, 24 top white varieties, 2000, 2010 and 2016 (ha and %)

Airén	**2000**	bearing area (ha)	% of world	**2010**	bearing area (ha)	% of world	**2016**	bearing area (ha)	% of world
	1 Spain	387978	100.0	1 Spain	252364	100.0	1 Spain	203276	99.7
	2			2			2 Morocco	440	0.2
	3			3			3 Tunisia	85	0.0
	4			4			4		
	5			5			5		
	6			6			6		
	7			7			7		
	8			8			8		
	9			9			9		
	10			10			10		
	11			11			11		
	12			12			12		
	13			13			13		
	14			14			14		
	15			15			15		
	16			16			16		
	17			17			17		
	18			18			18		
	19			19			19		
	20			20			20		
	Others			Others			Others		
	Total	387978	100.0	Total	252364	100.0	Total	203801	100.0

Chardonnay	**2000**	bearing area (ha)	% of world	**2010**	bearing area (ha)	% of world	**2016**	bearing area (ha)	% of world
	1 France	36496	25.1	1 France	45243	22.7	1 France	47451	23.5
	2 United States	35791	24.6	2 United States	40846	20.4	2 United States	41392	20.5
	3 Australia	17266	11.9	3 Australia	27773	13.9	3 Australia	21321	10.6
	5 Italy	11687	8.0	5 Italy	19709	9.9	5 Italy	19769	9.8
	6 Chile	7672	5.3	6 Chile	13082	6.5	6 Chile	11435	5.7
	7 South Africa	6067	4.2	7 South Africa	8278	4.1	7 Spain	6866	3.4
	9 Moldova	5134	3.5	9 Spain	6958	3.5	9 South Africa	6856	3.4
	4 Argentina	4682	3.2	4 Argentina	6584	3.3	4 Argentina	6227	3.1
	8 Hungary	2954	2.0	8 Moldova	5134	2.6	8 China	6100	3.0
	10 New Zealand	2787	1.9	10 New Zealand	3911	2.0	10 Moldova	4133	2.0
	11 Bulgaria	1862	1.3	11 Ukraine	2985	1.5	11 Russia	3481	1.7
	12 Spain	1814	1.2	12 Hungary	2757	1.4	12 New Zealand	3117	1.5
	13 Russia	1639	1.1	13 Bulgaria	2457	1.2	13 Bulgaria	3087	1.5
	14 Slovenia	1549	1.1	14 Russia	1981	1.0	14 Hungary	2464	1.2
	15 Romania	1376	0.9	15 Austria	1380	0.7	15 Romania	1878	0.9
	16 Canada	973	0.7	16 Germany	1228	0.6	16 Austria	1577	0.8
	17 Slovakia	623	0.4	17 Slovenia	1208	0.6	17 Ukraine	1500	0.7
	18 Czechia	567	0.4	18 Canada	1178	0.6	18 Germany	1485	0.7
	19 Germany	531	0.4	19 Romania	1067	0.5	19 Serbia	1455	0.7
	20 Brazil	330	0.2	20 Portugal	803	0.4	20 Canada	1417	0.7
	Others	3742	2.6	Others	5181	2.6	Others	8639	4.3
	Total	145543	100.0	Total	199743	100.0	Total	201649	100.0

Table 31 (cont.): National ranking of top 20 countries, 24 top white varieties, 2000, 2010 and 2016 (ha and %)

Sauvignon Blanc	**2000**	bearing area (ha)	% of world	**2010**	bearing area (ha)	% of world	**2016**	bearing area (ha)	% of world
	1 France	20933	32.1	1 France	27931	25.0	1 France	28084	22.5
	2 Moldova	8151	12.5	2 New Zealand	16205	14.5	2 New Zealand	20497	16.4
	3 Chile	6662	10.2	3 Chile	12159	10.9	3 Chile	14999	12.0
	4 South Africa	5436	8.3	4 South Africa	9551	8.6	4 South Africa	9246	7.4
	5 Romania	4613	7.1	5 Moldova	8151	7.3	5 Moldova	6909	5.5
	6 United States	4191	6.4	6 United States	6584	5.9	6 United States	6747	5.4
	7 Italy	3312	5.1	7 Australia	6467	5.8	7 Australia	6044	4.8
	8 Australia	2602	4.0	8 Romania	4157	3.7	8 Romania	5594	4.5
	9 New Zealand	2423	3.7	9 Spain	4011	3.6	9 Spain	4562	3.7
	10 Slovenia	1221	1.9	10 Italy	3744	3.4	10 Italy	3935	3.2
	11 Argentina	865	1.3	11 Ukraine	3123	2.8	11 Russia	2501	2.0
	12 Spain	467	0.7	12 Argentina	2297	2.1	12 Argentina	2148	1.7
	13 Bulgaria	405	0.6	13 Slovenia	1061	1.0	13 China	2000	1.6
	14 Hungary	324	0.5	14 Russia	951	0.9	14 Ukraine	1550	1.2
	15 Austria	314	0.5	15 Hungary	907	0.8	15 Austria	1170	0.9
	16 Israel	263	0.4	16 Austria	845	0.8	16 Slovenia	1121	0.9
	17 Greece	158	0.2	17 Czechia	804	0.7	17 Hungary	982	0.8
	18 Canada	148	0.2	18 Germany	518	0.5	18 Czechia	906	0.7
	19 Uruguay	142	0.2	19 Canada	320	0.3	19 Serbia	741	0.6
	20 Brazil	140	0.2	20 Israel	263	0.2	20 Germany	736	0.6
	Others	2420	3.7	Others	1502	1.3	Others	4227	3.4
	Total	65190	100.0	Total	111552	100.0	Total	124700	100.0

Trebbiano Toscano	**2000**	bearing area (ha)	% of world	**2010**	bearing area (ha)	% of world	**2016**	bearing area (ha)	% of world
	1 France	90341	65.8	1 France	83445	75.0	1 France	78842	65.5
	2 Italy	39447	28.8	2 Italy	22702	20.4	2 Italy	35441	29.4
	3 Argentina	2765	2.0	3 Argentina	2425	2.2	3 Argentina	1622	1.3
	5 Bulgaria	1821	1.3	5 Uruguay	762	0.7	5 China	1500	1.2
	6 Greece	746	0.5	6 Bulgaria	723	0.6	6 Bulgaria	738	0.6
	7 Brazil	688	0.5	7 Greece	298	0.3	7 Uruguay	682	0.6
	9 Australia	685	0.5	9 Portugal	212	0.2	9 India	300	0.2
	4 Portugal	382	0.3	4 Croatia	210	0.2	4 Moldova	277	0.2
	8 United States	151	0.1	8 Brazil	149	0.1	8 Brazil	231	0.2
	10 South Africa	147	0.1	10 Australia	86	0.1	10 Greece	211	0.2
	11 Spain	28	0.0	11 United States	80	0.1	11 South Africa	156	0.1
	12			12 South Africa	74	0.1	12 Portugal	122	0.1
	13			13 Russia	66	0.1	13 United States	88	0.1
	14			14 Spain	45	0.0	14 Russia	66	0.1
	15			15 Romania	11	0.0	15 Spain	49	0.0
	16			16 Canada	2	0.0	16 Australia	14	0.0
	17			17			17 Romania	3	0.0
	18			18			18 Canada	2	0.0
	19			19			19		
	20			20			20		
	Others			Others			Others		
	Total	137201	100.0	Total	111290	100.0	Total	120343	100.0

Table 31 (cont.): National ranking of top 20 countries, 24 top white varieties, 2000, 2010 and 2016 (ha and %)

			bearing	% of			bearing	% of			bearing	% of
Riesling		**2000**	area (ha)	world		**2010**	area (ha)	world		**2016**	area (ha)	world
	1	Germany	22350	51.6	1	Germany	22429	44.8	1	Germany	21540	36.0
	2	France	3407	7.9	2	United States	4852	9.7	2	Romania	6121	10.2
	3	Australia	3129	7.2	3	Australia	4114	8.2	3	United States	4952	8.3
	4	United States	1965	4.5	4	France	3513	7.0	4	France	4025	6.7
	5	Austria	1643	3.8	5	Ukraine	2702	5.4	5	Australia	3114	5.2
	6	Hungary	1619	3.7	6	Austria	1852	3.7	6	Russia	2232	3.7
	7	Russia	1376	3.2	7	Moldova	1343	2.7	7	Austria	2016	3.4
	8	Moldova	1343	3.1	8	Hungary	1304	2.6	8	Moldova	1701	2.8
	9	Czechia	793	1.8	9	Czechia	1181	2.4	9	China	1600	2.7
	10	Bulgaria	647	1.5	10	New Zealand	979	2.0	10	Italy	1461	2.4
	11	Italy	599	1.4	11	Russia	882	1.8	11	Serbia	1361	2.3
	12	New Zealand	490	1.1	12	Canada	871	1.7	12	Ukraine	1350	2.3
	13	Canada	482	1.1	13	Croatia	676	1.4	13	Hungary	1261	2.1
	14	South Africa	477	1.1	14	Slovenia	676	1.4	14	Canada	1188	2.0
	15	Chile	286	0.7	15	Slovakia	605	1.2	15	Czechia	1172	2.0
	16	Luxembourg	175	0.4	16	Italy	446	0.9	16	N. Macedonia	900	1.5
	17	Argentina	150	0.3	17	China	437	0.9	17	New Zealand	767	1.3
	18	Spain	97	0.2	18	Chile	367	0.7	18	Croatia	625	1.0
	19	Switzerland	8	0.0	19	South Africa	211	0.4	19	Slovakia	620	1.0
	20				20	Spain	161	0.3	20	Slovenia	607	1.0
		Others	2279	5.3		Others	412	0.8		Others	1192	2.0
		Total	43316	100.0		Total	50014	100.0		Total	59805	100.0

			bearing	% of			bearing	% of			bearing	% of
Rkatsiteli		**2000**	area (ha)	world		**2010**	area (ha)	world		**2016**	area (ha)	world
	1	Georgia	19741	29.3	1	Georgia	25324	43.2	1	Georgia	25324	49.3
	2	Russia	13152	19.5	2	Ukraine	11552	19.7	2	Russia	6477	12.6
	3	Moldova	11508	17.1	3	Moldova	11508	19.6	3	Ukraine	5775	11.2
	4	Bulgaria	9429	14.0	4	Kazakhstan	3552	6.1	4	Bulgaria	5415	10.5
	5	Armenia	2469	3.7	5	Bulgaria	3121	5.3	5	Moldova	3898	7.6
	6	Romania	506	0.8	6	Armenia	2469	4.2	6	Kazakhstan	3552	6.9
	7				7	Russia	702	1.2	7	N. Macedonia	460	0.9
	8				8	Romania	356	0.6	8	Romania	413	0.8
	9				9	Croatia	57	0.1	9	Serbia	60	0.1
	10				10				10			
	11				11				11			
	12				12				12			
	13				13				13			
	14				14				14			
	15				15				15			
	16				16				16			
	17				17				17			
	18				18				18			
	19				19				19			
	20				20				20			
		Others	10549	15.7		Others				Others		
		Total	67354	100.0		Total	58641	100.0		Total	51374	100.0

Table 31 (cont.): National ranking of top 20 countries, 24 top white varieties, 2000, 2010 and 2016 (ha and %)

Macabeo	2000	bearing area (ha)	% of world	2010	bearing area (ha)	% of world	2016	bearing area (ha)	% of world
	1 Spain	42902	89.1	1 Spain	38417	94.0	1 Spain	36963	95.7
	2 France	5223	10.9	2 France	2446	6.0	2 France	1657	4.3
	3 Argentina	3	0.0	3 Argentina	2	0.0	3 South Africa	4	0.0
	4			4			4		
	5			5			5		
	6			6			6		
	7			7			7		
	8			8			8		
	9			9			9		
	10			10			10		
	11			11			11		
	12			12			12		
	13			13			13		
	14			14			14		
	15			15			15		
	16			16			16		
	17			17			17		
	18			18			18		
	19			19			19		
	20			20			20		
	Others			Others			Others		
	Total	48128	100.0	Total	40864	100.0	Total	38625	100.0

Cayetana Blanca	2000	bearing area (ha)	% of world	2010	bearing area (ha)	% of world	2016	bearing area (ha)	% of world
	1 Spain	55527	99.6	1 Spain	39633	99.6	1 Spain	36252	99.6
	2 Australia	249	0.4	2 Portugal	148	0.4	2 Portugal	132	0.4
	3			3			3 Australia	17	0.0
	5			5			5		
	6			6			6		
	7			7			7		
	9			9			9		
	4			4			4		
	8			8			8		
	10			10			10		
	11			11			11		
	12			12			12		
	13			13			13		
	14			14			14		
	15			15			15		
	16			16			16		
	17			17			17		
	18			18			18		
	19			19			19		
	20			20			20		
	Others			Others			Others		
	Total	55776	100.0	Total	39781	100.0	Total	36401	100.0

Table 31 (cont.): National ranking of top 20 countries, 24 top white varieties, 2000, 2010 and 2016 (ha and %)

Muscat Alexandria	2000	bearing area (ha)	% of world	2010	bearing area (ha)	% of world	2016	bearing area (ha)	% of world
	1 Spain	6144	20.8	1 Spain	8237	29.8	1 Spain	9534	27.4
	2 Argentina	5515	18.6	2 Argentina	4035	14.6	2 Chile	5424	15.6
	3 South Africa	4047	13.7	3 Morocco	3669	13.3	3 China	3000	8.6
	4 Morocco	3669	12.4	4 France	2603	9.4	4 Argentina	2716	7.8
	5 France	3027	10.2	5 South Africa	2167	7.8	5 France	2462	7.1
	6 Australia	2495	8.4	6 Australia	2043	7.4	6 Australia	2179	6.3
	7 United States	2013	6.8	7 Italy	1521	5.5	7 Morocco	2093	6.0
	8 Italy	1157	3.9	8 United States	1285	4.6	8 United States	1987	5.7
	9 Brazil	809	2.7	9 Chile	1090	3.9	9 South Africa	1781	5.1
	10 Portugal	510	1.7	10 Portugal	647	2.3	10 Italy	1375	3.9
	11 Israel	202	0.7	11 Israel	202	0.7	11 Greece	773	2.2
	12			12 Cyprus	120	0.4	12 Portugal	509	1.5
	13			13 Russia	21	0.1	13 Tunisia	405	1.2
	14			14 Brazil	7	0.0	14 Israel	220	0.6
	15			15			15 Algeria	200	0.6
	16			16			16 India	100	0.3
	17			17			17 Uruguay	22	0.1
	18			18			18 Russia	21	0.1
	19			19			19 Brazil	6	0.0
	20			20			20		
	Others			Others			Others		
	Total	29590	100.0	Total	27648	100.0	Total	34805	100.0

Muscat Blanc à Petits Grains	2000	bearing area (ha)	% of world	2010	bearing area (ha)	% of world	2016	bearing area (ha)	% of world
	1 Italy	13016	43.4	1 Italy	11506	36.8	1 Italy	13334	39.5
	2 France	6935	23.1	2 France	7671	24.5	2 France	7333	21.7
	3 Greece	2232	7.4	3 Greece	2162	6.9	3 Romania	1579	4.7
	4 Hungary	1538	5.1	4 Spain	1291	4.1	4 Greece	1568	4.6
	5 Portugal	1419	4.7	5 Brazil	1005	3.2	5 Spain	1350	4.0
	6 Romania	1012	3.4	6 Romania	840	2.7	6 United States	1218	3.6
	7 South Africa	773	2.6	7 United States	733	2.3	7 Portugal	1031	3.1
	8 Armenia	526	1.8	8 Hungary	709	2.3	8 Australia	857	2.5
	9 United States	515	1.7	9 South Africa	689	2.2	9 South Africa	839	2.5
	10 Spain	223	0.7	10 Ukraine	674	2.2	10 Austria	823	2.4
	11 Australia	214	0.7	11 Portugal	535	1.7	11 Hungary	762	2.3
	12 Moldova	172	0.6	12 Australia	533	1.7	12 Slovenia	586	1.7
	13 Argentina	149	0.5	13 Armenia	526	1.7	13 Russia	483	1.4
	14 Austria	143	0.5	14 Austria	492	1.6	14 N. Macedonia	400	1.2
	15 Germany	87	0.3	15 Peru	361	1.2	15 Peru	361	1.1
	16 Switzerland	44	0.1	16 Slovenia	353	1.1	16 Ukraine	338	1.0
	17			17 Mexico	246	0.8	17 Mexico	246	0.7
	18			18 Germany	190	0.6	18 Germany	240	0.7
	19			19 Moldova	172	0.6	19 Turkey	129	0.4
	20			20 Russia	145	0.5	20 Argentina	94	0.3
	Others	981	3.3	Others	427	1.4	Others	168	0.5
	Total	29979	100.0	Total	31259	100.0	Total	33739	100.0

Table 31 (cont.): National ranking of top 20 countries, 24 top white varieties, 2000, 2010 and 2016 (ha and %)

Chenin Blanc	**2000**	bearing area (ha)	% of world		**2010**	bearing area (ha)	% of world		**2016**	bearing area (ha)	% of world
1	South Africa	22566	49.3	1	South Africa	18515	51.9	1	South Africa	17707	55.0
2	France	9837	21.5	2	France	9825	27.5	2	France	9432	29.3
3	United States	8433	18.4	3	United States	3221	9.0	3	Argentina	2157	6.7
4	Argentina	3445	7.5	4	Argentina	2851	8.0	4	United States	1969	6.1
5	Australia	841	1.8	5	Australia	541	1.5	5	Australia	406	1.3
6	New Zealand	146	0.3	6	Mexico	275	0.8	6	Mexico	275	0.9
7	Spain	105	0.2	7	Israel	101	0.3	7	Spain	106	0.3
8	Israel	101	0.2	8	Spain	100	0.3	8	Ethiopia	54	0.2
9	Chile	76	0.2	9	Chile	57	0.2	9	Chile	39	0.1
10	Switzerland	1	0.0	10	Ethiopia	54	0.2	10	New Zealand	24	0.1
11				11	New Zealand	50	0.1	11	Thailand	16	0.1
12				12	Italy	45	0.1	12	Switzerland	8	0.0
13				13	Brazil	18	0.0	13	Italy	7	0.0
14				14	Thailand	13	0.0	14	Brazil	7	0.0
15				15	China	10	0.0	15	Canada	6	0.0
16				16	Canada	7	0.0	16	Hungary	6	0.0
17				17	Uruguay	7	0.0	17	Uruguay	2	0.0
18				18	Switzerland	6	0.0	18	Peru	2	0.0
19				19	Hungary	6	0.0	19	Portugal	0	0.0
20				20	Peru	2	0.0	20			
	Others	210	0.5		Others	1	0.0		Others		
	Total	45761	100.0		Total	35703	100.0		Total	32221	100.0

Colombard	**2000**	bearing area (ha)	% of world		**2010**	bearing area (ha)	% of world		**2016**	bearing area (ha)	% of world
1	United States	18010	46.6	1	South Africa	11990	36.4	1	South Africa	11512	38.4
2	South Africa	11432	29.6	2	United States	10025	30.4	2	France	8441	28.1
3	France	6896	17.8	3	France	8173	24.8	3	United States	7991	26.6
4	Australia	1801	4.7	4	Australia	2205	6.7	4	Australia	1789	6.0
5	Israel	486	1.3	5	Israel	486	1.5	5	Israel	220	0.7
6				6	Brazil	46	0.1	6	Brazil	22	0.1
7				7	Thailand	11	0.0	7	Thailand	15	0.0
8				8	Uruguay	4	0.0	8	Spain	6	0.0
9				9	Spain	4	0.0	9			
10				10				10			
11				11				11			
12				12				12			
13				13				13			
14				14				14			
15				15				15			
16				16							
17				17				17			
18				18				18			
19				19				19			
20				20				20			
	Others	8	0.0		Others				Others		
	Total	38632	100.0		Total	32944	100.0		Total	29996	100.0

Table 31 (cont.): National ranking of top 20 countries, 24 top white varieties, 2000, 2010 and 2016 (ha and %)

Catarratto Bianco	**2000**	bearing area (ha)	% of world		**2010**	bearing area (ha)	% of world		**2016**	bearing area (ha)	% of world
1	Italy	50711	100.0	1	Italy	34794	99.8	1	Italy	28563	99.8
2				2	United States	68	0.2	2	United States	50	0.2
3				3				3			
4				4				4			
5				5				5			
6				6				6			
7				7				7			
8				8				8			
9				9				9			
10				10				10			
11				11				11			
12				12				12			
13				13				13			
14				14				14			
15				15				15			
16				16				16			
17				17				17			
18				18				18			
19				19				19			
20				20				20			
	Others				Others				Others		
	Total	50711	100.0		Total	34863	100.0		Total	28613	100.0

Aligoté	**2000**	bearing area (ha)	% of world		**2010**	bearing area (ha)	% of world		**2016**	bearing area (ha)	% of world
1	Moldova	15790	44.3	1	Moldova	15790	43.7	1	Moldova	7765	28.8
2	Romania	7608	21.3	2	Ukraine	9627	26.7	2	Russia	5843	21.7
3	Russia	1821	5.1	3	Romania	7297	20.2	3	Romania	5840	21.7
4	France	1756	4.9	4	France	1953	5.4	4	Ukraine	4814	17.9
5	Bulgaria	1659	4.7	5	Russia	1029	2.8	5	France	1927	7.2
6	Georgia	97	0.3	6	Kazakhstan	277	0.8	6	Bulgaria	285	1.1
7	Switzerland	20	0.1	7	Georgia	124	0.3	7	Kazakhstan	277	1.0
8				8	Switzerland	23	0.1	8	Georgia	124	0.5
9				9	Hungary	0	0.0	9	Canada	30	0.1
10				10	Brazil	0	0.0	10	Switzerland	24	0.1
11				11				11	Hungary	0	0.0
12				12				12			
13				13				13			
14				14				14			
15				15				15			
16				16				16			
17				17				17			
18				18				18			
19				19				19			
20				20				20			
	Others	6916	19.4		Others				Others		
	Total	35668	100.0		Total	36120	100.0		Total	26929	100.0

Table 31 (cont.): National ranking of top 20 countries, 24 top white varieties, 2000, 2010 and 2016 (ha and %)

Graševina	**2000**	bearing area (ha)	% of world		**2010**	bearing area (ha)	% of world		**2016**	bearing area (ha)	% of world
1	Serbia	33120	35.9	1	Serbia	33120	54.1	1	Croatia	4459	18.3
2	Croatia	16051	17.4	2	Romania	7530	12.3	2	Hungary	3933	16.1
3	Romania	15014	16.3	3	Croatia	4701	7.7	3	Austria	3233	13.3
4	Hungary	6677	7.2	4	Hungary	4664	7.6	4	China	3000	12.3
5	Austria	4323	4.7	5	Austria	3462	5.7	5	Serbia	2037	8.4
6	Slovakia	3895	4.2	6	Slovenia	2360	3.9	6	Slovenia	1935	7.9
7	Bulgaria	3602	3.9	7	Slovakia	1655	2.7	7	Romania	1437	5.9
8	Slovenia	3568	3.9	8	Italy	1568	2.6	8	Italy	1259	5.2
9	Italy	2007	2.2	9	Czechia	1148	1.9	9	Czechia	1114	4.6
10	Spain	1923	2.1	10	Spain	791	1.3	10	Spain	1064	4.4
11	Czechia	1246	1.4	11	Brazil	200	0.3	11	Slovakia	456	1.9
12	Brazil	880	1.0	12				12	N. Macedonia	270	1.1
13				13				13	Brazil	188	0.8
14				14				14			
15				15				15			
16				16				16			
17				17				17			
18				18				18			
19				19				19			
20				20				20			
	Others				Others				Others		
	Total	92306	100.0		Total	61200	100.0		Total	24384	100.0

Palomino Fino	**2000**	bearing area (ha)	% of world		**2010**	bearing area (ha)	% of world		**2016**	bearing area (ha)	% of world
1	Spain	27685	90.7	1	Spain	18836	83.0	1	Spain	20110	86.7
2	South Africa	1632	5.3	2	Portugal	3033	13.4	2	Portugal	2594	11.2
3	France	440	1.4	3	South Africa	270	1.2	3	South Africa	134	0.6
4	United States	319	1.0	4	Argentina	163	0.7	4	Mexico	109	0.5
5	Argentina	216	0.7	5	United States	135	0.6	5	Argentina	104	0.4
6	Australia	124	0.4	6	France	134	0.6	6	United States	70	0.3
7	New Zealand	21	0.1	7	Mexico	109	0.5	7	France	41	0.2
8				8	New Zealand	14	0.1	8	Australia	19	0.1
9				9				9	New Zealand	7	0.0
10				10				10	Switzerland	0	0.0
11				11				11			
12				12				12			
13				13				13			
14				14				14			
15				15				15			
16				16				16			
17				17				17			
18				18				18			
19				19				19			
20				20				20			
	Others	76	0.2		Others				Others		
	Total	30513	100.0		Total	22693	100.0		Total	23190	100.0

Table 31 (cont.): National ranking of top 20 countries, 24 top white varieties, 2000, 2010 and 2016 (ha and %)

Prosecco	2000	bearing area (ha)	% of world	2010	bearing area (ha)	% of world	2016	bearing area (ha)	% of world
	1 Italy	7498	99.9	1 Italy	18255	99.0	1 Italy	19730	98.1
	2 Argentina	9	0.1	2 Brazil	173	0.9	2 Brazil	207	1.0
	3			3 Argentina	10	0.1	3 Australia	160	0.8
	4			4			4 Argentina	11	0.1
	5			5			5		
	6			6			6		
	7			7			7		
	8			8			8		
	9			9			9		
	10			10			10		
	11			11			11		
	12			12			12		
	13			13			13		
	14			14			14		
	15			15			15		
	16			16			16		
	17			17			17		
	18			18			18		
	19			19			19		
	20			20			20		
	Others			Others			Others		
	Total	7507	100.0	Total	18437	100.0	Total	20109	100.0

Müller-Thurgau	2000	bearing area (ha)	% of world	2010	bearing area (ha)	% of world	2016	bearing area (ha)	% of world
	1 Germany	20706	61.6	1 Germany	13638	59.5	1 Germany	11664	59.8
	2 Austria	3289	9.8	2 Hungary	2098	9.2	2 Austria	1777	9.1
	3 Hungary	3278	9.8	3 Austria	2044	8.9	3 Hungary	1670	8.6
	5 Slovakia	1870	5.6	5 Czechia	1572	6.9	5 Czechia	1479	7.6
	6 Czechia	1586	4.7	6 Italy	1312	5.7	6 Italy	1296	6.6
	7 Italy	996	3.0	7 Slovakia	932	4.1	7 Slovakia	509	2.6
	9 Switzerland	686	2.0	9 Switzerland	493	2.1	9 Switzerland	465	2.4
	4 Luxembourg	459	1.4	4 Luxembourg	184	0.8	4 Luxembourg	316	1.6
	8 New Zealand	419	1.2	8 Moldova	173	0.8	8 Slovenia	128	0.7
	10 Moldova	173	0.5	10 Japan	172	0.7	10 Russia	106	0.5
	11 France	5	0.0	11 Russia	106	0.5	11 United States	39	0.2
	12			12 New Zealand	79	0.3	12 Japan	22	0.1
	13			13 Croatia	60	0.3	13 United Kingdom	15	0.1
	14			14 United Kingdom	43	0.2	14 Canada	6	0.0
	15			15 Canada	7	0.0	15 Moldova	4	0.0
	16			16 France	5	0.0	16 France	2	0.0
	17			17 Romania	0	0.0	17 New Zealand	2	0.0
	18			18			18 Australia	1	0.0
	19			19			19 Romania	0	0.0
	20			20			20		
	Others	120	0.4	Others			Others		
	Total	33587	100.0	Total	22917	100.0	Total	19501	100.0

Table 31 (cont.): National ranking of top 20 countries, 24 top white varieties, 2000, 2010 and 2016 (ha and %)

Grüner Veltiner	**2000**	bearing area (ha)	% of world	**2010**	bearing area (ha)	% of world	**2016**	bearing area (ha)	% of world
	1 Austria	17479	74.1	1 Austria	13519	71.8	1 Austria	14376	75.2
	2 Slovakia	2960	12.5	2 Slovakia	2091	11.1	2 Slovakia	1627	8.5
	3 Czechia	1700	7.2	3 Hungary	1533	8.1	3 Czechia	1538	8.0
	4 Hungary	1335	5.7	4 Czechia	1527	8.1	4 Hungary	1381	7.2
	5 Italy	129	0.5	5 Italy	165	0.9	5 United States	60	0.3
	6			6 Romania	0	0.0	6 Italy	55	0.3
	7			7			7 New Zealand	43	0.2
	8			8			8 Germany	14	0.1
	9			9			9 Australia	9	0.0
	10			10			10 Argentina	7	0.0
	11			11			11 South Africa	4	0.0
	12			12			12 Canada	3	0.0
	13			13			13 Switzerland	1	0.0
	14			14			14 Romania	0	0.0
	15			15			15		
	16			16			16		
	17			17			17		
	18			18			18		
	19			19			19		
	20			20			20		
	Others			Others			Others		
	Total	23604	100.0	Total	18834	100.0	Total	19118	100.0

Trebbiano Romagnolo	**2000**	bearing area (ha)	% of world	**2010**	bearing area (ha)	% of world	**2016**	bearing area (ha)	% of world
	1 Italy	19492	100.0	1 Italy	15893	100.0	1 Italy	19059	100.0
	2			2			2		
	3			3			3		
	5			5			5		
	6			6			6		
	7			7			7		
	9			9			9		
	4			4			4		
	8			8			8		
	10			10			10		
	11			11			11		
	12			12			12		
	13			13			13		
	14			14			14		
	15			15			15		
	16			16			16		
	17			17			17		
	18			18			18		
	19			19			19		
	20			20			20		
	Others			Others			Others		
	Total	19492	100.0	Total	15893	100.0	Total	19059	100.0

Table 31 (cont.): National ranking of top 20 countries, 24 top white varieties, 2000, 2010 and 2016 (ha and %)

Sémillon	**2000**	bearing area (ha)	% of world	**2010**	bearing area (ha)	% of world	**2016**	bearing area (ha)	% of world
	1 France	14015	53.4	1 France	11566	52.2	1 France	10234	54.7
	2 Australia	6528	24.9	2 Australia	6112	27.6	2 Australia	4556	24.4
	3 Chile	1893	7.2	3 South Africa	1192	5.4	3 South Africa	1121	6.0
	4 Argentina	1033	3.9	4 Argentina	956	4.3	4 Chile	849	4.5
	5 South Africa	1033	3.9	5 Chile	846	3.8	5 Argentina	767	4.1
	6 United States	709	2.7	6 Turkey	547	2.5	6 Turkey	529	2.8
	7 Brazil	384	1.5	7 United States	436	2.0	7 United States	340	1.8
	8 New Zealand	229	0.9	8 New Zealand	201	0.9	8 Portugal	76	0.4
	9 Greece	21	0.1	9 Portugal	89	0.4	9 New Zealand	63	0.3
	10 Italy	8	0.0	10 Hungary	58	0.3	10 Hungary	43	0.2
	11 Switzerland	3	0.0	11 Italy	31	0.1	11 Russia	25	0.1
	12			12 Russia	25	0.1	12 Canada	19	0.1
	13			13 Uruguay	24	0.1	13 Romania	18	0.1
	14			14 Brazil	24	0.1	14 Uruguay	14	0.1
	15			15 Canada	19	0.1	15 Italy	13	0.1
	16			16 Romania	18	0.1	16 Greece	11	0.1
	17			17 Greece	8	0.0	17 Brazil	6	0.0
	18			18 Switzerland	5	0.0	18 Switzerland	4	0.0
	19			19 China	1	0.0	19 Moldova	3	0.0
	20			20			20 Spain	2	0.0
	Others	382	1.5	Others			Others		
	Total	26239	100.0	Total	22157	100.0	Total	18693	100.0

Verdejo	**2000**	bearing area (ha)	% of world	**2010**	bearing area (ha)	% of world	**2016**	bearing area (ha)	% of world
	1 Spain	4453	100.0	1 Spain	16578	100.0	1 Spain	17923	100.0
	2			2			2 Australia	6	0.0
	3			3			3 Chile	2	0.0
	5			5			5		
	6			6			6		
	7			7			7		
	9			9			9		
	4			4			4		
	8			8			8		
	10			10			10		
	11			11			11		
	12			12			12		
	13			13			13		
	14			14			14		
	15			15			15		
	16			16			16		
	17			17			17		
	18			18			18		
	19			19			19		
	20			20			20		
	Others			Others			Others		
	Total	4453	100.0	Total	16578	100.0	Total	17931	100.0

Table 31 (cont.): National ranking of top 20 countries, 24 top white varieties, 2000, 2010 and 2016 (ha and %)

Viognier

	2000	bearing area (ha)	% of world		**2010**	bearing area (ha)	% of world		**2016**	bearing area (ha)	% of world
1	France	2360	74.7	1	France	4823	40.9	1	France	8823	54.9
2	United States	314	10.0	2	Australia	1402	11.9	2	Italy	1827	11.4
3	Argentina	151	4.8	3	United States	1374	11.7	3	United States	1481	9.2
4	Chile	128	4.1	4	Italy	1210	10.3	4	Chile	839	5.2
5	Australia	117	3.7	5	South Africa	892	7.6	5	South Africa	822	5.1
6	South Africa	50	1.6	6	Argentina	754	6.4	6	Argentina	773	4.8
7	Italy	27	0.8	7	Chile	753	6.4	7	Australia	753	4.7
8	Spain	1	0.0	8	New Zealand	163	1.4	8	Spain	213	1.3
9				9	Portugal	127	1.1	9	New Zealand	129	0.8
10				10	Canada	83	0.7	10	Portugal	125	0.8
11				11	Spain	78	0.7	11	Canada	101	0.6
12				12	Uruguay	45	0.4	12	Switzerland	44	0.3
13				13	Switzerland	31	0.3	13	Uruguay	41	0.3
14				14	Romania	19	0.2	14	Moldova	31	0.2
15				15	Turkey	16	0.1	15	Romania	20	0.1
16				16	Brazil	8	0.1	16	Turkey	15	0.1
17				17	Hungary	6	0.1	17	Hungary	13	0.1
18				18	Thailand	1	0.0	18	Brazil	11	0.1
19				19				19	Thailand	1	0.0
20				20				20			
	Others	12	0.4		Others				Others		
	Total	3160	100.0		Total	11785	100.0		Total	16063	100.0

Pedro Giménez

	2000	bearing area (ha)	% of world		**2010**	bearing area (ha)	% of world		**2016**	bearing area (ha)	% of world
1	Argentina	14862	100.0	1	Argentina	13384	100.0	1	Argentina	11197	100.0
	Total	14862	100.0		Total	13384	100.0		Total	11197	100.0

Pedro Ximénez (ranked 38)

	2000	bearing area (ha)	% of world		**2010**	bearing area (ha)	% of world		**2016**	bearing area (ha)	% of world
1	Spain	14803	85.7	1	Spain	9036	97.8	1	Spain	8528	96.8
2	Chile	2379	13.8	2	Portugal	197	2.1	2	Portugal	259	2.9
3	Australia	89	0.5	3	Uruguay	1	0.0	3	Australia	20	0.2
4	Others	1	0.0	4	Peru	1	0.0	4	South Africa	2	0.0
	Total	17272	100.0		Total	9235	100.0		Total	8809	100.0

Table 32: National ranking of top 20 countries, 8 top grey varieties, 2000, 2010 and 2016 (ha and %)

Pinot Gris	**2000**	bearing area (ha)	% of world	**2010**	bearing area (ha)	% of world	**2016**	bearing area (ha)	% of world
	1 Italy	6608	35.0	1 Italy	17281	39.5	1 Italy	18821	38.7
	2 Germany	2637	14.0	2 United States	5231	12.0	2 United States	7462	15.4
	3 Romania	2388	12.6	3 Germany	4514	10.3	3 Germany	4887	10.1
	4 Moldova	2042	10.8	4 Australia	3296	7.5	4 Australia	3652	7.5
	5 France	1969	10.4	5 France	2674	6.1	5 France	2867	5.9
	6 Hungary	890	4.7	6 Moldova	2042	4.7	6 New Zealand	2422	5.0
	7 United States	826	4.4	7 Hungary	1624	3.7	7 Hungary	1594	3.3
	8 Austria	293	1.5	8 New Zealand	1501	3.4	8 Romania	1561	3.2
	9 Canada	210	1.1	9 Romania	1301	3.0	9 Moldova	1208	2.5
	10 Luxembourg	155	0.8	10 Czechia	706	1.6	10 Czechia	826	1.7
	11 Switzerland	149	0.8	11 Ukraine	685	1.6	11 Canada	649	1.3
	12 New Zealand	127	0.7	12 Canada	549	1.3	12 Slovenia	508	1.0
	13 South Africa	104	0.5	13 Slovenia	501	1.1	13 Chile	437	0.9
	14 Argentina	13	0.1	14 Argentina	386	0.9	14 Argentina	401	0.8
	15 Chile	2	0.0	15 South Africa	261	0.6	15 South Africa	369	0.8
	16			16 Croatia	219	0.5	16 Switzerland	230	0.5
	17			17 Switzerland	216	0.5	17 Austria	224	0.5
	18			18 Austria	215	0.5	18 Luxembourg	196	0.4
	19			19 Slovakia	211	0.5	19 Serbia	112	0.2
	20			20 Luxembourg	146	0.3	20 Russia	78	0.2
	Others	480	2.5	Others	212	0.5	Others	66	0.1
	Total	18893	100.0	Total	43773	100.0	Total	48570	100.0

Cereza	**2000**	bearing area (ha)	% of world	**2010**	bearing area (ha)	% of world	**2016**	bearing area (ha)	% of world
	1 Argentina	31113	100.0	1 Argentina	29934	100.0	1 Argentina	28887	100.0
	Total	31113	100.0	Total	29934	100.0	Total	28887	100.0

Roditis	**2000**	bearing area (ha)	% of world	**2010**	bearing area (ha)	% of world	**2016**	bearing area (ha)	% of world
	1 Greece	299	100.0	1 Greece	4668	100.0	1 Greece	8463	100.0
	Total	299	100.0	Total	4668	100.0	Total	8463	100.0

Misket Cherve	**2000**	bearing area (ha)	% of world	**2010**	bearing area (ha)	% of world	**2016**	bearing area (ha)	% of world
	1			1 Bulgaria	4159	100.0	1 Bulgaria	4349	100.0
	Total	0	0.0	Total	4159	100.0	Total	4349	100.0

Cserszegi Fűsz	**2000**	bearing area (ha)	% of world	**2010**	bearing area (ha)	% of world	**2016**	bearing area (ha)	% of world
	1 Hungary	2185	100.0	1 Hungary	3609	100.0	1 Hungary	4299	100.0
	Total	2185	100.0	Total	3609	100.0	Total	4299	100.0

Garnacha Roj	**2000**	bearing area (ha)	% of world	**2010**	bearing area (ha)	% of world	**2016**	bearing area (ha)	% of world
	1 France	2635	95.4	1 France	1635	69.1	1 France	1253	85.7
	2 Spain	122	4.4	2 Greece	645	27.3	2 Greece	114	7.8
	3 Greece	1	0.0	3 Spain	77	3.3	3 Spain	87	5.9
	4			4 Thailand	6	0.2	4 Thailand	5	0.4
	5			5 South Africa	2	0.1	5 South Africa	3	0.2
	Others	4	0.1	Others			Others		
	Total	2761	100.0	Total	2366	100.0	Total	1462	100.0

Moschofilero	**2000**	bearing area (ha)	% of world	**2010**	bearing area (ha)	% of world	**2016**	bearing area (ha)	% of world
	1 Greece	718	100.0	1 Greece	1111	100.0	1 Greece	1088	100.0
	Total	718	100.0	Total	1111	100.0	Total	1088	100.0

Roditis (R)	**2000**	bearing area (ha)	% of world	**2010**	bearing area (ha)	% of world	**2016**	bearing area (ha)	% of world
	1 Greece	6945	100.0	1 Greece	3826	100.0	1 Greece	828	100.0
	Total	6945	100.0	Total	3826	100.0	Total	828	100.0

Table 33: National winegrape area for world's top 30 red varieties, 2000 (hectares)

Country	*Cabernet Sauvignon*	*Garnacha Tinta*	*Merlot*	*Mazuelo*	*Syrah*	*Bobal*	*Tempranillo*	*Monastrell*
Algeria	1510	6040	1510	7550	1510			
Argentina	13776	9	6263	57	8888		4720	0
Armenia								
Australia	24997	2139	7669	90	29295		41	948
Austria	312		112					
Brazil	587		469					
Bulgaria	10441		11169					
Canada	569		674					
Chile	35967		12825	641	2040		1	22
Croatia								
Cyprus								
Czechia								
France	53413	95717	101309	95745	50676		1549	7634
Georgia	223							
Germany								
Greece	688	22	183	14	39			
Hungary	1052		486					
Israel	607		647	971				
Italy	7682	6781	21861	1721	1025		7	
Korea Rep.								
Luxembourg								
Moldova	7590		8123					
Morocco		802		1692				
New Zealand	654		657		60			
Portugal	318						7356	
Romania	8620		7810					
Russia	1578							
Serbia								
Slovakia	156							
Slovenia			1197					
South Africa	8824	40	4888	71	5631		14	13
Spain	4519	98131	1186	8103	86	100128	79310	67160
Switzerland	34		848		54			
Taiwan								
Tunisia	337	2020		7576	337			337
United Kingdom								
United States	17573	4519	16875	3088	1509		201	187
Uruguay	675		1057		62			
"Missing 9"	20373	129	5549	374	1278		171	3
Total	**223074**	**216349**	**213368**	**127692**	**102490**	**100128**	**93370**	**76304**

Table 33 (cont.) National winegrape area for world's top 30 red varieties, 2000 (hectares)

Country	*Sangiovese*	*Pinot Noir*	*Cabernet Franc*	*Cinsaut*	*Gamay Noir*	*Alicante Henri Bouschet*	*Barbera*	*Montepulciano*
Algeria		1510		7550		3020		
Argentina	2490	1114	252	6	2	113	1055	49
Armenia								
Australia	372	3223	744				103	
Austria		409	27					
Brazil			3784					
Bulgaria		769						
Canada		457	567		263			
Chile	123	1614	689	195		2882		
Croatia								
Cyprus								
Czechia								
France	1564	26526	36094	31593	34537	8764		
Georgia								
Germany		8643						
Greece			40	108		21		
Hungary		243	526					
Israel								
Italy	62761	3287	6639	274	152	510	27175	28679
Korea Rep.								
Luxembourg		66			1			
Moldova		6521						
Morocco				3940		1098		
New Zealand		1098	118					
Portugal						675		
Romania		1740						
Russia								
Serbia								
Slovakia								
Slovenia								
South Africa	35	487	488	3533	36	8	15	
Spain		417	30	3	1	18321		
Switzerland		4601	16		1977			
Taiwan								
Tunisia	842			842		842		
United Kingdom								
United States	682	5343	1189	33	684	563	4693	
Uruguay			364					
"Missing 9"	8	741	409	351	144	341		
Total	**68877**	**68810**	**51974**	**48428**	**37798**	**37157**	**33041**	**28728**

Table 33 (cont.) National winegrape area for world's top 30 red varieties, 2000 (hectares)

Country	*Isabella*	*Tribidrag*	*Côt*	*Criolla Grande*	*Pamid*	*Douce Noire*	*Negroamaro*	*Doukkali*
Algeria								
Argentina	74	6	18230	24264		15659		
Armenia								
Australia			429					
Austria								
Brazil	14285							
Bulgaria					22581			
Canada								
Chile		91	929					
Croatia								
Cyprus								
Czechia								
France		1	6129			1		
Georgia								
Germany								
Greece					15			
Hungary					121			
Israel								
Italy		7828	251			2642	16619	
Korea Rep.								
Luxembourg								
Moldova	11401		39					
Morocco								16557
New Zealand			67					
Portugal								
Romania								
Russia								
Serbia								
Slovakia								
Slovenia								
South Africa		28	76					
Spain			23					
Switzerland			0					
Taiwan								
Tunisia		337						
United Kingdom								
United States	16	18630	97			21		
Uruguay								
"Missing 9"	1673		16					
Total	**27450**	**26922**	**26285**	**24264**	**22718**	**18323**	**16619**	**16557**

Table 33 (cont.) National winegrape area for world's top 30 red varieties, 2000 (hectares)

Country	*Listan Prieto*	*Prokupac*	*Castelão*	*Blaufränkisch*	*Mencía*	*Pinot Meunier*	*Other Red*	*Total Red*
Algeria								30200
Argentina	700					11	7513	105251
Armenia								0
Australia						107	7216	77372
Austria				2641			9477	12977
Brazil							12595	31721
Bulgaria							13313	58273
Canada							2211	4740
Chile	15181						13736	86936
Croatia				1189			13895	15084
Cyprus							14625	14625
Czechia				680			2719	3399
France						10621	34506	596380
Georgia							4340	4563
Germany				1118		2289	13130	25180
Greece							17772	18901
Hungary				6920			11591	20939
Israel							744	2970
Italy				111		10	132901	328918
Korea Rep.							5300	5300
Luxembourg								67
Moldova							2067	35741
Morocco							20098	44187
New Zealand							171	2825
Portugal			14424		1971		94557	119300
Romania							43406	61576
Russia							9128	10706
Serbia		15180					6210	21390
Slovakia				1091			1402	2649
Slovenia							4906	6103
South Africa						6	9320	33512
Spain	11				11166	7	71931	460533
Switzerland							382	7913
Taiwan							1530	1530
Tunisia							3367	16836
United Kingdom							200	200
United States	340			45		80	19787	96153
Uruguay							6397	8555
"Missing 9"				203			12095	43858
Total	**16232**	**15180**	**14424**	**13997**	**13138**	**13131**	**624537**	**2417365**

Table 34: National shares of global winegrape area for world's top 30 red varieties, 2000 (%)

Country	*Cabernet Sauvignon*	*Garnacha Tinta*	*Merlot*	*Mazuelo*	*Syrah*	*Bobal*	*Tempranillo*	*Monastrell*
Algeria	0.68	2.79	0.71	5.91	1.47			
Argentina	6.18	0.00	2.94	0.04	8.67		5.06	0.00
Armenia								
Australia	11.21	0.99	3.59	0.07	28.58		0.04	1.24
Austria	0.14		0.05					
Brazil	0.26		0.22					
Bulgaria	4.68		5.23					
Canada	0.25		0.32					
Chile	16.12		6.01	0.50	1.99		0.00	0.03
Croatia								
Cyprus								
Czechia								
France	23.94	44.24	47.48	74.98	49.45		1.66	10.00
Georgia	0.10							
Germany								
Greece	0.31	0.01	0.09	0.01	0.04			
Hungary	0.47		0.23					
Israel	0.27		0.30	0.76				
Italy	3.44	3.13	10.25	1.35	1.00		0.01	
Korea Rep.								
Luxembourg								
Moldova	3.40		3.81					
Morocco		0.37		1.33				
New Zealand	0.29		0.31		0.06			
Portugal	0.14						7.88	
Romania	3.86		3.66					
Russia	0.71							
Serbia								
Slovakia	0.07							
Slovenia			0.56					
South Africa	3.96	0.02	2.29	0.06	5.49		0.01	0.02
Spain	2.03	45.36	0.56	6.35	0.08	100.00	84.94	88.02
Switzerland	0.02		0.40		0.05			
Taiwan								
Tunisia	0.15	0.93		5.93	0.33			0.44
United Kingdom								
United States	7.88	2.09	7.91	2.42	1.47		0.21	0.25
Uruguay	0.30		0.50		0.06			
"Missing 9"	9.13	0.06	2.60	0.29	1.25		0.18	0.00
Total	**100**	**100**	**100**	**100**	**100**	**100**	**100**	**100**

Table 34 (cont.): National shares of global winegrape area for world's top 30 red varieties, 2000 (%)

Country	*Sangiovese*	*Pinot Noir*	*Cabernet Franc*	*Cinsaut*	*Gamay Noir*	*Alicante Henri Bouschet*	*Barbera*	*Montepulciano*
Algeria		2.19		15.59		8.13		
Argentina	3.62	1.62	0.48	0.01	0.00	0.31	3.19	0.17
Armenia								
Australia	0.54	4.68	1.43				0.31	
Austria		0.59	0.05					
Brazil			7.28					
Bulgaria		1.12						
Canada		0.66	1.09		0.70			
Chile	0.18	2.35	1.33	0.40		7.76		
Croatia								
Cyprus								
Czechia								
France	2.27	38.55	69.45	65.24	91.37	23.59		
Georgia								
Germany		12.56						
Greece			0.08	0.22		0.06		
Hungary		0.35	1.01					
Israel								
Italy	91.12	4.78	12.77	0.57	0.40	1.37	82.25	99.83
Korea Rep.								
Luxembourg		0.10			0.00			
Moldova		9.48						
Morocco				8.14		2.96		
New Zealand		1.60	0.23					
Portugal						1.82		
Romania		2.53						
Russia								
Serbia								
Slovakia								
Slovenia								
South Africa	0.05	0.71	0.94	7.30	0.10	0.02	0.05	
Spain		0.61	0.06	0.01	0.00	49.31		
Switzerland		6.69	0.03		5.23			
Taiwan								
Tunisia	1.22			1.74		2.27		
United Kingdom								
United States	0.99	7.76	2.29	0.07	1.81	1.51	14.20	
Uruguay			0.70					
"Missing 9"	0.01	1.08	0.79	0.73	0.38	0.92		
Total	**100**	**100**	**100**	**100**	**100**	**100**	**100**	**100**

Table 34 (cont.): National shares of global winegrape area for world's top 30 red varieties, 2000 (%)

Country	*Isabella*	*Tribidrag*	*Côt*	*Criolla Grande*	*Pamid*	*Douce Noire*	*Negroamaro*	*Doukkali*
Algeria								
Argentina	0.27	0.02	69.35	100.00		85.46		
Armenia								
Australia			1.63					
Austria								
Brazil	52.04							
Bulgaria					99.40			
Canada								
Chile		0.34	3.53					
Croatia								
Cyprus								
Czechia								
France		0.01	23.32			0.00		
Georgia								
Germany								
Greece					0.07			
Hungary					0.53			
Israel								
Italy		29.08	0.96			14.42	100.00	
Korea Rep.								
Luxembourg								
Moldova	41.53		0.15					
Morocco								100.00
New Zealand			0.25					
Portugal								
Romania								
Russia								
Serbia								
Slovakia								
Slovenia								
South Africa		0.10	0.29					
Spain			0.09					
Switzerland			0.00					
Taiwan								
Tunisia		1.25						
United Kingdom								
United States	0.06	69.20	0.37			0.11		
Uruguay								
"Missing 9"	6.10		0.06					
Total	**100**	**100**	**100**	**100**	**100**	**100**	**100**	**100**

Table 34 (cont.): National shares of global winegrape area for world's top 30 red varieties, 2000 (%)

Country	*Listan Prieto*	*Prokupac*	*Castelão*	*Blaufränkisch*	*Mencía*	*Pinot Meunier*	*Other Red*	*Total Red*
Algeria								1.25
Argentina	4.31					0.08	1.20	4.35
Armenia								0.00
Australia						0.82	1.16	3.20
Austria				18.87			1.52	0.54
Brazil							2.02	1.31
Bulgaria							2.13	2.41
Canada							0.35	0.20
Chile	93.53						2.20	3.60
Croatia				8.49			2.22	0.62
Cyprus							2.34	0.61
Czechia				4.86			0.44	0.14
France						80.89	5.53	24.67
Georgia							0.69	0.19
Germany				7.99		17.43	2.10	1.04
Greece							2.85	0.78
Hungary				49.44			1.86	0.87
Israel							0.12	0.12
Italy				0.79		0.07	21.28	13.61
Korea Rep.							0.85	0.22
Luxembourg								0.00
Moldova							0.33	1.48
Morocco							3.22	1.83
New Zealand							0.03	0.12
Portugal			100.00		15.01		15.14	4.94
Romania							6.95	2.55
Russia							1.46	0.44
Serbia		100.00					0.99	0.88
Slovakia				7.79			0.22	0.11
Slovenia							0.79	0.25
South Africa						0.05	1.49	1.39
Spain	0.07				84.99	0.05	11.52	19.05
Switzerland							0.06	0.33
Taiwan							0.24	0.06
Tunisia							0.54	0.70
United Kingdom							0.03	0.01
United States	2.09			0.32		0.61	3.17	3.98
Uruguay							1.02	0.35
"Missing 9"				1.45			1.94	1.81
Total	**100**	**100**	**100**	**100**	**100**	**100**	**100**	**100**

Table 35: Shares of world's top 30 red varieties in national winegrape area, by country, 2000 (%)

Country	*Cabernet Sauvignon*	*Garnacha Tinta*	*Merlot*	*Mazuelo*	*Syrah*	*Bobal*	*Tempranillo*	*Monastrell*
Algeria	5.00	20.00	5.00	25.00	5.00			
Argentina	6.98	0.00	3.17	0.03	4.50		2.39	0.00
Armenia								
Australia	19.14	1.64	5.87	0.07	22.43		0.03	0.73
Austria	0.64		0.23					
Brazil	1.11		0.89					
Bulgaria	10.88		11.64					
Canada	6.69		7.93					
Chile	31.56		11.25	0.56	1.79		0.00	0.02
Croatia								
Cyprus								
Czechia								
France	6.18	11.07	11.71	11.07	5.86		0.18	0.88
Georgia	0.60							
Germany								
Greece	1.35	0.04	0.36	0.03	0.08			
Hungary	1.21		0.56					
Israel	12.51		13.35	20.02				
Italy	1.21	1.07	3.43	0.27	0.16		0.00	
Korea Rep.								
Luxembourg								
Moldova	8.45		9.04					
Morocco		1.62		3.41				
New Zealand	6.58		6.61		0.60			
Portugal	0.16						3.59	
Romania	3.88		3.52					
Russia	2.80							
Serbia								
Slovakia	1.00							
Slovenia			5.10					
South Africa	9.42	0.04	5.22	0.08	6.01		0.01	0.01
Spain	0.38	8.30	0.10	0.69	0.01	8.47	6.71	5.68
Switzerland	0.23		5.64		0.36			
Taiwan								
Tunisia	2.00	12.00		45.00	2.00			2.00
United Kingdom								
United States	10.00	2.57	9.60	1.76	0.86		0.11	0.11
Uruguay	7.60		11.90		0.70			
"Missing 9"	25.40	0.16	6.92	0.47	1.59		0.21	0.00
Total	**4.56**	**4.43**	**4.37**	**2.61**	**2.10**	**2.05**	**1.91**	**1.56**

Table 35 (cont.): Shares of world's top 30 red varieties in national winegrape area, by country, 2000 (%)

Country	*Sangiovese*	*Pinot Noir*	*Cabernet Franc*	*Cinsaut*	*Gamay Noir*	*Alicante Henri Bouschet*	*Barbera*	*Montepulciano*
Algeria		5.00		25.00		10.00		
Argentina	1.26	0.56	0.13	0.00	0.00	0.06	0.53	0.02
Armenia								
Australia	0.28	2.47	0.57				0.08	
Austria		0.84	0.06					
Brazil			7.16					
Bulgaria		0.80						
Canada		5.38	6.67		3.10			
Chile	0.11	1.42	0.60	0.17		2.53		
Croatia								
Cyprus								
Czechia								
France	0.18	3.07	4.17	3.65	3.99	1.01		
Georgia								
Germany		8.29						
Greece			0.08	0.21		0.04		
Hungary		0.28	0.61					
Israel								
Italy	9.86	0.52	1.04	0.04	0.02	0.08	4.27	4.50
Korea Rep.								
Luxembourg		4.90			0.07			
Moldova		7.26						
Morocco				7.94		2.21		
New Zealand		11.04	1.19					
Portugal						0.33		
Romania		0.78						
Russia								
Serbia								
Slovakia								
Slovenia								
South Africa	0.04	0.52	0.52	3.77	0.04	0.01	0.02	
Spain		0.04	0.00	0.00	0.00	1.55		
Switzerland		30.59	0.10		13.14			
Taiwan								
Tunisia	5.00			5.00		5.00		
United Kingdom								
United States	0.39	3.04	0.68	0.02	0.39	0.32	2.67	
Uruguay			4.10					
"Missing 9"	0.01	0.92	0.51	0.44	0.18	0.42		
Total	**1.41**	**1.41**	**1.06**	**0.99**	**0.77**	**0.76**	**0.68**	**0.59**

Table 35 (cont.): Shares of world's top 30 red varieties in national winegrape area, by country, 2000 (%)

Country	*Isabella*	*Tribidrag*	*Côt*	*Criolla Grande*	*Pamid*	*Douce Noire*	*Negroamaro*	*Doukkali*
Algeria								
Argentina	0.04	0.00	9.23	12.29		7.93		
Armenia								
Australia			0.33					
Austria								
Brazil	27.04							
Bulgaria					23.52			
Canada								
Chile		0.08	0.82					
Croatia								
Cyprus								
Czechia								
France		0.00	0.71			0.00		
Georgia								
Germany								
Greece					0.03			
Hungary					0.14			
Israel								
Italy		1.23	0.04			0.42	2.61	
Korea Rep.								
Luxembourg								
Moldova	12.69		0.04					
Morocco								33.38
New Zealand			0.67					
Portugal								
Romania								
Russia								
Serbia								
Slovakia								
Slovenia								
South Africa		0.03	0.08					
Spain			0.00					
Switzerland			0.00					
Taiwan								
Tunisia		2.00						
United Kingdom								
United States	0.01	10.60	0.06			0.01		
Uruguay								
"Missing 9"	2.09		0.02					
Total	**0.56**	**0.55**	**0.54**	**0.50**	**0.46**	**0.37**	**0.34**	**0.34**

Table 35 (cont.): Shares of world's top 30 red varieties in national winegrape area, by country, 2000 (%)

Country	*Listan Prieto*	*Prokupac*	*Castelão*	*Blaufränkisch*	*Mencía*	*Pinot Meunier*	*Other Red*	*Total Red*	*Total*
Algeria								100.00	100
Argentina	0.35					0.01	3.81	53.31	100
Armenia								0.00	100
Australia						0.08	5.53	59.24	100
Austria				5.44			19.54	26.76	100
Brazil							23.84	60.03	100
Bulgaria							13.87	60.70	100
Canada							26.02	55.78	100
Chile	13.32						12.05	76.28	100
Croatia				2.00			23.37	25.37	100
Cyprus							80.00	80.00	100
Czechia				6.00			24.00	30.00	100
France						1.23	3.99	68.96	100
Georgia							11.60	12.19	100
Germany				1.07		2.20	12.60	24.16	100
Greece							34.91	37.12	100
Hungary				7.96			13.34	24.10	100
Israel							15.35	61.22	100
Italy				0.02		0.00	20.87	51.66	100
Korea Rep.							98.15	98.15	100
Luxembourg								4.97	100
Moldova							2.30	39.78	100
Morocco							40.52	89.09	100
New Zealand							1.72	28.41	100
Portugal			7.04		0.96		46.12	58.19	100
Romania							19.54	27.72	100
Russia							16.20	19.00	100
Serbia		22.00					9.00	31.00	100
Slovakia				7.00			9.00	17.00	100
Slovenia							20.90	26.00	100
South Africa						0.01	9.95	35.78	100
Spain	0.00				0.94	0.00	6.09	38.97	100
Switzerland							2.54	52.61	100
Taiwan							54.00	54.00	100
Tunisia							20.00	100.00	100
United Kingdom							22.91	22.91	100
United States	0.19			0.03		0.05	11.26	54.73	100
Uruguay							72.04	96.34	100
"Missing 9"				0.25			15.08	54.67	100
Total	**0.33**	**0.31**	**0.30**	**0.29**	**0.27**	**0.27**	**12.78**	**49.46**	**100**

Table 36: National winegrape area for world's top 30 red varieties, 2010 (hectares)

Country	*Cabernet Sauvignon*	*Merlot*	*Tempranillo*	*Syrah*	*Garnacha Tinta*	*Pinot Noir*	*Bobal*	*Sangiovese*
Algeria	1510	1510		1510	6040	1510		
Argentina	17674	6953	6565	13093	22	1689		2231
Armenia								
Australia	25967	10028	476	42675	1748	4690		589
Austria	592	644		137		646		
Brazil	914	766	16		1	145		26
Bulgaria	8436	10573						
Canada	542	999	6	274	2	640		3
Chile	40728	10041	48	6027	37	2884		100
China	22612	3560		223	11	40		
Croatia	646	780		187	103	180		
Cyprus	369	63		244	84			
Czechia	230	90				688		
Ethiopia								90
France	54434	114675	680	67382	90991	30086		1511
Georgia	286							
Germany	295	471				11724		
Greece	1550	1248	19	641				
Hungary	2863	1907		177		1091		1
Israel	607	647						
Italy	13724	28042	23	6739	6372	5046		71619
Japan	469	817				64		
Kazakhstan	20					180		
Korea Rep.								
Luxembourg						3		
Mexico	756	391	229	145	140			
Moldova	7590	8123				6521		
Morocco					802			
Myanmar	1		3	27		7		
New Zealand	517	1369	7	293	2	4776		6
Peru	48	2		2	1	1		
Portugal	1671	772	16706	3501	84	148		
Romania	3718	10988	70	470		1089		88
Russia	3593	1588				533		
Serbia								
Slovakia	570							
Slovenia	453	996						
South Africa	12325	6497	38	10136	187	962		61
Spain	23237	15540	207677	20000	70202	1044	80120	
Switzerland	63	1028		181		4402		
Taiwan								
Thailand	7		4	66		1		2
Tunisia	337			337	2020			842
Turkey	391	355	9	1367	33	3		9
Ukraine	4869	2820				767		
United Kingdom	1	2				233		
United States	34788	22729	414	9197	2666	16776		852
Uruguay	682	875		87	5	55		
Total	**290083**	**267888**	**232988**	**185117**	**181553**	**98623**	**80120**	**78030**

Table 36 (cont.) National winegrape area for world's top 30 red varieties, 2010 (hectares)

Country	*Mazuelo*	*Monastrell*	*Cabernet Franc*	*Alicante Henri Bouschet*	*Côt*	*Montepulciano*	*Cinsaut*	*Tribidrag*
Algeria	7550			3020			7550	
Argentina	30	1	592	257	28543	94	6	0
Armenia								
Australia		692	591		356			149
Austria			56					
Brazil			8516	129	37	2		0
Bulgaria								
Canada		2	664		39			8
Chile	477	59	1321	4228	1264		198	58
China			507				3	
Croatia	34		95					65
Cyprus	481	172	203					
Czechia								
Ethiopia								
France	47720	9257	36302	4322	6123		19505	2
Georgia								
Germany			15					
Greece	16		23	56			43	
Hungary			1352	21	0			
Israel	971							
Italy	2023		6314	645	260	34824	51	12234
Japan								
Kazakhstan			56					
Korea Rep.								
Luxembourg								
Mexico	448							
Moldova					39			
Morocco	1692			1098			3940	
Myanmar	4			0				
New Zealand			163		156	7		4
Peru					10			
Portugal	338		24	3322			17	
Romania		4	73	20	7			8
Russia			20					
Serbia								
Slovakia								
Slovenia								
South Africa	81	403	934	10	450		2052	34
Spain	4749	58406	849	19551	93			
Switzerland			54		10			
Taiwan								
Thailand								
Tunisia	7576	337		842			842	337
Turkey	84	4	23	488	13		500	
Ukraine								
United Kingd			1					
United States	1441	404	2215	431	717	29	45	19857
Uruguay		1	334	22	41			
Total	**75716**	**69742**	**61295**	**38462**	**38158**	**34956**	**34751**	**32755**

Table 36 (cont.) National winegrape area for world's top 30 red varieties, 2010 (hectares)

Country	*Isabella*	*Gamay Noir*	*Barbera*	*Criolla Grande*	*Douce Noire*	*Blaufränkisch*	*Nero d'Avola*	*Doukkali*
Algeria								
Argentina		2	733	20745	18775		4	
Armenia								
Australia			116					
Austria						3228		
Brazil	18279	16	5		0			
Bulgaria								
Canada		220	1			4		
Chile		0	4					
China								
Croatia						558		
Cyprus								
Czechia						1160		
Ethiopia								
France		29698			0			
Georgia								
Germany						1749		
Greece								
Hungary		3				7998	50	
Israel								
Italy		122	20524		821	59	16595	
Japan								
Kazakhstan								
Korea Rep.								
Luxembourg		98						
Mexico								
Moldova	11401							
Morocco								16557
Myanmar								
New Zealand		12						
Peru						290		
Portugal		0						
Romania			0			760		
Russia	162							
Serbia								
Slovakia						1378		
Slovenia			134			680		
South Africa		19	51					
Spain								
Switzerland		1521				3		
Taiwan								
Thailand								
Tunisia								
Turkey		206						
Ukraine	2396							
United Kingdo		1						
United States			2798		34	22		
Uruguay	256	9						
Total	**32494**	**31927**	**24366**	**20745**	**19630**	**17890**	**16649**	**16557**

Table 36 (cont.) National winegrape area for world's top 30 red varieties, 2010 (hectares)

Country	*Prokupac*	*Pinot Meunier*	*Concord*	*Touriga Franca*	*Negroamaro*	*Carmenère*	*Other Red*	*Total Red*
Algeria								30200
Argentina		11		0		33	7035	125090
Armenia								0
Australia							4124	92200
Austria							10835	16137
Brazil			3543			7	8476	40875
Bulgaria							16534	35543
Canada		5	252			2	887	4548
Chile						8827	4718	81017
China						1353	40	28350
Croatia						19	4418	7085
Cyprus							4091	5707
Czechia							3649	5817
Ethiopia							21	111
France		11087				29	37729	561535
Georgia							5567	5853
Germany		2301					20242	36797
Greece							20369	23965
Hungary						0	5625	21090
Israel							744	2970
Italy		14			11492	1074	117451	356067
Japan							864	2213
Kazakhstan							1084	1340
Korea Rep.							5300	5300
Luxembourg							0	101
Mexico							1300	3409
Moldova							2067	35741
Morocco							19564	43653
Myanmar							2	44
New Zealanc		19					358	7689
Peru							1938	2292
Portugal				11582			71660	109824
Romania				4			35524	52823
Russia							3899	9795
Serbia	15180						6210	21390
Slovakia							1987	3934
Slovenia							3019	5282
South Africa		13		3			9794	44049
Spain		1					57390	558859
Switzerland							1312	8574
Taiwan							1530	1530
Thailand							19	99
Tunisia							3367	16836
Turkey							5194	8677
Ukraine							5297	16149
United King		50					117	405
United State		66	8421			22	19152	143076
Uruguay			22				3763	6152
Total	**15180**	**13566**	**12238**	**11590**	**11492**	**11366**	**534267**	**2590193**

Table 37: National shares of global winegrape area for world's top 30 red varieties, 2010 (%)

Country	*Cabernet Sauvignon*	*Merlot*	*Tempranillo*	*Syrah*	*Garnacha Tinta*	*Pinot Noir*	*Bobal*	*Sangiovese*
Algeria	0.52	0.56		0.82	3.33	1.53		
Argentina	6.09	2.60	2.82	7.07	0.01	1.71		2.86
Armenia								
Australia	8.95	3.74	0.20	23.05	0.96	4.76		0.75
Austria	0.20	0.24		0.07		0.65		
Brazil	0.31	0.29	0.01		0.00	0.15		0.03
Bulgaria	2.91	3.95						
Canada	0.19	0.37	0.00	0.15	0.00	0.65		0.00
Chile	14.04	3.75	0.02	3.26	0.02	2.92		0.13
China	7.80	1.33		0.12	0.01	0.04		
Croatia	0.22	0.29		0.10	0.06	0.18		
Cyprus	0.13	0.02		0.13	0.05			
Czechia	0.08	0.03				0.70		
Ethiopia								0.12
France	18.77	42.81	0.29	36.40	50.12	30.51		1.94
Georgia	0.10							
Germany	0.10	0.18				11.89		
Greece	0.53	0.47	0.01	0.35				
Hungary	0.99	0.71		0.10		1.11		0.00
Israel	0.21	0.24						
Italy	4.73	10.47	0.01	3.64	3.51	5.12		91.78
Japan	0.16	0.30				0.06		
Kazakhstan	0.01					0.18		
Korea Rep.								
Luxembourg						0.00		
Mexico	0.26	0.15	0.10	0.08	0.08			
Moldova	2.62	3.03				6.61		
Morocco					0.44			
Myanmar	0.00		0.00	0.01		0.01		
New Zealand	0.18	0.51	0.00	0.16	0.00	4.84		0.01
Peru	0.02	0.00		0.00	0.00	0.00		
Portugal	0.58	0.29	7.17	1.89	0.05	0.15		
Romania	1.28	4.10	0.03	0.25		1.10		0.11
Russia	1.24	0.59				0.54		
Serbia								
Slovakia	0.20							
Slovenia	0.16	0.37						
South Africa	4.25	2.43	0.02	5.48	0.10	0.98		0.08
Spain	8.01	5.80	89.14	10.80	38.67	1.06	100.00	
Switzerland	0.02	0.38		0.10		4.46		
Taiwan								
Thailand	0.00		0.00	0.04		0.00		0.00
Tunisia	0.12			0.18	1.11			1.08
Turkey	0.13	0.13	0.00	0.74	0.02	0.00		0.01
Ukraine	1.68	1.05				0.78		
United Kingdom	0.00	0.00				0.24		
United States	11.99	8.48	0.18	4.97	1.47	17.01		1.09
Uruguay	0.24	0.33		0.05	0.00	0.06		
Total	**100**	**100**	**100**	**100**	**100**	**100**	**100**	**100**

Table 37 (cont.): National shares of global winegrape area for world's top 30 red varieties, 2010 (%)

Country	*Mazuelo*	*Monastrell*	*Cabernet Franc*	*Alicante Henri Bouschet*	*Côt*	*Montepulciano*	*Cinsaut*	*Tribidrag*
Algeria	9.97			7.85			21.73	
Argentina	0.04	0.00	0.97	0.67	74.80	0.27	0.02	0.00
Armenia								
Australia		0.99	0.96		0.93			0.46
Austria			0.09					
Brazil			13.89	0.33	0.10	0.00		0.00
Bulgaria								
Canada		0.00	1.08		0.10			0.02
Chile	0.63	0.08	2.15	10.99	3.31		0.57	0.18
China			0.83				0.01	
Croatia	0.05		0.16					0.20
Cyprus	0.64	0.25	0.33					
Czechia								
Ethiopia								
France	63.03	13.27	59.22	11.24	16.05		56.13	0.00
Georgia								
Germany			0.02					
Greece	0.02		0.04	0.15			0.12	
Hungary			2.21	0.06	0.00			
Israel	1.28							
Italy	2.67		10.30	1.68	0.68	99.62	0.15	37.35
Japan								
Kazakhstan			0.09					
Korea Rep.								
Luxembourg								
Mexico	0.59							
Moldova					0.10			
Morocco	2.23			2.85			11.34	
Myanmar	0.01			0.00				
New Zealand			0.27		0.41	0.02		0.01
Peru					0.03			
Portugal	0.45		0.04	8.64			0.05	
Romania		0.01	0.12	0.05	0.02			0.03
Russia			0.03					
Serbia								
Slovakia								
Slovenia								
South Africa	0.11	0.58	1.52	0.03	1.18		5.90	0.10
Spain	6.27	83.75	1.39	50.83	0.24			
Switzerland			0.09		0.03			
Taiwan								
Thailand								
Tunisia	10.01	0.48		2.19			2.42	1.03
Turkey	0.11	0.01	0.04	1.27	0.03		1.44	
Ukraine								
United Kingd			0.00					
United States	1.90	0.58	3.61	1.12	1.88	0.08	0.13	60.62
Uruguay		0.00	0.54	0.06	0.11			
Total	**100**	**100**	**100**	**100**	**100**	**100**	**100**	**100**

Table 37 (cont.): National shares of global winegrape area for world's top 30 red varieties, 2010 (%)

Country	*Isabella*	*Gamay Noir*	*Barbera*	*Criolla Grande*	*Douce Noire*	*Blaufränkisch*	*Nero d'Avola*	*Doukkali*
Algeria								
Argentina		0.01	3.01	100.00	95.65		0.02	
Armenia								
Australia			0.48					
Austria						18.04		
Brazil	56.25	0.05	0.02		0.00			
Bulgaria								
Canada		0.69	0.01			0.02		
Chile		0.00	0.01					
China								
Croatia						3.12		
Cyprus								
Czechia						6.48		
Ethiopia								
France		93.02			0.00			
Georgia								
Germany						9.78		
Greece								
Hungary		0.01				44.71	0.30	
Israel								
Italy		0.38	84.23		4.18	0.33	99.67	
Japan								
Kazakhstan								
Korea Rep.								
Luxembourg		0.31						
Mexico								
Moldova	35.09							
Morocco								100.00
Myanmar								
New Zealand		0.04						
Peru						1.62		
Portugal		0.00						
Romania			0.00			4.25		
Russia	0.50							
Serbia								
Slovakia						7.70		
Slovenia			0.55			3.80		
South Africa		0.06	0.21					
Spain								
Switzerland		4.76				0.02		
Taiwan								
Thailand								
Tunisia								
Turkey		0.65						
Ukraine	7.37							
United Kingdc		0.00						
United States			11.48		0.17	0.12		
Uruguay	0.79	0.03						
Total	**100**	**100**	**100**	**100**	**100**	**100**	**100**	**100**

Table 37 (cont.): National shares of global winegrape area for world's top 30 red varieties, 2010 (%)

Country	*Prokupac*	*Pinot Meunier*	*Concord*	*Touriga Franca*	*Negroamaro*	*Carmenère*	*Other Red*	*Total Red*
Algeria								1.17
Argentina		0.08		0.00		0.29	1.32	4.83
Armenia								0.00
Australia							0.77	3.56
Austria							2.03	0.62
Brazil			28.95			0.06	1.59	1.58
Bulgaria							3.09	1.37
Canada		0.03	2.06			0.02	0.17	0.18
Chile						77.66	0.88	3.13
China						11.90	0.01	1.09
Croatia						0.16	0.83	0.27
Cyprus							0.77	0.22
Czechia							0.68	0.22
Ethiopia							0.00	0.00
France		81.72				0.26	7.06	21.68
Georgia							1.04	0.23
Germany		16.96					3.79	1.42
Greece							3.81	0.93
Hungary						0.00	1.05	0.81
Israel							0.14	0.11
Italy		0.10			100.00	9.45	21.98	13.75
Japan							0.16	0.09
Kazakhstan							0.20	0.05
Korea Rep.							0.99	0.20
Luxembourg							0.00	0.00
Mexico							0.24	0.13
Moldova							0.39	1.38
Morocco							3.66	1.69
Myanmar							0.00	0.00
New Zealanc		0.14					0.07	0.30
Peru							0.36	0.09
Portugal				99.93			13.41	4.24
Romania				0.03			6.65	2.04
Russia							0.73	0.38
Serbia	100.00						1.16	0.83
Slovakia							0.37	0.15
Slovenia							0.57	0.20
South Africa		0.09		0.03			1.83	1.70
Spain		0.01					10.74	21.58
Switzerland							0.25	0.33
Taiwan							0.29	0.06
Thailand							0.00	0.00
Tunisia							0.63	0.65
Turkey							0.97	0.33
Ukraine							0.99	0.62
United King		0.37					0.02	0.02
United State		0.48	68.81			0.20	3.58	5.52
Uruguay			0.18				0.70	0.24
Total	**100**	**100**	**100**	**100**	**100**	**100**	**100**	**100**

Table 38: Shares of world's top 30 red varieties in national winegrape area, by country, 2010 (%)

Country	*Cabernet Sauvignon*	*Merlot*	*Tempranillo*	*Syrah*	*Garnacha Tinta*	*Pinot Noir*	*Bobal*	*Sangiovese*
Algeria	5.00	5.00		5.00	20.00	5.00		
Argentina	8.28	3.26	3.08	6.14	0.01	0.79		1.05
Armenia								
Australia	17.11	6.61	0.31	28.12	1.15	3.09		0.39
Austria	1.30	1.41		0.30		1.42		
Brazil	1.85	1.55	0.03		0.00	0.29		0.05
Bulgaria	15.03	18.84						
Canada	5.37	9.89	0.06	2.71	0.02	6.34		0.03
Chile	36.52	9.00	0.04	5.40	0.03	2.59		0.09
China	76.54	12.05		0.75	0.04	0.14		
Croatia	3.11	3.76		0.90	0.50	0.87		
Cyprus	4.29	0.73		2.83	0.98			
Czechia	1.42	0.55				4.24		
Ethiopia								53.20
France	6.51	13.72	0.08	8.06	10.89	3.60		0.18
Georgia	0.60							
Germany	0.29	0.46				11.48		
Greece	2.85	2.29	0.03	1.18				
Hungary	4.11	2.74		0.25		1.56		0.00
Israel	12.51	13.35						
Italy	2.19	4.48	0.00	1.08	1.02	0.81		11.45
Japan	12.63	21.98				1.71		
Kazakhstan	0.28					2.59		
Korea Rep.								
Luxembourg						0.23		
Mexico	13.83	7.15	4.19	2.65	2.56			
Moldova	8.45	9.04				7.26		
Morocco					1.64			
Myanmar	0.97		4.46	35.83		8.98		
New Zealand	1.62	4.28	0.02	0.92	0.01	14.94		0.02
Peru	1.25	0.05		0.05	0.03	0.03		
Portugal	1.02	0.47	10.22	2.14	0.05	0.09		
Romania	2.18	6.45	0.04	0.28		0.64		0.05
Russia	14.02	6.20				2.08		
Serbia								
Slovakia	4.51							
Slovenia	2.77	6.09						
South Africa	12.20	6.43	0.04	10.03	0.19	0.95		0.06
Spain	2.26	1.51	20.20	1.95	6.83	0.10	7.79	
Switzerland	0.42	6.94		1.22		29.71		
Taiwan								
Thailand	4.65		2.85	44.46		0.35		1.41
Tunisia	2.00			2.00	12.00			5.00
Turkey	3.04	2.76	0.07	10.63	0.25	0.02		0.07
Ukraine	9.31	5.39				1.47		
United Kingdom	0.08	0.17				19.45		
United States	15.26	9.97	0.18	4.03	1.17	7.36		0.37
Uruguay	8.91	11.43		1.14	0.07	0.72		
Total	**6.29**	**5.81**	**5.05**	**4.01**	**3.93**	**2.14**	**1.74**	**1.69**

Table 38 (cont.): Shares of world's top 30 red varieties in national winegrape area, by country, 2010 (%)

Country	*Mazuelo*	*Monastrell*	*Cabernet Franc*	*Alicante Henri Bouschet*	*Côt*	*Montepulciano*	*Cinsaut*	*Tribidrag*
Algeria	25.00			10.00			25.00	
Argentina	0.01	0.00	0.28	0.12	13.38	0.04	0.00	0.00
Armenia								
Australia		0.46	0.39		0.23			0.10
Austria			0.12					
Brazil			17.23	0.26	0.08	0.00		0.00
Bulgaria								
Canada		0.02	6.58		0.38			0.08
Chile	0.43	0.05	1.18	3.79	1.13		0.18	0.05
China			1.71				0.01	
Croatia	0.17		0.46					0.31
Cyprus	5.59	2.00	2.36					
Czechia								
Ethiopia								
France	5.71	1.11	4.34	0.52	0.73		2.33	0.00
Georgia								
Germany			0.01					
Greece	0.03		0.04	0.10			0.08	
Hungary			1.94	0.03	0.00			
Israel	20.02							
Italy	0.32		1.01	0.10	0.04	5.57	0.01	1.96
Japan								
Kazakhstan			0.81					
Korea Rep.								
Luxembourg								
Mexico	8.20							
Moldova					0.04			
Morocco	3.45			2.24			8.04	
Myanmar	5.87			0.31				
New Zealand			0.51		0.49	0.02		0.01
Peru					0.26			
Portugal	0.21		0.01	2.03			0.01	
Romania		0.00	0.04	0.01	0.00			0.00
Russia			0.08					
Serbia								
Slovakia								
Slovenia								
South Africa	0.08	0.40	0.92	0.01	0.45		2.03	0.03
Spain	0.46	5.68	0.08	1.90	0.01			
Switzerland			0.36		0.07			
Taiwan								
Thailand								
Tunisia	45.00	2.00		5.00			5.00	2.00
Turkey	0.65	0.03	0.18	3.79	0.10		3.89	
Ukraine								
United Kingd			0.08					
United States	0.63	0.18	0.97	0.19	0.31	0.01	0.02	8.71
Uruguay		0.01	4.36	0.29	0.54			
Total	**1.64**	**1.51**	**1.33**	**0.83**	**0.83**	**0.76**	**0.75**	**0.71**

Table 38 (cont.): Shares of world's top 30 red varieties in national winegrape area, by country, 2010 (%)

Country	*Isabella*	*Gamay Noir*	*Barbera*	*Criolla Grande*	*Douce Noire*	*Blaufränkisch*	*Nero d'Avola*	*Doukkali*
Algeria								
Argentina		0.00	0.34	9.72	8.80		0.00	
Armenia								
Australia			0.08					
Austria						7.09		
Brazil	36.99	0.03	0.01		0.00			
Bulgaria								
Canada		2.18	0.01			0.04		
Chile		0.00	0.00					
China								
Croatia						2.69		
Cyprus								
Czechia						7.14		
Ethiopia								
France		3.55			0.00			
Georgia								
Germany						1.71		
Greece								
Hungary		0.00				11.47	0.07	
Israel								
Italy		0.02	3.28		0.13	0.01	2.65	
Japan								
Kazakhstan								
Korea Rep.								
Luxembourg		7.52						
Mexico								
Moldova	12.69							
Morocco								33.79
Myanmar								
New Zealand		0.04						
Peru						7.57		
Portugal		0.00						
Romania			0.00			0.45		
Russia	0.63							
Serbia								
Slovakia						10.91		
Slovenia			0.82			4.16		
South Africa		0.02	0.05					
Spain								
Switzerland		10.26				0.02		
Taiwan								
Thailand								
Tunisia								
Turkey		1.60						
Ukraine	4.58							
United Kingdo		0.08						
United States			1.23		0.01	0.01		
Uruguay	3.34	0.12						
Total	**0.70**	**0.69**	**0.53**	**0.45**	**0.43**	**0.39**	**0.36**	**0.36**

Table 38 (cont.): Shares of world's top 30 red varieties in national winegrape area, by country, 2010 (%)

Country	*Prokupac*	*Pinot Meunier*	*Concord*	*Touriga Franca*	*Negroamaro*	*Carmenèr e*	*Other Red*	*Total Red*	*Total*
Algeria								100.00	100
Argentina		0.01		0.00		0.02	3.30	58.63	100
Armenia								0.00	100
Australia							2.72	60.74	100
Austria							23.79	35.44	100
Brazil			7.17			0.01	17.15	82.72	100
Bulgaria							29.46	63.32	100
Canada		0.05	2.50			0.02	8.79	45.05	100
Chile						7.91	4.23	72.65	100
China						4.58	0.14	95.95	100
Croatia						0.09	21.29	34.14	100
Cyprus							47.53	66.30	100
Czechia							22.47	35.81	100
Ethiopia							12.34	65.54	100
France		1.33				0.00	4.52	67.21	100
Georgia							11.60	12.19	100
Germany		2.25					19.82	36.03	100
Greece							37.45	44.06	100
Hungary						0.00	8.07	30.25	100
Israel							15.35	61.22	100
Italy		0.00			1.84	0.17	18.77	56.91	100
Japan							23.25	59.58	100
Kazakhstan							15.63	19.32	100
Korea Rep.							98.15	98.15	100
Luxembourg							0.01	7.75	100
Mexico							23.79	62.38	100
Moldova							2.30	39.78	100
Morocco							39.93	89.09	100
Myanmar							2.57	58.99	100
New Zealanc		0.06					1.12	24.05	100
Peru							50.59	59.83	100
Portugal				7.08			43.82	67.16	100
Romania				0.00			20.86	31.02	100
Russia							15.22	38.22	100
Serbia	22.00						9.00	31.00	100
Slovakia							15.72	31.13	100
Slovenia							18.46	32.30	100
South Africa		0.01		0.00			9.70	43.61	100
Spain		0.00					5.58	54.35	100
Switzerland							8.86	57.86	100
Taiwan							54.00	54.00	100
Thailand							12.67	66.40	100
Tunisia							20.00	100.00	100
Turkey							40.40	67.49	100
Ukraine							10.13	30.88	100
United King		4.17					9.77	33.81	100
United State		0.03	3.69			0.01	8.40	62.77	100
Uruguay			0.29				49.14	80.34	100
Total	**0.33**	**0.29**	**0.27**	**0.25**	**0.25**	**0.25**	**11.58**	**56.13**	**100**

Table 39: National winegrape area for world's top 30 red varieties, 2016 (hectares)

Country	*Cabernet Sauvignon*	*Merlot*	*Tempranillo*	*Syrah*	*Garnacha Tinta*	*Pinot Noir*	*Sangiovese*	*Bobal*
Algeria	1000	1000		1000	2000			
Argentina	15356	5632	6140	12707	22	1866	1837	
Armenia								
Australia	23987	8415	681	38942	1492	4806	430	
Austria	567	695		141		614		
Brazil	429	363	23			141	3	
Bulgaria	9327	10050		804		342		
Cambodia	2	2		3				
Canada	660	633	6	260	1	639	4	
Chile	42409	12057	127	7994	187	4091	152	
China	40300	16700		1000	4000	400		
Croatia	709	828						
Cyprus								
Czechia						697		
Ethiopia							90	
France	46555	108483	658	62211	78631	31602	1503	
Georgia	286							
Germany	329	553		46		11184		
Greece	1929	1393	22	1042				
Hungary	2677	1961		215		1092	1	
India	100			500		100		
Israel	990	715	55	385				
Italy	14240	24057	9	7693	5421	5057	68428	
Japan	42	197				20		
Kazakhstan	20					180		
Korea Rep.								
Lebanon	1000	500		300				
Luxembourg						121		
Mexico	756	391	229	145	140			
Moldova	8169	7689		87		2366		
Morocco	604	333		347	786			
Myanmar			4	27		7		
New Zealand	275	1239	18	436	1	5514	8	
N. Macedonia	1020	1240				500		
Norway								
Peru	48	2		2	1	1		
Portugal	2346	482	17014	4017	60	130		
Romania	5359	11647	67	504		1930	88	
Russia	8528	2988				918		
Serbia	2111	1968				633		
Slovakia	469							
Slovenia	423	817		18		202		
South Africa	10589	5558	92	9946	344	1153	70	
Spain	20139	12852	193597	19488	54606	969	2	59189
Switzerland	66	1124	0	194	1	4209	0	
Taiwan								
Thailand	7		4	74		1	2	
Tunisia	117	64		67	152			
Turkey	476	415	6	1439	33	10	18	
Ukraine	4935	1400				385		
United Kingdom						546		
United States	40837	21251	626	9083	2213	22998	827	
Uruguay	484	747		67	4	56		
Total	**310671**	**266440**	**219379**	**181185**	**150096**	**105480**	**73464**	**59189**

Table 39 (cont.) National winegrape area for world's top 30 red varieties, 2016 (hectares)

Country	*Cabernet Franc*	*Côt*	*Monastrell*	*Mazuelo*	*Alicante Henri Bouschet*	*Tribidrag*	*Montepul ciano*	*Gamay Noir*
Algeria				3000				
Argentina	929	40401	12	13	135	1	82	0
Armenia								
Australia	328	515	704	8	19	87	60	6
Austria	64							
Brazil	6834	30			101		1	5
Bulgaria	240							
Cambodia								
Canada	820	41	1		0	8		272
Chile	1578	2293	102	811	6908	66	2	0
China	600			100				
Croatia								
Cyprus								
Czechia								
Ethiopia								
France	32327	6100	8754	31760	2607	1		24095
Georgia								
Germany	32							
Greece	10			1	60			
Hungary	1368	3			14			3
India								
Israel	110	110	55	935				
Italy	5590	178		1686	286	13896	32724	64
Japan								
Kazakhstan	56							
Korea Rep.								
Lebanon								
Luxembourg								
Mexico				448				
Moldova	756	162						
Morocco				1230	919			
Myanmar								
New Zealand	109	129				4	8	7
N. Macedonia						1000		
Norway								
Peru		10						
Portugal	23			291	4547			0
Romania	72	7	3		20	9		
Russia	20							
Serbia	79							54
Slovakia								
Slovenia	27							4
South Africa	835	452	473	114	7	24		9
Spain	680	113	41303	5461	19294	1	0	
Switzerland	63	15	0		0	1		1349
Taiwan								
Thailand								
Tunisia				238	178			
Turkey	37	21	7	130	532			228
Ukraine								
United Kingd								
United States	2199	1610	515	1086	380	18551	58	123
Uruguay	266	43			24			1
Total	**56052**	**52233**	**51930**	**47312**	**36031**	**33649**	**32935**	**26221**

Table 39 (cont.) National winegrape area for world's top 30 red varieties, 2016 (hectares)

Country	*Cinsaut*	*Carmenère*	*Douce Noire*	*Barbera*	*Isabella*	*Blaufränkisch*	*Criolla Grande*	*Pinot Meunier*
Algeria								
Argentina	1	59	19072	444			15596	11
Armenia								
Australia	10	16		102	15	1		82
Austria						2808		
Brazil		10		2	11664			
Bulgaria								
Cambodia								
Canada		3		1		5		9
Chile	848	10503		5				2
China		11200						
Croatia						521		
Cyprus								
Czechia						1143		
Ethiopia								
France	15930	28	0					12130
Georgia								
Germany						1737		2002
Greece	4							
Hungary		0				7260		
India								
Israel								
Italy	4	635	630	15006		28		5
Japan								
Kazakhstan								
Korea Rep.								
Lebanon								
Luxembourg								
Mexico								
Moldova					3468			138
Morocco	3239							
Myanmar								
New Zealand								21
N. Macedonia								
Norway								
Peru						290		
Portugal	12							
Romania				0		729		
Russia					1362			
Serbia						727		
Slovakia						1216		
Slovenia				98		709		
South Africa	1767	8		35				14
Spain	10	0		0				2
Switzerland		0		0	1	4		0
Taiwan								
Thailand								
Tunisia	626							
Turkey	430							
Ukraine					1200			
United Kingdom								202
United States	45	24	31	2131		3		76
Uruguay					102			
Total	**22926**	**22486**	**19733**	**17824**	**17813**	**17180**	**15596**	**14695**

Table 39 (cont.) National winegrape area for world's top 30 red varieties, 2016 (hectares)

Country	*Nero d'Avola*	*Touriga Franca*	*Castelão*	*Moldova*	*Touriga Nacional*	*Negroamaro*	*Other Red*	*Total Red*
Algeria								8000
Argentina	0				27		9040	129384
Armenia							4405	4405
Australia	64				38	18	3314	84141
Austria							10418	15306
Brazil					6		8386	27999
Bulgaria							10433	31196
Cambodia							3	10
Canada							1909	5273
Chile					2		11139	101275
China							78053	152353
Croatia							2347	4405
Cyprus							3187	3187
Czechia							2552	4392
Ethiopia							21	111
France							70702	534077
Georgia							5567	5853
Germany							18298	34181
Greece							17031	21491
Hungary	86						4430	19111
India								700
Israel							793	4148
Italy	14129					11431	105241	326437
Japan							1510	1769
Kazakhstan							1074	1330
Korea Rep.							5300	5300
Lebanon							382	2182
Luxembourg							5	126
Mexico							1425	3534
Moldova				12375			7288	42497
Morocco							1371	8830
Myanmar							2	40
New Zealand							82	7851
N. Macedonia							11066	14826
Norway							4	4
Peru							1938	2292
Portugal		14217	12580		11411		51301	118432
Romania		4			11		41425	61876
Russia							5120	18936
Serbia							6397	11970
Slovakia							1378	3063
Slovenia							2626	4926
South Africa	2	3			105		10960	42561
Spain	0		0		4		41675	469385
Switzerland	0				0		1487	8515
Taiwan							94	94
Thailand							23	111
Tunisia							265	1707
Turkey							5573	9355
Ukraine							1250	9170
United Kingdom							118	867
United States					117		28334	153116
Uruguay							3682	5476
Total	**14281**	**14224**	**12580**	**12375**	**11722**	**11449**	**600423**	**2527573**

Table 40: National shares of global winegrape area for world's top 30 red varieties, 2016 (%)

Country	*Cabernet Sauvignon*	*Merlot*	*Tempranillo*	*Syrah*	*Garnacha Tinta*	*Pinot Noir*	*Sangiovese*	*Bobal*
Algeria	0.32	0.38		0.55	1.33			
Argentina	4.94	2.11	2.80	7.01	0.01	1.77	2.50	
Armenia								
Australia	7.72	3.16	0.31	21.49	0.99	4.56	0.58	
Austria	0.18	0.26		0.08		0.58		
Brazil	0.14	0.14	0.01			0.13	0.00	
Bulgaria	3.00	3.77		0.44		0.32		
Cambodia	0.00	0.00		0.00				
Canada	0.21	0.24	0.00	0.14	0.00	0.61	0.00	
Chile	13.65	4.53	0.06	4.41	0.12	3.88	0.21	
China	12.97	6.27		0.55	2.66	0.38		
Croatia	0.23	0.31						
Cyprus								
Czechia						0.66		
Ethiopia							0.12	
France	14.99	40.72	0.30	34.34	52.39	29.96	2.05	
Georgia	0.09							
Germany	0.11	0.21		0.03		10.60		
Greece	0.62	0.52	0.01	0.58				
Hungary	0.86	0.74		0.12		1.04	0.00	
India	0.03			0.28		0.09		
Israel	0.32	0.27	0.03	0.21				
Italy	4.58	9.03	0.00	4.25	3.61	4.79	93.15	
Japan	0.01	0.07				0.02		
Kazakhstan	0.01					0.17		
Korea Rep.								
Lebanon	0.32	0.19		0.17				
Luxembourg						0.11		
Mexico	0.24	0.15	0.10	0.08	0.09			
Moldova	2.63	2.89		0.05		2.24		
Morocco	0.19	0.13		0.19	0.52			
Myanmar			0.00	0.01		0.01		
New Zealand	0.09	0.46	0.01	0.24	0.00	5.23	0.01	
N. Macedonia	0.33	0.47				0.47		
Norway								
Peru	0.02	0.00		0.00	0.00	0.00		
Portugal	0.76	0.18	7.76	2.22	0.04	0.12		
Romania	1.72	4.37	0.03	0.28		1.83	0.12	
Russia	2.75	1.12				0.87		
Serbia	0.68	0.74				0.60		
Slovakia	0.15							
Slovenia	0.14	0.31		0.01		0.19		
South Africa	3.41	2.09	0.04	5.49	0.23	1.09	0.10	
Spain	6.48	4.82	88.25	10.76	36.38	0.92	0.00	100.00
Switzerland	0.02	0.42	0.00	0.11	0.00	3.99	0.00	
Taiwan								
Thailand	0.00		0.00	0.04		0.00	0.00	
Tunisia	0.04	0.02		0.04	0.10			
Turkey	0.15	0.16	0.00	0.79	0.02	0.01	0.02	
Ukraine	1.59	0.53				0.36		
United Kingdom						0.52		
United States	13.14	7.98	0.29	5.01	1.47	21.80	1.13	
Uruguay	0.16	0.28		0.04	0.00	0.05		
Total	**100**	**100**	**100**	**100**	**100**	**100**	**100**	**100**

Table 40 (cont.): National shares of global winegrape area for world's top 30 red varieties, 2016 (%)

Country	*Cabernet Franc*	*Côt*	*Monastrell*	*Mazuelo*	*Alicante Henri Bouschet*	*Tribidrag*	*Montepul ciano*	*Gamay Noir*
Algeria				6.34				
Argentina	1.66	77.35	0.02	0.03	0.38	0.00	0.25	0.00
Armenia								
Australia	0.58	0.99	1.36	0.02	0.05	0.26	0.18	0.02
Austria	0.11							
Brazil	12.19	0.06			0.28		0.00	0.02
Bulgaria	0.43							
Cambodia								
Canada	1.46	0.08	0.00		0.00	0.03		1.04
Chile	2.82	4.39	0.20	1.71	19.17	0.20	0.00	0.00
China	1.07			0.21				
Croatia								
Cyprus								
Czechia								
Ethiopia								
France	57.67	11.68	16.86	67.13	7.24	0.00		91.89
Georgia								
Germany	0.06							
Greece	0.02			0.00	0.17			
Hungary	2.44	0.01			0.04			0.01
India								
Israel	0.20	0.21	0.11	1.98				
Italy	9.97	0.34		3.56	0.79	41.30	99.36	0.24
Japan								
Kazakhstan	0.10							
Korea Rep.								
Lebanon								
Luxembourg								
Mexico				0.95				
Moldova	1.35	0.31						
Morocco				2.60	2.55			
Myanmar								
New Zealand	0.19	0.25				0.01	0.02	0.03
N. Macedonia						2.97		
Norway								
Peru		0.02						
Portugal	0.04			0.62	12.62			0.00
Romania	0.13	0.01	0.01		0.06	0.03		
Russia	0.04							
Serbia	0.14							0.21
Slovakia								
Slovenia	0.05							0.02
South Africa	1.49	0.87	0.91	0.24	0.02	0.07		0.03
Spain	1.21	0.22	79.54	11.54	53.55	0.00	0.00	
Switzerland	0.11	0.03	0.00		0.00	0.00		5.15
Taiwan								
Thailand								
Tunisia				0.50	0.49			
Turkey	0.07	0.04	0.01	0.27	1.48			0.87
Ukraine								
United Kingd								
United States	3.92	3.08	0.99	2.30	1.06	55.13	0.18	0.47
Uruguay	0.47	0.08			0.07			0.00
Total	**100**	**100**	**100**	**100**	**100**	**100**	**100**	**100**

Table 40 (cont.): National shares of global winegrape area for world's top 30 red varieties, 2016 (%)

Country	*Cinsaut*	*Carmenère*	*Douce Noire*	*Barbera*	*Isabella*	*Blaufränkisch*	*Criolla Grande*	*Pinot Meunier*
Algeria								
Argentina	0.01	0.26	96.65	2.49			100.00	0.08
Armenia								
Australia	0.04	0.07		0.57	0.09	0.00		0.56
Austria						16.34		
Brazil		0.04		0.01	65.48			
Bulgaria								
Cambodia								
Canada		0.01		0.01		0.03		0.06
Chile	3.70	46.71		0.03				0.01
China		49.81						
Croatia						3.03		
Cyprus								
Czechia						6.65		
Ethiopia								
France	69.48	0.13	0.00					82.54
Georgia								
Germany						10.11		13.62
Greece	0.02							
Hungary		0.00				42.26		
India								
Israel								
Italy	0.02	2.82	3.19	84.19		0.16		0.04
Japan								
Kazakhstan								
Korea Rep.								
Lebanon								
Luxembourg								
Mexico								
Moldova					19.47			0.94
Morocco	14.13							
Myanmar								
New Zealand								0.14
N. Macedonia								
Norway								
Peru						1.69		
Portugal	0.05							
Romania				0.00		4.24		
Russia					7.65			
Serbia						4.23		
Slovakia						7.08		
Slovenia				0.55		4.13		
South Africa	7.71	0.04		0.20				0.10
Spain	0.04	0.00		0.00				0.02
Switzerland		0.00		0.00	0.01	0.02		0.00
Taiwan								
Thailand								
Tunisia	2.73							
Turkey	1.87							
Ukraine					6.74			
United Kingdom								1.38
United States	0.20	0.11	0.16	11.95		0.02		0.52
Uruguay					0.57			
Total	**100**	**100**	**100**	**100**	**100**	**100**	**100**	**100**

Table 40 (cont.): National shares of global winegrape area for world's top 30 red varieties, 2016 (%)

Country	*Nero d'Avola*	*Touriga Franca*	*Castelão*	*Moldova*	*Touriga Nacional*	*Negroam aro*	*Other Red*	*Total Red*
Algeria								0.32
Argentina	0.00				0.23		1.51	5.12
Armenia							0.73	0.17
Australia	0.45				0.33	0.16	0.55	3.33
Austria							1.74	0.61
Brazil					0.05		1.40	1.11
Bulgaria							1.74	1.23
Cambodia							0.00	0.00
Canada							0.32	0.21
Chile					0.01		1.86	4.01
China							13.00	6.03
Croatia							0.39	0.17
Cyprus							0.53	0.13
Czechia							0.43	0.17
Ethiopia							0.00	0.00
France							11.78	21.13
Georgia							0.93	0.23
Germany							3.05	1.35
Greece							2.84	0.85
Hungary	0.60						0.74	0.76
India								0.03
Israel							0.13	0.16
Italy	98.93					99.84	17.53	12.92
Japan							0.25	0.07
Kazakhstan							0.18	0.05
Korea Rep.							0.88	0.21
Lebanon							0.06	0.09
Luxembourg							0.00	0.01
Mexico							0.24	0.14
Moldova				100.00			1.21	1.68
Morocco							0.23	0.35
Myanmar							0.00	0.00
New Zealand							0.01	0.31
N. Macedonia							1.84	0.59
Norway							0.00	0.00
Peru							0.32	0.09
Portugal		99.95	100.00		97.35		8.54	4.69
Romania		0.03			0.10		6.90	2.45
Russia							0.85	0.75
Serbia							1.07	0.47
Slovakia							0.23	0.12
Slovenia							0.44	0.19
South Africa	0.01	0.02			0.90		1.83	1.68
Spain	0.00		0.00		0.03		6.94	18.57
Switzerland	0.00				0.00		0.25	0.34
Taiwan							0.02	0.00
Thailand							0.00	0.00
Tunisia							0.04	0.07
Turkey							0.93	0.37
Ukraine							0.21	0.36
United Kingdom							0.02	0.03
United States					1.00		4.72	6.06
Uruguay							0.61	0.22
Total	**100**	**100**	**100**	**100**	**100**	**100**	**100**	**100**

Table 41: Shares of world's top 30 red varieties in national winegrape area, by country, 2016 (%)

Country	*Cabernet Sauvignon*	*Merlot*	*Tempranillo*	*Syrah*	*Garnacha Tinta*	*Pinot Noir*	*Sangiovese*	*Bobal*
Algeria	12.05	12.05		12.05	24.10			
Argentina	7.44	2.73	2.98	6.16	0.01	0.90	0.89	
Armenia								
Australia	18.11	6.35	0.51	29.40	1.13	3.63	0.32	
Austria	1.25	1.53		0.31		1.35		
Brazil	1.29	1.09	0.07			0.43	0.01	
Bulgaria	17.61	18.97		1.52		0.65		
Cambodia	20.00	20.00		30.00				
Canada	5.23	5.03	0.05	2.06	0.01	5.07	0.03	
Chile	29.07	8.27	0.09	5.48	0.13	2.80	0.10	
China	22.64	9.38		0.56	2.25	0.22		
Croatia	6.04	7.05						
Cyprus								
Czechia						5.13		
Ethiopia							53.20	
France	5.71	13.31	0.08	7.63	9.65	3.88	0.18	
Georgia	0.60							
Germany	0.35	0.59		0.05		11.83		
Greece	3.79	2.74	0.04	2.05				
Hungary	4.19	3.07		0.34		1.71	0.00	
India	3.70			18.52		3.70		
Israel	19.80	14.30	1.10	7.70				
Italy	2.36	3.98	0.00	1.27	0.90	0.84	11.32	
Japan	1.07	5.09				0.53		
Kazakhstan	0.28					2.59		
Korea Rep.								
Lebanon	25.00	12.50		7.50				
Luxembourg						9.31		
Mexico	13.83	7.15	4.19	2.65	2.56			
Moldova	9.89	9.31		0.11		2.86		
Morocco	3.44	1.90		1.97	4.47			
Myanmar			5.00	38.57		10.00		
New Zealand	0.78	3.49	0.05	1.23	0.00	15.55	0.02	
N. Macedonia	4.12	5.00				2.02		
Norway								
Peru	1.25	0.05		0.05	0.03	0.03		
Portugal	1.28	0.26	9.32	2.20	0.03	0.07		
Romania	2.93	6.37	0.04	0.28		1.06	0.05	
Russia	16.79	5.88				1.81		
Serbia	9.59	8.94				2.88		
Slovakia	6.05							
Slovenia	2.65	5.11		0.12		1.26		
South Africa	11.06	5.80	0.10	10.38	0.36	1.20	0.07	
Spain	2.28	1.45	21.91	2.21	6.18	0.11	0.00	6.70
Switzerland	0.45	7.60	0.00	1.31	0.01	28.45	0.00	
Taiwan								
Thailand	3.37		1.81	35.85		0.25	1.03	
Tunisia	3.44	1.90		1.97	4.47			
Turkey	3.48	3.03	0.05	10.50	0.24	0.07	0.13	
Ukraine	19.61	5.56				1.53		
United Kingdom						29.70		
United States	17.04	8.87	0.26	3.79	0.92	9.60	0.35	
Uruguay	7.18	11.08		0.99	0.06	0.83		
Total	**6.93**	**5.94**	**4.89**	**4.04**	**3.35**	**2.35**	**1.64**	**1.32**

Table 41 (cont.): Shares of world's top 30 red varieties in national winegrape area, by country, 2016 (%)

Country	*Cabernet Franc*	*Côt*	*Monastrell*	*Mazuelo*	*Alicante Henri Bouschet*	*Tribidrag*	*Montepulciano*	*Gamay Noir*
Algeria				36.14				
Argentina	0.45	19.58	0.01	0.01	0.07	0.00	0.04	0.00
Armenia								
Australia	0.25	0.39	0.53	0.01	0.01	0.07	0.05	0.00
Austria	0.14							
Brazil	20.58	0.09			0.30		0.00	0.02
Bulgaria	0.45							
Cambodia								
Canada	6.50	0.33	0.01		0.00	0.07		2.16
Chile	1.08	1.57	0.07	0.56	4.74	0.05	0.00	0.00
China	0.34			0.06				
Croatia								
Cyprus								
Czechia								
Ethiopia								
France	3.97	0.75	1.07	3.90	0.32	0.00		2.96
Georgia								
Germany	0.03							
Greece	0.02			0.00	0.12			
Hungary	2.14	0.00			0.02			0.00
India								
Israel	2.20	2.20	1.10	18.70				
Italy	0.92	0.03		0.28	0.05	2.30	5.41	0.01
Japan								
Kazakhstan	0.81							
Korea Rep.								
Lebanon								
Luxembourg								
Mexico				8.20				
Moldova	0.92	0.20						
Morocco				7.00	5.23			
Myanmar								
New Zealand	0.31	0.36				0.01	0.02	0.02
N. Macedonia						4.04		
Norway								
Peru		0.26						
Portugal	0.01			0.16	2.49			0.00
Romania	0.04	0.00	0.00		0.01	0.00		
Russia	0.04							
Serbia	0.36							0.25
Slovakia								
Slovenia	0.17							0.03
South Africa	0.87	0.47	0.49	0.12	0.01	0.03		0.01
Spain	0.08	0.01	4.67	0.62	2.18	0.00	0.00	
Switzerland	0.43	0.10	0.00		0.00	0.00		9.12
Taiwan								
Thailand								
Tunisia				7.00	5.23			
Turkey	0.27	0.15	0.05	0.95	3.88			1.66
Ukraine								
United Kingd								
United States	0.92	0.67	0.21	0.45	0.16	7.74	0.02	0.05
Uruguay	3.94	0.64			0.36			0.01
Total	**1.25**	**1.17**	**1.16**	**1.06**	**0.80**	**0.75**	**0.73**	**0.59**

Table 41 (cont.): Shares of world's top 30 red varieties in national winegrape area, by country, 2016 (%)

Country	*Cinsaut*	*Carmenère*	*Douce Noire*	*Barbera*	*Isabella*	*Blaufränkisch*	*Criolla Grande*	*Pinot Meunier*
Algeria								
Argentina	0.00	0.03	9.24	0.21			7.56	0.01
Armenia								
Australia	0.01	0.01		0.08	0.01	0.00		0.06
Austria						6.18		
Brazil		0.03		0.00	35.13			
Bulgaria								
Cambodia								
Canada		0.02		0.01		0.04		0.07
Chile	0.58	7.20		0.00				0.00
China		6.29						
Croatia						4.44		
Cyprus								
Czechia						8.40		
Ethiopia								
France	1.95	0.00	0.00					1.49
Georgia								
Germany						1.84		2.12
Greece	0.01							
Hungary		0.00				11.37		
India								
Israel								
Italy	0.00	0.10	0.10	2.48		0.00		0.00
Japan								
Kazakhstan								
Korea Rep.								
Lebanon								
Luxembourg								
Mexico								
Moldova					4.20			0.17
Morocco	18.42							
Myanmar								
New Zealand								0.06
N. Macedonia								
Norway								
Peru						7.57		
Portugal	0.01							
Romania				0.00		0.40		
Russia					2.68			
Serbia						3.30		
Slovakia						15.69		
Slovenia				0.61		4.44		
South Africa	1.84	0.01		0.04				0.02
Spain	0.00	0.00		0.00				0.00
Switzerland		0.00		0.00	0.01	0.02		0.00
Taiwan								
Thailand								
Tunisia	18.42							
Turkey	3.14							
Ukraine					4.77			
United Kingdom								11.00
United States	0.02	0.01	0.01	0.89		0.00		0.03
Uruguay					1.51			
Total	**0.51**	**0.50**	**0.44**	**0.40**	**0.40**	**0.38**	**0.35**	**0.33**

Table 41 (cont.): Shares of world's top 30 red varieties in national winegrape area, by country, 2016 (%)

Country	*Nero d'Avola*	*Touriga Franca*	*Castelão*	*Moldova*	*Touriga Nacional*	*Negroamaro*	*Other Red*	*Total Red*	*Total*
Algeria								96.39	100
Argentina	0.00				0.01		4.38	62.70	100
Armenia							29.96	29.96	100
Australia	0.05				0.03	0.01	2.50	63.53	100
Austria							22.93	33.68	100
Brazil					0.02		25.26	84.32	100
Bulgaria							19.70	58.89	100
Cambodia							30.00	100.00	100
Canada							15.15	41.85	100
Chile					0.00		7.64	69.43	100
China							43.85	85.59	100
Croatia							19.98	37.50	100
Cyprus							62.09	62.09	100
Czechia							18.76	32.29	100
Ethiopia							12.34	65.54	100
France							8.68	65.54	100
Georgia							11.60	12.19	100
Germany							19.36	36.17	100
Greece							33.50	42.27	100
Hungary	0.14						6.94	29.92	100
India								25.93	100
Israel							15.85	82.95	100
Italy	2.34					1.89	17.41	54.00	100
Japan							39.03	45.73	100
Kazakhstan							15.47	19.16	100
Korea Rep.							98.15	98.15	100
Lebanon							9.55	54.55	100
Luxembourg							0.42	9.73	100
Mexico							26.07	64.66	100
Moldova				14.98			8.82	51.45	100
Morocco							7.79	50.20	100
Myanmar							2.86	56.43	100
New Zealand							0.23	22.14	100
N. Macedonia							44.66	59.84	100
Norway							34.00	34.00	100
Peru							50.59	59.83	100
Portugal		7.78	6.89		6.25		28.09	64.84	100
Romania		0.00			0.01		22.67	33.86	100
Russia							10.08	37.28	100
Serbia							29.06	54.37	100
Slovakia							17.78	39.53	100
Slovenia							16.43	30.81	100
South Africa	0.00	0.00			0.11		11.44	44.44	100
Spain	0.00		0.00		0.00		4.72	53.12	100
Switzerland	0.00				0.00		10.05	57.56	100
Taiwan							63.24	63.24	100
Thailand							11.10	53.43	100
Tunisia							7.79	50.20	100
Turkey							40.67	68.26	100
Ukraine							4.97	36.44	100
United Kingdom							6.44	47.14	100
United States					0.05		11.82	63.90	100
Uruguay							54.61	81.21	100
Total	**0.32**	**0.32**	**0.28**	**0.28**	**0.26**	**0.26**	**13.40**	**56.39**	**100**

Table 42: Change in national winegrape area since 2000 for world's top 30 red varieties in 2016 (hectares)

Country	*Cabernet Sauvignon*	*Merlot*	*Tempranillo*	*Syrah*	*Garnacha Tinta*	*Pinot Noir*	*Sangiovese*	*Bobal*
Algeria	-510	-510		-510	-4040			
Argentina	1579	-631	1420	3819	13	752	-653	
Armenia								
Australia	-1009	746	640	9647	-647	1583	58	
Austria	255	583				205		
Brazil	-158	-106						
Bulgaria	-1114	-1120				-426		
Cambodia								
Canada	91	-41				182		
Chile	6442	-768	126	5954		2477	29	
China								
Croatia								
Cyprus								
Czechia								
Ethiopia								
France	-6859	7174	-890	11535	-17086	5076	-61	
Georgia	63							
Germany						2541		
Greece	1241	1210		1003				
Hungary	1624	1475				849		
India								
Israel	383	68						
Italy	6558	2196	2	6668	-1360	1770	5667	
Japan								
Kazakhstan								
Korea Rep.								
Lebanon								
Luxembourg						55		
Mexico								
Moldova	579	-434				-4155		
Morocco					-16			
Myanmar								
New Zealand	-379	582		376		4416		
N. Macedonia								
Norway								
Peru								
Portugal	2028		9658					
Romania	-3261	3837				190		
Russia	6950							
Serbia								
Slovakia	313							
Slovenia		-380						
South Africa	1765	670	79	4315	305	666	35	
Spain	15621	11666	114287	19402	-43524	552		-40940
Switzerland	32	276		139		-393		
Taiwan								
Thailand								
Tunisia	-220			-270	-1868			
Turkey								
Ukraine								
United Kingdom								
United States	23263	4376	425	7574	-2307	17655	145	
Uruguay	-191	-310		5				
Total	87597	53072	126009	78695	-66253	36670	4587	-40940

Table 42 (cont.): Change in national winegrape area since 2000 for world's top 30 red varieties in 2016 (hectares)

Country	*Cabernet Franc*	*Côt*	*Monastrell*	*Mazuelo*	*Alicante Henri Bouschet*	*Tribidrag*	*Montepulciano*	*Gamay Noir*
Algeria				-4550				
Argentina	677	22172	12	-44	22	-5	33	-1
Armenia								
Australia	-416	86	-244	-82				
Austria	37							
Brazil	3051							
Bulgaria								
Cambodia								
Canada	253							9
Chile	889	1364	80	170	4026	-25		
China								
Croatia								
Cyprus								
Czechia								
Ethiopia								
France	-3768	-29	1120	-63985	-6157	0		-10443
Georgia								
Germany								
Greece	-30			-13	39			
Hungary	841							
India								
Israel				-36				
Italy	-1049	-73		-34	-224	6068	4045	-88
Japan								
Kazakhstan								
Korea Rep.								
Lebanon								
Luxembourg								
Mexico								
Moldova		123						
Morocco				-462	-179			
Myanmar								
New Zealand	-9	62						
N. Macedonia								
Norway								
Peru								
Portugal					3872			
Romania								
Russia								
Serbia								
Slovakia								
Slovenia								
South Africa	347	377	460	43	-2	-4		-28
Spain	650	90	-25857	-2642	973			
Switzerland	47	15						-628
Taiwan								
Thailand								
Tunisia				-7338	-664			
Turkey								
Ukraine								
United Kingd								
United States	1011	1513	328	-2002	-182	-79		-562
Uruguay	-98							
Total	4078	25948	-24374	-80380	-1125	6728	4207	-11577

Table 42 (cont.): Change in national winegrape area since 2000 for world's top 30 red varieties in 2016 (hectares)

Country	*Cinsaut*	*Carmenère*	*Douce Noire*	*Barbera*	*Isabella*	*Blaufränkisch*	*Criolla Grande*	*Pinot Meunier*
Algeria								
Argentina	-5		3412	-611			-8668	1
Armenia								
Australia				-1				-25
Austria						167		
Brazil					-2621			
Bulgaria								
Cambodia								
Canada								
Chile	653	5784						
China								
Croatia						-668		
Cyprus								
Czechia						463		
Ethiopia								
France	-15663	24	-1					1508
Georgia								
Germany						619		-287
Greece	-104							
Hungary						340		
India								
Israel								
Italy	-270	592	-2013	-12169		-84		-5
Japan								
Kazakhstan								
Korea Rep.								
Lebanon								
Luxembourg								
Mexico								
Moldova					-7933			
Morocco	-701							
Myanmar								
New Zealand								
N. Macedonia								
Norway								
Peru								
Portugal								
Romania								
Russia								
Serbia								
Slovakia						125		
Slovenia								
South Africa	-1767			20				8
Spain	7							-5
Switzerland								
Taiwan								
Thailand								
Tunisia	-216							
Turkey								
Ukraine								
United Kingdom								
United States	12		10	-2563		-41		-4
Uruguay								
Total	-25502	16775	1409	-15217	-9637	3184	-8668	1564

Table 42 (cont.): Change in national winegrape area since 2000 for world's top 30 red varieties in 2016 (hectares)

Country	*Nero d'Avola*	*Touriga Franca*	*Castelão*	*Moldova*	*Touriga Nacional*	*Negroamaro*	*Other Red*	*Total Red*
Algeria								-22200
Argentina	-5						833	24132
Armenia								4405
Australia					-34		-3829	6768
Austria							941	2329
Brazil							-4209	-3722
Bulgaria							-25461	-27077
Cambodia								10
Canada							-302	533
Chile							-13059	14339
China								152353
Croatia							-11548	-10679
Cyprus							-11438	-11438
Czechia							-167	993
Ethiopia								111
France							36201	-62303
Georgia							1227	1290
Germany							5168	9001
Greece							-756	2590
Hungary							-7282	-1829
India								700
Israel							48	1177
Italy	2810					-5188	-16300	-2481
Japan								1769
Kazakhstan								1330
Korea Rep.							0	0
Lebanon								2182
Luxembourg								59
Mexico								3534
Moldova							5221	6756
Morocco							-35285	-35358
Myanmar								40
New Zealand							-89	5027
N. Macedonia								14826
Norway								4
Peru								2292
Portugal		7546	-1844		7262		-34407	-868
Romania							-1981	299
Russia							-4008	8230
Serbia							-14992	-9420
Slovakia							-25	414
Slovenia							-2279	-1177
South Africa		0			64		1685	9049
Spain							-41433	8853
Switzerland							1104	602
Taiwan							-1436	-1436
Thailand								111
Tunisia							-3102	-15129
Turkey								9355
Ukraine								9170
United Kingdom							-82	667
United States							8207	56963
Uruguay							-2715	-3079
Total	2958	7550	-1844	12375	7459	-5170	-79967	110208

Table 43: National winegrape area for world's top 30 white varieties, 2000 (hectares)

Country	*Airén*	*Chardonnay*	*Trebbiano Toscano*	*Graševina*	*Rkatsiteli*	*Sauvignon Blanc*	*Cayetana Blanca*	*Catarratto Bianco*
Algeria								
Argentina		4682	2765			865		
Armenia					2469			
Australia		17266	685			2602	249	
Austria				4323		314		
Brazil		330	688	880		140		
Bulgaria		1862	1821	3602	9429	405		
Canada		973				148		
Chile		7672				6662		
Croatia				16051				
Cyprus								
Czechia		567		1246				
France		36496	90341			20933		
Georgia					19741			
Germany		531						
Greece		36	746			158		
Hungary		2954		6677		324		
Israel		142				263		
Italy		11687	39447	2007		3312		50711
Korea Rep.								
Luxembourg		8						
Moldova		5134			11508	8151		
Morocco								
New Zealand		2787				2423		
Portugal			382					
Romania		1376		15014	506	4613		
Russia		1639			13152			
Serbia				33120				
Slovakia		623		3895				
Slovenia		1549		3568		1221		
South Africa		6067	147			5436		
Spain	387978	1814	28	1923		467	55527	
Switzerland		226				38		
Taiwan								
Tunisia								
United Kingdom		68						
United States		35791	151			4191		
Uruguay		142				142		
"Missing 9"		3120			10549	2383		
Total	**387978**	**145543**	**137201**	**92306**	**67354**	**65190**	**55776**	**50711**

Table 43 (cont.) National winegrape area for world's top 30 white varieties, 2000 (hectares)

Country	*Macabeo*	*Chenin Blanc*	*Riesling*	*Colombard*	*Aligoté*	*Müller-Thurgau*	*Palomino Fino*	*Blanc à Petits Grains*
Algeria								
Argentina	3	3445	150				216	149
Armenia								526
Australia		841	3129	1801			124	214
Austria			1643			3289		143
Brazil								
Bulgaria			647		1659			
Canada			482					
Chile		76	286					
Croatia								
Cyprus								
Czechia			793			1586		
France	5223	9837	3407	6896	1756	5	440	6935
Georgia					97			
Germany			22350			20706		87
Greece								2232
Hungary			1619			3278		1538
Israel		101		486				
Italy			599			996		13016
Korea Rep.								
Luxembourg			175			459		
Moldova			1343		15790	173		172
Morocco								
New Zealand		146	490			419	21	
Portugal								1419
Romania					7608			1012
Russia			1376		1821			
Serbia								
Slovakia						1870		
Slovenia								
South Africa		22566	477	11432			1632	773
Spain	42902	105	97				27685	223
Switzerland		1	8		20	686		44
Taiwan								
Tunisia								
United Kingdom								
United States		8433	1965	18010			319	515
Uruguay								
"Missing 9"		210	2279	8	6916	120	76	981
Total	**48128**	**45761**	**43316**	**38632**	**35668**	**33587**	**30513**	**29979**

Table 43 (cont.) National winegrape area for world's top 30 white varieties, 2000 (hectares)

Country	*Muscat of Alexandria*	*Sémillon*	*Fetească Albă*	*Grüner Veltliner*	*Trebbiano Romagnolo*	*Pedro Ximénez*	*Pinot Blanc*	*Garganega*
Algeria								
Argentina	5515	1033					40	4
Armenia								
Australia	2495	6528				89		
Austria				17479			2936	
Brazil	809	384						
Bulgaria								
Canada							146	
Chile		1893				2379	14	
Croatia								
Cyprus								
Czechia				1700				
France	3027	14015					1401	
Georgia							171	
Germany							2396	
Greece		21						
Hungary			971	1335				
Israel	202							
Italy	1157	8		129	19492		4993	16549
Korea Rep.								
Luxembourg							138	
Moldova			4334				350	
Morocco	3669							
New Zealand		229						
Portugal	510							
Romania			18211					
Russia							2995	
Serbia								
Slovakia			312	2960			623	
Slovenia								
South Africa	4047	1033					35	
Spain	6144					14803		
Switzerland		3					77	
Taiwan								
Tunisia								
United Kingdom								
United States	2013	709					431	
Uruguay								
"Missing 9"		382				1	238	
Total	**29590**	**26239**	**23828**	**23604**	**19492**	**17272**	**16983**	**16553**

Table 43 (cont.) National winegrape area for world's top 30 white varieties, 2000 (hectares)

Country	*Niagara*	*Pedro Giménez*	*Fernão Pires*	*Chasselas*	*Melon*	*Malvasia Bianca di Candia*	*Other White*	*Total White*
Algeria							0	0
Argentina		14862					26927	60656
Armenia							8211	11206
Australia							17206	53230
Austria							4712	34840
Brazil	13436						4452	21119
Bulgaria							18299	37724
Canada	461						1338	3548
Chile				404			7642	27028
Croatia							28313	44364
Cyprus							3656	3656
Czechia							2040	7932
France				943	13253		48693	263602
Georgia							12827	32836
Germany				1198			27973	75241
Greece							20859	24052
Hungary				1902			41059	61657
Israel							687	1881
Italy				22		11921	125055	301101
Korea Rep.								0
Luxembourg							346	1126
Moldova							5106	52061
Morocco							1744	5413
New Zealand				25			444	6984
Portugal			14206				69186	85703
Romania							109869	158209
Russia							24643	45626
Serbia				3450			11040	47609
Slovakia							2649	12932
Slovenia							11032	17369
South Africa			339				6058	60040
Spain				1			181455	721151
Switzerland				5373			503	6979
Taiwan							1303	1303
Tunisia								0
United Kingdom							605	673
United States	1357					968	3763	78616
Uruguay							41	325
"Missing 9"	89						8368	35720
Total	**15343**	**14862**	**14545**	**13318**	**13253**	**12889**	**838104**	**2403514**

Table 44: National shares of global winegrape area for world's top 30 white varieties, 2000 (%)

Country	*Airén*	*Chardonnay*	*Trebbiano Toscano*	*Graševina*	*Rkatsiteli*	*Sauvignon Blanc*	*Cayetana Blanca*	*Catarratto Bianco*
Algeria								
Argentina		3.22	2.01			1.33		
Armenia					3.67			
Australia		11.86	0.50			3.99	0.45	
Austria				4.68		0.48		
Brazil		0.23	0.50	0.95		0.21		
Bulgaria		1.28	1.33	3.90	14.00	0.62		
Canada		0.67				0.23		
Chile		5.27				10.22		
Croatia				17.39				
Cyprus								
Czechia		0.39		1.35				
France		25.08	65.85			32.11		
Georgia					29.31			
Germany		0.36						
Greece		0.02	0.54			0.24		
Hungary		2.03		7.23		0.50		
Israel		0.10				0.40		
Italy		8.03	28.75	2.17		5.08		100.00
Korea Rep.								
Luxembourg		0.01						
Moldova		3.53			17.09	12.50		
Morocco								
New Zealand		1.91				3.72		
Portugal			0.28					
Romania		0.95		16.27	0.75	7.08		
Russia		1.13			19.53			
Serbia				35.88				
Slovakia		0.43		4.22				
Slovenia		1.06		3.87		1.87		
South Africa		4.17	0.11			8.34		
Spain	100.00	1.25	0.02	2.08		0.72	99.55	
Switzerland		0.16				0.06		
Taiwan								
Tunisia								
United Kingdom		0.05						
United States		24.59	0.11			6.43		
Uruguay		0.10				0.22		
"Missing 9"		2.14			15.66	3.66		
Total	**100**	**100**	**100**	**100**	**100**	**100**	**100**	**100**

Table 44 (cont.): National shares of global winegrape area for world's top 30 white varieties, 2000 (%)

Country	*Macabeo*	*Chenin Blanc*	*Riesling*	*Colombard*	*Aligoté*	*Müller-Thurgau*	*Palomino Fino*	*Blanc à Petits Grains*
Algeria								
Argentina	0.01	7.53	0.35				0.71	0.50
Armenia								1.75
Australia		1.84	7.22	4.66			0.41	0.71
Austria			3.79			9.79		0.48
Brazil								
Bulgaria			1.49		4.65			
Canada			1.11					
Chile		0.17	0.66					
Croatia								
Cyprus								
Czechia			1.83			4.72		
France	10.85	21.50	7.87	17.85	4.92	0.01	1.44	23.13
Georgia					0.27			
Germany			51.60			61.65		0.29
Greece								7.45
Hungary			3.74			9.76		5.13
Israel		0.22		1.26				
Italy			1.38			2.97		43.42
Korea Rep.								
Luxembourg			0.40			1.37		
Moldova			3.10		44.27	0.52		0.57
Morocco								
New Zealand		0.32	1.13			1.25	0.07	
Portugal								4.73
Romania					21.33			3.37
Russia			3.18		5.11			
Serbia								
Slovakia						5.57		
Slovenia								
South Africa		49.31	1.10	29.59			5.35	2.58
Spain	89.14	0.23	0.22				90.73	0.74
Switzerland		0.00	0.02		0.06	2.04		0.15
Taiwan								
Tunisia								
United Kingdom								
United States		18.43	4.54	46.62			1.05	1.72
Uruguay								
"Missing 9"		0.46	5.26	0.02	19.39	0.36	0.25	3.27
Total	**100**	**100**	**100**	**100**	**100**	**100**	**100**	**100**

Table 44 (cont.): National shares of global winegrape area for world's top 30 white varieties, 2000 (%)

Country	*Muscat of Alexandria*	*Sémillon*	*Fetească Albă*	*Grüner Veltliner*	*Trebbiano Romagnolo*	*Pedro Ximénez*	*Pinot Blanc*	*Garganega*
Algeria								
Argentina	18.64	3.94					0.23	0.03
Armenia								
Australia	8.43	24.88				0.52		
Austria				74.05			17.29	
Brazil	2.74	1.47						
Bulgaria								
Canada							0.86	
Chile		7.21				13.77	0.08	
Croatia								
Cyprus								
Czechia				7.20				
France	10.23	53.41					8.25	
Georgia							1.01	
Germany							14.11	
Greece		0.08						
Hungary			4.08	5.66				
Israel	0.68							
Italy	3.91	0.03		0.55	100.00		29.40	99.97
Korea Rep.								
Luxembourg							0.81	
Moldova			18.19				2.06	
Morocco	12.40							
New Zealand		0.87						
Portugal	1.72							
Romania			76.43					
Russia							17.63	
Serbia								
Slovakia			1.31	12.54			3.67	
Slovenia								
South Africa	13.68	3.94					0.20	
Spain	20.77					85.71		
Switzerland		0.01					0.46	
Taiwan								
Tunisia								
United Kingdom								
United States	6.80	2.70					2.54	
Uruguay								
"Missing 9"		1.46				0.00	1.40	
Total	**100**	**100**	**100**	**100**	**100**	**100**	**100**	**100**

Table 44 (cont.): National shares of global winegrape area for world's top 30 white varieties, 2000 (%)

Country	*Niagara*	*Pedro Giménez*	*Fernão Pires*	*Chasselas*	*Melon*	*Malvasia Bianca di Candia*	*Other White*	*Total White*
Algeria								0.00
Argentina		100.00					3.21	2.52
Armenia							0.98	0.47
Australia							2.05	2.21
Austria							0.56	1.45
Brazil	87.57						0.53	0.88
Bulgaria							2.18	1.57
Canada	3.01						0.16	0.15
Chile				3.03			0.91	1.12
Croatia							3.38	1.85
Cyprus							0.44	0.15
Czechia							0.24	0.33
France				7.08	100.00		5.81	10.97
Georgia							1.53	1.37
Germany				9.00			3.34	3.13
Greece							2.49	1.00
Hungary				14.28			4.90	2.57
Israel							0.08	0.08
Italy				0.16		92.49	14.92	12.53
Korea Rep.								0.00
Luxembourg							0.04	0.05
Moldova							0.61	2.17
Morocco							0.21	0.23
New Zealand				0.19			0.05	0.29
Portugal			97.67				8.26	3.57
Romania							13.11	6.58
Russia							2.94	1.90
Serbia				25.90			1.32	1.98
Slovakia							0.32	0.54
Slovenia							1.32	0.72
South Africa			2.33				0.72	2.50
Spain				0.01			21.65	30.00
Switzerland				40.34			0.06	0.29
Taiwan							0.16	0.05
Tunisia								0.00
United Kingdom							0.07	0.03
United States	8.84					7.51	0.45	3.27
Uruguay							0.00	0.01
"Missing 9"	0.58						1.00	1.49
Total	**100**	**100**	**100**	**100**	**100**	**100**	**100**	**100**

Table 45: Shares of world's top 30 white varieties in national winegrape area, by country, 2000 (%)

Country	*Airén*	*Chardonnay*	*Trebbiano Toscano*	*Graševina*	*Rkatsiteli*	*Sauvignon Blanc*	*Cayetana Blanca*	*Catarratto Bianco*
Algeria								
Argentina		2.37	1.40			0.44		
Armenia					22.03			
Australia		13.22	0.52			1.99	0.19	
Austria				8.91		0.65		
Brazil		0.62	1.30	1.67		0.26		
Bulgaria		1.94	1.90	3.75	9.82	0.42		
Canada		11.45				1.74		
Chile		6.73				5.85		
Croatia				27.00				
Cyprus								
Czechia		5.00		11.00				
France		4.22	10.45			2.42		
Georgia					52.76			
Germany		0.51						
Greece		0.07	1.46			0.31		
Hungary		3.40		7.69		0.37		
Israel		2.92				5.42		
Italy		1.84	6.20	0.32		0.52		7.97
Korea Rep.								
Luxembourg		0.59						
Moldova		5.71			12.81	9.07		
Morocco								
New Zealand		28.03				24.37		
Portugal			0.19					
Romania		0.62		6.76	0.23	2.08		
Russia		2.91			23.35			
Serbia				48.00				
Slovakia		4.00		25.00				
Slovenia		6.60		15.20		5.20		
South Africa		6.48	0.16			5.80		
Spain	32.83	0.15	0.00	0.16		0.04	4.70	
Switzerland		1.50				0.25		
Taiwan								
Tunisia								
United Kingdom		7.79						
United States		20.37	0.09			2.39		
Uruguay		1.60				1.60		
"Missing 9"		3.89			13.15	2.97		
Total	**7.94**	**2.98**	**2.81**	**1.89**	**1.38**	**1.33**	**1.14**	**1.04**

Table 45 (cont.): Shares of world's top 30 white varieties in national winegrape area, by country, 2000 (%)

Country	*Macabeo*	*Chenin Blanc*	*Riesling*	*Colombard*	*Aligoté*	*Müller-Thurgau*	*Palomino Fino*	*Blanc à Petits Grains*
Algeria								
Argentina	0.00	1.75	0.08				0.11	0.08
Armenia								4.69
Australia		0.64	2.40	1.38			0.10	0.16
Austria			3.39			6.78		0.30
Brazil								
Bulgaria			0.67		1.73			
Canada			5.67					
Chile		0.07	0.25					
Croatia								
Cyprus								
Czechia			7.00			14.00		
France	0.60	1.14	0.39	0.80	0.20	0.00	0.05	0.80
Georgia					0.26			
Germany			21.44			19.87		0.08
Greece								4.38
Hungary			1.86			3.77		1.77
Israel		2.09		10.01				
Italy			0.09			0.16		2.04
Korea Rep.								
Luxembourg			12.98			34.05		
Moldova			1.49		17.57	0.19		0.19
Morocco								
New Zealand		1.47	4.93			4.21	0.21	
Portugal								0.69
Romania					3.42			0.46
Russia			2.44		3.23			
Serbia								
Slovakia						12.00		
Slovenia								
South Africa		24.09	0.51	12.21			1.74	0.83
Spain	3.63	0.01	0.01				2.34	0.02
Switzerland		0.01	0.05		0.13	4.56		0.29
Taiwan								
Tunisia								
United Kingdom								
United States		4.80	1.12	10.25			0.18	0.29
Uruguay								
"Missing 9"		0.26	2.84	0.01	8.62	0.15	0.09	1.22
Total	**0.98**	**0.94**	**0.89**	**0.79**	**0.73**	**0.69**	**0.62**	**0.61**

Table 45 (cont.): Shares of world's top 30 white varieties in national winegrape area, by country, 2000 (%)

Country	*Muscat of Alexandria*	*Sémillon*	*Feteascǎ Albǎ*	*Grüner Veltliner*	*Trebbiano Romagnolo*	*Pedro Ximénez*	*Pinot Blanc*	*Garganega*
Algeria								
Argentina	2.79	0.52					0.02	0.00
Armenia								
Australia	1.91	5.00				0.07		
Austria				36.04			6.05	
Brazil	1.53	0.73						
Bulgaria								
Canada							1.71	
Chile		1.66				2.09	0.01	
Croatia								
Cyprus								
Czechia				15.00				
France	0.35	1.62					0.16	
Georgia							0.46	
Germany							2.30	
Greece		0.04						
Hungary			1.12	1.54				
Israel	4.17							
Italy	0.18	0.00		0.02	3.06		0.78	2.60
Korea Rep.								
Luxembourg							10.24	
Moldova			4.82				0.39	
Morocco	7.40							
New Zealand		2.30						
Portugal	0.25							
Romania			8.20					
Russia							5.32	
Serbia								
Slovakia			2.00	19.00			4.00	
Slovenia								
South Africa	4.32	1.10					0.04	
Spain	0.52					1.25		
Switzerland		0.02					0.51	
Taiwan								
Tunisia								
United Kingdom								
United States	1.15	0.40					0.25	
Uruguay								
"Missing 9"		0.48				0.00	0.30	
Total	**0.61**	**0.54**	**0.49**	**0.48**	**0.40**	**0.35**	**0.35**	**0.34**

Table 45 (cont.): Shares of world's top 30 white varieties in national winegrape area, by country, 2000 (%)

Country	*Niagara*	*Pedro Giménez*	*Fernão Pires*	*Chasselas*	*Melon*	*Malvasia Bianca di Candia*	*Other White*	*Total White*	*Total*
Algeria							0.00	0.00	100
Argentina		7.53					13.64	30.72	100
Armenia							73.28	100.00	100
Australia							13.17	40.76	100
Austria							9.72	71.84	100
Brazil	25.43						8.43	39.97	100
Bulgaria							19.06	39.30	100
Canada	5.43						15.74	41.74	100
Chile				0.35			6.71	23.72	100
Croatia							47.63	74.63	100
Cyprus							20.00	20.00	100
Czechia							18.00	70.00	100
France				0.11	1.53		5.63	30.48	100
Georgia							34.28	87.75	100
Germany				1.15			26.84	72.19	100
Greece							40.97	47.24	100
Hungary				2.19			47.26	70.96	100
Israel							14.17	38.78	100
Italy				0.00		1.87	19.64	47.29	100
Korea Rep.								0.00	100
Luxembourg							25.67	83.53	100
Moldova							5.68	57.95	100
Morocco							3.52	10.91	100
New Zealand				0.25			4.47	70.25	100
Portugal			6.93				33.75	41.81	100
Romania							49.45	71.21	100
Russia							43.75	81.00	100
Serbia				5.00			16.00	69.00	100
Slovakia							17.00	83.00	100
Slovenia							47.00	74.00	100
South Africa			0.36				6.47	64.11	100
Spain				0.00			15.35	61.02	100
Switzerland				35.72			3.34	46.40	100
Taiwan							46.00	46.00	100
Tunisia								0.00	100
United Kingdom							69.30	77.09	100
United States	0.77					0.55	2.14	44.75	100
Uruguay							0.46	3.66	100
"Missing 9"	0.11						10.43	44.53	100
Total	**0.31**	**0.30**	**0.30**	**0.27**	**0.27**	**0.26**	**17.15**	**49.18**	**100**

Table 46: National winegrape area for world's top 30 white varieties, 2010 (hectares)

Country	*Airén*	*Chardonnay*	*Sauvignon Blanc*	*Trebbiano Toscano*	*Graševina*	*Rkatsiteli*	*Riesling*	*Macabeo*
Algeria								
Argentina		6584	2297	2425			110	2
Armenia						2469		
Australia		27773	6467	86			4114	
Austria		1380	845		3462		1852	
Brazil		377	45	149	200		9	
Bulgaria		2457		723		3121		
Canada		1178	320	2			871	
Chile		13082	12159				367	
China		738	1				437	
Croatia		668	249	210	4701	57	676	
Cyprus		128						
Czechia		766	804		1148		1181	
Ethiopia								
France		45243	27931	83445			3513	2446
Georgia						25324		
Germany		1228	518				22429	
Greece		586	256	298			2	
Hungary		2757	907		4664		1304	
Israel		142	263					
Italy		19709	3744	22702	1568		446	
Japan		602					11	
Kazakhstan						3552	111	
Korea Rep.								
Luxembourg		16					121	
Mexico			120					
Moldova		5134	8151			11508	1343	
Morocco								
Myanmar		1	22					
New Zealand		3911	16205				979	
Peru		1						
Portugal		803	171	212			18	
Romania		1067	4157	11	7530	356		
Russia		1981	951	66		702	882	
Serbia					33120			
Slovakia		310	208		1655		605	
Slovenia		1208	1061		2360		676	
South Africa		8278	9551	74			211	
Spain	252364	6958	4011	45	791		161	38417
Switzerland		321	134				12	
Taiwan								
Thailand								
Tunisia								
Turkey		142	146				3	
Ukraine		2985	3123			11552	2702	
United Kingdom		235	3				1	
United States		40846	6584	80			4852	
Uruguay		149	147	762			15	
Total	**252364**	**199743**	**111552**	**111290**	**61200**	**58641**	**50014**	**40864**

Table 46 (cont.) National winegrape area for world's top 30 white varieties, 2010 (hectares)

Country	*Cayetana Blanca*	*Aligoté*	*Chenin Blanc*	*Catarratto Bianco*	*Colombard*	*Muscat Blanc à Petits Grains*	*Muscat of Alexandria*	*Müller-Thurgau*
Algeria								
Argentina			2851			139	4035	
Armenia						526		
Australia			541		2205	533	2043	
Austria						492		2044
Brazil		0	18		46	1005	7	
Bulgaria								
Canada			7					7
Chile			57				1090	
China			10					
Croatia						56		60
Cyprus							120	
Czechia								1572
Ethiopia			54					
France		1953	9825		8173	7671	2603	5
Georgia		124						
Germany						190		13638
Greece						2162		
Hungary		0	6			709		2098
Israel			101		486		202	
Italy			45	34794		11506	1521	1312
Japan								172
Kazakhstan		277						
Korea Rep.								
Luxembourg								184
Mexico			275			246		
Moldova		15790				172		173
Morocco							3669	
Myanmar			1			7		
New Zealand			50					79
Peru			2			361		
Portugal	148		0			535	647	
Romania		7297				840		0
Russia		1029				145	21	106
Serbia								
Slovakia						48		932
Slovenia						353		
South Africa			18515		11990	689	2167	
Spain	39633		100		4	1291	8237	
Switzerland		23	6			49		493
Taiwan								
Thailand			13		11	3		
Tunisia								
Turkey						114		
Ukraine		9627				674		
United Kingdom								43
United States			3221	68	10025	733	1285	
Uruguay			7		4	10		
Total	**39781**	**36120**	**35703**	**34863**	**32944**	**31259**	**27648**	**22917**

Table 46 (cont.) National winegrape area for world's top 30 white varieties, 2010 (hectares)

Country	*Palomino Fino*	*Sémillon*	*Grüner Veltliner*	*Prosecco*	*Fetească Albă*	*Verdejo*	*Trebbiano Romagnolo*	*Garganega*
Algeria								
Argentina	163	956		10				21
Armenia								
Australia		6112						
Austria			13519					
Brazil		24		173				0
Bulgaria								
Canada		19						
Chile		846						
China		1						
Croatia								
Cyprus								
Czechia			1527					
Ethiopia								
France	134	11566						
Georgia								
Germany								
Greece		8						
Hungary		58	1533					
Israel								
Italy		31	165	18255			15893	15375
Japan								
Kazakhstan								
Korea Rep.								
Luxembourg								
Mexico	109							
Moldova					4334			
Morocco								
Myanmar								
New Zealand	14	201						
Peru								
Portugal	3033	89						
Romania		18	0		12916			
Russia		25						
Serbia								
Slovakia			2091		219			
Slovenia								
South Africa	270	1192						
Spain	18836					16578		
Switzerland		5						
Taiwan								
Thailand								
Tunisia								
Turkey		547						
Ukraine								
United Kingdom								
United States	135	436						
Uruguay		24						
Total	**22693**	**22157**	**18834**	**18437**	**17469**	**16578**	**15893**	**15397**

Table 46 (cont.) National winegrape area for world's top 30 white varieties, 2010 (hectares)

Country	*Pinot Blanc*	*Gewürztraminer*	*Pedro Giménez*	*Fetească Regală*	*Chasselas*	*Melon*	*Other White*	*Total White*
Algeria								0
Argentina	27		13384			1	24586	57591
Armenia							8211	11206
Australia		834					5583	56292
Austria	1914	309					3073	28890
Brazil	1	13			3		6454	8523
Bulgaria		747					9383	16431
Canada	125	404			5		2032	4971
Chile	10	316	118		88		2275	30408
China	2	5						1193
Croatia	188	234			21		6273	13394
Cyprus							2653	2901
Czechia	732	601					1190	9521
Ethiopia							4	58
France	1280	3168			2362	12305	45955	269578
Georgia	219						16455	42122
Germany	3939				1132		16758	59832
Greece							16862	20174
Hungary	237				1892		26171	42335
Israel							687	1881
Italy	3086	1408			34		100664	252260
Japan							498	1282
Kazakhstan							1658	5598
Korea Rep.								0
Luxembourg	162	20					554	1057
Mexico							1306	2056
Moldova	350	2731					2375	52061
Morocco							1678	5347
Myanmar								31
New Zealand	16	311			2		1004	22772
Peru							1168	1532
Portugal	22	0			76		47935	53689
Romania	0	385		12905			68417	115899
Russia	695				21		9131	15755
Serbia					3450		11040	47609
Slovakia	523	265		231			1374	8460
Slovenia	525						4388	10571
South Africa	14	122					3629	56704
Spain		301			20		81575	469322
Switzerland	105	49			4013		819	6029
Taiwan							1303	1303
Thailand							17	44
Tunisia								0
Turkey							3229	4179
Ukraine	338	961					3496	35458
United Kingdom	23	1					428	734
United States	269	1144					9888	79566
Uruguay	9	24					342	1493
Total	**14812**	**14355**	**13502**	**13136**	**13119**	**12306**	**552521**	**1928113**

Table 47: National shares of global winegrape area for world's top 30 white varieties, 2010 (%)

Country	*Airén*	*Chardonnay*	*Sauvignon Blanc*	*Trebbiano Toscano*	*Graševina*	*Rkatsiteli*	*Riesling*	*Macabeo*
Algeria								
Argentina		3.30	2.06	2.18			0.22	0.00
Armenia						4.21		
Australia		13.90	5.80	0.08			8.23	
Austria		0.69	0.76		5.66		3.70	
Brazil		0.19	0.04	0.13	0.33		0.02	
Bulgaria		1.23		0.65		5.32		
Canada		0.59	0.29	0.00			1.74	
Chile		6.55	10.90				0.73	
China		0.37	0.00				0.87	
Croatia		0.33	0.22	0.19	7.68	0.10	1.35	
Cyprus		0.06						
Czechia		0.38	0.72		1.88		2.36	
Ethiopia								
France		22.65	25.04	74.98			7.02	5.98
Georgia						43.18		
Germany		0.61	0.46				44.85	
Greece		0.29	0.23	0.27			0.00	
Hungary		1.38	0.81		7.62		2.61	
Israel		0.07	0.24					
Italy		9.87	3.36	20.40	2.56		0.89	
Japan		0.30					0.02	
Kazakhstan						6.06	0.22	
Korea Rep.								
Luxembourg		0.01					0.24	
Mexico			0.11					
Moldova		2.57	7.31			19.62	2.69	
Morocco								
Myanmar		0.00	0.02					
New Zealand		1.96	14.53				1.96	
Peru		0.00						
Portugal		0.40	0.15	0.19			0.04	
Romania		0.53	3.73	0.01	12.30	0.61		
Russia		0.99	0.85	0.06		1.20	1.76	
Serbia					54.12			
Slovakia		0.16	0.19		2.70		1.21	
Slovenia		0.60	0.95		3.86		1.35	
South Africa		4.14	8.56	0.07			0.42	
Spain	100.00	3.48	3.60	0.04	1.29		0.32	94.01
Switzerland		0.16	0.12				0.02	
Taiwan								
Thailand								
Tunisia								
Turkey		0.07	0.13				0.00	
Ukraine		1.49	2.80			19.70	5.40	
United Kingdom		0.12	0.00				0.00	
United States		20.45	5.90	0.07			9.70	
Uruguay		0.07	0.13	0.68			0.03	
Total	**100**	**100**	**100**	**100**	**100**	**100**	**100**	**100**

Table 47 (cont.) National shares of global winegrape area for world's top 30 white varieties, 2010 (%)

Country	*Cayetana Blanca*	*Aligoté*	*Chenin Blanc*	*Catarratto Bianco*	*Colombard*	*Muscat Blanc à Petits Grains*	*Muscat of Alexandria*	*Müller-Thurgau*
Algeria								
Argentina			7.98			0.45	14.60	
Armenia						1.68		
Australia			1.52		6.69	1.70	7.39	
Austria						1.57		8.92
Brazil		0.00	0.05		0.14	3.22	0.02	
Bulgaria								
Canada			0.02					0.03
Chile			0.16				3.94	
China			0.03					
Croatia						0.18		0.26
Cyprus							0.43	
Czechia								6.86
Ethiopia			0.15					
France		5.41	27.52		24.81	24.54	9.41	0.02
Georgia		0.34						
Germany						0.61		59.51
Greece						6.92		
Hungary		0.00	0.02			2.27		9.16
Israel			0.28		1.47		0.73	
Italy			0.13	99.80		36.81	5.50	5.73
Japan								0.75
Kazakhstan		0.77						
Korea Rep.								
Luxembourg								0.80
Mexico			0.77			0.79		
Moldova		43.72				0.55		0.75
Morocco							13.27	
Myanmar			0.00			0.02		
New Zealand			0.14					0.34
Peru			0.01			1.15		
Portugal	0.37		0.00			1.71	2.34	
Romania		20.20				2.69		0.00
Russia		2.85				0.46	0.08	0.46
Serbia								
Slovakia						0.15		4.07
Slovenia						1.13		
South Africa			51.86		36.40	2.20	7.84	
Spain	99.63		0.28		0.01	4.13	29.79	
Switzerland		0.06	0.02			0.16		2.15
Taiwan								
Thailand			0.04		0.03	0.01		
Tunisia								
Turkey						0.36		
Ukraine		26.65				2.16		
United Kingd								0.19
United States			9.02	0.20	30.43	2.34	4.65	
Uruguay			0.02		0.01	0.03		
Total	**100**	**100**	**100**	**100**	**100**	**100**	**100**	**100**

Table 47 (cont.) National shares of global winegrape area for world's top 30 white varieties, 2010 (%)

Country	*Palomino Fino*	*Sémillon*	*Grüner Veltliner*	*Prosecco*	*Fetească Albă*	*Verdejo*	*Trebbiano Romagnolo*	*Garganega*
Algeria								
Argentina	0.72	4.31		0.05				0.14
Armenia								
Australia		27.59						
Austria			71.78					
Brazil		0.11		0.94				0.00
Bulgaria								
Canada		0.09						
Chile		3.82						
China		0.00						
Croatia								
Cyprus								
Czechia			8.11					
Ethiopia								
France	0.59	52.20						
Georgia								
Germany								
Greece		0.04						
Hungary		0.26	8.14					
Israel								
Italy		0.14	0.88	99.01			100.00	99.86
Japan								
Kazakhstan								
Korea Rep.								
Luxembourg								
Mexico	0.48							
Moldova					24.81			
Morocco								
Myanmar								
New Zealand	0.06	0.91						
Peru								
Portugal	13.36	0.40						
Romania		0.08	0.00		73.94			
Russia		0.11						
Serbia								
Slovakia			11.10		1.25			
Slovenia								
South Africa	1.19	5.38						
Spain	83.00					100.00		
Switzerland		0.02						
Taiwan								
Thailand								
Tunisia								
Turkey		2.47						
Ukraine								
United Kingdc								
United States	0.59	1.97						
Uruguay		0.11						
Total	**100**	**100**	**100**	**100**	**100**	**100**	**100**	**100**

Table 47 (cont.) National shares of global winegrape area for world's top 30 white varieties, 2010 (%)

Country	*Pinot Blanc*	*Gewürztramin er*	*Pedro Giménez*	*Fetească Regală*	*Chasselas*	*Melon*	*Other White*	*Total White*
Algeria								0.00
Argentina	0.18		99.13			0.01	4.45	2.99
Armenia							1.49	0.58
Australia		5.81					1.01	2.92
Austria	12.93	2.15					0.56	1.50
Brazil	0.01	0.09			0.02		1.17	0.44
Bulgaria		5.20					1.70	0.85
Canada	0.85	2.81			0.04		0.37	0.26
Chile	0.07	2.20	0.87		0.67		0.41	1.58
China	0.01	0.04						0.06
Croatia	1.27	1.63			0.16		1.14	0.69
Cyprus							0.48	0.15
Czechia	4.94	4.19					0.22	0.49
Ethiopia							0.00	0.00
France	8.64	22.07			18.01	99.99	8.32	13.98
Georgia	1.48						2.98	2.18
Germany	26.59				8.63		3.03	3.10
Greece							3.05	1.05
Hungary	1.60				14.42		4.74	2.20
Israel							0.12	0.10
Italy	20.83	9.81			0.26		18.22	13.08
Japan							0.09	0.07
Kazakhstan							0.30	0.29
Korea Rep.								0.00
Luxembourg	1.09	0.14					0.10	0.05
Mexico							0.24	0.11
Moldova	2.36	19.02					0.43	2.70
Morocco							0.30	0.28
Myanmar								0.00
New Zealan(	0.11	2.17			0.02		0.18	1.18
Peru							0.21	0.08
Portugal	0.15	0.00			0.58		8.68	2.78
Romania	0.00	2.68		98.24			12.38	6.01
Russia	4.69				0.16		1.65	0.82
Serbia					26.30		2.00	2.47
Slovakia	3.53	1.85		1.76			0.25	0.44
Slovenia	3.54						0.79	0.55
South Africa	0.10	0.85					0.66	2.94
Spain		2.10			0.15		14.76	24.34
Switzerland	0.71	0.34			30.59		0.15	0.31
Taiwan							0.24	0.07
Thailand							0.00	0.00
Tunisia								0.00
Turkey							0.58	0.22
Ukraine	2.28	6.70					0.63	1.84
United King	0.16	0.01					0.08	0.04
United State	1.82	7.97					1.79	4.13
Uruguay	0.06	0.17					0.06	0.08
Total	**100**	**100**	**100**	**100**	**100**	**100**	**100**	**100**

Table 48: Shares of world's top 30 white varieties in national winegrape area, by country, 2010 (%)

Country	*Airén*	*Chardonnay*	*Sauvignon Blanc*	*Trebbiano Toscano*	*Graševina*	*Rkatsiteli*	*Riesling*	*Macabeo*
Algeria								
Argentina		3.09	1.08	1.14			0.05	0.00
Armenia						22.03		
Australia		18.30	4.26	0.06			2.71	
Austria		3.03	1.86		7.60		4.07	
Brazil		0.76	0.09	0.30	0.41		0.02	
Bulgaria		4.38		1.29		5.56		
Canada		11.67	3.17	0.02			8.63	
Chile		11.73	10.90				0.33	
China		2.50	0.00				1.48	
Croatia		3.22	1.20	1.01	22.65	0.27	3.26	
Cyprus		1.49						
Czechia		4.72	4.95		7.07		7.27	
Ethiopia								
France		5.41	3.34	9.99			0.42	0.29
Georgia						52.76		
Germany		1.20	0.51				21.96	
Greece		1.08	0.47	0.55			0.00	
Hungary		3.95	1.30		6.69		1.87	
Israel		2.92	5.42					
Italy		3.15	0.60	3.63	0.25		0.07	
Japan		16.19					0.30	
Kazakhstan						51.20	1.60	
Korea Rep.								
Luxembourg		1.23					9.28	
Mexico			2.20					
Moldova		5.71	9.07			12.81	1.49	
Morocco								
Myanmar		1.50	29.23					
New Zealand		12.24	50.70				3.06	
Peru		0.03						
Portugal		0.49	0.10	0.13			0.01	
Romania		0.63	2.44	0.01	4.42	0.21		
Russia		7.73	3.71	0.26		2.74	3.44	
Serbia					48.00			
Slovakia		2.45	1.65		13.10		4.79	
Slovenia		7.39	6.49		14.43		4.13	
South Africa		8.20	9.45	0.07			0.21	
Spain	24.54	0.68	0.39	0.00	0.08		0.02	3.74
Switzerland		2.17	0.90				0.08	
Taiwan								
Thailand								
Tunisia								
Turkey		1.10	1.13				0.02	
Ukraine		5.71	5.97			22.09	5.17	
United Kingdom		19.62	0.25				0.08	
United States		17.92	2.89	0.03			2.13	
Uruguay		1.95	1.92	9.95			0.20	
Total	**5.47**	**4.33**	**2.42**	**2.41**	**1.33**	**1.27**	**1.08**	**0.89**

Table 48 (cont.): Shares of world's top 30 white varieties in national winegrape area, by country, 2010 (%)

Country	*Cayetana Blanca*	*Aligoté*	*Chenin Blanc*	*Catarratto Bianco*	*Colombard*	*Muscat Blanc à Petits Grains*	*Muscat of Alexandria*	*Müller-Thurgau*
Algeria								
Argentina			1.34			0.07	1.89	
Armenia						4.69		
Australia			0.36		1.45	0.35	1.35	
Austria						1.08		4.49
Brazil		0.00	0.04		0.09	2.03	0.01	
Bulgaria								
Canada			0.07					0.07
Chile			0.05				0.98	
China			0.03					
Croatia						0.27		0.29
Cyprus							1.39	
Czechia								9.68
Ethiopia			31.92					
France		0.23	1.18		0.98	0.92	0.31	0.00
Georgia		0.26						
Germany						0.19		13.35
Greece						3.97		
Hungary		0.00	0.01			1.02		3.01
Israel			2.09		10.01		4.17	
Italy			0.01	5.56		1.84	0.24	0.21
Japan								4.62
Kazakhstan		3.99						
Korea Rep.								
Luxembourg								14.11
Mexico			5.03			4.50		
Moldova		17.57				0.19		0.19
Morocco							7.49	
Myanmar			0.81			9.46		
New Zealand			0.16					0.25
Peru			0.05			9.42		
Portugal	0.09		0.00			0.33	0.40	
Romania		4.28				0.49		0.00
Russia		4.02				0.57	0.08	0.41
Serbia								
Slovakia						0.38		7.37
Slovenia						2.16		
South Africa			18.33		11.87	0.68	2.15	
Spain	3.85		0.01		0.00	0.13	0.80	
Switzerland		0.15	0.04			0.33		3.32
Taiwan								
Thailand			8.86		7.48	2.19		
Tunisia								
Turkey						0.88		
Ukraine		18.41				1.29		
United Kingd								3.59
United States			1.41	0.03	4.40	0.32	0.56	
Uruguay			0.09		0.05	0.13		
Total	**0.86**	**0.78**	**0.77**	**0.76**	**0.71**	**0.68**	**0.60**	**0.50**

Table 48 (cont.): Shares of world's top 30 white varieties in national winegrape area, by country, 2010 (%)

Country	*Palomino Fino*	*Sémillon*	*Grüner Veltliner*	*Prosecco*	*Fetească Albă*	*Verdejo*	*Trebbiano Romagnolo*	*Garganega*
Algeria								
Argentina	0.08	0.45		0.00				0.01
Armenia								
Australia		4.03						
Austria			29.69					
Brazil		0.05		0.35				0.00
Bulgaria								
Canada		0.19						
Chile		0.76						
China		0.00						
Croatia								
Cyprus								
Czechia			9.40					
Ethiopia								
France	0.02	1.38						
Georgia								
Germany								
Greece		0.01						
Hungary		0.08	2.20					
Israel								
Italy		0.00	0.03	2.92			2.54	2.46
Japan								
Kazakhstan								
Korea Rep.								
Luxembourg								
Mexico	1.99							
Moldova					4.82			
Morocco								
Myanmar								
New Zealand	0.04	0.63						
Peru								
Portugal	1.85	0.05						
Romania		0.01	0.00		7.58			
Russia		0.10						
Serbia								
Slovakia			16.54		1.73			
Slovenia								
South Africa	0.27	1.18						
Spain	1.83					1.61		
Switzerland		0.03						
Taiwan								
Thailand								
Tunisia								
Turkey		4.25						
Ukraine								
United Kingdc								
United States	0.06	0.19						
Uruguay		0.31						
Total	**0.49**	**0.48**	**0.41**	**0.40**	**0.38**	**0.36**	**0.34**	**0.33**

Table 48 (cont.): Shares of world's top 30 white varieties in national winegrape area, by country, 2010 (%)

Country	*Pinot Blanc*	*Gewürztraminer*	*Pedro Giménez*	*Fetească Regală*	*Chasselas*	*Melon*	*Other White*	*Total White*	*Total*
Algeria								0.00	100
Argentina	0.01		6.27			0.00	11.52	26.99	100
Armenia							73.28	100.00	100
Australia		0.55					3.68	37.09	100
Austria	4.20	0.68					6.75	63.45	100
Brazil	0.00	0.03			0.01		13.06	17.25	100
Bulgaria		1.33					16.72	29.27	100
Canada	1.24	4.00			0.05		20.13	49.23	100
Chile	0.01	0.28	0.11		0.08		2.04	27.27	100
China	0.01	0.02						4.04	100
Croatia	0.90	1.13			0.10		30.23	64.54	100
Cyprus							30.82	33.70	100
Czechia	4.51	3.70					7.33	58.62	100
Ethiopia							2.54	34.46	100
France	0.15	0.38			0.28	1.47	5.50	32.26	100
Georgia	0.46						34.28	87.75	100
Germany	3.86				1.11		16.41	58.58	100
Greece							31.00	37.09	100
Hungary	0.34				2.71		37.54	60.73	100
Israel							14.17	38.78	100
Italy	0.49	0.23			0.01		16.09	40.32	100
Japan							13.39	34.51	100
Kazakhstan							23.90	80.68	100
Korea Rep.								0.00	100
Luxembourg	12.42	1.53					42.48	81.05	100
Mexico							23.89	37.62	100
Moldova	0.39	3.04					2.64	57.95	100
Morocco							3.43	10.91	100
Myanmar								41.01	100
New Zealanc	0.05	0.97			0.01		3.14	71.24	100
Peru							30.49	39.99	100
Portugal	0.01	0.00			0.05		29.31	32.83	100
Romania	0.00	0.23		7.58			40.18	68.06	100
Russia	2.71				0.08		35.63	61.47	100
Serbia					5.00		16.00	69.00	100
Slovakia	4.14	2.10		1.83			10.87	66.95	100
Slovenia	3.21						26.83	64.64	100
South Africa	0.01	0.12					3.59	56.13	100
Spain		0.03			0.00		7.93	45.64	100
Switzerland	0.71	0.33			27.08		5.53	40.68	100
Taiwan							46.00	46.00	100
Thailand							11.26	29.79	100
Tunisia								0.00	100
Turkey							25.11	32.51	100
Ukraine	0.65	1.84					6.69	67.81	100
United King	1.92	0.08					35.73	61.27	100
United State	0.12	0.50					4.34	34.91	100
Uruguay	0.12	0.31					4.47	19.50	100
Total	**0.32**	**0.31**	**0.29**	**0.28**	**0.28**	**0.27**	**11.97**	**41.79**	**100**

Table 49: National winegrape area for world's top 30 white varieties, 2016 (hectares)

Country	*Airén*	*Chardonnay*	*Sauvignon Blanc*	*Trebbiano Toscano*	*Riesling*	*Rkatsiteli*	*Macabeo*	*Cayetana Blanca*
Algeria								
Argentina		6227	2148	1622	93			
Armenia								
Australia		21321	6044	14	3114			17
Austria		1577	1170		2016			
Brazil		340	33	231	6			
Bulgaria		3087	637	738		5415		
Cambodia								
Canada		1417	285	2	1188			
Chile		11435	14999		413			
China		6100	2000	1500	1600			
Croatia		657			625			
Cyprus								
Czechia		820	906		1172			
Ethiopia								
France		47451	28084	78842	4025		1657	
Georgia						25324		
Germany		1485	736		21540			
Greece		673	727	211	1			
Hungary		2464	982		1261			
India		100	500	300				
Israel		165	110					
Italy		19769	3935	35441	1461			
Japan		137	15		22			
Kazakhstan					111	3552		
Korea Rep.								
Lebanon		1000	500					
Luxembourg		30			162			
Mexico			120					
Moldova		4133	6909	277	1701	3898		
Morocco	440	880	440					
Myanmar		2	22					
New Zealand		3117	20497		767			
N. Macedonia		750	185		900	460		
Norway								
Peru		1						
Portugal		547	102	122	13			132
Romania		1878	5594	3	6121	413		
Russia		3481	2501	66	2232	6477		
Serbia		1455	741		1361	60		
Slovakia					620			
Slovenia		1181	1121		607			
South Africa		6856	9246	156	152		4	
Spain	203276	6866	4562	49	184		36963	36252
Switzerland		359	170		19			
Taiwan								
Thailand								
Tunisia	85	170	85					
Turkey		177	153		3			
Ukraine		1500	1550		1350	5775		
United Kingdom		531						
United States		41392	6747	88	4952			
Uruguay		119	144	682	12			
Total	**203801**	**201649**	**124700**	**120343**	**59805**	**51374**	**38625**	**36401**

Table 49 (cont.) National winegrape area for world's top 30 white varieties, 2016 (hectares)

Country	*Muscat of Alexandria*	*Muscat Blanc à Petits Grains*	*Chenin Blanc*	*Colombard*	*Catarratto Bianco*	*Aligoté*	*Graševina*	*Palomino Fino*
Algeria	200							
Argentina	2716	94	2157					104
Armenia								
Australia	2179	857	406	1789				19
Austria		823					3233	
Brazil	6	32	7	22			188	
Bulgaria						285		
Cambodia								
Canada			6			30		
Chile	5424	1	39					
China	3000						3000	
Croatia							4459	
Cyprus								
Czechia							1114	
Ethiopia			54					
France	2462	7333	9432	8441		1927		41
Georgia						124		
Germany		240						
Greece	773	1568						
Hungary		762	6			0	3933	
India	100							
Israel	220			220				
Italy	1375	13334	7		28563		1259	
Japan								
Kazakhstan						277		
Korea Rep.								
Lebanon								
Luxembourg		1						
Mexico		246	275					109
Moldova		50				7765		
Morocco	2093							
Myanmar		7						
New Zealand		0	24					7
N. Macedonia		400					270	
Norway								
Peru		361	2					
Portugal	509	1031	0					2594
Romania		1579				5840	1437	
Russia	21	483				5843		
Serbia		31					2037	
Slovakia							456	
Slovenia		586					1935	
South Africa	1781	839	17707	11512				134
Spain	9534	1350	106	6			1064	20110
Switzerland		36	8			24		0
Taiwan								
Thailand			16	15				
Tunisia	405							
Turkey		129						
Ukraine		338				4814		
United Kingd								
United States	1987	1218	1969	7991	50			70
Uruguay	22	10	2					
Total	**34805**	**33739**	**32221**	**29996**	**28613**	**26929**	**24384**	**23190**

Table 49 (cont.) National winegrape area for world's top 30 white varieties, 2016 (hectares)

Country	*Prosecco*	*Müller-Thurgau*	*Grüner Veltliner*	*Trebbiano Romagnolo*	*Sémillon*	*Verdejo*	*Viognier*	*Pedro Giménez*
Algeria								
Argentina	11		7		767		773	11197
Armenia								
Australia	160	1	9		4556	6	753	
Austria		1777	14376					
Brazil	207				6		11	
Bulgaria								
Cambodia								
Canada		6	3		19		101	
Chile					849	2	839	4379
China								
Croatia								
Cyprus								
Czechia		1479	1538					
Ethiopia								
France		2			10234		8823	
Georgia								
Germany		11664	14					
Greece					11			
Hungary		1670	1381		43		13	
India								
Israel								
Italy	19730	1296	55	19059	13		1827	
Japan		22						
Kazakhstan								
Korea Rep.								
Lebanon								
Luxembourg		316						
Mexico								
Moldova		4			3		31	
Morocco								
Myanmar								
New Zealand		2	43		63		129	
N. Macedonia								
Norway								
Peru								
Portugal					76		125	
Romania		0	0		18		20	
Russia		106			25			
Serbia								
Slovakia		509	1627					
Slovenia		128						
South Africa			4		1121		822	
Spain					2	17923	213	
Switzerland		465	1		4		44	
Taiwan								
Thailand							1	
Tunisia								
Turkey					529		15	
Ukraine								
United Kingdom		15						
United States		39	60		340		1481	
Uruguay					14		41	
Total	**20109**	**19501**	**19118**	**19059**	**18693**	**17931**	**16063**	**15576**

Table 49 (cont.) National winegrape area for world's top 30 white varieties, 2016 (hectares)

Country	*Pinot Blanc*	*Fetească Albă*	*Fetească Regală*	*Gewürztraminer*	*Muscat Ottonel*	*Fernão Pires*	*Other White*	*Total White*
Algeria							100	300
Argentina	9						19748	47671
Armenia							10300	10300
Australia	5			252			3139	44641
Austria	1916				344		2391	29623
Brazil				9			4102	5199
Bulgaria				591	3679		2998	17429
Cambodia								0
Canada	109			398	32		3041	6637
Chile	18			371			5389	44160
China							8447	25647
Croatia							1600	7341
Cyprus							1946	1946
Czechia	762			591				8382
Ethiopia							4	58
France	1181			3320	172		63126	276553
Georgia	219						16454	42121
Germany	4323			824	12		14485	55323
Greece							14895	18860
Hungary	228			694	1256		23494	38187
India							1000	2000
Israel							137	852
Italy	2337			1321	0		108465	259247
Japan							959	1155
Kazakhstan							1669	5609
Korea Rep.								0
Lebanon							318	1818
Luxembourg	160			21			288	978
Mexico							1181	1931
Moldova	210	954	372	1099	1859		9631	38896
Morocco							4908	8760
Myanmar								31
New Zealand	12			277			248	25187
N. Macedonia							6986	9951
Norway							8	8
Peru							1168	1532
Portugal	15			0		12138	39591	56996
Romania	0	12428	12619	469	4779		65783	118982
Russia	865			500	34		9147	31780
Serbia				142	183		3923	9932
Slovakia	416						1057	4685
Slovenia	424				98		4474	10555
South Africa	9			106	9	73	2304	52834
Spain				373			73826	412658
Switzerland	111			51	5		4748	6045
Taiwan							55	55
Thailand							59	91
Tunisia							949	1693
Turkey							3344	4349
Ukraine	170			500				15996
United Kingdom	15						358	919
United States	263			897			9442	78987
Uruguay	2			17	2		191	1258
Total	**13779**	**13382**	**12991**	**12823**	**12464**	**12211**	**551875**	**1846150**

Table 50: National shares of global winegrape area for world's top 30 white varieties, 2016 (%)

Country	*Airén*	*Chardonnay*	*Sauvignon Blanc*	*Trebbiano Toscano*	*Riesling*	*Rkatsiteli*	*Macabeo*	*Cayetana Blanca*
Algeria								
Argentina		3.09	1.72	1.35	0.16			
Armenia								
Australia		10.57	4.85	0.01	5.21			0.05
Austria		0.78	0.94		3.37			
Brazil		0.17	0.03	0.19	0.01			
Bulgaria		1.53	0.51	0.61		10.54		
Cambodia								
Canada		0.70	0.23	0.00	1.99			
Chile		5.67	12.03		0.69			
China		3.03	1.60	1.25	2.68			
Croatia		0.33			1.05			
Cyprus								
Czechia		0.41	0.73		1.96			
Ethiopia								
France		23.53	22.52	65.51	6.73		4.29	
Georgia						49.29		
Germany		0.74	0.59		36.02			
Greece		0.33	0.58	0.18	0.00			
Hungary		1.22	0.79		2.11			
India		0.05	0.40	0.25				
Israel		0.08	0.09					
Italy		9.80	3.16	29.45	2.44			
Japan		0.07	0.01		0.04			
Kazakhstan					0.19	6.91		
Korea Rep.								
Lebanon		0.50	0.40					
Luxembourg		0.01			0.27			
Mexico			0.10					
Moldova		2.05	5.54	0.23	2.84	7.59		
Morocco	0.22	0.44	0.35					
Myanmar		0.00	0.02					
New Zealand		1.55	16.44		1.28			
N. Macedonia		0.37	0.15		1.50	0.90		
Norway								
Peru		0.00						
Portugal		0.27	0.08	0.10	0.02			0.36
Romania		0.93	4.49	0.00	10.23	0.80		
Russia		1.73	2.01	0.05	3.73	12.61		
Serbia		0.72	0.59		2.28	0.12		
Slovakia					1.04			
Slovenia		0.59	0.90		1.02			
South Africa		3.40	7.41	0.13	0.25		0.01	
Spain	99.74	3.41	3.66	0.04	0.31		95.70	99.59
Switzerland		0.18	0.14		0.03			
Taiwan								
Thailand								
Tunisia	0.04	0.08	0.07					
Turkey		0.09	0.12		0.01			
Ukraine		0.74	1.24		2.26	11.24		
United Kingdom		0.26						
United States		20.53	5.41	0.07	8.28			
Uruguay		0.06	0.12	0.57	0.02			
Total	**100**	**100**	**100**	**100**	**100**	**100**	**100**	**100**

Table 50 (cont.): National shares of global winegrape area for world's top 30 white varieties, 2016 (%)

Country	*Muscat of Alexandria*	*Muscat Blanc à Petits Grains*	*Chenin Blanc*	*Colombard*	*Catarratto Bianco*	*Aligoté*	*Graševina*	*Palomino Fino*
Algeria	0.57							
Argentina	7.80	0.28	6.69					0.45
Armenia								
Australia	6.26	2.54	1.26	5.96				0.08
Austria		2.44					13.26	
Brazil	0.02	0.09	0.02	0.07			0.77	
Bulgaria						1.06		
Cambodia								
Canada			0.02			0.11		
Chile	15.58	0.00	0.12					
China	8.62						12.30	
Croatia							18.29	
Cyprus								
Czechia							4.57	
Ethiopia			0.17					
France	7.07	21.73	29.27	28.14		7.16		0.18
Georgia						0.46		
Germany		0.71						
Greece	2.22	4.65						
Hungary		2.26	0.02			0.00	16.13	
India	0.29							
Israel	0.63			0.73				
Italy	3.95	39.52	0.02		99.82		5.16	
Japan								
Kazakhstan						1.03		
Korea Rep.								
Lebanon								
Luxembourg		0.00						
Mexico		0.73	0.85					0.47
Moldova		0.15				28.84		
Morocco	6.01							
Myanmar		0.02						
New Zealand		0.00	0.07					0.03
N. Macedonia		1.19					1.11	
Norway								
Peru		1.07	0.01					
Portugal	1.46	3.06	0.00					11.19
Romania		4.68				21.69	5.89	
Russia	0.06	1.43				21.70		
Serbia		0.09					8.35	
Slovakia							1.87	
Slovenia		1.74					7.93	
South Africa	5.12	2.49	54.95	38.38				0.58
Spain	27.39	4.00	0.33	0.02			4.36	86.72
Switzerland		0.11	0.02			0.09		0.00
Taiwan								
Thailand			0.05	0.05				
Tunisia	1.16							
Turkey		0.38						
Ukraine		1.00				17.87		
United Kingd								
United States	5.71	3.61	6.11	26.64	0.18			0.30
Uruguay	0.06	0.03	0.01					
Total	**100**	**100**	**100**	**100**	**100**	**100**	**100**	**100**

Table 50 (cont.): National shares of global winegrape area for world's top 30 white varieties, 2016 (%)

Country	*Prosecco*	*Müller-Thurgau*	*Grüner Veltliner*	*Trebbiano Romagnolo*	*Sémillon*	*Verdejo*	*Viognier*	*Pedro Giménez*
Algeria								
Argentina	0.06		0.04		4.10		4.81	71.89
Armenia								
Australia	0.80	0.00	0.05		24.37	0.04	4.69	
Austria		9.11	75.19					
Brazil	1.03				0.03		0.07	
Bulgaria								
Cambodia								
Canada		0.03	0.01		0.10		0.63	
Chile					4.54	0.01	5.23	28.11
China								
Croatia								
Cyprus								
Czechia		7.58	8.04					
Ethiopia								
France		0.01			54.74		54.93	
Georgia								
Germany		59.81	0.07					
Greece					0.06			
Hungary		8.56	7.22		0.23		0.08	
India								
Israel								
Italy	98.12	6.65	0.29	100.00	0.07		11.37	
Japan		0.11						
Kazakhstan								
Korea Rep.								
Lebanon								
Luxembourg		1.62						
Mexico								
Moldova		0.02			0.02		0.20	
Morocco								
Myanmar								
New Zealand		0.01	0.23		0.34		0.80	
N. Macedonia								
Norway								
Peru								
Portugal					0.41		0.78	
Romania		0.00	0.00		0.10		0.13	
Russia		0.54			0.13			
Serbia								
Slovakia		2.61	8.51					
Slovenia		0.66						
South Africa			0.02		6.00		5.12	
Spain					0.01	99.95	1.32	
Switzerland		2.39	0.00		0.02		0.28	
Taiwan								
Thailand							0.01	
Tunisia								
Turkey					2.83		0.09	
Ukraine								
United Kingdom		0.08						
United States		0.20	0.32		1.82		9.22	
Uruguay					0.07		0.26	
Total	**100**	**100**	**100**	**100**	**100**	**100**	**100**	**100**

Table 50 (cont.): National shares of global winegrape area for world's top 30 white varieties, 2016 (%)

Country	*Pinot Blanc*	*Fetească Albă*	*Fetească Regală*	*Gewürztraminer*	*Muscat Ottonel*	*Fernão Pires*	*Other White*	*Total White*
Algeria							0.02	0.02
Argentina	0.06						3.58	2.58
Armenia							1.87	0.56
Australia	0.04			1.96			0.57	2.42
Austria	13.91				2.76		0.43	1.60
Brazil				0.07			0.74	0.28
Bulgaria				4.61	29.51		0.54	0.94
Cambodia								0.00
Canada	0.79			3.11	0.26		0.55	0.36
Chile	0.13			2.90			0.98	2.39
China							1.53	1.39
Croatia							0.29	0.40
Cyprus							0.35	0.11
Czechia	5.53			4.61				0.45
Ethiopia							0.00	0.00
France	8.57			25.89	1.38		11.44	14.98
Georgia	1.59						2.98	2.28
Germany	31.37			6.43	0.10		2.62	3.00
Greece							2.70	1.02
Hungary	1.66			5.41	10.08		4.26	2.07
India							0.18	0.11
Israel							0.02	0.05
Italy	16.96			10.30	0.00		19.65	14.04
Japan							0.17	0.06
Kazakhstan							0.30	0.30
Korea Rep.								0.00
Lebanon							0.06	0.10
Luxembourg	1.16			0.16			0.05	0.05
Mexico							0.21	0.10
Moldova	1.52	7.13	2.86	8.57	14.91		1.75	2.11
Morocco							0.89	0.47
Myanmar								0.00
New Zealand	0.08			2.16			0.04	1.36
N. Macedonia							1.27	0.54
Norway							0.00	0.00
Peru							0.21	0.08
Portugal	0.11			0.00		99.40	7.17	3.09
Romania	0.00	92.87	97.14	3.66	38.34		11.92	6.44
Russia	6.28			3.90	0.27		1.66	1.72
Serbia				1.10	1.47		0.71	0.54
Slovakia	3.02						0.19	0.25
Slovenia	3.08				0.78		0.81	0.57
South Africa	0.06			0.83	0.08	0.60	0.42	2.86
Spain				2.91			13.38	22.35
Switzerland	0.81			0.40	0.04		0.86	0.33
Taiwan							0.01	0.00
Thailand							0.01	0.00
Tunisia							0.17	0.09
Turkey							0.61	0.24
Ukraine	1.23			3.90				0.87
United Kingdom	0.11						0.06	0.05
United States	1.91			6.99			1.71	4.28
Uruguay	0.01			0.13	0.02		0.03	0.07
Total	**100**	**100**	**100**	**100**	**100**	**100**	**100**	**100**

Table 51: Shares of world's top 30 white varieties in national winegrape area, by country, 2016 (%)

Country	*Airén*	*Chardonnay*	*Sauvignon Blanc*	*Trebbiano Toscano*	*Riesling*	*Rkatsiteli*	*Macabeo*	*Cayetana Blanca*
Algeria								
Argentina		3.02	1.04	0.79	0.05			
Armenia								
Australia		16.10	4.56	0.01	2.35			0.01
Austria		3.47	2.58		4.44			
Brazil		1.02	0.10	0.69	0.02			
Bulgaria		5.83	1.20	1.39		10.22		
Cambodia								
Canada		11.24	2.26	0.02	9.43			
Chile		7.84	10.28		0.28			
China		3.43	1.12	0.84	0.90			
Croatia		5.59			5.32			
Cyprus								
Czechia		6.03	6.66		8.62			
Ethiopia								
France		5.82	3.45	9.68	0.49		0.20	
Georgia						52.76		
Germany		1.57	0.78		22.79			
Greece		1.32	1.43	0.41	0.00			
Hungary		3.86	1.54		1.97			
India		3.70	18.52	11.11				
Israel		3.30	2.20					
Italy		3.27	0.65	5.86	0.24			
Japan		3.55	0.38		0.56			
Kazakhstan					1.60	51.20		
Korea Rep.								
Lebanon		25.00	12.50					
Luxembourg		2.31			12.46			
Mexico			2.20					
Moldova		5.00	8.36	0.34	2.06	4.72		
Morocco	2.50	5.00	2.50					
Myanmar		2.14	31.43					
New Zealand		8.79	57.80		2.16			
N. Macedonia		3.03	0.75		3.63	1.86		
Norway								
Peru		0.03						
Portugal		0.30	0.06	0.07	0.01			0.07
Romania		1.03	3.06	0.00	3.35	0.23		
Russia		6.85	4.92	0.13	4.39	12.75		
Serbia		6.61	3.36		6.18	0.27		
Slovakia					8.00			
Slovenia		7.39	7.01		3.80			
South Africa		7.16	9.65	0.16	0.16		0.00	
Spain	23.01	0.78	0.52	0.01	0.02		4.18	4.10
Switzerland		2.43	1.15		0.13			
Taiwan								
Thailand								
Tunisia	2.50	5.00	2.50					
Turkey		1.29	1.12		0.02			
Ukraine		5.96	6.16		5.36	22.95		
United Kingdom		28.90						
United States		17.27	2.82	0.04	2.07			
Uruguay		1.76	2.14	10.11	0.18			
Total	**4.55**	**4.50**	**2.78**	**2.68**	**1.33**	**1.15**	**0.86**	**0.81**

Table 51 (cont.): Shares of world's top 30 white varieties in national winegrape area, by country, 2016 (%)

Country	*Muscat of Alexandria*	*Muscat Blanc à Petits Grains*	*Chenin Blanc*	*Colombard*	*Catarratto Bianco*	*Aligoté*	*Graševina*	*Palomino Fino*
Algeria	2.74							
Argentina	1.32	0.05	1.05					0.05
Armenia								
Australia	1.65	0.65	0.31	1.35				0.01
Austria		1.81					7.11	
Brazil	0.02	0.10	0.02	0.07			0.57	
Bulgaria						0.54		
Cambodia								
Canada			0.05			0.24		
Chile	3.72	0.00	0.03					
China	1.69						1.69	
Croatia							37.96	
Cyprus								
Czechia							8.19	
Ethiopia			31.92					
France	0.30	0.90	1.16	1.04		0.24		0.01
Georgia						0.26		
Germany		0.25						
Greece	1.52	3.08						
Hungary		1.19	0.01			0.00	6.16	
India	3.70							
Israel	4.40			4.40				
Italy	0.23	2.21	0.00		4.72		0.21	
Japan								
Kazakhstan						3.99		
Korea Rep.								
Lebanon								
Luxembourg		0.08						
Mexico		4.50	5.03					1.99
Moldova		0.06				9.40		
Morocco	11.90							
Myanmar		10.00						
New Zealand		0.00	0.07					0.02
N. Macedonia		1.61					1.09	
Norway								
Peru		9.42	0.05					
Portugal	0.28	0.56	0.00					1.42
Romania		0.86				3.20	0.79	
Russia	0.04	0.95				11.50		
Serbia		0.14					9.25	
Slovakia							5.89	
Slovenia		3.67					12.10	
South Africa	1.86	0.88	18.49	12.02				0.14
Spain	1.08	0.15	0.01	0.00			0.12	2.28
Switzerland		0.24	0.05			0.16		0.00
Taiwan								
Thailand			7.80	7.02				
Tunisia	11.90							
Turkey		0.94						
Ukraine		1.34				19.13		
United Kingd								
United States	0.83	0.51	0.82	3.33	0.02			0.03
Uruguay	0.33	0.15	0.03					
Total	**0.78**	**0.75**	**0.72**	**0.67**	**0.64**	**0.60**	**0.54**	**0.52**

Table 51 (cont.): Shares of world's top 30 white varieties in national winegrape area, by country, 2016 (%)

Country	*Prosecco*	*Müller-Thurgau*	*Grüner Veltliner*	*Trebbiano Romagnolo*	*Sémillon*	*Verdejo*	*Viognier*	*Pedro Giménez*
Algeria								
Argentina	0.01		0.00		0.37		0.37	5.43
Armenia								
Australia	0.12	0.00	0.01		3.44	0.00	0.57	
Austria		3.91	31.64					
Brazil	0.62				0.02		0.03	
Bulgaria								
Cambodia								
Canada		0.05	0.02		0.15		0.80	
Chile					0.58	0.00	0.58	3.00
China								
Croatia								
Cyprus								
Czechia		10.88	11.31					
Ethiopia								
France		0.00			1.26		1.08	
Georgia								
Germany		12.34	0.01					
Greece					0.02			
Hungary		2.61	2.16		0.07		0.02	
India								
Israel								
Italy	3.26	0.21	0.01	3.15	0.00		0.30	
Japan		0.57						
Kazakhstan								
Korea Rep.								
Lebanon								
Luxembourg		24.31						
Mexico								
Moldova		0.00			0.00		0.04	
Morocco								
Myanmar								
New Zealand		0.01	0.12		0.18		0.36	
N. Macedonia								
Norway								
Peru								
Portugal					0.04		0.07	
Romania		0.00	0.00		0.01		0.01	
Russia		0.21			0.05			
Serbia								
Slovakia		6.57	21.00					
Slovenia		0.80						
South Africa			0.00		1.17		0.86	
Spain					0.00	2.03	0.02	
Switzerland		3.14	0.01		0.03		0.30	
Taiwan								
Thailand							0.60	
Tunisia								
Turkey					3.86		0.11	
Ukraine								
United Kingdom		0.80						
United States		0.02	0.03		0.14		0.62	
Uruguay					0.21		0.61	
Total	**0.45**	**0.44**	**0.43**	**0.43**	**0.42**	**0.40**	**0.36**	**0.35**

Table 51 (cont.): Shares of world's top 30 white varieties in national winegrape area, by country, 2016 (%)

Country	*Pinot Blanc*	*Fetească Albă*	*Fetească Regală*	*Gewürztraminer*	*Muscat Ottonel*	*Fernão Pires*	*Other White*	*Total White*	*Total*
Algeria							1.37	4.11	100
Argentina	0.00						9.57	23.10	100
Armenia							70.04	70.04	100
Australia	0.00			0.19			2.37	33.71	100
Austria	4.22				0.76		5.26	65.19	100
Brazil				0.03			12.35	15.66	100
Bulgaria				1.11	6.94		5.66	32.90	100
Cambodia								0.00	100
Canada	0.86			3.16	0.25		24.14	52.67	100
Chile	0.01			0.25			3.69	30.27	100
China							4.75	14.41	100
Croatia							13.62	62.50	100
Cyprus							37.91	37.91	100
Czechia	5.60			4.35				61.63	100
Ethiopia							2.54	34.46	100
France	0.14			0.41	0.02		7.75	33.94	100
Georgia	0.46						34.28	87.75	100
Germany	4.57			0.87	0.01		15.33	58.54	100
Greece							29.29	37.09	100
Hungary	0.36			1.09	1.97		36.78	59.78	100
India							37.04	74.07	100
Israel							2.75	17.05	100
Italy	0.39			0.22	0.00		17.94	42.88	100
Japan							24.79	29.86	100
Kazakhstan							24.05	80.84	100
Korea Rep.								0.00	100
Lebanon							7.95	45.45	100
Luxembourg	12.31			1.62			22.12	75.19	100
Mexico							21.61	35.34	100
Moldova	0.25	1.15	0.45	1.33	2.25		11.66	47.09	100
Morocco							27.90	49.80	100
Myanmar								43.57	100
New Zealand	0.03			0.78			0.70	71.02	100
N. Macedonia							28.19	40.16	100
Norway							66.00	66.00	100
Peru							30.49	39.99	100
Portugal	0.01			0.00		6.65	21.68	31.21	100
Romania	0.00	6.80	6.90	0.26	2.61		35.99	65.10	100
Russia	1.70			0.98	0.07		18.01	62.57	100
Serbia				0.64	0.83		17.82	45.12	100
Slovakia	5.37						13.65	60.47	100
Slovenia	2.65				0.61		27.98	66.02	100
South Africa	0.01			0.11	0.01	0.08	2.41	55.16	100
Spain				0.04			8.36	46.70	100
Switzerland	0.75			0.34	0.04		32.10	40.87	100
Taiwan							36.76	36.76	100
Thailand							28.66	44.08	100
Tunisia							27.90	49.80	100
Turkey							24.40	31.74	100
Ukraine	0.68			1.99				63.56	100
United Kingdom	0.80						19.46	49.96	100
United States	0.11			0.37			3.94	32.96	100
Uruguay	0.03			0.25	0.03		2.83	18.66	100
Total	**0.31**	**0.30**	**0.29**	**0.29**	**0.28**	**0.27**	**12.31**	**41.19**	**100**

Table 52: Change in national winegrape area since 2000 for world's top 30 white varieties in 2016 (hectares)

Country	*Airén*	*Chardonnay*	*Sauvignon Blanc*	*Trebbiano Toscano*	*Riesling*	*Rkatsiteli*	*Macabeo*	*Cayetana Blanca*
Algeria								
Argentina		1545	1283	-1143	-57			
Armenia								
Australia		4055	3442	-671	-15			-232
Austria			856		373			
Brazil		10	-107	-457				
Bulgaria		1226	232	-1083		-4014		
Cambodia								
Canada		443	137		706			
Chile		3763	8337		127			
China								
Croatia								
Cyprus								
Czechia		253			379			
Ethiopia								
France		10955	7151	-11499	618		-3566	
Georgia						5583		
Germany		954			-810			
Greece		637	569	-535				
Hungary		-491	659		-358			
India								
Israel		23	-153					
Italy		8082	622	-4006	861			
Japan								
Kazakhstan								
Korea Rep.								
Lebanon								
Luxembourg		22			-13			
Mexico								
Moldova		-1001	-1242		358	-7610		
Morocco								
Myanmar								
New Zealand		330	18074		277			
N. Macedonia								
Norway								
Peru								
Portugal				-260				
Romania		502	981			-93		
Russia		1842			856	-6675		
Serbia								
Slovakia								
Slovenia		-368	-99					
South Africa		788	3810	9	-325			
Spain	-184702	5052	4095	21	87		-5939	-19274
Switzerland		133	132		11			
Taiwan								
Thailand								
Tunisia								
Turkey								
Ukraine								
United Kingdom		463						
United States		5601	2556	-63	2987			
Uruguay		-23	2					
Total	**-184177**	**56106**	**59510**	**-16858**	**16489**	**-15980**	**-9504**	**-19374**

Table 52 (cont.): Change in national winegrape area since 2000 for world's top 30 white varieties in 2016 (hectares)

Country	*Muscat of Alexandria*	*Muscat Blanc à Petits Grains*	*Chenin Blanc*	*Colombard*	*Catarratto Bianco*	*Aligoté*	*Graševina*	*Palomino Fino*
Algeria								
Argentina	-2800	-55	-1288					-112
Armenia								
Australia	-316	643	-435	-13				-105
Austria		680					-1090	
Brazil	-804						-692	
Bulgaria						-1374		
Cambodia								
Canada								
Chile			-37					
China								
Croatia							-11592	
Cyprus								
Czechia							-132	
Ethiopia								
France	-566	398	-406	1546		171		-399
Georgia						27		
Germany		153						
Greece		-664						
Hungary		-776					-2745	
India								
Israel	18			-266				
Italy	217	319			-22148		-748	
Japan								
Kazakhstan								
Korea Rep.								
Lebanon								
Luxembourg								
Mexico								
Moldova		-122				-8025		
Morocco	-1576							
Myanmar								
New Zealand			-122					-14
N. Macedonia								
Norway								
Peru								
Portugal	-1	-388						
Romania		567				-1768	-13577	
Russia						4021		
Serbia							-31082	
Slovakia							-3439	
Slovenia							-1633	
South Africa	-2266	66	-4859	80				-1498
Spain	3389	1127	1				-859	-7574
Switzerland		-8	7			4		
Taiwan								
Thailand								
Tunisia								
Turkey								
Ukraine								
United Kingd(								
United States	-26	703	-6464	-10018				-249
Uruguay								
Total	**5215**	**3761**	**-13540**	**-8636**	**-22098**	**-8739**	**-67922**	**-7323**

Table 52 (cont.): Change in national winegrape area since 2000 for world's top 30 white varieties in 2016 (hectares)

Country	*Prosecco*	*Müller-Thurgau*	*Grüner Veltliner*	*Trebbiano Romagnolo*	*Sémillon*	*Verdejo*	*Viognier*	*Pedro Giménez*
Algeria								
Argentina	2				-266		621	-3665
Armenia								
Australia					-1973		637	
Austria		-1512	-3103					
Brazil					-378			
Bulgaria								
Cambodia								
Canada								
Chile					-1044		711	
China								
Croatia								
Cyprus								
Czechia		-107	-162					
Ethiopia								
France		-2			-3782		6463	
Georgia								
Germany		-9042						
Greece					-10			
Hungary		-1608	45					
India								
Israel								
Italy	12232	300	-74	-432	5		1800	
Japan								
Kazakhstan								
Korea Rep.								
Lebanon								
Luxembourg		-143						
Mexico								
Moldova		-169						
Morocco								
Myanmar								
New Zealand		-417			-166			
N. Macedonia								
Norway								
Peru								
Portugal								
Romania								
Russia								
Serbia								
Slovakia		-1361	-1333					
Slovenia								
South Africa					89		772	
Spain						13470	211	
Switzerland		-221			1			
Taiwan								
Thailand								
Tunisia								
Turkey								
Ukraine								
United Kingdom								
United States					-369		1166	
Uruguay								
Total	**12601**	**-14085**	**-4486**	**-432**	**-7546**	**13479**	**12903**	**714**

Table 52 (cont.): Change in national winegrape area since 2000 for world's top 30 white varieties in 2016 (hectares)

Country	*Pinot Blanc*	*Fetească Albă*	*Fetească Regală*	*Gewürztraminer*	*Muscat Ottonel*	*Fernão Pires*	*Other White*	*Total White*
Algeria								300
Argentina	-31						-7023	-12985
Armenia							2089	-906
Australia				-269			-13518	-8589
Austria	-1019				-74		-1540	-5217
Brazil				-130			-13646	-15921
Bulgaria				-381	765		-11416	-20295
Cambodia								
Canada	-37			162			1479	3089
Chile	4			250			-4786	17133
China								25647
Croatia							-26713	-37023
Cyprus							-1710	-1710
Czechia								450
Ethiopia								58
France	-220			562	10		5516	12951
Georgia	48						3626	9285
Germany	1927				5		-14679	-19918
Greece							-5964	-5191
Hungary					-160		-17484	-23470
India								2000
Israel							-550	-1029
Italy	-2656			780			-37016	-41854
Japan								1155
Kazakhstan								5609
Korea Rep.								
Lebanon								1818
Luxembourg	22			9			-46	-148
Mexico								1931
Moldova	-140	-3380		-1632	339		8776	-13165
Morocco							3164	3348
Myanmar								31
New Zealand				136			-80	18203
N. Macedonia								9951
Norway								8
Peru								1532
Portugal						-2068	-29595	-28707
Romania		-5783	10919	24	-1008		-36154	-39227
Russia	-2130						-15496	-13846
Serbia							-10567	-37677
Slovakia	-207						-1280	-8246
Slovenia							-6558	-6814
South Africa	-26			-104	-25	-266	-3460	-7205
Spain				327			-117932	-308493
Switzerland	34			20			-1097	-934
Taiwan							-1248	-1248
Thailand								91
Tunisia								1693
Turkey								4349
Ukraine								15996
United Kingdom							-247	246
United States	-169			169			4396	370
Uruguay							150	933
Total	**-3204**	**-10446**	**10413**	**2152**	**205**	**-2334**	**-334230**	**-557364**

Table 53: National winegrape area for world's top 6 grey varieties, 2010 (hectares)

Country	*Pinot Gris*	*Cereza*	*Roditis*	*Misket Cherven*	*Roditis (R)*	*Cserszegi Fűszeres*	*Other Grey*	*Total Grey*
Algeria								0
Argentina	386	29934					371	30691
Armenia								0
Australia	3296							3296
Austria	215						292	507
Brazil	7						7	14
Bulgaria				4159				4159
Canada	549						28	578
Chile	100							100
China	2							2
Croatia	219						56	275
Cyprus								0
Czechia	706						198	904
Ethiopia								0
France	2674						1767	4441
Georgia							26	26
Germany	4514						992	5506
Greece			4668		3826		1756	10251
Hungary	1624					3609	1057	6290
Israel								0
Italy	17281						93	17374
Japan							220	220
Kazakhstan								0
Korea Rep.							100	100
Luxembourg	146							146
Mexico								0
Moldova	2042							2042
Morocco								0
Myanmar								0
New Zealand	1501						2	1503
Peru							7	7
Portugal	5						5	9
Romania	1301						269	1570
Russia	78							78
Serbia								0
Slovakia	211						32	242
Slovenia	501							501
South Africa	261						2	264
Spain							77	77
Switzerland	216							216
Taiwan								0
Thailand							6	6
Tunisia								0
Turkey								0
Ukraine	685							685
United Kingdom	9						50	59
United States	5231						75	5307
Uruguay	12							12
Total	**43773**	**29934**	**4668**	**4159**	**3826**	**3609**	**7487**	**97455**

Table 54: National shares of global winegrape area for world's top 6 grey varieties, 2010 (%)

Country	*Pinot Gris*	*Cereza*	*Roditis*	*Misket Cherven*	*Roditis (R)*	*Cserszegi Fűszeres*	*Other Grey*	*Total Grey*
Algeria								0.00
Argentina	0.88	100.00					4.95	31.49
Armenia								0.00
Australia	7.53							3.38
Austria	0.49						3.90	0.52
Brazil	0.02						0.09	0.01
Bulgaria				100.00				4.27
Canada	1.25						0.38	0.59
Chile	0.23							0.10
China	0.00							0.00
Croatia	0.50						0.75	0.28
Cyprus								0.00
Czechia	1.61						2.64	0.93
Ethiopia								0.00
France	6.11						23.60	4.56
Georgia							0.34	0.03
Germany	10.31						13.25	5.65
Greece			100.00		100.00		23.46	10.52
Hungary	3.71					100.00	14.12	6.45
Israel								0.00
Italy	39.48						1.24	17.83
Japan							2.93	0.23
Kazakhstan								0.00
Korea Rep.							1.34	0.10
Luxembourg	0.33							0.15
Mexico								0.00
Moldova	4.67							2.10
Morocco								0.00
Myanmar								0.00
New Zealand	3.43						0.03	1.54
Peru							0.09	0.01
Portugal	0.01						0.06	0.01
Romania	2.97						3.59	1.61
Russia	0.18							0.08
Serbia								0.00
Slovakia	0.48						0.42	0.25
Slovenia	1.14							0.51
South Africa	0.60						0.03	0.27
Spain							1.03	0.08
Switzerland	0.49							0.22
Taiwan								0.00
Thailand							0.08	0.01
Tunisia								0.00
Turkey								0.00
Ukraine	1.57							0.70
United Kingdom	0.02						0.67	0.06
United States	11.95						1.01	5.45
Uruguay	0.03							0.01
Total	**100**	**100**	**100**	**100**	**100**	**100**	**100**	**100**

Table 55: National winegrape area for world's top 6 grey varieties, 2016 (hectares)

Country	*Pinot Gris*	*Cereza*	*Roditis*	*Misket Cherven*	*Cserszegi Fűszeres*	*Garnacha Roja (Gris)*	*Other Grey*	*Total Grey*
Algeria								0
Argentina	401	28887						29288
Armenia								0
Australia	3652						1	3653
Austria	224						286	510
Brazil	7						0	8
Bulgaria				4349				4349
Cambodia								0
Canada	649						41	690
Chile	437							437
China								0
Croatia								0
Cyprus								0
Czechia	826							826
Ethiopia								0
France	2867					1253	131	4252
Georgia							26	26
Germany	4887						110	4997
Greece			8463			114	1916	10493
Hungary	1594				4299		689	6583
India								0
Israel								0
Italy	18821						46	18867
Japan							945	945
Kazakhstan								0
Korea Rep.							100	100
Lebanon								0
Luxembourg	196							196
Mexico								0
Moldova	1208							1208
Morocco								0
Myanmar								0
New Zealand	2422						3	2425
N. Macedonia								0
Norway								0
Peru							7	7
Portugal	5						7216	7221
Romania	1561						344	1905
Russia	78							78
Serbia	112						1	112
Slovakia								0
Slovenia	508							508
South Africa	369					3	8	380
Spain						87	1428	1514
Switzerland	230						3	233
Taiwan								0
Thailand						5		5
Tunisia								0
Turkey								0
Ukraine								0
United Kingdom	44						9	53
United States	7462						66	7529
Uruguay	9							9
Total	**48570**	**28887**	**8463**	**4349**	**4299**	**1462**	**13375**	**109406**

Table 56: National winegrape area for world's top 6 grey varieties, 2016 (hectares)

Country	*Pinot Gris*	*Cereza*	*Roditis*	*Misket Cherven*	*Cserszegi Fűszeres*	*Garnacha Roja (Gris)*	*Other Grey*	*Total Grey*
Algeria								0.00
Argentina	0.83	100.00						26.77
Armenia								0.00
Australia	7.52						0.01	3.34
Austria	0.46						2.14	0.47
Brazil	0.02						0.00	0.01
Bulgaria				100.00				3.98
Cambodia								0.00
Canada	1.34						0.31	0.63
Chile	0.90							0.40
China								0.00
Croatia								0.00
Cyprus								0.00
Czechia	1.70							0.75
Ethiopia								0.00
France	5.90					85.71	0.98	3.89
Georgia							0.19	0.02
Germany	10.06						0.82	4.57
Greece			100.00			7.81	14.32	9.59
Hungary	3.28				100.00		5.15	6.02
India								0.00
Israel								0.00
Italy	38.75						0.34	17.24
Japan							7.06	0.86
Kazakhstan								0.00
Korea Rep.							0.75	0.09
Lebanon								0.00
Luxembourg	0.40							0.18
Mexico								0.00
Moldova	2.49							1.10
Morocco								0.00
Myanmar								0.00
New Zealand	4.99						0.02	2.22
N. Macedonia								0.00
Norway								0.00
Peru							0.05	0.01
Portugal	0.01						53.95	6.60
Romania	3.21						2.57	1.74
Russia	0.16							0.07
Serbia	0.23						0.01	0.10
Slovakia								0.00
Slovenia	1.05							0.46
South Africa	0.76					0.19	0.06	0.35
Spain						5.93	10.67	1.38
Switzerland	0.47						0.02	0.21
Taiwan								0.00
Thailand						0.35		0.00
Tunisia								0.00
Turkey								0.00
Ukraine								0.00
United Kingdom	0.09						0.07	0.05
United States	15.36						0.50	6.88
Uruguay	0.02							0.01
Total	**100**	**100**	**100**	**100**	**100**	**100**	**100**	**100**

IV. Winegrape areas and Varietal Intensity Indexes for national top varieties

Table 57: Winegrape areas and national and global shares (and Varietal Intensity Indexes) for national top 20 varieties, 2000 and 2016

Algeria

2000						2016					
Prime variety	Col	Area	Nat'l share	Glob'l share	VII*	Prime variety	Col	Area	Nat'l share	Glob'l share	VII*
Cinsaut	R	7550	25.0	15.6	25.2	Mazuelo	R	3000	36.1	6.3	34.2
Mazuelo	R	7550	25.0	5.9	9.6	Garnacha Tinta	R	2000	24.1	1.3	7.2
Garnacha Tinta	R	6040	20.0	2.8	4.5	Cabernet Sauvignon	R	1000	12.0	0.3	1.7
Alicante Henri Bouschet	R	3020	10.0	8.1	13.2	Merlot	R	1000	12.0	0.4	2.0
Cabernet Sauvignon	R	1510	5.0	0.7	1.1	Syrah	R	1000	12.0	0.6	3.0
Merlot	R	1510	5.0	0.7	1.1	Muscat of Alexandria	W	200	2.4	0.6	3.1
Pinot Noir	R	1510	5.0	2.2	3.6	Sultaniye	W	100	1.2	1.9	10.1
Syrah	R	1510	5.0	1.5	2.4						
Total of above		30200	100.0			Total of above		8300	100.0		

Argentina

2000						2016					
Prime variety	Col	Area	Nat'l share	Glob'l share	VII*	Prime variety	Col	Area	Nat'l share	Glob'l share	VII*
Cereza	G	31113	15.8	100.0	24.8	Côt	R	40401	19.6	77.3	16.8
Criolla Grande	R	24264	12.3	100.0	24.8	Cereza	G	28887	14.0	100.0	21.7
Côt	R	18230	9.2	69.4	17.2	Douce Noire	R	19072	9.2	96.6	21.0
Douce Noire	R	15659	7.9	85.5	21.2	Criolla Grande	R	15596	7.6	100.0	21.7
Pedro Giménez	W	14862	7.5	100.0	24.8	Cabernet Sauvignon	R	15356	7.4	4.9	1.1
Cabernet Sauvignon	R	13776	7.0	6.2	1.5	Syrah	R	12707	6.2	7.0	1.5
Muscat Blanc à Petits Grains (G)	W	10442	5.3	100.0	24.8	Pedro Giménez	W	11197	5.4	71.9	15.6
Syrah	R	8888	4.5	8.7	2.1	Torrontés Riojano	W	8208	4.0	92.6	20.1
Torrontés Riojano	W	8127	4.1	99.1	24.5	Muscat Blanc à Petits Grains (G)	W	6526	3.2	79.0	17.2
Merlot	R	6263	3.2	2.9	0.7	Chardonnay	W	6227	3.0	3.1	0.7
Muscat of Alexandria	W	5515	2.8	18.6	4.6	Tempranillo	R	6140	3.0	2.8	0.6
Tempranillo	R	4720	2.4	5.1	1.3	Merlot	R	5632	2.7	2.1	0.5
Chardonnay	W	4682	2.4	3.2	0.8	Aspiran Bouschet	R	4087	2.0	100.0	21.7
Chenin Blanc	W	3445	1.7	7.5	1.9	Muscat of Alexandria	W	2716	1.3	7.8	1.7
Torrontés Sanjuanino	W	3170	1.6	100.0	24.8	Chenin Blanc	W	2157	1.0	6.7	1.5
Trebbiano Toscano	W	2765	1.4	2.0	0.5	Sauvignon Blanc	W	2148	1.0	1.7	0.4
Sangiovese	R	2490	1.3	3.6	0.9	Torrontés Sanjuanino	W	1885	0.9	51.5	11.2
Red Globe	R	1940	1.0	91.8	22.7	Pinot Noir	R	1866	0.9	1.8	0.4
Gibi	W	1227	0.6	100.0	24.8	Sangiovese	R	1837	0.9	2.5	0.5
Pinot Noir	R	1114	0.6	1.6	0.4	Trebbiano Toscano	W	1622	0.8	1.3	0.3
Total of above		182692	92.5			Total of above		194264	94.1		

Table 57 (cont.): Winegrape areas and national and global shares (and Varietal Intensity Indexes) for national top 20 varieties, 2000 and 2016

Armenia

2000						2016					
Prime variety	Col	Area	Nat'l share	Glob'l share	VII*	Prime variety	Col	Area	Nat'l share	Glob'l share	VII*
Rkatsiteli	W	2469	22.0	3.7	16.0	white varieties	W	10300	70.0	100.0	304.9
Mskhali	W	1093	9.8	100.0	436.2	red varieties	R	4405	30.0	100.0	304.9
Garandmak	W	931	8.3	100.0	436.2						
Kangun	W	850	7.6	100.0	436.2						
Voskeat	W	809	7.2	100.0	436.2						
Muscat Blanc à Petits Grains	W	526	4.7	1.8	7.7						
Total of above		6677	59.6			Total of above		14705	100.0		

Australia

2000						2016					
Prime variety	Col	Area	Nat'l share	Glob'l share	VII*	Prime variety	Col	Area	Nat'l share	Glob'l share	VII*
Syrah	R	29295	22.4	28.6	10.7	Syrah	R	38942	29.4	21.5	7.3
Cabernet Sauvignon	R	24997	19.1	11.2	4.2	Cabernet Sauvignon	R	23987	18.1	7.7	2.6
Chardonnay	W	17266	13.2	11.9	4.4	Chardonnay	W	21321	16.1	10.6	3.6
Sultaniye	W	10298	7.9	85.0	31.8	Merlot	R	8415	6.4	3.2	1.1
Merlot	R	7669	5.9	3.6	1.3	Sauvignon Blanc	W	6044	4.6	4.8	1.6
Sémillon	W	6528	5.0	24.9	9.3	Pinot Noir	R	4806	3.6	4.6	1.5
Pinot Noir	R	3223	2.5	4.7	1.8	Sémillon	W	4556	3.4	24.4	8.2
Riesling	W	3129	2.4	7.2	2.7	Pinot Gris	G	3652	2.8	7.5	2.5
Sauvignon Blanc	W	2602	2.0	4.0	1.5	Riesling	W	3114	2.4	5.2	1.8
Muscat of Alexandria	W	2495	1.9	8.4	3.2	Muscat of Alexandria	W	2179	1.6	6.3	2.1
Ruby Cabernet	R	2424	1.9	32.7	12.2	Colombard	W	1789	1.4	6.0	2.0
Garnacha Tinta	R	2139	1.6	1.0	0.4	Garnacha Tinta	R	1492	1.1	1.0	0.3
Colombard	W	1801	1.4	4.7	1.7	Petit Verdot	R	1118	0.8	13.8	4.7
Verdelho	W	1293	1.0	78.7	29.4	Verdelho	W	1016	0.8	67.0	22.7
Monastrell	R	948	0.7	1.2	0.5	Savagnin Blanc	W	870	0.7	38.4	13.0
Chenin Blanc	W	841	0.6	1.8	0.7	Muscat Blanc à Petits Grains	W	857	0.6	2.5	0.9
Korinthiaki	R	778	0.6	93.3	34.9	Ruby Cabernet	R	849	0.6	16.0	5.4
Cabernet Franc	R	744	0.6	1.4	0.5	Viognier	W	753	0.6	4.7	1.6
Petit Verdot	R	721	0.6	44.0	16.5	Monastrell	R	704	0.5	1.4	0.5
Trebbiano Toscano	W	685	0.5	0.5	0.2	Tempranillo	R	681	0.5	0.3	0.1
Total of above		119877	91.8			Total of above		127144	96.0		

Table 57 (cont.): Winegrape areas and national and global shares (and Varietal Intensity Indexes) for national top 20 varieties, 2000 and 2016

Austria

2000						2016					
Prime variety	Col	Area	Nat'l share	Glob'l share	VII*	Prime variety	Col	Area	Nat'l share	Glob'l share	VII*
Grüner Veltliner	W	17479	36.0	74.1	74.6	Grüner Veltliner	W	14376	31.6	75.2	74.2
Zweigelt	R	4350	9.0	59.9	60.3	Zweigelt	R	6311	13.9	69.6	68.7
Graševina	W	4323	8.9	4.7	4.7	Graševina	W	3233	7.1	13.3	13.1
Müller-Thurgau	W	3289	6.8	9.8	9.9	Blaufränkisch	R	2808	6.2	16.3	16.1
Pinot Blanc	W	2936	6.1	17.3	17.4	Riesling	W	2016	4.4	3.4	3.3
Blaufränkisch	R	2641	5.4	18.9	19.0	Pinot Blanc	W	1916	4.2	13.9	13.7
Blauer Portugieser	R	2358	4.9	25.8	26.0	Müller-Thurgau	W	1777	3.9	9.1	9.0
Riesling	W	1643	3.4	3.8	3.8	Chardonnay	W	1577	3.5	0.8	0.8
Neuburger	W	1094	2.3	76.3	76.9	Blauer Portugieser	R	1265	2.8	19.2	18.9
Blauburger	R	884	1.8	88.2	88.9	Sauvignon Blanc	W	1170	2.6	0.9	0.9
Frühroter Veltliner	R	626	1.3	99.0	99.8	Muscat Blanc à Petits Grains	W	823	1.8	2.4	2.4
Scheurebe	W	529	1.1	14.5	14.6	Blauburger	R	750	1.7	61.3	60.5
Blauer Wildbacher	R	464	1.0	98.3	99.0	Sankt Laurent	R	724	1.6	22.1	21.8
Muscat Ottonel	W	418	0.9	3.4	3.4	Merlot	R	695	1.5	0.3	0.3
Sankt Laurent	R	415	0.9	16.2	16.4	Pinot Noir	R	614	1.4	0.6	0.6
Pinot Noir	R	409	0.8	0.6	0.6	Cabernet Sauvignon	R	567	1.2	0.2	0.2
Bouvier	W	365	0.8	100.0	100.8	Neuburger	W	507	1.1	87.7	86.5
Gewürztraminer	W	363	0.7	3.4	3.4	Blauer Wildbacher	R	434	1.0	99.3	98.0
Sauvignon Blanc	W	314	0.6	0.5	0.5	Frühroter Veltliner	R	369	0.8	95.0	93.7
Cabernet Sauvignon	R	312	0.6	0.1	0.1	Scheurebe	W	351	0.8	21.6	21.3
Total of above		45212	93.2			Total of above		42282	93.1		

Brazil

2000						2016					
Prime variety	Col	Area	Nat'l share	Glob'l share	VII*	Prime variety	Col	Area	Nat'l share	Glob'l share	VII*
Isabella	R	14285	27.0	52.0	48.1	Isabella	R	11664	35.1	65.5	88.4
Niagara	W	13436	25.4	87.6	81.0	Cabernet Franc	R	6834	20.6	12.2	16.5
Cabernet Franc	R	3784	7.2	7.3	6.7	Couderc Noir	R	1938	5.8	90.7	122.5
Concord	R	2509	4.7	21.2	19.6	Concord	R	1687	5.1	16.0	21.6
Seibel	R	1967	3.7	98.8	91.4	Niagara	W	1430	4.3	43.8	59.1
Herbemont	R	1453	2.7	100.0	92.5	Jacquez	R	1274	3.8	88.3	119.2
Graševina	W	880	1.7	1.0	0.9	Moscato Embrapa	W	683	2.1	100.0	135.0
Muscat of Alexandria	W	809	1.5	2.7	2.5	Muscat	W	671	2.0	90.2	121.8
Trebbiano Toscano	W	688	1.3	0.5	0.5	Violeta	R	636	1.9	100.0	135.0
Cabernet Sauvignon	R	587	1.1	0.3	0.2	Cora	R	570	1.7	100.0	135.0
Merlot	R	469	0.9	0.2	0.2	Lorena	W	500	1.5	100.0	135.0
Sémillon	W	384	0.7	1.5	1.4	Seibel	R	478	1.4	99.0	133.7
Chardonnay	W	330	0.6	0.2	0.2	Couderc 13	W	474	1.4	100.0	135.0
Couderc Noir	R	299	0.6	48.8	45.1	Niagara Red	R	469	1.4	100.0	135.0
Tannat	R	182	0.3	3.3	3.0	Cabernet Sauvignon	R	429	1.3	0.1	0.2
Jacquez	R	170	0.3	75.3	69.6	Merlot	R	363	1.1	0.1	0.2
Gewürztraminer	W	140	0.3	1.3	1.2	Chardonnay	W	340	1.0	0.2	0.2
Sauvignon Blanc	W	140	0.3	0.2	0.2	Carmem	R	328	1.0	100.0	135.0
						Trebbiano Toscano	W	231	0.7	0.2	0.3
						Prosecco	W	207	0.6	1.0	1.4
Total of above		42512	80.5			Total of above		31205	94.0		

Table 57 (cont.): Winegrape areas and national and global shares (and Varietal Intensity Indexes) for national top 20 varieties, 2000 and 2016

Bulgaria

2000						2016					
Prime variety	Col	Area	Nat'l share	Glob'l share	VII*	Prime variety	Col	Area	Nat'l share	Glob'l share	VII*
Pamid	R	22581	23.5	99.4	50.6	Merlot	R	10050	19.0	3.8	3.2
Merlot	R	11169	11.6	5.2	2.7	Cabernet Sauvignon	R	9327	17.6	3.0	2.5
Cabernet Sauvignon	R	10441	10.9	4.7	2.4	Pamid	R	6874	13.0	69.0	58.4
Rkatsiteli	W	9429	9.8	14.0	7.1	Rkatsiteli	W	5415	10.2	10.5	8.9
Dimyat	W	7649	8.0	98.8	50.3	Misket Cherven	G	4349	8.2	100.0	84.6
Shiroka Melnishka	R	3804	4.0	100.0	50.9	Muscat Ottonel	W	3679	6.9	29.5	25.0
Misket	W	3764	3.9	100.0	50.9	Chardonnay	W	3087	5.8	1.5	1.3
Graševina	W	3602	3.8	3.9	2.0	Dimyat	W	2998	5.7	30.9	26.2
Cardinal	R	3035	3.2	78.4	39.9	Shiroka Melnishka	R	1205	2.3	100.0	84.6
Muscat Ottonel	W	2914	3.0	23.8	12.1	Mavrud	R	1193	2.3	100.0	84.6
Chardonnay	W	1862	1.9	1.3	0.7	Kadarka	R	1161	2.2	71.5	60.5
Trebbiano Toscano	W	1821	1.9	1.3	0.7	Syrah	R	804	1.5	0.4	0.4
Aligoté	W	1659	1.7	4.7	2.4	Trebbiano Toscano	W	738	1.4	0.6	0.5
Kadarka	R	1619	1.7	61.5	31.3	Sauvignon Blanc	W	637	1.2	0.5	0.4
Gewürztraminer	W	971	1.0	9.1	4.6	Gewürztraminer	W	591	1.1	4.6	3.9
Pinot Noir	R	769	0.8	1.1	0.6	Pinot Noir	R	342	0.6	0.3	0.3
Mavrud	R	647	0.7	100.0	50.9	Aligoté	W	285	0.5	1.1	0.9
Riesling	W	647	0.7	1.5	0.8	Cabernet Franc	R	240	0.5	0.4	0.4
Königin der Weingärten	W	567	0.6	75.5	38.4						
Muscat of Hamburg	R	445	0.5	6.3	3.2						
Total of above		89395	93.1			Total of above		52974	100.0		

Cambodia

2000						2016					
Prime variety	Col	Area	Nat'l share	Glob'l share	VII*	Prime variety	Col	Area	Nat'l share	Glob'l share	VII*
						Black Queen	R	3	30.0	2.1	9379.7
						Syrah	R	3	30.0	0.0	7.4
						Cabernet Sauvignon	R	2	20.0	0.0	2.9
						Merlot	R	2	20.0	0.0	3.4
Total of above						Total of above		10	100.0		

Table 57 (cont.): Winegrape areas and national and global shares (and Varietal Intensity Indexes) for national top 20 varieties, 2000 and 2016

Canada

2000						2016					
Prime variety	Col	Area	Nat'l share	Glob'l share	VII*	Prime variety	Col	Area	Nat'l share	Glob'l share	VII*
Concord	R	977	11.5	8.3	47.6	Seyval Blanc	W	2259	17.9	83.7	297.8
Chardonnay	W	973	11.5	0.7	3.8	Chardonnay	W	1417	11.2	0.7	2.5
Merlot	R	674	7.9	0.3	1.8	Riesling	W	1188	9.4	2.0	7.1
Cabernet Sauvignon	R	569	6.7	0.3	1.5	Cabernet Franc	R	820	6.5	1.5	5.2
Cabernet Franc	R	567	6.7	1.1	6.3	Baco Noir	R	704	5.6	95.8	340.8
Vidal	W	514	6.0	84.1	483.9	Cabernet Sauvignon	R	660	5.2	0.2	0.8
Riesling	W	482	5.7	1.1	6.4	Pinot Gris	G	649	5.1	1.3	4.8
Niagara	W	461	5.4	3.0	17.3	Pinot Noir	R	639	5.1	0.6	2.2
Pinot Noir	R	457	5.4	0.7	3.8	Merlot	R	633	5.0	0.2	0.8
Baco Noir	R	279	3.3	70.3	404.1	Gewürztraminer	W	398	3.2	3.1	11.1
Gamay Noir	R	263	3.1	0.7	4.0	Arinarnoa	R	289	2.3	59.5	211.6
Gewürztraminer	W	237	2.8	2.2	12.8	Sauvignon Blanc	W	285	2.3	0.2	0.8
Pinot Gris	G	210	2.5	1.1	6.4	Gamay Noir	R	272	2.2	1.0	3.7
Sauvignon Blanc	W	148	1.7	0.2	1.3	Syrah	R	260	2.1	0.1	0.5
Pinot Blanc	W	146	1.7	0.9	4.9	Concord	R	183	1.4	1.7	6.2
Seyval Blanc	W	132	1.5	33.8	194.6	Millot-Foch	R	124	1.0	98.0	348.8
De Chaunac	R	119	1.4	64.1	368.8	Pinot Blanc	W	109	0.9	0.8	2.8
Maréchal Foch	R	109	1.3	63.1	362.8	Viognier	W	101	0.8	0.6	2.2
Fredonia	R	73	0.9	84.9	488.3	Maréchal Foch	R	94	0.7	41.1	146.2
Elvira	W	69	0.8	20.0	115.2	Niagara	W	87	0.7	2.7	9.5
Total of above		7458	87.8			Total of above		11170	88.7		

Chile

2000						2016					
Prime variety	Col	Area	Nat'l share	Glob'l share	VII*	Prime variety	Col	Area	Nat'l share	Glob'l share	VII*
Cabernet Sauvignon	R	35967	31.6	16.1	6.9	Cabernet Sauvignon	R	42409	29.1	13.7	4.2
Listan Prieto	R	15181	13.3	93.5	40.1	Sauvignon Blanc	W	14999	10.3	12.0	3.7
Merlot	R	12825	11.3	6.0	2.6	Merlot	R	12057	8.3	4.5	1.4
Chardonnay	W	7672	6.7	5.3	2.3	Chardonnay	W	11435	7.8	5.7	1.7
Sauvignon Blanc	W	6662	5.8	10.2	4.4	Carmenère	R	10503	7.2	46.7	14.4
Carmenère	R	4719	4.1	82.6	35.4	Listan Prieto	R	9693	6.6	94.4	29.0
Malvasia Fina	W	4305	3.8	60.6	26.0	Syrah	R	7994	5.5	4.4	1.4
Alicante Henri Bouschet	R	2882	2.5	7.8	3.3	Alicante Henri Bouschet	R	6908	4.7	19.2	5.9
Pedro Ximénez	W	2379	2.1	13.8	5.9	Muscat of Alexandria	W	5424	3.7	15.6	4.8
Syrah	R	2040	1.8	2.0	0.9	Pedro Giménez	W	4379	3.0	28.1	8.6
Sémillon	W	1893	1.7	7.2	3.1	Pinot Noir	R	4091	2.8	3.9	1.2
Pinot Noir	R	1614	1.4	2.3	1.0	Côt	R	2293	1.6	4.4	1.3
Côt	R	929	0.8	3.5	1.5	Torrontés Sanjuanino	W	1771	1.2	48.5	14.9
Cabernet Franc	R	689	0.6	1.3	0.6	Muscat Blanc à Petits Grains (C	W	1732	1.2	21.0	6.4
Mazuelo	R	641	0.6	0.5	0.2	Cabernet Franc	R	1578	1.1	2.8	0.9
Chasselas	W	404	0.4	3.0	1.3	Petit Verdot	R	863	0.6	10.6	3.3
Riesling	W	286	0.3	0.7	0.3	Sémillon	W	849	0.6	4.5	1.4
Cinsaut	R	195	0.2	0.4	0.2	Cinsaut	R	848	0.6	3.7	1.1
Sauvignonasse	W	132	0.1	2.4	1.0	Viognier	W	839	0.6	5.2	1.6
Viognier	W	128	0.1	4.1	1.7	Mazuelo	R	811	0.6	1.7	0.5
Total of above		101543	89.1			Total of above		141476	97.0		

Table 57 (cont.): Winegrape areas and national and global shares (and Varietal Intensity Indexes) for national top 20 varieties, 2000 and 2016

China											
2000						2016					
Prime variety	Col	Area	Nat'l share	Glob'l share	VII*	Prime variety	Col	Area	Nat'l share	Glob'l share	VII*
						Cabernet Sauvignon	R	40300	22.6	13.0	3.3
						Merlot	R	16700	9.4	6.3	1.6
						Carmenère	R	11200	6.3	49.8	12.5
						Chardonnay	W	6100	3.4	3.0	0.8
						Yan 73	R	4800	2.7	100.0	25.2
						Garnacha Tinta	R	4000	2.2	2.7	0.7
						Graševina	W	3000	1.7	12.3	3.1
						Muscat of Alexandria	W	3000	1.7	8.6	2.2
						Sauvignon Blanc	W	2000	1.1	1.6	0.4
						Beibinghong	R	1600	0.9	100.0	25.2
						Riesling	W	1600	0.9	2.7	0.7
						Trebbiano Toscano	W	1500	0.8	1.2	0.3
						Vidal	W	1500	0.8	77.5	19.5
						Longyan	R	1000	0.6	100.0	25.2
						Syrah	R	1000	0.6	0.6	0.1
						Cabernet Franc	R	600	0.3	1.1	0.3
						Pinot Noir	R	400	0.2	0.4	0.1
						Mazuelo	R	100	0.1	0.2	0.1
Total of above		0	0			Total of above		100400	56.4		

Croatia											
2000						2016					
Prime variety	Col	Area	Nat'l share	Glob'l share	VII*	Prime variety	Col	Area	Nat'l share	Glob'l share	VII*
Graševina	W	16051	27.0	17.4	14.3	Graševina	W	4459	38.0	18.3	69.8
Malvazija Istarska	W	7134	12.0	94.4	77.6	Plavac Mali	R	1664	14.2	97.1	370.5
Plavac Mali	R	6539	11.0	100.0	82.2	Malvazija Istarska	W	1600	13.6	57.4	219.0
Pošip Bijeli	W	6539	11.0	100.0	82.2	Merlot	R	828	7.0	0.3	1.2
Babić	R	1189	2.0	100.0	82.2	Cabernet Sauvignon	R	709	6.0	0.2	0.9
Blaufränkisch	R	1189	2.0	8.5	7.0	Plavina	R	683	5.8	100.0	381.7
Mondeuse Noire	R	1189	2.0	84.7	69.6	Chardonnay	W	657	5.6	0.3	1.2
						Riesling	W	625	5.3	1.0	4.0
						Blaufränkisch	R	521	4.4	3.0	11.6
Total of above		39830	67			Total of above		11746	100.0		

Table 57 (cont.): Winegrape areas and national and global shares (and Varietal Intensity Indexes) for national top 20 varieties, 2000 and 2016

Cyprus

2000						2016					
Prime variety	Col	Area	Nat'l share	Glob'l share	VII*	Prime variety	Col	Area	Nat'l share	Glob'l share	VII*
Mavro	R	10969	60.0	100.0	267.4	Mavro	R	3187	62.1	100.0	873.4
Xynisteri	W	2742	15.0	100.0	267.4	Xynisteri	W	1946	37.9	100.0	873.4
Total of above		13711	75.0			Total of above		5133	100.0		

Czechia

2000						2016					
Prime variety	Col	Area	Nat'l share	Glob'l share	VII*	Prime variety	Col	Area	Nat'l share	Glob'l share	VII*
Grüner Veltliner	W	1700	15.0	7.2	31.1	Grüner Veltliner	W	1538	11.3	8.0	26.5
Müller-Thurgau	W	1586	14.0	4.7	20.4	Müller-Thurgau	W	1479	10.9	7.6	25.0
Graševina	W	1246	11.0	1.4	5.8	Sankt Laurent	R	1183	8.7	36.2	119.2
Sankt Laurent	R	1020	9.0	39.9	172.2	Riesling	W	1172	8.6	2.0	6.5
Riesling	W	793	7.0	1.8	7.9	Blaufränkisch	R	1143	8.4	6.7	21.9
Blaufränkisch	R	680	6.0	4.9	21.0	Graševina	W	1114	8.2	4.6	15.1
Chardonnay	W	567	5.0	0.4	1.7	Sauvignon Blanc	W	906	6.7	0.7	2.4
Zweigelt	R	453	4.0	6.2	26.9	Pinot Gris	G	826	6.1	1.7	5.6
Blauer Portugieser	R	340	3.0	3.7	16.0	Chardonnay	W	820	6.0	0.4	1.3
Neuburger	W	340	3.0	23.7	102.3	Zweigelt	R	770	5.7	8.5	28.0
						Pinot Blanc	W	762	5.6	5.5	18.2
						Pinot Noir	R	697	5.1	0.7	2.2
						Blauer Portugieser	R	599	4.4	9.1	30.0
						Gewürztraminer	W	591	4.3	4.6	15.2
Total of above		8725	77.0			Total of above		13600	100.0		

Table 57 (cont.): Winegrape areas and national and global shares (and Varietal Intensity Indexes) for national top 20 varieties, 2000 and 2016

Ethiopia

2000						2016					
Prime variety	Col	Area	Nat'l share	Glob'l share	VII*	Prime variety	Col	Area	Nat'l share	Glob'l share	VII*
						Sangiovese	R	90	53.2	0.1	32.5
						Chenin Blanc	W	54	31.9	0.2	44.4
						Flame Seedless	R	13	7.8	23.9	6326.2
						Crimson Seedless	R	6	3.3	73.5	19471.4
						Sultaniye	W	4	2.5	0.1	21.3
						Red Globe	R	2	1.3	0.9	237.5
Total of above						Total of above		169	100.0		

France

2000						2016					
Prime variety	Col	Area	Nat'l share	Glob'l share	VII*	Prime variety	Col	Area	Nat'l share	Glob'l share	VII*
Merlot	R	101309	11.7	47.5	2.7	Merlot	R	108483	13.3	40.7	2.2
Mazuelo	R	95745	11.1	75.0	4.2	Trebbiano Toscano	W	78842	9.7	65.5	3.6
Garnacha Tinta	R	95717	11.1	44.2	2.5	Garnacha Tinta	R	78631	9.6	52.4	2.9
Trebbiano Toscano	W	90341	10.4	65.8	3.7	Syrah	R	62211	7.6	34.3	1.9
Cabernet Sauvignon	R	53413	6.2	23.9	1.4	Chardonnay	W	47451	5.8	23.5	1.3
Syrah	R	50676	5.9	49.4	2.8	Cabernet Sauvignon	R	46555	5.7	15.0	0.8
Chardonnay	W	36496	4.2	25.1	1.4	Cabernet Franc	R	32327	4.0	57.7	3.2
Cabernet Franc	R	36094	4.2	69.4	3.9	Mazuelo	R	31760	3.9	67.1	3.7
Gamay Noir	R	34537	4.0	91.4	5.2	Pinot Noir	R	31602	3.9	30.0	1.6
Cinsaut	R	31593	3.7	65.2	3.7	Sauvignon Blanc	W	28084	3.4	22.5	1.2
Pinot Noir	R	26526	3.1	38.5	2.2	Gamay Noir	R	24095	3.0	91.9	5.1
Sauvignon Blanc	W	20933	2.4	32.1	1.8	Cinsaut	R	15930	2.0	69.5	3.8
Sémillon	W	14015	1.6	53.4	3.0	Pinot Meunier	R	12130	1.5	82.5	4.5
Melon	W	13253	1.5	100.0	5.7	Sémillon	W	10234	1.3	54.7	3.0
Pinot Meunier	R	10621	1.2	80.9	4.6	Melon	W	9550	1.2	100.0	5.5
Chenin Blanc	W	9837	1.1	21.5	1.2	Chenin Blanc	W	9432	1.2	29.3	1.6
Aramon Noir	R	9157	1.1	100.0	5.7	Viognier	W	8823	1.1	54.9	3.0
Alicante Henri Bouschet	R	8764	1.0	23.6	1.3	Monastrell	R	8754	1.1	16.9	0.9
Monastrell	R	7634	0.9	10.0	0.6	Colombard	W	8441	1.0	28.1	1.5
Muscat Blanc à Petits Grains	W	6935	0.8	23.1	1.3	Muscat Blanc à Petits Grains	W	7333	0.9	21.7	1.2
Total of above		753598	87.1			Total of above		660665	81.1		

Table 57 (cont.): Winegrape areas and national and global shares (and Varietal Intensity Indexes) for national top 20 varieties, 2000 and 2016

Georgia

2000						2016					
Prime variety	Col	Area	Nat'l share	Glob'l share	VII*	Prime variety	Col	Area	Nat'l share	Glob'l share	VII*
Rkatsiteli	W	19741	52.8	29.3	38.3	Rkatsiteli	W	25324	52.8	49.3	46.0
Tsolikouri	W	6161	16.5	100.0	130.6	Tsolikouri	W	7903	16.5	100.0	93.4
Saperavi	R	3704	9.9	55.2	72.1	Saperavi	R	4751	9.9	73.3	68.5
Tsitska	W	2839	7.6	100.0	130.6	Tsitska	W	3642	7.6	100.0	93.4
Chinuri	W	955	2.6	100.0	130.6	Chinuri	W	1225	2.6	100.0	93.4
Mtsvane Kakhuri	W	249	0.7	100.0	130.6	Mtsvane Kakhuri	W	319	0.7	100.0	93.4
Goruli Mtsvane	W	224	0.6	100.0	130.6	Goruli Mtsvane	W	287	0.6	100.0	93.4
Cabernet Sauvignon	R	223	0.6	0.1	0.1	Cabernet Sauvignon	R	286	0.6	0.1	0.1
Aleksandrouli	R	219	0.6	100.0	130.6	Aleksandrouli	R	281	0.6	100.0	93.4
Pinot Blanc	W	171	0.5	1.0	1.3	Pinot Blanc	W	219	0.5	1.6	1.5
Tsulukidzis Tetra	W	152	0.4	100.0	130.6	Tsulukidzis Tetra	W	195	0.4	100.0	93.4
Aligoté	W	97	0.3	0.3	0.4	Aligoté	W	124	0.3	0.5	0.4
Aladasturi	R	46	0.1	100.0	130.6	Aladasturi	R	59	0.1	100.0	93.4
Krakhuna	W	36	0.1	100.0	130.6	Krakhuna	W	46	0.1	100.0	93.4
Tavkveri	R	29	0.1	100.0	130.6	Tavkveri	R	37	0.1	100.0	93.4
Ojaleshi	R	25	0.1	100.0	130.6	Ojaleshi	R	32	0.1	100.0	93.4
Chkhaveri	G	20	0.1	100.0	130.6	Chkhaveri	G	26	0.1	100.0	93.4
Kisi	W	20	0.1	100.0	130.6	Kisi	W	26	0.1	100.0	93.4
Usakhelouri	R	8	0.0	100.0	130.6	Usakhelouri	R	10	0.0	100.0	93.4
Khikhvi	W	5	0.0	100.0	130.6	Khikhvi	W	6	0.0	100.0	93.4
Total of above		34924	93.3			Total of above		44801	93.3		

Germany

2000						2016					
Prime variety	Col	Area	Nat'l share	Glob'l share	VII*	Prime variety	Col	Area	Nat'l share	Glob'l share	VII*
Riesling	W	22350	21.4	51.6	24.2	Riesling	W	21540	22.8	36.0	17.1
Müller-Thurgau	W	20706	19.9	61.6	28.9	Müller-Thurgau	W	11664	12.3	59.8	28.4
Pinot Noir	R	8643	8.3	12.6	5.9	Pinot Noir	R	11184	11.8	10.6	5.0
Silvaner	W	6859	6.6	62.1	29.1	Dornfelder	R	7761	8.2	98.6	46.8
Kerner	W	6846	6.6	96.0	45.0	Pinot Gris	G	4887	5.2	10.1	4.8
Blauer Portugieser	R	4878	4.7	53.3	25.0	Silvaner	W	4627	4.9	76.2	36.2
Dornfelder	R	3765	3.6	100.0	46.9	Pinot Blanc	W	4323	4.6	31.4	14.9
Bacchus	W	3262	3.1	96.7	45.3	Blauer Portugieser	R	3177	3.4	48.2	22.9
Scheurebe	W	3126	3.0	85.5	40.1	Kerner	W	2646	2.8	91.5	43.4
Pinot Gris	G	2637	2.5	14.0	6.5	Schiava Grossa	R	2197	2.3	97.4	46.2
Schiava Grossa	R	2530	2.4	66.8	31.3	Pinot Meunier	R	2002	2.1	13.6	6.5
Pinot Blanc	W	2396	2.3	14.1	6.6	Regent	R	1902	2.0	96.4	45.7
Pinot Meunier	R	2289	2.2	17.4	8.2	Blaufränkisch	R	1737	1.8	10.1	4.8
Faberrebe	W	1586	1.5	100.0	46.9	Bacchus	W	1610	1.7	91.5	43.4
Huxelrebe	W	1289	1.2	100.0	46.9	Chardonnay	W	1485	1.6	0.7	0.3
Chasselas	W	1198	1.1	9.0	4.2	Scheurebe	W	1266	1.3	77.9	36.9
Morio-Muskat	W	1167	1.1	98.2	46.1	Chasselas	W	1046	1.1	14.2	6.7
Blaufränkisch	R	1118	1.1	8.0	3.7	Gewürztraminer	W	824	0.9	6.4	3.0
Ortega	W	1054	1.0	100.0	46.9	Sauvignon Blanc	W	736	0.8	0.6	0.3
Elbling	W	1043	1.0	86.4	40.5	Sankt Laurent	R	633	0.7	19.3	9.2
Total of above		98742	94.7			Total of above		87247	92.3		

Table 57 (cont.): Winegrape areas and national and global shares (and Varietal Intensity Indexes) for national top 20 varieties, 2000 and 2016

Greece

2000						2016					
Prime variety	Col	Area	Nat'l share	Glob'l share	VII*	Prime variety	Col	Area	Nat'l share	Glob'l share	VII*
Savatiano	W	12747	25.0	100.0	96.0	Savatiano	W	10268	20.2	100.0	88.2
Roditis (R)	G	6945	13.6	100.0	96.0	Roditis	G	8463	16.6	100.0	88.2
Liatiko	R	2476	4.9	100.0	96.0	Agiorgitiko	R	3270	6.4	99.9	88.1
Agiorgitiko	R	2320	4.6	100.0	96.0	Liatiko	R	2633	5.2	100.0	88.2
Moschomavro	R	2295	4.5	100.0	96.0	Muscat of Hamburg	R	2288	4.5	29.8	26.3
Muscat Blanc à Petits Grains	W	2232	4.4	7.4	7.1	Xinomavro	R	2135	4.2	100.0	88.2
Xinomavro	R	1816	3.6	100.0	96.0	Cabernet Sauvignon	R	1929	3.8	0.6	0.5
Athiri	W	1350	2.7	100.0	96.0	Assyrtiko	W	1770	3.5	100.0	88.2
Kotsifali	R	1148	2.3	100.0	96.0	Mavrouda	R	1658	3.3	100.0	88.2
Assyrtiko	W	1106	2.2	100.0	96.0	Muscat Blanc à Petits Grains	W	1568	3.1	4.6	4.1
Mandilaria	R	845	1.7	100.0	96.0	Merlot	R	1393	2.7	0.5	0.5
Trebbiano Toscano	W	746	1.5	0.5	0.5	Kotsifali	R	1338	2.6	100.0	88.2
Moschofilero	G	718	1.4	100.0	96.0	Romeiko	R	1131	2.2	100.0	88.2
Cabernet Sauvignon	R	688	1.4	0.3	0.3	Moschofilero	G	1088	2.1	100.0	88.2
Mavrodafni	R	537	1.1	100.0	96.0	Syrah	R	1042	2.0	0.6	0.5
Vilana	W	506	1.0	100.0	96.0	Mandilaria	R	932	1.8	100.0	88.2
Vertzami	R	491	1.0	100.0	96.0	Roditis (R)	G	828	1.6	100.0	88.2
Debina	W	455	0.9	100.0	96.0	Muscat of Alexandria	W	773	1.5	2.2	2.0
Asprouda	W	433	0.9	100.0	96.0	Sauvignon Blanc	W	727	1.4	0.6	0.5
Monemvassia	W	418	0.8	100.0	96.0	Chardonnay	W	673	1.3	0.3	0.3
Total of above		40271	79.1			Total of above		45908	90.3		

Hungary

2000						2016					
Prime variety	Col	Area	Nat'l share	Glob'l share	VII*	Prime variety	Col	Area	Nat'l share	Glob'l share	VII*
Blaufränkisch	R	6920	8.0	49.4	27.8	Blaufränkisch	R	7260	11.4	42.3	29.7
Graševina	W	6677	7.7	7.2	4.1	Bianca	W	4898	7.7	50.2	35.2
Furmint	W	3480	4.0	100.0	56.2	Cserszegi Fűszeres	G	4299	6.7	100.0	70.2
Müller-Thurgau	W	3278	3.8	9.8	5.5	Graševina	W	3933	6.2	16.1	11.3
Ezerjó	W	3157	3.6	100.0	56.3	Furmint	W	3862	6.0	87.1	61.1
Chardonnay	W	2954	3.4	2.0	1.1	Cabernet Sauvignon	R	2677	4.2	0.9	0.6
Arany Sárfehér	W	2914	3.4	100.0	56.3	Chardonnay	W	2464	3.9	1.2	0.9
Zalagyöngye	W	2550	2.9	58.9	33.1	Merlot	R	1961	3.1	0.7	0.5
Zweigelt	R	2266	2.6	31.2	17.5	Zweigelt	R	1687	2.6	18.6	13.1
Cserszegi Fűszeres	G	2185	2.5	100.0	56.3	Aletta	W	1676	2.6	100.0	70.2
Chasselas	W	1902	2.2	14.3	8.0	Müller-Thurgau	W	1670	2.6	8.6	6.0
Riesling	W	1619	1.9	3.7	2.1	Hárslevelű	W	1603	2.5	99.0	69.5
Silvaner	W	1619	1.9	14.7	8.2	Pinot Gris	G	1594	2.5	3.3	2.3
Muscat Blanc à Petits Grains	W	1538	1.8	5.1	2.9	Irsai Olivér	W	1531	2.4	85.5	60.0
Muscat Ottonel	W	1416	1.6	11.6	6.5	Grüner Veltliner	W	1381	2.2	7.2	5.1
Kunleány	W	1376	1.6	100.0	56.3	Cabernet Franc	R	1368	2.1	2.4	1.7
Grüner Veltliner	W	1335	1.5	5.7	3.2	Riesling	W	1261	2.0	2.1	1.5
Blauer Portugieser	R	1255	1.4	13.7	7.7	Muscat Ottonel	W	1256	2.0	10.1	7.1
Kövidinka	G	1214	1.4	100.0	56.3	Chasselas	W	1159	1.8	15.7	11.0
Cabernet Sauvignon	R	1052	1.2	0.5	0.3	Pinot Noir	R	1092	1.7	1.0	0.7
Total of above		50707	58.4			Total of above		48631	76.1		

Table 57 (cont.): Winegrape areas and national and global shares (and Varietal Intensity Indexes) for national top 20 varieties, 2000 and 2016

India

2000						2016					
Prime variety	Col	Area	Nat'l share	Glob'l share	VII*	Prime variety	Col	Area	Nat'l share	Glob'l share	VII*
						Sultaniye	W	1000	37.0	18.8	311.8
						Sauvignon Blanc	W	500	18.5	0.4	6.7
						Syrah	R	500	18.5	0.3	4.6
						Trebbiano Toscano	W	300	11.1	0.2	4.1
						Cabernet Sauvignon	R	100	3.7	0.0	0.5
						Chardonnay	W	100	3.7	0.0	0.8
						Muscat of Alexandria	W	100	3.7	0.3	4.8
						Pinot Noir	R	100	3.7	0.1	1.6
Total of above						Total of above		2700	100.0		

Israel

2000						2016					
Prime variety	Col	Area	Nat'l share	Glob'l share	VII*	Prime variety	Col	Area	Nat'l share	Glob'l share	VII*
Mazuelo	R	971	20.0	0.8	7.7	Cabernet Sauvignon	R	990	19.8	0.3	2.9
Merlot	R	647	13.3	0.3	3.1	Mazuelo	R	935	18.7	2.0	17.7
Cabernet Sauvignon	R	607	12.5	0.3	2.7	Merlot	R	715	14.3	0.3	2.4
Colombard	W	486	10.0	1.3	12.7	Syrah	R	385	7.7	0.2	1.9
Emerald Riesling	W	344	7.1	36.7	369.9	Argaman	R	275	5.5	100.0	896.6
Sauvignon Blanc	W	263	5.4	0.4	4.1	Petit Verdot	R	275	5.5	3.4	30.4
Argaman	R	202	4.2	100.0	1007.5	Colombard	W	220	4.4	0.7	6.6
Muscat of Alexandria	W	202	4.2	0.7	6.9	Muscat of Alexandria	W	220	4.4	0.6	5.7
Chardonnay	W	142	2.9	0.1	1.0	Chardonnay	W	165	3.3	0.1	0.7
Chenin Blanc	W	101	2.1	0.2	2.2	Cabernet Franc	R	110	2.2	0.2	1.8
						Côt	R	110	2.2	0.2	1.9
						Durif	R	110	2.2	2.3	20.5
						Emerald Riesling	W	110	2.2	62.0	555.7
						Sauvignon Blanc	W	110	2.2	0.1	0.8
						Monastrell	R	55	1.1	0.1	0.9
						Tempranillo	R	55	1.1	0.0	0.2
Total of above		3966	81.7			Total of above		4840	96.8		

Table 57 (cont.): Winegrape areas and national and global shares (and Varietal Intensity Indexes) for national top 20 varieties, 2000 and 2016

Italy

2000						2016					
Prime variety	Col	Area	Nat'l share	Glob'l share	VII*	Prime variety	Col	Area	Nat'l share	Glob'l share	VII*
Sangiovese	R	62761	9.9	91.1	7.0	Sangiovese	R	68428	11.3	93.1	6.9
Catarratto Bianco	W	50711	8.0	100.0	7.7	Trebbiano Toscano	W	35441	5.9	29.4	2.2
Trebbiano Toscano	W	39447	6.2	28.8	2.2	Montepulciano	R	32724	5.4	99.4	7.4
Montepulciano	R	28679	4.5	99.8	7.7	Catarratto Bianco	W	28563	4.7	99.8	7.4
Barbera	R	27175	4.3	82.2	6.3	Merlot	R	24057	4.0	9.0	0.7
Merlot	R	21861	3.4	10.2	0.8	Chardonnay	W	19769	3.3	9.8	0.7
Trebbiano Romagnolo	W	19492	3.1	100.0	7.7	Prosecco	W	19730	3.3	98.1	7.3
Negroamaro	R	16619	2.6	100.0	7.7	Trebbiano Romagnolo	W	19059	3.2	100.0	7.4
Garganega	W	16549	2.6	100.0	7.7	Pinot Gris	G	18821	3.1	38.7	2.9
Muscat Blanc à Petits Grains	W	13016	2.0	43.4	3.3	Barbera	R	15006	2.5	84.2	6.2
Malvasia Bianca di Candia	W	11921	1.9	92.5	7.1	Cabernet Sauvignon	R	14240	2.4	4.6	0.3
Chardonnay	W	11687	1.8	8.0	0.6	Nero d'Avola	R	14129	2.3	98.9	7.3
Nero d'Avola	R	11318	1.8	100.0	7.7	Tribidrag	R	13896	2.3	41.3	3.1
Aglianico	R	9264	1.5	99.1	7.6	Muscat Blanc à Petits Grains	W	13334	2.2	39.5	2.9
Inzolia	W	9259	1.5	100.0	7.7	Negroamaro	R	11431	1.9	99.8	7.4
Trebbiano d'Abruzzo	W	8435	1.3	100.0	7.7	Aglianico	R	9627	1.6	98.9	7.3
Manzoni Bianco	W	8290	1.3	100.0	7.7	Malvasia Bianca di Candia	W	9028	1.5	93.2	6.9
Tribidrag	R	7828	1.2	29.1	2.2	Garganega	W	8522	1.4	99.6	7.4
Cabernet Sauvignon	R	7682	1.2	3.4	0.3	Syrah	R	7693	1.3	4.2	0.3
Prosecco	W	7498	1.2	99.9	7.7	Nebbiolo	R	7551	1.2	94.4	7.0
Total of above		389493	61.2			Total of above		391050	64.7		

Japan

2000						2016					
Prime variety	Col	Area	Nat'l share	Glob'l share	VII*	Prime variety	Col	Area	Nat'l share	Glob'l share	VII*
						Koshu	G	690	17.8	100.0	1158.8
						Niagara	W	551	14.2	16.9	195.7
						Muscat Bailey A	R	521	13.5	28.6	331.7
						Concord	R	292	7.6	2.8	32.1
						Delaware	G	254	6.6	60.5	700.6
						Campbell Early	R	238	6.1	100.0	1158.8
						Merlot	R	197	5.1	0.1	0.9
						Chardonnay	W	137	3.5	0.1	0.8
						Kerner	W	76	2.0	2.6	30.5
						Kyoho (4N)	R	62	1.6	2.2	26.0
						Zweigelt	R	59	1.5	0.6	7.5
						Cabernet Sauvignon	R	42	1.1	0.0	0.2
						Verdelet	W	39	1.0	98.5	1141.9
						Portland	W	39	1.0	100.0	1158.8
						Yamabudo	R	35	0.9	100.0	1158.8
						Black Queen	R	28	0.7	19.8	229.4
						Ryugan	W	27	0.7	100.0	1158.8
						Adirondac	R	24	0.6	100.0	1158.8
						Yama Sauvignon	R	24	0.6	100.0	1158.8
						Müller-Thurgau	W	22	0.6	0.1	1.3
Total of above						Total of above		3359	86.8		

Table 57 (cont.): Winegrape areas and national and global shares (and Varietal Intensity Indexes) for national top 20 varieties, 2000 and 2016

Kazakhstan

2000						2016					
Prime variety	Col	Area	Nat'l share	Glob'l share	VII*	Prime variety	Col	Area	Nat'l share	Glob'l share	VII*
						Rkatsiteli	W	3552	51.2	6.9	44.7
						Bayanshira	W	645	9.3	100.0	646.2
						Saperavi	R	428	6.2	6.6	42.7
						Kuldzhinskii	R	385	5.6	100.0	646.2
						Aligoté	W	277	4.0	1.0	6.6
						Muscat Blanc à Petits Grains (F	W	255	3.7	17.7	114.4
						Pinot Noir	R	180	2.6	0.2	1.1
						Riesling	W	111	1.6	0.2	1.2
						Maiskii Chernyi	R	110	1.6	100.0	646.2
						Cabernet Franc	R	56	0.8	0.1	0.7
						Madrasa	R	28	0.4	100.0	646.2
						Cabernet Sauvignon	R	20	0.3	0.0	0.0
						Ruby	R	9	0.1	100.0	646.2
						Rubinovy Magaracha	R	0	0.0	100.0	646.2
Total of above						Total of above		6056	87.3		

Korea, Rep.

2000						2016					
Prime variety	Col	Area	Nat'l share	Glob'l share	VII*	Prime variety	Col	Area	Nat'l share	Glob'l share	VII*
Kyoho (4N)	R	2700	50.0	67.4	610.5	Kyoho (4N)	R	2700	50.0	97.8	811.6
Muscat Bailey A	R	1300	24.1	94.8	857.8	Muscat Bailey A	R	1300	24.1	71.4	592.6
Sheridan	R	500	9.3	100.0	905.1	Sheridan	R	500	9.3	100.0	830.2
Delaware	G	100	1.9	42.8	387.1	Delaware	G	100	1.9	23.8	197.4
Total of above		4600	85.2			Total of above		4600	85.2		

Table 57 (cont.): Winegrape areas and national and global shares (and Varietal Intensity Indexes) for national top 20 varieties, 2000 and 2016

Lebanon

2000						2016					
Prime variety	Col	Area	Nat'l share	Glob'l share	VII*	Prime variety	Col	Area	Nat'l share	Glob'l share	VII*
						Cabernet Sauvignon	R	1000	25.0	0.3	3.6
						Chardonnay	W	1000	25.0	0.5	5.6
						Merlot	R	500	12.5	0.2	2.1
						Sauvignon Blanc	W	500	12.5	0.4	4.5
						Syrah	R	300	7.5	0.2	1.9
Total of above						Total of above		3300	82.5		

Luxembourg

2000						2016					
Prime variety	Col	Area	Nat'l share	Glob'l share	VII*	Prime variety	Col	Area	Nat'l share	Glob'l share	VII*
Müller-Thurgau	W	459	34.1	1.4	49.6	Müller-Thurgau	W	316	24.3	1.6	55.9
Riesling	W	175	13.0	0.4	14.6	Pinot Gris	G	196	15.1	0.4	13.9
Auxerrois	W	169	12.5	7.3	266.2	Auxerrois	W	190	14.6	6.7	229.7
Elbling	W	164	12.2	13.6	492.3	Riesling	W	162	12.5	0.3	9.3
Pinot Gris	G	155	11.5	0.8	29.7	Pinot Blanc	W	160	12.3	1.2	40.0
Pinot Blanc	W	138	10.2	0.8	29.5	Pinot Noir	R	121	9.3	0.1	4.0
Pinot Noir	R	66	4.9	0.1	3.5	Elbling	W	86	6.6	8.9	305.2
Gewürztraminer	W	12	0.9	0.1	4.1	Chardonnay	W	30	2.3	0.0	0.5
Chardonnay	W	8	0.6	0.0	0.2	Gewürztraminer	W	21	1.6	0.2	5.6
Gamay Noir	R	1	0.1	0.0	0.1	Sankt Laurent	R	4	0.3	0.1	4.2
Silvaner	W	1	0.1	0.0	0.3	Muscat Blanc à Petits Grains	W	1	0.1	0.0	0.1
Total of above		1348	100.0			Total of above		1287	99.0		

Table 57 (cont.): Winegrape areas and national and global shares (and Varietal Intensity Indexes) for national top 20 varieties, 2000 and 2016

Mexico

2000						2016					
Prime variety	Col	Area	Nat'l share	Glob'l share	VII*	Prime variety	Col	Area	Nat'l share	Glob'l share	VII*
						Sultaniye	W	841	15.4	15.8	129.6
						Cabernet Sauvignon	R	756	13.8	0.2	2.0
						Mazuelo	R	448	8.2	0.9	7.8
						Merlot	R	391	7.2	0.1	1.2
						Salvador	R	350	6.4	99.8	818.6
						Chenin Blanc	W	275	5.0	0.9	7.0
						Muscat Blanc à Petits Grains	W	246	4.5	0.7	6.0
						Fiesta	W	230	4.2	100.0	820.3
						Tempranillo	R	229	4.2	0.1	0.9
						Nebbiolo	R	180	3.3	2.3	18.5
						Cardinal	R	168	3.1	10.1	83.0
						Syrah	R	145	2.7	0.1	0.7
						Garnacha Tinta	R	140	2.6	0.1	0.8
						Durif	R	133	2.4	2.8	22.7
						Sauvignon Blanc	W	120	2.2	0.1	0.8
						Palomino Fino	W	109	2.0	0.5	3.9
						Jacquez	R	80	1.5	5.5	45.5
Total of above						Total of above		4841	88.6		

Moldova

2000						2016					
Prime variety	Col	Area	Nat'l share	Glob'l share	VII*	Prime variety	Col	Area	Nat'l share	Glob'l share	VII*
Aligoté	W	15790	17.6	44.3	24.1	Moldova	R	12375	15.0	100.0	54.3
Rkatsiteli	W	11508	12.8	17.1	9.3	Cabernet Sauvignon	R	8169	9.9	2.6	1.4
Isabella	R	11401	12.7	41.5	22.6	Aligoté	W	7765	9.4	28.8	15.7
Sauvignon Blanc	W	8151	9.1	12.5	6.8	Merlot	R	7689	9.3	2.9	1.6
Merlot	R	8123	9.0	3.8	2.1	Sauvignon Blanc	W	6909	8.4	5.5	3.0
Cabernet Sauvignon	R	7590	8.4	3.4	1.9	Chardonnay	W	4133	5.0	2.0	1.1
Pinot Noir	R	6521	7.3	9.5	5.2	Rkatsiteli	W	3898	4.7	7.6	4.1
Chardonnay	W	5134	5.7	3.5	1.9	Isabella	R	3468	4.2	19.5	10.6
Fetească Albă	W	4334	4.8	18.2	9.9	Pinot Noir	R	2366	2.9	2.2	1.2
Gewürztraminer	W	2731	3.0	25.6	13.9	Muscat Ottonel	W	1859	2.3	14.9	8.1
Pinot Gris	G	2042	2.3	10.8	5.9	Riesling	W	1701	2.1	2.8	1.5
Muscat Ottonel	W	1520	1.7	12.4	6.7	Bianca	W	1340	1.6	13.7	7.4
Riesling	W	1343	1.5	3.1	1.7	Pinot Gris	G	1208	1.5	2.5	1.3
Bastardo Magarachsky	R	1040	1.2	52.8	28.7	Kodryanka	R	1143	1.4	100.0	54.3
Saperavi	R	716	0.8	10.7	5.8	Gewürztraminer	W	1099	1.3	8.6	4.7
Sukholimansky Bely	W	599	0.7	36.7	20.0	Fetească Albă	W	954	1.2	7.1	3.9
Pinot Blanc	W	350	0.4	2.1	1.1	Magaracha Rannii	R	884	1.1	100.0	54.3
Müller-Thurgau	W	173	0.2	0.5	0.3	Alb de Suruceni	W	780	0.9	100.0	54.3
Muscat Blanc à Petits Grains	W	172	0.2	0.6	0.3	Cabernet Franc	R	756	0.9	1.3	0.7
Silvaner	W	98	0.1	0.9	0.5	Muscat Yantarnyi	W	683	0.8	100.0	54.3
Total of above		89336	99.4			Total of above		69178	83.8		

Table 57 (cont.): Winegrape areas and national and global shares (and Varietal Intensity Indexes) for national top 20 varieties, 2000 and 2016

Morocco

2000						2016					
Prime variety	Col	Area	Nat'l share	Glob'l share	VII*	Prime variety	Col	Area	Nat'l share	Glob'l share	VII*
Doukkali	R	16557	33.4	100.0	98.5	Italia	W	3333	19.0	64.3	163.8
Cinsaut	R	3940	7.9	8.1	8.0	Cinsaut	R	3239	18.4	14.1	36.0
Muscat of Alexandria	W	3669	7.4	12.4	12.2	Muscat of Alexandria	W	2093	11.9	6.0	15.3
Teneron	R	3488	7.0	100.0	98.5	Mazuelo	R	1230	7.0	2.6	6.6
Abbo	R	2375	4.8	100.0	98.5	Alicante Henri Bouschet	R	919	5.2	2.6	6.5
Mazuelo	R	1692	3.4	1.3	1.3	Chardonnay	W	880	5.0	0.4	1.1
Alicante Henri Bouschet	R	1098	2.2	3.0	2.9	Garnacha Tinta	R	786	4.5	0.5	1.3
Garnacha Tinta	R	802	1.6	0.4	0.4	Sultaniye	W	721	4.1	13.5	34.5
						Cardinal	R	624	3.6	37.6	95.9
						Cabernet Sauvignon	R	604	3.4	0.2	0.5
						Airén	W	440	2.5	0.2	0.5
						Sauvignon Blanc	W	440	2.5	0.4	0.9
						Syrah	R	347	2.0	0.2	0.5
						Merlot	R	333	1.9	0.1	0.3
						Clairette	W	113	0.6	4.7	11.9
Total of above		33621	67.8			Total of above		16104	91.6		

Myanmar

2000						2016					
Prime variety	Col	Area	Nat'l share	Glob'l share	VII*	Prime variety	Col	Area	Nat'l share	Glob'l share	VII*
						Syrah	R	27	38.6	0.0	9.5
						Sauvignon Blanc	W	22	31.4	0.0	11.3
						Muscat Blanc à Petits Grains	W	7	10.0	0.0	13.3
						Pinot Noir	R	7	10.0	0.0	4.3
						Tempranillo	R	4	5.0	0.0	1.0
						Petit Verdot	R	2	2.9	0.0	15.8
						Chardonnay	W	2	2.1	0.0	0.5
Total of above						Total of above		70	100.0		

Table 57 (cont.): Winegrape areas and national and global shares (and Varietal Intensity Indexes) for national top 20 varieties, 2000 and 2016

New Zealand

2000						2016					
Prime variety	Col	Area	Nat'l share	Glob'l share	VII*	Prime variety	Col	Area	Nat'l share	Glob'l share	VII*
Chardonnay	W	2787	28.0	1.9	9.4	Sauvignon Blanc	W	20497	57.8	16.4	20.8
Sauvignon Blanc	W	2423	24.4	3.7	18.3	Pinot Noir	R	5514	15.5	5.2	6.6
Pinot Noir	R	1098	11.0	1.6	7.8	Chardonnay	W	3117	8.8	1.5	2.0
Merlot	R	657	6.6	0.3	1.5	Pinot Gris	G	2422	6.8	5.0	6.3
Cabernet Sauvignon	R	654	6.6	0.3	1.4	Merlot	R	1239	3.5	0.5	0.6
Riesling	W	490	4.9	1.1	5.6	Riesling	W	767	2.2	1.3	1.6
Müller-Thurgau	W	419	4.2	1.2	6.1	Syrah	R	436	1.2	0.2	0.3
Sémillon	W	229	2.3	0.9	4.3	Gewürztraminer	W	277	0.8	2.2	2.7
Chenin Blanc	W	146	1.5	0.3	1.6	Cabernet Sauvignon	R	275	0.8	0.1	0.1
Gewürztraminer	W	141	1.4	1.3	6.5	Côt	R	129	0.4	0.2	0.3
Pinot Gris	G	127	1.3	0.7	3.3	Viognier	W	129	0.4	0.8	1.0
Cabernet Franc	R	118	1.2	0.2	1.1	Cabernet Franc	R	109	0.3	0.2	0.2
Pinotage	R	73	0.7	1.1	5.5	Sauvignon Blanc (G)	W	104	0.3	9.6	12.2
Côt	R	67	0.7	0.3	1.3	Sémillon	W	63	0.2	0.3	0.4
Reichensteiner	W	62	0.6	19.4	95.5	Grüner Veltliner	W	43	0.1	0.2	0.3
Syrah	R	60	0.6	0.1	0.3	Pinotage	R	38	0.1	0.5	0.7
Breidecker	W	28	0.3	100.0	491.6	Muscat	W	36	0.1	4.8	6.1
Chasselas	W	25	0.3	0.2	0.9	Arneis	W	33	0.1	2.8	3.6
Palomino Fino	W	21	0.2	0.1	0.3	Alvarinho	W	26	0.1	0.5	0.6
Blauburger	R	13	0.1	1.3	6.4	Chenin Blanc	W	24	0.1	0.1	0.1
Total of above		9638	96.9			Total of above		35279	99.5		

North Macedonia

2000						2016					
Prime variety	Col	Area	Nat'l share	Glob'l share	VII*	Prime variety	Col	Area	Nat'l share	Glob'l share	VII*
						Vranac	R	9500	38.3	100.0	180.9
						Dimyat	W	6500	26.2	67.0	121.3
						Merlot	R	1240	5.0	0.5	0.8
						Cabernet Sauvignon	R	1020	4.1	0.3	0.6
						Tribidrag	R	1000	4.0	3.0	5.4
						Riesling	W	900	3.6	1.5	2.7
						Chardonnay	W	750	3.0	0.4	0.7
						Pinot Noir	R	500	2.0	0.5	0.9
						Rkatsiteli	W	460	1.9	0.9	1.6
						Prokupac	R	445	1.8	32.7	59.2
						Muscat Blanc à Petits Grains	W	400	1.6	1.2	2.1
						Stanušina Crna	R	400	1.6	100.0	180.9
						Muscat of Hamburg	R	350	1.4	4.6	8.2
						Graševina	W	270	1.1	1.1	2.0
						Pamid	R	250	1.0	2.5	4.5
						Župljanka	W	250	1.0	49.5	89.5
						Sauvignon Blanc	W	185	0.7	0.1	0.3
						Žilavka	W	185	0.7	100.0	180.9
						Plavac Mali	R	50	0.2	2.9	5.3
Total of above						Total of above		24655	99.5		

Table 57 (cont.): Winegrape areas and national and global shares (and Varietal Intensity Indexes) for national top 20 varieties, 2000 and 2016

Norway

2000						2016					
Prime variety	Col	Area	Nat'l share	Glob'l share	VII*	Prime variety	Col	Area	Nat'l share	Glob'l share	VII*
						Solaris	W	8	60.0	6.3	22727.0
						Rondo	R	4	30.0	7.4	26610.8
Total of above						Total of above		11	90.0		

Peru

2000						2016					
Prime variety	Col	Area	Nat'l share	Glob'l share	VII*	Prime variety	Col	Area	Nat'l share	Glob'l share	VII*
						Negramoll	R	1252	32.7	41.6	486.6
						Italia	W	1011	26.4	19.5	228.1
						Muscat Blanc à Petits Grains	W	361	9.4	1.1	12.5
						Quebranta	R	330	8.6	100.0	1170.2
						Blaufränkisch	R	290	7.6	1.7	19.8
						Red Globe	R	240	6.3	99.1	1159.7
						Pecorino	W	114	3.0	6.5	76.6
						Cardinal	R	53	1.4	3.2	37.4
						Cabernet Sauvignon	R	48	1.3	0.0	0.2
						Flame Seedless	R	42	1.1	76.1	890.9
						Alphonse Lavallée	R	18	0.5	2.8	33.2
						Sultaniye	W	15	0.4	0.3	3.3
						Côt	R	10	0.3	0.0	0.2
						Superior Seedless	W	9	0.2	100.0	1170.2
						Torrontés Riojano	W	8	0.2	0.1	1.1
						Albillo Mayor	W	7	0.2	0.6	7.1
						Ar99	G	5	0.1	100.0	1170.2
						Chenin Blanc	W	2	0.1	0.0	0.1
						Crimson Seedless	R	2	0.1	26.5	310.4
						Crystal	W	2	0.1	1.1	13.4
Total of above						Total of above		3819	99.7		

Table 57 (cont.): Winegrape areas and national and global shares (and Varietal Intensity Indexes) for national top 20 varieties, 2000 and 2016

Portugal											
2000						2016					
Prime variety	Col	Area	Nat'l share	Glob'l share	VII*	Prime variety	Col	Area	Nat'l share	Glob'l share	VII*
Castelão	R	14424	7.0	100.0	23.8	Tempranillo	R	17014	9.3	7.8	1.9
Fernão Pires	W	14206	6.9	97.7	23.3	Touriga Franca	R	14217	7.8	100.0	24.5
Tempranillo	R	7356	3.6	7.9	1.9	Castelão	R	12580	6.9	100.0	24.5
Trincadeira	R	7264	3.5	100.0	23.8	Fernão Pires	W	12138	6.6	99.4	24.4
Baga	R	6730	3.3	100.0	23.8	Touriga Nacional	R	11411	6.2	97.3	23.9
Touriga Franca	R	6671	3.3	100.0	23.8	Trincadeira	R	10493	5.7	99.8	24.5
Tinta Barroca	R	5657	2.8	93.5	22.3	Baga	R	6750	3.7	100.0	24.5
Vinhao	R	5296	2.6	89.2	21.3	Síria	W	6438	3.5	91.5	22.5
Touriga Nacional	R	4149	2.0	97.3	23.2	Arinto de Bucelas	W	5409	3.0	100.0	24.5
Codega de Larinho	W	4058	2.0	100.0	23.8	Tinta Barroca	R	4733	2.6	96.1	23.6
Arinto de Bucelas	W	3966	1.9	100.0	23.8	Alicante Henri Bouschet	R	4547	2.5	12.6	3.1
Loureiro	W	3939	1.9	89.7	21.4	Loureiro	W	4402	2.4	93.7	23.0
Marufo	R	3512	1.7	55.4	13.2	Vinhao	R	4055	2.2	90.7	22.3
Azal	W	3302	1.6	100.0	23.8	Syrah	R	4017	2.2	2.2	0.5
Rufete	R	2338	1.1	68.8	16.4	Marufo	R	3367	1.8	71.9	17.6
Rabo de Ovelha	W	2330	1.1	100.0	23.8	Malvasia Fina	W	2922	1.6	89.0	21.9
Malvasia Fina	W	2328	1.1	32.8	7.8	Alvarelhão	R	2860	1.6	98.3	24.1
Vital	W	2246	1.1	100.0	23.8	Palomino Fino	W	2594	1.4	11.2	2.7
Malvasia Preta	R	2210	1.1	100.0	23.8	Mencía	R	2561	1.4	23.2	5.7
Borraçal	R	2035	1.0	76.7	18.3	Cabernet Sauvignon	R	2346	1.3	0.8	0.2
Total of above		#####	50.7			Total of above		134855	73.8		

Romania											
2000						2016					
Prime variety	Col	Area	Nat'l share	Glob'l share	VII*	Prime variety	Col	Area	Nat'l share	Glob'l share	VII*
Fetească Albă	W	18211	8.2	76.4	16.8	Fetească Regală	W	12619	6.9	97.1	23.8
Graševina	W	15014	6.8	16.3	3.6	Fetească Albă	W	12428	6.8	92.9	22.8
Cabernet Sauvignon	R	8620	3.9	3.9	0.9	Merlot	R	11647	6.4	4.4	1.1
Merlot	R	7810	3.5	3.7	0.8	Riesling	W	6121	3.3	10.2	2.5
Aligoté	W	7608	3.4	21.3	4.7	Aligoté	W	5840	3.2	21.7	5.3
Muscat Ottonel	W	5787	2.6	47.2	10.4	Sauvignon Blanc	W	5594	3.1	4.5	1.1
Sauvignon Blanc	W	4613	2.1	7.1	1.6	Cabernet Sauvignon	R	5359	2.9	1.7	0.4
Băbească Neagră	R	3642	1.6	97.9	21.5	Muscat Ottonel	W	4779	2.6	38.3	9.4
Pinot Gris	G	2388	1.1	12.6	2.8	Fetească Neagră	R	2845	1.6	87.6	21.5
Pinot Noir	R	1740	0.8	2.5	0.6	Pamid	R	2716	1.5	27.3	6.7
Fetească Regală	W	1700	0.8	65.9	14.5	Băbească Neagră	R	2696	1.5	100.0	24.5
Chardonnay	W	1376	0.6	0.9	0.2	Pinot Noir	R	1930	1.1	1.8	0.4
Fetească Neagră	R	1214	0.5	100.0	22.0	Chardonnay	W	1878	1.0	0.9	0.2
Muscat Blanc à Petits Grains	W	1012	0.5	3.4	0.7	Muscat Blanc à Petits Grains	W	1579	0.9	4.7	1.1
Grasă de Cotnari	W	850	0.4	100.0	22.0	Pinot Gris	G	1561	0.9	3.2	0.8
Galbenă de Odobești	W	546	0.2	100.0	22.0	Graševina	W	1437	0.8	5.9	1.4
Rkatsiteli	W	506	0.2	0.8	0.2	Blaufränkisch	R	729	0.4	4.2	1.0
Gewürztraminer	W	445	0.2	4.2	0.9	Grasă de Cotnari	W	571	0.3	90.3	22.2
						Syrah	R	504	0.3	0.3	0.1
						Gewürztraminer	W	469	0.3	3.7	0.9
Total of above		83082	37.4			Total of above		83302	45.6		

Table 57 (cont.): Winegrape areas and national and global shares (and Varietal Intensity Indexes) for national top 20 varieties, 2000 and 2016

Russia

2000						2016					
Prime variety	Col	Area	Nat'l share	Glob'l share	VII*	Prime variety	Col	Area	Nat'l share	Glob'l share	VII*
Rkatsiteli	W	13152	23.3	19.5	16.9	Cabernet Sauvignon	R	8528	16.8	2.7	2.4
Pinot Blanc	W	2995	5.3	17.6	15.3	Rkatsiteli	W	6477	12.8	12.6	11.1
Pervenets Magaracha	W	2388	4.2	84.2	73.0	Aligoté	W	5843	11.5	21.7	19.1
Bianca	W	2165	3.8	99.3	86.2	Bianca	W	3513	6.9	36.0	31.8
Aligoté	W	1821	3.2	5.1	4.4	Chardonnay	W	3481	6.9	1.7	1.5
Zalagyöngye	W	1781	3.2	41.1	35.7	Merlot	R	2988	5.9	1.1	1.0
Chardonnay	W	1639	2.9	1.1	1.0	Sauvignon Blanc	W	2501	4.9	2.0	1.8
Cabernet Sauvignon	R	1578	2.8	0.7	0.6	Pervenets Magaracha	W	2238	4.4	81.2	71.7
Riesling	W	1376	2.4	3.2	2.8	Riesling	W	2232	4.4	3.7	3.3
Agadai	W	1265	2.2	100.0	86.8	Isabella	R	1362	2.7	7.6	6.7
Saperavi	R	931	1.7	13.9	12.0	Odessky Cherny	R	1250	2.5	49.8	44.0
						Pinot Noir	R	918	1.8	0.9	0.8
						Levokumskij	R	890	1.8	100.0	88.3
						Pinot Blanc	W	865	1.7	6.3	5.5
						Saperavi	R	716	1.4	11.1	9.8
						Krasnostop Zolotovsky	R	562	1.1	100.0	88.3
						Gewürztraminer	W	500	1.0	3.9	3.4
						Dunavski Lazur	W	483	1.0	100.0	88.3
						Muscat Blanc à Petits Grains	W	483	0.9	1.4	1.3
						Tsimlyansky Cherny	R	451	0.9	100.0	88.3
Total of above		31090	55.2			Total of above		46280	91.1		

Serbia

2000						2016					
Prime variety	Col	Area	Nat'l share	Glob'l share	VII*	Prime variety	Col	Area	Nat'l share	Glob'l share	VII*
Graševina	W	33120	48.0	35.9	25.4	Cabernet Sauvignon	R	2111	9.6	0.7	1.4
Prokupac	R	15180	22.0	100.0	70.8	Graševina	W	2037	9.3	8.4	17.0
Chasselas	W	3450	5.0	25.9	18.3	Merlot	R	1968	8.9	0.7	1.5
Muscat of Hamburg	R	2760	4.0	39.1	27.7	Chardonnay	W	1455	6.6	0.7	1.5
						Riesling	W	1361	6.2	2.3	4.6
						Prokupac	R	916	4.2	67.3	137.1
						Sauvignon Blanc	W	741	3.4	0.6	1.2
						Blaufränkisch	R	727	3.3	4.2	8.6
						Pinot Noir	R	633	2.9	0.6	1.2
						Muscat of Hamburg	R	624	2.8	8.1	16.5
						Župljanka	W	255	1.2	50.5	102.9
						Dimyat	W	192	0.9	2.0	4.0
						Muscat Ottonel	W	183	0.8	1.5	3.0
						Gewürztraminer	W	142	0.6	1.1	2.2
						Muscat Fleur d'Oranger	W	116	0.5	38.7	78.9
						Pinot Gris	G	112	0.5	0.2	0.5
						Marselan	R	84	0.4	2.1	4.3
						Cabernet Franc	R	79	0.4	0.1	0.3
						Afus Ali	W	73	0.3	34.4	70.0
						Pamid	R	67	0.3	0.7	1.4
Total of above		54509	79.0			Total of above		13874	63.0		

Table 57 (cont.): Winegrape areas and national and global shares (and Varietal Intensity Indexes) for national top 20 varieties, 2000 and 2016

Slovakia

2000						2016					
Prime variety	Col	Area	Nat'l share	Glob'l share	VII*	Prime variety	Col	Area	Nat'l share	Glob'l share	VII*
Graševina	W	3895	25.0	4.2	13.2	Grüner Veltliner	W	1627	21.0	8.5	49.2
Grüner Veltliner	W	2960	19.0	12.5	39.3	Blaufränkisch	R	1216	15.7	7.1	41.0
Müller-Thurgau	W	1870	12.0	5.6	17.5	Sankt Laurent	R	717	9.3	21.9	126.8
Blaufränkisch	R	1091	7.0	7.8	24.4	Riesling	W	620	8.0	1.0	6.0
Sankt Laurent	R	935	6.0	36.6	114.8	Müller-Thurgau	W	509	6.6	2.6	15.1
Chardonnay	W	623	4.0	0.4	1.3	Cabernet Sauvignon	R	469	6.1	0.2	0.9
Pinot Blanc	W	623	4.0	3.7	11.5	Graševina	W	456	5.9	1.9	10.8
Fetească Albă	W	312	2.0	1.3	4.1	Pinot Blanc	W	416	5.4	3.0	17.5
Fetească Regală	W	312	2.0	12.1	37.9						
Cabernet Sauvignon	R	156	1.0	0.1	0.2						
Silvaner	W	156	1.0	1.4	4.4						
Total of above		12932	83.0			Total of above		6030	77.8		

Slovenia

2000						2016					
Prime variety	Col	Area	Nat'l share	Glob'l share	VII*	Prime variety	Col	Area	Nat'l share	Glob'l share	VII*
Graševina	W	3568	15.2	3.9	8.0	Graševina	W	1935	12.1	7.9	22.2
Chardonnay	W	1549	6.6	1.1	2.2	Refosco	R	1340	8.4	100.0	280.3
Terrano	R	1267	5.4	86.8	180.6	Chardonnay	W	1181	7.4	0.6	1.6
Sauvignon Blanc	W	1221	5.2	1.9	3.9	Sauvignon Blanc	W	1121	7.0	0.9	2.5
Merlot	R	1197	5.1	0.6	1.2	Malvazija Istarska	W	915	5.7	32.8	92.0
Ribolla Gialla	W	1127	4.8	80.1	166.8	Žametovka	R	822	5.1	100.0	280.4
						Merlot	R	817	5.1	0.3	0.9
						Blaufränkisch	R	709	4.4	4.1	11.6
						Riesling	W	607	3.8	1.0	2.8
						Ribolla Gialla	W	597	3.7	62.3	174.6
						Muscat Blanc à Petits Grains	W	586	3.7	1.7	4.9
						Furmint	W	546	3.4	12.3	34.5
						Pinot Gris	G	508	3.2	1.0	2.9
						Pinot Blanc	W	424	2.7	3.1	8.6
						Cabernet Sauvignon	R	423	2.6	0.1	0.4
						Sauvignonasse	W	231	1.4	6.0	16.8
						Savagnin Blanc	W	209	1.3	9.2	25.8
						Pinot Noir	R	202	1.3	0.2	0.5
						Kraljevina	W	199	1.2	100.0	280.4
						Müller-Thurgau	W	128	0.8	0.7	1.8
Total of above		9929	42.3			Total of above		13502	84.4		

Table 57 (cont.): Winegrape areas and national and global shares (and Varietal Intensity Indexes) for national top 20 varieties, 2000 and 2016

South Africa

2000						2016					
Prime variety	Col	Area	Nat'l share	Glob'l share	VII*	Prime variety	Col	Area	Nat'l share	Glob'l share	VII*
Chenin Blanc	W	22566	24.1	49.3	25.7	Chenin Blanc	W	17707	18.5	55.0	25.7
Colombard	W	11432	12.2	29.6	15.4	Colombard	W	11512	12.0	38.4	18.0
Cabernet Sauvignon	R	8824	9.4	4.0	2.1	Cabernet Sauvignon	R	10589	11.1	3.4	1.6
Pinotage	R	6501	6.9	98.9	51.6	Syrah	R	9946	10.4	5.5	2.6
Chardonnay	W	6067	6.5	4.2	2.2	Sauvignon Blanc	W	9246	9.7	7.4	3.5
Syrah	R	5631	6.0	5.5	2.9	Pinotage	R	7052	7.4	98.9	46.3
Sauvignon Blanc	W	5436	5.8	8.3	4.4	Chardonnay	W	6856	7.2	3.4	1.6
Merlot	R	4888	5.2	2.3	1.2	Merlot	R	5558	5.8	2.1	1.0
Muscat of Alexandria	W	4047	4.3	13.7	7.1	Ruby Cabernet	R	2306	2.4	43.4	20.3
Cinsaut	R	3533	3.8	7.3	3.8	Muscat of Alexandria	W	1781	1.9	5.1	2.4
Crouchen	W	2161	2.3	95.7	49.9	Cinsaut	R	1767	1.8	7.7	3.6
Ruby Cabernet	R	2050	2.2	27.6	14.4	Pinot Noir	R	1153	1.2	1.1	0.5
Palomino Fino	W	1632	1.7	5.3	2.8	Sémillon	W	1121	1.2	6.0	2.8
Sémillon	W	1033	1.1	3.9	2.1	Muscat Blanc à Petits Grains	W	839	0.9	2.5	1.2
Clairette	W	938	1.0	21.5	11.2	Cabernet Franc	R	835	0.9	1.5	0.7
Muscat Blanc à Petits Grains	W	773	0.8	2.6	1.3	Viognier	W	822	0.9	5.1	2.4
Cabernet Franc	R	488	0.5	0.9	0.5	Petit Verdot	R	749	0.8	9.2	4.3
Pinot Noir	R	487	0.5	0.7	0.4	Monastrell	R	473	0.5	0.9	0.4
Riesling	W	477	0.5	1.1	0.6	Côt	R	452	0.5	0.9	0.4
Servant	W	432	0.5	81.5	42.5	Nouvelle	W	428	0.4	100.0	46.8
Total of above		89394	95.4			Total of above		91193	95.2		

Spain

2000						2016					
Prime variety	Col	Area	Nat'l share	Glob'l share	VII*	Prime variety	Col	Area	Nat'l share	Glob'l share	VII*
Airén	W	387978	32.8	100.0	4.1	Airén	W	203276	23.0	99.7	5.1
Bobal	R	100128	8.5	100.0	4.1	Tempranillo	R	193597	21.9	88.2	4.5
Garnacha Tinta	R	98131	8.3	45.4	1.9	Bobal	R	59189	6.7	100.0	5.1
Tempranillo	R	79310	6.7	84.9	3.5	Garnacha Tinta	R	54606	6.2	36.4	1.8
Monastrell	R	67160	5.7	88.0	3.6	Monastrell	R	41303	4.7	79.5	4.0
Cayetana Blanca	W	55527	4.7	99.6	4.1	Macabeo	W	36963	4.2	95.7	4.9
Macabeo	W	42902	3.6	89.1	3.7	Cayetana Blanca	W	36252	4.1	99.6	5.1
Palomino Fino	W	27685	2.3	90.7	3.8	Cabernet Sauvignon	R	20139	2.3	6.5	0.3
Alicante Henri Bouschet	R	18321	1.6	49.3	2.0	Palomino Fino	W	20110	2.3	86.7	4.4
Pedro Ximénez	W	14803	1.3	85.7	3.5	Syrah	R	19488	2.2	10.8	0.5
Parellada	W	11188	0.9	100.0	4.1	Alicante Henri Bouschet	R	19294	2.2	53.5	2.7
Mencía	R	11166	0.9	85.0	3.5	Verdejo	W	17923	2.0	100.0	5.1
Chelva	W	10877	0.9	100.0	4.1	Merlot	R	12852	1.5	4.8	0.2
Xarello	W	10299	0.9	100.0	4.1	Muscat of Alexandria	W	9534	1.1	27.4	1.4
Calagrano	W	8229	0.7	100.0	4.1	Xarello	W	8534	1.0	100.0	5.1
Mazuelo	R	8103	0.7	6.3	0.3	Pedro Ximénez	W	8528	1.0	96.8	4.9
Tinto Velasco	R	7998	0.7	100.0	4.1	Mencía	R	8489	1.0	76.8	3.9
Merseguera	W	7460	0.6	100.0	4.1	Parellada	W	7137	0.8	100.0	5.1
Pardillo	W	7272	0.6	100.0	4.1	Chardonnay	W	6866	0.8	3.4	0.2
Muscat of Alexandria	W	6144	0.5	20.8	0.9	Mazuelo	R	5461	0.6	11.5	0.6
Total of above		980681	83.0			Total of above		789542	89.4		

Table 57 (cont.): Winegrape areas and national and global shares (and Varietal Intensity Indexes) for national top 20 varieties, 2000 and 2016

Switzerland

2000						2016					
Prime variety	Col	Area	Nat'l share	Glob'l share	VII*	Prime variety	Col	Area	Nat'l share	Glob'l share	VII*
Chasselas	W	5373	35.7	40.3	131.1	Pinot Noir	R	4209	28.4	4.0	12.1
Pinot Noir	R	4601	30.6	6.7	21.7	Chasselas	W	3838	25.9	52.0	157.7
Gamay Noir	R	1977	13.1	5.2	17.0	Gamay Noir	R	1349	9.1	5.1	15.6
Merlot	R	848	5.6	0.4	1.3	Merlot	R	1124	7.6	0.4	1.3
Müller-Thurgau	W	686	4.6	2.0	6.6	Müller-Thurgau	W	465	3.1	2.4	7.2
Chardonnay	W	226	1.5	0.2	0.5	Gamaret	R	425	2.9	96.3	292.0
Silvaner	W	208	1.4	1.9	6.1	Chardonnay	W	359	2.4	0.2	0.5
Pinot Gris	G	149	1.0	0.8	2.6	Silvaner	W	250	1.7	4.1	12.5
Cornalin	R	92	0.6	98.7	320.7	Pinot Gris	G	230	1.6	0.5	1.4
Pinot Blanc	W	77	0.5	0.5	1.5	Garanoir	R	225	1.5	98.2	297.7
Gamaret	R	60	0.4	84.8	275.4	Syrah	R	194	1.3	0.1	0.3
Arvine	W	57	0.4	93.0	302.2	Arvine	W	178	1.2	92.6	280.7
Syrah	R	54	0.4	0.1	0.2	Sauvignon Blanc	W	170	1.1	0.1	0.4
Garanoir	R	50	0.3	65.8	214.0	Cornalin	R	138	0.9	93.8	284.3
Muscat Blanc à Petits Grains	W	44	0.3	0.1	0.5	Rouge du Pays	R	136	0.9	100.0	303.1
Sauvignon Blanc	W	38	0.2	0.1	0.2	Savagnin Blanc	W	127	0.9	5.6	17.0
Cabernet Sauvignon	R	34	0.2	0.0	0.1	Diolinoir	R	120	0.8	98.3	297.8
Marsanne	W	33	0.2	2.2	7.1	Pinot Blanc	W	111	0.8	0.8	2.4
Gewürztraminer	W	31	0.2	0.3	0.9	Cabernet Sauvignon	R	66	0.4	0.0	0.1
Diolinoir	R	25	0.2	79.6	258.5	Cabernet Franc	R	63	0.4	0.1	0.3
Total of above		14664	97.5			Total of above		13776	93.1		

Taiwan

2000						2016					
Prime variety	Col	Area	Nat'l share	Glob'l share	VII*	Prime variety	Col	Area	Nat'l share	Glob'l share	VII*
Kyoho (4N)	R	1303	46.0	32.6	561.6	Black Queen	R	94	63.2	65.6	19773.7
Golden Muscat	W	1190	42.0	100.0	1725.4	Golden Muscat	W	50	33.4	99.2	29910.5
Black Queen	R	227	8.0	40.0	690.5	Musann Blanc	W	5	3.4	100.0	30163.0
Italia	W	113	4.0	10.5	181.7						
Total of above		2833	100.0			Total of above		149	100.0		

Table 57 (cont.): Winegrape areas and national and global shares (and Varietal Intensity Indexes) for national top 20 varieties, 2000 and 2016

Thailand

2000						2016					
Prime variety	Col	Area	Nat'l share	Glob'l share	VII*	Prime variety	Col	Area	Nat'l share	Glob'l share	VII*
						Syrah	R	74	35.9	0.0	8.9
						Malaga Blanc	W	54	26.0	100.0	21598.8
						Black Queen	R	18	8.7	12.6	2712.7
						Chenin Blanc	W	16	7.8	0.1	10.9
						Colombard	W	15	7.0	0.0	10.5
						Cabernet Sauvignon	R	7	3.4	0.0	0.5
						Garnacha Roja (Gris)	G	5	2.5	0.4	76.5
						Tempranillo	R	4	1.8	0.0	0.4
						Muscat	W	3	1.5	0.4	92.8
						Dornfelder	R	3	1.4	0.0	8.0
						Sangiovese	R	2	1.0	0.0	0.6
						Viognier	W	1	0.6	0.0	1.7
						Verdelho	W	1	0.5	0.1	15.0
						Durif	R	1	0.3	0.0	2.4
						Pinot Noir	R	1	0.3	0.0	0.1
Total of above						Total of above		205	98.7		

Tunisia

2000						2016					
Prime variety	Col	Area	Nat'l share	Glob'l share	VII*	Prime variety	Col	Area	Nat'l share	Glob'l share	VII*
Mazuelo	R	7576	45.0	5.9	17.2	Italia	W	644	19.0	12.4	163.8
Garnacha Tinta	R	2020	12.0	0.9	2.7	Cinsaut	R	626	18.4	2.7	36.0
Alicante Henri Bouschet	R	842	5.0	2.3	6.6	Muscat of Alexandria	W	405	11.9	1.2	15.3
Cinsaut	R	842	5.0	1.7	5.0	Mazuelo	R	238	7.0	0.5	6.6
Sangiovese	R	842	5.0	1.2	3.5	Alicante Henri Bouschet	R	178	5.2	0.5	6.5
Cabernet Sauvignon	R	337	2.0	0.2	0.4	Chardonnay	W	170	5.0	0.1	1.1
Monastrell	R	337	2.0	0.4	1.3	Garnacha Tinta	R	152	4.5	0.1	1.3
Syrah	R	337	2.0	0.3	1.0	Sultaniye	W	139	4.1	2.6	34.5
Tribidrag	R	337	2.0	1.3	3.6	Cardinal	R	121	3.6	7.3	95.9
						Cabernet Sauvignon	R	117	3.4	0.0	0.5
						Airén	W	85	2.5	0.0	0.5
						Sauvignon Blanc	W	85	2.5	0.1	0.9
						Syrah	R	67	2.0	0.0	0.5
						Merlot	R	64	1.9	0.0	0.3
						Clairette	W	22	0.6	0.9	11.9
Total of above		13469	80.0			Total of above		3113	91.6		

Table 57 (cont.): Winegrape areas and national and global shares (and Varietal Intensity Indexes) for national top 20 varieties, 2000 and 2016

Turkey											
2000						2016					
Prime variety	Col	Area	Nat'l share	Glob'l share	VII*	Prime variety	Col	Area	Nat'l share	Glob'l share	VII*
						Sultaniye	W	2461	18.0	46.2	151.2
						Öküzgözü	R	1601	11.7	100.0	327.2
						Syrah	R	1439	10.5	0.8	2.6
						Boğazkere	R	1436	10.5	100.0	327.2
						Çalkarası	R	806	5.9	100.0	327.2
						Narince	W	787	5.7	100.0	327.2
						Kalecik Karası	R	704	5.1	100.0	327.2
						Dimrit	R	704	5.1	100.0	327.2
						Alicante Henri Bouschet	R	532	3.9	1.5	4.8
						Sémillon	W	529	3.9	2.8	9.3
						Cabernet Sauvignon	R	476	3.5	0.2	0.5
						Cinsaut	R	430	3.1	1.9	6.1
						Merlot	R	415	3.0	0.2	0.5
						Gamay Noir	R	228	1.7	0.9	2.8
						Papazkarası	R	204	1.5	100.0	327.2
						Chardonnay	W	177	1.3	0.1	0.3
						Sauvignon Blanc	W	153	1.1	0.1	0.4
						Mazuelo	R	130	0.9	0.3	0.9
						Muscat Blanc à Petits Grains	W	129	0.9	0.4	1.3
						Adakarası	R	89	0.6	100.0	327.2
Total of above						Total of above		13430	98.0		

Ukraine											
2000						2016					
Prime variety	Col	Area	Nat'l share	Glob'l share	VII*	Prime variety	Col	Area	Nat'l share	Glob'l share	VII*
						Rkatsiteli	W	5775	22.9	11.2	20.0
						Cabernet Sauvignon	R	4935	19.6	1.6	2.8
						Aligoté	W	4814	19.1	17.9	31.8
						Sauvignon Blanc	W	1550	6.2	1.2	2.2
						Chardonnay	W	1500	6.0	0.7	1.3
						Merlot	R	1400	5.6	0.5	0.9
						Riesling	W	1350	5.4	2.3	4.0
						Odessky Cherny	R	1250	5.0	49.8	88.8
						Isabella	R	1200	4.8	6.7	12.0
						Gewürztraminer	W	500	2.0	3.9	6.9
						Pinot Noir	R	385	1.5	0.4	0.7
						Muscat Blanc à Petits Grains	W	338	1.3	1.0	1.8
						Pinot Blanc	W	170	0.7	1.2	2.2
Total of above						Total of above		25166	100.0		

Table 57 (cont.): Winegrape areas and national and global shares (and Varietal Intensity Indexes) for national top 20 varieties, 2000 and 2016

United Kingdom

2000						2016					
Prime variety	Col	Area	Nat'l share	Glob'l share	VII*	Prime variety	Col	Area	Nat'l share	Glob'l share	VII*
Bacchus	W	112	12.8	3.3	185.8	Pinot Noir	R	546	29.7	0.5	12.6
Blauburger	R	105	12.0	10.5	586.7	Chardonnay	W	531	28.9	0.3	6.4
Dalkauer	W	100	11.5	100.0	5598.7	Pinot Meunier	R	202	11.0	1.4	33.6
Seyval Blanc	W	85	9.7	21.9	1224.4	Bacchus	W	127	6.9	7.2	175.9
Chardonnay	W	68	7.8	0.0	2.6	Seyval Blanc	W	77	4.2	2.9	69.8
Auxerrois	W	58	6.6	2.5	141.1	Pinot Gris	G	44	2.4	0.1	2.2
Cabernet Dorsa	R	43	4.9	100.0	5598.7	Reichensteiner	W	42	2.3	35.4	862.7
Albalonga	W	41	4.7	71.9	4027.1	Madeleine × Angevine 7672	W	39	2.1	80.0	1949.5
Ehrenfelser	W	34	3.9	11.8	658.7	Rondo	R	37	2.0	72.8	1774.1
						Regent	R	26	1.4	1.3	31.8
						Ortega	W	24	1.3	4.5	109.5
						Müller-Thurgau	W	15	0.8	0.1	1.8
						Pinot Blanc	W	15	0.8	0.1	2.6
						Dornfelder	R	13	0.7	0.2	4.0
						Schönburger	G	9	0.5	26.1	636.6
						Huxelrebe	W	4	0.2	0.8	19.3
Total of above		646	74.0			Total of above		1751	95.2		

United States

2000						2016					
Prime variety	Col	Area	Nat'l share	Glob'l share	VII*	Prime variety	Col	Area	Nat'l share	Glob'l share	VII*
Chardonnay	W	35791	20.4	24.6	6.8	Chardonnay	W	41392	17.3	20.5	3.8
Tribidrag	R	18630	10.6	69.2	19.3	Cabernet Sauvignon	R	40837	17.0	13.1	2.5
Colombard	W	18010	10.3	46.6	13.0	Pinot Noir	R	22998	9.6	21.8	4.1
Cabernet Sauvignon	R	17573	10.0	7.9	2.2	Merlot	R	21251	8.9	8.0	1.5
Merlot	R	16875	9.6	7.9	2.2	Tribidrag	R	18551	7.7	55.1	10.3
Chenin Blanc	W	8433	4.8	18.4	5.1	Syrah	R	9083	3.8	5.0	0.9
Concord	R	8330	4.7	70.5	19.6	Concord	R	8349	3.5	79.2	14.8
Pinot Noir	R	5343	3.0	7.8	2.2	Colombard	W	7991	3.3	26.6	5.0
Barbera	R	4693	2.7	14.2	4.0	Pinot Gris	G	7462	3.1	15.4	2.9
Garnacha Tinta	R	4519	2.6	2.1	0.6	Sauvignon Blanc	W	6747	2.8	5.4	1.0
Sauvignon Blanc	W	4191	2.4	6.4	1.8	Riesling	W	4952	2.1	8.3	1.5
Rubired	R	4153	2.4	100.0	27.8	Rubired	R	4825	2.0	98.2	18.4
Mazuelo	R	3088	1.8	2.4	0.7	Durif	R	3698	1.5	76.9	14.4
Ruby Cabernet	R	2895	1.6	39.0	10.9	Garnacha Tinta	R	2213	0.9	1.5	0.3
Muscat of Alexandria	W	2013	1.1	6.8	1.9	Cabernet Franc	R	2199	0.9	3.9	0.7
Riesling	W	1965	1.1	4.5	1.3	Barbera	R	2131	0.9	12.0	2.2
Syrah	R	1509	0.9	1.5	0.4	Ruby Cabernet	R	2114	0.9	39.8	7.5
Niagara	W	1357	0.8	8.8	2.5	Muscat of Alexandria	W	1987	0.8	5.7	1.1
Cabernet Franc	R	1189	0.7	2.3	0.6	Chenin Blanc	W	1969	0.8	6.1	1.1
Malvasia Bianca di Candia	W	968	0.6	7.5	2.1	Côt	R	1610	0.7	3.1	0.6
Total of above		161524	91.9			Total of above		30200	100.0		

Table 57 (cont.): Winegrape areas and national and global shares (and Varietal Intensity Indexes) for national top 20 varieties, 2000 and 2016

Uruguay

2000						2016					
Prime variety	Col	Area	Nat'l share	Glob'l share	VII*	Prime variety	Col	Area	Nat'l share	Glob'l share	VII*
Muscat of Hamburg	R	2886	32.5	40.8	224.8	Tannat	R	1725	25.6	30.7	204.4
Tannat	R	2433	27.4	43.5	239.4	Muscat of Hamburg	R	1267	18.8	16.5	109.7
Merlot	R	1057	11.9	0.5	2.7	Merlot	R	747	11.1	0.3	1.9
Cabernet Sauvignon	R	675	7.6	0.3	1.7	Trebbiano Toscano	W	682	10.1	0.6	3.8
Cabernet Franc	R	364	4.1	0.7	3.9	Cabernet Sauvignon	R	484	7.2	0.2	1.0
Chardonnay	W	142	1.6	0.1	0.5	Cabernet Franc	R	266	3.9	0.5	3.2
Sauvignon Blanc	W	142	1.6	0.2	1.2	Sauvignon Blanc	W	144	2.1	0.1	0.8
Syrah	R	62	0.7	0.1	0.3	Marselan	R	120	1.8	3.0	20.2
						Chardonnay	W	119	1.8	0.1	0.4
						Isabella	R	102	1.5	0.6	3.8
						Syrah	R	67	1.0	0.0	0.2
						Pinot Noir	R	56	0.8	0.1	0.4
						Arinarnoa	R	45	0.7	9.3	61.5
						Côt	R	43	0.6	0.1	0.5
						Viognier	W	41	0.6	0.3	1.7
						Petit Verdot	R	26	0.4	0.3	2.1
						Nebbiolo	R	25	0.4	0.3	2.1
						Alicante Henri Bouschet	R	24	0.4	0.1	0.4
						Concord	R	24	0.4	0.2	1.5
						Muscat of Alexandria	W	22	0.3	0.1	0.4
Total of above		7762	87.4			Total of above		6029	89.4		

Missing 9

2000						2016					
Prime variety	Col	Area	Nat'l share	Glob'l share	VII*	Prime variety	Col	Area	Nat'l share	Glob'l share	VII*
Cabernet Sauvignon	R	20373	25.4	9.1	5.6						
Rkatsiteli	W	10549	13.1	15.7	9.5						
Aligoté	W	6916	8.6	19.4	11.8						
Merlot	R	5549	6.9	2.6	1.6						
Chardonnay	W	3120	3.9	2.1	1.3						
Sauvignon Blanc	W	2383	3.0	3.7	2.2						
Riesling	W	2279	2.8	5.3	3.2						
Sultaniye	W	1820	2.3	15.0	9.1						
Odessky Cherny	R	1694	2.1	100.0	60.9						
Isabella	R	1673	2.1	6.1	3.7						
Saperavi	R	1356	1.7	20.2	12.3						
Syrah	R	1278	1.6	1.2	0.8						
Öküzgözü	R	1033	1.3	100.0	60.9						
Sukholimansky Bely	W	1032	1.3	63.3	38.5						
Muscat Blanc à Petits Grains	W	981	1.2	3.3	2.0						
Carmenère	R	945	1.2	16.5	10.1						
Bastardo Magarachsky	R	929	1.2	47.2	28.7						
Negramoll	R	874	1.1	24.6	15.0						
Boğazkere	R	773	1.0	100.0	60.9						
Pinot Noir	R	741	0.9	1.1	0.7						
Total of above		66299	82.6			Total of above					

Table 58: Varietal Intensity Indexes (and winegrape areas and national shares) for national top 20 varieties, 2000 and 2016

Algeria											
2000						2016					
Prime variety	Col	Area	Nat'l share	NVII*	VII*	Prime variety	Col	Area	Nat'l share	NVII*	VII*
Cinsaut	R	7550	25.0	1.5	25.2	Mazuelo	R	3000	36.1	0.6	34.2
Mazuelo	R	7550	25.0	1.4	9.6	Garnacha Tinta	R	2000	24.1	0.4	7.2
Garnacha Tinta	R	6040	20.0	1.0	4.5	Syrah	R	1000	12.0	0.1	3.0
Alicante Henri Bouschet	R	3020	10.0	0.6	13.2	Merlot	R	1000	12.0	0.1	2.0
Pinot Noir	R	1510	5.0	0.2	3.6	Cabernet Sauvignon	R	1000	12.0	0.1	1.7
Syrah	R	1510	5.0	0.2	2.4	Muscat of Alexandria	W	200	2.4	0.0	3.1
Merlot	R	1510	5.0	0.0	1.1	Sultaniye	W	100	1.2	0.0	10.1
Cabernet Sauvignon	R	1510	5.0	0.0	1.1						
Total of above		30200	100.0			Total of above		8300	100.0		

Argentina											
2000						2016					
Prime variety	Col	Area	Nat'l share	NVII*	VII*	Prime variety	Col	Area	Nat'l share	NVII*	VII*
Cereza	G	31113	15.8	6.1	24.8	Côt	R	40401	19.6	8.5	16.8
Criolla Grande	R	24264	12.3	4.8	24.8	Cereza	G	28887	14.0	6.1	21.7
Côt	R	18230	9.2	3.5	17.2	Douce Noire	R	19072	9.2	4.1	21.0
Douce Noire	R	15659	7.9	3.1	21.2	Criolla Grande	R	15596	7.6	3.3	21.7
Pedro Giménez	W	14862	7.5	2.9	24.8	Pedro Giménez	W	11197	5.4	2.3	15.6
Muscat Blanc à Petits Grains (C	W	10442	5.3	2.1	24.8	Torrontés Riojano	W	8208	4.0	1.7	20.1
Torrontés Riojano	W	8127	4.1	1.6	24.5	Muscat Blanc à Petits Grains (C	W	6526	3.2	1.4	17.2
Cabernet Sauvignon	R	13776	7.0	1.0	1.5	Syrah	R	12707	6.2	1.0	1.5
Syrah	R	8888	4.5	1.0	2.1	Aspiran Bouschet	R	4087	2.0	0.9	21.7
Muscat of Alexandria	W	5515	2.8	0.9	4.6	Torrontés Sanjuanino	W	1885	0.9	0.4	11.2
Torrontés Sanjuanino	W	3170	1.6	0.6	24.8	Muscat of Alexandria	W	2716	1.3	0.2	1.7
Red Globe	R	1940	1.0	0.4	22.7	Cabernet Sauvignon	R	15356	7.4	0.2	1.1
Chenin Blanc	W	3445	1.7	0.3	1.9	Ancellotta	R	991	0.5	0.2	7.9
Gibi	W	1227	0.6	0.2	24.8	Gibi	W	785	0.4	0.2	21.7
Béquignol Noir	R	1082	0.5	0.2	24.7	Chenin Blanc	W	2157	1.0	0.2	1.5
Tempranillo	R	4720	2.4	0.2	1.3	Torrontes Mendocino	W	653	0.3	0.1	21.7
Torrontes Mendocino	W	780	0.4	0.2	24.8	Béquignol Noir	R	616	0.3	0.1	21.7
Damaschino	W	827	0.4	0.1	6.4	Tannat	R	837	0.4	0.1	3.2
Sauvignonasse	W	756	0.4	0.1	3.4	Damaschino	W	527	0.3	0.1	7.1
Cardinal	R	680	0.3	0.1	4.3	Petit Verdot	R	740	0.4	0.1	2.0
Total of above		169503	85.9			Total of above		173939	84.3		

Table 58 (cont.): Varietal Intensity Indexes (and winegrape areas and national shares) for national top 20 varieties, 2000 and 2016

Armenia

2000						2016					
Prime variety	Col	Area	Nat'l share	NVII*	VII*	Prime variety	Col	Area	Nat'l share	NVII*	VII*
Rkatsiteli	W	2469	22.0	0.5	16.0	white varieties	W	10300	70.0	2.3	304.9
Mskhali	W	1093	9.8	0.2	436.2	red varieties	R	4405	30.0	1.0	304.9
Garandmak	W	931	8.3	0.2	436.2						
Kangun	W	850	7.6	0.2	436.2						
Voskeat	W	809	7.2	0.2	436.2						
Muscat Blanc à Petits Grains	W	526	4.7	0.1	7.7						
Total of above		6677	59.6			Total of above		14705	100.0		

Australia

2000						2016					
Prime variety	Col	Area	Nat'l share	NVII*	VII*	Prime variety	Col	Area	Nat'l share	NVII*	VII*
Syrah	R	29295	22.4	5.4	10.7	Syrah	R	38942	29.4	7.5	7.3
Cabernet Sauvignon	R	24997	19.1	3.9	4.2	Chardonnay	W	21321	16.1	3.4	3.6
Chardonnay	W	17266	13.2	2.7	4.4	Cabernet Sauvignon	R	23987	18.1	3.3	2.6
Sultaniye	W	10298	7.9	2.0	31.8	Sémillon	W	4556	3.4	0.9	8.2
Sémillon	W	6528	5.0	1.2	9.3	Sauvignon Blanc	W	6044	4.6	0.5	1.6
Ruby Cabernet	R	2424	1.9	0.5	12.2	Pinot Gris	G	3652	2.8	0.5	2.5
Riesling	W	3129	2.4	0.4	2.7	Pinot Noir	R	4806	3.6	0.4	1.5
Merlot	R	7669	5.9	0.4	1.3	Riesling	W	3114	2.4	0.3	1.8
Muscat of Alexandria	W	2495	1.9	0.3	3.2	Muscat of Alexandria	W	2179	1.6	0.3	2.1
Pinot Noir	R	3223	2.5	0.3	1.8	Verdelho	W	1016	0.8	0.2	22.7
Verdelho	W	1293	1.0	0.3	29.4	Colombard	W	1789	1.4	0.2	2.0
Sauvignon Blanc	W	2602	2.0	0.2	1.5	Petit Verdot	R	1118	0.8	0.2	4.7
Colombard	W	1801	1.4	0.2	1.7	Savagnin Blanc	W	870	0.7	0.2	13.0
Korinthiaki	R	778	0.6	0.2	34.9	Ruby Cabernet	R	849	0.6	0.2	5.4
Petit Verdot	R	721	0.6	0.1	16.5	Merlot	R	8415	6.4	0.1	1.1
Muscat Blanc à Petits Grains (R	W	380	0.3	0.1	12.3	Durif	R	540	0.4	0.1	3.8
Afus Ali	W	328	0.3	0.1	6.7	Viognier	W	753	0.6	0.1	1.6
Gewürztraminer	W	521	0.4	0.0	1.8	Muscat Blanc à Petits Grains (R	W	240	0.2	0.0	5.7
Marsanne	W	216	0.2	0.0	5.3	Canada Muscat	W	120	0.1	0.0	33.9
Durif	R	181	0.1	0.0	5.7	Marsanne	W	161	0.1	0.0	3.0
Total of above		116147	88.9			Total of above		124472	94.0		

Table 58 (cont.): Varietal Intensity Indexes (and winegrape areas and national shares) for national top 20 varieties, 2000 and 2016

Austria

2000						2016					
Prime variety	Col	Area	Nat'l share	NVII*	VII*	Prime variety	Col	Area	Nat'l share	NVII*	VII*
Grüner Veltliner	W	17479	36.0	3.5	74.6	Grüner Veltliner	W	14376	31.6	3.2	74.2
Zweigelt	R	4350	9.0	0.9	60.3	Zweigelt	R	6311	13.9	1.4	68.7
Graševina	W	4323	8.9	0.7	4.7	Graševina	W	3233	7.1	0.7	13.1
Müller-Thurgau	W	3289	6.8	0.6	9.9	Blaufränkisch	R	2808	6.2	0.6	16.1
Pinot Blanc	W	2936	6.1	0.6	17.4	Pinot Blanc	W	1916	4.2	0.4	13.7
Blaufränkisch	R	2641	5.4	0.5	19.0	Müller-Thurgau	W	1777	3.9	0.4	9.0
Blauer Portugieser	R	2358	4.9	0.5	26.0	Riesling	W	2016	4.4	0.3	3.3
Riesling	W	1643	3.4	0.2	3.8	Blauer Portugieser	R	1265	2.8	0.3	18.9
Neuburger	W	1094	2.3	0.2	76.9	Blauburger	R	750	1.7	0.2	60.5
Blauburger	R	884	1.8	0.2	88.9	Sankt Laurent	R	724	1.6	0.2	21.8
Frühroter Veltliner	R	626	1.3	0.1	99.8	Neuburger	W	507	1.1	0.1	86.5
Scheurebe	W	529	1.1	0.1	14.6	Muscat Blanc à Petits Grains	W	823	1.8	0.1	2.4
Blauer Wildbacher	R	464	1.0	0.1	99.0	Blauer Wildbacher	R	434	1.0	0.1	98.0
Sankt Laurent	R	415	0.9	0.1	16.4	Frühroter Veltliner	R	369	0.8	0.1	93.7
Bouvier	W	365	0.8	0.1	100.8	Scheurebe	W	351	0.8	0.1	21.3
Goldburger	W	309	0.6	0.1	100.8	Savagnin Blanc	W	288	0.6	0.1	12.5
Muscat Ottonel	W	418	0.9	0.1	3.4	Muscat Ottonel	W	344	0.8	0.0	2.7
Gewürztraminer	W	363	0.7	0.1	3.4	Bouvier	W	216	0.5	0.0	95.3
Roter Veltliner	G	258	0.5	0.1	100.8	Rösler	R	216	0.5	0.0	98.2
Rotgipfler	W	118	0.2	0.0	100.8	Roter Veltliner	G	198	0.4	0.0	98.7
Total of above		44861	92.5			Total of above		38921	85.7		

Brazil

2000						2016					
Prime variety	Col	Area	Nat'l share	NVII*	VII*	Prime variety	Col	Area	Nat'l share	NVII*	VII*
Isabella	R	14285	27.0	2.9	48.1	Isabella	R	11664	35.1	2.6	88.4
Niagara	W	13436	25.4	2.7	81.0	Cabernet Franc	R	6834	20.6	1.4	16.5
Cabernet Franc	R	3784	7.2	0.7	6.7	Couderc Noir	R	1938	5.8	0.4	122.5
Concord	R	2509	4.7	0.5	19.6	Concord	R	1687	5.1	0.4	21.6
Seibel	R	1967	3.7	0.4	91.4	Niagara	W	1430	4.3	0.3	59.1
Herbemont	R	1453	2.7	0.3	92.5	Jacquez	R	1274	3.8	0.3	119.2
Muscat of Alexandria	W	809	1.5	0.1	2.5	Moscato Embrapa	W	683	2.1	0.2	135.0
Couderc Noir	R	299	0.6	0.1	45.1	Muscat	W	671	2.0	0.1	121.8
Jacquez	R	170	0.3	0.0	69.6	Violeta	R	636	1.9	0.1	135.0
Tannat	R	182	0.3	0.0	3.0	Cora	R	570	1.7	0.1	135.0
Sémillon	W	384	0.7	0.0	1.4	Lorena	W	500	1.5	0.1	135.0
Gewürztraminer	W	140	0.3	0.0	1.2	Seibel	R	478	1.4	0.1	133.7
Graševina	W	880	1.7	0.0	0.9	Couderc 13	W	474	1.4	0.1	135.0
Sauvignon Blanc	W	140	0.3	-0.1	0.2	Niagara Red	R	469	1.4	0.1	135.0
Trebbiano Toscano	W	688	1.3	-0.2	0.5	Carmem	R	328	1.0	0.1	135.0
Chardonnay	W	330	0.6	-0.3	0.2	Concord Clone 30	R	196	0.6	0.0	135.0
Cabernet Sauvignon	R	587	1.1	-0.4	0.2	Rúbea	R	181	0.5	0.0	135.0
Merlot	R	469	0.9	-0.4	0.2	Herbemont	R	112	0.3	0.0	135.0
						Malvasia Bianca di Candia	W	165	0.5	0.0	2.3
						Moscato Giallo	W	90	0.3	0.0	7.4
Total of above		42512	80.5			Total of above		30380	91.5		

Table 58 (cont.): Varietal Intensity Indexes (and winegrape areas and national shares) for national top 20 varieties, 2000 and 2016

Bulgaria

2000						2016					
Prime variety	Col	Area	Nat'l share	NVII*	VII*	Prime variety	Col	Area	Nat'l share	NVII*	VII*
Pamid	R	22581	23.5	4.5	50.6	Merlot	R	10050	19.0	1.5	3.2
Rkatsiteli	W	9429	9.8	1.7	7.1	Pamid	R	6874	13.0	1.5	58.4
Dimyat	W	7649	8.0	1.5	50.3	Cabernet Sauvignon	R	9327	17.6	1.3	2.5
Merlot	R	11169	11.6	1.4	2.7	Rkatsiteli	W	5415	10.2	1.1	8.9
Cabernet Sauvignon	R	10441	10.9	1.2	2.4	Misket Cherven	G	4349	8.2	1.0	84.6
Shiroka Melnishka	R	3804	4.0	0.8	50.9	Muscat Ottonel	W	3679	6.9	0.8	25.0
Misket	W	3764	3.9	0.8	50.9	Dimyat	W	2998	5.7	0.6	26.2
Cardinal	R	3035	3.2	0.6	39.9	Shiroka Melnishka	R	1205	2.3	0.3	84.6
Muscat Ottonel	W	2914	3.0	0.5	12.1	Mavrud	R	1193	2.3	0.3	84.6
Graševina	W	3602	3.8	0.4	2.0	Kadarka	R	1161	2.2	0.3	60.5
Kadarka	R	1619	1.7	0.3	31.3	Chardonnay	W	3087	5.8	0.2	1.3
Aligoté	W	1659	1.7	0.2	2.4	Gewürztraminer	W	591	1.1	0.1	3.9
Gewürztraminer	W	971	1.0	0.2	4.6	Aligoté	W	285	0.5	0.0	0.9
Mavrud	R	647	0.7	0.1	50.9	Cabernet Franc	R	240	0.5	-0.1	0.4
Königin der Weingärten	W	567	0.6	0.1	38.4	Trebbiano Toscano	W	738	1.4	-0.2	0.5
Muscat of Hamburg	R	445	0.5	0.1	3.2	Sauvignon Blanc	W	637	1.2	-0.2	0.4
Riesling	W	647	0.7	0.0	0.8	Pinot Noir	R	342	0.6	-0.2	0.3
Pinot Noir	R	769	0.8	-0.1	0.6	Syrah	R	804	1.5	-0.3	0.4
Trebbiano Toscano	W	1821	1.9	-0.2	0.7						
Sauvignon Blanc	W	405	0.4	-0.2	0.3						
Total of above		87938	91.6			Total of above		52974	100.0		

Cambodia

2000						2016					
Prime variety	Col	Area	Nat'l share	NVII*	VII*	Prime variety	Col	Area	Nat'l share	NVII*	VII*
						Black Queen	R	3	30.0	0.0	9379.7
						Syrah	R	3	30.0	0.0	7.4
						Merlot	R	2	20.0	0.0	3.4
						Cabernet Sauvignon	R	2	20.0	0.0	2.9
Total of above						Total of above		10	100.0		

Table 58 (cont.): Varietal Intensity Indexes (and winegrape areas and national shares) for national top 20 varieties, 2000 and 2016

Canada

2000						2016					
Prime variety	Col	Area	Nat'l share	NVII*	VII*	Prime variety	Col	Area	Nat'l share	NVII*	VII*
Concord	R	977	11.5	0.2	47.6	Seyval Blanc	W	2259	17.9	0.5	297.8
Chardonnay	W	973	11.5	0.1	3.8	Riesling	W	1188	9.4	0.2	7.1
Vidal	W	514	6.0	0.1	483.9	Chardonnay	W	1417	11.2	0.2	2.5
Cabernet Franc	R	567	6.7	0.1	6.3	Baco Noir	R	704	5.6	0.2	340.8
Niagara	W	461	5.4	0.1	17.3	Cabernet Franc	R	820	6.5	0.1	5.2
Riesling	W	482	5.7	0.1	6.4	Pinot Gris	G	649	5.1	0.1	4.8
Pinot Noir	R	457	5.4	0.1	3.8	Gewürztraminer	W	398	3.2	0.1	11.1
Merlot	R	674	7.9	0.1	1.8	Pinot Noir	R	639	5.1	0.1	2.2
Baco Noir	R	279	3.3	0.1	404.1	Arinarnoa	R	289	2.3	0.1	211.6
Gewürztraminer	W	237	2.8	0.0	12.8	Gamay Noir	R	272	2.2	0.0	3.7
Gamay Noir	R	263	3.1	0.0	4.0	Concord	R	183	1.4	0.0	6.2
Cabernet Sauvignon	R	569	6.7	0.0	1.5	Millot-Foch	R	124	1.0	0.0	348.8
Pinot Gris	G	210	2.5	0.0	6.4	Maréchal Foch	R	94	0.7	0.0	146.2
Seyval Blanc	W	132	1.5	0.0	194.6	Niagara	W	87	0.7	0.0	9.5
De Chaunac	R	119	1.4	0.0	368.8	Pinot Blanc	W	109	0.9	0.0	2.8
Pinot Blanc	W	146	1.7	0.0	4.9	L'Acadie Blanc	W	65	0.5	0.0	355.8
Maréchal Foch	R	109	1.3	0.0	362.8	Frontenac	R	64	0.5	0.0	107.0
Fredonia	R	73	0.9	0.0	488.3	Viognier	W	101	0.8	0.0	2.2
Elvira	W	69	0.8	0.0	115.2	Vidal	W	59	0.5	0.0	10.9
Sauvignon Blanc	W	148	1.7	0.0	1.3	De Chaunac	R	53	0.4	0.0	184.6
Total of above		7458	87.8			Total of above		9574	76.0		

Chile

2000						2016					
Prime variety	Col	Area	Nat'l share	NVII*	VII*	Prime variety	Col	Area	Nat'l share	NVII*	VII*
Cabernet Sauvignon	R	35967	31.6	6.3	6.9	Cabernet Sauvignon	R	42409	29.1	7.2	4.2
Listan Prieto	R	15181	13.3	3.0	40.1	Sauvignon Blanc	W	14999	10.3	2.4	3.7
Merlot	R	12825	11.3	1.6	2.6	Carmenère	R	10503	7.2	2.2	14.4
Sauvignon Blanc	W	6662	5.8	1.1	4.4	Listan Prieto	R	9693	6.6	2.1	29.0
Carmenère	R	4719	4.1	0.9	35.4	Alicante Henri Bouschet	R	6908	4.7	1.3	5.9
Chardonnay	W	7672	6.7	0.9	2.3	Chardonnay	W	11435	7.8	1.1	1.7
Malvasia Fina	W	4305	3.8	0.8	26.0	Muscat of Alexandria	W	5424	3.7	1.0	4.8
Alicante Henri Bouschet	R	2882	2.5	0.4	3.3	Pedro Giménez	W	4379	3.0	0.9	8.6
Pedro Ximénez	W	2379	2.1	0.4	5.9	Merlot	R	12057	8.3	0.8	1.4
Sémillon	W	1893	1.7	0.3	3.1	Syrah	R	7994	5.5	0.5	1.4
Côt	R	929	0.8	0.1	1.5	Torrontés Sanjuanino	W	1771	1.2	0.4	14.9
Blanca Ovoide	W	107	0.1	0.0	42.9	Muscat Blanc à Petits Grains (C	W	1732	1.2	0.3	6.4
Chasselas	W	404	0.4	0.0	1.3	Pinot Noir	R	4091	2.8	0.1	1.2
Viognier	W	128	0.1	0.0	1.7	Petit Verdot	R	863	0.6	0.1	3.3
Petit Verdot	R	84	0.1	0.0	2.2	Côt	R	2293	1.6	0.1	1.3
Pinot Noir	R	1614	1.4	0.0	1.0	Sauvignonasse	W	658	0.5	0.1	5.2
Sauvignon Blanc (G)	W	7	0.0	0.0	4.0	Torrontés Riojano	W	643	0.4	0.1	2.2
Sauvignonasse	W	132	0.1	0.0	1.0	Viognier	W	839	0.6	0.1	1.6
Marsanne	W	1	0.0	0.0	0.0	Sémillon	W	849	0.6	0.1	1.4
Albillo Real	W	2	0.0	0.0	0.0	Lacrima Christi	R	226	0.2	0.0	30.7
Total of above		97893	85.9			Total of above		139767	95.8		

Table 58 (cont.): Varietal Intensity Indexes (and winegrape areas and national shares) for national top 20 varieties, 2000 and 2016

China

2000						2016					
Prime variety	Col	Area	Nat'l share	NVII*	VII*	Prime variety	Col	Area	Nat'l share	NVII*	VII*
						Cabernet Sauvignon	R	40300	22.6	6.2	3.3
						Carmenère	R	11200	6.3	2.3	12.5
						Merlot	R	16700	9.4	1.4	1.6
						Yan 73	R	4800	2.7	1.0	25.2
						Graševina	W	3000	1.7	0.5	3.1
						Muscat of Alexandria	W	3000	1.7	0.4	2.2
						Beibinghong	R	1600	0.9	0.3	25.2
						Vidal	W	1500	0.8	0.3	19.5
						Longyan	R	1000	0.6	0.2	25.2
						Riesling	W	1600	0.9	0.2	0.7
						Cabernet Franc	R	600	0.3	-0.4	0.3
						Mazuelo	R	100	0.1	-0.4	0.1
						Chardonnay	W	6100	3.4	-0.4	0.8
						Garnacha Tinta	R	4000	2.2	-0.4	0.7
						Sauvignon Blanc	W	2000	1.1	-0.7	0.4
						Trebbiano Toscano	W	1500	0.8	-0.7	0.3
						Pinot Noir	R	400	0.2	-0.8	0.1
						Syrah	R	1000	0.6	-1.4	0.1
Total of above		0	0			Total of above		#####	56.4		

Croatia

2000						2016					
Prime variety	Col	Area	Nat'l share	NVII*	VII*	Prime variety	Col	Area	Nat'l share	NVII*	VII*
Graševina	W	16051	27.0	3.1	14.3	Graševina	W	4459	38.0	1.0	69.8
Malvazija Istarska	W	7134	12.0	1.4	77.6	Plavac Mali	R	1664	14.2	0.4	370.5
Plavac Mali	R	6539	11.0	1.3	82.2	Malvazija Istarska	W	1600	13.6	0.4	219.0
Pošip Bijeli	W	6539	11.0	1.3	82.2	Plavina	R	683	5.8	0.2	381.7
Babić	R	1189	2.0	0.2	82.2	Blaufränkisch	R	521	4.4	0.1	11.6
Mondeuse Noire	R	1189	2.0	0.2	69.6	Riesling	W	625	5.3	0.1	4.0
Blaufränkisch	R	1189	2.0	0.2	7.0	Merlot	R	828	7.0	0.0	1.2
						Chardonnay	W	657	5.6	0.0	1.2
						Cabernet Sauvignon	R	709	6.0	0.0	0.9
Total of above		39830	67			Total of above		11746	100.0		

Table 58 (cont.): Varietal Intensity Indexes (and winegrape areas and national shares) for national top 20 varieties, 2000 and 2016

Cyprus

2000						2016					
Prime variety	Col	Area	Nat'l share	NVII*	VII*	Prime variety	Col	Area	Nat'l share	NVII*	VII*
Mavro	R	10969	60.0	2.2	267.4	Mavro	R	3187	62.1	0.7	873.4
Xynisteri	W	2742	15.0	0.6	267.4	Xynisteri	W	1946	37.9	0.4	873.4
Total of above		13711	75.0			Total of above		5133	100.0		

Czechia

2000						2016					
Prime variety	Col	Area	Nat'l share	NVII*	VII*	Prime variety	Col	Area	Nat'l share	NVII*	VII*
Grüner Veltliner	W	1700	15.0	0.3	31.1	Grüner Veltliner	W	1538	11.3	0.3	26.5
Müller-Thurgau	W	1586	14.0	0.3	20.4	Müller-Thurgau	W	1479	10.9	0.3	25.0
Graševina	W	1246	11.0	0.2	5.8	Sankt Laurent	R	1183	8.7	0.3	119.2
Sankt Laurent	R	1020	9.0	0.2	172.2	Blaufränkisch	R	1143	8.4	0.2	21.9
Riesling	W	793	7.0	0.1	7.9	Graševina	W	1114	8.2	0.2	15.1
Blaufränkisch	R	680	6.0	0.1	21.0	Riesling	W	1172	8.6	0.2	6.5
Zweigelt	R	453	4.0	0.1	26.9	Zweigelt	R	770	5.7	0.2	28.0
Neuburger	W	340	3.0	0.1	102.3	Pinot Blanc	W	762	5.6	0.2	18.2
Blauer Portugieser	R	340	3.0	0.1	16.0	Pinot Gris	G	826	6.1	0.2	5.6
Chardonnay	W	567	5.0	0.0	1.7	Blauer Portugieser	R	599	4.4	0.1	30.0
						Gewürztraminer	W	591	4.3	0.1	15.2
						Sauvignon Blanc	W	906	6.7	0.1	2.4
						Pinot Noir	R	697	5.1	0.1	2.2
						Chardonnay	W	820	6.0	0.0	1.3
Total of above		8725	77.0			Total of above		13600	100.0		

Table 58 (cont.): Varietal Intensity Indexes (and winegrape areas and national shares) for national top 20 varieties, 2000 and 2016

Ethiopia

2000						2016					
Prime variety	Col	Area	Nat'l share	NVII*	VII*	Prime variety	Col	Area	Nat'l share	NVII*	VII*
						Sangiovese	R	90	53.2	0.0	32.5
						Chenin Blanc	W	54	31.9	0.0	44.4
						Flame Seedless	R	13	7.8	0.0	6326.2
						Crimson Seedless	R	6	3.3	0.0	19471.4
						Sultaniye	W	4	2.5	0.0	21.3
						Red Globe	R	2	1.3	0.0	237.5
Total of above						Total of above		169	100.0		

France

2000						2016					
Prime variety	Col	Area	Nat'l share	NVII*	VII*	Prime variety	Col	Area	Nat'l share	NVII*	VII*
Mazuelo	R	95745	11.1	15.0	4.2	Merlot	R	108483	13.3	13.4	2.2
Trebbiano Toscano	W	90341	10.4	13.5	3.7	Trebbiano Toscano	W	78842	9.7	12.7	3.6
Merlot	R	101309	11.7	13.0	2.7	Garnacha Tinta	R	78631	9.6	11.5	2.9
Garnacha Tinta	R	95717	11.1	11.8	2.5	Syrah	R	62211	7.6	6.5	1.9
Syrah	R	50676	5.9	6.7	2.8	Mazuelo	R	31760	3.9	5.2	3.7
Gamay Noir	R	34537	4.0	5.7	5.2	Cabernet Franc	R	32327	4.0	4.9	3.2
Cabernet Franc	R	36094	4.2	5.5	3.9	Gamay Noir	R	24095	3.0	4.3	5.1
Cinsaut	R	31593	3.7	4.7	3.7	Pinot Noir	R	31602	3.9	2.8	1.6
Pinot Noir	R	26526	3.1	2.9	2.2	Cinsaut	R	15930	2.0	2.6	3.8
Cabernet Sauvignon	R	53413	6.2	2.9	1.4	Chardonnay	W	47451	5.8	2.4	1.3
Melon	W	13253	1.5	2.2	5.7	Pinot Meunier	R	12130	1.5	2.1	4.5
Chardonnay	W	36496	4.2	2.2	1.4	Melon	W	9550	1.2	1.7	5.5
Sauvignon Blanc	W	20933	2.4	1.9	1.8	Sémillon	W	10234	1.3	1.5	3.0
Sémillon	W	14015	1.6	1.9	3.0	Viognier	W	8823	1.1	1.3	3.0
Pinot Meunier	R	10621	1.2	1.7	4.6	Sauvignon Blanc	W	28084	3.4	1.2	1.2
Aramon Noir	R	9157	1.1	1.5	5.7	Garnacha Blanca	W	5130	0.6	0.8	3.8
Garnacha Blanca	W	6461	0.7	0.9	3.4	Chenin Blanc	W	9432	1.2	0.8	1.6
Mauzac Blanc	W	3310	0.4	0.6	5.7	Colombard	W	8441	1.0	0.7	1.5
Clairette	W	3274	0.4	0.5	4.2	Marselan	R	3662	0.4	0.7	5.1
Grolleau Noir	R	3006	0.3	0.5	5.7	Vermentino	W	4642	0.6	0.6	2.2
Total of above		736479	85.2			Total of above		611458	75.0		

Table 58 (cont.): Varietal Intensity Indexes (and winegrape areas and national shares) for national top 20 varieties, 2000 and 2016

Georgia										
2000						2016				
Prime variety	Col	Area	Nat'l share	NVII*	VII*	Prime variety	Col	Area	Nat'l share	NVII*	VII*
Rkatsiteli	W	19741	52.8	3.9	38.3	Rkatsiteli	W	25324	52.8	5.5	46.0
Tsolikouri	W	6161	16.5	1.3	130.6	Tsolikouri	W	7903	16.5	1.7	93.4
Saperavi	R	3704	9.9	0.7	72.1	Saperavi	R	4751	9.9	1.0	68.5
Tsitska	W	2839	7.6	0.6	130.6	Tsitska	W	3642	7.6	0.8	93.4
Chinuri	W	955	2.6	0.2	130.6	Chinuri	W	1225	2.6	0.3	93.4
Mtsvane Kakhuri	W	249	0.7	0.1	130.6	Mtsvane Kakhuri	W	319	0.7	0.1	93.4
Goruli Mtsvane	W	224	0.6	0.0	130.6	Goruli Mtsvane	W	287	0.6	0.1	93.4
Aleksandrouli	R	219	0.6	0.0	130.6	Aleksandrouli	R	281	0.6	0.1	93.4
Tsulukidzis Tetra	W	152	0.4	0.0	130.6	Tsulukidzis Tetra	W	195	0.4	0.0	93.4
Aladasturi	R	46	0.1	0.0	130.6	Pinot Blanc	W	219	0.5	0.0	1.5
Pinot Blanc	W	171	0.5	0.0	1.3	Aladasturi	R	59	0.1	0.0	93.4
Krakhuna	W	36	0.1	0.0	130.6	Krakhuna	W	46	0.1	0.0	93.4
Tavkveri	R	29	0.1	0.0	130.6	Tavkveri	R	37	0.1	0.0	93.4
Ojaleshi	R	25	0.1	0.0	130.6	Ojaleshi	R	32	0.1	0.0	93.4
Chkhaveri	G	20	0.1	0.0	130.6	Chkhaveri	G	26	0.1	0.0	93.4
Kisi	W	20	0.1	0.0	130.6	Kisi	W	26	0.1	0.0	93.4
Usakhelouri	R	8	0.0	0.0	130.6	Usakhelouri	R	10	0.0	0.0	93.4
Khikhvi	W	5	0.0	0.0	130.6	Khikhvi	W	6	0.0	0.0	93.4
Otskhanuri Sapere	R	5	0.0	0.0	130.6	Otskhanuri Sapere	R	6	0.0	0.0	93.4
Aligoté	W	97	0.3	0.0	0.4	Aligoté	W	124	0.3	0.0	0.4
Total of above		34706	92.7			Total of above		44521	92.8		

Germany											
2000						2016					
Prime variety	Col	Area	Nat'l share	NVII*	VII*	Prime variety	Col	Area	Nat'l share	NVII*	VII*
Riesling	W	22350	21.4	4.4	24.2	Riesling	W	21540	22.8	4.5	17.1
Müller-Thurgau	W	20706	19.9	4.1	28.9	Müller-Thurgau	W	11664	12.3	2.5	28.4
Pinot Noir	R	8643	8.3	1.5	5.9	Pinot Noir	R	11184	11.8	2.0	5.0
Kerner	W	6846	6.6	1.4	45.0	Dornfelder	R	7761	8.2	1.7	46.8
Silvaner	W	6859	6.6	1.4	29.1	Silvaner	W	4627	4.9	1.0	36.2
Blauer Portugieser	R	4878	4.7	1.0	25.0	Pinot Blanc	W	4323	4.6	0.9	14.9
Dornfelder	R	3765	3.6	0.8	46.9	Pinot Gris	G	4887	5.2	0.9	4.8
Bacchus	W	3262	3.1	0.7	45.3	Blauer Portugieser	R	3177	3.4	0.7	22.9
Scheurebe	W	3126	3.0	0.6	40.1	Kerner	W	2646	2.8	0.6	43.4
Schiava Grossa	R	2530	2.4	0.5	31.3	Schiava Grossa	R	2197	2.3	0.5	46.2
Pinot Gris	G	2637	2.5	0.5	6.5	Regent	R	1902	2.0	0.4	45.7
Pinot Blanc	W	2396	2.3	0.4	6.6	Pinot Meunier	R	2002	2.1	0.4	6.5
Pinot Meunier	R	2289	2.2	0.4	8.2	Bacchus	W	1610	1.7	0.4	43.4
Faberrebe	W	1586	1.5	0.3	46.9	Blaufränkisch	R	1737	1.8	0.3	4.8
Huxelrebe	W	1289	1.2	0.3	46.9	Scheurebe	W	1266	1.3	0.3	36.9
Morio-Muskat	W	1167	1.1	0.2	46.1	Chasselas	W	1046	1.1	0.2	6.7
Ortega	W	1054	1.0	0.2	46.9	Sankt Laurent	R	633	0.7	0.1	9.2
Elbling	W	1043	1.0	0.2	40.5	Gewürztraminer	W	824	0.9	0.1	3.0
Chasselas	W	1198	1.1	0.2	4.2	Ortega	W	482	0.5	0.1	43.0
Savagnin Rose	G	848	0.8	0.2	45.0	Elbling	W	489	0.5	0.1	23.9
Total of above		98472	94.5			Total of above		85997	91.0		

Table 58 (cont.): Varietal Intensity Indexes (and winegrape areas and national shares) for national top 20 varieties, 2000 and 2016

Greece											
2000						2016					
Prime variety	Col	Area	Nat'l share	NVII*	VII*	Prime variety	Col	Area	Nat'l share	NVII*	VII*
Savatiano	W	12747	25.0	2.6	96.0	Savatiano	W	10268	20.2	2.3	88.2
Roditis (R)	G	6945	13.6	1.4	96.0	Roditis	G	8463	16.6	1.9	88.2
Liatiko	R	2476	4.9	0.5	96.0	Agiorgitiko	R	3270	6.4	0.7	88.1
Agiorgitiko	R	2320	4.6	0.5	96.0	Liatiko	R	2633	5.2	0.6	88.2
Moschomavro	R	2295	4.5	0.5	96.0	Muscat of Hamburg	R	2288	4.5	0.5	26.3
Muscat Blanc à Petits Grains	W	2232	4.4	0.4	7.1	Xinomavro	R	2135	4.2	0.5	88.2
Xinomavro	R	1816	3.6	0.4	96.0	Assyrtiko	W	1770	3.5	0.4	88.2
Athiri	W	1350	2.7	0.3	96.0	Mavrouda	R	1658	3.3	0.4	88.2
Kotsifali	R	1148	2.3	0.2	96.0	Kotsifali	R	1338	2.6	0.3	88.2
Assyrtiko	W	1106	2.2	0.2	96.0	Muscat Blanc à Petits Grains	W	1568	3.1	0.3	4.1
Mandilaria	R	845	1.7	0.2	96.0	Romeiko	R	1131	2.2	0.2	88.2
Moschofilero	G	718	1.4	0.1	96.0	Moschofilero	G	1088	2.1	0.2	88.2
Mavrodafni	R	537	1.1	0.1	96.0	Mandilaria	R	932	1.8	0.2	88.2
Vilana	W	506	1.0	0.1	96.0	Roditis (R)	G	828	1.6	0.2	88.2
Vertzami	R	491	1.0	0.1	96.0	Vilana	W	650	1.3	0.1	88.2
Debina	W	455	0.9	0.1	96.0	Athiri	W	577	1.1	0.1	88.2
Asprouda	W	433	0.9	0.1	96.0	Muscat of Alexandria	W	773	1.5	0.1	2.0
Monemvassia	W	418	0.8	0.1	96.0	Mavrodafni	R	324	0.6	0.1	88.2
Romeiko	R	382	0.8	0.1	96.0	Fokiano	R	212	0.4	0.0	88.2
Robola	W	359	0.7	0.1	96.0	Limnio	R	176	0.3	0.0	88.2
Total of above		39579	77.7			Total of above		42083	82.8		

Hungary											
2000						2016					
Prime variety	Col	Area	Nat'l share	NVII*	VII*	Prime variety	Col	Area	Nat'l share	NVII*	VII*
Blaufränkisch	R	6920	8.0	1.4	27.8	Blaufränkisch	R	7260	11.4	1.6	29.7
Graševina	W	6677	7.7	1.0	4.1	Bianca	W	4898	7.7	1.1	35.2
Furmint	W	3480	4.0	0.7	56.2	Cserszegi Fűszeres	G	4299	6.7	0.9	70.2
Ezerjó	W	3157	3.6	0.6	56.3	Furmint	W	3862	6.0	0.8	61.1
Arany Sárfehér	W	2914	3.4	0.6	56.3	Graševina	W	3933	6.2	0.8	11.3
Müller-Thurgau	W	3278	3.8	0.5	5.5	Aletta	W	1676	2.6	0.4	70.2
Zalagyöngye	W	2550	2.9	0.5	33.1	Hárslevelű	W	1603	2.5	0.4	69.5
Cserszegi Fűszeres	G	2185	2.5	0.4	56.3	Zweigelt	R	1687	2.6	0.3	13.1
Zweigelt	R	2266	2.6	0.4	17.5	Irsai Olivér	W	1531	2.4	0.3	60.0
Chasselas	W	1902	2.2	0.3	8.0	Müller-Thurgau	W	1670	2.6	0.3	6.0
Silvaner	W	1619	1.9	0.3	8.2	Grüner Veltliner	W	1381	2.2	0.2	5.1
Kunleány	W	1376	1.6	0.3	56.3	Muscat Ottonel	W	1256	2.0	0.2	7.1
Muscat Ottonel	W	1416	1.6	0.2	6.5	Chasselas	W	1159	1.8	0.2	11.0
Kövidinka	G	1214	1.4	0.2	56.3	Zalagyöngye	W	1065	1.7	0.2	59.4
Blauer Portugieser	R	1255	1.4	0.2	7.7	Kunleány	W	974	1.5	0.2	70.2
Muscat Blanc à Petits Grains	W	1538	1.8	0.2	2.9	Blauer Portugieser	R	1023	1.6	0.2	10.9
Hárslevelű	W	1012	1.2	0.2	43.9	Pinot Gris	G	1594	2.5	0.2	2.3
Kadarka	R	1012	1.2	0.2	21.6	Királyleányka	W	784	1.2	0.2	70.2
Grüner Veltliner	W	1335	1.5	0.2	3.2	Leányka	W	719	1.1	0.2	70.2
Riesling	W	1619	1.9	0.2	2.1	Kövidinka	G	658	1.0	0.1	70.1
Total of above		48724	56.1			Total of above		43031	67.4		

Table 58 (cont.): Varietal Intensity Indexes (and winegrape areas and national shares) for national top 20 varieties, 2000 and 2016

India

2000						2016					
Prime variety	Col	Area	Nat'l share	NVII*	VII*	Prime variety	Col	Area	Nat'l share	NVII*	VII*
						Sultaniye	W	1000	37.0	0.2	311.8
						Sauvignon Blanc	W	500	18.5	0.1	6.7
						Syrah	R	500	18.5	0.1	4.6
						Trebbiano Toscano	W	300	11.1	0.1	4.1
						Muscat of Alexandria	W	100	3.7	0.0	4.8
						Pinot Noir	R	100	3.7	0.0	1.6
						Chardonnay	W	100	3.7	0.0	0.8
						Cabernet Sauvignon	R	100	3.7	0.0	0.5
Total of above						Total of above		2700	100.0		

Israel

2000						2016					
Prime variety	Col	Area	Nat'l share	NVII*	VII*	Prime variety	Col	Area	Nat'l share	NVII*	VII*
Mazuelo	R	971	20.0	0.2	7.7	Mazuelo	R	935	18.7	0.2	17.7
Colombard	W	486	10.0	0.1	12.7	Cabernet Sauvignon	R	990	19.8	0.1	2.9
Merlot	R	647	13.3	0.1	3.1	Merlot	R	715	14.3	0.1	2.4
Cabernet Sauvignon	R	607	12.5	0.1	2.7	Argaman	R	275	5.5	0.1	896.6
Emerald Riesling	W	344	7.1	0.1	369.9	Petit Verdot	R	275	5.5	0.1	30.4
Argaman	R	202	4.2	0.0	1007.5	Colombard	W	220	4.4	0.0	6.6
Sauvignon Blanc	W	263	5.4	0.0	4.1	Syrah	R	385	7.7	0.0	1.9
Muscat of Alexandria	W	202	4.2	0.0	6.9	Muscat of Alexandria	W	220	4.4	0.0	5.7
Chenin Blanc	W	101	2.1	0.0	2.2	Emerald Riesling	W	110	2.2	0.0	555.7
Chardonnay	W	142	2.9	0.0	1.0	Durif	R	110	2.2	0.0	20.5
						Côt	R	110	2.2	0.0	1.9
						Cabernet Franc	R	110	2.2	0.0	1.8
						Monastrell	R	55	1.1	0.0	0.9
						Sauvignon Blanc	W	110	2.2	0.0	0.8
						Chardonnay	W	165	3.3	0.0	0.7
						Tempranillo	R	55	1.1	0.0	0.2
Total of above		3966	81.7			Total of above		4840	96.8		

Table 58 (cont.): Varietal Intensity Indexes (and winegrape areas and national shares) for national top 20 varieties, 2000 and 2016

Italy

2000						2016					
Prime variety	Col	Area	Nat'l share	NVII*	VII*	Prime variety	Col	Area	Nat'l share	NVII*	VII*
Sangiovese	R	62761	9.9	11.0	7.0	Sangiovese	R	68428	11.3	13.1	6.9
Catarratto Bianco	W	50711	8.0	9.0	7.7	Montepulciano	R	32724	5.4	6.3	7.4
Montepulciano	R	28679	4.5	5.1	7.7	Catarratto Bianco	W	28563	4.7	5.5	7.4
Barbera	R	27175	4.3	4.7	6.3	Trebbiano Toscano	W	35441	5.9	4.3	2.2
Trebbiano Toscano	W	39447	6.2	4.4	2.2	Prosecco	W	19730	3.3	3.8	7.3
Trebbiano Romagnolo	W	19492	3.1	3.5	7.7	Trebbiano Romagnolo	W	19059	3.2	3.7	7.4
Negroamaro	R	16619	2.6	3.0	7.7	Barbera	R	15006	2.5	2.8	6.2
Garganega	W	16549	2.6	2.9	7.7	Pinot Gris	G	18821	3.1	2.7	2.9
Malvasia Bianca di Candia	W	11921	1.9	2.1	7.1	Nero d'Avola	R	14129	2.3	2.7	7.3
Nero d'Avola	R	11318	1.8	2.0	7.7	Negroamaro	R	11431	1.9	2.2	7.4
Muscat Blanc à Petits Grains	W	13016	2.0	1.9	3.3	Tribidrag	R	13896	2.3	2.1	3.1
Inzolia	W	9259	1.5	1.6	7.7	Muscat Blanc à Petits Grains	W	13334	2.2	2.0	2.9
Aglianico	R	9264	1.5	1.6	7.6	Aglianico	R	9627	1.6	1.9	7.3
Trebbiano d'Abruzzo	W	8435	1.3	1.5	7.7	Malvasia Bianca di Candia	W	9028	1.5	1.7	6.9
Manzoni Bianco	W	8290	1.3	1.5	7.7	Garganega	W	8522	1.4	1.6	7.4
Prosecco	W	7498	1.2	1.3	7.7	Nebbiolo	R	7551	1.2	1.4	7.0
Dolcetto	R	7156	1.1	1.3	7.6	Grillo	W	7382	1.2	1.4	7.4
Pignoletto	W	6009	0.9	1.1	7.7	Lambrusco Salamino	R	6228	1.0	1.2	7.4
Verdicchio Bianco	W	5043	0.8	0.9	7.7	Corvina Veronese	R	6222	1.0	1.2	7.4
Tribidrag	R	7828	1.2	0.9	2.2	Vermentino	W	6703	1.1	1.1	4.3
Total of above		366470	57.6			Total of above		351824	58.2		

Japan

2000						2016					
Prime variety	Col	Area	Nat'l share	NVII*	VII*	Prime variety	Col	Area	Nat'l share	NVII*	VII*
						Koshu	G	690	17.8	0.2	1158.8
						Niagara	W	551	14.2	0.1	195.7
						Muscat Bailey A	R	521	13.5	0.1	331.7
						Concord	R	292	7.6	0.1	32.1
						Delaware	G	254	6.6	0.1	700.6
						Campbell Early	R	238	6.1	0.1	1158.8
						Kerner	W	76	2.0	0.0	30.5
						Kyoho (4N)	R	62	1.6	0.0	26.0
						Zweigelt	R	59	1.5	0.0	7.5
						Verdelet	W	39	1.0	0.0	1141.9
						Portland	W	39	1.0	0.0	1158.8
						Yamabudo	R	35	0.9	0.0	1158.8
						Black Queen	R	28	0.7	0.0	229.4
						Ryugan	W	27	0.7	0.0	1158.8
						Adirondac	R	24	0.6	0.0	1158.8
						Yama Sauvignon	R	24	0.6	0.0	1158.8
						Cascade	R	22	0.6	0.0	1158.8
						Yamasachi	R	20	0.5	0.0	1158.8
						Müller-Thurgau	W	22	0.6	0.0	1.3
						Riesling Forte	W	2	0.1	0.0	1158.8
Total of above						Total of above		3027	78.2		

Table 58 (cont.): Varietal Intensity Indexes (and winegrape areas and national shares) for national top 20 varieties, 2000 and 2016

Kazakhstan

2000						2016					
Prime variety	Col	Area	Nat'l share	NVII*	VII*	Prime variety	Col	Area	Nat'l share	NVII*	VII*
						Rkatsiteli	W	3552	51.2	0.8	44.7
						Bayanshira	W	645	9.3	0.1	646.2
						Saperavi	R	428	6.2	0.1	42.7
						Kuldzhinskii	R	385	5.6	0.1	646.2
						Muscat Blanc à Petits Grains (R	W	255	3.7	0.1	114.4
						Aligoté	W	277	4.0	0.1	6.6
						Maiskii Chernyi	R	110	1.6	0.0	646.2
						Madrasa	R	28	0.4	0.0	646.2
						Riesling	W	111	1.6	0.0	1.2
						Pinot Noir	R	180	2.6	0.0	1.1
						Ruby	R	9	0.1	0.0	646.2
						Rubinovy Magaracha	R	0	0.0	0.0	646.2
						Cabernet Franc	R	56	0.8	0.0	0.7
						Cabernet Sauvignon	R	20	0.3	-0.1	0.0
Total of above						Total of above		6056	87.3		

Korea, Rep.

2000						2016					
Prime variety	Col	Area	Nat'l share	NVII*	VII*	Prime variety	Col	Area	Nat'l share	NVII*	VII*
Kyoho (4N)	R	2700	50.0	0.6	610.5	Kyoho (4N)	R	2700	50.0	0.6	811.6
Muscat Bailey A	R	1300	24.1	0.3	857.8	Muscat Bailey A	R	1300	24.1	0.3	592.6
Sheridan	R	500	9.3	0.1	905.1	Sheridan	R	500	9.3	0.1	830.2
Delaware	G	100	1.9	0.0	387.1	Delaware	G	100	1.9	0.0	197.4
Total of above		4600	85.2			Total of above		4600	85.2		

Table 58 (cont.): Varietal Intensity Indexes (and winegrape areas and national shares) for national top 20 varieties, 2000 and 2016

Lebanon

2000						2016					
Prime variety	Col	Area	Nat'l share	NVII*	VII*	Prime variety	Col	Area	Nat'l share	NVII*	VII*
						Chardonnay	W	1000	25.0	0.2	5.6
						Cabernet Sauvignon	R	1000	25.0	0.2	3.6
						Sauvignon Blanc	W	500	12.5	0.1	4.5
						Merlot	R	500	12.5	0.1	2.1
						Syrah	R	300	7.5	0.0	1.9
Total of above						Total of above		3300	82.5		

Luxembourg

2000						2016					
Prime variety	Col	Area	Nat'l share	NVII*	VII*	Prime variety	Col	Area	Nat'l share	NVII*	VII*
Müller-Thurgau	W	459	34.1	0.1	49.6	Müller-Thurgau	W	316	24.3	0.1	55.9
Auxerrois	W	169	12.5	0.0	266.2	Auxerrois	W	190	14.6	0.0	229.7
Elbling	W	164	12.2	0.0	492.3	Pinot Gris	G	196	15.1	0.0	13.9
Riesling	W	175	13.0	0.0	14.6	Pinot Blanc	W	160	12.3	0.0	40.0
Pinot Gris	G	155	11.5	0.0	29.7	Riesling	W	162	12.5	0.0	9.3
Pinot Blanc	W	138	10.2	0.0	29.5	Pinot Noir	R	121	9.3	0.0	4.0
Pinot Noir	R	66	4.9	0.0	3.5	Elbling	W	86	6.6	0.0	305.2
Gewürztraminer	W	12	0.9	0.0	4.1	Gewürztraminer	W	21	1.6	0.0	5.6
Silvaner	W	1	0.1	0.0	0.3	Sankt Laurent	R	4	0.3	0.0	4.2
Gamay Noir	R	1	0.1	0.0	0.1	Muscat Blanc à Petits Grains	W	1	0.1	0.0	0.1
Chardonnay	W	8	0.6	0.0	0.2	Chardonnay	W	30	2.3	0.0	0.5
Total of above		1348	100.0			Total of above		1287	99.0		

Table 58 (cont.): Varietal Intensity Indexes (and winegrape areas and national shares) for national top 20 varieties, 2000 and 2016

Mexico

2000						2016					
Prime variety	Col	Area	Nat'l share	NVII*	VII*	Prime variety	Col	Area	Nat'l share	NVII*	VII*
						Sultaniye	W	841	15.4	0.2	129.6
						Mazuelo	R	448	8.2	0.1	7.8
						Cabernet Sauvignon	R	756	13.8	0.1	2.0
						Salvador	R	350	6.4	0.1	818.6
						Chenin Blanc	W	275	5.0	0.1	7.0
						Fiesta	W	230	4.2	0.1	820.3
						Muscat Blanc à Petits Grains	W	246	4.5	0.0	6.0
						Nebbiolo	R	180	3.3	0.0	18.5
						Cardinal	R	168	3.1	0.0	83.0
						Durif	R	133	2.4	0.0	22.7
						Palomino Fino	W	109	2.0	0.0	3.9
						Jacquez	R	80	1.5	0.0	45.5
						Merlot	R	391	7.2	0.0	1.2
						Sauvignon Blanc	W	120	2.2	0.0	0.8
						Tempranillo	R	229	4.2	0.0	0.9
						Garnacha Tinta	R	140	2.6	0.0	0.8
						Syrah	R	145	2.7	0.0	0.7
Total of above						Total of above		4841	88.6		

Moldova

2000						2016					
Prime variety	Col	Area	Nat'l share	NVII*	VII*	Prime variety	Col	Area	Nat'l share	NVII*	VII*
Aligoté	W	15790	17.6	3.1	24.1	Moldova	R	12375	15.0	2.7	54.3
Isabella	R	11401	12.7	2.2	22.6	Aligoté	W	7765	9.4	1.6	15.7
Rkatsiteli	W	11508	12.8	2.1	9.3	Sauvignon Blanc	W	6909	8.4	1.0	3.0
Sauvignon Blanc	W	8151	9.1	1.4	6.8	Isabella	R	3468	4.2	0.7	10.6
Pinot Noir	R	6521	7.3	1.1	5.2	Rkatsiteli	W	3898	4.7	0.7	4.1
Merlot	R	8123	9.0	0.9	2.1	Merlot	R	7689	9.3	0.6	1.6
Fetească Albă	W	4334	4.8	0.8	9.9	Cabernet Sauvignon	R	8169	9.9	0.5	1.4
Cabernet Sauvignon	R	7590	8.4	0.7	1.9	Muscat Ottonel	W	1859	2.3	0.4	8.1
Gewürztraminer	W	2731	3.0	0.5	13.9	Bianca	W	1340	1.6	0.3	7.4
Chardonnay	W	5134	5.7	0.5	1.9	Kodryanka	R	1143	1.4	0.3	54.3
Pinot Gris	G	2042	2.3	0.3	5.9	Magaracha Rannii	R	884	1.1	0.2	54.3
Muscat Ottonel	W	1520	1.7	0.3	6.7	Gewürztraminer	W	1099	1.3	0.2	4.7
Bastardo Magarachsky	R	1040	1.2	0.2	28.7	Alb de Suruceni	W	780	0.9	0.2	54.3
Saperavi	R	716	0.8	0.1	5.8	Fetească Albă	W	954	1.2	0.2	3.9
Sukholimansky Bely	W	599	0.7	0.1	20.0	Muscat Yantarnyi	W	683	0.8	0.1	54.3
Riesling	W	1343	1.5	0.1	1.7	Riesling	W	1701	2.1	0.1	1.5
Noah	W	71	0.1	0.0	54.4	Victoria	W	565	0.7	0.1	49.4
Pervomaisky	R	64	0.1	0.0	54.4	Pervenets Magaracha	W	517	0.6	0.1	10.2
Gamay Teinturier Freaux	R	55	0.1	0.0	22.7	Saperavi	R	573	0.7	0.1	4.8
Golubok	R	50	0.1	0.0	54.4	Cardinal	R	473	0.6	0.1	15.5
Total of above		88783	98.8			Total of above		62844	76.1		

Table 58 (cont.): Varietal Intensity Indexes (and winegrape areas and national shares) for national top 20 varieties, 2000 and 2016

Morocco

2000						2016					
Prime variety	Col	Area	Nat'l share	NVII*	VII*	Prime variety	Col	Area	Nat'l share	NVII*	VII*
Doukkali	R	16557	33.4	3.4	98.5	Italia	W	3333	19.0	0.7	163.8
Teneron	R	3488	7.0	0.7	98.5	Cinsaut	R	3239	18.4	0.7	36.0
Cinsaut	R	3940	7.9	0.7	8.0	Muscat of Alexandria	W	2093	11.9	0.4	15.3
Muscat of Alexandria	W	3669	7.4	0.7	12.2	Mazuelo	R	1230	7.0	0.2	6.6
Abbo	R	2375	4.8	0.5	98.5	Alicante Henri Bouschet	R	919	5.2	0.2	6.5
Alicante Henri Bouschet	R	1098	2.2	0.1	2.9	Sultaniye	W	721	4.1	0.2	34.5
Mazuelo	R	1692	3.4	0.1	1.3	Cardinal	R	624	3.6	0.1	95.9
Garnacha Tinta	R	802	1.6	-0.3	0.4	Garnacha Tinta	R	786	4.5	0.0	1.3
						Clairette	W	113	0.6	0.0	11.9
						Chardonnay	W	880	5.0	0.0	1.1
						Sauvignon Blanc	W	440	2.5	0.0	0.9
						Airén	W	440	2.5	-0.1	0.5
						Syrah	R	347	2.0	-0.1	0.5
						Cabernet Sauvignon	R	604	3.4	-0.1	0.5
						Merlot	R	333	1.9	-0.2	0.3
Total of above		33621	67.8			Total of above		16104	91.6		

Myanmar

2000						2016					
Prime variety	Col	Area	Nat'l share	NVII*	VII*	Prime variety	Col	Area	Nat'l share	NVII*	VII*
						Syrah	R	27	38.6	0.0	9.5
						Sauvignon Blanc	W	22	31.4	0.0	11.3
						Muscat Blanc à Petits Grains	W	7	10.0	0.0	13.3
						Pinot Noir	R	7	10.0	0.0	4.3
						Petit Verdot	R	2	2.9	0.0	15.8
						Tempranillo	R	4	5.0	0.0	1.0
						Chardonnay	W	2	2.1	0.0	0.5
Total of above						Total of above		70	100.0		

Table 58 (cont.): Varietal Intensity Indexes (and winegrape areas and national shares) for national top 20 varieties, 2000 and 2016

New Zealand

2000						2016					
Prime variety	Col	Area	Nat'l share	NVII*	VII*	Prime variety	Col	Area	Nat'l share	NVII*	VII*
Chardonnay	W	2787	28.0	0.5	9.4	Sauvignon Blanc	W	20497	57.8	4.4	20.8
Sauvignon Blanc	W	2423	24.4	0.5	18.3	Pinot Noir	R	5514	15.5	1.0	6.6
Pinot Noir	R	1098	11.0	0.2	7.8	Pinot Gris	G	2422	6.8	0.5	6.3
Riesling	W	490	4.9	0.1	5.6	Chardonnay	W	3117	8.8	0.3	2.0
Müller-Thurgau	W	419	4.2	0.1	6.1	Riesling	W	767	2.2	0.1	1.6
Merlot	R	657	6.6	0.0	1.5	Gewürztraminer	W	277	0.8	0.0	2.7
Cabernet Sauvignon	R	654	6.6	0.0	1.4	Sauvignon Blanc (G)	W	104	0.3	0.0	12.2
Sémillon	W	229	2.3	0.0	4.3	Muscat	W	36	0.1	0.0	6.1
Gewürztraminer	W	141	1.4	0.0	6.5	Arneis	W	33	0.1	0.0	3.6
Pinot Gris	G	127	1.3	0.0	3.3	Reichensteiner	W	14	0.0	0.0	15.0
Reichensteiner	W	62	0.6	0.0	95.5	Kolor	R	7	0.0	0.0	126.4
Pinotage	R	73	0.7	0.0	5.5	Flora	G	3	0.0	0.0	27.7
Chenin Blanc	W	146	1.5	0.0	1.6	Viognier	W	129	0.4	0.0	1.0
Breidecker	W	28	0.3	0.0	491.6	Osteiner	W	1	0.0	0.0	126.4
Côt	R	67	0.7	0.0	1.3	Ehrenfelser	W	1	0.0	0.0	1.8
Cabernet Franc	R	118	1.2	0.0	1.1	Breidecker	W	0	0.0	0.0	126.4
Blauburger	R	13	0.1	0.0	6.4	Lagrein	R	2	0.0	0.0	1.1
Flora	G	6	0.1	0.0	491.6	Würzer	W	0	0.0	0.0	1.0
Chasselas	W	25	0.3	0.0	0.9	Seibel	R	3	0.0	0.0	0.8
Silvaner	W	4	0.0	0.0	0.2	Chambourcin	R	3	0.0	0.0	0.4
Total of above		9567	96.2			Total of above		32931	92.9		

North Macedonia

2000						2016					
Prime variety	Col	Area	Nat'l share	NVII*	VII*	Prime variety	Col	Area	Nat'l share	NVII*	VII*
						Vranac	R	9500	38.3	2.1	180.9
						Dimyat	W	6500	26.2	1.4	121.3
						Tribidrag	R	1000	4.0	0.2	5.4
						Riesling	W	900	3.6	0.1	2.7
						Prokupac	R	445	1.8	0.1	59.2
						Stanušina Crna	R	400	1.6	0.1	180.9
						Muscat of Hamburg	R	350	1.4	0.1	8.2
						Župljanka	W	250	1.0	0.1	89.5
						Muscat Blanc à Petits Grains	W	400	1.6	0.0	2.1
						Pamid	R	250	1.0	0.0	4.5
						Žilavka	W	185	0.7	0.0	180.9
						Rkatsiteli	W	460	1.9	0.0	1.6
						Graševina	W	270	1.1	0.0	2.0
						Plavac Mali	R	50	0.2	0.0	5.3
						Pinot Noir	R	500	2.0	0.0	0.9
						Merlot	R	1240	5.0	-0.1	0.8
						Chardonnay	W	750	3.0	-0.1	0.7
						Sauvignon Blanc	W	185	0.7	-0.1	0.3
						Cabernet Sauvignon	R	1020	4.1	-0.2	0.6
Total of above						Total of above		24655	99.5		

Table 58 (cont.): Varietal Intensity Indexes (and winegrape areas and national shares) for national top 20 varieties, 2000 and 2016

Norway

2000						2016					
Prime variety	Col	Area	Nat'l share	NVII*	VII*	Prime variety	Col	Area	Nat'l share	NVII*	VII*
						Solaris	W	8	60.0	0.0	22727.0
						Rondo	R	4	30.0	0.0	26610.8
Total of above						Total of above		11	90.0		

Peru

2000						2016					
Prime variety	Col	Area	Nat'l share	NVII*	VII*	Prime variety	Col	Area	Nat'l share	NVII*	VII*
						Negramoll	R	1252	32.7	0.3	486.6
						Italia	W	1011	26.4	0.2	228.1
						Muscat Blanc à Petits Grains	W	361	9.4	0.1	12.5
						Quebranta	R	330	8.6	0.1	1170.2
						Blaufränkisch	R	290	7.6	0.1	19.8
						Red Globe	R	240	6.3	0.1	1159.7
						Pecorino	W	114	3.0	0.0	76.6
						Cardinal	R	53	1.4	0.0	37.4
						Flame Seedless	R	42	1.1	0.0	890.9
						Alphonse Lavallée	R	18	0.5	0.0	33.2
						Sultaniye	W	15	0.4	0.0	3.3
						Superior Seedless	W	9	0.2	0.0	1170.2
						Albillo Mayor	W	7	0.2	0.0	7.1
						Ar99	G	5	0.1	0.0	1170.2
						Crimson Seedless	R	2	0.1	0.0	310.4
						Crystal	W	2	0.1	0.0	13.4
						Ar110	G	1	0.0	0.0	1170.2
						Imperial Seedless	G	1	0.0	0.0	1170.2
						Rosa Arica	R	1	0.0	0.0	1170.2
						Perlette	W	1	0.0	0.0	522.5
Total of above						Total of above		3755	98.0		

Table 58 (cont.): Varietal Intensity Indexes (and winegrape areas and national shares) for national top 20 varieties, 2000 and 2016

Portugal

2000						2016					
Prime variety	Col	Area	Nat'l share	NVII*	VII*	Prime variety	Col	Area	Nat'l share	NVII*	VII*
Castelão	R	14424	7.0	2.8	23.8	Touriga Franca	R	14217	7.8	3.0	24.5
Fernão Pires	W	14206	6.9	2.8	23.3	Castelão	R	12580	6.9	2.7	24.5
Trincadeira	R	7264	3.5	1.4	23.8	Fernão Pires	W	12138	6.6	2.6	24.4
Baga	R	6730	3.3	1.3	23.8	Touriga Nacional	R	11411	6.2	2.4	23.9
Touriga Franca	R	6671	3.3	1.3	23.8	Trincadeira	R	10493	5.7	2.2	24.5
Tinta Barroca	R	5657	2.8	1.1	22.3	Tempranillo	R	17014	9.3	1.8	1.9
Vinhao	R	5296	2.6	1.0	21.3	Baga	R	6750	3.7	1.4	24.5
Touriga Nacional	R	4149	2.0	0.8	23.2	Síria	W	6438	3.5	1.4	22.5
Codega de Larinho	W	4058	2.0	0.8	23.8	Arinto de Bucelas	W	5409	3.0	1.2	24.5
Arinto de Bucelas	W	3966	1.9	0.8	23.8	Tinta Barroca	R	4733	2.6	1.0	23.6
Loureiro	W	3939	1.9	0.8	21.4	Loureiro	W	4402	2.4	0.9	23.0
Tempranillo	R	7356	3.6	0.7	1.9	Vinhao	R	4055	2.2	0.9	22.3
Marufo	R	3512	1.7	0.7	13.2	Marufo	R	3367	1.8	0.7	17.6
Azal	W	3302	1.6	0.6	23.8	Alicante Henri Bouschet	R	4547	2.5	0.7	3.1
Rabo de Ovelha	W	2330	1.1	0.5	23.8	Malvasia Fina	W	2922	1.6	0.6	21.9
Rufete	R	2338	1.1	0.4	16.4	Alvarelhão	R	2860	1.6	0.6	24.1
Vital	W	2246	1.1	0.4	23.8	Mencía	R	2561	1.4	0.5	5.7
Malvasia Preta	R	2210	1.1	0.4	23.8	Caladoc	R	2180	1.2	0.4	10.2
Malvasia Fina	W	2328	1.1	0.4	7.8	Rabigato	W	1969	1.1	0.4	24.5
Borraçal	R	2035	1.0	0.4	18.3	Malvasia Preta	R	1933	1.1	0.4	24.5
Total of above		#####	50.7			Total of above		131979	72.3		

Romania

2000						2016					
Prime variety	Col	Area	Nat'l share	NVII*	VII*	Prime variety	Col	Area	Nat'l share	NVII*	VII*
Fetească Albă	W	18211	8.2	3.5	16.8	Fetească Regală	W	12619	6.9	2.7	23.8
Graševina	W	15014	6.8	2.2	3.6	Fetească Albă	W	12428	6.8	2.7	22.8
Aligoté	W	7608	3.4	1.2	4.7	Aligoté	W	5840	3.2	1.1	5.3
Muscat Ottonel	W	5787	2.6	1.1	10.4	Muscat Ottonel	W	4779	2.6	1.0	9.4
Băbească Neagră	R	3642	1.6	0.7	21.5	Riesling	W	6121	3.3	0.8	2.5
Sauvignon Blanc	W	4613	2.1	0.3	1.6	Fetească Neagră	R	2845	1.6	0.6	21.5
Fetească Regală	W	1700	0.8	0.3	14.5	Băbească Neagră	R	2696	1.5	0.6	24.5
Pinot Gris	G	2388	1.1	0.3	2.8	Pamid	R	2716	1.5	0.5	6.7
Fetească Neagră	R	1214	0.5	0.2	22.0	Merlot	R	11647	6.4	0.2	1.1
Grasă de Cotnari	W	850	0.4	0.2	22.0	Grasă de Cotnari	W	571	0.3	0.1	22.2
Galbenă de Odobeşti	W	546	0.2	0.1	22.0	Sauvignon Blanc	W	5594	3.1	0.1	1.1
Gewürztraminer	W	445	0.2	0.0	0.9	Graševina	W	1437	0.8	0.1	1.4
Muscat Blanc à Petits Grains	W	1012	0.5	-0.1	0.7	Crimposie	W	450	0.2	0.1	24.5
Pinot Noir	R	1740	0.8	-0.3	0.6	Galbenă de Odobeşti	W	417	0.2	0.1	24.5
Cabernet Sauvignon	R	8620	3.9	-0.3	0.9	Frâncuşă	W	365	0.2	0.1	24.5
Merlot	R	7810	3.5	-0.4	0.8	Busuioacă de Bohotin	G	343	0.2	0.1	24.5
Rkatsiteli	W	506	0.2	-0.5	0.2	Iordan	W	311	0.2	0.1	24.5
Chardonnay	W	1376	0.6	-1.1	0.2	Băbească Neagră (G)	R	297	0.2	0.1	24.5
						Mustoasă de Măderat	W	282	0.2	0.1	24.5
						Şarbă	W	266	0.1	0.1	24.5
Total of above		83082	37.4			Total of above		72024	39.4		

Table 58 (cont.): Varietal Intensity Indexes (and winegrape areas and national shares) for national top 20 varieties, 2000 and 2016

Russia											
2000						2016					
Prime variety	Col	Area	Nat'l share	NVII*	VII*	Prime variety	Col	Area	Nat'l share	NVII*	VII*
Rkatsiteli	W	13152	23.3	2.5	16.9	Rkatsiteli	W	6477	12.8	1.3	11.1
Pinot Blanc	W	2995	5.3	0.6	15.3	Aligoté	W	5843	11.5	1.2	19.1
Pervenets Magaracha	W	2388	4.2	0.5	73.0	Cabernet Sauvignon	R	8528	16.8	1.1	2.4
Bianca	W	2165	3.8	0.4	86.2	Bianca	W	3513	6.9	0.8	31.8
Zalagyöngye	W	1781	3.2	0.4	35.7	Pervenets Magaracha	W	2238	4.4	0.5	71.7
Aligoté	W	1821	3.2	0.3	4.4	Riesling	W	2232	4.4	0.3	3.3
Agadai	W	1265	2.2	0.3	86.8	Odessky Cherny	R	1250	2.5	0.3	44.0
Riesling	W	1376	2.4	0.2	2.8	Chardonnay	W	3481	6.9	0.3	1.5
Saperavi	R	931	1.7	0.2	12.0	Isabella	R	1362	2.7	0.3	6.7
Chardonnay	W	1639	2.9	0.0	1.0	Sauvignon Blanc	W	2501	4.9	0.2	1.8
Cabernet Sauvignon	R	1578	2.8	-0.2	0.6	Levokumskij	R	890	1.8	0.2	88.3
						Pinot Blanc	W	865	1.7	0.2	5.5
						Saperavi	R	716	1.4	0.1	9.8
						Krasnostop Zolotovsky	R	562	1.1	0.1	88.3
						Dunavski Lazur	W	483	1.0	0.1	88.3
						Tsimlyansky Cherny	R	451	0.9	0.1	88.3
						Gewürztraminer	W	500	1.0	0.1	3.4
						Saperavi Severny	R	325	0.6	0.1	88.3
						Citronny Magarach	W	307	0.6	0.1	88.3
						Viorika	W	307	0.6	0.1	48.6
Total of above		31090	55.2			Total of above		42831	84.3		

Serbia											
2000						2016					
Prime variety	Col	Area	Nat'l share	NVII*	VII*	Prime variety	Col	Area	Nat'l share	NVII*	VII*
Graševina	W	33120	48.0	6.5	25.4	Graševina	W	2037	9.3	0.4	17.0
Prokupac	R	15180	22.0	3.1	70.8	Riesling	W	1361	6.2	0.2	4.6
Chasselas	W	3450	5.0	0.7	18.3	Prokupac	R	916	4.2	0.2	137.1
Muscat of Hamburg	R	2760	4.0	0.5	27.7	Merlot	R	1968	8.9	0.1	1.5
						Blaufränkisch	R	727	3.3	0.1	8.6
						Muscat of Hamburg	R	624	2.8	0.1	16.5
						Cabernet Sauvignon	R	2111	9.6	0.1	1.4
						Chardonnay	W	1455	6.6	0.1	1.5
						Župljanka	W	255	1.2	0.1	102.9
						Dimyat	W	192	0.9	0.0	4.0
						Sauvignon Blanc	W	741	3.4	0.0	1.2
						Muscat Ottonel	W	183	0.8	0.0	3.0
						Pinot Noir	R	633	2.9	0.0	1.2
						Muscat Fleur d'Oranger	W	116	0.5	0.0	78.9
						Gewürztraminer	W	142	0.6	0.0	2.2
						Afus Ali	W	73	0.3	0.0	70.0
						Marselan	R	84	0.4	0.0	4.3
						Victoria	W	55	0.3	0.0	18.1
						Morava	W	34	0.2	0.0	203.6
						Pamid	R	67	0.3	0.0	1.4
Total of above		54509	79.0			Total of above		13773	62.6		

Table 58 (cont.): Varietal Intensity Indexes (and winegrape areas and national shares) for national top 20 varieties, 2000 and 2016

Slovakia

2000						2016					
Prime variety	Col	Area	Nat'l share	NVII*	VII*	Prime variety	Col	Area	Nat'l share	NVII*	VII*
Graševina	W	3895	25.0	0.7	13.2	Grüner Veltliner	W	1627	21.0	0.4	49.2
Grüner Veltliner	W	2960	19.0	0.6	39.3	Blaufränkisch	R	1216	15.7	0.3	41.0
Müller-Thurgau	W	1870	12.0	0.4	17.5	Sankt Laurent	R	717	9.3	0.2	126.8
Blaufränkisch	R	1091	7.0	0.2	24.4	Riesling	W	620	8.0	0.1	6.0
Sankt Laurent	R	935	6.0	0.2	114.8	Müller-Thurgau	W	509	6.6	0.1	15.1
Pinot Blanc	W	623	4.0	0.1	11.5	Graševina	W	456	5.9	0.1	10.8
Fetească Regală	W	312	2.0	0.1	37.9	Pinot Blanc	W	416	5.4	0.1	17.5
Fetească Albă	W	312	2.0	0.0	4.1	Cabernet Sauvignon	R	469	6.1	0.0	0.9
Chardonnay	W	623	4.0	0.0	1.3						
Silvaner	W	156	1.0	0.0	4.4						
Cabernet Sauvignon	R	156	1.0	-0.1	0.2						
Total of above		12932	83.0			Total of above		6030	77.8		

Slovenia

2000						2016					
Prime variety	Col	Area	Nat'l share	NVII*	VII*	Prime variety	Col	Area	Nat'l share	NVII*	VII*
Graševina	W	3568	15.2	0.6	8.0	Graševina	W	1935	12.1	0.4	22.2
Terrano	R	1267	5.4	0.3	180.6	Refosco	R	1340	8.4	0.3	280.3
Ribolla Gialla	W	1127	4.8	0.2	166.8	Malvazija Istarska	W	915	5.7	0.2	92.0
Sauvignon Blanc	W	1221	5.2	0.2	3.9	Žametovka	R	822	5.1	0.2	280.4
Chardonnay	W	1549	6.6	0.2	2.2	Sauvignon Blanc	W	1121	7.0	0.2	2.5
Merlot	R	1197	5.1	0.0	1.2	Blaufränkisch	R	709	4.4	0.1	11.6
						Ribolla Gialla	W	597	3.7	0.1	174.6
						Furmint	W	546	3.4	0.1	34.5
						Muscat Blanc à Petits Grains	W	586	3.7	0.1	4.9
						Chardonnay	W	1181	7.4	0.1	1.6
						Riesling	W	607	3.8	0.1	2.8
						Pinot Blanc	W	424	2.7	0.1	8.6
						Pinot Gris	G	508	3.2	0.1	2.9
						Sauvignonasse	W	231	1.4	0.0	16.8
						Savagnin Blanc	W	209	1.3	0.0	25.8
						Kraljevina	W	199	1.2	0.0	280.4
						Kerner	W	111	0.7	0.0	10.8
						Plavec Žuti	W	82	0.5	0.0	280.4
						Zelen	W	75	0.5	0.0	280.4
						Pinella	W	61	0.4	0.0	135.0
Total of above		9929	42.3			Total of above		12262	76.7		

Table 58 (cont.): Varietal Intensity Indexes (and winegrape areas and national shares) for national top 20 varieties, 2000 and 2016

South Africa

2000						2016					
Prime variety	Col	Area	Nat'l share	NVII*	VII*	Prime variety	Col	Area	Nat'l share	NVII*	VII*
Chenin Blanc	W	22566	24.1	4.4	25.7	Chenin Blanc	W	17707	18.5	3.8	25.7
Colombard	W	11432	12.2	2.2	15.4	Colombard	W	11512	12.0	2.4	18.0
Pinotage	R	6501	6.9	1.3	51.6	Pinotage	R	7052	7.4	1.5	46.3
Cabernet Sauvignon	R	8824	9.4	0.9	2.1	Sauvignon Blanc	W	9246	9.7	1.5	3.5
Sauvignon Blanc	W	5436	5.8	0.9	4.4	Syrah	R	9946	10.4	1.4	2.6
Syrah	R	5631	6.0	0.8	2.9	Cabernet Sauvignon	R	10589	11.1	0.9	1.6
Muscat of Alexandria	W	4047	4.3	0.7	7.1	Chardonnay	W	6856	7.2	0.6	1.6
Chardonnay	W	6067	6.5	0.7	2.2	Ruby Cabernet	R	2306	2.4	0.5	20.3
Cinsaut	R	3533	3.8	0.5	3.8	Cinsaut	R	1767	1.8	0.3	3.6
Crouchen	W	2161	2.3	0.4	49.9	Muscat of Alexandria	W	1781	1.9	0.2	2.4
Ruby Cabernet	R	2050	2.2	0.4	14.4	Sémillon	W	1121	1.2	0.2	2.8
Palomino Fino	W	1632	1.7	0.2	2.8	Petit Verdot	R	749	0.8	0.1	4.3
Clairette	W	938	1.0	0.2	11.2	Viognier	W	822	0.9	0.1	2.4
Merlot	R	4888	5.2	0.2	1.2	Nouvelle	W	428	0.4	0.1	46.8
Sémillon	W	1033	1.1	0.1	2.1	Muscat Blanc à Petits Grains (R	W	404	0.4	0.1	13.2
Servant	W	432	0.5	0.1	42.5	Villard Blanc	W	321	0.3	0.1	20.2
Muscat Blanc à Petits Grains (R	W	360	0.4	0.1	16.3	Crouchen	W	271	0.3	0.1	39.8
Chenel	W	339	0.4	0.1	52.2	Roobernet	R	269	0.3	0.1	46.8
Emerald Riesling	W	340	0.4	0.1	19.0	Clairette	W	195	0.2	0.0	3.8
Tinta Barroca	R	395	0.4	0.1	3.4	Muscat Blanc à Petits Grains	W	839	0.9	0.0	1.2
Total of above		88604	94.6			Total of above		84181	87.9		

Spain

2000						2016					
Prime variety	Col	Area	Nat'l share	NVII*	VII*	Prime variety	Col	Area	Nat'l share	NVII*	VII*
Airén	W	387978	32.8	60.2	4.1	Airén	W	203276	23.0	36.4	5.1
Bobal	R	100128	8.5	15.5	4.1	Tempranillo	R	193597	21.9	33.5	4.5
Tempranillo	R	79310	6.7	11.6	3.5	Bobal	R	59189	6.7	10.6	5.1
Monastrell	R	67160	5.7	10.0	3.6	Monastrell	R	41303	4.7	6.9	4.0
Garnacha Tinta	R	98131	8.3	9.4	1.9	Macabeo	W	36963	4.2	6.5	4.9
Cayetana Blanca	W	55527	4.7	8.6	4.1	Cayetana Blanca	W	36252	4.1	6.5	5.1
Macabeo	W	42902	3.6	6.4	3.7	Garnacha Tinta	R	54606	6.2	5.6	1.8
Palomino Fino	W	27685	2.3	4.2	3.8	Palomino Fino	W	20110	2.3	3.5	4.4
Pedro Ximénez	W	14803	1.3	2.2	3.5	Verdejo	W	17923	2.0	3.2	5.1
Alicante Henri Bouschet	R	18321	1.6	1.9	2.0	Alicante Henri Bouschet	R	19294	2.2	2.7	2.7
Parellada	W	11188	0.9	1.7	4.1	Xarello	W	8534	1.0	1.5	5.1
Chelva	W	10877	0.9	1.7	4.1	Pedro Ximénez	W	8528	1.0	1.5	4.9
Mencía	R	11166	0.9	1.6	3.5	Mencía	R	8489	1.0	1.4	3.9
Xarello	W	10299	0.9	1.6	4.1	Parellada	W	7137	0.8	1.3	5.1
Calagrano	W	8229	0.7	1.3	4.1	Tinto Velasco	R	5369	0.6	1.0	5.1
Tinto Velasco	R	7998	0.7	1.2	4.1	Alvarinho	W	5393	0.6	1.0	4.9
Merseguera	W	7460	0.6	1.2	4.1	Chelva	W	5029	0.6	0.9	5.1
Pardillo	W	7272	0.6	1.1	4.1	Alarije	W	4407	0.5	0.8	5.1
Zalema	W	5969	0.5	0.9	4.1	Prieto Picudo	R	4293	0.5	0.8	5.1
Beba	W	4762	0.4	0.7	4.1	Zalema	W	4015	0.5	0.7	5.1
Total of above		977164	82.7			Total of above		743707	84.2		

Table 58 (cont.): Varietal Intensity Indexes (and winegrape areas and national shares) for national top 20 varieties, 2000 and 2016

Switzerland

2000						2016					
Prime variety	Col	Area	Nat'l share	NVII*	VII*	Prime variety	Col	Area	Nat'l share	NVII*	VII*
Chasselas	W	5373	35.7	1.1	131.1	Pinot Noir	R	4209	28.4	0.9	12.1
Pinot Noir	R	4601	30.6	0.9	21.7	Chasselas	W	3838	25.9	0.9	157.7
Gamay Noir	R	1977	13.1	0.4	17.0	Gamay Noir	R	1349	9.1	0.3	15.6
Müller-Thurgau	W	686	4.6	0.1	6.6	Gamaret	R	425	2.9	0.1	292.0
Merlot	R	848	5.6	0.0	1.3	Müller-Thurgau	W	465	3.1	0.1	7.2
Silvaner	W	208	1.4	0.0	6.1	Merlot	R	1124	7.6	0.1	1.3
Cornalin	R	92	0.6	0.0	320.7	Silvaner	W	250	1.7	0.1	12.5
Pinot Gris	G	149	1.0	0.0	2.6	Garanoir	R	225	1.5	0.0	297.7
Gamaret	R	60	0.4	0.0	275.4	Arvine	W	178	1.2	0.0	280.7
Arvine	W	57	0.4	0.0	302.2	Cornalin	R	138	0.9	0.0	284.3
Garanoir	R	50	0.3	0.0	214.0	Rouge du Pays	R	136	0.9	0.0	303.1
Marsanne	W	33	0.2	0.0	7.1	Diolinoir	R	120	0.8	0.0	297.8
Pinot Blanc	W	77	0.5	0.0	1.5	Savagnin Blanc	W	127	0.9	0.0	17.0
Diolinoir	R	25	0.2	0.0	258.5	Pinot Gris	G	230	1.6	0.0	1.4
Räuschling	W	23	0.2	0.0	324.9	Pinot Blanc	W	111	0.8	0.0	2.4
Amigne	W	21	0.1	0.0	324.9	Marsanne	W	48	0.3	0.0	7.8
Bondola	R	17	0.1	0.0	324.9	Amigne	W	42	0.3	0.0	303.1
Savagnin Blanc	W	17	0.1	0.0	12.4	Doral	W	35	0.2	0.0	303.1
Regent	R	16	0.1	0.0	15.7	Galotta	R	35	0.2	0.0	303.1
Humagne	W	9	0.1	0.0	324.9	Regent	R	38	0.3	0.0	5.9
Total of above		14340	95.3			Total of above		13121	88.7		

Taiwan

2000						2016					
Prime variety	Col	Area	Nat'l share	NVII*	VII*	Prime variety	Col	Area	Nat'l share	NVII*	VII*
Kyoho (4N)	R	1303	46.0	0.3	561.6	Black Queen	R	94	63.2	0.0	19773.7
Golden Muscat	W	1190	42.0	0.2	1725.4	Golden Muscat	W	50	33.4	0.0	29910.5
Black Queen	R	227	8.0	0.0	690.5	Musann Blanc	W	5	3.4	0.0	30163.0
Italia	W	113	4.0	0.0	181.7						
Total of above		2833	100.0			Total of above		149	100.0		

Table 58 (cont.): Varietal Intensity Indexes (and winegrape areas and national shares) for national top 20 varieties, 2000 and 2016

Thailand

2000						2016					
Prime variety	Col	Area	Nat'l share	NVII*	VII*	Prime variety	Col	Area	Nat'l share	NVII*	VII*
						Syrah	R	74	35.9	0.0	8.9
						Malaga Blanc	W	54	26.0	0.0	21598.8
						Black Queen	R	18	8.7	0.0	2712.7
						Chenin Blanc	W	16	7.8	0.0	10.9
						Colombard	W	15	7.0	0.0	10.5
						Garnacha Roja (Gris)	G	5	2.5	0.0	76.5
						Muscat	W	3	1.5	0.0	92.8
						Dornfelder	R	3	1.4	0.0	8.0
						Verdelho	W	1	0.5	0.0	15.0
						Viognier	W	1	0.6	0.0	1.7
						Durif	R	1	0.3	0.0	2.4
						Sangiovese	R	2	1.0	0.0	0.6
						Pinot Noir	R	1	0.3	0.0	0.1
						Tempranillo	R	4	1.8	0.0	0.4
						Cabernet Sauvignon	R	7	3.4	0.0	0.5
Total of above						Total of above		205	98.7		

Tunisia

2000						2016					
Prime variety	Col	Area	Nat'l share	NVII*	VII*	Prime variety	Col	Area	Nat'l share	NVII*	VII*
Mazuelo	R	7576	45.0	1.5	17.2	Italia	W	644	19.0	0.1	163.8
Garnacha Tinta	R	2020	12.0	0.3	2.7	Cinsaut	R	626	18.4	0.1	36.0
Alicante Henri Bouschet	R	842	5.0	0.1	6.6	Muscat of Alexandria	W	405	11.9	0.1	15.3
Cinsaut	R	842	5.0	0.1	5.0	Mazuelo	R	238	7.0	0.0	6.6
Sangiovese	R	842	5.0	0.1	3.5	Alicante Henri Bouschet	R	178	5.2	0.0	6.5
Tribidrag	R	337	2.0	0.0	3.6	Sultaniye	W	139	4.1	0.0	34.5
Monastrell	R	337	2.0	0.0	1.3	Cardinal	R	121	3.6	0.0	95.9
Syrah	R	337	2.0	0.0	1.0	Garnacha Tinta	R	152	4.5	0.0	1.3
Cabernet Sauvignon	R	337	2.0	-0.1	0.4	Clairette	W	22	0.6	0.0	11.9
						Chardonnay	W	170	5.0	0.0	1.1
						Sauvignon Blanc	W	85	2.5	0.0	0.9
						Airén	W	85	2.5	0.0	0.5
						Syrah	R	67	2.0	0.0	0.5
						Cabernet Sauvignon	R	117	3.4	0.0	0.5
						Merlot	R	64	1.9	0.0	0.3
Total of above		13469	80.0			Total of above		3113	91.6		

Table 58 (cont.): Varietal Intensity Indexes (and winegrape areas and national shares) for national top 20 varieties, 2000 and 2016

Turkey

2000						2016					
Prime variety	Col	Area	Nat'l share	NVII*	VII*	Prime variety	Col	Area	Nat'l share	NVII*	VII*
						Sultaniye	W	2461	18.0	0.5	151.2
						Öküzgözü	R	1601	11.7	0.4	327.2
						Boğazkere	R	1436	10.5	0.3	327.2
						Syrah	R	1439	10.5	0.2	2.6
						Çalkarası	R	806	5.9	0.2	327.2
						Narince	W	787	5.7	0.2	327.2
						Kalecik Karası	R	704	5.1	0.2	327.2
						Dimrit	R	704	5.1	0.2	327.2
						Sémillon	W	529	3.9	0.1	9.3
						Alicante Henri Bouschet	R	532	3.9	0.1	4.8
						Cinsaut	R	430	3.1	0.1	6.1
						Papazkarası	R	204	1.5	0.0	327.2
						Gamay Noir	R	228	1.7	0.0	2.8
						Adakarası	R	89	0.6	0.0	327.2
						Emir	W	89	0.6	0.0	327.2
						Muscat Blanc à Petits Grains	W	129	0.9	0.0	1.3
						Karalahna	R	4	0.0	0.0	327.2
						Karasakiz	R	4	0.0	0.0	327.2
						Vasilaki	W	4	0.0	0.0	327.2
						Çavuş	W	3	0.0	0.0	327.2
Total of above						Total of above		12183	88.9		

Ukraine

2000						2016					
Prime variety	Col	Area	Nat'l share	NVII*	VII*	Prime variety	Col	Area	Nat'l share	NVII*	VII*
						Rkatsiteli	W	5775	22.9	1.2	20.0
						Aligoté	W	4814	19.1	1.0	31.8
						Cabernet Sauvignon	R	4935	19.6	0.7	2.8
						Odessky Cherny	R	1250	5.0	0.3	88.8
						Isabella	R	1200	4.8	0.2	12.0
						Riesling	W	1350	5.4	0.2	4.0
						Sauvignon Blanc	W	1550	6.2	0.2	2.2
						Gewürztraminer	W	500	2.0	0.1	6.9
						Chardonnay	W	1500	6.0	0.1	1.3
						Muscat Blanc à Petits Grains	W	338	1.3	0.0	1.8
						Pinot Blanc	W	170	0.7	0.0	2.2
						Merlot	R	1400	5.6	0.0	0.9
						Pinot Noir	R	385	1.5	0.0	0.7
Total of above						Total of above		25166	100.0		

Table 58 (cont.): Varietal Intensity Indexes (and winegrape areas and national shares) for national top 20 varieties, 2000 and 2016

United Kingdom

2000						2016					
Prime variety	Col	Area	Nat'l share	NVII*	VII*	Prime variety	Col	Area	Nat'l share	NVII*	VII*
Bacchus	W	112	12.8	0.0	185.8	Pinot Noir	R	546	29.7	0.1	12.6
Blauburger	R	105	12.0	0.0	586.7	Chardonnay	W	531	28.9	0.1	6.4
Dalkauer	W	100	11.5	0.0	5598.7	Pinot Meunier	R	202	11.0	0.0	33.6
Seyval Blanc	W	85	9.7	0.0	1224.4	Bacchus	W	127	6.9	0.0	175.9
Auxerrois	W	58	6.6	0.0	141.1	Seyval Blanc	W	77	4.2	0.0	69.8
Cabernet Dorsa	R	43	4.9	0.0	5598.7	Reichensteiner	W	42	2.3	0.0	862.7
Chardonnay	W	68	7.8	0.0	2.6	Madeleine × Angevine 7672	W	39	2.1	0.0	1949.5
Albalonga	W	41	4.7	0.0	4027.1	Rondo	R	37	2.0	0.0	1774.1
Ehrenfelser	W	34	3.9	0.0	658.7	Regent	R	26	1.4	0.0	31.8
						Pinot Gris	G	44	2.4	0.0	2.2
						Ortega	W	24	1.3	0.0	109.5
						Dornfelder	R	13	0.7	0.0	4.0
						Schönburger	G	9	0.5	0.0	636.6
						Pinot Blanc	W	15	0.8	0.0	2.6
						Müller-Thurgau	W	15	0.8	0.0	1.8
						Huxelrebe	W	4	0.2	0.0	19.3
Total of above		646	74.0			Total of above		1751	95.2		

United States

2000						2016					
Prime variety	Col	Area	Nat'l share	NVII*	VII*	Prime variety	Col	Area	Nat'l share	NVII*	VII*
Chardonnay	W	35791	20.4	6.3	6.8	Chardonnay	W	41392	17.3	6.8	3.8
Tribidrag	R	18630	10.6	3.6	19.3	Cabernet Sauvignon	R	40837	17.0	5.4	2.5
Colombard	W	18010	10.3	3.4	13.0	Pinot Noir	R	22998	9.6	3.9	4.1
Cabernet Sauvignon	R	17573	10.0	2.0	2.2	Tribidrag	R	18551	7.7	3.7	10.3
Merlot	R	16875	9.6	1.9	2.2	Concord	R	8349	3.5	1.7	14.8
Concord	R	8330	4.7	1.6	19.6	Merlot	R	21251	8.9	1.6	1.5
Chenin Blanc	W	8433	4.8	1.4	5.1	Colombard	W	7991	3.3	1.4	5.0
Rubired	R	4153	2.4	0.8	27.8	Pinot Gris	G	7462	3.1	1.1	2.9
Barbera	R	4693	2.7	0.7	4.0	Rubired	R	4825	2.0	1.0	18.4
Pinot Noir	R	5343	3.0	0.6	2.2	Durif	R	3698	1.5	0.8	14.4
Ruby Cabernet	R	2895	1.6	0.5	10.9	Ruby Cabernet	R	2114	0.9	0.4	7.5
Sauvignon Blanc	W	4191	2.4	0.4	1.8	Riesling	W	4952	2.1	0.4	1.5
Muscat of Alexandria	W	2013	1.1	0.2	1.9	Barbera	R	2131	0.9	0.3	2.2
Durif	R	923	0.5	0.2	21.5	Niagara	W	1196	0.5	0.2	6.9
Niagara	W	1357	0.8	0.2	2.5	Petit Verdot	R	1219	0.5	0.2	2.8
Monbadon	W	656	0.4	0.1	27.8	Viognier	W	1481	0.6	0.1	1.7
Catawba	R	635	0.4	0.1	27.8	Symphony	W	647	0.3	0.1	18.7
Carnelian	R	625	0.4	0.1	27.8	Catawba	R	625	0.3	0.1	18.7
Malvasia Bianca di Candia	W	968	0.6	0.1	2.1	Triplett Blanc	W	412	0.2	0.1	18.7
Riesling	W	1965	1.1	0.1	1.3	Elbling	W	396	0.2	0.1	7.6
Total of above		154060	87.7			Total of above		30200	100.0		

Table 58 (cont.): Varietal Intensity Indexes (and winegrape areas and national shares) for national top 20 varieties, 2000 and 2016

Uruguay

2000						2016					
Prime variety	Col	Area	Nat'l share	NVII*	VII*	Prime variety	Col	Area	Nat'l share	NVII*	VII*
Muscat of Hamburg	R	2886	32.5	0.6	224.8	Tannat	R	1725	25.6	0.4	204.4
Tannat	R	2433	27.4	0.5	239.4	Muscat of Hamburg	R	1267	18.8	0.3	109.7
Merlot	R	1057	11.9	0.1	2.7	Trebbiano Toscano	W	682	10.1	0.1	3.8
Cabernet Franc	R	364	4.1	0.1	3.9	Merlot	R	747	11.1	0.1	1.9
Cabernet Sauvignon	R	675	7.6	0.1	1.7	Cabernet Franc	R	266	3.9	0.0	3.2
Sauvignon Blanc	W	142	1.6	0.0	1.2	Marselan	R	120	1.8	0.0	20.2
Chardonnay	W	142	1.6	0.0	0.5	Isabella	R	102	1.5	0.0	3.8
Syrah	R	62	0.7	0.0	0.3	Arinarnoa	R	45	0.7	0.0	61.5
						Viognier	W	41	0.6	0.0	1.7
						Cabernet Sauvignon	R	484	7.2	0.0	1.0
						Petit Verdot	R	26	0.4	0.0	2.1
						Nebbiolo	R	25	0.4	0.0	2.1
						Sauvignon Blanc (G)	W	10	0.1	0.0	6.2
						Concord	R	24	0.4	0.0	1.5
						Arriloba	W	6	0.1	0.0	73.6
						Canari Noir	R	4	0.1	0.0	31.8
						Egiodola	R	2	0.0	0.0	4.7
						Malvasia Fina	W	6	0.1	0.0	1.2
						Roussanne	W	2	0.0	0.0	0.6
						Marsanne	W	1	0.0	0.0	0.4
Total of above		7762	87.4			Total of above		5585	82.8		

Missing 9

2000						2016					
Prime variety	Col	Area	Nat'l share	NVII*	VII*	Prime variety	Col	Area	Nat'l share	NVII*	VII*
Cabernet Sauvignon	R	20373	25.4	3.4	5.6						
Rkatsiteli	W	10549	13.1	1.9	9.5						
Aligoté	W	6916	8.6	1.3	11.8						
Merlot	R	5549	6.9	0.4	1.6						
Odessky Cherny	R	1694	2.1	0.3	60.9						
Sultaniye	W	1820	2.3	0.3	9.1						
Riesling	W	2279	2.8	0.3	3.2						
Sauvignon Blanc	W	2383	3.0	0.3	2.2						
Saperavi	R	1356	1.7	0.3	12.3						
Isabella	R	1673	2.1	0.3	3.7						
Öküzgözü	R	1033	1.3	0.2	60.9						
Sukholimansky Bely	W	1032	1.3	0.2	38.5						
Bastardo Magarachsky	R	929	1.2	0.2	28.7						
Carmenère	R	945	1.2	0.2	10.1						
Negramoll	R	874	1.1	0.2	15.0						
Boğazkere	R	773	1.0	0.2	60.9						
Chardonnay	W	3120	3.9	0.1	1.3						
Italia	W	706	0.9	0.1	40.0						
Kokur Bely	W	641	0.8	0.1	60.9						
Dimrit	R	602	0.8	0.1	60.9						
Total of above		65248	81.3			Total of above					

V. National Varietal Intensity Indexes for world's top varieties

Table 59: National VIIs for the world's top 24 red varieties, 2000

Cabernet Sauvignon		**Garnacha Tinta**		**Merlot**		**Mazuelo**	
Country	*VII*	*Country*	*VII*	*Country*	*VII*	*Country*	*VII*
Chile	6.91	Algeria	4.52	Israel	3.06	Tunisia	17.22
Australia	4.19	Tunisia	2.71	Uruguay	2.73	Algeria	9.57
Israel	2.74	France	2.50	France	2.68	Israel	7.66
Bulgaria	2.38	Spain	1.88	Bulgaria	2.67	France	4.24
United States	2.19	United States	0.58	Chile	2.58	Morocco	1.31
South Africa	2.06	Australia	0.37	United States	2.20	United States	0.67
Moldova	1.85	Morocco	0.37	Moldova	2.07	Spain	0.26
Uruguay	1.67	Italy	0.24	Canada	1.82	Chile	0.22
Argentina	1.53	Greece	0.01	New Zealand	1.51	Italy	0.10
Canada	1.47	South Africa	0.01	Australia	1.35	South Africa	0.03
New Zealand	1.44	Argentina	0.00	Switzerland	1.29	Australia	0.03
France	1.35			South Africa	1.20	Argentina	0.01
Algeria	1.10			Slovenia	1.17	Greece	0.01
Romania	0.85			Algeria	1.15		
Russia	0.61			Romania	0.81		

Syrah		**Bobal**		**Tempranillo**		**Monastrell**	
Country	*VII*	*Country*	*VII*	*Country*	*VII*	*Country*	*VII*
Australia	10.70	Spain	4.14	Spain	3.51	Spain	3.64
South Africa	2.87			Portugal	1.88	Tunisia	1.28
France	2.79			Argentina	1.25	France	0.57
Algeria	2.38			France	0.09	Australia	0.46
Argentina	2.15			United States	0.06	United States	0.07
Tunisia	0.95			Australia	0.02	Chile	0.01
Chile	0.85			South Africa	0.01	South Africa	0.01
United States	0.41			Italy	0.00	Argentina	0.00
Uruguay	0.33			Chile	0.00		
New Zealand	0.29						
Switzerland	0.17						
Italy	0.08						
Greece	0.04						
Spain	0.00						

Table 59 (cont.): National VIIs for the world's top 24 red varieties, 2000

Sangiovese		Pinot Noir		Cabernet Franc		Cinsaut	
Country	*VII*	*Country*	*VII*	*Country*	*VII*	*Country*	*VII*
Italy	7.00	Switzerland	21.73	Brazil	6.73	Algeria	25.23
Tunisia	3.55	New Zealand	7.84	Canada	6.27	Morocco	8.02
Argentina	0.90	Germany	5.89	France	3.92	Tunisia	5.05
United States	0.28	Moldova	5.16	Uruguay	3.86	South Africa	3.81
Australia	0.20	Canada	3.82	New Zealand	1.12	France	3.69
France	0.13	Algeria	3.55	Italy	0.98	Greece	0.21
Chile	0.08	Luxembourg	3.48	United States	0.64	Chile	0.17
South Africa	0.03	France	2.18	Hungary	0.57	Italy	0.04
		United States	2.16	Chile	0.57	United States	0.02
		Australia	1.75	Australia	0.54	Argentina	0.00
		Chile	1.01	South Africa	0.49	Spain	0.00
		Austria	0.60	Argentina	0.12		
		Bulgaria	0.57	Switzerland	0.10		
		Romania	0.56	Greece	0.07		
		Argentina	0.40	Austria	0.05		

Gamay Noir		Alicante Henri Bouschet		Barbera		Montepulciano	
VII	*VII*	*Country*	*VII*	*Country*	*VII*	*Country*	*VII*
Switzerland	17.00	Algeria	13.15	Italy	6.31	Italy	7.66
France	5.16	Tunisia	6.58	United States	3.95	Argentina	0.04
Canada	4.00	Chile	3.33	Argentina	0.79		
United States	0.50	Morocco	2.91	Australia	0.12		
Luxembourg	0.10	Spain	2.04	South Africa	0.02		
South Africa	0.05	France	1.33				
Italy	0.03	Portugal	0.43				
Argentina	0.00	United States	0.42				
Spain	0.00	Italy	0.11				
		Argentina	0.08				
		Greece	0.05				
		South Africa	0.01				

Table 59 (cont.): National VIIs for the world's top 24 red varieties, 2000

Isabella		Tribidrag		Côt		Criolla Grande	
Country	*VII*	*Country*	*VII*	*Country*	*VII*	*Country*	*VII*
Brazil	48.14	United States	19.25	Argentina	17.17	Argentina	24.76
Moldova	22.59	Tunisia	3.63	Chile	1.52		
Argentina	0.07	Italy	2.23	France	1.32		
United States	0.02	Chile	0.14	New Zealand	1.25		
		South Africa	0.05	Australia	0.61		
		Argentina	0.01	South Africa	0.15		
		France	0.00	United States	0.10		
				Moldova	0.08		
				Italy	0.07		
				Spain	0.00		
				Switzerland	0.00		

Pamid		Douce Noire		Negroamaro		Doukkali	
Country	*VII*	*Country*	*VII*	*Country*	*VII*	*Country*	*VII*
Bulgaria	50.61	Argentina	21.16	Italy	7.68	Morocco	98.54
Hungary	0.30	Italy	1.11				
Greece	0.06	United States	0.03				
		France	0.00				

Table 60: National VIIs for the world's top 24 red varieties, 2016

Cabernet Sauvignon		**Merlot**		**Tempranillo**		**Syrah**	
Country	*VII*	*Country*	*VII*	*Country*	*VII*	*Country*	*VII*
Chile	4.20	Cambodia	3.37	Spain	4.48	Myanmar	9.54
Lebanon	3.61	Bulgaria	3.19	Portugal	1.90	Thailand	8.87
China	3.27	Israel	2.41	Myanmar	1.02	Cambodia	7.42
Cambodia	2.89	France	2.24	Mexico	0.86	Australia	7.28
Israel	2.86	Lebanon	2.10	Argentina	0.61	India	4.58
Ukraine	2.83	Algeria	2.03	Thailand	0.37	Algeria	2.98
Australia	2.61	Uruguay	1.86	Israel	0.22	Turkey	2.60
Bulgaria	2.54	China	1.58	Australia	0.11	South Africa	2.57
United States	2.46	Moldova	1.57	United States	0.05	Israel	1.91
Russia	2.42	Serbia	1.50	South Africa	0.02	France	1.89
Mexico	2.00	United States	1.49	Chile	0.02	Lebanon	1.86
Algeria	1.74	Chile	1.39	France	0.02	Argentina	1.52
South Africa	1.60	Switzerland	1.28	Brazil	0.01	Chile	1.36
Moldova	1.43	Mexico	1.20	New Zealand	0.01	United States	0.94
Serbia	1.38	Croatia	1.19	Canada	0.01	Mexico	0.66

Garnacha Tinta		**Pinot Noir**		**Sangiovese**		**Bobal**	
Country	*VII*	*Country*	*VII*	*Country*	*VII*	*Country*	*VII*
Algeria	7.20	United Kingdom	12.62	Ethiopia	32.47	Spain	5.07
France	2.88	Switzerland	12.09	Italy	6.91		
Spain	1.85	New Zealand	6.61	Thailand	0.63		
Tunisia	1.34	Germany	5.03	Argentina	0.54		
Morocco	1.34	Myanmar	4.25	United States	0.21		
Mexico	0.77	United States	4.08	Australia	0.20		
China	0.67	Luxembourg	3.96	France	0.11		
Australia	0.34	Czechia	2.18	Turkey	0.08		
United States	0.28	Canada	2.16	Chile	0.06		
Italy	0.27	France	1.65	South Africa	0.04		
South Africa	0.11	India	1.57	Romania	0.03		
Turkey	0.07	Australia	1.54	Canada	0.02		
Chile	0.04	Serbia	1.22	New Zealand	0.01		
Uruguay	0.02	Moldova	1.22	Brazil	0.01		
Portugal	0.01	Chile	1.19	Hungary	0.00		

Table 60 (cont.): National VIIs for the world's top 24 red varieties, 2016

Cabernet Franc		**Côt**		**Monastrell**		**Mazuelo**	
Country	*VII*	*Country*	*VII*	*Country*	*VII*	*Country*	*VII*
Brazil	16.46	Argentina	16.81	Spain	4.04	Algeria	34.25
Canada	5.20	Israel	1.89	Israel	0.95	Israel	17.72
France	3.17	Chile	1.35	France	0.93	Mexico	7.77
Uruguay	3.15	France	0.64	Australia	0.46	Morocco	6.63
Israel	1.76	United States	0.58	South Africa	0.43	Tunisia	6.63
Hungary	1.71	Uruguay	0.55	United States	0.19	France	3.69
Chile	0.87	South Africa	0.41	Chile	0.06	Turkey	0.90
Italy	0.74	Australia	0.33	Turkey	0.04	Spain	0.59
United States	0.73	New Zealand	0.31	Canada	0.01	Chile	0.53
Moldova	0.73	Canada	0.28	Argentina	0.01	United States	0.43
South Africa	0.70	Peru	0.22	Romania	0.00	Italy	0.26
Kazakhstan	0.65	Moldova	0.17	Switzerland	0.00	Portugal	0.15
Bulgaria	0.36	Turkey	0.13			South Africa	0.11
Argentina	0.36	Switzerland	0.09			China	0.05
Switzerland	0.34	Brazil	0.08			Argentina	0.01

Alicante Henri Bouschet		**Tribidrag**		**Montepulciano**		**Gamay Noir**	
Country	*VII*	*Country*	*VII*	*Country*	*VII*	*Country*	*VII*
Tunisia	6.50	United States	10.31	Italy	7.37	Switzerland	15.59
Morocco	6.50	North Macedonia	5.38	Australia	0.06	France	5.06
Chile	5.89	Italy	3.06	Argentina	0.05	Canada	3.70
Turkey	4.83	Canada	0.09	United States	0.03	Turkey	2.84
Portugal	3.10	Australia	0.09	New Zealand	0.03	Serbia	0.42
Spain	2.72	Chile	0.06	Brazil	0.01	United States	0.09
Uruguay	0.44	South Africa	0.03	Chile	0.00	Slovenia	0.05
France	0.40	New Zealand	0.02	Spain	0.00	New Zealand	0.04
Brazil	0.38	Romania	0.01			Brazil	0.03
United States	0.20	Switzerland	0.00			Uruguay	0.03
Greece	0.15	Argentina	0.00			Italy	0.02
Argentina	0.08	France	0.00			South Africa	0.02
Italy	0.06	Spain	0.00			Hungary	0.01
Hungary	0.03					Australia	0.01
Australia	0.02					Chile	0.00

Table 60 (cont.): National VIIs for the world's top 24 red varieties, 2016

Cinsaut		Carmenère		Douce Noire		Barbera	
Country	*VII*	*Country*	*VII*	*Country*	*VII*	*Country*	*VII*
Tunisia	36.01	Chile	14.36	Argentina	21.00	Italy	6.24
Morocco	36.01	China	12.55	Italy	0.24	United States	2.24
Turkey	6.13	Italy	0.21	United States	0.03	Slovenia	1.54
France	3.82	Brazil	0.06	France	0.00	Argentina	0.54
South Africa	3.61	Argentina	0.06			Australia	0.19
Chile	1.14	Canada	0.04			South Africa	0.09
United States	0.04	Australia	0.02			Canada	0.03
Greece	0.02	United States	0.02			Brazil	0.01
Australia	0.01	South Africa	0.02			Switzerland	0.01
Portugal	0.01	France	0.01			Chile	0.01
Spain	0.00	Hungary	0.00			Romania	0.00
Argentina	0.00	Switzerland	0.00			Spain	0.00
Italy	0.00	Spain	0.00				

Isabella		Blaufränkisch		Criolla Grande		Pinot Meunier	
Country	*VII*	*Country*	*VII*	*Country*	*VII*	*Country*	*VII*
Brazil	88.41	Slovakia	40.95	Argentina	21.73	United Kingdom	33.56
Ukraine	12.00	Hungary	29.66			Germany	6.46
Moldova	10.57	Czechia	21.93			France	4.54
Russia	6.75	Peru	19.75			Moldova	0.51
Uruguay	3.81	Austria	16.12			Canada	0.22
Australia	0.03	Croatia	11.57			Australia	0.19
Switzerland	0.02	Slovenia	11.57			New Zealand	0.18
		Serbia	8.62			United States	0.10
		Germany	4.80			South Africa	0.05
		Romania	1.04			Argentina	0.02
		Canada	0.10			Chile	0.00
		Switzerland	0.06			Switzerland	0.00
		Italy	0.01			Italy	0.00
		United States	0.00			Spain	0.00
		Australia	0.00				

Table 61: National VIIs for the world's top 24 white varieties, 2000

Airén		**Chardonnay**		**Trebbiano Toscano**		**Graševina**	
Country	*VII*	*Country*	*VII*	*Country*	*VII*	*Country*	*VII*
Spain	4.14	New Zealand	9.41	France	3.72	Serbia	25.42
		United States	6.84	Italy	2.21	Croatia	14.30
		Australia	4.44	Bulgaria	0.68	Slovakia	13.24
		Canada	3.85	Greece	0.52	Slovenia	8.05
		United Kingdom	2.62	Argentina	0.50	Czechia	5.82
		Chile	2.26	Brazil	0.46	Austria	4.72
		Slovenia	2.22	Australia	0.19	Hungary	4.07
		South Africa	2.18	Portugal	0.07	Romania	3.58
		Moldova	1.92	South Africa	0.06	Bulgaria	1.99
		Czechia	1.68	United States	0.03	Brazil	0.88
		France	1.42	Spain	0.00	Italy	0.17
		Slovakia	1.34			Spain	0.09
		Hungary	1.14				
		Israel	0.98				
		Russia	0.98				

Rkatsiteli		**Sauvignon Blanc**		**Cayetana Blanca**		**Catarratto Bianco**	
Country	*VII*	*Country*	*VII*	*Country*	*VII*	*Country*	*VII*
Georgia	38.28	New Zealand	18.27	Spain	4.12	Italy	7.68
Russia	16.94	Moldova	6.80	Australia	0.17		
Armenia	15.99	Chile	4.38				
Moldova	9.29	South Africa	4.35				
Bulgaria	7.13	Israel	4.07				
Romania	0.17	Slovenia	3.90				
		France	1.81				
		United States	1.79				
		Romania	1.56				
		Australia	1.49				
		Canada	1.30				
		Uruguay	1.20				
		Austria	0.49				
		Italy	0.39				
		Argentina	0.33				

Table 61 (cont.): National VIIs for the world's top 24 white varieties, 2000

Macabeo		**Chenin Blanc**		**Riesling**		**Colombard**	
Country	*VII*	*Country*	*VII*	*Country*	*VII*	*Country*	*VII*
Spain	3.69	South Africa	25.73	Germany	24.19	South Africa	15.44
France	0.61	United States	5.13	Luxembourg	14.65	United States	12.97
Argentina	0.00	Israel	2.23	Czechia	7.90	Israel	12.66
		Argentina	1.86	Canada	6.39	Australia	1.75
		New Zealand	1.57	New Zealand	5.56	France	1.01
		France	1.21	Austria	3.82		
		Australia	0.69	Russia	2.76		
		Chile	0.07	Australia	2.70		
		Spain	0.01	Hungary	2.10		
		Switzerland	0.01	Moldova	1.69		
				United States	1.26		
				Bulgaria	0.76		
				South Africa	0.57		
				France	0.44		
				Chile	0.28		

Aligoté		**Müller-Thurgau**		**Palomino Fino**		**Muscat Blanc à Petits Grains**	
Country	*VII*	*Country*	*VII*	*Country*	*VII*	*Country*	*VII*
Moldova	24.08	Luxembourg	49.55	Spain	3.75	Armenia	7.65
Romania	4.69	Germany	28.91	South Africa	2.79	Greece	7.15
Russia	4.43	Czechia	20.37	New Zealand	0.34	Italy	3.33
Bulgaria	2.37	Slovakia	17.46	United States	0.29	Hungary	2.89
Georgia	0.36	Austria	9.87	Argentina	0.18	South Africa	1.35
France	0.28	Switzerland	6.64	Australia	0.15	France	1.31
Switzerland	0.18	New Zealand	6.13	France	0.08	Portugal	1.13
		Hungary	5.49			Romania	0.74
		Moldova	0.28			Austria	0.48
		Italy	0.23			United States	0.48
		France	0.00			Switzerland	0.48
						Moldova	0.31
						Australia	0.27
						Germany	0.14
						Argentina	0.12

Table 61 (cont.): National VIIs for the world's top 24 white varieties, 2000

		Sémillon		**Fetească Albă**		**Grüner Veltliner**	
Country	*VII*	*Country*	*VII*	*Country*	*VII*	*Country*	*VII*
Morocco	12.22	Australia	9.31	Romania	16.81	Austria	74.63
South Africa	7.14	New Zealand	4.29	Moldova	9.89	Slovakia	39.34
Israel	6.89	Chile	3.09	Slovakia	4.10	Czechia	31.06
Argentina	4.61	France	3.02	Hungary	2.29	Hungary	3.18
Australia	3.16	South Africa	2.05			Italy	0.04
Brazil	2.53	Brazil	1.36				
United States	1.89	Argentina	0.97				
Spain	0.86	United States	0.75				
France	0.58	Greece	0.08				
Portugal	0.41	Switzerland	0.04				
Italy	0.30	Italy	0.00				

Trebbiano Romagnolo		**Pedro Ximénez**		**Pinot Blanc**		**Garganega**	
Country	*VII*	*Country*	*VII*	*Country*	*VII*	*Country*	*VII*
Italy	7.68	Chile	5.91	Luxembourg	29.46	Italy	7.67
		Spain	3.54	Austria	17.42	Argentina	0.01
		Australia	0.19	Russia	15.30		
				Slovakia	11.51		
				Germany	6.62		
				Canada	4.93		
				Italy	2.26		
				Switzerland	1.48		
				Georgia	1.32		
				Moldova	1.12		
				United States	0.71		
				France	0.47		
				South Africa	0.11		
				Argentina	0.06		
				Chile	0.04		

Table 62: National VIIs for the world's top 24 white varieties, 2016

Airen		Chardonnay		Sauvignon Blanc		Trebbiano Toscano	
Country	*VII*	*Country*	*VII*	*Country*	*VII*	*Country*	*VII*
Spain	5.06	United Kingdom	6.43	New Zealand	20.78	India	4.14
Tunisia	0.55	Lebanon	5.56	Myanmar	11.30	Uruguay	3.77
Morocco	0.55	United States	3.84	India	6.66	France	3.60
		Australia	3.58	Lebanon	4.49	Italy	2.18
		Canada	2.50	Chile	3.70	Bulgaria	0.52
		New Zealand	1.95	South Africa	3.47	China	0.31
		Chile	1.74	Moldova	3.01	Argentina	0.29
		Slovenia	1.64	Slovenia	2.52	Brazil	0.26
		South Africa	1.59	Czechia	2.39	Greece	0.15
		Russia	1.52	Ukraine	2.21	Moldova	0.13
		Serbia	1.47	Russia	1.77	South Africa	0.06
		Czechia	1.34	Australia	1.64	Russia	0.05
		Ukraine	1.33	France	1.24	Portugal	0.02
		Bulgaria	1.30	Serbia	1.21	United States	0.01
		France	1.29	Romania	1.10	Canada	0.01

Riesling		Rkatsiteli		Macabeo		Cayetana Blanca	
Country	*VII*	*Country*	*VII*	*Country*	*VII*	*Country*	*VII*
Germany	17.09	Georgia	46.04	Spain	4.86	Spain	5.05
Luxembourg	9.34	Kazakhstan	44.68	France	0.24	Portugal	0.09
Canada	7.07	Ukraine	20.03	South Africa	0.01	Australia	0.02
Czechia	6.46	Russia	11.13				
Slovakia	6.00	Bulgaria	8.92				
Serbia	4.63	Moldova	4.12				
Ukraine	4.02	North Macedonia	1.62				
Croatia	3.99	Serbia	0.24				
Austria	3.33	Romania	0.20				
Russia	3.29						
Slovenia	2.85						
North Macedonia	2.72						
Romania	2.51						
Australia	1.76						
New Zealand	1.62						

Table 62 (cont.): National VIIs for the world's top 24 white varieties, 2016

Muscat Alexandria		**Muscat Blanc à Petits Grains**		**Chenin Blanc**		**Colombard**	
Country	*VII*	*Country*	*VII*	*Country*	*VII*	*Country*	*VII*
Morocco	15.33	Myanmar	13.29	Ethiopia	44.41	South Africa	17.96
Tunisia	15.33	Peru	12.52	South Africa	25.72	Thailand	10.49
Israel	5.67	Mexico	5.98	Thailand	10.85	Israel	6.58
Chile	4.79	Slovenia	4.87	Mexico	7.00	United States	4.98
India	4.77	Greece	4.10	France	1.61	Australia	2.02
Algeria	3.10	Italy	2.93	Argentina	1.45	France	1.55
South Africa	2.40	Austria	2.41	United States	1.14	Brazil	0.10
China	2.17	North Macedonia	2.15	Australia	0.43	Spain	0.00
Australia	2.12	Ukraine	1.78	New Zealand	0.09		
Greece	1.96	Hungary	1.59	Switzerland	0.07		
Argentina	1.70	Russia	1.26	Peru	0.07		
Spain	1.39	Turkey	1.25	Canada	0.07		
United States	1.07	France	1.20	Uruguay	0.04		
Uruguay	0.42	South Africa	1.16	Chile	0.04		
France	0.39	Romania	1.15	Brazil	0.03		

Catarratto Bianco		**Aligoté**		**Graševina**		**Palomino Fino**	
Country	*VII*	*Country*	*VII*	*Country*	*VII*	*Country*	*VII*
Italy	7.40	Ukraine	31.84	Croatia	69.79	Spain	4.40
United States	0.03	Russia	19.15	Slovenia	22.25	Mexico	3.86
		Moldova	15.65	Serbia	17.01	Portugal	2.75
		Kazakhstan	6.64	Czechia	15.06	South Africa	0.27
		Romania	5.32	Austria	13.08	Argentina	0.10
		Bulgaria	0.90	Hungary	11.32	United States	0.06
		Georgia	0.43	Slovakia	10.82	New Zealand	0.04
		Canada	0.40	China	3.10	Australia	0.03
		France	0.39	North Macedonia	2.00	France	0.01
		Switzerland	0.27	Romania	1.45	Switzerland	0.00
		Hungary	0.00	Brazil	1.04		
				Italy	0.38		
				Spain	0.22		

Table 62 (cont.): National VIIs for the world's top 24 white varieties, 2016

Prosecco		**Müller-Thurgau**		**Grüner Veltliner**		**Trebbiano Romagnolo**	
Country	*VII*	*Country*	*VII*	*Country*	*VII*	*Country*	*VII*
Italy	7.28	Luxembourg	55.88	Austria	74.19	Italy	7.42
Brazil	1.39	Germany	28.37	Slovakia	49.24		
Australia	0.27	Czechia	25.00	Czechia	26.52		
Argentina	0.01	Slovakia	15.10	Hungary	5.07		
		Austria	8.99	New Zealand	0.29		
		Switzerland	7.23	United States	0.06		
		Hungary	6.01	Canada	0.05		
		Slovenia	1.85	Germany	0.03		
		United Kingdom	1.84	Italy	0.02		
		Japan	1.30	Australia	0.02		
		Italy	0.49	Switzerland	0.01		
		Russia	0.48	South Africa	0.01		
		Canada	0.11	Argentina	0.01		
		United States	0.04	Romania	0.00		
		New Zealand	0.01				

Sémillon		**Verdejo**		**Viognier**		**Pedro Giménez**	
Country	*VII*	*Country*	*VII*	*Country*	*VII*	*Country*	*VII*
Turkey	9.25	Spain	5.07	France	3.02	Argentina	15.62
Australia	8.25	Australia	0.01	South Africa	2.40	Chile	8.64
France	3.01	Chile	0.00	Canada	2.23		
South Africa	2.81			United States	1.72		
Chile	1.40			Uruguay	1.70		
Argentina	0.89			Thailand	1.69		
Uruguay	0.50			Chile	1.61		
New Zealand	0.43			Australia	1.59		
Canada	0.36			Argentina	1.05		
United States	0.34			New Zealand	1.02		
Hungary	0.16			Italy	0.84		
Russia	0.12			Switzerland	0.84		
Portugal	0.10			Turkey	0.30		
Switzerland	0.06			Portugal	0.19		
Greece	0.05			Moldova	0.11		

Table 63: National NVIIs for the world's top 24 red varieties, 2000

Cabernet Sauvignon		**Garnacha Tinta**		**Merlot**		**Mazuelo**	
Country	*VII*	*Country*	*VII*	*Country*	*VII*	*Country*	*VII*
Chile	6.29	France	11.75	France	13.00	France	14.97
Australia	3.89	Spain	9.37	United States	1.88	Tunisia	1.46
France	2.85	Algeria	0.96	Chile	1.61	Algeria	1.38
United States	1.95	Tunisia	0.26	Bulgaria	1.43	Israel	0.17
Bulgaria	1.24	Morocco	-0.29	Moldova	0.86	Morocco	0.08
Argentina	0.98	Greece	-0.46	Australia	0.40	Greece	-0.27
South Africa	0.93	United States	-0.67	South Africa	0.16	United States	-0.31
Moldova	0.71	Australia	-0.75	Uruguay	0.14	Chile	-0.48
Israel	0.08	South Africa	-0.84	Israel	0.09	South Africa	-0.49
Uruguay	0.06	Argentina	-1.79	Canada	0.06	Australia	-0.68
New Zealand	0.04	Italy	-4.38	New Zealand	0.05	Argentina	-1.04
Canada	0.04			Switzerland	0.04	Italy	-3.05
Algeria	0.03			Algeria	0.04	Spain	-4.66
Tunisia	-0.09			Slovenia	0.04		
Slovakia	-0.11			Brazil	-0.38		
Switzerland	-0.13			Romania	-0.39		

Syrah		**Bobal**		**Tempranillo**		**Monastrell**	
Country	*VII*	*Country*	*VII*	*Country*	*VII*	*Country*	*VII*
France	6.66	Spain	15.53	Spain	11.61	Spain	9.97
Australia	5.43			Portugal	0.70	Tunisia	0.02
Argentina	0.97			Argentina	0.19	Australia	-0.22
South Africa	0.75			South Africa	-0.36	South Africa	-0.30
Algeria	0.18			Chile	-0.45	Chile	-0.36
Tunisia	0.00			Australia	-0.50	United States	-0.52
Uruguay	-0.03			United States	-0.65	Argentina	-0.63
New Zealand	-0.03			Italy	-2.49	France	-1.20
Switzerland	-0.05			France	-3.06		
Chile	-0.07						
Greece	-0.21						
United States	-0.45						
Italy	-2.52						
Spain	-5.05						

Table 63 (cont.): National NVIIs for the world's top 24 red varieties, 2000

Sangiovese		Pinot Noir		Cabernet Franc		Cinsaut	
Country	*VII*	*Country*	*VII*	*Country*	*VII*	*Country*	*VII*
Italy	11.01	France	2.94	France	5.50	France	4.71
Tunisia	0.12	Germany	1.47	Brazil	0.66	Algeria	1.48
Argentina	-0.06	Moldova	1.08	Canada	0.10	Morocco	0.71
South Africa	-0.26	Switzerland	0.90	Uruguay	0.06	South Africa	0.53
Australia	-0.30	United States	0.59	New Zealand	0.00	Tunisia	0.14
Chile	-0.30	Australia	0.28	Italy	-0.03	Greece	-0.08
United States	-0.37	Algeria	0.22	Switzerland	-0.03	Chile	-0.19
France	-2.17	New Zealand	0.20	Hungary	-0.08	United States	-0.35
		Canada	0.07	Austria	-0.10	Argentina	-0.40
		Luxembourg	0.01	Greece	-0.10	Italy	-1.23
		Chile	0.00	South Africa	-0.10	Spain	-2.40
		Austria	-0.06	Chile	-0.11		
		Bulgaria	-0.12	Australia	-0.13		
		South Africa	-0.17	United States	-0.14		
		Hungary	-0.20	Argentina	-0.38		
		Romania	-0.28	Spain	-2.57		

Gamay Noir		Alicante Henri Bouschet		Barbera		Montepulciano	
Country	*VII*	*Country*	*VII*	*Country*	*VII*	*Country*	*VII*
France	5.70	Spain	1.91	Italy	4.68	Italy	5.10
Switzerland	0.38	Algeria	0.57	United States	0.72	Argentina	-0.23
Canada	0.04	France	0.45	Argentina	-0.06		
Luxembourg	0.00	Chile	0.41	South Africa	-0.13		
United States	-0.14	Morocco	0.15	Australia	-0.16		
South Africa	-0.14	Tunisia	0.15				
Argentina	-0.31	Greece	-0.07				
Italy	-0.98	South Africa	-0.14				
Spain	-1.87	United States	-0.16				
		Portugal	-0.18				
		Argentina	-0.28				
		Italy	-0.89				

Table 63 (cont.): National NVIIs for the world's top 24 red varieties, 2000

Isabella		**Tribidrag**		**Côt**		**Criolla Grande**	
Country	*VII*	*Country*	*VII*	*Country*	*VII*	*Country*	*VII*
Brazil	2.86	United States	3.61	Argentina	3.51	Argentina	4.76
Moldova	2.23	Italy	0.88	France	0.30		
United States	-0.20	Tunisia	0.05	Chile	0.06		
Argentina	-0.21	South Africa	-0.10	New Zealand	0.00		
		Chile	-0.11	Switzerland	-0.02		
		Argentina	-0.22	Australia	-0.06		
		France	-0.97	South Africa	-0.09		
				Moldova	-0.09		
				United States	-0.17		
				Italy	-0.65		
				Spain	-1.30		

Pamid		**Douce Noire**		**Negroamaro**		**Doukkali**	
Country	*VII*	*Country*	*VII*	*Country*	*VII*	*Country*	*VII*
Bulgaria	4.53	Argentina	3.05	Italy	2.96	Morocco	3.35
Greece	-0.05	Italy	0.05				
Hungary	-0.06	United States	-0.13				
		France	-0.66				

Table 64: National NVIIs for the world's top 24 red varieties, 2016

Cabernet Sauvignon		Merlot		Tempranillo		Syrah	
Country	*VII*	*Country*	*VII*	*Country*	*VII*	*Country*	*VII*
Chile	7.20	France	13.40	Spain	33.54	Australia	7.49
China	6.24	United States	1.56	Portugal	1.80	France	6.53
United States	5.40	Bulgaria	1.54	Myanmar	0.00	South Africa	1.36
Australia	3.30	China	1.37	Thailand	0.00	Argentina	0.97
Bulgaria	1.26	Chile	0.76	Mexico	-0.01	Chile	0.47
Russia	1.12	Moldova	0.62	Israel	-0.04	Turkey	0.20
South Africa	0.88	Romania	0.18	Canada	-0.14	Algeria	0.15
Ukraine	0.71	Serbia	0.15	Turkey	-0.15	India	0.09
Moldova	0.55	Australia	0.12	Switzerland	-0.16	Israel	0.04
Argentina	0.24	Algeria	0.11	Brazil	-0.36	Lebanon	0.03
Lebanon	0.16	Israel	0.09	New Zealand	-0.38	Thailand	0.01
Israel	0.14	Uruguay	0.08	Greece	-0.55	Myanmar	0.01
Serbia	0.13	Lebanon	0.06	Argentina	-0.88	Cambodia	0.00
Algeria	0.09	Switzerland	0.05	South Africa	-1.02	Tunisia	-0.02
Mexico	0.08	Croatia	0.03	Australia	-1.29	Mexico	-0.02

Garnacha Tinta		Pinot Noir		Sangiovese		Bobal	
Country	*VII*	*Country*	*VII*	*Country*	*VII*	*Country*	*VII*
France	11.45	United States	3.87	Italy	13.05	Spain	10.60
Spain	5.58	France	2.77	Ethiopia	0.02		
Algeria	0.38	Germany	2.00	Thailand	0.00		
Morocco	0.04	New Zealand	1.04	Canada	-0.05		
Tunisia	0.01	Switzerland	0.86	Turkey	-0.05		
Mexico	-0.01	Australia	0.38	Switzerland	-0.05		
Peru	-0.03	Chile	0.15	Brazil	-0.12		
Uruguay	-0.05	United Kingdom	0.11	New Zealand	-0.13		
Canada	-0.09	Moldova	0.09	Hungary	-0.23		
Turkey	-0.09	Czechia	0.08	South Africa	-0.33		
Switzerland	-0.11	Canada	0.08	Argentina	-0.34		
New Zealand	-0.26	Serbia	0.03	Australia	-0.39		
China	-0.44	Luxembourg	0.02	Chile	-0.50		
South Africa	-0.64	India	0.01	Romania	-0.65		
Australia	-0.66	Kazakhstan	0.00	United States	-0.69		

Table 64 (cont.): National NVIIs for the world's top 24 red varieties, 2016

Cabernet Franc		**Côt**		**Monastrell**		**Mazuelo**	
Country	*VII*	*Country*	*VII*	*Country*	*VII*	*Country*	*VII*
France	4.94	Argentina	8.48	Spain	6.93	France	5.17
Brazil	1.43	Chile	0.13	Israel	0.00	Algeria	0.65
Canada	0.15	Israel	0.01	Canada	-0.03	Morocco	0.23
Hungary	0.13	Peru	-0.01	Turkey	-0.03	Israel	0.20
Uruguay	0.04	Uruguay	-0.01	Switzerland	-0.04	Mexico	0.09
Israel	0.01	Canada	-0.02	South Africa	-0.14	Tunisia	0.05
Kazakhstan	-0.01	Turkey	-0.03	France	-0.15	Turkey	0.00
Switzerland	-0.03	Switzerland	-0.04	Australia	-0.19	Greece	-0.12
Turkey	-0.03	New Zealand	-0.06	Chile	-0.35	Chile	-0.16
Slovenia	-0.04	Brazil	-0.08	Romania	-0.47	South Africa	-0.20
Serbia	-0.04	South Africa	-0.15	United States	-0.50	Australia	-0.31
Chile	-0.05	Hungary	-0.17	Argentina	-0.53	United States	-0.32
Moldova	-0.06	Moldova	-0.18			Portugal	-0.37
New Zealand	-0.07	Australia	-0.23			China	-0.40
South Africa	-0.08	United States	-0.26			Argentina	-0.48

Alicante Henri Bouschet		**Tribidrag**		**Montepulciano**		**Gamay Noir**	
Country	*VII*	*Country*	*VII*	*Country*	*VII*	*Country*	*VII*
Spain	2.72	United States	3.74	Italy	6.31	France	4.31
Chile	1.28	Italy	2.09	Brazil	-0.05	Switzerland	0.28
Portugal	0.69	North Macedonia	0.18	New Zealand	-0.06	Canada	0.04
Morocco	0.17	Canada	-0.02	Australia	-0.20	Turkey	0.03
Turkey	0.09	Switzerland	-0.02	Chile	-0.24	Uruguay	-0.01
Tunisia	0.03	New Zealand	-0.06	Argentina	-0.32	Serbia	-0.02
Uruguay	-0.01	South Africa	-0.15	United States	-0.38	Slovenia	-0.02
Canada	-0.02	Australia	-0.20	Spain	-1.45	Brazil	-0.04
Switzerland	-0.03	Chile	-0.23			New Zealand	-0.04
Brazil	-0.04	Romania	-0.30			Hungary	-0.08
Greece	-0.08	Argentina	-0.35			South Africa	-0.12
Hungary	-0.11	France	-1.36			Australia	-0.17
South Africa	-0.17	Spain	-1.48			Chile	-0.19
Australia	-0.23					Portugal	-0.24
Romania	-0.32					Argentina	-0.27

Table 64 (cont.): National NVIIs for the world's top 24 red varieties, 2016

Cinsaut		Carmenère		Douce Noire		Barbera	
Country	*VII*	*Country*	*VII*	*Country*	*VII*	*Country*	*VII*
France	2.62	China	2.30	Argentina	4.05	Italy	2.81
Morocco	0.70	Chile	2.18	United States	-0.23	United States	0.26
South Africa	0.28	Canada	-0.01	Italy	-0.45	Slovenia	0.01
Tunisia	0.14	Switzerland	-0.02	France	-0.80	Canada	-0.01
Turkey	0.08	Brazil	-0.03			Switzerland	-0.01
Chile	0.02	Hungary	-0.07			Brazil	-0.03
Greece	-0.06	South Africa	-0.11			South Africa	-0.08
Australia	-0.15	Australia	-0.14			Argentina	-0.08
Portugal	-0.21	Argentina	-0.22			Australia	-0.09
Argentina	-0.24	United States	-0.26			Chile	-0.13
United States	-0.26	Italy	-0.53			Romania	-0.16
Italy	-0.69	France	-0.91			Spain	-0.78
Spain	-1.01	Spain	-0.99				

Isabella		Blaufränkisch		Criolla Grande		Pinot Meunier	
Country	*VII*	*Country*	*VII*	*Country*	*VII*	*Country*	*VII*
Brazil	2.57	Hungary	1.56	Argentina	3.32	France	2.11
Moldova	0.70	Austria	0.59			Germany	0.38
Russia	0.26	Germany	0.31			United Kingdom	0.04
Ukraine	0.25	Slovakia	0.26			Canada	-0.01
Uruguay	0.02	Czechia	0.24			Switzerland	-0.01
Switzerland	-0.01	Slovenia	0.14			New Zealand	-0.02
Australia	-0.11	Serbia	0.14			Moldova	-0.03
		Croatia	0.11			South Africa	-0.07
		Peru	0.06			Australia	-0.08
		Romania	0.01			Chile	-0.11
		Canada	-0.01			Argentina	-0.15
		Switzerland	-0.01			United States	-0.16
		Australia	-0.11			Italy	-0.44
		United States	-0.20			Spain	-0.65
		Italy	-0.51				

Table 65: National NVIIs for the world's top 24 white varieties, 2000

Airén		**Chardonnay**		**Trebbiano Toscano**		**Graševina**	
Country	*VII*	*Country*	*VII*	*Country*	*VII*	*Country*	*VII*
Spain	60.19	United States	6.25	France	13.52	Serbia	6.51
		Australia	2.74	Italy	4.41	Croatia	3.05
		France	2.20	Greece	-0.14	Romania	2.21
		Chile	0.88	Brazil	-0.16	Hungary	1.03
		South Africa	0.67	Bulgaria	-0.18	Slovakia	0.74
		New Zealand	0.51	South Africa	-0.51	Austria	0.70
		Moldova	0.50	Argentina	-0.57	Slovenia	0.64
		Slovenia	0.17	Australia	-0.61	Bulgaria	0.37
		Canada	0.15	United States	-0.98	Czechia	0.21
		Hungary	0.08	Portugal	-1.10	Brazil	-0.02
		Czechia	0.05	Spain	-6.78	Italy	-2.05
		Slovakia	0.03			Spain	-4.17
		United Kingdom	0.01				
		Israel	0.00				
		Luxembourg	-0.01				

Rkatsiteli		**Sauvignon Blanc**		**Cayetana Blanca**		**Catarratto Bianco**	
Country	*VII*	*Country*	*VII*	*Country*	*VII*	*Country*	*VII*
Georgia	3.93	France	1.92	Spain	8.60	Italy	9.02
Russia	2.53	Moldova	1.42	Australia	-0.25		
Moldova	2.10	Chile	1.05				
Bulgaria	1.66	South Africa	0.86				
Armenia	0.47	New Zealand	0.47				
Romania	-0.52	United States	0.38				
		Romania	0.34				
		Slovenia	0.19				
		Australia	0.18				
		Israel	0.04				
		Canada	0.01				
		Uruguay	0.00				
		Switzerland	-0.03				
		Austria	-0.07				
		Greece	-0.11				

Table 65 (cont.): National NVIIs for the world's top 24 white varieties, 2000

Macabeo		Chenin Blanc		Riesling		Colombard	
Country	*VII*	*Country*	*VII*	*Country*	*VII*	*Country*	*VII*
Spain	6.40	South Africa	4.44	Germany	4.38	United States	3.40
Argentina	-0.40	United States	1.39	Australia	0.40	South Africa	2.19
France	-0.67	France	0.36	Austria	0.25	Australia	0.16
		Argentina	0.33	Russia	0.18	Israel	0.09
		Israel	0.01	Hungary	0.17	France	0.01
		New Zealand	0.01	Czechia	0.14		
		Switzerland	-0.03	Moldova	0.11		
		Australia	-0.08	United States	0.08		
		Chile	-0.20	Canada	0.08		
		Spain	-2.24	New Zealand	0.08		
				Luxembourg	0.03		
				Switzerland	-0.03		
				Bulgaria	-0.04		
				South Africa	-0.07		
				Chile	-0.15		

Aligoté		Müller-Thurgau		Palomino Fino		Muscat Blanc à Petits Grains	
Country	*VII*	*Country*	*VII*	*Country*	*VII*	*Country*	*VII*
Moldova	3.10	Germany	4.09	Spain	4.15	Italy	1.86
Romania	1.22	Austria	0.60	South Africa	0.21	Greece	0.39
Russia	0.29	Hungary	0.55	New Zealand	-0.01	France	0.33
Bulgaria	0.20	Slovakia	0.36	Australia	-0.14	Hungary	0.21
Switzerland	-0.02	Czechia	0.31	United States	-0.16	Armenia	0.09
Georgia	-0.04	Switzerland	0.12	Argentina	-0.21	South Africa	0.04
France	-0.93	Luxembourg	0.09	France	-1.01	Portugal	0.03
		New Zealand	0.07			Switzerland	-0.01
		Moldova	-0.09			Austria	-0.03
		Italy	-0.69			Romania	-0.07
		France	-1.21			Moldova	-0.08
						Germany	-0.11
						United States	-0.12
						Australia	-0.12
						Argentina	-0.22

Table 65 (cont.): National NVIIs for the world's top 24 white varieties, 2000

Muscat of Alexandria		Sémillon		Fetească Albă		Grüner Veltliner	
Country	*VII*	*Country*	*VII*	*Country*	*VII*	*Country*	*VII*
Argentina	0.88	France	1.92	Romania	3.50	Austria	3.53
South Africa	0.71	Australia	1.19	Moldova	0.80	Slovakia	0.59
Morocco	0.69	Chile	0.26	Hungary	0.11	Czechia	0.34
Australia	0.35	South Africa	0.11	Slovakia	0.05	Hungary	0.19
United States	0.19	New Zealand	0.04			Italy	-0.60
Brazil	0.10	Brazil	0.02				
Israel	0.04	Argentina	-0.01				
Portugal	-0.15	Switzerland	-0.02				
Spain	-0.21	United States	-0.05				
France	-0.45	Greece	-0.05				
Italy	-0.55	Italy	-0.70				

Trebbiano Romagnolo		Pedro Ximénez		Pinot Blanc		Garganega	
Country	*VII*	*Country*	*VII*	*Country*	*VII*	*Country*	*VII*
Italy	3.47	Spain	2.17	Russia	0.57	Italy	2.94
		Chile	0.40	Italy	0.57	Argentina	-0.14
		Australia	-0.08	Austria	0.57		
				Germany	0.42		
				Slovakia	0.12		
				Luxembourg	0.03		
				Canada	0.02		
				Georgia	0.01		
				Moldova	0.01		
				Switzerland	0.01		
				United States	-0.04		
				South Africa	-0.06		
				Chile	-0.08		
				Argentina	-0.13		
				France	-0.33		

Table 66: National NVIIs for the world's top 24 white varieties, 2016

Airén		Chardonnay		Sauvignon Blanc		Trebbiano Toscano	
Country	*VII*	*Country*	*VII*	*Country*	*VII*	*Country*	*VII*
Spain	36.38	United States	6.83	New Zealand	4.35	France	12.71
Tunisia	-0.02	Australia	3.43	Chile	2.44	Italy	4.29
Morocco	-0.08	France	2.41	South Africa	1.47	Uruguay	0.11
		Chile	1.09	France	1.21	India	0.05
		South Africa	0.57	Moldova	1.03	Canada	-0.07
		New Zealand	0.34	Australia	0.53	Brazil	-0.15
		Russia	0.27	Russia	0.24	Bulgaria	-0.15
		Canada	0.19	Ukraine	0.19	Greece	-0.26
		Lebanon	0.18	Slovenia	0.15	Russia	-0.29
		Bulgaria	0.16	Czechia	0.12	Moldova	-0.43
		Serbia	0.10	Romania	0.11	South Africa	-0.54
		Slovenia	0.10	India	0.09	China	-0.73
		United Kingdom	0.10	Lebanon	0.09	Australia	-0.79
		Moldova	0.09	Serbia	0.03	Argentina	-0.87
		Ukraine	0.08	United States	0.02	Portugal	-1.07
Riesling		**Rkatsiteli**		**Macabeo**		**Cayetana Blanca**	
Country	*VII*	*Country*	*VII*	*Country*	*VII*	*Country*	*VII*
Germany	4.52	Georgia	5.53	Spain	6.55	Spain	6.49
Romania	0.82	Russia	1.31	South Africa	-0.18	Australia	-0.24
United States	0.39	Ukraine	1.22	France	-1.20	Portugal	-0.30
Russia	0.35	Bulgaria	1.07				
Austria	0.31	Kazakhstan	0.77				
Australia	0.30	Moldova	0.66				
Serbia	0.24	North Macedonia	0.04				
Canada	0.23	Serbia	-0.04				
Ukraine	0.23	Romania	-0.38				
Czechia	0.22						
Moldova	0.13						
North Macedonia	0.13						
Slovakia	0.12						
Croatia	0.10						

Table 66 (cont.): National NVIIs for the world's top 24 white varieties, 2016

Muscat Alexandria		**Muscat Blanc à Petits Grains**		**Chenin Blanc**		**Colombard**	
Country	*VII*	*Country*	*VII*	*Country*	*VII*	*Country*	*VII*
Chile	0.96	Italy	1.96	South Africa	3.80	South Africa	2.42
Spain	0.60	France	0.27	France	0.80	United States	1.42
Morocco	0.44	Greece	0.26	Argentina	0.15	France	0.67
China	0.36	Austria	0.11	United States	0.05	Australia	0.20
Australia	0.26	Slovenia	0.10	Mexico	0.05	Israel	0.04
Argentina	0.25	Peru	0.07	Ethiopia	0.01	Thailand	0.00
South Africa	0.23	Hungary	0.06	Thailand	0.00	Brazil	-0.04
Greece	0.08	North Macedonia	0.05	Peru	-0.01	Spain	-1.32
Tunisia	0.08	Mexico	0.05	Uruguay	-0.01		
Israel	0.04	Romania	0.05	Canada	-0.02		
Algeria	0.03	Ukraine	0.03	Switzerland	-0.02		
United States	0.03	South Africa	0.03	New Zealand	-0.05		
India	0.02	Russia	0.02	Brazil	-0.05		
Uruguay	-0.01	Turkey	0.01	Hungary	-0.10		
Brazil	-0.06	Myanmar	0.00	Australia	-0.12		
Catarratto Bianco		**Aligoté**		**Graševina**		**Palomino Fino**	
Country	*VII*	*Country*	*VII*	*Country*	*VII*	*Country*	*VII*
Italy	5.51	Moldova	1.62	Croatia	0.98	Spain	3.47
United States	-0.33	Russia	1.24	Hungary	0.80	Portugal	0.37
		Romania	1.06	Austria	0.67	Mexico	0.02
		Ukraine	1.04	China	0.45	Switzerland	-0.02
		Kazakhstan	0.05	Serbia	0.43	New Zealand	-0.04
		Bulgaria	-0.01	Slovenia	0.41	South Africa	-0.08
		Canada	-0.01	Czechia	0.23	Australia	-0.15
		Switzerland	-0.01	Romania	0.10	Argentina	-0.21
		Georgia	-0.04	Slovakia	0.09	United States	-0.26
		Hungary	-0.09	North Macedonia	0.03	France	-0.93
		France	-0.66	Brazil	0.00		
				Italy	-0.45		
				Spain	-0.83		

Table 66 (cont.): National NVIIs for the world's top 24 white varieties, 2016

Prosecco		Müller-Thurgau		Grüner Veltliner		Trebbiano Romagnolo	
Country	*VII*	*Country*	*VII*	*Country*	*VII*	*Country*	*VII*
Italy	3.80	Germany	2.51	Austria	3.16	Italy	3.68
Brazil	0.01	Austria	0.35	Slovakia	0.36		
Australia	-0.10	Czechia	0.32	Czechia	0.33		
Argentina	-0.20	Hungary	0.31	Hungary	0.25		
		Slovakia	0.11	Canada	-0.01		
		Switzerland	0.09	Switzerland	-0.01		
		Luxembourg	0.07	New Zealand	-0.02		
		Slovenia	0.01	Germany	-0.09		
		United Kingdom	0.00	South Africa	-0.09		
		Japan	0.00	Australia	-0.12		
		Canada	-0.01	Romania	-0.17		
		Russia	-0.03	Argentina	-0.19		
		New Zealand	-0.03	United States	-0.21		
		Moldova	-0.08	Italy	-0.56		
		Australia	-0.13				

Sémillon		Verdejo		Viognier		Pedro Giménez	
Country	*VII*	*Country*	*VII*	*Country*	*VII*	*Country*	*VII*
France	1.52	Spain	3.21	France	1.32	Argentina	2.34
Australia	0.89	Australia	-0.12	United States	0.14	Chile	0.86
South Africa	0.16	Chile	-0.13	South Africa	0.11		
Turkey	0.11			Chile	0.07		
Chile	0.05			Australia	0.06		
Uruguay	0.00			Canada	0.01		
Canada	-0.01			Argentina	0.01		
Switzerland	-0.01			Uruguay	0.00		
New Zealand	-0.02			New Zealand	0.00		
Argentina	-0.02			Thailand	0.00		
Brazil	-0.03			Switzerland	0.00		
Russia	-0.04			Turkey	-0.01		
Greece	-0.04			Brazil	-0.02		
Hungary	-0.05			Hungary	-0.05		

VI. Regional coverage of each country

Table 67: Winegrape area by region and region's national share, by country, 2000

Region	*Area (hectares)*	*National share (%)*	*Region*	*Area (hectares)*	*National share (%)*
Algeria	**30200**	**100**	**Argentina (cont.)**		
			Poman	29	0.01
Argentina			Puelen	122	0.06
Adolfo Alsina	10	0.00	Rawson	1584	0.80
Albardón	1238	0.63	Rivadavia - Mza	15919	8.06
Andalgala	7	0.00	Rivadavia - San Juan	480	0.24
Añelo	624	0.32	San Alberto	1	0.00
Angaco	2328	1.18	San Blas De Los Sauces	45	0.02
Arauco	26	0.01	San Carlos - Mza	4220	2.14
Avellaneda - Río Negro	443	0.22	San Carlos - Salta	495	0.25
Ayacucho	12	0.01	San Javier	1	0.00
Belén	129	0.07	San Martín - Mza	30344	15.37
Cachi	4	0.00	San Martín - San Juan	3408	1.73
Cafayatc	1334	0.68	San Rafael	16109	8.16
Calingasta	72	0.04	Sanagasta	3	0.00
Capital San Juan	17	0.01	Santa Lucía	1380	0.70
Capital Santiago del Esterc	2	0.00	Santa María - Catamarca	579	0.29
Castro Barros	219	0.11	Santa Rosa - Mza	10905	5.52
Caucete	6606	3.35	Sarmiento - San Juan	6533	3.31
Chilecito	5810	2.94	Tafí del Valle	18	0.01
Chimbas	823	0.42	Tinogasta	1414	0.72
Colón - Entre Ríos	224	0.11	Totoral	6	0.00
Conesa	50	0.03	Tunuyán	4041	2.05
Confluencia	39	0.02	Tupungato	5935	3.01
Coronel Felipe Varela	856	0.43	Ullum	889	0.45
Coronel Pringles	5	0.00	Valle Fértil	25	0.01
Cruz del Eje	60	0.03	Veinticinco de Mayo - San	7423	3.76
Cushamen	20	0.01	Vinchina	75	0.04
El Cuy	44	0.02	Zonda	998	0.51
Famatina	229	0.12	**Total**	**197418**	**100**
General Alvear	5675	2.87			
General Lamadrid	45	0.02	**Armenia**	**11206**	**100**
General Pueyrredón	2	0.00			
General Roca	2076	1.05	**Australia**		
Godoy Cruz	10	0.00	Adelaide Hills	1811	1.39
Guaymallén	1101	0.56	Alpine Valleys/Beechwortl	803	0.61
Iglesia	24	0.01	Australian Capital Territor	15	0.01
Ischilin	48	0.02	Barossa - other	249	0.19
Jachal	42	0.02	Barossa Valley	7673	5.88
Junín - Mza	11112	5.63	Beechworth	35	0.03
La Paz	369	0.19	Bendigo	607	0.46
La Rioja	21	0.01	Big Rivers - other	1015	0.78
Las Heras	1640	0.83	Blackwood Valley	501	0.38
Lavalle	12762	6.46	Canberra District	110	0.08
Luján de Cuyo	10809	5.48	Central Ranges - other	302	0.23
Maipú	11463	5.81	Central Victoria - other	876	0.67
Molinos	30	0.01	Central Western Australia	70	0.05
Nueve de Julio	2675	1.35	Clare Valley	3617	2.77
Pichi Mahuida	38	0.02	Cowra	1533	1.17
Pocito	3267	1.65	Currency Creek	940	0.72

Table 67 (cont.): Winegrape area by region and region's national share, by country, 2000

Region	*Area (hectares)*	*National share (%)*
Australia (cont.)		
Eastern Plains, Inland and	39	0.03
Eden Valley	1224	0.94
Far North - other	107	0.08
Fleurieu - other	510	0.39
Geelong	322	0.25
Geographe	480	0.37
Gippsland	174	0.13
Goulburn Valley	1090	0.83
Grampians	424	0.32
Granite Belt	433	0.33
Great Southern	2391	1.83
Greater Perth - other	395	0.30
Hastings River	118	0.09
Henty	183	0.14
Hilltops	383	0.29
Hunter	3669	2.81
Hunter Valley - other	278	0.21
Kangaroo Island	34	0.03
Langhorne Creek	3737	2.86
Limestone Coast - other	7529	5.76
Lower Murray - other	145	0.11
Margaret River	3401	2.60
McLaren Vale	4695	3.59
Mornington Peninsula	402	0.31
Mount Benson	299	0.23
Mount Lofty Ranges - othe	488	0.37
Mudgee	2152	1.65
Murray Darling (NSW)	5576	4.27
Murray Darling (VIC)	15663	11.99
North East Victoria - other	1254	0.96
North West Victoria - othe	3177	2.43
Northern Rivers - other	12	0.01
Northern Slopes - other	174	0.13
Northern Territory	302	0.23
Orange	995	0.76
Padthaway	3226	2.47
Perricoota	153	0.12
Perth Hills	316	0.24
Port Phillip - other	129	0.10
Pyrenees	428	0.33
Queensland - other	1207	0.92
Riverina	12398	9.49
Riverland	18336	14.04
Rutherglen	793	0.61
South Burnett	344	0.26
South Coast - other	112	0.09
South West Australia - oth	802	0.61
Southern Fleurieu	328	0.25
Southern NSW - other	914	0.70
Australia (cont.)		
Sunbury	79	0.06
Swan District	812	0.62
Swan Hill (NSW)	544	0.42
Swan Hill (VIC)	3725	2.85
Tasmania	680	0.52
The Peninsulas	50	0.04
Tumbarumba	278	0.21
Western Australia Southea	65	0.05
Western Plains - other	328	0.25
Western Victoria - other	110	0.08
Yarra Valley	2038	1.56
Total	**130602**	**100**
Austria		
Burgenland	14540	29.98
Niederosterreich	29975	61.81
Steiermark	3283	6.77
Wien and other regions	699	1.44
Total	**48496**	**100**
Brazil	**52840**	**100**
Bulgaria	**95997**	**100**
Canada	**8498**	**100**
Chile		
Araucania	5	0.00
Atacama	797	0.70
Coquimbo	11083	9.72
Del Bio Bio	13747	12.06
Del Maule	45053	39.53
Metropolitana	9453	8.29
O'Higgins	29044	25.48
Valparaiso	4783	4.20
Total	**113966**	**100**
Canada	**8498**	**100**
Croatia	**59448**	**100**
Cyprus	**18282**	**100**
Czechia	**11331**	**100**

Table 67 (cont.): Winegrape area by region and region's national share, by country, 2000

Region	*Area (hectares)*	*National share (%)*	*Region*	*Area (hectares)*	*National share (%)*
France			**Greece**		
Aisne	2340	0.27	Anatoliki Makedonia, Thra	384	0.01
Alpes-de-Haute-Provence,	1082	0.13	Attiki	6112	0.12
Alsace	15128	1.75	Dytiki Ellada	8031	0.16
Aquitaine except Gironde	25732	2.98	Dytiki Makedonia	1690	0.03
Ardeche	12295	1.42	Ionia Nisia	2707	0.05
Aude	85270	9.86	Ipeiros	632	0.01
Auvergne	1096	0.13	Kentriki Makedonia	2494	0.05
Bouches-du-Rhone	11089	1.28	Kriti	5312	0.10
Bourgogne	29941	3.46	Notio Aigaio	4124	0.08
Centre-Val de Loire	22316	2.58	Peloponnissos	8197	0.16
Champagne-Ardenne	28671	3.32	Sterea Ellada	5991	0.12
Charente	38514	4.45	Thessalia	3043	0.06
Charente-Maritime	40321	4.66	Voreio Aigaio	2196	0.04
Correze, Haute-Vienne	50	0.01	**Total**	**50915**	**100**
Corse	6992	0.81			
Deux-Sevres, Vienne	1973	0.23	**Hungary**	**86886**	**100**
Franche Comté	2020	0.23			
Gard	67133	7.76	**Israel**	**4851**	**100**
Gers	19913	2.30			
Gironde	124617	14.41	**Italy**		
Herault	105647	12.22	Agrigento	19624	3.08
Lorraine	184	0.02	Alessandria	14873	2.34
Midi-Pyrenees except Gers	18872	2.18	Ancona	4591	0.72
Pays de la Loire except Ma	37882	4.38	Arezzo	4887	0.77
Pyrenees-Orientales	37659	4.35	Ascoli Piceno	8211	1.29
Rhone-Alpes except Ardec	44861	5.19	Asti	18016	2.83
Seine-et-Marne	26	0.00	Avellino	6287	0.99
Var	31420	3.63	Bari	16954	2.66
Vaucluse	51801	5.99	Belluno	9	0.00
Total	**864846**	**100**	Benevento	10291	1.62
			Bergamo	881	0.14
Georgia	**37419**	**100**	Biella	289	0.05
			Bologna	7049	1.11
Germany			Bolzano-Bozen	4781	0.75
Ahr	520	0.50	Brescia	3904	0.61
Baden	15554	14.92	Brindisi	13498	2.12
Franken	5993	5.75	Cagliari	8398	1.32
Hessische Bergstraße	456	0.44	Caltanissetta	5526	0.87
Mittelrhein	572	0.55	Campobasso	4858	0.76
Mosel	11521	11.05	Caserta	2964	0.47
Nahe	4604	4.42	Catania	3875	0.61
Pfalz	23342	22.39	Catanzaro	1336	0.21
Rheingau	3217	3.09	Chieti	25986	4.08
Rheinhessen	26385	25.31	Como	18	0.00
Saale	618	0.59	Cosenza	5107	0.80
Sachsen	409	0.39	Cremona	39	0.01
Württemberg	11042	10.59	Crotone	3266	0.51
Total	**104233**	**100**	Cuneo	16272	2.56

Table 67 (cont.): Winegrape area by region and region's national share, by country, 2000

Region	*Area (hectares)*	*National share (%)*
Italy (cont.)		
Enna	586	0.09
Ferrara	569	0.09
Firenze	15940	2.50
Foggia	28797	4.52
Forlì-Cesena	6746	1.06
Frosinone	4162	0.65
Genova	147	0.02
Gorizia	3474	0.55
Grosseto	3843	0.60
Imperia	454	0.07
Isernia	563	0.09
La Spezia	833	0.13
L'Aquila	855	0.13
Latina	6045	0.95
Lecce	10021	1.57
Lecco	44	0.01
Livorno	1214	0.19
Lodi	22	0.00
Lucca	728	0.11
Macerata	1482	0.23
Mantova	1830	0.29
Massa-Carrara	275	0.04
Matera	1645	0.26
Messina	2253	0.35
Milano	238	0.04
Modena	7585	1.19
Napoli	1956	0.31
Novara	575	0.09
Nuoro	7208	1.13
Oristano	3250	0.51
Padova	6313	0.99
Palermo	16515	2.59
Parma	574	0.09
Pavia	13734	2.16
Perugia	8040	1.26
Pesaro e Urbino	1391	0.22
Pescara	3146	0.49
Piacenza	5568	0.87
Pisa	2386	0.37
Pistoia	633	0.10
Pordenone	6671	1.05
Potenza	5431	0.85
Prato	335	0.05
Ragusa	1520	0.24
Ravenna	16910	2.66
Reggio di Calabria	2101	0.33
Reggio nell'Emilia	8263	1.30
Rieti	1490	0.23
Rimini	2653	0.42
Italy (cont.)		
Roma	10519	1.65
Rovigo	258	0.04
Salerno	5222	0.82
Sassari	6232	0.98
Savona	399	0.06
Siena	15718	2.47
Siracusa	1583	0.25
Sondrio	987	0.16
Taranto	15156	2.38
Teramo	2785	0.44
Terni	5548	0.87
Torino	1777	0.28
Trapani	59078	9.28
Trento	8844	1.39
Treviso	24008	3.77
Trieste	182	0.03
Udine	7171	1.13
Valle d'Aosta	424	0.07
Varese	19	0.00
Venezia	6018	0.95
Verbano-Cusio-Ossola	29	0.00
Vercelli	169	0.03
Verona	23359	3.67
Vibo Valentia	845	0.13
Vicenza	6898	1.08
Viterbo	4628	0.73
Total	**636662**	**100**
Korea, Rep.	**5400**	**100**
Luxembourg	**1348**	**100**
Moldova	**89844**	**100**
Morocco	**49600**	**100**
New Zealand		
Auckland	392	3.95
Canterbury	232	2.33
Gisborne	1681	16.91
Hawkes Bay	2443	24.57
Marlborough	4054	40.77
Nelson	203	2.04
Otago	280	2.81
Waikato	119	1.20
Waipara	210	2.11
Wairarapa	328	3.30
Total	**9942**	**100**

Table 67 (cont.): Winegrape area by region and region's national share, by country, 2000

Region	*Area (hectares)*	*National share (%)*
Portugal		
Acores	1689	0.82
Alentejo	14905	7.27
Algarve	1900	0.93
Alto Tras-os-Montes	63371	30.91
Beira Interior	20851	10.17
Beira Litoral	23921	11.67
Entre Douro e Minho	29678	14.48
Madeira	1513	0.74
Ribatejo e Oeste	47173	23.01
Total	**205003**	**100**
Romania	**222173**	**100**
Russia	**56332**	**100**
Serbia	**68999**	**100**
Slovakia	**15580**	**100**
Slovenia	**23472**	**100**
South Africa		
Breedekloof	10385	11.09
Little Karoo	3168	3.38
Northern Cape	5025	5.36
Olifants River	9015	9.63
Paarl	17249	18.42
Robertson	12227	13.06
Stellenbosch	16112	17.20
Swartland	13670	14.60
Worcester	6805	7.27
Total	**93656**	**100**
Spain		
Alava	11338	0.96
Albacete	107563	9.10
Alicante	25466	2.15
Almeria, Granada, Jaen, S(	9635	0.82
Avila, Palencia, Salamanca	13995	1.18
Badajoz	81209	6.87
Barcelona	24223	2.05
Burgos	13264	1.12
Caceres	4310	0.36
Cadiz	9931	0.84
Canarias	13727	1.16
Cantabria	39	0.00
Castellon	1190	0.10
Ciudad Real	206366	17.46
Comunidad de Madrid	17475	1.48
Spain		
Comunidad Foral de Navar	23619	2.00
Cordoba	12777	1.08
Cuenca	97721	8.27
Galicia	31747	2.69
Girona, Lleida	7390	0.63
Guadalajara	2827	0.24
Guipuzcoa, Vizcaya	243	0.02
Huelva	6747	0.57
Huesca, Teruel	7599	0.64
Illes Balears	1718	0.15
La Rioja	39459	3.34
Leon	15964	1.35
Malaga	6467	0.55
Principado de Asturias	84	0.01
Region de Murcia	45058	3.81
Tarragona	32839	2.78
Toledo	172334	14.58
Valencia	64216	5.43
Valladolid	15837	1.34
Zamora	15883	1.34
Zaragoza	41546	3.52
Total	**1181806**	**100**
Switzerland		
Aargau	395	2.62
Basel Land	99	0.66
Bern	257	1.71
Fribourg	116	0.77
Geneva	1355	9.01
Graubünden	414	2.75
Jura	7	0.05
Lucerne	19	0.12
Neuchâtel	605	4.02
Schaffhausen	500	3.32
Schwyz	30	0.20
St. Gallen	217	1.44
Thurgau	274	1.82
Ticino	961	6.39
Valais	5255	34.94
Vaud	3879	25.79
Zürich	645	4.29
other regions	14	0.10
Total	**15042**	**100**
Taiwan	**2833**	**100**
Tunisia	**16836**	**100**
United Kingdom	**873**	**100**

Table 67 (cont.): Winegrape area by region and region's national share, by country, 2000

Region	Area (hectares)	National share (%)
United States		
Alameda	546	0.31
Amador	1014	0.58
Benton Co.	88	0.05
Butte	58	0.03
Calaveras	114	0.06
Chautauqua-Erie	8116	4.62
Columbia River	293	0.17
Colusa	539	0.31
Contra Costa	397	0.23
Douglas Co.	190	0.11
El Dorado	338	0.19
Finger Lakes	3692	2.10
Fresno	17606	10.02
Glenn	580	0.33
Humboldt	4	0.00
Josephine Co.	117	0.07
Kern	11198	6.37
Kings	949	0.54
Lake	1444	0.82
Lane Co.	254	0.14
Los Angeles	12	0.01
Madera	17427	9.92
Marin	33	0.02
Marion Co.	221	0.13
Mariposa	23	0.01
Mendocino	5050	2.87
Merced	5901	3.36
Michigan	526	0.30
Monterey	11688	6.65
Napa	12258	6.98
Nevada	76	0.04
New York - other	1544	0.88
Oregon - other	216	0.12
Placer	37	0.02
Polk Co.	383	0.22
Riverside	845	0.48
Sacramento	3611	2.06
San Benito	720	0.41
San Bernardino	558	0.32
San Diego	25	0.01
San Joaquin	20930	11.91
San Luis Obispo	5047	2.87
San Mateo	19	0.01
Santa Barbara	4043	2.30
Santa Clara	443	0.25
Santa Cruz	68	0.04
Shasta	15	0.01
Solano	698	0.40
Sonoma	14708	8.37
United States (cont.)		
Stanislaus	5358	3.05
Sutter	32	0.02
Tehama	53	0.03
Trinity	15	0.01
Tulare	4602	2.62
Valley - other	106	0.06
Ventura	3	0.00
Washington	6880	3.92
Washington Co.	393	0.22
Yamhill Co.	1016	0.58
Yolo	2446	1.39
Yuba	126	0.07
Total	**175693**	**100**
Uruguay	**8880**	**100**
Missing 9	**80221**	**100**

Table 68: Winegrape area by region and region's national share, by country, 2010

Region	*Area (hectares)*	*National share (%)*
Algeria	**30200**	**100**
Argentina		
Adolfo Alsina	67	0.03
Albardón	839	0.39
Ambato	2	0.00
Andalgala	8	0.00
Añelo	1504	0.70
Angaco	1961	0.92
Arauco	5	0.00
Avellaneda - Río Negro	224	0.10
Belén	152	0.07
Cachi	37	0.02
Cafayate	1608	0.75
Calamuchita	1	0.00
Calingasta	103	0.05
Capital San Juan	11	0.01
Castro Barros	206	0.10
Caucete	6443	3.02
Chilecito	6016	2.82
Chimbas	765	0.36
Chos Malal	2	0.00
Collon Cura	5	0.00
Colón - Cba	155	0.07
Colón - Entre Ríos	5	0.00
Concordia	3	0.00
Conesa	47	0.02
Confluencia	93	0.04
Coronel Felipe Varela	883	0.41
Coronel Suarez	13	0.01
Cruz del Eje	29	0.01
Cushamen	20	0.01
El Cuy	67	0.03
Famatina	322	0.15
General Alvear	5341	2.50
General Lamadrid	74	0.03
General Roca	1879	0.88
Godoy Cruz	1	0.00
Guaymallén	955	0.45
Iglesia	21	0.01
Ischilin	46	0.02
Jachal	44	0.02
Junín - Mza	12058	5.65
Junín - San Luis	26	0.01
La Paz	409	0.19
La Rioja	8	0.00
La Viña	4	0.00
Las Heras	1699	0.80
Argentina (cont.)		
Lavalle	13969	6.55
Leandro Alem	0	0.00
Luján de Cuyo	13990	6.56
Maipú	12586	5.90
Malargüe	1	0.00
Molinos	120	0.06
Nogoya	0	0.00
Nueve de Julio	2386	1.12
Pichi Mahuida	41	0.02
Picunches	25	0.01
Pocito	2670	1.25
Poman	30	0.01
Puelen	217	0.10
Rawson	1464	0.69
Rivadavia - Mza	16976	7.96
Rivadavia - San Juan	440	0.21
San Blas De Los Sauces	46	0.02
San Carlos - Mza	7030	3.29
San Carlos - Salta	522	0.24
San Javier	2	0.00
San Martín - Mza	31161	14.60
San Martín - San Juan	3361	1.57
San Rafael	15326	7.18
Sanagasta	11	0.01
Santa Lucía	1209	0.57
Santa María - Catamarca	618	0.29
Santa María - Cba	9	0.00
Santa Rosa - Mza	11545	5.41
Sarmiento - San Juan	7522	3.53
Tafí del Valle	47	0.02
Tandil	6	0.00
Tilcara	2	0.00
Tinogasta	1582	0.74
Tornquist	12	0.01
Totoral	2	0.00
Tulumba	0	0.00
Tumbaya	0	0.00
Tunuyán	7241	3.39
Tupungato	8149	3.82
Ullum	580	0.27
Valle Fértil	13	0.01
Veinticinco de Mayo - Mis	0	0.00
Veinticinco de Mayo - San	7416	3.48
Victoria	1	0.00
Villarino	25	0.01
Vinchina	75	0.03
Zonda	781	0.37
Total	**213372**	**100**

Table 68 (cont.): Winegrape area by region and region's national share, by country, 2010

Region	*Area (hectares)*	*National share (%)*	*Region*	*Area (hectares)*	*National share (%)*
Armenia	**11206**	**100**	**Australia (cont.)**		
			Macedon Ranges	224	0.15
Australia			Manjimup	179	0.12
Adelaide Hills	3861	2.54	Margaret River	4894	3.22
Adelaide Plains	880	0.58	McLaren Vale	6490	4.28
Alpine Valleys	705	0.46	Mornington Peninsula	752	0.50
Australian Capital Territor	4	0.00	Mount Benson	233	0.15
Barossa - other	91	0.06	Mount Lofty Ranges - othe	468	0.31
Barossa Valley	9763	6.43	Mudgee	3414	2.25
Beechworth	57	0.04	Murray Darling (NSW)	6533	4.30
Bendigo	771	0.51	Murray Darling (VIC)	8339	5.49
Big Rivers - other	629	0.41	New England Australia	123	0.08
Blackwood Valley	249	0.16	North East Victoria - other	74	0.05
Canberra District (ACT)	105	0.07	North West Victoria - othe	121	0.08
Canberra District (NSW)	378	0.25	Northern Rivers - other	41	0.03
Central Ranges - other	227	0.15	Northern Slopes - other	145	0.10
Central Victoria - other	56	0.04	Orange	1546	1.02
Central Western Australia	62	0.04	Padthaway	5028	3.31
Clare Valley	4801	3.16	Peel	96	0.06
Coonawarra	5985	3.94	Pemberton	622	0.41
Cowra	1427	0.94	Perricoota	671	0.44
Currency Creek	871	0.57	Perth Hills	295	0.19
Eastern Plains, Inland and	25	0.02	Port Phillip - other	68	0.04
Eden Valley	1933	1.27	Pyrenees	874	0.58
Far North - other	11	0.01	Queensland - other	187	0.12
Fleurieu - other	187	0.12	Riverina	20154	13.28
Geelong	515	0.34	Riverland	20009	13.18
Geographe	1181	0.78	Robe	644	0.42
Gippsland	236	0.16	Rutherglen	853	0.56
Glenrowan	203	0.13	Shoalhaven Coast	40	0.03
Goulburn Valley	1612	1.06	South Burnett	240	0.16
Grampians	506	0.33	South Coast - other	113	0.07
Granite Belt	331	0.22	South West Australia - oth	101	0.07
Great Southern	2804	1.85	Southern Fleurieu	414	0.27
Greater Perth - other	36	0.02	Southern Flinders Ranges	180	0.12
Gundagai	408	0.27	Southern Highlands	202	0.13
Hastings River	18	0.01	Southern NSW - other	119	0.08
Heathcote	1245	0.82	Strathbogie Ranges	369	0.24
Henty	183	0.12	Sunbury	129	0.09
Hilltops	484	0.32	Swan District	784	0.52
Hunter	3450	2.27	Swan Hill (NSW)	308	0.20
Hunter Valley - other	24	0.02	Swan Hill (VIC)	3869	2.55
Kangaroo Island	89	0.06	Tasmania	1251	0.82
King Valley	1320	0.87	The Peninsulas	93	0.06
Langhorne Creek	5957	3.92	Tumbarumba	254	0.17
Limestone Coast - other	476	0.31	Upper Goulburn	245	0.16
Lower Murray - other	260	0.17	Western Australia Southea	19	0.01

Table 68 (cont.): Winegrape area by region and region's national share, by country, 2010

Region	*Area (hectares)*	*National share (%)*
Australia (cont.)		
Western Plains - other	236	0.16
Western Victoria - other	73	0.05
Wrattonbully	2818	1.86
Yarra Valley	2440	1.61
Total	**151788**	**100**
Austria		
Burgenland	13842	30.40
Niederosterreich	27184	59.70
Steiermark	3867	8.49
Wien and other regions	640	1.40
Total	**45533**	**100**
Brazil	**49412**	**100**
Bulgaria		
North Central	3868	6.89
Northeast	5837	10.40
Northwest	5830	10.39
South Central	17466	31.12
Southeast	19533	34.80
Southwest	3599	6.41
Total	**56133**	**100**
Canada		
British Colombia	3995	39.56
Ontario	6102	60.44
Total	**10096**	**100**
Chile		
Araucania	12	0.01
Atacama	12	0.01
Coquimbo	2155	1.93
De Los Lagos	6	0.01
Del Bio Bio	3420	3.07
Del Maule	49014	43.95
Metropolitana	12214	10.95
O'Higgins	36170	32.43
Valparaiso	8522	7.64
Total	**111525**	**100**
China		
Beijing	3067	10.38
Gansu	4987	16.88
Ningxia	11152	37.74
Shandong	67	0.23
ShanXi	547	1.85
Sichuan	533	1.81
Tianjin	400	1.35
Xinjiang	3133	10.61
Yantai	4373	14.80
other regions	1287	4.35
Total	**29545**	**100**
Croatia		
Dalmatinska Zagora	602	2.90
Hrvatsko Primorje	210	1.01
Istra	3083	14.85
Moslavina	228	1.10
Plesivica	452	2.18
Podunavlje	3206	15.45
Pokuplje	41	0.20
Prigorje - Bilogora	791	3.81
Sjeverna Dalmacija	2333	11.24
Slavonija	3307	15.94
Srednja Juzna Dalmacija	2972	14.32
Zagorje-Medimurje	1266	6.10
other regions	2263	10.90
Total	**20754**	**100**
Cyprus	**8608**	**100**
Czechia		
Cechy	785	4.83
Morava	15457	95.17
Total	**16242**	**100**
Ethiopia	**169**	**100**

Table 68 (cont.): Winegrape area by region and region's national share, by country, 2010

Region	*Area (hectares)*	*National share (%)*
France		
Ain	954	0.11
Aisne	3229	0.39
Allier	717	0.09
Alpes-de-Haute-Provence,	843	0.10
Alpes-Maritimes	152	0.02
Ardeche	11960	1.43
Ariege	133	0.02
Aube	7870	0.94
Aude	72058	8.62
Aveyron	884	0.11
Bas-Rhin	6971	0.83
Bouches-du-Rhone	11663	1.40
Cantal	11	0.00
Charente	41093	4.92
Charente-Maritime	40790	4.88
Cher	4182	0.50
Correze	287	0.03
Corse	968	0.12
Cote-d'Or	9683	1.16
Deux-Sevres	1227	0.15
Dordogne	13820	1.65
Doubs	51	0.01
Drome	18507	2.21
Eure-et-Loire	0	0.00
Gard	59210	7.09
Gers	18897	2.26
Gironde	123023	14.72
Haute-Corse	5561	0.67
Haute-Garonne	1993	0.24
Haute-Loire	83	0.01
Haute-Marne	114	0.01
Hautes-Alpes	232	0.03
Haute-Saone	172	0.02
Haute-Savoie	286	0.03
Hautes-Pyrenees	680	0.08
Haut-Rhin	9208	1.10
Herault	93944	11.24
Indre	980	0.12
Indre-et-Loire	10421	1.25
Isere	264	0.03
Jura	2267	0.27
Landes	2107	0.25
Loire	1067	0.13
Loire-Atlantique	15618	1.87
Loiret	309	0.04
Loir-et-Cher	7474	0.89
France (cont.)		
Lot	5571	0.67
Lot-et-Garonne	6846	0.82
Lozere	11	0.00
Maine-et-Loire	20766	2.49
Marne	22615	2.71
Mayenne	1	0.00
Meurthe-et-Moselle	159	0.02
Meuse	45	0.01
Moselle	94	0.01
Nievre	1719	0.21
Puy-de-Dome	881	0.11
Pyrenees-Atlantiques	2635	0.32
Pyrenees-Orientales	28817	3.45
Rhone	19137	2.29
Saone-et-Loire	13509	1.62
Sarthe	321	0.04
Savoie	2092	0.25
Seine-et-Marne	82	0.01
Tarn	7087	0.85
Tarn-et-Garonne	3821	0.46
Var	31913	3.82
Vaucluse	54117	6.48
Vendee	1886	0.23
Vienne	2199	0.26
Vosges	42	0.01
Yonne	7224	0.86
Total	**835554**	**100**
Georgia	**48001**	**100**
Germany		
Ahr	562	0.55
Baden	15836	15.50
Franken	6104	5.98
Hessische Bergstraße	427	0.42
Mittelrhein	458	0.45
Mosel	8976	8.79
Nahe	4163	4.08
Pfalz	23467	22.98
Rheingau	3062	3.00
Rheinhessen	26480	25.93
Saale	704	0.69
Sachsen	461	0.45
Württemberg	11435	11.20
Total	**102135**	**100**

Table 68 (cont.): Winegrape area by region and region's national share, by country, 2010

Region	*Area (hectares)*	*National share (%)*
Greece		
Anatoliki Makedonia, Thra	1234	2.27
Attiki	5599	10.29
Dytiki Ellada	6484	11.92
Dytiki Makedonia	2083	3.83
Ionia Nisia	2422	4.45
Ipeiros	559	1.03
Kentriki Makedonia	4256	7.83
Kriti	7863	14.46
Notio Aigaio	3547	6.52
Peloponnissos	8134	14.96
Sterea Ellada	6457	11.87
Thessalia	3415	6.28
Voreio Aigaio	2337	4.30
Total	**54390**	**100**
Hungary		
Badacsony	1618	2.32
Balatonboglar	3305	4.74
Balatonfelvidek	1025	1.47
Balatonfured-Csopak	2180	3.13
Bukk	1055	1.51
Csongrad	1513	2.17
Eger	5509	7.90
Etyek-Budai	1717	2.46
Hajos-bajai	1982	2.84
Kunsag	22263	31.93
Matra	6294	9.03
Mor	730	1.05
Nagy-Somlo	598	0.86
Neszmely	1587	2.28
Pannonhalma	615	0.88
Pecs	777	1.11
Sopron	1919	2.75
Szekszard	2333	3.35
Tokaj	5994	8.60
Tolna	2526	3.62
Villany	2582	3.70
Zala	1592	2.28
Total	**69715**	**100**
Israel	**4851**	**100**

Region	*Area (hectares)*	*National share (%)*
Italy		
Agrigento	16781	2.68
Alessandria	12328	1.97
Ancona	4739	0.76
Arezzo	7031	1.12
Ascoli Piceno	6111	0.98
Asti	15559	2.49
Avellino	5711	0.91
Bari	7586	1.21
Barletta-Andria-Trani	16241	2.60
Belluno	52	0.01
Benevento	10488	1.68
Bergamo	804	0.13
Biella	289	0.05
Bologna	6845	1.09
Bolzano-Bozen	5282	0.84
Brescia	6016	0.96
Brindisi	9677	1.55
Cagliari	4563	0.73
Caltanissetta	4170	0.67
Campobasso	4715	0.75
Carbonia-Iglesias	1948	0.31
Caserta	2062	0.33
Catania	2990	0.48
Catanzaro	706	0.11
Chieti	25989	4.15
Como	23	0.00
Cosenza	4162	0.67
Cremona	65	0.01
Crotone	3229	0.52
Cuneo	16065	2.57
Enna	278	0.04
Fermo	1706	0.27
Ferrara	572	0.09
Firenze	18362	2.93
Foggia	25249	4.04
Forlì-Cesena	7017	1.12
Frosinone	1877	0.30
Genova	177	0.03
Gorizia	4064	0.65
Grosseto	7438	1.19
Imperia	414	0.07
Isernia	439	0.07
La Spezia	617	0.10
L'Aquila	440	0.07
Latina	3757	0.60

Table 68 (cont.): Winegrape area by region and region's national share, by country, 2010

Region	*Area (hectares)*	*National share (%)*	*Region*	*Area (hectares)*	*National share (%)*
Lecce	8379	1.34	**Italy (cont.)**		
Lecco	57	0.01	Taranto	15627	2.50
Livorno	2425	0.39	Teramo	2602	0.42
Lodi	20	0.00	Terni	4871	0.78
Lucca	1053	0.17	Torino	1313	0.21
Macerata	2185	0.35	Trapani	61649	9.85
Mantova	1803	0.29	Trento	10377	1.66
Massa-Carrara	759	0.12	Treviso	28580	4.57
Matera	1243	0.20	Trieste	205	0.03
Medio Campidano	638	0.10	Udine	7285	1.16
Messina	866	0.14	Valle d'Aosta	463	0.07
Milano	207	0.03	Varese	23	0.00
Modena	7874	1.26	Venezia	6604	1.06
Monza e della Brianza	3	0.00	Verbano-Cusio-Ossola	28	0.00
Napoli	1609	0.26	Vercelli	200	0.03
Novara	535	0.09	Verona	27750	4.44
Nuoro	2401	0.38	Vibo Valentia	344	0.06
Ogliastra	1605	0.26	Vicenza	8445	1.35
Olbia-Tempio	2136	0.34	Viterbo	2957	0.47
Oristano	2228	0.36	**Total**	**625700**	**100**
Padova	5850	0.93			
Palermo	14569	2.33	**Japan**		
Parma	736	0.12	Hokkaido	835	22.47
Pavia	13193	2.11	Nagano	754	20.30
Perugia	7620	1.22	Yamagata	392	10.56
Pesaro e Urbino	2004	0.32	Yamanashi	632	17.01
Pescara	3159	0.50	other regions	1102	29.66
Piacenza	5906	0.94	**Total**	**3715**	**100**
Pisa	3163	0.51			
Pistoia	776	0.12	**Kazakhstan**		
Pordenone	7695	1.23	Almaty	4553	65.62
Potenza	3620	0.58	East Kazakhstan	3	0.04
Prato	509	0.08	South Kazakhstan	2162	31.16
Ragusa	1381	0.22	West Kazakhstan	2	0.03
Ravenna	16388	2.62	Zhambyl	217	3.12
Reggio di Calabria	1343	0.21	other regions	2	0.03
Reggio nell'Emilia	8027	1.28	**Total**	**6938**	**100**
Rieti	750	0.12			
Rimini	2431	0.39	**Korea, Rep.**	**5400**	**100**
Roma	7061	1.13			
Rovigo	363	0.06	**Luxembourg**	**1304**	**100**
Salerno	3315	0.53			
Sassari	2946	0.47	**Mexico**		
Savona	329	0.05	Aguascalientes	850	15.55
Siena	18323	2.93	Baja California	2863	52.39
Siracusa	1383	0.22	Coahuila	279	5.11
Sondrio	874	0.14	Sonora	1164	21.30
			Zacatecas	309	5.65
			Total	**5465**	**100**

Table 68 (cont.): Winegrape area by region and region's national share, by country, 2010

Region	*Area (hectares)*	*National share (%)*
Moldova	**89844**	**100**
Morocco	**49000**	**100**
Myanmar	**75**	**100**
New Zealand		
Auckland	543	1.70
Canterbury	320	1.00
Gisborne	2149	6.72
Hawkes Bay	4921	15.40
Marlborough	18401	57.57
Nelson	813	2.54
Otago	1532	4.79
Waikato	147	0.46
Waipara	1442	4.51
Wairarapa	859	2.69
other regions	836	2.62
Total	**31964**	**100**
Peru		
Arequipa	1356	35.40
Lima	783	20.44
Moquegua	877	22.89
Tacna	815	21.27
Total	**3831**	**100**
Portugal		
Acores	176	0.11
Alentejo	21892	13.39
Algarve	1078	0.66
Alto Tras-os-Montes	59112	36.15
Beira Interior	15968	9.76
Beira Litoral	15241	9.32
Entre Douro e Minho	14288	8.74
Madeira	847	0.52
Ribatejo e Oeste	34920	21.36
Total	**163522**	**100**
Romania		
Bucuresti - Ilfov	443	0.26
Centru	5349	3.27
Nord-Est	28072	17.17
Nord-Vest	6667	4.08
Sud - Muntenia	27069	16.55
Sud-Est	68081	41.63
Sud-Vest Oltenia	31966	19.55
Vest	2645	1.62
Total	**170292**	**104**

Region	*Area (hectares)*	*National share (%)*
Russia		
Krasnodar Krai	21224	82.82
Rostov Oblast	4404	17.18
Total	**25628**	**100**
Serbia	**68999**	**100**
Slovakia		
Juznoslovenska	4141	32.77
Malokarpatska	3683	29.14
Nitrianska	2652	20.98
Stredné Slovensko	1155	9.14
Tokajska	453	3.59
Východné Slovensko	553	4.38
Total	**12637**	**100**
Slovenia		
Bela Krajina	365	2.23
Bizeljsko Sremic	907	5.54
Dolenjska	1476	9.02
Goriska brda	1898	11.61
Kras	593	3.62
Prekmurje	564	3.45
Slovenska Istra	1626	9.94
Stajerska Slovenija	6374	38.97
Vipavska dolina	2526	15.44
other regions	26	0.16
Total	**16354**	**100**
South Africa		
Breedekloof	12568	12.44
Little Karoo	2822	2.79
Northern Cape	5078	5.03
Olifants River	9997	9.90
Paarl	16568	16.40
Robertson	14004	13.86
Stellenbosch	17107	16.93
Swartland	14224	14.08
Worcester	8649	8.56
Total	**101016**	**100**

Table 68 (cont.): Winegrape area by region and region's national share, by country, 2010

Region	Area (hectares)	National share (%)	Region	Area (hectares)	National share (%)
Spain			**Switzerland (cont.)**		
Alava	13040	1.27	Lucerne	41	0.28
Albacete	96745	9.41	Neuchâtel	591	3.99
Alicante	14661	1.43	Schaffhausen	478	3.22
Almeria, Granada, Jaen, Se	6950	0.68	Schwyz	38	0.26
Avila, Palencia, Salamanca	9154	0.89	St. Gallen	215	1.45
Badajoz	82749	8.05	Thurgau	263	1.78
Barcelona	22339	2.17	Ticino	1069	7.21
Burgos	16276	1.58	Valais	5070	34.21
Caceres	3857	0.38	Vaud	3819	25.77
Cadiz	10156	0.99	Zürich	614	4.14
Canarias	8653	0.84	other regions	25	0.17
Cantabria	50	0.00	**Total**	**14820**	**100**
Castellon	1214	0.12			
Ciudad Real	175764	17.09	**Taiwan**	**2833**	**100**
Comunidad de Madrid	23963	2.33			
Comunidad Foral de Navar	11024	1.07	**Thailand**	**149**	**100**
Cordoba	8278	0.81			
Cuenca	94883	9.23	**Tunisia**	**16836**	**100**
Galicia	25457	2.48			
Girona, Lleida	7560	0.74	**Turkey**		
Guadalajara	2149	0.21	Aegean	6770	52.66
Guipuzcoa, Vizcaya	736	0.07	Central East	1895	14.74
Huelva	4230	0.41	Central North	428	3.33
Huesca, Teruel	9277	0.90	Central South	1553	12.08
Illes Balears	1544	0.15	Marmara	1745	13.57
La Rioja	44576	4.34	Mediterranean	28	0.21
Leon	12149	1.18	South East	438	3.40
Malaga	2079	0.20	**Total**	**12856**	**100**
Principado de Asturias	95	0.01			
Region de Murcia	35437	3.45	**Ukraine**	**52293**	**100**
Tarragona	29617	2.88			
Toledo	125760	12.23	**United Kingdom**	**1198**	**100**
Valencia	57559	5.60			
Valladolid	22081	2.15	**United States**		
Zamora	12906	1.26	Alameda	1145	0.50
Zaragoza	35294	3.43	Amador	1255	0.55
Total	**1028258**	**100**	Arizona	101	0.04
			Arkansas	243	0.11
Switzerland			Benton Co.	155	0.07
Aargau	399	2.69	Butte	58	0.03
Basel Land	114	0.77	Calaveras	253	0.11
Bern	242	1.63	Chautauqua-Erie	7561	3.32
Fribourg	117	0.79	Colorado	271	0.12
Geneva	1292	8.72	Columbia Gorge	159	0.07
Graubünden	421	2.84	Columbia River	610	0.27
Jura	14	0.09	Columbia Valley	3023	1.33

Table 68 (cont.): Winegrape area by region and region's national share, by country, 2010

Region	*Area (hectares)*	*National share (%)*	*Region*	*Area (hectares)*	*National share (%)*
United States (cont.)			**United States (cont.)**		
Colusa	646	0.28	San Benito	959	0.42
Contra Costa	675	0.30	San Bernardino	209	0.09
Douglas Co.	350	0.15	San Diego	78	0.03
El Dorado	660	0.29	San Joaquin	27146	11.91
Finger Lakes	3801	1.67	San Luis Obispo	11484	5.04
Fresno	16010	7.02	San Mateo	29	0.01
Georgia	567	0.25	Santa Barbara	6512	2.86
Glenn	329	0.14	Santa Clara	609	0.27
Horse Heaven Hills	4283	1.88	Santa Cruz	160	0.07
Humboldt	36	0.02	Shasta	41	0.02
Illinois	373	0.16	Siskiyou	8	0.00
Indiana	263	0.12	Snipes Mountain	285	0.12
Iowa	194	0.09	Solano	1231	0.54
Jackson Co.	536	0.24	Sonoma	22265	9.77
Josephine Co.	162	0.07	Stanislaus	3079	1.35
Kentucky	210	0.09	Sutter	54	0.02
Kern	8422	3.69	Tehama	59	0.03
Kings	615	0.27	Texas	1214	0.53
Lake	3122	1.37	Trinity	49	0.02
Lake Chelan	100	0.04	Tulare	3432	1.51
Lane Co.	341	0.15	Tuolumne	12	0.01
Los Angeles	53	0.02	Ventura	21	0.01
Madera	14273	6.26	Virginia	1065	0.47
Marin	62	0.03	Wahluke Slope	2689	1.18
Marion Co.	660	0.29	Walla Walla Valley	528	0.23
Mariposa	24	0.01	Washington Co.	670	0.29
Mendocino	6555	2.88	Willamette Valley - other	154	0.07
Merced	4418	1.94	Yakima Valley	5444	2.39
Michigan	1072	0.47	Yamhill Co.	2273	1.00
Minnesota	418	0.18	Yolo	4263	1.87
Missouri	647	0.28	Yuba	39	0.02
Monterey	15600	6.84	**Total**	**227949**	**100**
Napa	17768	7.79			
Nevada	159	0.07	**Uruguay**	**7657**	**100**
New York - other	1508	0.66			
North Carolina	728	0.32			
Ohio	436	0.19			
Orange	0	0.00			
Pennsylvania	1004	0.44			
Placer	70	0.03			
Polk Co.	928	0.41			
Puget Sound	72	0.03			
Rattlesnake Hills	647	0.28			
Red Mountain	515	0.23			
Riverside	333	0.15			
Sacramento	7406	3.25			

Table 69: Winegrape area by region and region's national share, by country, 2016

Region	*Area (hectares)*	*National share (%)*	*Region*	*Area (hectares)*	*National share (%)*
Algeria	**8300**	**100**	**Argentina (cont.)**		
			Famatina	344	0.17
Argentina			Futaleufu	5	0.00
Adolfo Alsina	98	0.05	General Alvear	4028	1.95
Albardón	711	0.34	General Belgrano	4	0.00
Andalgala	6	0.00	General Lamadrid	60	0.03
Añelo	1588	0.77	General Pueyrredón	11	0.01
Angaco	1774	0.86	General Roca	1174	0.57
Avellaneda - Río Negro	154	0.07	Godoy Cruz	1	0.00
Ayacucho	10	0.00	Gualeguaychu	1	0.00
Balcarce	9	0.00	Guaymallén	630	0.31
Bariloche	1	0.00	Humahuaca	5	0.00
Belén	185	0.09	Iglesia	7	0.00
Benito Juárez	0	0.00	Ischilin	49	0.02
Cachi	78	0.04	Junín - Bs. As.	0	0.00
Cafayate	2411	1.17	Junín - Mza	11709	5.67
Cainguas	7	0.00	Junín - San Luis	27	0.01
Calamuchita	18	0.01	La Paz	300	0.15
Calingasta	124	0.06	La Viña	5	0.00
Cañuelas	2	0.00	Lacar	1	0.00
Capayán	1	0.00	Languiñeo	0	0.00
Capital Misiones	2	0.00	Las Heras	1529	0.74
Capital San Juan	2	0.00	Lavalle	13763	6.67
Capital San Luis	60	0.03	Leandro Alem	1	0.00
Capital Santiago del Este	2	0.00	Luján de Cuyo	15830	7.67
Castro Barros	245	0.12	Maipú	12343	5.98
Caucete	5681	2.75	Molinos	124	0.06
Chilecito	5356	2.60	Nogoya	0	0.00
Chimbas	612	0.30	Ñorquin	1	0.00
Chos Malal	6	0.00	Nueve de Julio	2138	1.04
Collon Cura	5	0.00	Paraná	6	0.00
Colón - Cba	113	0.05	Pehuenches	3	0.00
Colón - Entre Ríos	6	0.00	Pichi Mahuida	38	0.02
Concordia	13	0.01	Picún Leufú	1	0.00
Conesa	22	0.01	Picunches	22	0.01
Confluencia	126	0.06	Pocito	2190	1.06
Coronel Felipe Varela	506	0.25	Poman	27	0.01
Coronel Pringles	2	0.00	Puelen	240	0.12
Coronel Suarez	19	0.01	Punilla	4	0.00
Cruz del Eje	27	0.01	Rawson	1156	0.56
Curaco	3	0.00	Rivadavia - Mza	15618	7.57
Cushamen	20	0.01	Rivadavia - San Juan	364	0.18
Daireaux	1	0.00	Saavedra	18	0.01
De La Costa	0	0.00	San Alberto	5	0.00
Diamante	4	0.00	San Blas De Los Sauces	41	0.02
El Carmen	3	0.00	San Carlos - Mza	8676	4.20
El Cuy	60	0.03	San Carlos - Salta	624	0.30

Table 69 (cont.): Winegrape area by region and region's national share, by country, 2016

Region	*Area (hectares)*	*National share (%)*	*Region*	*Area (hectares)*	*National share (%)*
Argentina (cont.)			**Australia (cont.)**		
San Javier	12	0.01	Blackwood Valley	304	0.23
San Martín - Mza	28869	13.99	Canberra District (ACT)	92	0.07
San Martín - San Juan	2940	1.42	Canberra District (NSW)	243	0.18
San Rafael	13620	6.60	Central Ranges - other	390	0.29
Sanagasta	12	0.01	Central Victoria - other	190	0.14
Santa Lucía	989	0.48	Central Western Australia	16	0.01
Santa María - Catamarca	708	0.34	Clare Valley	4200	3.17
Santa María - Cba	9	0.00	Coonawarra	4726	3.57
Santa Rosa - Catamarca	1	0.00	Cowra	643	0.49
Santa Rosa - Mza	10012	4.85	Currency Creek	864	0.65
Sarmiento - Chubut	38	0.02	Eastern Plains, Inland anc	2	0.00
Sarmiento - San Juan	7225	3.50	Eden Valley	1799	1.36
Tafí del Valle	105	0.05	Far North - other	2	0.00
Tandil	7	0.00	Fleurieu - other	323	0.24
Tilcara	13	0.01	Geelong	284	0.21
Tinogasta	1458	0.71	Geographe	338	0.25
Tornquist	12	0.01	Gippsland	177	0.13
Trancas	5	0.00	Glenrowan	179	0.13
Tulumba	5	0.00	Goulburn Valley	1011	0.76
Tumbaya	0	0.00	Grampians	511	0.39
Tunuyán	9320	4.52	Granite Belt	237	0.18
Tupungato	9852	4.77	Great Southern	1886	1.42
Ullum	525	0.25	Greater Perth - other	168	0.13
Uruguay	0	0.00	Gundagai	332	0.25
Valle Fértil	0	0.00	Hastings River	13	0.01
Valle Viejo	1	0.00	Heathcote	1159	0.88
Veinticinco de Mayo - M	0	0.00	Henty	158	0.12
Veinticinco de Mayo - Sa	6586	3.19	Hilltops	566	0.43
Victoria	7	0.00	Hunter	2309	1.74
Villa Gesell	0	0.00	Hunter Valley - other	52	0.04
Villarino	25	0.01	Kangaroo Island	95	0.07
Vinchina	48	0.02	King Valley	1336	1.01
Zonda	747	0.36	Langhorne Creek	5282	3.99
Total	**206342**	**100**	Limestone Coast - other	2113	1.60
			Lower Murray - other	856	0.65
Armenia	**14705**	**100**	Macedon Ranges	140	0.11
			Manjimup	54	0.04
Australia			Margaret River	4816	3.64
Adelaide Hills	2967	2.24	McLaren Vale	5995	4.53
Adelaide Plains	586	0.44	Mornington Peninsula	775	0.59
Alpine Valleys	258	0.19	Mount Benson	257	0.19
Barossa - other	27	0.02	Mount Gambier	186	0.14
Barossa Valley	8899	6.72	Mount Lofty Ranges - otl	470	0.35
Beechworth	107	0.08	Mudgee	1070	0.81
Bendigo	610	0.46	Murray Darling (NSW)	6298	4.76
Big Rivers - other	605	0.46	Murray Darling (VIC)	9214	6.96

Table 69 (cont.): Winegrape area by region and region's national share, by country, 2016

Region	*Area (hectares)*	*National share (%)*	*Region*	*Area (hectares)*	*National share (%)*
Australia (cont.)			**Austria**		
New England Australia	132	0.10	Bergland	140	0.31
North East Victoria - othe	76	0.06	Carnuntum	906	1.99
North West Victoria - oth	106	0.08	Kamptal	3907	8.60
Northern Rivers - other	7	0.01	Kremstal	2368	5.21
Northern Slopes	48	0.04	Mittelburgenland	1897	4.17
Orange	1098	0.83	Neusiedlersee	7098	15.62
Padthaway	3237	2.44	Neusiedlersee Hügelland	2835	6.24
Peel	36	0.03	Steirerland - other	0	0.00
Pemberton	332	0.25	Südburgenland	419	0.92
Perricoota	367	0.28	Südsteiermark	2162	4.76
Perth Hills	114	0.09	Thermenregion	2182	4.80
Port Phillip - other	520	0.39	Traisental	815	1.79
Pyrenees	493	0.37	Vulkanland Steiermark	1623	3.57
Queensland - other	83	0.06	Wachau	1344	2.96
Riverina	18610	14.05	Wagram	2720	5.99
Riverland	18774	14.18	Weinviertel	13858	30.50
Robe	650	0.49	Weststeiermark	538	1.18
Rutherglen	398	0.30	Wien	581	1.28
Shoalhaven Coast	33	0.03	other regions	45	0.10
South Burnett	215	0.16	**Total**	**45439**	**100**
South Coast - other	71	0.05			
South West Australia - ot	41	0.03	**Brazil**	**33205**	**100**
Southern Fleurieu	181	0.14			
Southern Flinders Ranges	175	0.13	**Bulgaria**		
Southern Highlands	124	0.09	North Central	3962	7.48
Southern NSW - other	46	0.04	Northeast	6858	12.95
Strathbogie Ranges	627	0.47	Northwest	4169	7.87
Sunbury	67	0.05	South Central	16720	31.56
Swan District	720	0.54	Southeast	18725	35.35
Swan Hill (NSW)	131	0.10	Southwest	2540	4.79
Swan Hill (VIC)	1717	1.30	**Total**	**52974**	**100**
Tasmania	1442	1.09			
The Peninsulas	69	0.05	**Cambodia**	**10**	**100**
Tumbarumba	210	0.16			
Upper Goulburn	384	0.29	**Canada**		
Western Australia Southe	20	0.02	British Colombia	4152	32.95
Western Plains	236	0.18	Nova Scotia	325	2.58
Western Victoria - other	62	0.05	Ontario	6530	51.82
Wrattonbully	2478	1.87	Quebec	384	3.05
Yarra Valley	2116	1.60	other regions	1209	9.59
Total	**132435**	**100**	**Total**	**12600**	**100**

Table 69 (cont.): Winegrape area by region and region's national share, by country, 2016

Region	*Area (hectares)*	*National share (%)*
Chile		
Antofagasta	5	0.00
Araucania	65	0.04
Arica	15	0.01
Atacama	567	0.39
Coquimbo	11076	7.59
De Los Lagos	27	0.02
Del Bio Bio	12093	8.29
Del Maule	52962	36.31
Metropolitana	12908	8.85
O'Higgins	46337	31.77
Tarapaca	2	0.00
Valparaiso	9816	6.73
Total	**145873**	**100**
China	**178000**	**100**
Croatia		
Jadranska Hrvatska	5308	45.19
Kontinentalna Hrvatska	6438	54.81
Total	**11746**	**100**
Cyprus	**5133**	**100**
Czechia		
Cechy	248	1.82
Jihovýchod	12800	94.12
Morava	268	1.97
Praha	10	0.07
Severozápad	274	2.01
Total	**13600**	**100**
Ethiopia	**169**	**100**
France		
Alsace	16804	2.06
Aquitaine	144314	17.71
Auvergne	1010	0.12
Bourgogne	32673	4.01
Centre-Val de Loire	22463	2.76
Champagne-Ardenne	34332	4.21
Corse	6660	0.82
Franche Comté	1819	0.22
Île de France	4	0.00
Languedoc Roussillon	245964	30.18
Limousin	83	0.01
Lorraine	211	0.03

Region	*Area (hectares)*	*National share (%)*
France (cont.)		
Midi Pyrénées	36827	4.52
Pays de la Loire	35844	4.40
Picardie	2242	0.28
Poitou Charentes	85808	10.53
Provence-Alpes-Cote d'A	96470	11.84
Rhône Alpes	51354	6.30
Total	**814882**	**100**
Georgia	**48000**	**100**
Germany		
Ahr	522	0.55
Baden	14666	15.52
Franken	5796	6.13
Hessische Bergstraße	414	0.44
Mittelrhein	429	0.45
Mosel	8137	8.61
Nahe	3897	4.12
Pfalz	21541	22.79
Rheingau	2962	3.13
Rheinhessen	24117	25.52
Saale	695	0.74
Sachsen	466	0.49
Württemberg	10859	11.49
Total	**94501**	**100**
Greece		
Anatoliki Makedonia, Th	1706	3.36
Attiki	5957	11.72
Dytiki Ellada	7043	13.85
Dytiki Makedonia	1824	3.59
Ionia Nisia	665	1.31
Ipeiros	153	0.30
Kentriki Makedonia	3115	6.13
Kriti	6767	13.31
Notio Aigaio	2581	5.08
Peloponnissos	8739	17.19
Sterea Ellada	6291	12.37
Thessalia	3426	6.74
Voreio Aigaio	2578	5.07
Total	**50845**	**100**

Table 69 (cont.): Winegrape area by region and region's national share, by country, 2016

Region	*Area (hectares)*	*National share (%)*
Hungary		
Badacsony	1382	2.16
Balatonboglar	3231	5.06
Balatonfelvidek	830	1.30
Balatonfured-Csopak	2032	3.18
Bukk	1015	1.59
Csongrad	1126	1.76
Eger	5353	8.38
Etyek-Budai	1513	2.37
Hajos-bajai	1889	2.96
Kunsag	20519	32.12
Matra	6248	9.78
Mor	573	0.90
Nagy-Somlo	553	0.87
Neszmely	1077	1.69
Pannonhalma	622	0.97
Pecs	673	1.05
Sopron	1650	2.58
Szekszard	2183	3.42
Tokaj	5709	8.94
Tolna	2399	3.76
Villany	2469	3.86
Zala	833	1.30
Total	**63881**	**100**
India	**2700**	**100**
Israel	**5000**	**100**
Italy		
Abruzzo	27138	4.49
Basilicata	3827	0.63
Bolzano-Bozen	4072	0.67
Calabria	6509	1.08
Campania	17217	2.85
Emilia-Romagna	58675	9.71
Friuli-Venezia Giulia	19879	3.29
Lazio	14968	2.48
Liguria	1004	0.17
Lombardia	23221	3.84
Marche	14766	2.44
Molise	3991	0.66
Piemonte	46327	7.66
Puglia	94348	15.61
Sardegna	17784	2.94
Sicilia	86725	14.35
Toscana	57373	9.49

Region	*Area (hectares)*	*National share (%)*
Italy (cont.)		
Trento	10079	1.67
Umbria	11768	1.95
Valle d'Aosta	375	0.06
Veneto	84505	13.98
Total	**604551**	**100**
Japan		
Hokkaido	868	22.43
Iwate	160	4.15
Nagano	754	19.49
Niigata	62	1.61
Yamagata	414	10.69
Yamanashi	1202	31.08
other regions	408	10.55
Total	**3869**	**100**
Kazakhstan		
Almaty	4553	65.62
East Kazakhstan	3	0.04
South Kazakhstan	2162	31.16
West Kazakhstan	2	0.03
Zhambyl	217	3.12
other regions	2	0.03
Total	**6938**	**100**
Korea, Rep.	**5400**	**100**
Lebanon	**4000**	**100**
Luxembourg	**1300**	**100**
Mexico		
Aguascalientes	850	15.55
Baja California	2863	52.39
Coahuila	279	5.11
Sonora	1164	21.30
Zacatecas	309	5.65
Total	**5465**	**100**
Moldova	**82600**	**100**
Morocco	**17590**	**100**
Myanmar	**70**	**100**

Table 69 (cont.): Winegrape area by region and region's national share, by country, 2016

Region	*Area (hectares)*	*National share (%)*
New Zealand		
Auckland	346	0.98
Canterbury	197	0.56
Gisborne	1440	4.06
Hawkes Bay	4638	13.08
Marlborough	23452	66.13
Nelson	1141	3.22
Northland	57	0.16
Otago	1942	5.48
Waikato	16	0.05
Waipara	1231	3.47
Wairarapa	1003	2.83
Total	**35463**	**100**
North Macedonia	**24777**	**100**
Norway	**13**	**100**
Peru		
Arequipa	1356	35.40
Lima	783	20.44
Moquegua	877	22.89
Tacna	815	21.27
Total	**3831**	**100**
Portugal		
Acores	176	0.10
Alentejo	29987	16.42
Algarve	1571	0.86
Centro	67240	36.81
Lisboa	7958	4.36
Madeira	847	0.46
Norte	74870	40.99
Total	**182649**	**100**
Romania	**182762**	**100**
Russia		
Crimea	25166	49.55
Krasnodar Krai	21224	41.78
Rostov Oblast	4404	8.67
Total	**50794**	**100**

Region	*Area (hectares)*	*National share (%)*
Serbia		
Bačka	23	0.10
Banat	98	0.45
Belgrade	1132	5.14
Čačak-Kraljevo	65	0.29
Knjaževac	1077	4.89
Leskovac	1451	6.59
Mlava	815	3.70
Negotinska Krajina	975	4.43
Niš	1316	5.98
Nišava	477	2.17
South Banat	1722	7.82
Srem	2142	9.73
Subotica	311	1.41
Šumadija	1118	5.08
Telečka	115	0.52
Tisa	261	1.19
Toplica	761	3.46
Tri Morave	7544	34.27
Valjevo	191	0.87
Vranje	419	1.90
Total	**22014**	**100**
Slovakia		
Bratislavský kraj	1571	20.28
Stredné Slovensko	780	10.07
Východné Slovensko	394	5.09
Západné Slovensko	5003	64.57
Total	**7748**	**100**
Slovenia		
Bela Krajina	369	2.31
Bizeljsko Sremic	878	5.49
Dolenjska	1600	10.01
Goriska brda	1803	11.28
Kras	582	3.64
Prekmurje	528	3.30
Slovenska Istra	1840	11.51
Stajerska Slovenija	6113	38.23
Vipavska dolina	2276	14.24
Total	**15989**	**100**

Table 69 (cont.): Winegrape area by region and region's national share, by country, 2016

Region	Area (hectares)	National share (%)	Region	Area (hectares)	National share (%)
South Africa			**Switzerland (cont.)**		
Breedekloof	12893	13.46	Neuchâtel	604	4.09
Cape South Coast	2657	2.77	Nidwalden	0	0.00
Little Karoo	2443	2.55	Owalden	2	0.01
Northern Cape	4360	4.55	Schaffhausen	483	3.26
Olifants River	10018	10.46	Schwyz	39	0.26
Paarl	15280	15.95	Solothurn	10	0.06
Robertson	13227	13.81	St. Gallen	211	1.42
Stellenbosch	15339	16.02	Thunersee	19	0.13
Swartland	12940	13.51	Thurgau	257	1.74
Worcester	6618	6.91	Ticino	1098	7.42
Total	**95775**	**100**	Uri	4	0.03
			Valais	4906	33.17
Spain			Vaud	3771	25.49
Andalucía	32054	3.63	Zug	2	0.01
Aragón	33729	3.82	Zürich	607	4.10
Canarias	10878	1.23	other regions	3	0.02
Cantabria	106	0.01	**Total**	**14793**	**100**
Castilla y León	72364	8.19			
Castilla-La Mancha	409969	46.40	**Taiwan**	**149**	**100**
Cataluña	51908	5.87			
Comunidad de Madrid	11255	1.27	**Thailand**	**208**	**100**
Comunidad Foral de Nav	17015	1.93			
Comunidad Valenciana	57677	6.53	**Tunisia**	**3400**	**100**
Extremadura	78323	8.86			
Galicia	30120	3.41	**Turkey**		
Illes Balears	1718	0.19	Aegean	8214	59.94
La Rioja	40081	4.54	Central East	2066	15.08
País Vasco	13481	1.53	Central North	214	1.56
Principado de Asturias	104	0.01	Central South	770	5.62
Región de Murcia	22774	2.58	Marmara	1737	12.68
Total	**883558**	**100**	Mediterranean	25	0.18
			South East	677	4.94
Switzerland			**Total**	**13704**	**100**
Aargau	385	2.60			
Appenzell Ausserrhoden	4	0.03	**Ukraine**	**25166**	**100**
Appenzell Innerrhoden	1	0.00			
Basel Land	114	0.77	**United Kingdom**	**1839**	**100**
Basel Stadt	5	0.03			
Fribourg	117	0.79			
Geneva	1411	9.54			
Glarus	2	0.01			
Graubünden - Mesolcina	30	0.20			
Graubünden - other	421	2.85			
Jura	15	0.10			
Lac de Bienne	221	1.49			
Lucerne	53	0.36			

Table 69 (cont.): Winegrape area by region and region's national share, by country, 2016

Region	*Area (hectares)*	*National share (%)*	*Region*	*Area (hectares)*	*National share (%)*
United States			**United States (cont.)**		
Alameda	1188	0.50	North Texas (DFW)	161	0.07
Amador	1458	0.61	North Willamette Valley	6072	2.53
Arizona	250	0.10	Ohio	607	0.25
Arkansas	251	0.10	Ontario	216	0.09
Butte	69	0.03	Orange	0	0.00
Calaveras	288	0.12	Pennsylvania	1040	0.43
Cattaraugus	139	0.06	Placer	76	0.03
Chautauqua	6740	2.81	Puget Sound	32	0.01
Colorado	248	0.10	Rattlesnake Hills	546	0.23
Columbia Gorge	85	0.04	Red Mountain	646	0.27
Columbia River	645	0.27	Riverside	437	0.18
Columbia Valley	2907	1.21	Rogue Valley	1186	0.49
Colusa	606	0.25	Sacramento	8031	3.35
Contra Costa	725	0.30	San Benito	1115	0.47
El Dorado	830	0.35	San Bernardino	193	0.08
Erie	682	0.28	San Diego	199	0.08
Fresno	15909	6.64	San Joaquin	28107	11.73
Georgia	586	0.24	San Luis Obispo	12392	5.17
Glenn	408	0.17	San Mateo	41	0.02
Hill Country	307	0.13	Santa Barbara	6113	2.55
Horse Heaven Hills	5345	2.23	Santa Clara	606	0.25
Humboldt	49	0.02	Santa Cruz	193	0.08
Illinois	386	0.16	Schuyler	459	0.19
Indiana	262	0.11	Seneca	540	0.23
Iowa	145	0.06	Shasta	38	0.02
Kentucky	217	0.09	Siskiyou	7	0.00
Kern	6913	2.88	Snipes Mountain	225	0.09
Kings	805	0.34	Solano	1433	0.60
Lake	3264	1.36	Sonoma	23528	9.82
Lake Chelan	42	0.02	South Texas and Gulf Co	123	0.05
Lassen	1	0.00	South Willamette Valley	1165	0.49
Los Angeles	84	0.04	Stanislaus	2765	1.15
Madera	13813	5.76	Steuben	422	0.18
Marin	66	0.03	Suffolk	826	0.34
Mariposa	27	0.01	Sutter	8	0.00
Mendocino	6660	2.78	Tehama	46	0.02
Merced	5265	2.20	Texas High Plains and Pa	1087	0.45
Michigan	1150	0.48	Trinity	53	0.02
Minnesota	432	0.18	Tulare	2890	1.21
Missouri	688	0.29	Tuolumne	11	0.00
Monterey	17909	7.47	Ulster	64	0.03
Naches Heights	6	0.00	Umpqua Valley	945	0.39
Napa	17914	7.48	Ventura	17	0.01
Nevada	179	0.07	Virginia	1284	0.54
Niagara	357	0.15	Wahluke Slope	2834	1.18
North Carolina	754	0.31	Walla Walla Valley	564	0.24

Table 69 (cont.): Winegrape area by region and region's national share, by country, 2016

Region	*Area (hectares)*	*National share (%)*	*Region*	*Area (hectares)*	*National share (%)*
United States (cont.)					
Wayne	15	0.01			
West Texas	160	0.07			
Yakima Valley	5806	2.42			
Yates	2164	0.90			
Yolo	5059	2.11			
Yuba	27	0.01			
Total	**239632**	**100**			
Uruguay					
Artigas	36	0.53			
Canelones	4269	63.31			
Colonia	522	7.74			
Durazno	47	0.70			
Florida	42	0.62			
Lavalleja	7	0.10			
Maldonado	326	4.83			
Montevideo	813	12.06			
Paysandu	160	2.37			
Rivera	41	0.61			
Rocha	12	0.18			
Salto	54	0.80			
San Jose	392	5.81			
Soriano	3	0.04			
Tacuarembo	19	0.29			
Total	**6743**	**100**			

Table 70: States/admin regions and wine regions within them in Australia, Italy and the United States

Australia			**Italy**		
Aus. Capital Terr.	Far North Other	***Western Australia***	***Abruzzo***	Imperia	***Sicilia***
Aus. Capital Terr.	Fleurieu other	Blackwood Valley	Chieti	La Spezia	Agrigento
Canberra ACT	Kangaroo Island	Central WA	L'Aquila	Savona	Caltanissetta
New South Wales	Langhorne Creek	Eastern Plain ect.	Pescara	***Lombardia***	Catania
Big Rivers Other	Limestone C. other	Geographe	Teramo	Bergamo	Enna
Canberra NSW	Lower Murray Oth	Great Southern	***Basilicata***	Brescia	Messina
Central Ranges other	McLaren Vale	Greater Perth Othe	Matera	Como	Palermo
Cowra	Mount Benson	Manjimup	Potenza	Cremona	Ragusa
Gundagai	Mt Lofty Rgs Othe	Margaret River	***Bolzano-Bozen***	Lecco	Siracusa
Hastings River	Padthaway	Peel	***Calabria***	Lodi	Trapani
Hilltops	Riverland	Pemberton	Catanzaro	Mantova	***Toscana***
Hunter	Robe	Perth Hills	Cosenza	Milano	Arezzo
Hunter Valley Other	SA - other	SE Coastal WA	Crotone	Pavia	Firenze
Mudgee	Southern Fleurieu	SW Australia Othe	Reggio di Calabri	Sondrio	Grosseto
Murray Darling NSW	Sthn Flinders	Swan District	Vibo Valentia	Varese	Livorno
New England	The Peninsulas	WA other	***Campania***	***Marche***	Lucca
NSW other	Wrattonbully	***Tasmania***	Avellino	Ancona	Massa-Carrara
Nthn Rivers other	***Victoria***		Benevento	Ascoli Piceno	Pisa
Nthn Slopes Other	Central Victoria		Caserta	Macerata	Pistoia
Orange	Alpine Valleys		Napoli	Pesaro e Urbino	Prato
Perricoota	Beechworth		Salerno	***Molise***	Siena
Riverina	Bendigo		***Emilia Romagna***	Campobasso	***Trentino AA***
Shoalhaven Coast	Central Vic Other		Bologna	Isernia	Bolzano-Bozen
South Coast Other	Geelong		Ferrara	***Piemonte***	Trento
Southern Highlands	Gippsland		Forli-Cesena	Alessandria	***Umbria***
Southern NSW other	Glenrowan		Modena	Asti	Perugia
Swan Hill NSW	Heathcote		Parma	Biella	Terni
Tumbarumba	Henty		Piacenza	Cuneo	***Valle d'Aosta***
Upper Goulburn	King Valley		Ravenna	Novara	***Veneto***
West Plains Other	Macedon Ranges		Reggio nell'Emilia	Verbano-C.-Os	Belluno
Queensland	Mornington Peninsula		Rimini	Torino	Padova
Granite Belt	Murray Darling Vic		***Friuli -Venezia G.***	Vercelli	Rovigo
Queensland Other	N. E. Vic Other		Gorizia	***Puglia***	Treviso
South Burnett	N. W. Vic Other		Pordenone	Bari	Venezia
South Australia	Port Phillip - other		Trieste	Brindisi	Verona
Adelaide Hills	Pyrenees		Udine	Foggia	Vicenza
Adelaide Plains	Rutherglen		***Lazio***	Lecce	
Barossa Other	Strathbogie Ranges		Frosinone	Taranto	
Barossa Valley	Sunbury		Latina	***Sardegna***	
Bordertown	Swan Hill Vic		Rieti	Cagliari	
Clare Valley	West Vic Other		Roma	Nuoro	
Coonawarra	Western Victoria		Viterbo	Oristano	
Currency Creek	Yarra Valley		***Liguria***	Sassari	
Eden Valley			Genova		

Table 70 (cont.): States/admin regions and wine regions within them in Australia, Italy and the United States

United States

California

Alameda
Amador
Butte
Calaveras
Colusa
Contra Costa
El Dorado
Fresno
Glenn
Humboldt
Kern
Kings
Lake
Los Angeles
Madera
Mariposa
Marin
Mendocino
Merced
Monterey
Napa
Nevada
Orange
Placer
Riverside
Sacramento
San Benito
San Bernardino
San Diego
San Joaquin
San Mateo
San Luis Obispo
Santa Barbara
Santa Clara
Santa Cruz
Shasta
Siskiyou
Solano
Stanislaus
Sutter
Sonoma
Tehama
Trinity
Tulare
Tuolumne
Ventura
Yolo
Yuba

New York

Chautauqua-Erie
Finger Lakes
Other New York

Oregon

Benton Co.
Columbia River
Douglas Co.
Jackson Co.
Josephine Co.
Lane Co.
Marion Co.
Other W. Valley
Other Valley
Polk Co.
Washington Co.
Yamhill Co.
Oregon-Other

Texas

Hill Country
North Texas
Southeast Texas and Gulf Coast
Texas High Plains and Panhandle
West Texas

Washington

Columbia Gorge
Columbia Valley
Horse Heaven Hills
Lake Chelan
Puget Sound
Rattlesnake Hills
Red Mountain
Snipes Mountain
Wahluke Slope
Walla Walla Valley
Yakima Valley

Arizona

Arkansas

Colorado

Georgia

Illinois

Indiana

Iowa

Kentucky

Michigan

Minnesota

Missouri

North Carolina

Ohio

Pennsylvania

Virginia

Table 71: Concordance over time between regions of selected countries

2000	2010	2016
Australia		
Australian Capital Territory	Australian Capital Territory + Canberra District (ACT)	Canberra District (ACT)
Canberra District	Canberra District (NSW)	Canberra District (NSW)
Northern Slopes - other	Northern Slopes - other + New England Australia	New England Australia + Northern Slopes
South Coast - other	South Coast - other + Shoalhaven Coast + Southern Highlands	South Coast - other + Shoalhaven Coast + Southern Highlands
Southern NSW - other	Southern NSW - other + Gundagai	Southern NSW - other + Gundagai
Western Plains - other	Western Plains - other	Western Plains
Far North - other	Far North - other + Southern Flinders Ranges	Far North - other + Southern Flinders Ranges
Limestone Coast - other	Limestone Coast - other + Coonawarra + Robe + Wrattonbully	Limestone Coast - other + Coonawarra + Mount Gambier + Robe + Wrattonbully
Mount Lofty Ranges - other	Mount Lofty Ranges - other + Adelaide Plains	Mount Lofty Ranges - other + Adelaide Plains
Alpine Valleys/Beechworth + Beechworth	Alpine Valleys + Beechworth	Alpine Valleys + Beechworth
Central Victoria - other	Central Victoria - other + Heathcote + Strathbogie Ranges + Upper Goulburn	Central Victoria - other + Heathcote + Strathbogie Ranges + Upper Goulburn
North East Victoria - other	North East Victoria - other + Glenrowan + King Valley	North East Victoria - other + Glenrowan + King Valley
Port Phillip - other	Port Phillip - other + Macedon Ranges	Port Phillip - other + Macedon Ranges
Greater Perth - other	Greater Perth - other + Peel	Greater Perth - other + Peel
South West Australia - other	South West Australia - other + Manjimup + Pemberton	South West Australia - other + Manjimup + Pemberton
Croatia		
	Dalmatinska Zagora + Hrvatsko Primorje + Istra + Sjeverna Dalmacija + Srednja Juzna Dalmacija	Jadranska Hrvatska
	Moslavina + Plesivica + Podunavlje + Pokuplje + Prigorje - Bilogora + Slavonija + Zagorje-	Kontinentalna Hrvatska
Czechia		
	Cechy	Cechy + Praha + Severozápad
	Morava	Morava + Jihovýchod
Italy		
Ascoli Piceno	Ascoli Piceno + Fermo	
Bari	Bari + (part of) Barletta-Andria-Trani	
Foggia	Foggia + (part of) Barletta-Andria-Trani	
Cagliari	Cagliari + Carbonia-Iglesias + Medio Campidano	

Table 71 (cont.): Concordance over time between regions of selected countries

2000	2010	2016
Italy (cont.)		
Nuoro	Nuoro + Ogliastra + (part of) Olbia-Tempio	
Sassari	Sassari + (part of) Olbia-Tempio	
Portugal		
Beira Interior + Beira Litoral + Ribatejo e Oeste	Beira Interior + Beira Litoral + Ribatejo e Oeste	Centro + Lisboa
Alto Tras-os-Montes + Entre Douro e Minho	Alto Tras-os-Montes + Entre Douro e Minho	Norte
Slovakia		
	Juznoslovenska + Malokarpatska + Nitrianska	Západné Slovensko + Bratislavský kraj
	Tokajska + Východné Slovensko	Východné Slovensko
Switzerland		
Graubünden	Graubünden	Graubünden - Mesolcina + Graubünden - other
United States		
Chautauqua-Erie	Chautauqua-Erie	Chautauqua + Erie
Finger Lakes	Finger Lakes	Ontario + Schuyler + Seneca + Steuben + Wayne + Yates
New York - other	New York - other	Niagara + Suffolk + Ulster
Marion Co. + Polk Co. + Washington Co. + Yamhill Co.	Marion Co. + Polk Co. + Washington Co. + Yamhill Co.	North Willamette Valley
Josephine Co.	Jackson Co. + Josephine Co.	Rogue Valley
Benton Co. + Lane Co.	Benton Co. + Lane Co.	South Willamette Valley
Douglas Co.	Douglas Co.	Umpqua Valley

Table 72: Concordance of this database's regions of France, Italy and Spain with those in the 2019 *World Atlas of Wine*

Database		2019 World Atlas of Wine			
Regions	Subregions	Regions	Subregions	Sub-subregions	Pages
France					
Bourgogne	Cote-d'Or, Yonne, Saone-et-Loire, Rhone	Burgundy			54-55
Bourgogne	Cote-d'Or, Saone-et-Loire		Cote d'Or		56-58
Bourgogne	Cote-d'Or			Southern Cote de Beaune	59
Bourgogne	Cote-d'Or			Central Cote de Beaune	60-61
Bourgogne	Cote-d'Or			Northern Cote de Beaune	62-63
Bourgogne	Cote-d'Or			Southern Cote de Nuits	64-65
Bourgogne	Cote-d'Or			Northern Cote de Nuits	66-67
Bourgogne	Saone-et-Loire		Cote Chalonnaise		68
Bourgogne	Saone-et-Loire			Maconnais (including Pouilly-Fuisse)	69-71
Bourgogne	Saone-et-Loire, Rhone		Beaujolais		72-73
Bourgogne	Saone-et-Loire, Rhone			The Crus of Beaujolais	74-75
Bourgogne	Yonne		Chablis		76-77
Bourgogne	Yonne			The Heart of Chablis	78-79
Champagne Ardenne, Île de France, Picardie	Marne, Aube, Seine-et-Marne, Aisne	Champagne			80-81
Champagne Ardenne	Marne		The heart of Champagne		82-83
Aquitaine	Gironde	Bordeaux			84-87
Aquitaine	Gironde		Northern Medoc		88-89
Aquitaine	Gironde		St-Estephe		90-91
Aquitaine	Gironde		Pauillac		92-93
Aquitaine	Gironde		St-Julien		94-95
Aquitaine	Gironde		Central Medoc		96-97
Aquitaine	Gironde		Margaux and Southern Medoc		98-99
Aquitaine	Gironde		Graves and Entre-Deux-Mers		100-101
Aquitaine	Gironde		Pessac-Leognan		102-103
Aquitaine	Gironde		Sauternes and Barsac		104-105
Aquitaine	Gironde		The Right Bank		106-107
Aquitaine	Gironde			Pomerol	108-109
Aquitaine	Gironde			St-Emilion	110-112
Aquitaine, Midi Pyrénées	Dordogne, Landes, Lot-et-Garonne, Pyrenees-Atlantiques, Haute-Garonne, Gers, Lot, Hautes-Pyrenees, Tarn, Tarn-et-Garonne	Wines of the Southwest			113-115
Bourgogne, Centre-Val de Loire, Pays de la Loire, Poitou Charentes	Nievre, Cher, Indre, Indre-et-Loire, Loiret, Loir-et-Cher, Loire-Atlantique, Maine-et-Loire, Sarthe, Vendee, Deux-Sevres, Vienne	The Loire Valley			116-117
Pays de la Loire	Maine-et-Loire		Anjou		118
Pays de la Loire	Maine-et-Loire		Saumur		119
Centre-Val de Loire	Indre-et-Loire		Chinon and Bourgueil		120
Centre-Val de Loire	Indre-et-Loire		Vouvray and Montlouis		121
Centre-Val de Loire, Bourgogne	Cher, Nievre		Sancerre and Pouilly		122-123
Alsace	Bas-Rhin, Haut-Rhin	Alsace			124-125
Alsace	Haut-Rhin		The heart of Alsace		125-127
Rhône Alpes	Ardeche, Drome, Isere, Loire, Rhone	Northern Rhone			128-129
Rhône Alpes	Loire, Rhone		Cote-Rotie and Condrieu		130-131
Rhône Alpes	Ardeche, Drome		Hermitage		132-133
Languedoc Roussillon, Provence-Alpes-Cote d'Azur, Rhône Alpes	Gard, Vaucluse, Ardeche, Drome	Southern Rhone			134-135
Languedoc Roussillon, Provence-Alpes-Cote d'Azur	Gard, Vaucluse		Heart of the Southern Rhone		136-137
Provence-Alpes-Cote d'Azur	Vaucluse			Chateauneuf-du-Pape	138-139
Languedoc Roussillon	Aude, Herault	Western Languedoc			140-141

Table 72 (cont.): Concordance of this database's regions of France, Italy and Spain with those in the 2019 *World Atlas of Wine*

Database		2019 World Atlas of Wine			
Regions	Subregions	Regions	Subregions	Sub-subregions	Pages
France (cont.)					
Languedoc Roussillon	Gard, Herault	Eastern Languedoc			142-143
Languedoc Roussillon	Pyrenees-Orientales	Rousillon			144-145
Provence-Alpes-Cote d'Azur	Bouches-du-Rhone, Var	Provence			146-147
Provence-Alpes-Cote d'Azur	Var		Bandol		148
Corse	Corse-du-Sud, Haute-Corse	Corsica			149
Franche Comté, Rhône Alpes	Jura, Ain, Savoie, Haute-Savoie	Jura, Savoie, and Bugey			150-152
Italy					
Piemonte, Valle d'Aosta, Liguria, Emilia-Romagna, Lombardia	*some sub-regions from Emilia-Romagna and Lombardia, and all sub-regions from other regions*	Northwest Italy			156-157
Piemonte	*all sub-regions from the region mentioned*		Piemonte		158-159
Piemonte	Cuneo			Barbaresco	160-161
Piemonte	Cuneo			Barolo	162-163
Bolzano-Bozen, Trento, Veneto, Friuli-Venezia Giulia, Lombardia, Emilia-Romagna	*some sub-regions from Emilia-Romagna and Lombardia, and all sub-regions from other regions*	Northeast Italy			164-165
Bolzano-Bozen, Trento	*all sub-regions from the regions mentioned*		Trentino and Alto Adige		166-167
Veneto, Lombardia	Verona, Vicenza, Brescia		Verona		168-169
Friuli-Venezia Giulia	*not specified*		Friuli		170-171
Toscana, Umbria, Lazio, Marche, Abruzzo, Emilia-Romagna	*some sub-regions from Emilia-Romagna, and all sub-regions from other regions*	Central Italy			172-173
Toscana	Livorno		Maremma		174-175
Toscana	Firenze, Siena		Chianti Classico		176-178
Toscana	Siena		Montalcino		179
Toscana	Siena		Montepulciano		180
Umbria, Lazio, Toscana	Perugia, Terni, Viterbo, Siena		Umbria		181
Sicilia, Sardegna, Molise, Puglia, Campania, Basilicata, Calabria	*all sub-regions from the regions mentioned*	Southern Italy			182-183
Sicilia	*all sub-regions from Sicilia*		Sicily		184-185
Sardegna	*all sub-regions from Sardegna*		Sardinia		186
Spain					
	not specified	Others			189-191
País Vasco, Castilla y León, Aragón	*not specified*		The north		189-190
Comunidad Valenciana, Región de Murcia, Castilla-La Mancha	*not specified*		The east		190-191
Community of Madrid, Castilla-La Mancha, Extremadura	*not specified*		South of Madrid		191
Canarias	*not specified*		The islands		191
Galicia	Galicia	Northwest Spain			192
Galicia	Galicia		Rias Baixas		193
Castilla y León	Burgos, Segovia, Soria, Valladolid	Ribera del Duero			194-195
Castilla y León	Valladolid, Zamora	Toro and Rueda			196
Comunidad Foral de Navarra	Comunidad Foral de Navarra	Navarra			197
La Rioja, País Vasco, Comunidad Foral de Navarra	La Rioja, Alava, Comunidad Foral de Navarra	Rioja			198-199
Cataluña	Barcelona, Girona, Lleida, Tarragona	Catalunya			200-201
Cataluña	Tarragona		Priorat		202
Andalucía	Almeria, Granada, Jaen, Sevilla, Cadiz, Cordoba, Huelva, Malaga	Andalucia			203-205

Table 73: Concordance of regions of France, Italy and Spain in the 2019 *World Atlas of Wine* with those in this book's database

2019 World Atlas of Wine				Database	
Regions	Subregions	Sub-subregions	Pages	Regions	Subregions
France					
Burgundy			54-55	Bourgogne	Cote-d'Or, Yonne, Saone-et-Loire, Rhone
	Cote d'Or		56-58	Bourgogne	Cote-d'Or, Saone-et-Loire
		Southern Cote de Beaune	59	Bourgogne	Cote-d'Or
		Central Cote de Beaune	60-61	Bourgogne	Cote-d'Or
		Northern Cote de Beaune	62-63	Bourgogne	Cote-d'Or
		Southern Cote de Nuits	64-65	Bourgogne	Cote-d'Or
		Northern Cote de Nuits	66-67	Bourgogne	Cote-d'Or
	Cote Chalonnaise		68	Bourgogne	Saone-et-Loire
		Maconnais (including Pouilly-Fu	69-71	Bourgogne	Saone-et-Loire
	Beaujolais		72-73	Bourgogne	Saone-et-Loire, Rhone
		The Crus of Beaujolais	74-75	Bourgogne	Saone-et-Loire, Rhone
	Chablis		76-77	Bourgogne	Yonne
		The Heart of Chablis	78-79	Bourgogne	Yonne
Champagne			80-81	Champagne Ardenne, Île de France, Picardie	Marne, Aube, Seine-et-Marne, Aisne
	The heart of Champagne		82-83	Champagne Ardenne	Marne
Bordeaux			84-87	Aquitaine	Gironde
	Northern Medoc		88-89	Aquitaine	Gironde
	St-Estephe		90-91	Aquitaine	Gironde
	Pauillac		92-93	Aquitaine	Gironde
	St-Julien		94-95	Aquitaine	Gironde
	Central Medoc		96-97	Aquitaine	Gironde
	Margaux and Southern Medoc		98-99	Aquitaine	Gironde
	Graves and Entre-Deux-Mers		100-101	Aquitaine	Gironde
	Pessac-Leognan		102-103	Aquitaine	Gironde
	Sauternes and Barsac		104-105	Aquitaine	Gironde
	The Right Bank		106-107	Aquitaine	Gironde
		Pomerol	108-109	Aquitaine	Gironde
		St-Emilion	110-112	Aquitaine	Gironde

Table 73 (cont.): Concordance of regions of France, Italy and Spain in the 2019 *World Atlas of Wine* with those in this book's database

2019 World Atlas of Wine				Database	
Regions	Subregions	Sub-subregions	Pages	Regions	Subregions
France (cont.)					
Wines of the Southwest			113-115	Aquitaine, Midi Pyrénées	Dordogne, Landes, Lot-et-Garonne, Pyrenees-Atlantiques, Haute-Garonne, Gers, Lot, Hautes-Pyrenees, Tarn, Tarn-et-Garonne
The Loire Valley			116-117	Bourgogne, Centre-Val de Loire, Pays de la Loire, Poitou Charentes	Nievre, Cher, Indre, Indre-et-Loire, Loiret, Loir-et-Cher, Loire-Atlantique, Maine-et-Loire, Sarthe, Vendee, Deux-Sevres, Vienne
	Anjou		118	Pays de la Loire	Maine-et-Loire
	Saumur		119	Pays de la Loire	Maine-et-Loire
	Chinon and Bourgueil		120	Centre-Val de Loire	Indre-et-Loire
	Vouvray and Montlouis		121	Centre-Val de Loire	Indre-et-Loire
	Sancerre and Pouilly		122-123	Centre-Val de Loire, Bourgogne	Cher, Nievre
Alsace			124-125	Alsace	Bas-Rhin, Haut-Rhin
	The heart of Alsace		125-127	Alsace	Haut-Rhin
Northern Rhone			128-129	Rhône Alpes	Ardeche, Drome, Isere, Loire, Rhone
	Cote-Rotie and Condrieu		130-131	Rhône Alpes	Loire, Rhone
	Hermitage		132-133	Rhône Alpes	Ardeche, Drome

Table 73 (cont.): Concordance of regions of France, Italy and Spain in the 2019 *World Atlas of Wine* with those in this book's database

2019 World Atlas of Wine				Database	
Regions	Subregions	Sub-subregions	Pages	Regions	Subregions
France (cont.)					
Southern Rhone			134-135	Languedoc Roussillon, Provence-Alpes-Cote d'Azur, Rhône Alpes	Gard, Vaucluse, Ardeche, Drome
	Heart of the Southern Rhone		136-137	Languedoc Roussillon, Provence-Alpes-Cote d'Azur	Gard, Vaucluse
		Chateauneuf-du-Pape	138-139	Provence-Alpes-Cote d'Azur	Vaucluse
Western Languedoc			140-141	Languedoc Roussillon	Aude, Herault
Eastern Languedoc			142-143	Languedoc Roussillon	Gard, Herault
Rousillon			144-145	Languedoc Roussillon	Pyrenees-Orientales
Provence			146-147	Provence-Alpes-Cote d'Azur	Bouches-du-Rhone, Var
	Bandol		148	Provence-Alpes-Cote d'Azur	Var
Corsica			149	Corse	Corse-du-Sud, Haute-Corse
Jura, Savoie, and Bugey			150-152	Franche Comté, Rhône Alpes	Jura, Ain, Savoie, Haute-Savoie
Italy					
Northwest Italy			156-157	Piemonte, Valle d'Aosta, Liguria, Emilia-Romagna, Lombardia	*some sub-regions from Emilia-Romagna and Lombardia, and all sub-regions from other regions*
	Piemonte		158-159	Piemonte	*all sub-regions from the region mentioned*
		Barbaresco	160-161	Piemonte	Cuneo
		Barolo	162-163	Piemonte	Cuneo

Table 73 (cont.): Concordance of regions of France, Italy and Spain in the 2019 *World Atlas of Wine* with those in this book's database

2019 World Atlas of Wine				Database	
Regions	Subregions	Sub-subregions	Pages	Regions	Subregions
Italy (cont.)					
Northeast Italy			164-165	Bolzano-Bozen, Trento, Veneto, Friuli-Venezia Giulia, Lombardia, Emilia-Romagna	*some sub-regions from Emilia-Romagna and Lombardia, and all sub-regions from other regions*
	Trentino and Alto Adige		166-167	Bolzano-Bozen, Trento	*all sub-regions from the regions mentioned*
	Verona		168-169	Veneto, Lombardia	Verona, Vicenza, Brescia
	Friuli		170-171	Friuli-Venezia Giulia	*not specified*
Central Italy			172-173	Toscana, Umbria, Lazio, Marche, Abruzzo, Emilia-Romagna	*some sub-regions from Emilia-Romagna, and all sub-regions from other regions*
	Maremma		174-175	Toscana	Livorno
	Chianti Classico		176-178	Toscana	Firenze, Siena
	Montalcino		179	Toscana	Siena
	Montepulciano		180	Toscana	Siena
	Umbria		181	Umbria, Lazio, Toscana	Perugia, Terni, Viterbo, Siena
Southern Italy			182-183	Sicilia, Sardegna, Molise, Puglia, Campania, Basilicata, Calabria	*all sub-regions from the regions mentioned*
	Sicily		184-185	Sicilia	*all sub-regions from Sicilia*
	Sardinia		186	Sardegna	*all sub-regions from Sardegna*

Table 73: Concordance of regions of France, Italy and Spain in the 2019 *World Atlas of Wine* with those in this book's database

2019 World Atlas of Wine				Database	
Regions	Subregions	Sub-subregions	Pages	Regions	Subregions
Spain					
Others			189-191		*not specified*
	The north		189-190	País Vasco, Castilla y León, Aragón	*not specified*
	The east		190-191	Comunidad Valenciana, Región de Murcia, Castilla-La Mancha	*not specified*
	South of Madrid		191	Community of Madrid, Castilla-La Mancha, Extremadura	*not specified*
	The islands		191	Canarias	*not specified*
Northwest Spain			192	Galicia	Galicia
	Rias Baixas		193	Galicia	Galicia
Ribera del Duero			194-195	Castilla y León	Burgos, Segovia, Soria, Valladolid
Toro and Rueda			196	Castilla y León	Valladolid, Zamora
Navarra			197	Comunidad Foral de Navarra	Comunidad Foral de Navarra
Rioja			198-199	La Rioja, País Vasco, Comunidad Foral de Navarra	La Rioja, Alava, Comunidad Foral de Navarra
Catalunya			200-201	Cataluña	Barcelona, Girona, Lleida, Tarragona
	Priorat		202	Cataluña	Tarragona
Andalucia			203-205	Andalucía	Almeria, Granada, Jaen, Sevilla, Cadiz, Cordoba, Huelva, Malaga

Table 74: World's wine regions' bearing areas and names in *World Atlas of Wine*, by country, 2000, 2010 and 2016
(final column indicates whether climate is cool, temperate, warm or hot)

Name	2000	2010	2016	ha 2000	ha 2010	ha 2016	Atlas name	Pages	Town	Zone
ALGERIA	**ALGERIA**	**ALGERIA**	**ALGERIA**	**30200**	**30200**	**8300**	**ALGERIA**	**48**	**Algiers**	**Hot**
ARGENTINA	**ARGENTINA**	**ARGENTINA**	**ARGENTINA**	**197418**	**213372**	**206342**	**ARGENTINA**	**338-341**		**Hot**
Buenos Aires				**7**	**55**	**110**	**Buenos Aires**	**341**		**Warm**
Balcarce			Balcarce			9			Balcarce	Warm
Benito Juárez			Benito Juárez			0			Benito Juárez	Warm
Cañuelas			Cañuelas			2			Cañuelas	Hot
Coronel Pringles	Coronel Pringles		Coronel Pringles	5		2			Coronel Pringles	Warm
Coronel Suarez		Coronel Suarez	Coronel Suarez		13	19			Coronel Suarez	Warm
Daireaux			Daireaux			1			Daireaux	Hot
De La Costa			De La Costa			0				Warm
General Belgrano			General Belgrano			4			General Belgrano	Hot
General Pueyrredón	General Pueyrredón		General Pueyrredón	2		11	Chapadmalal	341	General Pueyrredón	Warm
Junín - Bs. As.			Junín - Bs. As.			0			Junín - Bs. As.	Hot
Saavedra			Saavedra			18			Saavedra	Warm
Tandil		Tandil	Tandil		6	7	Tandil	map 330	Tandil	Warm
Tornquist		Tornquist	Tornquist		12	12			Tornquist	Warm
Villa Gesell			Villa Gesell			0			Villa Gesell	Warm
Villarino		Villarino	Villarino		25	25			Médanos	Hot
Catamarca				**2178**	**2413**	**2407**	**Catamarca**	**339**		**Hot**
Ambato		Ambato			2				El Bolsón	Hot
Andalgala	Andalgala	Andalgala	Andalgala	7	8	6			Andalgalá	Hot
Belén	Belén	Belén	Belén	129	152	185			Belén	Hot
Capayán			Capayán			1			Capayán	Hot
Cushamen	Cushamen	Cushamen	Cushamen	20	20	20			Lago Puelo	Cool
Poman	Poman	Poman	Poman	29	30	27			Pomán	Hot
Santa María - Catamarca	Santa María - Catamarca	Santa María - Catamarca	Santa María - Catamarca	579	618	708	Santa María	339	Santa María	Hot
Santa Rosa - Catamarca			Santa Rosa - Catamarca			1			Bañado de Ovanta	Hot
Tinogasta	Tinogasta	Tinogasta	Tinogasta	1414	1582	1458			Tinogasta	Hot
Valle Viejo			Valle Viejo			1			El Portezuelo	Hot
Chubut						**43**	**Chubut**	**341**		**Cool**
Futaleufu			Futaleufu			5	Trevelin	341	Esquel	Cool
Languiñeo			Languiñeo			0			Tecka	Cool
Sarmiento - Chubut			Sarmiento - Chubut			38			Sarmiento	Cool
Córdoba				**117**	**245**	**243**	**Córdoba**	**map 330**		**Hot**
Calamuchita		Calamuchita	Calamuchita		1	18	Valle de Calamucl	map 330	San Agustín	Hot
Colón - Cordoba		Colón - Cordoba	Colón - Cordoba		155	113			Jesús María	Hot
Cruz del Eje	Cruz del Eje	Cruz del Eje	Cruz del Eje	60	29	27			Cruz del Eje	Hot
Ischilin	Ischilin	Ischilin	Ischilin	48	46	49			Deán Funes	Hot
Punilla			Punilla			4			Cosquín	Warm
San Alberto	San Alberto		San Alberto	1		5			Villa Cura Brochero	Hot
San Javier	San Javier	San Javier	San Javier	1	2	12			Villa Dolores	Hot
Santa María - Cordoba		Santa María - Cordoba	Santa María - Cordoba		9	9			Alta Gracia	Hot
Totoral	Totoral	Totoral		6	2				Villa del Totoral	Hot
Tulumba		Tulumba	Tulumba		0	5			Villa Tulumba	Hot

Table 74 (cont.): World's wine regions' bearing areas and names in *World Atlas of Wine*, by country, 2000, 2010 and 2016

Name	2000	2010	2016	ha 2000	ha 2010	ha 2016	Atlas name	Pages	Town	Zone
Entre Ríos				**224**	**9**	**36**				**Hot**
Colón - Entre Ríos	Colón - Entre Ríos	Colón - Entre Ríos	Colón - Entre Ríos	224	5	6	Colón	map 330	Colón	Hot
Concordia		Concordia	Concordia		3	13	Concordia	map 330	Concordia	Hot
Diamante			Diamante			4			Diamante	Hot
Gualeguaychu			Gualeguaychu			1	Gualeguaychu	map 330	Gualeguaychú	Hot
Nogoya		Nogoya	Nogoya		0	0			Nogoyá	Hot
Paraná			Paraná			6	Paraná	map 330	Paraná	Hot
Uruguay			Uruguay			0			Concepción del Uruguay	Hot
Victoria		Victoria	Victoria		1	7	Victoria	map 330	Victoria	Hot
Jujuy					**3**	**21**	**Jujuy**	**339**		**Temp.**
El Carmen			El Carmen			3			El Carmen	Hot
Humahuaca			Humahuaca			5			Humahuaca	Temp.
Tilcara		Tilcara	Tilcara		2	13	Tilcara	map 330	Tilcara	Temp.
Tumbaya		Tumbaya	Tumbaya		0	0	Tumbaya	map 330	Tumbaya	Warm
La Pampa				**122**	**217**	**243**	**La Pampa**	**map 330**		**Hot**
Curaco			Curaco			3			Puelches	Hot
Puelen	Puelen	Puelen	Puelen	122	217	240			Puelén	Hot
La Rioja				**7328**	**7646**	**6611**	**La Rioja**	**339**		**Hot**
Arauco	Arauco	Arauco		26	5				Aimogasta	Hot
Capital La Rioja	Capital La Rioja	Capital La Rioja		21	8				La Rioja	Hot
Castro Barros	Castro Barros	Castro Barros	Castro Barros	219	206	245			Aminga	Hot
Chilecito	Chilecito	Chilecito	Chilecito	5810	6016	5356	Chilecito	339	Chilecito	Hot
Coronel Felipe Varela	Coronel Felipe Varela	Coronel Felipe Varela	Coronel Felipe Varela	856	883	506			Villa Unión	Hot
Famatina	Famatina	Famatina	Famatina	229	322	344	Famatina Valley	339	Famatina	Hot
General Lamadrid	General Lamadrid	General Lamadrid	General Lamadrid	45	74	60			Villa Castelli	Hot
San Blas De Los Sauces	San Blas De Los Sauces	San Blas De Los Sauces	San Blas De Los Sauces	45	46	41			San Blas de los Sauces	Hot
Sanagasta	Sanagasta	Sanagasta	Sanagasta	3	11	12			Sanagasta	Hot
Vinchina	Vinchina	Vinchina	Vinchina	75	75	48			Vinchina	Hot
Mendoza				**142414**	**158439**	**156099**	**Mendoza**	**341**		**Hot**
General Alvear	General Alvear	General Alvear	General Alvear	5675	5341	4028			General Alvear	Hot
Godoy Cruz	Godoy Cruz	Godoy Cruz	Godoy Cruz	10	1	1	Godoy Cruz	map 340	Godoy Cruz	Hot
Guaymallén	Guaymallén	Guaymallén	Guaymallén	1101	955	630			Guaymallén	Hot
Junín - Mendoza	Junín - Mendoza	Junín - Mendoza	Junín - Mendoza	11112	12058	11709			Junín	Hot
La Paz	La Paz	La Paz	La Paz	369	409	300			La Paz	Hot
Las Heras	Las Heras	Las Heras	Las Heras	1640	1699	1529	Las Heras	map 340	Las Heras	Hot
Lavalle	Lavalle	Lavalle	Lavalle	12762	13969	13763			Villa Tulumaya	Hot
Luján de Cuyo	Luján de Cuyo	Luján de Cuyo	Luján de Cuyo	10809	13990	15830	Luján de Cuyo	341	Luján de Cuyo	Hot
Maipú	Maipú	Maipú	Maipú	11463	12586	12343	Maipú	341	Maipú	Hot
Malargüe		Malargüe			1				Malargüe	Temp.
Rivadavia - Mendoza	Rivadavia - Mendoza	Rivadavia - Mendoza	Rivadavia - Mendoza	15919	16976	15618	Rivadavia	map 340	Rivadavia	Hot
San Carlos - Mendoza	San Carlos - Mendoza	San Carlos - Mendoza	San Carlos - Mendoza	4220	7030	8676	San Carlos	341	San Carlos	Warm
San Martín - Mendoza	San Martín - Mendoza	San Martín - Mendoza	San Martín - Mendoza	30344	31161	28869	San Martín	map 340	San Martín	Hot
San Rafael	San Rafael	San Rafael	San Rafael	16109	15326	13620	San Rafael	341	San Rafael	Warm
Santa Rosa - Mendoza	Santa Rosa - Mendoza	Santa Rosa - Mendoza	Santa Rosa - Mendoza	10905	11545	10012	Santa Rosa	map 340	Santa Rosa	Hot
Tunuyán	Tunuyán	Tunuyán	Tunuyán	4041	7241	9320	Tunuyán	341	Tunuyán	Warm
Tupungato	Tupungato	Tupungato	Tupungato	5935	8149	9852	Tupungato	341	Tupungato	Warm
Misiones					**1**	**10**				**Hot**

Table 74 (cont.): World's wine regions' bearing areas and names in *World Atlas of Wine*, by country, 2000, 2010 and 2016

Name	2000	2010	2016	ha 2000	ha 2010	ha 2016	Atlas name	Pages	Town	Zone
Cainguas			Cainguas			7			Campo Grande	Hot
Capital Misiones			Capital Misiones			2			Posadas	Hot
Leandro Alem		Leandro Alem	Leandro Alem		0	1			Leandro N. Alem	Hot
Veinticinco de Mayo - Misiones		Veinticinco de Mayo - Mis	Veinticinco de Mayo - Mis		0	0			Alba Posse	Hot
Neuquén				**663**	**1628**	**1752**	**Neuquén**	**341**		**Warm**
Añelo	Añelo	Añelo	Añelo	624	1504	1588			Añelo	Warm
Chos Malal		Chos Malal	Chos Malal		2	6			Chos Malal	Temp.
Collon Cura		Collon Cura	Collon Cura		5	5			Collón Cura	Hot
Confluencia	Confluencia	Confluencia	Confluencia	39	93	126			Neuquén	Hot
Lacar			Lacar			1			San Martín de los Andes	Cool
Ñorquin			Ñorquin			1			El Huecú	Cool
Pehuenches			Pehuenches			3			Rincón de los Sauces	Warm
Picún Leufú		Picún Leufú	Picún Leufú		25	22			Picún Leufú	Warm
Picunches			Picunches			1			Las Lajas	Temp.
Río Negro				**2660**	**2324**	**1546**	**Río Negro**	**341**		**Hot**
Adolfo Alsina	Adolfo Alsina	Adolfo Alsina	Adolfo Alsina	10	67	98			Viedma	Warm
Avellaneda - Río Negro	Avellaneda - Río Negro	Avellaneda - Río Negro	Avellaneda - Río Negro	443	224	154			Choele Choel	Hot
Bariloche			Bariloche			1			San Carlos de Bariloche	Cool
Conesa	Conesa	Conesa	Conesa	50	47	22			General Conesa	Hot
El Cuy	El Cuy	El Cuy	El Cuy	44	67	60			El Cuy	Temp.
General Roca	General Roca	General Roca	General Roca	2076	1879	1174			General Roca	Hot
Pichi Mahuida	Pichi Mahuida	Pichi Mahuida	Pichi Mahuida	38	41	38			Río Colorado	Hot
Salta				**1862**	**2292**	**3243**	**Salta**	**339**		**Hot**
Cachi	Cachi	Cachi	Cachi	4	37	78	Cachi	map 339	Cachi	Warm
Cafayate	Cafayate	Cafayate	Cafayate	1334	1608	2411	Cafayate	339	Cafayate	Hot
La Viña		La Viña	La Viña		4	5			La Viña	Hot
Molinos	Molinos	Molinos	Molinos	30	120	124	Molinos	map 339	Molinos	Warm
San Carlos - Salta	San Carlos - Salta	San Carlos - Salta	San Carlos - Salta	455	522	624	San Carlos	map 339	San Carlos	Hot
San Juan				**39812**	**38028**	**33771**	**San Juan**	**339**		**Hot**
Albardón	Albardón	Albardón	Albardón	1238	839	711			General San Martín	Hot
Angaco	Angaco	Angaco	Angaco	2328	1961	1774			Villa del Salvador	Hot
Calingasta	Calingasta	Calingasta	Calingasta	72	103	124	Calingasta Valley	339	Tamberías	Temp.
Capital San Juan	Capital San Juan	Capital San Juan	Capital San Juan	17	11	2			San Juan	Hot
Caucete	Caucete	Caucete	Caucete	6606	6443	5681			Caucete	Hot
Chimbas	Chimbas	Chimbas	Chimbas	823	765	612			Paula A. de Sarmiento	Hot
Iglesia	Iglesia	Iglesia	Iglesia	24	21	7			Rodeo	Temp.
Jachal	Jachal	Jachal		42	44				Huaco	Hot
Nueve de Julio	Nueve de Julio	Nueve de Julio	Nueve de Julio	2675	2386	2138			Nueve de Julio	Hot
Pocito	Pocito	Pocito	Pocito	3267	2670	2190			Villa Aberastain	Hot
Rawson	Rawson	Rawson	Rawson	1584	1464	1156			Villa Krause	Hot
Rivadavia - San Juan	Rivadavia - San Juan	Rivadavia - San Juan	Rivadavia - San Juan	480	440	364			Rivadavia	Hot
San Martín - San Juan	San Martín - San Juan	San Martín - San Juan	San Martín - San Juan	3408	3361	2940			San Martín	Hot
Santa Lucía	Santa Lucía	Santa Lucía	Santa Lucía	1380	1209	989			Santa Lucía	Hot
Sarmiento - San Juan	Sarmiento - San Juan	Sarmiento - San Juan	Sarmiento - San Juan	6533	7522	7225			Media Agua	Hot
Ullum	Ullum	Ullum	Ullum	889	580	525	Ullum		Villa Ibañez	Hot
Valle Fértil	Valle Fértil	Valle Fértil	Valle Fértil	25	13	0			San Agustín	Hot
Veinticinco de Mayo - San Juan	Veinticinco de Mayo - San	Veinticinco de Mayo - San	Veinticinco de Mayo - San	7423	7416	6586			Santa Rosa	Hot

Table 74 (cont.): World's wine regions' bearing areas and names in *World Atlas of Wine*, by country, 2000, 2010 and 2016

Name	2000	2010	2016	ha 2000	ha 2010	ha 2016	Atlas name	Pages	Town	Zone
Zonda	Zonda	Zonda	Zonda	998	781	747	Zonda Valley	339	Bsilio Nievas	Hot
San Luis				**12**	**26**	**98**				**Hot**
Ayacucho	Ayacucho		Ayacucho	12		10			San Francisco del Monte (	Hot
Capital San Luis			Capital San Luis			60			San Luis	Hot
Junín - San Luis		Junín - San Luis	Junín - San Luis		26	27			Santa Rosa	Hot
Santiago del Estero				**2**		**2**	**Santiago del Este**	**map 330**		**Hot**
Capital Santiago del Estero	Capital Santiago del Estero		Capital Santiago del Esterc	2		2			Santiago del Estero	Hot
Tucumán				**18**	**47**	**110**	**Tucuman**	**339**		**Warm**
Tafí del Valle	Tafí del Valle	Tafí del Valle	Tafí del Valle	18	47	105	Tafí del Valle	map 339	Tafí del Valle	Warm
Trancas			Trancas			5	Trancas	map 339	Trancas	Hot
ARMENIA	**ARMENIA**	**ARMENIA**	**ARMENIA**	**11206**	**11206**	**14705**	**ARMENIA**	**277**	**Yerevan**	**Warm**
AUSTRALIA	**AUSTRALIA**	**AUSTRALIA**	**AUSTRALIA**	**130602**	**151788**	**132435**	**AUSTRALIA**	**344-366**		**Hot**
Australian Capital Territory				**15**	**109**	**92**				**Temp.**
Australian Capital Territory	Australian Capital Territor	Australian Capital Territory		15	4		Canberra District	365		Warm
Canberra District (ACT)		Canberra District (ACT)	Canberra District (ACT)		105	92	Canberra District	365	Canberra	Warm
New South Wales				**31043**	**40944**	**33634**	**New South Wales**	**364-365**		**Hot**
Big Rivers - other	Big Rivers - other	Big Rivers - other	Big Rivers - other	1015	629	605				Hot
Canberra District	Canberra District			110			Canberra District	365		Warm
Canberra District (NSW)		Canberra District (NSW)	Canberra District (NSW)		378	243	Canberra District	365	Yass	Warm
Central Ranges - other	Central Ranges - other	Central Ranges - other	Central Ranges - other	302	227	390				Hot
Cowra	Cowra	Cowra	Cowra	1533	1427	643	Cowra	365	Cowra	Hot
Gundagai		Gundagai	Gundagai		408	332	Gundagai	map 345	Cootamundra	Hot
Hastings River	Hastings River	Hastings River	Hastings River	118	18	13	Hastings River	365	Port Macquarie	Hot
Hilltops	Hilltops	Hilltops	Hilltops	383	484	566	Hilltops	365	Young	Warm
Hunter	Hunter	Hunter	Hunter	3669	3450	2309	Hunter Valley	364	Cessnook	Hot
Hunter Valley - other	Hunter Valley - other	Hunter Valley - other	Hunter Valley - other	278	24	52	Hunter Valley	364		Hot
Mudgee	Mudgee	Mudgee	Mudgee	2152	3414	1070	Mudgee	365	Mudgee	Warm
Murray Darling (NSW)	Murray Darling (NSW)	Murray Darling (NSW)	Murray Darling (NSW)	5576	6533	6298	Murray Darling	345	Wentworth	Hot
New England Australia		New England Australia	New England Australia		123	132	New England	365	Glen Innes	Temp.
Northern Rivers - other	Northern Rivers - other	Northern Rivers - other	Northern Rivers - other	12	41	7				Hot
Northern Slopes			Northern Slopes			48			Armidale	Warm
Northern Slopes - other	Northern Slopes - other	Northern Slopes - other		174	145					Warm
Orange	Orange	Orange	Orange	995	1546	1098	Orange	365	Orange	Temp.
Perricoota	Perricoota	Perricoota	Perricoota	153	671	367	Perricoota	map 344	Moama	Hot
Riverina	Riverina	Riverina	Riverina	12398	20154	18610	Riverina	345	Griffith	Hot
Shoalhaven Coast		Shoalhaven Coast	Shoalhaven Coast		40	33	Shoalhaven Coast	365	Nowra	Hot
South Coast - other	South Coast - other	South Coast - other	South Coast - other	112	113	71				Hot
Southern Highlands		Southern Highlands	Southern Highlands		202	124	Southern Highlanc	map 345	Bowral	Temp.
Southern NSW - other	Southern NSW - other	Southern NSW - other	Southern NSW - other	914	119	46				Hot
Swan Hill (NSW)	Swan Hill (NSW)	Swan Hill (NSW)	Swan Hill (NSW)	544	308	131	Swan Hill	map 344	Koraleigh	Hot
Tumbarumba	Tumbarumba	Tumbarumba	Tumbarumba	278	254	210	Tumbarumba	365	Tumbarumba	Temp.
Western Plains			Western Plains			236			Bourke	Hot
Western Plains - other	Western Plains - other	Western Plains - other		328	236					Hot
Northern Territory	**Northern Territory**			**302**					**Darwin**	**Hot**
Queensland				**1984**	**758**	**535**	**Queensland**	**345**		**Hot**
Granite Belt	Granite Belt	Granite Belt	Granite Belt	433	331	237	Granite Belt	345	Stanthorpe	Warm
Queensland - other	Queensland - other	Queensland - other	Queensland - other	1207	187	83				Hot

Table 74 (cont.): World's wine regions' bearing areas and names in *World Atlas of Wine*, by country, 2000, 2010 and 2016

Name	2000	2010	2016	ha 2000	ha 2010	ha 2016	Atlas name	Pages	Town	Zone
South Burnett	South Burnett	South Burnett	South Burnett	344	240	215	South Burnett	map 345	Murgon	Hot
South Australia				**54997**	**71542**	**65211**	**South Australia**	**350-357**		**Warm**
Adelaide Hills	Adelaide Hills	Adelaide Hills	Adelaide Hills	1811	3861	2967	Adelaide Hills	356	Lenswood	Warm
Adelaide Plains		Adelaide Plains	Adelaide Plains		880	586			Elizabeth	Hot
Barossa Valley	Barossa Valley	Barossa Valley	Barossa Valley	7673	9763	8899	Barossa Valley	351-352	Nuriootpa	Warm
Barossa - other	Barossa - other	Barossa - other	Barossa - other	249	91	27				Warm
Clare Valley	Clare Valley	Clare Valley	Clare Valley	3617	4801	4200	Clare Valley	353	Clare	Warm
Coonawarra		Coonawarra	Coonawarra		5985	4726	Coonawarra	357	Coonawarra	Temp.
Currency Creek	Currency Creek	Currency Creek	Currency Creek	940	871	864	Currency Creek	355	Milang	Warm
Eden Valley	Eden Valley	Eden Valley	Eden Valley	1224	1933	1799	Eden Valley	352	Eden Valley	Warm
Far North - other	Far North - other	Far North - other	Far North - other	107	11	2				Hot
Fleurieu - other	Fleurieu - other	Fleurieu - other	Fleurieu - other	510	187	323				Warm
Kangaroo Island	Kangaroo Island	Kangaroo Island	Kangaroo Island	34	89	95	Kangaroo Island	354	Parndana	Warm
Langhorne Creek	Langhorne Creek	Langhorne Creek	Langhorne Creek	3737	5957	5282	Langhorne Creek	355	Langhorne Creek	Warm
Limestone Coast - other	Limestone Coast - other	Limestone Coast - other	Limestone Coast - other	7529	476	2113	Limestone Coast	346		Warm
Lower Murray - other	Lower Murray - other	Lower Murray - other	Lower Murray - other	145	260	856				Hot
McLaren Vale	McLaren Vale	McLaren Vale	McLaren Vale	4695	6490	5995	McLaren Vale	354-355	McLaren Vale	Warm
Mount Benson	Mount Benson	Mount Benson	Mount Benson	299	233	257	Mount Benson	346	Mount Benson	Temp.
Mount Gambier			Mount Gambier			186	Mount Gambier	346	Mount Gambier	Temp.
Mount Lofty Ranges - other	Mount Lofty Ranges - othe	Mount Lofty Ranges - othe	Mount Lofty Ranges - othe	488	468	470	Mount Lofty Rang	356		Warm
Padthaway	Padthaway	Padthaway	Padthaway	3226	5028	3237	Padthaway	346	Padthaway	Warm
Riverland	Riverland	Riverland	Riverland	18336	20009	18774	Riverland	345	Renmark	Hot
Robe		Robe	Robe		644	650	Robe	346	Robe	Temp.
Southern Fleurieu	Southern Fleurieu	Southern Fleurieu	Southern Fleurieu	328	414	181	Southern Fleurieu	354	Victor Harbor	Warm
Southern Flinders Ranges		Southern Flinders Ranges	Southern Flinders Ranges		180	175	Southern Flinders	map 344	Gladstone	Hot
The Peninsulas	The Peninsulas	The Peninsulas	The Peninsulas	50	93	69			Maitland	Hot
Wrattonbully		Wrattonbully	Wrattonbully		2818	2478	Wrattonbully	346	Naracoorte	Warm
Tasmania	**Tasmania**	**Tasmania**	**Tasmania**	**680**	**1251**	**1442**	**Tasmania**	**366**	**Launceston**	**Cool**
Victoria				**32310**	**25838**	**22674**	**Victoria**	**358-363**		**Hot**
Alpine Valleys		Alpine Valleys	Alpine Valleys		705	258	Alpine Valleys	358	Myrtleford	Warm
Alpine Valleys/Beechworth	Alpine Valleys/Beechworth			803						Warm
Beechworth	Beechworth	Beechworth	Beechworth	35	57	107	Beechworth	358	Beechworth	Temp.
Bendigo	Bendigo	Bendigo	Bendigo	607	771	610	Bendigo	360	Bendigo	Warm
Central Victoria - other	Central Victoria - other	Central Victoria - other	Central Victoria - other	876	56	190	Central Victoria	360		Hot
Geelong	Geelong	Geelong	Geelong	322	515	284	Geelong	360	Geelong	Warm
Gippsland	Gippsland	Gippsland	Gippsland	174	236	177	Gippsland	360	Lakes Entrance	Warm
Glenrowan		Glenrowan	Glenrowan		203	179	Glenrowan	358	Glenrowan	Warm
Goulburn Valley	Goulburn Valley	Goulburn Valley	Goulburn Valley	1090	1612	1011	Goulburn Valley	360	Shepparton	Hot
Grampians	Grampians	Grampians	Grampians	424	506	511	Grampians	359	Ararat	Temp.
Heathcote		Heathcote	Heathcote		1245	1159	Heathcote	360	Heathcote	Warm
Henty	Henty	Henty	Henty	183	183	158	Henty	359	Hamilton	Temp.
King Valley		King Valley	King Valley		1320	1336	King Valley	358	Whitfield	Warm
Macedon Ranges		Macedon Ranges	Macedon Ranges		224	140	Macedon Ranges	360	Woodend	Cool
Mornington Peninsula	Mornington Peninsula	Mornington Peninsula	Mornington Peninsula	402	752	775	Mornington Penin	361	Hastings	Temp.
Murray Darling (VIC)	Murray Darling (VIC)	Murray Darling (VIC)	Murray Darling (VIC)	15663	8339	9214	Murray Darling	345	Mildura	Hot
North East Victoria - other	North East Victoria - other	North East Victoria - other	North East Victoria - other	1254	74	76	North East Victori	358		Warm
North West Victoria - other	North West Victoria - othe	North West Victoria - othe	North West Victoria - othe	3177	121	106				Hot

Table 74 (cont.): World's wine regions' bearing areas and names in *World Atlas of Wine*, by country, 2000, 2010 and 2016

Name	2000	2010	2016	ha 2000	ha 2010	ha 2016	Atlas name	Pages	Town	Zone
Port Phillip - other	Port Phillip - other	Port Phillip - other	Port Phillip - other	129	68	520	Port Phillip	360		Temp.
Pyrenees	Pyrenees	Pyrenees	Pyrenees	428	874	493	Pyrenees	359	Moonambel	Temp.
Rutherglen	Rutherglen	Rutherglen	Rutherglen	793	853	398	Rutherglen	358	Rutherglen	Hot
Strathbogie Ranges		Strathbogie Ranges	Strathbogie Ranges		369	627	Strathbogie Range	360	Strathbogie	Temp.
Sunbury	Sunbury	Sunbury	Sunbury	79	129	67	Sunbury	360	Sunbury	Temp.
Swan Hill (VIC)	Swan Hill (VIC)	Swan Hill (VIC)	Swan Hill (VIC)	3725	3869	1717	Swan Hill	map 344	Swan Hill	Hot
Upper Goulburn		Upper Goulburn	Upper Goulburn		245	384	Upper Goulburn	map 344	Alexandra	Warm
Western Victoria - other	Western Victoria - other	Western Victoria - other	Western Victoria - other	110	73	62	Western Victoria	359		Temp.
Yarra Valley	Yarra Valley	Yarra Valley	Yarra Valley	2038	2440	2116	Yarra Valley	362-363	Healesville	Temp.
Western Australia				**9272**	**11346**	**8847**	**Western Australi**	**347-349**		**Warm**
Blackwood Valley	Blackwood Valley	Blackwood Valley	Blackwood Valley	501	249	304	Blackwood Valley	348	Nannup	Warm
Central Western Australia	Central Western Australia	Central Western Australia	Central Western Australia	70	62	16			Northam	Hot
Eastern Plains, Inland and North WA	Eastern Plains, Inland and	Eastern Plains, Inland and	Eastern Plains, Inland and	39	25	2			Nyabing	Hot
Geographe	Geographe	Geographe	Geographe	480	1181	338	Geographe	348	Bunbury	Hot
Great Southern	Great Southern	Great Southern	Great Southern	2391	2804	1886	Great Southern	347	Denmark	Warm
Greater Perth - other	Greater Perth - other	Greater Perth - other	Greater Perth - other	395	36	168				Hot
Manjimup		Manjimup	Manjimup		179	54	Manjimup	348	Manjimup	Warm
Margaret River	Margaret River	Margaret River	Margaret River	3401	4894	4816	Margaret River	349	Margaret River	Warm
Peel		Peel	Peel		96	36	Peel	map 347	Mandurah	Hot
Pemberton		Pemberton	Pemberton		622	332	Pemberton	348	Pemberton	Warm
Perth Hills	Perth Hills	Perth Hills	Perth Hills	316	295	114	Perth Hills	map 347	Mundaring	Hot
South West Australia - other	South West Australia - oth	South West Australia - oth	South West Australia - oth	802	101	41				Hot
Swan District	Swan District	Swan District	Swan District	812	784	720	Swan District	map 347	Upper Swan	Hot
Western Australia Southeast Coast	Western Australia Southea	Western Australia Southea	Western Australia Southea	65	19	20			Esperance	Warm
AUSTRIA	**AUSTRIA**	**AUSTRIA**	**AUSTRIA**	**48496**	**45533**	**45439**	**AUSTRIA**	**254-261**		**Temp.**
Burgenland	**Burgenland**	**Burgenland**		**14540**	**13842**	**12249**	**Burgenland**	**260-261**		**Temp.**
Mittelburgenland			Mittelburgenland			1897	Mittelburgenland	260	Horitschon	Temp.
Neusiedlersee			Neusiedlersee			7098	Neusiedlersee	260	Neusiedl am See	Temp.
Neusiedlersee Hügelland			Neusiedlersee Hügelland			2835	Neusiedlersee-Hüg	261	Rust	Temp.
Südburgenland			Südburgenland			419	Südburgenland	261	Eisenberg	Temp.
Niederösterreich	**Niederösterreich**	**Niederösterreich**		**29975**	**27184**	**28240**	**Niederösterreich**	**254**		**Temp.**
Bergland			Bergland			140			Melk	Temp.
Carnuntum			Carnuntum			906	Carnuntum	254	Stixneusiedl	Temp.
Kamptal			Kamptal			3907	Kamptal	258-259	Langenlois	Temp.
Kremstal			Kremstal			2368	Kremstal	258-259	Krems	Cool
Thermenregion			Thermenregion			2182	Thermenregion	254-255	Baden	Temp.
Traisental			Traisental			815	Traisental	254	Traismauer	Temp.
Wachau			Wachau			1344	Wachau	256-257	Dürnstein	Cool
Wagram			Wagram			2720	Wagram	254	Tulln	Temp.
Weinviertel			Weinviertel			13858	Weinviertel	254	Mailberg	Temp.
Steiermark	**Steiermark**	**Steiermark**		**3283**	**3867**	**4324**	**Steiermark**	**255**		**Temp.**
Steiermark - other			Steiermark - other			0				Temp.
Südsteiermark			Südsteiermark			2162	Südsteiermark	255	Kitzeck im Sausal	Cool
Vulkanland Steiermark			Vulkanland Steiermark			1623	Vulkanland Steier	255	Klöch	Temp.
Weststeiermark			Weststeiermark			538	Weststeiermark	255	Stainz	Temp.
Wien and other regions	**Wien and other regions**	**Wien and other regions**		**699**	**640**	**626**				**Temp.**
Wien			Wien			581	Wien	254	Wien	Temp.

Table 74 (cont.): World's wine regions' bearing areas and names in *World Atlas of Wine*, by country, 2000, 2010 and 2016

Name	2000	2010	2016	ha 2000	ha 2010	ha 2016	Atlas name	Pages	Town	Zone
Other regions			Other regions			45				Temp.
BRAZIL	**BRAZIL**	**BRAZIL**	**BRAZIL**	**52840**	**49412**	**33205**	**BRAZIL**	**331**	**Bento Gonçalves**	**Hot**
BULGARIA	**BULGARIA**	**BULGARIA**	**BULGARIA**	**95997**	**56133**	**52974**	**BULGARIA**	**274-275**		**Warm**
North Central		North Central	North Central		3868	3962	Danubian Plain (c	map 274	Svishtov	Warm
Northeast		Northeast	Northeast		5837	6858	Danubian Plain (e	map 274	Dobrich	Warm
Northwest		Northwest	Northwest		5830	4169	Danubian Plain (w	map 274	Vidin	Warm
South Central		South Central	South Central		17466	16720	Thracian Lowland	map 274	Plodvid	Warm
Southeast		Southeast	Southeast		19533	18725	Thracian Lowland	map 274	Melnik	Warm
Southwest		Southwest	Southwest		3599	2540	Thracian Lowland	map 274	Burgas	Warm
CAMBODIA			**CAMBODIA**			**10**			**Phnom Penh**	**Hot**
CANADA	**CANADA**	**CANADA**	**CANADA**	**8498**	**10096**	**12600**	**CANADA**	**291-293**		**Temp.**
British Columbia		British Columbia	British Columbia		3995	4152	British Columbia	292	Summerland	Cool
Nova Scotia			Nova Scotia			325	Nova Scotia	291	Port Williams	Cool
Ontario		Ontario	Ontario		6102	6530	Ontario	293	St Catherines	Temp.
Quebec			Quebec			384	Quebec	291	Quebec	Cool
Other regions			Other regions			1209				Temp.
CHILE	**CHILE**	**CHILE**	**CHILE**	**113966**	**111525**	**145873**	**CHILE**	**333-337**		**Warm**
Antofagasta			Antofagasta			5			Antofagasta	Hot
Araucania	Araucania	Araucania	Araucania	5	12	65	Araucania	map 333	Temuco	Cool
Arica			Arica			15			Arica	Hot
Atacama	Atacama	Atacama	Atacama	797	12	567	Atacama	map 333	Copiapó	Warm
Coquimbo	Coquimbo	Coquimbo	Coquimbo	11083	2155	11076	Coquimbo	map 333	Ovalle	Warm
De Los Lagos		De Los Lagos	De Los Lagos		6	27			Puerto Montt	Cool
Del Bio Bio	Del Bio Bio	Del Bio Bio	Del Bio Bio	13747	3420	12093	Bío Bío	map 333	Chillan	Temp.
Del Maule	Del Maule	Del Maule	Del Maule	45053	49014	52962	Maule	map 333	Curico	Warm
Metropolitana	Metropolitana	Metropolitana	Metropolitana	9453	12214	12908	Metropolitana	map 333	Santiago	Hot
O'Higgins	O'Higgins	O'Higgins	O'Higgins	29044	36170	46337	Libertador	map 333	San Fernando	Warm
Tarapaca			Tarapaca			2			Iquique	Hot
Valparaiso	Valparaiso	Valparaiso	Valparaiso	4783	8522	9816	Valparaiso	map 333	Casablanca	Warm
CHINA		**CHINA**	**CHINA**		**29545**	**178000**	**CHINA**	**388-390**		**Warm**
Beijing		Beijing			3067		Beijing	map 388	Beijing	Hot
Gansu		Gansu			4987		Gansu	390	Lanzhou	Temp.
Ningxia		Ningxia			11152		Ningxia	390	Yinchuan	Warm
Shandong		Shandong			67		Shandong	389	Qingdao	Hot
ShanXi		ShanXi			547		Shanxi	390	Taiyuan	Warm
Sichuan		Sichuan			533		Sichuan	map 388	Chengdu	Hot
Tianjin		Tianjin			400		Tianjin	map 388	Tianjin	Hot
Xinjiang		Xinjiang			3133		Xinjiang	390	Ürümqi	Warm
Yantai		Yantai			4373		Yantai	map 388	Yantai	Hot
Other regions		Other regions			1287					Warm
CROATIA	**CROATIA**	**CROATIA**	**CROATIA**	**59448**	**20754**	**11746**	**CROATIA**	**270-271**		**Warm**
Dalmatinska Zagora		Dalmatinska Zagora			602		Dalmatinska Zago	270	Sinj	Warm
Hrvatsko Primorje		Hrvatsko Primorje			210		Hrvatsko Primorje	270	Rijeka	Hot
Istra		Istra			3083		Hrvatska Istra	270	Novigrad	Hot
Jadranska Hrvatska			Jadranska Hrvatska			5308	Coastal Croatia	270		Warm
Kontinentalna Hrvatska			Kontinentalna Hrvatska			6438	Continental Croati	270		Warm
Moslavina		Moslavina			228		Moslavina	map 267	Kutina	Warm

Table 74 (cont.): World's wine regions' bearing areas and names in *World Atlas of Wine*, by country, 2000, 2010 and 2016

Name	2000	2010	2016	ha 2000	ha 2010	ha 2016	Atlas name	Pages	Town	Zone
Plesivica		Plesivica			452		Plesivica	270	Plešivica	Temp.
Podunavlje		Podunavlje			3206		Hrvatsko Podunav	map 267	Bilje	Warm
Pokuplje		Pokuplje			41		Pokuplje	map 267	Karlovac	Temp.
Prigorje - Bilogora		Prigorje - Bilogora			791		Prigorje - Bilogora	map 267	Bjelovar	Warm
Sjeverna Dalmacija		Sjeverna Dalmacija			2333		Sjeverna Dalmacij	270	Zadar	Hot
Slavonija		Slavonija			3307		Slavonija	270	Virovitica	Warm
Srednja Juzna Dalmacija		Srednja Juzna Dalmacija			2972		Srednja i Juzna Da	270	Split	Hot
Zagorje-Medimurje		Zagorje-Medimurje			1266		Zagorje-Medimurj	map 267	Krapina	Temp.
Other regions		Other regions			2263					Warm
CYPRUS	**CYPRUS**	**CYPRUS**	**CYPRUS**	**18282**	**8608**	**5133**	**CYPRUS**	**284**	**Doros**	**Hot**
CZECHIA	**CZECHIA**	**CZECHIA**	**CZECHIA**	**11331**	**16242**	**13600**	**CZECHIA**	**266**		**Temp.**
Cechy		Cechy	Cechy		785	248	Bohemia	266	Kutná Hora	Cool
Jihovýchod			Jihovýchod			12800			Znojmo	Temp.
Morava		Morava	Morava		15457	268	Moravia	266	Bzenec	Temp.
Praha			Praha			10	Prague	266	Prague	Temp.
Severozápad			Severozápad			274			Most	Cool
ETHIOPIA		**ETHIOPIA**	**ETHIOPIA**		**169**	**169**			**Addis Ababa**	**Temp.**
FRANCE	**FRANCE**	**FRANCE**	**FRANCE**	**864846**	**835554**	**814882**	**FRANCE**	**50-152**		**Warm**
Alsace	**Alsace**		**Alsace**	**15128**	**16179**	**16804**	**Alsace**	**124-127**		**Temp.**
Bas-Rhin		Bas-Rhin			6971		Bas-Rhin	map 53	Sélestat	Temp.
Haut-Rhin		Haut-Rhin			9208		Haut-Rhin	map 53	Colmar	Temp.
Aquitaine			**Aquitaine**	**150349**	**148431**	**144314**				**Temp.**
Aquitaine except Gironde	Aquitaine except Gironde			25732						Warm
Dordogne		Dordogne			13820		Dordogne	113	Bergerac	Temp.
Gironde	Gironde	Gironde		124617	123023		Bordeaux	84-112	Merignac	Temp.
Landes		Landes			2107		Landes	map 53	Villeneuve-de-Marsan	Temp.
Lot-et-Garonne		Lot-et-Garonne			6846		Lot-et-Garonne	map 53	Buzet	Warm
Pyrenees-Atlantiques		Pyrenees-Atlantiques			2635		Pyrenees-Atlantiq	map 53	Pau	Warm
Auvergne	**Auvergne**		**Auvergne**	**1096**	**1692**	**1010**				**Temp.**
Allier		Allier			717		Allier	123	St Pourcain sur Sioule	Temp.
Cantal		Cantal			11		Cantal	map 53	Aurillac	Cool
Haute-Loire		Haute-Loire			83		Haute-Loire	map 53	Le Puy	Cool
Puy-de-Dome		Puy-de-Dome			881		Puy-de-Dome	map 53	Clermont-Ferrand	Temp.
Bourgogne	**Bourgogne**		**Bourgogne**	**29941**	**32134**	**32673**				**Temp.**
Cote-d'Or		Cote-d'Or			9683		Cote d'Or	56-57	Dijon	Temp.
Nievre		Nievre			1719		Nievre	map 53	Cosne-Cours-sur-Loire	Temp.
Saone-et-Loire		Saone-et-Loire			13509		Saone-et-Loire	map 53	Mâcon	Temp.
Yonne		Yonne			7224		Yonne	map 53	Chablis	Temp.
Centre-Val de Loire	**Centre-Val de Loire**		**Centre-Val de Loire**	**22316**	**23365**	**22463**				**Temp.**
Cher		Cher			4182		Cher	map 53	Bourges	Temp.
Eure-et-Loire		Eure-et-Loire			0		Eure-et-Loire	map 53	Chartres	Cool
Indre		Indre			980		Indre	map 53	Châteauroux	Temp.
Indre-et-Loire		Indre-et-Loire			10421		Indre-et-Loire	map 53	Tours	Temp.
Loiret		Loiret			309		Loiret	map 53	Orléans	Temp.
Loir-et-Cher		Loir-et-Cher			7474		Loir-et-Cher	map 53	Blois	Temp.
Champagne-Ardenne	**Champagne-Ardenne**		**Champagne-Ardenne**	**28671**	**30599**	**34332**				**Cool**
Aube		Aube			7870		Aube	map 53	Fontette	Cool

Table 74 (cont.): World's wine regions' bearing areas and names in *World Atlas of Wine*, by country, 2000, 2010 and 2016

Name	2000	2010	2016	ha 2000	ha 2010	ha 2016	Atlas name	Pages	Town	Zone
Haute-Marne		Haute-Marne			114		Haute-Marne	map 53	Chaumont	Cool
Marne		Marne			22615		Marne	map 53	Reims	Cool
Corse	**Corse**		**Corse**	**6992**	**6529**	**6660**	**Corsica**	**149**		**Warm**
Corse-du-Sud		Corse-du-Sud			968				Ajaccio	Hot
Haute-Corse		Haute-Corse			5561				Bastia	Warm
Franche Comté	**Franche Comté**		**Franche Comté**	**2020**	**2490**	**1819**				**Temp.**
Doubs		Doubs			51		Doubs	map 53	Besançon	Temp.
Haute-Saone		Haute-Saone			172		Haute-Saone	map 53	Vesoul	Temp.
Jura		Jura			2267		Jura	150	Lons-le-Saunier	Temp.
Île de France			**Île de France**	**26**	**82**	**4**				**Cool**
Seine-et-Marne	Seine-et-Marne	Seine-et-Marne		26	82		Seine-et-Marne	map 53	La Ferté-sous-Jouarre	Cool
Languedoc Roussillon			**Languedoc Roussillon**	**295708**	**254039**	**245964**				**Warm**
Aude	Aude	Aude		85270	72058		Aude	map 53	Carcassonne	Warm
Gard	Gard	Gard		67133	59210		Gard	map 53	Nîmes	Hot
Herault	Herault	Herault		105647	93944		Herault	map 53	Beziers	Hot
Lozere		Lozere			11		Lozere	map 53	Mende	Cool
Pyrenees-Orientales	Pyrenees-Orientales	Pyrenees-Orientales		37659	28817		Pyrenees-Orientale	map 53	Perpignan	Hot
Limousin			**Limousin**	**50**	**287**	**83**				**Temp.**
Correze		Correze			287		Correze	map 53	Tulle	Temp.
Correze, Haute-Vienne	Correze, Haute-Vienne			50			Correze	map 53	Tulle	Temp.
Lorraine	**Lorraine**		**Lorraine**	**184**	**341**	**211**				**Cool**
Meurthe-et-Moselle		Meurthe-et-Moselle			159		Meurthe-et-Mosel	map 53	Nancy	Cool
Meuse		Meuse			45		Meuse	map 53	Toul	Cool
Moselle		Moselle			94		Moselle	map 53	Metz	Temp.
Vosges		Vosges			42		Vosges	map 53	Épinal	Cool
Midi Pyrénées			**Midi Pyrénées**	**38786**	**39065**	**36827**				**Warm**
Midi-Pyrenees except Gers	Midi-Pyrenees except Gers			18872						Warm
Ariege		Ariege			133		Ariege	map 53	Foix	Temp.
Aveyron		Aveyron			884		Aveyron	map 53	Estaing	Temp.
Gers	Gers	Gers		19913	18897		Gers	map 53	Auch	Warm
Haute-Garonne		Haute-Garonne			1993		Haute-Garonne	map 53	Toulouse	Warm
Hautes-Pyrenees		Hautes-Pyrenees			680		Hautes-Pyrenees	map 53	Tarbes	Warm
Lot		Lot			5571		Lot	map 53	Cahors	Temp.
Tarn		Tarn			7087		Tarn	map 53	Gaillac	Warm
Tarn-et-Garonne		Tarn-et-Garonne			3821		Tarn-et-Garonne	map 53	Montauban	Warm
Pays de la Loire			**Pays de la Loire**	**37882**	**38593**	**35844**				**Temp.**
Pays de la Loire except Mayenne	Pays de la Loire except Mayenne			37882						Temp.
Loire-Atlantique		Loire-Atlantique			15618		Loire-Atlantique	map 53	Nantes	Temp.
Maine-et-Loire		Maine-et-Loire			20766		Maine-et-Loire	map 53	Angers	Temp.
Mayenne		Mayenne			1		Mayenne	map 53	Laval	Temp.
Sarthe		Sarthe			321		Sarthe	map 53	Le Mans	Temp.
Vendee		Vendee			1886		Vendee	map 53	La Roche-sur-Yon	Temp.
Picardie			**Picardie**	**2340**	**3229**	**2242**				**Cool**
Aisne	Aisne	Aisne		2340	3229		Aisne	map 53	Soissons	Cool
Poitou Charentes			**Poitou Charentes**	**80809**	**85310**	**85808**				**Temp.**
Charente	Charente	Charente		38514	41093		Charente	map 53	Angoulême	Temp.
Charente-Maritime	Charente-Maritime	Charente-Maritime		4032[illegible]	40790		Charente-Maritim	map 53	La Rochelle	Temp.

Table 74 (cont.): World's wine regions' bearing areas and names in *World Atlas of Wine*, by country, 2000, 2010 and 2016

Name	2000	2010	2016	ha 2000	ha 2010	ha 2016	Atlas name	Pages	Town	Zone
Deux-Sevres		Deux-Sevres			1227		Deux-Sevres	map 53	Thouars	Temp.
Deux-Sevres, Vienne	Deux-Sevres, Vienne			1973						Temp.
Vienne		Vienne			2199		Vienne	map 53	Poitiers	Temp.
Provence-Alpes-Cote d'Azur			**Provence-Alpes-Cote d'A**	**95392**	**98921**	**96470**				**Hot**
Alpes-de-Haute-Provence		Alpes-de-Haute-Provence			843		Alpes-de-Haute-Pı	map 53	Pierrevert	Warm
Alpes-de-Haute-Provence, Hautes-Alpe:	Alpes-de-Haute-Provence, Hautes-Alpes, Alpes-Maritimes			1082						Warm
Alpes-Maritimes		Alpes-Maritimes			152		Alpes-Maritimes	map 53	Nice	Warm
Bouches-du-Rhone	Bouches-du-Rhone	Bouches-du-Rhone		11089	11663		Bouches-du-Rhon	map 53	Aix-en-Provence	Warm
Hautes-Alpes		Hautes-Alpes			232		Hautes-Alpes	map 53	Gap	Cool
Var	Var	Var		31420	31913		Var	map 53	Toulon	Hot
Vaucluse	Vaucluse	Vaucluse		51801	54117		Vaucluse	map 53	Avignon	Hot
Rhône Alpes			**Rhône Alpes**	**57156**	**54267**	**51354**				**Warm**
Rhône-Alpes except Ardeche	Rhône-Alpes except Ardeche			44861						Warm
Ain		Ain			954		Ain	map 53	Belley	Temp.
Ardeche	Ardeche	Ardeche		12295	11960		Ardeche	map 53	Tournon	Warm
Drome		Drome			18507		Drome	map 53	Valence	Warm
Haute-Savoie		Haute-Savoie			286		Haute-Savoie	map 53	Annecy	Temp.
Isere		Isere			264		Isere	map 53	Grenoble	Warm
Loire		Loire			1067		Loire	map 53	Boën-sur-Lignon	Temp.
Rhone		Rhone			19137		Rhone	map 53	Lyon	Warm
Savoie		Savoie			2092		Savoie	151	Chambery	Temp.
GEORGIA	**GEORGIA**	**GEORGIA**	**GEORGIA**	**37419**	**48001**	**48000**	**GEORGIA**	**278-279**	**Telavi**	**Temp.**
GERMANY	**GERMANY**	**GERMANY**	**GERMANY**	**104233**	**102135**	**94501**	**GERMANY**	**223-247**		**Cool**
Ahr	Ahr	Ahr	Ahr	520	562	522	Ahr	226	Ahrweiler	Cool
Baden	Baden	Baden	Baden	15554	15836	14666	Baden	244-245	Baden-Baden	Cool
Franken	Franken	Franken	Franken	5993	6104	5796	Franken	246-247	Würzburg	Cool
Hessische Bergstraße	Hessische Bergstraße	Hessische Bergstraße	Hessische Bergstraße	456	427	414	Hessische Bergstrı	map 223	Bensheim	Temp.
Mittelrhein	Mittelrhein	Mittelrhein	Mittelrhein	572	458	429	Mittelrhein	225	Bad Ems	Cool
Mosel	Mosel	Mosel	Mosel	11521	8976	8137	Mosel	227-233	Trier	Cool
Nahe	Nahe	Nahe	Nahe	4604	4163	3897	Nahe	234-235	Schloßböckelheim	Cool
Pfalz	Pfalz	Pfalz	Pfalz	23342	23467	21541	Pfalz	241-243	Bad Dürkheim	Temp.
Rheingau	Rheingau	Rheingau	Rheingau	3217	3062	2962	Rheingau	236-237	Geisenheim	Cool
Rheinhessen	Rheinhessen	Rheinhessen	Rheinhessen	26385	26480	24117	Rheinhessen	238-240	Worms	Temp.
Saale	Saale	Saale	Saale	618	704	695	Saale-Unstrut	225	Naumburg	Cool
Sachsen	Sachsen	Sachsen	Sachsen	409	461	466	Sachsen	225	Dresden	Temp.
Württemberg	Württemberg	Württemberg	Württemberg	11042	11435	10859	Württemberg	245	Stuttgart	Cool
GREECE	**GREECE**	**GREECE**	**GREECE**	**50915**	**54390**	**50845**	**GREECE**	**280-283**		**Hot**
Anatoliki Makedonia, Thraki	Anatoliki Makedonia, Thrı	Anatoliki Makedonia, Thrı	Anatoliki Makedonia, Thrı	384	1234	1706	Anatoliki Makedo	map 281	Maronia	Hot
Attiki	Attiki	Attiki	Attiki	6112	5599	5957	Attica	280	Athens	Hot
Dytiki Ellada	Dytiki Ellada	Dytiki Ellada	Dytiki Ellada	8031	6484	7043	Dytiki Ellada	map 281	Patras	Hot
Dytiki Makedonia	Dytiki Makedonia	Dytiki Makedonia	Dytiki Makedonia	1690	2083	1824	Dytiki Makedonia	map 281	Amindeo	Warm
Ionia Nisia	Ionia Nisia	Ionia Nisia	Ionia Nisia	2707	2422	665	Ionioi Nisia	map 281	Argostoli	Hot
Ipeiros	Ipeiros	Ipeiros	Ipeiros	632	559	153	Epirus	280	Zitsa	Warm
Kentriki Makedonia	Kentriki Makedonia	Kentriki Makedonia	Kentriki Makedonia	2494	4256	3115	Kentriki Makedon	map 281	Goumenissa	Hot
Kriti	Kriti	Kriti	Kriti	5312	7863	6767	Crete	282	Iraklion	Hot
Notio Aigaio	Notio Aigaio	Notio Aigaio	Notio Aigaio	4124	3547	2581	Notio Aigaio	map 281	Paros	Hot
Peloponnissos	Peloponnissos	Peloponnissos	Peloponnissos	8197	8134	8739	Peloponnissos	map 281	Nemea	Hot

Table 74 (cont.): World's wine regions' bearing areas and names in *World Atlas of Wine*, by country, 2000, 2010 and 2016

Name	2000	2010	2016	ha 2000	ha 2010	ha 2016	Atlas name	Pages	Town	Zone
Sterea Ellada	Sterea Ellada	Sterea Ellada	Sterea Ellada	5991	6457	6291	Sterea Ellada	map 281	Halkída	Hot
Thessalia	Thessalia	Thessalia	Thessalia	3043	3415	3426	Thessaly	280	Rapsani	Hot
Voreio Aigaio	Voreio Aigaio	Voreio Aigaio	Voreio Aigaio	2196	2337	2578	Voreio Aigaio	map 281	Limnos	Hot
HUNGARY	**HUNGARY**	**HUNGARY**	**HUNGARY**	**86886**	**69715**	**63881**	**HUNGARY**	**262-265**		**Temp.**
Badacsony		Badacsony	Badacsony		1618	1382	Badacsony	263	Badacsonytomaj	Temp.
Balatonboglar		Balatonboglar	Balatonboglar		3305	3231	Balatonboglar	263	Balatonboglár	Temp.
Balatonfelvidek		Balatonfelvidek	Balatonfelvidek		1025	830	Balatonfelvidek	map 263	Monostorapáti	Temp.
Balatonfured-Csopak		Balatonfured-Csopak	Balatonfured-Csopak		2180	2032	Balatonfured-Csop	263	Balatonfüred	Temp.
Bukk		Bukk	Bukk		1055	1015	Bukk	map 263	Miskolc	Temp.
Csongrad		Csongrad	Csongrad		1513	1126	Csongrad	262	Hódmezővásárhely	Temp.
Eger		Eger	Eger		5509	5353	Eger	262	Eger	Temp.
Etyek-Budai		Etyek-Budai	Etyek-Budai		1717	1513	Etyek-Buda	263	Etyek	Temp.
Hajos-Bajai		Hajos-bajai	Hajos-Bajai		1982	1889	Hajos-Baja	262	Baja	Warm
Kunsag		Kunsag	Kunsag		22263	20519	Kunsag	262	Kecskemét	Temp.
Matra		Matra	Matra		6294	6248	Matra	263	Gyöngyös	Temp.
Mor		Mor	Mor		730	573	Mor	263	Mór	Temp.
Nagy-Somlo		Nagy-Somlo	Nagy-Somlo		598	553	Nagy-Somlo	map 263	Somlóvásárhely	Temp.
Neszmely		Neszmely	Neszmely		1587	1077	Neszmely	263	Neszmély	Temp.
Pannonhalma		Pannonhalma	Pannonhalma		615	622	Pannonhalma	map 263	Pannonhalma	Temp.
Pecs		Pecs	Pecs		777	673	Pecs	263	Pécs	Warm
Sopron		Sopron	Sopron		1919	1650	Sopron	263	Sopron	Temp.
Szekszard		Szekszard	Szekszard		2333	2183	Szekszard	262	Szekszárd	Warm
Tokaj		Tokaj	Tokaj		5994	5709	Tokaj	264-265	Tokaj	Temp.
Tolna		Tolna	Tolna		2526	2399	Tolna	262	Paks	Warm
Villany		Villany	Villany		2582	2469	Villany	262	Villány	Warm
Zala		Zala	Zala		1592	833	Zala	263	Nagykanizsa	Temp.
INDIA			**INDIA**			**2700**	**INDIA**	**385**	**Nashik**	**Hot**
ISRAEL	**ISRAEL**	**ISRAEL**	**ISRAEL**	**4851**	**4851**	**5000**	**ISRAEL**	**287**	**Jerusalem**	**Hot**
ITALY	**ITALY**	**ITALY**	**ITALY**	**636662**	**625700**	**604551**	**ITALY**	**154-186**		**Hot**
Abruzzo			**Abruzzo**	**32773**	**32189**	**27138**	**Abruzzo**	**map 144**		**Hot**
Chieti	Chieti	Chieti		25986	25989		Chieti	map 173	Chieti	Hot
L'Aquila	L'Aquila	L'Aquila		855	440		L'Aquila	map 173	L'Aquila	Warm
Pescara	Pescara	Pescara		3145	3159		Pescara	map 173	Pescara	Hot
Teramo	Teramo	Teramo		2785	2602		Teramo	map 173	Teramo	Warm
Basilicata			**Basilicata**	**7076**	**4863**	**3827**	**Basilicata**	**183**		**Warm**
Matera	Matera	Matera		1645	1243		Matera	map 182	Matera	Hot
Potenza	Potenza	Potenza		5431	3620		Potenza	map 182	Potenza	Temp.
Calabria			**Calabria**	**12655**	**9785**	**6509**	**Calabria**	**183**		**Hot**
Catanzaro	Catanzaro	Catanzaro		1335	706		Catanzaro	map 182	Catanzaro	Hot
Cosenza	Cosenza	Cosenza		5107	4162		Terre di Cosenza	183	Cosenza	Hot
Crotone	Crotone	Crotone		3265	3229		Crotone	map 182	Crotone	Hot
Reggio di Calabria	Reggio di Calabria	Reggio di Calabria		2101	1343		Reggio di Calabri	map 182	Reggio di Calabria	Hot
Vibo Valentia	Vibo Valentia	Vibo Valentia		845	344		Vibo Valentia	map 182	Vibo Valentia	Hot
Campania			**Campania**	**26720**	**23185**	**17217**	**Campania**	**182**		**Hot**
Avellino	Avellino	Avellino		6287	5711		Avellino	map 182	Avellino	Warm
Benevento	Benevento	Benevento		10291	10488		Benevento	map 182	Benevento	Hot
Caserta	Caserta	Caserta		2964	2062		Caserta	map 182	Caserta	Hot

Table 74 (cont.): World's wine regions' bearing areas and names in *World Atlas of Wine*, by country, 2000, 2010 and 2016

Name	2000	2010	2016	ha 2000	ha 2010	ha 2016	Atlas name	Pages	Town	Zone
Napoli	Napoli	Napoli		1956	1609		Napoli	map 182	Napoli	Hot
Salerno	Salerno	Salerno		5222	3315		Salerno	map 182	Salerno	Hot
Emilia-Romagna			**Emilia-Romagna**	**55916**	**55796**	**58675**	**Emilia-Romagna**	**map 144**		**Hot**
Bologna	Bologna	Bologna		7049	6845		Bologna	map 165	Bologna	Hot
Ferrara	Ferrara	Ferrara		569	572		Ferrara	map 165	Ferrara	Hot
Forli-Cesena	Forli-Cesena	Forli-Cesena		6746	7017		Forlì	map 165	Forlì	Hot
Modena	Modena	Modena		7585	7874		Modena	map 165	Modena	Hot
Parma	Parma	Parma		574	736		Parma	map 165	Parma	Hot
Piacenza	Piacenza	Piacenza		5568	5906		Piacenza	map 157	Piacenza	Warm
Ravenna	Ravenna	Ravenna		16910	16388		Ravenna	map 165	Ravenna	Hot
Reggio nell'Emilia	Reggio nell'Emilia	Reggio nell'Emilia		8263	8027		Reggio nell'Emilia	map 165	Reggio nell'Emilia	Hot
Rimini	Rimini	Rimini		2653	2431		Rimini	map 173	Rimini	Hot
Friuli-Venezia Giulia			**Friuli-Venezia Giulia**	**17499**	**19250**	**19879**	**Friuli-Venezia Gi**	**map 144**		**Warm**
Gorizia	Gorizia	Gorizia		3474	4064		Gorizia	map 165	Gorizia	Warm
Pordenone	Pordenone	Pordenone		6671	7695		Pordenone	map 165	Pordenone	Warm
Trieste	Trieste	Trieste		182	205		Trieste	map 165	Trieste	Hot
Udine	Udine	Udine		7171	7285		Udine	171	Udine	Warm
Lazio			**Lazio**	**26844**	**16401**	**14968**	**Lazio**	**172**		**Hot**
Frosinone	Frosinone	Frosinone		4162	1877		Frosinone	map 173	Frosinone	Hot
Latina	Latina	Latina		6045	3757		Latina	map 173	Latina	Hot
Rieti	Rieti	Rieti		1490	750		Rieti	map 173	Rieti	Warm
Roma	Roma	Roma		10519	7061		Roma	map 173	Roma	Hot
Viterbo	Viterbo	Viterbo		4628	2957		Viterbo	map 173	Viterbo	Warm
Liguria			**Liguria**	**1833**	**1538**	**1004**	**Liguria**	**157**		**Hot**
Genova	Genova	Genova		147	177		Genova	map 157	Genova	Hot
Imperia	Imperia	Imperia		454	414		Imperia	map 157	Imperia	Hot
La Spezia	La Spezia	La Spezia		833	617		La Spezia	map 157	La Spezia	Hot
Savona	Savona	Savona		399	329		Savona	map 157	Savona	Hot
Lombardia			**Lombardia**	**21715**	**23089**	**23221**	**Lombardy**	**156**		**Warm**
Bergamo	Bergamo	Bergamo		881	804		Bergamo	map 157	Bergamo	Warm
Brescia	Brescia	Brescia		3904	6016		Brescia	map 157	Brescia	Warm
Como	Como	Como		18	23		Como	map 157	Como	Temp.
Cremona	Cremona	Cremona		39	65		Cremona	map 157	Cremona	Warm
Lecco	Lecco	Lecco		44	57		Lecco	map 157	Lecco	Temp.
Lodi	Lodi	Lodi		22	20		Lodi	map 157	Lodi	Warm
Mantova	Mantova	Mantova		1830	1803		Mantova	map 165	Mantova	Hot
Milano	Milano	Milano		238	207		Milano	map 157	Milano	Hot
Monza e della Brianza		Monza e della Brianza			3		Monza	map 157	Monza	Warm
Pavia	Pavia	Pavia		13734	13193		Pavia	map 157	Pavia	Warm
Sondrio	Sondrio	Sondrio		987	874		Sondrio	map 157	Sondrio	Cool
Varese	Varese	Varese		19	23		Varese	map 157	Varese	Temp.
Marche			**Marche**	**15676**	**16745**	**14766**	**Marche**	**172**		**Warm**
Ancona	Ancona	Ancona		4591	4739		Ancona	map 173	Ancona	Hot
Ascoli Piceno	Ascoli Piceno	Ascoli Piceno		8211	6111		Ascoli Piceno	map 173	Ascoli Piceno	Hot
Fermo		Fermo			1706		Fermo	map 173	Fermo	Warm
Macerata	Macerata	Macerata		1482	2185		Macerata	map 173	Macerata	Warm
Pesaro e Urbino	Pesaro e Urbino	Pesaro e Urbino		1391	2004		Urbino	map 173	Urbino	Warm

Table 74 (cont.): World's wine regions' bearing areas and names in *World Atlas of Wine*, by country, 2000, 2010 and 2016

Name	2000	2010	2016	ha 2000	ha 2010	ha 2016	Atlas name	Pages	Town	Zone
Molise			**Molise**	**5422**	**5154**	**3991**	**Molise**	**183**		**Warm**
Campobasso	Campobasso	Campobasso		4858	4715		Campobasso	map 182	Campobasso	Warm
Isernia	Isernia	Isernia		563	439		Isernia	map 182	Isernia	Warm
Piemonte			**Piemonte**	**51999**	**46317**	**46327**	**Piemonte**	**158-159**		**Warm**
Alessandria	Alessandria	Alessandria		14873	12328		Alessandria	map 157	Alessandria	Warm
Asti	Asti	Asti		18016	15559		Asti	158	Asti	Warm
Biella	Biella	Biella		289	289		Biella	map 157	Biella	Temp.
Cuneo	Cuneo	Cuneo		16272	16065		Cuneo	map 157	Cuneo	Warm
Novara	Novara	Novara		575	535		Novara	map 157	Novara	Warm
Torino	Torino	Torino		1777	1313		Torino	157	Torino	Warm
Verbano-Cusio-Ossola	Verbano-Cusio-Ossola	Verbano-Cusio-Ossola		29	28		Domodossola	map 157	Domodossola	Cool
Vercelli	Vercelli	Vercelli		169	200		Vercelli	map 157	Vercelli	Warm
Puglia			**Puglia**	**84426**	**82760**	**94348**	**Puglia**	**183**		**Hot**
Bari	Bari	Bari		16954	7586		Bari	map 182	Bari	Hot
Barletta-Andria-Trani		Barletta-Andria-Trani			16241		Barletta	map 182	Barletta	Hot
Brindisi	Brindisi	Brindisi		13498	9677		Brindisi	183	Brindisi	Hot
Foggia	Foggia	Foggia		28797	25249		Foggia	map 182	Foggia	Hot
Lecce	Lecce	Lecce		10021	8379		Lecce	map 182	Lecce	Hot
Taranto	Taranto	Taranto		15156	15627		Taranto	map 182	Taranto	Hot
Sardegna			**Sardegna**	**25087**	**18465**	**17784**	**Sardinia**	**186**		**Hot**
Cagliari	Cagliari	Cagliari		8398	4563		Cagliari	map 186	Cagliari	Hot
Carbonia-Iglesias		Carbonia-Iglesias			1948		Carbonia Iglesias	map 186	Carbonia	Hot
Medio Campidano		Medio Campidano			638		Medio Campidano		Sardara	Hot
Nuoro	Nuoro	Nuoro		7208	2401		Nuoro	map 186	Nuoro	Warm
Ogliastra		Ogliastra			1605		Oligastra		Baunei	Hot
Olbia-Tempio		Olbia-Tempio			2136		Olbia-Tempio	map 186	Olbia	Hot
Oristano	Oristano	Oristano		3250	2228		Oristano	map 186	Oristano	Hot
Sassari	Sassari	Sassari		6232	2946		Sassari	map 186	Sassari	Hot
Sicilia			**Sicilia**	**110562**	**104068**	**86725**	**Sicily**	**184-185**		**Hot**
Agrigento	Agrigento	Agrigento		19624	16781		Agrigento	map 184	Agrigento	Hot
Caltanissetta	Caltanissetta	Caltanissetta		5525	4170		Caltanissetta	map 184	Caltanissetta	Hot
Catania	Catania	Catania		3875	2990		Catania	map 184	Catania	Hot
Enna	Enna	Enna		585	278		Enna	map 184	Enna	Hot
Messina	Messina	Messina		2253	866		Messina	map 184	Messina	Hot
Palermo	Palermo	Palermo		16515	14569		Palermo	map 184	Palermo	Hot
Ragusa	Ragusa	Ragusa		1520	1381		Ragusa	map 184	Ragusa	Hot
Siracusa	Siracusa	Siracusa		1583	1383		Siracusa	map 184	Siracusa	Hot
Trapani	Trapani	Trapani		59073	61649		Trapani	map 184	Trapani	Hot
Toscana			**Toscana**	**45961**	**59839**	**57373**	**Toscana**	**178**		**Hot**
Arezzo	Arezzo	Arezzo		4887	7031		Arezzo	map 173	Arezzo	Warm
Firenze	Firenze	Firenze		15940	18362		Firenze	176	Firenze	Hot
Grosseto	Grosseto	Grosseto		3843	7438		Grosseto	map 173	Grosseto	Hot
Livorno	Livorno	Livorno		1214	2425		Livorno	map 173	Livorno	Hot
Lucca	Lucca	Lucca		728	1053		Lucca	map 173	Lucca	Hot
Massa-Carrara	Massa-Carrara	Massa-Carrara		275	759		Massa	map 157	Massa	Hot
Pisa	Pisa	Pisa		2386	3163		Pisa	map 173	Pisa	Hot
Pistoia	Pistoia	Pistoia		633	776		Pistoia	map 173	Pistoia	Hot

Table 74 (cont.): World's wine regions' bearing areas and names in *World Atlas of Wine*, by country, 2000, 2010 and 2016

Name	2000	2010	2016	ha 2000	ha 2010	ha 2016	Atlas name	Pages	Town	Zone
Prato	Prato	Prato		335	509		Prato	map 173	Prato	Hot
Siena	Siena	Siena		15718	18323		Siena	map 173	Siena	Warm
Trentino-Alto Adige				**13624**	**15658**	**14151**				**Temp.**
Bolzano-Bozen	Bolzano-Bozen	Bolzano-Bozen	Bolzano-Bozen	4781	5282	4072	Bolzano	167	Bolzano-Bozen	Cool
Trento	Trento	Trento	Trento	8844	10377	10079	Trento	166	Trento	Temp.
Umbria			**Umbria**	**13588**	**12491**	**11768**	**Umbria**	**181**		**Warm**
Perugia	Perugia	Perugia		8040	7620		Perugia	181	Perugia	Warm
Terni	Terni	Terni		5548	4871		Terni	map 173	Terni	Hot
Valle d'Aosta	**Valle d'Aosta**	**Valle d'Aosta**	**Valle d'Aosta**	**424**	**463**	**375**	**Valle d'Aosta**	**156**	**Aosta**	**Cool**
Veneto			**Veneto**	**66864**	**77644**	**84505**	**Veneto**	**168**		**Hot**
Belluno	Belluno	Belluno		9	52		Belluno	map 165	Belluno	Temp.
Padova	Padova	Padova		6313	5850		Padova	map 165	Padova	Hot
Rovigo	Rovigo	Rovigo		258	363		Rovigo	map 165	Rovigo	Hot
Treviso	Treviso	Treviso		24008	28580		Treviso	map 165	Treviso	Hot
Venezia	Venezia	Venezia		6018	6604		Venezia	map 165	Venezia	Hot
Verona	Verona	Verona		23359	27750		Verona	168-169	Verona	Hot
Vicenza	Vicenza	Vicenza		6898	8445		Vicenza		Vicenza	Hot
JAPAN		**JAPAN**	**JAPAN**		**3715**	**3869**	**JAPAN**	**386-387**		**Warm**
Hokkaido		Hokkaido	Hokkaido		835	868	Hokkaido	387	Sapporo	Temp.
Iwate			Iwate			160			Morioka	Warm
Nagano		Nagano	Nagano		754	754	Nagano	387	Nagano	Warm
Niigata			Niigata			62	Niigata	387	Niigata	Hot
Yamagata		Yamagata	Yamagata		392	414	Yamagata	387	Yamagata	Warm
Yamanashi		Yamanashi	Yamanashi		632	1202	Yamanashi	387	Kofu	Hot
Other regions		Other regions	Other regions		1102	408				Warm
KAZAKHSTAN		**KAZAKHSTAN**	**KAZAKHSTAN**		**6938**	**6938**				**Warm**
Almaty		Almaty	Almaty		4553	4553			Almaty	Warm
East Kazakhstan		East Kazakhstan	East Kazakhstan		3	3			Ust'-Kamenogorsk	Cool
South Kazakhstan		South Kazakhstan	South Kazakhstan		2162	2162			Turkistan	Hot
West Kazakhstan		West Kazakhstan	West Kazakhstan		2	2			Uralsk	Temp.
Zhambyl		Zhambyl	Zhambyl		217	217			Taraz	Hot
Other regions		Other regions	Other regions		2	2				Warm
KOREA, REP.	**KOREA, REP.**	**KOREA, REP.**	**KOREA, REP.**	**5400**	**5400**	**5400**	**SOUTH KOREA**	**385**	**Seoul**	**Hot**
LEBANON			**LEBANON**			**4000**	**LEBANON**	**286**	**Chtaura**	**Hot**
LUXEMBOURG	**LUXEMBOURG**	**LUXEMBOURG**	**LUXEMBOURG**	**1348**	**1304**	**1300**	**LUXEMBOURG**	**53**	**Luxembourg**	**Cool**
MEXICO		**MEXICO**	**MEXICO**		**5465**	**5465**	**MEXICO**	**327**		**Hot**
Aguascalientes		Aguascalientes	Aguascalientes		850	850	Aguascalientes	327	Aguascalientes	Hot
Baja California		Baja California	Baja California		2863	2863	Baja California	327	Guadalupe	Hot
Coahuila		Coahuila	Coahuila		279	279	Coahuila	327	San Buenaventura	Hot
Sonora		Sonora	Sonora		1164	1164			Hermosillo	Hot
Zacatecas		Zacatecas	Zacatecas		309	309	Zacatecas	327	Zacatecas	Warm
MOLDOVA	**MOLDOVA**	**MOLDOVA**	**MOLDOVA**	**89844**	**89844**	**82600**		**276**	**Chisinau**	**Warm**
MOROCCO	**MOROCCO**	**MOROCCO**	**MOROCCO**	**49600**	**49000**	**17590**	**MOROCCO**	**48**	**Meknes**	**Hot**
MYANMAR		**MYANMAR**	**MYANMAR**		**75**	**70**	**MYANMAR/BU**	**385**	**Mandalay**	**Hot**
NEW ZEALAND	**NEW ZEALAND**	**NEW ZEALAND**	**NEW ZEALAND**	**9942**	**31964**	**35463**	**NEW ZEALAND**	**367-375**		**Temp.**
Auckland	Auckland	Auckland	Auckland	392	543	346	Auckland	368	Auckland	Warm
Canterbury	Canterbury	Canterbury	Canterbury	232	320	197	Canterbury	371	Christchurch	Cool

Table 74 (cont.): World's wine regions' bearing areas and names in *World Atlas of Wine*, by country, 2000, 2010 and 2016

Name	2000	2010	2016	ha 2000	ha 2010	ha 2016	Atlas name	Pages	Town	Zone
Gisborne	Gisborne	Gisborne	Gisborne	1681	2149	1440	Gisborne	368	Gisborne	Temp.
Hawkes Bay	Hawkes Bay	Hawkes Bay	Hawkes Bay	2443	4921	4638	Hawke's Bay	369	Napier	Temp.
Marlborough	Marlborough	Marlborough	Marlborough	4054	18401	23452	Marlborough	372-374	Blenheim	Temp.
Nelson	Nelson	Nelson	Nelson	203	813	1141	Nelson	368	Nelson	Temp.
Northland			Northland			57	Northland	368	Whangarei	Warm
Otago	Otago	Otago	Otago	280	1532	1942	Otago	375	Cromwell	Cool
Waikato	Waikato	Waikato	Waikato	119	147	16	Waikato	map 367	Hamilton	Temp.
Waipara	Waipara	Waipara	Waipara	210	1442	1231	Waipara	371	Waipara	Cool
Wairarapa	Wairarapa	Wairarapa	Wairarapa	328	859	1003	Wairarapa	370	Martinborough	Temp.
Other regions		Other regions			836					Temp.
NORTH MACEDONIA			**NORTH MACEDONIA**			**24777**	**NORTH MACEI**	**267**	**Bitola**	**Temp.**
NORWAY			**NORWAY**			**13**			**Oslo**	**Cool**
PERU		**PERU**	**PERU**		**3831**	**3831**	**PERU**	**330**		**Warm**
Arequipa		Arequipa	Arequipa		1356	1356	Arequipa	330	Arequipa	Temp.
Lima		Lima	Lima		783	783	Lima	330	Lima	Hot
Moquegua		Moquegua	Moquegua		877	877	Moquegua	330	Moquegua	Hot
Tacna		Tacna	Tacna		815	815	Tacna	330	Tacna	Hot
PORTUGAL	**PORTUGAL**	**PORTUGAL**	**PORTUGAL**	**205003**	**163522**	**182649**	**PORTUGAL**	**207-221**		**Warm**
Acores	Acores	Acores	Acores	1689	176	176	Acores	map 208	Ponta Delgada	Hot
Alentejo	Alentejo	Alentejo	Alentejo	14905	21892	29987	Alentejo	218-219	Evora	Hot
Algarve	Algarve	Algarve	Algarve	1900	1078	1571	Algarve	208	Lagoa	Hot
Alto Tras-os-Montes	Alto Tras-os-Montes	Alto Tras-os-Montes		63371	59112		Tras-os-Montes	208	Chaves	Warm
Beira Interior	Beira Interior	Beira Interior		20851	15968		Beira Interior	208	Castelo Branco	Hot
Beira Litoral	Beira Litoral	Beira Litoral		23921	15241		Beira Atlantico	map 208	Coimbra	Hot
Centro			Centro			67240				Hot
Entre Douro e Minho	Entre Douro e Minho	Entre Douro e Minho		29678	14288		Douro Valley	210-213	Pinhao	Temp.
Lisboa			Lisboa			7958	Lisboa	215	Lisbon	Hot
Madeira	Madeira	Madeira	Madeira	1513	847	847	Madeira	220-221	Funchal	Hot
Norte			Norte			74870				Warm
Ribatejo e Oeste	Ribatejo e Oeste	Ribatejo e Oeste		47173	34920				Coruche	Hot
ROMANIA	**ROMANIA**	**ROMANIA**	**ROMANIA**	**222173**	**170292**	**182762**	**ROMANIA**	**272-273**		**Warm**
Bucuresti - Ilfov		Bucuresti - Ilfov			443		Bucuresti	map 273	Bucharest	Warm
Centru		Centru			5349		part of Transylvan	map 273	Blaj	Temp.
Nord-Est		Nord-Est			28072		part of Moldovan	map 273	Iași	Temp.
Nord-Vest		Nord-Vest			6667		part of Transylvan	map 273	Oradea	Temp.
Sud-Est		Sud-Est			68081		part of Moldovan	map 273	Constanța	Warm
Sud-Muntenia		Sud-Muntenia			27069		part of Oltenia anc	map 273	Pitești	Temp.
Sud-Vest Oltenia		Sud-Vest Oltenia			31966		part of Oltenia anc	map 273	Rogova	Warm
Vest		Vest			2645		part of Banat, part	map 273	Timișoara	Warm
RUSSIA	**RUSSIA**	**RUSSIA**	**RUSSIA**	**56332**	**25628**	**50794**	**RUSSIA**	**276**		**Warm**
Crimea			Crimea			25166	Crimea	276	Simferopol	Warm
Krasnodar Krai		Krasnodar Krai	Krasnodar Krai		21224	21224	Kuban	276	Krasnodar	Warm
Rostov Oblast		Rostov Oblast	Rostov Oblast		4404	4404			Rostov-on-Don	Warm
SERBIA	**SERBIA**	**SERBIA**	**SERBIA**	**68999**	**68999**	**22014**	**SERBIA**	**267**		**Warm**
Bačka			Bačka			23	Bačka	map 267	Bačka Topola	Warm
Banat			Banat			98	Banat	map 267	Kikinda	Warm
Belgrade			Belgrade			1132	Belgrade	267	Belgrade	Warm

Table 74 (cont.): World's wine regions' bearing areas and names in *World Atlas of Wine*, by country, 2000, 2010 and 2016

Name	2000	2010	2016	ha 2000	ha 2010	ha 2016	Atlas name	Pages	Town	Zone
Čačak-Kraljevo			Čačak-Kraljevo			65	Čačak-Kraljevo	map 267	Čačak	Warm
Knjaževac			Knjaževac			1077	Knjaževac	map 267	Knjaževac	Temp.
Leskovac			Leskovac			1451	Leskovac	map 267	Leskovac	Warm
Mlava			Mlava			815	Mlava	map 267	Požarevac	Warm
Negotinska Krajina			Negotinska Krajina			975	Negotinska Krajin	267	Negotin	Warm
Niš			Niš			1316	Niš	map 267	Niš	Warm
Nišava			Nišava			477	Nišava	map 267	Bela Palanka	Temp.
South Banat			South Banat			1722	Južni Banat	map 267	Bela Crkva	Warm
Srem			Srem			2142	Srem	map 267	Novi Sad	Warm
Subotica			Subotica			311	Subotica	267	Subotica	Warm
Šumadija			Šumadija			1118	Šumadija	map 267	Kragujevac	Temp.
Telečka			Telečka			115	Telečka	map 267	Sombor	Warm
Tisa			Tisa			261	Tisa	267	Senta	Warm
Toplica			Toplica			761	Toplica	map 267	Zitoradja	Warm
Tri Morave			Tri Morave			7544	Tri Morave	map 267	Kruševac	Warm
Valjevo			Valjevo			191	Cer-Valjevo	map 267	Koceljeva	Warm
Vranje			Vranje			419	Vranje	map 267	Vranje	Temp.
SLOVAKIA	**SLOVAKIA**	**SLOVAKIA**	**SLOVAKIA**	**15580**	**12637**	**7748**	**SLOVAKIA**	**266**		**Temp.**
Bratislavský kraj			Bratislavský kraj			1571	Bratislava	266	Bratislava	Temp.
Juznoslovenska		Juznoslovenska			4141		Juznoslovenska	map 266	Dunajská Streda	Temp.
Malokarpatska		Malokarpatska			3683		Malokarpatska	map 266	Modra	Temp.
Nitrianska		Nitrianska			2652		Nitrianska	map 266	Nitra	Temp.
Stredné Slovensko		Stredné Slovensko	Stredné Slovensko		1155	780	Stredoslovenska	map 266	Rimavská Sobota	Temp.
Tokajska		Tokajska			453		Tokaj	266	Čerhov	Temp.
Východné Slovensko		Východné Slovensko	Východné Slovensko		553	394	Východoslovensk	map 266	Kráľovský Chlmec	Temp.
Západné Slovensko			Západné Slovensko			5003			Nitra	Temp.
SLOVENIA	**SLOVENIA**	**SLOVENIA**	**SLOVENIA**	**23472**	**16354**	**15989**	**SLOVENIA**	**268-269**		**Temp.**
Bela Krajina		Bela Krajina	Bela Krajina		365	369	Bela Krajina	269	Črnomelj	Temp.
Bizeljsko Sremic		Bizeljsko Sremic	Bizeljsko Sremic		907	878	Bizeljsko Sremic	269	Bizeljsko	Temp.
Dolenjska		Dolenjska	Dolenjska		1476	1600	Dolenjska	269	Dobrovo	Warm
Goriska brda		Goriska brda	Goriska brda		1898	1803	Brda	268	Neblo	Warm
Kras		Kras	Kras		593	582	Kras	268	Križ	Temp.
Prekmurje		Prekmurje	Prekmurje		564	528	Prekmurje	269	Gornji Petrovci	Temp.
Slovenska Istra		Slovenska Istra	Slovenska Istra		1626	1840	Slovenska Istra	268	Koper	Hot
Stajerska Slovenija		Stajerska Slovenija	Stajerska Slovenija		6374	6113	Štajerska Slovenij	269	Slovenske Konjice	Temp.
Vipavska dolina		Vipavska dolina	Vipavska dolina		2526	2276	Vipava Valley	268	Nova Gorica	Warm
Other regions		Other regions			26					Temp.
SOUTH AFRICA	**SOUTH AFRICA**	**SOUTH AFRICA**	**SOUTH AFRICA**	**93656**	**101016**	**95775**	**SOUTH AFRICA**	**378-384**		**Hot**
Breedekloof	Breedekloof	Breedekloof	Breedekloof	10385	12568	12893	Breedekloof	379	Rawsonville	Hot
Cape South Coast			Cape South Coast			2657	Cape South Coast	384	Bot River	Warm
Little Karoo	Little Karoo	Little Karoo	Little Karoo	3168	2822	2443	Klein Karoo	378	Calitzdorp	Hot
Northern Cape	Northern Cape	Northern Cape	Northern Cape	5025	5078	4360	Northern Cape	378	Rand	Hot
Olifants River	Olifants River	Olifants River	Olifants River	9015	9997	10018	Olifants River	378	Klawer	Hot
Paarl	Paarl	Paarl	Paarl	17249	16568	15280	Paarl	382	Paarl	Hot
Robertson	Robertson	Robertson	Robertson	12227	14004	13227	Robertson	379	Robertson	Hot
Stellenbosch	Stellenbosch	Stellenbosch	Stellenbosch	16112	17107	15339	Stellenbosch	382	Stellenbosch	Hot
Swartland	Swartland	Swartland	Swartland	13670	14224	12940	Swartland	381	Malmesbury	Hot

Table 74 (cont.): World's wine regions' bearing areas and names in *World Atlas of Wine*, by country, 2000, 2010 and 2016

Name	2000	2010	2016	ha 2000	ha 2010	ha 2016	Atlas name	Pages	Town	Zone
Worcester	Worcester	Worcester	Worcester	6805	8649	6618	Worcester	379	Worcester	Hot
SPAIN	**SPAIN**	**SPAIN**	**SPAIN**	**1181806**	**1028258**	**883558**	**SPAIN**	**188-205**		**Hot**
Andalucía			**Andalucía**	**45557**	**31693**	**32054**	**Andalucia**	**203-205**		**Hot**
Almeria, Granada, Jaen, Sevilla	Almeria, Granada, Jaen, Se	Almeria, Granada, Jaen, Sevilla		9635	6950		Almeria, Granada,	map 188	Granada	Hot
Cadiz	Cadiz	Cadiz		9931	10156		Cadiz	204	Jerez de la Frontera	Hot
Cordoba	Cordoba	Cordoba		12777	8278		Cordoba	map 188	Córdoba	Hot
Huelva	Huelva	Huelva		6747	4230		Huelva	map 188	Huelva	Hot
Malaga	Malaga	Malaga		6467	2079		Malaga	203	Málaga	Hot
Aragón			**Aragón**	**49146**	**44571**	**33729**	**Aragón**	**map 188**		**Hot**
Huesca, Teruel	Huesca, Teruel	Huesca, Teruel		7599	9277		Huesca, Teruel	map 188	Huesca	Hot
Zaragoza	Zaragoza	Zaragoza		41545	35294		Zaragoza	map 188	Zaragoza	Hot
Canarias	**Canarias**	**Canarias**	**Canarias**	**13727**	**8653**	**10878**	**Canary Islands**	**191**	**Santa Cruz de Tenerife**	**Hot**
Cantabria	**Cantabria**	**Cantabria**	**Cantabria**	**39**	**50**	**106**	**Cantabria**	**map 188**	**Santander**	**Warm**
Castilla y León			**Castilla y León**	**74943**	**72566**	**72364**	**Castilla y León**	**map 188**		**Temp.**
Avila, Palencia, Salamanca, Segovia, So	Avila, Palencia, Salamanca	Avila, Palencia, Salamanca, Segovia, Soria		13995	9154		Avila, Palencia, Sa	map 188	Salamanca	Temp.
Burgos	Burgos	Burgos		13264	16276		Burgos	map 188	Burgos	Temp.
Leon	Leon	Leon		15964	12149		Tierra de Leon	189	León	Temp.
Valladolid	Valladolid	Valladolid		15837	22081		Valladolid	195	Valladolid	Warm
Zamora	Zamora	Zamora		15883	12906		Zamora	map 188	Zamora	Warm
Castilla-La Mancha			**Castilla-La Mancha**	**586811**	**495301**	**409969**	**Castilla-La Manc**	**map 188**		**Hot**
Albacete	Albacete	Albacete		107563	96745		Albacete	map 188	Albacete	Hot
Ciudad Real	Ciudad Real	Ciudad Real		206365	175764		Ciudad Real	map 188	Ciudad Real	Hot
Cuenca	Cuenca	Cuenca		97721	94883		Cuenca	map 188	Cuenca	Warm
Guadalajara	Guadalajara	Guadalajara		2827	2149		Guadalajara	map 188	Guadalajara	Warm
Toledo	Toledo	Toledo		172334	125760		Toledo	map 188	Toledo	Hot
Cataluña			**Cataluña**	**64452**	**59516**	**51908**	**Catalunya**	**200-201**		**Hot**
Barcelona	Barcelona	Barcelona		24223	22339		Barcelona	map 188	Barcelona	Hot
Girona, Lleida	Girona, Lleida	Girona, Lleida		7390	7560		Girona, Lleida	map 188	Lleida	Hot
Tarragona	Tarragona	Tarragona		32839	29617		Tarragona	200	Reus	Hot
Comunidad de Madrid	**Comunidad de Madrid**	**Comunidad de Madrid**	**Comunidad de Madrid**	**17475**	**23963**	**11255**	**Madrid**	**map 188**	**Madrid**	**Hot**
Comunidad Foral de Navarra	**Comunidad Foral de Nav**	**Comunidad Foral de Nav**	**Comunidad Foral de Nav**	**23619**	**11024**	**17015**	**Navarra**	**197**	**Estella**	**Temp.**
Comunidad Valenciana			**Comunidad Valenciana**	**90872**	**73434**	**57677**	**Valencia**	**map 188**		**Hot**
Alicante	Alicante	Alicante		25465	14661		Alicante	190	Alicante	Hot
Castellon	Castellon	Castellon		1190	1214			map 188	Castellón de la Plana	Hot
Valencia	Valencia	Valencia		64215	57559		Valencia	190	Valencia	Hot
Extremadura			**Extremadura**	**85519**	**86606**	**78323**	**Extremadura**	**map 188**		**Hot**
Badajoz	Badajoz	Badajoz		81209	82749		Badajoz	map 188	Badajoz	Hot
Caceres	Caceres	Caceres		4310	3857		Caceres	map 188	Cáceres	Hot
Galicia	**Galicia**	**Galicia**	**Galicia**	**31747**	**25457**	**30120**	**Galicia**	**map 188**	**Vigo**	**Warm**
Illes Balears	**Illes Balears**	**Illes Balears**	**Illes Balears**	**1718**	**1544**	**1718**	**Islas Baleares**	**map 188**	**Palma**	**Hot**
La Rioja	**La Rioja**	**La Rioja**	**La Rioja**	**39459**	**44576**	**40081**	**Rioja**	**198-199**	**Logrono**	**Warm**
País Vasco			**País Vasco**	**11581**	**13776**	**13481**	**País Vasco**	**map 188**		**Temp.**
Alava	Alava	Alava		11333	13040		Alava	map 199	Vitoria-Gasteiz	Temp.
Guipuzcoa, Vizcaya	Guipuzcoa, Vizcaya	Guipuzcoa, Vizcaya		243	736				Bilbao	Warm
Principado de Asturias	**Principado de Asturias**	**Principado de Asturias**	**Principado de Asturias**	**84**	**95**	**104**	**Asturias**	**map 188**	**Oviedo**	**Temp.**
Región de Murcia	**Región de Murcia**	**Región de Murcia**	**Región de Murcia**	**45058**	**35437**	**22774**	**Murcia**	**map 188**	**Murcia**	**Hot**
SWITZERLAND	**SWITZERLAND**	**SWITZERLAND**	**SWITZERLAND**	**15042**	**14820**	**14793**		**250-253**		**Cool**

Table 74 (cont.): World's wine regions' bearing areas and names in *World Atlas of Wine*, by country, 2000, 2010 and 2016

Name	2000	2010	2016	ha 2000	ha 2010	ha 2016	Atlas name	Pages	Town	Zone
Aargau	Aargau	Aargau	Aargau	395	399	385	Aargau	251	Hägglingen	Cool
Appenzell Ausserrhoden			Appenzell Ausserrhoden			4			Waldstatt	Cool
Appenzell Innerrhoden			Appenzell Innerrhoden			1	Appenzell Innerrh	251	Appenzell	Cool
Basel Land	Basel Land	Basel Land	Basel Land	99	114	114	Baselbiet	map 251	Liestal	Cool
Basel Stadt			Basel Stadt			5			Basel	Temp.
Bern	Bern	Bern	Bern	257	242	221	Bern	map 251	Bern	Cool
Fribourg	Fribourg	Fribourg	Fribourg	116	117	117	Fribourg	251	Fribourg	Cool
Geneva	Geneva	Geneva	Geneva	1355	1292	1411	Geneva	253	Geneva	Temp.
Glarus			Glarus			2	Glarus	map 251	Glarus	Cool
Graubünden	Graubünden	Graubünden		414	421		Graubünden	251	Chur	Cool
Graubünden - Mesolcina			Graubünden - Mesolcina			30	Graubünden	251	Cama	Cool
Graubünden - other			Graubünden - other			421	Graubünden	251		Cool
Jura	Jura	Jura	Jura	7	14	15			Delémont	Cool
Lucerne	Lucerne	Lucerne	Lucerne	19	41	53	Luzern	map 251	Lucerne	Cool
Neuchâtel	Neuchâtel	Neuchâtel	Neuchâtel	605	591	604	Neuchâtel	251	Neuchâtel	Cool
Nidwalden			Nidwalden			0			Dallenwil	Cool
Owalden			Owalden			2			Sarnen	Cool
Schaffhausen	Schaffhausen	Schaffhausen	Schaffhausen	500	478	483	Schaffhausen	251	Schaffhausen	Cool
Schwyz	Schwyz	Schwyz	Schwyz	30	38	39	Schwyz	map 251	Schwyz	Cool
Solothurn			Solothurn			10	Solothurn	map 251	Solothurn	Cool
St. Gallen	St. Gallen	St. Gallen	St. Gallen	217	215	211	St. Gallen	map 251	St. Gallen	Cool
Thunersee			Thunersee			19	Thunersee	map 251	Thun	Cool
Thurgau	Thurgau	Thurgau	Thurgau	274	263	257	Thurgau	250	Märstetten	Cool
Ticino	Ticino	Ticino	Ticino	961	1069	1098	Ticino	251	Frasco	Cool
Uri			Uri			4			Gurtnellen	Cool
Valais	Valais	Valais	Valais	5255	5070	4906	Valais	252	Valais	Cool
Vaud	Vaud	Vaud	Vaud	3879	3819	3771	Vaud	252	Crissier	Temp.
Zug			Zug			2	Zug	map 251	Zug	Cool
Zürich	Zürich	Zürich	Zürich	645	614	607	Zurich	251	Zürich	Cool
Other regions	Other regions	Other regions	Other regions	14	25	3				Cool
TAIWAN	**TAIWAN**	**TAIWAN**	**TAIWAN**	**2833**	**2833**	**149**	**TAIWAN**	**385**	**Taipei**	**Hot**
THAILAND		**THAILAND**	**THAILAND**		**149**	**208**	**THAILAND**	**385**	**Bangkok**	**Hot**
TUNISIA	**TUNISIA**	**TUNISIA**	**TUNISIA**	**16836**	**16836**	**3400**			**Kelibia**	**Hot**
TURKEY		**TURKEY**	**TURKEY**		**12856**	**13704**	**TURKEY**	**285**		**Hot**
Aegean		Aegean	Aegean		6770	8214	Aegean	285	İzmir	Hot
Central East		Central East	Central East		1895	2066	Eastern Anatolia	map 285	Elazığ	Hot
Central North		Central North	Central North		428	214	Black Sea	map 285	Tokat	Warm
Central South		Central South	Central South		1553	770	Central Anatolia	map 285	Ankara	Warm
Marmara		Marmara	Marmara		1745	1737	Thrace-Marmara	map 285	Tekirdağ	Warm
Mediterranean		Mediterranean	Mediterranean		28	25	Mediterranean	285	Antalya	Hot
South East		South East	South East		438	677	Southeastern Anat	map 285	Gaziantep	Hot
UKRAINE		**UKRAINE**	**UKRAINE**		**52293**	**25166**	**UKRAINE**	**276**	**Odessa**	**Warm**
UNITED KINGDOM	**UNITED KINGDOM**	**UNITED KINGDOM**	**UNITED KINGDOM**	**873**	**1198**	**1839**	**ENGLAND AND**	**249**	**Brighton**	**Cool**
UNITED STATES	**UNITED STATES**	**UNITED STATES**	**UNITED STATES**	**175693**	**227949**	**239632**	**UNITED STATE**	**294-326**		**Hot**
Arizona		**Arizona**	**Arizona**		**101**	**250**	**Arizona**	**326**	**Willcox**	**Hot**
Arkansas		**Arkansas**	**Arkansas**		**243**	**251**	**Arkansas**	**290**	**Little Rock**	**Hot**
California				**151657**	**181687**	**187816**	**California**	**302-322**		**Hot**

Table 74 (cont.): World's wine regions' bearing areas and names in *World Atlas of Wine*, by country, 2000, 2010 and 2016

Name	2000	2010	2016	ha 2000	ha 2010	ha 2016	Atlas name	Pages	Town	Zone
Alameda	Alameda	Alameda	Alameda	546	1145	1188	Alameda	map 303	Livermore	Hot
Amador	Amador	Amador	Amador	1014	1255	1458	Amador	map 303	Jackson	Hot
Butte	Butte	Butte	Butte	58	58	69	Butte	map 303	Chico	Hot
Calaveras	Calaveras	Calaveras	Calaveras	114	253	288	Calaveras	map 303	San Andreas	Hot
Colusa	Colusa	Colusa	Colusa	539	646	606	Colusa	map 303	Colusa	Hot
Contra Costa	Contra Costa	Contra Costa	Contra Costa	397	675	725	Contra Costa	map 303	Concord	Hot
El Dorado	El Dorado	El Dorado	El Dorado	338	660	830	El Dorado	318	Placerville	Hot
Fresno	Fresno	Fresno	Fresno	17606	16010	15909	Fresno	map 303	Fresno	Hot
Glenn	Glenn	Glenn	Glenn	580	329	408	Glenn	map 303	Orland	Hot
Humboldt	Humboldt	Humboldt	Humboldt	4	36	49			Eureka	Cool
Kern	Kern	Kern	Kern	11198	8422	6913	Kern	map 303	Bakersfield	Hot
Kings	Kings	Kings	Kings	949	615	805	Kings	map 303	Hanford	Hot
Lake	Lake	Lake	Lake	1444	3122	3264	Lake County	304	Clearlake	Warm
Lassen			Lassen			1			Susanville	Cool
Los Angeles	Los Angeles	Los Angeles	Los Angeles	12	53	84	Los Angeles	map 303	Los Angeles	Hot
Madera	Madera	Madera	Madera	17427	14273	13813	Madera	map 303	Madera	Hot
Marin	Marin	Marin	Marin	33	62	66	Marin	map 303	San Rafael	Warm
Mariposa	Mariposa	Mariposa	Mariposa	23	24	27	Mariposa	map 303	Mariposa	Hot
Mendocino	Mendocino	Mendocino	Mendocino	5050	6555	6660	Mendocino	304	Ukiah	Warm
Merced	Merced	Merced	Merced	5901	4418	5265	Merced	map 303	Merced	Hot
Monterey	Monterey	Monterey	Monterey	11688	15600	17909	Monterrey	316	Salinas	Temp.
Napa	Napa	Napa	Napa	12258	17768	17914	Napa Valley	310-315	St Helena	Warm
Nevada	Nevada	Nevada	Nevada	76	159	179	Nevada	map 303	Nevada City	Warm
Orange		Orange	Orange		0	0			Irvine	Hot
Placer	Placer	Placer	Placer	37	70	76	Placer	map 303	Auburn	Hot
Riverside	Riverside	Riverside	Riverside	845	333	437			Riverside	Hot
Sacramento	Sacramento	Sacramento	Sacramento	3611	7406	8031	Sacramento	map 303	Sacramento	Hot
San Benito	San Benito	San Benito	San Benito	720	959	1115	San Benito	map 303	Hollister	Warm
San Bernardino	San Bernardino	San Bernardino	San Bernardino	558	209	193			San Bernardino	Hot
San Diego	San Diego	San Diego	San Diego	25	78	199	San Diego	289	San Diego	Hot
San Joaquin	San Joaquin	San Joaquin	San Joaquin	20930	27146	28107	San Joaquin	map 303	Lodi	Hot
San Luis Obispo	San Luis Obispo	San Luis Obispo	San Luis Obispo	5047	11484	12392	San Luis Obispo	319	San Luis Obispo	Warm
San Mateo	San Mateo	San Mateo	San Mateo	19	29	41	San Mateo	map 303	San Mateo	Temp.
Santa Barbara	Santa Barbara	Santa Barbara	Santa Barbara	4043	6512	6113	Santa Barbara Cou	322	Santa Maria	Temp.
Santa Clara	Santa Clara	Santa Clara	Santa Clara	443	609	606	Santa Clara	map 303	San Jose	Hot
Santa Cruz	Santa Cruz	Santa Cruz	Santa Cruz	68	160	193	Santa Cruz	map 303	Santa Cruz	Temp.
Shasta	Shasta	Shasta	Shasta	15	41	38			Redding	Hot
Siskiyou		Siskiyou	Siskiyou		8	7			Montague	Temp.
Solano	Solano	Solano	Solano	698	1231	1433	Solano	map 303	Fairfield	Hot
Sonoma	Sonoma	Sonoma	Sonoma	14708	22265	23528	Sonoma	305-309	Sonoma	Warm
Stanislaus	Stanislaus	Stanislaus	Stanislaus	5358	3079	2765	Stanislaus	map 303	Modesto	Hot
Sutter	Sutter	Sutter	Sutter	32	54	8	Sutter	map 303	Yuba City	Hot
Tehama	Tehama	Tehama	Tehama	53	59	46	Tehama	map 303	Corning	Hot
Trinity	Trinity	Trinity	Trinity	15	49	53			Hayfork	Temp.
Tulare	Tulare	Tulare	Tulare	4602	3432	2890	Tulare	map 303	Visalia	Hot
Tuolumne		Tuolumne	Tuolumne		12	11	Tuolumne	map 303	Sonora	Hot
Ventura	Ventura	Ventura	Ventura	3	21	17	Ventura	map 303	Oxnard	Warm

Table 74 (cont.): World's wine regions' bearing areas and names in *World Atlas of Wine*, by country, 2000, 2010 and 2016

Name	2000	2010	2016	ha 2000	ha 2010	ha 2016	Atlas name	Pages	Town	Zone
Yolo	Yolo	Yolo	Yolo	2446	4263	5059	Yolo	map 303	Woodland	Hot
Yuba	Yuba	Yuba	Yuba	126	39	27	Yuba	map 303	Browns Valley	Hot
Colorado		**Colorado**	**Colorado**		**271**	**248**	**Colorado**	**326**	**Grand Junction**	**Warm**
Georgia		**Georgia**	**Georgia**		**567**	**586**	**Georgia**	**290**	**Atlanta**	**Hot**
Illinois		**Illinois**	**Illinois**		**373**	**386**	**Illinois**	**map 290**	**Springfield**	**Hot**
Indiana		**Indiana**	**Indiana**		**263**	**262**	**Indiana**	**map 290**	**Indianapolis**	**Warm**
Iowa		**Iowa**	**Iowa**		**194**	**145**	**Iowa**	**map 290**	**Des Moines**	**Warm**
Kentucky		**Kentucky**	**Kentucky**		**210**	**217**	**Kentucky**	**map 290**	**Frankfort**	**Hot**
Michigan	**Michigan**	**Michigan**	**Michigan**	**526**	**1072**	**1150**	**Michigan**	**290**	**Detroit**	**Warm**
Minnesota		**Minnesota**	**Minnesota**		**418**	**432**	**Minnesota**		**Minneapolis**	**Temp.**
Missouri		**Missouri**	**Missouri**		**647**	**688**	**Missuri**	**290**	**St. Louis**	**Hot**
New York				**13352**	**12870**	**12626**	**New York**	**324-325**		**Temp.**
Cattaraugus			Cattaraugus			139			Cattaraugus	Cool
Chautauqua			Chautauqua			6740			Chautauqua	Temp.
Chautauqua-Erie	Chautauqua-Erie	Chautauqua-Erie		8116	7561					Temp.
Erie			Erie			682	Lake Erie	324	Buffalo	Temp.
Finger Lakes	Finger Lakes	Finger Lakes		3692	3801		Finger Lakes	324		Temp.
New York - other	New York - other	New York - other		1544	1508					Temp.
Niagara			Niagara			357	Niagara Escarpme	325	Niagara Falls	Temp.
Ontario			Ontario			216	Ontario	map 325	Canandaigua	Temp.
Schuyler			Schuyler			459	Schuyler	map 325	Watkins Glen	Temp.
Seneca			Seneca			540	Seneca	map 325	Ovid	Temp.
Steuben			Steuben			422	Steuben	map 325	Bath	Cool
Suffolk			Suffolk			826			Riverhead	Warm
Ulster			Ulster			64			Ellenville	Temp.
Wayne			Wayne			15	Wayne	map 325	Newark	Temp.
Yates			Yates			2164	Yates	map 325	Penn Yan	Temp.
North Carolina		**North Carolina**	**North Carolina**		**728**	**754**	**North Carolina**	**map 290**	**Raleigh**	**Hot**
Ohio		**Ohio**	**Ohio**		**436**	**607**	**Ohio**	**290**	**Cleveland**	**Warm**
Oregon				**3278**	**6685**	**10013**	**Oregon**	**324**		**Temp.**
Benton Co.	Benton Co.	Benton Co.		88	155				Corvallis	Temp.
Columbia River	Columbia River	Columbia River	Columbia River	293	610	645			The Dalles	Warm
Douglas Co.	Douglas Co.	Douglas Co.		190	350				Waterville	Cool
Jackson Co.		Jackson Co.			536				Medford	Temp.
Josephine Co.	Josephine Co.	Josephine Co.		117	162				Grants Pass	Warm
Lane Co.	Lane Co.	Lane Co.		254	341				Eugene	Temp.
Marion Co.	Marion Co.	Marion Co.		221	660		Marion	map 297	Salem	Temp.
North Willamette Valley			North Willamette Valley			6072	Willamatte Valley	296	McMinnville	Temp.
Oregon - other	Oregon - other	Oregon - other		216	154					Temp.
Polk Co.	Polk Co.	Polk Co.		383	928		Polk	map 297	Dallas	Temp.
Rogue Valley			Rogue Valley			1186	Rogue Valley	294	Medford	Temp.
South Willamette Valley			South Willamette Valley			1165	Willamatte Valley	296-297	Eugene	Temp.
Umpqua Valley			Umpqua Valley			945	Umpqua Valley	294	Sutherlin	Temp.
Valley - other	Valley - other			106						Temp.
Washington Co.	Washington Co.	Washington Co.		393	670		Washington	map 297	Hilsboro	Temp.
Yamhill Co.	Yamhill Co.	Yamhill Co.		1016	2273		Yamhill	map 297	McMinnville	Temp.
Pennsylvania		**Pennsylvania**	**Pennsylvania**		**1004**	**1040**	**Pennsylvania**	**290**	**Erie**	**Temp.**

Table 74 (cont.): World's wine regions' bearing areas and names in *World Atlas of Wine*, by country, 2000, 2010 and 2016

Name	2000	2010	2016	ha 2000	ha 2010	ha 2016	Atlas name	Pages	Town	Zone
Texas		**Texas**			**1214**	**1838**	**Texas**	**326**		**Hot**
Hill Country			Hill Country			307	Texas Hill Countr	map 326	Austin	Hot
North Texas (DFW)			North Texas (DFW)			161			Dallas	Hot
South Texas and Gulf Coast			South Texas and Gulf Coa:			123			Houston	Hot
Texas High Plains and Panhandle			Texas High Plains and Pan			1087	Texas High Plains	map 326	Lubbock	Hot
West Texas			West Texas			160			El Paso	Hot
Virginia		**Virginia**	**Virginia**		**1065**	**1284**	**Virginia**	**323**	**Charlottesville**	**Hot**
Washington	**Washington**			**6880**	**17899**	**19038**	**Washington**	**298-301**		**Warm**
Columbia Gorge		Columbia Gorge	Columbia Gorge		159	85	Columbia Gorge	294	White Salmon	Temp.
Columbia Valley		Columbia Valley	Columbia Valley		3023	2907	Columbia Valley	299	Yakima	Temp.
Horse Heaven Hills		Horse Heaven Hills	Horse Heaven Hills		4283	5345	Horse Heaven Hil	300	Paterson	Warm
Lake Chelan		Lake Chelan	Lake Chelan		100	42	Lake Chelan	294	Chelan	Temp.
Naches Heights			Naches Heights			6	Naches Heights	300	Gleed	Temp.
Puget Sound		Puget Sound	Puget Sound		72	32	Puget Sound	294	Seattle	Cool
Rattlesnake Hills		Rattlesnake Hills	Rattlesnake Hills		647	546	Rattlesnake Hills	300	Granger	Temp.
Red Mountain		Red Mountain	Red Mountain		515	646	Red Mountain	300	Benton City	Warm
Snipes Mountain		Snipes Mountain	Snipes Mountain		285	225	Snipes Mountain	300	Granger	Temp.
Wahluke Slope		Wahluke Slope	Wahluke Slope		2689	2834	Wahluke Slope	300	Mattawa	Warm
Walla Walla Valley		Walla Walla Valley	Walla Walla Valley		528	564	Walla Walla Valle	300	Walla Walla	Warm
Yakima Valley		Yakima Valley	Yakima Valley		5444	5806	Yakima Valley	300	Prosser	Temp.
URUGUAY	**URUGUAY**	**URUGUAY**	**URUGUAY**	**8880**	**7657**	**6743**	**URUGUAY**	**332**		**Hot**
Artigas			Artigas			36	Artigas	map 330	Artigas	Hot
Canelones			Canelones			4269	Canelones	332	Canelones	Hot
Colonia			Colonia			522	Colonia	332	Carmelo	Hot
Durazno			Durazno			47			Durazno	Hot
Florida			Florida			42	Florida	map 330	Florida	Hot
Lavalleja			Lavalleja			7			Lavalleja	Hot
Maldonado			Maldonado			326	Maldonado	map 332	Maldonado	Hot
Montevideo			Montevideo			813	Montevideo	332	Montevideo	Hot
Paysandu			Paysandu			160	Paysandu	map 330	Paysandú	Hot
Rivera			Rivera			41	Rivera	332	Rivera	Hot
Rocha			Rocha			12	Rocha	map 332	Rocha	Hot
Salto			Salto			54	Salto	332	Salto	Hot
San Jose			San Jose			392	San Jose	332	San José	Hot
Soriano			Soriano			3	Soriano	map 330	Palmitas	Hot
Tacuarembo			Tacuarembo			19			Tacuarembó	Hot

VII. Location and climate indicators of the world's wine regions

Table 75: Geographic location, elevation and growing season average temperature and precipitation (GST, GSP) for nearest town to world's wine regions
(Final column indicates whether climate is cool, temperate, warm or hot)

Name	2000	2010	2016	Town	Latitude	Longitude	Elevation	GST	GSP	Zone
ALGERIA	**ALGERIA**	**ALGERIA**	**ALGERIA**	**Algiers**	**36.75**	**3.04**	**190**	**21.3**	**250**	**Hot**
ARGENTINA	**ARGENTINA**	**ARGENTINA**	**ARGENTINA**		**-32.83**	**-68.45**	**755**	**20.6**	**198**	**Hot**
Buenos Aires					**-37.86**	**-61.12**	**177**	**18.2**	**550**	**Warm**
Balcarce			Balcarce	Balcarce	-37.85	-58.26	115	17.3	528	Warm
Benito Juárez			Benito Juárez	Benito Juárez	-37.67	-59.80	217	17.2	539	Warm
Cañuelas			Cañuelas	Cañuelas	-35.05	-58.76	34	20.3	752	Hot
Coronel Pringles	Coronel Pringles		Coronel Pringles	Coronel Pringles	-37.99	-61.35	254	17.7	602	Warm
Coronel Suarez		Coronel Suarez	Coronel Suarez	Coronel Suarez	-37.46	-61.93	242	18.3	618	Warm
Daireaux			Daireaux	Daireaux	-36.60	-61.75	119	19.4	669	Hot
De La Costa			De La Costa		-37.86	-61.12	177	18.2	550	Warm
General Belgrano			General Belgrano	General Belgrano	-35.77	-58.50	21	19.3	703	Hot
General Pueyrredón	General Pueyrredón		General Pueyrredón	General Pueyrredón	-37.95	-57.78	88	17.2	558	Warm
Junín - Bs. As.			Junín - Bs. As.	Junín - Bs. As.	-34.59	-60.95	83	19.9	792	Hot
Saavedra			Saavedra	Saavedra	-37.76	-62.35	343	17.7	580	Warm
Tandil		Tandil	Tandil	Tandil	-37.33	-59.14	201	17.3	636	Warm
Tornquist		Tornquist	Tornquist	Tornquist	-38.10	-62.22	294	18.0	579	Warm
Villa Gesell			Villa Gesell	Villa Gesell	-37.26	-56.97	10	17.9	551	Warm
Villarino		Villarino	Villarino	Médanos	-38.83	-62.69	34	19.4	396	Hot
Catamarca					**-27.75**	**-67.09**	**1408**	**21.2**	**157**	**Hot**
Ambato		Ambato		El Bolsón	-27.90	-65.88	1214	21.2	467	Hot
Andalgala	Andalgala	Andalgala	Andalgala	Andalgalá	-27.58	-66.31	1081	22.9	281	Hot
Belén	Belén	Belén	Belén	Belén	-27.65	-67.03	1263	21.7	156	Hot
Capayán			Capayán	Capayán	-28.77	-66.05	419	24.5	385	Hot
Cushamen	Cushamen	Cushamen	Cushamen	Lago Puelo	-42.06	-71.60	216	13.9	376	Cool
Poman	Poman	Poman	Poman	Pomán	-28.40	-66.22	1181	22.8	338	Hot
Santa María - Catamarca	Santa María - Catamarca	Santa María - Catamarca	Santa María - Catamarca	Santa María	-26.70	-66.05	1896	19.5	183	Hot
Santa Rosa - Catamarca			Santa Rosa - Catamarca	Bañado de Ovanta	-28.10	-65.32	473	23.7	610	Hot
Tinogasta	Tinogasta	Tinogasta	Tinogasta	Tinogasta	-28.06	-67.57	1214	22.0	137	Hot
Valle Viejo			Valle Viejo	El Portezuelo	-28.47	-65.64	634	23.6	467	Hot
Chubut					**-45.29**	**-69.32**	**310**	**14.4**	**85**	**Cool**
Futaleufu			Futaleufu	Esquel	-42.91	-71.31	584	12.2	250	Cool
Languiñeo			Languiñeo	Tecka	-43.49	-70.81	716	12.5	149	Cool
Sarmiento - Chubut			Sarmiento - Chubut	Sarmiento	-45.59	-69.07	274	14.7	64	Cool
Córdoba					**-31.05**	**-64.43**	**531**	**20.1**	**576**	**Hot**
Calamuchita		Calamuchita	Calamuchita	San Agustín	-31.98	-64.38	551	19.8	594	Hot
Colón - Cordoba		Colón - Cordoba	Colón - Cordoba	Jesús María	-30.98	-64.10	535	19.1	637	Hot
Cruz del Eje	Cruz del Eje	Cruz del Eje	Cruz del Eje	Cruz del Eje	-30.72	-64.81	468	21.5	470	Hot
Ischilin	Ischilin	Ischilin	Ischilin	Deán Funes	-30.73	-64.79	472	21.2	473	Hot
Punilla			Punilla	Cosquín	-31.25	-64.47	718	18.8	607	Warm
San Alberto	San Alberto		San Alberto	Villa Cura Brochero	-31.67	-65.02	928	19.6	561	Hot
San Javier	San Javier	San Javier	San Javier	Villa Dolores	-31.95	-65.19	523	22.2	580	Hot
Santa María - Cordoba		Santa María - Cordoba	Santa María - Cordoba	Alta Gracia	-31.66	-64.43	553	20.0	630	Hot
Totoral	Totoral	Totoral		Villa del Totoral	-30.71	-64.07	551	19.5	607	Hot
Tulumba		Tulumba	Tulumba	Villa Tulumba	-30.40	-64.12	703	19.6	580	Hot
Entre Ríos					**-31.94**	**-59.13**	**46**	**21.9**	**852**	**Hot**
Colón - Entre Ríos	Colón - Entre Ríos	Colón - Entre Ríos	Colón - Entre Ríos	Colón	-32.22	-58.14	24	21.7	834	Hot
Concordia		Concordia	Concordia	Concordia	-31.39	-58.02	35	22.4	947	Hot
Diamante			Diamante	Diamante	-32.07	-60.64	81	21.6	823	Hot
Gualeguaychu			Gualeguaychu	Gualeguaychú	-33.01	-58.51	11	21.4	758	Hot
Nogoya		Nogoya	Nogoya	Nogoyá	-32.39	-59.79	47	21.5	751	Hot
Paraná			Paraná	Paraná	-31.74	-60.51	65	21.8	875	Hot
Uruguay			Uruguay	Concepción del Uruguay	-32.48	-58.23	20	21.6	779	Hot
Victoria		Victoria	Victoria	Victoria	-32.62	-60.16	55	21.5	714	Hot
Jujuy					**-23.62**	**-65.36**	**2390**	**16.7**	**313**	**Temp.**
El Carmen			El Carmen	El Carmen	-24.39	-65.26	1183	20.7	727	Hot
Humahuaca			Humahuaca	Humahuaca	-23.20	-65.35	2952	15.2	175	Temp.
Tilcara		Tilcara	Tilcara	Tilcara	-23.58	-65.39	2479	16.3	263	Temp.
Tumbaya		Tumbaya	Tumbaya	Tumbaya	-23.87	-65.47	2105	17.4	383	Warm
La Pampa					**-37.35**	**-67.60**	**447**	**19.3**	**198**	**Hot**
Curaco			Curaco	Puelches	-38.14	-65.92	234	19.8	239	Hot
Puelen	Puelen	Puelen	Puelen	Puelén	-37.34	-67.62	450	19.3	197	Hot
La Rioja					**-29.14**	**-67.54**	**1157**	**21.2**	**179**	**Hot**
Arauco	Arauco	Arauco		Aimogasta	-28.55	-66.82	845	23.3	267	Hot

Table 75 (cont.): Geographic location, elevation and growing season average temperature and precipitation (GST, GSP) for nearest town to world's wine regions

Name	2000	2010	2016	Town	Latitude	Longitude	Elevation	GST	GSP	Zone
Capital La Rioja	Capital La Rioja	Capital La Rioja		La Rioja	-29.41	-66.86	511	23.6	325	Hot
Castro Barros	Castro Barros	Castro Barros	Castro Barros	Aminga	-28.85	-66.93	1339	20.9	241	Hot
Chilecito	Chilecito	Chilecito	Chilecito	Chilecito	-29.16	-67.50	1122	21.5	183	Hot
Coronel Felipe Varela	Coronel Felipe Varela	Coronel Felipe Varela	Coronel Felipe Varela	Villa Unión	-29.32	-68.23	1154	20.0	121	Hot
Famatina	Famatina	Famatina	Famatina	Famatina	-28.93	-67.52	1560	19.3	168	Hot
General Lamadrid	General Lamadrid	General Lamadrid	General Lamadrid	Villa Castelli	-29.02	-68.23	1284	19.5	114	Hot
San Blas De Los Sauces	San Blas De Los Sauces	San Blas De Los Sauces	San Blas De Los Sauces	San Blas de los Sauces	-28.40	-67.09	1019	22.8	206	Hot
Sanagasta	Sanagasta	Sanagasta	Sanagasta	Sanagasta	-29.29	-67.02	1022	21.2	236	Hot
Vinchina	Vinchina	Vinchina	Vinchina	Vinchina	-28.75	-68.20	1472	19.2	114	Hot
Mendoza					**-33.33**	**-68.61**	**753**	**20.2**	**191**	**Hot**
General Alvear	General Alvear	General Alvear	General Alvear	General Alvear	-34.98	-67.69	478	20.2	237	Hot
Godoy Cruz	Godoy Cruz	Godoy Cruz	Godoy Cruz	Godoy Cruz	-32.93	-68.86	840	20.3	219	Hot
Guaymallén	Guaymallén	Guaymallén	Guaymallén	Guaymallén	-32.90	-68.80	747	20.8	208	Hot
Junín - Mendoza	Junín - Mendoza	Junín - Mendoza	Junín - Mendoza	Junín	-33.14	-68.48	672	20.8	168	Hot
La Paz	La Paz	La Paz	La Paz	La Paz	-33.46	-67.55	512	21.8	219	Hot
Las Heras	Las Heras	Las Heras	Las Heras	Las Heras	-32.84	-68.83	723	21.2	190	Hot
Lavalle	Lavalle	Lavalle	Lavalle	Villa Tulumaya	-32.72	-68.59	613	21.8	175	Hot
Luján de Cuyo	Luján de Cuyo	Luján de Cuyo	Luján de Cuyo	Luján de Cuyo	-33.03	-68.88	960	19.5	190	Hot
Maipú	Maipú	Maipú	Maipú	Maipú	-32.96	-68.79	804	20.5	190	Hot
Malargüe		Malargüe		Malargüe	-35.48	-69.58	1420	15.1	164	Temp.
Rivadavia - Mendoza	Rivadavia - Mendoza	Rivadavia - Mendoza	Rivadavia - Mendoza	Rivadavia	-33.19	-68.46	667	20.8	174	Hot
San Carlos - Mendoza	San Carlos - Mendoza	San Carlos - Mendoza	San Carlos - Mendoza	San Carlos	-33.77	-69.04	957	18.1	245	Warm
San Martín - Mendoza	San Martín - Mendoza	San Martín - Mendoza	San Martín - Mendoza	San Martín	-33.08	-68.47	664	21.2	172	Hot
San Rafael	San Rafael	San Rafael	San Rafael	San Rafael	-34.61	-68.34	708	18.7	218	Warm
Santa Rosa - Mendoza	Santa Rosa - Mendoza	Santa Rosa - Mendoza	Santa Rosa - Mendoza	Santa Rosa	-33.25	-68.15	611	21.3	187	Hot
Tunuyán	Tunuyán	Tunuyán	Tunuyán	Tunuyán	-33.58	-69.02	886	18.8	219	Warm
Tupungato	Tupungato	Tupungato	Tupungato	Tupungato	-33.37	-69.15	1076	18.1	194	Warm
Misiones					**-27.27**	**-55.21**	**352**	**23.5**	**1053**	**Hot**
Cainguas			Cainguas	Campo Grande	-27.21	-54.98	448	23.2	1013	Hot
Capital Misiones			Capital Misiones	Posadas	-27.36	-55.90	110	24.6	1164	Hot
Leandro Alem		Leandro Alem	Leandro Alem	Leandro N. Alem	-27.60	-55.32	296	23.4	1070	Hot
Veinticinco de Mayo - Misiones		Veinticinco de Mayo - Misiones	Veinticinco de Mayo - Misiones	Alba Posse	-27.57	-54.68	126	23.8	1041	Hot
Neuquén					**-38.40**	**-68.75**	**406**	**18.6**	**113**	**Warm**
Añelo	Añelo	Añelo	Añelo	Añelo	-38.35	-68.79	414	18.5	107	Warm
Chos Malal		Chos Malal	Chos Malal	Chos Malal	-37.38	-70.27	857	16.7	85	Temp.
Collon Cura		Collon Cura	Collon Cura	Collón Cura	-38.95	-68.10	282	19.0	202	Hot
Confluencia	Confluencia	Confluencia	Confluencia	Neuquén	-38.95	-68.06	281	19.0	196	Hot
Lacar			Lacar	San Martín de los Andes	-40.16	-71.35	650	12.0	416	Cool
Ñorquin			Ñorquin	El Huecú	-37.64	-70.58	1225	13.4	164	Cool
Pehuenches			Pehuenches	Rincón de los Sauces	-37.39	-68.93	601	18.4	113	Warm
Picún Leufú		Picún Leufú	Picún Leufú	Picún Leufú	-39.52	-69.29	393	17.2	54	Warm
Picunches			Picunches	Las Lajas	-38.52	-70.36	718	16.0	78	Temp.
Río Negro					**-39.22**	**-67.01**	**225**	**19.0**	**125**	**Hot**
Adolfo Alsina	Adolfo Alsina	Adolfo Alsina	Adolfo Alsina	Viedma	-40.81	-63.00	8	18.3	230	Warm
Avellaneda - Río Negro	Avellaneda - Río Negro	Avellaneda - Río Negro	Avellaneda - Río Negro	Choele Choel	-39.28	-65.66	136	19.6	179	Hot
Bariloche			Bariloche	San Carlos de Bariloche	-41.13	-71.31	851	11.6	333	Cool
Conesa	Conesa	Conesa	Conesa	General Conesa	-40.10	-64.46	62	19.1	152	Hot
El Cuy	El Cuy	El Cuy	El Cuy	El Cuy	-39.93	-68.35	718	16.0	102	Temp.
General Roca	General Roca	General Roca	General Roca	General Roca	-39.03	-67.59	237	19.1	106	Hot
Pichi Mahuida	Pichi Mahuida	Pichi Mahuida	Pichi Mahuida	Río Colorado	-38.99	-64.10	81	19.5	238	Hot
Salta					**-25.99**	**-65.99**	**1658**	**19.9**	**158**	**Hot**
Cachi	Cachi	Cachi	Cachi	Cachi	-25.12	-66.16	2345	17.1	167	Warm
Cafayate	Cafayate	Cafayate	Cafayate	Cafayate	-26.07	-65.98	1625	20.0	158	Hot
La Viña		La Viña	La Viña	La Viña	-25.47	-65.57	1194	21.2	373	Hot
Molinos	Molinos	Molinos	Molinos	Molinos	-25.44	-66.29	2045	18.5	143	Warm
San Carlos - Salta	San Carlos - Salta	San Carlos - Salta	San Carlos - Salta	San Carlos	-25.89	-65.93	1628	20.4	160	Hot
San Juan					**-31.69**	**-68.41**	**596**	**22.6**	**243**	**Hot**
Albardón	Albardón	Albardón	Albardón	General San Martín	-31.44	-68.52	641	22.5	226	Hot

Table 75 (cont.): Geographic location, elevation and growing season average temperature and precipitation (GST, GSP) for nearest town to world's wine regions

Name	2000	2010	2016	Town	Latitude	Longitude	Elevation	GST	GSP	Zone
Angaco	Angaco	Angaco	Angaco	Villa del Salvador	-31.45	-68.40	601	22.6	240	Hot
Calingasta	Calingasta	Calingasta	Calingasta	Tamberías	-31.46	-69.42	1469	16.7	91	Temp.
Capital San Juan	Capital San Juan	Capital San Juan	Capital San Juan	San Juan	-31.54	-68.54	651	22.4	257	Hot
Caucete	Caucete	Caucete	Caucete	Caucete	-31.65	-68.28	577	22.9	260	Hot
Chimbas	Chimbas	Chimbas	Chimbas	Paula A. de Sarmiento	-31.49	-68.53	644	22.5	235	Hot
Iglesia	Iglesia	Iglesia	Iglesia	Rodeo	-30.21	-69.14	1636	16.1	75	Temp.
Jachal	Jachal	Jachal		Huaco	-30.16	-68.48	974	20.4	138	Hot
Nueve de Julio	Nueve de Julio	Nueve de Julio	Nueve de Julio	Nueve de Julio	-31.65	-68.39	584	22.8	254	Hot
Pocito	Pocito	Pocito	Pocito	Villa Aberastain	-31.65	-68.58	617	22.4	233	Hot
Rawson	Rawson	Rawson	Rawson	Villa Krause	-31.57	-68.53	641	22.4	261	Hot
Rivadavia - San Juan	Rivadavia - San Juan	Rivadavia - San Juan	Rivadavia - San Juan	Rivadavia	-31.53	-68.59	691	22.1	220	Hot
San Martín - San Juan	San Martín - San Juan	San Martín - San Juan	San Martín - San Juan	San Martín	-31.52	-68.35	597	22.8	249	Hot
Santa Lucía	Santa Lucía	Santa Lucía	Santa Lucía	Santa Lucía	-31.53	-68.48	621	22.5	243	Hot
Sarmiento - San Juan	Sarmiento - San Juan	Sarmiento - San Juan	Sarmiento - San Juan	Media Agua	-31.98	-68.43	555	22.8	227	Hot
Ullum	Ullum	Ullum	Ullum	Villa Ibañez	-31.46	-68.72	789	21.4	197	Hot
Valle Fértil	Valle Fértil	Valle Fértil	Valle Fértil	San Agustin	-30.63	-67.47	854	21.4	269	Hot
Veinticinco de Mayo - San Juan	Veinticinco de Mayo - San Juan	Veinticinco de Mayo - San Juan	Veinticinco de Mayo - San Juan	Santa Rosa	-31.74	-68.31	571	22.8	254	Hot
Zonda	Zonda	Zonda	Zonda	Bsilio Nievas	-31.54	-68.73	778	21.4	204	Hot
San Luis					**-32.96**	**-66.00**	**698**	**21.5**	**432**	**Hot**
Ayacucho	Ayacucho		Ayacucho	San Francisco del Monte de Oro	-32.60	-66.13	798	22.1	430	Hot
Capital San Luis			Capital San Luis	San Luis	-33.30	-66.34	729	21.4	378	Hot
Junín - San Luis		Junín - San Luis	Junín - San Luis	Santa Rosa	-32.34	-65.20	592	21.5	553	Hot
Santiago del Estero					**-27.78**	**-64.26**	**192**	**24.5**	**575**	**Hot**
Capital Santiago del Estero	Capital Santiago del Estero		Capital Santiago del Estero	Santiago del Estero	-27.78	-64.26	192	24.5	575	Hot
Tucumán					**-26.82**	**-65.69**	**1960**	**18.1**	**336**	**Warm**
Tafí del Valle	Tafí del Valle	Tafí del Valle	Tafí del Valle	Tafí del Valle	-26.85	-65.71	2012	17.9	332	Warm
Trancas			Trancas	Trancas	-26.23	-65.28	779	22.4	421	Hot
ARMENIA	**ARMENIA**	**ARMENIA**	**ARMENIA**	**Yerevan**	**40.19**	**44.52**	**1014**	**17.8**	**229**	**Warm**
AUSTRALIA	**AUSTRALIA**	**AUSTRALIA**	**AUSTRALIA**		**-34.32**	**138.55**	**128**	**19.2**	**231**	**Hot**
Australian Capital Territory					**-33.81**	**142.91**	**554**	**16.4**	**391**	**Temp.**
Australian Capital Territory	Australian Capital Territory	Australian Capital Territory			-35.28	149.13	578	17.1	408	Warm
Canberra District (ACT)		Canberra District (ACT)	Canberra District (ACT)	Canberra	-35.28	149.13	578	17.1	408	Warm
New South Wales					**-34.01**	**146.02**	**185**	**21.0**	**274**	**Hot**
Big Rivers - other	Big Rivers - other	Big Rivers - other	Big Rivers - other		-34.11	141.92	37	21.5	155	Hot
Canberra District	Canberra District				-35.28	149.13	578	17.1	408	Warm
Canberra District (NSW)		Canberra District (NSW)	Canberra District (NSW)	Yass	-34.84	148.91	491	18.1	403	Warm
Central Ranges - other	Central Ranges - other	Central Ranges - other	Central Ranges - other		-33.83	148.68	318	20.2	369	Hot
Cowra	Cowra	Cowra	Cowra	Cowra	-33.83	148.68	318	20.2	369	Hot
Gundagai		Gundagai	Gundagai	Cootamundra	-34.64	148.03	375	19.5	327	Hot
Hastings River	Hastings River	Hastings River	Hastings River	Port Macquarie	-31.43	152.90	14	21.1	1019	Hot
Hilltops	Hilltops	Hilltops	Hilltops	Young	-34.30	148.30	490	18.7	368	Warm
Hunter	Hunter	Hunter	Hunter	Cessnook	-32.50	151.36	90	21.3	569	Hot
Hunter Valley - other	Hunter Valley - other	Hunter Valley - other	Hunter Valley - other		-32.50	151.36	90	21.3	569	Hot
Mudgee	Mudgee	Mudgee	Mudgee	Mudgee	-32.61	149.57	585	18.9	455	Warm
Murray Darling (NSW)	Murray Darling (NSW)	Murray Darling (NSW)	Murray Darling (NSW)	Wentworth	-34.11	141.92	37	21.5	155	Hot
New England Australia		New England Australia	New England Australia	Glen Innes	-29.75	151.74	1091	16.9	649	Temp.
Northern Rivers - other	Northern Rivers - other	Northern Rivers - other	Northern Rivers - other		-31.43	152.90	14	21.1	1019	Hot
Northern Slopes			Northern Slopes	Armidale	-30.50	151.67	1041	17.3	544	Warm
Northern Slopes - other	Northern Slopes - other	Northern Slopes - other			-30.50	151.67	1041	17.3	544	Warm
Orange	Orange	Orange	Orange	Orange	-33.28	149.10	889	16.7	499	Temp.
Perricoota	Perricoota	Perricoota	Perricoota	Moama	-36.11	144.76	104	19.6	215	Hot
Riverina	Riverina	Riverina	Riverina	Griffith	-34.28	146.05	148	21.5	238	Hot
Shoalhaven Coast		Shoalhaven Coast	Shoalhaven Coast	Nowra	-34.88	150.60	37	19.5	729	Hot

Table 75 (cont.): Geographic location, elevation and growing season average temperature and precipitation (GST, GSP) for nearest town to world's wine regions

Name	**2000**	**2010**	**2016**	**Town**	**Latitude**	**Longitude**	**Elevation**	**GST**	**GSP**	**Zone**
South Coast - other	South Coast - other	South Coast - other	South Coast - other		-34.88	150.60	37	19.5	729	Hot
Southern Highlands		Southern Highlands	Southern Highlands	Bowral	-34.48	150.42	680	16.4	664	Temp.
Southern NSW - other	Southern NSW - other	Southern NSW - other	Southern NSW - other		-34.64	148.03	375	19.5	327	Hot
Swan Hill (NSW)	Swan Hill (NSW)	Swan Hill (NSW)	Swan Hill (NSW)	Koraleigh	-35.16	143.42	67	20.7	176	Hot
Tumbarumba	Tumbarumba	Tumbarumba	Tumbarumba	Tumbarumba	-35.78	148.01	645	16.3	497	Temp.
Western Plains			Western Plains	Bourke	-30.09	145.94	108	25.2	200	Hot
Western Plains - other	Western Plains - other	Western Plains - other			-30.09	145.94	108	25.2	200	Hot
Northern Territory	**Northern Territory**			**Darwin**	**-12.46**	**130.85**	**8**	**28.5**	**1701**	**Hot**
Queensland					**-27.51**	**151.94**	**609**	**20.6**	**571**	**Hot**
Granite Belt	Granite Belt	Granite Belt	Granite Belt	Stanthorpe	-28.66	151.94	854	18.9	550	Warm
Queensland - other	Queensland - other	Queensland - other	Queensland - other		-27.51	151.94	609	20.6	571	Hot
South Burnett	South Burnett	South Burnett	South Burnett	Murgon	-26.24	151.94	339	22.6	595	Hot
South Australia					**-35.06**	**139.76**	**102**	**18.9**	**194**	**Warm**
Adelaide Hills	Adelaide Hills	Adelaide Hills	Adelaide Hills	Lenswood	-35.06	138.83	363	17.0	298	Warm
Adelaide Plains		Adelaide Plains	Adelaide Plains	Elizabeth	-34.72	138.67	39	19.6	212	Hot
Barossa Valley	Barossa Valley	Barossa Valley	Barossa Valley	Nuriootpa	-34.55	138.99	116	18.0	218	Warm
Barossa - other	Barossa - other	Barossa - other	Barossa - other		-34.55	138.99	116	18.0	218	Warm
Clare Valley	Clare Valley	Clare Valley	Clare Valley	Clare	-33.83	138.61	393	18.5	221	Warm
Coonawarra		Coonawarra	Coonawarra	Coonawarra	-37.75	140.86	63	16.1	283	Temp.
Currency Creek	Currency Creek	Currency Creek	Currency Creek	Milang	-35.41	138.96	11	19.0	162	Warm
Eden Valley	Eden Valley	Eden Valley	Eden Valley	Eden Valley	-34.64	139.10	399	17.7	215	Warm
Far North - other	Far North - other	Far North - other	Far North - other		-33.27	138.35	237	20.2	190	Hot
Fleurieu - other	Fleurieu - other	Fleurieu - other	Fleurieu - other		-35.55	138.62	21	18.3	201	Warm
Kangaroo Island	Kangaroo Island	Kangaroo Island	Kangaroo Island	Parndana	-35.79	137.26	160	17.6	206	Warm
Langhorne Creek	Langhorne Creek	Langhorne Creek	Langhorne Creek	Langhorne Creek	-35.30	139.04	26	19.0	176	Warm
Limestone Coast - other	Limestone Coast - other	Limestone Coast - other	Limestone Coast - other		-36.96	140.74	70	17.5	225	Warm
Lower Murray - other	Lower Murray - other	Lower Murray - other	Lower Murray - other		-34.17	140.74	23	21.0	133	Hot
McLaren Vale	McLaren Vale	McLaren Vale	McLaren Vale	McLaren Vale	-35.22	138.55	86	18.7	202	Warm
Mount Benson	Mount Benson	Mount Benson	Mount Benson	Mount Benson	-37.06	139.84	15	16.9	218	Temp.
Mount Gambier			Mount Gambier	Mount Gambier	-37.83	140.78	36	16.3	281	Temp.
Mount Lofty Ranges - other	Mount Lofty Ranges - other	Mount Lofty Ranges - other	Mount Lofty Ranges - other		-35.06	138.83	363	17.0	298	Warm
Padthaway	Padthaway	Padthaway	Padthaway	Padthaway	-36.60	140.49	39	17.8	201	Warm
Riverland	Riverland	Riverland	Riverland	Renmark	-34.17	140.74	23	21.0	133	Hot
Robe		Robe	Robe	Robe	-37.16	139.75	2	16.8	211	Temp.
Southern Fleurieu	Southern Fleurieu	Southern Fleurieu	Southern Fleurieu	Victor Harbor	-35.55	138.62	21	18.3	201	Warm
Southern Flinders Ranges		Southern Flinders Ranges	Southern Flinders Ranges	Gladstone	-33.27	138.35	237	20.2	190	Hot
The Peninsulas	The Peninsulas	The Peninsulas	The Peninsulas	Maitland	-34.37	137.67	195	19.6	169	Hot
Wrattonbully		Wrattonbully	Wrattonbully	Naracoorte	-36.96	140.74	70	17.5	225	Warm
Tasmania	**Tasmania**	**Tasmania**	**Tasmania**	**Launceston**	**-41.54**	**147.14**	**166**	**14.4**	**326**	**Cool**
Victoria					**-35.76**	**143.80**	**130**	**19.1**	**273**	**Hot**
Alpine Valleys		Alpine Valleys	Alpine Valleys	Myrtleford	-36.56	146.72	226	18.0	439	Warm
Alpine Valleys/Beechworth	Alpine Valleys/Beechworth				-36.77	146.41	259	17.2	469	Warm
Beechworth	Beechworth	Beechworth	Beechworth	Beechworth	-36.36	146.69	547	16.8	434	Temp.
Bendigo	Bendigo	Bendigo	Bendigo	Bendigo	-36.76	144.28	218	17.5	271	Warm
Central Victoria - other	Central Victoria - other	Central Victoria - other	Central Victoria - other		-36.38	145.40	118	19.2	267	Hot
Geelong	Geelong	Geelong	Geelong	Geelong	-38.15	144.36	19	17.2	300	Warm
Gippsland	Gippsland	Gippsland	Gippsland	Lakes Entrance	-37.88	147.99	5	17.4	410	Warm
Glenrowan		Glenrowan	Glenrowan	Glenrowan	-36.46	146.22	233	18.7	341	Warm
Goulburn Valley	Goulburn Valley	Goulburn Valley	Goulburn Valley	Shepparton	-36.38	145.40	118	19.2	267	Hot
Grampians	Grampians	Grampians	Grampians	Ararat	-37.28	142.93	325	16.1	287	Temp.
Heathcote		Heathcote	Heathcote	Heathcote	-36.92	144.70	236	17.3	275	Warm
Henty	Henty	Henty	Henty	Hamilton	-37.74	142.02	192	15.9	291	Temp.
King Valley		King Valley	King Valley	Whitfield	-36.77	146.41	259	17.2	469	Warm
Macedon Ranges		Macedon Ranges	Macedon Ranges	Woodend	-37.36	144.52	573	14.9	391	Cool
Mornington Peninsula	Mornington Peninsula	Mornington Peninsula	Mornington Peninsula	Hastings	-38.31	145.19	11	16.9	389	Temp.
Murray Darling (VIC)	Murray Darling (VIC)	Murray Darling (VIC)	Murray Darling (VIC)	Mildura	-34.21	142.14	55	21.1	159	Hot

Table 75 (cont.): Geographic location, elevation and growing season average temperature and precipitation (GST, GSP) for nearest town to world's wine regions

Name	2000	2010	2016	Town	Latitude	Longitude	Elevation	GST	GSP	Zone
North East Victoria - other	North East Victoria - other	North East Victoria - other	North East Victoria - other		-36.77	146.41	259	17.2	469	Warm
North West Victoria - other	North West Victoria - other	North West Victoria - other	North West Victoria - other		-34.21	142.14	55	21.1	159	Hot
Port Phillip - other	Port Phillip - other	Port Phillip - other	Port Phillip - other		-38.31	145.19	11	16.9	389	Temp.
Pyrenees	Pyrenees	Pyrenees	Pyrenees	Moonambel	-36.99	143.33	279	16.6	273	Temp.
Rutherglen	Rutherglen	Rutherglen	Rutherglen	Rutherglen	-36.05	146.46	171	19.3	285	Hot
Strathbogie Ranges		Strathbogie Ranges	Strathbogie Ranges	Strathbogie	-36.86	145.86	480	15.5	490	Temp.
Sunbury	Sunbury	Sunbury	Sunbury	Sunbury	-37.58	144.71	243	17.0	332	Temp.
Swan Hill (VIC)	Swan Hill (VIC)	Swan Hill (VIC)	Swan Hill (VIC)	Swan Hill	-35.34	143.55	70	20.4	183	Hot
Upper Goulburn		Upper Goulburn	Upper Goulburn	Alexandra	-37.19	145.71	223	17.3	366	Warm
Western Victoria - other	Western Victoria - other	Western Victoria - other	Western Victoria - other		-37.74	142.02	192	15.9	291	Temp.
Yarra Valley	Yarra Valley	Yarra Valley	Yarra Valley	Healesville	-37.81	145.51	130	16.8	532	Temp.
Western Australia					**-33.95**	**115.78**	**85**	**18.8**	**256**	**Warm**
Blackwood Valley	Blackwood Valley	Blackwood Valley	Blackwood Valley	Nannup	-33.98	115.77	109	18.7	222	Warm
Central Western Australia	Central Western Australia	Central Western Australia	Central Western Australia	Northam	-31.65	116.67	155	22.2	120	Hot
Eastern Plains, Inland and North WA	Eastern Plains, Inland and North WA	Eastern Plains, Inland and North WA	Eastern Plains, Inland and North WA	Nyabing	-33.54	118.15	334	19.1	143	Hot
Geographe	Geographe	Geographe	Geographe	Bunbury	-33.32	115.64	2	20.0	178	Hot
Great Southern	Great Southern	Great Southern	Great Southern	Denmark	-34.96	117.35	36	17.6	338	Warm
Greater Perth - other	Greater Perth - other	Greater Perth - other	Greater Perth - other		-32.54	115.74	6	20.8	179	Hot
Manjimup		Manjimup	Manjimup	Manjimup	-34.24	116.15	282	17.8	250	Warm
Margaret River	Margaret River	Margaret River	Margaret River	Margaret River	-33.95	115.07	109	18.8	245	Warm
Peel		Peel	Peel	Mandurah	-32.54	115.74	6	20.8	179	Hot
Pemberton		Pemberton	Pemberton	Pemberton	-34.44	116.04	117	17.7	324	Warm
Perth Hills	Perth Hills	Perth Hills	Perth Hills	Mundaring	-31.90	116.16	295	19.8	206	Hot
South West Australia - other	South West Australia - other	South West Australia - other	South West Australia - other		-31.77	116.03	23	21.5	154	Hot
Swan District	Swan District	Swan District	Swan District	Upper Swan	-31.77	116.03	23	21.5	154	Hot
Western Australia Southeast Coast	Western Australia Southeast Coast	Western Australia Southeast Coast	Western Australia Southeast Coast	Esperance	-33.86	121.89	9	18.9	231	Warm
AUSTRIA	**AUSTRIA**	**AUSTRIA**	**AUSTRIA**		**48.15**	**16.19**	**213**	**15.5**	**438**	**Temp.**
Burgenland	**Burgenland**	**Burgenland**			**47.83**	**16.75**	**149**	**15.9**	**426**	**Temp.**
Mittelburgenland			Mittelburgenland	Horitschon	47.59	16.55	243	15.5	475	Temp.
Neusiedlersee			Neusiedlersee	Neusiedl am See	47.95	16.85	129	15.9	409	Temp.
Neusiedlersee Hügelland			Neusiedlersee Hügelland	Rust	47.80	16.67	122	16.0	422	Temp.
Südburgenland			Südburgenland	Eisenberg	47.18	16.43	247	15.4	519	Temp.
Niederösterreich	**Niederösterreich**	**Niederösterreich**			**48.49**	**16.02**	**210**	**15.4**	**403**	**Temp.**
Bergland			Bergland	Melk	48.23	15.35	264	15.0	471	Temp.
Carnuntum			Carnuntum	Stixneusiedl	48.04	16.68	170	15.9	420	Temp.
Kamptal			Kamptal	Langenlois	48.47	15.69	212	15.2	410	Temp.
Kremstal			Kremstal	Krems	48.42	15.60	207	14.9	437	Cool
Thermenregion			Thermenregion	Baden	48.00	16.23	229	15.5	434	Temp.
Traisental			Traisental	Traismauer	48.35	15.74	195	15.3	410	Temp.
Wachau			Wachau	Dürnstein	48.40	15.52	240	14.5	458	Cool
Wagram			Wagram	Tulln	48.33	16.06	179	15.7	397	Temp.
Weinviertel			Weinviertel	Mailberg	48.67	16.18	213	15.6	384	Temp.
Steiermark	**Steiermark**	**Steiermark**			**46.79**	**15.62**	**420**	**15.0**	**708**	**Temp.**
Steiermark - other			Steiermark - other		46.79	15.62	420	15.0	708	Temp.
Südsteiermark			Südsteiermark	Kitzeck im Sausal	46.78	15.45	543	14.9	764	Cool
Vulkanland Steiermark			Vulkanland Steiermark	Klöch	46.76	15.97	281	15.1	620	Temp.
Weststeiermark			Weststeiermark	Stainz	46.89	15.26	342	15.3	746	Temp.
Wien and other regions	**Wien and other regions**	**Wien and other regions**			**48.21**	**16.37**	**182**	**16.0**	**407**	**Temp.**
Wien			Wien	Wien	48.21	16.37	182	16.0	407	Temp.
Other regions			Other regions		48.21	16.37	182	16.0	407	Temp.
BRAZIL	**BRAZIL**	**BRAZIL**	**BRAZIL**	**Bento Gonçalves**	**-29.17**	**-51.52**	**657**	**19.6**	**999**	**Hot**
BULGARIA	**BULGARIA**	**BULGARIA**	**BULGARIA**		**42.38**	**24.69**	**223**	**18.3**	**317**	**Warm**

Table 75 (cont.): Geographic location, elevation and growing season average temperature and precipitation (GST, GSP) for nearest town to world's wine regions

Name	2000	2010	2016	Town	Latitude	Longitude	Elevation	GST	GSP	Zone
North Central		North Central	North Central	Svishtov	43.62	25.35	3	18.7	364	Warm
Northeast		Northeast	Northeast	Dobrich	43.57	27.83	219	17.3	312	Warm
Northwest		Northwest	Northwest	Vidin	44.00	22.87	38	18.4	360	Warm
South Central		South Central	South Central	Plodvid	42.13	24.75	179	18.8	315	Warm
Southeast		Southeast	Southeast	Melnik	41.52	23.39	378	18.2	301	Warm
Southwest		Southwest	Southwest	Burgas	42.50	27.46	23	18.5	314	Warm
CAMBODIA			**CAMBODIA**	**Phnom Penh**	**11.56**	**104.93**	**8**	**29.0**	**1150**	**Hot**
CANADA	**CANADA**	**CANADA**	**CANADA**		**45.70**	**-93.29**	**206**	**15.1**	**431**	**Temp.**
British Columbia		British Columbia	British Columbia	Summerland	49.61	-119.68	434	14.3	189	Cool
Nova Scotia			Nova Scotia	Port Williams	45.10	-64.41	15	13.3	625	Cool
Ontario		Ontario	Ontario	St Catherines	43.18	-79.25	79	15.8	553	Temp.
Quebec			Quebec	Quebec	46.81	-71.21	59	13.9	817	Cool
Other regions			Other regions		45.70	-93.29	206	15.1	431	Temp.
CHILE	**CHILE**	**CHILE**	**CHILE**		**-34.38**	**-71.19**	**282**	**17.9**	**103**	**Warm**
Antofagasta			Antofagasta	Antofagasta	-23.65	-70.40	28	19.3	0	Hot
Araucania	Araucania	Araucania	Araucania	Temuco	-38.74	-72.59	116	14.3	414	Cool
Arica			Arica	Arica	-18.48	-70.31	26	20.4	0	Hot
Atacama	Atacama	Atacama	Atacama	Copiapó	-27.37	-70.33	394	17.0	0	Warm
Coquimbo	Coquimbo	Coquimbo	Coquimbo	Ovalle	-30.60	-71.20	219	17.7	7	Warm
De Los Lagos		De Los Lagos	De Los Lagos	Puerto Montt	-41.47	-72.94	26	13.3	798	Cool
Del Bio Bio	Del Bio Bio	Del Bio Bio	Del Bio Bio	Chillan	-36.60	-72.10	129	16.7	247	Temp.
Del Maule	Del Maule	Del Maule	Del Maule	Curico	-34.97	-71.25	228	17.7	109	Warm
Metropolitana	Metropolitana	Metropolitana	Metropolitana	Santiago	-33.49	-70.67	508	19.0	47	Hot
O'Higgins	O'Higgins	O'Higgins	O'Higgins	San Fernando	-34.59	-70.99	339	18.3	112	Warm
Tarapaca			Tarapaca	Iquique	-20.24	-70.14	31	19.5	0	Hot
Valparaiso	Valparaiso	Valparaiso	Valparaiso	Casablanca	-33.32	-71.41	266	17.6	38	Warm
CHINA		**CHINA**	**CHINA**		**38.40**	**107.47**	**835**	**18.1**	**341**	**Warm**
Beijing		Beijing		Beijing	39.90	116.41	49	20.7	541	Hot
Gansu		Gansu		Lanzhou	36.06	103.83	1526	16.9	325	Temp.
Ningxia		Ningxia		Yinchuan	38.28	106.23	1111	17.2	192	Warm
Shandong		Shandong		Qingdao	36.07	120.38	66	20.0	609	Hot
ShanXi		ShanXi		Taiyuan	37.85	112.57	802	18.3	411	Warm
Sichuan		Sichuan		Chengdu	30.57	104.07	485	22.3	940	Hot
Tianjin		Tianjin		Tianjin	39.08	117.20	5	21.2	513	Hot
Xinjiang		Xinjiang		Ürümqi	43.83	87.62	838	18.1	173	Warm
Yantai		Yantai		Yantai	37.46	121.45	28	19.4	618	Hot
Other regions		Other regions			38.40	107.47	835	18.1	341	Warm
CROATIA	**CROATIA**	**CROATIA**	**CROATIA**		**45.06**	**16.32**	**79**	**18.5**	**504**	**Warm**
Dalmatinska Zagora		Dalmatinska Zagora		Sinj	43.70	16.64	331	18.4	483	Warm
Hrvatsko Primorje		Hrvatsko Primorje		Rijeka	45.33	14.44	60	19.1	773	Hot
Istra		Istra		Novigrad	45.32	13.56	2	19.3	538	Hot
Jadranska Hrvatska			Jadranska Hrvatska		45.06	16.32	79	18.5	504	Warm
Kontinentalna Hrvatska			Kontinentalna Hrvatska		45.06	16.32	79	18.5	504	Warm
Moslavina		Moslavina		Kutina	45.48	16.78	112	17.4	591	Warm
Plesivica		Plesivica		Plešivica	45.73	15.65	381	15.2	740	Temp.
Podunavlje		Podunavlje		Bilje	45.60	18.74	85	17.2	422	Warm
Pokuplje		Pokuplje		Karlovac	45.49	15.37	196	16.1	771	Temp.
Prigorje - Bilogora		Prigorje - Bilogora		Bjelovar	45.90	16.84	135	17.4	529	Warm
Sjeverna Dalmacija		Sjeverna Dalmacija		Zadar	44.12	15.23	7	19.8	471	Hot
Slavonija		Slavonija		Virovitica	45.83	17.39	122	17.3	533	Warm
Srednja Juzna Dalmacija		Srednja Juzna Dalmacija		Split	43.51	16.44	1	21.3	396	Hot
Zagorje-Medimurje		Zagorje-Medimurje		Krapina	46.16	15.87	179	15.7	713	Temp.
Other regions		Other regions			45.06	16.32	79	18.5	504	Warm
CYPRUS	**CYPRUS**	**CYPRUS**	**CYPRUS**	**Doros**	**34.82**	**32.91**	**479**	**21.2**	**117**	**Hot**
CZECHIA	**CZECHIA**	**CZECHIA**	**CZECHIA**		**48.92**	**16.01**	**284**	**15.3**	**390**	**Temp.**
Cechy		Cechy	Cechy	Kutná Hora	49.95	15.27	307	14.2	382	Cool
Jihovýchod			Jihovýchod	Znojmo	48.86	16.05	285	15.3	387	Temp.
Morava		Morava	Morava	Bzenec	48.97	17.27	196	15.1	479	Temp.
Praha			Praha	Prague	50.08	14.44	276	15.7	384	Temp.
Severozápad			Severozápad	Most	50.50	13.63	282	14.1	464	Cool
ETHIOPIA		**ETHIOPIA**	**ETHIOPIA**	**Addis Ababa**	**8.98**	**38.76**	**2296**	**17.0**	**942**	**Temp.**
FRANCE	**FRANCE**	**FRANCE**	**FRANCE**		**44.75**	**2.40**	**78**	**17.6**	**409**	**Warm**
Alsace	**Alsace**		**Alsace**		**48.07**	**7.40**	**194**	**15.6**	**420**	**Temp.**
Bas-Rhin		Bas-Rhin		Sélestat	48.26	7.45	176	15.6	447	Temp.

Table 75 (cont.): Geographic location, elevation and growing season average temperature and precipitation (GST, GSP) for nearest town to world's wine regions

Name	2000	2010	2016	Town	Latitude	Longitude	Elevation	GST	GSP	Zone
Haut-Rhin		Haut-Rhin		Colmar	47.93	7.36	207	15.6	400	Temp.
Aquitaine			**Aquitaine**		**44.39**	**-0.50**	**50**	**16.9**	**459**	**Temp.**
Aquitaine except Gironde	Aquitaine except Gironde				44.45	0.28	63	17.0	524	Warm
Dordogne		Dordogne		Bergerac	44.85	0.48	39	16.9	490	Temp.
Gironde	Gironde	Gironde		Merignac	44.38	-0.66	47	16.9	445	Temp.
Landes		Landes		Villeneuve-de-Marsan	43.89	-0.31	89	16.9	543	Temp.
Lot-et-Garonne		Lot-et-Garonne		Buzet	44.25	0.30	47	17.0	507	Warm
Pyrenees-Atlantiques		Pyrenees-Atlantiques		Pau	43.30	-0.37	206	17.7	736	Warm
Auvergne	**Auvergne**		**Auvergne**		**45.96**	**3.21**	**358**	**15.5**	**459**	**Temp.**
Allier		Allier		St Pourcain sur Sioule	46.31	3.29	252	15.6	475	Temp.
Cantal		Cantal		Aurillac	44.93	2.44	630	14.1	487	Cool
Haute-Loire		Haute-Loire		Le Puy	45.04	3.88	637	14.1	487	Cool
Puy-de-Dome		Puy-de-Dome		Clermont-Ferrand	45.78	3.09	415	15.5	443	Temp.
Bourgogne	**Bourgogne**		**Bourgogne**		**47.00**	**4.56**	**193**	**15.7**	**470**	**Temp.**
Cote-d'Or		Cote-d'Or		Dijon	47.27	5.04	219	15.6	459	Temp.
Nievre		Nievre		Cosne-Cours-sur-Loire	47.38	2.93	179	15.4	424	Temp.
Saone-et-Loire		Saone-et-Loire		Mâcon	46.31	4.83	197	16.2	509	Temp.
Yonne		Yonne		Chablis	47.82	3.80	152	15.2	423	Temp.
Centre-Val de Loire	**Centre-Val de Loire**		**Centre-Val de Loire**		**47.40**	**1.26**	**109**	**15.7**	**386**	**Temp.**
Cher		Cher		Bourges	47.06	2.40	161	15.8	421	Temp.
Eure-et-Loire		Eure-et-Loire		Chartres	48.44	1.49	155	14.9	354	Cool
Indre		Indre		Châteauroux	46.81	1.69	155	15.8	438	Temp.
Indre-et-Loire		Indre-et-Loire		Tours	47.44	0.68	108	15.6	373	Temp.
Loiret		Loiret		Orléans	47.90	1.91	114	15.6	371	Temp.
Loir-et-Cher		Loir-et-Cher		Blois	47.59	1.34	75	15.6	377	Temp.
Champagne-Ardenne	**Champagne-Ardenne**		**Champagne-Ardenne**		**48.99**	**4.18**	**147**	**14.6**	**399**	**Cool**
Aube		Aube		Fontette	48.08	4.61	306	14.3	480	Cool
Haute-Marne		Haute-Marne		Chaumont	48.11	5.14	290	14.5	443	Cool
Marne		Marne		Reims	49.31	4.03	91	14.7	370	Cool
Corse	**Corse**		**Corse**		**42.27**	**9.34**	**11**	**17.8**	**318**	**Warm**
Corse-du-Sud		Corse-du-Sud		Ajaccio	41.92	8.74	16	19.1	282	Hot
Haute-Corse		Haute-Corse		Bastia	42.33	9.45	10	17.6	324	Warm
Franche Comté	**Franche Comté**		**Franche Comté**		**46.75**	**5.60**	**261**	**15.9**	**681**	**Temp.**
Doubs		Doubs		Besançon	47.23	6.02	354	15.3	666	Temp.
Haute-Saone		Haute-Saone		Vesoul	47.62	6.15	220	15.1	603	Temp.
Jura		Jura		Lons-le-Saunier	46.67	5.55	262	16.0	687	Temp.
Île de France			**Île de France**		**48.95**	**3.13**	**63**	**14.7**	**372**	**Cool**
Seine-et-Marne	Seine-et-Marne	Seine-et-Marne		La Ferté-sous-Jouarre	48.95	3.13	63	14.7	372	Cool
Languedoc Roussillon			**Languedoc Roussillon**		**43.34**	**3.20**	**53**	**18.9**	**348**	**Warm**
Aude	Aude	Aude		Carcassonne	43.21	2.35	115	18.2	362	Warm
Gard	Gard	Gard		Nîmes	43.84	4.36	45	19.2	435	Hot
Herault	Herault	Herault		Beziers	43.32	3.22	15	19.1	301	Hot
Lozere		Lozere		Mende	44.52	3.50	753	13.2	516	Cool
Pyrenees-Orientales	Pyrenees-Orientales	Pyrenees-Orientales		Perpignan	42.74	2.89	42	19.6	290	Hot
Limousin			**Limousin**		**45.27**	**1.77**	**248**	**15.8**	**510**	**Temp.**
Correze		Correze		Tulle	45.27	1.77	248	15.8	510	Temp.
Correze, Haute-Vienne	Correze, Haute-Vienne			Tulle	45.27	1.77	248	15.8	510	Temp.
Lorraine	**Lorraine**		**Lorraine**		**48.74**	**6.17**	**211**	**14.9**	**449**	**Cool**
Meurthe-et-Moselle		Meurthe-et-Moselle		Nancy	48.69	6.18	198	14.9	447	Cool
Meuse		Meuse		Toul	48.68	5.89	211	14.7	425	Cool
Moselle		Moselle		Metz	49.12	6.18	180	15.1	430	Temp.
Vosges		Vosges		Épinal	48.17	6.44	327	14.4	523	Cool
Midi Pyrénées			**Midi Pyrénées**		**43.85**	**1.11**	**153**	**17.4**	**493**	**Warm**
Midi-Pyrenees except Gers	Midi-Pyrenees except Gers				44.04	1.59	156	17.4	428	Warm
Ariege		Ariege		Foix	42.96	1.61	394	16.1	499	Temp.
Aveyron		Aveyron		Estaing	44.55	2.67	331	15.5	405	Temp.
Gers	Gers	Gers		Auch	43.65	0.59	151	17.4	561	Warm
Haute-Garonne		Haute-Garonne		Toulouse	43.60	1.44	145	18.0	396	Warm

Table 75 (cont.): Geographic location, elevation and growing season average temperature and precipitation (GST, GSP) for nearest town to world's wine regions

Name	**2000**	**2010**	**2016**	**Town**	**Latitude**	**Longitude**	**Elevation**	**GST**	**GSP**	**Zone**
Hautes-Pyrenees		Hautes-Pyrenees		Tarbes	43.23	0.08	309	17.4	663	Warm
Lot		Lot		Cahors	44.45	1.44	165	16.9	459	Temp.
Tarn		Tarn		Gaillac	43.90	1.90	145	17.8	397	Warm
Tarn-et-Garonne		Tarn-et-Garonne		Montauban	44.02	1.35	91	17.5	420	Warm
Pays de la Loire			**Pays de la Loire**		**47.31**	**-0.99**	**33**	**15.9**	**385**	**Temp.**
Pays de la Loire except Mayenne	Pays de la Loire except Mayenne				47.31	-0.99	33	15.9	385	Temp.
Loire-Atlantique		Loire-Atlantique		Nantes	47.15	-1.55	26	15.9	396	Temp.
Maine-et-Loire		Maine-et-Loire		Angers	47.47	-0.55	36	15.9	377	Temp.
Mayenne		Mayenne		Laval	48.08	-0.77	64	15.3	397	Temp.
Sarthe		Sarthe		Le Mans	48.01	-0.20	69	15.3	383	Temp.
Vendee		Vendee		La Roche-sur-Yon	46.67	-1.43	57	15.9	383	Temp.
Picardie			**Picardie**		**49.38**	**3.32**	**51**	**14.8**	**377**	**Cool**
Aisne	Aisne	Aisne		Soissons	49.38	3.32	51	14.8	377	Cool
Poitou Charentes			**Poitou Charentes**		**45.94**	**-0.47**	**58**	**16.6**	**403**	**Temp.**
Charente	Charente	Charente		Angoulême	45.65	0.16	103	16.5	431	Temp.
Charente-Maritime	Charente-Maritime	Charente-Maritime		La Rochelle	46.16	-1.15	8	16.8	376	Temp.
Deux-Sevres		Deux-Sevres		Thouars	46.98	-0.22	90	15.8	393	Temp.
Deux-Sevres, Vienne	Deux-Sevres, Vienne				46.72	0.14	105	15.8	386	Temp.
Vienne		Vienne		Poitiers	46.58	0.34	114	15.8	382	Temp.
Provence-Alpes-Cote d'Azur			**Provence-Alpes-Cote d'Azur**		**43.61**	**5.26**	**51**	**19.3**	**387**	**Hot**
Alpes-de-Haute-Provence		Alpes-de-Haute-Provence		Pierrevert	43.81	5.75	420	18.0	401	Warm
Alpes-de-Haute-Provence, Hautes-Alpes, Alpes-Maritimes	Alpes-de-Haute-Provence, Hautes-Alpes, Alpes-Maritimes				43.94	6.00	433	17.6	418	Warm
Alpes-Maritimes		Alpes-Maritimes		Nice	43.71	7.26	25	19.0	391	Warm
Bouches-du-Rhône	Bouches-du-Rhône	Bouches-du-Rhône		Aix-en-Provence	43.53	5.45	197	19.0	353	Warm
Hautes-Alpes		Hautes-Alpes		Gap	44.56	6.08	746	14.9	499	Cool
Var	Var	Var		Toulon	43.12	5.93	11	19.2	321	Hot
Vaucluse	Vaucluse	Vaucluse		Avignon	43.91	4.81	34	19.5	432	Hot
Rhône Alpes			**Rhône Alpes**		**45.31**	**4.91**	**169**	**17.2**	**542**	**Warm**
Rhône-Alpes except Ardeche	Rhône-Alpes except Ardeche				45.38	4.93	181	17.3	542	Warm
Ain		Ain		Belley	45.76	5.69	292	16.3	616	Temp.
Ardeche	Ardeche	Ardeche		Tournon	45.07	4.83	124	17.1	543	Warm
Drome		Drome		Valence	44.91	4.89	160	17.6	552	Warm
Haute-Savoie		Haute-Savoie		Annecy	45.90	6.13	455	15.5	562	Temp.
Isere		Isere		Grenoble	45.19	5.72	209	17.5	563	Warm
Loire		Loire		Boën-sur-Lignon	45.75	4.00	407	15.3	487	Temp.
Rhône		Rhône		Lyon	45.76	4.84	173	17.4	525	Warm
Savoie		Savoie		Chambery	45.64	5.92	235	15.7	594	Temp.
GEORGIA	**GEORGIA**	**GEORGIA**	**GEORGIA**	**Telavi**	**41.91**	**45.48**	**716**	**16.6**	**581**	**Temp.**
GERMANY	**GERMANY**	**GERMANY**	**GERMANY**		**49.41**	**8.34**	**187**	**15.0**	**443**	**Cool**
Ahr	Ahr	Ahr	Ahr	Ahrweiler	50.54	7.12	108	13.9	438	Cool
Baden	Baden	Baden	Baden	Baden-Baden	48.77	8.23	327	15.0	684	Cool
Franken	Franken	Franken	Franken	Würzburg	49.79	9.95	275	14.4	373	Cool
Hessische Bergstraße	Hessische Bergstraße	Hessische Bergstraße	Hessische Bergstraße	Bensheim	49.69	8.62	107	15.6	427	Temp.
Mittelrhein	Mittelrhein	Mittelrhein	Mittelrhein	Bad Ems	50.34	7.71	149	13.7	504	Cool
Mosel	Mosel	Mosel	Mosel	Trier	49.75	6.64	137	14.4	462	Cool
Nahe	Nahe	Nahe	Nahe	Schloßböckelheim	49.81	7.74	158	14.1	357	Cool
Pfalz	Pfalz	Pfalz	Pfalz	Bad Dürkheim	49.46	8.17	136	15.1	361	Temp.
Rheingau	Rheingau	Rheingau	Rheingau	Geisenheim	49.59	7.97	115	14.2	367	Cool
Rheinhessen	Rheinhessen	Rheinhessen	Rheinhessen	Worms	49.63	8.35	105	15.6	386	Temp.
Saale	Saale	Saale	Saale	Naumburg	51.15	11.81	131	14.4	378	Cool
Sachsen	Sachsen	Sachsen	Sachsen	Dresden	51.05	13.74	140	15.3	435	Temp.
Württemberg	Württemberg	Württemberg	Württemberg	Stuttgart	48.78	9.18	314	14.5	482	Cool
GREECE	**GREECE**	**GREECE**	**GREECE**		**38.27**	**23.36**	**148**	**21.2**	**194**	**Hot**
Anatoliki Makedonia, Thraki	Anatoliki Makedonia, Thraki	Anatoliki Makedonia, Thraki	Anatoliki Makedonia, Thraki	Maronia	40.90	25.52	181	19.2	248	Hot

Table 75 (cont.): Geographic location, elevation and growing season average temperature and precipitation (GST, GSP) for nearest town to world's wine regions

Name	2000	2010	2016	Town	Latitude	Longitude	Elevation	GST	GSP	Zone
Attiki	Attiki	Attiki	Attiki	Athens	37.98	23.73	69	23.0	139	Hot
Dytiki Ellada	Dytiki Ellada	Dytiki Ellada	Dytiki Ellada	Patras	38.25	21.73	1	21.4	223	Hot
Dytiki Makedonia	Dytiki Makedonia	Dytiki Makedonia	Dytiki Makedonia	Amindeo	40.69	21.68	594	18.0	330	Warm
Ionia Nisia	Ionia Nisia	Ionia Nisia	Ionia Nisia	Argostoli	38.17	20.49	9	21.4	262	Hot
Ipeiros	Ipeiros	Ipeiros	Ipeiros	Zitsa	39.75	20.65	684	18.8	414	Warm
Kentriki Makedonia	Kentriki Makedonia	Kentriki Makedonia	Kentriki Makedonia	Goumenissa	40.95	22.45	231	19.7	274	Hot
Kriti	Kriti	Kriti	Kriti	Iraklion	35.34	25.14	12	22.6	120	Hot
Notio Aigaio	Notio Aigaio	Notio Aigaio	Notio Aigaio	Paros	37.09	25.15	3	22.1	87	Hot
Peloponnissos	Peloponnissos	Peloponnissos	Peloponnissos	Nemea	37.82	22.66	309	20.3	230	Hot
Sterea Ellada	Sterea Ellada	Sterea Ellada	Sterea Ellada	Halkída	38.47	23.62	6	21.9	161	Hot
Thessalia	Thessalia	Thessalia	Thessalia	Rapsani	39.90	22.55	494	19.9	245	Hot
Voreio Aigaio	Voreio Aigaio	Voreio Aigaio	Voreio Aigaio	Limnos	39.92	25.14	144	20.1	184	Hot
HUNGARY	**HUNGARY**	**HUNGARY**	**HUNGARY**		**47.14**	**19.36**	**135**	**16.7**	**386**	**Temp.**
Badacsony		Badacsony	Badacsony	Badacsonytomaj	46.80	17.51	155	16.5	431	Temp.
Balatonboglar		Balatonboglar	Balatonboglar	Balatonboglár	46.77	17.66	138	16.8	415	Temp.
Balatonfelvidek		Balatonfelvidek	Balatonfelvidek	Monostorapáti	46.93	17.56	154	16.0	435	Temp.
Balatonfured-Csopak		Balatonfured-Csopak	Balatonfured-Csopak	Balatonfüred	46.96	17.89	139	16.2	412	Temp.
Bukk		Bukk	Bukk	Miskolc	48.10	20.78	215	15.9	396	Temp.
Csongrad		Csongrad	Csongrad	Hódmezővásárhely	46.42	20.33	76	17.0	350	Temp.
Eger		Eger	Eger	Eger	47.90	20.38	163	16.7	388	Temp.
Etyek-Budai		Etyek-Budai	Etyek-Budai	Etyek	47.45	18.75	178	16.8	367	Temp.
Hajos-Bajai		Hajos-bajai	Hajos-Bajai	Baja	46.18	18.95	99	17.2	382	Warm
Kunsag		Kunsag	Kunsag	Kecskemét	46.90	19.69	115	16.9	359	Temp.
Matra		Matra	Matra	Gyöngyös	47.78	19.93	164	16.8	374	Temp.
Mor		Mor	Mor	Mór	47.37	18.20	177	16.4	378	Temp.
Nagy-Somlo		Nagy-Somlo	Nagy-Somlo	Somlóvásárhely	47.12	17.38	152	16.1	422	Temp.
Neszmely		Neszmely	Neszmely	Neszmély	47.73	18.35	161	16.4	367	Temp.
Pannonhalma		Pannonhalma	Pannonhalma	Pannonhalma	47.55	17.76	173	16.2	385	Temp.
Pecs		Pecs	Pecs	Pécs	46.07	18.23	207	17.2	428	Warm
Sopron		Sopron	Sopron	Sopron	47.68	16.58	217	15.6	473	Temp.
Szekszard		Szekszard	Szekszard	Szekszárd	46.35	18.71	100	17.2	383	Warm
Tokaj		Tokaj	Tokaj	Tokaj	48.10	21.41	133	15.8	404	Temp.
Tolna		Tolna	Tolna	Paks	46.61	18.85	93	17.3	371	Warm
Villany		Villany	Villany	Villány	45.87	18.45	110	17.0	426	Warm
Zala		Zala	Zala	Nagykanizsa	46.46	16.99	157	16.0	510	Temp.
INDIA			**INDIA**	**Nashik**	**20.00**	**73.79**	**589**	**26.9**	**1137**	**Hot**
ISRAEL	**ISRAEL**	**ISRAEL**	**ISRAEL**	**Jerusalem**	**31.77**	**35.21**	**773**	**21.2**	**36**	**Hot**
ITALY	**ITALY**	**ITALY**	**ITALY**		**42.44**	**12.56**	**123**	**19.6**	**382**	**Hot**
Abruzzo			**Abruzzo**		**42.39**	**14.12**	**290**	**19.3**	**390**	**Hot**
Chieti	Chieti	Chieti		Chieti	42.35	14.16	320	19.4	382	Hot
L'Aquila	L'Aquila	L'Aquila		L'Aquila	42.35	13.40	712	17.0	574	Warm
Pescara	Pescara	Pescara		Pescara	42.46	14.22	7	19.6	354	Hot
Teramo	Teramo	Teramo		Teramo	42.66	13.70	270	18.5	480	Warm
Basilicata			**Basilicata**		**40.65**	**16.01**	**706**	**17.2**	**307**	**Warm**
Matera	Matera	Matera		Matera	40.67	16.60	393	19.0	296	Hot
Potenza	Potenza	Potenza		Potenza	40.64	15.81	813	16.6	311	Temp.
Calabria			**Calabria**		**39.01**	**16.48**	**150**	**20.9**	**313**	**Hot**
Catanzaro	Catanzaro	Catanzaro		Catanzaro	38.91	16.59	304	20.9	334	Hot
Cosenza	Cosenza	Cosenza		Cosenza	39.30	16.25	246	20.0	362	Hot
Crotone	Crotone	Crotone		Crotone	39.08	17.13	7	21.4	240	Hot
Reggio di Calabria	Reggio di Calabria	Reggio di Calabria		Reggio di Calabria	38.11	15.65	29	23.2	315	Hot
Vibo Valentia	Vibo Valentia	Vibo Valentia		Vibo Valentia	38.68	16.10	487	19.7	362	Hot
Campania			**Campania**		**40.99**	**14.71**	**173**	**19.5**	**398**	**Hot**
Avellino	Avellino	Avellino		Avellino	40.91	14.79	352	18.4	392	Warm
Benevento	Benevento	Benevento		Benevento	41.13	14.79	160	19.6	378	Hot
Caserta	Caserta	Caserta		Caserta	41.07	14.33	66	20.3	422	Hot
Napoli	Napoli	Napoli		Napoli	40.85	14.27	6	20.4	441	Hot
Salerno	Salerno	Salerno		Salerno	40.68	14.77	56	20.2	438	Hot
Emilia-Romagna			**Emilia-Romagna**		**44.54**	**11.39**	**34**	**19.4**	**422**	**Hot**
Bologna	Bologna	Bologna		Bologna	44.49	11.34	91	19.6	423	Hot
Ferrara	Ferrara	Ferrara		Ferrara	44.84	11.62	7	20.3	374	Hot
Forlì-Cesena	Forlì-Cesena	Forlì-Cesena		Forlì	44.22	12.04	27	19.5	434	Hot
Modena	Modena	Modena		Modena	44.65	10.93	37	19.5	434	Hot
Parma	Parma	Parma		Parma	44.80	10.33	57	19.2	439	Hot
Piacenza	Piacenza	Piacenza		Piacenza	45.05	9.69	59	18.6	481	Warm
Ravenna	Ravenna	Ravenna		Ravenna	44.42	12.20	-3	19.7	384	Hot
Reggio nell'Emilia	Reggio nell'Emilia	Reggio nell'Emilia		Reggio nell'Emilia	44.70	10.63	56	19.2	430	Hot

Table 75 (cont.): Geographic location, elevation and growing season average temperature and precipitation (GST, GSP) for nearest town to world's wine regions

Name	2000	2010	2016	Town	Latitude	Longitude	Elevation	GST	GSP	Zone
Rimini	Rimini	Rimini		Rimini	44.07	12.57	3	19.2	432	Hot
Friuli-Venezia Giulia			**Friuli-Venezia Giulia**		**45.99**	**13.09**	**70**	**18.5**	**857**	**Warm**
Gorizia	Gorizia	Gorizia		Gorizia	45.94	13.62	83	18.3	920	Warm
Pordenone	Pordenone	Pordenone		Pordenone	45.96	12.66	24	18.6	714	Warm
Trieste	Trieste	Trieste		Trieste	45.65	13.78	13	19.1	665	Hot
Udine	Udine	Udine		Udine	46.06	13.23	113	18.4	977	Warm
Lazio			**Lazio**		**41.89**	**12.64**	**128**	**19.7**	**332**	**Hot**
Frosinone	Frosinone	Frosinone		Frosinone	41.63	13.34	175	19.3	537	Hot
Latina	Latina	Latina		Latina	41.47	12.90	27	20.4	388	Hot
Rieti	Rieti	Rieti		Rieti	42.40	12.86	391	18.3	381	Warm
Roma	Roma	Roma		Roma	41.90	12.50	58	20.1	317	Hot
Viterbo	Viterbo	Viterbo		Viterbo	42.42	12.13	328	18.3	153	Warm
Liguria			**Liguria**		**44.12**	**8.95**	**5**	**19.4**	**453**	**Hot**
Genova	Genova	Genova		Genova	44.41	8.95	6	19.4	650	Hot
Imperia	Imperia	Imperia		Imperia	43.89	8.04	9	19.3	361	Hot
La Spezia	La Spezia	La Spezia		La Spezia	44.10	9.82	-3	19.3	455	Hot
Savona	Savona	Savona		Savona	44.30	8.46	10	19.8	459	Hot
Lombardia			**Lombardia**		**45.33**	**9.61**	**102**	**18.4**	**511**	**Warm**
Bergamo	Bergamo	Bergamo		Bergamo	45.70	9.68	257	18.1	731	Warm
Brescia	Brescia	Brescia		Brescia	45.54	10.21	138	18.8	593	Warm
Como	Como	Como		Como	45.81	9.09	206	16.3	888	Temp.
Cremona	Cremona	Cremona		Cremona	45.13	10.02	42	18.9	490	Warm
Lecco	Lecco	Lecco		Lecco	45.86	9.40	222	16.9	829	Temp.
Lodi	Lodi	Lodi		Lodi	45.31	9.50	80	18.4	534	Warm
Mantova	Mantova	Mantova		Mantova	45.16	10.79	21	19.3	414	Hot
Milano	Milano	Milano		Milano	45.46	9.19	110	19.0	632	Hot
Monza e della Brianza		Monza e della Brianza		Monza	45.58	9.27	168	18.4	678	Warm
Pavia	Pavia	Pavia		Pavia	45.18	9.16	74	18.4	463	Warm
Sondrio	Sondrio	Sondrio		Sondrio	46.17	9.88	299	13.5	602	Cool
Varese	Varese	Varese		Varese	45.82	8.83	386	17.0	862	Temp.
Marche			**Marche**		**43.26**	**13.44**	**180**	**18.9**	**433**	**Warm**
Ancona	Ancona	Ancona		Ancona	43.62	13.52	27	19.1	404	Hot
Ascoli Piceno	Ascoli Piceno	Ascoli Piceno		Ascoli Piceno	42.85	13.57	156	19.1	449	Hot
Fermo		Fermo		Fermo	43.16	13.72	227	19.0	399	Warm
Macerata	Macerata	Macerata		Macerata	43.30	13.45	289	18.5	408	Warm
Pesaro e Urbino	Pesaro e Urbino	Pesaro e Urbino		Urbino	43.74	12.64	452	18.3	507	Warm
Molise			**Molise**		**41.56**	**14.62**	**679**	**17.5**	**304**	**Warm**
Campobasso	Campobasso	Campobasso		Campobasso	41.56	14.66	698	17.4	297	Warm
Isernia	Isernia	Isernia		Isernia	41.60	14.23	474	18.5	382	Warm
Piemonte			**Piemonte**		**44.74**	**8.07**	**272**	**18.1**	**433**	**Warm**
Alessandria	Alessandria	Alessandria		Alessandria	44.91	8.61	96	18.5	452	Warm
Asti	Asti	Asti		Asti	44.90	8.21	124	18.4	392	Warm
Biella	Biella	Biella		Biella	45.56	8.06	416	16.8	577	Temp.
Cuneo	Cuneo	Cuneo		Cuneo	44.38	7.54	551	17.5	440	Warm
Novara	Novara	Novara		Novara	45.45	8.62	168	18.0	568	Warm
Torino	Torino	Torino		Torino	45.20	7.69	302	17.7	556	Warm
Verbano-Cusio-Ossola	Verbano-Cusio-Ossola	Verbano-Cusio-Ossola		Domodossola	46.11	8.29	277	15.0	727	Cool
Vercelli	Vercelli	Vercelli		Vercelli	45.32	8.42	131	18.0	516	Warm
Puglia			**Puglia**		**41.01**	**16.68**	**36**	**20.7**	**250**	**Hot**
Bari	Bari	Bari		Bari	41.12	16.87	6	20.5	276	Hot
Barletta-Andria-Trani		Barletta-Andria-Trani		Barletta	41.32	16.28	13	20.6	233	Hot
Brindisi	Brindisi	Brindisi		Brindisi	40.65	17.94	10	21.0	213	Hot
Foggia	Foggia	Foggia		Foggia	41.46	15.54	71	20.4	265	Hot
Lecce	Lecce	Lecce		Lecce	40.35	18.18	47	21.2	244	Hot
Taranto	Taranto	Taranto		Taranto	40.46	17.25	30	20.9	257	Hot
Sardegna			**Sardegna**		**39.96**	**9.01**	**168**	**20.4**	**209**	**Hot**
Cagliari	Cagliari	Cagliari		Cagliari	39.22	9.12	27	21.2	170	Hot
Carbonia-Iglesias		Carbonia-Iglesias		Carbonia	39.16	8.52	98	21.0	225	Hot
Medio Campidano		Medio Campidano		Sardara	39.61	8.82	147	21.0	213	Hot
Nuoro	Nuoro	Nuoro		Nuoro	40.32	9.33	536	18.9	247	Warm
Ogliastra		Ogliastra		Baunei	40.03	9.66	478	20.5	156	Hot
Olbia-Tempio		Olbia-Tempio		Olbia	40.92	9.50	7	20.4	209	Hot
Oristano	Oristano	Oristano		Oristano	39.91	8.59	7	20.8	210	Hot
Sassari	Sassari	Sassari		Sassari	40.73	8.56	205	19.5	254	Hot
Sicilia			**Sicilia**		**37.86**	**13.05**	**78**	**21.5**	**187**	**Hot**
Agrigento	Agrigento	Agrigento		Agrigento	37.31	13.59	242	21.2	169	Hot
Caltanissetta	Caltanissetta	Caltanissetta		Caltanissetta	37.49	14.06	576	20.1	195	Hot

Table 75 (cont.): Geographic location, elevation and growing season average temperature and precipitation (GST, GSP) for nearest town to world's wine regions

Name	2000	2010	2016	Town	Latitude	Longitude	Elevation	GST	GSP	Zone
Catania	Catania	Catania		Catania	37.51	15.08	30	21.5	205	Hot
Enna	Enna	Enna		Enna	37.57	14.28	913	19.0	212	Hot
Messina	Messina	Messina		Messina	38.19	15.55	9	22.7	330	Hot
Palermo	Palermo	Palermo		Palermo	38.12	13.36	9	21.5	197	Hot
Ragusa	Ragusa	Ragusa		Ragusa	36.93	14.73	530	20.0	196	Hot
Siracusa	Siracusa	Siracusa		Siracusa	37.08	15.29	23	21.9	208	Hot
Trapani	Trapani	Trapani		Trapani	38.02	12.54	6	21.7	186	Hot
Toscana			**Toscana**		**43.48**	**11.22**	**157**	**19.1**	**419**	**Hot**
Arezzo	Arezzo	Arezzo		Arezzo	43.46	11.88	278	18.5	451	Warm
Firenze	Firenze	Firenze		Firenze	43.80	11.26	44	19.3	475	Hot
Grosseto	Grosseto	Grosseto		Grosseto	42.76	11.11	12	19.9	312	Hot
Livorno	Livorno	Livorno		Livorno	43.55	10.31	6	19.4	383	Hot
Lucca	Lucca	Lucca		Lucca	43.84	10.50	26	19.3	501	Hot
Massa-Carrara	Massa-Carrara	Massa-Carrara		Massa	44.04	10.14	56	19.3	539	Hot
Pisa	Pisa	Pisa		Pisa	43.72	10.40	7	19.3	488	Hot
Pistoia	Pistoia	Pistoia		Pistoia	43.93	10.91	65	19.3	501	Hot
Prato	Prato	Prato		Prato	43.88	11.10	62	19.7	490	Hot
Siena	Siena	Siena		Siena	43.32	11.33	347	18.6	372	Warm
Trentino-Alto Adige					**46.18**	**11.19**	**212**	**15.4**	**542**	**Temp.**
Bolzano-Bozen	Bolzano-Bozen	Bolzano-Bozen	Bolzano-Bozen	Bolzano-Bozen	46.46	11.35	241	14.6	549	Cool
Trento	Trento	Trento	Trento	Trento	46.07	11.12	200	15.8	539	Temp.
Umbria			**Umbria**		**42.89**	**12.49**	**176**	**18.7**	**368**	**Warm**
Perugia	Perugia	Perugia		Perugia	43.10	12.39	208	18.2	403	Warm
Terni	Terni	Terni		Terni	42.56	12.64	126	19.4	314	Hot
Valle d'Aosta	**Valle d'Aosta**	**Valle d'Aosta**	**Valle d'Aosta**	**Aosta**	**45.73**	**7.31**	**580**	**14.4**	**449**	**Cool**
Veneto			**Veneto**		**45.51**	**11.69**	**38**	**19.3**	**563**	**Hot**
Belluno	Belluno	Belluno		Belluno	46.14	12.22	395	15.8	681	Temp.
Padova	Padova	Padova		Padova	45.41	11.88	-1	19.3	519	Hot
Rovigo	Rovigo	Rovigo		Rovigo	45.07	11.79	0	19.6	415	Hot
Treviso	Treviso	Treviso		Treviso	45.67	12.24	17	19.1	596	Hot
Venezia	Venezia	Venezia		Venezia	45.44	12.32	10	19.0	481	Hot
Verona	Verona	Verona		Verona	45.38	10.99	73	19.4	535	Hot
Vicenza	Vicenza	Vicenza		Vicenza	45.55	11.54	41	19.7	646	Hot
JAPAN		**JAPAN**	**JAPAN**		**38.27**	**139.53**	**215**	**18.0**	**846**	**Warm**
Hokkaido		Hokkaido	Hokkaido	Sapporo	43.06	141.35	20	15.6	769	Temp.
Iwate			Iwate	Morioka	39.70	141.15	150	17.3	948	Warm
Nagano		Nagano	Nagano	Nagano	36.65	138.20	389	18.7	795	Warm
Niigata			Niigata	Niigata	37.92	139.04	16	19.7	1003	Hot
Yamagata		Yamagata	Yamagata	Yamagata	38.26	140.34	168	18.6	831	Warm
Yamanashi		Yamanashi	Yamanashi	Kofu	35.67	138.57	281	19.1	918	Hot
Other regions		Other regions	Other regions		38.27	139.53	215	18.0	846	Warm
KAZAKHSTAN		**KAZAKHSTAN**	**KAZAKHSTAN**		**43.24**	**73.99**	**647**	**18.8**	**264**	**Warm**
Almaty		Almaty	Almaty	Almaty	43.22	76.85	852	17.7	357	Warm
East Kazakhstan		East Kazakhstan	East Kazakhstan	Ust'-Kamenogorsk	49.97	82.60	292	13.2	299	Cool
South Kazakhstan		South Kazakhstan	South Kazakhstan	Turkistan	43.31	68.23	217	21.1	78	Hot
West Kazakhstan		West Kazakhstan	West Kazakhstan	Uralsk	51.23	51.39	34	15.7	209	Temp.
Zhambyl		Zhambyl	Zhambyl	Taraz	42.90	71.40	625	19.0	162	Hot
Other regions		Other regions	Other regions		43.24	73.99	647	18.8	264	Warm
KOREA, REP.	**KOREA, REP.**	**KOREA, REP.**	**KOREA, REP.**	**Seoul**	**37.57**	**126.98**	**56**	**19.7**	**1192**	**Hot**
LEBANON			**LEBANON**	**Chtaura**	**33.82**	**35.85**	**910**	**21.7**	**102**	**Hot**
LUXEMBOURG	**LUXEMBOURG**	**LUXEMBOURG**	**LUXEMBOURG**	**Luxembourg**	**49.61**	**6.13**	**325**	**14.2**	**474**	**Cool**
MEXICO		**MEXICO**	**MEXICO**		**29.08**	**-111.60**	**682**	**22.2**	**176**	**Hot**
Aguascalientes		Aguascalientes	Aguascalientes	Aguascalientes	21.89	-102.29	1879	19.9	436	Hot
Baja California		Baja California	Baja California	Guadalupe	32.10	-116.57	334	20.6	49	Hot
Coahuila		Coahuila	Coahuila	San Buenaventura	27.06	-101.54	502	26.6	269	Hot
Sonora		Sonora	Sonora	Hermosillo	29.07	-110.96	229	28.1	221	Hot
Zacatecas		Zacatecas	Zacatecas	Zacatecas	22.77	-102.58	2491	17.1	376	Warm
MOLDOVA	**MOLDOVA**	**MOLDOVA**	**MOLDOVA**	**Chisinau**	**47.01**	**28.86**	**41**	**17.2**	**346**	**Warm**
MOROCCO	**MOROCCO**	**MOROCCO**	**MOROCCO**	**Meknes**	**33.87**	**-5.54**	**557**	**20.9**	**181**	**Hot**
MYANMAR		**MYANMAR**	**MYANMAR**	**Mandalay**	**21.98**	**96.08**	**790**	**29.6**	**784**	**Hot**
NEW ZEALAND	**NEW ZEALAND**	**NEW ZEALAND**	**NEW ZEALAND**		**-41.31**	**174.23**	**39**	**15.7**	**451**	**Temp.**
Auckland	Auckland	Auckland	Auckland	Auckland	-36.85	174.76	31	17.5	750	Warm
Canterbury	Canterbury	Canterbury	Canterbury	Christchurch	-43.53	172.64	31	14.9	404	Cool
Gisborne	Gisborne	Gisborne	Gisborne	Gisborne	-38.66	178.02	2	16.8	571	Temp.
Hawkes Bay	Hawkes Bay	Hawkes Bay	Hawkes Bay	Napier	-39.50	176.91	2	16.3	498	Temp.
Marlborough	Marlborough	Marlborough	Marlborough	Blenheim	-41.50	173.96	35	15.7	430	Temp.

Table 75 (cont.): Geographic location, elevation and growing season average temperature and precipitation (GST, GSP) for nearest town to world's wine regions

Name	2000	2010	2016	Town	Latitude	Longitude	Elevation	GST	GSP	Zone
Nelson	Nelson	Nelson	Nelson	Nelson	-41.27	173.28	23	15.2	652	Temp.
Northland			Northland	Whangarei	-35.73	174.32	6	17.3	818	Warm
Otago	Otago	Otago	Otago	Cromwell	-45.03	169.20	202	14.1	374	Cool
Waikato	Waikato	Waikato	Waikato	Hamilton	-37.79	175.28	42	16.9	733	Temp.
Waipara	Waipara	Waipara	Waipara	Waipara	-43.05	172.76	71	14.6	378	Cool
Wairarapa	Wairarapa	Wairarapa	Wairarapa	Martinborough	-41.22	175.46	29	15.3	444	Temp.
Other regions		Other regions			-41.31	174.23	39	15.7	451	Temp.
NORTH MACEDONIA			**NORTH MACEDONIA**	**Bitola**	**41.03**	**21.33**	**622**	**16.7**	**356**	**Temp.**
NORWAY			**NORWAY**	**Oslo**	**59.91**	**10.75**	**117**	**11.9**	**484**	**Cool**
PERU		**PERU**	**PERU**		**-16.04**	**-72.25**	**1303**	**18.4**	**33**	**Warm**
Arequipa		Arequipa	Arequipa	Arequipa	-16.41	-71.54	2325	15.6	76	Temp.
Lima		Lima	Lima	Lima	-12.05	-77.04	127	20.8	9	Hot
Moquegua		Moquegua	Moquegua	Moquegua	-17.19	-70.93	1431	19.2	17	Hot
Tacna		Tacna	Tacna	Tacna	-18.01	-70.25	595	19.8	3	Hot
PORTUGAL	**PORTUGAL**	**PORTUGAL**	**PORTUGAL**		**40.10**	**-7.98**	**230**	**18.8**	**410**	**Warm**
Acores	Acores	Acores	Acores	Ponta Delgada	37.74	-25.67	13	19.3	437	Hot
Alentejo	Alentejo	Alentejo	Alentejo	Evora	38.57	-7.91	309	19.8	243	Hot
Algarve	Algarve	Algarve	Algarve	Lagoa	37.13	-8.45	46	20.2	163	Hot
Alto Tras-os-Montes	Alto Tras-os-Montes	Alto Tras-os-Montes		Chaves	41.74	-7.47	365	17.6	642	Warm
Beira Interior	Beira Interior	Beira Interior		Castelo Branco	39.82	-7.50	398	20.1	278	Hot
Beira Litoral	Beira Litoral	Beira Litoral		Coimbra	40.20	-8.41	37	19.2	383	Hot
Centro			Centro		39.55	-8.13	128	19.8	296	Hot
Entre Douro e Minho	Entre Douro e Minho	Entre Douro e Minho		Pinhao	41.11	-7.55	120	16.6	470	Temp.
Lisboa			Lisboa	Lisbon	38.72	-9.14	77	19.8	261	Hot
Madeira	Madeira	Madeira	Madeira	Funchal	32.63	-16.92	58	19.9	218	Hot
Norte			Norte		41.62	-7.49	317	17.4	609	Warm
Ribatejo e Oeste	Ribatejo e Oeste	Ribatejo e Oeste		Coruche	38.96	-8.52	32	19.9	258	Hot
ROMANIA	**ROMANIA**	**ROMANIA**	**ROMANIA**		**45.03**	**26.24**	**113**	**17.7**	**344**	**Warm**
Bucuresti - Ilfov		Bucuresti - Ilfov		Bucharest	44.43	26.10	75	17.6	409	Warm
Centru		Centru		Blaj	46.18	23.93	284	15.9	434	Temp.
Nord-Est		Nord-Est		Iași	47.16	27.60	118	16.9	402	Temp.
Nord-Vest		Nord-Vest		Oradea	47.05	21.92	124	16.7	375	Temp.
Sud-Est		Sud-Est		Constanța	44.16	28.63	22	18.3	254	Warm
Sud-Muntenia		Sud-Muntenia		Pitești	44.86	24.87	305	16.9	432	Temp.
Sud-Vest Oltenia		Sud-Vest Oltenia		Rogova	44.47	22.81	112	18.4	382	Warm
Vest		Vest		Timişoara	45.75	21.21	90	17.2	398	Warm
RUSSIA	**RUSSIA**	**RUSSIA**	**RUSSIA**		**45.19**	**36.62**	**117**	**17.6**	**347**	**Warm**
Crimea			Crimea	Simferopol	44.95	34.10	205	17.1	298	Warm
Krasnodar Krai		Krasnodar Krai	Krasnodar Krai	Krasnodar	45.04	38.97	26	18.2	409	Warm
Rostov Oblast		Rostov Oblast	Rostov Oblast	Rostov-on-Don	47.24	39.70	58	17.5	331	Warm
SERBIA	**SERBIA**	**SERBIA**	**SERBIA**		**44.00**	**21.27**	**161**	**17.3**	**408**	**Warm**
Bačka			Bačka	Bačka Topola	45.81	19.64	102	17.2	380	Warm
Banat			Banat	Kikinda	45.83	20.46	78	17.6	366	Warm
Belgrade			Belgrade	Belgrade	44.79	20.45	123	17.8	442	Warm
Čačak-Kraljevo			Čačak-Kraljevo	Čačak	43.89	20.35	241	17.0	512	Warm
Knjaževac			Knjaževac	Knjaževac	43.57	22.25	246	17.0	392	Temp.
Leskovac			Leskovac	Leskovac	43.00	21.94	239	17.5	376	Warm
Mlava			Mlava	Požarevac	44.62	21.18	79	17.5	425	Warm
Negotinska Krajina			Negotinska Krajina	Negotin	44.23	22.53	46	18.1	369	Warm
Niš			Niš	Niš	43.32	21.90	201	17.4	384	Warm
Nišava			Nišava	Bela Palanka	43.21	22.32	329	16.6	389	Temp.
South Banat			South Banat	Bela Crkva	44.90	21.42	121	17.7	439	Warm
Srem			Srem	Novi Sad	45.27	19.83	80	17.7	406	Warm
Subotica			Subotica	Subotica	46.10	19.66	113	17.1	375	Warm
Šumadija			Šumadija	Kragujevac	44.01	20.91	184	17.0	458	Temp.
Telečka			Telečka	Sombor	45.77	19.12	86	17.3	402	Warm
Tisa			Tisa	Senta	45.92	20.08	81	17.3	359	Warm
Toplica			Toplica	Zitoradja	43.19	21.72	214	17.4	387	Warm
Tri Morave			Tri Morave	Kruševac	43.58	21.33	162	17.1	409	Warm
Valjevo			Valjevo	Koceljeva	44.47	19.81	125	17.1	504	Warm
Vranje			Vranje	Vranje	42.55	21.90	446	15.7	399	Temp.
SLOVAKIA	**SLOVAKIA**	**SLOVAKIA**	**SLOVAKIA**		**48.24**	**17.99**	**152**	**15.9**	**420**	**Temp.**
Bratislavský kraj			Bratislavský kraj	Bratislava	48.15	17.11	167	16.0	428	Temp.
Juznoslovenska		Juznoslovenska		Dunajská Streda	47.99	17.62	115	16.2	394	Temp.
Malokarpatska		Malokarpatska		Modra	48.33	17.31	201	15.7	514	Temp.
Nitrianska		Nitrianska		Nitra	48.31	18.08	148	16.0	390	Temp.
Stredné Slovensko		Stredné Slovensko	Stredné Slovensko	Rimavská Sobota	48.38	20.02	220	15.4	393	Temp.

Table 75 (cont.): Geographic location, elevation and growing season average temperature and precipitation (GST, GSP) for nearest town to world's wine regions

Name	2000	2010	2016	Town	Latitude	Longitude	Elevation	GST	GSP	Zone
Tokajska		Tokajska		Čerhov	48.46	21.64	121	15.5	427	Temp.
Východné Slovensko		Východné Slovensko	Východné Slovensko	Kráľovský Chlmec	48.42	21.98	11	15.8	419	Temp.
Západné Slovensko			Západné Slovensko	Nitra	48.31	18.08	148	16.0	390	Temp.
SLOVENIA	**SLOVENIA**	**SLOVENIA**	**SLOVENIA**		**46.08**	**14.55**	**202**	**16.8**	**849**	**Temp.**
Bela Krajina		Bela Krajina	Bela Krajina	Črnomelj	45.57	15.19	152	16.4	789	Temp.
Bizeljsko Sremic		Bizeljsko Sremic	Bizeljsko Sremic	Bizeljsko	46.02	15.69	176	15.4	728	Temp.
Dolenjska		Dolenjska	Dolenjska	Dobrovo	46.00	13.53	124	18.2	967	Warm
Goriska brda		Goriska brda	Goriska brda	Neblo	46.01	13.50	91	17.6	1105	Warm
Kras		Kras	Kras	Križ	45.74	13.87	330	16.7	948	Temp.
Prekmurje		Prekmurje	Prekmurje	Gornji Petrovci	46.80	16.22	282	15.4	576	Temp.
Slovenska Istra		Slovenska Istra	Slovenska Istra	Koper	45.55	13.73	9	19.5	624	Hot
Stajerska Slovenija		Stajerska Slovenija	Stajerska Slovenija	Slovenske Konjice	46.34	15.42	339	15.3	811	Temp.
Vipavska dolina		Vipavska dolina	Vipavska dolina	Nova Gorica	45.95	13.65	100	18.1	944	Warm
Other regions		Other regions			46.08	14.55	202	16.8	849	Temp.
SOUTH AFRICA	**SOUTH AFRICA**	**SOUTH AFRICA**	**SOUTH AFRICA**		**-33.28**	**19.27**	**182**	**21.0**	**173**	**Hot**
Breedekloof	Breedekloof	Breedekloof	Breedekloof	Rawsonville	-33.68	19.31	221	20.5	196	Hot
Cape South Coast			Cape South Coast	Bot River	-34.23	19.20	108	18.7	257	Warm
Little Karoo	Little Karoo	Little Karoo	Little Karoo	Calitzdorp	-33.53	21.68	262	20.6	150	Hot
Northern Cape	Northern Cape	Northern Cape	Northern Cape	Rand	-28.45	21.25	808	24.8	179	Hot
Olifants River	Olifants River	Olifants River	Olifants River	Klawer	-31.77	18.62	51	22.1	55	Hot
Paarl	Paarl	Paarl	Paarl	Paarl	-33.73	18.96	125	21.0	197	Hot
Robertson	Robertson	Robertson	Robertson	Robertson	-33.80	19.89	190	21.3	163	Hot
Stellenbosch	Stellenbosch	Stellenbosch	Stellenbosch	Stellenbosch	-33.90	18.87	146	19.7	222	Hot
Swartland	Swartland	Swartland	Swartland	Malmesbury	-33.47	18.72	115	20.9	147	Hot
Worcester	Worcester	Worcester	Worcester	Worcester	-33.65	19.46	226	21.0	180	Hot
SPAIN	**SPAIN**	**SPAIN**	**SPAIN**		**40.71**	**-3.35**	**465**	**20.1**	**249**	**Hot**
Andalucía			**Andalucía**		**37.11**	**-5.22**	**210**	**21.5**	**186**	**Hot**
Almeria, Granada, Jaen, Sevilla	Almeria, Granada, Jaen, Sevilla	Almeria, Granada, Jaen, Sevilla		Granada	37.18	-3.60	689	20.3	167	Hot
Cadiz	Cadiz	Cadiz		Jerez de la Frontera	36.45	-6.12	55	21.4	191	Hot
Cordoba	Cordoba	Cordoba		Córdoba	37.89	-4.78	126	22.4	213	Hot
Huelva	Huelva	Huelva		Huelva	37.26	-6.94	35	21.7	165	Hot
Malaga	Malaga	Malaga		Málaga	36.72	-4.43	54	21.5	165	Hot
Aragón			**Aragón**		**41.75**	**-0.79**	**266**	**19.9**	**248**	**Hot**
Huesca, Teruel	Huesca, Teruel	Huesca, Teruel		Huesca	42.13	-0.41	459	19.4	334	Hot
Zaragoza	Zaragoza	Zaragoza		Zaragoza	41.65	-0.89	215	20.0	226	Hot
Canarias	**Canarias**	**Canarias**	**Canarias**	**Santa Cruz de Tenerife**	**28.46**	**-16.25**	**18**	**22.1**	**58**	**Hot**
Cantabria	**Cantabria**	**Cantabria**	**Cantabria**	**Santander**	**43.46**	**-3.81**	**10**	**17.2**	**590**	**Warm**
Castilla y León			**Castilla y León**		**41.87**	**-4.94**	**806**	**16.5**	**264**	**Temp.**
Avila, Palencia, Salamanca, Segovia, Soria	Avila, Palencia, Salamanca, Segovia, Soria	Avila, Palencia, Salamanca, Segovia, Soria		Salamanca	40.97	-5.66	803	16.9	258	Temp.
Burgos	Burgos	Burgos		Burgos	42.34	-3.70	857	15.4	282	Temp.
Leon	Leon	Leon		León	42.60	-5.57	841	16.3	283	Temp.
Valladolid	Valladolid	Valladolid		Valladolid	41.70	-4.72	846	17.0	227	Warm
Zamora	Zamora	Zamora		Zamora	41.50	-5.75	645	17.1	289	Warm
Castilla-La Mancha			**Castilla-La Mancha**		**39.42**	**-3.20**	**659**	**19.6**	**215**	**Hot**
Albacete	Albacete	Albacete		Albacete	38.99	-1.86	683	19.5	224	Hot
Ciudad Real	Ciudad Real	Ciudad Real		Ciudad Real	38.98	-3.93	631	20.1	195	Hot
Cuenca	Cuenca	Cuenca		Cuenca	40.07	-2.14	924	17.7	278	Warm
Guadalajara	Guadalajara	Guadalajara		Guadalajara	40.63	-3.16	702	18.4	216	Warm
Toledo	Toledo	Toledo		Toledo	39.86	-4.03	479	20.5	187	Hot
Cataluña			**Cataluña**		**41.30**	**1.45**	**61**	**20.0**	**347**	**Hot**
Barcelona	Barcelona	Barcelona		Barcelona	41.39	2.17	8	20.0	390	Hot
Girona, Lleida	Girona, Lleida	Girona, Lleida		Lleida	41.62	0.62	175	20.0	228	Hot
Tarragona	Tarragona	Tarragona		Reus	41.15	1.11	71	20.0	345	Hot
Comunidad de Madrid	**Comunidad de Madrid**	**Comunidad de Madrid**	**Comunidad de Madrid**	**Madrid**	**40.42**	**-3.70**	**655**	**19.2**	**221**	**Hot**
Comunidad Foral de Navarra	**Comunidad Foral de Navarra**	**Comunidad Foral de Navarra**	**Comunidad Foral de Navarra**	**Estella**	**42.67**	**-2.03**	**430**	**16.1**	**383**	**Temp.**
Comunidad Valenciana			**Comunidad Valenciana**		**39.25**	**-0.40**	**9**	**21.5**	**237**	**Hot**

Table 75 (cont.): Geographic location, elevation and growing season average temperature and precipitation (GST, GSP) for nearest town to world's wine regions

Name	2000	2010	2016	Town	Latitude	Longitude	Elevation	GST	GSP	Zone
Alicante	Alicante	Alicante		Alicante	38.34	-0.49	17	21.8	183	Hot
Castellon	Castellon	Castellon		Castellón de la Plana	39.99	-0.05	36	21.2	284	Hot
Valencia	Valencia	Valencia		Valencia	39.47	-0.38	6	21.5	250	Hot
Extremadura			**Extremadura**		**38.91**	**-6.94**	**202**	**20.7**	**218**	**Hot**
Badajoz	Badajoz	Badajoz		Badajoz	38.88	-6.97	191	20.7	218	Hot
Caceres	Caceres	Caceres		Cáceres	39.48	-6.37	439	20.8	220	Hot
Galicia	**Galicia**	**Galicia**	**Galicia**	**Vigo**	**42.24**	**-8.72**	**261**	**17.2**	**640**	**Warm**
Illes Balears	**Illes Balears**	**Illes Balears**	**Illes Balears**	**Palma**	**39.57**	**2.65**	**31**	**20.7**	**229**	**Hot**
La Rioja	**La Rioja**	**La Rioja**	**La Rioja**	**Logrono**	**42.45**	**-2.45**	**353**	**17.7**	**249**	**Warm**
País Vasco			**País Vasco**		**42.87**	**-2.68**	**507**	**15.8**	**430**	**Temp.**
Alava	Alava	Alava		Vitoria-Gasteiz	42.85	-2.67	535	15.7	421	Temp.
Guipuzcoa, Vizcaya	Guipuzcoa, Vizcaya	Guipuzcoa, Vizcaya		Bilbao	43.26	-2.93	17	17.3	595	Warm
Principado de Asturias	**Principado de Asturias**	**Principado de Asturias**	**Principado de Asturias**	**Oviedo**	**43.36**	**-5.85**	**258**	**16.6**	**483**	**Temp.**
Región de Murcia	**Región de Murcia**	**Región de Murcia**	**Región de Murcia**	**Murcia**	**37.99**	**-1.13**	**37**	**22.4**	**157**	**Hot**
SWITZERLAND	**SWITZERLAND**	**SWITZERLAND**	**SWITZERLAND**		**46.58**	**7.38**	**489**	**14.4**	**607**	**Cool**
Aargau	Aargau	Aargau	Aargau	Hägglingen	47.39	8.25	462	14.6	713	Cool
Appenzell Ausserrhoden			Appenzell Ausserrhoden	Waldstatt	47.36	9.29	818	12.4	928	Cool
Appenzell Innerrhoden			Appenzell Innerrhoden	Appenzell	47.33	9.41	766	11.5	985	Cool
Basel Land	Basel Land	Basel Land	Basel Land	Liestal	47.49	7.73	322	14.4	653	Cool
Basel Stadt			Basel Stadt	Basel	47.56	7.59	251	15.6	514	Temp.
Bern	Bern	Bern	Bern	Bern	46.95	7.45	532	14.1	696	Cool
Fribourg	Fribourg	Fribourg	Fribourg	Fribourg	46.81	7.16	590	13.9	666	Cool
Geneva	Geneva	Geneva	Geneva	Geneva	46.20	6.14	377	15.4	546	Temp.
Glarus			Glarus	Glarus	47.04	9.07	480	12.7	1035	Cool
Graubünden	Graubünden	Graubünden		Chur	46.85	9.53	598	13.7	666	Cool
Graubünden - Mesolcina			Graubünden - Mesolcina	Cama	46.27	9.17	351	12.5	991	Cool
Graubünden - other			Graubünden - other		46.85	9.53	598	13.7	666	Cool
Jura	Jura	Jura	Jura	Delémont	47.37	7.35	426	14.4	602	Cool
Lucerne	Lucerne	Lucerne	Lucerne	Lucerne	47.05	8.31	443	14.9	841	Cool
Neuchâtel	Neuchâtel	Neuchâtel	Neuchâtel	Neuchâtel	46.99	6.93	432	14.9	593	Cool
Nidwalden			Nidwalden	Dallenwil	46.93	8.39	486	13.4	940	Cool
Owalden			Owalden	Sarnen	46.90	8.25	475	14.0	898	Cool
Schaffhausen	Schaffhausen	Schaffhausen	Schaffhausen	Schaffhausen	47.70	8.64	394	14.4	567	Cool
Schwyz	Schwyz	Schwyz	Schwyz	Schwyz	47.02	8.65	512	14.3	939	Cool
Solothurn			Solothurn	Solothurn	47.21	7.53	443	13.6	721	Cool
St. Gallen	St. Gallen	St. Gallen	St. Gallen	St. Gallen	47.42	9.38	673	13.2	874	Cool
Thunersee			Thunersee	Thun	46.76	7.63	556	13.8	714	Cool
Thurgau	Thurgau	Thurgau	Thurgau	Märstetten	47.59	9.06	425	14.3	628	Cool
Ticino	Ticino	Ticino	Ticino	Frasco	46.34	8.80	890	9.0	1001	Cool
Uri			Uri	Gurtnellen	46.74	8.63	950	9.3	1064	Cool
Valais	Valais	Valais	Valais	Valais	46.22	7.36	482	14.8	413	Cool
Vaud	Vaud	Vaud	Vaud	Crissier	46.55	6.57	448	15.2	711	Temp.
Zug			Zug	Zug	47.17	8.52	428	14.7	813	Cool
Zürich	Zürich	Zürich	Zürich	Zürich	47.48	8.54	404	14.6	650	Cool
Other regions	Other regions	Other regions	Other regions		46.58	7.38	489	14.4	607	Cool
TAIWAN	**TAIWAN**	**TAIWAN**	**TAIWAN**	**Taipei**	**25.05**	**121.55**	**21**	**25.5**	**1911**	**Hot**
THAILAND		**THAILAND**	**THAILAND**	**Bangkok**	**13.76**	**100.50**	**8**	**29.1**	**1233**	**Hot**
TUNISIA	**TUNISIA**	**TUNISIA**	**TUNISIA**	**Kelibia**	**36.85**	**11.10**	**9**	**22.2**	**193**	**Hot**
TURKEY		**TURKEY**	**TURKEY**		**38.83**	**29.99**	**284**	**21.6**	**185**	**Hot**
Aegean		Aegean	Aegean	İzmir	38.42	27.14	25	22.9	155	Hot
Central East		Central East	Central East	Elazığ	38.67	39.22	1069	20.2	253	Hot
Central North		Central North	Central North	Tokat	40.32	36.55	626	17.3	229	Warm
Central South		Central South	Central South	Ankara	39.93	32.86	880	17.6	197	Warm
Marmara		Marmara	Marmara	Tekirdağ	40.98	27.51	52	18.7	259	Warm
Mediterranean		Mediterranean	Mediterranean	Antalya	36.90	30.71	56	23.2	191	Hot
South East		South East	South East	Gaziantep	37.07	37.38	853	22.1	131	Hot
UKRAINE		**UKRAINE**	**UKRAINE**	**Odessa**	**46.48**	**30.72**	**40**	**17.1**	**272**	**Warm**
UNITED KINGDOM	**UNITED KINGDOM**	**UNITED KINGDOM**	**UNITED KINGDOM**	**Brighton**	**50.82**	**-0.13**	**12**	**13.5**	**399**	**Cool**
UNITED STATES	**UNITED STATES**	**UNITED STATES**	**UNITED STATES**		**38.87**	**-117.67**	**111**	**19.0**	**154**	**Hot**
Arizona		**Arizona**	**Arizona**	**Willcox**	**32.25**	**-109.83**	**1274**	**21.9**	**218**	**Hot**
Arkansas		**Arkansas**	**Arkansas**	**Little Rock**	**34.75**	**-92.29**	**105**	**22.8**	**745**	**Hot**

Table 75 (cont.): Geographic location, elevation and growing season average temperature and precipitation (GST, GSP) for nearest town to world's wine regions

Name	2000	2010	2016	Town	Latitude	Longitude	Elevation	GST	GSP	Zone
California					**37.43**	**-121.25**	**72**	**19.7**	**92**	**Hot**
Alameda	Alameda	Alameda	Alameda	Livermore	37.68	-121.77	148	19.5	70	Hot
Amador	Amador	Amador	Amador	Jackson	38.35	-120.77	368	20.2	173	Hot
Butte	Butte	Butte	Butte	Chico	39.73	-121.84	71	21.1	122	Hot
Calaveras	Calaveras	Calaveras	Calaveras	San Andreas	38.20	-120.68	312	20.3	151	Hot
Colusa	Colusa	Colusa	Colusa	Colusa	39.21	-122.01	18	21.1	74	Hot
Contra Costa	Contra Costa	Contra Costa	Contra Costa	Concord	37.98	-122.03	27	19.7	80	Hot
El Dorado	El Dorado	El Dorado	El Dorado	Placerville	38.73	-120.80	572	19.5	202	Hot
Fresno	Fresno	Fresno	Fresno	Fresno	36.74	-119.79	112	23.1	59	Hot
Glenn	Glenn	Glenn	Glenn	Orland	39.75	-122.20	78	21.2	102	Hot
Humboldt	Humboldt	Humboldt	Humboldt	Eureka	40.80	-124.16	17	14.1	242	Cool
Kern	Kern	Kern	Kern	Bakersfield	35.37	-119.02	122	23.8	38	Hot
Kings	Kings	Kings	Kings	Hanford	36.33	-119.65	77	22.6	43	Hot
Lake	Lake	Lake	Lake	Clearlake	38.96	-122.63	453	18.5	121	Warm
Lassen			Lassen	Susanville	40.42	-120.65	1279	14.8	118	Cool
Los Angeles	Los Angeles	Los Angeles	Los Angeles	Los Angeles	34.05	-118.24	128	21.4	56	Hot
Madera	Madera	Madera	Madera	Madera	36.96	-120.06	86	22.0	64	Hot
Marin	Marin	Marin	Marin	San Rafael	37.97	-122.53	14	17.8	160	Warm
Mariposa	Mariposa	Mariposa	Mariposa	Mariposa	37.49	-119.97	595	19.8	169	Hot
Mendocino	Mendocino	Mendocino	Mendocino	Ukiah	39.15	-123.21	193	17.3	183	Warm
Merced	Merced	Merced	Merced	Merced	37.30	-120.48	70	21.6	66	Hot
Monterey	Monterey	Monterey	Monterey	Salinas	36.68	-121.66	15	16.6	71	Temp.
Napa	Napa	Napa	Napa	St Helena	38.50	-122.47	69	18.1	142	Warm
Nevada	Nevada	Nevada	Nevada	Nevada City	39.26	-121.02	770	17.5	264	Warm
Orange		Orange	Orange	Irvine	33.68	-117.83	6	20.2	46	Hot
Placer	Placer	Placer	Placer	Auburn	38.90	-121.08	372	20.5	173	Hot
Riverside	Riverside	Riverside	Riverside	Riverside	33.98	-117.38	268	21.7	56	Hot
Sacramento	Sacramento	Sacramento	Sacramento	Sacramento	38.58	-121.49	26	21.1	83	Hot
San Benito	San Benito	San Benito	San Benito	Hollister	36.85	-121.40	96	18.7	64	Warm
San Bernardino	San Bernardino	San Bernardino	San Bernardino	San Bernardino	34.11	-117.29	349	22.6	57	Hot
San Diego	San Diego	San Diego	San Diego	San Diego	32.72	-117.16	16	20.1	47	Hot
San Joaquin	San Joaquin	San Joaquin	San Joaquin	Lodi	38.11	-121.27	12	20.4	83	Hot
San Luis Obispo	San Luis Obispo	San Luis Obispo	San Luis Obispo	San Luis Obispo	35.28	-120.66	68	17.5	86	Warm
San Mateo	San Mateo	San Mateo	San Mateo	San Mateo	37.56	-122.33	15	16.8	88	Temp.
Santa Barbara	Santa Barbara	Santa Barbara	Santa Barbara	Santa Maria	34.55	-120.44	77	15.8	82	Temp.
Santa Clara	Santa Clara	Santa Clara	Santa Clara	San Jose	37.34	-121.89	31	19.4	74	Hot
Santa Cruz	Santa Cruz	Santa Cruz	Santa Cruz	Santa Cruz	36.97	-122.04	8	16.5	121	Temp.
Shasta	Shasta	Shasta	Shasta	Redding	40.59	-122.39	199	22.2	238	Hot
Siskiyou		Siskiyou	Siskiyou	Montague	41.73	-122.30	775	15.5	161	Temp.
Solano	Solano	Solano	Solano	Fairfield	38.25	-122.04	6	19.3	83	Hot
Sonoma	Sonoma	Sonoma	Sonoma	Sonoma	38.30	-122.46	30	18.4	122	Warm
Stanislaus	Stanislaus	Stanislaus	Stanislaus	Modesto	37.64	-121.00	30	21.9	64	Hot
Sutter	Sutter	Sutter	Sutter	Yuba City	39.18	-121.69	18	22.0	97	Hot
Tehama	Tehama	Tehama	Tehama	Corning	39.93	-122.18	84	21.7	115	Hot
Trinity	Trinity	Trinity	Trinity	Hayfork	40.55	-123.18	725	16.3	204	Temp.
Tulare	Tulare	Tulare	Tulare	Visalia	36.33	-119.29	110	22.7	63	Hot
Tuolumne		Tuolumne	Tuolumne	Sonora	37.98	-120.38	549	19.4	168	Hot
Ventura	Ventura	Ventura	Ventura	Oxnard	34.20	-119.18	32	18.1	36	Warm
Yolo	Yolo	Yolo	Yolo	Woodland	38.68	-121.77	26	21.4	79	Hot
Yuba	Yuba	Yuba	Yuba	Browns Valley	39.24	-121.41	83	22.1	136	Hot
Colorado		**Colorado**	**Colorado**	**Grand Junction**	**39.06**	**-108.55**	**1403**	**18.9**	**144**	**Warm**
Georgia		**Georgia**	**Georgia**	**Atlanta**	**33.75**	**-84.39**	**316**	**21.9**	**718**	**Hot**
Illinois		**Illinois**	**Illinois**	**Springfield**	**39.78**	**-89.65**	**190**	**19.1**	**648**	**Hot**
Indiana		**Indiana**	**Indiana**	**Indianapolis**	**39.77**	**-86.16**	**226**	**18.3**	**670**	**Warm**
Iowa		**Iowa**	**Iowa**	**Des Moines**	**41.59**	**-93.62**	**269**	**18.3**	**699**	**Warm**
Kentucky		**Kentucky**	**Kentucky**	**Frankfort**	**38.20**	**-84.87**	**156**	**19.3**	**706**	**Hot**
Michigan	**Michigan**	**Michigan**	**Michigan**	**Detroit**	**42.33**	**-83.05**	**189**	**17.1**	**561**	**Warm**
Minnesota		**Minnesota**	**Minnesota**	**Minneapolis**	**44.98**	**-93.27**	**269**	**16.5**	**610**	**Temp.**
Missouri		**Missouri**	**Missouri**	**St. Louis**	**38.63**	**-90.20**	**149**	**20.1**	**640**	**Hot**
New York					**42.31**	**-78.21**	**314**	**15.4**	**670**	**Temp.**
Cattaraugus			Cattaraugus	Cattaraugus	42.33	-78.87	419	14.0	697	Cool
Chautauqua			Chautauqua	Chautauqua	42.21	-79.47	415	15.1	737	Temp.
Chautauqua-Erie	Chautauqua-Erie	Chautauqua-Erie			42.27	-79.42	394	15.1	723	Temp.
Erie			Erie	Buffalo	42.89	-78.88	184	15.8	589	Temp.
Finger Lakes	Finger Lakes	Finger Lakes			42.61	-77.04	237	15.4	580	Temp.
New York - other	New York - other	New York - other			41.59	-74.58	61	16.8	622	Temp.
Niagara			Niagara	Niagara Falls	43.10	-79.04	176	15.6	581	Temp.
Ontario			Ontario	Canandaigua	42.89	-77.28	225	15.5	581	Temp.
Schuyler			Schuyler	Watkins Glen	42.38	-76.87	144	15.2	583	Temp.
Seneca			Seneca	Ovid	42.68	-76.82	294	15.4	606	Temp.
Steuben			Steuben	Bath	42.34	-77.32	337	14.7	572	Cool

Table 75 (cont.): Geographic location, elevation and growing season average temperature and precipitation (GST, GSP) for nearest town to world's wine regions

Name	2000	2010	2016	Town	Latitude	Longitude	Elevation	GST	GSP	Zone
Suffolk			Suffolk	Riverhead	40.92	-72.66	7	17.4	632	Warm
Ulster			Ulster	Ellenville	41.72	-74.40	106	16.0	719	Temp.
Wayne			Wayne	Newark	43.05	-77.10	138	15.7	590	Temp.
Yates			Yates	Penn Yan	42.66	-77.05	225	15.5	575	Temp.
North Carolina		**North Carolina**	**North Carolina**	**Raleigh**	**35.78**	**-78.64**	**101**	**21.2**	**723**	**Hot**
Ohio		**Ohio**	**Ohio**	**Cleveland**	**41.50**	**-81.69**	**225**	**17.4**	**606**	**Warm**
Oregon					**44.68**	**-122.98**	**131**	**15.6**	**289**	**Temp.**
Benton Co.	Benton Co.	Benton Co.		Corvallis	44.56	-123.26	70	15.7	328	Temp.
Columbia River	Columbia River	Columbia River	Columbia River	The Dalles	45.59	-121.18	81	17.3	112	Warm
Douglas Co.	Douglas Co.	Douglas Co.		Waterville	47.65	-120.07	800	14.0	111	Cool
Jackson Co.		Jackson Co.		Medford	42.33	-122.88	446	16.8	159	Temp.
Josephine Co.	Josephine Co.	Josephine Co.		Grants Pass	42.44	-123.33	288	17.1	198	Warm
Lane Co.	Lane Co.	Lane Co.		Eugene	44.05	-123.09	173	16.0	349	Temp.
Marion Co.	Marion Co.	Marion Co.		Salem	44.94	-123.04	45	15.8	291	Temp.
North Willamette Valley			North Willamette Valley	McMinnville	45.13	-123.20	47	15.0	301	Temp.
Oregon - other	Oregon - other	Oregon - other			44.68	-122.98	131	15.6	289	Temp.
Polk Co.	Polk Co.	Polk Co.		Dallas	44.92	-123.32	101	15.0	356	Temp.
Rogue Valley			Rogue Valley	Medford	42.33	-122.88	422	16.8	159	Temp.
South Willamette Valley			South Willamette Valley	Eugene	44.05	-123.09	173	16.0	349	Temp.
Umpqua Valley			Umpqua Valley	Sutherlin	43.39	-123.31	157	16.3	323	Temp.
Valley - other	Valley - other				44.96	-123.18	67	15.2	309	Temp.
Washington Co.	Washington Co.	Washington Co.		Hilsboro	45.53	-122.40	98	15.9	510	Temp.
Yamhill Co.	Yamhill Co.	Yamhill Co.		McMinnville	45.21	-123.20	45	15.3	288	Temp.
Pennsylvania		**Pennsylvania**	**Pennsylvania**	**Erie**	**42.13**	**-80.09**	**203**	**16.2**	**666**	**Temp.**
Texas		**Texas**			**32.54**	**-100.70**	**733**	**23.4**	**449**	**Hot**
Hill Country			Hill Country	Austin	30.27	-97.74	199	25.5	561	Hot
North Texas (DFW)			North Texas (DFW)	Dallas	32.78	-96.80	180	25.0	624	Hot
South Texas and Gulf Coast			South Texas and Gulf Coast	Houston	29.76	-95.37	15	25.9	791	Hot
Texas High Plains and Panhandle			Texas High Plains and Panhandle	Lubbock	33.58	-101.86	985	22.3	394	Hot
West Texas			West Texas	El Paso	31.76	-106.49	1157	23.9	165	Hot
Virginia		**Virginia**	**Virginia**	**Charlottesville**	**38.13**	**-78.48**	**190**	**19.5**	**704**	**Hot**
Washington	**Washington**				**46.28**	**-119.83**	**222**	**17.2**	**81**	**Warm**
Columbia Gorge		Columbia Gorge	Columbia Gorge	White Salmon	45.73	-121.49	188	15.3	210	Temp.
Columbia Valley		Columbia Valley	Columbia Valley	Yakima	46.60	-120.51	331	17.0	76	Temp.
Horse Heaven Hills		Horse Heaven Hills	Horse Heaven Hills	Paterson	45.94	-119.60	124	17.6	80	Warm
Lake Chelan		Lake Chelan	Lake Chelan	Chelan	47.84	-120.02	342	15.8	98	Temp.
Naches Heights			Naches Heights	Gleed	46.66	-120.61	386	15.9	78	Temp.
Puget Sound		Puget Sound	Puget Sound	Seattle	47.61	-122.33	70	14.4	358	Cool
Rattlesnake Hills		Rattlesnake Hills	Rattlesnake Hills	Granger	46.34	-120.19	227	16.9	68	Temp.
Red Mountain		Red Mountain	Red Mountain	Benton City	46.26	-119.49	151	18.0	72	Warm
Snipes Mountain		Snipes Mountain	Snipes Mountain	Granger	46.34	-120.19	227	16.9	68	Temp.
Wahluke Slope		Wahluke Slope	Wahluke Slope	Mattawa	46.74	-119.90	234	17.4	66	Warm
Walla Walla Valley		Walla Walla Valley	Walla Walla Valley	Walla Walla	46.06	-118.34	286	17.3	178	Warm
Yakima Valley		Yakima Valley	Yakima Valley	Prosser	46.20	-119.77	253	17.0	82	Temp.
URUGUAY	**URUGUAY**	**URUGUAY**	**URUGUAY**		**-34.39**	**-56.44**	**29**	**20.2**	**670**	**Hot**
Artigas			Artigas	Artigas	-30.41	-56.48	126	22.6	975	Hot
Canelones			Canelones	Canelones	-34.53	-56.29	25	20.1	665	Hot
Colonia			Colonia	Carmelo	-34.00	-58.29	14	21.0	716	Hot
Durazno			Durazno	Durazno	-33.38	-56.53	82	21.1	787	Hot
Florida			Florida	Florida	-34.09	-56.22	66	20.4	737	Hot
Lavalleja			Lavalleja	Lavalleja	-34.36	-55.26	117	19.4	716	Hot
Maldonado			Maldonado	Maldonado	-34.90	-54.97	34	19.4	575	Hot
Montevideo			Montevideo	Montevideo	-34.82	-56.21	35	19.8	575	Hot
Paysandu			Paysandu	Paysandú	-32.31	-58.08	44	21.6	853	Hot
Rivera			Rivera	Rivera	-30.92	-55.55	195	21.6	955	Hot
Rocha			Rocha	Rocha	-34.48	-54.33	27	19.4	645	Hot
Salto			Salto	Salto	-31.39	-57.96	43	22.4	934	Hot
San Jose			San Jose	San José	-34.35	-56.71	32	20.3	743	Hot
Soriano			Soriano	Palmitas	-33.51	-57.80	93	20.9	782	Hot
Tacuarembo			Tacuarembo	Tacuarembó	-31.72	-55.99	141	21.6	796	Hot

Table 76: Key climate indicators of the world's wine regions

Data Source: TerraClimate, as compiled by Gregory Jones of Linfield University, Oregon with the asssistance of German Puga of the University of Adelaide, South Australia
Time Period: 1958-2019
AnnP - annual precipitation (mm)
GSP - growing season precipitation (mm); growing season Apr-Oct NH, Oct-Apr SH
RipeP - ripening period precipitation (mm); September NH, March SH
GST - growing season average temperature (°C); growing season Apr-Oct NH, Oct-Apr SH
GDD - growing degree days (C° units); growing season Apr-Oct NH, Oct-Apr SH; base temperature 10°C, no upper cut off
AnnT - annual average temperature (°C)
RPT - ripening period average temperature (°C); August-September NH, February-March SH
GSDTR - growing season diurnal temperature range (°C); growing season Apr-Oct NH, Oct-Apr SH
RPDTR - ripening period diurnal temperature range (°C); August-September NH, February-March SH

	AnnP	GSP	RipeP	GST	GDD	AnnT	RPT	GSDTR	RPDTR
Mean	713	403	59	18.3	1787	13.9	20.4	12.1	12.3
Median	681	383	52	18.3	1760	13.9	20.3	11.7	11.9
Mode	642	383	54	19.3	1551	9.7	19.6	11.4	13.0
Standard D	349.5	242.5	40.3	2.8	580.1	3.3	2.8	2.7	2.9
Kurtosis	1.89	2.71	6.51	0.64	0.61	0.72	0.01	-0.35	0.17
Skewness	0.80	1.12	1.68	0.35	0.47	0.34	0.11	0.34	0.49
Range	2888	1911	337	20.6	3992	25.3	19.8	14.6	17.8
Minimum	0	0	0	9.0	205	3.2	11.1	5.7	4.9
Maximum	2888	1911	337	29.6	4197	28.5	30.9	20.3	22.7
Count	813	813	813	813	813	813	813	813	813

Location	Latitude	Longitude	AnnP	GSP	RipeP	GST	GDD	AnnT	RPT	GSDTR	RPDTR
Addis Ababa	8.98	38.76	1078	942	145	17.0	1495	16.9	16.0	12.5	10.7
Agrigento	37.31	13.59	438	169	31	21.2	2392	17.6	24.1	6.9	6.6
Aguascalientes	21.89	-102.29	481	436	82	19.9	2121	17.6	19.8	15.9	13.9
Ahrweiler	50.54	7.12	739	438	57	13.9	883	9.7	15.8	9.4	9.4
Aimogasta	-28.55	-66.82	296	267	47	23.3	2821	19.0	23.8	15.3	14.6
Aix-en-Provence	43.53	5.45	632	353	76	19.0	1927	14.3	21.6	10.2	10.5
Ajaccio	41.92	8.74	663	282	52	19.1	1949	15.4	21.8	7.4	7.8
Alba Posse	-27.57	-54.68	1724	1041	136	23.8	2924	20.9	24.9	12.4	12.1
Albacete	38.99	-1.86	376	224	33	19.5	2033	14.5	22.8	13.0	13.6
Alessandria	44.91	8.61	755	452	64	18.5	1820	12.8	21.3	9.3	9.4
Alexandra	-37.19	145.71	772	366	40	17.3	1551	13.9	19.6	14.3	15.1
Algiers	36.75	3.04	777	250	36	21.3	2426	17.5	24.6	8.4	8.3
Alicante	38.34	-0.49	319	183	43	21.8	2525	18.0	24.9	10.9	10.9
Almaty	43.22	76.85	561	357	25	17.7	1655	9.7	20.0	11.7	13.0
Alta Gracia	-31.66	-64.43	718	630	100	20.0	2107	16.6	20.6	13.7	13.4
Amindeo	40.69	21.68	603	330	50	18.0	1707	12.5	20.7	13.1	14.4
Aminga	-28.85	-66.93	260	241	40	20.9	2318	16.9	21.5	15.3	14.6
Ancona	43.62	13.52	699	404	69	19.1	1944	14.5	21.5	7.7	7.8
Andalgalá	-27.58	-66.31	311	281	50	22.9	2724	18.9	23.3	14.7	13.5
Añelo	-38.35	-68.79	164	107	13	18.5	1809	14.1	19.8	16.4	17.0
Angers	47.47	-0.55	726	377	57	15.9	1271	12.1	18.0	10.7	11.0
Angoulême	45.65	0.16	826	431	65	16.5	1398	12.6	18.7	11.3	11.6

Table 76 (cont.): Key climate indicators of the world's wine regions

Location	Latitude	Longitude	AnnP	GSP	RipeP	GST	GDD	AnnT	RPT	GSDTR	RPDTR
Ankara	39.93	32.86	402	197	17	17.6	1627	11.6	20.6	14.6	15.6
Annecy	45.90	6.13	981	562	87	15.5	1192	10.7	17.6	11.2	11.6
Antalya	36.90	30.71	783	191	18	23.2	2830	18.9	26.4	11.2	11.3
Antofagasta	-23.65	-70.40	0	0	0	19.3	1975	17.5	20.7	7.0	7.4
Aosta	45.73	7.31	760	449	54	14.4	989	9.4	16.5	10.5	10.7
Appenzell	47.33	9.41	1432	985	127	11.5	510	6.9	13.8	9.1	9.1
Ararat	-37.28	142.93	602	287	33	16.1	1292	13.1	18.2	14.0	14.8
Arequipa	-16.41	-71.54	78	76	17	15.6	1189	15.2	15.6	14.9	13.5
Arezzo	43.46	11.88	828	451	72	18.5	1827	13.6	21.1	11.8	12.9
Argostoli	38.17	20.49	886	262	50	21.4	2446	17.7	24.4	9.6	9.9
Arica	-18.48	-70.31	1	0	0	20.4	2210	18.6	22.1	7.0	7.3
Armidale	-30.50	151.67	760	544	57	17.3	1535	13.7	18.7	13.6	12.9
Artigas	-30.41	-56.48	1482	975	149	22.6	2665	19.3	23.9	12.9	12.8
Ascoli Piceno	42.85	13.57	746	449	77	19.1	1951	14.1	21.9	9.6	10.2
Asti	44.90	8.21	636	392	49	18.4	1801	12.9	21.1	9.0	8.9
Athens	37.98	23.73	430	139	17	23.0	2790	18.3	25.9	9.7	10.2
Atlanta	33.75	-84.39	1299	718	92	21.9	2545	16.4	24.0	12.0	11.3
Auburn	38.90	-121.08	833	173	12	20.5	2250	15.9	23.5	15.5	16.8
Auch	43.65	0.59	1015	561	72	17.4	1577	13.3	19.8	11.4	11.6
Auckland	-36.85	174.76	1432	750	103	17.5	1585	15.3	19.2	8.3	8.6
Aurillac	44.93	2.44	795	487	71	14.1	928	10.0	16.1	11.8	12.1
Austin	30.27	-97.74	829	561	96	25.5	3329	20.3	27.9	12.1	12.4
Avellino	40.91	14.79	884	392	74	18.4	1798	13.9	21.3	9.0	9.1
Avignon	43.91	4.81	731	432	83	19.5	2030	14.7	22.0	11.3	11.5
Bačka Topola	45.81	19.64	580	380	44	17.2	1551	11.3	19.3	12.0	13.0
Bad Dürkheim	49.46	8.17	591	361	43	15.1	1098	10.3	16.9	10.6	10.8
Bad Ems	50.34	7.71	873	504	67	13.7	850	9.3	15.4	9.6	9.6
Badacsonytomaj	46.80	17.51	621	431	62	16.5	1391	10.8	18.5	11.4	11.8
Badajoz	38.88	-6.97	557	218	29	20.7	2298	16.7	23.6	13.5	14.4
Baden	48.00	16.23	603	434	53	15.5	1181	10.0	17.6	11.6	11.8
Baden-Baden	48.77	8.23	1150	684	84	15.0	1072	10.2	16.8	10.6	11.0
Baja	46.18	18.95	561	382	58	17.2	1551	11.4	19.3	11.7	12.5
Bakersfield	35.37	-119.02	148	38	4	23.8	2957	18.8	26.5	15.9	16.2
Balatonboglár	46.77	17.66	603	415	62	16.8	1465	11.0	18.8	11.4	11.8
Balatonfüred	46.96	17.89	609	412	64	16.2	1334	10.5	18.3	11.2	11.6
Balcarce	-37.85	-58.26	844	528	102	17.3	1537	13.9	18.9	13.4	13.1
Bañado de Ovanta	-28.10	-65.32	670	610	107	23.7	2905	19.9	24.0	12.6	11.6
Bangkok	13.76	100.50	1345	1233	293	29.1	4094	28.5	28.5	8.0	7.6
Barcelona	41.39	2.17	622	390	79	20.0	2153	16.4	22.7	7.5	7.1
Bari	41.12	16.87	558	276	57	20.5	2256	16.3	23.2	9.0	9.2
Barletta	41.32	16.28	464	233	47	20.6	2264	16.7	23.3	9.3	9.6
Basel	47.56	7.59	782	514	63	15.6	1202	10.8	17.7	10.3	10.6
Bastia	42.33	9.45	741	324	64	17.6	1631	13.5	20.5	6.4	6.7
Bath	42.34	-77.32	849	572	77	14.7	1111	7.9	17.6	13.9	13.9
Baunei	40.03	9.66	372	156	35	20.5	2241	16.4	23.5	7.9	8.1
Beechworth	-36.36	146.69	952	434	53	16.8	1443	13.1	19.2	13.6	14.1
Beijing	39.90	116.41	569	541	49	20.7	2300	12.6	22.8	11.2	10.3
Bela Crkva	44.90	21.42	683	439	54	17.7	1660	11.8	20.0	11.9	13.0
Bela Palanka	43.21	22.32	635	389	51	16.6	1425	10.9	18.7	12.4	13.6
Belén	-27.65	-67.03	169	156	26	21.7	2483	17.9	22.2	15.7	14.6
Belgrade	44.79	20.45	686	442	54	17.8	1664	12.1	19.8	11.2	11.8

Table 76 (cont.): Key climate indicators of the world's wine regions

Location	Latitude	Longitude	AnnP	GSP	RipeP	GST	GDD	AnnT	RPT	GSDTR	RPDTR
Belley	45.76	5.69	1028	616	98	16.3	1357	11.4	18.3	11.2	11.5
Belluno	46.14	12.22	1008	681	91	15.8	1255	10.6	18.0	11.4	11.6
Bendigo	-36.76	144.28	573	271	30	17.5	1583	14.1	19.5	13.8	14.1
Benevento	41.13	14.79	835	378	74	19.6	2062	14.8	22.5	9.9	9.9
Benito Juárez	-37.67	-59.80	790	539	100	17.2	1518	13.5	19.0	14.2	14.2
Bensheim	49.69	8.62	681	427	54	15.6	1208	10.7	17.6	10.8	10.8
Bento Gonçalves	-29.17	-51.52	1697	999	155	19.6	2026	17.1	20.9	9.7	9.3
Benton City	46.26	-119.49	184	72	8	18.0	1707	12.0	20.8	15.6	16.8
Bergamo	45.70	9.68	1076	731	109	18.1	1738	12.7	20.6	10.4	10.6
Bergerac	44.85	0.48	939	490	80	16.9	1485	13.0	19.0	11.9	12.3
Bern	46.95	7.45	1038	696	85	14.1	928	9.1	16.1	9.8	9.9
Besançon	47.23	6.02	1123	666	95	15.3	1146	10.7	17.4	10.6	10.7
Beziers	43.32	3.22	561	301	53	19.1	1946	15.0	21.5	10.6	10.7
Biella	45.56	8.06	842	577	66	16.8	1462	11.5	19.1	9.9	9.9
Bienne	47.14	7.25	970	636	80	14.4	976	9.6	16.4	10.0	10.2
Bilbao	43.26	-2.93	1195	595	78	17.3	1556	14.4	19.6	6.5	6.7
Bilje	45.60	18.74	655	422	54	17.2	1537	11.3	19.1	12.1	12.9
Bitola	41.03	21.33	686	356	50	16.7	1430	11.1	19.3	12.9	14.1
Bizeljsko	46.02	15.69	1110	728	118	15.4	1168	10.1	17.2	10.8	11.1
Bjelovar	45.90	16.84	825	529	80	17.4	1583	11.5	18.9	11.5	12.2
Blaj	46.18	23.93	587	434	46	15.9	1255	9.6	17.6	12.5	13.0
Blenheim	-41.50	173.96	779	430	64	15.7	1197	12.9	17.1	9.7	10.0
Blois	47.59	1.34	662	377	53	15.6	1207	11.5	17.8	10.7	10.9
Boën-sur-Lignon	45.75	4.00	710	487	72	15.3	1143	10.9	17.4	11.2	11.5
Bologna	44.49	11.34	742	423	67	19.6	2059	14.1	22.3	10.3	10.7
Bolzano-Bozen	46.46	11.35	758	549	71	14.6	1020	9.6	16.9	11.5	11.6
Bot River	-34.23	19.20	693	257	30	18.7	1848	16.4	20.0	10.6	10.6
Bourges	47.06	2.40	712	421	57	15.8	1242	11.5	18.0	11.0	11.4
Bourke	-30.09	145.94	306	200	37	25.2	3216	20.6	26.8	14.9	14.2
Bowral	-34.48	150.42	1021	664	116	16.4	1362	13.3	17.9	11.7	11.0
Bratislava	48.15	17.11	645	428	57	16.0	1296	10.4	18.0	11.7	11.9
Brescia	45.54	10.21	904	593	84	18.8	1875	13.2	21.1	10.5	10.3
Brighton	50.82	-0.13	808	399	65	13.5	786	10.3	15.4	9.2	9.6
Brindisi	40.65	17.94	485	213	43	21.0	2357	17.0	23.9	8.4	8.3
Browns Valley	39.24	-121.41	695	136	9	22.1	2589	17.2	24.8	15.9	17.1
Bsilio Nievas	-31.54	-68.73	232	204	29	21.4	2420	16.9	22.4	15.8	15.6
Bucharest	44.43	26.10	602	409	46	17.6	1637	11.2	19.8	12.8	13.4
Buffalo	42.89	-78.88	962	589	90	15.8	1337	9.0	19.1	10.6	10.6
Bunbury	-33.32	115.64	801	178	20	20.0	2125	17.7	22.2	13.8	14.4
Burgas	42.50	27.46	559	314	50	18.5	1822	13.1	21.5	9.3	10.1
Burgos	42.34	-3.70	522	282	37	15.4	1195	11.2	18.6	11.7	12.4
Buzet	44.25	0.30	960	507	75	17.0	1503	13.0	19.3	11.8	12.1
Bzenec	48.97	17.27	653	479	57	15.1	1104	9.4	16.9	11.4	11.7
Čačak	43.89	20.35	801	512	64	17.0	1505	11.4	19.0	11.6	12.5
Cáceres	39.48	-6.37	541	220	28	20.8	2307	16.2	24.2	12.8	13.6
Cachi	-25.12	-66.16	169	167	21	17.1	1501	14.4	17.3	13.7	12.5
Cafayate	-26.07	-65.98	160	158	20	20.0	2113	17.1	20.3	12.1	10.9
Cagliari	39.22	9.12	428	170	33	21.2	2406	17.3	24.3	9.5	9.5
Cahors	44.45	1.44	808	459	64	16.9	1481	12.8	19.0	11.9	12.1
Calitzdorp	-33.53	21.68	242	150	24	20.6	2240	17.8	22.2	15.4	15.1
Caltanissetta	37.49	14.06	486	195	38	20.1	2158	16.0	23.0	7.3	7.2
Cama	46.27	9.17	1388	991	141	12.5	663	7.9	14.6	8.2	8.1

Table 76 (cont.): Key climate indicators of the world's wine regions

Location	Latitude	Longitude	AnnP	GSP	RipeP	GST	GDD	AnnT	RPT	GSDTR	RPDTR
Campo Grande	-27.21	-54.98	1620	1013	132	23.2	2790	20.4	24.2	12.5	12.1
Campobasso	41.56	14.66	597	297	47	17.4	1580	12.7	20.4	8.6	8.9
Canandaigua	42.89	-77.28	878	581	83	15.5	1269	8.6	18.6	11.9	11.9
Canberra	-35.28	149.13	660	408	53	17.1	1509	13.4	19.0	14.3	14.1
Canelones	-34.53	-56.29	1112	665	109	20.1	2142	16.8	21.7	11.3	11.2
Cañuelas	-35.05	-58.76	1078	752	117	20.3	2180	16.8	21.7	13.0	12.9
Capayán	-28.77	-66.05	422	385	68	24.5	3064	20.3	24.7	13.8	13.0
Carbonia	39.16	8.52	642	225	40	21.0	2354	17.1	23.9	9.4	9.6
Carcassonne	43.21	2.35	691	362	54	18.2	1753	14.0	20.7	10.4	10.6
Carmelo	-34.00	-58.29	1050	716	136	21.0	2327	17.6	22.4	10.5	10.3
Casablanca	-33.32	-71.41	438	38	4	17.6	1615	15.1	18.6	12.8	13.1
Caserta	41.07	14.33	939	422	85	20.3	2215	16.1	23.1	10.2	10.6
Castellón de la Plana	39.99	-0.05	463	284	65	21.2	2395	17.3	24.1	8.3	8.1
Castelo Branco	39.82	-7.50	873	278	39	20.1	2161	15.9	23.0	11.4	12.4
Catania	37.51	15.08	536	205	45	21.5	2464	17.6	24.6	9.7	9.8
Catanzaro	38.91	16.59	907	334	48	20.9	2327	16.8	24.0	8.1	8.6
Cattaraugus	42.33	-78.87	1095	697	104	14.0	1007	7.3	17.0	12.7	12.4
Caucete	-31.65	-68.28	282	260	38	22.9	2722	18.0	23.7	15.6	15.1
Čerhov	48.46	21.64	621	427	51	15.5	1187	9.5	17.5	11.8	12.1
Cessnook	-32.50	151.36	798	569	93	21.3	2388	17.9	22.5	12.8	11.8
Chablis	47.82	3.80	723	423	57	15.2	1112	10.8	17.0	10.8	11.0
Chambery	45.64	5.92	1031	594	94	15.7	1224	10.8	17.8	11.3	11.6
Charlottesville	38.13	-78.48	1096	704	105	19.5	2038	13.3	22.0	12.9	12.4
Chartres	48.44	1.49	592	354	47	14.9	1058	10.9	17.1	10.3	10.5
Châteauroux	46.81	1.69	745	438	63	15.8	1236	11.6	18.0	10.6	10.8
Chaumont	48.11	5.14	745	443	56	14.5	997	10.0	16.8	10.3	10.5
Chautauqua	42.21	-79.47	1158	737	112	15.1	1186	8.1	18.1	11.0	10.8
Chaves	41.74	-7.47	1705	642	94	17.6	1628	14.0	20.5	12.1	13.0
Chelan	47.84	-120.02	283	98	11	15.8	1261	9.6	18.8	14.5	15.7
Chengdu	30.57	104.07	1019	940	141	22.3	2634	17.0	24.1	7.5	7.3
Chico	39.73	-121.84	621	122	10	21.1	2384	16.4	23.5	16.9	18.6
Chieti	42.35	14.16	702	382	66	19.4	2010	14.8	22.0	9.2	9.8
Chilecito	-29.16	-67.50	197	183	31	21.5	2427	17.3	22.2	15.2	14.9
Chillan	-36.60	-72.10	1072	247	25	16.7	1417	13.5	18.0	14.5	15.5
Chisinau	47.01	28.86	540	346	43	17.2	1536	10.5	19.3	10.3	10.8
Choele Choel	-39.28	-65.66	289	179	28	19.6	2041	15.3	20.8	17.2	17.6
Chos Malal	-37.38	-70.27	270	85	10	16.7	1424	12.8	18.0	18.2	19.3
Christchurch	-43.53	172.64	793	404	64	14.9	1026	12.3	16.3	9.6	9.6
Chtaura	33.82	35.85	783	102	2	21.7	2499	16.6	24.4	15.0	16.0
Chur	46.85	9.53	987	666	92	13.7	850	8.9	15.8	9.7	9.7
Ciudad Real	38.98	-3.93	413	195	26	20.1	2169	15.1	23.5	15.0	16.1
Clare	-33.83	138.61	552	221	23	18.5	1802	15.1	20.5	15.4	15.8
Clearlake	38.96	-122.63	722	121	10	18.5	1827	14.2	21.7	19.3	21.2
Clermont-Ferrand	45.78	3.09	610	443	60	15.5	1187	11.2	17.7	11.7	12.0
Cleveland	41.50	-81.69	953	606	87	17.4	1594	10.5	20.4	11.7	11.6
Coimbra	40.20	-8.41	982	383	47	19.2	1973	16.0	21.5	11.5	12.7
Collón Cura	-38.95	-68.10	280	202	28	19.0	1909	14.6	20.2	16.6	17.0
Colmar	47.93	7.36	623	400	50	15.6	1201	10.7	17.5	10.5	10.8
Colón	-32.22	-58.14	1205	834	128	21.7	2472	18.3	23.2	12.9	12.6
Colusa	39.21	-122.01	407	74	7	21.1	2373	16.5	23.2	17.8	19.4
Como	45.81	9.09	1277	888	129	16.3	1351	11.3	18.6	9.2	9.1

Table 76 (cont.): Key climate indicators of the world's wine regions

Location	Latitude	Longitude	AnnP	GSP	RipeP	GST	GDD	AnnT	RPT	GSDTR	RPDTR
Concepción del Uruguay	-32.48	-58.23	1147	779	119	21.6	2456	18.2	23.1	12.4	12.4
Concord	37.98	-122.03	462	80	5	19.7	2073	16.0	21.8	13.3	14.1
Concordia	-31.39	-58.02	1350	947	149	22.4	2620	18.9	23.6	13.0	12.7
Constanța	44.16	28.63	428	254	36	18.3	1782	12.4	21.0	7.7	7.9
Coonawarra	-37.75	140.86	728	283	35	16.1	1289	13.8	17.7	11.6	12.1
Cootamundra	-34.64	148.03	576	327	41	19.5	2014	15.2	21.8	15.4	15.3
Copiapó	-27.37	-70.33	8	0	0	17.0	1483	14.9	18.3	10.2	10.1
Córdoba	37.89	-4.78	574	213	29	22.4	2650	17.9	25.5	13.8	14.7
Corning	39.93	-122.18	562	115	11	21.7	2500	16.7	24.3	16.4	17.9
Coronel Pringles	-37.99	-61.35	814	602	97	17.7	1630	13.9	19.2	14.0	13.9
Coronel Suarez	-37.46	-61.93	815	618	99	18.3	1757	14.3	19.5	14.3	14.4
Coruche	38.96	-8.52	721	258	30	19.9	2126	16.6	22.5	10.1	10.9
Corvallis	44.56	-123.26	1168	328	39	15.7	1214	11.8	18.4	14.1	16.1
Cosenza	39.30	16.25	925	362	57	20.0	2136	15.6	23.0	7.4	7.6
Cosne-Cours-sur-Loire	47.38	2.93	720	424	55	15.4	1153	11.0	17.1	11.0	11.3
Cosquín	-31.25	-64.47	694	607	94	18.8	1867	15.5	19.5	13.8	13.3
Cowra	-33.83	148.68	625	369	45	20.2	2155	16.0	22.2	16.0	15.8
Cremona	45.13	10.02	808	490	73	18.9	1897	13.1	21.3	10.8	10.8
Crissier	46.55	6.57	1146	711	99	15.2	1120	10.4	17.1	10.6	10.9
Črnomelj	45.57	15.19	1244	789	128	16.4	1373	10.9	18.2	10.6	11.1
Cromwell	-45.03	169.20	728	374	59	14.1	855	10.4	15.7	11.9	12.4
Crotone	39.08	17.13	684	240	39	21.4	2446	17.7	24.4	9.2	9.3
Cruz del Eje	-30.72	-64.81	528	470	65	21.5	2445	18.1	22.0	13.6	12.9
Cuenca	40.07	-2.14	518	278	40	17.7	1651	13.0	21.1	14.0	15.0
Cuneo	44.38	7.54	733	440	68	17.5	1611	12.4	20.0	8.8	8.8
Curico	-34.97	-71.25	706	109	14	17.7	1630	14.2	19.0	16.7	17.9
Daireaux	-36.60	-61.75	873	669	111	19.4	1991	15.4	20.5	14.1	14.3
Dallas	44.92	-123.32	1367	356	44	15.0	1089	11.3	18.0	15.2	17.4
Dallas	32.78	-96.80	921	624	97	25.0	3221	19.1	27.7	11.7	11.9
Dallenwil	46.93	8.39	1425	940	106	13.4	810	8.6	15.3	8.8	8.6
Darwin	-12.46	130.85	1742	1701	324	28.5	3925	27.3	28.1	9.1	10.1
Deán Funes	-30.73	-64.79	532	473	66	21.2	2374	17.8	21.7	13.6	12.9
Delémont	47.37	7.35	934	602	76	14.4	973	9.5	16.4	10.0	10.2
Denmark	-34.96	117.35	981	338	42	17.6	1610	15.7	19.2	10.4	10.6
Des Moines	41.59	-93.62	887	699	89	18.3	1770	10.2	21.0	12.2	12.3
Detroit	42.33	-83.05	871	561	83	17.1	1551	10.0	20.1	10.7	10.6
Diamante	-32.07	-60.64	1035	823	152	21.6	2463	18.2	22.9	12.3	12.0
Dijon	47.27	5.04	754	459	64	15.6	1190	10.8	17.7	10.5	10.6
Dobrich	43.57	27.83	527	312	45	17.3	1567	11.4	19.9	9.0	9.6
Dobrovo	46.00	13.53	1569	967	170	18.2	1758	13.1	20.4	10.3	10.6
Domodossola	46.11	8.29	1074	727	84	15.0	1086	10.1	17.1	9.5	9.4
Doros	34.82	32.91	622	117	10	21.2	2400	17.2	23.9	12.3	12.8
Dresden	51.05	13.74	650	435	50	15.3	1141	10.1	17.3	9.7	9.8
Dunajská Streda	47.99	17.62	585	394	52	16.2	1330	10.4	18.1	11.9	12.2
Durazno	-33.38	-56.53	1310	787	133	21.1	2350	17.6	22.5	12.1	11.8
Dürnstein	48.40	15.52	617	458	51	14.5	992	9.2	16.0	11.5	11.8
Eden Valley	-34.64	139.10	539	215	26	17.7	1633	14.5	19.6	14.2	14.8
Eger	47.90	20.38	542	388	48	16.7	1444	10.5	18.7	12.3	12.8
Eisenberg	47.18	16.43	719	519	71	15.4	1159	9.8	17.3	11.4	11.7
El Bolsón	-27.90	-65.88	512	467	77	21.2	2368	17.4	21.6	13.3	12.1

Table 76 (cont.): Key climate indicators of the world's wine regions

Location	Latitude	Longitude	AnnP	GSP	RipeP	GST	GDD	AnnT	RPT	GSDTR	RPDTR
El Carmen	-24.39	-65.26	769	727	106	20.7	2268	17.8	20.8	12.0	10.7
El Cuy	-39.93	-68.35	161	102	18	16.0	1272	11.7	17.4	16.3	17.0
El Huecú	-37.64	-70.58	562	164	22	13.4	749	9.6	14.6	18.3	19.3
El Paso	31.76	-106.49	214	165	38	23.9	2984	18.1	26.0	16.3	15.0
El Portezuelo	-28.47	-65.64	506	467	80	23.6	2882	19.7	23.9	12.9	12.1
Elazığ	38.67	39.22	675	253	9	20.2	2195	13.1	24.2	12.1	13.4
Elizabeth	-34.72	138.67	544	212	24	19.6	2029	16.4	21.4	13.0	13.4
Ellenville	41.72	-74.40	1147	719	100	16.0	1337	9.1	18.8	12.8	12.6
Enna	37.57	14.28	530	212	43	19.0	1934	14.7	22.3	7.4	7.5
Épinal	48.17	6.44	918	523	70	14.4	980	9.8	16.5	10.7	11.0
Erie	42.13	-80.09	1056	666	106	16.2	1401	9.5	19.6	10.7	10.6
Esperance	-33.86	121.89	630	231	32	18.9	1872	16.8	20.3	10.7	10.4
Esquel	-42.91	-71.31	671	250	34	12.2	551	8.8	13.6	13.8	14.8
Estaing	44.55	2.67	660	405	63	15.5	1187	11.4	17.8	12.2	12.5
Estella	42.67	-2.03	715	383	46	16.1	1299	12.1	19.4	11.5	12.3
Etyek	47.45	18.75	562	367	48	16.8	1449	10.7	18.8	11.0	11.2
Eugene	44.05	-123.09	1234	349	42	16.0	1282	12.1	18.6	13.8	15.8
Eugene	44.05	-123.09	1234	349	42	16.0	1282	12.1	18.6	13.8	15.8
Eureka	40.80	-124.16	1037	242	21	14.1	867	12.2	15.5	7.2	7.3
Evora	38.57	-7.91	655	243	30	19.8	2106	16.2	22.6	11.7	12.9
Fairfield	38.25	-122.04	533	83	5	19.3	1999	15.6	21.3	15.3	16.4
Famatina	-28.93	-67.52	180	168	27	19.3	1971	15.5	20.0	15.5	14.9
Fermo	43.16	13.72	686	399	70	19.0	1918	14.3	21.5	8.5	8.8
Ferrara	44.84	11.62	627	374	55	20.3	2202	14.4	22.9	9.1	9.3
Firenze	43.80	11.26	911	475	78	19.3	2003	14.6	22.1	10.6	10.9
Florida	-34.09	-56.22	1241	737	123	20.4	2195	16.9	21.9	11.9	11.7
Foggia	41.46	15.54	507	265	49	20.4	2237	16.4	23.4	11.5	11.8
Foix	42.96	1.61	843	499	64	16.1	1303	12.1	18.6	10.2	10.4
Fontette	48.08	4.61	843	480	66	14.3	964	10.0	16.7	10.4	10.6
Forlì	44.22	12.04	748	434	74	19.5	2047	14.0	22.2	10.0	10.4
Frankfort	38.20	-84.87	1157	706	87	19.3	1995	12.8	22.1	13.1	13.1
Frasco	46.34	8.80	1445	1001	133	9.0	205	4.8	11.1	7.0	6.9
Fresno	36.74	-119.79	264	59	5	23.1	2809	18.0	25.9	17.4	18.3
Fribourg	46.81	7.16	1020	666	84	13.9	905	9.1	16.1	10.0	10.1
Frosinone	41.63	13.34	1170	537	96	19.3	1998	14.7	22.3	12.2	13.0
Funchal	32.63	-16.92	627	218	44	19.9	2111	18.4	22.1	5.9	6.2
Gaillac	43.90	1.90	695	397	62	17.8	1667	13.4	20.0	11.4	11.6
Gap	44.56	6.08	916	499	79	14.9	1091	10.2	17.5	11.5	12.0
Gaziantep	37.07	37.38	546	131	5	22.1	2598	15.6	25.9	14.5	15.2
Geelong	-38.15	144.36	541	300	33	17.2	1520	14.7	19.0	10.2	10.3
Geisenheim	49.59	7.97	589	367	44	14.2	940	9.6	16.3	10.2	10.4
General Alvear	-34.98	-67.69	295	237	34	20.2	2167	15.8	21.1	16.8	16.9
General Belgrano	-35.77	-58.50	1007	703	104	19.3	1963	15.7	20.7	13.1	13.1
General Conesa	-40.10	-64.46	256	152	27	19.1	1931	14.9	20.3	15.4	15.6
General Pueyrredón	-37.95	-57.78	875	558	107	17.2	1514	13.9	18.8	12.0	11.6
General Roca	-39.03	-67.59	181	106	15	19.1	1937	14.8	20.3	17.3	18.0
General San Martín	-31.44	-68.52	249	226	31	22.5	2637	17.7	23.4	15.9	15.6
Geneva	46.20	6.14	937	546	87	15.4	1164	10.6	17.5	10.7	11.0
Genova	44.41	8.95	1194	650	123	19.4	2024	15.4	22.2	8.0	8.0
Gisborne	-38.66	178.02	1200	571	99	16.8	1427	14.3	18.1	10.3	10.1

Table 76 (cont.): Key climate indicators of the world's wine regions

Location	Latitude	Longitude	AnnP	GSP	RipeP	GST	GDD	AnnT	RPT	GSDTR	RPDTR
Gladstone	-33.27	138.35	437	190	21	20.2	2149	16.5	22.1	14.6	14.9
Glarus	47.04	9.07	1539	1035	130	12.7	688	7.9	14.7	8.8	8.6
Gleed	46.66	-120.61	227	78	9	15.9	1269	10.2	18.7	16.7	18.3
Glen Innes	-29.75	151.74	912	649	74	16.9	1456	13.4	18.4	13.1	12.4
Glenrowan	-36.46	146.22	709	341	41	18.7	1845	14.9	21.0	14.4	14.9
Godoy Cruz	-32.93	-68.86	268	219	36	20.3	2189	16.0	21.1	15.3	14.9
Gorizia	45.94	13.62	1498	920	165	18.3	1789	13.2	20.8	10.0	10.3
Gornji Petrovci	46.80	16.22	806	576	81	15.4	1150	9.8	17.1	11.3	11.6
Goumenissa	40.95	22.45	465	274	38	19.7	2073	14.0	22.6	12.2	13.1
Granada	37.18	-3.60	425	167	23	20.3	2217	15.9	23.4	13.7	14.6
Grand Junction	39.06	-108.55	223	144	26	18.9	1912	11.9	21.9	17.4	17.5
Granger	46.34	-120.19	174	68	10	16.9	1469	11.1	19.4	16.3	17.6
Granger	46.34	-120.19	174	68	10	16.9	1469	11.1	19.4	16.3	17.6
Grants Pass	42.44	-123.33	823	198	21	17.1	1515	12.7	20.1	18.6	21.3
Grenoble	45.19	5.72	934	563	93	17.5	1597	12.5	19.6	11.3	11.5
Griffith	-34.28	146.05	418	238	40	21.5	2436	17.1	23.8	14.9	14.6
Grosseto	42.76	11.11	633	312	72	19.9	2111	15.5	22.7	10.4	10.8
Guadalajara	40.63	-3.16	427	216	32	18.4	1808	13.6	21.8	11.9	12.5
Guadalupe	32.10	-116.57	265	49	6	20.6	2272	17.6	23.8	14.6	14.9
Gualeguaychú	-33.01	-58.51	1079	758	116	21.4	2419	17.9	22.8	12.4	12.4
Guaymallén	-32.90	-68.80	251	208	35	20.8	2296	16.5	21.6	14.8	14.4
Gurtnellen	46.74	8.63	1589	1064	135	9.3	232	5.0	11.4	7.1	6.9
Gyöngyös	47.78	19.93	538	374	48	16.8	1465	10.6	18.9	12.0	12.4
Hägglingen	47.39	8.25	1169	713	84	14.6	1003	9.5	16.5	9.4	9.4
Halkída	38.47	23.62	494	161	28	21.9	2550	17.4	24.9	9.7	9.9
Hamilton	-37.74	142.02	647	291	36	15.9	1255	13.3	18.1	12.8	14.1
Hamilton	-37.79	175.28	1358	733	89	16.9	1463	14.3	18.6	9.6	10.1
Hanford	36.33	-119.65	203	43	4	22.6	2694	17.5	25.1	17.4	18.1
Hastings	-38.31	145.19	772	389	48	16.9	1465	14.5	18.8	9.9	10.2
Hayfork	40.55	-123.18	1035	204	20	16.3	1376	11.8	19.9	20.3	22.7
Healesville	-37.81	145.51	1028	532	62	16.8	1430	13.9	18.8	12.3	12.9
Heathcote	-36.92	144.70	575	275	31	17.3	1548	13.9	19.3	13.8	14.1
Hermosillo	29.07	-110.96	287	221	40	28.1	3881	23.6	30.9	15.2	12.9
Hilsboro	45.53	-122.40	1454	510	64	15.9	1259	11.8	18.5	12.6	14.1
Hódmező-vásárhely	46.42	20.33	531	350	44	17.0	1493	10.9	18.9	12.1	12.6
Hollister	36.85	-121.40	350	64	3	18.7	1868	15.5	20.9	15.8	16.5
Horitschon	47.59	16.55	662	475	66	15.5	1179	10.0	17.5	11.5	11.7
Houston	29.76	-95.37	1221	791	132	25.9	3396	21.2	27.9	9.7	9.7
Huaco	-30.16	-68.48	152	138	19	20.4	2200	16.4	21.3	15.8	15.6
Huelva	37.26	-6.94	491	165	16	21.7	2514	18.2	24.3	8.5	8.6
Huesca	42.13	-0.41	553	334	55	19.4	2015	14.5	22.3	11.3	11.8
Humahuaca	-23.20	-65.35	179	175	22	15.2	1093	12.6	15.3	17.0	16.0
Iași	47.16	27.60	562	402	47	16.9	1478	10.3	18.9	11.7	12.1
Imperia	43.89	8.04	737	361	72	19.3	2001	15.5	22.2	8.5	8.8
Indianapolis	39.77	-86.16	1045	670	75	18.3	1769	11.1	21.0	12.1	12.3
Iquique	-20.24	-70.14	0	0	0	19.5	2007	18.0	20.7	7.5	7.7
Iraklion	35.34	25.14	507	120	17	22.6	2690	19.0	25.0	5.7	4.9
Irvine	33.68	-117.83	309	46	7	20.2	2184	17.8	22.5	11.2	11.6
Isernia	41.60	14.23	766	382	68	18.5	1830	13.6	21.5	9.7	10.2
İzmir	38.42	27.14	704	155	17	22.9	2768	18.0	25.8	11.8	11.8
Jackson	38.35	-120.77	824	173	13	20.2	2178	15.6	23.5	17.6	19.0

Table 76 (cont.): Key climate indicators of the world's wine regions

Location	Latitude	Longitude	AnnP	GSP	RipeP	GST	GDD	AnnT	RPT	GSDTR	RPDTR
Jerez de la Frontera	36.45	-6.12	632	191	20	21.4	2433	18.1	24.0	7.5	7.3
Jerusalem	31.77	35.21	405	36	0	21.2	2407	17.0	23.4	13.4	13.0
Jesús María	-30.98	-64.10	727	637	100	19.1	1928	15.9	19.7	13.6	13.0
Junín	-33.14	-68.48	198	168	29	20.8	2292	16.3	21.6	15.6	15.3
Junín - Bs. As.	-34.59	-60.95	1010	792	146	19.9	2096	16.2	21.0	13.9	14.0
Karlovac	45.49	15.37	1229	771	120	16.1	1315	10.7	17.9	10.7	11.3
Kecskemét	46.90	19.69	550	359	49	16.9	1470	10.8	18.9	11.4	11.9
Kelibia	36.85	11.10	510	193	40	22.2	2613	18.7	25.4	14.0	14.5
Kikinda	45.83	20.46	572	366	46	17.6	1637	11.6	19.6	11.8	12.6
Kitzeck im Sausal	46.78	15.45	1043	764	104	14.9	1072	9.2	16.7	11.4	11.9
Klawer	-31.77	18.62	202	55	7	22.1	2561	19.3	23.6	15.4	15.8
Klöch	46.76	15.97	857	620	85	15.1	1106	9.5	16.9	11.3	11.6
Knjaževac	43.57	22.25	643	392	51	17.0	1491	11.2	19.0	12.1	13.1
Koceljeva	44.47	19.81	775	504	60	17.1	1522	11.5	19.1	11.6	12.4
Kofu	35.67	138.57	1169	918	157	19.1	1954	13.2	23.0	10.2	9.4
Koper	45.55	13.73	1017	624	109	19.5	2029	14.5	21.9	7.4	7.2
Koraleigh	-35.16	143.42	321	176	22	20.7	2264	16.9	22.5	15.2	15.4
Kragujevac	44.01	20.91	714	458	55	17.0	1494	11.4	19.0	11.9	13.1
Kráľovský Chlmec	48.42	21.98	635	419	52	15.8	1252	9.8	17.7	11.7	11.9
Krapina	46.16	15.87	1083	713	110	15.7	1214	10.2	17.3	11.0	11.3
Krasnodar	45.04	38.97	754	409	46	18.2	1755	12.1	20.6	9.2	9.4
Krems	48.42	15.60	583	437	47	14.9	1072	9.5	16.7	11.5	11.8
Križ	45.74	13.87	1634	948	156	16.7	1437	11.7	18.6	8.4	8.6
Kruševac	43.58	21.33	666	409	51	17.1	1514	11.4	19.0	12.5	13.6
Kutina	45.48	16.78	935	591	86	17.4	1575	11.7	18.9	11.8	12.6
Kutná Hora	49.95	15.27	504	382	40	14.2	960	8.8	16.1	10.7	10.9
La Ferté-sous-Jouarre	48.95	3.13	620	372	49	14.7	1020	10.6	16.7	10.7	10.9
La Paz	-33.46	-67.55	261	219	30	21.8	2492	17.0	22.5	15.9	15.6
La Rioja (Arg)	-29.41	-66.86	352	325	57	23.6	2886	19.3	24.1	14.3	13.6
La Rochelle	46.16	-1.15	763	376	60	16.8	1457	13.1	19.0	8.3	8.4
La Roche-sur-Yon	46.67	-1.43	774	383	59	15.9	1275	12.2	17.9	8.9	9.1
La Spezia	44.10	9.82	981	455	81	19.3	2001	14.9	22.0	8.4	8.6
La Viña	-25.47	-65.57	383	373	55	21.2	2370	18.3	21.4	13.1	11.8
Lago Puelo	-42.06	-71.60	1115	376	55	13.9	819	10.6	15.1	14.9	16.0
Lagoa	37.13	-8.45	540	163	18	20.2	2185	17.2	22.5	9.8	10.3
Lakes Entrance	-37.88	147.99	707	410	52	17.4	1560	15.1	18.8	10.4	10.4
Langenlois	48.47	15.69	543	410	44	15.2	1121	9.7	17.1	11.6	11.8
Langhorne Creek	-35.30	139.04	379	176	20	19.0	1894	16.2	20.5	12.4	12.6
Lanzhou	36.06	103.83	347	325	54	16.9	1483	9.5	18.5	12.8	11.8
L'Aquila	42.35	13.40	864	574	99	17.0	1508	12.0	20.0	11.9	13.1
Las Heras	-32.84	-68.83	230	190	32	21.2	2364	16.7	21.9	14.6	14.0
Las Lajas	-38.52	-70.36	233	78	10	16.0	1265	12.1	17.2	18.5	19.3
Latina	41.47	12.90	947	388	81	20.4	2223	16.1	23.4	11.2	11.7
Launceston	-41.54	147.14	687	326	40	14.4	918	11.7	16.0	11.8	12.4
Laval	48.08	-0.77	751	397	58	15.3	1132	11.7	17.4	10.2	10.5
Lavalleja	-34.36	-55.26	1260	716	114	19.4	1986	16.2	21.0	11.5	11.2
Le Mans	48.01	-0.20	735	383	58	15.3	1136	11.7	17.4	10.7	11.1

Table 76 (cont.): Key climate indicators of the world's wine regions

Location	Latitude	Longitude	AnnP	GSP	RipeP	GST	GDD	AnnT	RPT	GSDTR	RPDTR
Le Puy	45.04	3.88	721	487	73	14.1	931	9.7	16.4	11.2	11.8
Leandro N. Alem	-27.60	-55.32	1698	1070	143	23.4	2846	20.6	24.5	12.7	12.2
Lecce	40.35	18.18	579	244	53	21.2	2400	16.9	24.1	8.8	9.1
Lecco	45.86	9.40	1189	829	123	16.9	1470	11.8	19.1	9.9	9.9
Lenswood	-35.06	138.83	831	298	34	17.0	1482	14.2	18.8	13.2	14.0
León	42.60	-5.57	570	283	38	16.3	1352	11.9	19.4	13.6	14.3
Leskovac	43.00	21.94	637	376	53	17.5	1608	11.6	19.6	12.8	14.0
Liestal	47.49	7.73	1048	653	80	14.4	979	9.7	16.3	9.9	10.0
Lima	-12.05	-77.04	47	9	2	20.8	2276	19.2	22.8	9.5	9.8
Limnos	39.92	25.14	563	184	34	20.1	2172	15.5	22.9	8.4	8.4
Lisbon	38.72	-9.14	752	261	30	19.8	2098	16.9	22.0	9.0	9.8
Little Rock	34.75	-92.29	1305	745	99	22.8	2738	16.7	25.1	12.0	11.9
Livermore	37.68	-121.77	379	70	5	19.5	2027	15.7	21.9	16.6	18.1
Livorno	43.55	10.31	724	383	74	19.4	2007	15.3	22.1	9.9	10.4
Lleida	41.62	0.62	357	228	42	20.0	2144	15.1	22.9	12.8	13.0
Lodi	45.31	9.50	850	534	77	18.4	1794	12.6	20.8	10.6	10.6
Lodi	38.11	-121.27	444	83	6	20.4	2221	16.0	22.5	17.3	18.5
Logrono	42.45	-2.45	432	249	29	17.7	1658	13.5	20.5	11.7	12.3
Lons-le-Saunier	46.67	5.55	1157	687	99	16.0	1286	11.2	17.8	10.5	10.8
Los Angeles	34.05	-118.24	382	56	6	21.4	2443	18.9	23.8	11.4	11.9
Lubbock	33.58	-101.86	480	394	63	22.3	2625	16.2	24.4	14.9	14.1
Lucca	43.84	10.50	952	501	91	19.3	1992	14.6	22.1	10.5	10.9
Lucerne	47.05	8.31	1292	841	97	14.9	1057	9.9	16.8	9.2	9.1
Luján de Cuyo	-33.03	-68.88	240	190	33	19.5	2016	15.2	20.3	15.7	15.5
Luxembourg	49.61	6.13	843	474	63	14.2	929	9.8	16.2	9.9	9.6
Lyon	45.76	4.84	813	525	78	17.4	1581	12.6	19.4	11.1	11.4
Macerata	43.30	13.45	697	408	69	18.5	1812	13.7	21.1	8.1	8.2
Mâcon	46.31	4.83	830	509	78	16.2	1325	11.5	18.3	10.7	11.0
Madera	36.96	-120.06	272	64	4	22.0	2566	17.0	24.7	18.1	19.0
Madrid	40.42	-3.70	439	221	30	19.2	1972	14.5	22.5	12.0	12.1
Mailberg	48.67	16.18	530	384	44	15.6	1213	10.0	17.6	11.4	11.6
Maipú	-32.96	-68.79	228	190	33	20.5	2213	16.1	21.2	15.2	14.9
Maitland	-34.37	137.67	436	169	20	19.6	2023	16.5	21.4	12.7	12.9
Málaga	36.72	-4.43	559	165	20	21.5	2470	18.0	24.4	8.5	8.4
Malargüe	-35.48	-69.58	423	164	23	15.1	1073	11.2	16.2	16.5	16.4
Maldonado	-34.90	-54.97	1034	575	86	19.4	1982	16.5	21.1	7.6	7.5
Malmesbury	-33.47	18.72	496	147	19	20.9	2306	17.9	22.6	13.4	13.8
Mandalay	21.98	96.08	833	784	155	29.6	4197	27.3	29.0	8.9	7.6
Mandurah	-32.54	115.74	823	179	19	20.8	2275	18.3	22.9	12.0	12.6
Manjimup	-34.24	116.15	895	250	26	17.8	1639	15.2	19.7	12.5	13.4
Mantova	45.16	10.79	664	414	56	19.3	2002	13.6	21.6	10.4	10.6
Margaret River	-33.95	115.07	1085	245	26	18.8	1860	16.6	20.7	9.6	10.1
Mariposa	37.49	-119.97	799	169	9	19.8	2095	15.1	23.7	17.4	18.7
Maronia	40.90	25.52	607	248	34	19.2	1969	14.1	22.1	10.6	11.1
Märstetten	47.59	9.06	931	628	74	14.3	966	9.3	16.2	9.7	9.9
Martinborough	-41.22	175.46	937	444	74	15.3	1117	12.8	16.9	9.5	10.0
Massa	44.04	10.14	1116	539	89	19.3	1993	14.7	22.1	9.4	9.9
Matera	40.67	16.60	619	296	47	19.0	1930	14.5	21.9	9.6	10.2
Mattawa	46.74	-119.90	173	66	8	17.4	1579	11.4	20.2	15.1	16.0
McLaren Vale	-35.22	138.55	539	202	25	18.7	1848	16.0	20.4	10.8	11.0
McMinnville	45.13	-123.20	1134	301	38	15.0	1078	11.4	18.0	15.1	17.3

Table 76 (cont.): Key climate indicators of the world's wine regions

Location	Latitude	Longitude	AnnP	GSP	RipeP	GST	GDD	AnnT	RPT	GSDTR	RPDTR
McMinnville	45.21	-123.20	1070	288	36	15.3	1132	11.6	18.1	15.1	17.2
Médanos	-38.83	-62.69	558	396	66	19.4	1981	15.2	20.6	14.3	14.3
Medford	42.33	-122.88	516	159	18	16.8	1457	12.3	19.9	18.0	20.0
Medford	42.33	-122.88	516	159	18	16.8	1457	12.3	19.9	18.0	20.0
Media Agua	-31.98	-68.43	258	227	34	22.8	2704	18.0	23.6	15.4	15.0
Meknes	33.87	-5.54	545	181	16	20.9	2333	16.9	24.0	15.2	16.1
Melk	48.23	15.35	643	471	54	15.0	1097	9.5	16.7	11.8	12.1
Melnik	41.52	23.39	524	301	38	18.2	1756	12.4	20.9	11.8	12.5
Mende	44.52	3.50	882	516	80	13.2	766	9.0	15.6	11.2	11.6
Merced	37.30	-120.48	325	66	5	21.6	2496	16.8	24.2	17.4	18.5
Merignac	44.38	-0.66	886	445	76	16.9	1468	13.0	19.2	11.3	11.7
Messina	38.19	15.55	818	330	70	22.7	2712	19.0	26.0	7.5	7.4
Metz	49.12	6.18	747	430	57	15.1	1097	10.6	17.1	10.4	10.7
Milang	-35.41	138.96	368	162	19	19.0	1892	16.3	20.5	11.8	11.9
Milano	45.46	9.19	985	632	89	19.0	1936	13.5	21.4	10.1	9.6
Mildura	-34.21	142.14	283	159	16	21.1	2349	17.3	22.9	15.2	15.4
Minneapolis	44.98	-93.27	759	610	79	16.5	1455	7.7	19.3	11.8	11.6
Miskolc	48.10	20.78	547	396	45	15.9	1265	9.8	17.6	12.7	13.1
Moama	-36.11	144.76	412	215	27	19.6	2022	15.8	21.5	14.8	15.1
Modena	44.65	10.93	725	434	65	19.5	2028	13.7	22.1	10.6	10.9
Modesto	37.64	-121.00	306	64	6	21.9	2550	17.2	24.4	17.1	17.9
Modra	48.33	17.31	730	514	63	15.7	1227	10.1	17.7	11.7	11.9
Molinos	-25.44	-66.29	145	143	17	18.5	1805	15.7	18.8	13.3	12.1
Monostorapáti	46.93	17.56	628	435	64	16.0	1293	10.4	18.1	11.3	11.6
Montague	41.73	-122.30	527	161	15	15.5	1239	10.6	19.3	19.0	21.1
Montauban	44.02	1.35	732	420	58	17.5	1606	13.3	19.6	11.6	11.8
Montevideo	-34.82	-56.21	958	575	92	19.8	2072	16.5	21.5	10.7	10.6
Monza	45.58	9.27	1025	678	98	18.4	1806	13.0	20.8	10.0	9.9
Moonambel	-36.99	143.33	618	273	31	16.6	1401	13.4	18.6	14.5	15.1
Moquegua	-17.19	-70.93	17	17	0	19.2	1953	18.6	19.9	13.2	12.8
Mór	47.37	18.20	571	378	54	16.4	1366	10.5	18.4	11.4	11.6
Morioka	39.70	141.15	1355	948	159	17.3	1601	10.6	21.7	9.5	8.6
Most	50.50	13.63	691	464	52	14.1	957	8.9	16.3	9.4	9.4
Mount Benson	-37.06	139.84	642	218	27	16.9	1468	14.9	18.3	10.5	10.9
Mount Gambier	-37.83	140.78	732	281	35	16.3	1326	14.0	17.9	11.1	11.4
Mudgee	-32.61	149.57	729	455	59	18.9	1885	15.1	20.6	14.5	13.8
Mundaring	-31.90	116.16	924	206	21	19.8	2065	16.7	22.2	13.9	14.6
Murcia	37.99	-1.13	291	157	28	22.4	2646	18.3	25.4	12.6	13.1
Murgon	-26.24	151.94	778	595	73	22.6	2667	19.2	23.5	12.9	11.4
Myrtleford	-36.56	146.72	960	439	52	18.0	1696	14.3	20.3	14.0	14.6
Nagano	36.65	138.20	1160	795	138	18.7	1874	12.3	23.0	9.8	8.6
Nagykanizsa	46.46	16.99	759	510	76	16.0	1280	10.4	17.8	11.4	11.8
Nancy	48.69	6.18	763	447	60	14.9	1065	10.4	16.9	10.5	10.8
Nannup	-33.98	115.77	866	222	24	18.7	1837	16.0	20.8	12.8	13.4
Nantes	47.15	-1.55	794	396	61	15.9	1269	12.3	17.9	9.6	9.9
Napier	-39.50	176.91	915	498	77	16.3	1333	13.8	17.7	10.0	10.1
Napoli	40.85	14.27	995	441	91	20.4	2234	16.4	23.2	9.8	10.2
Naracoorte	-36.96	140.74	570	225	25	17.5	1593	14.7	19.4	14.6	15.6
Nashik	20.00	73.79	1167	1137	219	26.9	3607	25.1	25.0	9.6	7.1
Naumburg	51.15	11.81	562	378	45	14.4	980	9.5	16.6	10.2	10.4
Neblo	46.01	13.50	1814	1105	188	17.6	1636	12.7	19.7	10.2	10.6
Negotin	44.23	22.53	620	369	48	18.1	1740	11.9	20.3	12.0	13.1

Table 76 (cont.): Key climate indicators of the world's wine regions

Location	Latitude	Longitude	AnnP	GSP	RipeP	GST	GDD	AnnT	RPT	GSDTR	RPDTR
Nelson	-41.27	173.28	1228	652	79	15.2	1106	12.6	16.8	9.0	9.3
Nemea	37.82	22.66	690	230	35	20.3	2199	15.6	23.0	12.2	13.1
Neszmély	47.73	18.35	561	367	47	16.4	1377	10.6	18.3	11.7	12.1
Neuchâtel	46.99	6.93	920	593	79	14.9	1065	10.2	16.8	10.4	10.6
Neuquén	-38.95	-68.06	272	196	27	19.0	1909	14.6	20.2	16.7	17.3
Neusiedl am See	47.95	16.85	601	409	56	15.9	1274	10.3	18.0	11.8	11.9
Nevada City	39.26	-121.02	1295	264	16	17.5	1614	13.2	20.8	16.3	18.0
Newark	43.05	-77.10	911	590	84	15.7	1289	8.7	18.6	11.6	11.6
Niagara Falls	43.10	-79.04	950	581	87	15.6	1295	8.8	18.9	10.5	10.4
Nice	43.71	7.26	788	391	75	19.0	1920	15.0	21.7	8.5	8.6
Niigata	37.92	139.04	1978	1003	162	19.7	2077	13.7	24.4	7.9	7.6
Nîmes	43.84	4.36	758	435	77	19.2	1979	14.7	21.7	11.1	11.4
Niš	43.32	21.90	642	384	51	17.4	1591	11.7	19.6	12.7	13.8
Nitra	48.31	18.08	583	390	49	16.0	1280	10.2	17.9	12.1	12.5
Nitra	48.31	18.08	583	390	49	16.0	1280	10.2	17.9	12.1	12.5
Nogoyá	-32.39	-59.79	1017	751	144	21.5	2428	18.0	22.8	12.6	12.6
Northam	-31.65	116.67	410	120	19	22.2	2578	18.3	24.6	16.1	16.1
Nova Gorica	45.95	13.65	1547	944	169	18.1	1744	13.1	20.4	9.8	10.2
Novara	45.45	8.62	902	568	74	18.0	1707	12.4	20.2	11.1	11.4
Novi Sad	45.27	19.83	622	406	45	17.7	1642	11.9	19.7	12.4	13.6
Novigrad	45.32	13.56	894	538	97	19.3	2001	14.3	21.7	7.3	7.4
Nowra	-34.88	150.60	1149	729	130	19.5	2009	16.8	20.9	9.2	8.4
Nueve de Julio	-31.65	-68.39	278	254	37	22.8	2717	18.0	23.7	15.8	15.4
Nuoro	40.32	9.33	682	247	36	18.9	1910	14.6	21.9	8.8	9.3
Nuriootpa	-34.55	138.99	581	218	26	18.0	1684	14.8	19.8	14.0	14.5
Nyabing	-33.54	118.15	373	143	21	19.1	1927	15.9	21.0	14.4	14.3
Odessa	46.48	30.72	468	272	38	17.1	1517	10.7	19.5	7.6	8.1
Olbia	40.92	9.50	554	209	28	20.4	2229	16.7	23.4	8.1	8.4
Oradea	47.05	21.92	609	375	41	16.7	1439	10.7	18.8	12.0	12.4
Orange	-33.28	149.10	879	499	57	16.7	1409	12.8	18.5	14.0	13.8
Oristano	39.91	8.59	556	210	32	20.8	2311	17.0	23.7	9.0	9.3
Orland	39.75	-122.20	537	102	9	21.2	2399	16.4	23.5	17.0	18.5
Orléans	47.90	1.91	634	371	49	15.6	1202	11.5	17.8	10.6	10.8
Oslo	59.91	10.75	742	484	76	11.9	657	6.5	14.0	8.2	7.9
Ovalle	-30.60	-71.20	120	7	0	17.7	1631	15.5	19.0	9.6	9.8
Ovid	42.68	-76.82	912	606	86	15.4	1245	8.6	18.4	12.0	12.1
Oviedo	43.36	-5.85	967	483	54	16.6	1409	13.5	18.8	9.9	10.0
Oxnard	34.20	-119.18	328	36	3	18.1	1731	16.4	19.7	10.5	10.3
Paarl	-33.73	18.96	646	197	24	21.0	2335	18.0	22.8	13.6	14.1
Padova	45.41	11.88	845	519	65	19.3	2002	13.7	21.6	10.8	11.4
Padthaway	-36.60	140.49	512	201	21	17.8	1643	14.9	19.6	14.8	15.8
Paks	46.61	18.85	553	371	55	17.3	1555	11.3	19.3	11.3	11.9
Palermo	38.12	13.36	470	197	44	21.5	2472	18.1	24.6	7.0	6.8
Palma	39.57	2.65	442	229	54	20.7	2287	17.0	23.7	8.7	8.7
Palmitas	-33.51	-57.80	1163	782	137	20.9	2297	17.4	22.2	11.5	11.1
Pannonhalma	47.55	17.76	573	385	54	16.2	1326	10.4	18.2	11.7	12.0
Paraná	-31.74	-60.51	1088	875	152	21.8	2506	18.4	23.1	12.2	12.1
Parma	44.80	10.33	744	439	64	19.2	1976	13.5	21.8	10.4	10.6
Parndana	-35.79	137.26	658	206	29	17.6	1603	15.2	19.1	8.0	7.9
Paros	37.09	25.15	407	87	7	22.1	2583	18.4	24.4	5.9	5.3
Paterson	45.94	-119.60	203	80	9	17.6	1625	12.0	20.2	15.6	16.8
Patras	38.25	21.73	715	223	39	21.4	2446	17.1	24.4	10.9	11.8

Table 76 (cont.): Key climate indicators of the world's wine regions

Location	Latitude	Longitude	AnnP	GSP	RipeP	GST	GDD	AnnT	RPT	GSDTR	RPDTR
Pau	43.30	-0.37	1428	736	109	17.7	1649	14.0	20.3	11.0	11.0
Paula A. de Sarmiento	-31.49	-68.53	258	235	33	22.5	2650	17.7	23.4	15.8	15.6
Pavia	45.18	9.16	758	463	67	18.4	1802	12.7	20.9	10.2	10.3
Paysandú	-32.31	-58.08	1230	853	130	21.6	2454	18.2	23.2	13.0	12.8
Pécs	46.07	18.23	630	428	59	17.2	1542	11.4	19.2	11.2	11.6
Pemberton	-34.44	116.04	1126	324	32	17.7	1636	15.5	19.5	11.4	12.1
Penn Yan	42.66	-77.05	857	575	81	15.5	1266	8.7	18.5	12.6	12.6
Perpignan	42.74	2.89	539	290	49	19.6	2051	15.6	22.0	9.5	9.4
Perugia	43.10	12.39	732	403	69	18.2	1758	13.4	21.0	9.6	10.1
Pescara	42.46	14.22	658	354	61	19.6	2059	15.2	22.2	8.8	9.2
Phnom Penh	11.56	104.93	1348	1150	231	29.0	4054	28.3	28.5	7.8	7.0
Piacenza	45.05	9.69	810	481	65	18.6	1847	12.9	21.1	10.9	11.0
Picún Leufú	-39.52	-69.29	119	54	9	17.2	1524	13.1	18.5	15.6	16.4
Pierrevert	43.81	5.75	702	401	73	18.0	1726	13.3	20.8	10.4	10.6
Pinhao	41.11	-7.55	1299	470	59	16.6	1425	13.0	19.5	11.8	12.7
Pisa	43.72	10.40	914	488	93	19.3	1987	14.8	21.9	10.4	10.9
Pistoia	43.93	10.91	940	501	86	19.3	1985	14.5	22.0	10.6	11.0
Pitești	44.86	24.87	619	432	43	16.9	1476	10.7	19.1	12.4	13.0
Placerville	38.73	-120.80	940	202	16	19.5	2039	14.9	23.0	16.8	18.4
Plešivica	45.73	15.65	1161	740	117	15.2	1112	10.0	17.0	10.6	11.1
Plodvid	42.13	24.75	534	315	39	18.8	1886	12.8	21.1	11.7	12.4
Poitiers	46.58	0.34	694	382	53	15.8	1239	11.8	17.9	10.9	11.1
Pomán	-28.40	-66.22	362	338	58	22.8	2708	18.6	23.1	14.4	13.5
Ponta Delgada	37.74	-25.67	980	437	83	19.3	1987	17.6	21.8	6.4	6.9
Pordenone	45.96	12.66	1094	714	102	18.6	1837	13.2	20.9	10.9	11.1
Port Macquarie	-31.43	152.90	1487	1019	195	21.1	2340	18.4	22.4	9.1	8.6
Port Williams	45.10	-64.41	1170	625	97	13.3	894	6.9	16.9	11.4	11.7
Posadas	-27.36	-55.90	1762	1164	145	24.6	3098	21.7	25.6	12.2	11.9
Potenza	40.64	15.81	623	311	51	16.6	1425	12.1	19.5	8.5	9.0
Požarevac	44.62	21.18	665	425	52	17.5	1612	11.7	19.6	11.8	12.9
Prague	50.08	14.44	512	384	39	15.7	1218	10.4	17.7	11.2	11.3
Prato	43.88	11.10	923	490	82	19.7	2087	14.9	22.4	10.6	11.0
Prosser	46.20	-119.77	206	82	10	17.0	1493	11.0	19.9	15.9	17.2
Puelches	-38.14	-65.92	319	239	34	19.8	2078	15.3	21.1	17.3	17.7
Puelén	-37.34	-67.62	273	197	28	19.3	1972	14.8	20.6	17.6	18.0
Puerto Montt	-41.47	-72.94	1858	798	114	13.3	699	11.1	14.4	8.5	8.6
Qingdao	36.07	120.38	690	609	78	20.0	2138	13.1	23.9	6.8	6.4
Quebec	46.81	-71.21	1264	817	120	13.9	1078	5.5	17.1	10.7	11.1
Ragusa	36.93	14.73	513	196	39	20.0	2149	16.0	23.1	8.1	8.1
Raleigh	35.78	-78.64	1169	723	107	21.2	2403	15.5	23.7	12.4	11.6
Rand	-28.45	21.25	204	179	37	24.8	3131	20.6	26.1	16.7	15.9
Rapsani	39.90	22.55	472	245	39	19.9	2129	14.6	23.0	12.8	13.5
Ravenna	44.42	12.20	658	384	65	19.7	2082	14.1	22.4	9.0	9.3
Rawsonville	-33.68	19.31	615	196	24	20.5	2221	17.5	22.3	14.3	14.4
Redding	40.59	-122.39	1119	238	22	22.2	2610	17.0	25.6	16.4	18.1
Reggio di Calabria	38.11	15.65	794	315	64	23.2	2832	19.4	26.3	8.2	8.3
Reggio nell'Emilia	44.70	10.63	731	430	63	19.2	1976	13.5	21.8	10.4	10.7
Reims	49.31	4.03	613	370	49	14.7	1020	10.5	16.7	10.9	11.1
Renmark	-34.17	140.74	249	133	11	21.0	2333	17.4	22.7	15.7	15.9

Table 76 (cont.): Key climate indicators of the world's wine regions

Location	Latitude	Longitude	AnnP	GSP	RipeP	GST	GDD	AnnT	RPT	GSDTR	RPDTR
Reus	41.15	1.11	541	345	78	20.0	2143	16.2	22.9	8.1	8.2
Rieti	42.40	12.86	650	381	67	18.3	1780	13.5	21.2	11.5	12.4
Rijeka	45.33	14.44	1424	773	151	19.1	1952	14.4	21.7	8.7	9.1
Rimavská Sobota	48.38	20.02	549	393	44	15.4	1184	9.3	17.4	12.5	12.9
Rimini	44.07	12.57	725	432	73	19.2	1973	14.1	21.8	10.0	10.1
Rincón de los Sauces	-37.39	-68.93	209	113	14	18.4	1786	14.1	19.7	17.4	18.0
Río Colorado	-38.99	-64.10	344	238	40	19.5	2003	15.3	20.7	15.9	16.4
Rivadavia	-33.19	-68.46	206	174	30	20.8	2281	16.3	21.5	15.6	15.4
Rivadavia	-31.53	-68.59	244	220	31	22.1	2565	17.4	23.0	15.9	15.6
Rivera	-30.92	-55.55	1539	955	135	21.6	2454	18.4	22.9	12.6	12.6
Riverhead	40.92	-72.66	1142	632	90	17.4	1610	11.3	20.8	9.9	9.6
Riverside	33.98	-117.38	311	56	8	21.7	2514	18.4	24.6	16.4	17.2
Robe	-37.16	139.75	642	211	28	16.8	1444	14.9	18.1	9.6	9.9
Robertson	-33.80	19.89	396	163	22	21.3	2396	18.1	23.0	14.5	14.4
Rocha	-34.48	-54.33	1161	645	95	19.4	1984	16.4	21.1	10.5	10.5
Rodeo	-30.21	-69.14	94	75	9	16.1	1296	12.8	17.1	15.4	15.4
Rogova	44.47	22.81	629	382	47	18.4	1797	12.0	20.6	11.8	12.7
Roma	41.90	12.50	718	317	63	20.1	2172	15.8	23.0	11.6	11.9
Rostov-on-Don	47.24	39.70	586	331	41	17.5	1623	10.2	19.8	10.9	11.6
Rovigo	45.07	11.79	678	415	57	19.6	2068	13.9	22.0	9.6	9.9
Rust	47.80	16.67	601	422	58	16.0	1297	10.5	18.1	11.7	11.9
Rutherglen	-36.05	146.46	573	285	33	19.3	1971	15.2	21.8	14.5	14.8
Saavedra	-37.76	-62.35	767	580	92	17.7	1620	13.7	19.0	14.5	14.5
Sacramento	38.58	-121.49	452	83	6	21.1	2376	16.7	23.4	16.7	18.0
Salamanca	40.97	-5.66	549	258	35	16.9	1475	12.5	19.8	14.0	15.1
Salem	44.94	-123.04	1001	291	37	15.8	1234	11.9	18.4	14.4	16.4
Salerno	40.68	14.77	1050	438	85	20.2	2192	16.1	23.3	9.4	9.3
Salinas	36.68	-121.66	402	71	5	16.6	1420	14.6	18.1	11.9	12.1
Salto	-31.39	-57.96	1342	934	147	22.4	2615	18.9	23.5	12.8	12.6
San Agustín	-31.98	-64.38	681	594	97	19.8	2080	16.4	20.6	13.5	13.3
San Agustín	-30.63	-67.47	287	269	38	21.4	2422	17.2	22.1	14.6	14.0
San Andreas	38.20	-120.68	735	151	12	20.3	2206	15.7	23.5	18.0	19.4
San Bernardino	34.11	-117.29	289	57	8	22.6	2693	18.9	25.5	16.7	17.8
San Blas de los Sauces	-28.40	-67.09	221	206	35	22.8	2704	18.5	23.3	16.0	15.2
San Buenaventura	27.06	-101.54	335	269	72	26.6	3550	21.9	27.5	15.5	14.6
San Carlos	-33.77	-69.04	323	245	46	18.1	1722	13.8	19.0	17.6	17.8
San Carlos	-25.89	-65.93	162	160	20	20.4	2204	17.4	20.8	12.4	11.1
San Carlos de Bariloche	-41.13	-71.31	950	333	50	11.6	438	8.4	13.1	13.8	14.9
San Diego	32.72	-117.16	265	47	5	20.1	2163	18.0	22.4	8.0	8.0
San Fernando	-34.59	-70.99	732	112	11	18.3	1751	14.7	19.6	14.3	14.9
San Francisco del Monte de Oro	-32.60	-66.13	509	430	62	22.1	2569	18.7	22.9	14.2	13.8
San Jose	37.34	-121.89	386	74	4	19.4	2018	16.4	21.4	13.4	13.8
San José	-34.35	-56.71	1215	743	131	20.3	2189	17.0	21.9	11.9	11.8
San Juan	-31.54	-68.54	281	257	34	22.4	2620	17.6	23.3	15.9	15.6
San Luis	-33.30	-66.34	441	378	51	21.4	2422	17.5	22.1	14.7	14.4
San Luis Obispo	35.28	-120.66	560	86	8	17.5	1615	15.4	19.4	13.9	14.6

Table 76 (cont.): Key climate indicators of the world's wine regions

Location	Latitude	Longitude	AnnP	GSP	RipeP	GST	GDD	AnnT	RPT	GSDTR	RPDTR
San Martín	-33.08	-68.47	201	172	29	21.2	2363	16.6	21.9	15.4	15.1
San Martín	-31.52	-68.35	272	249	35	22.8	2700	17.9	23.6	15.7	15.3
San Martín de los Andes	-40.16	-71.35	1134	416	58	12.0	513	8.9	13.5	14.8	15.9
San Mateo	37.56	-122.33	503	88	5	16.8	1456	14.5	18.4	9.3	9.4
San Rafael	-34.61	-68.34	293	218	35	18.7	1841	14.3	19.7	16.1	16.0
San Rafael	37.97	-122.53	1047	160	7	17.8	1672	15.0	19.7	11.0	11.4
Sanagasta	-29.29	-67.02	253	236	40	21.2	2381	17.1	21.9	14.9	14.3
Santa Cruz	36.97	-122.04	744	121	7	16.5	1401	14.4	18.0	12.1	12.3
Santa Cruz de Tenerife	28.46	-16.25	275	58	7	22.1	2596	20.5	24.3	7.4	7.7
Santa Lucía	-31.53	-68.48	264	243	34	22.5	2644	17.7	23.4	15.8	15.5
Santa Maria	34.55	-120.44	478	82	6	15.8	1252	14.3	17.9	10.3	10.3
Santa María	-26.70	-66.05	187	183	23	19.5	2012	16.3	19.9	11.8	10.6
Santa Rosa	-31.74	-68.31	277	254	37	22.8	2706	18.0	23.6	15.6	15.1
Santa Rosa	-32.34	-65.20	673	553	85	21.5	2434	17.8	22.2	14.0	13.9
Santa Rosa	-33.25	-68.15	216	187	30	21.3	2394	16.6	22.0	15.7	15.5
Santander	43.46	-3.81	1207	590	92	17.2	1550	14.7	19.5	6.2	6.3
Santiago	-33.49	-70.67	342	47	5	19.0	1914	15.6	20.2	16.2	16.9
Santiago del Estero	-27.78	-64.26	618	575	87	24.5	3073	20.7	24.9	14.5	13.5
Sapporo	43.06	141.35	1268	769	141	15.6	1287	8.7	20.1	8.3	7.7
Sardara	39.61	8.82	566	213	37	21.0	2350	16.8	24.1	10.0	10.4
Sarmiento	-45.59	-69.07	139	64	11	14.7	996	11.0	15.9	12.8	12.9
Sarnen	46.90	8.25	1377	898	99	14.0	905	9.0	15.8	9.0	8.9
Sassari	40.73	8.56	649	254	39	19.5	2027	15.6	22.4	9.5	9.9
Savona	44.30	8.46	849	459	83	19.8	2089	15.6	22.5	8.4	8.6
Schaffhausen	47.70	8.64	897	567	62	14.4	974	9.4	16.5	9.9	10.2
Schloßböckel-heim	49.81	7.74	568	357	45	14.1	926	9.5	16.1	10.4	10.5
Schwyz	47.02	8.65	1397	939	112	14.3	962	9.3	16.2	9.0	8.9
Seattle	47.61	-122.33	1015	358	47	14.4	951	10.9	16.9	9.8	10.4
Sélestat	48.26	7.45	709	447	58	15.6	1199	10.8	17.6	10.4	10.8
Senta	45.92	20.08	544	359	42	17.3	1559	11.3	19.3	11.9	12.6
Seoul	37.57	126.98	1368	1192	152	19.7	2073	12.3	23.3	9.6	8.6
Shepparton	-36.38	145.40	522	267	33	19.2	1938	15.4	21.2	14.1	14.3
Siena	43.32	11.33	744	372	61	18.6	1854	14.1	21.2	10.6	11.3
Simferopol	44.95	34.10	539	298	41	17.1	1530	11.5	20.1	10.9	11.4
Sinj	43.70	16.64	933	483	82	18.4	1798	13.3	20.8	9.5	10.1
Siracusa	37.08	15.29	495	208	33	21.9	2559	18.3	25.0	8.7	8.8
Slovenske Konjice	46.34	15.42	1170	811	121	15.3	1145	10.0	17.1	11.1	11.4
Soissons	49.38	3.32	633	377	49	14.8	1028	10.7	16.7	10.4	10.4
Solothurn	47.21	7.53	1129	721	88	13.6	859	8.9	15.7	9.4	9.6
Sombor	45.77	19.12	619	402	50	17.3	1574	11.4	19.3	12.5	13.8
Somlóvásárhely	47.12	17.38	601	422	62	16.1	1311	10.5	18.2	11.5	11.8
Sondrio	46.17	9.88	838	602	88	13.5	822	8.6	15.6	10.0	10.0
Sonoma	38.30	-122.46	742	122	7	18.4	1808	15.1	20.6	17.5	19.0
Sonora	37.98	-120.38	813	168	11	19.4	2020	14.8	23.3	18.2	19.6
Sopron	47.68	16.58	661	473	66	15.6	1190	10.1	17.6	11.5	11.7
Split	43.51	16.44	803	396	71	21.3	2423	16.6	24.0	9.2	9.5
Springfield	39.78	-89.65	947	648	85	19.1	1942	11.7	21.6	12.4	12.7

Table 76 (cont.): Key climate indicators of the world's wine regions

Location	Latitude	Longitude	AnnP	GSP	RipeP	GST	GDD	AnnT	RPT	GSDTR	RPDTR
St Catherines	43.18	-79.25	864	553	84	15.8	1329	9.0	19.1	10.3	10.3
St Helena	38.50	-122.47	864	142	9	18.1	1744	14.6	20.5	17.3	18.6
St Pourcain sur Sioule	46.31	3.29	727	475	65	15.6	1196	11.3	17.7	11.8	12.1
St. Gallen	47.42	9.38	1245	874	111	13.2	795	8.5	15.3	9.5	9.5
St. Louis	38.63	-90.20	983	640	83	20.1	2164	13.0	22.7	12.1	12.3
Stainz	46.89	15.26	1002	746	99	15.3	1139	9.6	17.1	11.6	12.0
Stanthorpe	-28.66	151.94	766	550	71	18.9	1873	15.3	20.1	12.5	11.3
Stellenbosch	-33.90	18.87	701	222	26	19.7	2061	17.0	21.4	12.8	13.4
Stixneusiedl	48.04	16.68	609	420	56	15.9	1272	10.3	18.0	11.6	11.9
Strathbogie	-36.86	145.86	1148	490	54	15.5	1156	12.0	17.9	13.4	14.1
Stuttgart	48.78	9.18	710	482	54	14.5	983	9.7	16.4	10.4	10.6
Subotica	46.10	19.66	571	375	45	17.1	1525	11.1	19.1	11.6	12.5
Summerland	49.61	-119.68	327	189	23	14.3	1031	8.3	17.1	14.3	15.2
Sunbury	-37.58	144.71	573	332	37	17.0	1474	14.0	19.0	12.1	12.4
Susanville	40.42	-120.65	408	118	12	14.8	1101	9.5	18.1	20.0	22.1
Sutherlin	43.39	-123.31	1133	323	36	16.3	1358	12.4	19.1	15.8	18.0
Svishtov	43.62	25.35	572	364	42	18.7	1857	12.2	21.3	12.3	12.9
Swan Hill	-35.34	143.55	345	183	24	20.4	2207	16.6	22.4	15.0	15.2
Szekszárd	46.35	18.71	566	383	58	17.2	1549	11.3	19.3	11.4	11.9
Tacna	-18.01	-70.25	8	3	0	19.8	2074	18.0	21.4	11.2	11.4
Tacuarembó	-31.72	-55.99	1315	796	129	21.6	2465	18.3	22.9	12.8	12.6
Tafí del Valle	-26.85	-65.71	348	332	44	17.9	1682	14.9	18.3	11.3	10.2
Taipei	25.05	121.55	2888	1911	337	25.5	3315	22.0	27.1	6.2	6.4
Taiyuan	37.85	112.57	453	411	61	18.3	1786	10.5	20.0	13.8	12.6
Tamberías	-31.46	-69.42	134	91	11	16.7	1414	13.0	17.8	15.4	15.5
Tandil	-37.33	-59.14	909	636	104	17.3	1547	13.7	19.1	14.2	14.3
Taranto	40.46	17.25	570	257	45	20.9	2325	16.8	23.7	9.0	9.1
Taraz	42.90	71.40	324	162	9	19.0	1934	11.5	21.5	14.4	15.7
Tarbes	43.23	0.08	1227	663	80	17.4	1595	13.5	20.0	10.9	10.9
Tecka	-43.49	-70.81	374	149	24	12.5	609	8.9	13.9	13.9	14.7
Tekirdağ	40.98	27.51	642	259	45	18.7	1870	13.9	21.7	10.1	10.4
Telavi	41.91	45.48	760	581	63	16.6	1415	10.8	19.3	11.5	11.6
Temuco	-38.74	-72.59	1243	414	48	14.3	917	11.9	15.6	13.5	14.7
Teramo	42.66	13.70	784	480	81	18.5	1821	13.7	21.3	10.2	10.9
Terni	42.56	12.64	574	314	56	19.4	2015	14.6	22.2	11.3	11.9
The Dalles	45.59	-121.18	414	112	14	17.3	1573	12.2	20.0	15.0	16.4
Thouars	46.98	-0.22	750	393	58	15.8	1253	11.9	18.0	10.5	10.8
Thun	46.76	7.63	1109	714	83	13.8	884	9.0	15.9	9.3	9.3
Tianjin	39.08	117.20	546	513	42	21.2	2398	13.1	23.5	9.8	8.5
Tilcara	-23.58	-65.39	268	263	34	16.3	1331	13.7	16.4	15.5	14.4
Timişoara	45.75	21.21	605	398	46	17.2	1551	11.2	19.3	12.4	13.0
Tinogasta	-28.06	-67.57	147	137	21	22.0	2547	17.9	22.5	17.2	16.3
Tokaj	48.10	21.41	609	404	46	15.8	1251	9.7	17.5	12.1	12.4
Tokat	40.32	36.55	458	229	22	17.3	1570	12.0	20.1	13.4	14.4
Toledo	39.86	-4.03	365	187	24	20.5	2249	15.6	23.8	13.0	13.6
Torino	45.20	7.69	783	556	66	17.7	1644	12.2	20.0	10.2	10.0
Tornquist	-38.10	-62.22	776	579	93	18.0	1687	14.0	19.3	14.2	14.0
Toul	48.68	5.89	719	425	54	14.7	1025	10.2	16.7	10.3	10.6
Toulon	43.12	5.93	652	321	56	19.2	1979	15.6	21.8	8.4	8.5
Toulouse	43.60	1.44	671	396	52	18.0	1725	13.8	20.3	11.0	11.3
Tournon	45.07	4.83	854	543	97	17.1	1514	12.4	19.3	11.5	11.8

Table 76 (cont.): Key climate indicators of the world's wine regions

Location	Latitude	Longitude	AnnP	GSP	RipeP	GST	GDD	AnnT	RPT	GSDTR	RPDTR
Tours	47.44	0.68	685	373	50	15.6	1210	11.7	17.8	10.7	10.9
Traismauer	48.35	15.74	551	410	46	15.3	1145	9.8	17.0	11.7	11.9
Trancas	-26.23	-65.28	435	421	70	22.4	2625	19.6	22.8	11.6	10.4
Trapani	38.02	12.54	471	186	42	21.7	2509	18.2	24.9	7.8	7.7
Trento	46.07	11.12	780	539	69	15.8	1244	10.6	18.0	11.4	11.6
Treviso	45.67	12.24	932	596	80	19.1	1942	13.5	21.4	10.3	10.6
Trier	49.75	6.64	804	462	60	14.4	960	9.8	16.3	10.1	10.4
Trieste	45.65	13.78	1087	665	113	19.1	1961	14.3	21.8	7.7	7.4
Tulle	45.27	1.77	863	510	75	15.8	1239	11.7	17.8	12.1	12.3
Tulln	48.33	16.06	552	397	47	15.7	1213	10.1	17.6	11.6	11.9
Tumbarumba	-35.78	148.01	1037	497	58	16.3	1335	12.4	18.9	15.3	15.6
Tumbaya	-23.87	-65.47	394	383	53	17.4	1558	14.8	17.5	14.5	13.4
Tunuyán	-33.58	-69.02	291	219	41	18.8	1863	14.5	19.6	17.1	17.2
Tupungato	-33.37	-69.15	288	194	33	18.1	1723	14.0	19.0	16.8	16.9
Turkistan	43.31	68.23	205	78	2	21.1	2377	12.7	23.1	16.0	17.5
Udine	46.06	13.23	1540	977	148	18.4	1808	13.2	21.0	10.5	10.6
Ukiah	39.15	-123.21	1070	183	14	17.3	1568	13.8	20.3	17.6	19.5
Upper Swan	-31.77	116.03	700	154	16	21.5	2432	18.5	24.1	13.8	14.3
Uralsk	51.23	51.39	340	209	27	15.7	1398	6.1	17.8	13.1	13.9
Urbino	43.74	12.64	894	507	84	18.3	1784	13.2	20.9	9.0	9.2
Ürümqi	43.83	87.62	235	173	20	18.1	1774	7.7	20.5	12.0	12.3
Ust'-Kamenogorsk	49.97	82.60	466	299	34	13.2	990	3.2	15.0	13.4	14.3
Valais	46.22	7.36	726	413	50	14.8	1047	9.7	16.8	10.4	10.4
Valence	44.91	4.89	872	552	103	17.6	1625	12.9	19.9	11.6	11.9
Valencia	39.47	-0.38	451	250	55	21.5	2454	17.8	24.5	8.9	8.6
Valladolid	41.70	-4.72	437	227	28	17.0	1508	12.6	20.0	13.7	14.4
Varese	45.82	8.83	1273	862	115	17.0	1490	11.6	19.1	10.0	10.0
Venezia	45.44	12.32	770	481	68	19.0	1936	13.6	21.5	8.6	8.9
Vercelli	45.32	8.42	809	516	64	18.0	1717	12.4	20.3	10.5	10.6
Verona	45.38	10.99	825	535	68	19.4	2015	13.6	21.8	10.0	10.1
Vesoul	47.62	6.15	1043	603	85	15.1	1094	10.4	17.0	11.0	11.3
Vibo Valentia	38.68	16.10	918	362	65	19.7	2076	15.9	22.7	7.7	7.9
Vicenza	45.55	11.54	1097	646	78	19.7	2080	13.9	22.1	9.8	9.9
Victor Harbor	-35.55	138.62	536	201	25	18.3	1748	16.0	19.6	9.3	9.1
Victoria	-32.62	-60.16	965	714	147	21.5	2426	18.1	22.7	12.7	12.8
Vidin	44.00	22.87	594	360	45	18.4	1798	12.1	20.5	11.7	12.7
Viedma	-40.81	-63.00	363	230	44	18.3	1760	14.5	19.8	13.5	13.4
Vigo	42.24	-8.72	1626	640	92	17.2	1533	14.5	19.1	7.1	7.0
Villa Aberastain	-31.65	-68.58	257	233	33	22.4	2626	17.7	23.4	15.9	15.5
Villa Castelli	-29.02	-68.23	131	114	19	19.5	2017	15.9	20.4	15.5	15.1
Villa Cura Brochero	-31.67	-65.02	669	561	87	19.6	2028	16.1	20.2	14.0	13.6
Villa del Salvador	-31.45	-68.40	262	240	33	22.6	2665	17.8	23.5	15.7	15.4
Villa del Totoral	-30.71	-64.07	692	607	97	19.5	2008	16.2	20.1	13.7	13.1
Villa Dolores	-31.95	-65.19	717	580	89	22.2	2590	18.5	22.8	14.1	13.9
Villa Gesell	-37.26	-56.97	926	551	89	17.9	1672	14.6	19.7	10.3	10.4
Villa Ibañez	-31.46	-68.72	224	197	27	21.4	2412	16.9	22.3	15.9	15.6
Villa Krause	-31.57	-68.53	286	261	35	22.4	2628	17.6	23.3	15.9	15.6
Villa Tulumaya	-32.72	-68.59	207	175	30	21.8	2504	17.3	22.6	14.7	14.1
Villa Tulumba	-30.40	-64.12	659	580	96	19.6	2038	16.4	20.3	13.9	13.3
Villa Unión	-29.32	-68.23	136	121	20	20.0	2108	16.2	20.9	15.5	15.2

Table 76 (cont.): Key climate indicators of the world's wine regions

Location	Latitude	Longitude	AnnP	GSP	RipeP	GST	GDD	AnnT	RPT	GSDTR	RPDTR
Villány	45.87	18.45	641	426	57	17.0	1506	11.2	19.0	11.5	12.1
Villeneuve-de-Marsan	43.89	-0.31	1054	543	85	16.9	1483	13.1	19.3	11.8	12.1
Vinchina	-28.75	-68.20	131	114	20	19.2	1943	15.6	20.0	15.6	15.1
Virovitica	45.83	17.39	815	533	75	17.3	1571	11.7	19.1	11.7	12.4
Visalia	36.33	-119.29	266	63	5	22.7	2713	17.7	25.1	16.9	17.6
Viterbo	42.42	12.13	292	153	26	18.3	1783	13.6	21.3	10.4	10.8
Vitoria-Gasteiz	42.85	-2.67	835	421	49	15.7	1225	11.9	18.5	9.8	10.4
Vranje	42.55	21.90	644	399	56	15.7	1228	10.2	18.0	12.2	13.3
Waipara	-43.05	172.76	655	378	62	14.6	982	11.9	16.2	9.0	8.9
Waldstatt	47.36	9.29	1340	928	118	12.4	649	7.8	14.4	9.2	9.2
Walla Walla	46.06	-118.34	407	178	17	17.3	1556	11.8	20.2	15.5	17.1
Waterville	47.65	-120.07	293	111	13	14.0	989	7.6	17.1	14.8	16.0
Watkins Glen	42.38	-76.87	866	583	80	15.2	1193	8.5	18.4	13.3	13.4
Wentworth	-34.11	141.92	276	155	17	21.5	2428	17.6	23.3	15.2	15.4
Whangarei	-35.73	174.32	1595	818	127	17.3	1540	15.2	19.0	8.3	8.4
White Salmon	45.73	-121.49	814	210	26	15.3	1149	10.7	18.2	14.7	16.5
Whitfield	-36.77	146.41	1071	469	54	17.2	1525	13.7	19.5	14.2	14.9
Wien	48.21	16.37	596	407	49	16.0	1283	10.5	18.0	11.6	11.8
Willcox	32.25	-109.83	314	218	40	21.9	2550	16.6	24.4	18.8	16.6
Woodend	-37.36	144.52	814	391	45	14.9	1037	11.8	17.3	12.8	13.3
Woodland	38.68	-121.77	464	79	6	21.4	2451	16.8	23.9	18.1	19.7
Worcester	-33.65	19.46	538	180	22	21.0	2329	17.8	22.8	14.6	14.7
Worms	49.63	8.35	591	386	44	15.6	1205	10.8	17.7	10.6	10.6
Würzburg	49.79	9.95	616	373	42	14.4	982	9.6	16.2	10.7	10.9
Yakima	46.60	-120.51	205	76	9	17.0	1498	11.0	19.9	16.9	18.4
Yamagata	38.26	140.34	1295	831	133	18.6	1842	11.9	22.9	9.7	8.9
Yantai	37.46	121.45	709	618	81	19.4	2019	12.4	23.1	7.3	6.0
Yass	-34.84	148.91	715	403	53	18.1	1702	14.1	20.1	15.1	14.9
Yerevan	40.19	44.52	349	229	14	17.8	1674	11.1	21.3	14.0	14.8
Yinchuan	38.28	106.23	205	192	30	17.2	1540	9.2	19.0	13.2	12.1
Young	-34.30	148.30	666	368	45	18.7	1845	14.6	21.0	16.0	15.9
Yuba City	39.18	-121.69	516	97	8	22.0	2576	17.2	24.3	17.1	18.7
Zacatecas	22.77	-102.58	426	376	82	17.1	1529	15.2	16.6	16.0	14.4
Zadar	44.12	15.23	885	471	100	19.8	2090	15.2	22.2	11.2	12.3
Zamora	41.50	-5.75	668	289	40	17.1	1526	12.9	20.0	13.6	14.2
Zaragoza	41.65	-0.89	361	226	33	20.0	2146	15.5	22.8	12.0	12.3
Zitoradja	43.19	21.72	648	387	52	17.4	1588	11.6	19.5	12.7	13.8
Zitsa	39.75	20.65	1142	414	78	18.8	1892	14.0	21.5	11.9	12.6
Znojmo	48.86	16.05	526	387	43	15.3	1150	9.7	17.3	11.3	11.5
Zug	47.17	8.52	1228	813	95	14.7	1032	9.6	16.6	9.1	8.9
Zürich	47.48	8.54	1026	650	75	14.6	1011	9.6	16.6	9.4	9.4

Table 77: Shares of Old World, New World and world winegrape area of each variety in cool, temperate, warm and hot climates, and average growing season temperature, top 40 varieties, 2000 and 2016 (% and °C)

New World	2000						2016					
Variety	Area (ha)	Cool	Temp.	Warm	Hot	av. GST	Area (ha)	Cool	Temp.	Warm	Hot	av. GST
Elvira	344	0	100	0	0	15.3	231	7	93	0	0	15.4
Aurore	299	0	100	0	0	15.4	255	13	86	0	0	15.4
Pinot Meunier	198	4	19	66	12	17.8	418	54	14	26	6	15.4
Baco Noir	391	0	100	0	0	15.2	729	3	97	0	0	15.7
Seyval Blanc	383	22	69	8	0	15.2	2584	5	90	1	3	15.7
Catawba	635	0	100	0	0	15.4	625	6	82	5	7	15.9
Concord	11816	0	79	0	21	16.1	10535	2	76	5	17	16.1
Campbell Early	0	0	0	0	0	0.0	238	0	72	28	0	16.1
Arinarnoa	1	0	0	0	100	21.3	339	0	85	1	14	16.3
Pinot Blanc	631	0	70	23	7	16.8	440	29	32	26	13	16.5
Gamay Noir	949	0	52	32	16	17.5	424	17	54	15	15	16.5
Sauvignon Blanc (G)	13	0	0	88	12	18.4	253	0	41	52	7	17.1
Pinot Noir	13311	4	38	46	13	17.4	42339	9	43	36	12	17.1
Cardinal	680	0	0	0	100	22.2	223	0	10	75	15	17.4
Gewürztraminer	1885	1	37	33	28	17.8	2328	14	40	23	23	17.4
Negramoll	0	0	0	0	0	0.0	1258	0	52	0	48	17.5
Riesling	6504	3	28	54	15	17.6	12319	6	38	45	11	17.5
Niagara	15253	0	12	0	88	19.1	3264	1	45	10	44	17.5
Listan Prieto	16221	0	35	59	6	17.6	10267	0	37	57	5	17.6
Sauvignon Blanc	22567	1	15	45	39	18.7	62799	1	35	39	25	17.9
Vidal	611	0	93	7	0	15.3	1936	3	5	83	8	18.0
Lacrima Christi	0	0	0	0	0	0.0	226	0	0	100	0	18.1
Yan 73	0	0	0	0	0	0.0	4800	0	0	100	0	18.1
Beibinghong	0	0	0	0	0	0.0	1600	0	0	100	0	18.1
Longyan	0	0	0	0	0	0.0	1000	0	0	100	0	18.1
Carmenère	4719	0	1	91	8	18.1	21822	0	1	95	5	18.1
Delaware	198	0	49	0	51	17.5	421	3	18	36	43	18.2
Graševina	880	0	0	0	100	19.6	3188	0	0	94	6	18.2
Alicante Henri Bouschet	3558	0	0	75	24	18.8	7575	0	1	84	15	18.3
Traminette	5	0	0	100	0	17.1	239	0	20	26	54	18.3
Vignoles	68	0	49	51	0	16.3	241	0	21	21	58	18.5
Pinot Gris	1178	5	73	18	4	16.2	15454	7	33	15	45	18.5
Cabernet Sauvignon	103622	0	4	55	41	19.0	176280	0	6	66	28	18.7
Merlot	51325	0	7	49	44	19.2	73187	1	8	57	33	18.8
Sauvignonasse	935	0	0	22	77	20.2	1126	0	0	65	35	18.9
Prosecco	9	0	0	93	7	18.8	379	0	1	35	64	18.9
Cabernet Franc	8193	0	12	23	65	18.9	14499	2	8	27	64	18.9
Chardonnay	75704	1	19	37	44	19.0	99094	2	18	38	42	19.0
Koshu	0	0	0	0	0	0.0	690	0	0	14	86	19.0
Roussanne	21	0	29	49	22	17.6	291	3	14	38	46	19.0
Red Globe	1940	0	0	2	98	21.9	242	0	26	0	74	19.0
Chambourcin	28	0	59	41	0	16.4	373	0	14	19	67	19.0
Garnacha Tinta	6667	0	1	19	81	21.4	8405	0	3	67	30	19.2
Marsanne	233	0	15	12	73	19.5	259	1	11	34	54	19.3
Muscat	0	0	0	0	0	0.0	723	2	5	0	93	19.4
Viognier	760	0	5	55	40	19.2	4951	2	8	35	55	19.4
Muscat Bailey A	1300	0	0	0	100	19.7	1821	0	0	9	91	19.4
Durif	1104	0	8	38	54	19.5	4697	0	3	39	58	19.5
Côt	19751	0	1	30	69	19.9	45524	0	1	43	55	19.5
Petit Verdot	1060	0	2	38	60	19.8	4764	1	3	39	57	19.5
Cinsaut	3767	0	3	2	94	20.6	2671	0	30	2	68	19.5
Italia	370	0	0	0	100	23.0	1013	0	5	0	94	19.5
Seibel	1967	0	0	0	100	19.6	481	0	0	1	99	19.6
Couderc Noir	299	0	0	0	100	19.6	1938	0	0	0	100	19.6
Moscato Embrapa	0	0	0	0	0	0.0	683	0	0	0	100	19.6
Violeta	0	0	0	0	0	0.0	636	0	0	0	100	19.6
Cora	0	0	0	0	0	0.0	570	0	0	0	100	19.6
Lorena	0	0	0	0	0	0.0	500	0	0	0	100	19.6
Couderc 13	0	0	0	0	0	0.0	474	0	0	0	100	19.6
Niagara Red	0	0	0	0	0	0.0	469	0	0	0	100	19.6
Carmem	0	0	0	0	0	0.0	328	0	0	0	100	19.6
Isabella	14375	0	0	0	100	19.6	11781	0	0	0	100	19.6
Syrah	47432	0	3	37	61	19.9	81186	0	7	41	52	19.6
Monastrell	1157	0	1	20	79	20.2	1808	0	3	39	59	19.6

Table 77 (cont.): Shares of national winegrape area of each variety in cool, temperate, warm and hot climates, and average growing season temperature, top 40 varieties, 2000 and 2016 (% and °C)

New World (cont.)	2000						2016					
Variety	Area (ha)	Cool	Temp.	Warm	Hot	av. GST	Area (ha)	Cool	Temp.	Warm	Hot	av. GST
Kyoho (4N)	4003	0	0	0	100	21.6	2762	0	0	1	99	19.7
Garnacha Blanca	0	0	0	0	0	0.0	218	0	14	26	61	19.7
Sheridan	500	0	0	0	100	19.7	500	0	0	0	100	19.7
Nebbiolo	349	0	6	18	76	21.1	443	1	10	22	66	19.7
Muscat of Alexandria	14758	0	0	0	100	21.8	17214	0	23	27	51	19.7
Norton	0	0	0	0	0	0.0	328	0	0	1	99	19.7
Mazuelo	3875	0	2	24	74	20.6	2581	0	3	42	55	19.7
Moscato Giallo	197	0	0	2	98	20.8	526	0	1	31	68	19.9
Salvador	320	0	0	0	100	23.6	350	0	0	0	100	19.9
Muscat Blanc à Petits Grains (	10442	0	0	23	77	20.4	8258	0	0	37	63	19.9
Sémillon	11772	0	4	43	53	19.5	7736	0	3	35	62	19.9
Sangiovese	3668	0	3	26	71	20.2	3423	0	6	28	66	19.9
Torrontés Sanjuanino	3170	0	0	4	95	22.2	3656	0	0	49	50	20.1
Quebranta	0	0	0	0	0	0.0	330	0	5	0	95	20.2
Tempranillo	4963	0	0	31	69	20.2	7950	0	2	29	69	20.2
Pedro Giménez	14862	0	0	12	88	21.1	15576	0	0	34	66	20.2
Tribidrag	18727	0	2	20	78	20.4	18742	0	1	26	74	20.2
Touriga Nacional	73	0	0	39	60	19.8	296	0	1	17	82	20.3
Savagnin Blanc	11	9	0	45	47	19.3	892	2	3	24	72	20.3
Jacquez	170	0	0	0	100	19.6	1443	0	0	0	100	20.3
Muscat of Hamburg	2905	0	0	0	100	20.2	1415	0	0	1	99	20.3
Blaufränkisch	45	0	0	100	0	17.2	299	2	4	0	94	20.3
Trebbiano Toscano	4289	0	0	3	96	21.0	4595	0	0	33	67	20.4
Muscat Blanc à Petits Grains	1636	0	3	14	83	20.9	3666	0	13	12	76	20.5
Tannat	2790	0	0	0	100	20.2	3072	0	1	3	97	20.5
Malvasia Bianca di Candia	968	0	7	0	93	21.6	651	0	9	1	90	20.6
Torrontés Riojano	8192	0	0	8	92	21.0	8859	0	1	12	87	20.8
Crouchen	98	0	0	20	80	20.3	318	0	0	0	100	20.8
Verdelho	1297	0	3	22	75	20.4	1156	0	1	13	86	20.8
Pinotage	6506	0	1	0	99	20.8	7131	0	0	1	98	20.8
Maticha	354	0	0	9	91	20.9	257	0	0	15	85	20.8
Criolla Grande	24264	0	0	11	89	20.8	15596	0	0	9	91	20.8
Roobernet	0	0	0	0	0	0.0	269	0	0	0	100	20.9
Douce Noire	15680	0	0	18	82	20.8	19103	0	0	16	84	20.9
Muscat Blanc à Petits Grains (	380	0	1	14	85	19.9	644	0	0	6	93	20.9
Damaschino	827	0	0	9	91	20.8	527	0	0	5	95	21.0
Nouvelle	0	0	0	0	0	0.0	428	0	0	1	99	21.0
Ancellotta	13	0	0	7	93	21.7	1011	0	0	6	94	21.0
Chenin Blanc	35353	0	2	5	93	21.2	22664	0	1	4	95	21.2
Béquignol Noir	1082	0	0	4	96	21.1	616	0	0	3	97	21.2
Aspiran Bouschet	432	0	0	0	100	21.3	4088	0	0	4	96	21.2
Gibi	1227	0	0	1	99	21.2	785	0	0	1	99	21.4
Torrontes Mendocino	780	0	0	11	89	20.2	653	0	1	6	93	21.5
Cereza	31113	0	0	6	94	21.6	28887	0	0	4	96	21.6
Ruby Cabernet	7301	0	0	1	99	21.8	5297	0	0	0	100	21.9
Barbera	5851	0	0	4	96	22.3	2719	0	1	8	91	21.9
Symphony	184	0	0	0	100	22.2	647	0	0	0	100	22.1
Colombard	31014	0	0	1	99	22.2	21329	0	0	0	100	22.1
Elbling	0	0	0	0	0	0.0	396	0	0	0	100	22.3
Palomino Fino	680	0	1	19	80	21.2	443	0	0	12	88	22.6
Greco Nero	513	0	0	0	100	22.6	356	0	0	0	99	22.6
Rubired	4153	0	0	0	100	22.9	4916	0	0	0	100	22.8
Triplett Blanc	0	0	0	0	0	0.0	412	0	0	0	100	22.8
Villard Blanc	0	0	0	0	0	0.0	321	0	0	0	100	23.4
Sultaniye	10298	0	0	1	99	21.1	1902	0	1	0	99	27.2
Fiesta	0	0	0	0	0	0.0	230	0	0	0	100	28.1
Top 120	**737405**	**0**	**8**	**28**	**64**	**19.9**	**1011090**	**1**	**12**	**39**	**48**	**19.2**
All	**800601**	**0**	**9**	**27**	**64**	**19.8**	**1109790**	**1**	**11**	**43**	**45**	**19.2**

Table 77 (cont.): Shares of Old World, New World and world winegrape area of each variety in cool, temperate, warm and hot climates, and average growing season temperature, top 40 varieties, 2000 and 2016 (% and °C)

Old World	2000						2016					
Variety	Area (ha)	Cool	Temp.	Warm	Hot	av. GST	Area (ha)	Cool	Temp.	Warm	Hot	av. GST
Pinot Meunier	12927	98	2	0	0	14.6	14277	97	2	1	0	14.6
Silvaner	11041	24	75	1	0	15.5	6061	35	65	0	0	15.2
Dornfelder	3766	20	80	0	0	15.1	7812	17	83	0	0	15.2
Müller-Thurgau	33033	37	62	1	0	15.3	19417	38	59	2	1	15.4
Sankt Laurent	2554	1	99	0	0	15.6	3271	5	95	0	0	15.5
Grüner Veltliner	23604	0	99	0	0	15.6	18992	11	87	2	0	15.6
Chasselas	12889	26	41	29	3	16.1	7170	32	55	12	1	15.7
Blauer Portugieser	9155	7	90	3	0	15.6	6587	7	78	14	1	15.7
Furmint	3481	0	100	0	0	16.7	4434	0	100	0	0	15.8
Zweigelt	7107	1	99	0	0	15.9	8990	7	87	6	0	15.8
Riesling	34057	43	45	11	1	15.3	47486	27	42	30	1	15.9
Pinot Noir	54726	42	31	22	5	15.9	63141	41	33	22	4	15.9
Melon	13253	0	100	0	0	15.9	9550	0	100	0	0	15.9
Pinot Blanc	16071	13	40	30	17	16.8	13338	25	51	16	9	16.1
Chenin Blanc	10044	0	92	5	3	16.0	9558	0	93	6	1	16.1
Blaufränkisch	13748	8	83	9	0	16.4	16882	10	67	23	0	16.3
Gewürztraminer	7899	3	40	55	2	16.7	10495	9	50	36	5	16.5
Tsolikouri	6161	0	100	0	0	16.6	7903	0	100	0	0	16.6
Tsitska	2839	0	100	0	0	16.6	3642	0	100	0	0	16.6
Prieto Picudo	3256	0	88	10	1	16.4	4293	0	99	0	1	16.6
Vranac	0	0	0	0	0	0.0	9503	0	100	0	0	16.7
Gamay Noir	36669	3	30	67	0	16.8	25797	3	27	70	0	16.8
Cserszegi Fűszeres	2185	0	100	0	0	16.7	4299	0	86	14	0	16.9
Saperavi	5351	0	69	31	0	16.8	6468	0	73	24	2	17.0
Sémillon	14047	0	67	32	2	17.0	10957	0	91	6	2	17.0
Cabernet Franc	43372	0	68	18	13	17.0	41553	0	69	18	12	17.1
Rkatsiteli	56805	0	35	65	0	17.3	51374	0	50	48	2	17.1
Dimyat	7740	0	0	99	1	18.4	9696	0	68	32	0	17.1
Graševina	91426	0	22	76	2	17.5	21196	2	47	46	6	17.2
Isabella	11401	0	0	100	0	17.2	6031	0	0	100	0	17.2
Moldova	0	0	0	0	0	0.0	12375	0	0	100	0	17.2
Alvarinho	0	0	0	0	0	0.0	5393	0	0	99	1	17.2
Aligoté	28752	0	6	94	0	17.3	26899	0	8	92	0	17.2
Bianca	2180	0	0	100	0	17.6	9766	0	49	51	0	17.3
Côt	6443	0	24	73	4	17.3	6709	0	29	65	6	17.4
Sauvignon Blanc	40199	0	38	48	14	17.3	61900	2	35	50	13	17.5
Muscat Ottonel	12225	0	16	84	0	17.6	12421	0	15	85	0	17.5
Mencía	13138	0	48	29	23	17.3	11050	0	47	30	23	17.5
Chardonnay	66644	14	38	29	19	17.0	102554	12	29	39	21	17.5
Pinot Gris	17129	13	30	42	14	17.0	33116	10	30	30	30	17.5
Colombard	7381	0	16	77	7	17.5	8667	0	21	76	3	17.6
Loureiro	4392	0	90	10	0	16.7	4696	0	0	92	8	17.6
Fetească Albă	23828	0	5	95	0	17.5	13382	0	0	100	0	17.7
Verdejo	4453	0	20	66	14	17.5	17923	0	64	0	35	17.7
Fetească Regală	2578	0	34	66	0	17.3	12991	0	0	100	0	17.7
Trebbiano Toscano	132765	0	57	12	31	17.8	115748	0	64	7	29	17.8
white varieties	0	0	0	0	0	0.0	10300	0	0	100	0	17.8
red varieties	0	0	0	0	0	0.0	4405	0	0	100	0	17.8
Vinhao	5886	0	89	10	1	16.7	4388	0	0	79	21	17.9
Tinta Barroca	5657	0	0	94	6	17.8	4733	0	0	79	21	17.9
Merlot	156443	1	43	34	22	17.7	193253	1	42	39	19	17.9
Pamid	22718	0	1	99	0	18.3	9961	0	3	97	0	18.1
Dolcetto	7156	0	0	97	3	18.0	4381	0	0	98	2	18.1
Touriga Franca	6671	0	0	100	0	17.6	14221	0	0	63	37	18.3
Cabernet Sauvignon	99080	0	35	41	23	17.9	134391	0	25	45	30	18.3
Muscat Blanc à Petits Grains	27345	0	8	56	36	18.5	30074	2	8	70	20	18.3
Malvasia Fina	2797	0	0	43	57	19.0	3276	0	0	64	36	18.3
Nebbiolo	4778	18	1	75	6	17.0	7554	1	0	80	19	18.4
Barbera	27175	0	1	85	14	18.5	15105	0	0	80	20	18.4
Misket Cherven	0	0	0	0	0	0.0	4349	0	0	100	0	18.5
Touriga Nacional	4149	0	0	63	37	18.4	11426	0	0	50	50	18.6
Marufo	6339	0	0	53	47	18.8	4683	0	0	48	52	18.6
Bonarda Piemontese	23	0	0	69	31	18.6	5926	0	0	68	32	18.7
Arinto de Bucelas	3966	0	70	0	30	17.6	5409	0	0	46	54	18.7

Table 77 (cont.): Shares of national winegrape area of each variety in cool, temperate, warm and hot climates, and average growing season temperature, top 40 varieties, 2000 and 2016 (% and °C)

Old World (cont.)	2000						2016					
Variety	Area (ha)	Cool	Temp.	Warm	Hot	av. GST	Area (ha)	Cool	Temp.	Warm	Hot	av. GST
Viognier	2388	0	0	44	56	18.5	11112	0	0	74	25	18.7
Tempranillo	88222	0	27	42	32	17.9	211429	0	26	18	56	18.8
Marselan	176	0	0	50	50	18.8	3761	0	0	87	13	18.8
Verdicchio Bianco	5043	0	0	19	81	19.2	4674	0	0	91	9	18.8
Síria	2791	0	6	21	73	19.2	7037	0	1	36	63	18.9
Muscat of Hamburg	4162	0	1	88	11	17.7	6265	1	10	19	70	19.0
Aglianico	9264	0	26	35	39	18.5	9627	0	0	27	73	19.0
Trincadeira	7264	0	0	14	86	19.5	10493	0	0	31	69	19.1
Garnacha Tinta	209513	0	7	20	73	19.2	141691	0	6	34	60	19.1
Alarije	1686	0	0	0	100	20.7	4407	0	40	0	60	19.2
Garnacha Blanca	10794	0	0	11	89	19.5	7191	0	0	55	44	19.2
Alicante Henri Bouschet	33250	0	2	34	64	19.1	28457	0	2	30	68	19.2
Prosecco	7498	0	0	1	99	19.1	19730	0	0	2	98	19.3
Syrah	53728	0	0	32	68	18.9	99998	0	1	50	49	19.3
Corvina Veronese	4781	0	0	1	99	19.4	6222	0	0	0	100	19.3
Caladoc	1427	0	0	71	29	18.6	5242	0	0	41	59	19.3
Mazuelo	123372	0	1	30	69	19.3	44732	0	1	59	40	19.3
Petit Verdot	462	0	84	3	13	17.4	3360	0	17	10	73	19.4
Fernão Pires	14206	0	0	3	97	19.7	12138	0	0	17	83	19.4
Lambrusco Salamino	4147	0	0	0	100	19.4	6228	0	0	3	97	19.4
Cinsaut	44310	0	0	18	82	19.6	20255	0	0	47	53	19.4
Trebbiano Romagnolo	19492	0	0	1	99	19.7	19059	0	0	0	100	19.4
Sangiovese	65167	0	0	30	70	19.4	70041	0	0	11	89	19.5
Baga	6730	0	0	0	100	19.3	6750	0	0	12	88	19.5
Montepulciano	28679	0	0	19	81	19.4	32724	0	0	17	83	19.5
Falanghina Flegrea	0	0	0	0	0	0.0	3634	0	0	3	97	19.6
Airén	387978	0	0	11	89	19.9	203801	0	0	0	100	19.6
Tinto Velasco	7998	0	0	0	100	20.4	5369	0	0	0	100	19.6
Pardillo	7272	0	0	13	87	19.3	3283	0	0	0	100	19.6
Vermentino	5835	0	0	25	75	19.2	11345	0	0	22	78	19.6
Castelão	14424	0	0	0	100	19.9	12580	0	0	4	96	19.7
Macabeo	48125	0	7	17	76	19.3	38620	0	5	10	85	19.8
Malvasia Bianca di Candia	11921	0	0	15	85	19.8	9034	0	0	7	93	19.9
Parellada	11188	0	0	0	100	20.0	7137	0	0	0	100	20.0
Xarello	10299	0	0	0	100	20.0	8534	0	0	0	100	20.0
Palomino Fino	28125	0	10	30	60	19.8	22746	0	10	20	71	20.2
Garganega	16549	0	0	1	99	20.2	8522	0	0	1	99	20.3
Gaglioppo	3592	0	0	0	99	21.0	4626	0	0	25	75	20.3
Tribidrag	8167	0	0	0	99	20.8	14908	0	7	1	92	20.4
Bobal	100128	0	0	41	58	19.5	59189	0	0	0	100	20.4
Chelva	10877	0	2	0	98	20.7	5029	0	3	0	97	20.5
Lambrusco Maestri	1362	0	0	0	100	19.6	5610	0	0	3	97	20.5
Cayetana Blanca	55527	0	0	0	100	20.7	36385	0	0	0	100	20.7
Negroamaro	16619	0	0	0	100	21.1	11431	0	0	0	100	20.7
Monastrell	75130	0	0	3	97	21.0	50122	0	0	10	90	20.8
Muscat of Alexandria	14710	0	0	1	99	20.9	17591	0	0	15	85	20.8
Italia	0	0	0	0	0	0.0	4174	0	0	3	96	20.9
Roditis	299	0	0	0	100	20.4	8463	0	0	0	100	21.0
Pedro Ximénez	14803	0	0	0	99	22.1	8787	0	0	0	100	21.3
Inzolia	9259	0	0	2	98	21.3	4740	0	0	0	100	21.4
Nero d'Avola	11318	0	0	1	99	20.9	14215	0	0	0	99	21.5
Zalema	5969	0	0	0	100	21.7	4015	0	0	0	100	21.5
Grillo	1803	0	0	0	100	21.7	7382	0	0	0	100	21.5
Catarratto Bianco	50711	0	0	0	100	21.6	28563	0	0	0	100	21.5
Savatiano	12747	0	0	0	100	22.3	10268	0	0	0	100	22.4
Sultaniye	4	0	0	12	88	21.6	3423	0	0	0	100	22.4
Top 120	**2984182**	**3**	**19**	**28**	**50**	**18.6**	**2819746**	**3**	**22**	**28**	**46**	**18.5**
All	**4006807**	**3**	**18**	**31**	**48**	**18.6**	**3373338**	**3**	**21**	**32**	**44**	**18.5**

Table 77 (cont.): Shares of Old World, New World and world winegrape area of each variety in cool, temperate, warm and hot climates, and average growing season temperature, top 40 varieties, 2000 and 2016 (% and °C)

World	2000						2016					
Variety	Area (ha)	Cool	Temp.	Warm	Hot	av. GST	Area (ha)	Cool	Temp.	Warm	Hot	av. GST
Pinot Meunier	13125	97	2	1	0	14.7	14695	96	2	2	0	14.7
Silvaner	11048	24	75	1	0	15.5	6072	35	65	0	0	15.2
Dornfelder	3766	20	80	0	0	15.1	7871	17	82	0	0	15.2
Müller-Thurgau	33453	36	63	1	0	15.3	19501	38	59	2	1	15.4
Grüner Veltliner	23604	0	99	0	0	15.6	19118	11	87	2	0	15.6
Chasselas	13318	26	43	28	3	16.1	7377	31	56	11	1	15.7
Blauer Portugieser	9155	7	90	3	0	15.6	6590	7	78	14	1	15.7
Zweigelt	7107	1	99	0	0	15.9	9068	8	87	6	0	15.8
Melon	13253	0	100	0	0	15.9	9551	0	100	0	0	15.9
Pinot Blanc	16702	12	41	30	17	16.8	13779	25	50	16	9	16.1
Concord	11816	0	79	0	21	16.1	10544	2	76	5	17	16.1
Riesling	40561	37	42	18	3	15.7	59805	23	41	33	3	16.2
Blaufränkisch	13793	8	82	10	0	16.4	17180	10	66	23	2	16.4
Pinot Noir	68037	35	32	27	6	16.2	105480	28	37	28	7	16.4
Tsolikouri	6161	0	100	0	0	16.6	7903	0	100	0	0	16.6
Gewürztraminer	9784	3	40	51	7	16.9	12823	10	48	34	8	16.6
Vranac	0	0	0	0	0	0.0	9503	0	100	0	0	16.7
Gamay Noir	37618	3	30	66	1	16.8	26221	3	27	69	0	16.8
Saperavi	5351	0	69	31	0	16.8	6478	0	73	24	2	17.0
Rkatsiteli	56805	0	35	65	0	17.3	51374	0	50	48	2	17.1
Dimyat	7740	0	0	99	1	18.4	9696	0	68	32	0	17.1
Moldova	0	0	0	0	0	0.0	12375	0	0	100	0	17.2
Aligoté	28752	0	6	94	0	17.3	26929	0	8	92	0	17.2
Alvarinho	0	0	0	0	0	0.0	5545	0	1	97	2	17.2
Graševina	92306	0	21	75	3	17.5	24384	2	41	52	6	17.3
Bianca	2180	0	0	100	0	17.6	9766	0	49	51	0	17.3
Muscat Ottonel	12225	0	16	84	0	17.6	12464	0	15	85	0	17.5
Mencia	13138	0	48	29	23	17.3	11052	0	47	30	23	17.5
Cabernet Franc	51565	0	59	19	22	17.3	56052	1	53	20	26	17.5
Listan Prieto	16232	0	35	59	6	17.6	10267	0	37	57	5	17.6
Loureiro	4392	0	90	10	0	16.7	4696	0	0	92	8	17.6
Fetească Albă	23828	0	5	95	0	17.5	13382	0	0	100	0	17.7
Sauvignon Blanc	62765	0	30	47	23	17.8	124700	2	35	45	19	17.7
Verdejo	4453	0	20	66	14	17.5	17931	0	64	0	35	17.7
Fetească Regală	2578	0	34	66	0	17.3	12991	0	0	100	0	17.7
white varieties	0	0	0	0	0	0.0	10300	0	0	100	0	17.8
Pinot Gris	18307	12	33	41	14	17.0	48570	9	31	25	35	17.8
Trebbiano Toscano	137054	0	55	12	33	17.9	120343	0	61	8	31	17.9
Tinta Barroca	5657	0	0	94	6	17.8	4926	0	0	76	24	18.0
Pamid	22718	0	1	99	0	18.3	9961	0	3	97	0	18.1
Yan 73	0	0	0	0	0	0.0	4800	0	0	100	0	18.1
Carmenère	4766	0	1	90	9	18.1	22486	0	1	92	7	18.2
Merlot	207768	0	34	38	28	18.1	266440	1	33	44	23	18.2
Chardonnay	142348	7	28	33	32	18.0	201649	7	23	39	31	18.2
Sémillon	25819	0	38	37	25	18.1	18693	0	55	18	27	18.2
Touriga Franca	6671	0	0	100	0	17.6	14224	0	0	63	37	18.3
Nebbiolo	5127	16	2	72	10	17.3	7997	1	1	77	22	18.4
Cabernet Sauvignon	202702	0	19	48	32	18.5	310671	0	14	57	29	18.5
Muscat Blanc à Petits Grains	28981	0	7	54	38	18.6	33739	1	9	64	26	18.6
Marufo	6339	0	0	53	47	18.8	4683	0	0	48	52	18.6
Touriga Nacional	4221	0	0	62	38	18.4	11722	0	0	50	50	18.6
Bonarda Piemontese	23	0	0	69	31	18.6	5926	0	0	68	32	18.7
Arinto de Bucelas	3966	0	70	0	30	17.6	5409	0	0	46	54	18.7
Isabella	25776	0	0	44	56	18.5	17813	0	0	34	66	18.8
Verdicchio Bianco	5043	0	0	19	81	19.2	4682	0	0	91	9	18.8
Tempranillo	93185	0	25	41	34	18.1	219379	0	25	19	56	18.8
Viognier	3148	0	1	47	52	18.7	16063	1	3	62	34	18.9
Síria	2791	0	6	21	73	19.2	7037	0	1	36	63	18.9
Barbera	33026	0	1	70	28	19.2	17824	0	0	69	31	18.9
Tannat	5594	0	0	49	51	18.8	5611	0	14	32	54	19.0
Alicante Henri Bouschet	36808	0	2	38	60	19.1	36031	0	2	41	57	19.1
Aglianico	9346	0	26	35	39	18.5	9734	0	0	27	73	19.1
Trincadeira	7264	0	0	14	86	19.5	10510	0	0	31	69	19.1
Garnacha Tinta	216180	0	7	20	73	19.3	150096	0	6	36	58	19.1

Table 77 (cont.): Shares of national winegrape area of each variety in cool, temperate, warm and hot climates, and average growing season temperature, top 40 varieties, 2000 and 2016 (% and °C)

World (cont.)	2000						2016					
Variety	Area (ha)	Cool	Temp.	Warm	Hot	av. GST	Area (ha)	Cool	Temp.	Warm	Hot	av. GST
Côt	26194	0	6	41	53	19.2	52233	0	5	46	49	19.2
Garnacha Blanca	10794	0	0	11	89	19.5	7409	0	1	54	45	19.2
Muscat of Hamburg	7068	0	0	52	48	18.8	7680	1	8	15	75	19.2
Prosecco	7507	0	0	1	99	19.1	20109	0	0	3	97	19.3
Corvina Veronese	4800	0	0	1	99	19.4	6240	0	0	0	100	19.3
Caladoc	1427	0	0	71	29	18.6	5258	0	0	41	59	19.3
Mazuelo	127247	0	1	30	69	19.3	47312	0	1	58	41	19.4
Lambrusco Salamino	4147	0	0	0	100	19.4	6228	0	0	3	97	19.4
Syrah	101160	0	1	34	64	19.4	181185	0	4	46	51	19.4
Fernão Pires	14206	0	0	3	97	19.7	12211	0	0	17	83	19.4
Cinsaut	48077	0	0	16	83	19.7	22926	0	4	42	55	19.4
Trebbiano Romagnolo	19492	0	0	1	99	19.7	19059	0	0	0	100	19.4
Petit Verdot	1522	0	27	27	46	19.1	8124	0	9	27	64	19.5
Durif	1105	0	8	38	54	19.5	4807	0	3	38	59	19.5
Sangiovese	68834	0	0	30	70	19.5	73464	0	0	12	88	19.5
Baga	6730	0	0	0	100	19.3	6750	0	0	12	88	19.5
Montepulciano	28728	0	0	19	81	19.4	32935	0	0	17	83	19.5
Airén	387978	0	0	11	89	19.9	203801	0	0	0	100	19.6
Tinto Velasco	7998	0	0	0	100	20.4	5369	0	0	0	100	19.6
Vermentino	5838	0	0	25	75	19.2	11483	0	0	22	78	19.6
Chenin Blanc	45397	0	22	5	73	20.1	32221	0	28	4	67	19.7
Castelão	14424	0	0	0	100	19.9	12580	0	0	4	96	19.7
Macabeo	48128	0	7	17	76	19.3	38625	0	5	10	85	19.8
Muscat Blanc à Petits Grains	10442	0	0	23	77	20.4	8258	0	0	37	63	19.9
Malvasia Bianca di Candia	12889	0	1	14	86	20.0	9685	0	1	6	93	20.0
Parellada	11188	0	0	0	100	20.0	7137	0	0	0	100	20.0
Xarello	10299	0	0	0	100	20.0	8534	0	0	0	100	20.0
Pedro Giménez	14862	0	0	12	88	21.1	15576	0	0	34	66	20.2
Garganega	16553	0	0	1	99	20.2	8554	0	0	1	99	20.3
Palomino Fino	28805	0	10	29	60	19.8	23190	0	10	20	71	20.3
Muscat of Alexandria	29468	0	0	1	99	21.3	34805	0	11	21	68	20.3
Gaglioppo	3592	0	0	0	99	21.0	4626	0	0	25	75	20.3
Tribidrag	26894	0	1	14	84	20.5	33649	0	3	15	82	20.3
Bobal	100128	0	0	41	58	19.5	59189	0	0	0	100	20.4
Chelva	10877	0	2	0	98	20.7	5029	0	3	0	97	20.5
Lambrusco Maestri	1513	0	0	2	98	19.8	5657	0	0	3	97	20.5
Cayetana Blanca	55776	0	0	0	100	20.7	36401	0	0	0	100	20.7
Italia	370	0	0	0	100	23.0	5188	0	1	3	96	20.7
Negroamaro	16619	0	0	0	100	21.1	11449	0	0	0	100	20.7
Monastrell	76288	0	0	3	97	21.0	51930	0	0	11	89	20.8
Torrontés Riojano	8192	0	0	8	92	21.0	8859	0	1	12	87	20.8
Douce Noire	18323	0	0	28	71	20.4	19733	0	0	17	83	20.8
Pinotage	6506	0	1	0	99	20.8	7132	0	0	1	98	20.8
Colombard	38395	0	3	15	82	21.3	29996	0	6	22	72	20.8
Criolla Grande	. 24264	0	0	11	89	20.8	15596	0	0	9	91	20.8
Roditis	299	0	0	0	100	20.4	8463	0	0	0	100	21.0
Pedro Ximénez	17271	0	0	14	85	21.5	8810	0	0	0	100	21.3
Inzolia	9259	0	0	2	98	21.3	4740	0	0	0	100	21.4
Nero d'Avola	11323	0	0	1	99	20.9	14281	0	0	0	99	21.5
Catarratto Bianco	50711	0	0	0	100	21.6	28613	0	0	0	100	21.5
Grillo	1803	0	0	0	100	21.7	7383	0	0	0	100	21.5
Cereza	31113	0	0	6	94	21.6	28887	0	0	4	96	21.6
Ruby Cabernet	7323	0	0	1	99	21.8	5309	0	0	0	100	21.9
Savatiano	12747	0	0	0	100	22.3	10268	0	0	0	100	22.4
Rubired	4153	0	0	0	100	22.9	4916	0	0	0	100	22.8
Sultaniye	10302	0	0	1	99	21.1	5325	0	0	0	100	24.1
Top 120	**3644585**	**2**	**17**	**28**	**53**	**18.9**	**3725659**	**3**	**19**	**31**	**47**	**18.7**
All	**4807408**	**2**	**16**	**31**	**51**	**18.8**	**4483128**	**3**	**18**	**35**	**44**	**18.6**

Table 78: Shares of national winegrape area of each variety in cool, temperate, warm and hot climates, and average growing season temperature, top 40 varieties, 2000 and 2016 (% and°C)

Argentina

Variety	2000 Area (ha)	Cool	Temp.	Warm	Hot	av. GST	2016 Area (ha)	Cool	Temp.	Warm	Hot	av. GST
Pinot Noir	1114	1	0	58	41	19.2	1866	2	0	76	23	18.7
Fer	348	0	0	78	22	19.1	183	0	0	81	19	19.0
Sémillon	1033	0	1	47	53	19.3	767	0	0	46	54	19.4
Côt	18230	0	0	26	74	20.0	40401	0	0	42	58	19.6
Cabernet Franc	252	0	0	15	85	20.4	929	0	0	42	58	19.7
Merlot	6263	0	0	37	62	19.9	5632	0	0	39	60	19.9
Sauvignon Blanc	865	0	0	23	77	20.5	2148	0	0	40	59	19.9
Petit Verdot	159	0	0	19	81	20.3	740	0	0	26	73	19.9
Chardonnay	4682	0	0	34	66	20.1	6227	0	0	40	60	19.9
Cabernet Sauvignon	13776	0	0	23	77	20.4	15356	0	0	31	69	20.1
Tempranillo	4720	0	0	32	68	20.1	6140	0	0	28	72	20.3
Bonamico	253	0	0	28	72	20.3	137	0	0	28	72	20.3
Muscat Blanc à Petits Grains (G)	10442	0	0	23	77	20.4	6526	0	0	21	79	20.5
Chenin Blanc	3445	0	0	29	71	20.5	2157	0	0	27	73	20.6
Sauvignonasse	756	0	0	10	90	20.7	424	0	0	16	84	20.6
Sangiovese	2490	0	0	13	87	20.7	1837	0	0	10	89	20.7
Maticha	354	0	0	9	91	20.9	257	0	0	15	85	20.8
Criolla Grande	24264	0	0	11	89	20.8	15596	0	0	9	91	20.8
Douce Noire	15659	0	0	17	83	20.8	19072	0	0	15	85	20.9
Barbera	1055	0	0	14	86	20.9	444	0	0	17	83	20.9
Damaschino	827	0	0	9	91	20.8	527	0	0	5	95	21.0
Viognier	151	0	0	28	72	20.7	773	0	0	17	83	21.0
Pinot Gris	13	0	0	80	20	19.0	401	0	1	29	71	21.0
Torrontés Riojano	8127	0	0	7	93	21.0	8208	0	0	6	94	21.0
Syrah	8888	0	0	13	87	21.1	12707	0	0	16	84	21.0
Ancellotta	13	0	0	7	93	21.7	991	0	0	7	93	21.1
Tannat	174	0	0	6	94	20.6	837	0	0	5	95	21.2
Alicante Henri Bouschet	113	0	0	2	98	21.0	135	0	1	5	94	21.2
Béquignol Noir	1082	0	0	4	96	21.1	616	0	0	3	97	21.2
Aspiran Bouschet	432	0	0	0	100	21.3	4087	0	0	4	96	21.2
Pedro Giménez	14862	0	0	12	88	21.1	11197	0	0	8	92	21.2
Listan Prieto	700	0	0	7	93	21.2	374	0	0	6	94	21.3
Gibi	1227	0	0	1	99	21.2	785	0	0	1	99	21.4
Trebbiano Toscano	2765	0	0	4	96	21.2	1622	0	0	1	99	21.4
Torrontes Mendocino	780	0	0	11	89	20.2	653	0	1	6	93	21.5
Cereza	31113	0	0	6	94	21.6	28887	0	0	4	96	21.6
Muscat of Alexandria	5515	0	0	0	99	22.3	2716	0	0	0	100	22.2
Torrontés Sanjuanino	3170	0	0	4	95	22.2	1885	0	0	2	97	22.4
Fintendo	144	0	0	0	100	22.5	185	0	0	0	100	22.6
Greco Nero	513	0	0	0	100	22.6	356	0	0	0	99	22.6
Top 40	**190770**	**0**	**0**	**16**	**84**	**20.8**	**204775**	**0**	**0**	**21**	**79**	**20.6**

Australia

Variety	2000 Area (ha)	Cool	Temp.	Warm	Hot	av. GST	2016 Area (ha)	Cool	Temp.	Warm	Hot	av. GST
Pinot Meunier	107	7	35	46	12	17.1	82	17	48	35	0	16.2
Pinot Noir	3223	6	29	46	19	17.8	4806	14	34	31	21	17.6
Nebbiolo	50	0	5	43	53	19.1	107	2	33	49	16	18.0
Prosecco	0	0	0	0	0	0.0	160	0	2	76	22	18.1
Cabernet Franc	744	2	7	55	36	18.9	328	2	17	66	15	18.2
Riesling	3129	2	3	73	22	18.6	3114	4	7	73	17	18.3
Tribidrag	0	0	0	0	0	0.0	87	0	8	73	18	18.6
Côt	429	0	2	65	32	18.9	515	0	5	75	20	18.7
Cabernet Sauvignon	24997	0	4	55	41	19.2	23987	0	17	48	34	18.9
Sauvignon Blanc	2602	2	11	59	28	18.6	6044	3	10	47	39	18.9
Sangiovese	372	0	4	41	55	19.7	430	0	10	53	37	18.9
Garnacha Tinta	2139	0	0	56	44	19.5	1492	0	1	74	25	19.0
Monastrell	948	0	0	20	80	20.4	704	0	2	60	38	19.1
Barbera	103	0	3	50	47	19.4	102	0	9	47	44	19.2
Syrah	29295	0	4	48	49	19.5	38942	0	7	50	43	19.3
Tempranillo	41	0	5	42	52	19.6	681	0	3	51	46	19.4
Viognier	117	0	3	62	35	18.9	753	0	6	43	50	19.4
Vermentino	0	0	0	0	0	0.0	118	0	4	43	53	19.6
Merlot	7669	0	5	44	51	19.5	8415	0	9	36	55	19.6
Dolcetto	0	0	0	0	0	0.0	100	0	3	38	59	19.6
Pinot Gris	0	0	0	0	0	0.0	3652	3	9	28	60	19.7
Marsanne	216	0	14	10	76	19.6	161	0	9	24	67	19.7
Fiano	0	0	0	0	0	0.0	95	0	2	42	56	19.9
Chardonnay	17266	1	6	34	59	19.7	21321	2	8	24	66	19.9
Muscat Blanc à Petits Grains (R)	380	0	1	14	85	19.9	240	0	0	15	85	20.1
Muscadelle	198	0	1	43	56	19.1	92	0	1	14	86	20.3
Savagnin Blanc	0	0	0	0	0	0.0	870	2	3	23	72	20.3
Sémillon	6528	0	1	36	62	20.1	4556	0	2	30	68	20.3
Petit Verdot	721	0	3	29	68	20.1	1118	0	3	19	78	20.4
Chenin Blanc	841	0	1	24	75	20.3	406	0	0	25	75	20.6
Durif	181	0	1	8	91	20.6	540	0	0	19	81	20.6
Gewürztraminer	521	1	2	27	70	20.3	252	1	1	12	86	20.6
Verdelho	1293	0	3	22	75	20.3	1016	0	1	14	85	20.9
Moscato Giallo	0	0	0	0	0	0.0	94	0	0	6	94	20.9
Rubired	0	0	0	0	0	0.0	90	0	0	0	100	21.0
Muscat Blanc à Petits Grains	214	0	0	38	62	19.8	857	0	0	11	89	21.0
Muscat of Alexandria	2495	0	0	1	99	21.1	2179	0	0	1	98	21.1
Colombard	1801	0	0	1	99	21.2	1789	0	1	0	99	21.2
Canada Muscat	49	0	0	3	97	21.0	120	0	0	0	100	21.3
Ruby Cabernet	2424	0	0	2	98	21.1	849	0	0	1	99	21.6
Top 40	**111094**	**1**	**5**	**43**	**51**	**19.5**	**131263**	**1**	**10**	**40**	**49**	**19.4**

Table 78 (cont.): Shares of national winegrape area of each variety in cool, temperate, warm and hot climates, and average growing season temperature, top 40 varieties, 2000 and 2016 (% and°C)

Austria	2000						2016					
Variety	Area (ha)	Cool	Temp.	Warm	Hot	av. GST	Area (ha)	Cool	Temp.	Warm	Hot	av. GST
Blauer Wildbacher	464	0	100	0	0	15.0	434	13	87	0	0	15.2
Sauvignon Blanc	314	0	100	0	0	15.3	1170	40	60	0	0	15.3
Riesling	1643	0	100	0	0	15.5	2016	26	74	0	0	15.3
Muscat Blanc à Petits Grains	143	0	100	0	0	15.2	823	34	66	0	0	15.3
Pinot Gris	293	0	100	0	0	15.6	224	25	75	0	0	15.4
Müller-Thurgau	3289	0	100	0	0	15.5	1777	17	83	0	0	15.4
Silvaner	53	0	100	0	0	15.4	38	12	88	0	0	15.4
Grüner Veltliner	17479	0	100	0	0	15.5	14376	14	86	0	0	15.5
Frühroter Veltliner	626	0	100	0	0	15.5	369	10	90	0	0	15.5
Scheurebe	529	0	100	0	0	15.5	351	24	76	0	0	15.5
Pinot Blanc	2936	0	100	0	0	15.5	1916	16	84	0	0	15.5
Neuburger	1094	0	100	0	0	15.6	507	14	86	0	0	15.5
Rotgipfler	118	0	100	0	0	15.5	123	0	100	0	0	15.5
Zierfandler	98	0	100	0	0	15.5	82	0	100	0	0	15.5
Roter Veltliner	258	0	100	0	0	15.5	198	7	93	0	0	15.5
Chardonnay	0	0	0	0	0	0.0	1577	16	84	0	0	15.6
Blauer Portugieser	2358	0	100	0	0	15.5	1265	3	97	0	0	15.6
Savagnin Blanc	0	0	0	0	0	0.0	288	13	87	0	0	15.6
Blauburger	884	0	100	0	0	15.5	750	4	96	0	0	15.6
Graševina	4323	0	100	0	0	15.6	3233	11	89	0	0	15.6
Zweigelt	4350	0	100	0	0	15.6	6311	8	92	0	0	15.6
Goldburger	309	0	100	0	0	15.7	98	4	96	0	0	15.6
Pinot Noir	409	0	100	0	0	15.7	614	5	95	0	0	15.7
Rösler	0	0	0	0	0	0.0	216	4	96	0	0	15.7
Jubiläumsrebe	30	0	100	0	0	15.8	6	11	89	0	0	15.7
Sankt Laurent	415	0	100	0	0	15.6	724	4	96	0	0	15.7
Merlot	112	0	100	0	0	15.6	695	3	97	0	0	15.7
Cabernet Sauvignon	312	0	100	0	0	15.7	567	3	97	0	0	15.7
Blaufränkisch	2641	0	100	0	0	15.9	2808	0	100	0	0	15.8
Syrah	0	0	0	0	0	0.0	141	1	99	0	0	15.8
Rathay	0	0	0	0	0	0.0	32	1	99	0	0	15.8
Cabernet Franc	27	0	100	0	0	15.7	64	2	98	0	0	15.8
Muscat Ottonel	418	0	100	0	0	15.8	344	5	95	0	0	15.8
Bouvier	365	0	100	0	0	15.8	216	3	97	0	0	15.9
Furmint	1	0	100	0	0	15.7	9	0	100	0	0	16.0
Top 35	**46291**	**0**	**100**	**0**	**0**	**15.6**	**44361**	**13**	**87**	**0**	**0**	**15.5**

Bulgaria	2000						2016					
Variety	Area (ha)	Cool	Temp.	Warm	Hot	av. GST	Area (ha)	Cool	Temp.	Warm	Hot	av. GST
Trebbiano Toscano	1821	0	0	100	0	18.3	738	0	0	100	0	17.9
Gewürztraminer	971	0	0	100	0	18.3	591	0	0	100	0	17.9
Muscat Ottonel	2914	0	0	100	0	18.3	3679	0	0	100	0	18.0
Dimyat	7649	0	0	100	0	18.3	2998	0	0	100	0	18.0
Chardonnay	1862	0	0	100	0	18.3	3087	0	0	100	0	18.0
Sauvignon Blanc	405	0	0	100	0	18.3	637	0	0	100	0	18.2
Pamid	22581	0	0	100	0	18.3	6874	0	0	100	0	18.3
Pinot Noir	769	0	0	100	0	18.3	342	0	0	100	0	18.3
Rkatsiteli	9429	0	0	100	0	18.3	5415	0	0	100	0	18.3
Aligoté	1659	0	0	100	0	18.3	285	0	0	100	0	18.4
Misket Cherven	0	0	0	0	0	0.0	4349	0	0	100	0	18.5
Cabernet Sauvignon	10441	0	0	100	0	18.3	9327	0	0	100	0	18.5
Shiroka Melnishka	3804	0	0	100	0	18.3	1205	0	0	100	0	18.5
Merlot	11169	0	0	100	0	18.3	10050	0	0	100	0	18.5
Cabernet Franc	0	0	0	0	0	0.0	240	0	0	100	0	18.5
Kadarka	1619	0	0	100	0	18.3	1161	0	0	100	0	18.6
Syrah	0	0	0	0	0	0.0	804	0	0	100	0	18.6
Mavrud	647	0	0	100	0	18.3	1193	0	0	100	0	18.7
Top 18	**77740**	**0**	**0**	**100**	**0**	**18.3**	**52974**	**0**	**0**	**100**	**0**	**18.3**

Table 78 (cont.): Shares of national winegrape area of each variety in cool, temperate, warm and hot climates, and average growing season temperature, top 40 varieties, 2000 and 2016 (% and °C)

Canada	2000						2016					
Variety	Area (ha)	Cool	Temp.	Warm	Hot	av. GST	Area (ha)	Cool	Temp.	Warm	Hot	av. GST
L'Acadie Blanc	0	0	0	0	0	0.0	65	100	0	0	0	13.3
Léon Millot	0	0	0	0	0	0.0	20	100	0	0	0	13.5
Lucie Kuhlmann	0	0	0	0	0	0.0	21	100	0	0	0	13.6
Marquette	0	0	0	0	0	0.0	44	100	0	0	0	13.8
Frontenac (W)	0	0	0	0	0	0.0	26	100	0	0	0	13.9
Frontenac (G)	0	0	0	0	0	0.0	25	100	0	0	0	13.9
Vidal	514	0	100	0	0	15.1	59	100	0	0	0	13.9
Maréchal Foch	109	0	100	0	0	15.1	94	100	0	0	0	14.0
Ortega	0	0	0	0	0	0.0	26	100	0	0	0	14.2
Siegerrebe	0	0	0	0	0	0.0	22	100	0	0	0	14.2
Merlot	674	0	100	0	0	15.1	633	100	0	0	0	14.3
Pinot Blanc	146	0	100	0	0	15.1	109	100	0	0	0	14.3
Muscat Swenson	0	0	0	0	0	0.0	36	100	0	0	0	14.3
Petit Verdot	0	0	0	0	0	0.0	26	100	0	0	0	14.3
Ehrenfelser	0	0	0	0	0	0.0	32	100	0	0	0	14.3
Bacchus	0	0	0	0	0	0.0	20	100	0	0	0	14.3
Côt	0	0	0	0	0	0.0	41	100	0	0	0	14.3
Sémillon	0	0	0	0	0	0.0	19	100	0	0	0	14.3
Viognier	0	0	0	0	0	0.0	101	90	10	0	0	14.4
Frontenac	0	0	0	0	0	0.0	64	70	30	0	0	14.4
Syrah	0	0	0	0	0	0.0	260	83	17	0	0	14.5
Gewürztraminer	237	0	100	0	0	15.1	398	75	25	0	0	14.6
Pinot Noir	457	0	100	0	0	15.1	639	70	30	0	0	14.7
Pinot Gris	210	0	100	0	0	15.1	649	70	30	0	0	14.7
Sauvignon Blanc	148	0	100	0	0	15.1	285	57	43	0	0	14.9
Auxerrois	0	0	0	0	0	0.0	38	52	48	0	0	14.9
Cabernet Sauvignon	569	0	100	0	0	15.1	660	46	54	0	0	15.0
Chardonnay	973	0	100	0	0	15.1	1417	30	70	0	0	15.2
Cabernet Franc	567	0	100	0	0	15.1	820	27	73	0	0	15.3
Gamay Noir	263	0	100	0	0	15.1	272	26	74	0	0	15.3
Riesling	482	0	100	0	0	15.1	1188	18	82	0	0	15.4
De Chaunac	119	0	100	0	0	15.1	53	12	88	0	0	15.5
Muscat Ottonel	0	0	0	0	0	0.0	32	2	98	0	0	15.7
Seyval Blanc	132	0	100	0	0	15.1	2259	2	98	0	0	15.7
Baco Noir	279	0	100	0	0	15.1	704	2	98	0	0	15.7
Niagara	461	0	100	0	0	15.1	87	0	100	0	0	15.7
Arinarnoa	0	0	0	0	0	0.0	289	0	100	0	0	15.7
Concord	977	0	100	0	0	15.1	183	0	100	0	0	15.7
Millot-Foch	0	0	0	0	0	0.0	124	0	100	0	0	15.7
Aligoté	0	0	0	0	0	0.0	30	0	100	0	0	15.7
Top 40	**7317**	**0**	**100**	**0**	**0**	**15.1**	**11870**	**37**	**63**	**0**	**0**	**15.1**

Chile	2000						2016					
Variety	Area (ha)	Cool	Temp.	Warm	Hot	av. GST	Area (ha)	Cool	Temp.	Warm	Hot	av. GST
Chasselas	404	0	93	7	0	16.8	197	0	100	0	0	16.7
Korinthiaki	0	0	0	0	0	0.0	89	0	100	0	0	16.7
Cinsaut	195	0	59	41	0	17.1	848	0	94	6	0	16.8
Muscat of Alexandria	0	0	0	0	0	0.0	5424	0	72	28	0	17.0
Listan Prieto	15181	0	37	63	0	17.3	9693	0	39	61	0	17.3
Pinot Noir	1614	0	28	65	7	17.5	4091	1	15	81	3	17.6
Torrontés Riojano	65	0	0	100	0	17.6	643	0	11	89	0	17.6
Riesling	286	0	12	68	20	17.9	413	0	16	80	4	17.6
Moscato Giallo	0	0	0	0	0	0.0	162	0	4	96	0	17.7
Torrontés Sanjuanino	0	0	0	0	0	0.0	1771	0	0	100	0	17.7
Pedro Giménez	0	0	0	0	0	0.0	4379	0	0	100	0	17.7
Mazuelo	641	0	15	85	0	17.6	811	0	9	89	2	17.7
Gewürztraminer	121	0	6	79	16	17.9	371	0	18	78	5	17.7
Blanca Ovoide	107	0	4	96	0	17.7	44	0	2	98	0	17.7
Muscat Blanc à Petits Grains (G)	0	0	0	0	0	0.0	1732	0	0	99	1	17.7
Sauvignonasse	132	0	0	100	0	17.8	658	0	0	99	1	17.7
Chenin Blanc	76	0	1	93	5	18.2	39	0	0	99	1	17.8
Pinot Gris	2	0	0	100	0	17.6	437	0	6	94	0	17.8
Sauvignon Blanc	6662	0	1	93	6	17.8	14999	0	2	93	5	17.8
Roussanne	0	0	0	0	0	0.0	25	1	0	99	0	17.8
Sauvignon Blanc (G)	7	0	0	100	0	17.9	134	0	0	96	4	17.8
Sémillon	1893	0	5	93	3	17.9	849	0	5	93	3	17.9
Chardonnay	7672	0	3	84	13	18.0	11435	0	5	88	7	17.9
Marselan	0	0	0	0	0	0.0	24	0	0	93	7	17.9
Tempranillo	1	0	0	100	0	18.3	127	0	0	97	3	17.9
Viognier	128	0	0	100	0	18.0	839	0	0	97	3	18.0
Garnacha Tinta	0	0	0	0	0	0.0	187	0	1	89	11	18.0
Alicante Henri Bouschet	2882	0	0	92	7	18.0	6908	0	1	92	7	18.1
Monastrell	22	0	0	59	41	18.3	102	0	1	87	12	18.1
Lacrima Christi	0	0	0	0	0	0.0	226	0	0	100	0	18.1
Côt	929	0	2	94	5	18.1	2293	0	4	86	10	18.1
Merlot	12825	0	1	91	9	18.0	12057	0	2	89	9	18.1
Durif	0	0	0	0	0	0.0	208	0	1	79	19	18.1
Syrah	2040	0	0	88	12	18.1	7994	0	1	84	15	18.1
Carmenère	4719	0	1	91	8	18.1	10503	0	2	90	8	18.1
Cabernet Sauvignon	35967	0	1	84	15	18.1	42409	0	2	82	16	18.2
Petit Verdot	84	0	0	90	10	18.2	863	0	0	89	11	18.2
Sangiovese	123	0	0	91	9	18.1	152	0	0	83	16	18.2
Cabernet Franc	689	0	1	84	15	18.1	1578	0	1	80	19	18.2
Tribidrag	91	0	0	63	37	18.5	66	0	0	59	41	18.6
Top 40	**95558**	**0**	**8**	**82**	**10**	**17.9**	**145782**	**0**	**8**	**83**	**9**	**17.9**

Table 78 (cont.): Shares of national winegrape area of each variety in cool, temperate, warm and hot climates, and average growing season temperature, top 40 varieties, 2000 and 2016 (% and°C)

France	2000						2016					
Variety	Area (ha)	Cool	Temp.	Warm	Hot	av. GST	Area (ha)	Cool	Temp.	Warm	Hot	av. GST
Pinot Meunier	10621	100	0	0	0	14.6	12130	100	0	0	0	14.6
Auxerrois	1997	2	98	0	0	15.6	2409	2	98	0	0	15.6
Pinot Noir	26526	45	51	3	1	15.3	31602	43	45	12	1	15.6
Riesling	3407	0	100	0	0	15.6	4025	0	99	1	0	15.6
Gewürztraminer	2759	0	99	0	1	15.6	3320	0	97	3	0	15.7
Pinot Gris	1969	1	99	0	0	15.6	2867	1	96	3	0	15.7
Aligoté	1756	0	97	3	0	15.8	1927	0	98	2	0	15.8
Melon	13253	0	100	0	0	15.9	9550	0	100	0	0	15.9
Grolleau Noir	3006	0	100	0	0	15.9	1949	0	100	0	0	15.9
Chenin Blanc	9837	0	94	5	1	16.0	9432	0	94	6	0	16.0
Chardonnay	36497	23	45	15	16	16.4	47451	22	41	35	1	16.6
Cabernet Franc	36094	0	80	17	3	16.6	32327	0	86	14	0	16.6
Trebbiano Toscano	90341	0	83	10	6	16.9	78842	0	93	4	2	16.7
Gamay Noir	34537	0	29	71	0	16.8	24095	0	26	74	0	16.9
Sémillon	14015	0	67	32	2	16.9	10234	0	98	1	2	16.9
Sauvignon Blanc	20933	0	63	19	17	16.9	28084	0	60	40	1	17.2
Côt	6129	0	25	75	1	17.2	6100	0	32	68	0	17.2
Gros Manseng	2160	0	0	100	0	17.2	3046	0	27	73	0	17.2
Tannat	2760	0	0	99	1	17.3	2513	0	31	69	0	17.2
Merlot	101309	0	63	20	17	17.4	108483	0	69	29	2	17.4
Colombard	6896	0	18	82	0	17.2	8441	0	22	78	0	17.5
Cabernet Sauvignon	53413	0	59	20	22	17.5	46555	0	57	37	6	17.7
Viognier	2360	0	0	44	56	18.5	8823	0	0	92	8	18.2
Muscat Blanc à Petits Grains	6935	0	1	22	77	18.9	7333	0	1	93	6	18.6
Roussanne	760	0	0	27	73	18.8	1831	0	0	82	18	18.7
Syrah	50676	0	0	34	66	18.7	62211	0	0	76	24	18.8
Clairette	3274	0	0	26	74	18.9	2042	0	0	53	46	18.8
Marselan	176	0	0	50	50	18.8	3662	0	0	86	14	18.8
Alicante Henri Bouschet	8764	0	0	33	66	18.8	2607	0	1	89	10	18.9
Muscat of Alexandria	3027	0	0	6	94	19.5	2462	0	0	100	0	18.9
Garnacha Blanca	6461	0	0	17	83	19.2	5130	0	0	77	23	18.9
Garnacha Tinta	95717	0	0	26	74	18.9	78631	0	0	56	44	18.9
Macabeo	5223	0	0	10	90	19.4	1657	0	0	100	0	18.9
Piquepoul Blanc	975	0	0	1	99	19.1	1564	0	0	99	1	18.9
Mazuelo	95745	0	0	38	62	18.8	31760	0	0	80	20	18.9
Vermentino	2634	0	0	43	57	18.7	4642	0	0	48	52	19.0
Caladoc	1427	0	0	71	29	18.6	3062	0	0	70	30	19.0
Cinsaut	31593	0	0	24	75	18.9	15930	0	0	57	43	19.0
Monastrell	7634	0	0	23	77	19.0	8754	0	0	58	42	19.0
Muscat of Hamburg	752	0	0	54	46	18.6	2325	0	2	14	84	19.1
Top 40	**804349**	**4**	**38**	**25**	**34**	**17.6**	**719807**	**5**	**43**	**40**	**12**	**17.5**

Georgia	2000						2016					
Variety	Area (ha)	Cool	Temp.	Warm	Hot	av. GST	Area (ha)	Cool	Temp.	Warm	Hot	av. GST
Rkatsiteli	19741	0	100	0	0	16.6	25324	0	100	0	0	16.6
Tsolikouri	6161	0	100	0	0	16.6	7903	0	100	0	0	16.6
Saperavi	3704	0	100	0	0	16.6	4751	0	100	0	0	16.6
Tsitska	2839	0	100	0	0	16.6	3642	0	100	0	0	16.6
Chinuri	955	0	100	0	0	16.6	1225	0	100	0	0	16.6
Mtsvane Kakhuri	249	0	100	0	0	16.6	319	0	100	0	0	16.6
Goruli Mtsvane	224	0	100	0	0	16.6	287	0	100	0	0	16.6
Cabernet Sauvignon	223	0	100	0	0	16.6	286	0	100	0	0	16.6
Aleksandrouli	219	0	100	0	0	16.6	281	0	100	0	0	16.6
Pinot Blanc	171	0	100	0	0	16.6	219	0	100	0	0	16.6
Tsulukidzis Tetra	152	0	100	0	0	16.6	195	0	100	0	0	16.6
Aligoté	97	0	100	0	0	16.6	124	0	100	0	0	16.6
Aladasturi	46	0	100	0	0	16.6	59	0	100	0	0	16.6
Krakhuna	36	0	100	0	0	16.6	46	0	100	0	0	16.6
Tavkveri	29	0	100	0	0	16.6	37	0	100	0	0	16.6
Ojaleshi	25	0	100	0	0	16.6	32	0	100	0	0	16.6
Chkhaveri	20	0	100	0	0	16.6	26	0	100	0	0	16.6
Kisi	20	0	100	0	0	16.6	26	0	100	0	0	16.6
Usakhelouri	8	0	100	0	0	16.6	10	0	100	0	0	16.6
Khikhvi	5	0	100	0	0	16.6	6	0	100	0	0	16.6
Otskhanuri Sapere	5	0	100	0	0	16.6	6	0	100	0	0	16.6
Top 21	**34929**	**0**	**100**	**0**	**0**	**16.6**	**44807**	**0**	**100**	**0**	**0**	**16.6**

Table 78 (cont.): Shares of national winegrape area of each variety in cool, temperate, warm and hot climates, and average growing season temperature, top 40 varieties, 2000 and 2016 (% and°C)

Germany

	2000						2016					
Variety	Area (ha)	Cool	Temp.	Warm	Hot	av. GST	Area (ha)	Cool	Temp.	Warm	Hot	av. GST
Elbling	1042	98	2	0	0	14.4	489	100	0	0	0	14.4
Domina	185	90	10	0	0	14.5	374	94	6	0	0	14.5
Schiava Grossa	2529	100	0	0	0	14.5	2197	100	0	0	0	14.5
Zweigelt	38	97	3	0	0	14.5	108	95	5	0	0	14.5
Muscat of Hamburg	0	0	0	0	0	0.0	86	97	3	0	0	14.5
Blaufränkisch	1117	98	2	0	0	14.5	1737	97	3	0	0	14.5
Pinot Meunier	2289	92	8	0	0	14.6	2002	89	11	0	0	14.6
Acolon	0	0	0	0	0	0.0	469	65	35	0	0	14.8
Riesling	22349	65	35	0	0	14.7	21540	56	44	0	0	14.8
Bacchus	3280	38	62	0	0	15.1	1610	60	40	0	0	14.8
Pinot Noir	8637	79	21	0	0	14.9	11184	73	27	0	0	14.9
Pinot Noir Précoce	79	65	35	0	0	14.7	250	40	60	0	0	14.9
Johanniter	0	0	0	0	0	0.0	92	61	39	0	0	14.9
Auxerrois	77	77	23	0	0	14.9	213	55	45	0	0	14.9
Chasselas	1198	99	1	0	0	15.0	1046	100	0	0	0	15.0
Muscat Blanc à Petits Grains	86	62	38	0	0	15.0	240	56	44	0	0	15.0
Cabernet Dorsa	0	0	0	0	0	0.0	242	44	56	0	0	15.0
Kerner	6846	36	64	0	0	15.0	2646	37	63	0	0	15.0
Solaris	0	0	0	0	0	0.0	91	58	42	0	0	15.0
Pinot Blanc	2388	55	45	0	0	15.0	4323	51	49	0	0	15.0
Müller-Thurgau	20691	52	48	0	0	15.0	11664	50	50	0	0	15.0
Gewürztraminer	0	0	0	0	0	0.0	824	37	63	0	0	15.0
Dunkelfelder	281	36	64	0	0	15.0	265	26	74	0	0	15.1
Cabernet Mitos	0	0	0	0	0	0.0	308	56	44	0	0	15.1
Silvaner	6860	34	66	0	0	15.1	4627	40	60	0	0	15.1
Pinot Gris	2635	64	36	0	0	15.0	4887	49	51	0	0	15.1
Sauvignon Blanc	0	0	0	0	0	0.0	736	31	69	0	0	15.1
Regent	324	46	54	0	0	15.0	1902	32	68	0	0	15.1
Heroldrebe	199	13	87	0	0	15.1	112	16	84	0	0	15.1
Merlot	0	0	0	0	0	0.0	553	22	78	0	0	15.1
Cabernet Sauvignon	0	0	0	0	0	0.0	329	18	82	0	0	15.2
Chardonnay	530	30	70	0	0	15.1	1485	24	76	0	0	15.2
Dornfelder	3766	20	80	0	0	15.1	7761	17	83	0	0	15.2
Blauer Portugieser	4877	12	88	0	0	15.2	3177	12	88	0	0	15.2
Scheurebe	3125	15	85	0	0	15.3	1266	22	78	0	0	15.2
Sankt Laurent	184	8	92	0	0	15.2	633	9	91	0	0	15.3
Morio-Muskat	1165	4	96	0	0	15.3	362	3	97	0	0	15.3
Ortega	1053	12	88	0	0	15.3	482	9	91	0	0	15.3
Huxelrebe	1291	4	96	0	0	15.4	462	4	96	0	0	15.4
Faberrebe	1585	8	92	0	0	15.4	331	7	93	0	0	15.5
Top 40	**100706**	**52**	**48**	**0**	**0**	**14.9**	**93105**	**51**	**49**	**0**	**0**	**15.0**

Greece

	2000						2016					
Variety	Area (ha)	Cool	Temp.	Warm	Hot	av. GST	Area (ha)	Cool	Temp.	Warm	Hot	av. GST
Xinomavro	1816	0	0	62	38	18.6	2135	0	0	60	40	18.7
Pamid	15	0	0	100	0	18.0	23	0	0	0	100	19.2
Negoska	96	0	0	0	100	19.7	17	0	0	21	79	19.4
Limnio	95	0	0	0	100	19.8	176	0	0	0	100	19.7
Moschomavro	2295	0	0	3	97	19.9	113	0	0	40	60	19.7
Sauvignon Blanc	158	0	0	0	100	19.3	727	0	0	10	90	19.9
Muscat of Hamburg	0	0	0	0	0	0.0	2288	0	0	1	99	19.9
Tempranillo	0	0	0	0	0	0.0	22	0	0	0	100	20.0
Muscat of Alexandria	0	0	0	0	0	0.0	773	0	0	2	98	20.0
Merlot	183	0	0	0	100	19.6	1393	0	0	8	92	20.1
Trebbiano Toscano	746	0	0	0	100	20.2	211	0	0	0	100	20.2
Malagousia	23	0	0	0	100	20.0	126	0	0	0	100	20.2
Alicante Henri Bouschet	21	0	0	0	100	20.3	60	0	0	0	100	20.3
Agiorgitiko	2320	0	0	0	100	20.3	3270	0	0	0	100	20.3
Cabernet Sauvignon	688	0	0	12	88	20.2	1929	0	0	13	87	20.3
Muscat Blanc à Petits Grains	2232	0	0	0	100	20.3	1568	0	0	0	100	20.4
Moschofilero	718	0	0	0	100	20.5	1088	0	0	0	100	20.4
Chardonnay	36	0	0	0	100	20.2	673	0	0	6	94	20.4
Asprouda	433	0	0	0	100	20.3	120	0	0	0	100	20.5
Fokiano	162	0	0	0	100	20.1	212	0	0	0	100	20.5
Syrah	39	0	0	9	91	19.7	1042	0	0	10	90	20.7
Roditis (R)	6945	0	0	1	99	21.0	828	0	0	0	100	20.8
Mavrouda	349	0	0	0	100	20.6	1658	0	0	1	99	20.9
Roditis	299	0	0	0	100	20.4	8463	0	0	0	100	21.0
Garnacha Roja (Gris)	1	0	0	0	100	19.2	114	0	0	0	100	21.2
Robola	359	0	0	0	100	21.4	152	0	0	0	100	21.3
Vertzami	491	0	0	1	99	21.4	60	0	0	1	99	21.4
Mavrodafni	537	0	0	0	100	21.4	324	0	0	0	100	21.4
Kakotrygis	0	0	0	0	0	0.0	28	0	0	0	100	21.4
Goustolidi	112	0	0	0	100	21.4	19	0	0	0	100	21.4
Athiri	1350	0	0	0	100	21.9	577	0	0	0	100	21.6
Assyrtiko	1106	0	0	0	100	21.7	1770	0	0	0	100	21.7
Monemvassia	418	0	0	0	100	22.1	81	0	0	0	100	21.9
Mandilaria	845	0	0	0	100	22.3	932	0	0	0	100	22.1
Savatiano	12747	0	0	0	100	22.3	10268	0	0	0	100	22.4
Vilana	506	0	0	0	100	22.6	650	0	0	0	100	22.5
Liatiko	2476	0	0	0	100	22.6	2633	0	0	0	100	22.5
Kotsifali	1148	0	0	0	100	22.6	1338	0	0	0	100	22.6
Romeiko	382	0	0	0	100	22.6	1131	0	0	0	100	22.6
Thrapsathiri	0	0	0	0	0	0.0	27	0	0	0	100	22.6
Top 40	**42148**	**0**	**0**	**3**	**97**	**21.3**	**49021**	**0**	**0**	**4**	**96**	**21.2**

Table 78 (cont.): Shares of national winegrape area of each variety in cool, temperate, warm and hot climates, and average growing season temperature, top 40 varieties, 2000 and 2016 (% and°C)

Hungary	2000						2016					
Variety	Area (ha)	Cool	Temp.	Warm	Hot	av. GST	Area (ha)	Cool	Temp.	Warm	Hot	av. GST
Furmint	3480	0	100	0	0	16.7	3862	0	100	0	0	15.8
Muscat Blanc à Petits Grains	1538	0	100	0	0	16.7	762	0	100	0	0	16.1
Hárslevelű	1012	0	100	0	0	16.7	1603	0	96	4	0	16.1
Graševina	6677	0	100	0	0	16.7	3933	0	87	13	0	16.5
Zengő	0	0	0	0	0	0.0	226	0	98	2	0	16.6
Leányka	0	0	0	0	0	0.0	719	0	97	3	0	16.6
Zenit	405	0	100	0	0	16.7	639	0	93	7	0	16.6
Gewürztraminer	0	0	0	0	0	0.0	694	0	87	13	0	16.6
Ezerfürtű	405	0	100	0	0	16.7	264	0	96	4	0	16.6
Pinot Gris	890	0	100	0	0	16.7	1594	0	91	9	0	16.7
Blauburger	0	0	0	0	0	0.0	453	0	93	7	0	16.7
Sauvignon Blanc	324	0	100	0	0	16.7	982	0	89	11	0	16.7
Pinot Blanc	0	0	0	0	0	0.0	228	0	65	35	0	16.7
Ezerjó	3157	0	100	0	0	16.7	636	0	100	0	0	16.7
Blaufränkisch	6920	0	100	0	0	16.7	7260	0	78	22	0	16.7
Syrah	0	0	0	0	0	0.0	215	0	60	40	0	16.7
Grüner Veltliner	1335	0	100	0	0	16.7	1381	0	80	20	0	16.7
Zweigelt	2266	0	100	0	0	16.7	1687	0	76	24	0	16.7
Müller-Thurgau	3278	0	100	0	0	16.7	1670	0	90	10	0	16.7
Chardonnay	2954	0	100	0	0	16.7	2464	0	77	23	0	16.7
Pinot Noir	243	0	100	0	0	16.7	1092	0	74	26	0	16.8
Királyleányka	0	0	0	0	0	0.0	784	0	81	19	0	16.8
Irsai Olivér	0	0	0	0	0	0.0	1531	0	91	9	0	16.8
Riesling	1619	0	100	0	0	16.7	1261	0	84	16	0	16.8
Chasselas	1902	0	100	0	0	16.7	1159	0	98	2	0	16.8
Muscat Ottonel	1416	0	100	0	0	16.7	1256	0	90	10	0	16.8
Cabernet Sauvignon	1052	0	100	0	0	16.7	2677	0	62	38	0	16.8
Merlot	486	0	100	0	0	16.7	1961	0	58	42	0	16.8
Zalagyöngye	2550	0	100	0	0	16.7	1065	0	98	2	0	16.9
Generosa	0	0	0	0	0	0.0	325	0	94	6	0	16.9
Arany Sárfehér	2914	0	100	0	0	16.7	586	0	100	0	0	16.9
Villard Blanc	0	0	0	0	0	0.0	199	0	99	1	0	16.9
Bianca	0	0	0	0	0	0.0	4898	0	97	3	0	16.9
Aletta	0	0	0	0	0	0.0	1676	0	97	3	0	16.9
Cserszegi Fűszeres	2185	0	100	0	0	16.7	4299	0	86	14	0	16.9
Kövidinka	1214	0	100	0	0	16.7	658	0	98	2	0	16.9
Kunleány	1376	0	100	0	0	16.7	974	0	94	6	0	16.9
Blauer Portugieser	1255	0	100	0	0	16.7	1023	0	49	51	0	16.9
Cabernet Franc	526	0	100	0	0	16.7	1368	0	50	50	0	16.9
Kadarka	1012	0	100	0	0	16.7	351	0	57	43	0	17.0
Top 40	**54390**	**0**	**100**	**0**	**0**	**16.7**	**60414**	**0**	**85**	**15**	**0**	**16.7**

Italy	2000						2016					
Variety	Area (ha)	Cool	Temp.	Warm	Hot	av. GST	Area (ha)	Cool	Temp.	Warm	Hot	av. GST
Dolcetto	7156	0	0	97	3	18.0	4381	0	0	98	2	18.1
Pinot Noir	3287	9	5	72	14	18.1	5057	8	5	68	20	18.2
Nebbiolo	4778	18	1	75	6	17.0	7551	1	0	80	19	18.4
Barbera	27175	0	1	85	14	18.5	15006	0	0	80	20	18.4
Muscat Blanc à Petits Grains	13016	0	2	84	15	18.3	13334	0	0	84	16	18.5
Bonarda Piemontese	23	0	0	69	31	18.6	5926	0	0	68	32	18.7
Pinot Gris	6608	4	16	43	37	18.2	18821	3	10	34	52	18.7
Sauvignonasse	4517	0	0	46	54	18.9	2503	0	0	64	36	18.8
Sauvignon Blanc	3312	3	4	49	45	18.7	3935	9	2	42	46	18.8
Verdicchio Bianco	5043	0	0	19	81	19.2	4674	0	0	91	9	18.8
Croatina	3116	0	1	93	6	18.6	2678	0	0	37	63	19.0
Aglianico	9264	0	26	35	39	18.5	9627	0	0	27	73	19.0
Cabernet Franc	6639	1	2	28	69	19.0	5590	1	2	23	74	19.1
Chardonnay	11687	4	21	33	42	18.3	19769	2	12	30	55	19.1
Prosecco	7498	0	0	1	99	19.1	19730	0	0	2	98	19.3
Rondinella	2797	0	0	1	99	19.4	2683	0	0	1	99	19.3
Corvina Veronese	4781	0	0	1	99	19.4	6222	0	0	0	100	19.3
Trebbiano d'Abruzzo	8435	0	0	16	84	19.6	2630	0	0	3	97	19.3
Lambrusco Salamino	4147	0	0	0	100	19.4	6228	0	0	3	97	19.4
Trebbiano Romagnolo	19492	0	0	1	99	19.7	19059	0	0	0	100	19.4
Merlot	21861	1	4	27	68	19.0	24057	1	3	22	74	19.5
Montepulciano	28679	0	0	19	81	19.4	32724	0	0	17	83	19.5
Sangiovese	62761	0	0	29	71	19.4	68428	0	0	9	91	19.5
Cabernet Sauvignon	7682	1	5	26	68	19.1	14240	1	2	17	81	19.6
Falanghina Flegrea	0	0	0	0	0	0.0	3634	0	0	3	97	19.6
Malvasia Bianca di Candia	11921	0	0	15	85	19.8	9028	0	0	7	93	19.9
Trebbiano Toscano	39447	0	0	12	87	19.9	35441	0	0	10	89	20.0
Vermentino	3201	0	0	10	90	19.6	6703	0	0	4	96	20.1
Garganega	16549	0	0	1	99	20.2	8522	0	0	1	99	20.3
Gaglioppo	3592	0	0	0	99	21.0	4626	0	0	25	75	20.3
Garnacha Tinta	6781	0	0	68	31	19.3	5421	0	1	0	98	20.3
Lambrusco Maestri	1362	0	0	0	100	19.6	5610	0	0	3	97	20.5
Tribidrag	7828	0	0	0	99	20.7	13896	0	0	1	99	20.6
Nero di Troia	1765	0	0	1	99	20.5	2512	0	0	0	100	20.7
Negroamaro	16619	0	0	0	100	21.1	11431	0	0	0	100	20.7
Syrah	1025	0	0	12	88	21.0	7693	0	0	2	98	20.9
Inzolia	9259	0	0	2	98	21.3	4740	0	0	0	100	21.4
Nero d'Avola	11318	0	0	1	99	20.9	14129	0	0	0	100	21.5
Grillo	1803	0	0	0	100	21.7	7382	0	0	0	100	21.5
Catarratto Bianco	50711	0	0	0	100	21.6	28563	0	0	0	100	21.5
Top 40	**456936**	**1**	**2**	**24**	**73**	**19.7**	**484184**	**0**	**1**	**19**	**79**	**19.7**

Table 78 (cont.): Shares of national winegrape area of each variety in cool, temperate, warm and hot climates, and average growing season temperature, top 40 varieties, 2000 and 2016 (% and °C)

Japan	2000						2016					
Variety	Area (ha)	Cool	Temp.	Warm	Hot	av. GST	Area (ha)	Cool	Temp.	Warm	Hot	av. GST
Cascade							22	0	100	0	0	15.6
Pinot Noir							20	0	100	0	0	15.6
Yamasachi							20	0	100	0	0	15.6
Portland							39	0	95	5	0	15.7
Müller-Thurgau							22	0	92	8	0	15.7
Zweigelt							59	0	95	1	4	15.7
Kerner							76	0	94	1	5	15.8
Campbell Early							238	0	72	28	0	16.1
Niagara							551	0	45	54	1	17.2
Riesling							22	0	0	100	0	17.4
Yamabudo							35	0	0	94	6	17.5
Yama Sauvignon							24	0	0	100	0	18.1
Delaware							254	0	11	58	31	18.4
Concord							292	0	0	100	0	18.6
Black Queen							28	0	0	100	0	18.6
Ryugan							27	0	0	100	0	18.6
Sauvignon Blanc							15	0	0	100	0	18.6
Merlot							197	0	0	77	23	18.7
Chardonnay							137	0	0	82	18	18.7
Verdelet							39	0	0	81	19	18.8
Kyoho (4N)							62	0	0	65	35	18.8
Cabernet Sauvignon							42	0	0	49	51	18.8
Muscat Bailey A							521	0	0	33	67	18.9
Koshu							690	0	0	14	86	19.0
Riesling Forte							2	0	0	14	86	19.0
Adirondac							24	0	0	14	86	19.0
Rose Ciotat							2	0	0	14	86	19.5
Red Millennium							2	0	0	14	86	19.5
Top 28							**3464**	**0**	**20**	**46**	**34**	**18.1**

Moldova	2000						2016					
Variety	Area (ha)	Cool	Temp.	Warm	Hot	av. GST	Area (ha)	Cool	Temp.	Warm	Hot	av. GST
Moldova	0	0	0	0	0	0.0	12375	0	0	100	0	17.2
Cabernet Sauvignon	7590	0	0	100	0	17.2	8169	0	0	100	0	17.2
Aligoté	15790	0	0	100	0	17.2	7765	0	0	100	0	17.2
Merlot	8123	0	0	100	0	17.2	7689	0	0	100	0	17.2
Sauvignon Blanc	8151	0	0	100	0	17.2	6909	0	0	100	0	17.2
Chardonnay	5134	0	0	100	0	17.2	4133	0	0	100	0	17.2
Rkatsiteli	11508	0	0	100	0	17.2	3898	0	0	100	0	17.2
Isabella	11401	0	0	100	0	17.2	3468	0	0	100	0	17.2
Pinot Noir	6521	0	0	100	0	17.2	2366	0	0	100	0	17.2
Muscat Ottonel	1520	0	0	100	0	17.2	1859	0	0	100	0	17.2
Riesling	1343	0	0	100	0	17.2	1701	0	0	100	0	17.2
Bianca	15	0	0	100	0	17.2	1340	0	0	100	0	17.2
Pinot Gris	2042	0	0	100	0	17.2	1208	0	0	100	0	17.2
Kodryanka	0	0	0	0	0	0.0	1143	0	0	100	0	17.2
Gewürztraminer	2731	0	0	100	0	17.2	1099	0	0	100	0	17.2
Fetească Albă	4334	0	0	100	0	17.2	954	0	0	100	0	17.2
Magaracha Rannii	0	0	0	0	0	0.0	884	0	0	100	0	17.2
Alb de Suruceni	0	0	0	0	0	0.0	780	0	0	100	0	17.2
Cabernet Franc	0	0	0	0	0	0.0	756	0	0	100	0	17.2
Muscat Yantarnyi	0	0	0	0	0	0.0	683	0	0	100	0	17.2
Saperavi	716	0	0	100	0	17.2	573	0	0	100	0	17.2
Victoria	0	0	0	0	0	0.0	565	0	0	100	0	17.2
Pervenets Magaracha	0	0	0	0	0	0.0	517	0	0	100	0	17.2
Cardinal	0	0	0	0	0	0.0	473	0	0	100	0	17.2
Alb de Onițcani	0	0	0	0	0	0.0	424	0	0	100	0	17.2
Fetească Neagră	0	0	0	0	0	0.0	402	0	0	100	0	17.2
Fetească Regală	0	0	0	0	0	0.0	372	0	0	100	0	17.2
Chasselas	0	0	0	0	0	0.0	329	0	0	100	0	17.2
Sukholimansky Bely	599	0	0	100	0	17.2	325	0	0	100	0	17.2
Riton	2	0	0	100	0	17.2	313	0	0	100	0	17.2
Nastea	0	0	0	0	0	0.0	303	0	0	100	0	17.2
Trebbiano Toscano	0	0	0	0	0	0.0	277	0	0	100	0	17.2
Muscat of Hamburg	0	0	0	0	0	0.0	254	0	0	100	0	17.2
Viorika	40	0	0	100	0	17.2	251	0	0	100	0	17.2
Kodrinskii	5	0	0	100	0	17.2	229	0	0	100	0	17.2
Prezentabil	0	0	0	0	0	0.0	215	0	0	100	0	17.2
Pinot Blanc	350	0	0	100	0	17.2	210	0	0	100	0	17.2
Noah	71	0	0	100	0	17.2	200	0	0	100	0	17.2
Irsai Olivér	0	0	0	0	0	0.0	180	0	0	100	0	17.2
Bastardo Magarachsky	1040	0	0	100	0	17.2	180	0	0	100	0	17.2
Top 40	**89026**	**0**	**0**	**100**	**0**	**17.2**	**75771**	**0**	**0**	**100**	**0**	**17.2**

Table 78 (cont.): Shares of national winegrape area of each variety in cool, temperate, warm and hot climates, and average growing season temperature, top 40 varieties, 2000 and 2016 (% and°C)

New Zealand

Variety	2000 Area (ha)	Cool	Temp.	Warm	Hot	av. GST	2016 Area (ha)	Cool	Temp.	Warm	Hot	av. GST
Sankt Laurent	0	0	0	0	0	0.0	1	85	15	0	0	14.6
Müller-Thurgau	420	2	96	1	0	16.4	2	78	22	0	0	15.0
Riesling	492	21	78	0	0	15.5	767	49	51	0	0	15.1
Pinot Noir	1098	23	75	2	0	15.5	5514	35	65	0	0	15.1
Pinot Blanc	0	0	0	0	0	0.0	12	30	70	0	0	15.2
Lagrein	0	0	0	0	0	0.0	2	0	100	0	0	15.5
Grüner Veltliner	0	0	0	0	0	0.0	43	4	96	0	0	15.6
Nebbiolo	0	0	0	0	0	0.0	1	37	34	29	0	15.6
Sauvignon Blanc	2424	3	97	0	0	15.7	20497	2	98	0	0	15.7
Sauvignon Blanc (G)	0	0	0	0	0	0.0	104	0	100	0	0	15.7
Pinot Gris	127	33	58	9	0	15.5	2422	18	81	2	0	15.7
Pinot Meunier	0	0	0	0	0	0.0	21	8	92	0	0	15.9
Gewürztraminer	139	9	91	0	0	16.0	277	13	85	3	0	15.9
Chenin Blanc	146	0	98	2	0	16.5	24	10	90	0	0	16.0
Sémillon	229	2	94	4	0	16.2	63	8	91	1	0	16.0
Tempranillo	0	0	0	0	0	0.0	18	10	83	7	0	16.0
Gamay Noir	0	0	0	0	0	0.0	7	11	89	0	0	16.1
Chardonnay	2786	6	91	3	0	16.1	3117	4	93	3	0	16.1
Kolor	0	0	0	0	0	0.0	7	0	84	16	0	16.1
Tribidrag	0	0	0	0	0	0.0	4	1	99	0	0	16.3
Alvarinho	0	0	0	0	0	0.0	26	5	85	10	0	16.3
Viognier	0	0	0	0	0	0.0	129	2	89	8	0	16.3
Merlot	656	3	87	10	0	16.3	1239	1	95	4	0	16.3
Syrah	60	7	75	19	0	16.3	436	3	84	13	0	16.4
Montepulciano	0	0	0	0	0	0.0	8	0	70	30	0	16.4
Cabernet Sauvignon	654	2	87	12	0	16.3	275	2	87	11	0	16.4
Côt	67	2	83	15	0	16.4	129	1	87	12	0	16.4
Arneis	0	0	0	0	0	0.0	33	0	100	0	0	16.4
Verdelho	0	0	0	0	0	0.0	7	7	71	22	0	16.5
Cabernet Franc	118	3	72	25	0	16.5	109	2	78	20	0	16.5
Muscat	0	0	0	0	0	0.0	36	3	96	2	0	16.6
Pinotage	73	1	72	27	0	16.5	38	2	76	23	0	16.6
Reichensteiner	62	1	99	0	0	16.7	14	0	100	0	0	16.8
Sangiovese	0	0	0	0	0	0.0	8	0	51	49	0	16.8
Petit Verdot	0	0	0	0	0	0.0	9	0	48	52	0	16.9
Tannat	0	0	0	0	0	0.0	2	0	31	69	0	17.1
Seibel	0	0	0	0	0	0.0	3	0	13	87	0	17.2
Chambourcin	0	0	0	0	0	0.0	3	0	12	88	0	17.2
Flora	6	0	100	0	0	16.8	3	0	8	92	0	17.3
Palomino Fino	21	0	38	62	0	17.3	7	0	0	100	0	17.5
Top 40	**9578**	**7**	**89**	**4**	**0**	**16.0**	**35420**	**9**	**89**	**1**	**0**	**15.7**

Portugal

Variety	2000 Area (ha)	Cool	Temp	Warm	Hot	av. GST	2016 Area (ha)	Cool	Temp.	Warm	Hot	av. GST
Avesso	636	0	100	0	0	16.6	699	0	0	100	0	17.4
Rufete	2338	0	0	0	100	19.9	1145	0	0	99	1	17.4
Trajadura	1788	0	100	0	0	16.6	1550	0	0	98	2	17.5
Loureiro	3939	0	100	0	0	16.6	4402	0	0	91	9	17.6
Rabigato	1133	0	0	100	0	17.6	1969	0	0	88	12	17.7
Alvarelhão	1109	0	96	0	4	16.8	2860	0	0	87	13	17.7
Trousseau	1754	0	0	67	33	18.4	1166	0	0	80	20	17.9
Tinta Barroca	5657	0	0	94	6	17.8	4733	0	0	79	21	17.9
Azal	3302	0	100	0	0	16.6	1443	0	0	77	23	18.0
Vinhao	5296	0	99	0	1	16.7	4055	0	0	77	23	18.0
Tinta Carvalha	1920	0	0	100	0	17.6	1113	0	0	69	31	18.2
Malvasia Fina	2328	0	0	46	54	18.9	2922	0	0	68	32	18.2
Marufo	3512	0	0	49	51	18.9	3367	0	0	67	33	18.2
Touriga Franca	6671	0	0	100	0	17.6	14217	0	0	63	37	18.3
Viosinho	17	0	0	0	100	20.1	916	0	0	62	38	18.3
Muscat Blanc à Petits Grains	1419	0	0	87	13	17.9	1031	0	0	59	41	18.4
Malvasia Preta	2210	0	0	52	48	18.7	1933	0	0	51	49	18.6
Palomino Fino	0	0	0	0	0	0.0	2594	0	0	50	50	18.6
Touriga Nacional	4149	0	0	63	37	18.4	11411	0	0	50	50	18.6
Arinto de Bucelas	3966	0	70	0	30	17.6	5409	0	0	46	54	18.7
Malvasia	0	0	0	0	0	0.0	801	0	0	46	54	18.7
Diagalves	1088	0	0	43	57	18.9	1090	0	0	38	62	18.9
Tempranillo	7356	0	0	73	27	18.2	17014	0	0	37	63	18.9
Siria	1682	0	0	0	100	19.9	6438	0	0	33	67	19.0
Trincadeira	7264	0	0	14	86	19.5	10493	0	0	31	69	19.1
Vital	2246	0	0	0	100	19.9	659	0	0	22	78	19.3
Alicante Henri Bouschet	675	0	0	0	100	19.9	4547	0	0	18	82	19.4
Fernão Pires	14206	0	0	3	97	19.7	12138	0	0	17	83	19.4
Damaschino	1530	0	0	0	100	19.9	880	0	0	13	87	19.5
Mencia	1971	0	0	0	100	19.4	2561	0	0	12	88	19.5
Baga	6730	0	0	0	100	19.3	6750	0	0	12	88	19.5
Syrah	0	0	0	0	0	0.0	4017	0	0	11	89	19.6
Bical	912	0	0	0	100	19.4	1076	0	0	5	95	19.7
Castelão	14424	0	0	0	100	19.9	12580	0	0	4	96	19.7
Alfrocheiro	518	0	0	0	100	19.6	1206	0	0	1	99	19.8
Cabernet Sauvignon	318	0	0	0	100	19.9	2346	0	0	1	99	19.8
Santarena	0	0	0	0	0	0.0	724	0	0	0	100	19.8
Caladoc	0	0	0	0	0	0.0	2180	0	0	0	100	19.8
Antao Vaz	376	0	0	0	100	19.8	1768	0	0	0	100	19.8
Negramoll	1040	0	31	0	69	19.0	605	0	0	1	99	20.0
Top 40	**115478**	**0**	**17**	**27**	**56**	**18.7**	**158808**	**0**	**0**	**40**	**60**	**18.9**

Table 78 (cont.): Shares of national winegrape area of each variety in cool, temperate, warm and hot climates, and average growing season temperature, top 40 varieties, 2000 and 2016 (% and°C)

Romania	2000						2016					
Variety	Area (ha)	Cool	Temp.	Warm	Hot	av. GST	Area (ha)	Cool	Temp.	Warm	Hot	av. GST
Fetească Regală	1700	0	0	100	0	17.7	12619	0	0	100	0	17.7
Fetească Albă	18211	0	0	100	0	17.7	12428	0	0	100	0	17.7
Merlot	7810	0	0	100	0	17.7	11647	0	0	100	0	17.7
Riesling	0	0	0	0	0	0.0	6121	0	0	100	0	17.7
Aligoté	7608	0	0	100	0	17.7	5840	0	0	100	0	17.7
Sauvignon Blanc	4613	0	0	100	0	17.7	5594	0	0	100	0	17.7
Cabernet Sauvignon	8620	0	0	100	0	17.7	5359	0	0	100	0	17.7
Muscat Ottonel	5787	0	0	100	0	17.7	4779	0	0	100	0	17.7
Fetească Neagră	1214	0	0	100	0	17.7	2845	0	0	100	0	17.7
Pamid	0	0	0	0	0	0.0	2716	0	0	100	0	17.7
Băbească Neagră	3642	0	0	100	0	17.7	2696	0	0	100	0	17.7
Pinot Noir	1740	0	0	100	0	17.7	1930	0	0	100	0	17.7
Chardonnay	1376	0	0	100	0	17.7	1878	0	0	100	0	17.7
Muscat Blanc à Petits Grains	1012	0	0	100	0	17.7	1579	0	0	100	0	17.7
Pinot Gris	2388	0	0	100	0	17.7	1561	0	0	100	0	17.7
Graševina	15014	0	0	100	0	17.7	1437	0	0	100	0	17.7
Blaufränkisch	0	0	0	0	0	0.0	729	0	0	100	0	17.7
Grasă de Cotnari	850	0	0	100	0	17.7	571	0	0	100	0	17.7
Syrah	0	0	0	0	0	0.0	504	0	0	100	0	17.7
Gewürztraminer	445	0	0	100	0	17.7	469	0	0	100	0	17.7
Crimposie	0	0	0	0	0	0.0	450	0	0	100	0	17.7
Galbenă de Odobeşti	546	0	0	100	0	17.7	417	0	0	100	0	17.7
Rkatsiteli	506	0	0	100	0	17.7	413	0	0	100	0	17.7
Frâncuşă	0	0	0	0	0	0.0	365	0	0	100	0	17.7
Busuioacă de Bohotin	0	0	0	0	0	0.0	343	0	0	100	0	17.7
Blauer Portugieser	0	0	0	0	0	0.0	329	0	0	100	0	17.7
Iordan	0	0	0	0	0	0.0	311	0	0	100	0	17.7
Băbească Neagră (G)	0	0	0	0	0	0.0	297	0	0	100	0	17.7
Mustoasă de Măderat	0	0	0	0	0	0.0	282	0	0	100	0	17.7
Şarbă	0	0	0	0	0	0.0	266	0	0	100	0	17.7
Plavay	0	0	0	0	0	0.0	152	0	0	100	0	17.7
Chasselas	0	0	0	0	0	0.0	127	0	0	100	0	17.7
Coarnă Neagră	0	0	0	0	0	0.0	104	0	0	100	0	17.7
Kadarka	0	0	0	0	0	0.0	97	0	0	100	0	17.7
Sangiovese	0	0	0	0	0	0.0	88	0	0	100	0	17.7
Zweigelt	0	0	0	0	0	0.0	86	0	0	100	0	17.7
Novac	0	0	0	0	0	0.0	74	0	0	100	0	17.7
Cabernet Franc	0	0	0	0	0	0.0	72	0	0	100	0	17.7
Neuburger	0	0	0	0	0	0.0	71	0	0	100	0	17.7
Tempranillo	0	0	0	0	0	0.0	67	0	0	100	0	17.7
Top 40	**83082**	**0**	**0**	**100**	**0**	**17.7**	**87713**	**0**	**0**	**100**	**0**	**17.7**

Serbia	2000						2016					
Variety	Area (ha)	Cool	Temp.	Warm	Hot	av. GST	Area (ha)	Cool	Temp.	Warm	Hot	av. GST
Vranac	0	0	0	0	0	0.0	3	0	100	0	0	16.6
Cardinal	0	0	0	0	0	0.0	3	0	100	0	0	16.6
Muscat Ottonel	0	0	0	0	0	0.0	183	0	49	51	0	16.8
Pamid	0	0	0	0	0	0.0	67	0	40	60	0	17.0
Morava	0	0	0	0	0	0.0	34	0	0	100	0	17.1
Kadarka	0	0	0	0	0	0.0	15	0	0	100	0	17.1
Blaufränkisch	0	0	0	0	0	0.0	727	0	1	99	0	17.2
Pinot Noir	0	0	0	0	0	0.0	633	0	9	91	0	17.2
Prokupac	15180	0	0	100	0	17.3	916	0	5	95	0	17.2
Chardonnay	0	0	0	0	0	0.0	1455	0	23	77	0	17.2
Panonia	0	0	0	0	0	0.0	2	0	0	100	0	17.2
Kosmopolita	0	0	0	0	0	0.0	1	0	0	100	0	17.2
Cabernet Franc	0	0	0	0	0	0.0	79	0	36	64	0	17.2
Merlot	0	0	0	0	0	0.0	1968	0	9	91	0	17.3
Muscat Fleur d'Oranger	0	0	0	0	0	0.0	116	0	0	100	0	17.3
Cabernet Sauvignon	0	0	0	0	0	0.0	2111	0	17	83	0	17.3
Sauvignon Blanc	0	0	0	0	0	0.0	741	0	25	75	0	17.4
Muscat of Hamburg	2760	0	0	100	0	17.3	624	0	39	61	0	17.4
Pinot Gris	0	0	0	0	0	0.0	112	0	0	100	0	17.4
Marselan	0	0	0	0	0	0.0	84	0	0	100	0	17.5
Afus Ali	0	0	0	0	0	0.0	73	0	0	100	0	17.5
Riesling	0	0	0	0	0	0.0	1361	0	10	90	0	17.5
Graševina	33120	0	0	100	0	17.3	2037	0	6	94	0	17.5
Dimyat	0	0	0	0	0	0.0	192	0	24	76	0	17.6
Gewürztraminer	0	0	0	0	0	0.0	142	0	0	100	0	17.6
Župljanka	0	0	0	0	0	0.0	255	0	1	99	0	17.7
Muscat Blanc à Petits Grains	0	0	0	0	0	0.0	31	0	0	100	0	17.7
Rkatsiteli	0	0	0	0	0	0.0	60	0	3	97	0	17.7
Gamay Noir	0	0	0	0	0	0.0	54	0	0	100	0	17.8
Victoria	0	0	0	0	0	0.0	55	0	0	100	0	17.8
Muscat	0	0	0	0	0	0.0	14	0	0	100	0	18.1
Top 31	**51059**	**0**	**0**	**100**	**0**	**17.3**	**14146**	**0**	**13**	**87**	**0**	**17.3**

Table 78 (cont.): Shares of national winegrape area of each variety in cool, temperate, warm and hot climates, and average growing season temperature, top 40 varieties, 2000 and 2016 (% and°C)

Slovenia	2000						2016					
Variety	Area (ha)	Cool	Temp.	Warm	Hot	av. GST	Area (ha)	Cool	Temp.	Warm	Hot	av. GST
Müller-Thurgau	0	0	0	0	0	0.0	128	0	100	0	0	15.3
Muscat Ottonel	0	0	0	0	0	0.0	98	0	100	0	0	15.3
Savagnin Blanc	0	0	0	0	0	0.0	209	0	100	0	0	15.3
Furmint	0	0	0	0	0	0.0	546	0	100	0	0	15.3
Zweigelt	0	0	0	0	0	0.0	5	0	100	0	0	15.4
Silvaner	0	0	0	0	0	0.0	81	0	100	0	0	15.4
Riesling	0	0	0	0	0	0.0	607	0	100	0	0	15.4
Kerner	0	0	0	0	0	0.0	111	0	100	0	0	15.4
Blauer Portugieser	0	0	0	0	0	0.0	80	0	100	0	0	15.4
Graševina	3568	0	100	0	0	16.8	1935	0	83	17	0	15.9
Pinot Noir	0	0	0	0	0	0.0	202	0	72	28	0	16.1
Pinot Blanc	0	0	0	0	0	0.0	424	0	68	32	0	16.2
Sauvignon Blanc	1221	0	100	0	0	16.8	1121	0	68	32	0	16.2
Muscat Blanc à Petits Grains	0	0	0	0	0	0.0	586	0	76	10	14	16.2
Sankt Laurent	0	0	0	0	0	0.0	6	0	100	0	0	16.4
Gamay Noir	0	0	0	0	0	0.0	4	0	100	0	0	16.4
Plavec Žuti	0	0	0	0	0	0.0	82	0	63	37	0	16.5
Chardonnay	1549	0	100	0	0	16.8	1181	0	55	41	4	16.6
Pinot Gris	0	0	0	0	0	0.0	508	0	49	47	4	16.6
Blaufränkisch	0	0	0	0	0	0.0	709	0	58	42	0	16.7
Vitovska Grganja	0	0	0	0	0	0.0	33	0	66	34	0	17.2
Kraljevina	0	0	0	0	0	0.0	199	0	32	68	0	17.5
Žametovka	0	0	0	0	0	0.0	822	0	23	77	0	17.6
Cabernet Franc	0	0	0	0	0	0.0	27	0	0	100	0	17.6
Sauvignonasse	0	0	0	0	0	0.0	231	0	0	100	0	17.7
Ribolla Gialla	1127	0	100	0	0	16.8	597	0	0	100	0	17.8
Merlot	1197	0	100	0	0	16.8	817	0	1	89	10	18.1
Barbera	0	0	0	0	0	0.0	98	0	0	100	0	18.1
Zelen	0	0	0	0	0	0.0	75	0	0	100	0	18.1
Pinella	0	0	0	0	0	0.0	61	0	0	100	0	18.1
Cabernet Sauvignon	0	0	0	0	0	0.0	423	0	3	75	22	18.3
Refosco	0	0	0	0	0	0.0	1340	0	32	6	61	18.5
Malvazija Istarska	0	0	0	0	0	0.0	915	0	5	32	63	18.9
Syrah	0	0	0	0	0	0.0	18	0	0	0	100	19.5
Top 34	**8661**	**0**	**100**	**0**	**0**	**16.8**	**14283**	**0**	**51**	**37**	**12**	**16.9**

South Africa	2000						2016					
Variety	Area (ha)	Cool	Temp.	Warm	Hot	av. GST	Area (ha)	Cool	Temp.	Warm	Hot	av. GST
Riesling	0	0	0	0	0	0.0	152	0	0	23	77	20.1
Pinot Noir	461	0	0	0	100	20.7	1153	0	0	27	73	20.2
Cabernet Franc	488	0	0	0	100	20.4	835	0	0	5	95	20.3
Sauvignon Blanc	5393	0	0	0	100	20.7	9246	0	0	9	91	20.5
Sémillon	995	0	0	0	100	20.7	1121	0	0	6	94	20.5
Côt	0	0	0	0	0	0.0	452	0	0	5	95	20.6
Gewürztraminer	0	0	0	0	0	0.0	106	0	0	7	93	20.6
Mazuelo	0	0	0	0	0	0.0	114	0	0	0	100	20.6
Merlot	4837	0	0	0	100	20.7	5558	0	0	3	97	20.6
Roussanne	0	0	0	0	0	0.0	76	0	0	9	91	20.6
Pinot Gris	0	0	0	0	0	0.0	369	0	0	2	98	20.6
Petit Verdot	0	0	0	0	0	0.0	749	0	0	3	97	20.7
Touriga Nacional	0	0	0	0	0	0.0	105	0	0	3	97	20.7
Cabernet Sauvignon	8824	0	0	0	100	20.7	10589	0	0	2	98	20.7
Crouchen	0	0	0	0	0	0.0	271	0	0	0	100	20.7
Monastrell	0	0	0	0	0	0.0	473	0	0	4	96	20.7
Syrah	5578	0	0	0	100	20.8	9946	0	0	3	97	20.7
Chardonnay	5995	0	0	0	100	20.9	6856	0	0	5	95	20.8
Viognier	50	0	0	0	100	20.9	822	0	0	4	96	20.8
Cinsaut	3533	0	0	0	100	20.8	1767	0	0	0	100	20.8
Tinta Barroca	0	0	0	0	0	0.0	190	0	0	2	98	20.8
Roobernet	0	0	0	0	0	0.0	269	0	0	0	100	20.9
Pinotage	6433	0	0	0	100	20.8	7052	0	0	1	99	20.9
Tempranillo	0	0	0	0	0	0.0	92	0	0	5	95	20.9
Garnacha Blanca	0	0	0	0	0	0.0	110	0	0	10	90	20.9
Garnacha Tinta	0	0	0	0	0	0.0	344	0	0	5	95	20.9
Clairette	0	0	0	0	0	0.0	195	0	0	0	100	21.0
Durif	0	0	0	0	0	0.0	114	0	0	0	100	21.0
Nouvelle	0	0	0	0	0	0.0	428	0	0	1	99	21.0
Trebbiano Toscano	0	0	0	0	0	0.0	156	0	0	0	100	21.1
Tannat	0	0	0	0	0	0.0	114	0	0	1	99	21.2
Ruby Cabernet	1980	0	0	0	100	21.4	2306	0	0	0	100	21.2
Chenin Blanc	22412	0	0	0	100	21.2	17707	0	0	0	100	21.2
Muscat of Alexandria	3926	0	0	0	100	21.4	1781	0	0	1	99	21.3
(R)	0	0	0	0	0	0.0	404	0	0	2	98	21.3
Muscat Blanc à Petits Grains	757	0	0	0	100	21.3	839	0	0	0	100	21.3
Palomino Fino	0	0	0	0	0	0.0	134	0	0	1	99	21.6
Morio-Muskat	0	0	0	0	0	0.0	78	0	0	0	100	21.9
Colombard	11203	0	0	0	100	22.0	11512	0	0	0	100	21.9
Villard Blanc	0	0	0	0	0	0.0	321	0	0	0	100	23.4
Top 40	**82866**	**0**	**0**	**0**	**100**	**21.1**	**94908**	**0**	**0**	**3**	**97**	**21.0**

Table 78 (cont.): Shares of national winegrape area of each variety in cool, temperate, warm and hot climates, and average growing season temperature, top 40 varieties, 2000 and 2016 (% and°C)

Spain	2000						2016					
Variety	Area (ha)	Cool	Temp.	Warm	Hot	av. GST	Area (ha)	Cool	Temp.	Warm	Hot	av. GST
Juan Garcia	2077	0	38	62	0	17.0	1409	0	91	9	0	16.6
Prieto Picudo	3256	0	88	10	1	16.4	4293	0	99	0	1	16.6
Mencia	11166	0	56	34	10	17.0	8489	0	62	35	4	16.9
Alvarinho	0	0	0	0	0	0.0	5393	0	0	99	1	17.2
Verdejo	4453	0	20	66	14	17.5	17923	0	64	0	35	17.7
Graciano	435	0	45	46	9	17.1	2080	0	25	31	45	18.5
Tempranillo	79310	0	30	39	31	17.9	193597	0	28	16	55	18.7
Alicante Henri Bouschet	18321	0	4	46	50	18.5	19294	0	3	27	70	19.0
Garnacha Tinta	98131	0	16	12	72	19.2	54606	0	15	8	77	19.1
Alarije	1686	0	0	0	100	20.7	4407	0	40	0	60	19.2
Sauvignon Blanc	467	0	10	70	20	17.6	4562	0	16	1	83	19.3
Mazuelo	8103	0	11	15	74	19.3	5461	0	8	16	76	19.3
Muscat Blanc à Petits Grains	223	0	26	14	60	19.1	1350	0	12	0	88	19.5
Merlot	1186	0	36	6	58	18.5	12852	0	14	0	86	19.6
Airén	387978	0	0	11	89	19.9	203276	0	0	0	100	19.6
Tinto Velasco	7998	0	0	0	100	20.4	5369	0	0	0	100	19.6
Pardillo	7272	0	0	13	87	19.3	3283	0	0	0	100	19.6
Marufo	2827	0	0	57	43	18.6	1316	0	0	0	100	19.6
Chardonnay	1814	0	9	1	90	19.6	6866	0	9	1	90	19.7
Cabernet Sauvignon	4519	0	27	7	66	18.8	20139	0	10	0	90	19.7
Macabeo	42902	0	8	18	75	19.3	36963	0	6	6	88	19.8
Petit Verdot	0	0	0	0	0	0.0	1804	0	1	0	99	19.9
Syrah	86	0	5	3	92	20.0	19488	0	2	0	98	19.9
Garnacha Blanca	4333	0	1	2	97	19.9	2061	0	1	1	98	19.9
Parellada	11188	0	0	0	100	20.0	7137	0	0	0	100	20.0
Xarello	10299	0	0	0	100	20.0	8534	0	0	0	100	20.0
Trepat	1763	0	0	0	100	20.4	1199	0	0	0	100	20.2
Palomino Fino	27685	0	10	29	61	19.8	20110	0	11	16	73	20.4
Bobal	100128	0	0	41	58	19.5	59189	0	0	0	100	20.4
Chelva	10877	0	2	0	98	20.7	5029	0	3	0	97	20.5
Cayetana Blanca	55527	0	0	0	100	20.7	36252	0	0	0	100	20.7
Beba	4762	0	0	0	100	20.7	2556	0	0	0	100	20.7
Monastrell	67160	0	0	0	100	21.3	41303	0	0	0	100	21.2
Muscat of Alexandria	6144	0	0	0	100	21.5	9534	0	0	1	99	21.3
Pedro Ximénez	14803	0	0	0	99	22.1	8528	0	0	0	100	21.3
Zalema	5969	0	0	0	100	21.7	4015	0	0	0	100	21.5
Negramoll	1642	0	5	1	95	21.2	1149	0	0	0	100	21.5
Merseguera	7460	0	0	0	100	21.5	2373	0	0	0	100	21.5
Malvasia	0	0	0	0	0	0.0	1362	0	1	4	95	21.6
Listan Negro	3291	0	0	0	100	22.1	2847	0	0	0	100	22.1
Top 40	**1017240**	**0**	**6**	**16**	**78**	**19.7**	**847399**	**0**	**12**	**7**	**82**	**19.6**

Switzerland	2000						2016					
Variety	Area (ha)	Cool	Temp.	Warm	Hot	av. GST	Area (ha)	Cool	Temp.	Warm	Hot	av. GST
Merlot	848	99	1	0	0	9.3	1124	92	8	0	0	10.2
Cabernet Sauvignon	34	74	26	0	0	13.6	66	67	33	0	0	13.7
Kerner	8	96	4	0	0	13.0	25	78	22	0	0	13.8
Cabernet Franc	16	70	30	0	0	12.1	63	55	45	0	0	13.9
Chardonnay	226	61	39	0	0	14.2	359	60	40	0	0	14.1
Solaris	0	0	0	0	0	0.0	20	95	5	0	0	14.1
Sauvignon Blanc	38	53	47	0	0	13.9	170	70	30	0	0	14.1
Cabernet Jura	0	0	0	0	0	0.0	27	97	3	0	0	14.2
Zweigelt	0	0	0	0	0	0.0	19	100	0	0	0	14.2
Regent	16	100	0	0	0	14.4	38	97	3	0	0	14.3
Cabernet Dorsa	0	0	0	0	0	0.0	28	98	2	0	0	14.4
Müller-Thurgau	686	90	10	0	0	14.5	465	94	6	0	0	14.4
Dornfelder	0	0	0	0	0	0.0	22	90	10	0	0	14.4
Räuschling	23	100	0	0	0	14.5	23	100	0	0	0	14.5
Riesling	8	87	13	0	0	14.7	19	94	6	0	0	14.5
Viognier	0	0	0	0	0	0.0	44	40	60	0	0	14.5
Pinot Noir	4601	87	13	0	0	14.6	4209	85	15	0	0	14.6
Ancellotta	2	98	2	0	0	14.2	28	96	4	0	0	14.6
Diolinoir	25	95	5	0	0	14.4	120	90	10	0	0	14.7
Gewürztraminer	31	66	34	0	0	14.7	51	62	38	0	0	14.7
Pinot Gris	149	74	26	0	0	14.7	230	74	26	0	0	14.7
Muscat Blanc à Petits Grains	44	99	1	0	0	14.7	36	100	0	0	0	14.7
Syrah	54	98	2	0	0	14.6	194	90	10	0	0	14.7
Amigne	21	100	0	0	0	14.8	42	100	0	0	0	14.8
Arvine	57	100	0	0	0	14.8	178	100	0	0	0	14.8
Humagne	9	100	0	0	0	14.8	29	100	0	0	0	14.8
Cornalin	92	100	0	0	0	14.8	138	100	0	0	0	14.8
Rouge du Pays	0	0	0	0	0	0.0	136	100	0	0	0	14.8
Marsanne	33	100	0	0	0	14.8	48	99	1	0	0	14.8
Silvaner	208	98	2	0	0	14.8	250	99	1	0	0	14.8
Savagnin Blanc	17	100	0	0	0	14.8	127	96	4	0	0	14.8
Doral	3	48	52	0	0	12.5	35	13	87	0	0	14.8
Pinot Blanc	77	37	63	0	0	15.0	111	50	50	0	0	14.8
Gamaret	60	20	80	0	0	14.8	425	37	63	0	0	14.9
Galotta	0	0	0	0	0	0.0	35	28	72	0	0	15.0
Dunkelfelder	0	0	0	0	0	0.0	23	44	56	0	0	15.0
Chasselas	5373	41	59	0	0	15.0	3838	33	67	0	0	15.0
Gamay Noir	1977	49	51	0	0	15.0	1349	45	55	0	0	15.1
Garanoir	50	47	53	0	0	14.9	225	26	74	0	0	15.1
Aligoté	20	5	95	0	0	15.4	24	2	98	0	0	15.4
Top 40	**14807**	**65**	**35**	**0**	**0**	**14.5**	**14391**	**65**	**35**	**0**	**0**	**14.4**

Table 78 (cont.): Shares of national winegrape area of each variety in cool, temperate, warm and hot climates, and average growing season temperature, top 40 varieties, 2000 and 2016 (% and°C)

Turkey	2000						2016					
Variety	Area (ha)	Cool	Temp.	Warm	Hot	av. GST	Area (ha)	Cool	Temp.	Warm	Hot	av. GST
Emir							89	0	0	100	0	17.6
Dimrit							704	0	0	82	18	18.5
Sémillon							529	0	0	100	0	18.7
Cinsaut							430	0	0	100	0	18.7
Gamay Noir							228	0	0	100	0	18.7
Papazkarası							204	0	0	100	0	18.7
Adakarası							89	0	0	100	0	18.7
Riesling							3	0	0	100	0	18.7
Narince							787	0	0	0	100	20.3
Öküzgözü							1601	0	0	1	99	20.9
Tempranillo							6	0	0	25	75	21.6
Sauvignon Blanc							153	0	0	31	69	21.6
Kalecik Karası							704	0	0	22	78	21.7
Viognier							15	0	0	28	72	21.8
Merlot							415	0	0	22	78	22.0
Boğazkere							1436	0	0	3	97	22.2
Cabernet Sauvignon							476	0	0	12	88	22.4
Syrah							1439	0	0	9	91	22.5
Cabernet Franc							37	0	0	10	90	22.5
Chardonnay							177	0	0	8	92	22.5
Côt							21	0	0	4	96	22.7
Sultaniye							2461	0	0	0	100	22.9
Çalkarası							806	0	0	0	100	22.9
Alicante Henri Bouschet							532	0	0	0	100	22.9
Mazuelo							130	0	0	0	100	22.9
Muscat Blanc à Petits Grains							129	0	0	0	100	22.9
Garnacha Tinta							33	0	0	0	100	22.9
Petit Verdot							19	0	0	0	100	22.9
Sangiovese							18	0	0	0	100	22.9
Monastrell							7	0	0	0	100	22.9
Karalahna							4	0	0	0	100	22.9
Karasakiz							4	0	0	0	100	22.9
Vasilaki							4	0	0	0	100	22.9
Çavuş							3	0	0	0	100	22.9
Pinot Noir							10	0	0	0	100	22.9
Top 35							**13704**	**0**	**0**	**20**	**80**	**21.6**

United States	2000						2016					
Variety	Area (ha)	Cool	Temp.	Warm	Hot	av. GST	Area (ha)	Cool	Temp.	Warm	Hot	av. GST
Concord	8330	0	100	0	0	15.2	8349	3	93	3	1	15.3
Niagara	1357	0	100	0	0	15.5	1196	3	94	2	1	15.4
Aurore	299	0	100	0	0	15.4	255	13	86	0	0	15.4
Catawba	635	0	100	0	0	15.4	625	6	82	5	7	15.9
Pinot Noir	5343	1	44	54	1	17.1	22998	0	52	37	11	17.2
Pinot Blanc	431	0	67	25	8	17.2	263	0	50	36	14	17.3
Riesling	1965	1	42	53	4	17.0	4952	1	61	25	13	17.4
Gewürztraminer	728	1	44	55	0	17.0	897	0	59	21	19	17.6
Vidal	97	0	55	45	0	16.3	366	0	28	30	42	17.9
Cabernet Franc	1189	0	22	70	9	17.8	2199	0	17	60	24	18.3
Chambourcin	28	0	59	41	0	16.4	315	0	17	20	64	18.7
Sauvignon Blanc	4191	0	11	55	34	18.8	6747	0	14	58	29	18.7
Cabernet Sauvignon	17573	0	10	56	33	18.9	40837	0	11	59	30	18.8
Petit Verdot	96	0	5	84	11	18.4	1219	0	9	55	36	18.8
Chardonnay	35791	0	26	38	36	18.7	41392	0	28	33	39	18.8
Syrah	1509	0	11	36	53	19.6	9083	0	26	36	38	18.9
Merlot	16875	0	10	46	44	19.3	21251	0	18	42	39	19.0
Pinot Gris	826	3	70	22	5	16.5	7462	0	35	10	55	19.0
Côt	97	0	4	71	25	18.9	1610	0	15	40	45	19.1
Viognier	314	0	11	57	32	18.8	1481	0	15	28	57	19.3
Durif	923	0	10	44	47	19.3	3698	0	4	43	53	19.3
Sangiovese	682	0	12	53	36	19.1	827	0	9	46	44	19.4
Sémillon	709	0	8	61	31	18.8	340	0	12	40	48	19.6
Monastrell	187	0	4	18	78	19.6	515	0	7	31	62	19.6
Tempranillo	201	0	0	1	99	21.8	626	0	17	21	61	20.2
Tribidrag	18630	0	2	20	78	20.4	18551	0	0	25	74	20.2
Mazuelo	3088	0	0	12	88	21.2	1086	0	0	21	79	20.9
Malvasia Bianca di Candia	968	0	7	0	93	21.6	486	0	12	2	87	20.9
Garnacha Tinta	4519	0	1	1	98	22.3	2213	0	10	15	75	21.0
Muscat Blanc à Petits Grains	515	0	8	30	62	20.6	1218	0	9	17	74	21.2
Chenin Blanc	8433	0	6	5	90	21.8	1969	0	4	5	91	21.6
Alicante Henri Bouschet	563	0	0	4	96	22.6	380	0	0	7	93	21.7
Symphony	184	0	0	0	100	22.2	647	0	0	0	100	22.1
Barbera	4693	0	0	1	99	22.7	2131	0	1	4	95	22.3
Elbling	0	0	0	0	0	0.0	396	0	0	0	100	22.3
Colombard	18010	0	0	1	99	22.5	7991	0	0	0	100	22.7
Ruby Cabernet	2895	0	0	0	100	22.8	2114	0	0	0	100	22.8
Muscat of Alexandria	2013	0	0	0	100	23.1	1987	0	0	0	100	22.8
Triplett Blanc	0	0	0	0	0	0.0	412	0	0	0	100	22.8
Rubired	4153	0	0	0	100	22.9	4825	0	0	0	100	22.8
Top 40	**169039**	**0**	**18**	**27**	**55**	**19.8**	**225907**	**0**	**24**	**34**	**42**	**19.0**

Table 78 (cont.): Shares of national winegrape area of each variety in cool, temperate, warm and hot climates, and average growing season temperature, top 40 varieties, 2000 and 2016 (% and °C)

Uruguay	2000						2016					
Variety	Area (ha)	Cool	Temp.	Warm	Hot	av. GST	Area (ha)	Cool	Temp.	Warm	Hot	av. GST
Pinot Gris	0	0	0	0	0	0.0	9	0	0	0	100	19.5
Petit Verdot	0	0	0	0	0	0.0	26	0	0	0	100	19.8
Pinot Blanc	0	0	0	0	0	0.0	2	0	0	0	100	19.8
Arriloba	0	0	0	0	0	0.0	6	0	0	0	100	20.0
Concord	0	0	0	0	0	0.0	24	0	0	0	100	20.1
Marselan	0	0	0	0	0	0.0	120	0	0	0	100	20.1
Sauvignon Blanc	142	0	0	0	100	20.2	144	0	0	0	100	20.1
Arinarnoa	0	0	0	0	0	0.0	45	0	0	0	100	20.1
Sauvignon Blanc (G)	0	0	0	0	0	0.0	10	0	0	0	100	20.1
Malvasia Fina	0	0	0	0	0	0.0	6	0	0	0	100	20.1
Chenin Blanc	0	0	0	0	0	0.0	2	0	0	0	100	20.1
Egiodola	0	0	0	0	0	0.0	2	0	0	0	100	20.1
Ruby Cabernet	0	0	0	0	0	0.0	2	0	0	0	100	20.1
Muscat of Hamburg	2886	0	0	0	100	20.2	1267	0	0	0	100	20.1
Trebbiano Toscano	0	0	0	0	0	0.0	682	0	0	0	100	20.2
Merlot	1057	0	0	0	100	20.2	747	0	0	0	100	20.2
Tannat	2433	0	0	0	100	20.2	1725	0	0	0	100	20.2
Alicante Henri Bouschet	0	0	0	0	0	0.0	24	0	0	0	100	20.2
Chardonnay	142	0	0	0	100	20.2	119	0	0	0	100	20.2
Pinot Noir	0	0	0	0	0	0.0	56	0	0	0	100	20.2
Isabella	0	0	0	0	0	0.0	102	0	0	0	100	20.3
Viognier	0	0	0	0	0	0.0	41	0	0	0	100	20.3
Cabernet Franc	364	0	0	0	100	20.2	266	0	0	0	100	20.3
Muscat Blanc à Petits Grains	0	0	0	0	0	0.0	10	0	0	0	100	20.3
Cabernet Sauvignon	675	0	0	0	100	20.2	484	0	0	0	100	20.3
Syrah	62	0	0	0	100	20.2	67	0	0	0	100	20.4
Sémillon	0	0	0	0	0	0.0	14	0	0	0	100	20.4
Gewürztraminer	0	0	0	0	0	0.0	17	0	0	0	100	20.5
Côt	0	0	0	0	0	0.0	43	0	0	0	100	20.5
Muscat of Alexandria	0	0	0	0	0	0.0	22	0	0	0	100	20.6
Muscat Ottonel	0	0	0	0	0	0.0	2	0	0	0	100	20.6
Canari Noir	0	0	0	0	0	0.0	4	0	0	0	100	20.7
Riesling	0	0	0	0	0	0.0	12	0	0	0	100	20.9
Nebbiolo	0	0	0	0	0	0.0	25	0	0	0	100	20.9
Roussanne	0	0	0	0	0	0.0	2	0	0	0	100	21.0
Gamay Noir	0	0	0	0	0	0.0	1	0	0	0	100	21.0
Marsanne	0	0	0	0	0	0.0	1	0	0	0	100	21.0
Garnacha Tinta	0	0	0	0	0	0.0	4	0	0	0	100	21.2
Top 38	**7762**	**0**	**0**	**0**	**100**	**20.2**	**6135**	**0**	**0**	**0**	**100**	**20.2**

Table 79: Shares of national winegrape area in cool, temperate, warm and hot climates, and average growing season temperature, by country, 2000 and 2016

Country	2000					2016				
	Cool	Temp.	Warm	Hot	GST (°C)	Cool	Temp.	Warm	Hot	GST (°C)
Algeria	0	0	0	100	21.3	0	0	0	100	21.3
Argentina	0	0	16	84	20.8	0	0	21	79	20.6
Armenia	0	0	100	0	17.8	0	0	100	0	17.8
Australia	1	4	37	58	19.8	1	10	40	49	19.4
Austria	0	100	0	0	15.6	13	87	0	0	15.5
Brazil	0	0	0	100	19.6	0	0	0	100	19.6
Bulgaria	0	0	100	0	18.3	0	0	100	0	18.3
Cambodia						0	0	0	100	29.0
Canada	0	100	0	0	15.1	39	61	0	0	15.1
Chile	0	12	80	8	17.8	0	8	83	9	17.9
China						0	0	100	0	18.1
Croatia	0	0	100	0	18.5	0	0	100	0	18.5
Cyprus	0	0	0	100	21.2	0	0	0	100	21.2
Czechia	0	100	0	0	15.3	4	96	0	0	15.3
Ethiopia						0	100	0	0	17.0
France	4	36	26	34	17.7	5	42	42	12	17.6
Georgia	0	100	0	0	16.6	0	100	0	0	16.6
Germany	51	49	0	0	14.9	51	49	0	0	15.0
Greece	0	0	5	95	21.3	0	0	4	96	21.2
Hungary	0	100	0	0	16.7	0	85	15	0	16.7
India						0	0	0	100	26.9
Israel	0	0	0	100	21.2	0	0	0	100	21.2
Italy	1	2	24	73	19.6	1	2	20	77	19.6
Japan						0	22	45	33	18.0
Kazakhstan						0	0	66	34	18.8
Korea, Rep.	0	0	0	100	19.7	0	0	0	100	19.7
Lebanon						0	0	0	100	21.7
Luxembourg	100	0	0	0	14.2	100	0	0	0	14.2
Mexico						0	0	6	94	22.2
Moldova	0	0	100	0	17.2	0	0	100	0	17.2
Morocco	0	0	0	100	20.9	0	0	0	100	20.9
Myanmar						0	0	0	100	29.6
New Zealand	7	89	4	0	16.0	10	89	1	0	15.7
N. Macedonia						0	100	0	0	16.7
Norway						100	0	0	0	11.9
Peru						0	35	0	65	18.4
Portugal	0	14	31	55	18.7	0	0	41	59	18.8
Romania	0	0	100	0	17.7	0	0	100	0	17.7
Russia	0	0	100	0	17.6	0	0	100	0	17.6
Serbia	0	0	100	0	17.3	0	14	86	0	17.3
Slovakia	0	100	0	0	15.9	0	100	0	0	15.9
Slovenia	0	100	0	0	16.8	0	53	36	12	16.8
South Africa	0	0	0	100	21.1	0	0	3	97	21.0
Spain	0	7	17	76	19.7	0	12	8	80	19.5
Switzerland	65	35	0	0	14.5	65	35	0	0	14.4
Taiwan	0	0	0	100	25.5	0	0	0	100	25.5
Thailand						0	0	0	100	29.1
Tunisia	0	0	0	100	22.2	0	0	0	100	22.2
Turkey						0	0	20	80	21.6
Ukraine						0	0	100	0	17.1
United Kingdom	100	0	0	0	13.5	100	0	0	0	13.5
United States	0	18	27	55	19.8	0	23	33	44	19.1
Uruguay	0	0	0	100	20.2	0	0	0	100	20.2
New World	**0**	**9**	**27**	**64**	**19.8**	**1**	**11**	**43**	**45**	**19.2**
Old World	**3**	**18**	**31**	**48**	**18.6**	**3**	**21**	**32**	**44**	**18.5**
World	**2**	**16**	**31**	**51**	**18.8**	**3**	**18**	**35**	**44**	**18.6**

VIII. Regional Varietal Intensity Indexes for world's top varieties

Table 80: VIIs for the top 25 regions for the world's top 24 red varieties, 2000

Cabernet Sauvignon		**Garnacha Tinta**		**Merlot**		**Mazuelo**	
Region	*Varietal Intensity Indexes*	*Region*	*Varietal Intensity Indexes*	*Region*	*Varietal Intensity Indexes*	*Region*	*Varietal Intensity Indexes*
CL Metropolitana	12.51	IT Nuoro	14.16	CH Ticino	19.18	TN Tunisia	17.23
AU Limestone Coast - other	11.38	ES Zaragoza	13.72	IT Rovigo	13.07	FR Aude	14.46
CL O'Higgins	10.98	FR Vaucluse	12.61	AR General Pueyrredón	11.46	FR Herault	10.87
AR General Pueyrredón	10.96	ES Comunidad de Madrid	9.88	FR Gironde	11.44	DZ Algeria	9.57
US Yuba	10.81	ES Avila, Palencia, Salama	9.17	AR Puelen	9.39	FR Pyrenees-Orientales	7.79
AR Calingasta	10.52	ES Huesca, Teruel	9.11	US Oregon - other	9.35	IL Israel	7.67
US Butte	10.04	ES Comunidad Foral de Na	8.33	AR Cushamen	9.17	FR Var	7.55
AU Currency Creek	9.23	US Glenn	7.92	US Marin	8.66	ES Girona, Lleida	7.27
AU Langhorne Creek	8.73	FR Bouches-du-Rhone	6.79	IT Padova	8.60	IT Cagliari	6.73
AR Santa María - Catamarca	8.69	FR Gard	6.00	IT Venezia	7.23	FR Gard	6.17
AU Lower Murray - other	8.34	ES La Rioja	5.14	IT Pordenone	7.10	FR Bouches-du-Rhone	5.48
AU Kangaroo Island	7.96	FR Var	4.95	US Trinity	7.00	FR Vaucluse	3.75
AU Hilltops	7.46	FR Rhone-Alpes except Ar	4.94	US Mariposa	6.55	ES Tarragona	3.67
AU Mount Benson	7.14	FR Pyrenees-Orientales	4.61	IT Bergamo	5.81	US Contra Costa	3.48
CL Del Maule	6.99	DZ Algeria	4.52	US Sacramento	5.66	US Placer	2.53
AU Western Victoria - other	6.79	FR Ardeche	4.49	FR Aquitaine except Gi	5.62	US Madera	2.45
AU Beechworth	6.67	US San Bernardino	4.10	US Columbia River	5.50	FR Alpes-de-Haute-Proven	2.15
US Napa	6.67	FR Alpes-de-Haute-Proven	4.06	US Washington	5.45	US Mendocino	2.07
AU Mudgee	6.12	ES Toledo	3.82	AR Añelo	5.30	FR Ardeche	2.04
AU Orange	6.10	ES Castellon	2.79	IT Udine	5.00	US Santa Clara	1.75
US San Luis Obispo	5.90	IT Sassari	2.77	AU Alpine Valleys/Bee	4.72	US San Joaquin	1.41
AU Margaret River	5.80	TN Tunisia	2.71	AR Coronel Pringles	4.58	EL Anatoliki Makedonia, T	1.36
AU Clare Valley	5.74	ES Zamora	2.67	IT Treviso	4.49	MA Morocco	1.31
M9 Missing 9	5.56	US Kings	2.61	US Tehama	4.41	US Stanislaus	1.26
AU Perricoota	5.48	FR Corse	2.45	AR El Cuy	4.22	ES Comunidad Foral de Na	1.21

Table 80 (cont.): VIIs for the top 25 regions for the world's top 24 red varieties, 2000

Syrah		Bobal		Tempranillo		Monastrell	
Region	*Varietal Intensity Indexes*	*Region*	*Varietal Intensity Indexes*	*Region*	*Varietal Intensity Indexes*	*Region*	*Varietal Intensity Indexes*
US Ventura	29.82	ES Valencia	28.76	ES Alava	39.23	ES Region de Murcia	48.59
AU Far North - other	28.91	ES Cuenca	20.71	ES Burgos	31.00	ES Alicante	23.33
AU Bendigo	25.69	ES Albacete	8.84	ES Guadalajara	27.48	US Contra Costa	15.43
AU Grampians	21.60	ES Caceres	0.93	ES La Rioja	25.70	ES Albacete	13.13
AU Kangaroo Island	20.45	ES Avila, Palencia, Sal	0.81	ES Valladolid	12.52	AU Perricoota	9.63
US Shasta	19.88	ES Huesca, Teruel	0.52	ES Zamora	12.40	US Placer	7.74
AU Barossa - other	19.63	ES Zaragoza	0.52	ES Comunidad Foral de	12.07	FR Var	4.37
AU McLaren Vale	18.98	ES Almeria, Granada, .	0.48	ES Castellon	10.38	ES Illes Balears	4.26
AU Pyrenees	18.06	ES Alicante	0.21	AR San Carlos - Mza	8.03	AU Barossa - other	3.17
AU Rutherglen	17.94	ES Illes Balears	0.20	ES Avila, Palencia, Salai	6.41	US San Diego	2.10
AU Australian Capital Ter	17.78	ES Castellon	0.19	AR El Cuy	5.26	AU Alpine Valleys/Beechwort	2.10
AU Barossa Valley	17.67	ES Tarragona	0.08	AR Tunuyán	4.61	AU Riverland	1.69
AU Southern NSW - other	17.65	ES Comunidad de Mad	0.04	PT Alto Tras-os-Montes	4.45	ES Castellon	1.64
AU Northern Slopes - othe	17.18	ES Guadalajara	0.04	PT Alentejo	4.26	AU Eastern Plains, Inland and	1.63
AU Southern Fleurieu	16.98	ES Girona, Lleida	0.04	ES Comunidad de Madri	3.44	ES Girona, Lleida	1.53
AU Mudgee	16.33	ES Region de Murcia	0.03	ES Huesca, Teruel	3.29	FR Vaucluse	1.48
AU Currency Creek	16.15	ES Toledo	0.03	ES Ciudad Real	3.02	AU Lower Murray - other	1.46
AU Langhorne Creek	15.92	ES Alava	0.02	ES Zaragoza	2.95	TN Tunisia	1.28
AU Sunbury	15.36	ES Zamora	0.02	AR Guaymallén	2.86	FR Pyrenees-Orientales	1.17
AU Orange	15.15	ES Ciudad Real	0.02	ES Girona, Lleida	2.69	ES Almeria, Granada, Jaen, Se	1.14
AU The Peninsulas	15.14	ES Comunidad Foral de	0.01	ES Illes Balears	2.61	FR Bouches-du-Rhone	1.07
AU Mount Lofty Ranges -	14.94	ES Valladolid	0.01	AR Maipú	2.03	FR Aude	1.00
AU Hilltops	14.84	ES Badajoz	0.00	AR San Martín - Mza	1.96	US Riverside	0.98
AU Clare Valley	14.70	ES Burgos	0.00	AR Tupungato	1.95	AU Barossa Valley	0.97
AU Central Victoria - othe	14.63	ES Canarias	0.00	ES Valencia	1.61	US San Bernardino	0.93

Table 80 (cont.): VIIs for the top 25 regions for the world's top 24 red varieties, 2000

Sangiovese		Pinot Noir		Cabernet Franc		Cinsaut	
Region	*Varietal Intensity Indexes*	*Region*	*Varietal Intensity Indexes*	*Region*	*Varietal Intensity Indexes*	*Region*	*Varietal Intensity Indexes*
IT Siena	49.34	CH Graubünden	54.60	FR Deux-Sevres, Vienne	28.37	DZ Algeria	25.23
IT Firenze	46.77	CH Schaffhausen	54.50	FR Centre-Val de Loire	22.89	FR Var	16.66
IT Rimini	45.85	CH St. Gallen	52.80	FR Pays de la Loire except M	20.87	FR Bouches-du-Rhone	9.90
IT Arezzo	44.29	US Benton Co.	47.59	FR Correze, Haute-Vienne	19.41	FR Herault	9.86
IT Prato	37.10	CH Thurgau	46.28	IT Venezia	13.08	FR Alpes-de-Haute-Proven	9.77
IT Pistoia	36.70	CH Basel Land	45.10	FR Aquitaine except Gironde	12.64	ZA Paarl	8.63
IT Forlì-Cesena	34.77	US Yamhill Co.	43.01	FR Gironde	10.71	FR Gard	8.57
IT Grosseto	27.63	CH Zürich	42.71	AR Puelen	9.27	MA Morocco	8.02
IT Pisa	27.26	CH Aargau	42.36	IT Udine	8.10	ZA Breedekloof	7.92
IT Ascoli Piceno	25.68	DE Ahr	40.44	US Nevada	7.54	FR Ardeche	6.68
IT Pesaro e Urbino	21.81	US Polk Co.	39.47	IT Treviso	7.42	FR Aude	5.69
IT Lucca	19.46	CH Other regions	37.27	IT Gorizia	7.28	FR Corse	5.27
IT Matera	19.11	US Humboldt	35.53	IT Pordenone	7.12	TN Tunisia	5.05
IT Foggia	16.94	AR Colón - Entre Ríos	35.44	NZ Auckland	7.09	ZA Swartland	5.04
IT Isernia	16.12	NZ Otago	34.63	BR Brazil	6.73	FR Vaucluse	4.78
IT Bari	16.11	CH Schwyz	32.14	IT Padova	6.67	EL Dytiki Makedonia	1.98
IT Livorno	16.06	US Marion Co.	31.76	CA Canada	6.27	IT Brindisi	1.93
FR Corse	15.88	US Valley - other	30.65	FR Midi-Pyrenees except Ger	5.58	ZA Worcester	1.93
IT Perugia	15.16	CH Neuchâtel	30.53	AU Australian Capital Territo	5.53	EL Thessalia	1.54
IT Macerata	10.73	US Washington Co.	28.69	IT Brescia	5.45	ZA Stellenbosch	1.53
IT Caserta	9.73	FR Champagne-Ardenne	28.60	IT Vicenza	5.12	FR Rhone-Alpes except Ar	1.17
IT Rieti	9.01	AU Mornington Peninsula	27.46	US New York - other	5.05	EL Kentriki Makedonia	1.15
IT Taranto	8.79	CH Lucerne	26.53	IT Lecco	5.02	ZA Robertson	1.03
US San Diego	8.15	US Lane Co.	25.69	IT Bergamo	4.40	CL Del Bio Bio	0.85
IT Caltanissetta	7.47	US Josephine Co.	25.08	AU Hastings River	4.32	ZA Olifants River	0.66

Table 80 (cont.): VIIs for the top 25 regions for the world's top 24 red varieties, 2000

Gamay Noir		Alicante Henri Bouschet		Barbera		Montepulciano	
Varietal Intensity Indexes	*Varietal Intensity Indexes*	*Region*	*Varietal Intensity Indexes*	*Region*	*Varietal Intensity Indexes*	*Region*	*Varietal Intensity Indexes*
FR Auvergne	85.30	ES Galicia	32.04	IT Asti	72.69	IT L'Aquila	114.17
FR Rhone-Alpes except Ardeche	65.01	DZ Algeria	13.16	IT Verbano-Cusio-Osso	57.08	IT Pescara	113.93
FR Lorraine	49.96	IT Reggio di Calabria	9.92	IT Alessandria	52.50	IT Teramo	98.22
CH Geneva	45.01	ES Albacete	7.46	IT Piacenza	43.07	IT Campobasso	80.93
CH Valais	23.64	TN Tunisia	6.58	IT Salerno	42.87	IT Chieti	73.94
FR Centre-Val de Loire	18.63	ES Huesca, Teruel	5.80	IT Torino	42.38	IT Ascoli Piceno	36.76
CH Vaud	18.15	ES Leon	5.24	AR Andalgala	40.27	IT Isernia	35.85
FR Correze, Haute-Vienne	15.32	CL O'Higgins	5.17	IT Milano	39.89	IT Bari	22.42
FR Bourgogne	14.77	US San Bernardino	4.87	IT Pavia	36.64	IT Ancona	17.63
FR Deux-Sevres, Vienne	14.70	FR Herault	4.48	IT Lodi	36.44	IT Foggia	16.57
US San Benito	9.60	IT Catania	3.93	IT Biella	33.71	IT Macerata	16.38
FR Midi-Pyrenees except Gers	8.82	CL Del Maule	3.92	IT Varese	26.18	IT Rieti	10.78
FR Pays de la Loire except Maye	7.47	US Tulare	3.75	IT Cuneo	22.32	IT Caserta	9.84
FR Ardeche	7.12	FR Aude	3.54	US Fresno	21.56	IT Taranto	8.94
US Solano	5.86	PT Alentejo	3.12	IT Lecco	20.18	IT Pesaro e Urbino	7.62
IT Valle d'Aosta	5.57	FR Gard	3.02	IT Parma	15.90	IT Benevento	6.87
CA Canada	4.01	CL Metropolitana	2.96	IT Vercelli	14.59	IT Matera	6.16
US Monterey	2.42	MA Morocco	2.91	AR Ischilin	14.55	IT Viterbo	5.35
US San Diego	2.12	FR Bouches-du-Rhone	2.89	IT Brescia	13.36	IT Frosinone	4.52
US Tehama	1.99	ES Alicante	2.82	IT Caserta	11.85	IT Brindisi	4.43
US Lake	1.92	ES Castellon	2.73	AU The Peninsulas	11.82	IT Avellino	3.61
CH Fribourg	1.83	US Kings	2.30	IT Como	11.01	IT Roma	3.23
IT Perugia	1.13	ES Zamora	2.09	AR Poman	10.14	IT Terni	2.55
US Mendocino	1.11	CL Coquimbo	1.84	IT Cremona	10.09	AR Ullum	2.40
FR Franche Comté	1.09	US Fresno	1.69	US Stanislaus	8.93	IT Salerno	2.31

Table 80 (cont.): VIIs for the top 25 regions for the world's top 24 red varieties, 2000

Isabella		**Tribidrag**		**Côt**		**Criolla Grande**	
Region	*Varietal Intensity Indexes*	*Region*	*Varietal Intensity Indexes*	*Region*	*Varietal Intensity Indexes*	*Region*	*Varietal Intensity Indexes*
AR Colón - Entre Ríos	58.51	US Colusa	154.88	AR Godoy Cruz	129.32	AR Ayacucho	69.99
BR Brazil	48.14	US Amador	130.97	AR Molinos	85.08	AR General Alvear	52.94
MD Moldova	22.59	US San Bernardino	92.58	AR Totoral	83.43	AR San Martín - Mza	52.37
AR Totoral	18.62	US Butte	82.03	AR San Carlos - Mza	72.25	AR Junín - Mza	52.22
M9 Missing 9	3.71	US San Joaquin	64.77	AR Luján de Cuyo	67.36	AR Rivadavia - Mza	51.16
US Finger Lakes	0.76	IT Taranto	62.44	AR Añelo	58.34	AR San Rafael	33.21
		US Placer	61.91	AR Calingasta	48.86	AR Santa Rosa - Mza	33.21
		US Contra Costa	54.15	AR Tunuyán	42.19	AR Godoy Cruz	28.73
		US El Dorado	46.96	AR Maipú	41.85	AR Lavalle	21.21
		US Sutter	45.43	AR Castro Barros	40.42	AR Vinchina	18.63
		US Glenn	44.52	FR Midi-Pyrenees excep	37.69	AR La Paz	17.18
		US Mariposa	35.70	AR Coronel Pringles	37.30	AR Capital San Juan	16.23
		US Calaveras	28.46	AR Guaymallén	35.02	AR Guaymallén	15.26
		US Mendocino	25.31	AR Adolfo Alsina	33.72	AR Las Heras	13.97
		US Ventura	22.72	AR Tupungato	33.33	AR El Cuy	11.50
		US San Luis Obispo	21.58	AR Cafayate	32.66	AR San Blas De Los Sauces	9.92
		US Lake	21.44	AR Santa María - Catama	27.69	AR Iglesia	9.41
		US Kings	19.45	AR Poman	25.26	AR Confluencia	8.79
		US Sonoma	19.42	AR San Carlos - Salta	23.23	AR Maipú	8.76
		US Sacramento	16.45	AR El Cuy	22.32	AR Jachal	7.20
		US Stanislaus	15.51	AR Jachal	22.22	AR Luján de Cuyo	6.86
		US Merced	15.23	AR Junín - Mza	22.11	AR Veinticinco de Mayo - S	6.72
		US Alameda	14.97	AR General Roca	19.84	AR Avellaneda - Río Negro	6.32
		IT Bari	14.93	AR Belén	17.59	AR Sarmiento - San Juan	5.87
		US Kern	14.66	AR Rivadavia - Mza	16.98	AR Caucete	5.80

Table 80 (cont.): VIIs for the top 25 regions for the world's top 24 red varieties, 2000

Pamid			**Douce Noire**			**Negroamaro**			**Doukkali**		
	Region	*Varietal Intensity Indexes*		*Region*	*Varietal Intensity Indexes*		*Region*	*Varietal Intensity Indexes*		*Region*	*Varietal Intensity Indexes*
BG	Bulgaria	50.61	AR	Capital San Juan	63.42	IT	Lecce	242.71	MA	Morocco	98.54
EL	Dytiki Makedonia	1.92	IT	Varese	44.13	IT	Brindisi	162.06			
HU	Hungary	0.30	IT	Pavia	40.15	IT	Taranto	15.86			
			AR	Lavalle	34.34	IT	Matera	1.99			
			AR	Santa Rosa - Mza	33.24	IT	Reggio di Calabria	1.05			
			AR	Tupungato	33.03	IT	Cosenza	0.52			
			AR	La Paz	31.23	IT	Potenza	0.41			
			AR	San Rafael	29.83	IT	Enna	0.28			
			AR	San Martín - Mza	28.17	IT	Agrigento	0.25			
			AR	Rivadavia - Mza	26.58	IT	Messina	0.22			
			AR	Castro Barros	26.02	IT	Asti	0.18			
			AR	Chilecito	24.34	IT	Cuneo	0.15			
			AR	Las Heras	23.11	IT	Catania	0.13			
			AR	General Alvear	22.22	IT	Bari	0.10			
			IT	Biella	21.56	IT	Salerno	0.08			
			AR	Guaymallén	21.43	IT	Valle d'Aosta	0.06			
			AR	Sarmiento - San Jua	20.61	IT	Foggia	0.05			
			AR	Junín - Mza	19.96	IT	Catanzaro	0.04			
			IT	Parma	18.38	IT	Pavia	0.04			
			IT	Lodi	16.82	IT	Torino	0.03			
			AR	Maipú	15.74	IT	Pordenone	0.03			
			AR	Confluencia	14.92	IT	Caserta	0.02			
			AR	Nueve de Julio	14.40	IT	Nuoro	0.02			
			AR	Luján de Cuyo	13.98	IT	Campobasso	0.02			
			AR	Chimbas	12.84	IT	Trento	0.02			

Table 81: VIIs for the top 25 regions for the world's top 24 red varieties, 2016

Cabernet Sauvignon		Merlot		Tempranillo		Syrah	
Region	*Varietal Intensity Indexes*	*Region*	*Varietal Intensity Indexes*	*Region*	*Varietal Intensity Indexes*	*Region*	*Varietal Intensity Indexes*
AR Tumbaya	9.62	AR Nogoya	16.83	ES País Vasco	17.43	AU Eastern Plains, Inla	24.74
US Red Mountain	9.18	AR Ñorquin	14.02	ES La Rioja	16.14	AU Southern Flinders F	19.94
AU Coonawarra	8.92	CH Graubünden - Mesolcina	13.84	ES Comunidad Foral de Na	10.55	AU Grampians	17.07
EL Ipeiros	7.92	CH Ticino	13.46	ES Castilla y León	9.80	AU Central Victoria - o	16.71
US Walla Walla Valley	7.66	AR Bariloche	9.35	AR Valle Viejo	5.84	AU Gundagai	15.41
US Columbia Valley	7.61	FR Aquitaine	8.53	ES Extremadura	5.47	AU Heathcote	15.33
CL Metropolitana	7.51	US Shasta	8.15	PT Alentejo	3.74	AU Barossa Valley	14.98
AR Daireaux	7.22	AR Picunches	7.99	ES Aragón	3.56	AU Barossa - other	14.77
US Butte	6.88	AR Cushamen	6.80	ES Castilla-La Mancha	3.41	AU Mount Lofty Range	14.02
US Wahluke Slope	6.85	AR Uruguay	5.61	ES Comunidad Valenciana	2.39	AU McLaren Vale	13.64
US Napa	6.35	US Suffolk	5.42	AR Junín - San Luis	2.25	AU Fleurieu - other	13.16
AU Wrattonbully	6.31	AR El Cuy	5.42	US Hill Country	2.10	AU Bendigo	12.96
US Horse Heaven Hills	6.23	US Rattlesnake Hills	5.32	US North Texas (DFW)	2.05	AU Far North - other	12.37
AU Robe	5.88	BG South Central	5.14	US Sutter	1.95	AR Valle Fértil	12.37
AU Limestone Coast - otl	5.80	US Wahluke Slope	4.63	AR San Carlos - Mza	1.86	AU Queensland - other	11.33
CL O'Higgins	5.76	US Snipes Mountain	4.54	US West Texas	1.86	AU North East Victoria	11.26
AU Northern Slopes	5.75	AR Humahuaca	4.21	PT Norte	1.70	AU North West Victori	11.14
US Lake	5.71	AR Saavedra	4.21	MX Baja California	1.63	US Ventura	10.86
AR Santa María - Catama	5.28	AU New England Australia	4.15	US Texas High Plains and P	1.52	TR Central North	10.74
AR El Carmen	5.12	AU Western Plains	3.83	PT Lisboa	1.45	AU Glenrowan	10.58
US San Luis Obispo	5.08	NZ Hawkes Bay	3.80	PT Centro	1.43	AU Pyrenees	9.76
AU Langhorne Creek	4.90	AU South West Australia - o	3.78	AR La Paz	1.19	AU Currency Creek	9.66
AU Southern NSW - othe	4.89	US Humboldt	3.59	ES Comunidad de Madrid	1.17	MM Myanmar	9.54
AR Uruguay	4.81	AR Villarino	3.56	ES Cataluña	1.10	AU Langhorne Creek	9.53
US Los Angeles	4.79	JP Niigata	3.50	AR San Martín - Mza	1.03	AU Hunter Valley - oth	9.51

Table 81 (cont.): VIIs for the top 25 regions for the world's top 24 red varieties, 2016

Garnacha Tinta		Pinot Noir		Sangiovese		Bobal	
Region	*Varietal Intensity Indexes*	*Region*	*Varietal Intensity Indexes*	*Region*	*Varietal Intensity Indexes*	*Region*	*Varietal Intensity Indexes*
ES Aragón	11.93	US Marin	33.12	IT Toscana	39.61	ES Comunidad Valenci	33.43
FR Provence-Alpes-Cote d	10.76	AR Collon Cura	32.96	ET Ethiopia	32.47	ES Castilla-La Mancha	6.18
IT Sardegna	8.48	NZ Otago	32.73	IT Umbria	12.99	ES Aragón	0.26
DZ Algeria	7.20	US Umpqua Valley	32.41	FR Corse	12.88	ES Extremadura	0.11
ES Comunidad de Madrid	6.96	CH Graubünden - othe	31.58	IT Marche	12.26	ES Illes Balears	0.08
ES Comunidad Foral de N:	5.05	US North Willamette '	31.23	IT Puglia	10.79	ES Comunidad de Madı	0.03
FR Rhône Alpes	4.91	CH Schaffhausen	28.15	US Tuolumne	9.04	ES Andalucía	0.01
US Glenn	4.65	CH Glarus	27.79	AU Central Western Australia	7.66	ES Castilla y León	0.01
FR Languedoc Roussillon	4.23	DE Ahr	26.79	IT Basilicata	5.47	ES Región de Murcia	0.01
AU Central Western Austra	4.12	CH St. Gallen	26.37	AR Colón - Entre Ríos	4.36	ES Comunidad Foral de	0.01
ES La Rioja	3.07	US South Willamette '	26.35	IT Emilia-Romagna	4.15	ES Canarias	0.00
US Kings	2.85	CH Basel Land	23.84	US Mariposa	3.64		
FR Corse	2.75	CH Aargau	23.59	US San Diego	3.60		
US Tehama	2.64	CH Neuchâtel	23.07	IT Campania	3.34		
AU Perth Hills	2.30	CH Basel Stadt	23.00	AR Pichi Mahuida	3.25		
ES Castilla y León	2.25	CH Thurgau	22.98	AR Coronel Suarez	3.23		
ES Cataluña	2.15	AR Sarmiento - Chubu	22.82	US Texas High Plains and Panha	3.18		
AU McLaren Vale	1.92	US Puget Sound	22.47	IT Lazio	3.12		
AU Barossa Valley	1.74	US Santa Cruz	22.23	AU Beechworth	3.07		
US San Bernardino	1.63	CH Zürich	22.07	US Riverside	3.05		
MX Baja California	1.46	FR Île de France	21.25	AR General Roca	3.01		
ES Castilla-La Mancha	1.39	NZ Wairarapa	21.22	IT Liguria	2.98		
TN Tunisia	1.34	CH Appenzell Ausserr	21.00	US Los Angeles	2.93		
MA Morocco	1.34	CH Uri	20.90	AU Kangaroo Island	2.70		
US Fresno	1.19	AU Mornington Penins	20.59	IT Molise	2.66		

Table 81 (cont.): VIIs for the top 25 regions for the world's top 24 red varieties, 2016

Cabernet Franc		**Côt**		**Monastrell**		**Mazuelo**	
Region	*Varietal Intensity Indexes*	*Region*	*Varietal Intensity Indexes*	*Region*	*Varietal Intensity Indexes*	*Region*	*Varietal Intensity Indexes*
AR Gualeguaychu	39.99	AR General Belgrano	85.83	ES Región de Murcia	71.29	MX Aguascalientes	41.80
AR Tornquist	31.47	AR Godoy Cruz	85.83	ES Comunidad Valenciana	9.73	DZ Algeria	34.25
FR Centre-Val de Loire	20.50	AR Benito Juárez	85.83	US Contra Costa	9.73	IL Israel	17.72
FR Pays de la Loire	20.12	AR Villa Gesell	85.83	AU Barossa - other	8.85	FR Languedoc Roussill	9.40
BR Brazil	16.46	AR De La Costa	77.32	ES Illes Balears	6.55	IT Sardegna	8.71
UY Artigas	15.55	AR Cachi	64.92	US Texas High Plains and Pa	5.14	TN Tunisia	6.63
UY Rocha	13.33	AR Punilla	61.88	FR Provence-Alpes-Cote d'A:	3.33	MA Morocco	6.63
AR Victoria	11.43	AR Diamante	59.60	ES Castilla-La Mancha	3.28	MX Zacatecas	6.13
HU Villany	10.66	AR Molinos	57.36	US El Dorado	3.07	FR Provence-Alpes-Co	6.11
US Virginia	9.63	AR Chos Malal	54.75	US Placer	2.77	MX Sonora	4.31
AR Punilla	9.30	AR La Viña	53.85	AU Swan Hill (VIC)	2.39	ES Cataluña	3.88
AR Tilcara	8.82	AR San Carlos - Mza	53.05	US Hill Country	2.16	US San Bernardino	3.58
US Nevada	8.69	AR Humahuaca	48.28	US Ventura	2.11	US Madera	3.45
HU Szekszard	8.51	AR Luján de Cuyo	45.75	AU Barossa Valley	2.08	US Contra Costa	3.33
US Suffolk	8.43	AR Tilcara	44.60	AU Adelaide Plains	1.85	ES La Rioja	2.12
FR Limousin	7.20	AR Pehuenches	44.30	AU Limestone Coast - other	1.72	US Mendocino	2.08
FR Aquitaine	6.72	AR Tunuyán	43.01	FR Languedoc Roussillon	1.69	US San Diego	1.93
UY San Jose	6.33	AR Junín - Bs. As.	42.91	AU Peel	1.53	FR Rhône Alpes	1.85
US Kentucky	6.18	AR Ayacucho	42.07	AU Granite Belt	1.50	ES Aragón	1.76
AU Kangaroo Island	6.07	AR San Alberto	41.05	AR Chos Malal	1.49	ES Comunidad Foral d	1.75
CA Ontario	5.96	AR Calamuchita	38.25	AU Heathcote	1.41	US Santa Clara	1.64
US Colorado	5.86	AR San Javier	37.91	AU Western Victoria - other	1.40	FR Corse	1.61
US Lake Chelan	5.56	AR Coronel Pringles	37.32	US Calaveras	1.33	TR Aegean	1.49
US Placer	5.56	AR Picunches	37.23	US Mariposa	1.29	CL Del Maule	1.19
UY Colonia	5.36	AR Picún Leufú	35.76	ZA Swartland	1.25	US Kern	0.84

Table 81 (cont.): VIIs for the top 25 regions for the world's top 24 red varieties, 2016

Alicante Henri Bouschet		**Tribidrag**		**Montepulciano**		**Gamay Noir**	
Region	*Varietal Intensity Indexes*	*Region*	*Varietal Intensity Indexes*	*Region*	*Varietal Intensity Indexes*	*Region*	*Varietal Intensity Indexes*
ES Galicia	21.79	US San Bernardino	98.04	IT Molise	80.73	FR Auvergne	83.62
AR Poman	17.71	US Amador	78.78	IT Abruzzo	79.36	FR Rhône Alpes	55.76
AR Santa María - Cba	9.47	US Colusa	71.86	IT Marche	26.13	CH Geneva	42.91
PT Alentejo	8.79	US Sutter	38.07	IT Puglia	13.19	FR Lorraine	37.27
CL O'Higgins	8.18	US San Joaquin	35.09	IT Lazio	5.29	TR Marmara	22.43
TR Aegean	8.06	US Tehama	31.83	US Hill Country	4.67	CH Valais	21.20
CL Del Maule	7.01	US Mariposa	31.82	IT Basilicata	3.84	CH Vaud	17.45
TN Tunisia	6.50	US Contra Costa	30.63	AR Ullum	2.98	FR Centre-Val de Lo	14.38
MA Morocco	6.50	US El Dorado	29.64	AR Ischilin	2.77	FR Bourgogne	10.92
CL Metropolitana	4.98	US Glenn	26.43	IT Campania	2.71	FR Pays de la Loire	7.34
US Riverside	3.57	US Placer	26.36	IT Emilia-Romagna	2.00	RS Toplica	6.31
ES Castilla-La Mancha	3.57	IT Puglia	19.16	US Texas High Plain:	1.67	IT Valle d'Aosta	6.26
ES Región de Murcia	3.40	US Butte	18.03	IT Umbria	1.48	FR Midi Pyrénées	5.04
CL Coquimbo	3.15	US Stanislaus	17.53	IT Sardegna	1.34	FR Limousin	4.68
PT Lisboa	2.97	US Kings	16.82	NZ Auckland	0.91	RS Negotinska Krajii	4.59
PT Centro	2.63	US Mendocino	15.58	US Riverside	0.88	CA Ontario	4.31
US Tulare	2.21	US Calaveras	14.97	US Marin	0.84	CA Other regions	3.23
ES Comunidad Valenciana	1.96	US Tuolumne	14.80	IT Toscana	0.74	US Solano	2.90
AR Calingasta	1.71	US Trinity	14.35	AR Veinticinco de M	0.62	CA British Colombia	2.86
AR Zonda	1.37	US Nevada	12.96	IT Liguria	0.55	AU Beechworth	2.20
PT Norte	1.34	US Sonoma	11.83	US San Diego	0.55	SI Bela Krajina	2.05
US Merced	1.31	US Merced	11.72	AR San Martín - San	0.51	CH Fribourg	1.99
PT Algarve	1.19	US Fresno	11.01	AU Queensland - oth	0.51	FR Franche Comté	0.84
FR Languedoc Roussillon	1.10	US San Luis Obispo	10.43	IT Valle d'Aosta	0.47	US Lake	0.81
AR Santa Lucía	0.97	US Tulare	10.39	IT Calabria	0.46	US Calaveras	0.48

Table 81 (cont.): VIIs for the top 25 regions for the world's top 24 red varieties, 2016

Cinsaut		Carmenère		Douce Noire		Barbera	
Region	*Varietal Intensity Indexes*	*Region*	*Varietal Intensity Indexes*	*Region*	*Varietal Intensity Indexes*	*Region*	*Varietal Intensity Indexes*
TR Marmara	48.37	CL O'Higgins	25.49	AR Pehuenches	109.93	IT Piemonte	52.84
MA Morocco	36.01	CL Metropolitana	13.78	AR Capayán	103.27	AR Andalgala	34.10
TN Tunisia	36.01	CN China	12.55	AR Valle Viejo	64.91	AR Ischilin	24.03
FR Provence-Alpes-Cote d'Az	13.86	CL Del Maule	11.97	AR Poman	58.87	IT Lombardia	22.63
CL Del Bio Bio	12.83	CL Valparaiso	4.61	AR La Paz	38.90	AR Coronel Suarez	19.96
ZA Paarl	8.10	AR Tandil	2.83	AR Lavalle	36.52	US Fresno	19.89
ZA Breedekloof	8.08	CL Del Bio Bio	2.75	AR Santa Rosa - Mza	35.64	US Amador	14.03
FR Languedoc Roussillon	6.75	CL Coquimbo	2.05	AR Castro Barros	34.01	US El Dorado	12.27
FR Corse	4.93	CL Tarapaca	1.90	AR Junín - San Luis	32.46	US Placer	12.11
ZA Swartland	4.18	AR Chilecito	0.83	AR San Martín - Mza	30.09	US Sutter	11.98
US Texas High Plains and Pan	2.91	IT Sardegna	0.81	AR San Rafael	29.52	US Kings	11.89
ZA Worcester	2.82	IT Lazio	0.73	AR Las Heras	29.43	IT Campania	11.00
AU Rutherglen	1.97	IT Veneto	0.56	AR Rivadavia - Mza	28.21	SI Vipavska dolina	10.82
US Riverside	1.81	IT Molise	0.50	AR General Alvear	24.38	US Tuolumne	9.32
FR Rhône Alpes	1.58	IT Basilicata	0.49	AR Guaymallén	24.35	IT Emilia-Romagna	8.74
ZA Robertson	1.38	IT Friuli-Venezia Giul	0.46	AR Chilecito	23.18	US Tulare	7.64
ZA Stellenbosch	1.31	AU Granite Belt	0.29	AR Sarmiento - San Juan	22.91	US Butte	7.40
ZA Olifants River	0.57	US Madera	0.22	AR Capital San Juan	22.72	US Madera	6.88
US Santa Clara	0.39	AU Heathcote	0.22	AR Junín - Mza	22.06	IT Basilicata	6.38
US Contra Costa	0.33	AR Luján de Cuyo	0.21	AR Tupungato	21.06	US Ventura	6.13
EL Kentriki Makedonia	0.28	CL Atacama	0.19	AR Santa María - Cba	19.76	AR Pichi Mahuida	6.02
US Calaveras	0.27	US Riverside	0.18	AR Rivadavia - San Juan	19.41	AR Chilecito	5.18
ZA Cape South Coast	0.16	AU Beechworth	0.15	AR San Javier	18.93	AR Poman	4.59
CL Del Maule	0.16	IT Umbria	0.15	AR Nueve de Julio	18.22	AU South Coast - other	4.09
US Santa Barbara	0.13	AU Swan Hill (VIC)	0.15	AR Maipú	17.41	US Nevada	3.98

Table 81 (cont.): VIIs for the top 25 regions for the world's top 24 red varieties, 2016

Isabella		Blaufränkisch		Criolla Grande		Pinot Meunier	
Region	*Varietal Intensity Indexes*	*Region*	*Varietal Intensity Indexes*	*Region*	*Varietal Intensity Indexes*	*Region*	*Varietal Intensity Indexes*
BR Brazil	88.41	AT Mittelburgenland	141.92	AR San Martín - Mza	54.06	FR Picardie	175.13
UY Tacuarembo	64.99	HU Sopron	139.50	AR General Alvear	52.27	FR Île de France	152.54
RU Crimea	12.00	AT Südburgenland	97.97	AR Junín - Mza	50.38	FR Champagne-Ardenne	95.81
UA Ukraine	12.00	PE Lima	85.32	AR Rivadavia - Mza	48.90	DE Württemberg	40.71
MD Moldova	10.57	HU Szekszard	70.94	AR Santa Rosa - Mza	35.47	UK UnitedKingdom	33.56
UY Paysandu	9.44	SK Východné Slovensko	64.24	AR San Rafael	28.47	AU Henty	20.83
UY Canelones	4.54	HU Csongrad	62.26	AR Vinchina	22.25	AU Tumbarumba	5.74
UY Montevideo	3.10	AT Neusiedlersee Hügell	61.83	AR Lavalle	20.40	AU Southern Highlands	5.49
RU Krasnodar Krai	1.92	HU Hajos-bajai	54.02	AR Iglesia	15.97	DE Baden	4.87
AU Alpine Valleys	1.57	SI Bela Krajina	51.93	AR La Paz	13.34	DE Franken	4.05
UY San Jose	1.28	SI Bizeljsko Sremic	51.15	AR Las Heras	12.51	CL De Los Lagos	3.43
UY Maldonado	0.77	HU Eger	50.34	AR Guaymallén	10.54	AU King Valley	3.41
UY Colonia	0.48	HU Bukk	49.31	AR Maipú	6.39	AU Strathbogie Ranges	3.32
CH Ticino	0.32	SI Dolenjska	48.01	AR Veinticinco de May	6.11	AU Tasmania	3.00
AU Riverina	0.18	SK Stredné Slovensko	46.17	AR Albardón	6.02	CA Nova Scotia	2.31
CH Graubünden - Mesol	0.03	SK Západné Slovensko	42.77	AR Sarmiento - San Jua	5.67	DE Pfalz	2.00
AU Margaret River	0.00	DE Württemberg	38.81	AR Rivadavia - San Jua	5.05	US Riverside	1.41
		HU Tolna	35.85	AR Luján de Cuyo	4.40	DE Sachsen	1.31
		HU Kunsag	29.58	AR Caucete	3.64	AU Yarra Valley	1.25
		AT Neusiedlersee	28.36	AR Angaco	3.24	AU Adelaide Hills	1.19
		SK Bratislavský kraj	26.74	AR Confluencia	3.20	DE Rheinhessen	0.96
		HU Villany	26.28	AR Nueve de Julio	3.20	AU Grampians	0.91
		AT Carnuntum	24.11	AR Chimbas	3.05	AU Pyrenees	0.82
		CZ Morava	23.37	AR Coronel Felipe Vare	2.44	FR Centre-Val de Loire	0.79
		CZ Jihovýchod	22.79	AR Santa Lucía	2.38	NZ Hawkes Bay	0.74

Table 82: VIIs for the top 25 regions for the world's top 24 white varieties, 2000

Airén		**Chardonnay**		**Trebbiano Toscano**		**Graševina**	
Region	*Varietal Intensity Indexes*	*Region*	*Varietal Intensity Indexes*	*Region*	*Varietal Intensity Indexes*	*Region*	*Varietal Intensity Indexes*
ES Ciudad Real	10.40	CL Araucania	33.60	FR Charente	34.36	RS Serbia	25.42
ES Toledo	9.27	US Santa Barbara	21.12	FR Charente-Maritime	32.94	ES Caceres	16.34
ES Cuenca	5.45	US Santa Cruz	20.88	US Los Angeles	15.45	HR Croatia	14.30
ES Comunidad de Madrid	4.74	US Riverside	19.84	FR Gers	12.63	SK Slovakia	13.24
ES Albacete	4.36	US Tehama	19.64	IT Pisa	10.02	AT Steiermark	11.23
ES Guadalajara	4.31	US Shasta	19.60	IT Terni	9.63	SI Slovenia	8.05
ES Almeria, Granada, Jaen,	0.65	US Yolo	19.19	IT Latina	9.36	AT Burgenland	7.19
ES Region de Murcia	0.63	US San Mateo	17.87	IT Chieti	8.99	IT Pavia	6.56
ES Malaga	0.45	US Sutter	16.80	IT Viterbo	8.94	CZ Czechia	5.82
ES Cordoba	0.34	US Monterey	16.75	IT Roma	8.43	HU Hungary	4.07
ES Alicante	0.09	NZ Gisborne	16.25	IT Foggia	8.21	RO Romania	3.58
ES Huelva	0.02	AU Tumbarumba	15.51	IT Livorno	8.18	AT Niederosterreich	2.86
ES Caceres	0.02	FR Franche Comté	15.25	IT Lucca	7.09	AT Wien and other regions	2.54
ES Badajoz	0.02	FR Bourgogne	14.63	IT Pistoia	7.03	BG Bulgaria	1.99
ES Canarias	0.01	CL Valparaiso	14.03	IT Grosseto	6.91	IT Lecco	1.49
ES Valencia	0.01	US Alameda	12.99	IT Perugia	6.42	IT Milano	1.14
ES Zamora	0.01	US Santa Clara	12.65	IT Prato	6.11	BR Brazil	0.88
ES Zaragoza	0.01	AU Hunter Valley - other	12.53	IT Arezzo	5.04	IT Rovigo	0.40
ES Valladolid	0.00	AU Cowra	11.92	IT Agrigento	4.96	ES Badajoz	0.37
ES Castellon	0.00	AU Hunter	11.80	IT Firenze	4.21	IT Gorizia	0.37
ES Burgos	0.00	US Sacramento	11.79	IT Frosinone	3.85	IT Bologna	0.26
ES Avila, Palencia, Salaman	0.00	US Nevada	11.50	IT Ascoli Piceno	3.80	IT Piacenza	0.24
ES Girona, Lleida	0.00	US Sonoma	11.40	IT Benevento	3.71	IT Venezia	0.23
ES Comunidad Foral de Nav	0.00	US Mendocino	11.26	IT Palermo	3.52	IT Pordenone	0.16
ES La Rioja	0.00	US Solano	11.15	IT Campobasso	3.36	IT Padova	0.15

Table 82 (cont.): VIIs for the top 25 regions for the world's top 24 white varieties, 2000

Rkatsiteli		Sauvignon Blanc		Cayetana Blanca		Catarratto Bianco	
Region	*Varietal Intensity Indexes*	*Region*	*Varietal Intensity Indexes*	*Region*	*Varietal Intensity Indexes*	*Region*	*Varietal Intensity Indexes*
GE Georgia	38.28	NZ Marlborough	35.16	ES Badajoz	59.18	IT Trapani	61.30
RU Russia	16.94	EL Anatoliki Makedon	23.38	ES Caceres	1.48	IT Palermo	53.74
AM Armenia	15.99	FR Centre-Val de Loire	19.12	ES Zaragoza	1.16	IT Agrigento	17.29
M9 Missing 9	9.54	NZ Nelson	17.43	AU Riverland	0.87	IT Catania	3.13
MD Moldova	9.29	US Yuba	16.11	AU Mount Lofty Ranges	0.50	IT Messina	2.67
BG Bulgaria	7.13	US Lake	16.00	AU Big Rivers - other	0.35	IT La Spezia	1.64
RO Romania	0.17	NZ Waipara	14.38	ES Malaga	0.33	IT Caltanissetta	1.63
		AU Western Australia S	12.66	AU Murray Darling (NSV	0.27	IT Sassari	0.63
		NZ Wairarapa	11.70	AU Swan District	0.12	IT Genova	0.62
		NZ Waikato	10.05	AU Murray Darling (VIC	0.12	IT Imperia	0.48
		US Calaveras	9.61	AU Riverina	0.11	IT Massa-Carrara	0.24
		IT Gorizia	9.52	AU Adelaide Hills	0.09	IT Savona	0.22
		FR Deux-Sevres, Vienr	9.26	ES Valladolid	0.07	IT Napoli	0.12
		AU Adelaide Hills	8.92	ES Guadalajara	0.04	IT Valle d'Aosta	0.08
		AU Mount Benson	8.54	AU Hilltops	0.02	IT Oristano	0.08
		NZ Hawkes Bay	8.47	AU Swan Hill (VIC)	0.02	IT Cagliari	0.08
		IT Lecco	8.25	AU Barossa Valley	0.02	IT Brescia	0.07
		US Mariposa	8.04	AU Langhorne Creek	0.01	IT Caserta	0.06
		AU Margaret River	7.92	ES Zamora	0.01	IT Pistoia	0.06
		CL Valparaiso	7.89	ES Avila, Palencia, Salai	0.01	IT Campobasso	0.06
		CL Del Maule	7.83	ES Ciudad Real	0.00	IT Torino	0.06
		AU Greater Perth - othe	7.65	AU McLaren Vale	0.00	IT Pisa	0.06
		ZA Stellenbosch	7.59	ES Cuenca	0.00	IT Lucca	0.05
		AU Pyrenees	7.30	ES Canarias	0.00	IT Cosenza	0.05
		MD Moldova	6.81	ES Galicia	0.00	IT Matera	0.05

Table 82 (cont.): VIIs for the top 25 regions for the world's top 24 white varieties, 2000

Macabeo		Chenin Blanc		Riesling		Colombard	
Region	*Varietal Intensity Indexes*	*Region*	*Varietal Intensity Indexes*	*Region*	*Varietal Intensity Indexes*	*Region*	*Varietal Intensity Indexes*
ES Tarragona	33.42	ZA Olifants River	37.56	DE Rheingau	90.36	ZA Northern Cape	66.60
ES Girona, Lleida	24.52	ZA Worcester	31.98	DE Mittelrhein	83.35	US Madera	40.20
ES Barcelona	24.21	ZA Swartland	30.64	DE Hessische Bergstraße	63.04	US Kern	35.66
ES Zaragoza	20.55	ZA Breedekloof	30.27	DE Mosel	61.84	US Fresno	34.04
ES La Rioja	14.92	ZA Northern Cape	28.88	DE Nahe	29.12	US Glenn	33.75
ES Huesca, Teruel	14.90	ZA Paarl	28.04	AU Eden Valley	28.28	ZA Little Karoo	32.15
ES Alava	13.13	US Sutter	26.79	DE Württemberg	25.74	FR Gers	31.86
FR Pyrenees-Orientales	12.41	ZA Little Karoo	22.90	FR Alsace	25.69	ZA Olifants River	30.66
ES Castellon	9.35	AR San Javier	20.67	NZ Canterbury	25.23	US Tulare	30.27
ES Valladolid	7.37	FR Deux-Sevres, Vienne	20.62	DE Pfalz	24.12	US Stanislaus	27.69
ES Comunidad Foral d(	7.31	ZA Robertson	17.80	US Columbia River	23.60	US Merced	22.05
AR El Cuy	3.56	AU Perth Hills	17.32	NZ Nelson	19.20	ZA Robertson	21.42
ES Valencia	1.62	FR Centre-Val de Loire	16.91	DE Sachsen	18.41	ZA Breedekloof	19.61
ES Illes Balears	1.57	FR Pays de la Loire excep	15.19	US Michigan	17.90	US Kings	17.20
ES Badajoz	1.52	AR San Alberto	15.08	NZ Waipara	17.18	ZA Worcester	17.10
ES Albacete	1.31	AU Swan District	13.87	US Yuba	16.09	IL Israel	12.74
FR Aude	0.63	US Kern	13.09	AU Clare Valley	15.90	US San Joaquin	5.03
ES Burgos	0.62	US Madera	12.57	LU Luxembourg	14.81	AU Riverina	4.59
ES Cuenca	0.58	ZA Stellenbosch	12.01	US Douglas Co.	14.56	US Calaveras	4.53
ES Caceres	0.33	AU Greater Perth - other	10.91	AT Wien and other regions	14.43	AU Riverland	4.38
ES Guipuzcoa, Vizcaya	0.27	US Fresno	10.68	AU Henty	13.09	ZA Paarl	4.22
ES Avila, Palencia, Sal	0.21	US Stanislaus	10.59	AU Canberra District	12.57	AU Murray Darling (NSW)	3.55
ES Region de Murcia	0.20	US Merced	9.64	US Washington Co.	12.22	AU Murray Darling (VIC)	3.35
ES Almeria, Granada, J	0.15	AU Central Western Austr	6.76	US Washington	11.95	US Alameda	3.30
ES Ciudad Real	0.13	US Tulare	6.51	US Humboldt	11.41	ZA Swartland	3.19

Table 82 (cont.): VIIs for the top 25 regions for the world's top 24 white varieties, 2000

Aligoté		**Müller-Thurgau**		**Palomino Fino**		**Muscat Blanc à Petits Grains**	
Varietal Intensity Indexes	*Varietal Intensity Indexes*	*Region*	*Varietal Intensity Indexes*	*Region*	*Varietal Intensity Indexes*	*Region*	*Varietal Intensity Indexes*
MD Moldova	24.08	CH Lucerne	75.82	ES Cadiz	161.52	EL Voreio Aigaio	140.32
M9 Missing 9	11.81	DE Franken	59.00	ES Canarias	83.26	IT Cuneo	43.29
FR Bourgogne	7.78	CH Other regions	57.87	ES Cantabria	77.25	IT Asti	34.83
RO Romania	4.69	LU Luxembourg	49.57	ES Leon	28.70	EL Anatoliki Makedonia	31.43
RU Russia	4.43	CH Aargau	47.31	ES Galicia	22.75	US Los Angeles	21.75
BG Bulgaria	2.37	CH Schwyz	45.33	ES Valladolid	19.98	IT Alessandria	17.32
CH Geneva	1.81	CH Zürich	40.32	ES Zamora	19.19	FR Pyrenees-Orientales	12.97
FR Auvergne	0.76	CH Thurgau	39.82	AR Puelen	13.85	IT Padova	10.30
GE Georgia	0.36	DE Baden	39.21	US San Bernardino	7.74	IT Matera	9.18
FR Rhone-Alpes except Ardeche	0.13	DE Sachsen	33.81	NZ Waikato	6.52	IT Pavia	8.56
CH Vaud	0.04	DE Saale	32.52	ES Almeria, Granada, Ja	6.38	US San Diego	8.02
CH Valais	0.03	DE Rheinhessen	30.94	NZ Auckland	5.52	AM Armenia	7.66
FR Alpes-de-Haute-Provence, Ha	0.01	DE Mosel	30.15	ES Principado de Asturia	3.34	IT Sassari	6.22
FR Gard	0.00	DE Nahe	29.03	US Riverside	3.32	IT Potenza	5.43
FR Bouches-du-Rhone	0.00	DE Pfalz	25.77	ES Huelva	3.05	ZA Robertson	5.26
FR Centre-Val de Loire	0.00	CH Schaffhausen	25.61	US Contra Costa	2.42	FR Corse	4.63
FR Herault	0.00	CH Jura	24.96	FR Gers	2.37	AT Steiermark	4.26
FR Vaucluse	0.00	CH Basel Land	24.09	ES Avila, Palencia, Salar	1.80	IT Salerno	4.01
FR Midi-Pyrenees except Gers	0.00	CH St. Gallen	21.19	US Madera	1.32	ZA Little Karoo	3.98
FR Ardeche	0.00	CZ Czechia	20.38	AU Lower Murray - othe	1.28	EL Dytiki Ellada	3.72
FR Pays de la Loire except Maye	0.00	SK Slovakia	17.47	ES Malaga	1.14	IT Bari	3.59
FR Charente	0.00	NZ Gisborne	16.92	AU Beechworth	0.97	FR Rhone-Alpes except	3.44
		DE Hessische Bergstraße	15.96	US Fresno	0.93	IT Bergamo	3.40
		AT Steiermark	15.55	AR San Rafael	0.84	IT Parma	3.32
		CH Graubünden	12.02	AR Las Heras	0.77	IT Piacenza	3.28

Table 82 (cont.): VIIs for the top 25 regions for the world's top 24 white varieties, 2000

Muscat of Alexandria		Sémillon		Feteascā Albā		Grüner Veltliner	
Region	*Varietal Intensity Indexes*	*Region*	*Varietal Intensity Indexes*	*Region*	*Varietal Intensity Indexes*	*Region*	*Varietal Intensity Indexes*
AR Poman	67.68	AU Northern Rivers -	33.92	RO Romania	16.81	AT Niederosterreich	100.43
AR Zonda	60.07	AU Hunter	33.88	MD Moldova	9.89	AT Wien and other regions	59.08
AR Rivadavia - San Juaı	57.23	FR Aquitaine except '	31.68	SK Slovakia	4.10	SK Slovakia	39.34
AR Albardón	52.53	AU Riverina	29.83	HU Hungary	2.29	AT Burgenland	38.96
AR Chimbas	38.66	AR Confluencia	28.75			CZ Czechia	31.06
AR Ischilin	36.76	AU Hunter Valley - oı	28.47			HU Hungary	3.18
AR Capital San Juan	34.48	AU Hastings River	27.60			IT Pescara	3.02
US Kings	32.75	AU Margaret River	23.57			IT La Spezia	0.88
AR Ullum	26.60	AR Conesa	23.52			IT Oristano	0.82
ES Malaga	24.53	AR El Cuy	23.38			IT Bolzano-Bozen	0.57
ZA Little Karoo	23.31	AU Southern Fleurieu	20.59			AT Steiermark	0.35
AR Veinticinco de May(	21.19	AU Barossa Valley	19.22			IT Genova	0.21
ZA Breedekloof	18.74	AU Southern NSW - (	17.02			IT Sassari	0.19
AR Santa Lucía	16.93	AU South Coast - oth(	15.04			IT Cagliari	0.14
AR Caucete	15.93	AU Cowra	14.17			IT Chieti	0.14
AR Angaco	15.78	AU Northern Slopes -	14.16			IT Viterbo	0.11
ZA Olifants River	15.64	FR Gironde	13.90			IT Frosinone	0.11
US Tulare	15.26	AU Fleurieu - other	13.63			IT Latina	0.10
ES Alicante	14.87	AU Big Rivers - other	12.11			IT Imperia	0.10
AR San Martín - San Jua	14.82	AU South Burnett	11.54			IT Foggia	0.05
ZA Northern Cape	14.70	AU Mudgee	11.06			IT Teramo	0.05
AR Ayacucho	14.06	AU Hilltops	10.92			IT Sondrio	0.03
AR Chilecito	13.78	AU Clare Valley	10.39			IT Ancona	0.03
AR Sarmiento - San Juaı	12.61	AR Adolfo Alsina	10.23			IT Ascoli Piceno	0.02
AR Nueve de Julio	12.56	AU Sunbury	10.21			IT Nuoro	0.02

Table 82 (cont.): VIIs for the top 25 regions for the world's top 24 white varieties, 2000

Trebbiano Romagnolo		Pedro Ximénez		Pinot Blanc		Garganega	
Region	*Varietal Intensity Indexes*	*Region*	*Varietal Intensity Indexes*	*Region*	*Varietal Intensity Indexes*	*Region*	*Varietal Intensity Indexes*
IT Ravenna	190.79	ES Cordoba	269.73	AT Steiermark	49.85	IT Verona	104.04
IT Ferrara	128.58	CL Coquimbo	56.71	AT Wien and other region:	36.86	IT Vicenza	97.53
IT Bologna	92.07	CL Atacama	56.10	DE Sachsen	36.68	IT Trapani	22.66
IT Forlì-Cesena	86.69	ES Malaga	31.01	DE Saale	31.76	IT Padova	11.65
IT Rimini	46.45	ES Almeria, Granada, J	9.03	IT Bolzano-Bozen	29.75	IT Agrigento	5.27
IT Rovigo	9.39	ES Caceres	7.52	LU Luxembourg	29.54	IT Palermo	4.29
IT Piacenza	6.30	ES Badajoz	3.41	US Oregon - other	25.39	IT Mantova	2.90
IT Mantova	4.50	AU Fleurieu - other	2.22	IT Brescia	24.59	IT Foggia	2.33
IT Pesaro e Urbino	2.98	AU Lower Murray - oth	2.15	FR Alsace	22.60	IT Bari	1.79
IT Parma	2.89	AU Rutherglen	1.93	AT Burgenland	20.71	IT Matera	1.55
IT Modena	2.18	AU Swan District	1.81	IT Gorizia	19.42	AR Ullum	1.42
IT Macerata	2.07	AU Greater Perth - othe	1.79	IT Lecco	17.46	IT Taranto	1.23
IT Rieti	1.74	ES Valencia	1.50	DE Baden	17.31	IT Perugia	1.19
IT Padova	1.34	AU Barossa Valley	1.07	IT Pordenone	16.08	IT Ascoli Piceno	1.10
IT Agrigento	1.18	AU Clare Valley	0.75	IT Treviso	15.50	IT Campobasso	0.92
IT Terni	1.06	ES Castellon	0.63	RU Russia	15.34	IT Brindisi	0.76
IT Taranto	0.80	ES Avila, Palencia, Sal	0.45	IT Venezia	13.09	IT Brescia	0.75
IT Frosinone	0.76	ES Principado de Astur	0.41	IT Udine	13.00	IT Sassari	0.73
IT Salerno	0.72	AU Sunbury	0.36	AT Niederosterreich	11.89	IT Modena	0.61
IT Cremona	0.70	AU Currency Creek	0.33	SK Slovakia	11.54	IT Lecce	0.54
IT Savona	0.68	AU Eden Valley	0.28	IT Vicenza	11.19	IT Bergamo	0.40
IT Foggia	0.66	AU Riverland	0.27	IT Bergamo	10.99	IT Udine	0.39
IT Catania	0.65	ES Tarragona	0.24	IT Padova	9.70	IT Rovigo	0.35
IT Ascoli Piceno	0.60	AU McLaren Vale	0.16	DE Nahe	8.08	IT Benevento	0.26
IT Caserta	0.60	ES Cuenca	0.15	CH Geneva	7.92	IT Latina	0.20

Table 83: VIIs for the top 25 regions for the world's top 24 white varieties, 2016

Airén		Chardonnay		Sauvignon Blanc		Trebbiano Toscano	
Region	*Varietal Intensity Indexes*	*Region*	*Varietal Intensity Indexes*	*Region*	*Varietal Intensity Indexes*	*Region*	*Varietal Intensity Indexes*
ES Comunidad de Madri	14.07	AR Languiñeo	22.23	NZ Marlborough	27.95	FR Poitou Charentes	31.79
ES Castilla-La Mancha	10.46	US Lassen	14.82	NZ Nelson	17.36	IT Lazio	10.64
TN Tunisia	0.55	AR Picún Leufú	12.97	AU Mount Gambier	16.83	IT Abruzzo	6.51
MA Morocco	0.55	AU Cowra	11.12	AU Greater Perth - other	13.81	IT Puglia	6.40
ES Región de Murcia	0.47	AU Tumbarumba	11.11	CL Valparaiso	13.66	IT Umbria	6.36
ES Andalucía	0.33	FR Franche Comté	10.94	NZ Wairarapa	11.51	UY Canelones	4.57
ES Comunidad Valencia	0.03	NZ Gisborne	10.86	MM Myanmar	11.30	UY Paysandu	4.19
ES Canarias	0.02	FR Bourgogne	10.65	AU Manjimup	11.21	IN India	4.14
ES Extremadura	0.01	AR Coronel Pringles	9.67	AU Pemberton	11.10	UY Montevideo	4.08
ES Castilla y León	0.01	AR Sarmiento - Chubut	9.07	ZA Cape South Coast	10.99	IT Molise	4.05
ES Cataluña	0.00	CL Araucania	8.74	AR Coronel Suarez	10.84	IT Marche	2.98
		US Yolo	8.59	AR Gualeguaychu	10.79	UY San Jose	2.76
		US Monterey	8.56	FR Centre-Val de Loire	10.52	FR Midi Pyrénées	2.43
		AR Daireaux	8.34	NZ Waipara	9.86	US Los Angeles	2.33
		AU Big Rivers - other	8.33	AR General Pueyrredón	9.74	IT Toscana	1.89
		US Santa Barbara	7.78	AU Adelaide Hills	9.16	IT Campania	1.49
		AR Uruguay	7.41	AR Balcarce	8.99	UY Colonia	1.36
		RS Vranje	7.36	US Naches Heights	8.99	BG Northeast	1.33
		AU Pemberton	7.25	AR Saavedra	8.99	EL Anatoliki Makedon	1.07
		AR Curaco	7.17	CL De Los Lagos	8.88	IT Sicilia	1.03
		AR General Pueyrredón	6.65	US Lake	8.62	BG Southeast	0.98
		FR Champagne Ardenne	6.65	NZ Canterbury	7.60	ZA Little Karoo	0.95
		NZ Northland	6.62	RS Subotica	7.58	AR Las Heras	0.94
		AU Perricoota	6.60	AT Südsteiermark	7.18	UY Rivera	0.91
		US Santa Cruz	6.54	NZ Hawkes Bay	7.03	UY Florida	0.89

Table 83 (cont.): VIIs for the top 25 regions for the world's top 24 white varieties, 2016

Riesling		Rkatsiteli		Macabeo		Cayetana Blanca	
Region	*Varietal Intensity Indexes*	*Region*	*Varietal Intensity Indexes*	*Region*	*Varietal Intensity Indexes*	*Region*	*Varietal Intensity Indexes*
AR Veinticinco de Mayo -	74.96	KZ Almaty	49.89	ES Cataluña	25.38	ES Extremadura	51.99
AR Capital Misiones	74.96	GE Georgia	46.04	ES Aragón	7.75	ES Comunidad de Madı	4.01
DE Rheingau	59.22	KZ Zhambyl	35.49	ES Extremadura	6.86	ES Andalucía	3.06
DE Mittelrhein	50.85	KZ South Kazakhstan	34.76	ES País Vasco	6.70	ES Castilla-La Mancha	0.58
DE Mosel	45.83	BG North Central	32.06	ES La Rioja	6.39	ES Aragón	0.31
AR Leandro Alem	44.98	UA Ukraine	20.03	ES Comunidad Valenciana	5.10	PT Norte	0.18
AU Far North - other	37.48	RU Crimea	20.03	ES Castilla-La Mancha	3.32	AU Riverland	0.11
DE Hessische Bergstraße	34.22	BG Northeast	12.45	ES Comunidad Foral de Navarra	2.27	PT Centro	0.03
RS Tisa	24.58	RU Rostov Oblast	11.27	ES Castilla y León	1.54	PT Alentejo	0.03
CZ Praha	22.49	BG Northwest	10.45	ES Illes Balears	1.36	ES La Rioja	0.03
RS Negotinska Krajina	22.40	KZ Other regions	9.94	FR Languedoc Roussillon	0.78	ES Castilla y León	0.01
DE Nahe	21.16	BG Southeast	7.51	ES Región de Murcia	0.61	ES País Vasco	0.01
AU Eden Valley	19.76	MD Moldova	4.12	ES Cantabria	0.37	ES Comunidad Valenci	0.01
DE Pfalz	18.22	BG South Central	4.08	ES Comunidad de Madrid	0.19	ES Canarias	0.00
FR Alsace	17.86	BG Southwest	3.04	ES Andalucía	0.17	ES Comunidad Foral de	0.00
US Rattlesnake Hills	17.64	RS South Banat	2.96	ZA Paarl	0.02	PT Algarve	0.00
US Michigan	16.83	MK North Macedonia	1.62	ZA Swartland	0.01		
US Yakima Valley	15.77	RU Krasnodar Krai	0.55	FR Provence-Alpes-Cote d'Azur	0.00		
NZ Waipara	15.72	KZ East Kazakhstan	0.52	FR Rhône Alpes	0.00		
AR Lacar	14.99	RS Nišava	0.36	ES Galicia	0.00		
US Seneca	14.54	RO Romania	0.20	ES Canarias	0.00		
AU Clare Valley	14.41						
RS Leskovac	14.28						
DE Württemberg	13.81						
AT Wachau	13.08						

Table 83 (cont.): VIIs for the top 25 regions for the world's top 24 white varieties, 2016

Muscat Alexandria		Muscat Blanc à Petits Grains		Chenin Blanc		Colombard	
Region	*Varietal Intensity Indexes*	*Region*	*Varietal Intensity Indexes*	*Region*	*Varietal Intensity Indexes*	*Region*	*Varietal Intensity Indexes*
AR Capital San Juan	57.96	EL Voreio Aigaio	65.25	ET Ethiopia	44.41	ZA Northern Cape	75.92
AR Poman	41.84	AT Steirerland - other	63.98	ZA Worcester	40.71	ZA Little Karoo	48.34
CL Del Bio Bio	41.64	MX Zacatecas	52.03	ZA Olifants River	39.68	ZA Olifants River	38.67
AR Rivadavia - San Juan	37.40	PE Arequipa	33.90	ZA Northern Cape	36.51	US Fresno	35.19
EL Voreio Aigaio	30.98	IT Piemonte	29.77	AU Swan District	33.64	ZA Worcester	27.26
AR Zonda	30.64	MX Aguascalientes	19.54	ZA Breedekloof	30.94	US Madera	25.87
AR Ischilin	27.75	MM Myanmar	13.29	ZA Swartland	27.81	ZA Robertson	23.48
CL Atacama	24.81	AT Südsteiermark	12.94	ZA Little Karoo	27.66	ZA Breedekloof	22.84
US Glenn	18.27	HU Tokaj	12.40	ZA Paarl	25.52	US Tulare	21.95
US Colusa	17.20	SI Stajerska Slovenija	8.37	FR Centre-Val de Loire	21.18	FR Midi Pyrénées	21.91
AR Albardón	17.18	AU Glenrowan	7.59	FR Pays de la Loire	19.87	US Kern	18.96
ES Andalucía	15.59	IT Valle d'Aosta	7.09	ZA Robertson	16.88	US Kings	18.57
TN Tunisia	15.33	US Tulare	6.55	MX Baja California	13.36	TH Thailand	10.49
MA Morocco	15.33	SI Bela Krajina	6.51	ZA Stellenbosch	12.22	IL Israel	6.58
UY Artigas	14.31	AT Vulkanland Steiermark	6.40	TH Thailand	10.85	AU Riverland	5.77
AR Ullum	13.70	RS Čačak-Kraljevo	6.26	US Kings	9.80	AU Lower Murray - oth	5.23
CL Coquimbo	13.06	PT Algarve	5.92	AU Central Western Australia	8.64	AU Murray Darling (VI	5.22
AR Veinticinco de Mayo	10.58	SI Slovenska Istra	5.71	US Yolo	6.99	AU Murray Darling (NS	5.18
CL Antofagasta	10.37	UY Durazno	5.65	US Fresno	5.84	US Stanislaus	5.01
ES Comunidad Valencian	9.51	SI Bizeljsko Sremic	5.58	AR San Rafael	5.65	ZA Paarl	3.71
AR Vinchina	9.43	FR Corse	4.99	AU Peel	5.43	US Ventura	3.65
AR Chimbas	9.25	US Texas High Plains and	4.95	AR Conesa	5.13	AU Riverina	3.52
AU Swan Hill (VIC)	8.83	US Tuolumne	4.92	AU Perth Hills	5.12	US Glenn	3.41
AU Upper Goulburn	8.80	AU Big Rivers - other	4.85	US Solano	4.44	AU Pyrenees	3.28
ZA Little Karoo	8.62	AT Bergland	4.78	ZA Cape South Coast	3.96	US Merced	3.20

Table 83 (cont.): VIIs for the top 25 regions for the world's top 24 white varieties, 2016

Catarratto Bianco		Aligoté		Graševina		Palomino Fino	
Region	*Varietal Intensity Indexes*	*Region*	*Varietal Intensity Indexes*	*Region*	*Varietal Intensity Indexes*	*Region*	*Varietal Intensity Indexes*
IT Sicilia	51.58	RU Crimea	31.84	HR Kontinentalna Hr	127.22	ES Canarias	98.64
US Merced	0.42	UA Ukraine	31.84	HU Balatonfelvidek	82.45	ES Andalucía	55.34
US San Joaquin	0.14	KZ East Kazakhstan	26.46	HU Badacsony	79.01	MX Coahuila	47.81
US Napa	0.05	RU Rostov Oblast	23.44	RS South Banat	70.05	ES Cantabria	32.35
US Sonoma	0.04	KZ Other regions	19.52	HU Balatonfured-Cso	67.92	ES Galicia	20.08
IT Friuli-Venezia Giulia	0.02	MD Moldova	15.65	SK Východné Sloven	66.73	AR Puelen	8.07
IT Veneto	0.01	FR Bourgogne	9.56	SI Prekmurje	62.54	US Riverside	6.99
IT Piemonte	0.01	KZ Almaty	9.25	HU Nagy-Somlo	49.60	MX Sonora	6.64
IT Emilia-Romagna	0.00	KZ West Kazakhstan	7.16	HU Pannonhalma	45.47	ES Castilla y León	5.87
IT Calabria	0.00	RO Romania	5.32	AT Vulkanland Steiei	36.90	NZ Auckland	3.68
IT Marche	0.00	BG North Central	4.39	SI Stajerska Sloveni	36.59	PT Centro	3.46
IT Puglia	0.00	RU Krasnodar Krai	3.21	AT Südburgenland	33.61	AU Rutherglen	3.44
IT Toscana	0.00	CH Geneva	2.63	SI Bizeljsko Sremic	33.09	PT Norte	3.37
IT Campania	0.00	KZ South Kazakhstan	1.70	AT Südsteiermark	30.20	NZ Northland	2.36
IT Molise	0.00	BG Southeast	1.61	RS Srem	28.81	ZA Olifants River	1.46
		CA Ontario	0.65	SI Bela Krajina	28.50	ZA Little Karoo	1.32
		CA Other regions	0.65	SK Bratislavský kraj	28.09	US Contra Costa	1.08
		KZ Zhambyl	0.59	RS Negotinska Krajii	23.65	PT Lisboa	0.92
		GE Georgia	0.43	HU Pecs	20.68	AR Las Heras	0.67
		FR Rhône Alpes	0.15	AT Neusiedlersee	20.65	AR Guaymallén	0.64
		FR Auvergne	0.06	RS Knjaževac	20.08	AR San Rafael	0.44
		CH Vaud	0.06	AT Neusiedlersee Hü	19.49	AR Veinticinco de M.	0.35
		CH Valais	0.02	SI Dolenjska	17.28	US Fresno	0.33
		FR Lorraine	0.01	SK Stredné Slovensk	17.21	AR Rawson	0.32
		FR Centre-Val de Loire	0.01	RS Leskovac	17.15	ZA Swartland	0.32

Table 83 (cont.): VIIs for the top 25 regions for the world's top 24 white varieties, 2016

Prosecco		Müller-Thurgau		Grüner Veltliner		Trebbiano Romagnolo	
Region	*Varietal Intensity Indexes*	*Region*	*Varietal Intensity Indexes*	*Region*	*Varietal Intensity Indexes*	*Region*	*Varietal Intensity Indexes*
IT Veneto	50.94	CH Appenzell Innerrho	223.52	AT Traisental	137.36	IT Emilia-Romagna	76.20
AU King Valley	17.38	CH Nidwalden	145.78	AT Wachau	133.56	IT Lombardia	0.12
AU Alpine Valleys	7.20	CH Thunersee	82.10	AT Kremstal	129.01	IT Marche	0.12
IT Friuli-Venezia Giulia	4.52	CH Zug	74.53	AT Kamptal	118.97	IT Umbria	0.09
AU Tumbarumba	2.65	CZ Severozápad	68.80	AT Wagram	114.67	IT Molise	0.06
AU Central Victoria - other	1.91	DE Franken	63.14	AT Weinviertel	112.89	IT Puglia	0.05
AU Gippsland	1.50	LU Luxembourg	55.88	AT Other regions	97.25	IT Liguria	0.04
BR Brazil	1.39	CH Thurgau	52.84	SK Bratislavský kraj	73.74	IT Lazio	0.04
AU Murray Darling (NSW)	0.82	CZ Praha	45.98	AT Wien	69.14	IT Calabria	0.03
AU Beechworth	0.67	CH Zürich	45.54	SK Západné Slovensko	52.35	IT Veneto	0.01
AU Hilltops	0.66	CH Glarus	44.23	AT Carnuntum	49.82	IT Abruzzo	0.01
AU Goulburn Valley	0.56	CH Appenzell Ausserrh	43.12	AT Neusiedlersee Hügellai	34.28	IT Basilicata	0.00
AU Macedon Ranges	0.54	CH Aargau	42.87	AT Bergland	29.42	IT Toscana	0.00
AU Adelaide Hills	0.30	CH Schwyz	41.32	CZ Jihovýchod	28.07	IT Campania	0.00
AR San Rafael	0.18	CH Lucerne	40.67	AT Neusiedlersee	24.73	IT Sicilia	0.00
AU Orange	0.12	CZ Morava	37.74	HU Etyek-Budai	23.28	IT Sardegna	0.00
AU McLaren Vale	0.08	CH Basel Stadt	37.54	AT Thermenregion	19.24		
AU Murray Darling (VIC)	0.05	DE Rheinhessen	36.37	HU Tolna	19.16		
AU Riverland	0.05	DE Baden	36.24	HU Balatonboglar	17.77		
IT Emilia-Romagna	0.03	CH Solothurn	36.20	HU Sopron	14.02		
IT Trento	0.02	CZ Cechy	36.15	AT Südburgenland	12.62		
IT Marche	0.02	DE Saale	35.39	HU Neszmely	11.20		
IT Molise	0.02	DE Sachsen	35.03	HU Pecs	10.24		
AU Currency Creek	0.02	CH Schaffhausen	32.87	HU Zala	10.04		
IT Basilicata	0.01	DE Nahe	28.96	SK Východné Slovensko	9.52		

Table 83 (cont.): VIIs for the top 25 regions for the world's top 24 white varieties, 2016

Sémillon		Verdejo		Viognier		Pedro Giménez	
Region	*Varietal Intensity Indexes*	*Region*	*Varietal Intensity Indexes*	*Region*	*Varietal Intensity Indexes*	*Region*	*Varietal Intensity Indexes*
TR Marmara	72.99	ES Castilla y León	39.47	AR Gualeguaychu	55.82	CL Atacama	142.79
AU Hunter	46.90	ES Castilla-La Mancha	3.44	AR Junín - San Luis	52.14	CL Coquimbo	103.06
AU Manjimup	44.09	ES Extremadura	2.07	AU Canberra District (A	50.72	AR Pichi Mahuida	49.76
AU Hunter Valley - othe	32.17	ES Comunidad Foral de	1.39	AR San Javier	23.26	AR Las Heras	41.11
AU Margaret River	31.68	ES Andalucía	0.46	US Virginia	22.79	AR Conesa	35.81
AR Conesa	30.95	ES La Rioja	0.33	FR Midi Pyrénées	22.48	AR Rawson	31.70
AU Swan Hill (NSW)	28.19	AU Eden Valley	0.20	AR Capital San Luis	17.10	AR Lavalle	29.88
AU Hastings River	26.80	AU Swan Hill (VIC)	0.18	AU South West Australi	16.87	AR Veinticinco de Mayo - S	28.11
AU Shoalhaven Coast	23.66	AU Murray Darling (NSV	0.09	AU Hastings River	15.59	AR Guaymallén	27.55
AU South Burnett	23.64	CL Metropolitana	0.04	AU Perth Hills	14.06	AR Nueve de Julio	27.19
AU Riverina	20.18	ES Aragón	0.04	US Ventura	13.61	AR San Martín - San Juan	26.64
AU Blackwood Valley	19.07	AU Barossa Valley	0.03	US San Diego	13.61	AR Rivadavia - Mza	24.90
AU Geographe	17.03	ES País Vasco	0.03	US Riverside	12.67	AR Sarmiento - San Juan	24.32
FR Aquitaine	16.50	ES Región de Murcia	0.01	US Hill Country	12.52	AR Avellaneda - Río Negro	23.77
AU Central Western Aus	16.25	ES Comunidad Valencia	0.01	US Texas High Plains a	12.47	AR Junín - Mza	22.92
AU New England Austra	15.57	AU Riverland	0.00	US Rogue Valley	12.35	AR Angaco	20.30
AU Great Southern	14.54			US Placer	11.94	AR San Martín - Mza	20.17
AU South West Australia	14.38			AU Central Victoria - ot	10.28	AR Santa Rosa - Mza	19.15
AU Southern Fleurieu	13.25			AU North East Victoria	9.27	AR La Paz	18.16
AR El Cuy	12.81			US Colorado	9.09	AR Caucete	15.83
AU Greater Perth - other	11.60			AU South Coast - other	9.09	AR General Roca	14.89
AU Pemberton	11.38			US Shasta	8.81	AR San Rafael	14.34
AU The Peninsulas	11.17			AU The Peninsulas	8.67	AR Maipú	14.11
AU Murray Darling (VIC	11.14			US Calaveras	8.62	AR Santa Lucía	10.74
AU Granite Belt	10.38			US Mariposa	8.33	AR General Alvear	10.11

Table 84: NVIIs for the top 25 regions for the world's top 24 red varieties, 2000

Cabernet Sauvignon		Garnacha Tinta		Merlot		Mazuelo	
Region	*NVII*	*Region*	*NVII*	*Region*	*NVII*	*Region*	*NVII*
FR Gironde	4.85	FR Vaucluse	5.44	FR Gironde	11.62	FR Aude	6.13
M9 Missing 9	3.42	ES Zaragoza	4.79	BG Bulgaria	1.43	FR Herault	5.57
CL O'Higgins	2.71	ES Toledo	4.40	FR Aude	1.13	FR Gard	1.85
CL Del Maule	2.52	FR Gard	3.04	FR Aquitaine except Gi	1.06	TN Tunisia	1.46
BG Bulgaria	1.24	FR Rhone-Alpes except Ar	1.60	FR Herault	0.96	DZ Algeria	1.38
CL Metropolitana	1.02	ES Comunidad Foral de Na	1.57	MD Moldova	0.86	FR Pyrenees-Orientales	1.37
AU Limestone Coast - other	0.73	ES La Rioja	1.48	CL Del Maule	0.79	FR Var	1.10
MD Moldova	0.71	ES Comunidad de Madrid	1.40	CL O'Higgins	0.76	FR Vaucluse	0.76
US Napa	0.65	FR Pyrenees-Orientales	1.23	IT Treviso	0.75	ES Tarragona	0.47
AU Riverland	0.54	FR Var	1.12	FR Gard	0.66	FR Bouches-du-Rhone	0.27
FR Aquitaine except Girond	0.41	ES Avila, Palencia, Salama	1.04	IT Padova	0.43	IT Cagliari	0.26
US Sonoma	0.39	FR Herault	0.99	M9 Missing 9	0.42	ES Girona, Lleida	0.25
ZA Stellenbosch	0.37	DZ Algeria	0.96	IT Pordenone	0.36	IL Israel	0.17
US San Joaquin	0.32	FR Aude	0.94	IT Venezia	0.33	US Madera	0.13
ZA Paarl	0.30	IT Nuoro	0.86	US Napa	0.33	MA Morocco	0.08
AU Langhorne Creek	0.27	FR Bouches-du-Rhone	0.58	US Sonoma	0.32	FR Ardeche	0.07
AU Murray Darling (VIC)	0.25	ES Huesca, Teruel	0.56	US Washington	0.27	US San Joaquin	0.05
AR Luján de Cuyo	0.23	FR Ardeche	0.39	IT Udine	0.26	FR Rhone-Alpes except Ar	0.03
US San Luis Obispo	0.23	TN Tunisia	0.26	US San Joaquin	0.23	US Mendocino	0.03
ZA Swartland	0.22	ES Zamora	0.24	IT Vicenza	0.18	ES La Rioja	0.03
AR Maipú	0.21	US Madera	0.13	ZA Stellenbosch	0.18	ES Comunidad Foral de Na	0.03
AU Barossa Valley	0.17	US Fresno	0.10	CH Ticino	0.16	US Stanislaus	0.01
FR Gard	0.17	IT Sassari	0.10	US Sacramento	0.15	FR Alpes-de-Haute-Proven	0.01
US Monterey	0.17	FR Corse	0.09	CL Metropolitana	0.15	US Contra Costa	0.01
AU McLaren Vale	0.17	IT Cagliari	0.09	US Monterey	0.14	US Santa Clara	0.00

Table 84 (cont.): NVIIs for the top 25 regions for the world's top 24 red varieties, 2000

Syrah		Bobal		Tempranillo		Monastrell	
Region	*NVII*	*Region*	*NVII*	*Region*	*NVII*	*Region*	*NVII*
FR Herault	1.73	ES Cuenca	8.07	ES La Rioja	3.81	ES Region de Murcia	6.85
FR Gard	1.62	ES Valencia	7.47	ES Alava	1.69	ES Albacete	4.17
FR Vaucluse	1.18	ES Albacete	3.54	ES Ciudad Real	1.63	ES Alicante	1.82
FR Aude	1.07	ES Caceres	0.00	ES Burgos	1.55	FR Var	0.34
AU Riverland	0.90	ES Castellon	0.00	ES Comunidad Foral de	1.02	FR Vaucluse	0.08
FR Rhone-Alpes except A	0.73	ES Illes Balears	-0.01	PT Alto Tras-os-Montes	0.85	AU Riverland	0.04
AU Riverina	0.57	ES Guadalajara	-0.01	ES Valladolid	0.71	FR Pyrenees-Orientales	0.02
AU Barossa Valley	0.55	ES Avila, Palencia, Sal	-0.01	ES Zamora	0.71	US Contra Costa	0.02
FR Pyrenees-Orientales	0.54	ES Huesca, Teruel	-0.02	ES Zaragoza	0.32	ES Illes Balears	0.02
FR Var	0.51	ES Almeria, Granada,	-0.02	ES Avila, Palencia, Sala	0.30	TN Tunisia	0.02
FR Ardeche	0.49	ES Girona, Lleida	-0.03	ES Guadalajara	0.29	ES Girona, Lleida	0.01
AU McLaren Vale	0.36	ES Alava	-0.05	PT Alentejo	0.19	ES Almeria, Granada, Jaen, S	0.00
AU Limestone Coast - oth	0.32	ES Burgos	-0.06	ES Comunidad de Madri	0.17	AU Perricoota	0.00
AU Langhorne Creek	0.24	ES Canarias	-0.06	ES Cuenca	0.16	AU Alpine Valleys/Beechwort	0.00
AU Murray Darling (VIC)	0.24	ES Zamora	-0.07	ES Valencia	0.15	FR Bouches-du-Rhone	0.00
AU Clare Valley	0.21	ES Valladolid	-0.07	AR San Carlos - Mza	0.12	ES Castellon	0.00
ZA Paarl	0.21	ES Leon	-0.07	AR San Martín - Mza	0.11	AU Barossa - other	0.00
FR Midi-Pyrenees except	0.20	ES Comunidad de Mad	-0.07	ES Huesca, Teruel	0.07	US Placer	0.00
ZA Stellenbosch	0.18	ES Zaragoza	-0.08	AR Tunuyán	0.06	AU Lower Murray - other	0.00
DZ Algeria	0.18	ES Alicante	-0.08	ES Girona, Lleida	0.05	US San Diego	0.00
ZA Swartland	0.18	ES Comunidad Foral d	-0.10	AR Maipú	0.05	AU Eastern Plains, Inland and	0.00
FR Bouches-du-Rhone	0.16	ES Barcelona	-0.10	ES Castellon	0.04	FR Aude	0.00
AU Padthaway	0.14	ES Tarragona	-0.13	ES Barcelona	0.03	US Riverside	0.00
AU Mudgee	0.14	ES Galicia	-0.13	AR Tupungato	0.02	US San Bernardino	0.00
AU Hunter	0.14	ES La Rioja	-0.17	AR Lavalle	0.02	AU Canberra District	0.00

Table 84 (cont.): NVIIs for the top 25 regions for the world's top 24 red varieties, 2000

Sangiovese		Pinot Noir		Cabernet Franc		Cinsaut	
Region	*NVII*	*Region*	*NVII*	*Region*	*NVII*	*Region*	*NVII*
IT Siena	2.19	FR Champagne-Ardenne	2.28	FR Gironde	2.63	FR Herault	1.90
IT Firenze	2.10	FR Bourgogne	2.04	FR Pays de la Loire except M	1.64	DZ Algeria	1.48
IT Foggia	1.32	MD Moldova	1.08	FR Centre-Val de Loire	1.06	FR Gard	1.03
IT Bari	0.74	DE Baden	0.95	BR Brazil	0.66	FR Var	1.00
IT Forlì-Cesena	0.66	CH Valais	0.36	FR Aquitaine except Gironde	0.65	FR Aude	0.81
IT Arezzo	0.61	IT Pavia	0.34	IT Treviso	0.34	MA Morocco	0.71
IT Ascoli Piceno	0.58	US Sonoma	0.26	FR Midi-Pyrenees except Ger	0.19	FR Vaucluse	0.40
IT Rimini	0.34	FR Alsace	0.24	IT Venezia	0.16	ZA Paarl	0.27
IT Taranto	0.34	DZ Algeria	0.22	FR Deux-Sevres, Vienne	0.12	FR Bouches-du-Rhone	0.20
IT Perugia	0.33	FR Centre-Val de Loire	0.18	IT Udine	0.11	ZA Breedekloof	0.15
FR Corse	0.30	DE Pfalz	0.14	CA Canada	0.10	FR Ardeche	0.14
IT Grosseto	0.29	US Yamhill Co.	0.12	IT Pordenone	0.09	TN Tunisia	0.14
IT Ravenna	0.25	US Napa	0.12	IT Padova	0.08	ZA Swartland	0.11
IT Pisa	0.18	CL Valparaiso	0.11	IT Vicenza	0.06	FR Corse	0.06
IT Chieti	0.14	AU Yarra Valley	0.10	UY Uruguay	0.06	IT Brindisi	0.03
IT Bologna	0.12	DE Württemberg	0.10	IT Gorizia	0.05	FR Alpes-de-Haute-Proven	0.02
TN Tunisia	0.12	NZ Marlborough	0.10	US Napa	0.04	ZA Stellenbosch	0.02
IT Benevento	0.11	CH Vaud	0.09	IT Brescia	0.04	FR Rhone-Alpes except Ar	0.02
IT Caltanissetta	0.10	DE Rheinhessen	0.08	FR Gers	0.03	ZA Worcester	0.01
IT Terni	0.10	CH Zürich	0.08	US Washington	0.03	EL Dytiki Makedonia	0.00
IT Matera	0.09	CH Schaffhausen	0.08	US New York - other	0.01	EL Thessalia	0.00
IT Salerno	0.08	US Santa Barbara	0.07	ZA Stellenbosch	0.01	EL Kentriki Makedonia	0.00
IT Pesaro e Urbino	0.08	US Monterey	0.07	IT Trento	0.01	ZA Robertson	0.00
IT Viterbo	0.08	CA Canada	0.07	US Sonoma	0.01	IT Trieste	0.00
IT Caserta	0.07	DE Rheingau	0.07	AU Cowra	0.01	US Contra Costa	0.00

Table 84 (cont.): NVIIs for the top 25 regions for the world's top 24 red varieties, 2000

Gamay Noir		Alicante Henri Bouschet		Barbera		Montepulciano	
NVII	*NVII*	*Region*	*NVII*	*Region*	*NVII*	*Region*	*NVII*
FR Rhone-Alpes except Ardeche	4.54	ES Galicia	1.53	IT Asti	1.79	IT Chieti	2.28
FR Bourgogne	0.65	ES Albacete	1.08	IT Alessandria	1.06	IT Foggia	0.54
FR Centre-Val de Loire	0.62	FR Herault	0.57	IT Pavia	0.68	IT Campobasso	0.47
FR Pays de la Loire except Maye	0.39	DZ Algeria	0.57	US Fresno	0.50	IT Bari	0.44
FR Midi-Pyrenees except Gers	0.23	FR Aude	0.34	IT Cuneo	0.48	IT Pescara	0.43
CH Valais	0.19	FR Gard	0.21	IT Piacenza	0.32	IT Ascoli Piceno	0.35
FR Auvergne	0.15	CL Del Maule	0.20	IT Salerno	0.30	IT Teramo	0.33
FR Ardeche	0.12	CL O'Higgins	0.19	US Madera	0.14	IT Taranto	0.14
CH Vaud	0.11	MA Morocco	0.15	IT Benevento	0.11	IT L'Aquila	0.12
CH Geneva	0.09	TN Tunisia	0.15	IT Torino	0.10	IT Ancona	0.09
FR Deux-Sevres, Vienne	0.04	ES Leon	0.11	IT Brescia	0.07	IT Benevento	0.07
CA Canada	0.04	ES Alicante	0.07	US Kern	0.06	IT Brindisi	0.06
US Monterey	0.03	ES Huesca, Teruel	0.06	US Stanislaus	0.06	IT Caserta	0.03
FR Lorraine	0.01	PT Alentejo	0.05	US Merced	0.05	IT Roma	0.03
US San Benito	0.01	FR Bouches-du-Rhone	0.03	IT Caserta	0.04	IT Macerata	0.03
US Solano	0.01	IT Reggio di Calabria	0.03	IT Avellino	0.04	IT Viterbo	0.02
IT Valle d'Aosta	0.00	CL Metropolitana	0.03	IT Potenza	0.04	IT Isernia	0.02
US Lake	0.00	ES Zamora	0.03	IT Bologna	0.03	IT Avellino	0.02
IT Perugia	0.00	US Tulare	0.02	US Tulare	0.03	IT Frosinone	0.02
FR Correze, Haute-Vienne	0.00	US Fresno	0.02	IT Vicenza	0.03	IT Rieti	0.02
US Mendocino	0.00	IT Catania	0.02	AR Chilecito	0.03	IT Pesaro e Urbino	0.01
FR Franche Comté	0.00	CL Coquimbo	0.01	IT Biella	0.01	IT Terni	0.01
CH Fribourg	0.00	FR Var	0.01	IT Milano	0.01	IT Matera	0.01
US Tehama	0.00	ES Valencia	0.01	IT Parma	0.01	IT Salerno	0.01
US San Diego	0.00	IT Cosenza	0.00	AR Tupungato	0.01	IT Perugia	0.01

Table 84 (cont.): NVIIs for the top 25 regions for the world's top 24 red varieties, 2000

Isabella		**Tribidrag**		**Côt**		**Criolla Grande**	
Region	*NVII*	*Region*	*NVII*	*Region*	*NVII*	*Region*	*NVII*
BR Brazil	2.86	US San Joaquin	1.50	AR Luján de Cuyo	0.79	AR San Martín - Mza	1.58
MD Moldova	2.23	IT Taranto	1.05	FR Midi-Pyrenees except	0.76	AR Rivadavia - Mza	0.81
M9 Missing 9	0.25	US Sonoma	0.31	AR Maipú	0.51	AR Junín - Mza	0.58
AR Colón - Entre Ríos	0.01	IT Bari	0.27	AR San Carlos - Mza	0.33	AR San Rafael	0.53
AR Totoral	0.00	US Madera	0.25	AR Rivadavia - Mza	0.28	AR Santa Rosa - Mza	0.36
US Finger Lakes	0.00	US Fresno	0.18	AR Junín - Mza	0.26	AR General Alvear	0.30
		US Kern	0.17	AR Tupungato	0.21	AR Lavalle	0.26
		US Amador	0.15	AR Tunuyán	0.18	AR Maipú	0.09
		US Mendocino	0.14	AR San Rafael	0.18	AR Luján de Cuyo	0.06
		US Napa	0.13	AR San Martín - Mza	0.11	AR Veinticinco de Mayo - S	0.04
		US San Luis Obispo	0.12	FR Gironde	0.07	AR Sarmiento - San Juan	0.03
		IT Brindisi	0.10	FR Centre-Val de Loire	0.07	AR Caucete	0.03
		US Merced	0.09	AR Santa Rosa - Mza	0.07	AR Las Heras	0.02
		US Colusa	0.09	CL O'Higgins	0.07	AR Guaymallén	0.02
		US Stanislaus	0.09	AR Veinticinco de Mayo	0.06	AR San Martín - San Juan	0.01
		US Tulare	0.07	FR Aquitaine except Gir	0.06	AR Angaco	0.01
		US Sacramento	0.06	AR Chilecito	0.06	AR San Carlos - Mza	0.01
		US San Bernardino	0.06	AR Sarmiento - San Juan	0.05	AR La Paz	0.01
		US Monterey	0.06	AR Lavalle	0.05	AR Nueve de Julio	0.01
		TN Tunisia	0.05	AR Cafayate	0.05	AR Santa Lucía	0.00
		US Lake	0.03	AR General Roca	0.04	AR Tupungato	0.00
		US Glenn	0.03	AR Guaymallén	0.04	AR Avellaneda - Río Negro	0.00
		IT Lecce	0.03	AR Añelo	0.04	AR Chimbas	0.00
		IT Salerno	0.03	AR Caucete	0.03	AR Rawson	0.00
		US Contra Costa	0.02	CL Del Maule	0.02	AR Rivadavia - San Juan	0.00

Table 84 (cont.): NVIIs for the top 25 regions for the world's top 24 red varieties, 2000

Pamid		Douce Noire		Negroamaro		Doukkali	
Region	*NVII*	*Region*	*NVII*	*Region*	*NVII*	*Region*	*NVII*
BG Bulgaria	4.53	AR San Martín - Mza	0.63	IT Lecce	1.68	MA Morocco	3.35
EL Dytiki Makedonia	0.00	IT Pavia	0.41	IT Brindisi	1.51		
HU Hungary	-0.06	AR San Rafael	0.36	IT Taranto	0.16		
		AR Lavalle	0.33	IT Matera	0.00		
		AR Rivadavia - Mza	0.31	IT Reggio di Calabria	0.00		
		AR Santa Rosa - Mza	0.27	IT Valle d'Aosta	0.00		
		AR Junín - Mza	0.16	IT Enna	0.00		
		AR Tupungato	0.15	IT Catanzaro	0.00		
		AR Maipú	0.13	IT Torino	0.00		
		AR Luján de Cuyo	0.11	IT Messina	0.00		
		AR Chilecito	0.10	IT Cosenza	0.00		
		AR Sarmiento - San Jua	0.10	IT Caserta	0.00		
		AR General Alvear	0.09	IT Potenza	0.00		
		AR Caucete	0.06	IT Catania	0.00		
		AR Veinticinco de May	0.04	IT Brescia	0.00		
		AR Las Heras	0.03	IT Frosinone	0.00		
		AR Nueve de Julio	0.03	IT Viterbo	0.00		
		AR Tunuyán	0.02	IT Campobasso	0.00		
		AR San Martín - San Ju	0.02	IT Salerno	0.00		
		AR Guaymallén	0.02	IT Sassari	0.00		
		IT Asti	0.01	IT Padova	0.00		
		AR Pocito	0.01	IT Pordenone	0.00		
		AR Tinogasta	0.01	IT Nuoro	0.00		
		IT Torino	0.01	IT Perugia	-0.01		
		AR San Carlos - Mza	0.01	IT Trento	-0.01		

Table 85: NVIIs for the top 25 regions for the world's top 24 red varieties, 2016

Cabernet Sauvignon		**Merlot**		**Tempranillo**		**Syrah**	
Region	*NVII*	*Region*	*NVII*	*Region*	*NVII*	*Region*	*NVII*
CN China	6.24	FR Aquitaine	14.40	ES Castilla-La Mancha	10.78	FR Languedoc Roussil	6.12
CL O'Higgins	3.41	FR Languedoc Roussillon	2.58	ES Castilla y León	6.95	FR Provence-Alpes-Co	2.45
FR Aquitaine	3.30	CN China	1.37	ES La Rioja	6.62	FR Rhône Alpes	1.37
CL Del Maule	2.70	BG South Central	0.92	ES Extremadura	3.82	AU Barossa Valley	1.12
US Napa	1.48	MD Moldova	0.62	ES País Vasco	2.42	AU Riverland	1.01
CL Metropolitana	1.30	CL O'Higgins	0.58	ES Comunidad Foral de Na	1.77	AU Riverina	0.82
US San Luis Obispo	0.78	BG Southeast	0.48	ES Aragón	0.94	AU McLaren Vale	0.68
RU Crimea	0.71	CL Del Maule	0.40	PT Alentejo	0.90	AU Langhorne Creek	0.41
UA Ukraine	0.71	US San Joaquin	0.29	ES Comunidad Valenciana	0.87	AU Murray Darling (VI	0.38
US Sonoma	0.68	IT Friuli-Venezia Giulia	0.27	PT Norte	0.57	ZA Paarl	0.37
US San Joaquin	0.61	US Napa	0.24	PT Centro	0.31	IT Sicilia	0.35
AU Coonawarra	0.58	US Monterey	0.23	AR San Carlos - Mza	0.08	ZA Swartland	0.34
MD Moldova	0.55	US San Luis Obispo	0.20	ES Cataluña	0.06	CL O'Higgins	0.33
BG South Central	0.54	ZA Stellenbosch	0.20	PT Lisboa	0.04	ZA Stellenbosch	0.32
RU Krasnodar Krai	0.46	IT Toscana	0.20	ES Comunidad de Madrid	0.02	AU Clare Valley	0.30
ZA Stellenbosch	0.43	US Sacramento	0.18	MX Baja California	0.02	TR Aegean	0.22
US Horse Heaven Hills	0.43	CH Ticino	0.18	AR San Martín - Mza	0.01	AR Lavalle	0.21
AU Riverland	0.42	RO Romania	0.18	US Texas High Plains and P	0.01	AU Padthaway	0.20
BG Southeast	0.42	NZ Hawkes Bay	0.17	US Hill Country	0.00	ES Aragón	0.20
ES Aragón	0.33	US Sonoma	0.16	US North Texas (DFW)	0.00	AU Coonawarra	0.19
AU Langhorne Creek	0.32	IT Umbria	0.15	US West Texas	0.00	AU Murray Darling (NS	0.19
ZA Paarl	0.31	US Wahluke Slope	0.14	AR La Paz	0.00	AU Hunter	0.15
US Columbia Valley	0.30	US Horse Heaven Hills	0.13	AR Junín - San Luis	0.00	AU Heathcote	0.15
AR Luján de Cuyo	0.28	ES Comunidad Foral de Nav	0.12	US Sutter	0.00	DZ Algeria	0.15
US Wahluke Slope	0.26	BG Northwest	0.12	AR Valle Viejo	0.00	CL Metropolitana	0.15

Table 85 (cont.): NVIIs for the top 25 regions for the world's top 24 red varieties, 2016

Garnacha Tinta		Pinot Noir		Sangiovese		Bobal	
Region	*NVII*	*Region*	*NVII*	*Region*	*NVII*	*Region*	*NVII*
FR Provence-Alpes-Cote d	7.03	FR Champagne-Arden	2.75	IT Toscana	8.10	ES Castilla-La Mancha	6.26
FR Languedoc Roussillon	5.93	FR Bourgogne	2.12	IT Puglia	3.38	ES Comunidad Valenci	5.51
ES Aragón	2.75	DE Baden	1.10	IT Emilia-Romagna	0.67	ES Illes Balears	0.00
FR Rhône Alpes	1.50	US Sonoma	0.97	IT Marche	0.61	ES Canarias	-0.03
ES Castilla-La Mancha	1.18	US North Willamette	0.96	IT Umbria	0.52	ES Comunidad de Mad	-0.03
IT Sardegna	0.99	US Monterey	0.66	FR Corse	0.29	ES Comunidad Foral de	-0.05
ES Castilla y León	0.68	IT Lombardia	0.52	IT Campania	0.15	ES Región de Murcia	-0.07
ES La Rioja	0.62	NZ Marlborough	0.44	IT Lazio	0.12	ES Aragón	-0.07
ES Comunidad Foral de N	0.52	US Santa Barbara	0.40	IT Sardegna	0.08	ES Andalucía	-0.09
ES Comunidad de Madrid	0.50	CL Valparaiso	0.35	IT Basilicata	0.06	ES Extremadura	-0.21
ES Cataluña	0.45	NZ Otago	0.32	AR Maipú	0.03	ES Castilla y León	-0.21
DZ Algeria	0.38	CH Valais	0.32	IT Calabria	0.03		
FR Corse	0.09	FR Alsace	0.28	IT Molise	0.02		
AU Barossa Valley	0.05	DE Pfalz	0.24	ET Ethiopia	0.02		
MA Morocco	0.04	DE Württemberg	0.23	IT Abruzzo	0.02		
AU McLaren Vale	0.04	US Mendocino	0.20	US Texas High Plains and Panha	0.01		
US Fresno	0.02	FR Centre-Val de Loi	0.19	AR General Roca	0.01		
US Madera	0.02	DE Rheinhessen	0.18	IT Liguria	0.01		
US Kings	0.01	US Napa	0.16	AR Lavalle	0.01		
US Glenn	0.01	US Umpqua Valley	0.16	US Amador	0.01		
MX Baja California	0.01	US South Willamette	0.15	AU King Valley	0.00		
TN Tunisia	0.01	AU Yarra Valley	0.14	US Riverside	0.00		
AU Perth Hills	0.00	AU Tasmania	0.13	AR Junín - Mza	0.00		
AU Swan District	0.00	UK UnitedKingdom	0.11	US San Diego	0.00		
US San Bernardino	0.00	US San Luis Obispo	0.11	US Hill Country	0.00		

Table 85 (cont.): NVIIs for the top 25 regions for the world's top 24 red varieties, 2016

Cabernet Franc		**Côt**		**Monastrell**		**Mazuelo**	
Region	*NVII*	*Region*	*NVII*	*Region*	*NVII*	*Region*	*NVII*
FR Aquitaine	2.30	AR Luján de Cuyo	1.84	ES Región de Murcia	4.14	FR Languedoc Roussill	4.87
FR Pays de la Loire	1.91	AR San Carlos - Mza	1.17	ES Castilla-La Mancha	2.41	FR Provence-Alpes-Co	1.16
BR Brazil	1.43	AR Tunuyán	1.02	ES Comunidad Valenciana	1.30	DZ Algeria	0.65
FR Centre-Val de Loire	1.22	AR Maipú	0.85	FR Provence-Alpes-Cote d'A:	0.58	ES Cataluña	0.35
IT Veneto	0.43	AR Tupungato	0.85	FR Languedoc Roussillon	0.44	IT Sardegna	0.32
IT Friuli-Venezia Giulia	0.18	FR Midi Pyrénées	0.73	AU Barossa Valley	0.02	MA Morocco	0.23
FR Midi Pyrénées	0.13	AR Rivadavia - Mza	0.44	ES Illes Balears	0.02	IL Israel	0.20
CA Ontario	0.09	AR San Rafael	0.41	US Contra Costa	0.02	ES La Rioja	0.11
HU Villany	0.07	AR Junín - Mza	0.34	US Texas High Plains and Pa	0.01	FR Rhône Alpes	0.10
US Napa	0.05	AR Cafayate	0.20	ZA Swartland	0.01	MX Aguascalientes	0.08
FR Languedoc Roussillor	0.05	AR Sarmiento - San Juan	0.16	AU Swan Hill (VIC)	0.01	US Madera	0.08
HU Szekszard	0.05	AR San Martín - Mza	0.14	US El Dorado	0.00	ES Aragón	0.06
ZA Stellenbosch	0.04	AR Añelo	0.13	AU Limestone Coast - other	0.00	TN Tunisia	0.05
CL O'Higgins	0.04	CL O'Higgins	0.12	AU McLaren Vale	0.00	ES Comunidad Foral d	0.03
CA British Colombia	0.04	AR Lavalle	0.10	AU Adelaide Plains	0.00	CL Del Maule	0.02
US Virginia	0.03	AR Chilecito	0.09	AU Heathcote	0.00	US Mendocino	0.02
CL Metropolitana	0.03	AR Santa Rosa - Mza	0.08	US Hill Country	0.00	FR Corse	0.01
HU Eger	0.03	AR Veinticinco de Mayo -	0.07	AU Barossa - other	0.00	TR Aegean	0.01
UY Canelones	0.02	FR Centre-Val de Loire	0.06	US Placer	0.00	MX Sonora	0.01
US Suffolk	0.02	AR Caucete	0.05	AU Granite Belt	0.00	US Contra Costa	0.00
CA Other regions	0.02	AR General Roca	0.05	US Calaveras	0.00	MX Zacatecas	0.00
AR Tunuyán	0.01	CL Del Maule	0.04	US Riverside	0.00	US San Bernardino	0.00
US Michigan	0.01	AR San Carlos - Salta	0.04	US Nevada	0.00	US Santa Clara	0.00
IL Israel	0.01	AR San Martín - San Juan	0.03	AU Western Victoria - other	0.00	US San Diego	0.00
RS Toplica	0.01	AR Guaymallén	0.03	AU Peel	0.00	US Calaveras	0.00

Table 85 (cont.): NVIIs for the top 25 regions for the world's top 24 red varieties, 2016

Alicante Henri Bouschet		Tribidrag		Montepulciano		Gamay Noir	
Region	*NVII*	*Region*	*NVII*	*Region*	*NVII*	*Region*	*NVII*
ES Castilla-La Mancha	1.89	IT Puglia	2.87	IT Abruzzo	3.48	FR Rhône Alpes	3.67
ES Galicia	1.12	US San Joaquin	1.60	IT Puglia	1.89	FR Bourgogne	0.42
CL O'Higgins	0.60	US Sonoma	0.43	IT Marche	0.61	FR Centre-Val de Lo	0.39
CL Del Maule	0.57	US Fresno	0.27	IT Molise	0.52	FR Pays de la Loire	0.30
PT Alentejo	0.42	US Madera	0.21	IT Lazio	0.11	FR Midi Pyrénées	0.19
PT Centro	0.20	US San Luis Obispo	0.20	IT Emilia-Romagna	0.10	CH Valais	0.13
MA Morocco	0.17	US Amador	0.19	IT Campania	0.05	FR Auvergne	0.11
TR Aegean	0.10	MK NorthMacedon	0.18	IT Basilicata	0.02	CH Vaud	0.08
ES Comunidad Valenciana	0.10	US Mendocino	0.16	IT Sardegna	0.01	CH Geneva	0.08
ES Región de Murcia	0.10	US Napa	0.10	IT Umbria	0.01	TR Marmara	0.05
CL Metropolitana	0.09	US Merced	0.09	US Hill Country	0.00	CA Ontario	0.03
FR Languedoc Roussillon	0.05	US Sacramento	0.09	AR Ullum	0.00	CA British Colombia	0.01
PT Norte	0.05	US Stanislaus	0.08	US Texas High Plain	0.00	FR Lorraine	0.01
CL Coquimbo	0.04	US Colusa	0.07	AR Ischilin	0.00	RS Toplica	0.01
TN Tunisia	0.03	US Kern	0.06	US Marin	0.00	CA Other regions	0.00
PT Lisboa	0.03	US Tulare	0.05	NZ Auckland	0.00	RS Negotinska Kraji	0.00
US Tulare	0.01	US El Dorado	0.04	AU Queensland - oth	0.00	US Solano	0.00
US Merced	0.00	US Contra Costa	0.04	US Riverside	0.00	IT Valle d'Aosta	0.00
US Riverside	0.00	US Lake	0.04	US San Diego	0.00	SI Bela Krajina	0.00
AR Poman	0.00	US San Bernardino	0.03	IT Valle d'Aosta	0.00	FR Limousin	0.00
PT Algarve	0.00	IT Basilicata	0.02	AU Alpine Valleys	0.00	AU Beechworth	0.00
AR Zonda	0.00	US Kings	0.02	AU Mount Lofty Ran	0.00	CH Fribourg	0.00
AR Calingasta	0.00	US Glenn	0.02	IT Liguria	0.00	AU Tumbarumba	0.00
AR Santa María - Cba	0.00	US Alameda	0.01	US Contra Costa	0.00	US Calaveras	0.00
US Marin	0.00	US Solano	0.01	US El Dorado	0.00	CH Lac de Bienne	0.00

Table 85 (cont.): NVIIs for the top 25 regions for the world's top 24 red varieties, 2016

Cinsaut		**Carmenère**		**Douce Noire**		**Barbera**	
Region	*NVII*	*Region*	*NVII*	*Region*	*NVII*	*Region*	*NVII*
FR Languedoc Roussillon	1.61	CN China	2.30	AR San Martín - Mza	0.82	IT Piemonte	2.13
FR Provence-Alpes-Cote d'Az	1.41	CL O'Higgins	1.27	AR Lavalle	0.48	IT Lombardia	0.45
MA Morocco	0.70	CL Del Maule	0.65	AR Rivadavia - Mza	0.42	IT Emilia-Romagna	0.40
CL Del Bio Bio	0.16	CL Metropolitana	0.18	AR San Rafael	0.38	US Fresno	0.27
TN Tunisia	0.14	CL Valparaiso	0.04	AR Santa Rosa - Mza	0.34	IT Campania	0.15
ZA Paarl	0.12	CL Del Bio Bio	0.02	AR Junín - Mza	0.24	US Madera	0.07
ZA Breedekloof	0.10	CL Coquimbo	0.01	AR Maipú	0.20	AR Chilecito	0.02
TR Marmara	0.09	AR Tandil	0.00	AR Tupungato	0.19	SI Vipavska dolina	0.02
ZA Swartland	0.05	CL Tarapaca	0.00	AR Sarmiento - San Juan	0.16	IT Basilicata	0.02
FR Rhône Alpes	0.03	AU Beechworth	0.00	AR Luján de Cuyo	0.14	US Tulare	0.02
FR Corse	0.03	AU Granite Belt	0.00	AR Chilecito	0.12	US Amador	0.02
ZA Worcester	0.01	AR Rivadavia - San Ju	0.00	AR General Alvear	0.09	US El Dorado	0.01
ZA Robertson	0.01	US Riverside	0.00	AR Veinticinco de Mayo -	0.09	US Kings	0.01
ZA Stellenbosch	0.01	CL Atacama	0.00	AR Caucete	0.08	AR San Martín - San Ju	0.00
US Texas High Plains and Pan	0.00	US El Dorado	0.00	AR Las Heras	0.04	AR Lavalle	0.00
AU Rutherglen	0.00	AU Heathcote	0.00	AR Nueve de Julio	0.04	US Alameda	0.00
US Riverside	0.00	AR Chilecito	0.00	AR Tunuyán	0.03	AU King Valley	0.00
US Calaveras	0.00	IT Liguria	0.00	AR San Martín - San Juan	0.03	AR Ischilin	0.00
AU Canberra District (NSW)	0.00	US San Benito	0.00	AR Tinogasta	0.02	AU Central Ranges - ot	0.00
US Santa Clara	0.00	CH Ticino	0.00	AR Pocito	0.02	US Placer	0.00
US Contra Costa	0.00	AU Swan Hill (VIC)	0.00	AR Santa Lucía	0.01	US Riverside	0.00
US El Dorado	0.00	IT Basilicata	0.00	AR Guaymallén	0.01	US Calaveras	0.00
AU Yarra Valley	0.00	IT Molise	0.00	IT Piemonte	0.01	AR Coronel Felipe Var	0.00
ZA Little Karoo	0.00	HU Villany	0.00	AR Angaco	0.01	AU South Burnett	0.00
ZA Cape South Coast	0.00	AR San Martín - San Ju	0.00	AR La Paz	0.01	US Nevada	0.00

Table 85 (cont.): NVIIs for the top 25 regions for the world's top 24 red varieties, 2016

Isabella		Blaufränkisch		Criolla Grande		Pinot Meunier	
Region	*NVII*	*Region*	*NVII*	*Region*	*NVII*	*Region*	*NVII*
BR Brazil	2.57	HU Kunsag	0.50	AR San Martín - Mza	1.19	FR Champagne-Ardenne	2.38
MD Moldova	0.70	DE Württemberg	0.35	AR Rivadavia - Mza	0.58	DE Württemberg	0.32
UA Ukraine	0.25	CZ Jihovýchod	0.24	AR Junín - Mza	0.45	FR Picardie	0.29
RU Crimea	0.25	AT Mittelburgenland	0.23	AR San Rafael	0.29	UK UnitedKingdom	0.04
RU Krasnodar Krai	0.02	HU Eger	0.23	AR Santa Rosa - Mza	0.27	DE Baden	0.04
UY Canelones	0.01	HU Sopron	0.20	AR Lavalle	0.21	DE Pfalz	0.02
UY Montevideo	0.00	SK Západné Slovensko	0.18	AR General Alvear	0.16	DE Franken	0.01
UY Paysandu	0.00	AT Neusiedlersee	0.17	AR Maipú	0.05	AU King Valley	0.00
UY Tacuarembo	0.00	AT Neusiedlersee Hügell	0.15	AR Luján de Cuyo	0.04	AU Henty	0.00
AU Alpine Valleys	0.00	HU Szekszard	0.13	AR Sarmiento - San Jua	0.03	AU Tasmania	0.00
UY San Jose	0.00	RS Tri Morave	0.11	AR Veinticinco de May	0.03	AU Strathbogie Ranges	0.00
CH Graubünden - Mesol	0.00	HU Matra	0.11	AR Las Heras	0.01	AU Tumbarumba	0.00
UY Maldonado	0.00	HR Kontinentalna Hrvats	0.10	AR Caucete	0.01	FR Île de France	0.00
UY Colonia	0.00	HU Hajos-bajai	0.09	AR Guaymallén	0.00	AU Adelaide Hills	0.00
CH Ticino	0.00	HU Tolna	0.07	AR Nueve de Julio	0.00	AU Southern Highlands	0.00
AU Margaret River	0.00	SI Dolenjska	0.06	AR Angaco	0.00	AU Yarra Valley	0.00
AU Riverina	-0.01	HU Csongrad	0.06	AR La Paz	0.00	CA Nova Scotia	0.00
		PE Lima	0.06	AR Albardón	0.00	US Riverside	0.00
		HU Villany	0.05	AR San Martín - San Ju	0.00	DE Sachsen	0.00
		HU Bukk	0.04	AR Rawson	0.00	CL De Los Lagos	0.00
		SI Bizeljsko Sremic	0.04	AR Rivadavia - San Jua	0.00	AU Grampians	0.00
		HU Balatonboglar	0.04	AR Santa Lucía	0.00	AU Pyrenees	0.00
		AT Südburgenland	0.03	AR Chimbas	0.00	DE Hessische Bergstraße	0.00
		SK Bratislavský kraj	0.03	AR Vinchina	0.00	DE Mittelrhein	0.00
		SK Stredné Slovensko	0.03	AR Tinogasta	0.00	NZ Canterbury	0.00

Table 86: NVIIs for the top 25 regions for the world's top 24 white varieties, 2000

Airén		**Chardonnay**		**Trebbiano Toscano**		**Graševina**	
Region	*NVII*	*Region*	*NVII*	*Region*	*NVII*	*Region*	*NVII*
ES Ciudad Real	31.51	FR Bourgogne	2.49	FR Charente-Maritime	7.39	RS Serbia	6.51
ES Toledo	23.14	FR Champagne-Ardenne	1.52	FR Charente	7.37	HR Croatia	3.05
ES Cuenca	7.07	US Monterey	1.12	FR Gers	1.33	RO Romania	2.21
ES Albacete	5.88	US Sonoma	0.93	IT Chieti	1.19	HU Hungary	1.03
ES Comunidad de Madrid	1.06	US San Joaquin	0.82	IT Foggia	1.19	SK Slovakia	0.74
ES Guadalajara	0.15	US Napa	0.52	IT Roma	0.45	SI Slovenia	0.64
ES Castellon	-0.02	MD Moldova	0.50	IT Agrigento	0.45	BG Bulgaria	0.37
ES Almeria, Granada, Jaen,	-0.05	US Santa Barbara	0.50	FR Var	0.33	AT Burgenland	0.35
ES Malaga	-0.06	IT Trento	0.44	IT Firenze	0.29	IT Pavia	0.29
ES Caceres	-0.07	CL Valparaiso	0.38	IT Latina	0.29	ES Caceres	0.26
ES Huelva	-0.11	US Washington	0.37	IT Terni	0.27	AT Niederosterreich	0.22
ES Girona, Lleida	-0.12	AU Riverland	0.35	IT Perugia	0.25	CZ Czechia	0.21
ES Cordoba	-0.14	AU Murray Darling (VIC)	0.33	IT Palermo	0.24	AT Steiermark	0.13
ES Cadiz	-0.16	US Mendocino	0.32	IT Viterbo	0.21	AT Wien and other regions	0.00
ES Alava	-0.18	US Yolo	0.27	IT Benevento	0.16	IT Milano	0.00
ES Burgos	-0.22	US San Luis Obispo	0.27	IT Ascoli Piceno	0.13	IT Lecco	0.00
ES Canarias	-0.22	CL Del Maule	0.24	IT Grosseto	0.13	IT Rovigo	0.00
ES Avila, Palencia, Salaman	-0.23	AU Hunter	0.24	IT Pisa	0.12	IT Trieste	0.00
ES Valladolid	-0.26	US Sacramento	0.24	IT Arezzo	0.11	IT Savona	0.00
ES Zamora	-0.26	ZA Robertson	0.24	FR Aquitaine except Gi	0.10	IT Valle d'Aosta	0.00
ES Leon	-0.26	AU Riverina	0.22	EL Peloponnissos	0.09	IT L'Aquila	0.00
ES Region de Murcia	-0.27	US Merced	0.21	IT Frosinone	0.07	IT Bergamo	0.00
ES Alicante	-0.38	CL O'Higgins	0.18	IT Siena	0.07	IT Livorno	0.00
ES Comunidad Foral de Na\	-0.38	FR Aude	0.18	IT Campobasso	0.07	IT Pesaro e Urbino	-0.01
ES Galicia	-0.52	FR Franche Comté	0.18	IT Bari	0.06	IT Rieti	-0.01

Table 86 (cont.): NVIIs for the top 25 regions for the world's top 24 white varieties, 2000

Rkatsiteli		**Sauvignon Blanc**		**Cayetana Blanca**		**Catarratto Bianco**	
Region	*NVII*	*Region*	*NVII*	*Region*	*NVII*	*Region*	*NVII*
GE Georgia	3.93	MD Moldova	1.42	ES Badajoz	11.03	IT Trapani	7.56
RU Russia	2.53	FR Centre-Val de Loire	1.10	ES Zaragoza	0.02	IT Palermo	1.85
MD Moldova	2.10	CL Del Maule	0.84	ES Caceres	0.00	IT Agrigento	0.68
M9 Missing 9	1.93	FR Gironde	0.76	AU Mount Lofty Ranges	0.00	IT Catania	0.02
BG Bulgaria	1.66	NZ Marlborough	0.38	AU Hilltops	0.00	IT Messina	0.01
AM Armenia	0.47	RO Romania	0.34	AU Big Rivers - other	0.00	IT Caltanissetta	0.01
RO Romania	-0.52	FR Aquitaine except G	0.31	AU Swan District	0.00	IT La Spezia	0.00
		ZA Stellenbosch	0.29	AU Adelaide Hills	0.00	IT Genova	0.00
		M9 Missing 9	0.27	AU Riverland	-0.01	IT Massa-Carrara	0.00
		FR Herault	0.21	ES Guadalajara	-0.01	IT Imperia	0.00
		SI Slovenia	0.19	AU Swan Hill (VIC)	-0.01	IT Savona	0.00
		FR Bourgogne	0.16	AU Langhorne Creek	-0.01	IT Valle d'Aosta	0.00
		ZA Robertson	0.16	AU Murray Darling (NSW	-0.01	IT Novara	0.00
		ZA Swartland	0.12	ES Malaga	-0.01	IT Pistoia	0.00
		CL O'Higgins	0.11	AU McLaren Vale	-0.01	IT Lucca	0.00
		US Napa	0.11	AU Barossa Valley	-0.02	IT Vibo Valentia	0.00
		ZA Paarl	0.10	AU Riverina	-0.03	IT Livorno	0.00
		CL Valparaiso	0.09	ES Alava	-0.03	IT Catanzaro	0.00
		IT Udine	0.09	ES Burgos	-0.03	IT Rieti	0.00
		ZA Worcester	0.08	ES Canarias	-0.03	IT Siracusa	0.00
		IT Gorizia	0.08	AU Murray Darling (VIC	-0.03	IT Matera	0.00
		ZA Breedekloof	0.07	ES Avila, Palencia, Salar	-0.03	IT Torino	0.00
		US Sonoma	0.07	ES Valladolid	-0.03	IT Napoli	0.00
		US San Joaquin	0.06	ES Zamora	-0.04	IT Reggio di Calabria	0.00
		AU Margaret River	0.06	ES Leon	-0.04	IT Pisa	0.00

Table 86 (cont.): NVIIs for the top 25 regions for the world's top 24 white varieties, 2000

Macabeo		Chenin Blanc		Riesling		Colombard	
Region	*NVII*	*Region*	*NVII*	*Region*	*NVII*	*Region*	*NVII*
ES Tarragona	2.14	FR Pays de la Loire excep	1.03	DE Mosel	1.26	US Madera	1.10
ES Zaragoza	1.64	ZA Paarl	0.89	DE Pfalz	0.97	FR Gers	0.99
ES Barcelona	1.13	ZA Swartland	0.77	FR Alsace	0.67	US Fresno	0.94
ES La Rioja	1.11	FR Centre-Val de Loire	0.68	DE Rheingau	0.52	US Kern	0.62
FR Pyrenees-Orientales	0.87	ZA Olifants River	0.63	DE Württemberg	0.49	ZA Northern Cape	0.53
ES Girona, Lleida	0.35	ZA Breedekloof	0.58	DE Rheinhessen	0.48	ZA Olifants River	0.43
ES Comunidad Foral de	0.30	ZA Worcester	0.40	M9 Missing 9	0.32	ZA Robertson	0.40
ES Alava	0.28	ZA Robertson	0.39	DE Baden	0.24	ZA Breedekloof	0.31
ES Huesca, Teruel	0.21	US Madera	0.39	DE Nahe	0.23	US Stanislaus	0.23
ES Valladolid	0.20	ZA Stellenbosch	0.34	AT Niederosterreich	0.22	US Tulare	0.22
ES Badajoz	0.09	US Fresno	0.33	RU Russia	0.18	US Merced	0.20
ES Valencia	0.08	ZA Northern Cape	0.27	HU Hungary	0.18	ZA Worcester	0.18
ES Albacete	0.07	US Kern	0.26	CZ Czechia	0.14	ZA Little Karoo	0.16
ES Castellon	0.02	AR San Rafael	0.15	US Washington	0.14	US San Joaquin	0.14
ES Illes Balears	0.00	ZA Little Karoo	0.13	MD Moldova	0.11	AU Riverland	0.10
AR El Cuy	0.00	US Stanislaus	0.10	AU Clare Valley	0.10	IL Israel	0.09
ES Guipuzcoa, Vizcaya	0.00	US San Joaquin	0.10	DE Mittelrhein	0.08	ZA Paarl	0.09
AR General Roca	0.00	US Merced	0.10	CA Canada	0.08	FR Aquitaine except Giron	0.08
ES Guadalajara	-0.01	FR Deux-Sevres, Vienne	0.07	AU Barossa Valley	0.08	AU Riverina	0.07
ES Caceres	-0.01	US Monterey	0.06	US Monterey	0.06	FR Charente-Maritime	0.06
ES Burgos	-0.01	US Tulare	0.05	AU Eden Valley	0.06	AU Murray Darling (VIC)	0.06
ES Almeria, Granada, J	-0.02	AR Lavalle	0.04	AU Padthaway	0.05	ZA Swartland	0.05
FR Bouches-du-Rhone	-0.02	AR Rivadavia - Mza	0.04	DE Hessische Bergstraße	0.05	US Glenn	0.03
ES Avila, Palencia, Sal	-0.02	AR Maipú	0.03	NZ Marlborough	0.05	US Kings	0.02
FR Ardeche	-0.02	US Sacramento	0.03	DE Franken	0.04	AU Murray Darling (NSW)	0.02

Table 86 (cont.): NVIIs for the top 25 regions for the world's top 24 white varieties, 2000

Aligoté		Müller-Thurgau		Palomino Fino		Muscat Blanc à Petits Grains	
NVII	*NVII*	*Region*	*NVII*	*Region*	*NVII*	*Region*	*NVII*
MD Moldova	3.10	DE Rheinhessen	1.11	ES Cadiz	1.93	IT Cuneo	0.86
M9 Missing 9	1.30	DE Baden	0.84	ES Canarias	1.37	IT Asti	0.76
RO Romania	1.22	DE Pfalz	0.81	ES Galicia	0.83	FR Pyrenees-Orientales	0.57
FR Bourgogne	0.30	HU Hungary	0.55	ES Leon	0.53	EL Voreio Aigaio	0.38
RU Russia	0.29	DE Franken	0.49	ES Valladolid	0.36	IT Alessandria	0.30
BG Bulgaria	0.20	DE Mosel	0.47	ES Zamora	0.35	FR Herault	0.24
CH Geneva	0.00	AT Niederosterreich	0.39	ES Almeria, Granada, Ja	0.06	HU Hungary	0.21
FR Auvergne	0.00	SK Slovakia	0.36	FR Gers	0.03	PT Alto Tras-os-Montes	0.17
FR Alpes-de-Haute-Provence, H:	0.00	CZ Czechia	0.31	ES Huelva	0.02	FR Rhone-Alpes except	0.14
CH Vaud	-0.01	DE Nahe	0.18	ES Avila, Palencia, Salaı	0.01	IT Pavia	0.13
CH Valais	-0.01	AT Burgenland	0.14	US Madera	0.01	M9 Missing 9	0.10
FR Bouches-du-Rhone	-0.02	DE Württemberg	0.11	US San Bernardino	0.00	AM Armenia	0.09
FR Ardeche	-0.02	IT Trento	0.10	ES Cantabria	0.00	IT Padova	0.07
FR Midi-Pyrenees except Gers	-0.03	LU Luxembourg	0.09	US Riverside	0.00	ZA Robertson	0.07
FR Centre-Val de Loire	-0.03	AT Steiermark	0.07	NZ Auckland	0.00	IT Bari	0.06
GE Georgia	-0.04	NZ Gisborne	0.04	AR Puelen	0.00	IT Sassari	0.04
FR Pays de la Loire except May‹	-0.06	CH Zürich	0.04	ES Malaga	0.00	FR Corse	0.03
FR Charente	-0.06	IT Bolzano-Bozen	0.03	NZ Waikato	0.00	FR Vaucluse	0.03
FR Rhone-Alpes except Ardeche	-0.06	DE Saale	0.03	US Contra Costa	0.00	IT Potenza	0.03
FR Vaucluse	-0.08	NZ Hawkes Bay	0.03	ES Principado de Asturi:	0.00	EL Dytiki Ellada	0.03
FR Gard	-0.10	CH Aargau	0.03	AU Lower Murray - othe	0.00	IT Salerno	0.02
FR Herault	-0.16	DE Sachsen	0.02	AU Beechworth	0.00	IT Matera	0.02
		CH Schaffhausen	0.02	AU Sunbury	0.00	IT Piacenza	0.02
		CH Thurgau	0.01	AU Greater Perth - other	0.00	EL Anatoliki Makedonia	0.01
		DE Rheingau	0.01	AU Rutherglen	0.00	AT Steiermark	0.01

Table 86 (cont.): NVIIs for the top 25 regions for the world's top 24 white varieties, 2000

Muscat of Alexandria		Sémillon		Fetească Albă		Grüner Veltliner	
Region	*NVII*	*Region*	*NVII*	*Region*	*NVII*	*Region*	*NVII*
MA Morocco	0.69	FR Gironde	1.76	RO Romania	3.50	AT Niederosterreich	2.94
FR Pyrenees-Orientales	0.52	FR Aquitaine except (	0.87	MD Moldova	0.80	SK Slovakia	0.59
ES Alicante	0.44	AU Riverina	0.39	HU Hungary	0.11	AT Burgenland	0.55
ES Valencia	0.37	CL Del Maule	0.20	SK Slovakia	0.05	CZ Czechia	0.34
ZA Breedekloof	0.23	AU Barossa Valley	0.15			HU Hungary	0.19
AU Riverland	0.20	AU Hunter	0.13			AT Wien and other regions	0.04
ES Malaga	0.19	AU Margaret River	0.08			IT Pescara	0.01
AR Veinticinco de May(	0.18	CL O'Higgins	0.07			IT La Spezia	0.00
ZA Olifants River	0.16	AU Murray Darling (V	0.06			IT Genova	0.00
IT Trapani	0.14	AU Riverland	0.06			IT Imperia	0.00
US Fresno	0.12	ZA Breedekloof	0.05			IT Oristano	0.00
AR Caucete	0.12	US Washington	0.04			IT Sondrio	0.00
US Kern	0.11	AR Tupungato	0.04			IT Rieti	0.00
BR Brazil	0.10	AU Clare Valley	0.04			IT Bolzano-Bozen	0.00
AR Sarmiento - San Juai	0.09	AR Luján de Cuyo	0.03			AT Steiermark	0.00
AR Chilecito	0.09	AU McLaren Vale	0.03			IT Teramo	0.00
ZA Little Karoo	0.09	AU Mudgee	0.02			IT Caserta	0.00
ZA Northern Cape	0.08	AR San Carlos - Mza	0.02			IT Frosinone	0.00
AU Murray Darling (VI(	0.08	AU Cowra	0.02			IT Grosseto	0.00
US Tulare	0.08	AU Great Southern	0.02			IT Viterbo	0.00
AR Albardón	0.08	BR Brazil	0.02			IT Ancona	0.00
AR Zonda	0.07	NZ Marlborough	0.02			IT Campobasso	0.00
AR San Martín - San Jua	0.06	AU Adelaide Hills	0.02			IT Sassari	0.00
ZA Worcester	0.06	AU Southern NSW - (	0.02			IT Cosenza	0.00
AU Murray Darling (NS	0.05	ZA Stellenbosch	0.02			IT Latina	-0.01

Table 86 (cont.): NVIIs for the top 25 regions for the world's top 24 white varieties, 2000

Trebbiano Romagnolo		**Pedro Ximénez**		**Pinot Blanc**		**Garganega**	
Region	*NVII*	*Region*	*NVII*	*Region*	*NVII*	*Region*	*NVII*
IT Ravenna	2.62	ES Cordoba	2.48	RU Russia	0.57	IT Verona	1.67
IT Bologna	0.52	CL Coquimbo	0.45	IT Treviso	0.25	IT Trapani	0.89
IT Forlì-Cesena	0.47	ES Badajoz	0.14	FR Alsace	0.23	IT Vicenza	0.46
IT Rimini	0.10	ES Malaga	0.14	AT Niederosterreich	0.23	IT Agrigento	0.06
IT Ferrara	0.06	ES Almeria, Granada, J	0.06	AT Burgenland	0.20	IT Padova	0.05
IT Piacenza	0.02	CL Atacama	0.03	DE Baden	0.18	IT Palermo	0.04
IT Modena	0.01	ES Valencia	0.02	SK Slovakia	0.12	IT Foggia	0.03
IT Mantova	0.01	ES Caceres	0.02	AT Steiermark	0.11	IT Bari	0.01
IT Agrigento	0.00	AU Rutherglen	0.00	DE Pfalz	0.11	IT Taranto	0.00
IT Pesaro e Urbino	0.00	AU Swan District	0.00	IT Bolzano-Bozen	0.10	IT Mantova	0.00
IT Rovigo	0.00	AU Fleurieu - other	0.00	IT Pordenone	0.07	IT Perugia	0.00
IT Padova	0.00	AU Barossa Valley	0.00	IT Brescia	0.07	IT Matera	0.00
IT Macerata	0.00	AU Greater Perth - othe	0.00	DE Rheinhessen	0.07	IT Ascoli Piceno	0.00
IT Rieti	0.00	AU Lower Murray - oth	0.00	IT Udine	0.06	AR Ullum	0.00
IT Parma	0.00	ES Principado de Astur	0.00	IT Venezia	0.05	IT Rovigo	0.00
IT Terni	0.00	AU Sunbury	0.00	IT Vicenza	0.05	IT Milano	0.00
IT Cremona	0.00	ES Castellon	0.00	IT Gorizia	0.05	IT Massa-Carrara	0.00
IT Lecco	0.00	AU Tasmania	0.00	IT Padova	0.04	IT Campobasso	0.00
IT Genova	0.00	AU Currency Creek	0.00	US Monterey	0.03	IT Imperia	0.00
IT Savona	0.00	AU Eden Valley	0.00	IT Trento	0.03	IT Valle d'Aosta	0.00
IT Massa-Carrara	0.00	AU Clare Valley	0.00	LU Luxembourg	0.03	IT Bergamo	0.00
IT Milano	0.00	AU Adelaide Hills	0.00	IT Bologna	0.03	IT Isernia	0.00
IT Imperia	0.00	ES Guadalajara	0.00	CA Canada	0.02	IT Lucca	0.00
IT Isernia	0.00	AU Swan Hill (VIC)	0.00	DE Nahe	0.02	IT Sondrio	0.00
IT Novara	0.00	AU McLaren Vale	0.00	IT Campobasso	0.02	IT Brescia	0.00

Table 87: NVIIs for the top 25 regions for the world's top 24 white varieties, 2016

Airén		Chardonnay		Sauvignon Blanc		Trebbiano Toscano	
Region	*NVII*	*Region*	*NVII*	*Region*	*NVII*	*Region*	*NVII*
ES Castilla-La Mancha	39.33	FR Bourgogne	3.16	NZ Marlborough	3.92	FR Poitou Charentes	15.82
ES Comunidad de Madri	1.49	FR Champagne-Ardenne	1.95	CL Del Maule	1.33	IT Puglia	3.05
TN Tunisia	-0.02	US Monterey	1.36	FR Centre-Val de Loire	1.33	IT Abruzzo	0.90
MA Morocco	-0.08	US Sonoma	1.16	MD Moldova	1.03	IT Lazio	0.86
ES Canarias	-0.11	US San Joaquin	0.94	CL Valparaiso	0.77	IT Umbria	0.38
ES Región de Murcia	-0.12	AU Riverland	0.77	FR Aquitaine	0.72	FR Midi Pyrénées	0.32
ES Andalucía	-0.22	AU Riverina	0.64	ZA Stellenbosch	0.49	IT Toscana	0.30
ES Cataluña	-0.53	FR Languedoc Roussillon	0.51	FR Midi Pyrénées	0.37	IT Marche	0.18
ES Comunidad Valencia	-0.57	IT Lombardia	0.45	FR Languedoc Roussillon	0.27	UY Canelones	0.09
ES Castilla y León	-0.73	US Napa	0.45	CL O'Higgins	0.26	IT Molise	0.07
ES Extremadura	-0.78	AU Murray Darling (VIC)	0.45	ZA Robertson	0.25	IN India	0.05
		IT Trento	0.43	UA Ukraine	0.19	IT Campania	0.05
		US Santa Barbara	0.42	RU Crimea	0.19	IT Sicilia	0.02
		CL Valparaiso	0.39	NZ Hawkes Bay	0.17	UY Montevideo	0.01
		US Madera	0.39	ZA Swartland	0.17	BG Northeast	0.01
		US Yolo	0.39	ZA Cape South Coast	0.16	UY San Jose	0.00
		CL Del Maule	0.34	ZA Breedekloof	0.16	UY Paysandu	0.00
		US Mendocino	0.34	IT Friuli-Venezia Giulia	0.16	UY Colonia	0.00
		US Sacramento	0.33	US Lake	0.15	EL Anatoliki Makedon	0.00
		US Yakima Valley	0.27	AU Adelaide Hills	0.15	US Los Angeles	0.00
		ZA Robertson	0.26	AU Margaret River	0.15	UY Rivera	0.00
		AU Murray Darling (NSW)	0.25	US Napa	0.14	UY Florida	0.00
		CL O'Higgins	0.25	FR Bourgogne	0.14	US Tehama	0.00
		RU Krasnodar Krai	0.23	ZA Paarl	0.12	AR Colón - Cba	0.00
		LB Lebanon	0.18	CZ Jihovýchod	0.12	AU Rutherglen	0.00

Table 87 (cont.): NVIIs for the top 25 regions for the world's top 24 white varieties, 2016

Riesling		Rkatsiteli		Macabeo		Cayetana Blanca	
Region	*NVII*	*Region*	*NVII*	*Region*	*NVII*	*Region*	*NVII*
DE Pfalz	1.10	GE Georgia	5.52	ES Cataluña	2.43	ES Extremadura	7.23
DE Mosel	1.09	RU Crimea	1.22	ES Castilla-La Mancha	1.82	ES Andalucía	0.12
FR Alsace	0.84	UA Ukraine	1.22	ES Extremadura	0.88	ES Comunidad de Madı	0.06
RO Romania	0.82	MD Moldova	0.66	ES Comunidad Valenciana	0.45	PT Algarve	0.00
DE Rheinhessen	0.80	KZ Almaty	0.57	ES Aragón	0.44	ES Canarias	-0.02
DE Rheingau	0.51	BG North Central	0.31	ES La Rioja	0.41	ES País Vasco	-0.02
DE Württemberg	0.41	BG Southeast	0.31	ES País Vasco	0.15	AU Riverland	-0.03
US Yakima Valley	0.26	BG Northeast	0.20	ES Castilla y León	0.08	ES Comunidad Foral de	-0.03
DE Nahe	0.23	KZ South Kazakhstan	0.19	ES Comunidad Foral de Navarra	0.04	ES Aragón	-0.04
UA Ukraine	0.23	BG South Central	0.13	ES Illes Balears	0.00	PT Alentejo	-0.05
RU Crimea	0.23	RU Rostov Oblast	0.12	ES Cantabria	0.00	ES La Rioja	-0.07
CZ Jihovýchod	0.20	BG Northwest	0.10	ES Región de Murcia	-0.02	ES Comunidad Valenci	-0.10
DE Baden	0.19	MK NorthMacedon	0.04	ES Comunidad de Madrid	-0.02	PT Norte	-0.11
AU Clare Valley	0.17	KZ Zhambyl	0.02	ES Canarias	-0.02	PT Centro	-0.12
CA Ontario	0.16	BG Southwest	0.01	ZA Swartland	-0.02	ES Castilla y León	-0.13
MD Moldova	0.13	RS South Banat	0.01	ZA Paarl	-0.03	ES Castilla-La Mancha	-0.31
RU Krasnodar Krai	0.13	KZ East Kazakhstan	0.00	ES Andalucía	-0.05		
MK NorthMacedon	0.13	RS Nišava	0.00	ES Galicia	-0.06		
HR Kontinentalna Hrvatska	0.12	KZ Other regions	0.00	FR Rhône Alpes	-0.10		
US Monterey	0.12	RU Krasnodar Krai	-0.02	FR Languedoc Roussillon	-0.10		
US Horse Heaven Hills	0.10	RO Romania	-0.38	FR Provence-Alpes-Cote d'Azur	-0.18		
SI Stajerska Slovenija	0.10						
AU Eden Valley	0.10						
AT Weinviertel	0.08						
AT Kamptal	0.08						

Table 87 (cont.): NVIIs for the top 25 regions for the world's top 24 white varieties, 2016

Muscat Alexandria		**Muscat Blanc à Petits Grains**		**Chenin Blanc**		**Colombard**	
Region	*NVII*	*Region*	*NVII*	*Region*	*NVII*	*Region*	*NVII*
CL Del Bio Bio	0.85	IT Piemonte	2.24	FR Pays de la Loire	1.08	FR Midi Pyrénées	1.15
ES Comunidad Valenciai	0.85	FR Languedoc Roussillon	0.77	FR Centre-Val de Loire	0.73	US Fresno	0.81
ES Andalucía	0.81	EL Voreio Aigaio	0.28	ZA Olifants River	0.62	ZA Olifants River	0.56
MA Morocco	0.44	FR Rhône Alpes	0.19	ZA Breedekloof	0.62	US Madera	0.51
CN China	0.36	HU Tokaj	0.11	ZA Paarl	0.60	ZA Northern Cape	0.49
CL Coquimbo	0.23	IT Lombardia	0.10	ZA Swartland	0.56	ZA Robertson	0.44
US Fresno	0.19	AU Riverina	0.09	ZA Worcester	0.42	ZA Breedekloof	0.42
AU Riverland	0.16	IT Puglia	0.09	ZA Robertson	0.34	ZA Worcester	0.26
IT Sicilia	0.15	SI Stajerska Slovenija	0.08	ZA Stellenbosch	0.28	US Kern	0.19
EL Voreio Aigaio	0.13	PE Arequipa	0.07	ZA Northern Cape	0.25	ZA Little Karoo	0.17
ZA Breedekloof	0.12	ZA Robertson	0.07	US Fresno	0.12	AU Riverland	0.13
AR Veinticinco de Mayo	0.11	US Fresno	0.05	ZA Little Karoo	0.10	FR Poitou Charentes	0.13
FR Languedoc Roussillor	0.11	MK NorthMacedon	0.05	AR San Rafael	0.10	US Tulare	0.09
AU Murray Darling (VIC	0.08	RO Romania	0.05	MX Baja California	0.06	AU Riverina	0.07
TN Tunisia	0.08	FR Corse	0.04	US Yolo	0.05	ZA Paarl	0.06
PT Lisboa	0.08	AT Südsteiermark	0.04	US Madera	0.04	AU Murray Darling (VI	0.06
ES Cataluña	0.08	RU Crimea	0.03	AR Lavalle	0.04	IL Israel	0.04
ZA Olifants River	0.07	UA Ukraine	0.03	AU Swan District	0.04	AU Murray Darling (NS	0.04
AR Chilecito	0.06	EL Dytiki Ellada	0.03	US Sacramento	0.02	ZA Swartland	0.03
AU Murray Darling (NSV	0.06	US Tulare	0.03	AR Rivadavia - Mza	0.02	US Kings	0.02
AR Sarmiento - San Juan	0.06	MX Zacatecas	0.03	AR Maipú	0.01	US Merced	0.02
US Kern	0.06	MX Aguascalientes	0.03	ZA Cape South Coast	0.01	US Stanislaus	0.02
AR Caucete	0.05	TR Aegean	0.01	ET Ethiopia	0.01	AU Lower Murray - oth	0.01
AU Riverina	0.04	AT Vulkanland Steiermark	0.01	US Kings	0.01	AU Swan Hill (VIC)	0.00
IL Israel	0.04	AT Weinviertel	0.01	AR Santa Rosa - Mza	0.01	US San Joaquin	0.00

Table 87 (cont.): NVIIs for the top 25 regions for the world's top 24 white varieties, 2016

Catarratto Bianco			Aligoté			Graševina			Palomino Fino		
	Region	*NVII*		*Region*	*NVII*		*Region*	*NVII*		*Region*	*NVII*
IT	Sicilia	6.25	MD	Moldova	1.62	HR	Kontinentalna Hr	0.99	ES	Andalucía	2.01
US	Merced	0.00	RO	Romania	1.06	CN	China	0.45	ES	Canarias	1.23
IT	Molise	-0.01	UA	Ukraine	1.04	SI	Stajerska Sloveni	0.26	ES	Galicia	0.66
IT	Calabria	-0.01	RU	Crimea	1.04	CZ	Jihovýchod	0.23	ES	Castilla y León	0.41
IT	Marche	-0.02	FR	Bourgogne	0.37	AT	Weinviertel	0.22	PT	Norte	0.20
US	Napa	-0.02	RU	Rostov Oblast	0.13	IT	Lombardia	0.21	PT	Centro	0.19
IT	Campania	-0.02	RU	Krasnodar Krai	0.06	AT	Neusiedlersee	0.17	MX	Coahuila	0.02
IT	Friuli-Venezia Giulia	-0.03	KZ	Almaty	0.05	HU	Balatonfured-Cso	0.16	MX	Sonora	0.01
US	Sonoma	-0.03	BG	North Central	0.02	RS	South Banat	0.14	ZA	Olifants River	0.01
US	San Joaquin	-0.03	BG	Southeast	0.02	ES	Extremadura	0.14	ES	Cantabria	0.00
IT	Piemonte	-0.07	CH	Geneva	0.00	HU	Badacsony	0.13	US	Riverside	0.00
IT	Toscana	-0.08	KZ	South Kazakhstan	0.00	RS	Tri Morave	0.12	AR	Puelen	0.00
IT	Emilia-Romagna	-0.08	KZ	East Kazakhstan	0.00	RO	Romania	0.10	AU	Rutherglen	0.00
IT	Veneto	-0.12	KZ	West Kazakhstan	0.00	HU	Matra	0.08	NZ	Auckland	0.00
IT	Puglia	-0.13	KZ	Zhambyl	0.00	HU	Balatonfelvidek	0.08	ZA	Little Karoo	0.00
			FR	Lorraine	0.00	AT	Südsteiermark	0.08	NZ	Northland	0.00
			CA	Other regions	0.00	RS	Srem	0.07	US	Contra Costa	0.00
			FR	Auvergne	0.00	AT	Vulkanland Steie	0.07	CH	Graubünden - Me	0.00
			KZ	Other regions	0.00	AT	Neusiedlersee Hü	0.06	AR	Guaymallén	0.00
			FR	Franche Comté	0.00	SK	Bratislavský kraj	0.05	AR	Las Heras	0.00
			CA	Ontario	0.00	HU	Eger	0.04	AR	Albardón	0.00
			CH	Vaud	0.00	HU	Villany	0.04	US	Santa Clara	0.00
			CH	Valais	-0.01	SI	Prekmurje	0.04	PT	Lisboa	0.00
			HU	Matra	-0.01	SI	Vipavska dolina	0.04	AU	Lower Murray - o	0.00
			FR	Centre-Val de Loire	-0.03	HU	Balatonboglar	0.04	AR	Rawson	0.00

Table 87 (cont.): NVIIs for the top 25 regions for the world's top 24 white varieties, 2016

Prosecco		Müller-Thurgau		Grüner Veltliner		Trebbiano Romagnolo	
Region	*NVII*	*Region*	*NVII*	*Region*	*NVII*	*Region*	*NVII*
IT Veneto	4.22	DE Rheinhessen	0.83	AT Weinviertel	1.47	IT Emilia-Romagna	4.18
IT Friuli-Venezia Giulia	0.07	DE Baden	0.50	AT Kamptal	0.44	IT Liguria	0.00
AU King Valley	0.02	DE Pfalz	0.40	CZ Jihovýchod	0.33	IT Molise	0.00
BR Brazil	0.01	DE Franken	0.35	AT Wagram	0.29	IT Basilicata	0.00
AU Alpine Valleys	0.00	CZ Jihovýchod	0.28	AT Kremstal	0.29	IT Calabria	-0.01
AU Tumbarumba	0.00	DE Mosel	0.22	SK Západné Slovensko	0.24	IT Umbria	-0.01
AU Central Victoria - other	0.00	IT Trento	0.16	AT Wachau	0.17	IT Marche	-0.01
AU Gippsland	0.00	HU Matra	0.13	AT Neusiedlersee	0.16	IT Lazio	-0.01
AU Beechworth	0.00	AT Weinviertel	0.11	SK Bratislavský kraj	0.11	IT Campania	-0.02
AU Macedon Ranges	0.00	DE Nahe	0.11	AT Traisental	0.11	IT Sardegna	-0.02
AU Hilltops	0.00	SK Západné Slovensko	0.10	AT Neusiedlersee Hügellai	0.09	IT Lombardia	-0.02
AU Goulburn Valley	0.00	LU Luxembourg	0.07	HU Balatonboglar	0.05	IT Abruzzo	-0.03
AU Currency Creek	0.00	DE Württemberg	0.06	AT Carnuntum	0.04	IT Toscana	-0.05
AU Orange	0.00	AT Kamptal	0.05	HU Tolna	0.04	IT Veneto	-0.08
AU Murray Darling (NSW)	0.00	IT Bolzano-Bozen	0.04	AT Thermenregion	0.04	IT Sicilia	-0.08
AU Adelaide Hills	0.00	AT Wagram	0.04	AT Wien	0.04	IT Puglia	-0.08
IT Basilicata	0.00	AT Neusiedlersee	0.03	HU Matra	0.03		
IT Molise	0.00	HU Kunsag	0.03	HU Etyek-Budai	0.03		
AU McLaren Vale	-0.01	CH Zürich	0.03	HU Sopron	0.02		
IT Calabria	-0.01	AT Kremstal	0.03	HU Kunsag	0.02		
AU Murray Darling (VIC)	-0.01	AT Südsteiermark	0.02	AT Mittelburgenland	0.01		
IT Trento	-0.01	HU Eger	0.02	HU Neszmely	0.01		
AR San Rafael	-0.01	DE Saale	0.02	HU Zala	0.01		
IT Umbria	-0.01	SI Stajerska Slovenija	0.02	HU Bukk	0.01		
AR Maipú	-0.01	AT Vulkanland Steiern	0.02	HU Csongrad	0.01		

Table 87 (cont.): NVIIs for the top 25 regions for the world's top 24 white varieties, 2016

Sémillon		**Verdejo**		**Viognier**		**Pedro Giménez**	
Region	*NVII*	*Region*	*NVII*	*Region*	*NVII*	*Region*	*NVII*
FR Aquitaine	2.08	ES Castilla y León	2.48	FR Languedoc Roussill	0.67	CL Coquimbo	0.88
AU Riverina	0.33	ES Castilla-La Mancha	0.89	FR Midi Pyrénées	0.63	AR San Martín - Mza	0.43
AU Margaret River	0.14	ES Extremadura	0.07	IT Sicilia	0.27	AR Lavalle	0.31
TR Marmara	0.12	ES Comunidad Foral de	0.01	FR Rhône Alpes	0.23	AR Rivadavia - Mza	0.29
AU Hunter	0.10	AU Swan Hill (VIC)	0.00	FR Provence-Alpes-Cot	0.09	AR Junín - Mza	0.20
AU Murray Darling (VIC	0.09	AU Eden Valley	0.00	US San Joaquin	0.05	AR San Rafael	0.14
ZA Breedekloof	0.08	AU Murray Darling (NSV	-0.01	CL O'Higgins	0.05	AR Santa Rosa - Mza	0.14
AU Riverland	0.07	AU Barossa Valley	-0.01	ZA Paarl	0.04	AR Veinticinco de Mayo - S	0.14
CL Del Maule	0.06	CL Metropolitana	-0.01	CL Del Maule	0.03	AR Sarmiento - San Juan	0.13
AU Barossa Valley	0.05	ES País Vasco	-0.01	US Virginia	0.02	AR Maipú	0.13
AR Tupungato	0.03	ES Andalucía	-0.02	US San Luis Obispo	0.02	AR Caucete	0.07
AU Great Southern	0.02	AU Riverland	-0.02	ZA Swartland	0.02	CL Atacama	0.06
ZA Stellenbosch	0.02	ES Región de Murcia	-0.02	US Yolo	0.02	AR San Martín - San Juan	0.06
AR Luján de Cuyo	0.02	ES La Rioja	-0.02	ZA Stellenbosch	0.02	AR Luján de Cuyo	0.05
AU Clare Valley	0.02	ES Aragón	-0.03	CA British Colombia	0.02	AR Las Heras	0.05
ZA Swartland	0.02	ES Comunidad Valencia	-0.05	US Santa Barbara	0.02	AR Nueve de Julio	0.04
CL O'Higgins	0.02			US Rogue Valley	0.01	AR General Alvear	0.03
ZA Paarl	0.02			IT Umbria	0.01	AR Rawson	0.03
ZA Cape South Coast	0.01			US Texas High Plains a	0.01	AR Angaco	0.03
ZA Worcester	0.01			ZA Worcester	0.01	AR Pocito	0.01
AU Murray Darling (NS'	0.01			NZ Hawkes Bay	0.01	AR San Carlos - Mza	0.01
AR Maipú	0.01			AU Riverland	0.01	AR Guaymallén	0.01
AR San Carlos - Mza	0.01			US Mendocino	0.01	AR General Roca	0.01
AU Adelaide Hills	0.01			AU Murray Darling (NS	0.01	AR Tunuyán	0.01
AU Mudgee	0.01			AU Riverina	0.01	AR Santa Lucía	0.01

IX. Varietal Regional Similarity Indexes, by country and region

Table 88: VSI of each country and region relative of the world, 2000

Region	Varietal Similarity Index	Region	Varietal Similarity Index
Algeria	**0.36**	**Argentina (cont.)**	
		Poman	0.06
Argentina	**0.24**	Puelen	0.37
Adolfo Alsina	0.08	Rawson	0.08
Albardón	0.09	Rivadavia - Mza	0.17
Andalgala	0.05	Rivadavia - San Juan	0.11
Añelo	0.34	San Alberto	0.05
Angaco	0.10	San Blas De Los Sauces	0.05
Arauco	0.00	San Carlos - Mza	0.24
Avellaneda - Río Negro	0.22	San Carlos - Salta	0.05
Ayacucho	0.05	San Javier	0.05
Belén	0.04	San Martín - Mza	0.15
Cachi	0.02	San Martín - San Juan	0.14
Cafayate	0.19	San Rafael	0.20
Calingasta	0.27	Sanagasta	0.04
Capital San Juan	0.06	Santa Lucía	0.08
Capital Santiago del Ester	0.00	Santa María - Catamarca	0.31
Castro Barros	0.34	Santa Rosa - Mza	0.22
Caucete	0.15	Sarmiento - San Juan	0.20
Chilecito	0.17	Tafí del Valle	0.03
Chimbas	0.09	Tinogasta	0.07
Colón - Entre Ríos	0.13	Totoral	0.12
Conesa	0.09	Tunuyán	0.39
Confluencia	0.15	Tupungato	0.38
Coronel Felipe Varela	0.06	Ullum	0.15
Coronel Pringles	0.33	Valle Fértil	0.03
Cruz del Eje	0.03	Veinticinco de Mayo - Sai	0.17
Cushamen	0.31	Vinchina	0.04
El Cuy	0.32	Zonda	0.16
Famatina	0.04		
General Alvear	0.12	**Armenia**	**0.40**
General Lamadrid	0.09		
General Pueyrredón	0.40	**Australia**	**0.46**
General Roca	0.27	Adelaide Hills	0.45
Godoy Cruz	0.04	Alpine Valleys/Beechwori	0.49
Guaymallén	0.19	Australian Capital Territoi	0.34
Iglesia	0.05	Barossa - other	0.33
Ischilin	0.17	Barossa Valley	0.36
Jachal	0.22	Beechworth	0.44
Junín - Mza	0.17	Bendigo	0.29
La Paz	0.24	Big Rivers - other	0.55
La Rioja	0.02	Blackwood Valley	0.44
Las Heras	0.18	Canberra District	0.42
Lavalle	0.18	Central Ranges - other	0.56
Luján de Cuyo	0.24	Central Victoria - other	0.38
Maipú	0.30	Central Western Australia	0.39
Molinos	0.14	Clare Valley	0.38
Nueve de Julio	0.15	Cowra	0.35
Pichi Mahuida	0.09	Currency Creek	0.38
Pocito	0.11	Eastern Plains, Inland and	0.37

Table 88 (cont.): VSI of each country and region relative of the world, 2000

Region	Varietal Similarity Index
Australia (cont.)	
Eden Valley	0.31
Far North - other	0.27
Fleurieu - other	0.43
Geelong	0.32
Geographe	0.44
Gippsland	0.37
Goulburn Valley	0.40
Grampians	0.32
Granite Belt	0.55
Great Southern	0.43
Greater Perth - other	0.39
Hastings River	0.42
Henty	0.33
Hilltops	0.38
Hunter	0.30
Hunter Valley - other	0.35
Kangaroo Island	0.35
Langhorne Creek	0.38
Limestone Coast - other	0.38
Lower Murray - other	0.40
Margaret River	0.43
McLaren Vale	0.38
Mornington Peninsula	0.34
Mount Benson	0.44
Mount Lofty Ranges - oth	0.41
Mudgee	0.38
Murray Darling (NSW)	0.41
Murray Darling (VIC)	0.30
North East Victoria - othe	0.50
North West Victoria - oth	0.28
Northern Rivers - other	0.44
Northern Slopes - other	0.36
Northern Territory	0.47
Orange	0.40
Padthaway	0.40
Perricoota	0.43
Perth Hills	0.30
Port Phillip - other	0.38
Pyrenees	0.37
Queensland - other	0.52
Riverina	0.36
Riverland	0.42
Rutherglen	0.27
South Burnett	0.59
South Coast - other	0.46
South West Australia - otl	0.47
Southern Fleurieu	0.37
Southern NSW - other	0.34
Sunbury	0.36

Region	Varietal Similarity Index
Australia (cont.)	
Swan District	0.53
Swan Hill (NSW)	0.35
Swan Hill (VIC)	0.39
Tasmania	0.34
The Peninsulas	0.38
Tumbarumba	0.28
Western Australia Southe	0.39
Western Plains - other	0.60
Western Victoria - other	0.41
Yarra Valley	0.39
Austria	**0.12**
Burgenland	0.15
Niederosterreich	0.09
Steiermark	0.15
Wien and other regions	0.23
Brazil	**0.23**
Bulgaria	**0.35**
Canada	**0.47**
Chile	**0.47**
Araucania	0.19
Atacama	0.35
Coquimbo	0.40
Del Bio Bio	0.26
Del Maule	0.39
Metropolitana	0.39
O'Higgins	0.40
Valparaiso	0.37
Croatia	**0.38**
Cyprus	**0.12**
Czechia	**0.38**

Table 88 (cont.): VSI of each country and region relative of the world, 2000

Region	*Varietal Similarity Index*	*Region*	*Varietal Similarity Index*
France	**0.59**	**Greece**	**0.20**
Aisne	0.05	Anatoliki Makedonia, Thr.	0.34
Alpes-de-Haute-Provence,	0.50	Attiki	0.02
Alsace	0.08	Dytiki Ellada	0.14
Aquitaine except Gironde	0.37	Dytiki Makedonia	0.08
Ardeche	0.49	Ionia Nisia	0.42
Aude	0.38	Ipeiros	0.15
Auvergne	0.11	Kentriki Makedonia	0.25
Bouches-du-Rhone	0.47	Kriti	0.03
Bourgogne	0.22	Notio Aigaio	0.22
Centre-Val de Loire	0.17	Peloponnissos	0.17
Champagne-Ardenne	0.16	Sterea Ellada	0.02
Charente	0.18	Thessalia	0.04
Charente-Maritime	0.19	Voreio Aigaio	0.05
Correze, Haute-Vienne	0.42		
Corse	0.39	**Hungary**	**0.51**
Deux-Sevres, Vienne	0.20		
Franche Comté	0.19	**Israel**	**0.54**
Gard	0.48		
Gers	0.21	**Italy**	**0.37**
Gironde	0.39	Agrigento	0.16
Herault	0.41	Alessandria	0.08
Lorraine	0.11	Ancona	0.04
Midi-Pyrenees except Ger:	0.22	Arezzo	0.14
Pays de la Loire except M:	0.10	Ascoli Piceno	0.19
Pyrenees-Orientales	0.36	Asti	0.07
Rhone-Alpes except Arde	0.20	Avellino	0.04
Seine-et-Marne	0.11	Bari	0.15
Var	0.43	Belluno	0.38
Vaucluse	0.35	Benevento	0.18
		Bergamo	0.43
Georgia	**0.14**	Biella	0.07
		Bologna	0.09
Germany	**0.11**	Bolzano-Bozen	0.14
Ahr	0.11	Brescia	0.36
Baden	0.12	Brindisi	0.06
Franken	0.06	Cagliari	0.16
Hessische Bergstraße	0.08	Caltanissetta	0.07
Mittelrhein	0.07	Campobasso	0.11
Mosel	0.07	Caserta	0.26
Nahe	0.09	Catania	0.02
Pfalz	0.09	Catanzaro	0.16
Rheingau	0.07	Chieti	0.14
Rheinhessen	0.08	Como	0.10
Saale	0.12	Cosenza	0.17
Sachsen	0.09	Cremona	0.05
Württemberg	0.09	Crotone	0.02
		Cuneo	0.06

Table 88 (cont.): VSI of each country and region relative of the world, 2000

Region	*Varietal Similarity Index*	*Region*	*Varietal Similarity Index*
Italy (cont.)		**Italy (cont.)**	
Enna	0.31	Roma	0.15
Ferrara	0.14	Rovigo	0.30
Firenze	0.14	Salerno	0.15
Foggia	0.19	Sassari	0.22
Forli-Cesena	0.11	Savona	0.09
Frosinone	0.34	Siena	0.13
Genova	0.05	Siracusa	0.02
Gorizia	0.29	Sondrio	0.02
Grosseto	0.23	Taranto	0.21
Imperia	0.05	Teramo	0.06
Isernia	0.18	Terni	0.37
La Spezia	0.37	Torino	0.10
L'Aquila	0.05	Trapani	0.09
Latina	0.28	Trento	0.31
Lecce	0.03	Treviso	0.26
Lecco	0.39	Trieste	0.06
Livorno	0.39	Udine	0.32
Lodi	0.08	Valle d'Aosta	0.08
Lucca	0.26	Varese	0.17
Macerata	0.14	Venezia	0.35
Mantova	0.29	Verbano-Cusio-Ossola	0.08
Massa-Carrara	0.19	Vercelli	0.02
Matera	0.14	Verona	0.09
Messina	0.26	Vibo Valentia	0.25
Milano	0.11	Vicenza	0.28
Modena	0.04	Viterbo	0.21
Napoli	0.14		
Novara	0.03	**Korea, Rep.**	**0.07**
Nuoro	0.28		
Oristano	0.23	**Luxembourg**	**0.08**
Padova	0.33		
Palermo	0.15	**Moldova**	**0.32**
Parma	0.10		
Pavia	0.15	**Morocco**	**0.25**
Perugia	0.39		
Pesaro e Urbino	0.11	**New Zealand**	**0.34**
Pescara	0.06	Auckland	0.48
Piacenza	0.07	Canterbury	0.24
Pisa	0.22	Gisborne	0.31
Pistoia	0.15	Hawkes Bay	0.45
Pordenone	0.37	Marlborough	0.21
Potenza	0.11	Nelson	0.27
Prato	0.19	Otago	0.19
Ragusa	0.08	Waikato	0.59
Ravenna	0.04	Waipara	0.28
Reggio di Calabria	0.22	Wairarapa	0.30
Reggio nell'Emilia	0.02		
Rieti	0.19		
Rimini	0.11		

Table 88 (cont.): VSI of each country and region relative of the world, 2000

Region	*Varietal Similarity Index*	*Region*	*Varietal Similarity Index*
Portugal	**0.44**	**Spain**	
Acores	0.26	Cuenca	0.49
Alentejo	0.23	Galicia	0.28
Algarve	0.25	Girona, Lleida	0.41
Alto Tras-os-Montes	0.45	Guadalajara	0.41
Beira Interior	0.43	Guipuzcoa, Vizcaya	0.03
Beira Litoral	0.28	Huelva	0.03
Entre Douro e Minho	0.23	Huesca, Teruel	0.38
Madeira	0.26	Illes Balears	0.34
Ribatejo e Oeste	0.28	La Rioja	0.29
		Leon	0.16
Romania	**0.55**	Malaga	0.45
		Principado de Asturias	0.20
Russia	**0.48**	Region de Murcia	0.19
		Tarragona	0.19
Serbia	**0.26**	Toledo	0.56
		Valencia	0.23
Slovakia	**0.32**	Valladolid	0.42
		Zamora	0.47
Slovenia	**0.55**	Zaragoza	0.33
South Africa	**0.29**	**Switzerland**	**0.13**
Breedekloof	0.22	Aargau	0.11
Little Karoo	0.32	Basel Land	0.14
Northern Cape	0.11	Bern	0.12
Olifants River	0.18	Fribourg	0.05
Paarl	0.41	Geneva	0.10
Robertson	0.44	Graubünden	0.12
Stellenbosch	0.59	Jura	0.30
Swartland	0.36	Lucerne	0.10
Worcester	0.34	Neuchâtel	0.08
		Schaffhausen	0.10
Spain	**0.69**	Schwyz	0.18
Alava	0.17	St. Gallen	0.12
Albacete	0.55	Thurgau	0.11
Alicante	0.41	Ticino	0.30
Almeria, Granada, Jaen, S	0.50	Valais	0.10
Avila, Palencia, Salamanc	0.41	Vaud	0.05
Badajoz	0.10	Zürich	0.12
Barcelona	0.20	other regions	0.10
Burgos	0.24		
Caceres	0.21	**Taiwan**	**0.00**
Cadiz	0.05		
Canarias	0.13	**Tunisia**	**0.35**
Cantabria	0.06		
Castellon	0.48	**United Kingdom**	**0.35**
Ciudad Real	0.55		
Comunidad de Madrid	0.56		
Comunidad Foral de Nava	0.46		
Cordoba	0.04		

Table 88 (cont.): VSI of each country and region relative of the world, 2000

Region	*Varietal Similarity Index*	*Region*	*Varietal Similarity Index*
United States	**0.42**	**United States (cont.)**	
Alameda	0.36	Stanislaus	0.33
Amador	0.08	Sutter	0.19
Benton Co.	0.15	Tehama	0.29
Butte	0.27	Trinity	0.42
Calaveras	0.38	Tulare	0.25
Chautauqua-Erie	0.02	Valley - other	0.20
Columbia River	0.46	Ventura	0.13
Colusa	0.07	Washington	0.44
Contra Costa	0.25	Washington Co.	0.23
Douglas Co.	0.41	Yamhill Co.	0.16
El Dorado	0.39	Yolo	0.31
Finger Lakes	0.11	Yuba	0.33
Fresno	0.21		
Glenn	0.30	**Uruguay**	**0.21**
Humboldt	0.21		
Josephine Co.	0.25	**Missing 9**	**0.41**
Kern	0.22		
Kings	0.20		
Lake	0.40		
Lane Co.	0.18		
Los Angeles	0.27		
Madera	0.27		
Marin	0.41		
Marion Co.	0.18		
Mariposa	0.43		
Mendocino	0.38		
Merced	0.37		
Michigan	0.41		
Monterey	0.33		
Napa	0.45		
Nevada	0.42		
New York - other	0.31		
Oregon - other	0.42		
Placer	0.15		
Polk Co.	0.18		
Riverside	0.27		
Sacramento	0.41		
San Benito	0.43		
San Bernardino	0.15		
San Diego	0.42		
San Joaquin	0.29		
San Luis Obispo	0.43		
San Mateo	0.28		
Santa Barbara	0.26		
Santa Clara	0.42		
Santa Cruz	0.26		
Shasta	0.23		
Solano	0.42		
Sonoma	0.40		

Table 89: VSI of each country and region relative of the world, 2010

Region	*Varietal Similarity Index*	*Region*	*Varietal Similarity Index*
Algeria	**0.36**	**Argentina (cont.)**	
		Lavalle	0.33
Argentina	**0.39**	Leandro Alem	0.06
Adolfo Alsina	0.48	Luján de Cuyo	0.31
Albardón	0.13	Maipú	0.43
Ambato	0.45	Malargüe	0.14
Andalgala	0.05	Molinos	0.24
Añelo	0.44	Nogoya	0.37
Angaco	0.13	Nueve de Julio	0.23
Arauco	0.04	Pichi Mahuida	0.22
Avellaneda - Río Negro	0.41	Picunches	0.37
Belén	0.05	Pocito	0.19
Cachi	0.23	Poman	0.08
Cafayate	0.31	Puelen	0.57
Calamuchita	0.63	Rawson	0.13
Calingasta	0.43	Rivadavia - Mza	0.28
Capital San Juan	0.06	Rivadavia - San Juan	0.19
Castro Barros	0.50	San Blas De Los Sauces	0.09
Caucete	0.25	San Carlos - Mza	0.31
Chilecito	0.30	San Carlos - Salta	0.07
Chimbas	0.13	San Javier	0.32
Chos Malal	0.39	San Martín - Mza	0.25
Collon Cura	0.18	San Martín - San Juan	0.23
Colón - Cba	0.26	San Rafael	0.36
Colón - Entre Ríos	0.45	Sanagasta	0.02
Concordia	0.25	Santa Lucía	0.10
Conesa	0.11	Santa María - Catamarca	0.45
Confluencia	0.45	Santa María - Cba	0.44
Coronel Felipe Varela	0.10	Santa Rosa - Mza	0.32
Coronel Suarez	0.34	Sarmiento - San Juan	0.32
Cruz del Eje	0.03	Tafí del Valle	0.35
Cushamen	0.44	Tandil	0.36
El Cuy	0.40	Tilcara	0.30
Famatina	0.18	Tinogasta	0.10
General Alvear	0.21	Tornquist	0.34
General Lamadrid	0.22	Totoral	0.14
General Roca	0.41	Tulumba	0.17
Godoy Cruz	0.05	Tumbaya	0.41
Guaymallén	0.31	Tunuyán	0.45
Iglesia	0.05	Tupungato	0.48
Ischilin	0.30	Ullum	0.37
Jachal	0.35	Valle Fértil	0.04
Junín - Mza	0.27	Veinticinco de Mayo - Mis	0.07
Junín - San Luis	0.26	Veinticinco de Mayo - San	0.28
La Paz	0.40	Victoria	0.38
La Rioja	0.06	Villarino	0.50
La Viña	0.29	Vinchina	0.04
Las Heras	0.31	Zonda	0.26

Table 89 (cont.): VSI of each country and region relative of the world, 2010

Region	*Varietal Similarity Index*	*Region*	*Varietal Similarity Index*
Armenia	**0.19**	**Australia (cont.)**	
		Macedon Ranges	0.44
Australia	**0.62**	Manjimup	0.63
Adelaide Hills	0.58	Margaret River	0.63
Adelaide Plains	0.60	McLaren Vale	0.48
Alpine Valleys	0.68	Mornington Peninsula	0.35
Australian Capital Territor	0.48	Mount Benson	0.65
Barossa - other	0.34	Mount Lofty Ranges - othe	0.46
Barossa Valley	0.45	Mudgee	0.62
Beechworth	0.67	Murray Darling (NSW)	0.61
Bendigo	0.47	Murray Darling (VIC)	0.59
Big Rivers - other	0.54	New England Australia	0.31
Blackwood Valley	0.60	North East Victoria - other	0.69
Canberra District (ACT)	0.54	North West Victoria - othe	0.62
Canberra District (NSW)	0.61	Northern Rivers - other	0.44
Central Ranges - other	0.63	Northern Slopes - other	0.66
Central Victoria - other	0.43	Orange	0.65
Central Western Australia	0.57	Padthaway	0.63
Clare Valley	0.51	Peel	0.63
Coonawarra	0.55	Pemberton	0.48
Cowra	0.46	Perricoota	0.58
Currency Creek	0.58	Perth Hills	0.48
Eastern Plains, Inland and	0.37	Port Phillip - other	0.46
Eden Valley	0.46	Pyrenees	0.49
Far North - other	0.45	Queensland - other	0.56
Fleurieu - other	0.52	Riverina	0.54
Geelong	0.41	Riverland	0.59
Geographe	0.65	Robe	0.61
Gippsland	0.46	Rutherglen	0.41
Glenrowan	0.54	Shoalhaven Coast	0.54
Goulburn Valley	0.61	South Burnett	0.56
Grampians	0.39	South Coast - other	0.58
Granite Belt	0.70	South West Australia - oth	0.60
Great Southern	0.60	Southern Fleurieu	0.56
Greater Perth - other	0.40	Southern Flinders Ranges	0.33
Gundagai	0.47	Southern Highlands	0.69
Hastings River	0.44	Southern NSW - other	0.48
Heathcote	0.41	Strathbogie Ranges	0.62
Henty	0.37	Sunbury	0.54
Hilltops	0.58	Swan District	0.56
Hunter	0.44	Swan Hill (NSW)	0.60
Hunter Valley - other	0.37	Swan Hill (VIC)	0.60
Kangaroo Island	0.54	Tasmania	0.34
King Valley	0.71	The Peninsulas	0.58
Langhorne Creek	0.59	Tumbarumba	0.40
Limestone Coast - other	0.63	Upper Goulburn	0.60
Lower Murray - other	0.56	Western Australia Southea	0.51

Table 89 (cont.): VSI of each country and region relative of the world, 2010

Region	*Varietal Similarity Index*	*Region*	*Varietal Similarity Index*
Australia (cont.)		**China**	**0.47**
Western Plains - other	0.48	Beijing	0.41
Western Victoria - other	0.62	Gansu	0.53
Wrattonbully	0.61	Ningxia	0.45
Yarra Valley	0.52	Shandong	0.41
		ShanXi	0.58
Austria	**0.15**	Sichuan	0.41
Burgenland	0.20	Tianjin	0.41
Niederosterreich	0.09	Xinjiang	0.53
Steiermark	0.27	Yantai	0.41
Wien and other regions	0.27	other regions	0.44
Brazil	**0.14**	**Croatia**	**0.32**
		Dalmatinska Zagora	0.12
Bulgaria	**0.55**	Hrvatsko Primorje	0.09
North Central	0.53	Istra	0.15
Northeast	0.39	Moslavina	0.21
Northwest	0.52	Plesivica	0.26
South Central	0.52	Podunavlje	0.18
Southeast	0.51	Pokuplje	0.16
Southwest	0.23	Prigorje - Bilogora	0.17
		Sjeverna Dalmacija	0.33
Canada	**0.51**	Slavonija	0.16
British Colombia	0.61	Srednja Juzna Dalmacija	0.06
Ontario	0.35	Zagorje-Medimurje	0.22
		other regions	0.22
Chile	**0.60**		
Araucania	0.33	**Cyprus**	**0.11**
Atacama	0.32		
Coquimbo	0.57	**Czechia**	**0.26**
De Los Lagos	0.33	Cechy	0.22
Del Bio Bio	0.37	Morava	0.26
Del Maule	0.58		
Metropolitana	0.55	**Ethiopia**	**0.12**
O'Higgins	0.58		
Valparaiso	0.42		

Table 89 (cont.): VSI of each country and region relative of the world, 2010

Region	*Varietal Similarity Index*	*Region*	*Varietal Similarity Index*
France	**0.73**	**France (cont.)**	
Ain	0.22	Lot	0.16
Aisne	0.09	Lot-et-Garonne	0.55
Allier	0.18	Lozere	0.38
Alpes-de-Haute-Provence,	0.55	Maine-et-Loire	0.17
Alpes-Maritimes	0.25	Marne	0.25
Ardeche	0.61	Mayenne	0.31
Ariege	0.45	Meurthe-et-Moselle	0.13
Aube	0.18	Meuse	0.23
Aude	0.62	Moselle	0.23
Aveyron	0.12	Nievre	0.19
Bas-Rhin	0.12	Puy-de-Dome	0.10
Bouches-du-Rhone	0.52	Pyrenees-Atlantiques	0.08
Cantal	0.33	Pyrenees-Orientales	0.43
Charente	0.17	Rhone	0.06
Charente-Maritime	0.18	Saone-et-Loire	0.31
Cher	0.20	Sarthe	0.07
Correze	0.28	Savoie	0.08
Corse	0.16	Seine-et-Marne	0.23
Cote-d'Or	0.22	Tarn	0.32
Deux-Sevres	0.20	Tarn-et-Garonne	0.15
Dordogne	0.45	Var	0.46
Doubs	0.37	Vaucluse	0.37
Drome	0.40	Vendee	0.31
Eure-et-Loire	0.26	Vienne	0.34
Gard	0.60	Vosges	0.00
Gers	0.27	Yonne	0.30
Gironde	0.51		
Haute-Corse	0.48	**Georgia**	**0.11**
Haute-Garonne	0.21		
Haute-Loire	0.16	**Germany**	**0.16**
Haute-Marne	0.20	Ahr	0.15
Hautes-Alpes	0.37	Baden	0.18
Haute-Saone	0.37	Franken	0.07
Haute-Savoie	0.06	Hessische Bergstraße	0.13
Hautes-Pyrenees	0.19	Mittelrhein	0.10
Haut-Rhin	0.12	Mosel	0.09
Herault	0.65	Nahe	0.13
Indre	0.32	Pfalz	0.16
Indre-et-Loire	0.12	Rheingau	0.10
Isere	0.20	Rheinhessen	0.13
Jura	0.28	Saale	0.10
Landes	0.26	Sachsen	0.12
Loire	0.18	Württemberg	0.11
Loire-Atlantique	0.05		
Loiret	0.33		
Loir-et-Cher	0.24		

Table 89 (cont.): VSI of each country and region relative of the world, 2010

Region	*Varietal Similarity Index*	*Region*	*Varietal Similarity Index*
Greece	**0.22**	**Italy**	**0.45**
Anatoliki Makedonia, Thr:	0.65	Agrigento	0.30
Attiki	0.02	Alessandria	0.08
Dytiki Ellada	0.07	Ancona	0.10
Dytiki Makedonia	0.09	Arezzo	0.18
Ionia Nisia	0.20	Ascoli Piceno	0.19
Ipeiros	0.18	Asti	0.08
Kentriki Makedonia	0.49	Avellino	0.04
Kriti	0.11	Bari	0.12
Notio Aigaio	0.15	Barletta-Andria-Trani	0.16
Peloponnissos	0.11	Belluno	0.16
Sterea Ellada	0.06	Benevento	0.13
Thessalia	0.16	Bergamo	0.61
Voreio Aigaio	0.09	Biella	0.06
		Bologna	0.18
Hungary	**0.35**	Bolzano-Bozen	0.31
Badacsony	0.20	Brescia	0.40
Balatonboglar	0.49	Brindisi	0.09
Balatonfelvidek	0.18	Cagliari	0.26
Balatonfured-Csopak	0.23	Caltanissetta	0.09
Bukk	0.19	Campobasso	0.18
Csongrad	0.13	Carbonia-Iglesias	0.13
Eger	0.39	Caserta	0.17
Etyek-Budai	0.40	Catania	0.07
Hajos-bajai	0.33	Catanzaro	0.11
Kunsag	0.13	Chieti	0.15
Matra	0.28	Como	0.35
Mor	0.19	Cosenza	0.13
Nagy-Somlo	0.19	Cremona	0.20
Neszmely	0.34	Crotone	0.03
Pannonhalma	0.25	Cuneo	0.08
Pecs	0.41	Enna	0.20
Sopron	0.13	Fermo	0.20
Szekszard	0.37	Ferrara	0.15
Tokaj	0.02	Firenze	0.17
Tolna	0.44	Foggia	0.22
Villany	0.49	Forlì-Cesena	0.14
Zala	0.24	Frosinone	0.29
		Genova	0.13
Israel	**0.52**	Gorizia	0.39
		Grosseto	0.23
		Imperia	0.07
		Isernia	0.23
		La Spezia	0.16
		L'Aquila	0.07
		Latina	0.30
		Lecce	0.05

Table 89 (cont.): VSI of each country and region relative of the world, 2010

Region	*Varietal Similarity Index*	*Region*	*Varietal Similarity Index*
Lecce	0.05	**Italy (cont.)**	
Lecco	0.57	Taranto	0.12
Livorno	0.59	Teramo	0.08
Lodi	0.11	Terni	0.44
Lucca	0.33	Torino	0.09
Macerata	0.25	Trapani	0.18
Mantova	0.36	Trento	0.37
Massa-Carrara	0.35	Treviso	0.15
Matera	0.25	Trieste	0.11
Medio Campidano	0.12	Udine	0.42
Messina	0.18	Valle d'Aosta	0.21
Milano	0.23	Varese	0.42
Modena	0.02	Venezia	0.41
Monza e della Brianza	0.14	Verbano-Cusio-Ossola	0.21
Napoli	0.06	Vercelli	0.03
Novara	0.05	Verona	0.10
Nuoro	0.27	Vibo Valentia	0.09
Ogliastra	0.26	Vicenza	0.50
Olbia-Tempio	0.11	Viterbo	0.27
Oristano	0.26		
Padova	0.44	**Japan**	**0.52**
Palermo	0.21	Hokkaido	0.03
Parma	0.19	Nagano	0.46
Pavia	0.18	Yamagata	0.46
Perugia	0.36	Yamanashi	0.26
Pesaro e Urbino	0.10	other regions	0.51
Pescara	0.08		
Piacenza	0.11	**Kazakhstan**	**0.14**
Pisa	0.22	Almaty	0.14
Pistoia	0.17	East Kazakhstan	0.27
Pordenone	0.33	South Kazakhstan	0.10
Potenza	0.07	West Kazakhstan	0.22
Prato	0.22	Zhambyl	0.20
Ragusa	0.12	other regions	0.25
Ravenna	0.07		
Reggio di Calabria	0.18	**Korea, Rep.**	**0.04**
Reggio nell'Emilia	0.03		
Rieti	0.32	**Luxembourg**	**0.07**
Rimini	0.18		
Roma	0.21	**Mexico**	**0.52**
Rovigo	0.34	Aguascalientes	0.09
Salerno	0.19	Baja California	0.65
Sassari	0.26	Coahuila	0.38
Savona	0.06	Sonora	0.01
Siena	0.17	Zacatecas	0.04
Siracusa	0.05		
Sondrio	0.02		

Table 89 (cont.): VSI of each country and region relative of the world, 2010

Region	*Varietal Similarity Index*	*Region*	*Varietal Similarity Index*
Moldova	**0.44**	**Russia**	**0.48**
		Krasnodar Krai	0.51
Morocco	**0.15**	Rostov Oblast	0.12
Myanmar	**0.38**	**Serbia**	**0.15**
New Zealand	**0.31**	**Slovakia**	**0.22**
Auckland	0.61	Juznoslovenska	0.26
Canterbury	0.30	Malokarpatska	0.16
Gisborne	0.39	Nitrianska	0.19
Hawkes Bay	0.61	Stredné Slovensko	0.21
Marlborough	0.21	Tokajska	0.03
Nelson	0.32	Východné Slovensko	0.21
Otago	0.17		
Waikato	0.65	**Slovenia**	**0.42**
Waipara	0.25	Bela Krajina	0.23
Wairarapa	0.26	Bizeljsko Sremic	0.20
other regions	0.19	Dolenjska	0.10
		Goriska brda	0.44
Peru	**0.03**	Kras	0.05
Arequipa	0.03	Prekmurje	0.26
Lima	0.02	Slovenska Istra	0.13
Moquegua	0.01	Stajerska Slovenija	0.31
Tacna	0.04	Vipavska dolina	0.54
		other regions	0.19
Portugal	**0.29**		
Acores	0.01	**South Africa**	0.51
Alentejo	0.33	Breedekloof	0.39
Algarve	0.11	Little Karoo	0.30
Alto Tras-os-Montes	0.19	Northern Cape	0.13
Beira Interior	0.14	Olifants River	0.29
Beira Litoral	0.14	Paarl	0.58
Entre Douro e Minho	0.02	Robertson	0.52
Madeira	0.01	Stellenbosch	0.66
Ribatejo e Oeste	0.21	Swartland	0.53
		Worcester	0.44
Romania	**0.33**		
Bucuresti - Ilfov	0.21		
Centru	0.20		
Nord-Est	0.22		
Nord-Vest	0.21		
Sud - Muntenia	0.31		
Sud-Est	0.39		
Sud-Vest Oltenia	0.27		
Vest	0.47		

Table 89 (cont.): VSI of each country and region relative of the world, 2010

Region	*Varietal Similarity Index*	*Region*	*Varietal Similarity Index*
Spain	**0.63**	**Switzerland (cont.)**	
Alava	0.46	Lucerne	0.17
Albacete	0.53	Neuchâtel	0.15
Alicante	0.21	Schaffhausen	0.16
Almeria, Granada, Jaen, S(	0.43	Schwyz	0.20
Avila, Palencia, Salamanc:	0.42	St. Gallen	0.18
Badajoz	0.25	Thurgau	0.16
Barcelona	0.26	Ticino	0.41
Burgos	0.33	Valais	0.19
Caceres	0.54	Vaud	0.07
Cadiz	0.05	Zürich	0.17
Canarias	0.08	other regions	0.18
Cantabria	0.18		
Castellon	0.49	**Taiwan**	**0.01**
Ciudad Real	0.45		
Comunidad de Madrid	0.56	**Thailand**	**0.31**
Comunidad Foral de Nava	0.48		
Cordoba	0.05	**Tunisia**	**0.26**
Cuenca	0.47		
Galicia	0.16	**Turkey**	**0.24**
Girona, Lleida	0.67	Aegean	0.27
Guadalajara	0.48	Central East	0.00
Guipuzcoa, Vizcaya	0.03	Central North	0.08
Huelva	0.01	Central South	0.01
Huesca, Teruel	0.69	Marmara	0.15
Illes Balears	0.56	Mediterranean	0.55
La Rioja	0.38	South East	0.00
Leon	0.07		
Malaga	0.08	**Ukraine**	**0.36**
Principado de Asturias	0.18		
Region de Murcia	0.15	**United Kingdom**	**0.27**
Tarragona	0.35		
Toledo	0.46	**United States**	**0.67**
Valencia	0.23	Alameda	0.64
Valladolid	0.28	Amador	0.15
Zamora	0.40	Arizona	0.21
Zaragoza	0.47	Arkansas	0.21
		Benton Co.	0.17
Switzerland	**0.22**	Butte	0.48
Aargau	0.17	Calaveras	0.70
Basel Land	0.17	Chautauqua-Erie	0.02
Bern	0.16	Colorado	0.67
Fribourg	0.11	Columbia Gorge	0.38
Geneva	0.20	Columbia River	0.64
Graubünden	0.17	Columbia Valley	0.66
Jura	0.25		

Table 89 (cont.): VSI of each country and region relative of the world, 2010

Region	*Varietal Similarity Index*	*Region*	*Varietal Similarity Index*
United States (cont.)		**United States (cont.)**	
Colusa	0.22	San Benito	0.51
Contra Costa	0.42	San Bernardino	0.08
Douglas Co.	0.28	San Diego	0.63
El Dorado	0.58	San Joaquin	0.48
Finger Lakes	0.15	San Luis Obispo	0.66
Fresno	0.26	San Mateo	0.36
Georgia	0.21	Santa Barbara	0.45
Glenn	0.34	Santa Clara	0.68
Horse Heaven Hills	0.69	Santa Cruz	0.37
Humboldt	0.63	Shasta	0.54
Illinois	0.13	Siskiyou	0.55
Indiana	0.15	Snipes Mountain	0.68
Iowa	0.00	Solano	0.52
Jackson Co.	0.56	Sonoma	0.60
Josephine Co.	0.29	Stanislaus	0.60
Kentucky	0.28	Sutter	0.21
Kern	0.39	Tehama	0.35
Kings	0.20	Texas	0.71
Lake	0.54	Trinity	0.53
Lake Chelan	0.40	Tulare	0.40
Lane Co.	0.15	Tuolumne	0.43
Los Angeles	0.60	Ventura	0.33
Madera	0.49	Virginia	0.56
Marin	0.19	Wahluke Slope	0.67
Marion Co.	0.17	Walla Walla Valley	0.61
Mariposa	0.62	Washington Co.	0.18
Mendocino	0.59	Willamette Valley - other	0.21
Merced	0.58	Yakima Valley	0.57
Michigan	0.36	Yamhill Co.	0.17
Minnesota	0.03	Yolo	0.46
Missouri	0.05	Yuba	0.48
Monterey	0.52		
Napa	0.62	**Uruguay**	**0.35**
Nevada	0.65		
New York - other	0.48		
North Carolina	0.21		
Ohio	0.07		
Orange	0.14		
Pennsylvania	0.12		
Placer	0.25		
Polk Co.	0.17		
Puget Sound	0.25		
Rattlesnake Hills	0.54		
Red Mountain	0.57		
Riverside	0.69		
Sacramento	0.66		

Table 90: VSI of each country and region relative of the world, 2016

Region	Varietal Similarity Index	Region	Varietal Similarity Index
Algeria	**0.43**	**Argentina (cont.)**	
		Famatina	0.19
Argentina	**0.37**	Futaleufu	0.32
Adolfo Alsina	0.44	General Alvear	0.24
Albardón	0.12	General Belgrano	0.07
Andalgala	0.08	General Lamadrid	0.17
Añelo	0.45	General Pueyrredón	0.41
Angaco	0.11	General Roca	0.39
Avellaneda - Río Negro	0.38	Godoy Cruz	0.07
Ayacucho	0.25	Gualeguaychu	0.16
Balcarce	0.35	Guaymallén	0.34
Bariloche	0.30	Humahuaca	0.27
Belén	0.08	Iglesia	0.07
Benito Juárez	0.07	Ischilin	0.35
Cachi	0.14	Junín - Bs. As.	0.06
Cafayate	0.26	Junín - Mza	0.28
Cainguas	0.02	Junín - San Luis	0.31
Calamuchita	0.38	La Paz	0.42
Calingasta	0.41	La Viña	0.24
Cañuelas	0.10	Lacar	0.40
Capayán	0.07	Languiñeo	0.28
Capital Misiones	0.08	Las Heras	0.30
Capital San Juan	0.04	Lavalle	0.32
Capital San Luis	0.38	Leandro Alem	0.07
Capital Santiago del Ester(	0.00	Luján de Cuyo	0.26
Castro Barros	0.46	Maipú	0.38
Caucete	0.23	Molinos	0.20
Chilecito	0.30	Nogoya	0.38
Chimbas	0.12	Ñorquin	0.38
Chos Malal	0.26	Nueve de Julio	0.21
Collon Cura	0.19	Paraná	0.44
Colón - Cba	0.38	Pehuenches	0.07
Colón - Entre Ríos	0.42	Pichi Mahuida	0.19
Concordia	0.30	Picún Leufú	0.27
Conesa	0.06	Picunches	0.35
Confluencia	0.37	Pocito	0.18
Coronel Felipe Varela	0.12	Poman	0.10
Coronel Pringles	0.28	Puelen	0.39
Coronel Suarez	0.40	Punilla	0.17
Cruz del Eje	0.17	Rawson	0.11
Curaco	0.46	Rivadavia - Mza	0.28
Cushamen	0.46	Rivadavia - San Juan	0.21
Daireaux	0.51	Saavedra	0.64
De La Costa	0.07	San Alberto	0.26
Diamante	0.13	San Blas De Los Sauces	0.13
El Carmen	0.58	San Carlos - Mza	0.24
El Cuy	0.35	San Carlos - Salta	0.11

Table 90 (cont.): VSI of each country and region relative of the world, 2016

Region	Varietal Similarity Index	Region	Varietal Similarity Index
Argentina (cont.)		**Australia (cont.)**	
San Javier	0.29	Blackwood Valley	0.62
San Martín - Mza	0.25	Canberra District (ACT)	0.54
San Martín - San Juan	0.22	Canberra District (NSW)	0.57
San Rafael	0.40	Central Ranges - other	0.61
Sanagasta	0.05	Central Victoria - other	0.38
Santa Lucía	0.09	Central Western Australia	0.45
Santa María - Catamarca	0.46	Clare Valley	0.53
Santa María - Cba	0.48	Coonawarra	0.56
Santa Rosa - Catamarca	0.04	Cowra	0.49
Santa Rosa - Mza	0.29	Currency Creek	0.59
Sarmiento - Chubut	0.31	Eastern Plains, Inland and	0.26
Sarmiento - San Juan	0.31	Eden Valley	0.49
Tafí del Valle	0.30	Far North - other	0.24
Tandil	0.46	Fleurieu - other	0.50
Tilcara	0.21	Geelong	0.38
Tinogasta	0.12	Geographe	0.69
Tornquist	0.35	Gippsland	0.47
Trancas	0.07	Glenrowan	0.51
Tulumba	0.37	Goulburn Valley	0.62
Tumbaya	0.43	Grampians	0.36
Tunuyán	0.32	Granite Belt	0.65
Tupungato	0.37	Great Southern	0.62
Ullum	0.41	Greater Perth - other	0.52
Uruguay	0.63	Gundagai	0.44
Valle Fértil	0.24	Hastings River	0.32
Valle Viejo	0.23	Heathcote	0.40
Veinticinco de Mayo - Mis	0.08	Henty	0.36
Veinticinco de Mayo - Sar	0.25	Hilltops	0.60
Victoria	0.37	Hunter	0.41
Villa Gesell	0.07	Hunter Valley - other	0.32
Villarino	0.53	Kangaroo Island	0.56
Vinchina	0.05	King Valley	0.67
Zonda	0.24	Langhorne Creek	0.59
		Limestone Coast - other	0.64
Armenia	**0.02**	Lower Murray - other	0.58
		Macedon Ranges	0.46
Australia	**0.63**	Manjimup	0.43
Adelaide Hills	0.53	Margaret River	0.64
Adelaide Plains	0.57	McLaren Vale	0.46
Alpine Valleys	0.68	Mornington Peninsula	0.33
Barossa - other	0.41	Mount Benson	0.64
Barossa Valley	0.41	Mount Gambier	0.37
Beechworth	0.63	Mount Lofty Ranges - othe	0.47
Bendigo	0.49	Mudgee	0.65
Big Rivers - other	0.56	Murray Darling (NSW)	0.64
		Murray Darling (VIC)	0.64

Table 90 (cont.): VSI of each country and region relative of the world, 2016

Region	*Varietal Similarity Index*	*Region*	*Varietal Similarity Index*
Australia (cont.)		**Austria**	**0.15**
New England Australia	0.58	Bergland	0.43
North East Victoria - other	0.47	Carnuntum	0.21
North West Victoria - othe	0.56	Kamptal	0.08
Northern Rivers - other	0.15	Kremstal	0.07
Northern Slopes	0.64	Mittelburgenland	0.11
Orange	0.69	Neusiedlersee	0.20
Padthaway	0.59	Neusiedlersee Hügelland	0.22
Peel	0.72	Steirerland - other	0.05
Pemberton	0.46	Südburgenland	0.19
Perricoota	0.60	Südsteiermark	0.29
Perth Hills	0.68	Thermenregion	0.24
Port Phillip - other	0.63	Traisental	0.07
Pyrenees	0.55	Vulkanland Steiermark	0.24
Queensland - other	0.46	Wachau	0.07
Riverina	0.57	Wagram	0.09
Riverland	0.61	Weinviertel	0.08
Robe	0.63	Weststeiermark	0.06
Rutherglen	0.40	Wien	0.21
Shoalhaven Coast	0.53	other regions	0.10
South Burnett	0.56		
South Coast - other	0.63	**Brazil**	**0.10**
South West Australia - oth	0.69		
Southern Fleurieu	0.58	**Bulgaria**	**0.55**
Southern Flinders Ranges	0.34	North Central	0.33
Southern Highlands	0.60	Northeast	0.30
Southern NSW - other	0.57	Northwest	0.55
Strathbogie Ranges	0.56	South Central	0.56
Sunbury	0.57	Southeast	0.53
Swan District	0.38	Southwest	0.25
Swan Hill (NSW)	0.62		
Swan Hill (VIC)	0.57	**Cambodia**	**0.47**
Tasmania	0.35		
The Peninsulas	0.61	**Canada**	**0.42**
Tumbarumba	0.35	British Colombia	0.63
Upper Goulburn	0.64	Nova Scotia	0.24
Western Australia Southea	0.48	Ontario	0.25
Western Plains	0.55	Quebec	0.12
Western Victoria - other	0.65	other regions	0.25
Wrattonbully	0.63		
Yarra Valley	0.53		

Table 90 (cont.): VSI of each country and region relative of the world, 2016

Region	*Varietal Similarity Index*	*Region*	*Varietal Similarity Index*
Chile	**0.64**	**France (cont.)**	
Antofagasta	0.29	Midi Pyrénées	0.39
Araucania	0.34	Pays de la Loire	0.19
Arica	0.01	Picardie	0.16
Atacama	0.05	Poitou Charentes	0.20
Coquimbo	0.15	Provence-Alpes-Cote d'Az	0.43
De Los Lagos	0.34	Rhône Alpes	0.36
Del Bio Bio	0.18		
Del Maule	0.62	**Georgia**	**0.10**
Metropolitana	0.58		
O'Higgins	0.61	**Germany**	**0.18**
Tarapaca	0.05	Ahr	0.16
Valparaiso	0.43	Baden	0.19
		Franken	0.07
China	**0.57**	Hessische Bergstraße	0.16
		Mittelrhein	0.11
Croatia	**0.20**	Mosel	0.11
Jadranska Hrvatska	0.20	Nahe	0.15
Kontinentalna Hrvatska	0.11	Pfalz	0.19
		Rheingau	0.11
Cyprus	**0.01**	Rheinhessen	0.15
		Saale	0.12
Czechia	**0.22**	Sachsen	0.14
Cechy	0.18	Württemberg	0.13
Jihovýchod	0.22		
Morava	0.21	**Greece**	**0.19**
Praha	0.16	Anatoliki Makedonia, Thr	0.67
Severozápad	0.13	Attiki	0.03
		Dytiki Ellada	0.06
Ethiopia	**0.11**	Dytiki Makedonia	0.12
		Ionia Nisia	0.20
France	**0.78**	Ipeiros	0.50
Alsace	0.15	Kentriki Makedonia	0.44
Aquitaine	0.54	Kriti	0.06
Auvergne	0.19	Notio Aigaio	0.03
Bourgogne	0.35	Peloponnissos	0.10
Centre-Val de Loire	0.26	Sterea Ellada	0.08
Champagne-Ardenne	0.26	Thessalia	0.07
Corse	0.45	Voreio Aigaio	0.14
Franche Comté	0.28		
Île de France	0.12		
Languedoc Roussillon	0.69		
Limousin	0.40		
Lorraine	0.15		

Table 90 (cont.): VSI of each country and region relative of the world, 2016

Region	*Varietal Similarity Index*
Hungary	**0.32**
Badacsony	0.15
Balatonboglar	0.50
Balatonfelvidek	0.16
Balatonfured-Csopak	0.21
Bukk	0.18
Csongrad	0.14
Eger	0.39
Etyek-Budai	0.41
Hajos-bajai	0.22
Kunsag	0.08
Matra	0.27
Mor	0.25
Nagy-Somlo	0.14
Neszmely	0.36
Pannonhalma	0.22
Pecs	0.44
Sopron	0.15
Szekszard	0.40
Tokaj	0.02
Tolna	0.44
Villany	0.51
Zala	0.39
India	**0.29**
Israel	**0.61**
Italy	**0.50**
Abruzzo	0.13
Basilicata	0.07
Bolzano-Bozen	0.32
Calabria	0.07
Campania	0.07
Emilia-Romagna	0.16
Friuli-Venezia Giulia	0.40
Lazio	0.26
Liguria	0.15
Lombardia	0.37
Marche	0.19
Molise	0.13
Piemonte	0.16
Puglia	0.26
Sardegna	0.26
Sicilia	0.22
Toscana	0.22
Italy (cont.)	
Trento	0.43
Umbria	0.40
Valle d'Aosta	0.28
Veneto	0.35
Japan	**0.20**
Hokkaido	0.13
Iwate	0.17
Nagano	0.21
Niigata	0.54
Yamagata	0.22
Yamanashi	0.06
other regions	0.19
Kazakhstan	**0.13**
Almaty	0.13
East Kazakhstan	0.35
South Kazakhstan	0.10
West Kazakhstan	0.34
Zhambyl	0.20
other regions	0.30
Korea, Rep.	**0.07**
Lebanon	**0.73**
Luxembourg	**0.14**
Mexico	
Aguascalientes	**0.06**
Baja California	0.72
Coahuila	0.40
Sonora	0.01
Zacatecas	0.04
Moldova	**0.54**
Morocco	**0.32**
Myanmar	**0.38**

Table 90 (cont.): VSI of each country and region relative of the world, 2016

Region	*Varietal Similarity Index*	*Region*	*Varietal Similarity Index*
New Zealand	**0.29**	**Serbia**	**0.61**
Auckland	0.65	Bačka	0.54
Canterbury	0.29	Banat	0.57
Gisborne	0.35	Belgrade	0.60
Hawkes Bay	0.61	Čačak-Kraljevo	0.31
Marlborough	0.22	Knjaževac	0.55
Nelson	0.29	Leskovac	0.59
Northland	0.54	Mlava	0.54
Otago	0.18	Negotinska Krajina	0.47
Waikato	0.49	Niš	0.52
Waipara	0.30	Nišava	0.32
Wairarapa	0.27	South Banat	0.15
		Srem	0.58
North Macedonia	**0.14**	Subotica	0.60
		Šumadija	0.62
Norway	**0.03**	Telečka	0.66
		Tisa	0.14
Peru	**0.04**	Toplica	0.48
Arequipa	0.03	Tri Morave	0.55
Lima	0.02	Valjevo	0.55
Moquegua	0.01	Vranje	0.49
Tacna	0.05		
		Slovakia	**0.28**
Portugal	**0.27**	Bratislavský kraj	0.20
Acores	0.01	Stredné Slovensko	0.27
Alentejo	0.34	Východné Slovensko	0.13
Algarve	0.18	Západné Slovensko	0.27
Centro	0.22		
Lisboa	0.14	**Slovenia**	**0.43**
Madeira	0.01	Bela Krajina	0.22
Norte	0.19	Bizeljsko Sremic	0.19
		Dolenjska	0.12
Romania	**0.41**	Goriska brda	0.45
		Kras	0.05
Russia	**0.52**	Prekmurje	0.24
Crimea	0.44	Slovenska Istra	0.12
Krasnodar Krai	0.54	Stajerska Slovenija	0.30
Rostov Oblast	0.12	Vipavska dolina	0.53

Table 90 (cont.): VSI of each country and region relative of the world, 2016

Region	*Varietal Similarity Index*	*Region*	*Varietal Similarity Index*
South Africa	**0.50**	**Switzerland (cont.)**	
Breedekloof	0.36	Neuchâtel	0.17
Cape South Coast	0.52	Nidwalden	0.10
Little Karoo	0.22	Owalden	0.00
Northern Cape	0.10	Schaffhausen	0.17
Olifants River	0.25	Schwyz	0.20
Paarl	0.57	Solothurn	0.23
Robertson	0.51	St. Gallen	0.20
Stellenbosch	0.66	Thunersee	0.13
Swartland	0.52	Thurgau	0.17
Worcester	0.29	Ticino	0.41
		Uri	0.20
Spain	**0.56**	Valais	0.20
Andalucía	0.15	Vaud	0.07
Aragón	0.53	Zug	0.10
Canarias	0.04	Zürich	0.18
Cantabria	0.19	other regions	0.13
Castilla y León	0.35		
Castilla-La Mancha	0.45	**Taiwan**	**0.00**
Cataluña	0.38		
Comunidad de Madrid	0.37	**Thailand**	**0.26**
Comunidad Foral de Nava	0.51		
Comunidad Valenciana	0.24	**Tunisia**	**0.32**
Extremadura	0.26		
Galicia	0.23	**Turkey**	**0.23**
Illes Balears	0.57	Aegean	0.25
La Rioja	0.34	Central East	0.00
País Vasco	0.32	Central North	0.20
Principado de Asturias	0.32	Central South	0.01
Región de Murcia	0.12	Marmara	0.15
		Mediterranean	0.59
Switzerland	**0.24**	South East	0.00
Aargau	0.19		
Appenzell Ausserrhoden	0.18	**Ukraine**	**0.44**
Appenzell Innerrhoden	0.03		
Basel Land	0.18	**United Kingdom**	**0.32**
Basel Stadt	0.20		
Fribourg	0.13		
Geneva	0.21		
Glarus	0.19		
Graubünden - Mesolcina	0.41		
Graubünden - other	0.18		
Jura	0.10		
Lac de Bienne	0.17		
Lucerne	0.20		

Table 90 (cont.): VSI of each country and region relative of the world, 2016

Region	*Varietal Similarity Index*	*Region*	*Varietal Similarity Index*
United States	**0.70**	**United States (cont.)**	
Alameda	0.65	North Texas (DFW)	0.50
Amador	0.16	North Willamette Valley	0.19
Arizona	0.32	Ohio	0.15
Arkansas	0.32	Ontario	0.18
Butte	0.54	Orange	0.26
Calaveras	0.72	Pennsylvania	0.13
Cattaraugus	0.02	Placer	0.39
Chautauqua	0.02	Puget Sound	0.27
Colorado	0.70	Rattlesnake Hills	0.59
Columbia Gorge	0.44	Red Mountain	0.57
Columbia River	0.73	Riverside	0.75
Columbia Valley	0.61	Rogue Valley	0.44
Colusa	0.21	Sacramento	0.69
Contra Costa	0.42	San Benito	0.60
El Dorado	0.56	San Bernardino	0.08
Erie	0.02	San Diego	0.74
Fresno	0.26	San Joaquin	0.53
Georgia	0.32	San Luis Obispo	0.68
Glenn	0.28	San Mateo	0.46
Hill Country	0.71	Santa Barbara	0.44
Horse Heaven Hills	0.66	Santa Clara	0.66
Humboldt	0.65	Santa Cruz	0.34
Illinois	0.19	Schuyler	0.38
Indiana	0.22	Seneca	0.38
Iowa	0.04	Shasta	0.57
Kentucky	0.30	Siskiyou	0.63
Kern	0.49	Snipes Mountain	0.69
Kings	0.21	Solano	0.55
Lake	0.57	Sonoma	0.61
Lake Chelan	0.50	South Texas and Gulf Coa	0.10
Lassen	0.32	South Willamette Valley	0.19
Los Angeles	0.69	Stanislaus	0.68
Madera	0.56	Steuben	0.19
Marin	0.22	Suffolk	0.64
Mariposa	0.60	Sutter	0.31
Mendocino	0.60	Tehama	0.43
Merced	0.64	Texas High Plains and Pan	0.71
Michigan	0.41	Trinity	0.54
Minnesota	0.04	Tulare	0.40
Missouri	0.09	Tuolumne	0.63
Monterey	0.54	Ulster	0.24
Naches Heights	0.33	Umpqua Valley	0.21
Napa	0.64	Ventura	0.52
Nevada	0.67	Virginia	0.58
Niagara	0.04	Wahluke Slope	0.65
North Carolina	0.32	Walla Walla Valley	0.60

Table 90 (cont.): VSI of each country and region relative of the world, 2016

Region	*Varietal Similarity Index*	*Region*	*Varietal Similarity Index*
United States (cont.)			
Wayne	0.20		
West Texas	0.47		
Yakima Valley	0.60		
Yates	0.11		
Yolo	0.48		
Yuba	0.47		
Uruguay	**0.36**		
Artigas	0.48		
Canelones	0.33		
Colonia	0.48		
Durazno	0.51		
Florida	0.25		
Lavalleja	0.53		
Maldonado	0.36		
Montevideo	0.31		
Paysandu	0.42		
Rivera	0.56		
Rocha	0.31		
Salto	0.32		
San Jose	0.43		
Soriano	0.30		
Tacuarembo	0.29		

Table 91: VSI of each region in 2016 relative to that region in 2000

Region	*Varietal Similarity Index*	*Region*	*Varietal Similarity Index*
Algeria	**0.75**	**Argentina (cont.)**	
		Coronel Pringles	0.31
Argentina	**0.91**	Colón - Entre Ríos	0.12
Capital Santiago del Estero	1.00	San Javier	0.08
Cushamen	0.99	General Pueyrredón	0.08
Tinogasta	0.99	Cachi	0.05
Rawson	0.99	San Alberto	0.00
Vinchina	0.99	Ayacucho	0.00
Coronel Felipe Varela	0.99	Valle Fértil	0.00
Santa María - Catamarca	0.99		
Santa Lucía	0.99	**Armenia**	**0.00**
Chilecito	0.99		
General Lamadrid	0.98	**Australia**	**0.95**
Nueve de Julio	0.98	Bendigo	1.00
Luján de Cuyo	0.97	Eden Valley	0.99
Castro Barros	0.97	Mudgee	0.99
Angaco	0.97	Langhorne Creek	0.99
Pocito	0.97	Tumbarumba	0.99
Pichi Mahuida	0.97	Far North - other	0.98
San Martín - Mza	0.97	Pyrenees	0.98
Caucete	0.97	Hilltops	0.98
Añelo	0.97	Margaret River	0.98
Godoy Cruz	0.97	Clare Valley	0.98
General Alvear	0.97	Limestone Coast - other	0.98
Sarmiento - San Juan	0.97	Goulburn Valley	0.98
Veinticinco de Mayo - San Juan	0.96	Great Southern	0.97
Santa Rosa - Mza	0.96	Cowra	0.97
San Martín - San Juan	0.96	Geographe	0.97
Rivadavia - San Juan	0.96	Padthaway	0.96
San Blas De Los Sauces	0.96	McLaren Vale	0.96
Las Heras	0.96	Currency Creek	0.96
Maipú	0.96	Yarra Valley	0.96
Junín - Mza	0.96	Southern NSW - other	0.96
Lavalle	0.95	Rutherglen	0.96
San Carlos - Mza	0.95	Mount Benson	0.96
San Carlos - Salta	0.95	Mount Lofty Ranges - other	0.96
Rivadavia - Mza	0.94	Riverland	0.95
Conesa	0.94	Gippsland	0.95
Guaymallén	0.93	Barossa Valley	0.94
Belén	0.92	The Peninsulas	0.94
Chimbas	0.92	Grampians	0.94
La Paz	0.91	Fleurieu - other	0.94
Molinos	0.91	Sunbury	0.93
Ischilin	0.90	Henty	0.93
San Rafael	0.90	Tasmania	0.93
Calingasta	0.90	Kangaroo Island	0.93
Cafayate	0.87	Mornington Peninsula	0.93
Tupungato	0.86	Orange	0.93
Albardón	0.85	Geelong	0.92
Tunuyán	0.83	Riverina	0.91
General Roca	0.80	Hunter	0.91
Cruz del Eje	0.79	Northern Slopes	0.91
Andalgala	0.78	Western Victoria - other	0.90
Poman	0.77	Barossa - other	0.90
Zonda	0.74	Central Victoria - other	0.90
Sanagasta	0.72	Southern Fleurieu	0.90
El Cuy	0.70	Blackwood Valley	0.89
Ullum	0.62	Adelaide Hills	0.86
Famatina	0.56	Western Australia Southeast Coast	0.85
Avellaneda - Río Negro	0.49	North East Victoria - other	0.85
Adolfo Alsina	0.48	Perth Hills	0.84
Capital San Juan	0.45	Alpine Valleys	0.84
Puelen	0.40	Perricoota	0.84
Iglesia	0.37	Canberra District (ACT)	0.81
Tafí del Valle	0.37	Lower Murray - other	0.81
Confluencia	0.32	Big Rivers - other	0.79

Table 91 (cont.): VSI of each region in 2016 relative to that region in 2000

Region	*Varietal Similarity Index*
Australia (cont.)	
Beechworth	0.78
Granite Belt	0.71
Hunter Valley - other	0.69
Port Phillip - other	0.68
South West Australia - other	0.67
Swan District	0.64
South Coast - other	0.64
Murray Darling (NSW)	0.64
South Burnett	0.59
Hastings River	0.57
Murray Darling (VIC)	0.51
Greater Perth - other	0.49
Swan Hill (VIC)	0.45
Western Plains	0.42
Central Western Australia	0.40
Swan Hill (NSW)	0.37
North West Victoria - other	0.28
Central Ranges - other	0.19
Northern Rivers - other	0.14
Queensland - other	0.09
Eastern Plains, Inland and North WA	0.05
Austria	**0.97**
Wien and other regions	0.94
Burgenland	0.90
Steiermark	0.88
Brazil	**0.72**
Bulgaria	**0.83**
Canada	**0.63**
Chile	**0.92**
Metropolitana	0.99
O'Higgins	0.98
Del Maule	0.96
Valparaiso	0.72
Araucania	0.63
Del Bio Bio	0.51
Coquimbo	0.02
Atacama	0.00
Croatia	**0.71**
Cyprus	**0.91**
Czechia	**0.74**
France	**0.92**
Champagne-Ardenne	1.00
Poitou Charentes	1.00
Bourgogne	0.99
Aquitaine	0.99
Centre-Val de Loire	0.98
Alsace	0.98
Pays de la Loire	0.98
Franche Comté	0.97
Rhône Alpes	0.97
Provence-Alpes-Cote d'Azur	0.96
Picardie	0.95
Auvergne	0.95
Corse	0.93
Lorraine	0.92
Limousin	0.90
Midi Pyrénées	0.82
Languedoc Roussillon	0.81
Georgia	**1.00**
Germany	**0.93**
Rheingau	1.00
Mittelrhein	1.00
Ahr	0.98
Mosel	0.98
Hessische Bergstraße	0.98
Franken	0.96
Württemberg	0.96
Baden	0.96
Nahe	0.93
Saale	0.92
Sachsen	0.92
Pfalz	0.89
Rheinhessen	0.85
Greece	**0.71**
Attiki	1.00
Ionia Nisia	0.97
Kriti	0.97
Sterea Ellada	0.96
Dytiki Makedonia	0.95
Voreio Aigaio	0.87
Peloponnissos	0.78
Notio Aigaio	0.74
Anatoliki Makedonia, Thraki	0.68
Kentriki Makedonia	0.67
Ipeiros	0.31
Dytiki Ellada	0.12
Thessalia	0.01
Hungary	**0.47**
Israel	**0.85**
Toscana	0.99
Abruzzo	0.98
Molise	0.97
Marche	0.97
Lazio	0.95
Emilia-Romagna	0.94
Liguria	0.94
Sicilia	0.93
Trento	0.92
Piemonte	0.89
Puglia	0.89
Sardegna	0.86
Umbria	0.85
Calabria	0.85
Campania	0.81
Friuli-Venezia Giulia	0.78
Valle d'Aosta	0.78
Veneto	0.65
Lombardia	0.61
Bolzano-Bozen	0.40
Basilicata	0.30
Korea, Rep.	**1.00**
Luxembourg	**0.96**
Moldova	**0.72**
Morocco	**0.27**
New Zealand	**0.76**
Otago	0.92
Marlborough	0.92

Île de France 0.77 Wairarapa 0.87

Table 91 (cont.): VSI of each region in 2016 relative to that region in 2000

Region	*Varietal Similarity Index*
New Zealand (cont.)	
Waipara	0.86
Gisborne	0.86
Hawkes Bay	0.85
Auckland	0.83
Canterbury	0.80
Nelson	0.73
Waikato	0.40
Portugal	**0.45**
Alentejo	0.67
Algarve	0.64
Madeira	0.19
Acores	0.08
Romania	**0.97**
Russia	**0.40**
Serbia	**0.49**
Slovakia	**0.77**
Slovenia	**0.59**
South Africa	**0.96**
Northern Cape	0.99
Olifants River	0.98
Breedekloof	0.96
Worcester	0.95
Robertson	0.92
Swartland	0.92
Paarl	0.91
Little Karoo	0.90
Stellenbosch	0.80
Spain	**0.85**
Región de Murcia	0.99
País Vasco	0.99
Canarias	0.97
Castilla-La Mancha	0.95
Cataluña	0.93
La Rioja	0.93
Andalucía	0.93
Comunidad Valenciana	0.90
Aragón	0.89
Illes Balears	0.89
Comunidad de Madrid	0.87
Extremadura	0.85
Castilla y León	0.79
Galicia	0.78
Comunidad Foral de Navarra	0.71
Cantabria	0.69
Principado de Asturias	0.53
Switzerland	**0.98**
Ticino	1.00
Graubünden	1.00
Schaffhausen	1.00
Thurgau	1.00
St. Gallen	1.00
Vaud	1.00
Zürich	0.99
Basel Land	0.99
Aargau	0.98
Valais	0.97

Region	*Varietal Similarity Index*
Switzerland (cont.)	
Neuchâtel	0.93
Lucerne	0.86
Other regions	0.80
Jura	0.60
Taiwan	**0.42**
Tunisia	**0.35**
United Kingdom	**0.28**
United States	**0.89**
Chautauqua-Erie	1.00
Amador	0.99
North Willamette Valley	0.97
South Willamette Valley	0.97
San Joaquin	0.95
San Bernardino	0.95
Colusa	0.95
Yolo	0.94
Finger Lakes	0.94
Napa	0.94
Sonoma	0.94
Mendocino	0.94
Fresno	0.93
Sacramento	0.93
Contra Costa	0.93
Monterey	0.93
El Dorado	0.92
Rogue Valley	0.92
Michigan	0.92
New York - other	0.90
San Benito	0.89
Mariposa	0.89
Ventura	0.88
Santa Clara	0.87
Lake	0.87
Solano	0.85
Butte	0.85
San Luis Obispo	0.84
Alameda	0.84
Santa Barbara	0.83
Kings	0.83
San Mateo	0.82
Tulare	0.82
Washington	0.81
Calaveras	0.78
Nevada	0.77
Madera	0.77
Columbia River	0.76
Placer	0.75
Trinity	0.73
Merced	0.73
Santa Cruz	0.71
San Diego	0.68
Humboldt	0.66
Stanislaus	0.58
Glenn	0.56
Marin	0.56
Kern	0.52
Yuba	0.51
Riverside	0.46
Tehama	0.40
Los Angeles	0.38
Sutter	0.32

Schwyz	0.96	Shasta	0.14
Geneva	0.95		
Fribourg	0.95	**Uruguay**	**0.93**

Table 92: Each country’s 10 most-similar winegrape countries in the world according to the VSI, 2000

Algeria	0.70 TN	0.66 FR	0.43 IL	0.19 AU	0.17 MA	0.17 US	0.16 CL	0.15 M9	0.15 ZA	0.12 ES
Argentina	0.31 AU	0.28 CL	0.25 M9	0.21 FR	0.20 ZA	0.20 US	0.17 IL	0.15 CA	0.15 NZ	0.14 BG
Armenia	0.84 RU	0.74 RO	0.71 SI	0.68 HU	0.50 GE	0.49 UK	0.48 HR	0.45 PT	0.38 CZ	0.30 SK
Australia	0.63 CL	0.59 M9	0.55 US	0.48 FR	0.47 NZ	0.44 ZA	0.43 CA	0.37 IL	0.31 AR	0.30 MD
Austria	0.71 SK	0.69 CZ	0.26 HU	0.21 RS	0.20 LU	0.20 DE	0.19 HR	0.14 SI	0.11 RO	0.08 RU
Brazil	0.37 CA	0.30 PT	0.29 RO	0.28 MD	0.28 SI	0.25 HU	0.24 RU	0.20 HR	0.19 MA	0.18 CZ
Bulgaria	0.50 M9	0.42 CL	0.37 MD	0.33 IL	0.29 RU	0.29 GE	0.28 AU	0.28 CA	0.28 FR	0.27 US
Canada	0.61 US	0.55 NZ	0.47 CL	0.43 AU	0.40 IL	0.38 M9	0.37 FR	0.37 BR	0.34 SI	0.33 MD
Chile	0.77 M9	0.63 AU	0.55 IL	0.50 US	0.47 CA	0.42 BG	0.41 NZ	0.40 FR	0.38 MD	0.36 ZA
Croatia	0.76 SI	0.73 RS	0.70 HU	0.68 RO	0.65 SK	0.54 CZ	0.48 RU	0.48 AM	0.45 PT	0.38 UK
Cyprus	0.25 PT	0.20 MA	0.18 RO	0.16 SI	0.16 RU	0.14 HU	0.12 TN	0.12 IL	0.11 CZ	0.10 HR
Czechia	0.87 SK	0.70 HU	0.69 AT	0.60 SI	0.55 RO	0.54 HR	0.45 RS	0.44 RU	0.43 DE	0.41 PT
France	0.66 DZ	0.57 IL	0.52 TN	0.48 AU	0.45 US	0.40 CL	0.37 CA	0.36 M9	0.33 NZ	0.32 IT
Georgia	0.60 RU	0.50 AM	0.40 M9	0.38 MD	0.29 BG	0.10 RO	0.09 SI	0.09 HU	0.06 PT	0.06 UK
Germany	0.74 LU	0.43 CZ	0.24 CH	0.22 NZ	0.22 CA	0.21 SK	0.20 AT	0.16 HU	0.10 MD	0.07 M9
Greece	0.29 PT	0.27 RO	0.25 SI	0.24 HU	0.23 RU	0.18 MA	0.17 AM	0.16 HR	0.16 UK	0.15 CZ
Hungary	0.88 SI	0.88 RO	0.70 HR	0.70 CZ	0.70 RU	0.68 AM	0.62 PT	0.61 SK	0.56 UK	0.48 RS
Israel	0.66 TN	0.57 FR	0.55 CL	0.48 US	0.43 DZ	0.43 M9	0.40 CA	0.40 ZA	0.37 AU	0.37 RO
Italy	0.32 FR	0.23 US	0.19 CA	0.17 IL	0.16 CL	0.15 NZ	0.14 AU	0.14 BG	0.13 SI	0.12 M9
Korea, Rep.	0.63 TW	0.17 PT	0.16 MA	0.10 TN	0.09 RO	0.09 RU	0.08 IL	0.08 SI	0.08 CY	0.08 CA
Luxembourg	0.74 DE	0.40 CZ	0.27 SK	0.20 AT	0.17 NZ	0.15 CH	0.14 CA	0.13 HU	0.07 MD	0.06 UK
Moldova	0.69 M9	0.46 NZ	0.38 CL	0.38 GE	0.37 BG	0.35 US	0.33 CA	0.30 RU	0.30 AU	0.29 FR
Morocco	0.459 PT	0.33 TN	0.28 IL	0.28 RO	0.27 RU	0.26 SI	0.22 HU	0.20 CA	0.20 CY	0.19 BR
New Zealand	0.66 US	0.55 CA	0.47 AU	0.46 MD	0.41 CL	0.36 ZA	0.33 FR	0.33 M9	0.30 IL	0.23 CH
Portugal	0.72 RO	0.68 SI	0.62 RU	0.62 HU	0.46 MA	0.45 HR	0.45 AM	0.44 UK	0.41 CZ	0.34 IL
Romania	0.94 SI	0.88 HU	0.78 RU	0.74 AM	0.72 PT	0.68 HR	0.60 UK	0.55 CZ	0.46 SK	0.41 RS
Russia	0.84 AM	0.78 RO	0.74 SI	0.70 HU	0.62 PT	0.60 GE	0.50 UK	0.48 HR	0.44 CZ	0.34 M9
Serbia	0.73 HR	0.67 SK	0.54 SI	0.48 HU	0.45 CZ	0.41 RO	0.24 RU	0.24 AM	0.21 PT	0.21 AT
Slovakia	0.87 CZ	0.71 AT	0.67 RS	0.65 HR	0.61 HU	0.56 SI	0.46 RO	0.31 RU	0.30 AM	0.27 LU
Slovenia	0.94 RO	0.88 HU	0.76 HR	0.74 RU	0.71 AM	0.68 PT	0.60 CZ	0.60 UK	0.56 SK	0.54 RS

Table 92 (cont.): Each country's 10 most-similar winegrape countries in the world according to the VSI, 2000

South Africa	0.56 US	0.44 AU	0.40 IL	0.36 NZ	0.36 CL	0.33 M9	0.31 FR	0.24 CA	0.22 MD	0.20 AR
Spain	0.20 RO	0.19 SI	0.18 HU	0.17 PT	0.16 RU	0.16 AM	0.13 HR	0.13 UK	0.12 DZ	0.12 FR
Switzerland	0.24 DE	0.23 NZ	0.22 CA	0.19 MD	0.18 FR	0.15 LU	0.13 US	0.09 HU	0.09 DZ	0.08 AU
Taiwan	0.63 KR	0.00 M9	0.00 AR							
Tunisia	0.70 DZ	0.66 IL	0.52 FR	0.33 MA	0.26 PT	0.14 RO	0.14 RU	0.13 CL	0.13 CA	0.12 SI
United Kingdc	0.60 RO	0.60 SI	0.56 HU	0.50 RU	0.49 AM	0.44 PT	0.38 HR	0.35 CZ	0.28 CA	0.26 SK
United States	0.66 NZ	0.61 CA	0.56 ZA	0.55 AU	0.50 CL	0.48 IL	0.45 FR	0.44 M9	0.35 MD	0.27 BG
Uruguay	0.27 CL	0.26 IL	0.24 CA	0.20 M9	0.19 FR	0.19 BG	0.18 PT	0.18 AU	0.17 US	0.16 MA
Missing 9	0.77 CL	0.69 MD	0.59 AU	0.50 BG	0.44 US	0.43 IL	0.40 GE	0.38 CA	0.36 FR	0.34 RU

Algeria(DZ), Argentina(AR), Armenia(AM), Australia(AU), Austria(AT), Brazil(BR), Bulgaria(BG), Canada(CA), Chile(CL), Croatia(HR), Cyprus(CY), Czechia(CZ), France(FR), Georgia(GE), Germany(DE), Greece(EL), Hungary(HU), Israel(IL), Italy(IT), Korea, Rep.(KR), Moldova(MD), Morocco(MA), New Zealand(NZ), Portugal(PT), Romania(RO), Russia(RU), Serbia(RS), Slovakia(SK), Slovenia(SI), South Africa(ZA), Spain(ES), Switzerland(CH), Taiwan(TW), Tunisia(TN), United Kingdom(UK), United States(US), Uruguay(UY)

Table 93: Each country's 10 most-similar winegrape countries in the world according to the VSI, 2010

Algeria	0.70 TN	0.53 FR	0.43 IL	0.31 MX	0.20 TR	0.18 MM	0.18 AU	0.18 US	0.18 CL	0.17 MA
Argentina	0.38 AU	0.37 CL	0.32 CN	0.31 US	0.29 ZA	0.28 FR	0.25 TH	0.23 RU	0.23 JP	0.23 MX
Armenia	0.71 RO	0.59 KZ	0.50 GE	0.32 SI	0.30 UA	0.26 HR	0.24 RS	0.23 BG	0.22 HU	0.18 MD
Australia	0.72 US	0.70 TH	0.67 CL	0.64 MM	0.62 ZA	0.58 FR	0.51 CA	0.48 CN	0.48 JP	0.45 RU
Austria	0.79 SK	0.71 CZ	0.43 HU	0.26 SI	0.26 HR	0.20 RS	0.20 DE	0.16 JP	0.14 CA	0.11 US
Brazil	0.37 MD	0.17 UY	0.16 CA	0.14 UA	0.11 FR	0.09 US	0.08 RO	0.08 HU	0.07 BG	0.06 AM
Bulgaria	0.59 JP	0.56 CL	0.56 IL	0.55 CN	0.55 US	0.49 MX	0.49 FR	0.46 RU	0.42 RO	0.40 MD
Canada	0.66 US	0.54 JP	0.51 AU	0.50 FR	0.45 CL	0.45 UK	0.43 RU	0.40 BG	0.39 DE	0.38 MD
Chile	0.90 CN	0.75 US	0.67 AU	0.64 RU	0.58 ZA	0.57 JP	0.56 BG	0.51 MX	0.49 FR	0.48 IL
China	0.90 CL	0.59 RU	0.58 US	0.55 BG	0.53 MX	0.48 AU	0.46 JP	0.43 IL	0.43 ZA	0.35 FR
Croatia	0.78 RS	0.78 SI	0.50 HU	0.50 SK	0.45 RO	0.39 CZ	0.26 AM	0.26 AT	0.26 BG	0.23 IL
Cyprus	0.15 MX	0.13 IL	0.12 TN	0.10 AU	0.10 FR	0.10 CL	0.09 DZ	0.09 CN	0.09 US	0.08 TR
Czechia	0.85 SK	0.71 AT	0.57 DE	0.56 HU	0.44 SI	0.39 HR	0.35 CA	0.33 LU	0.33 NZ	0.29 UK
Ethiopia	0.53 IT	0.30 ZA	0.12 TH	0.11 MX	0.08 TN	0.06 AR	0.04 US	0.03 IL	0.03 FR	0.02 TR
France	0.58 AU	0.58 US	0.53 DZ	0.53 JP	0.50 IL	0.50 CA	0.49 CL	0.49 BG	0.47 ZA	0.43 MX
Georgia	0.91 KZ	0.63 UA	0.50 AM	0.38 MD	0.19 BG	0.13 RU	0.10 RO	0.04 HR	0.04 SI	0.03 RS
Germany	0.57 CZ	0.41 LU	0.39 CA	0.35 UK	0.33 CH	0.30 SK	0.24 HU	0.20 AT	0.19 US	0.18 NZ
Greece	0.35 RO	0.27 BG	0.26 MA	0.26 IL	0.24 MX	0.21 SI	0.18 AM	0.16 TN	0.16 HR	0.15 US
Hungary	0.61 SK	0.58 SI	0.56 CZ	0.50 HR	0.43 AT	0.37 RS	0.36 RU	0.33 US	0.32 RO	0.30 CA
Israel	0.66 TN	0.60 MX	0.56 BG	0.50 FR	0.49 US	0.48 CL	0.46 ZA	0.43 DZ	0.43 CN	0.42 JP
Italy	0.53 ET	0.36 FR	0.35 US	0.29 BG	0.28 JP	0.27 CA	0.25 AU	0.23 IL	0.23 CL	0.22 SI
Japan	0.68 US	0.59 BG	0.57 CL	0.54 CA	0.53 FR	0.49 RU	0.48 AU	0.46 CN	0.42 IL	0.39 ZA
Kazakhstan	0.91 GE	0.68 UA	0.59 AM	0.44 MD	0.21 BG	0.20 RO	0.14 RU	0.09 SI	0.08 HR	0.06 HU
Korea, Rep.	0.63 TW	0.15 MA	0.10 EL	0.10 TN	0.10 RO	0.08 IL	0.08 BG	0.06 MX	0.05 SI	0.04 HR
Luxembourg	0.41 DE	0.33 CZ	0.20 SK	0.19 CA	0.12 HU	0.12 SI	0.11 AT	0.10 UK	0.09 CH	0.08 RU
Mexico	0.60 IL	0.53 CN	0.51 CL	0.49 BG	0.43 FR	0.41 US	0.41 ZA	0.40 TR	0.40 TN	0.36 AU
Moldova	0.86 UA	0.48 RU	0.44 US	0.44 KZ	0.43 CL	0.41 NZ	0.40 BG	0.38 GE	0.38 CA	0.37 BR
Morocco	0.33 TN	0.29 RO	0.28 IL	0.26 EL	0.21 BG	0.19 MX	0.17 DZ	0.15 KR	0.14 SI	0.12 HR
Myanmar	0.69 TH	0.64 AU	0.63 NZ	0.45 ZA	0.38 FR	0.32 TR	0.29 CL	0.24 US	0.23 MD	0.21 AR

Table 93 (cont.): Each country's 10 most-similar winegrape countries in the world according to the VSI, 2010

New Zealand	0.63 MM	0.41 MD	0.38 ZA	0.36 CL	0.36 CA	0.35 US	0.35 SI	0.33 CZ	0.31 UK	0.28 AU
Peru	0.11 HU	0.07 SK	0.06 MX	0.05 CZ	0.05 SI	0.04 MM	0.04 AT	0.04 TW	0.03 EL	0.03 CN
Portugal	0.32 ES	0.14 MX	0.14 MM	0.13 TH	0.12 AU	0.10 AR	0.08 CL	0.08 FR	0.07 ZA	0.07 US
Romania	0.71 AM	0.52 SI	0.45 HR	0.42 BG	0.40 IL	0.37 RS	0.35 EL	0.32 HU	0.30 MX	0.29 MA
Russia	0.64 CL	0.59 US	0.59 CN	0.49 JP	0.48 MD	0.48 UA	0.46 BG	0.45 AU	0.43 CA	0.39 ZA
Serbia	0.78 HR	0.60 SI	0.43 SK	0.37 HU	0.37 RO	0.27 CZ	0.24 AM	0.20 AT	0.09 EL	0.09 IL
Slovakia	0.85 CZ	0.79 AT	0.61 HU	0.50 HR	0.47 SI	0.43 RS	0.30 DE	0.20 LU	0.18 CL	0.17 CA
Slovenia	0.78 HR	0.60 RS	0.58 HU	0.52 RO	0.47 SK	0.44 CZ	0.38 US	0.36 CA	0.35 BG	0.35 NZ
South Africa	0.62 AU	0.60 US	0.58 CL	0.49 TH	0.47 FR	0.46 IL	0.45 MM	0.43 CN	0.41 MX	0.39 RU
Spain	0.32 PT	0.17 MX	0.16 FR	0.13 DZ	0.11 MM	0.11 AR	0.10 AU	0.09 TH	0.09 CL	0.09 US
Switzerland	0.47 UK	0.33 DE	0.30 CA	0.28 US	0.26 FR	0.24 NZ	0.23 MD	0.20 HU	0.19 CZ	0.17 JP
Taiwan	0.63 KR	0.05 JP	0.04 PE	0.03 TH	0.00 AR	0.00 FR	0.00 US	0.00 ES	0.00 BR	0.00 MM
Thailand	0.70 AU	0.69 MM	0.49 ZA	0.36 TR	0.33 FR	0.25 AR	0.21 US	0.20 CL	0.18 MX	0.13 PT
Tunisia	0.70 DZ	0.66 IL	0.40 MX	0.34 FR	0.33 MA	0.16 EL	0.15 IT	0.15 RO	0.14 BG	0.14 MM
Turkey	0.40 MX	0.39 AU	0.36 TH	0.32 MM	0.26 FR	0.23 ZA	0.20 CL	0.20 DZ	0.18 US	0.14 AR
Ukraine	0.86 MD	0.68 KZ	0.63 GE	0.48 RU	0.37 CL	0.36 BG	0.35 US	0.31 CN	0.30 AM	0.30 CA
United Kingdom	0.53 US	0.47 CH	0.45 CA	0.35 DE	0.34 AU	0.33 JP	0.31 NZ	0.29 CZ	0.26 MD	0.24 RU
United States	0.75 CL	0.72 AU	0.68 JP	0.66 CA	0.60 ZA	0.59 RU	0.58 CN	0.58 FR	0.55 BG	0.53 UK
Uruguay	0.41 FR	0.36 BG	0.32 CL	0.31 JP	0.30 US	0.30 CN	0.28 IL	0.26 RU	0.25 CA	0.25 MX

Algeria(DZ), Argentina(AR), Armenia(AM), Australia(AU), Austria(AT), Brazil(BR), Bulgaria(BG), Canada(CA), Chile(CL), China(CN), Croatia(HR), Cyprus(CY), Czechia(CZ), Ethiopia(ET), France(FR), Georgia(GE), Germany(DE), Greece(EL), Hungary(HU), Israel (IL), Italy(IT), Japan(JP), Kazakhstan(KZ), Korea, Rep.(KR), Luxembourg(LU), Mexico(MX), Moldova(MD), Morocco(MA), Myanmar(MM), New Zealand(NZ), Peru(PE), Portugal(PT), Romania(RO), Russia(RU), Serbia(RS), Slovakia(SK), Slovenia(SI), South Africa(ZA), Spain(ES), Switzerland(CH), Taiwan(TW), Thailand(TH), Tunisia(TN), Turkey(TR), Ukraine(UA), United Kingdom(UK), United States(US), Uruguay(UY)

Table 94: Each country's 10 most-similar winegrape countries in the world according to the VSI, 2016

Algeria	0.73 IL	0.59 FR	0.48 MX	0.36 AU	0.34 KH	0.32 CL	0.31 MA	0.31 TN	0.28 BG	0.28 US
Argentina	0.35 CL	0.34 AU	0.29 IL	0.28 US	0.28 LB	0.25 KH	0.24 ZA	0.24 FR	0.21 BG	0.20 RU
Armenia	0.00 AR	0.00 AT	0.00 AU	0.00 BG	0.00 BR	0.00 CA	0.00 CH	0.00 CL	0.00 CN	0.00 CY
Australia	0.73 LB	0.70 US	0.68 KH	0.66 CL	0.66 MM	0.61 TH	0.60 ZA	0.58 FR	0.57 IL	0.44 RU
Austria	0.72 SK	0.70 CZ	0.36 HU	0.25 SI	0.21 HR	0.19 DE	0.18 RS	0.16 LU	0.12 LB	0.12 CA
Brazil	0.18 MD	0.15 CA	0.13 UA	0.11 FR	0.11 UY	0.11 RU	0.08 JP	0.07 US	0.07 HU	0.06 IL
Bulgaria	0.62 RU	0.61 CL	0.60 LB	0.58 US	0.57 UA	0.57 IL	0.52 MD	0.51 FR	0.45 KH	0.44 AU
Cambodia	0.68 AU	0.59 TH	0.53 IL	0.52 TW	0.50 CL	0.49 FR	0.45 LB	0.45 BG	0.44 MM	0.41 US
Canada	0.54 US	0.47 UK	0.47 LB	0.40 AU	0.39 DE	0.38 FR	0.35 CL	0.34 RU	0.33 RS	0.33 SI
Chile	0.80 LB	0.74 US	0.68 IL	0.66 AU	0.64 RU	0.61 BG	0.54 ZA	0.54 UA	0.54 MD	0.50 KH
China	0.81 RS	0.60 LB	0.57 MX	0.50 RO	0.49 CL	0.45 FR	0.44 IL	0.44 US	0.39 MD	0.37 BG
Croatia	0.61 SI	0.39 HU	0.38 RS	0.34 CZ	0.23 SK	0.21 AT	0.20 US	0.20 LB	0.18 BG	0.17 CL
Cyprus	0.00 AM	0.00 HU	0.00 TN	0.00 NZ	0.00 NO	0.00 ES	0.00 RO	0.00 BR	0.00 BG	0.00 ZA
Czechia	0.70 AT	0.68 SK	0.57 DE	0.56 LU	0.52 HU	0.46 SI	0.34 NZ	0.34 HR	0.31 CA	0.29 UK
Ethiopia	0.52 IT	0.30 ZA	0.11 MX	0.10 TH	0.04 AR	0.03 IN	0.03 FR	0.03 TR	0.02 US	0.01 AU
France	0.60 LB	0.59 IL	0.59 DZ	0.58 US	0.58 AU	0.51 BG	0.49 KH	0.49 CL	0.47 RS	0.46 ZA
Georgia	0.91 KZ	0.56 UA	0.43 RU	0.28 BG	0.19 MD	0.10 RO	0.07 RS	0.05 SK	0.04 SI	0.04 MK
Germany	0.69 LU	0.57 CZ	0.39 CA	0.34 CH	0.34 UK	0.30 SK	0.23 SI	0.23 HU	0.23 US	0.21 RS
Greece	0.18 LB	0.17 CL	0.15 UY	0.15 AU	0.15 RS	0.15 US	0.15 IL	0.15 FR	0.13 MX	0.13 BG
Hungary	0.52 CZ	0.50 SI	0.47 SK	0.39 HR	0.36 AT	0.36 RS	0.33 US	0.33 RU	0.31 LB	0.27 CL
India	0.64 TR	0.54 MM	0.53 MX	0.42 AU	0.41 NZ	0.32 FR	0.31 TH	0.31 ZA	0.28 LB	0.27 CL
Israel	0.73 DZ	0.68 CL	0.61 MX	0.61 LB	0.59 FR	0.59 US	0.57 BG	0.57 AU	0.53 KH	0.47 ZA
Italy	0.52 ET	0.43 FR	0.35 US	0.33 LB	0.32 RS	0.28 RO	0.26 SI	0.25 AU	0.25 CN	0.24 BG
Japan	0.27 KR	0.27 RS	0.26 RO	0.21 LB	0.20 CN	0.20 US	0.18 FR	0.16 SI	0.14 MD	0.14 CA
Kazakhstan	0.91 GE	0.61 UA	0.47 RU	0.29 BG	0.24 MD	0.20 RO	0.13 RS	0.09 SK	0.08 SI	0.05 LB
Korea, Rep.	0.27 JP	0.21 CN	0.17 RS	0.11 RO	0.08 MX	0.06 SK	0.06 LB	0.05 FR	0.05 UY	0.03 TN
Lebanon	0.84 US	0.80 CL	0.73 AU	0.64 RS	0.61 RU	0.61 IL	0.60 FR	0.60 CN	0.60 BG	0.59 MD
Luxembourg	0.69 DE	0.56 CZ	0.28 CA	0.25 SK	0.24 CH	0.24 UK	0.20 SI	0.20 HU	0.18 US	0.16 AT
Mexico	0.61 IL	0.57 CN	0.53 IN	0.50 CL	0.49 LB	0.48 DZ	0.47 RS	0.46 TR	0.46 FR	0.42 US

Table 94 (cont.): Each country's 10 most-similar winegrape countries in the world according to the VSI, 2016

Moldova	0.69 RU	0.65 UA	0.59 LB	0.54 CL	0.52 US	0.52 BG	0.44 RS	0.44 FR	0.42 IL	0.39 NZ
Morocco	1.00 MA	0.35 PE	0.31 DZ	0.30 FR	0.30 IL	0.28 MX	0.26 LB	0.24 CL	0.23 RS	0.22 TR
Myanmar	0.66 AU	0.64 NZ	0.57 TH	0.54 IN	0.46 ZA	0.44 KH	0.36 FR	0.34 LB	0.31 CL	0.30 TR
New Zealand	0.64 MM	0.41 IN	0.40 LB	0.39 MD	0.37 SI	0.36 ZA	0.35 CL	0.34 CZ	0.29 US	0.27 UK
North Macedonia	0.23 BG	0.16 US	0.13 LB	0.13 RU	0.12 RS	0.12 CL	0.11 UA	0.11 MD	0.11 FR	0.10 IL
Norway	0.10 RO	0.08 RS	0.06 CN	0.05 SK	0.04 SI	0.03 LB	0.03 UK	0.03 MX	0.03 JP	0.02 IT
Peru	0.35 MA	0.35 TN	0.11 HU	0.08 SK	0.07 SI	0.06 MX	0.05 CZ	0.04 MM	0.04 AT	0.03 IT
Portugal	0.32 ES	0.13 MM	0.13 MX	0.13 AU	0.11 TH	0.09 CL	0.09 KH	0.09 IL	0.09 AR	0.08 LB
Romania	0.76 RS	0.50 CN	0.47 SK	0.43 SI	0.38 LB	0.32 MD	0.29 FR	0.29 MX	0.28 IT	0.26 JP
Russia	0.92 UA	0.69 MD	0.64 CL	0.62 BG	0.61 LB	0.59 US	0.47 KZ	0.47 IL	0.44 AU	0.43 GE
Serbia	0.81 CN	0.76 RO	0.64 LB	0.59 SI	0.52 SK	0.48 US	0.47 FR	0.47 MX	0.44 MD	0.39 CL
Slovakia	0.72 AT	0.68 CZ	0.52 RS	0.47 HU	0.47 RO	0.40 SI	0.33 CN	0.30 DE	0.25 LU	0.23 HR
Slovenia	0.61 HR	0.59 RS	0.50 HU	0.48 LB	0.46 CZ	0.43 RO	0.40 SK	0.40 US	0.37 NZ	0.35 MD
South Africa	0.60 AU	0.55 LB	0.54 US	0.54 CL	0.47 IL	0.46 FR	0.46 MM	0.44 TH	0.41 KH	0.39 MX
Spain	0.32 PT	0.18 MX	0.16 FR	0.15 DZ	0.13 MA	0.13 TN	0.12 MM	0.12 IL	0.11 AU	0.10 AR
Switzerland	0.51 UK	0.34 DE	0.33 US	0.28 FR	0.24 LU	0.23 NZ	0.23 CA	0.19 CZ	0.18 MD	0.18 MM
Taiwan	0.52 KH	0.16 TH	0.02 JP	0.00 US	0.00 MX	0.00 AU	0.00 ES	0.00 ET	0.00 CZ	0.00 GE
Thailand	0.61 AU	0.59 KH	0.57 MM	0.44 ZA	0.31 IN	0.28 TR	0.28 FR	0.24 IL	0.21 DZ	0.19 LB
Tunisia	1.00 TN	0.35 PE	0.31 DZ	0.30 FR	0.30 IL	0.28 MX	0.26 LB	0.24 CL	0.23 RS	0.22 TR
Turkey	0.64 IN	0.46 MX	0.37 AU	0.30 KH	0.30 MM	0.28 TH	0.24 FR	0.22 IL	0.22 CL	0.22 TN
Ukraine	0.92 RU	0.65 MD	0.61 KZ	0.57 BG	0.56 GE	0.54 CL	0.49 LB	0.46 US	0.39 IL	0.35 AU
United Kingdom	0.61 US	0.51 CH	0.47 CA	0.41 LB	0.34 DE	0.33 AU	0.29 FR	0.29 CZ	0.27 NZ	0.27 SI
United States	0.84 LB	0.74 CL	0.70 AU	0.61 UK	0.59 IL	0.59 RU	0.58 FR	0.58 BG	0.54 ZA	0.54 CA
Uruguay	0.42 FR	0.35 RS	0.32 CN	0.31 LB	0.29 BG	0.28 IL	0.27 CL	0.27 US	0.26 MD	0.24 MX

Algeria(DZ), Argentina(AR), Armenia(AM), Australia(AU), Austria(AT), Brazil(BR), Bulgaria(BG), Cambodia(KH), Canada(CA), Chile(CL), China(CN), Croatia(HR), Cyprus(CY), Czechia(CZ), Ethiopia(ET), France(FR), Georgia(GE), Germany(DE), Greece(EL), Hungary(HU), India(IN), Israel(IL), Italy(IT), Japan(JP), Kazakhstan(KZ), Korea, Rep.(KR), Lebanon(LB), Luxembourg(LU), Mexico(MX), Moldova(MD), Morocco(MA), Myanmar(MM), New Zealand(NZ), North Macedonia(MK), Norway(NO), Peru(PE), Portugal(PT), Romania(RO), Russia(RU), Serbia(RS), Slovakia(SK), Slovenia(SI), South Africa(ZA), Spain(ES), Switzerland(CH), Taiwan(TW), Thailand(TH), Tunisia(TN), Turkey(TR), Ukraine(UA), United Kingdom(UK), United States(US), Uruguay(UY)

Table 95: Each region's 3 most similar winegrape regions in the world according to the VSI, 2000

DZ	Algeria	0.89	FR	Var	0.84	FR	Herault	0.82	FR	Gard
AR	Adolfo Alsina	0.78	AR	Pichi Mahuida	0.70	AR	Las Heras	0.70	AR	Guaymallén
AR	Albardón	0.96	AR	Chimbas	0.90	AR	Rivadavia - San Juan	0.86	AR	Veinticinco de Mayo - San Juan
AR	Andalgala	0.69	AR	Tinogasta	0.69	AR	San Alberto	0.68	AR	Pocito
AR	Añelo	0.90	AR	Tunuyán	0.85	AR	Luján de Cuyo	0.84	AR	Tupungato
AR	Angaco	0.98	AR	San Martín - San Juan	0.97	AR	Caucete	0.91	AR	Santa Lucía
AR	Arauco	0.52	AR	Avellaneda - Río Negro	0.51	AR	Ullum	0.29	AR	Zonda
AR	Avellaneda - Río Negro	0.58	AR	Ullum	0.52	AR	Arauco	0.52	AR	General Roca
AR	Ayacucho	0.86	AR	Santa Lucía	0.84	AR	Tinogasta	0.84	AR	San Alberto
AR	Belén	0.98	AR	Famatina	0.97	AR	Coronel Felipe Varela	0.96	AR	General Lamadrid
AR	Cachi	0.98	AR	Tafí del Valle	0.94	AR	San Carlos - Salta	0.91	AR	San Blas De Los Sauces
AR	Cafayate	0.90	AR	Chilecito	0.89	AR	San Carlos - Salta	0.87	AR	Santa María - Catamarca
AR	Calingasta	0.84	AR	Santa María - Catamarca	0.84	AR	Castro Barros	0.82	CL	Metropolitana
AR	Capital San Juan	0.84	AR	Chimbas	0.80	AR	Caucete	0.78	AR	Albardón
AR	Capital Santiago del Estero	0.94	AR	La Rioja	0.76	AR	Valle Fértil	0.34	AR	Ischilin
AR	Castro Barros	0.88	AR	Tunuyán	0.85	AR	Luján de Cuyo	0.84	AR	Calingasta
AR	Caucete	0.98	AR	San Martín - San Juan	0.97	AR	Angaco	0.95	AR	Nueve de Julio
AR	Chilecito	0.91	AR	General Lamadrid	0.90	AR	San Carlos - Salta	0.90	AR	Cafayate
AR	Chimbas	0.96	AR	Albardón	0.94	AR	Santa Lucía	0.91	AR	Nueve de Julio
AR	Colón - Entre Ríos	0.83	CH	Graubünden	0.81	CH	St. Gallen	0.81	CH	Schaffhausen
AR	Conesa	0.82	AR	Pichi Mahuida	0.67	AR	Adolfo Alsina	0.61	AR	Belén
AR	Confluencia	0.66	AR	General Roca	0.65	AR	El Cuy	0.64	AU	Riverina
AR	Coronel Felipe Varela	0.99	AR	Famatina	0.97	AR	Vinchina	0.97	AR	Belén
AR	Coronel Pringles	0.75	AR	Castro Barros	0.73	AR	Tunuyán	0.73	AR	Añelo
AR	Cruz del Eje	0.93	AR	Belén	0.92	AR	Famatina	0.92	AR	Coronel Felipe Varela
AR	Cushamen	0.94	US	Marin	0.81	US	Oregon - other	0.79	US	Humboldt
AR	El Cuy	0.75	AR	General Roca	0.65	AR	Confluencia	0.62	AR	Puelen
AR	Famatina	0.99	AR	Coronel Felipe Varela	0.98	AR	Vinchina	0.98	AR	Belén
AR	General Alvear	0.95	AR	San Martín - Mza	0.92	AR	Santa Rosa - Mza	0.90	AR	San Rafael
AR	General Lamadrid	0.97	AR	Famatina	0.97	AR	Coronel Felipe Varela	0.96	AR	Belén
AR	General Pueyrredón	0.91	FR	Gironde	0.88	CL	O'Higgins	0.85	US	Oregon - other
AR	General Roca	0.75	AR	El Cuy	0.66	AR	Confluencia	0.59	AR	Tunuyán
AR	Godoy Cruz	0.89	AR	Luján de Cuyo	0.85	AR	San Carlos - Mza	0.80	AR	Molinos
AR	Guaymallén	0.90	AR	Maipú	0.84	AR	Las Heras	0.80	AR	Junín - Mza
AR	Iglesia	0.89	AR	Sanagasta	0.89	AR	Angaco	0.84	AR	Santa Lucía
AR	Ischilin	0.70	CH	Graubünden	0.67	CH	St. Gallen	0.67	US	Benton Co.
AR	Jachal	0.82	AR	Veinticinco de Mayo - San Juan	0.82	AR	Sarmiento - San Juan	0.82	AR	Caucete
AR	Junín - Mza	0.98	AR	Rivadavia - Mza	0.94	AR	San Martín - Mza	0.91	AR	Santa Rosa - Mza
AR	La Paz	0.86	AR	Las Heras	0.86	AR	Santa Rosa - Mza	0.86	AR	Lavalle
AR	La Rioja	0.94	AR	Capital Santiago del Estero	0.86	AR	Valle Fértil	0.49	AR	Cruz del Eje
AR	Las Heras	0.94	AR	Lavalle	0.87	AR	Santa Rosa - Mza	0.86	AR	La Paz
AR	Lavalle	0.94	AR	Santa Rosa - Mza	0.94	AR	Las Heras	0.87	AR	San Martín - Mza
AR	Luján de Cuyo	0.95	AR	Maipú	0.94	AR	San Carlos - Mza	0.89	AR	Godoy Cruz
AR	Maipú	0.95	AR	Luján de Cuyo	0.90	AR	Guaymallén	0.88	AR	Tunuyán
AR	Molinos	0.86	AR	Luján de Cuyo	0.83	AR	San Carlos - Mza	0.80	AR	Godoy Cruz
AR	Nueve de Julio	0.97	AR	Sarmiento - San Juan	0.96	AR	Rawson	0.96	AR	Veinticinco de Mayo - San Juan
AR	Pichi Mahuida	0.82	AR	Conesa	0.78	AR	Adolfo Alsina	0.71	AR	Las Heras
AR	Pocito	0.97	AR	Santa Lucía	0.96	AR	Rawson	0.96	AR	Tinogasta
AR	Poman	0.80	AR	Rivadavia - San Juan	0.77	AR	Zonda	0.64	AR	Albardón
AR	Puelen	0.91	CH	Ticino	0.90	IT	Rovigo	0.88	FR	Gironde
AR	Rawson	0.98	AR	Santa Lucía	0.97	AR	Tinogasta	0.96	AR	Pocito
AR	Rivadavia - Mza	0.98	AR	Junín - Mza	0.94	AR	San Martín - Mza	0.93	AR	Santa Rosa - Mza
AR	Rivadavia - San Juan	0.90	AR	Albardón	0.85	AR	Chimbas	0.83	AR	Zonda
AR	San Alberto	1.00	AR	San Javier	0.98	AR	Tinogasta	0.95	AR	Santa Lucía
AR	San Blas De Los Sauces	0.97	AR	Tafí del Valle	0.96	AR	San Carlos - Salta	0.95	AR	Belén
AR	San Carlos - Mza	0.94	AR	Luján de Cuyo	0.89	AR	Tunuyán	0.88	AR	Maipú
AR	San Carlos - Salta	0.98	AR	Tafí del Valle	0.96	AR	San Blas De Los Sauces	0.94	AR	Belén
AR	San Javier	1.00	AR	San Alberto	0.96	AR	Tinogasta	0.94	AR	Santa Lucía
AR	San Martín - Mza	0.96	AR	Santa Rosa - Mza	0.95	AR	General Alvear	0.94	AR	Junín - Mza
AR	San Martín - San Juan	0.98	AR	Angaco	0.98	AR	Caucete	0.94	AR	Veinticinco de Mayo - San Juan
AR	San Rafael	0.94	AR	Santa Rosa - Mza	0.91	AR	Rivadavia - Mza	0.90	AR	General Alvear

Table 95 (cont.): Each region's 3 most similar winegrape regions in the world according to the VSI, 2000

AR	Sanagasta	0.92	AR	Vinchina	0.89	AR	Iglesia	0.88	AR	Coronel Felipe Varela
AR	Santa Lucía	0.98	AR	Rawson	0.97	AR	Tinogasta	0.97	AR	Pocito
AR	Santa María - Catamarca	0.87	AR	Cafayate	0.84	AR	Calingasta	0.83	AR	Castro Barros
AR	Santa Rosa - Mza	0.96	AR	San Martín - Mza	0.94	AR	Lavalle	0.94	AR	San Rafael
AR	Sarmiento - San Juan	0.97	AR	Nueve de Julio	0.96	AR	Veinticinco de Mayo - San Juan	0.92	AR	Caucete
AR	Tafí del Valle	0.98	AR	Cachi	0.98	AR	San Carlos - Salta	0.97	AR	San Blas De Los Sauces
AR	Tinogasta	0.98	AR	San Alberto	0.97	AR	Santa Lucía	0.97	AR	Rawson
AR	Totoral	0.78	AR	Godoy Cruz	0.76	AR	Añelo	0.75	AR	Luján de Cuyo
AR	Tunuyán	0.90	AR	Añelo	0.89	AR	San Carlos - Mza	0.89	AR	Luján de Cuyo
AR	Tupungato	0.88	AR	Tunuyán	0.84	AR	Añelo	0.81	AR	Castro Barros
AR	Ullum	0.81	AR	Zonda	0.71	AR	Albardón	0.68	AR	Veinticinco de Mayo - San Juan
AR	Valle Fértil	0.86	AR	La Rioja	0.76	AR	Capital Santiago del Estero	0.68	AR	Pocito
AR	Veinticinco de Mayo - San Juan	0.96	AR	Nueve de Julio	0.96	AR	Sarmiento - San Juan	0.95	AR	Caucete
AR	Vinchina	0.98	AR	Famatina	0.97	AR	Coronel Felipe Varela	0.94	AR	Belén
AR	Zonda	0.83	AR	Rivadavia - San Juan	0.81	AR	Ullum	0.77	AR	Poman
AM	Armenia	0.84	RU	Russia	0.81	AU	Northern Territory	0.79	ES	Malaga
AU	Adelaide Hills	0.94	AU	Yarra Valley	0.93	AU	Gippsland	0.92	AU	Alpine Valleys/Beechworth
AU	Alpine Valleys/Beechworth	0.94	AU	Blackwood Valley	0.93	AU	South West Australia - other	0.93	AU	North East Victoria - other
AU	Australian Capital Territory	0.94	AU	Pyrenees	0.93	AU	Barossa - other	0.90	AU	Central Victoria - other
AU	Barossa - other	0.94	AU	Far North - other	0.94	AU	Barossa Valley	0.93	AU	Australian Capital Territory
AU	Barossa Valley	0.97	AU	McLaren Vale	0.97	AU	Southern NSW - other	0.96	AU	Mount Lofty Ranges - other
AU	Beechworth	0.95	AU	Western Victoria - other	0.88	AU	Yarra Valley	0.88	AU	Canberra District
AU	Bendigo	0.99	AU	Far North - other	0.97	AU	Grampians	0.97	AU	McLaren Vale
AU	Big Rivers - other	0.85	AU	South Coast - other	0.84	AU	Hunter Valley - other	0.84	AU	Granite Belt
AU	Blackwood Valley	0.98	AU	South West Australia - other	0.97	AU	Geographe	0.96	AU	Great Southern
AU	Canberra District	0.95	AU	Padthaway	0.95	AU	Sunbury	0.95	AU	Central Victoria - other
AU	Central Ranges - other	0.97	AU	Northern Territory	0.97	AU	Queensland - other	0.96	RO	Romania
AU	Central Victoria - other	0.98	AU	Sunbury	0.96	AU	Goulburn Valley	0.96	AU	Padthaway
AU	Central Western Australia	0.87	AU	Eastern Plains, Inland and North WA	0.84	PT	Acores	0.83	PT	Madeira
AU	Clare Valley	0.96	AU	Hilltops	0.96	AU	Great Southern	0.95	AU	Mudgee
AU	Cowra	0.96	AU	Hunter Valley - other	0.96	AU	Hunter	0.92	US	Shasta
AU	Currency Creek	0.99	AU	Langhorne Creek	0.98	AU	Lower Murray - other	0.97	AU	Hilltops
AU	Eastern Plains, Inland and North WA	0.99	IT	Belluno	0.97	PT	Acores	0.95	PT	Madeira
AU	Eden Valley	0.92	AU	Clare Valley	0.89	AU	Canberra District	0.88	AU	Sunbury
AU	Far North - other	0.99	AU	Bendigo	0.97	AU	Grampians	0.96	AU	Pyrenees
AU	Fleurieu - other	0.99	AU	Mudgee	0.98	AU	Southern Fleurieu	0.98	AU	Mount Lofty Ranges - other
AU	Geelong	0.97	AU	Gippsland	0.97	AU	Port Phillip - other	0.96	AU	Mornington Peninsula
AU	Geographe	0.98	AU	Great Southern	0.97	AU	Mudgee	0.97	AU	Orange
AU	Gippsland	0.98	AU	Yarra Valley	0.97	AU	Geelong	0.96	AU	Port Phillip - other
AU	Goulburn Valley	0.96	AU	Mudgee	0.96	AU	Central Victoria - other	0.96	AU	Mount Lofty Ranges - other
AU	Grampians	0.97	AU	McLaren Vale	0.97	AU	Bendigo	0.97	AU	Far North - other
AU	Granite Belt	0.92	AU	South Burnett	0.87	AU	Northern Rivers - other	0.87	AU	South Coast - other
AU	Great Southern	0.98	AU	Geographe	0.97	AU	South West Australia - other	0.97	AU	Padthaway
AU	Greater Perth - other	0.75	AU	Swan District	0.69	AU	Perth Hills	0.63	AU	Central Western Australia
AU	Hastings River	0.91	AU	Northern Rivers - other	0.89	AU	South Coast - other	0.82	AU	Granite Belt
AU	Henty	0.94	AU	Gippsland	0.92	AU	Tasmania	0.92	AU	Geelong
AU	Hilltops	0.98	AU	Langhorne Creek	0.98	AU	Kangaroo Island	0.98	AU	Mudgee
AU	Hunter	0.98	AU	Hunter Valley - other	0.96	AU	Cowra	0.88	US	Shasta
AU	Hunter Valley - other	0.98	AU	Hunter	0.96	AU	Cowra	0.87	US	Shasta
AU	Kangaroo Island	0.98	AU	Langhorne Creek	0.98	AU	Mudgee	0.98	AU	Hilltops
AU	Langhorne Creek	0.99	AU	Currency Creek	0.98	AU	Hilltops	0.98	AU	Kangaroo Island
AU	Limestone Coast - other	0.96	AU	Langhorne Creek	0.96	AU	Lower Murray - other	0.96	AU	Currency Creek
AU	Lower Murray - other	0.98	AU	Currency Creek	0.98	AU	Langhorne Creek	0.97	AU	Hilltops
AU	Margaret River	0.95	AU	Geographe	0.94	AU	Blackwood Valley	0.94	AU	South West Australia - other
AU	McLaren Vale	0.99	AU	Mount Lofty Ranges - other	0.98	AU	Fleurieu - other	0.97	AU	Grampians
AU	Mornington Peninsula	0.96	AU	Tasmania	0.96	AU	Geelong	0.95	AU	Port Phillip - other
AU	Mount Benson	0.96	AU	North East Victoria - other	0.95	AU	Orange	0.94	AU	Geographe

Table 95 (cont.): Each region's 3 most similar winegrape regions in the world according to the VSI, 2000

AU	Mount Lofty Ranges - other	0.99	AU	McLaren Vale	0.98	AU	Fleurieu - other	0.98	AU	Mudgee
AU	Mudgee	0.99	AU	Orange	0.99	AU	Fleurieu - other	0.98	AU	Kangaroo Island
AU	Murray Darling (NSW)	0.95	AU	Murray Darling (VIC)	0.89	AU	Swan Hill (NSW)	0.88	AU	Swan Hill (VIC)
AU	Murray Darling (VIC)	0.97	AU	North West Victoria - other	0.95	AU	Murray Darling (NSW)	0.93	AU	Swan Hill (NSW)
AU	North East Victoria - other	0.96	AU	Mount Benson	0.96	AU	South West Australia - other	0.95	AU	Great Southern
AU	North West Victoria - other	0.97	AU	Swan Hill (NSW)	0.97	AU	Murray Darling (VIC)	0.95	AU	Swan Hill (VIC)
AU	Northern Rivers - other	0.91	AU	Hastings River	0.89	AU	South Coast - other	0.87	AU	Granite Belt
AU	Northern Slopes - other	0.96	AU	Mount Lofty Ranges - other	0.96	AU	Southern NSW - other	0.96	AU	Southern Fleurieu
AU	Northern Territory	0.97	AU	Central Ranges - other	0.95	AU	Queensland - other	0.95	RO	Romania
AU	Orange	0.99	AU	Mudgee	0.98	AU	Fleurieu - other	0.98	AU	Kangaroo Island
AU	Padthaway	0.97	AU	Great Southern	0.96	AU	Central Victoria - other	0.96	AU	Blackwood Valley
AU	Perricoota	0.91	AU	Lower Murray - other	0.91	AU	Fleurieu - other	0.91	AU	Riverland
AU	Perth Hills	0.80	AU	Cowra	0.77	AU	Central Victoria - other	0.77	AU	Hunter
AU	Port Phillip - other	0.97	AU	Geelong	0.96	AU	Gippsland	0.95	AU	Mornington Peninsula
AU	Pyrenees	0.96	AU	Mount Lofty Ranges - other	0.96	AU	Far North - other	0.96	AU	McLaren Vale
AU	Queensland - other	0.97	AU	Central Ranges - other	0.95	AU	Northern Territory	0.95	RO	Romania
AU	Riverina	0.93	AU	Northern Slopes - other	0.92	AU	Southern NSW - other	0.90	AU	Southern Fleurieu
AU	Riverland	0.97	AU	Mount Lofty Ranges - other	0.96	AU	Mudgee	0.95	AU	Fleurieu - other
AU	Rutherglen	0.92	AU	Bendigo	0.91	AU	Far North - other	0.90	AU	Grampians
AU	South Burnett	0.92	AU	Granite Belt	0.82	AU	Big Rivers - other	0.81	AU	Central Western Australia
AU	South Coast - other	0.89	AU	Northern Rivers - other	0.89	AU	Hastings River	0.87	AU	Granite Belt
AU	South West Australia - other	0.98	AU	Blackwood Valley	0.97	AU	Geographe	0.97	AU	Great Southern
AU	Southern Fleurieu	0.98	AU	Fleurieu - other	0.98	AU	Mudgee	0.98	AU	Mount Lofty Ranges - other
AU	Southern NSW - other	0.98	AU	Mount Lofty Ranges - other	0.98	AU	Southern Fleurieu	0.97	AU	McLaren Vale
AU	Sunbury	0.98	AU	Central Victoria - other	0.96	AU	Mount Lofty Ranges - other	0.96	AU	Mudgee
AU	Swan District	0.80	AU	South Burnett	0.76	AU	Central Western Australia	0.75	AU	Greater Perth - other
AU	Swan Hill (NSW)	0.97	AU	Swan Hill (VIC)	0.97	AU	North West Victoria - other	0.93	AU	Murray Darling (VIC)
AU	Swan Hill (VIC)	0.97	AU	Swan Hill (NSW)	0.95	AU	North West Victoria - other	0.90	AU	Murray Darling (VIC)
AU	Tasmania	0.96	AU	Mornington Peninsula	0.95	NZ	Wairarapa	0.94	AU	Gippsland
AU	The Peninsulas	0.93	AU	Orange	0.92	AU	Kangaroo Island	0.92	AU	Mudgee
AU	Tumbarumba	0.96	US	San Mateo	0.95	FR	Bourgogne	0.95	US	Santa Cruz
AU	Western Australia Southeast Coast	0.95	AU	Geographe	0.94	AU	Blackwood Valley	0.94	AU	South West Australia - other
AU	Western Plains - other	0.88	AU	Central Ranges - other	0.82	AU	Northern Territory	0.81	AU	Queensland - other
AU	Western Victoria - other	0.95	AU	Beechworth	0.94	AU	North East Victoria - other	0.93	AU	Hilltops
AU	Yarra Valley	0.98	AU	Gippsland	0.94	AU	Adelaide Hills	0.94	AU	Geelong
AT	Burgenland	0.75	SK	Slovakia	0.70	AT	Wien and other regions	0.69	AT	Niederosterreich
AT	Niederosterreich	0.88	AT	Wien and other regions	0.69	AT	Burgenland	0.62	CZ	Czechia
AT	Steiermark	0.57	SK	Slovakia	0.54	AT	Burgenland	0.54	RS	Serbia
AT	Wien and other regions	0.88	AT	Niederosterreich	0.71	CZ	Czechia	0.70	AT	Burgenland
BR	Brazil	0.38	US	New York - other	0.37	CA	Canada	0.36	AR	Colón - Entre Ríos
BG	Bulgaria	0.50	M9	Missing 9	0.49	AR	General Pueyrredón	0.45	FR	Gironde
CA	Canada	0.82	US	New York - other	0.70	US	Michigan	0.67	NZ	Auckland
CL	Araucania	0.98	US	Riverside	0.98	US	Santa Barbara	0.95	US	Santa Cruz
CL	Atacama	0.95	CL	Coquimbo	0.81	ES	Malaga	0.77	AU	Northern Territory
CL	Coquimbo	0.95	CL	Atacama	0.68	ES	Malaga	0.67	AU	Central Ranges - other
CL	Del Bio Bio	0.67	AU	Eastern Plains, Inland and North WA	0.67	PT	Acores	0.66	IT	Belluno
CL	Del Maule	0.86	CL	O'Higgins	0.84	CL	Metropolitana	0.79	US	Yuba
CL	Metropolitana	0.99	CL	O'Higgins	0.93	AU	Limestone Coast - other	0.90	US	Yuba
CL	O'Higgins	0.99	CL	Metropolitana	0.91	AU	Limestone Coast - other	0.88	AR	General Pueyrredón
CL	Valparaiso	0.96	US	Monterey	0.94	US	San Mateo	0.94	US	Santa Barbara
HR	Croatia	0.76	SI	Slovenia	0.73	RS	Serbia	0.70	HU	Hungary
CY	Cyprus	0.32	AU	Eastern Plains, Inland and North WA	0.31	IT	Belluno	0.31	PT	Acores
CZ	Czechia	0.87	SK	Slovakia	0.71	AT	Wien and other regions	0.70	HU	Hungary
FR	Aisne	0.95	FR	Seine-et-Marne	0.68	FR	Champagne-Ardenne	0.43	DE	Württemberg
FR	Alpes-de-Haute-Provence, Hautes-Alpes, Alpes-Maritimes	0.91	FR	Ardeche	0.87	FR	Gard	0.85	FR	Bouches-du-Rhone
FR	Alsace	0.67	DE	Hessische Bergstraße	0.62	DE	Rheingau	0.61	DE	Mittelrhein
FR	Aquitaine except Gironde	0.88	FR	Gironde	0.75	IT	Venezia	0.74	AR	Puelen

Table 95 (cont.): Each region's 3 most similar winegrape regions in the world according to the VSI, 2000

FR	Ardeche	0.91	FR	Alpes-de-Haute-Provence, Hautes-Alpes, Alpes-Maritimes	0.89	FR	Gard	0.82	FR	Bouches-du-Rhone
FR	Aude	0.97	FR	Herault	0.86	TN	Tunisia	0.78	DZ	Algeria
FR	Auvergne	0.87	FR	Rhone-Alpes except Ardeche	0.83	FR	Lorraine	0.70	CH	Geneva
FR	Bouches-du-Rhone	0.96	FR	Gard	0.93	FR	Var	0.91	FR	Vaucluse
FR	Bourgogne	0.95	AU	Tumbarumba	0.91	AU	Mornington Peninsula	0.91	US	San Mateo
FR	Centre-Val de Loire	0.90	FR	Deux-Sevres, Vienne	0.59	NZ	Marlborough	0.47	FR	Pays de la Loire except Mayenne
FR	Champagne-Ardenne	0.82	AU	Mornington Peninsula	0.82	FR	Seine-et-Marne	0.81	AU	Henty
FR	Charente	1.00	FR	Charente-Maritime	0.86	US	Los Angeles	0.81	IT	Terni
FR	Charente-Maritime	1.00	FR	Charente	0.87	US	Los Angeles	0.82	IT	Terni
FR	Correze, Haute-Vienne	0.77	AU	Eastern Plains, Inland and North WA	0.77	PT	Acores	0.76	PT	Madeira
FR	Corse	0.69	IT	Siena	0.67	IT	Firenze	0.66	IT	Arezzo
FR	Deux-Sevres, Vienne	0.90	FR	Centre-Val de Loire	0.59	FR	Pays de la Loire except Mayenne	0.51	ZA	Worcester
FR	Franche Comté	0.88	US	Santa Cruz	0.87	US	Santa Barbara	0.86	US	San Mateo
FR	Gard	0.96	FR	Bouches-du-Rhone	0.90	FR	Var	0.89	FR	Ardeche
FR	Gers	0.79	FR	Charente-Maritime	0.78	FR	Charente	0.69	US	Los Angeles
FR	Gironde	0.91	AR	General Pueyrredón	0.89	IT	Padova	0.88	IT	Bergamo
FR	Herault	0.97	FR	Aude	0.84	DZ	Algeria	0.82	FR	Var
FR	Lorraine	0.83	FR	Auvergne	0.68	FR	Rhone-Alpes except Ardeche	0.65	CH	Valais
FR	Midi-Pyrenees except Gers	0.71	AR	Luján de Cuyo	0.68	AR	San Carlos - Mza	0.67	AR	Godoy Cruz
FR	Pays de la Loire except Mayenne	0.59	FR	Deux-Sevres, Vienne	0.47	FR	Centre-Val de Loire	0.29	ZA	Worcester
FR	Pyrenees-Orientales	0.78	FR	Gard	0.75	FR	Bouches-du-Rhone	0.73	FR	Var
FR	Rhone-Alpes except Ardeche	0.87	FR	Auvergne	0.68	FR	Lorraine	0.62	CH	Geneva
FR	Seine-et-Marne	0.95	FR	Aisne	0.82	FR	Champagne-Ardenne	0.54	AU	Henty
FR	Var	0.93	FR	Bouches-du-Rhone	0.90	FR	Gard	0.89	DZ	Algeria
FR	Vaucluse	0.93	IT	Nuoro	0.91	FR	Bouches-du-Rhone	0.90	ES	Zaragoza
GE	Georgia	0.60	RU	Russia	0.50	AM	Armenia	0.40	M9	Missing 9
DE	Ahr	0.95	CH	Graubünden	0.95	CH	St. Gallen	0.95	CH	Schaffhausen
DE	Baden	0.94	CH	Other regions	0.92	CH	Schwyz	0.92	CH	Aargau
DE	Franken	0.88	DE	Rheinhessen	0.84	DE	Saale	0.72	CH	Lucerne
DE	Hessische Bergstraße	0.98	DE	Mittelrhein	0.97	DE	Rheingau	0.96	DE	Mosel
DE	Mittelrhein	0.99	DE	Rheingau	0.98	DE	Hessische Bergstraße	0.95	DE	Mosel
DE	Mosel	0.96	DE	Hessische Bergstraße	0.95	DE	Mittelrhein	0.92	DE	Rheingau
DE	Nahe	0.95	DE	Pfalz	0.88	DE	Rheinhessen	0.87	DE	Mosel
DE	Pfalz	0.95	DE	Nahe	0.88	DE	Rheinhessen	0.82	DE	Sachsen
DE	Rheingau	0.99	DE	Mittelrhein	0.97	DE	Hessische Bergstraße	0.92	DE	Mosel
DE	Rheinhessen	0.88	DE	Franken	0.88	DE	Saale	0.88	DE	Nahe
DE	Saale	0.88	DE	Rheinhessen	0.86	DE	Sachsen	0.84	DE	Franken
DE	Sachsen	0.86	DE	Saale	0.86	LU	Luxembourg	0.84	DE	Nahe
DE	Württemberg	0.62	DE	Hessische Bergstraße	0.61	DE	Mittelrhein	0.61	DE	Mosel
EL	Anatoliki Makedonia, Thraki	0.70	NZ	Marlborough	0.63	US	Lake	0.59	ZA	Stellenbosch
EL	Attiki	1.00	EL	Sterea Ellada	0.30	EL	Peloponnissos	0.05	EL	Kentriki Makedonia
EL	Dytiki Ellada	0.55	EL	Kentriki Makedonia	0.50	EL	Peloponnissos	0.41	AU	Eastern Plains, Inland and North WA
EL	Dytiki Makedonia	0.84	EL	Kentriki Makedonia	0.27	PT	Acores	0.26	AU	Eastern Plains, Inland and North WA
EL	Ionia Nisia	0.92	AU	Northern Territory	0.89	ES	Malaga	0.89	AU	Central Ranges - other
EL	Ipeiros	0.35	PT	Acores	0.35	AU	Eastern Plains, Inland and North WA	0.35	PT	Madeira
EL	Kentriki Makedonia	0.84	EL	Dytiki Makedonia	0.55	EL	Dytiki Ellada	0.50	AU	Eastern Plains, Inland and North WA
EL	Kriti	0.08	AU	Eastern Plains, Inland and North WA	0.08	IT	Belluno	0.08	PT	Acores
EL	Notio Aigaio	0.44	AU	Northern Territory	0.43	AU	Central Ranges - other	0.42	ES	Malaga
EL	Peloponnissos	0.50	EL	Dytiki Ellada	0.31	EL	Kentriki Makedonia	0.30	EL	Sterea Ellada
EL	Sterea Ellada	1.00	EL	Attiki	0.30	EL	Peloponnissos	0.06	EL	Kentriki Makedonia
EL	Thessalia	0.28	EL	Kentriki Makedonia	0.21	EL	Dytiki Ellada	0.11	EL	Peloponnissos

Table 95 (cont.): Each region's 3 most similar winegrape regions in the world according to the VSI, 2000

EL	Voreio Aigaio	0.61	IT	Cuneo	0.46	EL	Anatoliki Makedonia, Thraki	0.39	IT	Asti
HU	Hungary	0.88	SI	Slovenia	0.88	RO	Romania	0.85	AU	Central Ranges - other
IL	Israel	0.68	FR	Aude	0.66	TN	Tunisia	0.65	FR	Herault
IT	Agrigento	0.64	IT	Palermo	0.60	IT	Caltanissetta	0.60	IT	Trapani
IT	Alessandria	0.90	IT	Asti	0.75	IT	Salerno	0.69	IT	Torino
IT	Ancona	0.91	IT	Macerata	0.34	IT	Brescia	0.21	IT	Ascoli Piceno
IT	Arezzo	1.00	IT	Firenze	0.99	IT	Prato	0.98	IT	Siena
IT	Ascoli Piceno	0.86	IT	Bari	0.85	IT	Isernia	0.84	IT	Arezzo
IT	Asti	0.90	IT	Alessandria	0.81	IT	Salerno	0.73	IT	Verbano-Cusio-Ossola
IT	Avellino	0.94	IT	Potenza	0.75	IT	Benevento	0.42	IT	Salerno
IT	Bari	0.86	IT	Ascoli Piceno	0.76	IT	Isernia	0.76	IT	Matera
IT	Belluno	0.99	AU	Eastern Plains, Inland and North WA	0.94	PT	Acores	0.93	PT	Madeira
IT	Benevento	0.77	IT	Potenza	0.75	IT	Avellino	0.55	IT	Rieti
IT	Bergamo	0.88	FR	Gironde	0.87	US	Oregon - other	0.84	AR	General Pueyrredón
IT	Biella	0.82	IT	Verbano-Cusio-Ossola	0.81	IT	Torino	0.77	IT	Varese
IT	Bologna	0.89	IT	Ravenna	0.83	IT	Ferrara	0.71	IT	Forlì-Cesena
IT	Bolzano-Bozen	0.47	DE	Württemberg	0.43	IT	Trento	0.35	US	Josephine Co.
IT	Brescia	0.69	US	Nevada	0.67	US	Santa Clara	0.67	US	Trinity
IT	Brindisi	0.99	IT	Lecce	0.25	IT	Taranto	0.09	IT	Bari
IT	Cagliari	0.61	IT	Oristano	0.45	TN	Tunisia	0.44	FR	Aude
IT	Caltanissetta	0.96	IT	Siracusa	0.95	IT	Ragusa	0.60	IT	Agrigento
IT	Campobasso	0.98	IT	Pescara	0.97	IT	L'Aquila	0.96	IT	Teramo
IT	Caserta	0.71	IT	Isernia	0.67	IT	Matera	0.62	IT	Ascoli Piceno
IT	Catania	0.61	IT	Messina	0.45	IT	Enna	0.20	IT	Catanzaro
IT	Catanzaro	0.68	IT	Cosenza	0.61	IT	Reggio di Calabria	0.52	IT	Vibo Valentia
IT	Chieti	0.94	IT	Campobasso	0.89	IT	Pescara	0.88	IT	Teramo
IT	Como	0.65	IT	Messina	0.59	IT	Reggio di Calabria	0.48	IT	Valle d'Aosta
IT	Cosenza	0.74	IT	Crotone	0.68	IT	Catanzaro	0.66	IT	Reggio di Calabria
IT	Cremona	0.75	IT	Reggio nell'Emilia	0.49	IT	Mantova	0.37	IT	Potenza
IT	Crotone	0.74	IT	Cosenza	0.30	IT	Reggio di Calabria	0.24	IT	Catanzaro
IT	Cuneo	0.66	IT	Alessandria	0.61	EL	Voreio Aigaio	0.60	IT	Asti
IT	Enna	0.81	IT	Messina	0.81	AU	Eastern Plains, Inland and North WA	0.80	IT	Belluno
IT	Ferrara	0.95	IT	Ravenna	0.83	IT	Bologna	0.55	IT	Forlì-Cesena
IT	Firenze	1.00	IT	Arezzo	0.99	IT	Siena	0.98	IT	Prato
IT	Foggia	0.89	IT	Pisa	0.87	IT	Lucca	0.86	IT	Grosseto
IT	Forlì-Cesena	0.94	IT	Rimini	0.81	IT	Siena	0.80	IT	Firenze
IT	Frosinone	0.65	IT	Latina	0.64	IT	Rieti	0.63	IT	Viterbo
IT	Genova	0.77	IT	La Spezia	0.65	IT	Massa-Carrara	0.51	IT	Savona
IT	Gorizia	0.89	IT	Udine	0.82	IT	Pordenone	0.75	IT	Venezia
IT	Grosseto	0.97	IT	Pisa	0.96	IT	Pistoia	0.95	IT	Prato
IT	Imperia	0.78	IT	Savona	0.56	IT	Massa-Carrara	0.53	IT	Sassari
IT	Isernia	0.85	IT	Ascoli Piceno	0.76	IT	Bari	0.74	IT	Matera
IT	La Spezia	0.77	IT	Genova	0.66	IT	Massa-Carrara	0.58	IT	Savona
IT	L'Aquila	0.99	IT	Pescara	0.97	IT	Campobasso	0.96	IT	Teramo
IT	Latina	0.90	IT	Viterbo	0.86	IT	Roma	0.77	FR	Charente-Maritime
IT	Lecce	0.99	IT	Brindisi	0.17	IT	Taranto	0.05	IT	Matera
IT	Lecco	0.75	US	Lake	0.74	US	Napa	0.72	NZ	Hawkes Bay
IT	Livorno	0.85	IT	Lucca	0.84	IT	Pisa	0.81	IT	Grosseto
IT	Lodi	0.81	IT	Milano	0.74	IT	Piacenza	0.69	IT	Pavia
IT	Lucca	0.97	IT	Pisa	0.95	IT	Grosseto	0.91	IT	Pistoia
IT	Macerata	0.91	IT	Ancona	0.52	IT	Ascoli Piceno	0.45	IT	Bari
IT	Mantova	0.49	IT	Cremona	0.48	US	Oregon - other	0.48	IT	Rovigo
IT	Massa-Carrara	0.72	IT	Savona	0.66	IT	La Spezia	0.65	IT	Genova
IT	Matera	0.83	IT	Siena	0.82	IT	Firenze	0.82	IT	Arezzo
IT	Messina	0.81	IT	Enna	0.68	IT	Reggio di Calabria	0.65	IT	Como
IT	Milano	0.85	IT	Piacenza	0.81	IT	Lodi	0.68	IT	Pavia
IT	Modena	0.44	IT	Reggio nell'Emilia	0.32	IT	Cremona	0.23	IT	Mantova
IT	Napoli	0.35	IT	Caserta	0.31	IT	Benevento	0.26	AU	Northern Territory
IT	Novara	0.86	IT	Vercelli	0.80	IT	Sondrio	0.65	IT	Biella
IT	Nuoro	0.93	FR	Vaucluse	0.92	ES	Zaragoza	0.88	ES	Huesca, Teruel
IT	Oristano	0.61	IT	Cagliari	0.45	AU	Eastern Plains, Inland and North WA	0.45	IT	Belluno

Table 95 (cont.): Each region's 3 most similar winegrape regions in the world according to the VSI, 2000

	Region			Most similar 1			Most similar 2			Most similar 3
IT	Padova	0.94	IT	Rovigo	0.91	CH	Ticino	0.90	IT	Venezia
IT	Palermo	0.98	IT	Trapani	0.64	IT	Agrigento	0.17	FR	Charente-Maritime
IT	Parma	0.47	IT	Piacenza	0.40	IT	Pavia	0.39	AR	Andalgala
IT	Pavia	0.69	IT	Lodi	0.69	IT	Asti	0.68	IT	Milano
IT	Perugia	0.88	IT	Pisa	0.86	IT	Lucca	0.86	IT	Grosseto
IT	Pesaro e Urbino	0.63	IT	Rimini	0.59	IT	Siena	0.58	IT	Firenze
IT	Pescara	0.99	IT	L'Aquila	0.99	IT	Teramo	0.98	IT	Campobasso
IT	Piacenza	0.85	IT	Milano	0.74	IT	Lodi	0.67	IT	Pavia
IT	Pisa	0.97	IT	Lucca	0.97	IT	Grosseto	0.95	IT	Pistoia
IT	Pistoia	0.98	IT	Arezzo	0.98	IT	Prato	0.97	IT	Firenze
IT	Pordenone	0.96	IT	Venezia	0.95	IT	Udine	0.86	IT	Padova
IT	Potenza	0.94	IT	Avellino	0.77	IT	Benevento	0.45	IT	Salerno
IT	Prato	0.99	IT	Arezzo	0.98	IT	Firenze	0.98	IT	Pistoia
IT	Ragusa	0.98	IT	Siracusa	0.95	IT	Caltanissetta	0.51	IT	Agrigento
IT	Ravenna	0.95	IT	Ferrara	0.89	IT	Bologna	0.66	IT	Forlì-Cesena
IT	Reggio di Calabria	0.69	IT	Enna	0.68	IT	Messina	0.66	IT	Cosenza
IT	Reggio nell'Emilia	0.75	IT	Cremona	0.46	IT	Mantova	0.44	IT	Modena
IT	Rieti	0.82	IT	Roma	0.79	IT	Viterbo	0.69	IT	Foggia
IT	Rimini	0.95	IT	Siena	0.94	IT	Firenze	0.94	IT	Forlì-Cesena
IT	Roma	0.86	IT	Latina	0.83	IT	Viterbo	0.82	IT	Rieti
IT	Rovigo	0.98	CH	Ticino	0.94	IT	Padova	0.90	AR	Puelen
IT	Salerno	0.81	IT	Asti	0.75	IT	Alessandria	0.71	IT	Verbano-Cusio-Ossola
IT	Sassari	0.68	IT	Savona	0.61	IT	Massa-Carrara	0.53	IT	Imperia
IT	Savona	0.78	IT	Imperia	0.72	IT	Massa-Carrara	0.68	IT	Sassari
IT	Siena	0.99	IT	Firenze	0.98	IT	Arezzo	0.96	IT	Prato
IT	Siracusa	0.98	IT	Ragusa	0.96	IT	Caltanissetta	0.50	IT	Agrigento
IT	Sondrio	0.97	IT	Vercelli	0.80	IT	Novara	0.57	IT	Verbano-Cusio-Ossola
IT	Taranto	0.88	US	Amador	0.87	US	Colusa	0.78	US	San Bernardino
IT	Teramo	0.99	IT	Pescara	0.96	IT	L'Aquila	0.96	IT	Campobasso
IT	Terni	0.84	IT	Viterbo	0.83	IT	Perugia	0.82	FR	Charente-Maritime
IT	Torino	0.81	IT	Biella	0.73	IT	Asti	0.69	IT	Verbano-Cusio-Ossola
IT	Trapani	0.98	IT	Palermo	0.60	IT	Agrigento	0.11	IT	Vicenza
IT	Trento	0.85	US	Yolo	0.85	US	Monterey	0.83	US	Tehama
IT	Treviso	0.67	IT	Padova	0.64	IT	Venezia	0.63	IT	Pordenone
IT	Trieste	0.23	HR	Croatia	0.20	IT	Gorizia	0.13	NZ	Marlborough
IT	Udine	0.95	IT	Pordenone	0.92	IT	Venezia	0.89	IT	Gorizia
IT	Valle d'Aosta	0.52	IT	Vercelli	0.51	IT	Novara	0.48	IT	Como
IT	Varese	0.77	IT	Biella	0.72	IT	Verbano-Cusio-Ossola	0.61	IT	Vercelli
IT	Venezia	0.96	IT	Pordenone	0.92	IT	Udine	0.90	IT	Padova
IT	Verbano-Cusio-Ossola	0.82	IT	Biella	0.73	IT	Asti	0.72	IT	Varese
IT	Vercelli	0.97	IT	Sondrio	0.86	IT	Novara	0.67	IT	Biella
IT	Verona	0.74	IT	Vicenza	0.14	IT	Padova	0.12	IT	Mantova
IT	Vibo Valentia	0.78	AU	Eastern Plains, Inland and North WA	0.78	PT	Acores	0.77	IT	Belluno
IT	Vicenza	0.74	IT	Verona	0.56	IT	Padova	0.50	IT	Venezia
IT	Viterbo	0.90	IT	Latina	0.84	IT	Terni	0.83	IT	Roma
KR	KoreaRep	0.63	TW	Taiwan	0.25	PT	Acores	0.25	PT	Madeira
LU	Luxembourg	0.86	DE	Sachsen	0.74	DE	Saale	0.71	CH	Lucerne
M9	Missing 9	0.84	CL	Metropolitana	0.84	CL	O'Higgins	0.79	AU	Limestone Coast - other
MD	Moldova	0.69	M9	Missing 9	0.51	AU	Adelaide Hills	0.49	AU	Alpine Valleys/Beechworth
MA	Morocco	0.62	PT	Acores	0.62	AU	Eastern Plains, Inland and North WA	0.61	PT	Madeira
NZ	Auckland	0.94	US	Napa	0.93	US	Nevada	0.93	NZ	Hawkes Bay
NZ	Canterbury	0.91	AU	Tasmania	0.91	NZ	Waipara	0.88	NZ	Otago
NZ	Gisborne	0.94	US	Riverside	0.94	US	Santa Barbara	0.94	CL	Araucania
NZ	Hawkes Bay	0.96	US	Solano	0.93	US	Nevada	0.93	NZ	Auckland
NZ	Marlborough	0.86	NZ	Nelson	0.80	NZ	Waipara	0.73	NZ	Wairarapa
NZ	Nelson	0.96	NZ	Waipara	0.88	NZ	Wairarapa	0.86	AU	Tasmania
NZ	Otago	0.98	US	Polk Co.	0.98	US	Humboldt	0.97	US	Yamhill Co.
NZ	Waikato	0.81	ZA	Stellenbosch	0.74	AU	Big Rivers - other	0.72	NZ	Hawkes Bay
NZ	Waipara	0.97	NZ	Wairarapa	0.96	NZ	Nelson	0.92	AU	Tasmania
NZ	Wairarapa	0.97	NZ	Waipara	0.95	AU	Tasmania	0.92	NZ	Otago
PT	Acores	0.98	PT	Madeira	0.97	AU	Eastern Plains, Inland and North WA	0.94	IT	Belluno
PT	Alentejo	0.64	PT	Ribatejo e Oeste	0.55	PT	Algarve	0.42	PT	Alto Tras-os-Montes

Table 95 (cont.): Each region's 3 most similar winegrape regions in the world according to the VSI, 2000

PT	Algarve	0.55	PT	Alentejo	0.54	PT	Ribatejo e Oeste	0.53	PT	Madeira
PT	Alto Tras-os-Montes	0.84	PT	Beira Interior	0.80	AU	Queensland - other	0.78	IT	Belluno
PT	Beira Interior	0.84	PT	Alto Tras-os-Montes	0.82	IT	Belluno	0.82	AU	Queensland - other
PT	Beira Litoral	0.60	PT	Beira Interior	0.57	IT	Belluno	0.56	AU	Eastern Plains, Inland and North WA
PT	Entre Douro e Minho	0.45	IT	Belluno	0.44	AU	Eastern Plains, Inland and North WA	0.43	PT	Beira Interior
PT	Madeira	0.98	PT	Acores	0.95	AU	Eastern Plains, Inland and North WA	0.93	IT	Belluno
PT	Ribatejo e Oeste	0.64	PT	Alentejo	0.54	PT	Algarve	0.53	PT	Beira Litoral
RO	Romania	0.96	AU	Central Ranges - other	0.95	AU	Queensland - other	0.95	AU	Northern Territory
RU	Russia	0.84	AM	Armenia	0.79	AU	Queensland - other	0.78	AU	Central Ranges - other
RS	Serbia	0.73	HR	Croatia	0.68	ES	Caceres	0.67	SK	Slovakia
SK	Slovakia	0.87	CZ	Czechia	0.75	AT	Burgenland	0.67	RS	Serbia
SI	Slovenia	0.94	RO	Romania	0.92	AU	Central Ranges - other	0.91	AU	Queensland - other
ZA	Breedekloof	0.96	ZA	Olifants River	0.94	ZA	Worcester	0.86	ZA	Little Karoo
ZA	Little Karoo	0.90	ZA	Olifants River	0.89	ZA	Northern Cape	0.87	ZA	Worcester
ZA	Northern Cape	0.89	ZA	Little Karoo	0.89	US	Kern	0.88	US	Madera
ZA	Olifants River	0.96	ZA	Breedekloof	0.94	ZA	Worcester	0.90	ZA	Little Karoo
ZA	Paarl	0.99	ZA	Swartland	0.88	ZA	Worcester	0.84	ZA	Breedekloof
ZA	Robertson	0.91	ZA	Worcester	0.87	ZA	Little Karoo	0.83	ZA	Olifants River
ZA	Stellenbosch	0.81	NZ	Waikato	0.80	ZA	Paarl	0.79	ZA	Swartland
ZA	Swartland	0.99	ZA	Paarl	0.88	ZA	Worcester	0.83	ZA	Breedekloof
ZA	Worcester	0.94	ZA	Olifants River	0.94	ZA	Breedekloof	0.91	ZA	Robertson
ES	Alava	0.95	ES	Burgos	0.91	ES	La Rioja	0.82	ES	Guadalajara
ES	Albacete	0.84	ES	Cuenca	0.77	ES	Ciudad Real	0.75	ES	Toledo
ES	Alicante	0.74	ES	Region de Murcia	0.73	ES	Malaga	0.70	AU	Northern Territory
ES	Almeria, Granada, Jaen, Sevilla	0.73	AU	Queensland - other	0.71	PT	Beira Interior	0.71	IT	Belluno
ES	Avila, Palencia, Salamanca, Segovia, Soria	0.88	ES	Zaragoza	0.88	ES	Comunidad Foral de Navarra	0.88	IT	Nuoro
ES	Badajoz	0.07	ES	Caceres	0.06	ES	Almeria, Granada, Jaen, Sevilla	0.05	ES	Malaga
ES	Barcelona	0.76	ES	Tarragona	0.61	ES	Girona, Lleida	0.26	ES	Castellon
ES	Burgos	0.95	ES	Alava	0.88	ES	La Rioja	0.80	ES	Guadalajara
ES	Caceres	0.68	RS	Serbia	0.54	HR	Croatia	0.53	SK	Slovakia
ES	Cadiz	0.88	ES	Canarias	0.68	ES	Cantabria	0.38	ES	Galicia
ES	Canarias	0.88	ES	Cadiz	0.60	ES	Cantabria	0.39	ES	Galicia
ES	Cantabria	0.84	ES	Leon	0.68	ES	Cadiz	0.60	ES	Canarias
ES	Castellon	0.77	ES	Comunidad Foral de Navarra	0.75	ES	La Rioja	0.74	ES	Valladolid
ES	Ciudad Real	0.97	ES	Toledo	0.77	ES	Albacete	0.72	ES	Cuenca
ES	Comunidad de Madrid	0.80	ES	Toledo	0.73	IT	Nuoro	0.72	ES	Zaragoza
ES	Comunidad Foral de Navarra	0.88	ES	Avila, Palencia, Salamanca, Segovia, Soria	0.85	ES	Huesca, Teruel	0.83	ES	Zaragoza
ES	Cordoba	0.44	CL	Coquimbo	0.33	CL	Atacama	0.16	ES	Malaga
ES	Cuenca	0.84	ES	Albacete	0.72	ES	Ciudad Real	0.70	ES	Toledo
ES	Galicia	0.50	ES	Leon	0.47	ES	Almeria, Granada, Jaen, Sevilla	0.44	IT	Belluno
ES	Girona, Lleida	0.85	ES	Tarragona	0.64	FR	Pyrenees-Orientales	0.61	ES	Barcelona
ES	Guadalajara	0.82	ES	Alava	0.80	ES	Burgos	0.74	ES	La Rioja
ES	Guipuzcoa, Vizcaya	0.06	AU	Northern Territory	0.06	AU	Central Ranges - other	0.06	ES	Malaga
ES	Huelva	0.07	ES	Almeria, Granada, Jaen, Sevilla	0.06	IT	Belluno	0.06	AU	Eastern Plains, Inland and North WA
ES	Huesca, Teruel	0.98	ES	Zaragoza	0.88	IT	Nuoro	0.87	FR	Vaucluse
ES	Illes Balears	0.40	ZA	Stellenbosch	0.39	AU	South Coast - other	0.39	AU	Granite Belt
ES	La Rioja	0.91	ES	Alava	0.88	ES	Burgos	0.81	ES	Comunidad Foral de Navarra
ES	Leon	0.84	ES	Cantabria	0.75	ES	Principado de Asturias	0.50	ES	Galicia
ES	Malaga	0.94	AU	Northern Territory	0.90	AU	Central Ranges - other	0.89	EL	Ionia Nisia
ES	Principado de Asturias	0.75	ES	Leon	0.66	PT	Acores	0.65	PT	Madeira
ES	Region de Murcia	0.74	ES	Alicante	0.51	ES	Albacete	0.51	US	Contra Costa
ES	Tarragona	0.85	ES	Girona, Lleida	0.76	ES	Barcelona	0.53	FR	Pyrenees-Orientales
ES	Toledo	0.97	ES	Ciudad Real	0.80	ES	Comunidad de Madrid	0.75	ES	Albacete
ES	Valencia	0.69	ES	Cuenca	0.42	ES	Albacete	0.17	AU	Queensland - other
ES	Valladolid	0.84	ES	Zamora	0.74	ES	Castellon	0.72	ES	Burgos

Table 95 (cont.): Each region's 3 most similar winegrape regions in the world according to the VSI, 2000

ES	Zamora	0.84	ES	Valladolid	0.73	ES	Castellon	0.70	ES	La Rioja
ES	Zaragoza	0.98	ES	Huesca, Teruel	0.92	IT	Nuoro	0.90	FR	Vaucluse
CH	Aargau	1.00	CH	Zürich	0.99	CH	Thurgau	0.99	CH	Other regions
CH	Basel Land	0.99	CH	St. Gallen	0.99	CH	Schaffhausen	0.98	CH	Thurgau
CH	Bern	0.98	CH	Neuchâtel	0.95	CH	Fribourg	0.90	CH	Valais
CH	Fribourg	0.97	CH	Vaud	0.95	CH	Bern	0.92	CH	Neuchâtel
CH	Geneva	0.84	CH	Vaud	0.80	CH	Valais	0.73	CH	Fribourg
CH	Graubünden	0.99	CH	St. Gallen	0.99	CH	Schaffhausen	0.98	CH	Basel Land
CH	Jura	0.61	CH	Schwyz	0.59	CH	Lucerne	0.55	CH	Aargau
CH	Lucerne	0.95	CH	Other regions	0.93	CH	Schwyz	0.91	DE	Baden
CH	Neuchâtel	0.98	CH	Bern	0.92	CH	Valais	0.92	CH	Fribourg
CH	Other regions	0.99	CH	Aargau	0.98	CH	Schwyz	0.97	CH	Zürich
CH	Schaffhausen	1.00	CH	St. Gallen	0.99	CH	Basel Land	0.99	CH	Graubünden
CH	Schwyz	0.98	CH	Other regions	0.98	CH	Aargau	0.98	CH	Zürich
CH	St. Gallen	1.00	CH	Schaffhausen	0.99	CH	Basel Land	0.99	CH	Graubünden
CH	Thurgau	1.00	CH	Zürich	0.99	CH	Aargau	0.98	CH	Schaffhausen
CH	Ticino	0.98	IT	Rovigo	0.91	IT	Padova	0.91	AR	Puelen
CH	Valais	0.92	CH	Neuchâtel	0.90	CH	Bern	0.82	CH	Fribourg
CH	Vaud	0.97	CH	Fribourg	0.89	CH	Bern	0.84	CH	Neuchâtel
CH	Zürich	1.00	CH	Thurgau	1.00	CH	Aargau	0.98	CH	Basel Land
TW	Taiwan	0.63	KR	KoreaRep	0.01	AR	Avellaneda - Río Negro	0.00	AR	Ullum
TN	Tunisia	0.86	FR	Aude	0.81	FR	Herault	0.70	DZ	Algeria
UK	UnitedKingdom	0.62	AU	Central Ranges - other	0.62	AU	Northern Territory	0.61	AU	Queensland - other
US	Alameda	0.96	US	Santa Clara	0.95	US	Solano	0.95	US	Nevada
US	Amador	0.99	US	Colusa	0.89	US	San Bernardino	0.88	IT	Taranto
US	Benton Co.	0.99	US	Yamhill Co.	0.98	US	Polk Co.	0.97	CH	Graubünden
US	Butte	0.85	US	El Dorado	0.76	US	San Joaquin	0.72	US	Amador
US	Calaveras	0.94	US	Mendocino	0.93	US	Solano	0.92	US	Sacramento
US	Chautauqua-Erie	0.85	US	Finger Lakes	0.56	US	New York - other	0.47	CA	Canada
US	Columbia River	0.82	AU	North East Victoria - other	0.82	US	Oregon - other	0.81	AR	General Pueyrredón
US	Colusa	0.99	US	Amador	0.89	US	San Bernardino	0.87	IT	Taranto
US	Contra Costa	0.80	US	San Joaquin	0.72	US	Mendocino	0.71	US	Sutter
US	Douglas Co.	0.92	US	Washington Co.	0.92	US	Josephine Co.	0.90	US	Valley - other
US	El Dorado	0.93	US	San Joaquin	0.91	US	Mariposa	0.85	US	Calaveras
US	Finger Lakes	0.85	US	Chautauqua-Erie	0.65	US	New York - other	0.59	CA	Canada
US	Fresno	0.93	US	Kern	0.91	US	Madera	0.88	US	Stanislaus
US	Glenn	0.71	US	Madera	0.68	US	Stanislaus	0.67	IT	Nuoro
US	Humboldt	0.98	NZ	Otago	0.97	US	Polk Co.	0.97	US	Yamhill Co.
US	Josephine Co.	0.97	US	Valley - other	0.97	US	Washington Co.	0.95	US	Lane Co.
US	Kern	0.95	US	Madera	0.93	US	Tulare	0.93	US	Fresno
US	Kings	0.75	US	Tulare	0.72	US	Kern	0.64	US	Fresno
US	Lake	0.90	US	San Luis Obispo	0.88	US	Napa	0.88	US	Calaveras
US	Lane Co.	0.98	US	Marion Co.	0.95	US	Josephine Co.	0.95	US	Washington Co.
US	Los Angeles	0.87	FR	Charente-Maritime	0.86	FR	Charente	0.72	IT	Terni
US	Madera	0.95	US	Kern	0.91	US	Stanislaus	0.91	US	Fresno
US	Marin	0.94	AR	Cushamen	0.92	US	Oregon - other	0.86	US	Trinity
US	Marion Co.	0.98	US	Lane Co.	0.97	US	Washington Co.	0.97	US	Valley - other
US	Mariposa	0.91	US	El Dorado	0.86	US	Calaveras	0.85	US	San Luis Obispo
US	Mendocino	0.97	US	Sonoma	0.96	US	Santa Clara	0.95	US	Alameda
US	Merced	0.85	US	Stanislaus	0.81	US	Tulare	0.79	US	Sacramento
US	Michigan	0.73	NZ	Canterbury	0.71	US	Douglas Co.	0.70	CA	Canada
US	Monterey	0.98	US	Yolo	0.98	US	Riverside	0.98	US	Santa Barbara
US	Napa	0.97	US	San Luis Obispo	0.94	US	San Benito	0.94	NZ	Auckland
US	Nevada	0.96	US	Santa Clara	0.96	US	Sonoma	0.96	US	Solano
US	New York - other	0.82	CA	Canada	0.70	US	Washington	0.69	US	Trinity
US	Oregon - other	0.92	US	Marin	0.90	US	Trinity	0.88	FR	Gironde
US	Placer	0.86	US	San Bernardino	0.75	US	Amador	0.74	US	Colusa
US	Polk Co.	1.00	US	Yamhill Co.	0.98	NZ	Otago	0.98	US	Benton Co.
US	Riverside	0.98	US	Yolo	0.98	US	Santa Barbara	0.98	CL	Araucania
US	Sacramento	0.97	US	Solano	0.96	US	Santa Clara	0.95	US	Nevada
US	San Benito	0.95	US	Sonoma	0.94	US	San Luis Obispo	0.94	US	Napa
US	San Bernardino	0.89	US	Amador	0.89	US	Colusa	0.86	US	Placer
US	San Diego	0.71	AU	Australian Capital Territory	0.69	AU	Alpine Valleys/Beechworth	0.69	US	Washington
US	San Joaquin	0.93	US	El Dorado	0.84	US	Colusa	0.84	US	Calaveras

Table 95 (cont.): Each region's 3 most similar winegrape regions in the world according to the VSI, 2000

US	San Luis Obispo	0.97	US	Napa	0.95	US	Sonoma	0.94	US	San Benito
US	San Mateo	0.98	US	Santa Cruz	0.97	US	Santa Barbara	0.96	AU	Tumbarumba
US	Santa Barbara	0.99	US	Santa Cruz	0.98	US	Riverside	0.98	CL	Araucania
US	Santa Clara	0.97	US	Sonoma	0.96	US	Nevada	0.96	US	Alameda
US	Santa Cruz	0.99	US	Santa Barbara	0.98	US	San Mateo	0.96	US	Monterey
US	Shasta	0.92	AU	Cowra	0.88	AU	Hunter	0.87	AU	Hunter Valley - other
US	Solano	0.97	US	Sacramento	0.96	US	Sonoma	0.96	US	Santa Clara
US	Sonoma	0.97	US	Santa Clara	0.97	US	Mendocino	0.96	US	Solano
US	Stanislaus	0.91	US	Madera	0.88	US	Fresno	0.87	US	Kern
US	Sutter	0.83	US	Mendocino	0.82	US	Riverside	0.82	US	Yolo
US	Tehama	0.98	US	Yolo	0.95	US	Riverside	0.94	US	Monterey
US	Trinity	0.94	US	Washington	0.90	US	Oregon - other	0.90	US	Sacramento
US	Tulare	0.93	US	Kern	0.89	US	Madera	0.88	US	Fresno
US	Valley - other	0.98	US	Washington Co.	0.97	US	Josephine Co.	0.97	US	Marion Co.
US	Ventura	0.86	AU	Far North - other	0.83	AU	Bendigo	0.83	AU	Barossa - other
US	Washington	0.95	US	Sacramento	0.94	US	Trinity	0.92	US	Nevada
US	Washington Co.	0.98	US	Valley - other	0.97	US	Josephine Co.	0.97	US	Marion Co.
US	Yamhill Co.	1.00	US	Polk Co.	0.99	US	Benton Co.	0.97	NZ	Otago
US	Yolo	0.98	US	Monterey	0.98	US	Riverside	0.98	US	Tehama
US	Yuba	0.90	CL	Metropolitana	0.86	CL	O'Higgins	0.82	AU	Limestone Coast - other
UY	Uruguay	0.46	AR	Coronel Pringles	0.34	FR	Correze, Haute-Vienne	0.33	IT	Bergamo

Table 96: Each region's 3 most similar winegrape regions in the world according to the VSI, 2010

DZ	Algeria	0.84	FR	Var	0.74	FR	Bouches-du-Rhone	0.72	FR	Herault
AR	Adolfo Alsina	0.91	AR	Puelen	0.89	AR	Añelo	0.88	AR	Tunuyán
AR	Albardón	0.97	AR	Chimbas	0.93	AR	La Rioja	0.90	AR	Santa Lucía
AR	Ambato	0.86	AU	Bendigo	0.85	AU	Central Western Australia	0.84	AU	Central Victoria - other
AR	Andalgala	0.61	AR	Arauco	0.60	AR	Tinogasta	0.60	AR	Santa Lucía
AR	Añelo	0.95	AR	Tunuyán	0.91	AR	Tupungato	0.91	AR	Luján de Cuyo
AR	Angaco	0.96	AR	San Martín - San Juan	0.96	AR	Caucete	0.95	AR	Valle Fértil
AR	Arauco	0.98	AR	Tinogasta	0.97	AR	Santa Lucía	0.96	AR	Rawson
AR	Avellaneda - Río Negro	0.81	AR	Maipú	0.79	AR	Tunuyán	0.78	AR	Confluencia
AR	Belén	0.96	AR	Famatina	0.96	AR	Cruz del Eje	0.95	AR	Coronel Felipe Varela
AR	Cachi	0.98	AR	Molinos	0.95	AR	Luján de Cuyo	0.94	AR	San Carlos - Mza
AR	Cafayate	0.97	AR	Tafí del Valle	0.91	AR	Santa María - Catamarca	0.91	AR	La Viña
AR	Calamuchita	0.90	AU	Mount Benson	0.87	AU	Blackwood Valley	0.87	AU	Geographe
AR	Calingasta	0.97	AR	Chos Malal	0.94	AR	Tumbaya	0.92	AR	Castro Barros
AR	Capital San Juan	0.65	AR	Rivadavia - San Juan	0.63	AR	Caucete	0.62	AR	Albardón
AR	Castro Barros	0.92	AR	Calingasta	0.89	AR	Santa María - Cba	0.88	AR	Maipú
AR	Caucete	0.98	AR	San Martín - San Juan	0.96	AR	Nueve de Julio	0.96	AR	Angaco
AR	Chilecito	0.93	AR	General Lamadrid	0.92	AR	Famatina	0.89	AR	San Carlos - Salta
AR	Chimbas	0.97	AR	Santa Lucía	0.97	AR	Albardón	0.96	AR	Pocito
AR	Chos Malal	0.98	AR	Tumbaya	0.97	AR	Calingasta	0.89	CN	Sichuan
AR	Collon Cura	0.99	FR	Aube	0.99	US	Marin	0.98	AR	Malargüe
AR	Colón - Cba	0.97	AR	Malargüe	0.97	US	Marin	0.97	CH	Graubünden
AR	Colón - Entre Ríos	0.85	AR	Tilcara	0.83	AR	Victoria	0.81	AR	Villarino
AR	Concordia	0.80	AR	Tandil	0.73	AR	Victoria	0.67	AR	Calingasta
AR	Conesa	0.84	AR	Pichi Mahuida	0.66	AR	Avellaneda - Río Negro	0.61	AR	Belén
AR	Confluencia	0.89	AR	Añelo	0.87	AR	Tunuyán	0.86	AR	Santa María - Cba
AR	Coronel Felipe Varela	0.96	AR	Vinchina	0.95	AR	Famatina	0.95	AR	Cruz del Eje
AR	Coronel Suarez	0.93	NZ	Nelson	0.91	NZ	Marlborough	0.91	AU	Pemberton
AR	Cruz del Eje	0.99	AR	Sanagasta	0.96	AR	Belén	0.95	AR	San Blas De Los Sauces
AR	Cushamen	0.80	JP	Nagano	0.79	US	Humboldt	0.75	AR	Colón - Cba
AR	El Cuy	0.97	AR	Picunches	0.87	AR	Puelen	0.81	AR	San Javier
AR	Famatina	0.96	AR	Belén	0.95	AR	Coronel Felipe Varela	0.93	AR	General Lamadrid
AR	General Alvear	0.95	AR	San Martín - Mza	0.91	AR	Santa Rosa - Mza	0.87	AR	Junín - Mza
AR	General Lamadrid	0.93	AR	Famatina	0.93	AR	Chilecito	0.92	AR	Coronel Felipe Varela
AR	General Roca	0.79	AR	El Cuy	0.77	AR	Confluencia	0.77	AR	Avellaneda - Río Negro
AR	Godoy Cruz	0.97	FR	Lot	0.92	AR	San Carlos - Mza	0.92	AR	Molinos
AR	Guaymallén	0.90	AR	Maipú	0.84	AR	San Rafael	0.82	AR	Las Heras
AR	Iglesia	0.92	AR	Angaco	0.92	AR	Valle Fértil	0.85	AR	Santa Lucía
AR	Ischilin	0.75	AR	Colón - Cba	0.74	AR	Totoral	0.74	US	Marin
AR	Jachal	0.88	AR	Veinticinco de Mayo - San Juan	0.87	AR	Caucete	0.87	AR	Sarmiento - San Juan
AR	Junín - Mza	0.98	AR	Rivadavia - Mza	0.94	AR	San Martín - Mza	0.90	AR	Santa Rosa - Mza
AR	Junín - San Luis	0.60	US	Ventura	0.59	AU	Heathcote	0.58	AU	Central Victoria - other
AR	La Paz	0.91	AR	Santa Rosa - Mza	0.89	AR	Lavalle	0.89	AR	Las Heras
AR	La Rioja	0.93	AR	Albardón	0.88	AR	Chimbas	0.85	AR	Rivadavia - San Juan
AR	La Viña	0.91	AR	Tafí del Valle	0.91	AR	Cafayate	0.89	AR	Santa María - Catamarca
AR	Las Heras	0.93	AR	Lavalle	0.89	AR	Santa Rosa - Mza	0.89	AR	La Paz
AR	Lavalle	0.95	AR	Santa Rosa - Mza	0.93	AR	Las Heras	0.89	AR	La Paz
AR	Leandro Alem	0.78	AR	Veinticinco de Mayo - Misiones	0.77	DE	Rheingau	0.77	DE	Mittelrhein
AR	Luján de Cuyo	0.98	AR	Molinos	0.97	AR	San Carlos - Mza	0.97	AR	Tunuyán
AR	Maipú	0.95	AR	Luján de Cuyo	0.94	AR	Tunuyán	0.91	AR	San Carlos - Mza
AR	Malargüe	0.99	AR	Totoral	0.99	CH	Graubünden	0.99	US	Marin
AR	Molinos	0.98	AR	Luján de Cuyo	0.98	AR	Cachi	0.97	AR	San Carlos - Mza
AR	Nogoya	1.00	CH	Ticino	0.98	CN	Other regions	0.92	FR	Gironde
AR	Nueve de Julio	0.97	AR	Pocito	0.97	AR	Sarmiento - San Juan	0.96	AR	Caucete
AR	Pichi Mahuida	0.84	AR	Conesa	0.73	AR	Las Heras	0.71	AR	Avellaneda - Río Negro
AR	Picunches	0.97	AR	El Cuy	0.85	AR	Nogoya	0.85	CH	Ticino
AR	Pocito	0.98	AR	Rawson	0.98	AR	Tinogasta	0.98	AR	Santa Lucía
AR	Poman	0.82	ES	Malaga	0.81	AR	Zonda	0.80	CL	Atacama
AR	Puelen	0.91	AR	Adolfo Alsina	0.90	AR	Añelo	0.87	AR	El Cuy
AR	Rawson	0.98	AR	Santa Lucía	0.98	AR	Pocito	0.97	AR	Tinogasta
AR	Rivadavia - Mza	0.98	AR	Junín - Mza	0.94	AR	San Martín - Mza	0.92	AR	Santa Rosa - Mza
AR	Rivadavia - San Juan	0.90	AR	Ullum	0.89	AR	Zonda	0.86	AR	Albardón
AR	San Blas De Los Sauces	0.95	AR	Cruz del Eje	0.95	AR	San Carlos - Salta	0.95	AR	Sanagasta
AR	San Carlos - Mza	0.97	AR	Luján de Cuyo	0.97	AR	Molinos	0.96	AR	Tunuyán
AR	San Carlos - Salta	0.95	AR	San Blas De Los Sauces	0.95	AR	Belén	0.95	AR	Sanagasta

Table 96 (cont.): Each region's 3 most similar winegrape regions in the world according to the VSI, 2010

AR	San Javier	0.86	AR	Tilcara	0.84	FR	Lot	0.82	AR	Luján de Cuyo
AR	San Martín - Mza	0.96	AR	Santa Rosa - Mza	0.95	AR	General Alvear	0.94	AR	Rivadavia - Mza
AR	San Martín - San Juan	0.98	AR	Caucete	0.96	AR	Veinticinco de Mayo - San	0.96	AR	Angaco
AR	San Rafael	0.90	AR	Santa Rosa - Mza	0.90	AR	Rivadavia - Mza	0.89	AR	Junín - Mza
AR	Sanagasta	0.99	AR	Cruz del Eje	0.95	AR	Belén	0.95	AR	San Carlos - Salta
AR	Santa Lucía	0.98	AR	Rawson	0.98	AR	Tinogasta	0.98	AR	Pocito
AR	Santa María - Catamarca	0.94	AR	Tafí del Valle	0.91	AR	Cafayate	0.89	AR	La Viña
AR	Santa María - Cba	0.92	AR	Tunuyán	0.90	AR	Luján de Cuyo	0.89	AR	Castro Barros
AR	Santa Rosa - Mza	0.96	AR	San Martín - Mza	0.95	AR	Lavalle	0.92	AR	Rivadavia - Mza
AR	Sarmiento - San Juan	0.97	AR	Nueve de Julio	0.97	AR	Veinticinco de Mayo - San Juan	0.93	AR	Caucete
AR	Tafí del Valle	0.97	AR	Cafayate	0.94	AR	Santa María - Catamarca	0.91	AR	La Viña
AR	Tandil	0.86	AR	Calingasta	0.85	AR	Victoria	0.81	AR	Villarino
AR	Tilcara	0.93	AR	Cachi	0.91	AR	Molinos	0.91	FR	Lot
AR	Tinogasta	0.98	AR	Santa Lucía	0.98	AR	Arauco	0.98	AR	Pocito
AR	Tornquist	0.73	FR	Indre-et-Loire	0.73	FR	Deux-Sevres	0.72	FR	Maine-et-Loire
AR	Totoral	0.99	AR	Malargüe	0.99	CH	Graubünden	0.99	US	Marin
AR	Tulumba	0.81	FR	Hautes-Pyrenees	0.65	UY	Uruguay	0.57	AR	Leandro Alem
AR	Tumbaya	0.98	AR	Chos Malal	0.94	AR	Calingasta	0.90	CN	Sichuan
AR	Tunuyán	0.97	AR	Luján de Cuyo	0.96	AR	San Carlos - Mza	0.95	AR	Añelo
AR	Tupungato	0.94	AR	Tunuyán	0.91	AR	Añelo	0.90	AR	Maipú
AR	Ullum	0.90	AR	Rivadavia - San Juan	0.87	AR	Veinticinco de Mayo - San Juan	0.85	AR	Zonda
AR	Valle Fértil	0.95	AR	Angaco	0.92	AR	Iglesia	0.91	AR	Arauco
AR	Veinticinco de Mayo - Misiones	0.99	DE	Rheingau	0.98	DE	Mittelrhein	0.96	DE	Mosel
AR	Veinticinco de Mayo - San Juan	0.97	AR	Sarmiento - San Juan	0.96	AR	San Martín - San Juan	0.96	AR	Nueve de Julio
AR	Victoria	0.94	AR	Villarino	0.85	AR	Tandil	0.83	AR	Colón - Entre Ríos
AR	Villarino	0.94	AR	Victoria	0.81	AR	Tandil	0.81	AR	Colón - Entre Ríos
AR	Vinchina	0.96	AR	Coronel Felipe Varela	0.91	AR	Sanagasta	0.91	AR	Cruz del Eje
AR	Zonda	0.89	AR	Rivadavia - San Juan	0.85	AR	Ullum	0.84	ES	Malaga
AM	Armenia	0.93	KZ	Zhambyl	0.81	HU	Zala	0.81	RO	Nord-Vest
AU	Adelaide Hills	0.96	AU	Strathbogie Ranges	0.92	AU	Pemberton	0.92	CL	Valparaiso
AU	Adelaide Plains	0.98	AU	Goulburn Valley	0.97	AU	Southern Fleurieu	0.97	AU	Western Australia Southeast Coast
AU	Alpine Valleys	0.93	AU	King Valley	0.90	AU	Strathbogie Ranges	0.89	AU	Central Ranges - other
AU	Australian Capital Territory	0.76	US	Colorado	0.75	US	Rattlesnake Hills	0.74	US	Michigan
AU	Barossa - other	1.00	AU	Southern Flinders Ranges	0.99	AU	Grampians	0.99	US	Ventura
AU	Barossa Valley	0.99	AU	Mount Lofty Ranges - other	0.99	AU	McLaren Vale	0.99	AU	Bendigo
AU	Beechworth	0.97	AU	Orange	0.96	AU	Mudgee	0.96	AU	Northern Slopes - other
AU	Bendigo	0.99	AU	McLaren Vale	0.99	AU	Mount Lofty Ranges - other	0.99	AU	Barossa Valley
AU	Big Rivers - other	0.97	AU	Murray Darling (NSW)	0.92	AU	Murray Darling (VIC)	0.91	AU	Cowra
AU	Blackwood Valley	0.97	AU	Geographe	0.97	AU	Margaret River	0.97	AU	Great Southern
AU	Canberra District (ACT)	0.89	AU	Eden Valley	0.89	AU	Canberra District (NSW)	0.89	AU	Goulburn Valley
AU	Canberra District (NSW)	0.97	AU	The Peninsulas	0.95	AU	Currency Creek	0.95	AU	Glenrowan
AU	Central Ranges - other	0.93	AU	Murray Darling (VIC)	0.93	AU	Murray Darling (NSW)	0.92	AU	Beechworth
AU	Central Victoria - other	0.99	AU	Grampians	0.99	AU	Heathcote	0.99	AU	Bendigo
AU	Central Western Australia	0.96	AU	McLaren Vale	0.95	AU	Bendigo	0.95	AU	Mount Lofty Ranges - other
AU	Clare Valley	0.95	AU	Canberra District (NSW)	0.94	AU	Eden Valley	0.91	AU	The Peninsulas
AU	Coonawarra	0.98	AU	Wrattonbully	0.97	US	Red Mountain	0.97	CL	Metropolitana
AU	Cowra	0.98	AU	Hunter Valley - other	0.95	US	Tehama	0.93	AU	Murray Darling (VIC)
AU	Currency Creek	0.99	AU	Kangaroo Island	0.98	AU	Langhorne Creek	0.98	AU	The Peninsulas
AU	Eastern Plains, Inland and North WA	0.84	AU	Swan District	0.75	AU	Perth Hills	0.65	AU	Queensland - other
AU	Eden Valley	0.94	AU	Clare Valley	0.89	AU	Canberra District (ACT)	0.85	AU	Canberra District (NSW)
AU	Far North - other	0.98	AU	Kangaroo Island	0.97	AU	Glenrowan	0.96	AU	Fleurieu - other
AU	Fleurieu - other	0.98	AU	Langhorne Creek	0.98	AU	Hilltops	0.98	AU	Kangaroo Island
AU	Geelong	0.97	AU	Macedon Ranges	0.95	AU	Port Phillip - other	0.94	AU	Gippsland
AU	Geographe	0.98	AU	Margaret River	0.97	AU	Blackwood Valley	0.97	AU	Great Southern
AU	Gippsland	0.99	AU	Macedon Ranges	0.99	AU	Port Phillip - other	0.98	AU	Yarra Valley
AU	Glenrowan	0.97	AU	The Peninsulas	0.97	AU	Currency Creek	0.97	AU	Far North - other
AU	Goulburn Valley	0.98	AU	Adelaide Plains	0.98	AU	Orange	0.97	AU	Southern Fleurieu
AU	Grampians	0.99	AU	Heathcote	0.99	AU	Barossa - other	0.99	AU	Central Victoria - other
AU	Granite Belt	0.94	AU	Northern Slopes - other	0.94	AU	Beechworth	0.94	AU	Swan Hill (NSW)
AU	Great Southern	0.97	AU	Blackwood Valley	0.97	AU	Geographe	0.95	AU	Margaret River
AU	Greater Perth - other	0.92	AU	Barossa Valley	0.92	AU	Mount Lofty Ranges - other	0.92	AU	Barossa - other

Table 96 (cont.): Each region's 3 most similar winegrape regions in the world according to the VSI, 2010

AU	Gundagai	0.98	AU	McLaren Vale	0.97	AU	Bendigo	0.97	AU	Southern Fleurieu
AU	Hastings River	0.89	AU	Shoalhaven Coast	0.81	AU	Northern Rivers - other	0.72	NZ	Auckland
AU	Heathcote	0.99	AU	Grampians	0.99	AU	Central Victoria - other	0.98	AU	Barossa - other
AU	Henty	0.97	AU	Tasmania	0.96	AU	Mornington Peninsula	0.95	NZ	Canterbury
AU	Hilltops	0.99	AU	Langhorne Creek	0.98	AU	Fleurieu - other	0.97	AU	Currency Creek
AU	Hunter	0.96	AU	Riverina	0.94	AU	South Coast - other	0.91	AU	Western Plains - other
AU	Hunter Valley - other	0.98	AU	Cowra	0.97	US	Tehama	0.94	FR	Yonne
AU	Kangaroo Island	0.99	AU	Currency Creek	0.98	AU	Far North - other	0.98	AU	Fleurieu - other
AU	King Valley	0.93	AU	Alpine Valleys	0.89	AU	Strathbogie Ranges	0.87	CA	British Colombia
AU	Langhorne Creek	0.99	AU	Hilltops	0.98	AU	Fleurieu - other	0.98	AU	Currency Creek
AU	Limestone Coast - other	0.97	AU	Wrattonbully	0.96	AU	Coonawarra	0.94	AU	Hilltops
AU	Lower Murray - other	0.98	AU	Perricoota	0.98	AU	Riverland	0.96	AU	Gundagai
AU	Macedon Ranges	0.99	AU	Gippsland	0.99	AU	Port Phillip - other	0.98	US	Santa Cruz
AU	Manjimup	0.88	US	Sacramento	0.88	AU	South West Australia - other	0.87	AU	Margaret River
AU	Margaret River	0.98	AU	Geographe	0.97	AU	Blackwood Valley	0.95	AU	Great Southern
AU	McLaren Vale	0.99	AU	Mount Lofty Ranges - other	0.99	AU	Barossa Valley	0.99	AU	Bendigo
AU	Mornington Peninsula	0.97	US	Santa Cruz	0.96	AU	Tasmania	0.96	AU	Henty
AU	Mount Benson	0.97	AU	Geographe	0.96	AU	Orange	0.96	AU	Blackwood Valley
AU	Mount Lofty Ranges - other	0.99	AU	Barossa Valley	0.99	AU	McLaren Vale	0.99	AU	Bendigo
AU	Mudgee	0.98	AU	Padthaway	0.98	AU	Orange	0.98	AU	Langhorne Creek
AU	Murray Darling (NSW)	0.97	AU	Murray Darling (VIC)	0.97	AU	Big Rivers - other	0.95	AU	Swan Hill (VIC)
AU	Murray Darling (VIC)	0.97	AU	Murray Darling (NSW)	0.95	AU	Riverland	0.95	AU	Swan Hill (VIC)
AU	New England Australia	0.79	AU	Hunter	0.74	AU	Riverina	0.72	AU	Cowra
AU	North East Victoria - other	0.96	US	Horse Heaven Hills	0.94	US	Wahluke Slope	0.93	US	San Luis Obispo
AU	North West Victoria - other	0.93	AU	Swan Hill (NSW)	0.93	AU	Riverland	0.92	AU	Mudgee
AU	Northern Rivers - other	0.87	NZ	Gisborne	0.87	AU	Hunter Valley - other	0.87	FR	Doubs
AU	Northern Slopes - other	0.97	AU	Mudgee	0.96	AU	Orange	0.96	AU	Padthaway
AU	Orange	0.98	AU	Padthaway	0.98	AU	Goulburn Valley	0.98	AU	Mudgee
AU	Padthaway	0.98	AU	Mudgee	0.98	AU	Orange	0.98	AU	Langhorne Creek
AU	Peel	0.96	AU	Hilltops	0.95	AU	Padthaway	0.95	AU	Langhorne Creek
AU	Pemberton	0.96	CL	Valparaiso	0.92	AU	Adelaide Hills	0.91	AR	Coronel Suarez
AU	Perricoota	0.98	AU	Riverland	0.98	AU	Lower Murray - other	0.96	AU	Padthaway
AU	Perth Hills	0.95	AU	Swan District	0.87	AU	Western Plains - other	0.86	AU	South Coast - other
AU	Port Phillip - other	0.99	AU	Macedon Ranges	0.99	AU	Gippsland	0.96	US	Santa Cruz
AU	Pyrenees	0.96	AU	Bendigo	0.95	AU	McLaren Vale	0.95	AU	Mount Lofty Ranges - other
AU	Queensland - other	0.95	AU	South Burnett	0.94	AU	Swan Hill (NSW)	0.94	AU	Southern Fleurieu
AU	Riverina	0.96	AU	Hunter	0.94	AU	South Coast - other	0.94	AU	Swan Hill (VIC)
AU	Riverland	0.98	AU	Perricoota	0.98	AU	Lower Murray - other	0.96	AU	Swan Hill (VIC)
AU	Robe	0.97	AU	Padthaway	0.96	AU	Perricoota	0.96	AU	Riverland
AU	Rutherglen	0.95	AU	Mount Lofty Ranges - other	0.95	AU	Grampians	0.95	AU	Bendigo
AU	Shoalhaven Coast	0.89	AU	Hastings River	0.81	AU	South Coast - other	0.77	AU	Northern Rivers - other
AU	South Burnett	0.97	AU	Mudgee	0.96	AU	Southern Fleurieu	0.95	AU	Queensland - other
AU	South Coast - other	0.94	AU	Hunter	0.94	AU	Riverina	0.91	AU	Swan Hill (VIC)
AU	South West Australia - other	0.92	AU	Blackwood Valley	0.92	AU	Margaret River	0.91	AU	Pemberton
AU	Southern Fleurieu	0.97	AU	Adelaide Plains	0.97	AU	Gundagai	0.97	AU	McLaren Vale
AU	Southern Flinders Ranges	1.00	AU	Barossa - other	0.99	US	Ventura	0.99	AU	Grampians
AU	Southern Highlands	0.90	AU	Strathbogie Ranges	0.89	AU	Adelaide Hills	0.88	AU	Upper Goulburn
AU	Southern NSW - other	0.86	AU	Adelaide Plains	0.83	AU	Goulburn Valley	0.82	AU	Southern Fleurieu
AU	Strathbogie Ranges	0.97	AU	Upper Goulburn	0.96	AU	Adelaide Hills	0.90	AU	Alpine Valleys
AU	Sunbury	0.94	AU	Adelaide Plains	0.93	AU	Southern Fleurieu	0.92	AU	Western Australia Southeast Coast
AU	Swan District	0.95	AU	Perth Hills	0.84	AU	Eastern Plains, Inland and North WA	0.82	AU	Queensland - other
AU	Swan Hill (NSW)	0.96	AU	Southern Fleurieu	0.95	AU	Mudgee	0.95	AU	South Burnett
AU	Swan Hill (VIC)	0.96	AU	Riverland	0.95	AU	Murray Darling (NSW)	0.95	AU	Murray Darling (VIC)
AU	Tasmania	0.98	NZ	Canterbury	0.97	AU	Henty	0.96	AU	Mornington Peninsula
AU	The Peninsulas	0.98	AU	Currency Creek	0.98	AU	Langhorne Creek	0.97	AU	Glenrowan
AU	Tumbarumba	0.95	US	Santa Barbara	0.94	AU	Macedon Ranges	0.94	AU	Gippsland
AU	Upper Goulburn	0.97	AU	Yarra Valley	0.97	AU	Strathbogie Ranges	0.93	AU	Gippsland
AU	Western Australia Southeast Coast	0.97	AU	Adelaide Plains	0.96	AU	Goulburn Valley	0.96	AU	Western Plains - other
AU	Western Plains - other	0.96	AU	Western Australia Southeast Coast	0.95	AU	Lower Murray - other	0.95	AU	Gundagai
AU	Western Victoria - other	0.94	AU	Canberra District (NSW)	0.93	AU	Langhorne Creek	0.93	AU	Hilltops
AU	Wrattonbully	0.98	AU	Coonawarra	0.97	AU	Limestone Coast - other	0.95	AU	Hilltops

Table 96 (cont.): Each region's 3 most similar winegrape regions in the world according to the VSI, 2010

AU	Yarra Valley	0.98	AU	Gippsland	0.97	AU	Upper Goulburn	0.96	AU	Macedon Ranges
AT	Burgenland	0.81	HU	Tolna	0.76	HU	Sopron	0.74	HU	Csongrad
AT	Niederosterreich	0.88	AT	Wien and other regions	0.77	SK	Malokarpatska	0.75	SK	Nitrianska
AT	Steiermark	0.74	SI	Stajerska Slovenija	0.72	SI	Prekmurje	0.67	HU	Pannonhalma
AT	Wien and other regions	0.88	AT	Niederosterreich	0.76	SK	Malokarpatska	0.74	SK	Nitrianska
BR	Brazil	0.37	MD	Moldova	0.35	FR	Deux-Sevres	0.35	AR	Tornquist
BG	North Central	0.90	BG	Northwest	0.77	US	Columbia River	0.77	MX	Baja California
BG	Northeast	0.73	BG	North Central	0.70	BG	Northwest	0.69	BG	Southeast
BG	Northwest	0.90	BG	North Central	0.76	FR	Ariege	0.75	RO	Sud-Vest Oltenia
BG	South Central	0.88	BG	Southeast	0.86	IT	Bergamo	0.84	FR	Gironde
BG	Southeast	0.88	BG	South Central	0.76	BG	North Central	0.74	BG	Northwest
BG	Southwest	0.40	BG	South Central	0.38	BG	Northwest	0.36	BG	Southeast
CA	British Colombia	0.87	AU	King Valley	0.85	AU	Alpine Valleys	0.82	US	Colorado
CA	Ontario	0.63	US	Michigan	0.63	US	Virginia	0.61	US	Yakima Valley
CL	Araucania	0.97	CL	De Los Lagos	0.90	AU	Tasmania	0.89	CL	Valparaiso
CL	Atacama	0.84	AR	Zonda	0.80	AR	Poman	0.77	ES	Malaga
CL	Coquimbo	0.92	AU	Swan Hill (VIC)	0.92	AU	Robe	0.92	AU	Riverland
CL	De Los Lagos	0.97	CL	Araucania	0.93	CL	Valparaiso	0.90	NZ	Nelson
CL	Del Bio Bio	0.77	CL	Atacama	0.77	AR	Zonda	0.76	ES	Malaga
CL	Del Maule	0.95	CL	Metropolitana	0.95	CL	O'Higgins	0.94	US	Napa
CL	Metropolitana	0.98	CL	O'Higgins	0.97	CN	Gansu	0.97	CN	Ningxia
CL	O'Higgins	0.98	CL	Metropolitana	0.95	CL	Del Maule	0.94	CN	Gansu
CL	Valparaiso	0.96	AU	Pemberton	0.93	CL	De Los Lagos	0.92	AU	Adelaide Hills
CN	Beijing	1.00	CN	Tianjin	1.00	CN	Sichuan	1.00	CN	Shandong
CN	Gansu	0.99	CN	Xinjiang	0.97	CL	Metropolitana	0.97	CN	Ningxia
CN	Ningxia	0.98	CN	Beijing	0.98	CN	Shandong	0.98	CN	Sichuan
CN	Other regions	0.98	AR	Nogoya	0.98	CH	Ticino	0.96	FR	Gironde
CN	Sichuan	1.00	CN	Tianjin	1.00	CN	Beijing	1.00	CN	Shandong
CN	ShanXi	0.91	US	Napa	0.90	CN	Gansu	0.89	US	Horse Heaven Hills
CN	Beijing	1.00	CN	Shandong	1.00	CN	Tianjin	1.00	CN	Sichuan
CN	Tianjin	1.00	CN	Sichuan	1.00	CN	Beijing	1.00	CN	Shandong
CN	Xinjiang	0.99	CN	Gansu	0.95	CL	Metropolitana	0.95	CN	Ningxia
CN	Yantai	1.00	CN	Beijing	1.00	CN	Sichuan	1.00	CN	Tianjin
HR	Dalmatinska Zagora	0.26	HR	Sjeverna Dalmacija	0.20	IT	Lecco	0.19	CH	Ticino
HR	Hrvatsko Primorje	0.12	US	Alameda	0.12	US	Horse Heaven Hills	0.12	US	Santa Clara
HR	Istra	0.64	SI	Slovenska Istra	0.60	IT	Trieste	0.45	SI	Vipavska dolina
HR	Moslavina	0.95	HR	Pokuplje	0.72	HR	Slavonija	0.72	HR	Podunavlje
HR	Other regions	0.98	RO	Bucuresti - Ilfov	0.92	RO	Sud - Muntenia	0.92	RO	Nord-Vest
HR	Plesivica	0.91	HR	Zagorje-Medimurje	0.89	HR	Prigorje - Bilogora	0.84	SI	Prekmurje
HR	Podunavlje	0.99	HR	Slavonija	0.96	HU	Balatonfured-Csopak	0.96	HU	Balatonfelvidek
HR	Pokuplje	0.95	HR	Moslavina	0.69	HR	Podunavlje	0.69	HR	Slavonija
HR	Prigorje - Bilogora	0.89	HR	Plesivica	0.82	HR	Slavonija	0.82	HR	Podunavlje
HR	Sjeverna Dalmacija	0.36	US	San Diego	0.36	FR	Ardeche	0.34	EL	Anatoliki Makedonia, Thraki
HR	Slavonija	0.99	HR	Podunavlje	0.97	HU	Balatonfured-Csopak	0.97	HU	Balatonfelvidek
HR	Srednja Juzna Dalmacija	0.27	HR	Sjeverna Dalmacija	0.16	HR	Dalmatinska Zagora	0.10	FR	Ariege
HR	Zagorje-Medimurje	0.91	HR	Plesivica	0.85	SI	Prekmurje	0.84	SI	Stajerska Slovenija
CY	Cyprus	0.14	FR	Aude	0.13	FR	Herault	0.13	IL	Israel
CZ	Cechy	0.82	CZ	Morava	0.76	DE	Sachsen	0.74	DE	Saale
CZ	Morava	0.86	SK	Stredné Slovensko	0.85	SK	Malokarpatska	0.84	SK	Nitrianska
ET	Ethiopia	0.84	IT	Siena	0.84	IT	Firenze	0.84	IT	Arezzo
FR	Ain	0.91	FR	Allier	0.86	FR	Puy-de-Dome	0.81	FR	Rhone
FR	Aisne	0.83	FR	Seine-et-Marne	0.80	FR	Marne	0.56	FR	Loiret
FR	Allier	0.95	FR	Puy-de-Dome	0.91	FR	Ain	0.89	FR	Rhone
FR	Alpes-de-Haute-Provence, Hautes-Alpes, Alpes-Maritimes	0.92	FR	Gard	0.90	FR	Ardeche	0.88	FR	Drome
FR	Alpes-Maritimes	0.71	US	North Carolina	0.71	US	Georgia	0.71	US	Arizona
FR	Ardeche	0.93	FR	Gard	0.90	FR	Alpes-de-Haute-Provence, Hautes-Alpes, Alpes-Maritimes	0.84	FR	Herault
FR	Ariege	0.89	FR	Correze	0.86	NZ	Other regions	0.86	US	Orange
FR	Aube	1.00	US	Marin	0.99	FR	Haute-Marne	0.99	US	Yamhill Co.
FR	Aude	0.97	FR	Herault	0.83	FR	Gard	0.78	FR	Ardeche
FR	Aveyron	0.45	FR	Tarn	0.44	FR	Cantal	0.33	FR	Ariege
FR	Bas-Rhin	0.95	FR	Haut-Rhin	0.65	DE	Nahe	0.64	DE	Hessische Bergstraße

Table 96 (cont.): Each region's 3 most similar winegrape regions in the world according to the VSI, 2010

FR	Bouches-du-Rhone	0.94	FR	Gard	0.94	FR	Var	0.93	FR	Vaucluse
FR	Cantal	0.85	FR	Correze	0.83	FR	Haute-Saone	0.80	US	Arkansas
FR	Charente	1.00	FR	Charente-Maritime	0.63	IT	Terni	0.59	IT	Foggia
FR	Charente-Maritime	1.00	FR	Charente	0.64	IT	Terni	0.59	IT	Foggia
FR	Cher	0.98	NZ	Marlborough	0.97	FR	Nievre	0.91	NZ	Waipara
FR	Correze	0.98	NZ	Other regions	0.97	US	North Carolina	0.97	US	Arkansas
FR	Corse	0.67	FR	Haute-Corse	0.44	IT	Olbia-Tempio	0.43	IT	Sassari
FR	Cote-d'Or	0.99	FR	Haute-Marne	0.98	FR	Aube	0.97	US	Marin
FR	Deux-Sevres	0.91	FR	Maine-et-Loire	0.90	FR	Indre-et-Loire	0.79	FR	Vienne
FR	Dordogne	0.83	FR	Lot-et-Garonne	0.82	FR	Gironde	0.72	CN	Other regions
FR	Doubs	0.91	NZ	Gisborne	0.87	AU	Northern Rivers - other	0.86	FR	Haute-Saone
FR	Drome	0.97	FR	Vaucluse	0.92	FR	Gard	0.91	FR	Bouches-du-Rhone
FR	Eure-et-Loire	0.92	FR	Yonne	0.91	US	Tehama	0.88	AU	Hunter Valley - other
FR	Gard	0.94	FR	Bouches-du-Rhone	0.93	FR	Ardeche	0.92	FR	Alpes-de-Haute-Provence, Hautes-Alpes, Alpes-Maritimes
FR	Gers	0.86	FR	Landes	0.70	ZA	Northern Cape	0.65	ZA	Little Karoo
FR	Gironde	0.96	CN	Other regions	0.95	FR	Lot-et-Garonne	0.92	CH	Ticino
FR	Haute-Corse	0.77	IT	Grosseto	0.74	IT	Lucca	0.74	IT	Siena
FR	Haute-Garonne	0.42	FR	Tarn-et-Garonne	0.36	AU	Pyrenees	0.33	FR	Deux-Sevres
FR	Haute-Loire	0.97	US	Orange	0.95	NZ	Other regions	0.92	FR	Correze
FR	Haute-Marne	0.99	FR	Aube	0.99	FR	Cote-d'Or	0.98	US	Marin
FR	Hautes-Alpes	0.84	FR	Correze	0.83	NZ	Other regions	0.82	FR	Ariege
FR	Haute-Saone	0.86	FR	Doubs	0.83	FR	Cantal	0.80	US	Puget Sound
FR	Haute-Savoie	0.95	CH	Vaud	0.85	FR	Tarn-et-Garonne	0.84	CH	Fribourg
FR	Hautes-Pyrenees	0.81	AR	Tulumba	0.69	UY	Uruguay	0.65	FR	Pyrenees-Atlantiques
FR	Haut-Rhin	0.95	FR	Bas-Rhin	0.64	US	Lake Chelan	0.62	US	Columbia Gorge
FR	Herault	0.97	FR	Aude	0.87	FR	Gard	0.84	FR	Ardeche
FR	Indre	0.77	FR	Loiret	0.74	FR	Cantal	0.73	FR	Loir-et-Cher
FR	Indre-et-Loire	0.98	FR	Maine-et-Loire	0.90	FR	Deux-Sevres	0.73	AR	Tornquist
FR	Isere	0.88	FR	Savoie	0.42	AU	Western Plains - other	0.41	AU	Western Australia Southeast Coast
FR	Jura	0.84	FR	Yonne	0.81	FR	Saone-et-Loire	0.81	US	Tehama
FR	Landes	0.86	FR	Gers	0.51	FR	Charente-Maritime	0.50	FR	Charente
FR	Loire	0.88	FR	Rhone	0.85	FR	Puy-de-Dome	0.79	FR	Allier
FR	Loire-Atlantique	0.20	FR	Vendee	0.13	FR	Maine-et-Loire	0.08	FR	Ain
FR	Loiret	0.77	FR	Indre	0.70	FR	Vienne	0.67	FR	Loir-et-Cher
FR	Loir-et-Cher	0.87	FR	Nievre	0.86	NZ	Marlborough	0.86	FR	Cher
FR	Lot	0.97	AR	Godoy Cruz	0.92	AR	Cachi	0.92	AR	San Carlos - Mza
FR	Lot-et-Garonne	0.95	FR	Gironde	0.87	CN	Other regions	0.83	CH	Ticino
FR	Lozere	0.93	AU	Geelong	0.92	AU	Tumbarumba	0.90	AU	Macedon Ranges
FR	Maine-et-Loire	0.98	FR	Indre-et-Loire	0.91	FR	Deux-Sevres	0.72	AR	Tornquist
FR	Marne	0.97	FR	Seine-et-Marne	0.80	FR	Aisne	0.74	US	Santa Barbara
FR	Mayenne	0.92	FR	Yonne	0.91	US	Santa Barbara	0.89	FR	Saone-et-Loire
FR	Meurthe-et-Moselle	0.82	FR	Allier	0.80	FR	Puy-de-Dome	0.74	FR	Rhone
FR	Meuse	0.80	FR	Haute-Marne	0.78	AU	Mornington Peninsula	0.78	US	Santa Cruz
FR	Moselle	0.86	US	Puget Sound	0.82	US	Douglas Co.	0.80	US	Willamette Valley - other
FR	Nievre	0.99	NZ	Marlborough	0.97	FR	Cher	0.88	AR	Coronel Suarez
FR	Puy-de-Dome	0.96	FR	Rhone	0.95	FR	Allier	0.86	FR	Ain
FR	Pyrenees-Atlantiques	0.65	FR	Hautes-Pyrenees	0.51	AR	Tulumba	0.42	UY	Uruguay
FR	Pyrenees-Orientales	0.86	FR	Drome	0.83	FR	Gard	0.81	FR	Vaucluse
FR	Rhone	0.96	FR	Puy-de-Dome	0.89	FR	Allier	0.88	FR	Loire
FR	Saone-et-Loire	0.92	FR	Yonne	0.92	US	Santa Barbara	0.89	US	San Benito
FR	Sarthe	0.59	ZA	Worcester	0.58	ZA	Breedekloof	0.58	ZA	Olifants River
FR	Savoie	0.88	FR	Isere	0.44	FR	Ain	0.35	FR	Allier
FR	Seine-et-Marne	0.97	FR	Marne	0.83	FR	Aisne	0.70	FR	Yonne
FR	Tarn	0.49	AU	Central Western Australia	0.47	AU	Goulburn Valley	0.47	AU	Adelaide Plains
FR	Tarn-et-Garonne	0.87	CH	Vaud	0.85	FR	Haute-Savoie	0.76	CH	Fribourg
FR	Var	0.94	FR	Bouches-du-Rhone	0.87	FR	Gard	0.85	FR	Vaucluse
FR	Vaucluse	0.97	FR	Drome	0.93	FR	Bouches-du-Rhone	0.93	IT	Ogliastra
FR	Vendee	0.70	FR	Vienne	0.67	FR	Cantal	0.63	FR	Indre
FR	Vienne	0.79	FR	Deux-Sevres	0.70	FR	Vendee	0.70	FR	Loiret
FR	Vosges	0.40	FR	Meurthe-et-Moselle	0.21	FR	Meuse	0.06	US	Iowa
FR	Yonne	0.97	US	Tehama	0.94	AU	Hunter Valley - other	0.93	US	Yolo
GE	Georgia	0.93	KZ	Almaty	0.77	KZ	South Kazakhstan	0.63	UA	Ukraine
DE	Ahr	0.98	AR	Malargüe	0.97	CH	Graubünden	0.97	NZ	Otago

Table 96 (cont.): Each region's 3 most similar winegrape regions in the world according to the VSI, 2010

DE	Baden	0.93	CH	Other regions	0.93	CH	Zürich	0.93	CH	Thurgau
DE	Franken	0.78	DE	Saale	0.75	DE	Rheinhessen	0.60	DE	Sachsen
DE	Hessische Bergstraße	0.98	DE	Mittelrhein	0.97	DE	Rheingau	0.96	DE	Mosel
DE	Mittelrhein	0.99	DE	Rheingau	0.98	AR	Veinticinco de Mayo - Misiones	0.98	DE	Mosel
DE	Mosel	0.98	DE	Mittelrhein	0.97	DE	Rheingau	0.96	DE	Hessische Bergstraße
DE	Nahe	0.96	DE	Pfalz	0.91	DE	Rheinhessen	0.88	DE	Mosel
DE	Pfalz	0.96	DE	Nahe	0.91	DE	Rheinhessen	0.83	DE	Hessische Bergstraße
DE	Rheingau	0.99	DE	Mittelrhein	0.99	AR	Veinticinco de Mayo - Misiones	0.97	DE	Hessische Bergstraße
DE	Rheinhessen	0.91	DE	Nahe	0.91	DE	Pfalz	0.87	DE	Saale
DE	Saale	0.87	DE	Rheinhessen	0.86	DE	Sachsen	0.78	DE	Franken
DE	Sachsen	0.86	DE	Saale	0.83	DE	Nahe	0.80	DE	Rheinhessen
DE	Württemberg	0.56	DE	Hessische Bergstraße	0.54	DE	Mittelrhein	0.54	DE	Rheingau
EL	Anatoliki Makedonia, Thraki	0.87	IT	Lecco	0.84	IT	Bergamo	0.81	US	Snipes Mountain
EL	Attiki	0.99	EL	Sterea Ellada	0.13	EL	Peloponnissos	0.06	EL	Ionia Nisia
EL	Dytiki Ellada	0.44	EL	Peloponnissos	0.42	EL	Thessalia	0.35	EL	Kentriki Makedonia
EL	Dytiki Makedonia	0.62	EL	Kentriki Makedonia	0.44	EL	Thessalia	0.33	US	Orange
EL	Ionia Nisia	0.88	US	Arizona	0.88	US	Georgia	0.88	US	Arkansas
EL	Ipeiros	0.75	NZ	Other regions	0.74	FR	Correze	0.73	US	North Carolina
EL	Kentriki Makedonia	0.69	IT	Bergamo	0.67	EL	Anatoliki Makedonia, Thraki	0.63	IT	Lecco
EL	Kriti	0.35	HR	Other regions	0.34	RO	Bucuresti - Ilfov	0.33	RO	Sud - Muntenia
EL	Notio Aigaio	0.45	HR	Other regions	0.45	RO	Sud-Vest Oltenia	0.45	RO	Sud - Muntenia
EL	Peloponnissos	0.44	EL	Dytiki Ellada	0.32	EL	Kentriki Makedonia	0.32	EL	Thessalia
EL	Sterea Ellada	0.99	EL	Attiki	0.18	EL	Peloponnissos	0.15	EL	Dytiki Ellada
EL	Thessalia	0.67	US	Orange	0.65	NZ	Other regions	0.65	FR	Haute-Loire
EL	Voreio Aigaio	0.60	IT	Cuneo	0.57	MX	Zacatecas	0.47	IT	Asti
HU	Badacsony	0.96	HU	Balatonfelvidek	0.95	HU	Balatonfured-Csopak	0.95	HR	Slavonija
HU	Balatonboglar	0.80	HU	Etyek-Budai	0.77	HU	Tolna	0.71	HU	Pecs
HU	Balatonfelvidek	0.97	HU	Balatonfured-Csopak	0.97	HR	Slavonija	0.96	HR	Podunavlje
HU	Balatonfured-Csopak	0.97	HR	Slavonija	0.97	HU	Balatonfelvidek	0.96	HR	Podunavlje
HU	Bukk	0.78	HU	Eger	0.70	AT	Burgenland	0.70	HU	Tolna
HU	Csongrad	0.83	HU	Sopron	0.77	HU	Hajos-bajai	0.76	HU	Szekszard
HU	Eger	0.89	HU	Szekszard	0.78	HU	Bukk	0.78	HU	Sopron
HU	Etyek-Budai	0.82	HU	Neszmely	0.80	HU	Balatonboglar	0.77	HU	Pecs
HU	Hajos-bajai	0.78	HU	Tolna	0.77	HU	Szekszard	0.77	HU	Csongrad
HU	Kunsag	0.69	HU	Csongrad	0.68	HU	Hajos-bajai	0.50	HU	Sopron
HU	Matra	0.70	HU	Neszmely	0.67	HU	Etyek-Budai	0.67	HU	Tolna
HU	Mor	0.55	HU	Neszmely	0.54	HU	Pecs	0.50	HU	Zala
HU	Nagy-Somlo	0.81	SI	Prekmurje	0.80	SI	Stajerska Slovenija	0.72	HU	Badacsony
HU	Neszmely	0.82	HU	Etyek-Budai	0.72	IT	Trento	0.70	HU	Matra
HU	Pannonhalma	0.90	HU	Balatonfured-Csopak	0.88	HU	Balatonfelvidek	0.87	HR	Slavonija
HU	Pecs	0.77	HU	Etyek-Budai	0.72	SI	Stajerska Slovenija	0.72	HU	Zala
HU	Sopron	0.86	HU	Szekszard	0.83	HU	Csongrad	0.78	HU	Eger
HU	Szekszard	0.89	HU	Eger	0.86	HU	Sopron	0.77	HU	Hajos-bajai
HU	Tokaj	0.97	SK	Tokajska	0.31	HR	Zagorje-Medimurje	0.31	HU	Nagy-Somlo
HU	Tolna	0.81	AT	Burgenland	0.78	HU	Hajos-bajai	0.77	HU	Balatonboglar
HU	Villany	0.73	HU	Szekszard	0.72	HU	Eger	0.71	CN	ShanXi
HU	Zala	0.98	RO	Nord-Vest	0.96	RO	Bucuresti - Ilfov	0.95	RO	Nord-Est
IL	Israel	0.66	TN	Tunisia	0.65	FR	Aude	0.62	IT	Bergamo
IT	Agrigento	0.88	IT	Caltanissetta	0.85	IT	Siracusa	0.85	IT	Ragusa
IT	Alessandria	0.89	IT	Asti	0.61	IT	Salerno	0.61	IT	Cuneo
IT	Ancona	0.80	IT	Macerata	0.38	IT	Chieti	0.37	IT	Fermo
IT	Arezzo	1.00	IT	Firenze	1.00	IT	Siena	0.99	IT	Grosseto
IT	Ascoli Piceno	0.98	IT	Fermo	0.94	IT	Barletta-Andria-Trani	0.90	IT	Isernia
IT	Asti	0.89	IT	Alessandria	0.66	IT	Salerno	0.64	IT	Piacenza
IT	Avellino	0.94	IT	Potenza	0.74	IT	Benevento	0.63	IT	Caserta
IT	Bari	0.80	IT	Taranto	0.74	US	Amador	0.73	US	San Bernardino
IT	Barletta-Andria-Trani	0.94	IT	Fermo	0.94	IT	Ascoli Piceno	0.89	IT	Isernia
IT	Belluno	0.87	US	Amador	0.87	US	San Bernardino	0.87	IT	Taranto
IT	Benevento	0.80	IT	Caserta	0.77	IT	Potenza	0.74	IT	Avellino
IT	Bergamo	0.87	FR	Gironde	0.86	IT	Lecco	0.86	BG	South Central
IT	Biella	0.89	IT	Novara	0.88	IT	Vercelli	0.87	IT	Verbano-Cusio-Ossola
IT	Bologna	0.85	IT	Ravenna	0.75	IT	Ferrara	0.70	IT	Forlì-Cesena

Table 96 (cont.): Each region's 3 most similar winegrape regions in the world according to the VSI, 2010

IT	Bolzano-Bozen	0.64	US	Columbia Gorge	0.64	IT	Trento	0.63	CA	British Colombia
IT	Brescia	0.92	FR	Yonne	0.90	US	Tehama	0.89	US	Yolo
IT	Brindisi	0.97	IT	Lecce	0.40	IT	Taranto	0.31	US	Amador
IT	Cagliari	0.83	IT	Medio Campidano	0.71	IT	Oristano	0.65	IT	Sassari
IT	Caltanissetta	0.99	IT	Siracusa	0.93	IT	Ragusa	0.88	IT	Agrigento
IT	Campobasso	0.98	IT	Pescara	0.98	IT	L'Aquila	0.97	IT	Teramo
IT	Carbonia-Iglesias	0.89	TN	Tunisia	0.70	MX	Aguascalientes	0.61	IL	Israel
IT	Caserta	0.80	IT	Benevento	0.67	IT	Potenza	0.63	IT	Napoli
IT	Catania	0.71	IT	Messina	0.44	IT	Enna	0.35	IT	Reggio di Calabria
IT	Catanzaro	0.51	IT	Cosenza	0.42	IT	Reggio di Calabria	0.40	IT	Vibo Valentia
IT	Chieti	0.93	IT	Pescara	0.93	IT	Campobasso	0.93	IT	Teramo
IT	Como	0.79	FR	Ariege	0.78	IT	Lecco	0.74	US	Orange
IT	Cosenza	0.88	IT	Crotone	0.57	IT	Reggio di Calabria	0.51	IT	Catanzaro
IT	Cremona	0.75	US	North Carolina	0.75	US	Arkansas	0.75	US	Georgia
IT	Crotone	0.88	IT	Cosenza	0.32	IT	Reggio di Calabria	0.27	IT	Catanzaro
IT	Cuneo	0.62	IT	Asti	0.61	IT	Alessandria	0.60	EL	Voreio Aigaio
IT	Enna	0.88	IT	Messina	0.87	IT	Reggio di Calabria	0.84	US	Orange
IT	Fermo	0.98	IT	Ascoli Piceno	0.94	IT	Barletta-Andria-Trani	0.89	IT	Isernia
IT	Ferrara	0.92	IT	Ravenna	0.75	IT	Bologna	0.43	IT	Forlì-Cesena
IT	Firenze	1.00	IT	Arezzo	1.00	IT	Siena	0.99	IT	Grosseto
IT	Foggia	0.83	IT	Barletta-Andria-Trani	0.83	IT	Fermo	0.77	IT	Ascoli Piceno
IT	Forlì-Cesena	0.95	IT	Rimini	0.89	IT	Siena	0.89	IT	Firenze
IT	Frosinone	0.71	RO	Sud - Muntenia	0.70	HR	Other regions	0.70	IT	Rieti
IT	Genova	0.77	IT	La Spezia	0.66	IT	Massa-Carrara	0.56	IT	Olbia-Tempio
IT	Gorizia	0.90	IT	Udine	0.81	IT	Pordenone	0.77	IT	Venezia
IT	Grosseto	0.99	IT	Arezzo	0.99	IT	Firenze	0.99	IT	Siena
IT	Imperia	0.81	IT	Savona	0.71	IT	Olbia-Tempio	0.65	IT	La Spezia
IT	Isernia	0.90	IT	Ascoli Piceno	0.89	IT	Fermo	0.89	IT	Barletta-Andria-Trani
IT	La Spezia	0.85	IT	Olbia-Tempio	0.83	IT	Savona	0.79	IT	Massa-Carrara
IT	L'Aquila	1.00	IT	Pescara	0.99	IT	Teramo	0.98	IT	Campobasso
IT	Latina	0.90	IT	Roma	0.84	IT	Viterbo	0.66	IT	Rieti
IT	Lecce	0.97	IT	Brindisi	0.17	IT	Como	0.16	IT	Taranto
IT	Lecco	0.87	EL	Anatoliki Makedonia, Thraki	0.86	IT	Bergamo	0.81	FR	Lot-et-Garonne
IT	Livorno	0.89	US	Red Mountain	0.88	CN	Gansu	0.88	US	Walla Walla Valley
IT	Lodi	0.91	IT	Milano	0.87	IT	Piacenza	0.76	IT	Pavia
IT	Lucca	0.95	IT	Pisa	0.93	IT	Grosseto	0.92	IT	Pistoia
IT	Macerata	0.80	IT	Ancona	0.70	IT	Fermo	0.67	IT	Barletta-Andria-Trani
IT	Mantova	0.56	IT	Cremona	0.56	JP	Nagano	0.55	US	Shasta
IT	Massa-Carrara	0.79	IT	La Spezia	0.68	IT	Olbia-Tempio	0.66	IT	Genova
IT	Matera	0.75	IT	Isernia	0.75	IT	Lucca	0.74	IT	Barletta-Andria-Trani
IT	Medio Campidano	0.83	IT	Cagliari	0.69	IT	Oristano	0.34	IT	Sassari
IT	Messina	0.88	IT	Enna	0.80	IT	Reggio di Calabria	0.71	US	North Carolina
IT	Milano	0.91	IT	Lodi	0.85	IT	Piacenza	0.84	IT	Pavia
IT	Modena	0.47	IT	Reggio nell'Emilia	0.32	IT	Mantova	0.25	IT	Cremona
IT	Monza e della Brianza	0.43	FR	Yonne	0.43	US	Tehama	0.42	AU	Hunter Valley - other
IT	Napoli	0.63	IT	Caserta	0.60	IT	Benevento	0.28	IT	Avellino
IT	Novara	0.95	IT	Vercelli	0.92	IT	Sondrio	0.89	IT	Biella
IT	Nuoro	1.00	IT	Ogliastra	0.93	FR	Vaucluse	0.90	ES	Zaragoza
IT	Ogliastra	1.00	IT	Nuoro	0.93	FR	Vaucluse	0.90	ES	Zaragoza
IT	Olbia-Tempio	0.95	IT	Savona	0.86	IT	Sassari	0.85	IT	La Spezia
IT	Oristano	0.71	IT	Cagliari	0.69	IT	Medio Campidano	0.59	TN	Tunisia
IT	Padova	0.85	CN	Other regions	0.84	FR	Gironde	0.84	CH	Ticino
IT	Palermo	0.98	IT	Trapani	0.51	IT	Agrigento	0.26	IT	Caltanissetta
IT	Parma	0.59	IT	Roma	0.48	IT	Salerno	0.47	IT	Cremona
IT	Pavia	0.84	IT	Milano	0.76	IT	Piacenza	0.76	IT	Lodi
IT	Perugia	0.85	IT	Lucca	0.83	IT	Pisa	0.82	IT	Grosseto
IT	Pesaro e Urbino	0.53	IT	Rimini	0.50	IT	Arezzo	0.50	IT	Firenze
IT	Pescara	1.00	IT	L'Aquila	1.00	IT	Teramo	0.98	IT	Campobasso
IT	Piacenza	0.87	IT	Lodi	0.85	IT	Milano	0.76	IT	Pavia
IT	Pisa	0.99	IT	Pistoia	0.99	IT	Grosseto	0.98	IT	Arezzo
IT	Pistoia	0.99	IT	Pisa	0.97	IT	Firenze	0.97	IT	Arezzo
IT	Pordenone	0.90	IT	Venezia	0.85	IT	Udine	0.81	IT	Gorizia
IT	Potenza	0.94	IT	Avellino	0.77	IT	Benevento	0.67	IT	Caserta
IT	Prato	0.97	IT	Arezzo	0.96	IT	Firenze	0.96	IT	Grosseto
IT	Ragusa	0.93	IT	Siracusa	0.93	IT	Caltanissetta	0.85	IT	Agrigento

Table 96 (cont.): Each region's 3 most similar winegrape regions in the world according to the VSI, 2010

IT	Ravenna	0.92	IT	Ferrara	0.85	IT	Bologna	0.60	IT	Forlì-Cesena
IT	Reggio di Calabria	0.87	IT	Enna	0.85	NZ	Other regions	0.83	US	Orange
IT	Reggio nell'Emilia	0.47	IT	Modena	0.47	IT	Mantova	0.43	IT	Cremona
IT	Rieti	0.76	IT	Viterbo	0.75	IT	Roma	0.70	IT	Frosinone
IT	Rimini	0.98	IT	Siena	0.98	IT	Firenze	0.98	IT	Arezzo
IT	Roma	0.90	IT	Latina	0.75	IT	Rieti	0.71	IT	Viterbo
IT	Rovigo	0.80	CH	Ticino	0.80	AR	Nogoya	0.79	CN	Other regions
IT	Salerno	0.66	IT	Como	0.66	IT	Torino	0.66	IT	Asti
IT	Sassari	0.86	IT	Olbia-Tempio	0.76	IT	Savona	0.72	IT	La Spezia
IT	Savona	0.95	IT	Olbia-Tempio	0.83	IT	La Spezia	0.81	IT	Imperia
IT	Siena	1.00	IT	Firenze	1.00	IT	Arezzo	0.99	IT	Grosseto
IT	Siracusa	0.99	IT	Caltanissetta	0.93	IT	Ragusa	0.85	IT	Agrigento
IT	Sondrio	0.99	IT	Vercelli	0.92	IT	Novara	0.88	IT	Verbano-Cusio-Ossola
IT	Taranto	0.97	US	Amador	0.96	US	San Bernardino	0.91	US	Colusa
IT	Teramo	1.00	IT	Pescara	0.99	IT	L'Aquila	0.97	IT	Campobasso
IT	Terni	0.74	IT	Perugia	0.68	IT	Foggia	0.66	IT	Viterbo
IT	Torino	0.66	IT	Salerno	0.63	IT	Biella	0.62	IT	Asti
IT	Trapani	0.98	IT	Palermo	0.46	IT	Agrigento	0.20	IT	Ragusa
IT	Trento	0.84	US	Solano	0.82	US	Yolo	0.80	AU	Big Rivers - other
IT	Treviso	0.64	IT	Padova	0.45	IT	Venezia	0.35	IT	Belluno
IT	Trieste	0.68	SI	Slovenska Istra	0.60	HR	Istra	0.54	SI	Kras
IT	Udine	0.92	IT	Venezia	0.90	IT	Gorizia	0.85	IT	Pordenone
IT	Valle d'Aosta	0.68	FR	Haute-Loire	0.68	US	Orange	0.67	KZ	West Kazakhstan
IT	Varese	0.80	IT	Verbano-Cusio-Ossola	0.72	IT	Lecco	0.66	US	Shasta
IT	Venezia	0.92	IT	Udine	0.90	IT	Pordenone	0.77	IT	Gorizia
IT	Verbano-Cusio-Ossola	0.92	IT	Vercelli	0.88	IT	Sondrio	0.87	IT	Novara
IT	Vercelli	0.99	IT	Sondrio	0.95	IT	Novara	0.92	IT	Verbano-Cusio-Ossola
IT	Verona	0.49	IT	Vicenza	0.42	IT	Rovigo	0.19	IT	Pordenone
IT	Vibo Valentia	0.51	IT	Cosenza	0.51	IT	Reggio di Calabria	0.48	US	Orange
IT	Vicenza	0.80	FR	Gironde	0.79	CN	Other regions	0.79	IT	Padova
IT	Viterbo	0.84	IT	Latina	0.76	IT	Rieti	0.72	IT	Foggia
JP	Hokkaido	0.47	DE	Saale	0.36	DE	Sachsen	0.35	DE	Franken
JP	Nagano	0.95	US	Shasta	0.94	JP	Yamagata	0.91	CH	Ticino
JP	Other regions	0.78	US	Santa Clara	0.77	US	Horse Heaven Hills	0.76	AU	North East Victoria - other
JP	Yamagata	0.94	JP	Nagano	0.90	US	Shasta	0.83	US	Merced
JP	Yamanashi	0.71	JP	Other regions	0.63	CN	Shandong	0.63	CN	Sichuan
KZ	Almaty	0.93	GE	Georgia	0.83	KZ	South Kazakhstan	0.71	KZ	Zhambyl
KZ	East Kazakhstan	0.87	KZ	West Kazakhstan	0.85	RO	Sud - Muntenia	0.85	RO	Sud-Vest Oltenia
KZ	Other regions	0.84	KZ	East Kazakhstan	0.71	HR	Other regions	0.69	RO	Bucuresti - Ilfov
KZ	South Kazakhstan	0.83	KZ	Almaty	0.77	GE	Georgia	0.61	KZ	Other regions
KZ	West Kazakhstan	0.94	NZ	Other regions	0.92	FR	Correze	0.92	US	North Carolina
KZ	Zhambyl	0.93	AM	Armenia	0.81	HU	Zala	0.80	RO	Nord-Vest
KR	KoreaRep	0.63	TW	Taiwan	0.25	US	Orange	0.25	FR	Haute-Loire
LU	Luxembourg	0.55	DE	Sachsen	0.46	DE	Saale	0.44	FR	Haut-Rhin
MX	Aguascalientes	0.70	IT	Carbonia-Iglesias	0.62	TN	Tunisia	0.43	FR	Aude
MX	Baja California	0.85	US	Walla Walla Valley	0.82	US	Texas	0.81	US	Columbia River
MX	Coahuila	0.58	US	Red Mountain	0.58	CN	Gansu	0.58	CN	Xinjiang
MX	Sonora	0.68	TR	Aegean	0.13	AU	North West Victoria - other	0.09	CY	Cyprus
MX	Zacatecas	0.57	EL	Voreio Aigaio	0.35	IT	Cuneo	0.29	PE	Arequipa
MD	Moldova	0.86	UA	Ukraine	0.50	NZ	Hawkes Bay	0.49	AU	Strathbogie Ranges
MA	Morocco	0.60	US	Orange	0.60	NZ	Other regions	0.59	FR	Haute-Loire
MM	Myanmar	0.81	AU	Sunbury	0.79	AU	Central Western Australia	0.77	AU	Western Australia Southeast Coast
NZ	Auckland	0.88	NZ	Hawkes Bay	0.85	US	Snipes Mountain	0.84	AU	Northern Rivers - other
NZ	Canterbury	0.98	AU	Tasmania	0.95	AU	Henty	0.92	AU	Mornington Peninsula
NZ	Gisborne	0.91	FR	Yonne	0.91	FR	Doubs	0.90	US	Yolo
NZ	Hawkes Bay	0.88	NZ	Auckland	0.87	AU	Strathbogie Ranges	0.85	AU	Alpine Valleys
NZ	Marlborough	0.99	FR	Nievre	0.98	FR	Cher	0.91	AR	Coronel Suarez
NZ	Nelson	0.94	NZ	Waipara	0.93	AR	Coronel Suarez	0.90	FR	Cher
NZ	Otago	1.00	US	Yamhill Co.	1.00	US	Polk Co.	0.99	US	Benton Co.
NZ	Other regions	0.98	FR	Correze	0.97	US	North Carolina	0.97	US	Arkansas
NZ	Waikato	0.83	US	Texas	0.81	NZ	Auckland	0.80	NZ	Hawkes Bay
NZ	Waipara	0.94	NZ	Nelson	0.91	FR	Cher	0.90	NZ	Marlborough
NZ	Wairarapa	0.93	NZ	Otago	0.93	AU	Tasmania	0.92	US	Yamhill Co.
PE	Arequipa	0.79	PT	Madeira	0.69	PE	Moquegua	0.55	PT	Algarve
PE	Lima	0.62	HU	Sopron	0.52	HU	Csongrad	0.50	HU	Szekszard

Table 96 (cont.): Each region's 3 most similar winegrape regions in the world according to the VSI, 2010

PE	Moquegua	0.94	PE	Tacna	0.69	PE	Arequipa	0.65	PT	Madeira	
PE	Tacna	0.94	PE	Moquegua	0.47	PE	Arequipa	0.41	PT	Madeira	
PT	Acores	0.46	PT	Entre Douro e Minho	0.13	PT	Ribatejo e Oeste	0.12	PT	Alentejo	
PT	Alentejo	0.61	ES	Burgos	0.60	ES	Alava	0.60	ES	La Rioja	
PT	Algarve	0.60	PT	Madeira	0.56	PT	Ribatejo e Oeste	0.55	PE	Arequipa	
PT	Alto Tras-os-Montes	0.49	PT	Alentejo	0.45	ES	Burgos	0.44	ES	La Rioja	
PT	Beira Interior	0.44	PT	Alto Tras-os-Montes	0.32	PT	Alentejo	0.26	ES	Avila, Palencia, Salamanca, Segovia, Soria	
PT	Beira Litoral	0.33	PT	Ribatejo e Oeste	0.28	ES	Zamora	0.28	ES	Leon	
PT	Entre Douro e Minho	0.46	PT	Acores	0.24	ES	Galicia	0.14	ES	Cantabria	
PT	Madeira	0.79	PE	Arequipa	0.65	PE	Moquegua	0.60	PT	Algarve	
PT	Ribatejo e Oeste	0.56	PT	Algarve	0.53	PT	Alentejo	0.33	PT	Beira Litoral	
RO	Bucuresti - Ilfov	0.98	HR	Other regions	0.98	RO	Nord-Vest	0.96	HU	Zala	
RO	Centru	0.77	RO	Vest	0.66	RO	Sud-Est	0.51	SI	Stajerska Slovenija	
RO	Nord-Est	0.98	RO	Nord-Vest	0.95	HU	Zala	0.95	RO	Bucuresti - Ilfov	
RO	Nord-Vest	0.98	HU	Zala	0.98	RO	Nord-Est	0.98	RO	Bucuresti - Ilfov	
RO	Sud - Muntenia	0.97	RO	Sud-Vest Oltenia	0.94	US	North Carolina	0.94	US	Arizona	
RO	Sud-Est	0.85	RO	Nord-Est	0.82	RO	Sud - Muntenia	0.81	RO	Nord-Vest	
RO	Sud-Vest Oltenia	0.97	RO	Sud - Muntenia	0.95	US	North Carolina	0.95	US	Arkansas	
RO	Vest	0.77	RO	Centru	0.70	RO	Sud-Est	0.61	SI	Vipavska dolina	
RU	Krasnodar Krai	0.75	US	Horse Heaven Hills	0.75	US	Napa	0.72	US	Riverside	
RU	Rostov Oblast	0.57	UA	Ukraine	0.48	KZ	Almaty	0.46	RU	Krasnodar Krai	
RS	Serbia	0.85	HU	Badacsony	0.85	HR	Slavonija	0.85	HR	Podunavlje	
SK	Juznoslovenska	0.94	SK	Nitrianska	0.92	SK	Malokarpatska	0.85	SK	Stredné Slovensko	
SK	Malokarpatska	0.98	SK	Nitrianska	0.92	SK	Juznoslovenska	0.89	SK	Stredné Slovensko	
SK	Nitrianska	0.98	SK	Malokarpatska	0.94	SK	Juznoslovenska	0.86	SK	Stredné Slovensko	
SK	Stredné Slovensko	0.89	SK	Malokarpatska	0.86	CZ	Morava	0.86	SK	Nitrianska	
SK	Tokajska	0.97	HU	Tokaj	0.32	HU	Nagy-Somlo	0.29	HR	Zagorje-Medimurje	
SK	Východné Slovensko	0.83	HU	Balatonfured-Csopak	0.83	SK	Juznoslovenska	0.81	HR	Slavonija	
SI	Bela Krajina	0.87	SI	Bizeljsko Sremic	0.82	SI	Other regions	0.72	SI	Dolenjska	
SI	Bizeljsko Sremic	0.90	SI	Other regions	0.87	SI	Bela Krajina	0.86	SI	Dolenjska	
SI	Dolenjska	0.90	SI	Other regions	0.86	SI	Bizeljsko Sremic	0.72	SI	Bela Krajina	
SI	Goriska brda	0.70	SI	Vipavska dolina	0.69	IT	Gorizia	0.68	IT	Udine	
SI	Kras	0.91	SI	Slovenska Istra	0.54	IT	Trieste	0.28	HR	Istra	
SI	Other regions	0.90	SI	Bizeljsko Sremic	0.90	SI	Dolenjska	0.82	SI	Bela Krajina	
SI	Prekmurje	0.94	SI	Stajerska Slovenija	0.88	HU	Badacsony	0.87	HR	Slavonija	
SI	Slovenska Istra	0.91	SI	Kras	0.68	IT	Trieste	0.64	HR	Istra	
SI	Stajerska Slovenija	0.94	SI	Prekmurje	0.84	HR	Zagorje-Medimurje	0.81	HR	Plesivica	
SI	Vipavska dolina	0.70	SI	Goriska brda	0.70	NZ	Hawkes Bay	0.69	IT	Lecco	
ZA	Breedekloof	0.97	ZA	Worcester	0.95	ZA	Olifants River	0.88	ZA	Little Karoo	
ZA	Little Karoo	0.94	ZA	Northern Cape	0.94	ZA	Olifants River	0.88	ZA	Breedekloof	
ZA	Northern Cape	0.94	ZA	Little Karoo	0.87	ZA	Olifants River	0.77	US	Fresno	
ZA	Olifants River	0.95	ZA	Breedekloof	0.94	ZA	Little Karoo	0.92	ZA	Worcester	
ZA	Paarl	0.99	ZA	Swartland	0.88	ZA	Stellenbosch	0.85	ZA	Worcester	
ZA	Robertson	0.91	ZA	Worcester	0.88	ZA	Breedekloof	0.81	ZA	Olifants River	
ZA	Stellenbosch	0.89	AU	Mount Benson	0.88	AU	Margaret River	0.88	ZA	Paarl	
ZA	Swartland	0.99	ZA	Paarl	0.87	ZA	Stellenbosch	0.85	ZA	Worcester	
ZA	Worcester	0.97	ZA	Breedekloof	0.92	ZA	Olifants River	0.91	ZA	Robertson	
ES	Alava	0.99	ES	La Rioja	0.97	ES	Zamora	0.97	ES	Comunidad de Madrid	
ES	Albacete	0.78	ES	Cuenca	0.60	ES	Guadalajara	0.59	ES	Alicante	
ES	Alicante	0.96	ES	Region de Murcia	0.59	ES	Albacete	0.29	CL	Atacama	
ES	Almeria, Granada, Jaen, Sevilla	0.64	HR	Other regions	0.62	RO	Bucuresti - Ilfov	0.62	RO	Sud - Muntenia	
ES	Avila, Palencia, Salamanca, Segovia, Soria	0.85	ES	Comunidad de Madrid	0.83	ES	Zaragoza	0.81	ES	Zamora	
ES	Badajoz	0.48	ES	La Rioja	0.47	ES	Burgos	0.47	ES	Alava	
ES	Barcelona	0.75	ES	Tarragona	0.53	ES	Girona, Lleida	0.30	ES	Huesca, Teruel	
ES	Burgos	0.98	ES	La Rioja	0.96	ES	Alava	0.96	ES	Zamora	
ES	Caceres	0.67	ES	Valladolid	0.66	ES	Comunidad de Madrid	0.64	ES	Alava	
ES	Cadiz	0.75	ES	Canarias	0.56	ES	Cantabria	0.51	MX	Coahuila	
ES	Canarias	0.75	ES	Cadiz	0.47	ES	Cantabria	0.39	ES	Galicia	
ES	Cantabria	0.69	ES	Galicia	0.60	ES	Leon	0.56	ES	Cadiz	
ES	Castellon	0.78	ES	Comunidad de Madrid	0.77	ES	Alava	0.76	ES	La Rioja	
ES	Ciudad Real	0.96	ES	Toledo	0.77	ES	Comunidad Foral de Navarra	0.69	ES	Cuenca	
ES	Comunidad de Madrid	0.97	ES	Alava	0.92	ES	La Rioja	0.92	ES	Zamora	

Table 96 (cont.): Each region's 3 most similar winegrape regions in the world according to the VSI, 2010

ES	Comunidad Foral de Navarra	0.88	ES	Toledo	0.77	ES	Ciudad Real	0.63	IT	Ogliastra
ES	Cordoba	0.23	ES	Almeria, Granada, Jaen, Sevilla	0.16	ES	Malaga	0.06	ES	Caceres
ES	Cuenca	0.78	ES	Albacete	0.75	ES	Valencia	0.69	ES	Ciudad Real
ES	Galicia	0.69	ES	Cantabria	0.47	RO	Sud - Muntenia	0.47	RO	Sud-Vest Oltenia
ES	Girona, Lleida	0.73	ES	Huesca, Teruel	0.72	ES	Tarragona	0.71	US	Santa Clara
ES	Guadalajara	0.88	ES	Burgos	0.86	ES	La Rioja	0.85	ES	Zamora
ES	Guipuzcoa, Vizcaya	0.13	ES	Cantabria	0.08	HU	Zala	0.07	RO	Nord-Vest
ES	Huelva	0.04	ES	Almeria, Granada, Jaen, Sevilla	0.01	AU	Southern Flinders Ranges	0.01	US	Ventura
ES	Huesca, Teruel	0.87	ES	Zaragoza	0.78	FR	Bouches-du-Rhone	0.77	FR	Gard
ES	Illes Balears	0.66	US	Horse Heaven Hills	0.66	AU	North East Victoria - other	0.66	US	Walla Walla Valley
ES	La Rioja	0.99	ES	Alava	0.98	ES	Burgos	0.97	ES	Zamora
ES	Leon	0.60	ES	Cantabria	0.54	ES	Principado de Asturias	0.34	ES	Galicia
ES	Malaga	0.84	AR	Zonda	0.82	AR	Poman	0.78	AR	Rivadavia - San Juan
ES	Principado de Asturias	0.64	US	Orange	0.62	NZ	Other regions	0.62	FR	Haute-Loire
ES	Region de Murcia	0.96	ES	Alicante	0.59	ES	Albacete	0.26	US	Contra Costa
ES	Tarragona	0.75	ES	Barcelona	0.72	ES	Girona, Lleida	0.57	ES	Huesca, Teruel
ES	Toledo	0.96	ES	Ciudad Real	0.88	ES	Comunidad Foral de Navarra	0.65	ES	Cuenca
ES	Valencia	0.75	ES	Cuenca	0.54	ES	Albacete	0.23	ES	Alava
ES	Valladolid	0.71	ES	Burgos	0.71	ES	Zamora	0.70	ES	La Rioja
ES	Zamora	0.97	ES	La Rioja	0.97	ES	Alava	0.96	ES	Burgos
ES	Zaragoza	0.90	IT	Ogliastra	0.90	IT	Nuoro	0.88	FR	Vaucluse
CH	Aargau	1.00	CH	Thurgau	1.00	CH	Zürich	0.99	CH	Basel Land
CH	Basel Land	1.00	CH	Schaffhausen	1.00	CH	St. Gallen	0.99	CH	Graubünden
CH	Bern	0.98	CH	Neuchâtel	0.95	CH	Fribourg	0.90	CH	Valais
CH	Fribourg	0.95	CH	Bern	0.94	CH	Vaud	0.90	CH	Neuchâtel
CH	Geneva	0.80	FR	Allier	0.77	FR	Puy-de-Dome	0.75	FR	Ain
CH	Graubünden	1.00	CH	St. Gallen	0.99	AR	Malargüe	0.99	CH	Schaffhausen
CH	Jura	0.78	HR	Other regions	0.75	RO	Sud - Muntenia	0.74	RO	Bucuresti - Ilfov
CH	Lucerne	0.96	CH	Other regions	0.93	CH	Schwyz	0.93	CH	Thurgau
CH	Neuchâtel	0.98	CH	Bern	0.92	CH	Valais	0.90	CH	Fribourg
CH	Other regions	0.98	CH	Thurgau	0.98	CH	Zürich	0.98	CH	Aargau
CH	Schaffhausen	1.00	CH	St. Gallen	1.00	CH	Basel Land	0.99	CH	Graubünden
CH	Schwyz	0.98	CH	Zürich	0.98	CH	Aargau	0.98	CH	Thurgau
CH	St. Gallen	1.00	CH	Schaffhausen	1.00	CH	Basel Land	1.00	CH	Graubünden
CH	Thurgau	1.00	CH	Aargau	1.00	CH	Zürich	0.98	CH	Schaffhausen
CH	Ticino	1.00	AR	Nogoya	0.98	CN	Other regions	0.92	FR	Gironde
CH	Valais	0.92	CH	Neuchâtel	0.90	CH	Bern	0.82	CH	Fribourg
CH	Vaud	0.95	FR	Haute-Savoie	0.94	CH	Fribourg	0.87	FR	Tarn-et-Garonne
CH	Zürich	1.00	CH	Aargau	1.00	CH	Thurgau	0.98	CH	Basel Land
TW	Taiwan	0.63	KR	KoreaRep	0.08	JP	Yamanashi	0.07	JP	Other regions
TH	Thailand	0.91	AU	Southern Flinders Ranges	0.91	US	Ventura	0.91	AU	Barossa - other
TN	Tunisia	0.89	IT	Carbonia-Iglesias	0.70	DZ	Algeria	0.66	IL	Israel
TR	Aegean	0.68	MX	Sonora	0.55	AU	North West Victoria - other	0.53	AU	Swan Hill (NSW)
TR	Central East	0.23	TR	Mediterranean	0.13	TR	Aegean	0.06	TR	Central North
TR	Central North	0.42	TR	Mediterranean	0.38	TR	Aegean	0.28	AU	Southern Flinders Ranges
TR	Central South	0.07	TR	Central North	0.05	TR	Aegean	0.03	TR	Mediterranean
TR	Marmara	0.53	AU	New England Australia	0.52	FR	Dordogne	0.39	DZ	Algeria
TR	Mediterranean	0.78	AU	Strathbogie Ranges	0.77	AU	Upper Goulburn	0.76	US	Sacramento
TR	South East	0.27	TR	Mediterranean	0.22	TR	Aegean	0.17	TR	Central North
UA	Ukraine	0.86	MD	Moldova	0.70	KZ	Almaty	0.63	GE	Georgia
UK	UnitedKingdom	0.84	US	San Mateo	0.83	US	Santa Cruz	0.81	AU	Tumbarumba
US	Alameda	0.94	US	Santa Clara	0.92	US	Riverside	0.92	US	Napa
US	Amador	0.97	US	San Bernardino	0.97	IT	Taranto	0.93	US	Colusa
US	Arizona	1.00	US	Arkansas	1.00	US	North Carolina	1.00	US	Georgia
US	Arkansas	1.00	US	Georgia	1.00	US	North Carolina	1.00	US	Arizona
US	Benton Co.	1.00	US	Washington Co.	1.00	US	Polk Co.	0.99	US	Yamhill Co.
US	Butte	0.90	US	Los Angeles	0.89	US	San Luis Obispo	0.88	AU	Coonawarra
US	Calaveras	0.92	AU	Mount Benson	0.90	AU	Geographe	0.90	US	Nevada
US	Chautauqua-Erie	0.99	US	Ohio	0.96	US	Pennsylvania	0.79	US	Finger Lakes
US	Colorado	0.91	US	Snipes Mountain	0.89	US	Columbia Valley	0.88	US	Wahluke Slope
US	Columbia Gorge	0.76	IT	Trento	0.70	NZ	Auckland	0.70	AU	Northern Rivers - other
US	Columbia River	0.88	US	Walla Walla Valley	0.87	US	Jackson Co.	0.86	US	Colorado

Table 96 (cont.): Each region's 3 most similar winegrape regions in the world according to the VSI, 2010

US	Columbia Valley	0.98	US	Wahluke Slope	0.96	US	Horse Heaven Hills	0.93	US	Yakima Valley
US	Colusa	0.93	US	Amador	0.91	US	San Bernardino	0.91	IT	Taranto
US	Contra Costa	0.87	US	San Joaquin	0.81	US	Merced	0.79	US	Yolo
US	Douglas Co.	0.98	US	Willamette Valley - other	0.96	US	Washington Co.	0.96	US	Polk Co.
US	El Dorado	0.87	US	Mariposa	0.86	US	Calaveras	0.86	US	San Luis Obispo
US	Finger Lakes	0.85	US	Pennsylvania	0.83	US	Ohio	0.79	US	Chautauqua-Erie
US	Fresno	0.87	US	Tulare	0.85	US	Madera	0.77	ZA	Northern Cape
US	Georgia	1.00	US	North Carolina	1.00	US	Arizona	1.00	US	Georgia
US	Glenn	0.78	US	Colusa	0.74	US	Kings	0.63	US	San Joaquin
US	Horse Heaven Hills	0.98	US	Wahluke Slope	0.96	US	Columbia Valley	0.96	AU	North East Victoria - other
US	Humboldt	0.85	AU	Western Victoria - other	0.85	AU	North East Victoria - other	0.80	AU	Alpine Valleys
US	Illinois	0.81	US	Indiana	0.70	US	Missouri	0.55	US	Arizona
US	Indiana	0.81	US	Illinois	0.64	HR	Other regions	0.63	RO	Bucuresti - Ilfov
US	Iowa	0.44	US	Minnesota	0.41	US	Illinois	0.21	US	Missouri
US	Jackson Co.	0.87	US	Columbia River	0.81	FR	Ariege	0.76	US	Colorado
US	Josephine Co.	0.97	US	Washington Co.	0.97	US	Willamette Valley - other	0.95	US	Douglas Co.
US	Kentucky	0.68	US	Missouri	0.62	US	Virginia	0.60	CA	Ontario
US	Kern	0.71	US	Tulare	0.71	ZA	Little Karoo	0.69	US	Fresno
US	Kings	0.74	US	Glenn	0.68	US	Tulare	0.61	US	Fresno
US	Lake	0.91	CL	Del Maule	0.88	CL	Metropolitana	0.87	US	Napa
US	Lake Chelan	0.75	US	Michigan	0.74	US	Josephine Co.	0.71	AU	Eden Valley
US	Lane Co.	0.95	US	Marion Co.	0.79	US	Washington Co.	0.79	US	Josephine Co.
US	Los Angeles	0.95	AU	Coonawarra	0.95	US	Red Mountain	0.94	CL	Metropolitana
US	Madera	0.87	US	Tulare	0.85	US	Fresno	0.82	US	Stanislaus
US	Marin	1.00	FR	Aube	0.99	AR	Malargüe	0.99	US	Yamhill Co.
US	Marion Co.	0.95	US	Lane Co.	0.94	US	Washington Co.	0.93	US	Benton Co.
US	Mariposa	0.91	US	San Luis Obispo	0.88	US	San Joaquin	0.87	US	El Dorado
US	Mendocino	0.98	US	Sonoma	0.94	US	Sacramento	0.92	US	San Benito
US	Merced	0.92	US	Sacramento	0.89	US	Mendocino	0.88	US	Santa Clara
US	Michigan	0.77	DE	Hessische Bergstraße	0.75	US	Lake Chelan	0.75	DE	Rheingau
US	Minnesota	0.44	US	Iowa	0.31	US	Illinois	0.13	US	Arizona
US	Missouri	0.70	US	Illinois	0.68	US	Kentucky	0.62	US	Indiana
US	Monterey	0.99	US	San Benito	0.96	US	Santa Barbara	0.94	US	Solano
US	Napa	0.95	US	San Luis Obispo	0.95	CL	Metropolitana	0.95	CN	Gansu
US	Nevada	0.90	AU	Peel	0.90	US	San Luis Obispo	0.90	US	Calaveras
US	New York - other	0.77	JP	Nagano	0.73	US	Virginia	0.73	JP	Yamagata
US	North Carolina	1.00	US	Arizona	1.00	US	Arkansas	1.00	US	Georgia
US	Ohio	0.99	US	Chautauqua-Erie	0.97	US	Pennsylvania	0.83	US	Finger Lakes
US	Orange	0.97	FR	Haute-Loire	0.96	NZ	Other regions	0.92	FR	Correze
US	Pennsylvania	0.97	US	Ohio	0.96	US	Chautauqua-Erie	0.85	US	Finger Lakes
US	Placer	0.76	US	Amador	0.74	US	San Bernardino	0.70	IT	Taranto
US	Polk Co.	1.00	US	Yamhill Co.	1.00	NZ	Otago	1.00	US	Benton Co.
US	Puget Sound	0.87	US	Douglas Co.	0.86	FR	Moselle	0.85	US	Willamette Valley - other
US	Rattlesnake Hills	0.90	US	Columbia Valley	0.88	US	Wahluke Slope	0.87	US	Yakima Valley
US	Red Mountain	0.97	US	Walla Walla Valley	0.97	AU	Coonawarra	0.97	CN	Gansu
US	Riverside	0.93	US	Horse Heaven Hills	0.92	US	Alameda	0.92	US	Santa Clara
US	Sacramento	0.95	US	Santa Clara	0.94	US	Mendocino	0.92	US	Merced
US	San Benito	0.99	US	Monterey	0.96	US	Santa Barbara	0.93	US	Trinity
US	San Bernardino	0.97	US	Amador	0.96	IT	Taranto	0.91	US	Colusa
US	San Diego	0.90	AU	The Peninsulas	0.90	AU	Glenrowan	0.89	AU	Canberra District (NSW)
US	San Joaquin	0.88	US	Mariposa	0.87	US	Contra Costa	0.86	US	Stanislaus
US	San Luis Obispo	0.95	US	Napa	0.93	AU	North East Victoria - other	0.92	US	Alameda
US	San Mateo	0.97	US	Santa Cruz	0.95	AU	Gippsland	0.94	AU	Mornington Peninsula
US	Santa Barbara	0.96	US	Monterey	0.96	US	San Benito	0.95	AU	Tumbarumba
US	Santa Clara	0.95	US	Sacramento	0.94	US	Alameda	0.92	AU	North East Victoria - other
US	Santa Cruz	0.98	AU	Macedon Ranges	0.97	AU	Gippsland	0.97	US	San Mateo
US	Shasta	0.95	JP	Nagano	0.90	JP	Yamagata	0.90	CH	Ticino
US	Siskiyou	0.83	US	Sonoma	0.83	US	Mendocino	0.79	US	Solano
US	Snipes Mountain	0.91	US	Colorado	0.90	US	Columbia Valley	0.89	US	Wahluke Slope
US	Solano	0.94	US	Yolo	0.94	US	Monterey	0.93	US	San Benito
US	Sonoma	0.98	US	Mendocino	0.92	US	Sacramento	0.91	AU	Yarra Valley
US	Stanislaus	0.88	US	Merced	0.86	US	San Joaquin	0.83	US	Sacramento
US	Sutter	0.71	US	Contra Costa	0.67	US	San Joaquin	0.64	US	Colusa
US	Tehama	0.97	FR	Yonne	0.97	AU	Hunter Valley - other	0.95	US	Yolo
US	Texas	0.88	US	Snipes Mountain	0.87	US	Riverside	0.86	US	Alameda
US	Trinity	0.93	US	San Benito	0.92	US	Monterey	0.91	US	Santa Barbara

Table 96 (cont.): Each region's 3 most similar winegrape regions in the world according to the VSI, 2010

US	Tulare	0.87	US	Fresno	0.87	US	Madera	0.73	US	Stanislaus
US	Tuolumne	0.97	AU	Bendigo	0.96	AU	McLaren Vale	0.96	AU	Mount Lofty Ranges - other
US	Ventura	0.99	AU	Southern Flinders Ranges	0.99	AU	Barossa - other	0.98	AU	Grampians
US	Virginia	0.77	US	Santa Clara	0.76	US	Colorado	0.75	US	Sacramento
US	Wahluke Slope	0.98	US	Horse Heaven Hills	0.98	US	Columbia Valley	0.94	AU	North East Victoria - other
US	Walla Walla Valley	0.97	US	Red Mountain	0.92	CN	Gansu	0.92	AU	Coonawarra
US	Washington Co.	1.00	US	Benton Co.	0.99	US	Willamette Valley - other	0.99	US	Polk Co.
US	Willamette Valley - other	0.99	US	Washington Co.	0.99	US	Polk Co.	0.99	NZ	Otago
US	Yakima Valley	0.93	US	Columbia Valley	0.87	US	Rattlesnake Hills	0.86	US	Colorado
US	Yamhill Co.	1.00	US	Polk Co.	1.00	NZ	Otago	0.99	US	Benton Co.
US	Yolo	0.95	US	Tehama	0.94	US	Solano	0.93	US	Monterey
US	Yuba	0.74	US	Alameda	0.74	US	Nevada	0.73	AU	Northern Slopes - other
UY	Uruguay	0.69	FR	Hautes-Pyrenees	0.65	AR	Tulumba	0.64	AR	Villarino

Table 97: Each region's 3 most similar winegrape regions in the world according to the VSI, 2016

DZ	Algeria	0.79	FR	Languedoc Roussillon	0.73	IL	Israel	0.67	FR	Provence-Alpes-Cote d'Azur
AR	AdolfoAlsina	0.95	AR	Calamuchita	0.93	AR	Avellaneda - Río Negro	0.93	AR	Puelen
AR	Albardón	0.99	AR	Chimbas	0.98	AR	Santa Lucía	0.97	AR	Pocito
AR	Andalgala	0.61	AR	Sarmiento - San Juan	0.60	AR	Famatina	0.58	AR	Maipú
AR	Añelo	0.96	AR	Puelen	0.96	AR	Confluencia	0.96	AR	Calamuchita
AR	Angaco	0.95	AR	Chimbas	0.95	AR	Caucete	0.95	AR	San Martín - San Juan
AR	AvellanedaRíoNegro	0.93	AR	Adolfo Alsina	0.91	AR	Puelen	0.89	AR	Tunuyán
AR	Ayacucho	0.94	AR	Capital San Luis	0.94	AR	San Alberto	0.91	AR	San Javier
AR	Balcarce	0.79	AU	Mount Gambier	0.79	CL	Valparaiso	0.76	NZ	Nelson
AR	Bariloche	0.78	AR	Nogoya	0.78	CH	Ticino	0.78	CH	Graubünden - Mesolcina
AR	Belén	0.97	AR	San Carlos - Salta	0.96	AR	Cruz del Eje	0.89	AR	Famatina
AR	BenitoJuárez	1.00	AR	Benito Juárez	1.00	AR	Godoy Cruz	1.00	AR	General Belgrano
AR	Cachi	0.99	AR	Villa Gesell	0.99	AR	General Belgrano	0.99	AR	Godoy Cruz
AR	Cafayate	0.96	AR	Famatina	0.96	AR	Tafí del Valle	0.87	AR	San Carlos - Salta
AR	Cainguas	0.96	AR	Cañuelas	0.89	UY	Florida	0.83	AR	Concordia
AR	Calamuchita	0.97	AR	Puelen	0.96	AR	Tunuyán	0.96	AR	Añelo
AR	Calingasta	0.94	AR	Calamuchita	0.92	AR	La Viña	0.92	AR	Santa María - Cba
AR	Cañuelas	0.96	AR	Cainguas	0.90	UY	Florida	0.87	AR	Concordia
AR	Capayán	0.94	AR	Pehuenches	0.74	AR	Valle Viejo	0.71	AR	Guaymallén
AR	CapitalMisiones	1.00	AR	Veinticinco de Mayo -	0.99	DE	Rheingau	0.98	DE	Mittelrhein
AR	CapitalSanJuan	0.57	AR	Poman	0.55	AR	Rivadavia - San Juan	0.48	CL	Del Bio Bio
AR	CapitalSanLuis	0.94	AR	Ayacucho	0.92	AR	San Javier	0.89	AR	Calingasta
AR	CapitalSantiagodelEstero	0.81	MX	Zacatecas	0.11	MA	Morocco	0.11	TN	Tunisia
AR	CastroBarros	0.91	AR	Santa María - Cba	0.90	AR	Maipú	0.89	AR	Calingasta
AR	Caucete	0.98	AR	San Martín - San Juan	0.97	AR	Nueve de Julio	0.97	AR	Veinticinco de Mayo - San
AR	Chilecito	0.92	AR	General Lamadrid	0.89	AR	San Blas De Los Sauces	0.89	AR	Cruz del Eje
AR	Chimbas	0.99	AR	Santa Lucía	0.99	AR	Albardón	0.99	AR	Tinogasta
AR	ChosMalal	0.99	AR	Luján de Cuyo	0.98	AR	Tunuyán	0.98	AR	Molinos
AR	CollonCura	0.99	US	Marin	0.98	CH	Graubünden - other	0.97	US	Umpqua Valley
AR	ColónCba	0.88	AR	Collon Cura	0.87	US	Umpqua Valley	0.86	US	Marin
AR	ColónEntreRíos	0.87	AR	Victoria	0.87	AR	Puelen	0.86	AR	San Javier
AR	Concordia	0.87	AR	Cañuelas	0.84	UY	Florida	0.83	AR	Cainguas
AR	Conesa	0.82	AR	Pichi Mahuida	0.63	AR	Belén	0.59	AR	San Carlos - Salta
AR	Confluencia	0.96	AR	Tupungato	0.96	AR	Añelo	0.96	AR	Tunuyán
AR	CoronelFelipeVarela	0.97	AR	General Lamadrid	0.95	AR	Vinchina	0.90	AR	Sanagasta
AR	CoronelPringles	0.96	AR	Picún Leufú	0.87	AR	Paraná	0.83	AR	Tupungato
AR	CoronelSuarez	0.86	CL	Valparaiso	0.86	AU	Mount Gambier	0.85	NZ	Nelson
AR	CruzdelEje	0.96	AR	Belén	0.91	AR	San Carlos - Salta	0.89	AR	General Lamadrid
AR	Curaco	0.82	AR	Tandil	0.81	AR	Calingasta	0.80	AR	Picún Leufú
AR	Cushamen	0.81	AR	Colón - Cba	0.80	US	Siskiyou	0.79	US	Humboldt
AR	Daireaux	0.89	US	Alameda	0.89	US	Napa	0.83	US	Horse Heaven Hills
AR	DeLaCosta	1.00	AR	Godoy Cruz	1.00	AR	General Belgrano	1.00	AR	Villa Gesell
AR	Diamante	0.97	AR	Cachi	0.96	AR	Benito Juárez	0.96	AR	Godoy Cruz
AR	ElCarmen	0.89	AR	Adolfo Alsina	0.85	AR	Tumbaya	0.83	AR	Añelo
AR	ElCuy	0.96	AR	Picunches	0.92	AR	General Roca	0.90	AR	Humahuaca
AR	Famatina	0.96	AR	Cafayate	0.93	AR	Tafí del Valle	0.89	AR	Belén
AR	Futaleufu	0.99	AU	Tasmania	0.97	CL	De Los Lagos	0.95	CL	Araucania
AR	GeneralAlvear	0.94	AR	San Martín - Mza	0.93	AR	Santa Rosa - Mza	0.87	AR	Junín - Mza
AR	GeneralBelgrano	1.00	AR	Villa Gesell	1.00	AR	Benito Juárez	1.00	AR	General Belgrano
AR	GeneralLamadrid	0.97	AR	Coronel Felipe Varela	0.93	AR	Sanagasta	0.93	AR	San Blas De Los Sauces
AR	GeneralPueyrredón	0.96	CL	De Los Lagos	0.94	CL	Valparaiso	0.93	AU	Adelaide Hills
AR	GeneralRoca	0.92	AR	El Cuy	0.88	AR	Avellaneda - Río Negro	0.84	AR	Picunches
AR	GodoyCruz	1.00	AR	General Belgrano	1.00	AR	Villa Gesell	1.00	AR	Benito Juárez
AR	Gualeguaychu	0.81	FR	Centre-Val de Loire	0.77	AR	Tornquist	0.53	FR	Pays de la Loire
AR	Guaymallén	0.93	AR	Maipú	0.88	AR	San Rafael	0.87	AR	San Javier
AR	Humahuaca	0.92	AR	Cachi	0.92	AR	Confluencia	0.92	AR	Añelo
AR	Iglesia	0.66	AR	Cachi	0.66	AR	Villa Gesell	0.66	AR	General Belgrano
AR	Ischilin	0.74	AR	Colón - Cba	0.72	US	Marin	0.71	AR	Collon Cura
AR	JunínBsAs	0.71	AR	Avellaneda - Río Negro	0.71	AR	Benito Juárez	0.71	AR	General Belgrano
AR	JunínMza	0.98	AR	Rivadavia - Mza	0.92	AR	San Martín - Mza	0.91	AR	Santa Rosa - Mza
AR	JunínSanLuis	0.64	AU	Canberra District (ACT)	0.61	AU	North East Victoria - other	0.61	AU	Central Victoria - other
AR	LaPaz	0.92	AR	Lavalle	0.91	AR	Santa Rosa - Mza	0.89	AR	Las Heras
AR	LaViña	0.99	AR	Molinos	0.98	AR	Luján de Cuyo	0.97	AR	Chos Malal
AR	Lacar	0.77	AR	Cushamen	0.71	US	Monterey	0.70	FR	Franche Comté
AR	Languiñeo	0.91	US	Yolo	0.89	US	Lassen	0.89	AU	Cowra
AR	LasHeras	0.93	AR	Lavalle	0.91	AR	Santa Rosa - Mza	0.89	AR	Sarmiento - San Juan
AR	Lavalle	0.96	AR	Santa Rosa - Mza	0.93	AR	Las Heras	0.92	AR	La Paz
AR	LeandroAlem	0.83	AR	Veinticinco de Mayo -	0.83	AR	Capital Misiones	0.82	DE	Rheingau
AR	LujándeCuyo	0.99	AR	Tunuyán	0.99	AR	Molinos	0.99	AR	Chos Malal
AR	Maipú	0.96	AR	Luján de Cuyo	0.95	AR	Tunuyán	0.94	AR	Tupungato
AR	Molinos	0.99	AR	Luján de Cuyo	0.99	AR	San Carlos - Mza	0.99	AR	La Viña
AR	Nogoya	1.00	CH	Graubünden - Mesolcina	1.00	CH	Ticino	0.98	AR	Ñorquin
AR	Ñorquin	0.98	AR	Nogoya	0.98	CH	Graubünden - Mesolcina	0.98	CH	Ticino
AR	NuevedeJulio	0.98	AR	Pocito	0.98	AR	Veinticinco de Mayo - San	0.97	AR	Caucete
AR	Paraná	0.91	AR	Confluencia	0.89	AR	Victoria	0.89	AR	Tupungato
AR	Pehuenches	0.94	AR	Capayán	0.79	AR	Tupungato	0.78	AR	Valle Viejo
AR	PichiMahuida	0.82	AR	Conesa	0.67	AR	Las Heras	0.66	CL	Coquimbo
AR	PicúnLeufú	0.96	AR	Coronel Pringles	0.82	AR	Paraná	0.81	AR	Languiñeo
AR	Picunches	0.96	AR	El Cuy	0.88	AR	Humahuaca	0.85	AR	Ñorquin
AR	Pocito	0.99	AR	Tinogasta	0.98	AR	Chimbas	0.98	AR	Santa Lucía
AR	Poman	0.68	AR	Rivadavia - San Juan	0.66	AR	Zonda	0.59	AR	Pehuenches

Table 97: Each region's 3 most similar winegrape regions in the world according to the VSI, 2016

Table 97: Each region's 3 most similar winegrape regions in the world according to the VSI, 2016

AR	Puelen	0.97	AR	Tunuyán	0.97	AR	Calamuchita	0.96	AR	Añelo
AR	Punilla	0.98	AR	Cachi	0.98	AR	Villa Gesell	0.98	AR	General Belgrano
AR	Rawson	0.98	AR	Santa Lucía	0.98	AR	Tinogasta	0.98	AR	Chimbas
AR	RivadaviaMza	0.98	AR	Junín - Mza	0.92	AR	San Martín - Mza	0.91	AR	Santa Rosa - Mza
AR	RivadaviaSanJuan	0.83	AR	Ullum	0.77	AR	Veinticinco de Mayo - San	0.77	AR	Zonda
AR	Saavedra	0.89	AU	South West Australia - other	0.89	LB	Lebanon	0.87	US	Sacramento
AR	SanAlberto	0.94	AR	Ayacucho	0.88	CL	Antofagasta	0.87	AR	Capital San Luis
AR	SanBlasDeLosSauces	0.96	AR	Trancas	0.94	AR	Sanagasta	0.93	AR	General Lamadrid
AR	SanCarlosMza	0.99	AR	Molinos	0.99	AR	Tunuyán	0.98	AR	Luján de Cuyo
AR	SanCarlosSalta	0.97	AR	Belén	0.91	AR	Cruz del Eje	0.88	AR	Famatina
AR	SanJavier	0.95	AR	Luján de Cuyo	0.95	AR	Tunuyán	0.94	AR	Molinos
AR	SanMartínMza	0.97	AR	Santa Rosa - Mza	0.94	AR	General Alvear	0.92	AR	Junín - Mza
AR	SanMartínSanJuan	0.98	AR	Caucete	0.97	AR	Veinticinco de Mayo - San	0.95	AR	Nueve de Julio
AR	SanRafael	0.91	AR	Rivadavia - Mza	0.89	AR	Junín - Mza	0.88	AR	La Paz
AR	Sanagasta	0.95	AR	Trancas	0.94	AR	San Blas De Los Sauces	0.93	AR	General Lamadrid
AR	SantaLucía	0.99	AR	Chimbas	0.98	AR	Rawson	0.98	AR	Tinogasta
AR	SantaMaríaCatamarca	0.86	AR	Tumbaya	0.86	AR	Cafayate	0.83	AR	Tafí del Valle
AR	SantaMaríaCba	0.92	AR	Calamuchita	0.92	AR	Calingasta	0.91	AR	Castro Barros
AR	SantaRosaCatamarca	0.98	AR	Santa Lucía	0.97	AR	Tinogasta	0.97	AR	Chimbas
AR	SantaRosaMza	0.97	AR	San Martín - Mza	0.96	AR	Lavalle	0.93	AR	General Alvear
AR	SarmientoChubut	0.99	CL	Araucania	0.98	US	Santa Cruz	0.95	US	Santa Barbara
AR	SarmientoSanJuan	0.96	AR	Veinticinco de Mayo - San	0.95	AR	Nueve de Julio	0.93	AR	Caucete
AR	TafídelValle	0.96	AR	Cafayate	0.94	AR	La Viña	0.93	AR	Famatina
AR	Tandil	0.87	AR	Villarino	0.83	AR	Calingasta	0.82	AR	Curaco
AR	Tilcara	0.95	AR	Punilla	0.95	AR	Cachi	0.93	AR	Tunuyán
AR	Tinogasta	0.99	AR	Chimbas	0.99	AR	Pocito	0.98	AR	Santa Lucía
AR	Tornquist	0.77	AR	Gualeguaychu	0.67	FR	Centre-Val de Loire	0.66	AR	Victoria
AR	Trancas	0.96	AR	San Blas De Los Sauces	0.95	AR	Sanagasta	0.91	AR	General Lamadrid
AR	Tulumba	0.94	AR	Calamuchita	0.93	AR	Añelo	0.89	AR	Confluencia
AR	Tumbaya	0.89	AR	Calingasta	0.86	CL	Metropolitana	0.86	US	Red Mountain
AR	Tunuyán	0.99	AR	Luján de Cuyo	0.99	AR	San Carlos - Mza	0.98	AR	Chos Malal
AR	Tupungato	0.97	AR	Tunuyán	0.96	AR	Confluencia	0.96	AR	Chos Malal
AR	Ullum	0.91	AR	Sarmiento - San Juan	0.88	AR	Veinticinco de Mayo - San	0.86	AR	Caucete
AR	Uruguay	0.91	US	Snipes Mountain	0.91	US	Sacramento	0.90	AU	South West Australia - other
AR	ValleFértil	0.89	CL	Antofagasta	0.82	AU	Eastern Plains, Inland and	0.80	AU	Southern Flinders Ranges
AR	ValleViejo	0.78	AR	Pehuenches	0.74	AR	Capayán	0.70	AR	Guaymallén
AR	VeinticincodeMayoMisiones	1.00	AR	Veinticinco de Mayo -	0.99	DE	Rheingau	0.98	DE	Mittelrhein
AR	VeinticincodeMayoSanJuan	0.98	AR	Nueve de Julio	0.97	AR	San Martín - San Juan	0.97	AR	Caucete
AR	Victoria	0.89	AR	Paraná	0.88	AR	Confluencia	0.87	AR	Colón - Entre Ríos
AR	VillaGesell	1.00	AR	Villa Gesell	1.00	AR	General Belgrano	1.00	AR	Godoy Cruz
AR	Villarino	0.87	AR	Tandil	0.80	AR	Concordia	0.78	AR	Colón - Entre Ríos
AR	Vinchina	0.95	AR	Coronel Felipe Varela	0.88	AR	General Lamadrid	0.86	AR	Sanagasta
AR	Zonda	0.77	AR	Rivadavia - San Juan	0.75	AR	Ullum	0.68	CL	Antofagasta
AM	Armenia	0.00	SI	Bela Krajina	0.00	US	Sacramento	0.00	US	Umpqua Valley
AU	AdelaideHills	0.96	CL	Valparaiso	0.94	AU	Pemberton	0.93	AR	General Pueyrredón
AU	AdelaidePlains	0.96	AU	Bendigo	0.96	AU	Pyrenees	0.95	AU	Goulburn Valley
AU	AlpineValleys	0.92	NZ	Hawkes Bay	0.92	AU	King Valley	0.87	AU	Southern Highlands
AU	Barossaother	0.98	AU	Mount Lofty Ranges - other	0.98	AU	McLaren Vale	0.97	AU	Barossa Valley
AU	BarossaValley	0.99	AU	McLaren Vale	0.99	AU	Southern Flinders Ranges	0.99	AU	Grampians
AU	Beechworth	0.93	AU	Sunbury	0.91	NZ	Waikato	0.90	US	Lake Chelan
AU	Bendigo	0.99	AU	Fleurieu - other	0.99	AU	Gundagai	0.99	AU	Mount Lofty Ranges - other
AU	BigRiversother	0.96	AU	Cowra	0.94	AU	Murray Darling (VIC)	0.94	AU	Murray Darling (NSW)
AU	BlackwoodValley	0.97	AU	Mount Benson	0.96	AU	Swan Hill (NSW)	0.96	AU	Great Southern
AU	CanberraDistrictACT	0.86	AU	Canberra District (NSW)	0.85	AU	The Peninsulas	0.85	AU	Goulburn Valley
AU	CanberraDistrictNSW	0.96	AU	Clare Valley	0.96	AU	The Peninsulas	0.95	AU	Bendigo
AU	CentralRangesother	0.99	AU	Currency Creek	0.98	AU	Port Phillip - other	0.98	AU	Langhorne Creek
AU	CentralVictoriaother	1.00	AU	Heathcote	0.99	AU	Grampians	0.99	AU	Southern Flinders Ranges
AU	CentralWesternAustralia	0.90	AU	Barossa Valley	0.90	AU	McLaren Vale	0.87	AU	Southern Flinders Ranges
AU	ClareValley	0.96	AU	Canberra District (NSW)	0.94	AU	The Peninsulas	0.92	AU	Eden Valley
AU	Coonawarra	0.98	US	Walla Walla Valley	0.97	US	Red Mountain	0.97	US	Columbia Valley
AU	Cowra	0.96	AU	Big Rivers - other	0.92	AU	Murray Darling (VIC)	0.90	NZ	Northland
AU	CurrencyCreek	0.99	AU	Langhorne Creek	0.99	AU	Central Ranges - other	0.98	AU	Hilltops
AU	EasternPlainsInlandandNor	0.99	AU	Southern Flinders Ranges	0.98	AU	Grampians	0.98	AU	Central Victoria - other
AU	EdenValley	0.92	AU	Clare Valley	0.90	AU	Canberra District (NSW)	0.90	AU	Far North - other
AU	FarNorthother	0.90	AU	Eden Valley	0.82	AU	Clare Valley	0.79	AU	Canberra District (NSW)
AU	Fleurieuother	0.99	AU	Bendigo	0.99	AU	Mount Lofty Ranges - other	0.99	AU	Gundagai
AU	Geelong	0.97	AU	Macedon Ranges	0.97	AU	Mornington Peninsula	0.96	AU	Gippsland
AU	Geographe	0.97	AU	Swan Hill (NSW)	0.96	AU	Goulburn Valley	0.95	AU	Great Southern
AU	Gippsland	0.99	AU	Macedon Ranges	0.99	AU	Yarra Valley	0.96	AU	Geelong
AU	Glenrowan	0.96	AU	Mount Lofty Ranges - other	0.96	AU	Central Ranges - other	0.96	AU	Currency Creek
AU	GoulburnValley	0.96	AU	Geographe	0.96	AU	Pyrenees	0.95	AU	Adelaide Plains
AU	Grampians	0.99	AU	Southern Flinders Ranges	0.99	AU	Heathcote	0.99	AU	Central Victoria - other
AU	GraniteBelt	0.95	AU	Geographe	0.94	AU	Goulburn Valley	0.94	AU	South Burnett
AU	GreatSouthern	0.97	AU	Swan Hill (NSW)	0.96	AU	Mount Benson	0.96	AU	Blackwood Valley
AU	GreaterPerthother	0.95	ZA	Cape South Coast	0.94	CL	Valparaiso	0.93	AU	Mount Gambier
AU	Gundagai	0.99	AU	Bendigo	0.99	AU	Fleurieu - other	0.99	AU	McLaren Vale
AU	HastingsRiver	0.82	AU	Shoalhaven Coast	0.68	AU	Cowra	0.68	NZ	Northland
AU	Heathcote	1.00	AU	Central Victoria - other	0.99	AU	Grampians	0.99	AU	Barossa Valley
AU	Henty	0.96	US	Santa Cruz	0.95	AU	Mornington Peninsula	0.95	AU	Tasmania
AU	Hilltops	0.99	AU	Langhorne Creek	0.98	AU	Currency Creek	0.98	AU	Central Ranges - other
AU	Hunter	0.93	AU	Hunter Valley - other	0.91	AU	Riverina	0.89	AU	South Burnett
AU	HunterValleyother	0.93	AU	Hunter	0.83	AU	South Burnett	0.82	AU	Queensland - other

Table 97: Each region's 3 most similar winegrape regions in the world according to the VSI, 2016

AU	KangarooIsland	0.95	AU	Adelaide Plains	0.94	AU	Bendigo	0.94	AU	Currency Creek
AU	KingValley	0.92	AU	Alpine Valleys	0.85	CA	British Colombia	0.85	NZ	Hawkes Bay
AU	LanghorneCreek	0.99	AU	Currency Creek	0.99	AU	Hilltops	0.98	AU	Central Ranges - other
AU	LimestoneCoastother	0.99	AU	Wrattonbully	0.99	AU	Northern Slopes	0.98	AU	Robe
AU	LowerMurrayother	0.99	AU	Riverland	0.98	AU	Swan Hill (VIC)	0.96	AU	Padthaway
AU	MacedonRanges	0.99	AU	Gippsland	0.98	AU	Yarra Valley	0.97	AU	Geelong
AU	Manjimup	0.90	ZA	Cape South Coast	0.89	AU	Greater Perth - other	0.88	CL	Valparaiso
AU	MargaretRiver	0.93	AU	Blackwood Valley	0.92	AU	Swan Hill (NSW)	0.90	AU	Geographe
AU	McLarenVale	0.99	AU	Barossa Valley	0.99	AU	Bendigo	0.99	AU	Gundagai
AU	MorningtonPeninsula	0.97	US	Santa Cruz	0.97	AU	Geelong	0.96	AU	Tasmania
AU	MountBenson	0.97	AU	Blackwood Valley	0.97	AU	Swan Hill (NSW)	0.96	AU	Great Southern
AU	MountGambier	0.98	CL	Valparaiso	0.95	NZ	Nelson	0.93	AU	Greater Perth - other
AU	MountLoftyRangesother	0.99	AU	Bendigo	0.99	AU	Fleurieu - other	0.99	AU	McLaren Vale
AU	Mudgee	0.98	AU	Currency Creek	0.97	AU	Central Ranges - other	0.97	AU	Langhorne Creek
AU	MurrayDarlingNSW	0.96	AU	Murray Darling (VIC)	0.94	AU	Big Rivers - other	0.93	NZ	Auckland
AU	MurrayDarlingVIC	0.98	AU	Riverland	0.96	AU	Perricoota	0.96	AU	Lower Murray - other
AU	NewEnglandAustralia	0.90	AU	North West Victoria - other	0.86	AU	The Peninsulas	0.85	AU	Port Phillip - other
AU	NorthEastVictoriaother	0.98	AU	Bendigo	0.97	AU	Fleurieu - other	0.97	AU	Mount Lofty Ranges - other
AU	NorthWestVictoriaother	0.95	AU	The Peninsulas	0.95	AU	Fleurieu - other	0.95	AU	Central Ranges - other
AU	NorthernRiversother	0.61	AU	Hastings River	0.53	US	Illinois	0.51	US	Indiana
AU	NorthernSlopes	0.99	AU	Limestone Coast - other	0.98	AU	Wrattonbully	0.98	AU	Robe
AU	Orange	0.96	AU	Great Southern	0.95	AU	Goulburn Valley	0.95	AU	Geographe
AU	Padthaway	0.97	AU	Perricoota	0.97	AU	Riverland	0.96	AU	Lower Murray - other
AU	Peel	0.96	AU	Mudgee	0.94	AU	Geographe	0.93	AU	Langhorne Creek
AU	Pemberton	0.94	CL	Valparaiso	0.94	AU	Adelaide Hills	0.92	AR	General Pueyrredón
AU	Perricoota	0.98	AU	Riverland	0.97	AU	Padthaway	0.96	AU	Murray Darling (VIC)
AU	PerthHills	0.91	AU	Granite Belt	0.90	AU	Riverland	0.90	AU	South Burnett
AU	PortPhillipother	0.98	AU	Central Ranges - other	0.97	AU	Currency Creek	0.97	AU	Mudgee
AU	Pyrenees	0.96	AU	Bendigo	0.96	AU	Swan Hill (VIC)	0.96	AU	Fleurieu - other
AU	Queenslandother	0.98	AU	Central Victoria - other	0.97	AU	Heathcote	0.97	AU	Bendigo
AU	Riverina	0.95	AU	Murray Darling (VIC)	0.94	AU	Riverland	0.94	AU	Lower Murray - other
AU	Riverland	0.99	AU	Lower Murray - other	0.98	AU	Murray Darling (VIC)	0.98	AU	Perricoota
AU	Robe	0.99	AU	Wrattonbully	0.98	AU	Limestone Coast - other	0.98	AU	Northern Slopes
AU	Rutherglen	0.84	AU	North East Victoria - other	0.83	AU	Queensland - other	0.80	AU	Heathcote
AU	ShoalhavenCoast	0.82	AU	Margaret River	0.82	AU	Hastings River	0.77	AU	South West Australia - other
AU	SouthBurnett	0.94	AU	Mudgee	0.94	AU	Granite Belt	0.94	AU	Swan Hill (NSW)
AU	SouthCoastother	0.88	NZ	Auckland	0.85	NZ	Northland	0.85	AU	Murray Darling (NSW)
AU	SouthWestAustraliaother	0.90	AR	Uruguay	0.89	US	Snipes Mountain	0.89	AR	Saavedra
AU	SouthernFleurieu	0.94	AU	Orange	0.94	AU	Pyrenees	0.94	AU	Goulburn Valley
AU	SouthernFlindersRanges	0.99	AU	Grampians	0.99	AU	Barossa Valley	0.99	AU	Central Victoria - other
AU	SouthernHighlands	0.89	AU	Upper Goulburn	0.88	AU	Adelaide Hills	0.87	AU	Alpine Valleys
AU	SouthernNSWother	0.97	AU	Hilltops	0.97	AU	Langhorne Creek	0.96	AU	Currency Creek
AU	StrathbogieRanges	0.96	AU	Yarra Valley	0.95	US	Santa Barbara	0.95	AU	Gippsland
AU	Sunbury	0.95	AU	Pyrenees	0.93	AU	Beechworth	0.92	AU	Southern Fleurieu
AU	SwanDistrict	0.72	ZA	Paarl	0.70	ZA	Swartland	0.70	AU	Perth Hills
AU	SwanHillNSW	0.97	AU	Geographe	0.97	AU	Mount Benson	0.97	AU	Great Southern
AU	SwanHillVIC	0.98	AU	Lower Murray - other	0.97	AU	Riverland	0.96	AU	Pyrenees
AU	Tasmania	0.99	AR	Futaleufu	0.96	CL	De Los Lagos	0.96	AU	Mornington Peninsula
AU	ThePeninsulas	0.98	AU	Central Ranges - other	0.97	AU	Mudgee	0.97	AU	Currency Creek
AU	Tumbarumba	0.99	FR	Bourgogne	0.98	US	Lassen	0.98	US	Santa Barbara
AU	UpperGoulburn	0.93	AU	Strathbogie Ranges	0.92	AU	Western Victoria - other	0.89	AU	Yarra Valley
AU	WesternAustraliaSoutheastCc	0.87	AU	Southern Fleurieu	0.84	AU	Sunbury	0.84	AU	Upper Goulburn
AU	WesternPlains	0.84	AU	Central Ranges - other	0.84	AU	Port Phillip - other	0.83	AU	The Peninsulas
AU	WesternVictoriaother	0.95	AU	Yarra Valley	0.92	AU	Upper Goulburn	0.91	AU	Strathbogie Ranges
AU	Wrattonbully	0.99	AU	Robe	0.99	AU	Limestone Coast - other	0.98	AU	Northern Slopes
AU	YarraValley	0.99	AU	Gippsland	0.98	AU	Macedon Ranges	0.96	AU	Strathbogie Ranges
AT	Bergland	0.76	AT	Wien	0.76	HU	Etyek-Budai	0.70	CZ	Jihovýchod
AT	Carnuntum	0.92	AT	Neusiedlersee	0.78	AT	Neusiedlersee Hügelland	0.76	AT	Wagram
AT	Kamptal	1.00	AT	Kremstal	0.99	AT	Wagram	0.99	AT	Traisental
AT	Kremstal	1.00	AT	Kamptal	0.99	AT	Traisental	0.99	AT	Wagram
AT	Mittelburgenland	0.97	HU	Sopron	0.87	AT	Südburgenland	0.83	AT	Neusiedlersee Hügelland
AT	Neusiedlersee	0.92	AT	Carnuntum	0.83	AT	Neusiedlersee Hügelland	0.73	AT	Thermenregion
AT	NeusiedlerseeHügelland	0.89	AT	Südburgenland	0.83	AT	Neusiedlersee	0.83	AT	Mittelburgenland
AT	Otherregions	0.99	AT	Kamptal	0.98	AT	Kremstal	0.98	AT	Wagram
AT	Steirerlandother	0.58	EL	Voreio Aigaio	0.45	AT	Vulkanland Steiermark	0.44	AT	Südsteiermark
AT	Südburgenland	0.89	HU	Sopron	0.89	AT	Neusiedlersee Hügelland	0.87	AT	Mittelburgenland
AT	Südsteiermark	0.91	AT	Vulkanland Steiermark	0.79	SI	Stajerska Slovenija	0.68	SI	Prekmurje
AT	Thermenregion	0.73	AT	Neusiedlersee	0.72	AT	Carnuntum	0.66	CZ	Jihovýchod
AT	Traisental	0.99	AT	Kremstal	0.99	AT	Kamptal	0.99	AT	Wagram
AT	VulkanlandSteiermark	0.91	AT	Südsteiermark	0.75	SI	Prekmurje	0.74	SI	Stajerska Slovenija
AT	Wachau	0.99	AT	Kremstal	0.98	AT	Kamptal	0.98	AT	Traisental
AT	Wagram	0.99	AT	Kamptal	0.99	AT	Kremstal	0.99	AT	Traisental
AT	Weinviertel	0.98	AT	Wagram	0.98	AT	Traisental	0.98	AT	Kamptal
AT	Weststeiermark	0.26	AT	Südsteiermark	0.23	AT	Vulkanland Steiermark	0.13	AT	Bergland
AT	Wien	0.94	AT	Other regions	0.94	AT	Wachau	0.93	AT	Kamptal
BR	Brazil	0.42	AR	Tornquist	0.41	UY	Tacuarembo	0.40	AR	Gualeguaychu
BG	NorthCentral	0.76	KZ	Almaty	0.75	UA	Ukraine	0.75	RU	Crimea
BG	Northeast	0.74	BG	Southeast	0.61	RS	Vranje	0.53	BG	North Central
BG	Northwest	0.85	BG	Southeast	0.82	BG	South Central	0.81	US	Wahluke Slope
BG	SouthCentral	0.89	BG	Southeast	0.83	FR	Aquitaine	0.82	BG	Northwest
BG	Southeast	0.89	BG	South Central	0.85	BG	Northwest	0.74	BG	Northeast
BG	Southwest	0.45	BG	Northwest	0.42	BG	South Central	0.42	BG	Southeast

Table 97: Each region's 3 most similar winegrape regions in the world according to the VSI, 2016

KH	Cambodia	0.77	AU	Central Ranges - other	0.77	AU	Port Phillip - other	0.76	AU	North West Victoria - other
CA	BritishColombia	0.85	AU	King Valley	0.84	US	Siskiyou	0.82	NZ	Auckland
CA	NovaScotia	0.51	RO	Romania	0.50	US	Arizona	0.49	RS	Nišava
CA	Ontario	1.00	CA	Other regions	0.52	US	Seneca	0.50	US	Michigan
CA	Otherregions	1.00	CA	Ontario	0.52	US	Seneca	0.50	US	Michigan
CA	Quebec	0.62	US	Minnesota	0.59	US	Iowa	0.44	US	Illinois
CL	Antofagasta	0.91	AR	Ayacucho	0.89	AR	Valle Fértil	0.88	AR	San Alberto
CL	Araucania	0.99	AR	Sarmiento - Chubut	0.97	US	Santa Barbara	0.97	US	Santa Cruz
CL	Arica	0.67	CL	Del Bio Bio	0.38	US	Placer	0.35	CL	Tarapaca
CL	Atacama	0.96	CL	Coquimbo	0.54	AR	Pichi Mahuida	0.47	AR	Las Heras
CL	Coquimbo	0.96	CL	Atacama	0.66	AR	Pichi Mahuida	0.52	AR	Las Heras
CL	DeLosLagos	0.97	AR	Futaleufu	0.96	AU	Tasmania	0.96	AR	General Pueyrredón
CL	DelBioBio	0.67	CL	Arica	0.53	AR	Poman	0.52	AR	Rivadavia - San Juan
CL	DelMaule	0.91	CL	O'Higgins	0.90	US	Lake	0.89	CL	Metropolitana
CL	Metropolitana	0.98	CL	O'Higgins	0.98	US	Columbia Valley	0.98	US	Red Mountain
CL	OHiggins	0.98	CL	Metropolitana	0.94	US	Columbia Valley	0.93	US	Red Mountain
CL	Tarapaca	0.37	AR	Sanagasta	0.35	CL	Arica	0.34	AR	San Blas De Los Sauces
CL	Valparaiso	0.98	AU	Mount Gambier	0.96	AU	Adelaide Hills	0.94	AU	Pemberton
CN	China	0.94	US	West Texas	0.94	UY	Lavalleja	0.87	FR	Limousin
HR	JadranskaHrvatska	0.46	SI	Vipavska dolina	0.40	SI	Slovenska Istra	0.31	FR	Aquitaine
HR	KontinentalnaHrvatska	0.95	HU	Balatonfured-Csopak	0.94	HU	Balatonfelvidek	0.93	HU	Badacsony
CY	Cyprus	0.00	HU	Badacsony	0.00	AR	Cushamen	0.00	AR	San Carlos - Salta
CZ	Cechy	0.91	CZ	Severozápad	0.89	CZ	Praha	0.80	DE	Sachsen
CZ	Jihovýchod	0.78	CZ	Morava	0.73	AT	Neusiedlersee Hügelland	0.71	AT	Wien
CZ	Morava	0.78	CZ	Jihovýchod	0.78	CZ	Cechy	0.78	CZ	Severozápad
CZ	Praha	0.89	CZ	Cechy	0.86	DE	Nahe	0.84	DE	Hessische Bergstraße
CZ	Severozápad	0.91	CZ	Cechy	0.80	LU	Luxembourg	0.79	CZ	Praha
ET	Ethiopia	0.84	IT	Toscana	0.55	FR	Corse	0.52	IT	Umbria
FR	Alsace	0.74	DE	Hessische Bergstraße	0.67	CZ	Cechy	0.66	DE	Mittelrhein
FR	Aquitaine	0.95	US	Shasta	0.93	CH	Ticino	0.93	CH	Graubünden - Mesolcina
FR	Auvergne	0.76	CH	Geneva	0.72	FR	Lorraine	0.70	FR	Rhône Alpes
FR	Bourgogne	0.99	AU	Tumbarumba	0.98	US	Lassen	0.95	US	Santa Barbara
FR	CentreValdeLoire	0.81	AR	Gualeguaychu	0.69	NZ	Marlborough	0.67	AR	Tornquist
FR	ChampagneArdenne	0.92	UK	UnitedKingdom	0.86	AU	Henty	0.85	FR	Île de France
FR	Corse	0.69	IT	Toscana	0.55	ET	Ethiopia	0.53	US	Tuolumne
FR	FrancheComté	0.87	US	Lassen	0.84	AR	Languiñeo	0.84	FR	Bourgogne
FR	ÎledeFrance	0.85	FR	Champagne-Ardenne	0.79	FR	Picardie	0.70	US	Umpqua Valley
FR	LanguedocRoussillon	0.81	FR	Provence-Alpes-Cote d'Azur	0.79	DZ	Algeria	0.73	AU	Perth Hills
FR	Limousin	0.95	US	Orange	0.94	US	West Texas	0.89	KZ	West Kazakhstan
FR	Lorraine	0.73	US	North Willamette Valley	0.72	NZ	Otago	0.72	FR	Auvergne
FR	MidiPyrénées	0.56	ZA	Little Karoo	0.56	ZA	Robertson	0.53	AR	Colón - Entre Ríos
FR	PaysdelaLoire	0.56	FR	Centre-Val de Loire	0.55	AR	Tornquist	0.53	AR	Gualeguaychu
FR	Picardie	0.79	FR	Île de France	0.73	FR	Champagne-Ardenne	0.51	UK	UnitedKingdom
FR	PoitouCharentes	0.67	IT	Lazio	0.51	IT	Puglia	0.50	IT	Umbria
FR	ProvenceAlpesCotedAzur	0.83	ES	Aragón	0.81	FR	Languedoc Roussillon	0.71	IT	Sardegna
FR	RhôneAlpes	0.70	FR	Auvergne	0.58	CH	Geneva	0.55	FR	Provence-Alpes-Cote d'Azur
GE	Georgia	0.93	KZ	Almaty	0.77	KZ	South Kazakhstan	0.73	BG	North Central
DE	Ahr	0.98	CH	Graubünden - other	0.97	NZ	Otago	0.97	US	Umpqua Valley
DE	Baden	0.92	CH	Basel Stadt	0.92	CH	Zürich	0.92	CH	Thurgau
DE	Franken	0.73	DE	Saale	0.71	DE	Rheinhessen	0.70	CH	Appenzell Innerrhoden
DE	HessischeBergstraße	0.97	DE	Mittelrhein	0.96	DE	Rheingau	0.96	DE	Mosel
DE	Mittelrhein	1.00	DE	Rheingau	0.99	DE	Mosel	0.98	AR	Veinticinco de Mayo -
DE	Mosel	0.99	DE	Mittelrhein	0.97	DE	Rheingau	0.97	AR	Veinticinco de Mayo -
DE	Nahe	0.96	DE	Pfalz	0.92	DE	Rheinhessen	0.91	DE	Hessische Bergstraße
DE	Pfalz	0.96	DE	Nahe	0.92	DE	Rheinhessen	0.87	DE	Hessische Bergstraße
DE	Rheingau	1.00	DE	Mittelrhein	0.99	AR	Capital Misiones	0.99	AR	Veinticinco de Mayo -
DE	Rheinhessen	0.92	DE	Nahe	0.92	DE	Pfalz	0.86	DE	Saale
DE	Saale	0.87	DE	Sachsen	0.86	DE	Rheinhessen	0.77	DE	Nahe
DE	Sachsen	0.87	DE	Saale	0.84	CZ	Praha	0.84	DE	Nahe
DE	Württemberg	0.59	DE	Hessische Bergstraße	0.56	DE	Mittelrhein	0.56	DE	Rheingau
EL	AnatolikiMakedoniaThraki	0.85	AR	Saavedra	0.82	ZA	Stellenbosch	0.81	AU	Alpine Valleys
EL	Attiki	0.97	EL	Sterea Ellada	0.15	EL	Peloponnissos	0.10	EL	Ionia Nisia
EL	DytikiEllada	0.53	EL	Peloponnissos	0.34	EL	Kentriki Makedonia	0.33	EL	Thessalia
EL	DytikiMakedonia	0.76	EL	Kentriki Makedonia	0.17	AU	Northern Slopes	0.16	AU	Limestone Coast - other
EL	IoniaNisia	0.85	US	Wayne	0.84	HU	Zala	0.77	RO	Romania
EL	Ipeiros	0.94	US	Red Mountain	0.93	CL	Metropolitana	0.93	US	Columbia Valley
EL	KentrikiMakedonia	0.76	EL	Dytiki Makedonia	0.58	US	Shasta	0.58	AU	South West Australia - other
EL	Kriti	0.09	HU	Zala	0.09	EL	Ionia Nisia	0.08	US	Wayne
EL	NotioAigaio	0.13	EL	Kentriki Makedonia	0.12	EL	Anatoliki Makedonia, Thraki	0.11	EL	Sterea Ellada
EL	Peloponnissos	0.53	EL	Dytiki Ellada	0.26	EL	Sterea Ellada	0.23	EL	Kentriki Makedonia
EL	StereaEllada	0.97	EL	Attiki	0.26	EL	Peloponnissos	0.26	EL	Dytiki Ellada
EL	Thessalia	0.62	UY	Montevideo	0.52	UY	Canelones	0.51	UY	San Jose
EL	VoreioAigaio	0.58	AT	Steirerland - other	0.54	IT	Piemonte	0.50	MX	Zacatecas
HU	Badacsony	0.97	HU	Balatonfelvidek	0.95	HU	Balatonfured-Csopak	0.93	HR	Kontinentalna Hrvatska
HU	Balatonboglar	0.84	HU	Etyek-Budai	0.81	HU	Tolna	0.80	HU	Pecs
HU	Balatonfelvidek	0.97	HU	Badacsony	0.96	HU	Balatonfured-Csopak	0.94	HR	Kontinentalna Hrvatska
HU	BalatonfuredCsopak	0.96	HU	Balatonfelvidek	0.95	HU	Badacsony	0.95	HR	Kontinentalna Hrvatska
HU	Bukk	0.80	HU	Eger	0.74	AT	Südburgenland	0.73	AT	Neusiedlersee Hügelland
HU	Csongrad	0.81	AT	Südburgenland	0.80	HU	Sopron	0.79	AT	Mittelburgenland
HU	Eger	0.90	HU	Szekszard	0.81	HU	Sopron	0.80	HU	Bukk
HU	EtyekBudai	0.84	HU	Balatonboglar	0.81	HU	Pecs	0.81	HU	Neszmely
HU	Hajosbajai	0.79	HU	Csongrad	0.74	HU	Kunsag	0.70	HU	Tolna

Table 97: Each region's 3 most similar winegrape regions in the world according to the VSI, 2016

HU	Kunsag	0.74	HU	Hajos-bajai	0.59	HU	Csongrad	0.43	RU	Rostov Oblast
HU	Matra	0.73	HU	Pecs	0.70	HU	Tolna	0.66	HU	Balatonboglar
HU	Mor	0.72	HU	Neszmely	0.54	HU	Etyek-Budai	0.49	HU	Pecs
HU	NagySomlo	0.78	SI	Prekmurje	0.72	SI	Stajerska Slovenija	0.71	RS	South Banat
HU	Neszmely	0.81	HU	Etyek-Budai	0.80	NZ	Gisborne	0.79	IT	Trento
HU	Pannonhalma	0.86	HU	Balatonfured-Csopak	0.84	HU	Balatonfelvidek	0.83	HR	Kontinentalna Hrvatska
HU	Pecs	0.81	HU	Etyek-Budai	0.80	HU	Balatonboglar	0.78	HU	Tolna
HU	Sopron	0.97	AT	Mittelburgenland	0.89	AT	Südburgenland	0.84	HU	Szekszard
HU	Szekszard	0.90	HU	Eger	0.84	HU	Sopron	0.81	AT	Mittelburgenland
HU	Tokaj	0.36	HU	Nagy-Somlo	0.27	SI	Stajerska Slovenija	0.15	SI	Prekmurje
HU	Tolna	0.82	AT	Neusiedlersee Hügelland	0.81	HU	Balatonboglar	0.78	HU	Pecs
HU	Villany	0.77	HU	Szekszard	0.73	HU	Eger	0.66	US	Wahluke Slope
HU	Zala	0.87	US	Wayne	0.84	EL	Ionia Nisia	0.83	RO	Romania
IN	India	0.80	TR	Aegean	0.75	MX	Sonora	0.54	MM	Myanmar
IL	Israel	0.74	US	Wahluke Slope	0.74	US	San Diego	0.73	DZ	Algeria
IT	Abruzzo	0.98	IT	Molise	0.58	IT	Marche	0.44	IT	Puglia
IT	Basilicata	0.86	IT	Campania	0.15	IT	Puglia	0.14	IT	Toscana
IT	BolzanoBozen	0.65	CA	British Colombia	0.61	HU	Neszmely	0.60	IT	Trento
IT	Calabria	0.19	IT	Lombardia	0.11	IT	Campania	0.09	IT	Basilicata
IT	Campania	0.86	IT	Basilicata	0.19	IT	Lazio	0.17	IT	Puglia
IT	EmiliaRomagna	0.20	ES	Principado de Asturias	0.20	US	Arizona	0.20	RS	Nišava
IT	FriuliVeneziaGiulia	0.68	US	Naches Heights	0.66	AU	King Valley	0.66	CA	British Colombia
IT	Lazio	0.67	FR	Poitou Charentes	0.51	IT	Umbria	0.51	IT	Puglia
IT	Liguria	0.57	IT	Sardegna	0.38	FR	Corse	0.37	US	Wayne
IT	Lombardia	0.64	US	Santa Barbara	0.62	US	Monterey	0.62	AU	Tumbarumba
IT	Marche	0.59	IT	Molise	0.58	IT	Abruzzo	0.57	IT	Toscana
IT	Molise	0.98	IT	Abruzzo	0.59	IT	Marche	0.42	IT	Puglia
IT	Piemonte	0.54	EL	Voreio Aigaio	0.42	AT	Steirerland - other	0.39	IT	Lombardia
IT	Puglia	0.63	IT	Umbria	0.58	IT	Toscana	0.56	IT	Marche
IT	Sardegna	0.71	FR	Provence-Alpes-Cote d'Azur	0.64	ES	Aragón	0.57	IT	Liguria
IT	Sicilia	0.23	AU	Murray Darling (NSW)	0.23	NZ	Auckland	0.23	AU	Murray Darling (VIC)
IT	Toscana	0.84	ET	Ethiopia	0.70	IT	Umbria	0.69	FR	Corse
IT	Trento	0.89	NZ	Gisborne	0.81	US	Yolo	0.79	HU	Neszmely
IT	Umbria	0.70	IT	Toscana	0.63	IT	Puglia	0.52	ET	Ethiopia
IT	ValledAosta	0.44	US	Columbia Gorge	0.41	RS	Mlava	0.41	US	Rogue Valley
IT	Veneto	0.49	IT	Friuli-Venezia Giulia	0.48	RS	Belgrade	0.48	US	Arizona
JP	Hokkaido	0.67	US	Ontario	0.66	JP	Iwate	0.54	US	Niagara
JP	Iwate	0.66	JP	Hokkaido	0.39	US	Ontario	0.36	RS	Negotinska Krajina
JP	Nagano	0.91	US	Niagara	0.81	US	Chautauqua	0.81	US	Pennsylvania
JP	Niigata	0.77	US	Suffolk	0.73	AR	Uruguay	0.67	RS	Telečka
JP	Otherregions	0.87	JP	Yamanashi	0.61	JP	Yamagata	0.51	JP	Nagano
JP	Yamagata	0.61	JP	Other regions	0.46	JP	Niigata	0.45	JP	Hokkaido
JP	Yamanashi	0.87	JP	Other regions	0.35	JP	Yamagata	0.26	JP	Niigata
KZ	Almaty	0.93	GE	Georgia	0.83	KZ	South Kazakhstan	0.76	BG	North Central
KZ	EastKazakhstan	0.87	KZ	West Kazakhstan	0.86	ES	Principado de Asturias	0.86	US	Arizona
KZ	Otherregions	0.84	KZ	East Kazakhstan	0.70	RO	Romania	0.69	US	Arizona
KZ	SouthKazakhstan	0.83	KZ	Almaty	0.77	GE	Georgia	0.63	BG	North Central
KZ	WestKazakhstan	0.92	UY	Soriano	0.92	US	North Carolina	0.92	US	Arkansas
KZ	Zhambyl	0.76	US	Wayne	0.75	EL	Ionia Nisia	0.75	HU	Zala
KR	KoreaRep	0.29	JP	Other regions	0.25	US	Orange	0.25	JP	Yamanashi
LB	Lebanon	0.89	AR	Saavedra	0.88	US	Sacramento	0.87	US	Alameda
LU	Luxembourg	0.80	DE	Sachsen	0.80	CZ	Severozápad	0.77	CZ	Praha
MX	Aguascalientes	0.53	DZ	Algeria	0.39	IL	Israel	0.26	FR	Languedoc Roussillon
MX	BajaCalifornia	0.81	CN	China	0.80	UY	Lavalleja	0.80	US	Texas High Plains and
MX	Coahuila	0.60	US	Wahluke Slope	0.58	US	Red Mountain	0.58	US	Walla Walla Valley
MX	Sonora	0.76	TR	Aegean	0.75	IN	India	0.13	TN	Tunisia
MX	Zacatecas	0.81	AR	Capital Santiago del Estero	0.50	EL	Voreio Aigaio	0.40	AT	Steirerland - other
MD	Moldova	0.65	RU	Crimea	0.65	UA	Ukraine	0.62	AR	Saavedra
MA	Morocco	1.00	MA	Morocco	0.53	PE	Tacna	0.40	PE	Moquegua
MM	Myanmar	0.80	AU	Adelaide Plains	0.80	ZA	Cape South Coast	0.79	AU	Western Australia Southeast
NZ	Auckland	0.93	AU	Murray Darling (NSW)	0.91	NZ	Northland	0.91	AU	Big Rivers - other
NZ	Canterbury	0.95	NZ	Waipara	0.95	NZ	Wairarapa	0.91	AU	Tasmania
NZ	Gisborne	0.94	US	Yolo	0.89	IT	Trento	0.89	AR	Languiñeo
NZ	HawkesBay	0.92	AU	Alpine Valleys	0.87	AR	Saavedra	0.85	AU	King Valley
NZ	Marlborough	0.94	NZ	Nelson	0.91	AU	Mount Gambier	0.86	AU	Greater Perth - other
NZ	Nelson	0.95	AU	Mount Gambier	0.94	NZ	Marlborough	0.91	CL	Valparaiso
NZ	Northland	0.93	AU	Big Rivers - other	0.91	NZ	Auckland	0.90	AU	Cowra
NZ	Otago	1.00	US	North Willamette Valley	0.99	US	Umpqua Valley	0.99	CH	Graubünden - other
NZ	Waikato	0.95	AU	Macedon Ranges	0.93	AU	Gippsland	0.93	AU	Strathbogie Ranges
NZ	Waipara	0.95	NZ	Canterbury	0.87	NZ	Wairarapa	0.86	NZ	Nelson
NZ	Wairarapa	0.95	NZ	Canterbury	0.91	CL	De Los Lagos	0.88	AU	Tasmania
MK	NorthMacedon	0.32	BG	Northeast	0.23	BG	Southeast	0.18	BG	South Central
NO	Norway	0.44	CH	Owalden	0.34	CH	Zug	0.25	CH	Lucerne
PE	Arequipa	0.79	PT	Madeira	0.69	PE	Moquegua	0.47	PE	Tacna
PE	Lima	0.62	HU	Sopron	0.58	AT	Mittelburgenland	0.55	AT	Südburgenland
PE	Moquegua	0.94	PE	Tacna	0.69	PE	Arequipa	0.65	PT	Madeira
PE	Tacna	0.94	PE	Moquegua	0.53	TN	Tunisia	0.53	MA	Morocco
PT	Acores	0.17	PT	Centro	0.16	PT	Norte	0.15	PT	Lisboa
PT	Alentejo	0.70	PT	Centro	0.62	ES	País Vasco	0.61	ES	La Rioja
PT	Algarve	0.64	PT	Lisboa	0.60	PT	Alentejo	0.58	PT	Centro
PT	Centro	0.70	PT	Alentejo	0.70	PT	Norte	0.58	PT	Algarve
PT	Lisboa	0.64	PT	Algarve	0.53	PT	Centro	0.51	PT	Alentejo

Table 97: Each region's 3 most similar winegrape regions in the world according to the VSI, 2016

PT	Madeira	0.79	PE	Arequipa	0.65	PE	Moquegua	0.42	PT	Algarve
PT	Norte	0.70	PT	Centro	0.52	PT	Alentejo	0.39	ES	País Vasco
RO	Romania	0.90	US	Arizona	0.88	RS	Nišava	0.87	ES	Principado de Asturias
RU	Crimea	1.00	UA	Ukraine	0.75	BG	North Central	0.65	MD	Moldova
RU	KrasnodarKrai	0.74	US	Horse Heaven Hills	0.74	US	Napa	0.71	US	Wahluke Slope
RU	RostovOblast	0.51	UA	Ukraine	0.51	RU	Crimea	0.48	KZ	Almaty
RS	Bačka	0.79	RS	Srem	0.76	RS	Telečka	0.75	RS	Vranje
RS	Banat	0.88	RS	Belgrade	0.87	RS	Knjaževac	0.87	RS	Mlava
RS	Belgrade	0.89	RS	Šumadija	0.88	RS	Banat	0.87	RS	Niš
RS	ČačakKraljevo	0.79	RS	Toplica	0.64	RS	Niš	0.62	RS	Tri Morave
RS	Knjaževac	0.89	RS	Srem	0.87	RS	Banat	0.83	RS	Tri Morave
RS	Leskovac	0.85	RS	Negotinska Krajina	0.78	US	Snipes Mountain	0.77	US	Colorado
RS	Mlava	0.87	RS	Banat	0.86	RS	Belgrade	0.82	RS	Srem
RS	NegotinskaKrajina	0.85	RS	Leskovac	0.80	RS	Knjaževac	0.75	US	Michigan
RS	Niš	0.91	RS	Tri Morave	0.87	RS	Belgrade	0.84	US	North Carolina
RS	Nišava	1.00	ES	Principado de Asturias	1.00	US	Arizona	0.97	US	North Carolina
RS	SouthBanat	0.89	SI	Prekmurje	0.89	HR	Kontinentalna Hrvatska	0.84	HU	Badacsony
RS	Srem	0.89	RS	Knjaževac	0.87	RS	Telečka	0.85	RS	Belgrade
RS	Subotica	0.89	RS	Telečka	0.80	RS	Srem	0.76	EL	Anatoliki Makedonia, Thraki
RS	Šumadija	0.89	RS	Belgrade	0.84	RS	Banat	0.81	RS	Mlava
RS	Telečka	0.89	RS	Subotica	0.87	RS	Srem	0.85	US	Colorado
RS	Tisa	0.59	AR	Veinticinco de Mayo -	0.59	AR	Capital Misiones	0.58	DE	Rheingau
RS	Toplica	0.79	RS	Čačak-Kraljevo	0.60	US	Napa	0.60	US	Horse Heaven Hills
RS	TriMorave	0.91	RS	Niš	0.87	RS	Banat	0.86	CN	China
RS	Valjevo	0.80	RS	Belgrade	0.77	RS	Niš	0.77	RS	Telečka
RS	Vranje	0.75	LB	Lebanon	0.75	RS	Bačka	0.75	US	Yolo
SK	Bratislavskýkraj	0.83	SK	Západné Slovensko	0.78	AT	Weinviertel	0.77	AT	Other regions
SK	StrednéSlovensko	0.76	RS	Knjaževac	0.75	RO	Romania	0.74	SI	Bela Krajina
SK	VýchodnéSlovensko	0.85	HU	Balatonfured-Csopak	0.84	HR	Kontinentalna Hrvatska	0.80	AT	Südburgenland
SK	ZápadnéSlovensko	0.83	SK	Bratislavský kraj	0.71	AT	Other regions	0.69	AT	Neusiedlersee Hügelland
SI	BelaKrajina	0.86	SI	Bizeljsko Sremic	0.78	AT	Südburgenland	0.74	SK	Stredné Slovensko
SI	BizeljskoSremic	0.86	SI	Bela Krajina	0.84	SI	Dolenjska	0.75	AT	Südburgenland
SI	Dolenjska	0.84	SI	Bizeljsko Sremic	0.66	SI	Bela Krajina	0.46	AT	Südburgenland
SI	Goriskabrda	0.75	SI	Vipavska dolina	0.70	NZ	Hawkes Bay	0.68	US	Sacramento
SI	Kras	0.86	SI	Slovenska Istra	0.15	SI	Vipavska dolina	0.08	SI	Goriska brda
SI	Prekmurje	0.92	SI	Stajerska Slovenija	0.89	RS	South Banat	0.87	HR	Kontinentalna Hrvatska
SI	SlovenskaIstra	0.86	SI	Kras	0.40	HR	Jadranska Hrvatska	0.37	SI	Vipavska dolina
SI	StajerskaSlovenija	0.92	SI	Prekmurje	0.79	AT	Südsteiermark	0.75	RS	South Banat
SI	Vipavskadolina	0.75	SI	Goriska brda	0.73	AR	Saavedra	0.70	NZ	Hawkes Bay
ZA	Breedekloof	0.97	ZA	Worcester	0.95	ZA	Olifants River	0.89	ZA	Robertson
ZA	CapeSouthCoast	0.95	AU	Greater Perth - other	0.94	CL	Valparaiso	0.93	AU	Adelaide Hills
ZA	LittleKaroo	0.96	ZA	Northern Cape	0.94	ZA	Olifants River	0.88	ZA	Worcester
ZA	NorthernCape	0.96	ZA	Little Karoo	0.90	ZA	Olifants River	0.81	ZA	Worcester
ZA	OlifantsRiver	0.98	ZA	Worcester	0.95	ZA	Breedekloof	0.94	ZA	Little Karoo
ZA	Paarl	0.99	ZA	Swartland	0.88	ZA	Stellenbosch	0.81	ZA	Breedekloof
ZA	Robertson	0.89	ZA	Breedekloof	0.84	ZA	Worcester	0.82	ZA	Olifants River
ZA	Stellenbosch	0.88	ZA	Paarl	0.88	AU	Blackwood Valley	0.88	ZA	Swartland
ZA	Swartland	0.99	ZA	Paarl	0.88	ZA	Stellenbosch	0.81	ZA	Breedekloof
ZA	Worcester	0.98	ZA	Olifants River	0.97	ZA	Breedekloof	0.88	ZA	Little Karoo
ES	Andalucía	0.61	ES	Canarias	0.37	ES	Cantabria	0.37	MX	Coahuila
ES	Aragón	0.83	FR	Provence-Alpes-Cote d'Azur	0.68	ES	Comunidad Foral de Navarra	0.64	FR	Languedoc Roussillon
ES	Canarias	0.61	ES	Andalucía	0.45	MX	Coahuila	0.41	ES	Cantabria
ES	Cantabria	0.70	ES	Galicia	0.41	ES	Canarias	0.37	ES	Andalucía
ES	CastillayLeón	0.93	ES	La Rioja	0.92	ES	País Vasco	0.90	ES	Comunidad Foral de Navarra
ES	CastillaLaMancha	0.92	ES	Comunidad de Madrid	0.34	ES	Comunidad Foral de Navarra	0.33	ES	La Rioja
ES	Cataluña	0.43	ES	Aragón	0.36	FR	Languedoc Roussillon	0.31	AU	Perth Hills
ES	ComunidaddeMadrid	0.92	ES	Castilla-La Mancha	0.33	ES	Aragón	0.30	FR	Provence-Alpes-Cote d'Azur
ES	ComunidadForaldeNavarra	0.95	ES	La Rioja	0.92	ES	País Vasco	0.90	ES	Castilla y León
ES	ComunidadValenciana	0.26	ES	Comunidad Foral de Navarra	0.26	ES	Castilla-La Mancha	0.25	ES	La Rioja
ES	Extremadura	0.53	ES	La Rioja	0.53	ES	País Vasco	0.51	ES	Comunidad Foral de Navarra
ES	Galicia	0.70	ES	Cantabria	0.60	US	Arizona	0.59	RS	Nišava
ES	IllesBalears	0.65	US	Wahluke Slope	0.65	AU	Limestone Coast - other	0.64	US	Snipes Mountain
ES	LaRioja	0.99	ES	País Vasco	0.95	ES	Comunidad Foral de Navarra	0.93	ES	Castilla y León
ES	PaísVasco	0.99	ES	La Rioja	0.92	ES	Comunidad Foral de Navarra	0.92	ES	Castilla y León
ES	PrincipadodeAsturias	1.00	US	Arizona	1.00	RS	Nišava	0.98	US	North Carolina
ES	RegióndeMurcia	0.31	US	Contra Costa	0.27	ES	Illes Balears	0.26	US	Texas High Plains and
CH	Aargau	1.00	CH	Zürich	1.00	CH	Thurgau	0.99	CH	Glarus
CH	AppenzellAusserrhoden	0.97	CH	Aargau	0.97	CH	Thurgau	0.97	CH	Zürich
CH	AppenzellInnerrhoden	0.87	CH	Nidwalden	0.76	CZ	Severozápad	0.70	DE	Franken
CH	BaselLand	1.00	CH	Schaffhausen	1.00	CH	St. Gallen	0.99	CH	Aargau
CH	BaselStadt	0.98	CH	Glarus	0.97	CH	Aargau	0.97	CH	Schaffhausen
CH	Fribourg	0.96	CH	Lac de Bienne	0.91	CH	Vaud	0.87	CH	Neuchâtel
CH	Geneva	0.76	FR	Auvergne	0.75	CH	Vaud	0.74	CH	Valais
CH	Glarus	0.99	CH	Aargau	0.99	CH	St. Gallen	0.99	CH	Schaffhausen
CH	GraubündenMesolcina	1.00	CH	Ticino	1.00	AR	Nogoya	0.98	AR	Ñorquin
CH	Graubündenother	0.99	CH	St. Gallen	0.99	CH	Schaffhausen	0.99	US	Umpqua Valley
CH	Jura	0.43	LU	Luxembourg	0.39	DE	Sachsen	0.38	DE	Saale
CH	LacdeBienne	0.97	CH	Neuchâtel	0.96	CH	Fribourg	0.91	CH	Valais
CH	Lucerne	0.89	CH	Thunersee	0.89	CH	Solothurn	0.89	CH	Thurgau
CH	Neuchâtel	0.97	CH	Lac de Bienne	0.92	CH	Valais	0.88	CH	Basel Land
CH	Nidwalden	0.94	CH	Thunersee	0.87	CH	Appenzell Innerrhoden	0.86	CH	Zug
CH	Otherregions	0.88	CH	Solothurn	0.84	CH	Thurgau	0.84	CH	Basel Land

Table 97: Each region's 3 most similar winegrape regions in the world according to the VSI, 2016

CH	Owalden	0.46	CH	Other regions	0.44	NO	Norway	0.22	CH	Lucerne
CH	Schaffhausen	1.00	CH	St. Gallen	1.00	CH	Basel Land	0.99	CH	Graubünden - other
CH	Schwyz	0.98	CH	Zürich	0.97	CH	Thurgau	0.97	CH	Glarus
CH	Solothurn	0.94	CH	Thurgau	0.94	CH	Zürich	0.93	CH	Aargau
CH	StGallen	1.00	CH	Schaffhausen	1.00	CH	Basel Land	0.99	CH	Graubünden - other
CH	Thunersee	0.94	CH	Nidwalden	0.92	CH	Thurgau	0.90	CH	Zürich
CH	Thurgau	1.00	CH	Zürich	1.00	CH	Aargau	0.98	CH	Glarus
CH	Ticino	1.00	CH	Graubünden - Mesolcina	1.00	AR	Nogoya	0.98	AR	Ñorquin
CH	Uri	0.96	NZ	Otago	0.96	US	North Willamette Valley	0.95	US	Umpqua Valley
CH	Valais	0.92	CH	Neuchâtel	0.91	CH	Lac de Bienne	0.84	CH	Fribourg
CH	Vaud	0.91	CH	Fribourg	0.79	CH	Lac de Bienne	0.75	CH	Geneva
CH	Zug	0.87	CH	Lucerne	0.86	CH	Nidwalden	0.86	CH	Thunersee
CH	Zürich	1.00	CH	Aargau	1.00	CH	Thurgau	0.99	CH	Glarus
TW	Taiwan	0.52	KH	Cambodia	0.16	TH	Thailand	0.07	JP	Nagano
TH	Thailand	0.77	AU	Southern Flinders Ranges	0.77	AU	Eastern Plains, Inland and	0.77	AU	Grampians
TN	Tunisia	1.00	TN	Tunisia	0.53	PE	Tacna	0.40	PE	Moquegua
TR	Aegean	0.80	IN	India	0.76	MX	Sonora	0.49	TR	Central North
TR	CentralEast	0.20	TR	Mediterranean	0.12	TR	Aegean	0.12	TR	Central North
TR	CentralNorth	0.77	AU	Eastern Plains, Inland and	0.75	AU	Southern Flinders Ranges	0.75	AU	Grampians
TR	CentralSouth	0.07	TR	Central North	0.06	TR	Aegean	0.06	TR	Mediterranean
TR	Marmara	0.38	AU	Manjimup	0.35	AU	Hunter	0.34	TN	Tunisia
TR	Mediterranean	0.83	AR	Saavedra	0.80	LB	Lebanon	0.78	AU	Margaret River
TR	SouthEast	0.40	TR	Central North	0.29	TR	Mediterranean	0.19	TR	Aegean
UA	Ukraine	1.00	RU	Crimea	0.75	BG	North Central	0.65	MD	Moldova
UK	UnitedKingdom	0.93	AR	Sarmiento - Chubut	0.93	CL	Araucania	0.92	AU	Tumbarumba
US	Alameda	0.94	US	Napa	0.93	US	San Luis Obispo	0.90	US	Horse Heaven Hills
US	Amador	0.97	US	San Bernardino	0.95	US	Colusa	0.82	US	El Dorado
US	Arizona	1.00	ES	Principado de Asturias	1.00	RS	Nišava	0.96	US	Georgia
US	Arkansas	1.00	US	Georgia	1.00	US	Arkansas	0.98	ES	Principado de Asturias
US	Butte	0.93	US	Red Mountain	0.93	AU	Coonawarra	0.93	US	San Luis Obispo
US	Calaveras	0.93	US	Nevada	0.89	US	San Diego	0.88	US	Riverside
US	Cattaraugus	1.00	US	Erie	0.99	US	Chautauqua	0.95	US	Pennsylvania
US	Chautauqua	1.00	US	Erie	0.99	US	Cattaraugus	0.96	US	Pennsylvania
US	Colorado	0.85	RS	Telečka	0.82	RS	Tri Morave	0.81	RS	Knjaževac
US	ColumbiaGorge	0.87	AU	Henty	0.85	AU	Mornington Peninsula	0.84	US	Santa Barbara
US	ColumbiaRiver	0.88	US	Humboldt	0.82	AU	Robe	0.81	US	San Luis Obispo
US	ColumbiaValley	0.98	US	Red Mountain	0.98	CL	Metropolitana	0.97	US	Horse Heaven Hills
US	Colusa	0.95	US	Amador	0.92	US	San Bernardino	0.80	US	El Dorado
US	ContraCosta	0.86	US	San Joaquin	0.75	US	Mendocino	0.74	US	Stanislaus
US	ElDorado	0.88	US	Mariposa	0.88	US	Calaveras	0.87	US	San Joaquin
US	Erie	1.00	US	Cattaraugus	1.00	US	Chautauqua	0.95	US	Pennsylvania
US	Fresno	0.86	US	Tulare	0.80	US	Madera	0.76	ZA	Little Karoo
US	Georgia	1.00	US	Georgia	1.00	US	Arkansas	0.98	ES	Principado de Asturias
US	Glenn	0.62	US	Colusa	0.62	US	Tehama	0.56	US	Contra Costa
US	HillCountry	0.82	US	Texas High Plains and	0.79	US	North Texas (DFW)	0.74	MX	Baja California
US	HorseHeavenHills	0.97	US	Columbia Valley	0.97	US	Wahluke Slope	0.96	US	Napa
US	Humboldt	0.88	US	Columbia River	0.87	AU	Western Victoria - other	0.81	AU	Upper Goulburn
US	Illinois	0.71	US	Missouri	0.69	US	Indiana	0.56	ES	Principado de Asturias
US	Indiana	0.77	RO	Romania	0.74	US	Wayne	0.74	US	Arizona
US	Iowa	0.83	US	Minnesota	0.59	CA	Quebec	0.32	US	Illinois
US	Kentucky	0.68	US	Missouri	0.63	US	Virginia	0.54	US	Illinois
US	Kern	0.77	US	Arizona	0.77	ES	Principado de Asturias	0.77	RS	Nišava
US	Kings	0.79	US	Tulare	0.74	US	Fresno	0.56	US	Madera
US	Lake	0.90	CL	Del Maule	0.89	US	Napa	0.88	CL	Metropolitana
US	LakeChelan	0.90	AU	Beechworth	0.89	AU	Southern Fleurieu	0.89	AU	Sunbury
US	Lassen	0.98	AU	Tumbarumba	0.98	FR	Bourgogne	0.94	US	Santa Barbara
US	LosAngeles	0.95	AU	Wrattonbully	0.95	US	Walla Walla Valley	0.95	AU	Limestone Coast - other
US	Madera	0.81	US	Tulare	0.80	US	Fresno	0.79	US	Stanislaus
US	Marin	0.99	CH	Graubünden - other	0.99	AR	Collon Cura	0.98	US	Umpqua Valley
US	Mariposa	0.91	US	San Joaquin	0.88	US	El Dorado	0.85	US	San Luis Obispo
US	Mendocino	0.98	US	Sonoma	0.96	US	Santa Clara	0.95	US	San Benito
US	Merced	0.90	US	Sacramento	0.87	US	Mendocino	0.87	US	Santa Clara
US	Michigan	0.90	US	Seneca	0.80	DE	Hessische Bergstraße	0.75	RS	Negotinska Krajina
US	Minnesota	0.83	US	Iowa	0.62	CA	Quebec	0.31	US	Illinois
US	Missouri	0.71	US	Illinois	0.68	US	Kentucky	0.49	US	Indiana
US	Monterey	0.96	US	San Benito	0.95	US	Solano	0.93	US	Lassen
US	NachesHeights	0.68	IT	Friuli-Venezia Giulia	0.66	MM	Myanmar	0.64	AU	Southern Fleurieu
US	Napa	0.96	US	San Luis Obispo	0.96	US	Horse Heaven Hills	0.95	US	Columbia Valley
US	Nevada	0.93	US	Calaveras	0.92	AU	Peel	0.91	US	San Diego
US	Niagara	0.91	JP	Nagano	0.77	US	Yates	0.77	US	Chautauqua
US	NorthCarolina	1.00	US	Arkansas	1.00	US	North Carolina	0.98	ES	Principado de Asturias
US	NorthTexasDFW	0.79	US	Hill Country	0.79	US	South Texas and Gulf Coast	0.54	US	Texas High Plains and
US	NorthWillametteValley	1.00	NZ	Otago	0.99	US	Umpqua Valley	0.98	CH	Graubünden - other
US	Ohio	0.98	US	Pennsylvania	0.95	US	Chautauqua	0.94	US	Erie
US	Ontario	0.74	US	Niagara	0.67	JP	Hokkaido	0.65	US	Steuben
US	Orange	0.95	FR	Limousin	0.94	US	West Texas	0.90	KZ	West Kazakhstan
US	Pennsylvania	0.98	US	Ohio	0.96	US	Chautauqua	0.95	US	Erie
US	Placer	0.81	US	El Dorado	0.76	US	Tehama	0.72	US	Amador
US	PugetSound	0.85	US	North Willamette Valley	0.84	US	Umpqua Valley	0.84	NZ	Otago
US	RattlesnakeHills	0.87	US	Snipes Mountain	0.82	US	Shasta	0.77	FR	Aquitaine
US	RedMountain	0.99	US	Walla Walla Valley	0.98	US	Columbia Valley	0.98	CL	Metropolitana
US	Riverside	0.89	US	San Diego	0.89	US	San Luis Obispo	0.88	US	Calaveras

Table 97: Each region's 3 most similar winegrape regions in the world according to the VSI, 2016

US	RogueValley	0.93	US	South Willamette Valley	0.92	US	North Willamette Valley	0.92	AU	Mornington Peninsula
US	Sacramento	0.92	US	Santa Clara	0.92	US	Mendocino	0.91	AR	Uruguay
US	SanBenito	0.98	US	Sonoma	0.96	US	Monterey	0.95	US	Mendocino
US	SanBernardino	0.97	US	Amador	0.92	US	Colusa	0.77	US	Sutter
US	SanDiego	0.91	US	Los Angeles	0.91	US	Nevada	0.91	AU	Limestone Coast - other
US	SanJoaquin	0.91	US	Mariposa	0.88	US	Stanislaus	0.87	US	El Dorado
US	SanLuisObispo	0.96	US	Napa	0.95	US	Columbia Valley	0.94	US	Los Angeles
US	SanMateo	0.93	US	San Benito	0.93	AU	Yarra Valley	0.93	US	Sonoma
US	SantaBarbara	0.98	AU	Tumbarumba	0.97	CL	Araucania	0.96	AU	Gippsland
US	SantaClara	0.98	US	Sonoma	0.96	US	Mendocino	0.95	US	San Benito
US	SantaCruz	0.98	AR	Sarmiento - Chubut	0.97	AU	Mornington Peninsula	0.97	CL	Araucania
US	Schuyler	0.85	US	Seneca	0.83	US	Steuben	0.78	US	Yates
US	Seneca	0.90	US	Michigan	0.85	US	Schuyler	0.75	RS	Negotinska Krajina
US	Shasta	0.95	FR	Aquitaine	0.95	CH	Graubünden - Mesolcina	0.95	CH	Ticino
US	Siskiyou	0.84	CA	British Colombia	0.81	US	Sacramento	0.81	US	San Benito
US	SnipesMountain	0.93	US	Wahluke Slope	0.91	US	Horse Heaven Hills	0.91	AR	Uruguay
US	Solano	0.95	US	Monterey	0.94	US	San Benito	0.93	US	Yolo
US	Sonoma	0.98	US	Mendocino	0.98	US	Santa Clara	0.98	US	San Benito
US	SouthTexasandGulfCoast	0.79	US	North Texas (DFW)	0.42	US	Hill Country	0.41	MX	Coahuila
US	SouthWillametteValley	0.97	US	North Willamette Valley	0.96	NZ	Otago	0.94	US	Umpqua Valley
US	Stanislaus	0.88	US	San Joaquin	0.85	US	Sacramento	0.84	US	Mendocino
US	Steuben	0.90	US	Yates	0.83	US	Schuyler	0.83	US	Ulster
US	Suffolk	0.83	RS	Telečka	0.82	AR	Uruguay	0.81	FR	Aquitaine
US	Sutter	0.80	US	Amador	0.77	US	San Bernardino	0.73	US	Colusa
US	Tehama	0.86	US	El Dorado	0.79	US	Colusa	0.78	US	Calaveras
US	TexasHighPlainsandPanhand	0.82	US	Hill Country	0.80	MX	Baja California	0.77	UY	Lavalleja
US	Trinity	0.92	US	Mendocino	0.92	US	Monterey	0.90	US	San Benito
US	Tulare	0.86	US	Fresno	0.81	US	Madera	0.79	US	Kings
US	Tuolumne	0.89	US	San Diego	0.86	US	Los Angeles	0.83	US	Nevada
US	Ulster	0.83	US	Steuben	0.74	US	Orange	0.73	US	North Carolina
US	UmpquaValley	0.99	NZ	Otago	0.99	US	North Willamette Valley	0.99	CH	Graubünden - other
US	Ventura	0.98	AU	McLaren Vale	0.98	AU	Bendigo	0.97	AU	Mount Lofty Ranges - other
US	Virginia	0.76	US	Suffolk	0.73	NZ	Auckland	0.72	US	Sacramento
US	WahlukeSlope	0.97	US	Horse Heaven Hills	0.95	US	Columbia Valley	0.94	US	San Luis Obispo
US	WallaWallaValley	0.99	US	Red Mountain	0.98	AU	Coonawarra	0.97	US	Columbia Valley
US	Wayne	0.87	HU	Zala	0.85	EL	Ionia Nisia	0.79	RO	Romania
US	WestTexas	0.94	CN	China	0.94	US	Orange	0.94	FR	Limousin
US	YakimaValley	0.83	US	Snipes Mountain	0.82	AU	Big Rivers - other	0.81	AU	King Valley
US	Yates	0.90	US	Steuben	0.88	US	Pennsylvania	0.87	US	Ohio
US	Yolo	0.94	NZ	Gisborne	0.93	US	Solano	0.92	US	Monterey
US	Yuba	0.82	AU	Glenrowan	0.79	AU	North East Victoria - other	0.77	AU	Currency Creek
UY	Artigas	0.76	UY	Rivera	0.75	UY	Colonia	0.72	UY	San Jose
UY	Canelones	0.99	UY	Montevideo	0.97	UY	San Jose	0.95	UY	Paysandu
UY	Colonia	0.94	UY	San Jose	0.93	UY	Paysandu	0.89	UY	Canelones
UY	Durazno	0.89	UY	Rivera	0.89	UY	Salto	0.83	UY	Florida
UY	Florida	0.92	UY	Salto	0.90	AR	Cañuelas	0.89	AR	Cainguas
UY	Lavalleja	0.94	CN	China	0.90	US	West Texas	0.87	UY	Rivera
UY	Maldonado	0.82	UY	Florida	0.82	UY	Colonia	0.80	UY	Paysandu
UY	Montevideo	0.99	UY	Canelones	0.96	UY	San Jose	0.93	UY	Paysandu
UY	Paysandu	0.95	UY	Canelones	0.94	UY	San Jose	0.93	UY	Colonia
UY	Rivera	0.89	UY	Durazno	0.87	UY	Lavalleja	0.86	UY	Salto
UY	Rocha	0.85	UY	Salto	0.77	UY	Durazno	0.76	UY	Florida
UY	Salto	0.92	UY	Florida	0.89	UY	Durazno	0.86	UY	Rivera
UY	SanJose	0.97	UY	Canelones	0.96	UY	Montevideo	0.94	UY	Paysandu
UY	Soriano	0.92	KZ	West Kazakhstan	0.89	US	Orange	0.86	FR	Limousin
UY	Tacuarembo	0.82	US	Orange	0.79	US	West Texas	0.78	FR	Limousin

X. Climatic National Similarity Index and a premium climate indicator

Table 98: Climatic National Similarity Index, 2016
(based on similarity of national shares of regions falling into the 4 climate zones: cool, temperate, warm, hot)

	Algeria	Argentina	Armenia	Australia	Austria	Brazil	Bulgaria	Cambodia	Canada
Algeria	1.00								
Argentina	0.97	1.00							
Armenia	0.00	0.26	1.00						
Australia	0.77	0.90	0.63	1.00					
Austria	0.00	0.00	0.00	0.15	1.00				
Brazil	1.00	0.97	0.00	0.77	0.00	1.00			
Bulgaria	0.00	0.26	1.00	0.63	0.00	0.00	1.00		
Cambodia	1.00	0.97	0.00	0.77	0.00	1.00	0.00	1.00	
Canada	0.00	0.00	0.00	0.14	0.92	0.00	0.00	0.00	1.00
Chile	0.11	0.36	0.99	0.71	0.10	0.11	0.99	0.11	0.08
China	0.00	0.26	1.00	0.63	0.00	0.00	1.00	0.00	0.00
Croatia	0.00	0.26	1.00	0.63	0.00	0.00	1.00	0.00	0.00
Cyprus	1.00	0.97	0.00	0.77	0.00	1.00	0.00	1.00	0.00
Czechia	0.00	0.00	0.00	0.15	0.99	0.00	0.00	0.00	0.87
Ethiopia	0.00	0.00	0.00	0.15	0.99	0.00	0.00	0.00	0.85
France	0.20	0.37	0.69	0.69	0.69	0.20	0.69	0.20	0.63
Georgia	0.00	0.00	0.00	0.15	0.99	0.00	0.00	0.00	0.85
Germany	0.00	0.00	0.00	0.12	0.79	0.00	0.00	0.00	0.97
Greece	1.00	0.98	0.04	0.79	0.00	1.00	0.04	1.00	0.00
Hungary	0.00	0.05	0.17	0.26	0.97	0.00	0.17	0.00	0.83
India	1.00	0.97	0.00	0.77	0.00	1.00	0.00	1.00	0.00
Israel	1.00	0.97	0.00	0.77	0.00	1.00	0.00	1.00	0.00
Italy	0.97	1.00	0.26	0.90	0.02	0.97	0.26	0.97	0.02
Japan	0.55	0.72	0.75	0.94	0.37	0.55	0.75	0.55	0.32
Kazakhstan	0.46	0.68	0.89	0.91	0.00	0.46	0.89	0.46	0.00
Korea, Rep.	1.00	0.97	0.00	0.77	0.00	1.00	0.00	1.00	0.00
Lebanon	1.00	0.97	0.00	0.77	0.00	1.00	0.00	1.00	0.00
Luxembourg	0.00	0.00	0.00	0.02	0.15	0.00	0.00	0.00	0.53
Mexico	1.00	0.98	0.06	0.80	0.00	1.00	0.06	1.00	0.00
Moldova	0.00	0.26	1.00	0.63	0.00	0.00	1.00	0.00	0.00
Morocco	1.00	0.97	0.00	0.77	0.00	1.00	0.00	1.00	0.00
Myanmar	1.00	0.97	0.00	0.77	0.00	1.00	0.00	1.00	0.00
New Zealand	0.00	0.00	0.01	0.16	1.00	0.00	0.01	0.00	0.90
N Macedonia	0.00	0.00	0.00	0.15	0.99	0.00	0.00	0.00	0.85
Norway	0.00	0.00	0.00	0.02	0.15	0.00	0.00	0.00	0.53
Peru	0.88	0.85	0.00	0.74	0.48	0.88	0.00	0.88	0.41
Portugal	0.82	0.94	0.57	0.99	0.00	0.82	0.57	0.82	0.00
Romania	0.00	0.26	1.00	0.63	0.00	0.00	1.00	0.00	0.00
Russia	0.00	0.26	1.00	0.63	0.00	0.00	1.00	0.00	0.00
Serbia	0.00	0.26	0.99	0.64	0.16	0.00	0.99	0.00	0.14
Slovakia	0.00	0.00	0.00	0.15	0.99	0.00	0.00	0.00	0.85
Slovenia	0.18	0.31	0.55	0.60	0.81	0.18	0.55	0.18	0.69
South Africa	1.00	0.97	0.03	0.78	0.00	1.00	0.03	1.00	0.00
Spain	0.98	0.98	0.10	0.84	0.14	0.98	0.10	0.98	0.12
Switzerland	0.00	0.00	0.00	0.09	0.60	0.00	0.00	0.00	0.87
Taiwan	1.00	0.97	0.00	0.77	0.00	1.00	0.00	1.00	0.00
Thailand	1.00	0.97	0.00	0.77	0.00	1.00	0.00	1.00	0.00
Tunisia	1.00	0.97	0.00	0.77	0.00	1.00	0.00	1.00	0.00
Turkey	0.97	1.00	0.24	0.89	0.00	0.97	0.24	0.97	0.00
Ukraine	0.00	0.26	1.00	0.63	0.00	0.00	1.00	0.00	0.00
United Kingdo	0.00	0.00	0.00	0.02	0.15	0.00	0.00	0.00	0.53
United States	0.74	0.85	0.55	0.97	0.39	0.74	0.55	0.74	0.34
Uruguay	1.00	0.97	0.00	0.77	0.00	1.00	0.00	1.00	0.00
New World	**0.71**	**0.86**	**0.68**	**1.00**	**0.18**	**0.71**	**0.68**	**0.71**	**0.16**
Old World	**0.76**	**0.87**	**0.55**	**0.97**	**0.36**	**0.76**	**0.55**	**0.76**	**0.33**
World	**0.75**	**0.87**	**0.59**	**0.99**	**0.31**	**0.75**	**0.59**	**0.75**	**0.29**

Table 98 (cont.): Climatic National Similarity Index, 2016
(based on similarity of national shares of regions falling into the 4 climate zones: cool, temperate, warm, hot)

	Chile	China	Croatia	Cyprus	Czechia	Ethiopia	France	Georgia	Germany	Greece
Algeria										
Argentina										
Armenia										
Australia										
Austria										
Brazil										
Bulgaria										
Cambodia										
Canada										
Chile	1.00									
China	0.99	1.00								
Croatia	0.99	1.00	1.00							
Cyprus	0.11	0.00	0.00	1.00						
Czechia	0.10	0.00	0.00	0.00	1.00					
Ethiopia	0.10	0.00	0.00	0.00	1.00	1.00				
France	0.77	0.69	0.69	0.20	0.69	0.69	1.00			
Georgia	0.10	0.00	0.00	0.00	1.00	1.00	0.69	1.00		
Germany	0.07	0.00	0.00	0.00	0.72	0.70	0.54	0.70	1.00	
Greece	0.15	0.04	0.04	1.00	0.00	0.00	0.22	0.00	0.00	1.00
Hungary	0.27	0.17	0.17	0.00	0.98	0.98	0.80	0.98	0.69	0.01
India	0.11	0.00	0.00	1.00	0.00	0.00	0.20	0.00	0.00	1.00
Israel	0.11	0.00	0.00	1.00	0.00	0.00	0.20	0.00	0.00	1.00
Italy	0.36	0.26	0.26	0.97	0.02	0.02	0.38	0.02	0.02	0.98
Japan	0.84	0.75	0.75	0.55	0.37	0.37	0.88	0.37	0.26	0.58
Kazakhstan	0.93	0.89	0.89	0.46	0.00	0.00	0.70	0.00	0.00	0.50
Korea, Rep.	0.11	0.00	0.00	1.00	0.00	0.00	0.20	0.00	0.00	1.00
Lebanon	0.11	0.00	0.00	1.00	0.00	0.00	0.20	0.00	0.00	1.00
Luxembourg	0.00	0.00	0.00	0.00	0.04	0.00	0.07	0.00	0.72	0.00
Mexico	0.16	0.06	0.06	1.00	0.00	0.00	0.24	0.00	0.00	1.00
Moldova	0.99	1.00	1.00	0.00	0.00	0.00	0.69	0.00	0.00	0.04
Morocco	0.11	0.00	0.00	1.00	0.00	0.00	0.20	0.00	0.00	1.00
Myanmar	0.11	0.00	0.00	1.00	0.00	0.00	0.20	0.00	0.00	1.00
New Zealand	0.11	0.01	0.01	0.00	1.00	0.99	0.70	0.99	0.77	0.00
N Macedonia	0.10	0.00	0.00	0.00	1.00	1.00	0.69	1.00	0.70	0.00
Norway	0.00	0.00	0.00	0.00	0.04	0.00	0.07	0.00	0.72	0.00
Peru	0.14	0.00	0.00	0.88	0.48	0.48	0.50	0.48	0.33	0.88
Portugal	0.65	0.57	0.57	0.82	0.00	0.00	0.56	0.00	0.00	0.84
Romania	0.99	1.00	1.00	0.00	0.00	0.00	0.69	0.00	0.00	0.04
Russia	0.99	1.00	1.00	0.00	0.00	0.00	0.69	0.00	0.00	0.04
Serbia	0.99	0.99	0.99	0.00	0.16	0.16	0.79	0.16	0.11	0.04
Slovakia	0.10	0.00	0.00	0.00	1.00	1.00	0.69	1.00	0.70	0.00
Slovenia	0.64	0.55	0.55	0.18	0.82	0.82	0.98	0.82	0.57	0.20
South Africa	0.13	0.03	0.03	1.00	0.00	0.00	0.22	0.00	0.00	1.00
Spain	0.21	0.10	0.10	0.98	0.14	0.14	0.36	0.14	0.10	0.99
Switzerland	0.05	0.00	0.00	0.00	0.51	0.48	0.39	0.48	0.96	0.00
Taiwan	0.11	0.00	0.00	1.00	0.00	0.00	0.20	0.00	0.00	1.00
Thailand	0.11	0.00	0.00	1.00	0.00	0.00	0.20	0.00	0.00	1.00
Tunisia	0.11	0.00	0.00	1.00	0.00	0.00	0.20	0.00	0.00	1.00
Turkey	0.34	0.24	0.24	0.97	0.00	0.00	0.36	0.00	0.00	0.98
Ukraine	0.99	1.00	1.00	0.00	0.00	0.00	0.69	0.00	0.00	0.04
United Kingdo	0.00	0.00	0.00	0.00	0.04	0.00	0.07	0.00	0.72	0.00
United States	0.66	0.55	0.55	0.74	0.39	0.39	0.80	0.39	0.28	0.76
Uruguay	0.11	0.00	0.00	1.00	0.00	0.00	0.20	0.00	0.00	1.00
New World	**0.77**	**0.68**	**0.68**	**0.71**	**0.18**	**0.18**	**0.73**	**0.18**	**0.14**	**0.73**
Old World	**0.66**	**0.55**	**0.55**	**0.76**	**0.36**	**0.36**	**0.78**	**0.36**	**0.29**	**0.78**
World	**0.69**	**0.59**	**0.59**	**0.75**	**0.31**	**0.31**	**0.77**	**0.31**	**0.25**	**0.77**

Table 98 (cont.): Climatic National Similarity Index, 2016
(based on similarity of national shares of regions falling into the 4 climate zones: cool, temperate, warm, hot)

	Hungary	India	Israel	Italy	Japan	Kazakhstan	Korea, Rep.	Lebanon	Luxembourg
Algeria									
Argentina									
Armenia									
Australia									
Austria									
Brazil									
Bulgaria									
Cambodia									
Canada									
Chile									
China									
Croatla									
Cyprus									
Czechia									
Ethiopia									
France									
Georgia									
Germany									
Greece									
Hungary	1.00								
India	0.00	1.00							
Israel	0.00	1.00	1.00						
Italy	0.07	0.97	0.97	1.00					
Japan	0.50	0.55	0.55	0.73	1.00				
Kazakhstan	0.15	0.46	0.46	0.67	0.92	1.00			
Korea, Rep.	0.00	1.00	1.00	0.97	0.55	0.46	1.00		
Lebanon	0.00	1.00	1.00	0.97	0.55	0.46	1.00	1.00	
Luxembourg	0.00	0.00	0.00	0.01	0.00	0.00	0.00	0.00	1.00
Mexico	0.01	1.00	1.00	0.98	0.59	0.52	1.00	1.00	0.00
Moldova	0.17	0.00	0.00	0.26	0.75	0.89	0.00	0.00	0.00
Morocco	0.00	1.00	1.00	0.97	0.55	0.46	1.00	1.00	0.00
Myanmar	0.00	1.00	1.00	0.97	0.55	0.46	1.00	1.00	0.00
New Zealand	0.98	0.00	0.00	0.02	0.38	0.01	0.00	0.00	0.11
N Macedonia	0.98	0.00	0.00	0.02	0.37	0.00	0.00	0.00	0.00
Norway	0.00	0.00	0.00	0.01	0.00	0.00	0.00	0.00	1.00
Peru	0.47	0.88	0.88	0.86	0.66	0.41	0.88	0.88	0.00
Portugal	0.10	0.82	0.82	0.94	0.88	0.89	0.82	0.82	0.00
Romania	0.17	0.00	0.00	0.26	0.75	0.89	0.00	0.00	0.00
Russia	0.17	0.00	0.00	0.26	0.75	0.89	0.00	0.00	0.00
Serbia	0.33	0.00	0.00	0.26	0.80	0.87	0.00	0.00	0.00
Slovakia	0.98	0.00	0.00	0.02	0.37	0.00	0.00	0.00	0.00
Slovenia	0.90	0.18	0.18	0.33	0.81	0.57	0.18	0.18	0.00
South Africa	0.00	1.00	1.00	0.97	0.57	0.49	1.00	1.00	0.00
Spain	0.16	0.98	0.98	0.98	0.66	0.54	0.98	0.98	0.00
Switzerland	0.47	0.00	0.00	0.02	0.18	0.00	0.00	0.00	0.88
Taiwan	0.00	1.00	1.00	0.97	0.55	0.46	1.00	1.00	0.00
Thailand	0.00	1.00	1.00	0.97	0.55	0.46	1.00	1.00	0.00
Tunisia	0.00	1.00	1.00	0.97	0.55	0.46	1.00	1.00	0.00
Turkey	0.04	0.97	0.97	1.00	0.71	0.66	0.97	0.97	0.00
Ukraine	0.17	0.00	0.00	0.26	0.75	0.89	0.00	0.00	0.00
United Kingdo	0.00	0.00	0.00	0.01	0.00	0.00	0.00	0.00	1.00
United States	0.48	0.74	0.74	0.86	0.96	0.83	0.74	0.74	0.00
Uruguay	0.00	1.00	1.00	0.97	0.55	0.46	1.00	1.00	0.00
New World	**0.29**	**0.71**	**0.71**	**0.86**	**0.96**	**0.93**	**0.71**	**0.71**	**0.02**
Old World	**0.45**	**0.76**	**0.76**	**0.88**	**0.96**	**0.83**	**0.76**	**0.76**	**0.05**
World	**0.41**	**0.75**	**0.75**	**0.88**	**0.96**	**0.86**	**0.75**	**0.75**	**0.04**

Table 98 (cont.): Climatic National Similarity Index, 2016
(based on similarity of national shares of regions falling into the 4 climate zones: cool, temperate, warm, hot)

	Mexico	Moldova	Morocco	Myanmar	New Zealand	N Macedonia	Norway	Peru	Portugal
Algeria									
Argentina									
Armenia									
Australia									
Austria									
Brazil									
Bulgaria									
Cambodia									
Canada									
Chile									
China									
Croatia									
Cyprus									
Czechia									
Ethiopia									
France									
Georgia									
Germany									
Greece									
Hungary									
India									
Israel									
Italy									
Japan									
Kazakhstan									
Korea, Rep.									
Lebanon									
Luxembourg									
Mexico	1.00								
Moldova	0.06	1.00							
Morocco	1.00	0.00	1.00						
Myanmar	1.00	0.00	1.00	1.00					
New Zealand	0.00	0.01	0.00	0.00	1.00				
N Macedonia	0.00	0.00	0.00	0.00	0.99	1.00			
Norway	0.00	0.00	0.00	0.00	0.11	0.00	1.00		
Peru	0.88	0.00	0.88	0.88	0.48	0.48	0.00	1.00	
Portugal	0.85	0.57	0.82	0.82	0.01	0.00	0.00	0.72	1.00
Romania	0.06	1.00	0.00	0.00	0.01	0.00	0.00	0.00	0.57
Russia	0.06	1.00	0.00	0.00	0.01	0.00	0.00	0.00	0.57
Serbia	0.06	0.99	0.00	0.00	0.17	0.16	0.00	0.08	0.56
Slovakia	0.00	0.00	0.00	0.00	0.99	1.00	0.00	0.48	0.00
Slovenia	0.21	0.55	0.18	0.18	0.82	0.82	0.00	0.55	0.46
South Africa	1.00	0.03	1.00	1.00	0.00	0.00	0.00	0.88	0.84
Spain	0.99	0.10	0.98	0.98	0.14	0.14	0.00	0.93	0.86
Switzerland	0.00	0.00	0.00	0.00	0.57	0.48	0.88	0.23	0.00
Taiwan	1.00	0.00	1.00	1.00	0.00	0.00	0.00	0.88	0.82
Thailand	1.00	0.00	1.00	1.00	0.00	0.00	0.00	0.88	0.82
Tunisia	1.00	0.00	1.00	1.00	0.00	0.00	0.00	0.88	0.82
Turkey	0.98	0.24	0.97	0.97	0.00	0.00	0.00	0.85	0.93
Ukraine	0.06	1.00	0.00	0.00	0.01	0.00	0.00	0.00	0.57
United Kingdo	0.00	0.00	0.00	0.00	0.11	0.00	1.00	0.00	0.00
United States	0.77	0.55	0.74	0.74	0.40	0.39	0.00	0.83	0.92
Uruguay	1.00	0.00	1.00	1.00	0.00	0.00	0.00	0.88	0.82
New World	**0.75**	**0.68**	**0.71**	**0.71**	**0.19**	**0.18**	**0.02**	**0.71**	**0.97**
Old World	**0.79**	**0.55**	**0.76**	**0.76**	**0.37**	**0.36**	**0.05**	**0.83**	**0.93**
World	**0.78**	**0.59**	**0.75**	**0.75**	**0.32**	**0.31**	**0.04**	**0.81**	**0.95**

Table 98 (cont.): Climatic National Similarity Index, 2016
(based on similarity of national shares of regions falling into the 4 climate zones: cool, temperate, warm, hot)

	Romania	Russia	Serbia	Slovakia	Slovenia	South Africa	Spain	Switzerland	Taiwan
Algeria									
Argentina									
Armenia									
Australia									
Austria									
Brazil									
Bulgaria									
Cambodia									
Canada									
Chile									
China									
Croatia									
Cyprus									
Czechia									
Ethiopia									
France									
Georgia									
Germany									
Greece									
Hungary									
India									
Israel									
Italy									
Japan									
Kazakhstan									
Korea, Rep.									
Lebanon									
Luxembourg									
Mexico									
Moldova									
Morocco									
Myanmar									
New Zealand									
N Macedonia									
Norway									
Peru									
Portugal									
Romania	1.00								
Russia	1.00	1.00							
Serbia	0.99	0.99	1.00						
Slovakia	0.00	0.00	0.16	1.00					
Slovenia	0.55	0.55	0.67	0.82	1.00				
South Africa	0.03	0.03	0.03	0.00	0.19	1.00			
Spain	0.10	0.10	0.12	0.14	0.35	0.99	1.00		
Switzerland	0.00	0.00	0.08	0.48	0.39	0.00	0.07	1.00	
Taiwan	0.00	0.00	0.00	0.00	0.18	1.00	0.98	0.00	1.00
Thailand	0.00	0.00	0.00	0.00	0.18	1.00	0.98	0.00	1.00
Tunisia	0.00	0.00	0.00	0.00	0.18	1.00	0.98	0.00	1.00
Turkey	0.24	0.24	0.24	0.00	0.30	0.98	0.98	0.00	0.97
Ukraine	1.00	1.00	0.99	0.00	0.55	0.03	0.10	0.00	0.00
United Kingdo	0.00	0.00	0.00	0.00	0.00	0.00	0.00	0.88	0.00
United States	0.55	0.55	0.61	0.39	0.75	0.75	0.83	0.19	0.74
Uruguay	0.00	0.00	0.00	0.00	0.18	1.00	0.98	0.00	1.00
New World	**0.68**	**0.68**	**0.70**	**0.18**	**0.64**	**0.73**	**0.79**	**0.10**	**0.71**
Old World	**0.55**	**0.55**	**0.60**	**0.36**	**0.73**	**0.77**	**0.85**	**0.22**	**0.76**
World	**0.59**	**0.59**	**0.63**	**0.31**	**0.71**	**0.76**	**0.84**	**0.19**	**0.75**

Table 98 (cont.): Climatic National Similarity Index, 2016
(based on similarity of national shares of regions falling into the 4 climate zones: cool, temperate, warm, hot)

	Thailand	Tunisia	Turkey	Ukraine	United Kingdom	United States	Uruguay	**New World**	**Old World**	**World**
Algeria										
Argentina										
Armenia										
Australia										
Austria										
Brazil										
Bulgaria										
Cambodia										
Canada										
Chile										
China										
Croatia										
Cyprus										
Czechia										
Ethiopia										
France										
Georgia										
Germany										
Greece										
Hungary										
India										
Israel										
Italy										
Japan										
Kazakhstan										
Korea, Rep.										
Lebanon										
Luxembourg										
Mexico										
Moldova										
Morocco										
Myanmar										
New Zealand										
N Macedonia										
Norway										
Peru										
Portugal										
Romania										
Russia										
Serbia										
Slovakia										
Slovenia										
South Africa										
Spain										
Switzerland										
Taiwan										
Thailand	1.00									
Tunisia	1.00	1.00								
Turkey	0.97	0.97	1.00							
Ukraine	0.00	0.00	0.24	1.00						
United Kingdo	0.00	0.00	0.00	0.00	1.00					
United States	0.74	0.74	0.85	0.55	0.00	1.00				
Uruguay	1.00	1.00	0.97	0.00	0.00	0.74	1.00			
New World	**0.71**	**0.71**	**0.85**	**0.68**	**0.02**	**0.97**	**0.71**	**1.00**		
Old World	**0.76**	**0.76**	**0.86**	**0.55**	**0.05**	**1.00**	**0.76**	**0.97**	**1.00**	
World	**0.75**	**0.75**	**0.87**	**0.59**	**0.04**	**1.00**	**0.75**	**0.98**	**1.00**	**1.00**

Table 99: Shares of selected varieties' bearing area within the ideal GST range for premium winegrapes, by country and globally, 2000 and 2016 (%)

2000	Cabernet Franc	Cabernet Sauvignon	Chard-onnay	Côt	Dolcetto	Garnacha Tinta	Gewürz-traminer	Mazuelo	Merlot	Müller-Thurgau	Nebbiolo
Algeria		0				0		0	0		
Argentina	15	24	0	26	0	9		35	37		5
Armenia											
Australia	62	60	10	68		57	1	13	42		69
Austria	98	0					46		0	11	
Brazil	0	0	0				0		0		
Bulgaria		100	0				0		100		
Cambodia											
Canada	0	0	100				100		0		
Chile	85	100	3	95			0	85	91		100
China											
Croatia											
Cyprus											
Czechia			100							0	
Ethiopia											
France	97	96	70	91	0	92	99	91	82	91	
Georgia		100									
Germany			100							52	
Greece	0	17	0			3		100	0		
Hungary	100	100	100						100	0	
India											
Israel		0	0					0	0		
Italy	30	76	25	24	97	7	48	1	26	1	94
Japan											
Kazakhstan											
Korea, Rep.											
Lebanon											
Luxembourg			100				100			100	
Mexico											
Moldova		100	100	100			0		100	0	
Morocco						0		0			
Myanmar											
New Zealand	91	16	97	82			45		80	2	
N. Macedonia											
Norway											
Peru											
Portugal		0									
Romania		100	0				0		100		
Russia		100	0								
Serbia											
Slovakia		0	100							0	
Slovenia			100						100		
South Africa	0	0	0						0		
Spain	12	7	9	89		30	0	16	39		
Switzerland	26	0	82	0			99		0	90	0
Taiwan											
Thailand											
Tunisia		0				0		0			
Turkey											
Ukraine											
United Kingdom			0								
United States	91	66	26	74	30	2	7	12	54		19
Uruguay	0	0	0						0		
New World	**27.8**	**63.6**	**19.9**	**30.8**	**25.7**	**19.6**	**18.8**	**24.8**	**52.5**	**2.3**	**19.3**
Old World	**86.4**	**88.0**	**60.3**	**88.8**	**97.1**	**56.1**	**40.4**	**71.9**	**75.7**	**37.1**	**94.2**
World	**77.1**	**75.5**	**38.8**	**45.1**	**96.6**	**55.0**	**36.3**	**70.5**	**70.0**	**36.6**	**89.1**

Table 99 (cont.): Shares of selected varieties' bearing area within the ideal GST range for premium winegrapes, by country and globally, 2000 and 2016 (%)

2000	Pinot Gris	Pinot Noir	Riesling	Sangiovese	Sauvignon Blanc	Sémillon	Syrah	Tempranillo	Tribidrag	Viognier
Algeria		0					0			
Argentina	0	0	1	13	0	36	13	18	0	16
Armenia										
Australia		9	6	43	38	21	50	28		55
Austria	23	100	100		100					
Brazil					0	0				
Bulgaria		0	0		0					
Cambodia										
Canada	100	100	100		100					
Chile	0	0	12	100	80	70	100	100	100	100
China										
Croatia										
Cyprus										
Czechia			100							
Ethiopia										
France	1	96	100	100	78	98	55	16	100	44
Georgia										
Germany	83	96	100							
Greece					0	0	9			
Hungary	0	0	100		100					
India										
Israel					0					
Italy	0	14	10	56	7	0	16	13	4	1
Japan										
Kazakhstan										
Korea, Rep.										
Lebanon										
Luxembourg	100	100	100							
Mexico										
Moldova	0	0	0		100					
Morocco										
Myanmar										
New Zealand	35	84	100		99	98	79			
N Macedonia										
Norway										
Peru										
Portugal								73		
Romania	0	0			100					
Russia			0							
Serbia										
Slovakia										
Slovenia					100					
South Africa		0			0	0	0			0
Spain		6	1		80		3	48		11
Switzerland	87	89	98		67	0	0			
Taiwan										
Thailand										
Tunisia				0			0		0	
Turkey										
Ukraine										
United Kingdom										
United States	11	33	44	53	30	58	41	1	18	60
Uruguay					0		0			
New World	**29.2**	**25.8**	**31.4**	**26.7**	**44.9**	**31.4**	**39.3**	**17.9**	**17.9**	**53.5**
Old World	**14.9**	**71.0**	**88.2**	**56.6**	**78.4**	**98.1**	**52.2**	**49.9**	**3.5**	**43.6**
World	**15.8**	**62.1**	**79.1**	**55.0**	**66.4**	**67.7**	**46.1**	**48.2**	**13.5**	**46.0**

Table 99 (cont.): Shares of selected varieties' bearing area within the ideal GST range for premium winegrapes, by country and globally, 2000 and 2016 (%)

2016	Cabernet Franc	Cabernet Sauvignon	Chard-onnay	Côt	Dolcetto	Garnacha Tinta	Gewürz-traminer	Mazuelo	Merlot	Müller-Thurgau	Nebbiolo
Algeria		0				0		0	0		
Argentina	42	32	0	42	0	22		27	39		14
Armenia											
Australia	83	53	13	78	33	75	1	68	38	0	36
Austria	90	0	100						11	18	
Brazil	0	0	0	0			0		0		100
Bulgaria	100	100	0				0		100		
Cambodia		0							0		
Canada	61	0	98	0		0	79		0	100	0
Chile	81	100	5	90	100	100	0	91	91		98
China	100	100	0			100		100	100		
Croatia		100	0						100		
Cyprus											
Czechia			100				100			8	
Ethiopia											
France	100	97	64	91		100	96	100	73	72	100
Georgia		100									
Germany	6	0	100				100		0	50	
Greece	9	25	0					0	8		
Hungary	100	87	80	51			2		94	0	
India		0	0								
Israel	0	0	0	0				0	0		
Italy	25	58	14	12	98	4	41	3	21	17	98
Japan		88	0						77	0	
Kazakhstan	100	26									
Korea, Rep.											
Lebanon		0	0						0		
Luxembourg			100				100			100	
Mexico		0				0		4	0		0
Moldova	100	100	100	100			0		100	0	
Morocco		0	0			0		0	0		
Myanmar			0								
New Zealand	94	11	97	94		4	56		96	78	0
N Macedonia		100	100						100		
Norway											
Peru		6	0	0		0			0		
Portugal	71	1	0			0	0	0	3		
Romania	100	100	0	100			0		100	0	100
Russia	100	100	43				0		100	0	
Serbia	100	98	48				0		99		
Slovakia		0								0	
Slovenia	100	78	55						90	0	
South Africa	5	2	0	5		5	0	0	3		52
Spain	0	2	9	12		13	0	18	14		100
Switzerland	24	0	79	0	0	0	98		0	94	0
Taiwan											
Thailand		0									
Tunisia		0	0			0		0	0		
Turkey	10	12	0	4		0		0	22		
Ukraine		100	100				0		100		
United Kingdom			0							100	
United States	75	70	28	53	43	21	9	23	60	80	33
Uruguay	0	1	0	0		0	0		0		0
New World	**33.4**	**73.3**	**19.6**	**44.6**	**35.3**	**69.0**	**23.8**	**43.1**	**64.5**	**63.2**	**19.5**
Old World	**87.5**	**72.7**	**47.8**	**85.6**	**98.1**	**60.5**	**49.9**	**73.3**	**65.6**	**37.6**	**98.0**
World	**73.5**	**73.1**	**33.9**	**49.8**	**95.9**	**61.0**	**45.1**	**71.7**	**65.3**	**37.7**	**93.6**

Table 99 (cont.): Shares of selected varieties' bearing area within the ideal GST range for premium winegrapes, by country and globally, 2000 and 2016 (%)

2016	Pinot Gris	Pinot Noir	Riesling	Sangiovese	Sauvignon Blanc	Sémillon	Syrah	Tempranillo	Tribidrag	Viognier
Algeria							0			
Argentina	0	1	1	14	1	34	16	19	100	11
Armenia										
Australia	3	19	12	57	37	13	54	37	63	42
Austria	47	100	100		99		0			
Brazil	0	0	0	0	0	0		0		0
Bulgaria		0			16		100			
Cambodia							0			
Canada	75	98	100	0	43	0	0	0	0	0
Chile	0	1	17	100	79	66	100	97	100	97
China		0	0		0		100			
Croatia			0							
Cyprus										
Czechia	9	100	100		100					
Ethiopia				100						
France	1	88	99	100	70	98	76	1	100	47
Georgia										
Germany	73	97	100		83		0			
Greece			0		0	0	10	0		
Hungary	0	9	87	0	100	100	84			83
India		0			0		0			
Israel					0		0	0		
Italy	3	12	33	63	3	0	17	0	1	5
Japan		100	0		0					
Kazakhstan		0	0							
Korea, Rep.										
Lebanon					0		0			
Luxembourg	100	100	100							
Mexico					0		0	0		
Moldova	0	0	0		100	100	100			100
Morocco					0		0			
Myanmar		0			0		0	0		
New Zealand	23	94	100	49	100	92	91	68	0	35
N Macedonia		0	100		100				0	
Norway										
Peru		0					0			
Portugal	0	0	0		30	28	11	37		14
Romania	0	0	0	100	100	100	100	100	100	100
Russia	0	0	0		62	0				
Serbia	0	0	38		96					
Slovakia			100							
Slovenia	0	72	100		78		0			
South Africa	0	0	0	9	0	0	3	0	0	4
Spain		1	0	98	17	0	1	45	0	1
Switzerland	88	88	99	0	57	21	0	0	0	0
Taiwan										
Thailand		0		0			0	0		0
Tunisia					0		0			
Turkey		0	0	0	1	0	9	25		28
Ukraine		0	100		100					
United Kingdom	100	0								
United States	10	31	62	49	27	36	56	36	23	35
Uruguay	0	0	0		0	0	0			0
New World	**12.8**	**33.1**	**44.3**	**33.7**	**58.0**	**20.8**	**46.4**	**22.1**	**23.1**	**36.4**
Old World	**14.2**	**70.5**	**72.6**	**64.2**	**67.7**	**92.3**	**51.1**	**44.0**	**1.3**	**39.2**
World	**13.8**	**55.5**	**66.8**	**62.8**	**62.8**	**62.7**	**49.0**	**43.2**	**13.4**	**38.4**

XI. National winegrape area, top 300 varieties

Table 100: National winegrape area for each of the world's top 300 varieties, 2016

	Agiorgitiko	Aglianico	Airén	Alarije	Alb de Suruceni	Albana	Albillo Mayor	Aletta	Alfrocheiro	Alicante Henri Bouschet	Aligoté	Alvarelhão	Alvarinho	Ancellotta	Antao Vaz	Aramon Noir	Arinto de Bucelas
Algeria																	
Argentina		51								135		6		991			
Australia		25								19		2					
Austria																	
Brazil										101		2		20			
Bulgaria											285						
Cambodia																	
Canada										0	30						
Chile							0			6908			1				
China																	
Croatia																	
Cyprus																	
Czechia																	
Ethiopia																	
France						0				2607	1927					1167	
Georgia											124						
Germany																	
Greece	3270									60							
Hungary								1676		14	0						
India																	
Israel																	
Italy		9627				782				286				1700			
Japan																	
Kazakhstan											277						
Korea, Rep.																	
Lebanon																	
Luxembourg																	
Mexico																	
Moldova					780						7765						
Morocco			440							919							
Myanmar																	
New Zealand													26				
N. Macedonia																	
Peru							7										
Portugal									1206	4547		2860			1768	14	5409
Romania										20	5840						
Russia											5843						
Serbia																	
Slovakia																	
Slovenia																	
South Africa	2									7		12					
Spain			203276	4407			1145		10	19294		28	5393	0			
Switzerland										0	24			28			
Thailand																	
Tunisia			85							178							
Turkey										532							
Ukraine											4814						
United Kingdom																	
United States		31								380			126				
Uruguay										24							
Old World	**3270**	**9627**	**203801**	**4407**	**780**	**782**	**1145**	**1676**	**1216**	**28457**	**26899**	**2888**	**5393**	**1728**	**1768**	**1181**	**5409**
New World	**2**	**107**					**7**			**7575**	**30**	**22**	**153**	**1011**			
World	**3272**	**9734**	**203801**	**4407**	**780**	**782**	**1152**	**1676**	**1216**	**36031**	**26929**	**2910**	**5545**	**2739**	**1768**	**1181**	**5409**

Table 100 (cont.): National winegrape area for each of the world's top 300 varieties, 2016

	Arneis	Aspiran Bouschet	Assyrtiko	Auxerrois	Avesso	Azal	Băbească Neagră	Bacchus	Baco Noir	Baga	Barbera	Beba	Beibinghong	Bianca	Biancame	Bical	Blauburger
Algeria																	
Argentina		4087									444						
Australia	37		0	0							102						
Austria																	750
Brazil		1									2						
Bulgaria																	
Cambodia																	
Canada				38				20	704		1						18
Chile											5						
China													1600				
Croatia																	
Cyprus																	
Czechia																	
Ethiopia																	
France				2409					5								
Georgia																	
Germany				213				1610									2
Greece			1770														
Hungary								0						4898			453
India																	
Israel																	
Italy	1108										15006				1336		
Japan																	
Kazakhstan																	
Korea, Rep.																	
Lebanon																	
Luxembourg				190													
Mexico																	
Moldova														1340			
Morocco																	
Myanmar																	
New Zealand	33																
N. Macedonia																	
Peru																	
Portugal					699	1443				6750		0				1076	
Romania							2696				0			13			
Russia														3513			
Serbia																	
Slovakia																	
Slovenia											98						
South Africa											35						
Spain										0	0	2556					
Switzerland	0			3				1	1		0			2			
Thailand																	
Tunisia																	
Turkey																	
Ukraine																	
United Kingdom								127									
United States									25		2131						
Uruguay																	
Old World	**1108**		**1770**	**2815**	**699**	**1443**	**2696**	**1612**	**6**	**6750**	**15105**	**2556**		**9766**	**1336**	**1076**	**1205**
New World	**71**	**4088**	**0**	**38**				**147**	**729**		**2719**		**1600**				**18**
World	**1179**	**4088**	**1770**	**2853**	**699**	**1443**	**2696**	**1759**	**735**	**6750**	**17824**	**2556**	**1600**	**9766**	**1336**	**1076**	**1223**

Table 100 (cont.): National winegrape area for each of the world's top 300 varieties, 2016

	Blauer Portugieser	Blaufränkisch	Bobal	Boğazkere	Bombino Bianco	Bombino Nero	Bonarda Piemontese	Brachetto del Piemonte	Cabernet Franc	Cabernet Sauvignon	Caladoc	Çalkarası	Canaiolo Nero	Cardinal	Carmenère
Algeria										1000					
Argentina					9				929	15356	15			2	59
Australia		1						2	328	23987			1		16
Austria	1265	2808							64	567					
Brazil									6834	429	0				10
Bulgaria									240	9327					
Cambodia										2					
Canada		5							820	660					3
Chile	3								1578	42409					10503
China									600	40300					11200
Croatia		521								709					
Cyprus															
Czechia	599	1143													
Ethiopia															
France	26								32327	46555	3062			202	28
Georgia										286					
Germany	3177	1737							32	329					
Greece									10	1929					
Hungary	1023	7260							1368	2677					0
India										100					
Israel									110	990					
Italy	51	28			1138	865	5926	1692	5590	14240			1031		635
Japan										42					
Kazakhstan									56	20					
Korea, Rep.															
Lebanon										1000					
Luxembourg															
Mexico										756				168	
Moldova									756	8169				473	
Morocco										604				624	
Myanmar															
New Zealand									109	275					
N. Macedonia										1020					
Peru		290								48				53	
Portugal	36								23	2346	2180				
Romania	329	729							72	5359					
Russia									20	8528					
Serbia		727							79	2111				3	
Slovakia		1216								469					
Slovenia	80	709							27	423					
South Africa									835	10589					8
Spain			59189						680	20139				14	0
Switzerland		4							63	66	0				0
Thailand										7					
Tunisia										117				121	
Turkey				1436					37	476		806			
Ukraine										4935					
United Kingdom															
United States		3							2199	40837					24
Uruguay									266	484					
Old World	**6587**	**16882**	**59189**	**1436**	**1138**	**865**	**5926**	**1692**	**41553**	**134391**	**5242**	**806**	**1031**	**1437**	**663**
New World	**3**	**299**			**9**			**2**	**14499**	**176280**	**16**		**1**	**223**	**21822**
World	**6590**	**17180**	**59189**	**1436**	**1147**	**865**	**5926**	**1694**	**56052**	**310671**	**5258**	**806**	**1033**	**1660**	**22486**

Table 100 (cont.): National winegrape area for each of the world's top 300 varieties, 2016

	Castelão	Catarratto Bianco	Cayetana Blanca	Cereza	Chambourcin	Chardonnay	Chasselas	Chelva	Chenin Blanc	Chinuri	Ciliegiolo	Cinsaut	Clairette	Cococciola	Colombard	Concord
Algeria																
Argentina				28887		6227			2157			1				
Australia			17		41	21321			406			10	0		1789	
Austria						1577										
Brazil						340			7						22	1687
Bulgaria						3087										
Cambodia																
Canada						1417	11		5							183
Chile						11435	197		39			848				
China						6100										
Croatia						657										
Cyprus																
Czechia						820										
Ethiopia									54							
France					592	47451	541		9432			15930	2042		8441	
Georgia										1225						
Germany						1485	1046									
Greece						673						4				
Hungary						2464	1159		6							
India						100										
Israel						165									220	
Italy		28563				19769	11		7		897	4	25	1671		
Japan						137										292
Kazakhstan																
Korea, Rep.																
Lebanon						1000										
Luxembourg						30										
Mexico									275							
Moldova						4133	329									9
Morocco						880						3239	113			
Myanmar						2										
New Zealand					3	3117			24							
N. Macedonia						750										
Peru						1			2							
Portugal	12580		132			547	72		0			12				
Romania					0	1878	127									
Russia						3481	21						22			
Serbia						1455										
Slovakia																
Slovenia						1181										
South Africa					13	6856			17707			1767	195		11512	
Spain	0		36252			6866	24	5029	106			10			6	
Switzerland					2	359	3838		8							
Thailand									16						15	
Tunisia						170						626	22			
Turkey						177						430				
Ukraine						1500										
United Kingdom						531										
United States		50			315	41392			1969			45			7991	8349
Uruguay						119			2							24
Old World	**12580**	**28563**	**36385**		**595**	**102554**	**7170**	**5029**	**9558**	**1225**	**897**	**20255**	**2225**	**1671**	**8667**	**9**
New World		**50**	**17**	**28887**	**373**	**99094**	**207**		**22664**			**2671**	**196**		**21329**	**10535**
World	**12580**	**28613**	**36401**	**28887**	**968**	**201649**	**7377**	**5029**	**32221**	**1225**	**897**	**22926**	**2420**	**1671**	**29996**	**10544**

Table 100 (cont.): National winegrape area for each of the world's top 300 varieties, 2016

	Cortese	Corvina Veronese	Corvinone	Côt	Couderc Noir	Criolla Grande	Croatina	Cserszegi Füszeres	Damaschino	Diagalves	Dimrit	Dimyat	Dolcetto	Dornfelder	Douce Noire	Duras	Durif
Algeria																	
Argentina		17	0	40401		15596	16		527				7		19072		
Australia		0		515									100	0			540
Austria																	
Brazil		0		30	1938												3
Bulgaria												2998					
Cambodia																	
Canada		1		41										13			
Chile		1		2293									1				208
China																	
Croatia																	
Cyprus																	
Czechia																	
Ethiopia																	
France				6100	198										0	785	
Georgia																	
Germany														7761			
Greece												6					
Hungary				3				4299						29			
India																	
Israel				110													110
Italy	2405	6222	1140	178			2678		52				4381		630		
Japan																	
Kazakhstan																	
Korea, Rep.																	
Lebanon																	
Luxembourg																	
Mexico																	133
Moldova				162													
Morocco																	
Myanmar																	
New Zealand				129													1
N. Macedonia												6500					
Peru				10													
Portugal									880	1090							
Romania				7										1			
Russia																	
Serbia												192					
Slovakia																	
Slovenia																	
South Africa				452													114
Spain				113					163								
Switzerland		0		15									0	22			
Thailand														3			1
Tunisia																	
Turkey				21							704						
Ukraine																	
United Kingdom														13			
United States				1610									56	29	31		3698
Uruguay				43													
Old World	**2405**	**6222**	**1140**	**6709**	**198**		**2678**	**4299**	**1095**	**1090**	**704**	**9696**	**4381**	**7812**	**630**	**785**	**110**
New World		**19**	**0**	**45524**	**1938**	**15596**	**16**		**527**				**164**	**59**	**19103**		**4697**
World	**2405**	**6240**	**1140**	**52233**	**2136**	**15596**	**2695**	**4299**	**1622**	**1090**	**704**	**9696**	**4545**	**7871**	**19733**	**785**	**4807**

Table 100 (cont.): National winegrape area for each of the world's top 300 varieties, 2016

	Elbling	Falanghina Flegrea	Fer	Fernão Pires	Fetească Albă	Fetească Neagră	Fetească Regală	Fiano	Folle Blanche	Furmint	Gaglioppo	Gamay Noir	Garganega	Garnacha Blanca	Garnacha Peluda	Garnacha Roja (Gris)	Garnacha Tinta
Algeria																	2000
Argentina			183					5				0	30	2			22
Australia			1					95		0		6	1				1492
Austria										9							
Brazil										0		5					
Bulgaria																	
Cambodia																	
Canada												272					1
Chile												0	1				187
China																	4000
Croatia																	
Cyprus																	
Czechia																	
Ethiopia																	
France	0		1502						1554			24095		5130	394	1253	78631
Georgia																	
Germany	489																
Greece																114	
Hungary						1				3862		3					
India																	
Israel																	
Italy		3634						2087			4626	64	8522				5421
Japan																	
Kazakhstan																	
Korea, Rep.																	
Lebanon																	
Luxembourg	86																
Mexico																	140
Moldova					954	402	372										
Morocco																	786
Myanmar																	
New Zealand												7					1
N. Macedonia																	
Peru																	1
Portugal				12138								0					60
Romania					12428	2845	12619			18							
Russia																	
Serbia												54					
Slovakia																	
Slovenia										546		4					
South Africa				73								9		110		3	344
Spain									20					2061	503	87	54606
Switzerland	1											1349	0				1
Thailand																5	
Tunisia																	152
Turkey												228					33
Ukraine																	
United Kingdom																	
United States	396											123		106			2213
Uruguay												1					4
Old World	**576**	**3634**	**1502**	**12138**	**13382**	**3248**	**12991**	**2087**	**1574**	**4434**	**4626**	**25797**	**8522**	**7191**	**898**	**1455**	**141691**
New World	**396**		**184**	**73**				**100**		**0**		**424**	**32**	**218**		**8**	**8405**
World	**972**	**3634**	**1686**	**12211**	**13382**	**3248**	**12991**	**2187**	**1574**	**4435**	**4626**	**26221**	**8554**	**7409**	**898**	**1462**	**150096**

Table 100 (cont.): National winegrape area for each of the world's top 300 varieties, 2016

	Gewürztraminer	Gibi	Godello	Graciano	Grand Noir	Graševina	Grechetto di Orvieto	Greco Bianco	Grignolino	Grillo	Grolleau Noir	Gros Manseng	Grüner Veltliner	Hárslevelű	Inzolia	Irsai Olivér	Isabella
Algeria																	
Argentina		785		18									7				
Australia	252			31	0					1		10	9	1			15
Austria						3233							14376				
Brazil	9					188											11664
Bulgaria	591																
Cambodia																	
Canada	398												3				
Chile	371			1													
China						3000											
Croatia						4459											
Cyprus																	
Czechia	591					1114							1538				
Ethiopia																	
France	3320			10	0						1949	3046					
Georgia																	
Germany	824												14				
Greece																	
Hungary	694					3933							1381	1603		1531	
India																	
Israel																	
Italy	1321			437		1259	1824	2050	911	7382			55		4740		
Japan																	
Kazakhstan																	
Korea, Rep.																	
Lebanon																	
Luxembourg	21																
Mexico																	
Moldova	1099															180	3468
Morocco																	
Myanmar																	
New Zealand	277												43				
N. Macedonia						270											
Peru																	
Portugal	0		584	326	167												
Romania	469					1437							0				
Russia	500															24	1362
Serbia	142					2037											
Slovakia						456							1627				
Slovenia						1935											
South Africa	106			8	3								4	14		48	
Spain	373		822	2080	536	1064						13					
Switzerland	51												1		0	7	1
Thailand																	
Tunisia																	
Turkey																	
Ukraine	500																1200
United Kingdom																	
United States	897												60				
Uruguay	17																102
Old World	**10495**		**1406**	**2852**	**703**	**21196**	**1824**	**2050**	**911**	**7382**	**1949**	**3059**	**18992**	**1603**	**4740**	**1742**	**6031**
New World	**2328**	**785**		**58**	**3**	**3188**				**1**		**10**	**127**	**16**		**48**	**11781**
World	**12823**	**785**	**1406**	**2910**	**707**	**24384**	**1824**	**2050**	**911**	**7383**	**1949**	**3069**	**19118**	**1618**	**4740**	**1790**	**17813**

Table 100 (cont.): National winegrape area for each of the world's top 300 varieties, 2016

	Italia	Jacquez	Juan García	Kadarka	Kalecik Karası	Kerner	Királyleányka	Kodryanka	Koshu	Kotsifali	Kunleány	Kyoho (4N)	Lambrusco di Sorbara	Lambrusco Grasparossa	Lambrusco Maestri	Lambrusco Marani
Algeria																
Argentina														11	48	
Australia	0	1				0										
Austria																
Brazil		1274														
Bulgaria				1161												
Cambodia																
Canada	2					11										
Chile																
China																
Croatia																
Cyprus																
Czechia																
Ethiopia																
France	75															
Georgia																
Germany						2646										
Greece										1338						
Hungary				351		2	784				974					
India																
Israel																
Italy						16							858	943	5610	1074
Japan						76			690			62				
Kazakhstan																
Korea, Rep.												2700				
Lebanon																
Luxembourg																
Mexico		80														
Moldova	118							1143								
Morocco	3333															
Myanmar																
New Zealand																
N. Macedonia																
Peru	1011															
Portugal			135													
Romania				97		3										
Russia																
Serbia				15												
Slovakia																
Slovenia						111										
South Africa						1										
Spain	4		1409													
Switzerland						25										
Thailand																
Tunisia	644															
Turkey					704											
Ukraine																
United Kingdom																
United States		88														
Uruguay																
Old World	**4174**		**1545**	**1625**	**704**	**2803**	**784**	**1143**		**1338**	**974**		**858**	**943**	**5610**	**1074**
New World	**1013**	**1443**				**89**			**690**			**2762**		**11**	**48**	
World	**5188**	**1443**	**1545**	**1625**	**704**	**2891**	**784**	**1143**	**690**	**1338**	**974**	**2762**	**858**	**954**	**5657**	**1074**

Table 100 (cont.): National winegrape area for each of the world's top 300 varieties, 2016

	Lambrusco Salamino	Leányka	Levokumskij	Liatiko	Listan Negro	Listan Prieto	Longyan	Loureiro	Macabeo	Magaracha Rannii	Malvasia	Malvasia Bianca di	Malvasia Bianca Lunga	Malvasia di Candia	Malvasia Fina	Malvasia Nera di
Algeria																
Argentina						374					21					
Australia																
Austria																
Brazil												165	0		0	
Bulgaria																
Cambodia																
Canada																
Chile						9693										
China							1000									
Croatia																
Cyprus																
Czechia																
Ethiopia																
France									1657							
Georgia																
Germany																
Greece				2633												
Hungary		719														
India																
Israel																
Italy	6228											9028	1247	1087		1247
Japan																
Kazakhstan																
Korea, Rep.																
Lebanon																
Luxembourg																
Mexico																
Moldova										884						
Morocco																
Myanmar																
New Zealand																
N. Macedonia																
Peru																
Portugal								4402			801	6			2922	
Romania																
Russia			890													
Serbia																
Slovakia																
Slovenia																
South Africa									4							
Spain					2847			294	36963		1362			121	354	16
Switzerland																
Thailand																
Tunisia																
Turkey																
Ukraine																
United Kingdom																
United States						200						486				
Uruguay															6	
Old World	**6228**	**719**	**890**	**2633**	**2847**			**4696**	**38620**	**884**	**2163**	**9034**	**1247**	**1208**	**3276**	**1264**
New World						**10267**	**1000**		**4**		**21**	**651**	**0**		**6**	
World	**6228**	**719**	**890**	**2633**	**2847**	**10267**	**1000**	**4696**	**38625**	**884**	**2184**	**9685**	**1247**	**1208**	**3282**	**1264**

Table 100 (cont.): National winegrape area for each of the world's top 300 varieties, 2016

	Malvasia Preta	Malvazija Istarska	Mammolo	Mandilaria	Marsanne	Marselan	Marufo	Marzemino	Mauzac Blanc	Mavro	Mavrouda	Mavrud	Mazuelo	Melon	Mencía	Merlot	Merseguera
Algeria													3000			1000	
Argentina					2	10							13	1		5632	
Australia		1	0		161			0					8		2	8415	
Austria																695	
Brazil						23										363	
Bulgaria												1193				10050	
Cambodia																2	
Canada					3											633	
Chile					18	24							811		1	12057	
China													100			16700	
Croatia		1600														828	
Cyprus										3187							
Czechia																	
Ethiopia																	
France			885		1484	3662			1526				31760	9550		108483	
Georgia																	
Germany																553	
Greece				932							1658		1			1393	
Hungary						2										1961	
India																	
Israel													935			715	
Italy		272	26		47			785					1686			24057	
Japan																197	
Kazakhstan																	
Korea, Rep.																	
Lebanon																500	
Luxembourg																	
Mexico													448			391	
Moldova																7689	
Morocco													1230			333	
Myanmar																	
New Zealand					0											1239	
N. Macedonia																1240	
Peru																2	
Portugal	1933						3367						291		2561	482	
Romania						11										11647	
Russia																2988	
Serbia						84										1968	
Slovakia																	
Slovenia		915														817	
South Africa					21	3							114			5558	
Spain						1	1316						5461		8489	12852	2373
Switzerland					48	2		0						0	0	1124	
Thailand																	
Tunisia													238			64	
Turkey													130			415	
Ukraine																1400	
United Kingdom																	
United States					52								1086			21251	
Uruguay					1	120										747	
Old World	**1933**	**2788**	**911**	**932**	**1580**	**3761**	**4683**	**785**	**1526**	**3187**	**1658**	**1193**	**44732**	**9550**	**11050**	**193253**	**2373**
New World		**1**	**0**		**259**	**180**		**0**					**2581**	**1**	**2**	**73187**	
World	**1933**	**2788**	**911**	**932**	**1838**	**3941**	**4683**	**785**	**1526**	**3187**	**1658**	**1193**	**47312**	**9551**	**11052**	**266440**	**2373**

Table 100 (cont.): National winegrape area for each of the world's top 300 varieties, 2016

	Misket Cherven	Moldova	Monastrell	Monica Nera	Montepulciano	Moscato Giallo	Moschofilero	Müller-Thurgau	Muscadelle	Muscat	Muscat Bailey A	Muscat Blanc à Petits Grains	Muscat Blanc à Petits Grains (G)	Muscat Blanc à Petits Grains (R)
Algeria														
Argentina			12		82	119						94	6526	
Australia			704		60	94		1	92			857		240
Austria								1777				823		
Brazil					1	90				671		32		
Bulgaria	4349													
Cambodia														
Canada			1					6		11				
Chile			102		2	162				1		1	1732	
China														
Croatia														
Cyprus														
Czechia								1479						
Ethiopia														
France			8754					2	1412			7333		419
Georgia														
Germany								11664				240		5
Greece							1088					1568		
Hungary								1670				762		
India														
Israel			55											
Italy				1203	32724	1108		1296				13334		
Japan								22			521			
Kazakhstan														255
Korea, Rep.											1300			
Lebanon														
Luxembourg								316				1		
Mexico												246		
Moldova		12375						4				50		
Morocco														
Myanmar												7		
New Zealand					8			2		36		0		
N. Macedonia												400		
Peru												361		
Portugal												1031		115
Romania			3					0	5			1579		
Russia								106				483		
Serbia										14		31		
Slovakia								509						
Slovenia								128				586		
South Africa			473									839		404
Spain			41303		0							1350		
Switzerland			0			0		465		7		36		
Thailand										3				
Tunisia														
Turkey			7									129		
Ukraine												338		
United Kingdom								15						
United States			515		58	61		39				1218		
Uruguay												10		
Old World	**4349**	**12375**	**50122**	**1203**	**32724**	**1108**	**1088**	**19417**	**1417**	**21**		**30074**		**793**
New World			**1808**		**211**	**526**		**85**	**92**	**723**	**1821**	**3666**	**8258**	**644**
World	**4349**	**12375**	**51930**	**1203**	**32935**	**1634**	**1088**	**19501**	**1509**	**744**	**1821**	**33739**	**8258**	**1438**

Table 100 (cont.): National winegrape area for each of the world's top 300 varieties, 2016

	Muscat of Alexandria	Muscat of Hamburg	Muscat Ottonel	Narince	Nebbiolo	Negramoll	Négrette	Negroamaro	Nerello Mascalese	Nero d'Avola	Nero di Troia	Niagara	Nuragus	Odessky Cherny	Öküzgözü	Ortrugo	Palomino Fino
Algeria	200																
Argentina	2716				32					0							104
Australia	2179	35			107	6		18		64							19
Austria			344														
Brazil	6				0							1430					
Bulgaria			3679														
Cambodia																	
Canada			32		0							87					
Chile	5424				11												
China	3000																
Croatia																	
Cyprus																	
Czechia																	
Ethiopia																	
France	2462	2325	172		1		1112										41
Georgia																	
Germany		86	12														
Greece	773	2288															
Hungary		12	1256							86				8			
India	100																
Israel	220																
Italy	1375	22	0		7551			11431	1805	14129	2512		1008			709	
Japan												551					
Kazakhstan																	
Korea, Rep.																	
Lebanon																	
Luxembourg																	
Mexico					180												109
Moldova		254	1859														
Morocco	2093																
Myanmar																	
New Zealand					1												7
N. Macedonia		350															
Peru						1252											
Portugal	509					606											2594
Romania			4779		0												
Russia	21	180	34											1250			
Serbia		624	183														
Slovakia																	
Slovenia			98														
South Africa	1781		9		24					2							134
Spain	9534	125			0	1149				0							20110
Switzerland		0	5		2					0							0
Thailand																	
Tunisia	405																
Turkey				787											1601		
Ukraine														1250			
United Kingdom																	
United States	1987	113			63							1196					70
Uruguay	22	1267	2		25												
Old World	**17591**	**6265**	**12421**	**787**	**7554**	**1755**	**1112**	**11431**	**1805**	**14215**	**2512**		**1008**	**2508**	**1601**	**709**	**22746**
New World	**17214**	**1415**	**43**		**443**	**1258**		**18**		**66**		**3264**					**443**
World	**34805**	**7680**	**12464**	**787**	**7997**	**3013**	**1112**	**11449**	**1805**	**14281**	**2512**	**3264**	**1008**	**2508**	**1601**	**709**	**23190**

Table 100 (cont.): National winegrape area for each of the world's top 300 varieties, 2016

	Pamid	Pardillo	Parellada	Passerina	Pecorino	Pedro Giménez	Pedro Ximénez	Perruno	Pervenets Magaracha	Petit Manseng	Petit Verdot	Pignoletto	Pinot Blanc	Pinot Gris	Pinot Meunier	Pinot Noir	Pinotage
Algeria																	
Argentina						11197				12	740		9	401	11	1866	
Australia							20				1118		5	3652	82	4806	1
Austria													1916	224		614	
Brazil											13			7		141	13
Bulgaria	6874															342	
Cambodia																	
Canada											26		109	649	9	639	6
Chile						4379				0	863		18	437	2	4091	
China																400	
Croatia																	
Cyprus																	
Czechia													762	826		697	
Ethiopia																	
France										1247	870		1181	2867	12130	31602	
Georgia													219				
Germany													4323	4887	2002	11184	
Greece	23																
Hungary	32										4		228	1594		1092	
India																100	
Israel											275						
Italy				933	1628					4	296	1174	2337	18821	5	5057	
Japan																20	
Kazakhstan																180	
Korea, Rep.																	
Lebanon																	
Luxembourg													160	196		121	
Mexico																	
Moldova									517				210	1208	138	2366	
Morocco																	
Myanmar											2					7	
New Zealand										1	9		12	2422	21	5514	38
N. Macedonia	250															500	
Peru					114		1									1	
Portugal							259				90		15	5		130	
Romania	2716												0	1561		1930	
Russia									2238				865	78		918	
Serbia	67													112		633	
Slovakia													416				
Slovenia													424	508		202	
South Africa							2				749		9	369	14	1153	7052
Spain		3283	7137				8528	745		3	1804				2	969	
Switzerland				0							2		111	230	0	4209	1
Thailand																1	
Tunisia																	
Turkey											19					10	
Ukraine													170			385	
United Kingdom													15	44	202	546	
United States										32	1219		263	7462	76	22998	20
Uruguay											26		2	9		56	
Old World	**9961**	**3283**	**7137**	**933**	**1628**		**8787**	**745**	**2755**	**1255**	**3360**	**1174**	**13338**	**33116**	**14277**	**63141**	**1**
New World					**114**	**15576**	**23**			**45**	**4764**		**440**	**15454**	**418**	**42339**	**7131**
World	**9961**	**3283**	**7137**	**933**	**1742**	**15576**	**8810**	**745**	**2755**	**1299**	**8124**	**1174**	**13779**	**48570**	**14695**	**105480**	**7132**

Table 100 (cont.): National winegrape area for each of the world's top 300 varieties, 2016

	Piquepoul Blanc	Planta Nova	Plavac Mali	Prieto Picudo	Prokupac	Prosecco	Prosecco Lungo	Rabigato	Refosco	Refosco dal Peduncolo Rosso	Regent	Ribolla Gialla	Riesling	Rkatsiteli	Roditis	Roditis (R)
Algeria																
Argentina						11							93			
Australia	1					160			0				3114			
Austria													2016			
Brazil						207			0				6			
Bulgaria														5415		
Cambodia																
Canada													1188			
Chile													413			
China													1600			
Croatia			1664										625			
Cyprus																
Czechia													1172			
Ethiopia																
France	1564												4025			
Georgia														25324		
Germany											1902		21540			
Greece										5			1		8463	828
Hungary											4		1261			
India																
Israel																
Italy						19730	1450			1267	1	362	1461			
Japan													22			
Kazakhstan													111	3552		
Korea, Rep.																
Lebanon																
Luxembourg													162			
Mexico																
Moldova													1701	3898		
Morocco																
Myanmar																
New Zealand													767			
N. Macedonia			50		445								900	460		
Peru																
Portugal		24						1969					13			
Romania											1		6121	413		
Russia													2232	6477		
Serbia					916								1361	60		
Slovakia													620			
Slovenia									1340			597	607			
South Africa	0												152			
Spain		864		4293									184			
Switzerland									0		38		19			
Thailand																
Tunisia																
Turkey													3			
Ukraine													1350	5775		
United Kingdom											26					
United States											2		4952			
Uruguay													12			
Old World	**1564**	**888**	**1714**	**4293**	**1361**	**19730**	**1450**	**1969**	**1341**	**1272**	**1946**	**959**	**47486**	**51374**	**8463**	**828**
New World	**1**					**379**			**0**		**28**		**12319**			
World	**1565**	**888**	**1714**	**4293**	**1361**	**20109**	**1450**	**1969**	**1341**	**1272**	**1974**	**959**	**59805**	**51374**	**8463**	**828**

Table 100 (cont.): National winegrape area for each of the world's top 300 varieties, 2016

	Rojal Tinta	Romeiko	Rondinella	Roussanne	Rubired	Ruby Cabernet	Rufete	Sagrantino	Sangiovese	Sankt Laurent	Santarena	Saperavi	Sauvignon Blanc	Sauvignon Blanc (G)	Sauvignonasse	Savagnin Blanc
Algeria																
Argentina			0			2			1837				2148	6	424	22
Australia			1	23	90	849		22	430			10	6044		1	870
Austria										724			1170			288
Brazil			0			24			3				33			
Bulgaria													637			
Cambodia																
Canada				8					4	0			285			1
Chile				25				1	152				14999	134	658	
China													2000			
Croatia																
Cyprus																
Czechia										1183			906			
Ethiopia									90							
France				1831					1503				28084	816		484
Georgia												4751				
Germany										633			736			
Greece		1131											727			
Hungary								1	1	1			982			
India													500			
Israel													110			
Italy			2683	12				998	68428				3935		2503	
Japan													15			
Kazakhstan												428				
Korea, Rep.																
Lebanon													500			
Luxembourg										4						
Mexico													120			
Moldova												573	6909			
Morocco													440			
Myanmar													22			
New Zealand				0					8	1			20497	104		
N. Macedonia													185			
Peru																
Portugal							1145				724		102			
Romania									88				5594			53
Russia												716	2501			214
Serbia													741			
Slovakia										717						
Slovenia										6			1121		231	209
South Africa				76		2306			70				9246			
Spain	736					12	714	0	2				4562			
Switzerland				3					0	2			170	7	0	127
Thailand									2							
Tunisia													85			
Turkey									18				153			
Ukraine													1550			
United Kingdom																
United States				156	4825	2114		4	827				6747		44	
Uruguay				2		2							144	10		
Old World	**736**	**1131**	**2683**	**1846**		**12**	**1859**	**999**	**70041**	**3271**	**724**	**6468**	**61900**	**823**	**2735**	**1375**
New World			**1**	**291**	**4916**	**5297**		**27**	**3423**	**2**		**10**	**62799**	**253**	**1126**	**892**
World	**736**	**1131**	**2684**	**2137**	**4916**	**5309**	**1859**	**1026**	**73464**	**3272**	**724**	**6478**	**124700**	**1076**	**3861**	**2267**

Table 100 (cont.): National winegrape area for each of the world's top 300 varieties, 2016

	Savatiano	Scheurebe	Schiava Grossa	Schiava Lombarda	Sémillon	Seyval Blanc	Shiroka Melnishka	Silvaner	Síria	Sultaniye	Syrah	Tannat	Tempranillo	Teroldego	Terret	Tinta Barroca	Tinta Carvalha
Algeria										100	1000						
Argentina					767			3			12707	837	6140				
Australia		1	0		4556			6		42	38942	30	681	1		1	
Austria		351						38			141						
Brazil					6	41						109	23	4			
Bulgaria							1205				804						
Cambodia											3						
Canada		1			19	2259		0			260	0	6	2			
Chile					849						7994	7	127			2	
China											1000						
Croatia																	
Cyprus																	
Czechia																	
Ethiopia										4							
France					10234	106		1027		1	62211	2513	658		872		
Georgia																	
Germany		1266	2197					4627			46						
Greece	10268				11			0			1042		22				
Hungary					43			7			215	1					
India										1000	500						
Israel											385		55				
Italy			58	701	13			18			7693	15	9	730			
Japan																	
Kazakhstan																	
Korea, Rep.																	
Lebanon											300						
Luxembourg																	
Mexico										841	145		229				
Moldova					3			11			87						
Morocco										721	347						
Myanmar											27		4				
New Zealand					63						436	2	18				
N. Macedonia																	
Peru										15	2						
Portugal					76				6438		4017	10	17014			4733	1113
Romania					18	1		1			504		67				
Russia					25												
Serbia																	
Slovakia																	
Slovenia								81			18						
South Africa					1121	0		1			9946	114	92			190	
Spain					2				599		19488	0	193597				
Switzerland		7	0		4	8		250			194	0	0			0	
Thailand											74		4				
Tunisia										139	67						
Turkey					529					2461	1439		6				
Ukraine																	
United Kingdom						77											
United States					340	207					9083	247	626	36			
Uruguay					14						67	1725					
Old World	**10268**	**1624**	**2255**	**701**	**10957**	**115**	**1205**	**6061**	**7037**	**3423**	**99998**	**2539**	**211429**	**730**	**872**	**4733**	**1113**
New World		**2**	**0**		**7736**	**2584**		**11**		**1902**	**81186**	**3072**	**7950**	**42**		**193**	
World	**10268**	**1626**	**2256**	**701**	**18693**	**2699**	**1205**	**6072**	**7037**	**5325**	**181185**	**5611**	**219379**	**772**	**872**	**4926**	**1113**

Table 100 (cont.): National winegrape area for each of the world's top 300 varieties, 2016

	Tinto Velasco	Torrontés Riojano	Torrontés Sanjuanino	Touriga Franca	Touriga Nacional	Trajadura	Trebbiano d'Abruzzo	Trebbiano Giallo	Trebbiano Romagnolo	Trebbiano Toscano	Trepat	Tribidrag	Trincadeira	Trousseau	Tsitska	Tsolikouri	Verdeca
Algeria																	
Argentina		8208	1885		27					1622		1	0	34			
Australia					38					14		87	1	0			
Austria																	
Brazil					6					231							
Bulgaria										738							
Cambodia																	
Canada										2		8					
Chile		643	1771		2							66					
China										1500							
Croatia																	
Cyprus																	
Czechia																	
Ethiopia																	
France										78842		1		45			
Georgia															3642	7903	
Germany																	
Greece										211							0
Hungary																	
India										300							
Israel																	
Italy							2630	2275	19059	35441		13896					912
Japan																	
Kazakhstan																	
Korea, Rep.																	
Lebanon																	
Luxembourg																	
Mexico																	
Moldova										277							
Morocco																	
Myanmar																	
New Zealand												4					
N. Macedonia												1000					
Peru		8															
Portugal				14217	11411	1550				122			10493	1166			
Romania				4	11					3		9					
Russia										66							
Serbia																	
Slovakia																	
Slovenia																	
South Africa				3	105					156		24	15	0			
Spain	5369				4	942				49	1199	1	0	18			
Switzerland					0							1		0			
Thailand																	
Tunisia																	
Turkey																	
Ukraine																	
United Kingdom																	
United States					117					88		18551					
Uruguay										682							
Old World	**5369**			**14221**	**11426**	**2492**	**2630**	**2275**	**19059**	**115748**	**1199**	**14908**	**10493**	**1229**	**3642**	**7903**	**913**
New World		**8859**	**3656**	**3**	**296**					**4595**		**18742**	**17**	**34**			
World	**5369**	**8859**	**3656**	**14224**	**11722**	**2492**	**2630**	**2275**	**19059**	**120343**	**1199**	**33649**	**10510**	**1263**	**3642**	**7903**	**913**

Table 100 (cont.): National winegrape area for each of the world's top 300 varieties, 2016

	Verdejo	Verdelho	Verdicchio Bianco	Verduzzo Friulano	Vermentino	Vernaccia di San	Vidal	Villard Blanc	Villard Noir	Vinhao	Viognier	Viosinho	Vranac	Xarello	Xinomavro	Xynisteri	Yan 73
Algeria																	
Argentina		18	7	2	2						773						
Australia	6	1016	0	1	118					1	753						
Austria																	
Brazil			1								11						
Bulgaria																	
Cambodia																	
Canada							59				101						
Chile	2				0						839						
China							1500										4800
Croatia																	
Cyprus																1946	
Czechia																	
Ethiopia																	
France					4642			223	763		8823						
Georgia																	
Germany																	
Greece															2135		
Hungary								199			13		0				
India																	
Israel																	
Italy			4674	687	6703	884					1827						
Japan																	
Kazakhstan																	
Korea, Rep.																	
Lebanon																	
Luxembourg																	
Mexico																	
Moldova											31						
Morocco																	
Myanmar																	
New Zealand		7			0						129						
N. Macedonia													9500				
Peru																	
Portugal		360								4055	125	916					
Romania											20						
Russia																	
Serbia													3				
Slovakia																	
Slovenia																	
South Africa		69			2		10	321	14	44	822						
Spain	17923									333	213			8534			
Switzerland							1				44						
Thailand		1									1						
Tunisia																	
Turkey											15						
Ukraine																	
United Kingdom																	
United States		45			17		366			36	1481						
Uruguay											41						
Old World	**17923**	**360**	**4674**	**687**	**11345**	**884**	**1**	**422**	**763**	**4388**	**11112**	**916**	**9503**	**8534**	**2135**	**1946**	
New World	**9**	**1156**	**8**	**2**	**138**		**1936**	**321**	**14**	**81**	**4951**						**4800**
World	**17931**	**1516**	**4682**	**690**	**11483**	**884**	**1936**	**743**	**777**	**4468**	**16063**	**916**	**9503**	**8534**	**2135**	**1946**	**4800**

Table 100 (cont.): National winegrape area for each of the world's top 300 varieties, 2016

	Zalagyöngye	Zalema	Žametovka	Zweigelt	Other varieties	Total
Algeria					**0**	**8300**
Argentina					**2693**	**206342**
Australia				1	**602**	**132435**
Austria				6311	**3359**	**45439**
Brazil					**4978**	**33205**
Bulgaria					**0**	**52974**
Cambodia					**3**	**10**
Canada				19	**1507**	**12600**
Chile					**370**	**145873**
China					**77600**	**178000**
Croatia					**683**	**11746**
Cyprus					**0**	**5133**
Czechia				770	**0**	**13600**
Ethiopia					**21**	**169**
France					**69890**	**814882**
Georgia					**4525**	**48000**
Germany	1			108	**5124**	**94501**
Greece					**4468**	**50845**
Hungary	1065			1687	**6526**	**63881**
India					**0**	**2700**
Israel					**545**	**5000**
Italy				4	**54810**	**604551**
Japan				59	**1162**	**3869**
Kazakhstan					**2060**	**6938**
Korea, Rep.					**1400**	**5400**
Lebanon					**700**	**4000**
Luxembourg					**13**	**1300**
Mexico					**1204**	**5465**
Moldova					**9903**	**82600**
Morocco					**1486**	**17590**
Myanmar					**0**	**70**
New Zealand					**69**	**35463**
N. Macedonia					**957**	**24777**
Peru					**652**	**3831**
Portugal					**19654**	**182649**
Romania				86	**98551**	**182762**
Russia	193				**5006**	**50794**
Serbia					**8403**	**22014**
Slovakia					**1718**	**7748**
Slovenia			822	5	**2157**	**15989**
South Africa					**1331**	**95775**
Spain		4015			**23136**	**883558**
Switzerland	0			19	**1781**	**14793**
Thailand					**16**	**149**
Tunisia					**287**	**3400**
Turkey					**396**	**13704**
Ukraine					**0**	**25166**
United Kingdom					**243**	**1839**
United States					**13080**	**239632**
Uruguay					**665**	**6743**
Old World	**1259**	**4015**	**822**	**8990**	**326140**	**3358633**
New World				**78**	**107596**	**1109570**
World	**1259**	**4015**	**822**	**9068**	**433735**	**4468203**

About the Wine Economics Research Centre

The Wine Economics Research Centre was established in 2010 by the School of Economics and the Wine2030 Research Network of the University of Adelaide, South Australia, having been previously a research program in the University's Centre for International Economic Studies. Its purpose is to promote and foster its growing research strength in the area of wine economics, and to complement the university's long-established strength in viticulture and oenology research.

The University of Adelaide is the Southern Hemisphere's premier wine research and teaching university and is part of the adjacent Wine Innovation Cluster which includes the University's School of Agriculture, Food and Wine and the Australian Wine Research Institute (established in 1955).

Adelaide is the capital of the state of South Australia, where nearly half of Australia's winegrapes are crushed and from where more than half of Australia's wine exports are shipped. Adelaide has four major wine regions and more than 200 cellar doors within an hour's drive (Adelaide Hills, Barossa Valley, McLaren Vale and Southern Fleurieu/Langhorne Creek), in addition to South Australia's three other key wine regions (Clare Valley, Coonawarra/Limestone Coast and Riverland). The University of Adelaide also is home to the National Wine Centre of Australia.

The Wine Economics Research Centre is unique in Australia and one of few similar centres around the world. It has close links with the Center for Wine Economics at the Robert Mondavi Institute for Wine and Food Science at the University of California, Davis, and with the American Association of Wine Economists' *Journal of Wine Economics*.

The key objectives of the Wine Economics Research Centre are to:

- publish wine economics research outputs and disseminate them to industry and government as well as academia
- contribute to economics journals, wine industry journals and related publications
- promote collaboration and sharing of information, data and analyses between industry and government agencies as well as research institutions
- sponsor wine economics seminars, workshops and conferences and contribute to other grape and wine meetings

The founding Executive Director of the Wine Economics Research Centre is Professor Kym Anderson. Contact details are as follows:

Wine Economics Research Centre
School of Economics
University of Adelaide
Adelaide SA 5005
Australia
Email: kym.anderson@adelaide.edu.au
Website: https://economics.adelaide.edu.au/wine-economics/

CPSIA information can be obtained
at www.ICGtesting.com
Printed in the USA
LVHW071517280920
667086LV00011B/110

9 781925 261868